Lecture Notes in Electrical Engineering

Volume 554

Series Editors

Leopoldo Angrisani, Department of Electrical and Information Technologies Engineering, University of Napoli Federico II, Napoli, Italy

Marco Arteaga, Departament de Control y Robótica, Universidad Nacional Autónoma de México, Coyoacán, Mexico

Bijaya Ketan Panigrahi, Electrical Engineering, Indian Institute of Technology Delhi, New Delhi, Delhi, India

Samarjit Chakraborty, Fakultät für Elektrotechnik und Informationstechnik, TU München, München, Germany

Jiming Chen, Zhejiang University, Hangzhou, Zhejiang, China

Shanben Chen, Materials Science & Engineering, Shanghai Jiao Tong University, Shanghai, China

Tan Kay Chen, Department of Electrical and Computer Engineering, National University of Singapore, Singapore, Singapore

Rüdiger Dillmann, Humanoids and Intelligent Systems Lab, Karlsruhe Institute for Technology, Karlsruhe, Baden-Württemberg, Germany

Haibin Duan, Beijing University of Aeronautics and Astronautics, Beijing, China

Gianluigi Ferrari, Università di Parma, Parma, Italy

Manuel Ferre, Centre for Automation and Robotics CAR (UPM-CSIC), Universidad Politécnica de Madrid, Madrid, Madrid, Spain

Sandra Hirche, Department of Electrical Engineering and Information Science, Technische Universität München, München, Germany

Faryar Jabbari, Department of Mechanical and Aerospace Engineering, University of California, Irvine, CA, USA

Limin Jia, State Key Laboratory of Rail Traffic Control and Safety, Beijing Jiaotong University, Beijing, China

Janusz Kacprzyk, Systems Research Institute, Polish Academy of Sciences, Warsaw, Poland

Alaa Khamis, German University in Egypt El Tagamoa El Khames, New Cairo City, Egypt

Torsten Kroeger, Stanford University, Stanford, CA, USA

Qilian Liang, Department of Electrical Engineering, University of Texas at Arlington, Arlington, TX, USA

Ferran Martin, Departament d'Enginyeria Electrònica, Universitat Autònoma de Barcelona, Bellaterra, Barcelona, Spain

Tan Cher Ming, College of Engineering, Nanyang Technological University, Singapore, Singapore

Wolfgang Minker, Institute of Information Technology, University of Ulm, Ulm, Germany

Pradeep Misra, Department of Electrical Engineering, Wright State University, Dayton, OH, USA

Sebastian Möller, Quality and Usability Lab, TU Berlin, Berlin, Germany

Subhas Mukhopadhyay, School of Engineering & Advanced Technology, Massey University, Palmerston North, Manawatu-Wanganui, New Zealand

Cun-Zheng Ning, Electrical Engineering, Arizona State University, Tempe, AZ, USA

Toyoaki Nishida, Graduate School of Informatics, Kyoto University, Kyoto, Kyoto, Japan

Federica Pascucci, Dipartimento di Ingegneria, Università degli Studi "Roma Tre", Rome, Italy

Yong Qin, State Key Laboratory of Rail Traffic Control and Safety, Beijing Jiaotong University, Beijing, China

Gan Woon Seng, School of Electrical & Electronic Engineering, Nanyang Technological University, Singapore, Singapore

Joachim Speidel, Institute of Telecommunications, Universität Stuttgart, Stuttgart, Baden-Württemberg, Germany

Germano Veiga, Campus da FEUP, INESC Porto, Porto, Portugal

Haitao Wu, Academy of Opto-electronics, Chinese Academy of Sciences, Beijing, China

Junjie James Zhang, Charlotte, NC, USA

The book series *Lecture Notes in Electrical Engineering* (LNEE) publishes the latest developments in Electrical Engineering - quickly, informally and in high quality. While original research reported in proceedings and monographs has traditionally formed the core of LNEE, we also encourage authors to submit books devoted to supporting student education and professional training in the various fields and applications areas of electrical engineering. The series cover classical and emerging topics concerning:

- Communication Engineering, Information Theory and Networks
- Electronics Engineering and Microelectronics
- Signal, Image and Speech Processing
- Wireless and Mobile Communication
- Circuits and Systems
- Energy Systems, Power Electronics and Electrical Machines
- Electro-optical Engineering
- Instrumentation Engineering
- Avionics Engineering
- Control Systems
- Internet-of-Things and Cybersecurity
- Biomedical Devices, MEMS and NEMS

For general information about this book series, comments or suggestions, please contact leontina.dicecco@springer.com.

To submit a proposal or request further information, please contact the Publishing Editor in your country:

China
Jasmine Dou, Associate Editor (jasmine.dou@springer.com)

India
Swati Meherishi, Executive Editor (swati.meherishi@springer.com)
Aninda Bose, Senior Editor (aninda.bose@springer.com)

Japan
Takeyuki Yonezawa, Editorial Director (takeyuki.yonezawa@springer.com)

South Korea
Smith (Ahram) Chae, Editor (smith.chae@springer.com)

Southeast Asia
Ramesh Nath Premnath, Editor (ramesh.premnath@springer.com)

USA, Canada:
Michael Luby, Senior Editor (michael.luby@springer.com)

All other Countries:
Leontina Di Cecco, Senior Editor (leontina.dicecco@springer.com)
Christoph Baumann, Executive Editor (christoph.baumann@springer.com)

** Indexing: The books of this series are submitted to ISI Proceedings, EI-Compendex, SCOPUS, MetaPress, Web of Science and Springerlink **

More information about this series at http://www.springer.com/series/7818

Ivan Zelinka · Pavel Brandstetter ·
Tran Trong Dao · Vo Hoang Duy ·
Sang Bong Kim
Editors

AETA 2018 - Recent Advances in Electrical Engineering and Related Sciences: Theory and Application

Editors
Ivan Zelinka
Department of Computer Science,
Faculty of Electrical Engineering
VŠB-TUO
Ostrava-Poruba, Czech Republic

Pavel Brandstetter
Department of Electronics,
Faculty of Electrical Engineering
VŠB-TUO
Ostrava, Czech Republic

Tran Trong Dao
International Cooperation,
Research and Training Institute
Ton Duc Thang University
Ho Chi Minh, Vietnam

Vo Hoang Duy
Department of Automatic Control
Ton Duc Thang University
Ho Chi Minh City, Vietnam

Sang Bong Kim
Department of Mechanical Design Engineering
Pukyong National University
Busan, Korea (Republic of)

ISSN 1876-1100   ISSN 1876-1119   (electronic)
Lecture Notes in Electrical Engineering
ISBN 978-3-030-14906-2   ISBN 978-3-030-14907-9   (eBook)
https://doi.org/10.1007/978-3-030-14907-9

Library of Congress Control Number: 2019933407

© Springer Nature Switzerland AG 2020
This work is subject to copyright. All rights are reserved by the Publisher, whether the whole or part of the material is concerned, specifically the rights of translation, reprinting, reuse of illustrations, recitation, broadcasting, reproduction on microfilms or in any other physical way, and transmission or information storage and retrieval, electronic adaptation, computer software, or by similar or dissimilar methodology now known or hereafter developed.
The use of general descriptive names, registered names, trademarks, service marks, etc. in this publication does not imply, even in the absence of a specific statement, that such names are exempt from the relevant protective laws and regulations and therefore free for general use.
The publisher, the authors and the editors are safe to assume that the advice and information in this book are believed to be true and accurate at the date of publication. Neither the publisher nor the authors or the editors give a warranty, express or implied, with respect to the material contained herein or for any errors or omissions that may have been made. The publisher remains neutral with regard to jurisdictional claims in published maps and institutional affiliations.

This Springer imprint is published by the registered company Springer Nature Switzerland AG
The registered company address is: Gewerbestrasse 11, 6330 Cham, Switzerland

Foreword

The modern world is based on vitally important technologies that merge electronics, cybernetics, computer science, telecommunication, and physics together. Since the beginning of our technologies, we have been confronted with numerous technological challenges such as finding the optimal solution of various problems including controlling technologies, power sources construction, and robotics. Technology development of those and related areas has had and continues to have profound impact on our civilization and our future lifestyle.

Therefore, this proceeding book containing articles of international conference AETA 2018 is a timely volume to be welcomed by the community focused on telecommunication, power control, and optimization as well as computational science community and beyond.

This proceeding book consists of the hottest topic areas of selected papers like telecommunication, power systems, digital signal processing, robotics, control system, renewable energy, power electronics, soft computing, and more. All selected papers represent interesting ideas and state-of-the-art overview.

Participations were carefully selected and reviewed; hence, this proceeding book certainly is one of the few discussing the benefit from the intersection of those modern and fruitful scientific fields of research. We hope that the proceeding book will be an instructional material for senior undergraduate and entry-level graduate students working in the area of electronics, power technologies, energy distribution, control and robotics, etc. The proceeding book will also be a resource and material for practitioners who want to apply discussed topics to solve real-life problems in their challenging applications. The important part of proceeding book is participation of four keynote speakers from the Russia, USA, and two from India.

The decision to organize AETA conference and to create this proceeding book was based on facts that technologies mentioned above, their use, and impact on life are interesting areas, which are under intensive research from many other branches of science today. This proceeding book is written to contain simplified versions of experiments with the aim to show how, in principle, problems about power systems can be solved.

It is obvious that this proceeding book does not encompass all aspects of discussed topics due to limited space and time of the conference. Only the main ideas and results of selected papers are reported here. The authors and editors hope that the readers will be inspired to do their own experiments and simulations, based on information reported in this proceeding book, thereby moving beyond the scope of it.

This proceeding book is devoted to the studies of common and related subjects in intensive research fields of modern electric, electronic, and related technologies. For these reasons, we believe that this proceeding book will be useful for scientists and engineers working in the above-mentioned fields of research and applications.

At the end, we **would like to thank** Ton Duc Thang University (Ho Chi Minh City, Vietnam) and VŠB-Technical University (Ostrava, Czech Republic) for interest and strong support in AETA conference organization. Also, **many thanks** belong to Springer Publishing Company for its highly professional, precise, and quick production process. Without all of this, it would be impossible to organize successful conference joining participants from the whole world.

September 2018   Editors

This conference was supported by the Ton Duc Thang University (Ho Chi Minh City, Vietnam) and VŠB - Technical University (Ostrava, Czech Republic).

Contents

Computer Science

New Neuromorphic AI NM500 and Its ADAS Application 3
Jungyun Kim

Analyzing *l*1-loss and *l*2-loss Support Vector Machines Implemented
in PERMON Toolbox 13
Marek Pecha and David Horák

Hybrid Fuzzy Neural Model Based Dempster-Shafer System
for Processing of Diagnostic Information 24
Alexander I. Dolgiy, Sergey M. Kovalev, Andrey V. Sukhanov,
and Vitezslav Styskala

ANFIS and Fuzzy Tuning of PID Controller for STATCOM
to Enhance Power Quality in Multi-machine System
Under Large Disturbance 34
Huu Vinh Nguyen, Hung Nguyen, and Kim Hung Le

Proposal of Electrode System for Measuring Level of Glucose
in the Blood 45
Klara Fiedorova and Martin Augustynek

Substitution Rules with Respect to a Context 55
Michal Fait and Marie Duží

Fuzzy Model Predictive Control for Discrete-Time System
with Input Delays 67
Sofiane Bououden, Ilyes Boulkaibet, Mohammed Chadli, and Ivan Zelinka

An Improvement of Fuzzy-Based Control Strategy for a Series
Hydraulic Hybrid Truck 78
Tri-Vien Vu, Bach H. Dinh, Anh-Minh Duc Tran, Chih-Keng Chen,
and Trung-Hieu Vu

A New Approach Newton-Raphson Load Flow Analysis in Power
System Networks with STATCOM............................ 88
Dung Vo Tien, Radomir Gono, and Zbigniew Leonowicz

Neural Network for Smart Adjustment of Industrial Camera - Study
of Deployed Application..................................... 101
Petr Dolezel and Daniel Honc

Risk Assessment Approach to Estimate Security of Cryptographic
Keys in Quantum Cryptography 114
Marcin Niemiec, Miralem Mehic, and Miroslav Voznak

Wavelet Transform Decomposition for Fetal Phonocardiogram
Extraction from Composite Abdominal Signal................... 125
Radana Kahankova and Radek Martinek

A CUDA Approach for Scenario Reduction in Hedging Models...... 134
Donald Davendra, Chin-mei Chueh, and Emmanuel Hamel

Using a Strain Gauge Load Cell for Analysis of Round Punch....... 144
Dora Lapkova

Geometrical Computational Method to Locate Hypocenter
by Signal Readings from a Three Receivers..................... 154
Alexander D. Krutas, Tatyana A. Smaglichenko, Alexander Smaglichenko, and Maria Sayankina

An Intelligent Question-Answer System over
Natural-Language Texts...................................... 162
Marie Duží and Bjørn Jespersen

An Efficient Reduced Basis Construction for Stochastic Galerkin
Matrix Equations Using Deflated Conjugate Gradients 175
Michal Béreš

An Investigation on Signal Comparison by Measuring of Numerical
Strings Similarity 185
Alexander Smaglichenko, Tatyana A. Smaglichenko, Arkady Genkin, and Boris Melnikov

Optimization

A Lightweight SHADE-Based Algorithm for Global
Optimization - liteSHADE 197
Adam Viktorin, Roman Senkerik, Michal Pluhacek, Tomas Kadavy, and Roman Jasek

Pupil Localization Using Self-organizing Migrating Algorithm 207
Radovan Fusek and Petr Dobeš

Differential Evolution Algorithms Used to Optimize Weights of Neural Network Solving Pole-Balancing Problem 217
Jan Vargovsky and Lenka Skanderova

The Use of Radial Basis Function Surrogate Models for Sampling Process Acceleration in Bayesian Inversion 228
Simona Domesová

An Optimised Hybrid Group Method in Data Handling (GMDH) Network 239
Donald Davendra and Petr Martinek

A Better Indexing Method for Closest Open Location Policy in Forklift Warehouse Operation 251
Duy Anh Nguyen, Truong Thinh Pham, and Viet Duong Nguyen

On-Line Efficiency-Optimization Control of Induction Motor Drives Using Particle Swarm Optimization Algorithm 261
Sang Dang Ho, Pavel Brandstetter, Cuong Tran Dinh, Thinh Cong Tran, Minh Chau Huu Nguyen, and Bach Hoang Dinh

Introducing the Run Support Strategy for the Bison Algorithm 272
Anezka Kazikova, Michal Pluhacek, Tomas Kadavy, and Roman Senkerik

Optimizing Automated Storage and Retrieval Algorithm in Cold Warehouse by Combining Dynamic Routing and Continuous Cluster Method 283
Ngoc Cuong Truong, Truong Giang Dang, and Duy Anh Nguyen

Dependency of GPA-ES Algorithm Efficiency on ES Parameters Optimization Strength 294
Tomas Brandejsky

A Modified Bat Algorithm to Improve the Search Performance Applying for the Optimal Combined Heat and Power Generations 303
Bach H. Dinh, An H. Ngo, and Thang T. Nguyen

Prediction of Hourly Vehicle Flows by Optimized Evolutionary Fuzzy Rules 313
Pavel Krömer, Jana Nowaková, Martin Hasal, and Jan Platoš

A New Simple, Fast and Robust Total Least Square Error Computation in E2: Experimental Comparison 325
Michal Smolik, Vaclav Skala, and Zuzana Majdisova

On the Self-organizing Migrating Algorithm Comparison by Means of Centrality Measures 335
Lukas Tomaszek, Patrik Lycka, and Ivan Zelinka

A Brief Overview of the Synergy Between Metaheuristics and Unconventional Dynamics 344
Roman Senkerik

Telecommunications

An Examination of Outage Performance for Selected Relay and Fixed Relay in Cognitive Radio-Aided NOMA 359
Tam Nguyen Kieu, Hong Nhu Nguyen, Long Nguyen Ngoc, Tu-Trinh Thi Nguyen, Jaroslav Zdralek, and Miroslav Voznak

Throughput Analysis of Power Beacon-Aided Multi-hop Relaying Networks Employing Non-orthogonal Multiple Access with Hardware Impairments 371
Phu Tran Tin, Pham Minh Nam, Tran Trung Duy, Phuong T. Tran, Tam Nguyen Kieu, and Miroslav Voznak

Optimum Selection of the Reference Signal for Correlation Receiver Applied to Marker Localization 382
Martin Vestenický and Peter Vestenický

Comparing of Transfer Process Data in PLC and MCU Based on IoT 390
Antonin Gavlas, Jiri Koziorek, and Robert Rakay

Protecting Gateway from ABP Replay Attack on LoRaWAN 400
Erik Gresak and Miroslav Voznak

Development of a Distributed VoIP Honeypot System with Advanced Malicious Traffic Detection 409
Ladislav Behan, Lukas Sevcik, and Miroslav Voznak

Proposal and Implementation of Probe for Sigfox Technology 420
Jakub Jalowiczor and Miroslav Voznak

IoT Approach to Street Lighting Control Using MQTT Protocol 429
Radim Kuncicky, Jakub Kolarik, Lukas Soustek, Lumir Kuncicky, and Radek Martinek

Materials

Temperature Dependence of Microstructure in Liquid Aluminosilicate 441
Mai Van Dung, Le The Vinh, Vo Hoang Duy, Nguyen Kieu Tam, Tran Thanh Nam, Nguyen Manh Tuan, Truong Duc Quynh, and Nguyen Van Yen

Study on Effect of Parameters on Friction Stir Welding Process of 6061 Aluminum Alloy Tubes 450
Van Vu Nguyen, Hoang Linh Nguyen, Tan Tien Nguyen,
Thien Phuc Tran, and Sang Bong Kim

Convergence Study of Different Approaches of Solving the Hartree-Fock Equation on the Potential Curve of the Hydrogen Fluoride 461
Martin Mrovec

Control Systems

Network Traffic Anomaly Detection in Railway Intelligent Control Systems Using Nonlinear Dynamics Approach 475
Maria A. Butakova, Andrey V. Chernov, Sergey M. Kovalev,
Andrey V. Sukhanov, and Stanislav Zajaczek

Advanced Methods of Detection of the Steganography Content 484
Jakub Hendrych and Lačezar Ličev

Robust Servo Controller Design Based on Linear Shift Invariant Differential Operator 494
Dae Hwan Kim and Sang Bong Kim

Servo Controller Design and Fault Detection Algorithm for Speed Control of a Conveyor System 505
Trong Hai Nguyen, Nguyen Thanh Phuong, and Hung Nguyen

A Control System for Power Electronics with an NXP Kinetis Series Microcontroller 514
Daniel Kouřil, Martin Sobek, and Petr Chamrád

A MIMO Robust Servo Controller Design Method for Omnidirectional Automated Guided Vehicles Using Polynomial Differential Operator 521
Van Lanh Nguyen, Sung Won Kim, Choong Hwan Lee, Dae Hwan Kim,
Hak Kyeong Kim, and Sang Bong Kim

Model Reference Adaptive Control Strategy for Application to Robot Manipulators 533
Manh Son Tran, Suk Ho Jung, Nhat Binh Le, Huy Hung Nguyen,
Dac Chi Dang, Anh Minh Duc Tran, and Young Bok Kim

Stabilization of Time-Varying Systems Subject to Actuator Saturation: A Takagi-Sugeno Approach 548
Sabrina Aouaouda and Mohammed Chadli

Observer Based Control for Systems with Mismatched Uncertainties in Output Matrix 561
Van Van Huynh, Tran Thanh Phong, and Bach Hoang Dinh

Nonlinear Disturbance Observer with Recurrent Neural Network Compensator 569
Shihono Yamada and Jun Ishikawa

Parameters Estimation for Sensorless Control of Induction Motor Drive Using Modify GA and CSA Algorithm 580
Thinh Cong Tran, Pavel Brandstetter, Cuong Dinh Tran, Sang Dang Ho, Minh Chau Huu Nguyen, and Pham Nhat Phuong

Study on Algorithms and Path-Optimization for USV's Obstacle Avoidance 592
Ngoc-Huy Tran and Nguyen Nhut-Thanh Pham

Visual Servoing Controller Design Based on Barrier Lyapunov Function for a Picking System 605
Jong Min Oh, Jotje Rantung, Sung Rak Kim, Sang Kwun Jeong, Hak Kyeong Kim, Sea June Oh, and Sang Bong Kim

Designing a PID Controller for Ship Autopilot System 618
Dinh Due Vo, Viet Anh Pham, Phung Hung Nguyen, and Duy Anh Nguyen

The Rotor Initial Position Determination of the Hi-Speed Switch-Reluctance Electrical Generator for the Steam-Microturbine ... 628
Pavel G. Kolpakhchyan, Vladimir I. Parshukov, Boris N. Lobov, Nikolay N. Efimov, and Vadim V. Kopitza

Stability and Chaotic Attractors of Memristor-Based Circuit with a Line of Equilibria 639
N. V. Kuznetsov, T. N. Mokaev, E. V. Kudryashova, O. A. Kuznetsova, R. N. Mokaev, M. V. Yuldashev, and R. V. Yuldashev

Mechanical Engineering

Behavior of Five-Pad Tilting–Pad Journal Bearings with Different Pivot Stiffness 647
Phuoc Vinh Dang, Steven Chatterton, and Paolo Pennacchi

Dynamic Characteristics of a Non-symmetric Tilting Pad Journal Bearing 658
Phuoc Vinh Dang, Steven Chatterton, and Paolo Pennacchi

Energy

DCM Boost Converter in CPM Operation for Tuning Piezoelectric Energy Harvesters 673
Andrés Gomez-Casseres, David Florez, and Darío Cortes

Effect of Weighting Coefficients on Behavior of the DTC Method with Direct Calculation of Voltage Vector 683
Jakub Baca, Martin Kuchar, and Petr Palacky

A New Protocol for Energy Harvesting Decode-and-Forward Relaying Networks 693
Duy-Hung Ha, Dac-Binh Ha, Jaroslav Zdralek, Miroslav Voznak, and Tan N. Nguyen

Average Bit Error Probability Analysis for Cooperative DF Relaying in Wireless Energy Harvesting Networks 705
Hoang-Sy Nguyen, Thanh-Sang Nguyen, Tan N. Nguyen, and Miroslav Voznak

LCCT vs. LLC Converter - Analysis of Operational Characteristics During Critical Modes of Operation 715
Michal Pridala, Michal Frivaldsky, and Pavol Spanik

Control Renewable Energy System and Optimize Performance by Using Weather Data 725
Duy Tan Nguyen, Duy Anh Nguyen, and Lien Son Chau Hoang

Analysis of Efficiency and THD in 7-Level Voltage Inverters with Reduced Number of Switches 736
Ales Havel, Martin Sobek, and Petr Chamrad

Waste Management - Weighing-Machine Automation 747
Zdenek Slanina, Rostislav Pokorny, and Jan Dedek

Optimization of Voltage Model for MRAS Based Sensorless Control of Induction Motor 758
Ondrej Lipcak and Jan Bauer

Capability of Predictive Torque Control Method to Control DC-Link Voltage Level in Small Autonomous Power System with Induction Generator 769
Pavel Karlovsky and Jiri Lettl

Feasibility Structural Analysis of Engineering Plastic Reel Module for Carrying Wound High-Voltage Electric Transmission Line 778
Jungyun Kim, Ho-Young Kang, Young-Geon Song, and Chan-Jung Kim

Improving Fault Tolerant Control to the One Current Sensor Failures for Induction Motor Drives 789
Cuong Dinh Tran, Pavel Brandstetter, Sang Dang Ho, Thinh Cong Tran, Minh Chau Huu Nguyen, Huy Xuan Phan, and Bach Hoang Dinh

Impact of Parameter Variation on Sensorless Indirect Field Oriented Control of Induction Machine 799
Andrej Kacenka, Pavol Makys, and Lubos Struharnansky

Validation the FEM Model of Asynchronous Motor by Analysis of External Radial Stray Field 810
Petr Kacor and Petr Bernat

Outage and Intercept Probability Analysis for Energy-Harvesting-Based Half-Duplex Relay Networks Assisted by Power Beacon Under the Existence of Eavesdropper 821
Tan N. Nguyen, Phuong T. Tran, Nguyen Dao, and Miroslav Voznak

Design of Electrical Regulated Drainage with Energy Harvesting 835
Vaclav Kolar, Roman Hrbac, Tomas Mlcak, and Jiri Placek

Analysis of Appliance Impact on Total Harmonic Distortion in Off-Grid System 844
Michal Petružela, Vojtěch Blažek, and Jan Vysocký

Influencing of Current Sensors by an External Magnetic Field of a Nearby Busbar 850
Tadeusz Sikora and Jan Hurta

A Model for Predicting Energy Savings Attainable by Using Lighting Systems Dimmable to a Constant Illuminance Level 860
T. Novak, J. Sumpich, J. Vanus, K. Sokansky, R. Gono, J. Latal, and P. Valicek

Strategy of Metropolis Electrical Energy Supply 870
Valery Beley, Andrey Nikishin, and Dmitriy Gorbatov

Robotics

Attitude Control of Jumping Robot with Bending-Stretching Mechanism 883
Chea Xin Ong, Yurika Nomura, and Jun Ishikawa

Geometric Foot Location Determination Algorithm for Façade Cleaning Robot 894
Shunsuke Nansai and Hiroshi Itoh

Smart Manipulation Approach for Assistant Robot 904
Yeyson Becerra, Jaime Leon, Santiago Orjuela, Mario Arbulu,
Fernando Matinez, and Fredy Martinez

**Computational Study on Upward Force Generation of Gymnotiform
Undulating Fin** 914
Van Hien Nguyen, Canh An Tien Pham, Van Dong Nguyen,
Hoang Long Phan, and Tan Tien Nguyen

Modular Design of Gymnotiform Undulating Fin 924
Van Dong Nguyen, Canh An Tien Pham, Van Hien Nguyen,
Thien Phuc Tran, and Tan Tien Nguyen

**Path Following Control of Automated Guide Vehicle
Using Camera Sensor** 932
Dae Hwan Kim and Sang Bong Kim

**Binary Classification of Terrains Using Energy Consumption
of Hexapod Robots** 939
Valeriia Iegorova and Sebastián Basterrech

**The Movement of Swarm Robots in an Unknown
Complex Environment** 949
Quoc Bao Diep and Ivan Zelinka

Image Processing

**Contour Detection Method of 3D Fish Using a Local Kernel
Regression Method** 963
Jong Min Oh, Sung Rak Kim, Sung Won Kim, Nam Soo Jeong,
Min Saeng Shin, Hak Kyeong Kim, and Sang Bong Kim

**Camera Based Tests of Dimensions, Shapes and Presence Based
on Virtual Instrumentation** 973
Lukas Soustek, Radek Martinek, Lukas Snajdr, and Petr Bilik

**A 3D Scanner Based on Virtual Instrumentation Implemented by a 1D
Laser Triangulation Method** 982
Jindrich Brablik, Radek Martinek, Marek Haluska, and Petr Bilík

Author Index 991

Computer Science

New Neuromorphic AI NM500 and Its ADAS Application

Jungyun Kim[(✉)]

School of Mechanical and Automotive Engineering,
Catholic University of Daegu, 13-13, Hayang-ro, Gyeongsan-si,
Gyeongsangbuk-do 38430, Korea
kjungyun@cu.ac.kr

Abstract. This article deals with an ADAS (Advanced Driver Assistance System) application using newly developed neuromorphic artificial intelligent chip NM500. Neuromorphic artificial intelligence is distinguished from other AI by its particular hardware structure and parallel algorithms of learning and recognition. Thus, neurons of NM500 can learn and recognize patterns extracted from any data sources with less energy and complexity than modern microprocessors. The proposed application can control the vehicle speed by recognizing the traffic information images marked on road. We have built a small-scaled vehicle model to discuss the real-time performance as well as hardware implementation with NM500. Taking advantages of NM500, the system simply consists of a low-priced surveillance camera attached in the front windshield of a vehicle and an Arduino kit, which processes the video signal from the camera and speed control signal.

Keywords: NM500 · Neuromorphic chip · Artificial intelligence ·
Edge computing · ADAS application

1 Introduction

Last year, Nepes, a highly reliable semiconductor manufacturer in Korea, has announced the newly developed neuromorphic artificial intelligent chip named NM500 and NeuroShield: a trainable board for IoT and smart appliances featuring NM500 chip, which can interface with Arduino boards or a PC (Fig. 1). NM500 is a hardware artificial intelligence opening new frontiers for smart sensors, IoT (Internet of Things), machine learning and cognitive computing. Its neurons can learn and recognize patterns extracted from any data sources such as images, audio waveform, bio signals, text and more, with less energy and complexity than modern microprocessors.

Furthermore, NM500 is a product that enables Edge computing, an extension of IoT, which is currently spreading throughout the industry. Edge computing is a machine that judges a machine to suit each situation and takes action accordingly [1]. While IoT provides information about each situation to the user and actions taken according to the judgment of the user, Edge computing judges and learns on its own, and judges appropriate situation. In short, each machine is equipped with artificial intelligence.

Neuromorphic computing has emerged in recent years as a complementary architecture to von Neumann systems. The term neuromorphic computing has been come out 1990 by Carver Mead [2]. This biologically inspired approach has created highly connected synthetic neurons and synapses that can be used to model neuroscience theories as well as solve machine-learning problems. The promise of the technology is to create a brain-like ability to learn and adapt, and is notable for being highly connected and parallel, requiring low-power, and collocating memory and processing [3, 4].

Many people think artificial intelligence in terms of software. As a representative example, Alpha Go and Alexa use 'deep learning' technology which learn through experience. These artificial intelligences require a large-capacity server storing data and a supercomputer processing the data. However, neurons of neuromorphic chip can learn and recognize patterns extracted from any data sources with less energy and complexity than modern microprocessors. Thus, it can make decisions based on situations without a large server or supercomputer, and integrate artificial intelligence into various industries.

Fig. 1. NM500 and NeuroShield.

In this paper, we have described an ADAS application using newly developed neuromorphic chip NM500. The proposed application can control the vehicle speed by recognizing the traffic information images marked on road such as crosswalk, school zone, and road-bump, etc. First, we briefly reviewed the neuromorphic artificial intelligence regarding its special features, neural network structure, and distinguished algorithms of learning and recognition. Then the hardware architecture of NM500 was introduced focused on the neuron connections and data flows. Finally, we have built a small-scaled vehicle model to discuss the real-time performance as well as hardware implementation with NM500. The system consists of a camera, a NeuroShield, and an Arduino kit, which processes the video signal from the camera and speed control signal.

2 Neuromorphic Artificial Intelligence

Much of work in neuromorphic computing have been driven by the development of hardware that could perform parallel operations, inspired by observed parallelism in biological brains, but on a single chip. The popular motivations have been inherent parallelism, extremely low power operation, real-time performance, speed in both operation and training, and scalability. Because of their fault tolerance characteristics or reliability in the face of hardware errors, neuromorphic computing has been implemented in neural network-style architectures (*i.e.,* architectures made up of neuron and synapse-like components). Moreover, many neuromorphic systems perform learning tasks in an unsupervised, low-power manner that is called as on-line learning defined as the ability to adapt to changes in a task as they occur.

Neural network supported by NM500 is FFNN (Feed Forward Neural Network) using RBF (radial basis function). It has a 3-layer logical structure with one input layer, one hidden layer, and one output layer (Fig. 2). The hidden layer can be extended to the number of physical neurons supported by NM500. For a single chip, 576 neurons are available and can scale up to several hundred million with the same processing time [5]. Each neuron in hidden layer has 256 bytes RAM and consists synapses. The neurons are capable of ranking similarities between input vectors and the reference patterns they hold in memory, but also reporting conflicting responses or cases of uncertainty, reporting unknown responses or cases of anomaly or novelty.

NM500 calculates two norm values using L_1 (Manhattan) or L_{sup} (Supremum) method as a criterion for determining the similarity between learned data and input data in a neuron. The recognition modes supported by NM500 provide RBF and KNN (k-Nearest Neighbor) algorithms [6, 7]. The neurons can learn and recognize input vectors autonomously and in parallel. If several neurons recognize a pattern, *i.e.* "firing", their responses can be retrieved automatically in increasing order of distance (equivalent to a decreasing order of confidence). The information from a firing neuron includes its distance, category, and neuron identifier. If the response is polled, this data can be consolidated to make a more sophisticated decision weighing the cost of uncertainty or else.

Learning is initiated by simply broadcasting a category, which is a label identifying the data to be learned, after an input pattern. If it represents novelty, the next neuron available in the chain automatically stores the pattern and its category. If some firing neurons recognize the pattern but with a category other than the category to learn, they auto-correct their influence fields. As the NM500 network broadcasts a new input pattern, all the neurons update their distance simultaneously by using parallel bus communication. They are ready to respond to a query as soon as the last component received. The neurons reacting to an input pattern autonomously order themselves per decreasing confidence. This unique feature pertains to the parallel architecture of NM500 network, which allows a winner-takes-all among the reacting neurons.

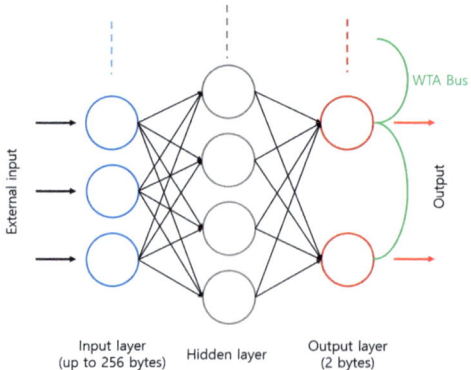

Fig. 2. 3-layer FFNN in NM500.

Table 1. Electrical and I/Os specifications of NM500.

Part	Specification
Clock frequency	37 MHz for single chip
	18 MHz for chain of multiple chips
I/O	Parallel bus (26lines)
Electrical	3.3 V I/O operation, 1.2 V core
Power consumption	< 153mWatt in active mode at 1.2 V and 3.3 V
Package	64-pin CSP 4.5 × 4.5 × 0.5 mm package

3   NM500 Architecture

The NM500 chip is composed of the following modules: neuron interconnect part and chain of neurons: daisy-chained and interconnected (Fig. 3). Inter-module and inter-neuron communications are made through a bi-directional parallel bus. The neuron cell is composed of a memory and a set of six. The detailed specifications of NM500 are listed in Table 1.

A neuron can have three states in the chain: Idle, RTL(Ready-To-Learn) or Committed. It becomes committed as soon as it learns a pattern and its category register is written with a value different from zero. Its DCO(Daisy-Chain-Out) control line automatically rises, changing its status from RTL to Committed. The next neuron in the chain becomes the RTL. It has its DCI(Daisy-Chain-In) high and DCO low. The transfer of the DCI-DCO from one neuron to the next is activated in the same way whether the two consecutive neurons belong to a same cluster or not.

One of the benefits of the NM500 architecture is that one can cascade multiple chips in parallel to expand the size of the neural network by increment of 576 neurons. The behavior of the neurons in a single-chip or multiple-chips configuration remains the same.

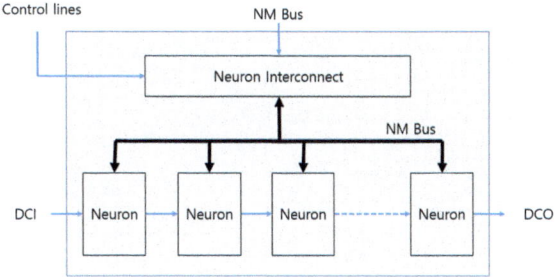

Fig. 3. NM500 architecture.

4 ADAS Application Using Image Learning and Recognition

Among various applications of NM500, we deal with an ADAS (Advanced Driver Assistance System) application using image learning and recognition function. It works like one of conventional ADAS functions such as cruise control or intelligent speed assist system in passenger cars. It can control the vehicle speed by recognizing the traffic information images marked on road such as crosswalk, school zone, and road-bump, *etc.* The proposed system is motivated by making cheaper and simplifying the current system (Fig. 4). Taking advantages of NM500, it consists of a low-priced surveillance camera attached in the front windshield of a vehicle, an NM500, and an Arduino kit, which processes the video signal from the camera and speed control signal.

4.1 Learning and Recognition of Traffic Information Images

There have been a myriad of popular theoretic and practical results on the image-processing field including learning and recognition solutions. In NM500 applications, there also need typical image processing steps like other computer vision applications (Fig. 5) [8–12]. Fortunately, one can carry out these pre-processing, learning, and recognition steps easily by using Knowledge Studio: a novel development tool supplied by Nepes. Though not skilled in a machine learning technology, one can develop various AI applications with NM500 by using the built-in functions only.

Knowledge Studio is a GUI (Graphic User Interface) and Show-and-Tell based development environment program. It can pre-process the data from various sensors and extract the genetic feature of learned data by using selective algorithms of sub-sampling, histogram, and composite profile. And it can decide, *e.g.*, classification, detection, and segmentation by interacting with neurons in NM500. Moreover, there has open-source libraries of Arduino kit, thus it can communicate with external sensors and actuators as like NM500 artificial intelligence works on-line.

In this study, we consider crosswalk and bump signs marked on the road for the traffic information images. First, a video of 150 s has been recorded using a surveillance camera in good driving conditions: bright sunshine weather in the afternoon, average vehicle speed about 40 km/hr. Image pre-processing through the sub-sampling algorithm has been carried out in average gray scale mode with Knowledge Studio

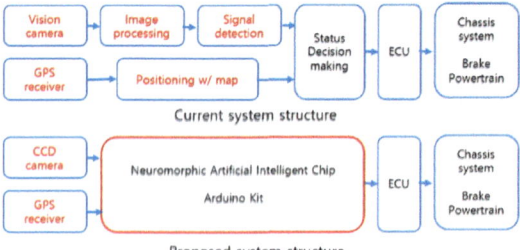

(a) Comparison of the proposed method with conventional ADAS application.

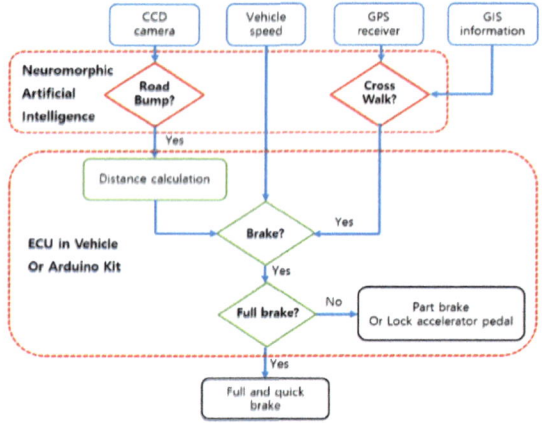

(b) Proposed ADAS algorithm.

Fig. 4. Structure and algorithm of proposed system.

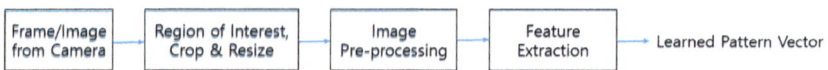

Fig. 5. Image processing steps.

(Fig. 6(a)). The test videos have been recorded with the resolution of 1024 × 768 pixels and frame rates of 30 frames/sec. The ROI (Region of Interest) for learning and recognition of road signs has been set to 270 × 70 pixels on the center of image area.

On the completing the training, only three neurons participate in learning the road signs; one for the road-bump and two for the crosswalk (Fig. 6(b)). Because of their relatively simple features, the learning and recognition of road signs can be accomplished even with a small number of neurons in NM500. Moreover, once new neurons are learned, previously learned neurons coordinate themselves by interacting with each other. Thus if the features of input data are similar to those of previously learned neurons, the learning does not proceed since it can be judged by already learned neurons. The evaluation of recognition performance has been resulted 88% using the test video; 22 signs have been detected successfully from all 25 signs.

 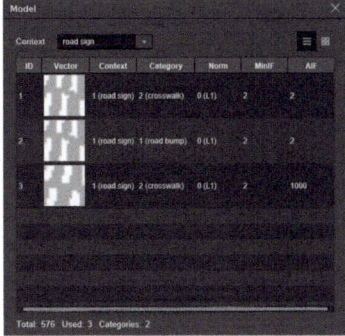

(a) Learning and pre-processing of crosswalk image.   (b) Resulted learning models.

Fig. 6. Learning and recognition with Knowledge Studio.

4.2 Real-Time Performance and Hardware Implementation

In order to discuss whether it is applicable to an ADAS application, the proposed system must guarantee the real-time performance as well as hardware implementation with NM500. Since it is beyond the scope of this study to build a test vehicle, we have made a small-scaled vehicle model with dc motors, which can emulate the vehicle movements.

The vehicle model is designed to validate the learning and recognition performance in real-time. Therefore, it consists of a camera, an image processor, and a NeuroShield (Fig. 7). And a micro-controller (Arduino Mega board) is added to manage the data and control signals with motor driver and NM500. Arduino is an open-source electronics platform based on easy-to-use hardware and software [9]. The camera and image processor are chosen among available Arduino peripherals; the detailed specifications of components are shown in Table 2.

(a) Rear view.   (b) Front view.

Fig. 7. Small-scaled vehicle model with the proposed system.

Table 2. Specifications of the prototype components.

Part	Specification	
Camera	OV2640 2Mega pixels: 1600x1200 UXGA 1/4" sensor size 15~60 fps image transfer rate	
Image processor	3.2 inch TFT LCD for ROI set SPI speed: 8MHz Resolution support: 0.3MP ~ 5MP	
Motor driver	Half-bridged DC motor controller 2 I/O channels Available motor driving voltage: 4.7~24V	
DC motors	5V DC, Installed a reduction gear	
Micro-controller	Arduino ATmega2560 I/O: Digital 54, Analog input 16 Clock speed: 16MHz Flash memory: 256 KB	

The test consists of following procedures. First, NM500 in NeuroShield has learned the road-bump sign by using Knowledge Studio or the embedded camera and image processor (Fig. 8). Only one neuron of NM500 has been found to learn and detect this image throughout the test. While moving on the flat road in constant speed (about 0.6 m/s), the miniature vehicle slows down its speed to 0.2 m/s, immediately on detecting the sign (Fig. 9). From the repeated test results, NM500 finished the learning process and surely detected the road sign in every case even with minimal number of neurons. After recognizing the road-bump image, it took almost 0.5 s to the target speed; it is short enough to guarantee the real-time performance in a real vehicle application (Fig. 10).

(a) Learning with Knowledge Studio.   (b) Learning using embedded camera.

Fig. 8. Learning the road bump sign.

Fig. 9. Miniature test: the miniature vehicle, road, and road-bump.

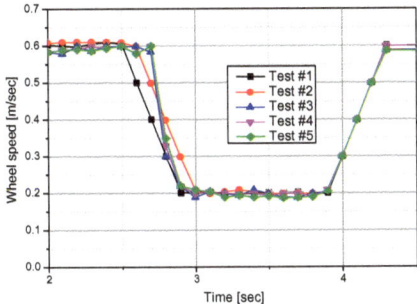

Fig. 10. Test results.

5 Conclusions

In this paper, we have described an ADAS application using newly developed neuromorphic chip NM500. Its neurons can learn and recognize patterns extracted from any data sources with less energy and complexity than modern microprocessors. The proposed application can control the vehicle speed by recognizing the traffic information images marked on road. Taking advantages of NM500, the system consists of a low-priced surveillance camera attached in the front windshield of a vehicle, an NM500, and an Arduino kit, which processes the video signal from the camera and speed control signal. Finally, in order to discuss the real-time performance as well as hardware implementation, we have made a small-scaled vehicle model. From the feasibility test results, NM500 finished the learning process and surely detected the road sign in every case even with minimal number of neurons. The major conclusions of this paper are as follows.

- NM500 is a hardware artificial intelligence and its neurons can learn and recognize patterns extracted from any data sources. Neuromorphic computing has been characterized by inherent parallelism, extremely low power operation, real-time performance, speed in both operation and training, and scalability.

- The proposed ADAS application using image learning and recognition function of NM500 can control the vehicle speed by recognizing the traffic information images marked on road such as crosswalk, school zone, and road-bump, etc. Motivated by making cheaper and simplifying the current system, it consists of a low-priced surveillance camera attached in the front windshield of a vehicle, an NM500, and an Arduino kit.
- In case of learning the crosswalk and bump signs marked on the road in bright sunshine weather, three neurons participate in learning the road signs; one for the road-bump and two for the crosswalk. And the evaluation of recognition performance has been resulted 88% using the test video.
- A small-scaled vehicle model has been built using NM500 and Arduino peripherals in order to discuss the real-time performance as well as hardware implementation. From the feasibility test results, NM500 finished the learning process and surely detected the road sign in every case even with minimal number of neurons.

Acknowledgement. This work was supported by research grants from the Catholic University of Daegu in 2017.

References

1. Ai, Y., Peng, M., Zhang, K.: Edge computing technologies for internet of things: a primer. Digit. Commun. Netw. **4**(2), 77–86 (2018)
2. Mead, C.: Neuromorphic electronic systems. Proc. IEEE **78**(10), 1629–1636 (1990)
3. Schuman, C.D., Potok, T.E., Patton, R.M., Birdwell, D., Dean, M.E., Rose, G.S., Plank, J.S.: A Survey of Neuromorphic Computing and Neural Networks in Hardware, ArXiv: 1705.06963, pp. 1–88 (2017)
4. Gerven, M.: Computational foundations of natural intelligence. Front. Comput. Neurosci. **11**, 112–136 (2017)
5. General Vision, NeuroMem Technology Reference Guide (2018)
6. Chen, S., Cowan, C.F.N., Grant, P.M.: Orthogonal least squares learning algorithm for radial basis function networks. IEEE Trans. Neural Networks **2**(2) (1991)
7. Keller, J.M., Gray, M.R., Givens Jr., J.A.: A fuzzy K-nearest neighbor algorithm. IEEE Trans. Syst. Man Cybern. **4**, 580–585 (1985). SMC-15
8. Sardar, S., Babu, K.A.: Hardware implementation of real-time, high performance, RCE-NN based face recognition system. In: 27th International Conference on VLSI Design and 13th International Conference on Embedded Systems, pp. 174–179 (2014)
9. Sinha, P., Balas, B., Ostrovsky, Y., Russell, R.: Face recognition by humans: nineteen results all computer vision researchers should know about. Proc. IEEE **94**(11) (2006)
10. Sardar, S., Tewari, G., Babu, K.A.: A hardware/software codesign model for face recognition using cognimem neural network chip. In: International Conference on Image Information Processing (ICIIP 2011) (2011)
11. Sui, C., Kwok, N.M., Ren, T.: A restricted coulomb energy (RCE) neural network system for hand image segmentation. In: 2011 Canadian Conference on Computer and Robot Vision (2011)
12. Batur, A.U., Flinchbaugh, B.E.: Performance analysis of face recognition algorithms on tms320c64x, Texas Instruments Application Report-SPRA874, pp. 1–12 (2002)

Analyzing $l1$-loss and $l2$-loss Support Vector Machines Implemented in PERMON Toolbox

Marek Pecha[1,2(✉)] and David Horák[1,2]

[1] Department of Applied Mathematics, VŠB – Technical University of Ostrava, Ostrava, Czech Republic
marek.pecha@vsb.cz
[2] Institute of Geonics CAS, Ostrava, Czech Republic

Abstract. This paper deals with investigating $l1$-loss and $l2$-loss $l2$-regularized Support Vector Machines implemented in PermonSVM – a part of our PERMON toolbox. The loss functions quantify error between predicted and correct classifications of samples in cases of non-perfectly linearly separable classifications. In numerical experiments, we study properties of Hessians related to performance score of models and analyze convergence rate on 4 public available datasets. The Modified Proportioning and Reduced Gradient Projection algorithm is used as a solver for the dual Quadratic Programming problem resulting from Support Vector Machines formulations.

Keywords: Support Vector Machines · SVM · PermonSVM · Hinge loss functions · Quadratic Programming · QP · MPRGP

1 Support Vector Machines for Classifications

In the last two decades, the Support Vector Machines (SVMs) [4], due to their accuracy and obliviousness to dimensionality [15], have become a popular machine learning technique with applications including genetics [3], image processing [7], and weather forecasting [14]. In this paper, we are only interested in SVMs for classification.

SVM is originally designed as a supervised binary classifier, i.e. a classifier that decides whether a sample falls into either Class A (label 1) or Class B (label -1) by means of a model. The model is determined from the already categorised training samples in the training phase of the classifier. Unless otherwise stated, let us assume the training samples are linearly separable. The essential idea of the SVM classifier training is to find the maximal-margin hyperplane that divides the Class A from the Class B samples by the widest possible empty strip, which is called the functional margin. The samples contributing to the definition of such hyperplane are called the support vectors – see the circled samples lying on the dashed hyperplanes depicted in Fig. 1.

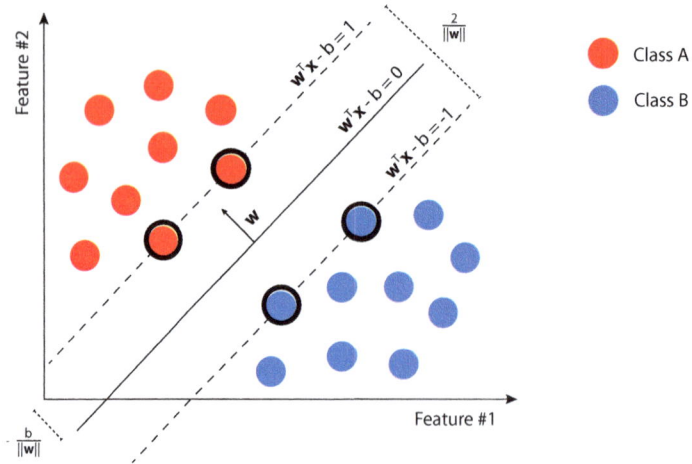

Fig. 1. An example of a two-class classification problem solved by the linear hard-margin SVM.

Let us denote the training samples as a set of ordered pairs such that

$$T := \{(\boldsymbol{x}_1, y_1), (\boldsymbol{x}_2, y_2), \ldots, (\boldsymbol{x}_m, y_m)\},$$

where m is the number of samples, $\boldsymbol{x}_i \in \mathbb{R}^n (n \in \mathbb{N}$ represents a number of attributes) is the i-th sample and $y_i \in \{-1, 1\}$ denotes the label of the i-th sample, $i \in \{1, 2, \ldots, m\}$. Let H be the maximal-margin hyperplane $\boldsymbol{w}^T \boldsymbol{x} - b = 0$, where $\boldsymbol{w}$ is a normal vector; $\frac{b}{\|\boldsymbol{w}\|}$ determines the offset of the hyperplane H from the origin along its normal vector $\boldsymbol{w}$. The problem of finding the hyperplane H can be formulated as a constrained optimization problem in the following hard-margin primal SVM formulation

$$\min_{\boldsymbol{w}, b} \frac{1}{2} \boldsymbol{w}^T \boldsymbol{w} \text{ s.t. } y_i \left(\boldsymbol{w}^T \boldsymbol{x}_i - b \right) \geq 1. \tag{1}$$

For the case of the non-perfectly linearly separable classifications, the soft-margin SVM was designed. We introduce hinge loss function $\xi_i = \max\left(0, \ 1 - y_i \left(\mathbf{w}^T \mathbf{x}_i - b\right)\right), i \in \{1, 2, \ldots, m\}$, which quantifies error between predicted and correct classification of sample $\mathbf{x}_i$. If sample x_i is correctly classified, a value of the hinge loss function equals 0. For the case of a sample misclassification, a value of hinge loss function is distance between the respective hyperplane and a misclassified sample. We can observe that if $0 \leq \xi_i \leq 1$, then the i-th sample lies somewhere between the margin and their respective hyperplane (illustrated in Fig. 2); if $\xi_i > 1$, the i-th sample is misclassified (illustrated in Fig. 3). Using the hinge loss function defined above, we can modify the hard-margin primal SVM formulation (1) into the soft-margin primal SVM formulation as follows

$$\min_{\boldsymbol{w},\, b,\, \xi_i} \frac{1}{2}\boldsymbol{w}^T\boldsymbol{w} + C\sum_{i=1}^{m}\xi_i \quad \text{s.t.} \quad \begin{cases} y_i\left(\boldsymbol{w}^T\boldsymbol{x}_i - b\right) \geq 1 - \xi_i, \\ \xi_i \geq 0, \end{cases} \tag{2}$$

where C is a user-specified penalty that penalizes misclassification error. Higher value of C increases the importance of minimising the hinge loss functions ξ_i and the importance of minimising $\|\boldsymbol{w}\|$ at the expense of satisfying the margin constraint for fewer samples. Formulation (2) is called $l1$-loss $l2$-regularized SVM.

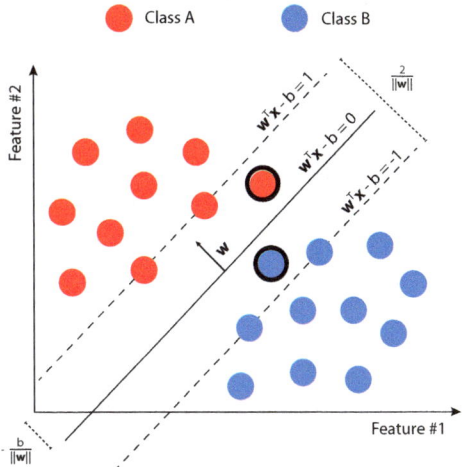

Fig. 2. Soft-margin SVM example: the encircled samples are correctly classified, but they are on the wrong side of their respective hyperplane

The primal formulation of the soft-margin SVM (2) can be modified by exploiting the Lagrange duality with the Lagrange multipliers $\boldsymbol{\alpha} = [\alpha_1, \alpha_2, \ldots, \alpha_m]^T$, $\boldsymbol{\beta} = [\beta_1, \beta_2, \ldots, \beta_m]^T$. Evaluating the Karush-Kuhn-Tucker conditions, the problem results into the dual formulation with inequality (box) and equality constraints [4]

$$\min_{\boldsymbol{\alpha}} \frac{1}{2}\boldsymbol{\alpha}^T\boldsymbol{Y}^T\boldsymbol{K}\boldsymbol{Y}\boldsymbol{\alpha} - \boldsymbol{\alpha}^T\boldsymbol{e} \quad \text{s.t.} \quad \begin{cases} \boldsymbol{o} \leq \boldsymbol{\alpha} \leq C\boldsymbol{e}, \\ \boldsymbol{B}_e\boldsymbol{\alpha} = 0, \end{cases} \tag{3}$$

where $\boldsymbol{e} = [1, 1, \ldots, 1]^T$, $\boldsymbol{o} = [0, 0, \ldots, 0]^T$, $\boldsymbol{X} = [\boldsymbol{x}_1, \boldsymbol{x}_2, \ldots, \boldsymbol{x}_m]$, $\boldsymbol{y} = [y_1, y_2, \ldots, y_m]^T$, $\boldsymbol{Y} = diag(\boldsymbol{y})$, $\boldsymbol{B}_e = [\boldsymbol{y}^T]$, and $\boldsymbol{K} \in \mathbb{R}^{n \times n}$ is symmetric positive semi-definite (SPSD) matrix such that $\boldsymbol{K} := \boldsymbol{X}^T\boldsymbol{X}$ called the Gram matrix or the kernel matrix; the Hessian of (3) is defined as follows $\boldsymbol{H} := \boldsymbol{Y}^T\boldsymbol{X}^T\boldsymbol{X}\boldsymbol{Y}$, which is symmetric SPSD either.

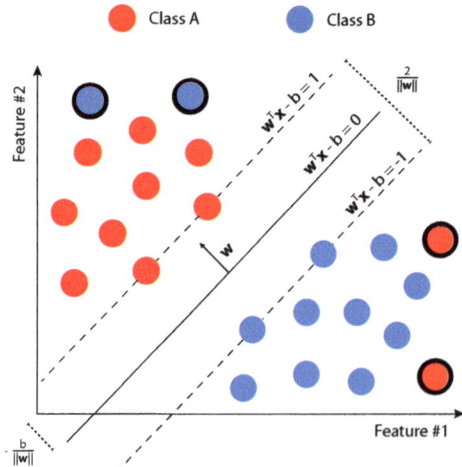

Fig. 3. Soft-margin SVM example: the encircled samples are misclassified.

Further, we introduce dual to primal reconstruction formulas for the normal vector

$$w = XY\alpha, \qquad (4)$$

and the bias

$$b = \frac{1}{|I^{SV}|} \sum_{i \in I^{SV}} \left(x_i^T w - y_i \right), \qquad (5)$$

where I^{SV} denotes the support vector index set, i.e. $I^{SV} := \{i \mid 0 < \alpha_i < C, \, i = 1, 2, \ldots, m\}$, and $|I^{SV}|$ is the cardinality of I^{SV}. From the normal vector w and bias b, we can easily set up the decision rule

$$\operatorname{sgn}\left(w^T x_i + b\right) = \begin{cases} +1 \ldots x_i \text{ belongs to Class A,} \\ -1 \ldots x_i \text{ belongs to Class B.} \end{cases} \qquad (6)$$

2 Hessian Regularization

In previous $l1$-loss $l2$-regularized SVM formulation (3), the Hessian matrix is SPSD. Instead of linear sum of the loss functions ξ_i, we can use the square sum of the loss functions in the objective function as follows

$$\min_{w,\, b,\, \xi_i} \frac{1}{2}\|w\|^2 + \frac{C}{2}\sum_{i=1}^{n}\xi_i^2 \text{ s.t. } \begin{cases} y_i\left(w^T x_i + b\right) \geq 1 - \xi_i, i \in \{1, 2, \ldots, n\}, \\ \xi_i \geq 0, i \in \{1, 2, \ldots, n\}. \end{cases} \qquad (7)$$

The formulation (7) is called $l2$-loss $l2$-regularized SVM. As for $l1$-loss $l2$-regularization SVM, we derive dual formulation by the Lagrange duality and, evaluating the Karush-Kuhn-Tucker conditions, the primal formulation (7) results into the dual formulation as follows

$$\min_{\alpha} \frac{1}{2}\alpha^T \left(H + C^{-1}I\right)\alpha - \alpha^T e \text{ s.t. } \begin{cases} 0 \leq \alpha, \\ B_e \alpha = 0. \end{cases} \quad (8)$$

Since the Hessian is regularized by matrix $C^{-1}I$, it becomes symmetric positive definite (SPD). Mathematically, the associated optimization problem would be more computationally stable than in a case of the $l1$-loss $l2$-regularized SVM.

3 No-Bias Data Classifications

In high dimensional space, we do not need the bias term in the primal formulation for sparse data [11], therefore the equality constraints vanished in the dual formulations. Mathematically, the soft-margin problems (both $l1$-loss and $l2$-loss $l2$-regularized SVM) reduce into solving rotation of the separating hyperplane that best fits the classification problems. Therefore, $l1$-loss and $l2$-loss functions become $\max\left(0, 1 - y_i w^T x_i\right)$ and $\max\left(0, 1 - y_i w^T x_i\right)^2$, respectively. For these loss functions, the primal formulations (2) and (7) take following forms

$$\min_{w,\,b,\,\xi_i} \frac{1}{2}\|w\|^2 + C\sum_{i=1}^{n} \xi_i \text{ s.t. } \begin{cases} y_i \left(w^T x_i\right) \geq 1 - \xi_i, i \in \{1,2,\ldots,n\}, \\ \xi_i \geq 0, i \in \{1,2,\ldots,n\}, \end{cases} \quad (9)$$

$$\min_{w,\,b,\,\xi_i} \frac{1}{2}\|w\|^2 + \frac{C}{2}\sum_{i=1}^{n} \xi_i^2 \text{ s.t. } \begin{cases} y_i \left(w^T x_i\right) \geq 1 - \xi_i, i \in \{1,2,\ldots,n\}, \\ \xi_i \geq 0, i \in \{1,2,\ldots,n\}, \end{cases} \quad (10)$$

which are called the no-bias formulations. Instead of the duals with the box or bound and equality constraints (3), (8), we solve following optimization problems with box or bound constraints

$$\min_{\alpha} \frac{1}{2}\alpha^T H\alpha - \alpha^T e \text{ s.t. } 0 \leq \alpha \leq C, \quad (11)$$

$$\min_{\alpha} \frac{1}{2}\alpha^T \left(H + C^{-1}I\right)\alpha - \alpha^T e \text{ s.t. } 0 \leq \alpha, \quad (12)$$

respectively. However, the no-bias formulations may be a cause of poor classification results in some applications. Solving the problem with equality constraint should be computationally expensive or Quadratic Programming (QP) solver diverges in some cases for large dimensional data. Alternatively, we can increase the dimension of samples by regularizing the bias term. Standard approach is modification of the data by appending each sample with an additional feature. Let us denote $\widetilde{\boldsymbol{x}}_i \in \mathbb{R}^{n+1}$ as the modified sample $\mathbf{x}_i$ in the following way $\mathbf{x}_i \leftarrow [\mathbf{x}_i, c]$, where $c \in \mathbb{R}^+$. In many classification problems, the value of c is typically set to 1 [11]. Further, let $\widetilde{\boldsymbol{w}} := [w_1, w_2, \ldots, w_n, w_{n+1}]$ be a hyperplane normal that solves the classification problem in the higher space. Let us write the separating hyperplane equation in a component-wise form such that

$$\underbrace{w_1 x_1 + w_2 x_2 + \cdots + w_m x_m}_{= \boldsymbol{w}^T \boldsymbol{x}} + \underbrace{1 w_{m+1}}_{=:b} = 0. \tag{13}$$

From (13), we can easily see the hyperplanes resulting from classification problem for modified samples $\widetilde{\boldsymbol{x}}_i \in \mathbb{R}^{n+1}, i \in \{1, 2 \ldots, m\}$ are equivalent to hyperplanes related to classification problems with the bias formulations (3), (7) in space $\mathbb{R}^n$.

4 PermonSVM: SVM Implementation on Top of PETSc

PermonSVM [8, 12] is a part of the PERMON toolbox [10]. PermonSVM is build on top of PETSc [2] and PermonQP [9]. PermonQP is a PETSc based package for the solution of large scale QP problems.

PermonSVM provides an implementations of the linear binary classification via soft-margin SVM. In the training procedure, PermonSVM takes advantage of an implicit representation of the Gram matrix (i.e. the matrix product $\boldsymbol{X}^T \boldsymbol{X}$ is not assembled), which saves memory and CPU time during Hessian assembling. The resulting QP problem with an implicit Hessian matrix is solved by the QP solver implemented in the PermonQP package.

Additional features include: model scores (accuracy, sensitivity, specifity, F_1, and confusion matrix), stratified cross validation + grid search, user defined penalty for unbalanced datasets ($C+$ a $C-$), probabilistic SVM related to logistic regression on top of soft margin SVM, bias and no-bias formulations, LIBSVM and HDF5 file loaders.

PermonSVM provides an executable for SVM classification as well as C API designed to be PETSc-like. Its typical usage is presented in Code 1.

```c
MPI_comm     comm = PETSC_COMM_WORLD;
SVM          svm;
PetscViewer  viewer;

Mat          Xt_test,Y_pred,Y_pred;

/*PETSC_DEFAULT means find best C by means grid-search and cross-validation*/
PetscReal  C = PETSC_DEFAULT;
PetscReal *params = NULL;

char       file_training[PETSC_MAX_PATH_LEN] = "examples/heart_scale.h5";
char       file_test[PETSC_MAX_PATH_LEN] = "examples/heart_scale.t.h5";

TRY( SVMCreate(comm,&svm) );
TRY( SVMSetType(svm,SVM_BINARY) );
TRY( SVMSetFromOptions(svm) );

TRY( PetscViewerHDF5Open(comm,file_training,FILE_MODE_READ,&viewer) );
TRY( SVMLoadTrainingDataset(svm,viewer) );
TRY( PetscViewerDestroy(&viewer) );

TRY( PetscViewerHDF5Open(comm,file_test,FILE_MODE_READ,&viewer) );
TRY( SVMLoadTestDataset(svm,viewer) );
TRY( PetscViewerDestroy(&viewer) );

TRY( SVMSetC(svm,C) );
/* Run grid search and cross validation */
TRY( SVMFindBestParams(svm,&params) );
TRY( SVMSetC(svm,params[0]) );

TRY( SVMTrain(svm) );
TRY( SVMGetTestDataset(svm,&Xt_test,&Y_test) );
TRY( SVMClassify(svm,Xt_test,Y_test,&Y_pred,NULL) );
```

Code 1. Calling PermonSVM API.

5 Numerical Experiments

In this section, we analyze $l1$-loss and $l2$-loss functions related to the no-bias classifications on 4 public available datasets, namely Australian, Diabetes, Heart, Ionosphere, downloaded from LIBSVM dataset webpages [1]. The Australian dataset (Australian Credit Approval) concerns credit card applications. The objective of the Diabetes is to diagnostically predict whether or not a patient has diabetes. The Heart dataset is based on data from the Cleveland Clinic Foundation and samples belonging to two classes: the presence or absence of heart disease. Ionosphere dataset contains radar data collected by a system in Goose Bay, Labrador. The targets determines if some structures are or not in the ionosphere (Table 1).

We evaluate performances of models on test datasets by means of accuracy and F_1 scores. The accuracy is defined as

$$\text{accuracy} := \frac{TP+TN}{TP+FP+FN+TN} * 100 \ [\%], \qquad (14)$$

Table 1. Australian, Diabetes, Heart, Ionosphere: the training and test descriptions of datasets.

Dataset	#samples	#samples+	#samples-	#features
Australian (training)	462	216	246	14
Australian (test)	228	91	137	
Diabetes (training)	514	332	182	8
Diabetes (test)	254	168	86	
Heart (training)	180	84	96	13
Heart (test)	90	36	54	
Ionosphere (training)	235	154	81	34
Ionosphere (test)	116	71	45	

where TP, FP, FN, TN are true positive, false positive, false negative and true negative samples, respectively, which are:

- TP samples are such samples, which are labeled as Class A (actual class) and predicted as Class A,
- FP are cases when the actual class of sample is Class A and predicted as Class B, i.e type I error,
- FN means the sample is labeled as Class B and predicted as Class A, i.e. type II error,
- TN sample is when sample is labeled as Class B and predicted as Class B.

$$F_1 = \frac{2}{\frac{1}{\text{precision}} + \frac{1}{\text{sensitivity}}} = 2 * \frac{\text{precision} * \text{sensitivity}}{\text{precision} + \text{sensitivity}} \quad (15)$$

where precision $:= \frac{TP}{TP+FP}$ and sensitivity $:= \frac{TP}{TP+FN}$. Mathematically, F_1 score is harmonic mean of the precision and sensitivity.

All initial guess components are set to $0.99 * C$, the relative norm of projected gradient being smaller than $1e-1$ [13] is used as stopping criterion for the MPRGP (Modified Proportioning and Reduced Gradient Projection) [6] algorithm. The expansion step-size is fixed and determined such as $\alpha = 2.0/\|\boldsymbol{H}\|_2$, where $\|\boldsymbol{H}\|_2 = \sqrt{\lambda_{max}(\boldsymbol{H}^T\boldsymbol{H})}$. The values of penalty C are chosen from the set $\{1.0, 5.0, 10.0\}$. We would like to note, the number of proportioning steps were 0 or 1, therefore they are not presented in tables.

Looking at Tables 2 and 3, the performance scores of models are slightly higher in cases of $l1$-loss for the Diabetes, Heart, and Ionosphere datasets – summarized in Table 4. For the Australian dataset, difference of accuracy between $l2$-loss and $l1$-loss SVM goes from 0.93% to 2.36% and F_1 from 0.03 to 0.04.

Table 2. $l1$-loss $l2$-regularized SVM: Comparison of the number of iterations, CG steps, expansion steps, and Hessian multiplications obtained after solver converged and evaluating performance scores of models.

Dataset	C	Hessian mult.	CG steps	Exp. steps	Accuracy [%]	F_1
Australian	1.0	137	1	67	84.65	0.83
Australian	5.0	212	29	90	84.65	0.83
Australian	10.0	251	41	103	84.65	0.83
Diabetes	1.0	365	1	181	73.62	0.80
Diabetes	5.0	563	38	261	73.23	0.80
Diabetes	10.0	680	107	285	73.23	0.80
Heart	1.0	130	0	64	80.00	0.74
Heart	5.0	273	69	99	83.33	0.78
Heart	10.0	327	98	112	82.22	0.76
Ionosphere	1.0	262	10	124	81.90	0.87
Ionosphere	5.0	321	47	135	82.76	0.87
Ionosphere	10.0	392	91	148	81.90	0.87

The theory described by Dostál et al. [5] guarantees the convergence of MPRGP for both SPSD and SPD matrices. We observe the convergence rate is slower for all tested datasets in case of SPSD Hessian, it corresponds to our remark about computation stability of the $l2$-loss $l2$-regularized SVM.

Table 3. $l2$-loss $l2$-regularized SVM: Comparison of the number of iterations, CG steps, expansion steps, and Hessian multiplication obtained after solver converged and evaluating performance scores of models.

Dataset	C	Hessian mult.	CG steps	Exp. steps	Accuracy [%]	F_1
Australian	1.0	99	14	42	87.01	0.87
Australian	5.0	136	27	54	85.58	0.86
Australian	10.0	133	26	53	86.15	0.86
Diabetes	1.0	68	3	32	72.83	0.79
Diabetes	5.0	79	8	35	72.44	0.79
Diabetes	10.0	83	10	36	72.83	0.80
Heart	1.0	32	1	15	77.78	0.71
Heart	5.0	36	5	15	77.78	0.71
Heart	10.0	38	7	15	80.00	0.74
Ionosphere	1.0	133	26	53	77.59	0.84
Ionosphere	5.0	204	39	82	75.86	0.83
Ionosphere	10.0	220	49	85	76.72	0.83

Table 4. Diabetes, Heart, Ionosphere: Comparing differences of model performance scores between $l2$-loss and $l1$-loss $l2$-regularized SVM.

	Minimum	Maximum	Mean	Median
Accuracy [%]	0.4	6.9	3.15	2.22
F_1	0.00	0.07	0.03	0.03

The maximum value of Hessian multiplication speed up is 8.61, minimum, mean, and median are 1.38, 4,25, 3.01, respectively. The number of expansion steps is approximately 3 to 5 times higher than CG steps for $l1$-loss and $l2$-loss, either. Standard implementation of the expansion step is more expensive than CG step, it needs one more Hessian multiplication, therefore reduction of expansion steps is required. One idea would be using non-fixed expansion step-size, i.e. the adaptive expansion step-size.

It seems, the $l1$-loss function in the SVM problem formulation is more robust (catching up outliers during training phase) than $l2$-loss, however $l1$-loss formulation produces SPSD Hessian in dual causing slower convergence rate. Therefore, we propose to use real power of the loss function for our future work. The exponent would be somewhere between 1 and 2. We assume, this approach regularizes Hessian to be SPD and should keep robustness of $l1$-loss approach.

6 Conclusions

In this paper, we analyzed $l1$-loss and $l2$-loss functions related to the no-bias SVM classifications implemented in PermonSVM. We benchmarked our implementation on 4 public available datasets, namely Australian, Diabetes, Heart, and Ionosphere. We evaluate performance score of models by accuracy and F_1 scores. For all tests, the MPRGP algorithm was used as a solver for the QP problem arising from the SVM dual formulations. We observe, the $l1$-loss is more robust than $l2$-loss, however, the formulation produces SPSD Hessian in the dual, which causes a slower convergence rate than in case of $l2$-loss. Therefore, for our related work, we assume using real exponent of the loss function between 1 and 2, which regularizes Hessian in the dual and should keep robustness of $l1$-loss. Further, we propose to use the adaptive expansion step-size to reduce the number of the expansion steps for $l1$-loss and $l2$-loss either.

Acknowledgments. This work has been supported by The Ministry of Education, Youth and Sports from the National Programme of Sustainability (NPU II) project IT4Innovations excellence in science - LQ1602; by Grant of SGS No. SP2018/165, VŠB - Technical University of Ostrava, Czech Republic and by the grant of the Czech Science Foundation (GACR) project no. GA17-22615S. The work has been also performed under the Project HPC-EUROPA3 (INFRAIA-2016-1-730897), with the support of the EC Research Innovation Action under the H2020 Programme; in particular, the author gratefully acknowledges the support of School of Mathematics, The University of Edinburgh, United Kingdom and the computer resources and technical support provided by Edinburgh Parallel Computing Centre (EPCC).

References

1. Libsvm data: Classification (binary class). https://www.csie.ntu.edu.tw/~cjlin/libsvmtools/datasets/binary.html
2. Balay, S., Abhyankar, S., Adams, M.F., Brown, J., Brune, P., Buschelman, K., Eijkhout, V., Gropp, W.D., Kaushik, D., Knepley, M.G., McInnes, L.C., Rupp, K., Smith, B.F., Zhang, H.: PETSc – Portable, Extensible Toolkit for Scientific Computation. http://www.mcs.anl.gov/petsc
3. Brown, M., Grundy, W., Lin, D., Cristianini, N., Sugnet, C., Furey, T., Ares Jr., M., Haussler, D.: Knowledge-based analysis of microarray gene expression data by using support vector machines. Proc. Nat. Acad. Sci. U.S.A. **97**(1), 262–267 (2000)
4. Cortes, C., Vapnik, V.: Support-vector networks. Mach. Learn. **20**(3), 273–297 (1995)
5. Dostál, Z., Pospíšil, L.: Minimizing quadratic functions with semidefinite hessian subject to bound constraints. Comput. Math. Appl. **70**(8), 2014–2028 (2015)
6. Dostál, Z.: Optimal Quadratic Programming Algorithms, with Applications to Variational Inequalities, vol. 23. SOIA. Springer, New York (2009)
7. Foody, G.M., Mathur, A.: The use of small training sets containing mixed pixels for accurate hard image classification: training on mixed spectral responses for classification by a SVM. Remote Sens. Environ. **103**(2), 179–189 (2006)
8. Hapla, V., Horák, D., Pecha, M.: PermonSVM (2017). http://permon.vsb.cz/permonsvm.htm
9. Hapla, V., Horák, D., Čermák, M., Kružík, J., Pospíšil, L., Sojka, R.: PermonQP (2015). http://permon.it4i.cz/qp/
10. Hapla, V., et al.: PERMON (Parallel, Efficient, Robust, Modular, Object-oriented, Numerical). http://permon.vsb.cz
11. Hsieh, C.J., Chang, K.W., Lin, C.J., Keerthi, S.S., Sundararajan, S.: A dual coordinate descent method for large-scale linear SVM. In: Proceedings of the 25th International Conference on Machine Learning, pp. 408–415. ACM (2008)
12. Kružík, J., Pecha, M., Hapla, V., Horák, D., Čermák, M.: Investigating Convergence of Linear SVM Implemented in PermonSVM Employing MPRGP Algorithm. Lecture Notes in Computer Science (including subseries Lecture Notes in Artificial Intelligence and Lecture Notes in Bioinformatics), vol. 11087 LNCS, pp. 115–129 (2018)
13. Pecha, M., Hapla, V., Horák, D., Čermák, M.: Notes on the preliminary results of a linear two-class classifier in the permon toolbox. In: AIP Conference Proceedings, vol. 1978 (2018)
14. Shi, J., Lee, W.J., Liu, Y., Yang, Y., Wang, P.: Forecasting power output of photovoltaic systems based on weather classification and support vector machines. IEEE Trans. Ind. Appl. **48**(3), 1064–1069 (2012)
15. Vishnu, A., Narasimhan, J., Holder, L., Kerbyson, D., Hoisie, A.: Fast and accurate support vector machines on large scale systems. In: 2015 IEEE International Conference on Cluster Computing, pp. 110–119, September 2015

Hybrid Fuzzy Neural Model Based Dempster-Shafer System for Processing of Diagnostic Information

Alexander I. Dolgiy[1], Sergey M. Kovalev[1], Andrey V. Sukhanov[1,2(✉)], and Vitezslav Styskala[2]

[1] JSC "NIIAS", Rostov Branch, Rostov-on-Don, Russia
a.suhanov@rfniias.ru
[2] VSB - Technical University of Ostrava, Ostrava, Czech Republic

Abstract. The paper presents a technique of data fusion obtained from the various sensors. The technique is proposed as important in reliable control over state of technical objects. It is based on adaptive models, which are combinations of fuzzy neural systems and network models implementing Dempster-Shafer computational methodology of evidence combining. The structure of fuzzy neural network is presented to compute basic belief assignments. The structure of hybrid network model based on synthesis of the fuzzy neural system and adaptive net of evidence combining is proposed. It is shown that presented idea provides probabilistic inference based on multi-sensor data even when belief assignments are missed. This advantage is achieved due to experimental based training. As well, linear convergence is proved for training.

Keywords: Dempster-Shafer theory · Fuzzy neural inference · Technical objects control · Hybrid classification models

1 Introduction

Automation of control and diagnostics over technical object is one of the key ways for increasing of its reliability and survivability, during which the tasks of fault detection and technical state prediction are decided utilizing various information obtained from different sources. Increasing number of sensors' types giving heterogeneous diagnostic data leads to necessity of the implementation of data fusion techniques. Data fusion is considered as merging, aggregation, integration, union and splicing of data under processing [1–4].

Sensor data fusion (or sensor fusion) is the sample of synergistic phenomena, in result of which data merging brings additional information than simple sum of these data [1]. The advantages of sensor fusion are stable system reliability when sensor falls, accuracy increasing of fault prediction and detection, reduction of false alarm rate due to more complete information resulted from various sensors.

Sensor fusion is usually used in integrated systems for security and monitoring over technical objects, for control over complex technological processes, for remote sensing, etc.

In sensor fusion area, the decision of complex tasks requires interdisciplinary approach, which includes integration of various algorithms and techniques. Dempster - Shaffer theory (DST) is one of the most efficient tools for data modeling and processing in uncertainty conditions [5]. The most promising thing seems to apply data fusion based on DST together with adaptive neural models. Neural models implement the possibility of experimental data training for DST, and DST implements fuzzy reasoning processing in strict mathematical descriptions for neural models.

The main objects and techniques of DST are belief assignments to hypotheses, techniques of their computation and techniques of their combining based on evident sets. However, in practice, there are serious problems with assignment computations because of incompleteness of initial information about modeled objects. Nevertheless there is an opportunity of assignment for part of hypotheses, where expert information or statistical one are available. To share this part for the whole hypothesis set, the appropriate approaches are required.

The considered approach is based on parameter identification for DST models on the basis of statistical and empirical information obtained from human experts. To achieve this aim, fuzzy neural system is combined with network implementation of Dempster-Shaffer theory.

2 Elements of Dempster-Shaffer Theory

The hypotheses set characterizing decision making situation is matched to belief interval, which shows belief degree for each hypothesis. Belief degree $Bel(P)$ is 0, if hypothesis P has no evidences, and 1 in opposite case. Plausibility degree $Pl(P)$ is defined via belief degree:

$$Pl(P) = 1 - Bel(not(P)) \qquad (1)$$

Let Ω be the universal hypotheses set and 2^{Ω} be the set of all subsets from Ω (power set). In DST, key notion is probability mass (or mass, Basic Belief Assignment, BBA) marked by $m(A)$ for element A from power set. It expresses the relation between all relevant and accessible evidences, which support the statement that hypothesis from Ω belongs to A. It should be noted that the value of $m(A)$ is related to set A and brings no additional information about other subsets of A (every subset has its own BBA).

According to assigned BBAs, thresholds can be given for interval of possibilities as belief and plausibility on set A. The first threshold is defined as sum of BBAs for all subsets of a considered set:

$$Bel(A) = \sum_{B \subseteq A} m(B). \qquad (2)$$

The second one is sum of BBAs for all subsets having at least one joint element with A:

$$Pl(A) = \sum_{B \cap A \neq \emptyset} m(B). \tag{3}$$

The main statement of DST is the combining rule (or evidence combining). Initial Dempster's rule of combination [5] is the generalization of Bayesian rule. Combination (so-called associated BBA) is computed for A as addition of BBAs for those two sets, which give A in their intersection:

$$\begin{array}{l} m_{1,2}(\emptyset) = 0, \\ m_{1,2}(A) = \frac{1}{1-K} \sum_{B \cap C = A \neq \emptyset} m_1(B) m_2(C), \end{array} \tag{4}$$

where K is the normalizing parameter characterizing the degree of conflict between two BBA:

$$K = \sum_{B \cap C = \emptyset} m_1(B) m_2(C). \tag{5}$$

3 Practical Task Specifications

Technical state of many railway objects (eg. railway automation and remote control, monitoring devices, etc.) is characterized by the set of states reflecting various degrees of their efficiency [6]. Railway staff maintaining these objects always check whether it is in serviceable or in fault condition. Another interesting object state is intermediate condition (or pre-failure). Each of the mentioned states is considered as some hypothesis (or state hypothesis).

State hypothesizing is made analyzing the set of parameters $X_1, X_2, \ldots, X_n$ (defined on some numerical scale) under control. Let X be the numerical parameter characterizing the state of controlled object. According to three possible conditions, the scale of X can be divided by three fuzzy intervals $\alpha_1, \alpha_2, \alpha_3$, which are matched to hypotheses in such manner that membership function (MF) $\mu_{\alpha_i}(x)$ characterizes the belief of hypothesis a_i, i.e. $m(a_i) = \mu_{\alpha_i}(x)$. $\alpha_1, \alpha_2, \alpha_3$ are called elementary hypotheses (or evidences), which are combined into complex hypotheses characterizing complex states of controlled object. Therefore, $2^3 = 8$ hypotheses can be made. BBA of complex hypotheses can be computed as conjunction of MS of the corresponding elementary hypotheses:

$$m(\alpha_{i_1}, \ldots, \alpha_{i_k}) = \bigwedge_{j=1}^{k} \mu_{\alpha_{i_j}}(x) \tag{6}$$

where $\wedge$ is the fuzzy conjunction computed as T-norm.

Therefore, $\mu_{\alpha_{i_j}}(x)$ divides numerical scale X by fuzzy intervals of efficiency degree. MF must be limited be the set of limits characterized by specifications of the considered task. Firstly, fuzzy sets $\alpha_1, \alpha_2, \alpha_3$ must form the ordered sequence of fuzzy sets corresponding to the natural order of technical object's states. Particularly, intermediate state α_2 is obtained after serviceable one (α_1).

Then, α_3 follows. Secondly, railway objects cannot be in $[\alpha_1, \alpha_3]$ state simultaneously because of railway specifics. However, it can be in other complex states. Based on these limitations, hypotheses focal set is $\{\alpha_1, \alpha_2, \alpha_3, [\alpha_1, \alpha_2], [\alpha_2, \alpha_3]\}$.

Formally, the above mentioned limits for the fuzzy division can be set as follows:

$$\forall \alpha_i \exists x' \Big(\big(x < x' \to \mu_{\alpha_{i-1}}(x) > \mu_{\alpha_i}(x) \big) \wedge \big(x > x' \to \mu_{\alpha_{i-1}(x)} < \mu_{\alpha_i}(x) \big) \Big) \quad (7)$$

$$\forall x \Big(\big(\mu_{\alpha_1}(x) \neq 0 \to \mu_{\alpha_3}(x) = 0 \big) \wedge \big((\mu_{\alpha_3}(x) \neq 0 \to \mu_{\alpha_1}(x) = 0) \big) \Big) \quad (8)$$

4 Adaptive Fuzzy Dempster-Shaffer Model

Situation of decision making is characterized by the certain value of $x \in X$. This value can be considered as experimental one, which is used for belief assignments to the state of controlled object. Here, BBA of elementary hypothesis is computed based on natural assumption that belief of α_i is as higher as x is closer to the supremum of the corresponding hypothesis on fuzzy interval α_i. It is expressed numerically via

$$\mu_{\alpha_i}(x, \vec{a_i}),$$

where $\vec{a_i}$ is the vector of parameters characterizing the form of MF.

It should be noted that $\mu_{\alpha_i}(x, \vec{a_i})$ simultaneously equals to $Bel(\alpha_i)$. $Pl(\alpha_i)$ is computed by the joining with BBAs of adjacent complex hypotheses $[\alpha_{i-1}, \alpha_i]$ and $[\alpha_i, \alpha_{i+1}]$. BBAs of complex hypotheses are calculated by multiplication of BBAs of the belonging elementary hypotheses. Multiplication is provided by fuzzy conjunction in form of T-norm. It is efficient to use parametric representation of T-norm, which makes the conjunction more flexible and gives adaptiveness to the model [7]:

$$T(\mu_1, \mu_2, c) = (\mu_1 \cdot \mu_2)(1 - c) + (\mu_1 + \mu_2)c,$$

where c is the disjunction parameter.

Therefore, the following equation system is given:

$$m(\alpha_i|x) = \mu_{\alpha_i}(x, \vec{a_i})(i = 1, 2, 3) \quad (9)$$

$$m([\alpha_i, \alpha_{i+1}]|x) = T(\mu_{\alpha_i}(x, \vec{a_i}), \mu_{\alpha_{i+1}}(x, \vec{a}_{i+1}), c)(i = 1, 2). \quad (10)$$

In practice, it is not allowed to priorly define both the form of MF for fuzzy divisions and the combination way for belief assignments, when complex BBA is computed. Adaptive DST models with experimentally adjusted parameters can be used to solve this problem and increase the efficiency of DST implementation.

To achieve this, fuzzy neural network (FNN) is used to compute belief assignment to hypotheses. It is presented in Fig. 1.

The FNN has some similarity with adaptive neural fuzzy models of Mamdani type [8]. It has 2 layers of neurons (L_1 and L_2). Neurons of L_1 compute $m(a_i)$

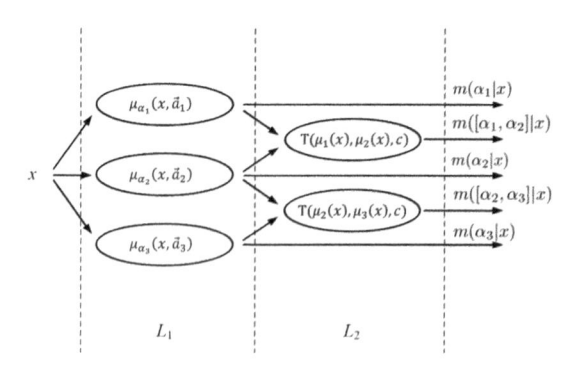

Fig. 1. Fuzzy neural network for BBA computing

for elementary hypotheses based on (9). Neurons of L_2 compute $m(\alpha_1, \alpha_2)$ and $m(\alpha_2, \alpha_3)$ for complex hypotheses based on (10). Adaptive parameters of the network are $\vec{\alpha_i}$ from $\mu_{\alpha_i}(x, \vec{\alpha_i})$ and c from $T(\mu_{\alpha_{i-1}}, \mu_{\alpha_i}, c)$. Training of the FNN is based on experimental data of the following form:

$$t = \Big[x, m(\alpha_1|x), m(\alpha_2|x), m(\alpha_3|x), m([\alpha_1, \alpha_2]|x), m([\alpha_2, \alpha_3]|x)\Big].$$

The first element of t is experimentally obtained value of x and the following ones are the real belief assignments to α_i, when x is obtained. Training algorithm is based on the backpropagation procedure.

5 Hybrid Fuzzy Neural Model Based on Evidence Theory

Above mentioned discussion refers to the analysis of one attribute. In this section, it is shown how the universal value is built for applications with many attributes.

Let $X_1, X_2, \ldots, X_n$ be the set of attributes describing technical state of controlled object. Let these attributes be independent variables. The mapping of evidences for each variable into the universal value is realized by their combination using (4).

It should be noted that every above mentioned rule is usable for decision making using criterion of "fault minimization" (or maximization of the belief of fault detection), because they leads to the same results.

To combine many evidences, the hybrid fuzzy neural network (HFNN) is considered in this work. HFNN is based on idea presented in [9]. HFNN is built by changing the form of neural network classifier based on conditions of the considered problem.

HFNN is obtained based on cascading of one presented in Fig. 1 and network implementing combination of belief assignment using DST. As a result, input part of HFNN provides fuzzy inference of BBA and the output one provides combination of particular belief assignment of hypotheses into vector of universal values according to DS. Unlike existing approach [9], HFNN computes belief assignments for complete hypotheses set taking into account limitations from (7) and (8).

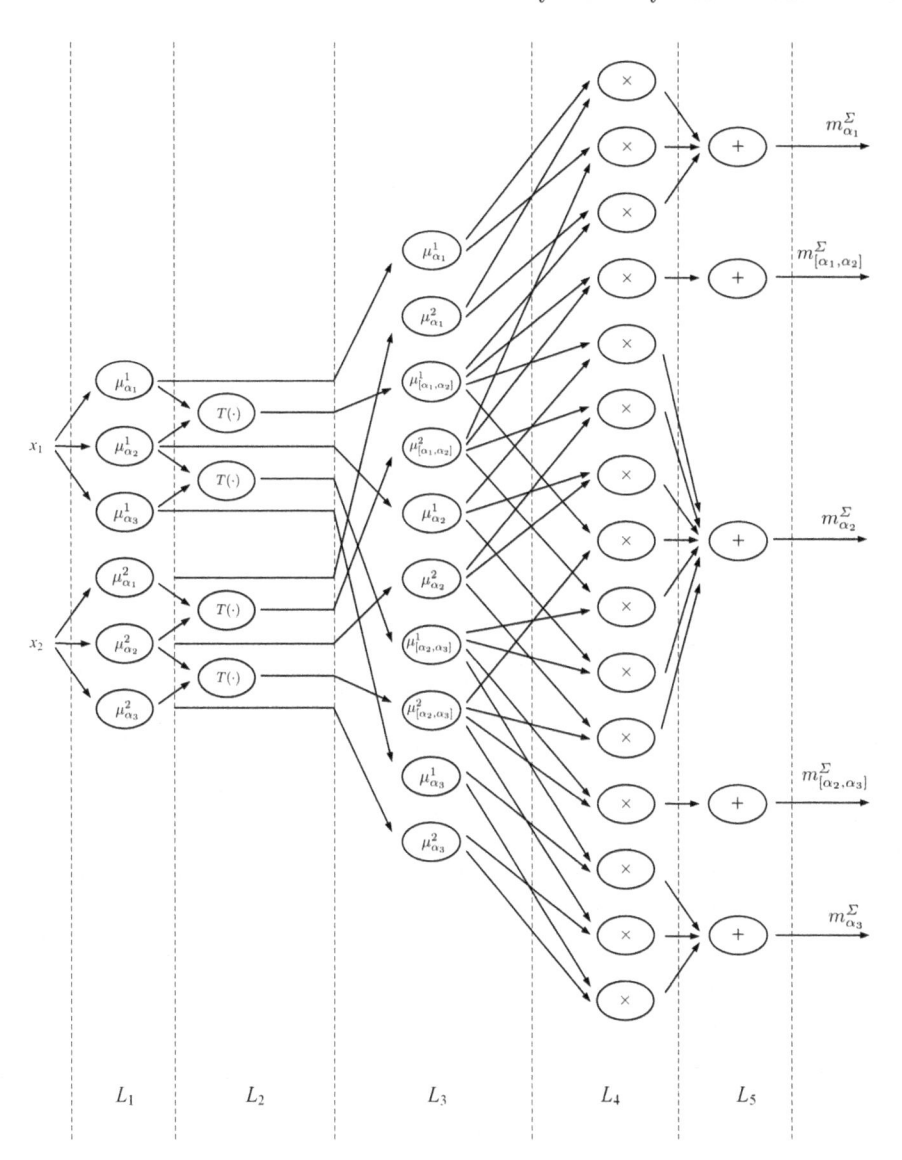

Fig. 2. Fuzzy neural network for combination of two evidences into universal value

HFNN is presented for two attributes (X_1 and X_2) in Fig. 2.

The first two layers L_1 and L_2 calculate the BBA of hypotheses for X_1 and X_2. The BBAs are inputs for L_3, which performs them for the following evidences combining. It should be noted that L_3 is used only for visual comprehension of the HFNN structure. Layer L_4 consists of neurons multiplying BBAs of X_1 and X_2. Layer L_5 summarizes beliefs of combined hypotheses according to DST and outputs the following vector:

$$\vec{m}^{\Sigma} = \left[m^{\Sigma}(\alpha_1), m^{\Sigma}(\alpha_2), m^{\Sigma}(\alpha_3), m^{\Sigma}([\alpha_1,\alpha_2]), m^{\Sigma}([\alpha_2,\alpha_3])\right]$$

This vector is conjunctive combination of m_1 and m_2:

$$\begin{aligned}
m^{\Sigma}(\alpha_1) &= & \mu_{\alpha_1}^{\Sigma} &= & \mu_{\alpha_1}^1 \mu_{\alpha_1}^2 + \mu_{\alpha_1}^1 \mu_{[\alpha_1,\alpha_2]}^2 + \mu_{[\alpha_1,\alpha_2]}^1 \mu_{\alpha_1}^2 \\
m^{\Sigma}(\alpha_3) &= & \mu_{\alpha_3}^{\Sigma} &= & \mu_{\alpha_3}^1 \mu_{\alpha_3}^2 + \mu_{\alpha_3}^1 \mu_{[\alpha_2,\alpha_3]}^2 + \mu_{[\alpha_2,\alpha_3]}^1 \mu_{\alpha_3}^2 \\
m^{\Sigma}(\alpha_2) &= & \mu_{\alpha_2}^{\Sigma} &= & \mu_{\alpha_2}^1 \mu_{\alpha_2}^2 + \mu_{\alpha_2}^1 \mu_{[\alpha_1,\alpha_2]}^2 + \mu_{[\alpha_1,\alpha_2]}^1 \mu_{\alpha_2}^2 + \\
& & & & + \mu_{\alpha_2}^1 \mu_{[\alpha_2,\alpha_3]}^2 + \mu_{[\alpha_2,\alpha_3]}^1 \mu_{\alpha_2}^2 + \\
& & & & + \mu_{[\alpha_1,\alpha_2]}^1 \mu_{[\alpha_2,\alpha_3]}^2 + \mu_{[\alpha_2,\alpha_3]}^1 \mu_{[\alpha_1,\alpha_2]}^2
\end{aligned} \quad (11)$$

$$m^{\Sigma}([\alpha_1,\alpha_2]) = \mu_{[\alpha_1,\alpha_2]}^{\Sigma} = \mu_{[\alpha_1,\alpha_2]}^1 \mu_{[\alpha_1,\alpha_2]}^2$$

$$m^{\Sigma}([\alpha_2,\alpha_3]) = \mu_{[\alpha_2,\alpha_3]}^{\Sigma} = \mu_{[\alpha_2,\alpha_3]}^1 \mu_{[\alpha_2,\alpha_3]}^2$$

Equations (11) are obtained based on connections between neurons from Fig. 2.

Structure of HFNN for more than two evidences is presented in Fig. 3.

It is represented as fixed cascade from $n-1$ identical modules implementing L_3, L_4 and L_5. Each module implements combination of belief assignments for the pair of attributes using (11). Output vector summarizes the beliefs for the whole set of evidences.

It should be noted that output does not depend on order of evidences because of commutativity and associativity of conjunction.

Although the presented structure seems to be more complicated than presented in [9], it requires the same number of calculations. Both networks require $O(nk)$ operations at activations of first hidden layer and $O(nP)$ ones for the output summarizing of first hidden layer (k is the number of features and P is the number of hypotheses). Therefore, the general complexity is $O(n(k+P))$.

6 HFNN Training

Proposed HFNN is trained based on optimization of an efficiency criterion as conventional fuzzy neural networks.

Let $t(\vec{x}) = [t_{\alpha_1}(\vec{x}), t_{\alpha_2}(\vec{x}), t_{\alpha_3}(\vec{x})]$ be the aim vector of beliefs for hypotheses α_1, α_2 and α_3 on training set Θ obtained from input vector $\vec{x} \in \vec{X}$ ($\vec{x} = (x_1, x_2, \ldots, x_n)$, $\vec{X} = X_1 \times X_2 \times \cdots \times X_n$). BBAs of hypotheses $t_{\alpha_i}(\vec{x})$ can be obtained based on human expert data or statistical one. Output error can be obtained by comparison between $t(\vec{x})$ and $\vec{m}^{\Sigma}$ calculated by HFNN. The normalization of dimensions must be applied because $\vec{m}^{\Sigma}$ has more attributes. It can be done by distribution of $m^{\Sigma}\left([\alpha_1,\alpha_2]\right)$ and $m^{\Sigma}\left([\alpha_2,\alpha_3]\right)$ between pairs

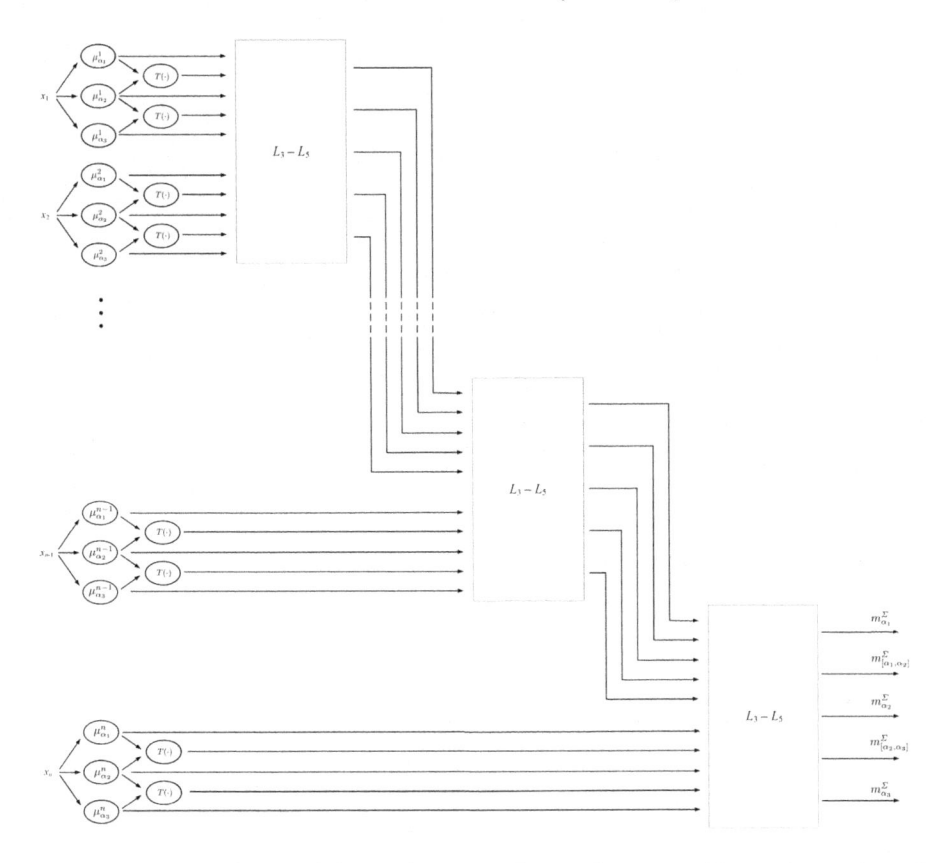

Fig. 3. Fuzzy neural network for combination of more than two evidences into universal value

$m^\Sigma(\alpha_1)$, $m^\Sigma(\alpha_2)$ and $m^\Sigma(\alpha_2)$, $m^\Sigma(\alpha_3)$, respectively. If the distribution is uniform, than resulting vector is one of pignistic probability:

$$\vec{m}'^\Sigma = \left[m'^\Sigma(\alpha_1), m'^\Sigma(\alpha_2), m'^\Sigma(\alpha_3)\right], \tag{12}$$

where

$$m'^\Sigma(\alpha_1) = m^\Sigma(\alpha_1) + \frac{m^\Sigma\Big([\alpha_1, \alpha_2]\Big)}{2},$$

$$m'^\Sigma(\alpha_3) = m^\Sigma(\alpha_3) + \frac{m^\Sigma\Big([\alpha_2, \alpha_3]\Big)}{2}.$$

$$m'^\Sigma(\alpha_2) = m^\Sigma(\alpha_2) + \frac{m^\Sigma\Big([\alpha_1, \alpha_2]\Big)}{2} + \frac{m^\Sigma\Big([\alpha_2, \alpha_3]\Big)}{2}.$$

Output error is computed as follows:

$$E_{\vec{x}} = ||t(\vec{x}) - \vec{m}'^{\Sigma}|| = \sum_{i=1}^{3}(t_{\alpha_i}(\vec{x}) - m'^{\Sigma}(\alpha_i))^2.$$

Mean output error for the complete training set is computed as follows:

$$\widehat{E}_{\vec{x}} = \frac{1}{L}\sum_{\vec{x}\in\Theta} E_{\vec{x}}. \qquad (13)$$

During training, parameters $\vec{a_i}$ and c_i are adjusted to minimize error from (13). The training is performed based on conventional backpropagation algorithm. The form of this algorithm proposed in [10] provides it convergence to the local minimum. Derivatives for all parameters are computed using chain rule. Complexity of the error gradient for each iteration is linear function from the total number of hypotheses and number of attributes.

7 Conclusions

The paper considers an approach for multisensor data fusion in technical diagnostics task, which is based on a new hybrid adaptive models. To implement the approach, hybrid fuzzy neural model is developed for various multidimensioned data processing. Hybrid model is based on combination of fuzzy neural network for computing the belief of object states and network model of BBA combination based on Dempster-Shafer theory.

The general advantage of the presented model is the possibility of computing the complex states for controlled objects, which allows to analyze the process dynamics and realize the state prediction. Moreover it allows to implement uncertainty degree for the basic hypotheses about state of technical object and predict the anomalous behavior.

Adjusted parameters of fuzzy neural model are parameters of membership function that provides the model to be sensibly interpreted. It allows to make a posterior estimation for the model.

As the future work, illustrative experimental results will be performed.

Acknowledgement. The work was supported Grant No. SP2018/163 "Diagnostics, reliability and efficiency of electrical machines and devices, problems of antenna systems" and by RFBR (Grants No. 17-20-01040 ofi_m_RZD, No. 16-07-00032-a and No. 16-07-00086-a).

References

1. Liggins II, M., Hall, D., Llinas, J.: Handbook of Multisensor Data Fusion: Theory and Practice. CRC press (2017)
2. Bleiholder, J., Naumann, F.: Data fusion and conflict resolution in integrated information systems. Ph.D. thesis, University of Potsdam (2010)
3. Valet, L., Mauris, G., Bolon, P.: A statistical overview of recent literature in information fusion. IEEE Aerosp. Electron. Syst. Mag. **16**(3), 7–14 (2001)
4. Goodman, I.R., Mahler, R.P., Nguyen, H.T.: Mathematics of Data Fusion, vol. 37. Springer Science & Business Media (2013)
5. Shafer, G.: A Mathematical Theory of Evidence, vol. 42. Princeton university press (1976)
6. Kovalev, S.M., Tarassov, V.B., Dolgiy, A.I., Dolgiy, I.D., Koroleva, M.N., Khatlamadzhiyan, A.E.: Towards intelligent measurement in railcar on-line monitoring: from measurement ontologies to hybrid information granulation system. In: International Conference on Intelligent Information Technologies for Industry, pp. 169–181. Springer (2017)
7. Batyrshin, I., Kaynak, O.: Parametric classes of generalized conjunction and disjunction operations for fuzzy modeling. IEEE Trans. Fuzzy Syst. **7**(5), 586–596 (1999)
8. Chai, Y., Jia, L., Zhang, Z.: Mamdani model based adaptive neural fuzzy inference system and its application. Int. J. Comput. Intell. **5**(1), 22–29 (2009)
9. Denoeux, T.: A neural network classifier based on dempster-shafer theory. IEEE Trans. Syst. Man Cybern. -Part A: Syst. Hum. **30**(2), 131–150 (2000)
10. Bishop, C., Bishop, C.M., et al.: Neural Networks for Pattern Recognition. Oxford university press (1995)

ANFIS and Fuzzy Tuning of PID Controller for STATCOM to Enhance Power Quality in Multi-machine System Under Large Disturbance

Huu Vinh Nguyen[1](✉), Hung Nguyen[2], and Kim Hung Le[3]

[1] Hochiminh City Power Company, Ho Chi Minh City, Vietnam
nguyenhuuvinhdlhcm@gmail.com
[2] Hochiminh City University of Technology (HUTECH),
Ho Chi Minh City, Vietnam
n.hung@hutech.edu.vn
[3] Danang University of Technology, Da Nang, Vietnam
lekimhung@dut.udn.vn

Abstract. STATCOM is one of the FACTS devices that are used in power systems. The algorithms used to control the STATCOM often use PID controller. However, there are a lot of elements in the network and have complex configurations and their dynamic model is highly non-linear, and convention PID controller are not robust for their stability control. In this paper, we propose the intelligent controllers for STATCOM based on dynamic model of the system and two control schemes have been developed: (i) Fuzzy-PID self-tuning controller (Hybrid F-PID); (ii) Adaptive neuro-fuzzy inference system – PID (ANFIS-PID) controller. The operating performance of the studied system is using the popular benchmark three-machine nine-bus system. The two-axis four-order model of synchronous generator (SG) is used. Time-domain scheme based on a nonlinear system model subject to a three-phase short-circuit fault at the load connected bus is utilized to examine the effectiveness of the proposed control schemes. It can be concluded from the simulation results that ANFIS has provide the best results for controlling STATCOM to enhance power quality in power system as compared to the conventional control strategies under large disturbance.

Keywords: STATCOM · Hybrid F-PID · ANFIS-PID · Power quality

1 Introduction

Flexible AC transmission systems (FACTS) devices have been proposed and installed in a lots of power systems. In which, static synchronous compensator (STATCOM) has an important role to enhance the power quality. Among them, STATCOM is installed at the point of common coupling (PCC) to maintain stable voltage in range to improve the power quality by protecting DFIG-based wind farm interconnected to weak grid from going offline during and after the disturbances [1]. In [2], the authors presented design procedure for STATCOM with constantly updates the parameters of PI controller for voltage regulator to enhance the voltage profile of the multi-machine system under dynamic disturbances. STATCOM can combine with Power System Stabilizer (PSS) in multi-machine power system connected to PV generation for eliminating transient stability is studied in [3]. In [4], the performance of the wind energy PMG-based system employing the dynamic voltage regulator (DVR) is compared to the performance of the system employing the STATCOM. This work recommends using STATCOM in systems with large loads where reactive power consumption from the grid could cause serious effects on connected loads. In [5], A state-based controlled STATCOM have suggested to support DFIG-based wind energy conversion systems during supply grid voltage sags. Each controller has its advantages and disadvantages, a STATCOM has been used to enhance the power stability using fuzzy logic controller (FLC) in the two-area four-generator interconnected power system [6]. In another work, such as [7], a cooperating PI controller and FL controller to enhance dynamic and steady state performance of a speed controller based interior permanent magnet synchronous motor are presented. In the distributed network, a distributed STATCOM (D-STATCOM) is proposed and different topologies of fuzzy controller with proportional plus integral control are implemented to control and maintain THD of the grid side current within the IEEE standards [8].

The PID controllers can be described by robust performances across a wide range of operating conditions and their functional simplicity. However, the high nonlinear of the power system means that a PID controller cannot perform well at all operating range, it can be robust performance at a particular operating range. There are two categories of PID controller. Firstly, the PID parameters are hold throughout control process; however, it is hard to meet good performances when the control system is nonlinear and heavily coupled. Secondly, in self-tuning PID, the parameters can be changed based on the parameters estimation [10]. For meeting the good global results, it is necessary to re-tune the PID controller when the operating range is changed, and different techniques from nonlinear control theory are required [11]. The fuzzy controller can be implemented without knowing the mathematical model of the system, and instead the through processes of human operators through linguistic rules. In the other hand, artificial networks are good at recognizing patterns and have ability to train the parameters of a control system, but they aren't good at explaining how they get their decisions. The neuro-adaptive learning techniques supply a procedure for the fuzzy modeling procedure to acquire information about a data set. This technique gives the fuzzy logic capability to compute the membership function parameters that effectively allow the associated fuzzy inference system to track the given input and output data [12]. Under the conditions of uncertainly, a method to identify the model parameters of parallel manipulators is to use the ANFIS control algorithm. Such an algorithm can be performed in a real-time control application [12].

This paper is mainly concerned with the applications of Fuzzy and ANFIS that are contained within the AI techniques to control a STATCOM to enhance power quality in power system. The STATCOM is connected to one of the load buses of a benchmark three-machine nine-bus system to absorb/inject reactive power to maintain the voltage at connected bus as well as to damp out the oscillations of the system. The paper is organized as below. System configuration of the studied system and mathematical models are introduced in Sect. 2. A Fuzzy-PID controller for STATCOM is depicted in Sect. 3.

In Sect. 4, the ANFIS controller is also designed for the same reason. Transient responses of the studied system with and without the designed controller are described in Sect. 5. Finally, specific important conclusions of this paper are drawn in last section.

2 Configuration of the Studied System

Figure 1 shows the configuration of the studied system containing three Synchronous Generators (SGs) supply power to three loads at bus 8, bus 9 and bus 10. A ±50-MVAR STATCOM is proposed and connected to bus 9 of the three-machine nine-bus system. The employed mathematical models are described as below.

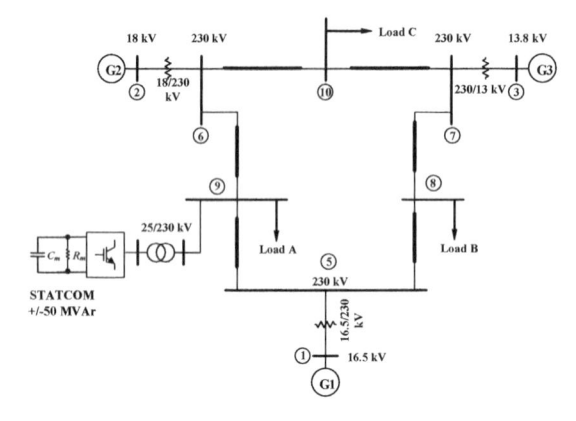

Fig. 1. Configuration of the benchmark three-machine nine-bus system with STATCOM

2.1 SG Model

In this paper, the generating units are modeled by its fourth order dynamic model, including a static excitation system [9]. The set of equations for each SG becomes:

$$\frac{d\delta}{dt} = \omega - \omega_0 \quad (1)$$

$$\frac{d\omega}{dt} = \frac{1}{M}(T_m - T_e - D\omega) \quad (2)$$

$$\frac{dE'_d}{dt} = \frac{1}{T'_{q0}}\left(-E'_d + \left(x_q - x'_q\right)i_q\right) \quad (3)$$

$$\frac{dE'_q}{dt} = \frac{1}{T'_{d0}}\left(E_{fd} - E'_q - \left(x_d - x'_d\right)i_d\right) \quad (4)$$

$$\frac{dE_{fd}}{dt} = \frac{1}{T_A}\left(-E_{fd} + K_A\left(V_{ref} - V_t\right)\right) \quad (5)$$

where δ (rad) and ω (rad/s) represent the rotor angular position and angular velocity; E'_d (pu) and E'_q (pu) are the internal transient voltages of the synchronous generator; E_{fd} (pu) is the excitation voltage; i_d (pu) and i_q (pu) are the d- and q-axis currents; $T'_{d0}(s)$ and $T'_{q0}(s)$ are the d- and q-open-circuit transient time constants; x'_d (pu) and x'_q (pu) are the d- and q- transient reactance; x_d (pu) and x_q (pu) are the d- and q- synchronous reactance; T_m (pu) and T_e (pu) are the mechanical and electromagnetic nominal torque; M is the inertia constant; D is the damping factor; K_A and T_A (s) are the system excitation gain and time constant; V_{ref} is the voltage reference; V_t is the terminal voltage magnitude.

2.2 STATCOM Model

For analyzing the STATCOM, the mathematical model is used. In which, the output voltage is separated into two components represented in d and q axes as follow [13]:

$$v_{dsta} = V_{dcsta} \cdot km_{sta} \cdot \sin(\theta_{bus} + \alpha_{sta}) \tag{6}$$

$$v_{qsta} = V_{dcsta} \cdot km_{sta} \cdot \cos(\theta_{bus} + \alpha_{sta}) \tag{7}$$

where: v_{dsta} and v_{qsta} are the voltages of d and q axes at the output terminals of the STATCOM, respectively; km_{sta} is the modulation index of the STATCOM; α_{sta} is phase angle of the STATCOM; θ_{bus} is the voltage phase angle of the common AC bus; V_{dcsta} is the DC voltage of the DC capacitor C_m. The relationship between DC voltage and current of the DC capacitor can be described as:

$$(C_m)p(V_{dcsta}) = \omega_b[I_{dcsta} - (V_{dcsta}/R_m)] \tag{8}$$

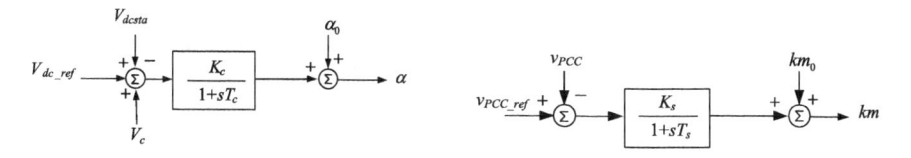

Fig. 2. Control scheme of STATCOM

In which: I_{dcsta} is the pu DC current flowing into the positive terminal of V_{dcsta}; R_m is the pu equivalent resistance considering the equivalent electrical losses of the STATCOM; i_{qsta} and i_{dsta} are the currents in q and d axes flowing into the terminals of the STATCOM, respectively. The fundamental control block diagram of the STATCOM including damping controller is shown in Fig. 2. The DC voltage V_{dcsta} is controlled by the phase angle α_{sta} while the voltage v_{sta} is varied by changing the modulation index km_{sta}.

3 Design of Fuzzy-PID Self-tuning Controller

According to [10], a fuzzy PID controller provides a control of the system with excellent performance in reliability, and accuracy. The basic approach is to try to detect inputs when the controller is not properly tuned and then seek to adjust the PID gains to improve the performance. The schematic structure of the fuzzy PID controller is given in Fig. 3. The basic schematic structure of the fuzzy PID controller is shown in general in Fig. 3, consisting of four basic segments: fuzzification (FZ); rule base (RB) or knowledge base (a set of If-Then rules); decision (DC); and de-fuzzification (DF).

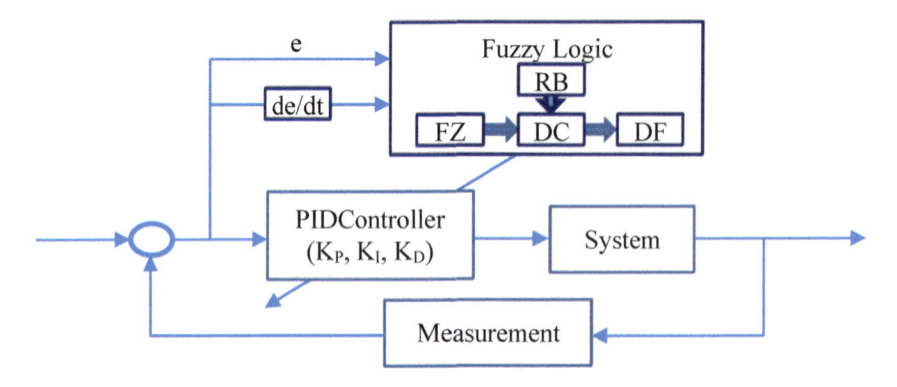

Fig. 3. Self-tuning-parameter fuzzy PID controller structure [10]

Table 1. Control rules of the FL

$\Delta\omega_{rs}$	ω_{rs}						
	NB	NM	NS	ZR	PS	PM	PB
PB	IB	IM	IS	KV	IS	IM	IB
PM	IM	IM	IS	KV	IS	IM	IB
PS	IS	IS	IS	KV	IS	IS	IS
ZR	DS	DS	DS	KV	IS	IS	IS
NS	DS	DS	DS	KV	DS	DS	DS
NM	DM	DM	DS	KV	DS	DM	DM
NB	DB	DM	DS	KV	DS	DM	DB

The fuzzy PID controller is based on a conventional PID controller: in Fig. 3, the input is the reference value and the output is the actual value. The parameters Kp, Ki and Kd are proportional, integral and differential coefficients of the PID controller. The inputs for the fuzzy block are the difference between the output and reference value, *e*, and its derivative, *de/dt*. These parameters generate the adjustment values ΔK_p, ΔK_i and ΔK_d for the PID parameters. The basic principle of the self-tuning-parameter PID parameters is as follows:

- First, the input variables are transformed into fuzzy quantities by membership functions, shown in Fig. 4.
- Then, refreshed fuzzy variables are generated by using a rule base. A rule-base contains a fuzzy logic quantification of the expert's linguistic description of how to achieve good control. In this research, seven linguistic variables for each input variable are used and denoted as NB, NM, NS, ZR, PS, PM and PB. There are also seven linguistic variables for output variable, namely, IB, IM, IS, KV, DS, DM and DB. The rule base for the tuning of the control system is shown in Table 1.
- After that, the values are formed through the de-fuzzification process by using the weighted average method, where μ is a weight co-efficient:

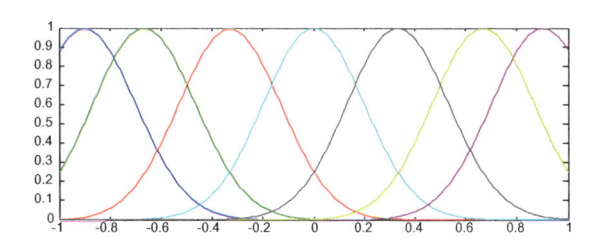

$$\mu(z) = min\{\mu(x); \mu(y)\} \quad (9)$$

Fig. 4. Inputs membership function

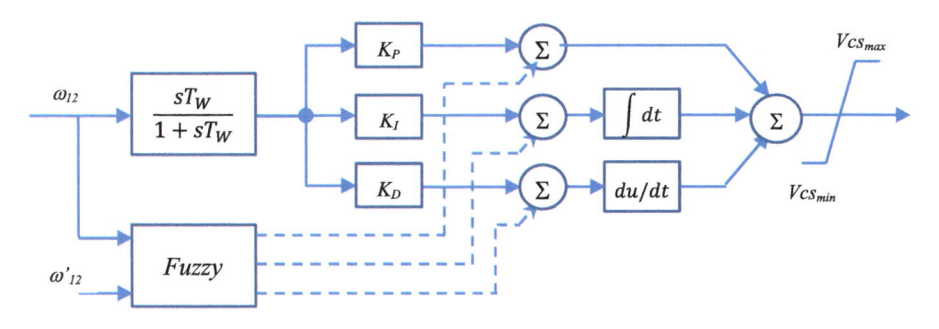

Fig. 5. Block diagram of the hybrid PI + FL controller [10]

- Finally, by the second scale transformation, the control signal was obtained, which is the input of the system in Fig. 3. The final parameter tuning results are as follows [10]: $K_i = K_{imin} + \mu_{K_i}(e; \Delta e).(K_{imax} - K_{imin})$, in which i = P, I, D.

The initial values for Fuzzy-PID were chosen the same as for the conventional PID controller. The idea for designing the Fuzzy - PID controller in this research is to compensate for the weakness of the PI controller in cases of large variation to the working equivalent point. The inputs for the fuzzy block are the difference between rotor speed of SG_1 and SG_2, ω_{12}, and its derivative, ω'_{12}. These parameters generate the adjustment values ΔK_p, ΔK_i and ΔK_d for the PID parameters (Fig. 5).

4 Design of ANFIS Controller for Tuning Gain PID

A typical architecture of used ANFIS is shown in Fig. 6; here in which a circle indicates a fixed node, whereas a square indicates an adaptive node. For simplicity, we consider two inputs x, y and one output f. For each model, a common rule set with two fuzzy if-then rules can be expressed as [14]. *Rule i: if x is Ai and y is Bi then $f_i = p_i x + q_i y + r_i$*, where A_i and B_i are fuzzy sets in the antecedent and $z = f(x, y)$ is a crisp function in the consequent; p_i, q_i, r_i are the updating parameters of rule. In this research, each ANFIS consists of five layers as follow [14]:

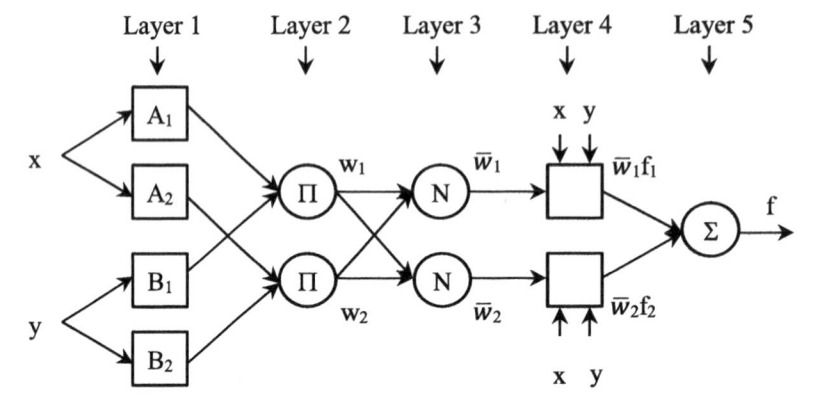

Fig. 6. Configuration of ANFIS [14]

Layer 1: In this layer input fuzzification takes place. Mathematically, this function can be expressed as:

$$O_{ij}^{(1)} = \mu_j\left(I_{ij}^{(1)}\right), \text{ in which } \mu_j(xi) = \frac{1}{1+\left|\frac{x_i-c_{ij}}{a_{ij}}\right|^{b_{ij}}} \quad (10)$$

where $O_{ij}^{(1)}$ is the Layer 1 node's output which corresponds to the j-th linguistic term of the i-th input variable $I_{ij}^{(1)}$. i is number of input variable and j is number of linguistic term of each input. In this research, $i = 2, j = 5$ and the triplet of parameters (a_{ij}, b_{ij}, c_{ij}) are referred to as premise parameters or non-linear parameters and they adjust the shape and the location of the membership function.

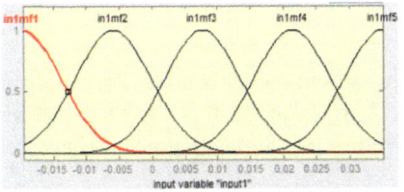

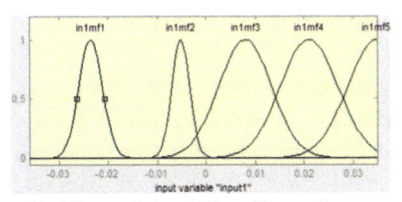

a. Initial MFs of input 1   b. After training MFs of input 1

Fig. 7. Initial and after training membership functions (MFs) for input 1

Layer 2: The total number of rules is 25 in this layer. Each node output represents the activation level of a rule:

$$o_k^{(2)} = w_k = \prod_{i=1}^{q} o_{ij}^{(1)} \qquad (11)$$

Layer 3: The output of the *k*-th node is the firing strength of each rule divided by the total sum of the activation values of all the fuzzy rules. This results in the normalization of the activation value for each fuzzy rule. This operation is simply written as:

$$o_k^{(3)} = \overline{w_k} = \frac{o_k^{(2)}}{\sum_{m=1}^{y^2} o_m^{(2)}} \qquad (12)$$

Layer 4: Each node *k* in this layer is accompanied by a set of adjustable parameters $d_{1k}, d_{2k}, \ldots, d_{N_{input}k}, d_{yk}, d_0$, and implements the linear function:

$$o_k^{(4)} = \overline{w_k} f_k = \overline{\omega_k}(d_{1k} I_1^{(1)} + d_{2k} I_2^{(1)} + \ldots\ldots + d_{yk} I_y^{(1)} + d_0) \qquad (13)$$

Layer 5: The single node in this layer computes the overall output as the summation of all incoming signals, which is expressed as:

$$o^{(5)} = \sum_{k=1}^{y^2} o_k^{(4)} = \sum_{k=1}^{y^2} \overline{w_k} f_k = \frac{\sum_{k=1}^{y^2} w_k f_k}{\sum_{k=1}^{y^2} w_k} \qquad (14)$$

The overall output is a linear combination of the modifiable parameters. The training algorithm requires a training set defined between inputs and outputs. Figure 7 shows optimized membership function for difference between rotor speed of SG_1 and SG_2, ω_{12}, and its derivative, ω'_{12}. The generalized bell MF is used for each input variables. It is clear from (13) that bell MF is specified by three parameters.

5 Simulation Results

For simulating the studied system, the nonlinear system model is used. The results are performed in MATLAB/SIMULINK toolbox. It is assumed that the three-machine system operates under stable condition as referred to [6]. Simulation results have been

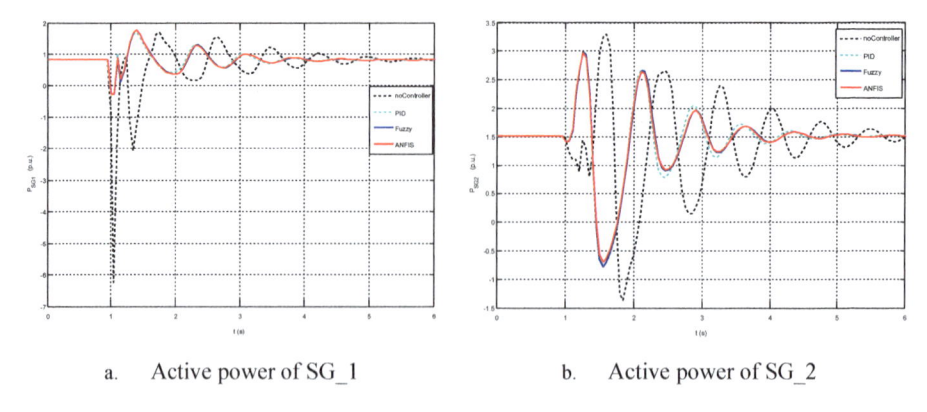

a. Active power of SG_1

b. Active power of SG_2

Fig. 8. Active power of SG_1, SG_2

a. Voltage of SG_1

b. Voltage of SG_2

c. Voltage of SG_3

d. Voltage at PCC

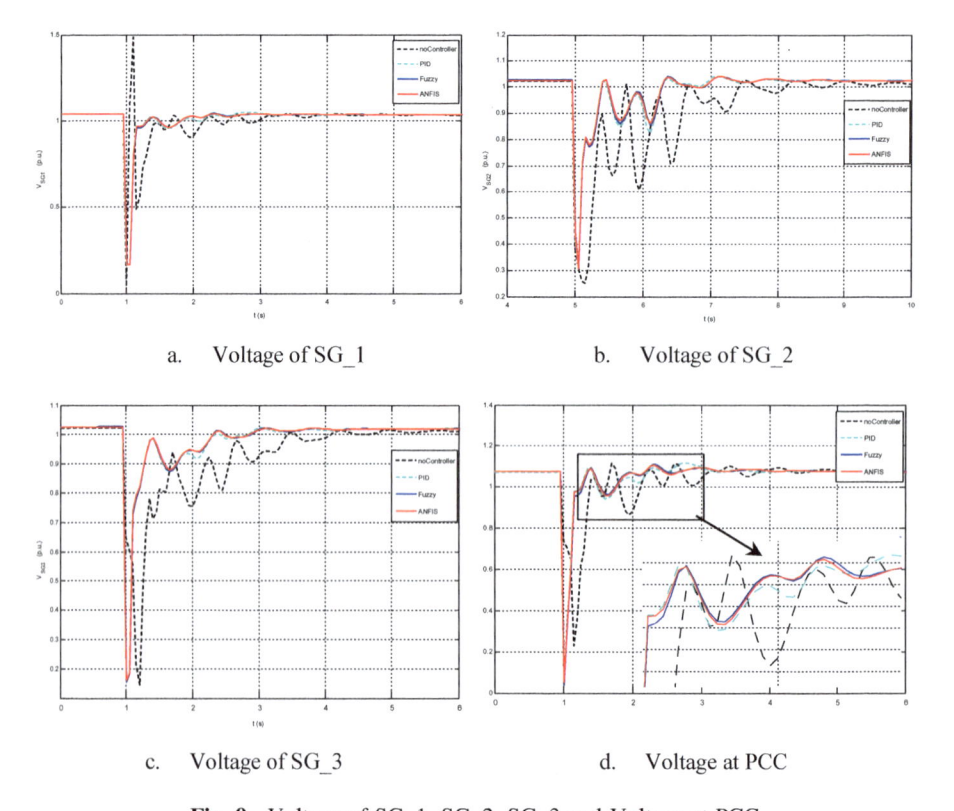

Fig. 9. Voltage of SG_1, SG_2, SG_3 and Voltage at PCC

presented in Figs. 8 and 9. These figure plots the comparative transient responses of the studied system with the proposed STATCOM in cases of without controller (black dotted lines), with PID controller (cyan dotted lines), with Fuzzy-PID controller (blue lines) and with ANFIS controller (red lines) subject to a 3-phase fault at bus 9. In which, Figs. 8a to b are represented for active power of SG_1, SG_2 respectively. While Figs. 9a to d are represented for voltage at SG_1, SG_2, SG_3 and PCC.

The simulation results also show that in the case of without controller, the waveforms of voltage at the PCC are recovered to the prefault steady-state operating conditions around 5 s with big oscillations. In the case of Fuzzy-PID and ANFIS controller, these waveforms are recovered to the prefault steady-state operating conditions around 3 s with small oscillations. It is clearly observed from the comparative transient simulation results shown in Figs. 8 and 9 that the proposed STATCOM with the different controllers can offer better damping characteristic, and enhance power quality in power system. All the results demonstrate that the ANFIS controller is better than Fuzzy PID and more effective than conventional PID controller.

6 Conclusion

The controller of STATCOM used in this system by ANFIS method has a good response without prior knowledge of the process. Also, by this method, more good responses than by the Fuzzy PID or only PID controllers are obtained. This control method is very useful to apply the process control system and helpful to select the most appropriate range for STATCOM operation. The proposed STATCOM joined with the designed controllers on suppressing inherent oscillations of the studied system and improving system stability as well as the power quality of the system under different operating conditions. It can be concluded that the proposed STATCOM joined with the designed damping controller has better damping characteristics to improve the performance of the studied multi-machine system under large disturbance.

References

1. Pokharel, B., Gao, W.: Mitigation of disturbances in DFIG-based wind farm connected to weak distribution system using STATCOM. In: Proceedings North American Power Symposium (NAPS), pp. 1–7, 26–28 September 2010
2. Ganesh, A., Dahiya, R., Singh, G.K.: Development of simple technique for STATCOM for voltage regulation and power quality improvement. In: 2016 IEEE International Conference on Power Electronics, PEDES, Trivandrum, India, pp. 1–6 (2016)
3. Selwa, F., Djamel, L., Imen, L., Hassiba, S.: Impact of PSS and STATCOM on transient stability of multi-machine power system connected to PV generation. In: 2015 International Conference on Renewable Energy Research and Applications, Palermo, pp. 1416–1421 (2015)
4. Eskander, M.N., Amer, S.I.: Mitigation of voltage dips and swells in grid-connected wind energy conversion systems. In: Proceedings ICCAS-SICE (2009)

5. Chhor, J., Tourou, P., Sourkounis, C.: Evaluation of state-based controlled STATCOM for DFIG-based WECS during voltage sags. In: 2016 IEEE International Conference on Renewable Energy Research and Applications (ICRERA), Birmingham, pp. 463–471 (2016)
6. Mak, L.O., Ni, Y.X., Shen, C.M.: STATCOM with fuzzy controllers for interconnected power systems. Electr. Power Syst. Res. **55**(2), 87–95 (2000)
7. Uddin, M.N., Rebeiro, R.S.: Improved dynamic and steady state performance of a hybrid speed controller based IPMSM drive. In: Proceedings IEEE Industry Applications Society Annual Meeting (IAS), pp. 1–8, 9–13 October 2011
8. Varshney, A., Garg, R.: Comparison of different topologies of fuzzy logic controller to control D-STATCOM. In: 2016 3rd International Conference on Computing for Sustainable Global Development (INDIACom), New Delhi, pp. 2492–2497 (2016)
9. Anderson, P.M., Fouad, A.A.: Power System Control and Stability, 2nd ed. IEEE Press (2003)
10. Tian, L.: Intelligent self-tuning of PID control for the robotic testing system for human musculoskeletal joints test. Ann. Biomed. Eng. **32**(6) (2004)
11. Adhyaru, D., Patel, J., Gianchandani, R.: Adaptive neurofuzzy inference system based control of robotic manipulators. In: Proceedings 2010 IEEE International Conference on Mechanical and Electrical Technology, pp. 353–358 (2010)
12. Bachir, O., Zoubir, A.F.: Adaptive neuro-fuzzy inference system based control of puma 600 robot manipulator. Int. J. Electr. Comput. Eng. **2**(1), 90–97 (2012)
13. Molinas, M., Suul, J.A., Undeland, T.: LVRT of wind farms with cage generators: STATCOM versus SVC. IEEE Trans. Power Electron. **23**(3), 1104–1117 (2008)
14. Jang, J., Sun, C., Mizutani, E.: Neuro-Fuzzy and Soft Computing. Prentice Hall, Upper Saddle River (1997)

Proposal of Electrode System for Measuring Level of Glucose in the Blood

Klara Fiedorova[✉] and Martin Augustynek

FEECS, VSB–Technical University of Ostrava,
K450, 17. listopadu 15, 708 33 Ostrava–Poruba, Czech Republic
{klara.fiedorova,martin.augustynek}@vsb.cz

Abstract. The paper describes the development of electrode system and blood glucose measuring circuit for glucose analysis. Measuring blood glucose levels is today one of the standards of medical procedures that determine the physiological state of a person. The task of the work was to implement the measuring system, to test the functionality of the whole system, the measurement tests were carried out and the measured data were subsequently processed statistically. In designing and constructing the electrode system, the resistive principle of measurement of non-electrical quantities was applied because the system focuses on the measurement of the concentration of a certain substance in the liquid. These are electrolytic sensors using gold electrodes. The part that deals with the construction of the electrical measuring circuit has determined the conditions that the circuit must meet. For this reason, a micro-ammeter was designed to provide the source voltage, measure the current and convert it back to a suitable magnitude. The current measured by the circuit is proportional to the concentration of glucose in the solution. Glucose oxidase enzyme was used to distinguish glucose molecules in the blood and determine its concentration, which, together with the supply of the required voltage, provides useful chemical reactions. Based on the processed data, it is possible to evaluate the constructed system and lay the foundations for further development of the issue.

Keywords: Measurement of blood glucose concentration · Glucose measurement sensors · Continuous glucose measurement

1 Introduction

Generally, blood concentration sensors are used in healthcare. It is used in diabetology to diagnose diabetes mellitus and for control of blood glucose level, depending on the individual's physical condition, the time of day, the diet, and the type of blood being analyzed. The blood glucose level is about 3.3–6.6 mmol/l in capillary blood, 6.9–5.5 mmol/l in venous blood, and 4.2–6.4 mmol/l in blood plasma [2, 4].

There are several procedures and principles to measure blood glucose. One of the most important facts by which glucose measurement methods can be divided is the invasiveness. Invasive and non-invasive methods are distinguished [4, 12]. Nowadays, invasive methods can be considered as standard in this issue because they are more accurate. To this group belongs laboratory tests made in laboratories, glucometer devices, and continuous methods using subcutaneous electrodes. The principle of measuring glucose by invasive methods is to evaluate the chemical reaction between glucose and the enzyme (glucose oxidase, hexokinase) [3, 7, 8]. Electrochemical or photometric methods based on amperometric, coulometric or resistance principles are used. Although invasive blood glucose measurement methods are the most appropriate in terms of accuracy [1, 2, 6], noninvasive techniques are developed to minimize pain and discomfort for the patient. They represent an alternative approach to measuring blood glucose. These include sensors placed on the patient's skin using microdialysis to obtain interstitial fluid. The most notorious noninvasive methods are NIR methods, reverse iontophoresis or raman spectroscopy [4, 10, 12].

Based on extensive literature research, the work has focused on the development of a new continuous technique for measuring blood glucose levels. The proposed biosensor should be inserted into a patient's body using a catheter and used to measure the glucose level, which would then be counted as the heart rate.

The following chapter, entitled Methods, deals with the design of the electrode system together with the measuring circuit and the measuring principle. Section 3 includes the measurement results together with their evaluation and statistical processing.

2 Methods

The design of the electrode system was influenced by several conditions and assumptions. Primary requirements and conditions include the fact that the conductivity of the solution will be measured to match the glucose concentration. Taking into account these facts, the system should function as an electrolytic sensor. More detailed information about electrical sensors is provided in the literature [1, 6]. Another important fact was using of an enzyme that has very specific chemical reactions with the help of electric energy from which the construction of the measuring circuit was developed. The Eqs. 1–3 describe the chemical reactions [3, 7, 8].

In principle, the proposed electrode system is based on resistance technology. The function of these sensors is based on the non-electric magnitude conversion to change the resistance in the circuit [1, 6] (Fig. 1).

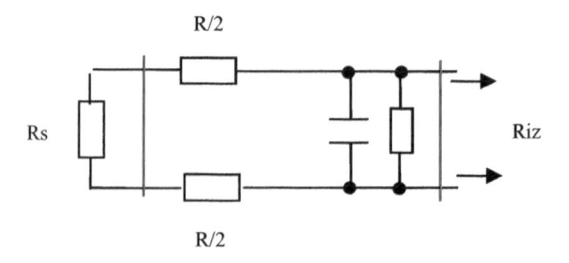

Fig. 1. Design scheme of the proposed system design

The resistive technique uses an electrolytic transducer that is proposed by this paper. These sensors work with electrolyte conduction. The electrolyte is composed of positive and negative ions, therefore it is a second-class conductor [1, 6].

The sensor is used to determine the concentration of ions. Because the measured glucose concentration parameter is evaluated from the analyzed blood, there is a problem with the separation of glucose ions. There is a large amount of free ions in the blood that belong to other substances. Glucose is used to refer to a substance called an enzyme that is specific to its reaction with one substance [2, 5, 7–9]. Thanks to this reaction, it is possible to transfer electrical current and distinguish glucose concentration. The electrical charge generated by the oxidation of hydrogen peroxide during a given reaction is led to the electrode. In the work sensors based on chemical reactions with oxygen consumption to produce hydrogen peroxide are used. In reactions, enzymes serve to achieve equilibrium between starters (substrates) and products and serve as the catalysis of the reaction [7, 8]. An important factor in the work is velocity of the chemical reaction, which is a function of the concentration of the substrate and it is solved by the Michaelis and Mentel equations [13, 14]. Electrochemical reactions applied in research:

$$GLUCOSE + GOx(FAD) \rightarrow GLUCAGON + GOx(FADH_2) \qquad (1)$$

$$GOx(FADH_2) + O_2 \rightarrow GOx(FAD) + H_2O_2 \qquad (2)$$

Due to oxidation of hydrogen peroxide, the potential on the working electrode must be 0.6 V.

$$H_2O_2 \xrightarrow[\text{exotherm, s=4MJ/kg}]{} 2H^+ + 2e^- \qquad (3)$$

Once the measurement principle has been determined, the choice of the electrode system material has come into play. Materials have been chosen that cannot influence the chemical reactions between the enzyme and glucose [11]. The first chosen material for the electrode was platinum, but it was replaced with gold-plated material because of its availability [11]. During the work, electrodes, which are characterized by a diameter of 0.6 mm, were used. In addition, the material constituting the container into which the electrodes are inserted must be chosen and used for the application of the test specimens. The most suitable material was a silicone like a tube. Silicone is electrically non-conductive, therefore it does not affect the function of the electrode system [11]. The silicone container has a volume of $25.3 * 10^{-8}$ ml. For connection the electrode system with the measuring circuit, the electrodes were taken out of the silicone vessel using copper conductors connected to the pinheads enabling connection to the proposed measuring circuit. The electrode system also contains copper ground electrode. This electrode is also part of the proposed measuring circuit and should eliminate the interference [1, 6].

The field of matter that deals with the measurement of the electrical conductivity of solutions in relation to the concentration of the substance is called conductometry [1, 6]. The parameters of the biosensor, which includes the cross-sectional area and distance, are important for calculating the resistance of the system (Tables 1 and 2).

Table 1. Parameters of the electrode system 1

Cross section [mm]	Distance [mm]	Specify resistance [μΩcm]	Length [mm]
0.6	0.47	2.35	4

Table 2. Parameters of the electrode system 2

Resistance [Ω]	Conductivity [S]
$0.02 * 10^{-6}$	$50 * 10^6$

$$R = \rho * 1/S \qquad (4)$$

$$G = 1/R \; [S] \qquad (5)$$

ρ = Specific Resistance, l = Distance of Electrode; S = Cross Section of Electrode;

$G = k * C$, where $C = S/l$ = Specific Conductivity

The realization of the measuring circuit was more demanding in terms of conditions and requirements for the functionality of the whole system. The basic element of the measuring circuit provides 0.6 V for supply to the working electrode, due to this voltage the decomposition of hydrogen peroxide and the formation of free electrons occurs [1, 8]. An important assumption during the implementation of the measuring circuit is the generation of a relatively small current, so the system should be able to detect small currents and process them. The measured current should also be converted to an output voltage that is more advantageous for purposes of this work [1, 8]. Designed circuit therefore works as a micro-ammeter. The diagram of the individual parts of the circuit together with the description of the parts is shown in the Fig. 2. The measuring system can be divided into two parts. The first part serves to supply the required voltage to the working electrode, i.e. the transformation of the source voltage to the desired value. The voltage source consists of two batteries connected in series. The second part serves to detect the generated current in the electrode system and to transfer the current to voltage.

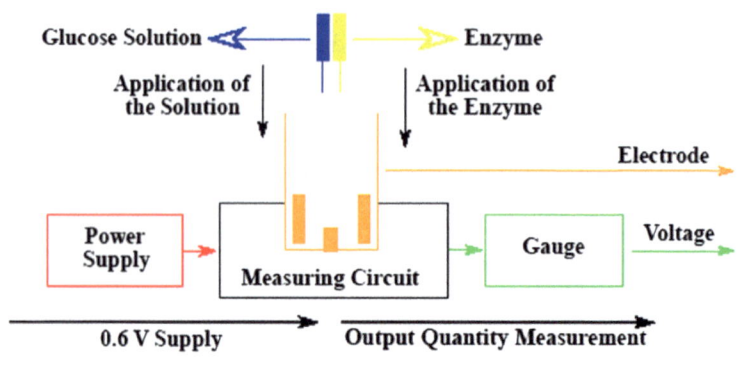

Fig. 2. Design of the measuring system

3 Results

The glucose concentration was measured in two variants, with the same number of measurements being performed for each variant. Measurements were made under steady conditions and the measured values were read at the same time. The first variation was done only with glucose solutions of different concentrations. These measurements were used to verify the functionality of the measuring system and to obtain basic processing data. The second variant was based on the using of the enzyme together with the glucose solution, resulting in an electrochemical reaction that provides the basis for further development of the technique and the calculation of the reaction kinetics. The subsequent measurement is based on the use of pure solution, made from distilled water so there are no other ions capable of transmitting the current in the solution. It can be assumed that the measured electrical current is proportional to the concentration of glucose in the solution.

Solutions with concentrations of 3 mM, 7 mM, 12 mM, 18 mM were used, each concentration was measured 10 times so that the data obtained had a predictive value and a statistical analysis could be performed. Based on this, measurements are made in two phases. In the first step, a 0.3 ml pure solution was used without the use of an enzyme where the time values 0 s, 20 s, 40 s, 60 s, 80 s, 100 s, 120 s, 140 s and 3 min were recorded. In the next part, 0.01 g of the enzyme was poured into the solution after ten second break. In this part of the measurement the value at the beginning of the measurement and during the addition of the enzyme was measured. The addition of the enzyme caused a change in voltage, due to the chemical reaction there was a sudden increase in voltage, it was important to measure the time of this change for the measurement. Once the chemical reaction had stabilized, the voltage was read again and then at 2:20 and 2:40 min. The recording of time helped to understand of the nature of the measurement and the principle of ongoing chemical reactions.

From the measured data, one representative was selected to represent and characterize the measured group. Tables 3, 4, 5 and 6 include the representatives, where Tables 3 and 4 are for pure glucose solution and Tables 5 and 6 are for measurements with enzyme. From both tables the primary assumption can be read, increasing the glucose concentration of the analyzed solution caused increasing of output voltage. In the first case, the value of the solution was 3 mM and voltage was 0.175 V, when higher concentrations of solutions were used, the measured value increased. For a 12 mM sample, the initial value of the 0.245 V voltage and the 18 mM sample gave a value of 0.348 V. Exceptions were samples with a concentration of 7 mM, where the initial measured value was higher and subsequently decreased. From 20 s the voltage value began to increase again. In the second case, the results show the same results, i.e. at the 3 mM sample, the initial voltage value was 0.173 V, the 7 mM solution was 0.241 V, the sample 12 mM had 0.261 V and the 0.296 V voltage at the 18 mM sample was measured. For measurement to provide the basic for further development of the techniques, the voltage samples were measured at the specified frequency for three minutes.

If we analyze the readings read in sixty seconds, higher concentration of pure glucose solution can measure a higher voltage value. A 3 mM sample gives a voltage value of 0.204 V, a 7 mM sample gives a value of 0.271 V, a voltage of 0.0.314 V for a 12 mM sample, and a sample of 18 mM giving 0.411 V. The course of measurement with addition of enzyme is characteristic of the specific chemical reaction between glucose oxidase and glucose. At this point, the chemical reaction time was measured, and then the voltage value was determined after the stabilization of chemical reaction. For the 3 mM sample, the chemical reaction started after 20 s and stabilized after 1:55 min. In this time, the value of voltage was 0.387 V. The response of 7 mM sample was observed at 16 s, stabilization was occurred at 1:17 min at a voltage of 0.492 V. For samples of 12 mM and 18 mM, the time of the reaction is 15 s and 13 s, stabilization of the reaction occurred at 1:18 min and at 1:21 min with measured voltages of 0.603 V and 0.649 V. Again, the fact that voltage increases with increasing concentration is confirmed.

Table 3. Measured values – pure solution 1

Solution	[V]	20 s/[V]	40 s/[V]	60 s/[V]	80 s/[V]
3 mM	0.175	0.182	0.199	0.204	0.21
7 mM	0.32	0.296	0.284	0.271	0.275
12 mM	0.245	0.274	0.293	0.314	0.339
18 mM	0.348	0.358	0.379	0.411	0.451

Table 4. Measured values – pure solution 2

Solution	100 s/[V]	120 s/[V]	140 s/[V]	3 min/[V]
3 mM	0.216	0.221	0.225	0.23
7 mM	0.276	0.279	0.28	0.285
12 mM	0.351	0.368	0.382	0.403
18 mM	0.348	0.358	0.379	0.411

Table 5. Measured values – solution with enzyme 1

Solution	[V]	Enzyme 10 s/[V]	Reaction [s]
3 mM	0.173	0.183	20
7 mM	0.241	0.246	16
12 mM	0.261	0.301	15
18 mM	0.296	0.302	13

Table 6. Measured values – solution with enzyme 2

Solution	Change [V] [min]	2:20 min/[V]	2:40/[V]
3 mM	0.387(1:55)	0.392	0.385
7 mM	0.492(1:17)	0.478	0.443
12 mM	0.603(1:18)	0.651	0.659
18 mM	0.649(1:21)	0.72	0.716

The graph in Fig. 3 presents that the measured voltage varies according to the concentration of the solution. The plots confirm the assumption, that the increasing concentration of the solutions caused increase of output voltage.

In the graph, there are two types of lines, where the full lines correspond to pure glucose solutions and the dotted lines represent the enzyme samples. Each color represents different concentration of glucose in solution. The color distribution is shown in Table 7.

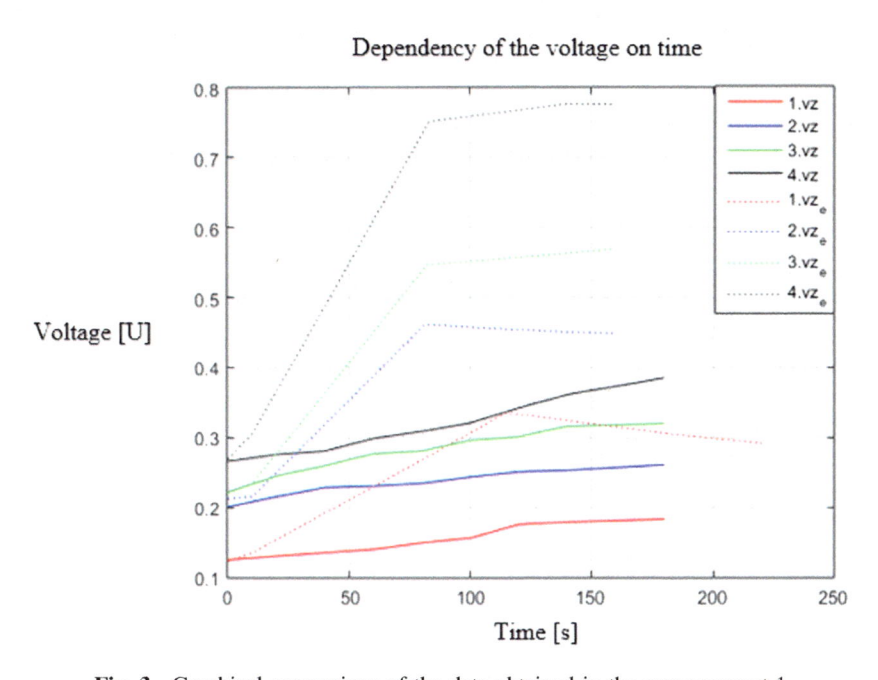

Fig. 3. Graphical comparison of the data obtained in the measurement 1

Table 7. The color distribution

3 mM	7 mM	12 mM	18 mM
Red	Blue	Green	Black

The lowest values answer the measured values which were obtained by using 3 mM glucose solution, it applies for the pure solution and for measurement with enzyme too. When analyzing the blue and red curves, where the blue curves include the voltage measured in the 7 mM solution, the higher concentration of the solution cause increase of the output voltage. These conclusions are confirmed by the analysis of further curves. The black lines correspond to a solution with a concentration of 18 mM glucose, thanks to the highest concentration the value of output voltage is the highest. It is clear from the graph in Fig. 3, and again it is applying for both measurement variants. The green lines represent the voltage of the 12 mM solution, and values are lower in comparison with samples of 18 mM. From the graph, it is also possible to see that the output voltage values generated using only pure glucose solution are lower. While the voltage in the second variant, when the enzyme was added to the solution, is more interesting. The curves have a characteristic shape that proves the chemical reaction between the enzyme and glucose. This fact is influenced by the number of glucose molecules in the solution, i.e. lower concentrations, does not provide many reagents for enzyme.

In order to support the conclusions and the fulfillment of the assumptions, the measured data were analyzed in terms of measured voltage in dependence on the concentration of glucose solution. See Fig. 4 for this dependency.

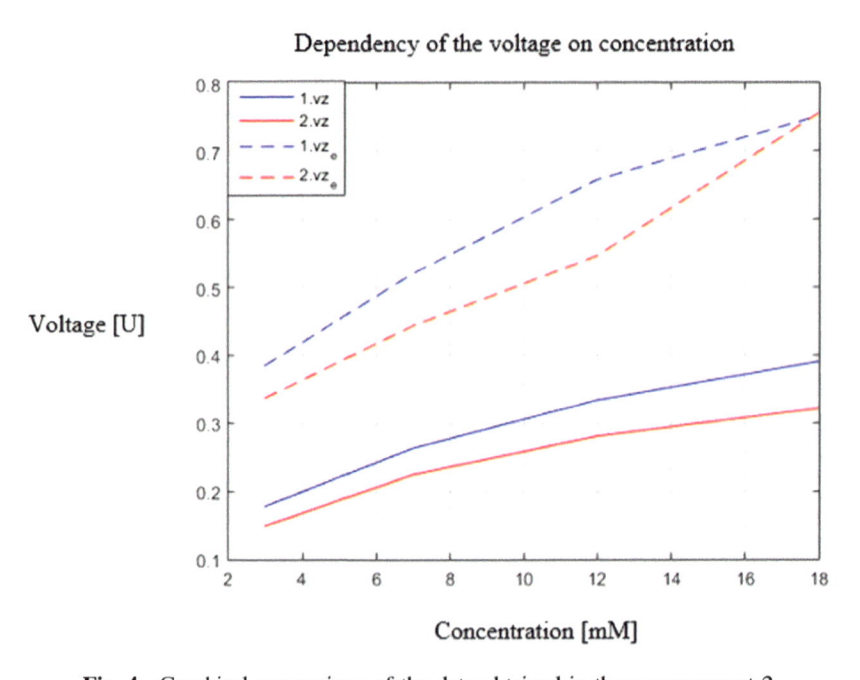

Fig. 4. Graphical comparison of the data obtained in the measurement 2

The graph shows the curve of four samples where the full lines carry the measured data pure glucose solution and the dashed line corresponds to the samples solution with the addition of the enzyme. For each possible case, ten samples were measured, which were randomly selected from one of the ten samples at one time. The values used in the graph were selected for the same time of measurement. These values were ranked based on the concentration from which they were obtained. The data were sorted according to the concentration, from the lowest values to the highest values.

All the results, obtained by measuring and processing in the form of tables and graphs, confirm the statistical analysis. The exact process of statistical analysis is shown in the literature [15].

4 Conclusion

The aim of the thesis was to create a measuring unit for the determination of the concentration of glucose in the solution. In the work, a simple electrode system has been created, which can be connected to the designed electrical circuit. Prior to the development of the technique, the requirements that the system should meet and the assumptions about how the measurement will evolve have been determined. This laid the foundation for the design and construction of the system. During the construction, we needed to consider the types of materials used to preserve all the necessary chemical reactions that we use at work. After making an existing measuring circuit along with the electrode system, the whole concept had to be subjected to a functional test and the correct values of the individual components had to be set. After verification of functionality, it was possible to approach a set of experimental tests.

The measurement was carried out in two variants where it was measured with pure glucose solution and then with pure solution and subsequent addition of enzyme. Even during the first set of measurements, the first assumption that the increasing concentration of the glucose solution increases the output voltage is confirmed, due to the higher number of free ions capable of transmitting the electric current. This finding has confirmed that if we measure the output voltage from the solution to which the enzyme is added, we will measure higher voltages than during the clear solution measurement. This was confirmed during the experimental measurement. The course of measurement with enzyme addition was characteristic of its course as we recorded the values before adding the enzyme and subsequently also during the addition of the enzyme. This is related to the saturation of the reaction. Since the course of the chemical reaction between enzyme and glucose is an essential part of the work and the basis for further development of the technique, due to the precise resolution of the generated signal from the chemical reaction. Therefore, there is an easy procedure for determining the kinetic properties of the glucose-enzyme reaction.

This work is the first step in more extensive research that should be to integrate the biosensor into the catheter to measure glucose concentration in the heart for subsequent calculation of cardiac output.

Acknowledgment. The work and the contributions were supported by the project SV4508811/2101 'Bi- medicínské inženýrské systémy XIV'. This study was supported by the research project The Czech Science Foundation (GACR) 2017 No. 15330/16/AGR Investment evaluation of medical device development run at the Faculty of Informatics and Management, University of Hradec Kralove, Czech Republic. This study was supported by the research project The Czech Science Foundation (TACR) ETA No. TL01000302 Medical devices development as an effective investment for public and private entities.

References

1. Comeaux, R., Novotny, P.: Biosensors: Properties, Materials and Applications. Biotechnology in Agriculture, Industry and Medicine Series. Nova Science Publishers, New York (2009). ISBN 978-1-60741-617-3
2. Clarke, S.F., Foster, J.R.: A history of blood glucose meters and their role in self-monitoring of diabetes mellitus. Br. J. Biomed. Sci. **69**(2), 83 (2012)
3. Chang, T.M.S.: Biomedical Applications of Immobilized Enzymes and Proteins. Springer Science & Business Media (2013)
4. Holzinger, U., et al.: Real-time continuous glucose monitoring in critically ill patients: a prospective randomized trial. Diabetes care **33**(3), 467–472 (2010)
5. Pickup, J.C., ed.: Insulin Pump Therapy and Continuous Glucose Monitoring. Oxford University Press, Oxford (2009). ISBN 978-0-19-956860-4
6. Eggins, B.R.: Chemical Sensors and Biosensors. Wiley, Chichester (2002). ISBN 0-471-89913-5
7. Hermanson, G.T.: Bioconjugate Techniques, 2nd edn. Academic Press, San Diego (2008). ISBN 978-0-12-370501-3
8. Bugg, T.D.H.: Introduction to Enzyme and Coenzyme Chemistry, 3rd ed. Wiley, Hoboken (2013). ISBN 9781118348994
9. Langendam, M., et al.: Continuous glucose monitoring systems for type 1 diabetes mellitus. The Cochrane Library (2012)
10. Rodbard, D.: Continuous glucose monitoring: a review of successes, challenges, and opportunities. Diab. Technol. Ther. **18**(S2), S2-3–S2-13 (2016)
11. Nichols, S.P., et al.: Biocompatible materials for continuous glucose monitoring devices. Chem. Rev. **113**(4), 2528–2549 (2013)
12. Wagner, J., Tennen, H., Wolpert, H.: Continuous glucose monitoring: a review for behavioral researchers. Psychosom. Med. **74**(4), 356 (2012)
13. English, B.P., et al.: Ever-fluctuating single enzyme molecules: Michaelis-Menten equation revisited. Nat. Chem. Biol. **2**(2), 87–94 (2006)
14. Purich, D.L.: Enzyme Kinetics: Catalysis & Control: A Reference of Theory and Best-Practice Methods. Elsevier/Academic Press, Amsterdam (2010). ISBN 9780123809247
15. Fiedorová, K.: Návrh elektrody pro měření koncentrace glukózy v krvi. Ostrava. Dostupné z: http://hdl.handle.net/10084/119180. Diplomová práce. Vysoká škola báňská - Technická univerzita Ostrava (2017). [cit. 2018-06-25]

Substitution Rules with Respect to a Context

Michal Fait$^{(\boxtimes)}$ and Marie Duží

Department of Computer Science FEI, VŠB-Technical University of Ostrava,
17. listopadu 15, 708 33 Ostrava, Czech Republic
{michal.fait,marie.duzi}@vsb.cz

Abstract. In this paper, we deal with Leibniz's rule of substitution of identicals, and describe how the rule can be applied in the TIL-Script language. The main goal is to introduce the algorithm of valid application of the substitution rules in all the three kinds of context that we distinguish in the TIL-Script language. The language is a computational variant of TIL, which is a hyperintensional, partial typed λ-calculus. Hyperintensional, because the meaning of TIL λ-terms are procedures producing functions rather than the denoted functions themselves. Partial, because TIL is a logic of partial functions, and typed, because all the entities of TIL ontology receive a type. Based on the results of context recognition the algorithm makes it possible to validly apply the substitution rules and derive relevant new pieces of analytic information.

Keywords: TIL · Substitution rules · Three kinds of context · TIL-Script · Algorithm of valid substitution

1 Introduction

Leibniz's Law is a fundamental law in any extensional logic. It actually consists of two principles; (i) identity of indiscernibles and (ii) indiscernibility of identicals. In this paper we deal with the principle (ii), which is formally defined in the second order predicate logic as follows:

$$\forall x \forall y (x = y \rightarrow \forall P(Px \leftrightarrow Py))$$

This rule is generally valid in direct (extensional) contexts, but in indirect (or opaque) contexts the rule ostensibly fails. Here is an example of such a failure:

Miloš Zeman is the president of the Czech Republic
Jiří Drahoš wants to become the president of the Czech Republic
―――
Jiří Drahoš wants to become Miloš Zeman.

The argument is obviously invalid. The problem is this. In the first premise, 'the president of Czech Republic' occurs *extensionally*, i.e., the *occupant* of the office denoted by the term rather than the office itself is identified with Miloš Zeman. For sure, Zeman is not the office, he just happens to occupy this office right now. While in the second premise the same term 'the president of Czech Republic' occurs *intensionally*. Jiří Drahoš wanted to occupy the office and is thus related to the office itself. Thus, in order the substitution be valid, we can substitute merely the same office rather than its current accidental occupant.

Here is another simple example:

$$\frac{\text{Tom calculates } 5+2}{\text{Tom calculates } \sqrt{49}.}$$

Again, the argument is obviously invalid due to mixing different contexts, this time an extensional and hyperintensional context. In the second premise both the terms referring to functions '+' and '$\sqrt{}$' occur *extensionally*. The *values* of two different functions at different arguments are identified. However, in the first premise the term '5+2' occurs *hyperintensionally*. Tom is not related to the value 7 denoted by the term '5 + 2'; if he were there would be nothing to compute. Rather, he is related to the very *procedure*, i.e. the *meaning* of the term '5 + 2'. He wants to find out what the *procedure* of applying the function plus to the arguments 5 and 2 produces.

Hence, we need a logical system that makes it possible to distinguish the three kinds of context, extensional, intensional and hyperintensional. And we have such a system at hand; it is Tichý's Transparent Intensional Logic (TIL).[1] TIL is a hyperintensional, typed λ-calculus of partial functions. A main feature of the λ-calculus is its ability to systematically distinguish between functions and functional values. An additional feature of TIL is its ability to systematically distinguish between functions and *modes of presentation* of functions and modes of presentation of functional values. These modes of presentation are *procedures* defined in TIL as six kinds of *constructions*.[2]

The goal of this paper is to demonstrate that Leibniz's Law is valid in all the three kinds of these contexts. The validity of the Law is not jeopardized by the context in which a given construction occurs. What depends on the context is the *type* of an object that can be validly substituted. In an extensional context, v-congruent constructions that happen to produce the same value for

---

[1] See [2,15].

[2] Recent years have seen a significant rise of interest in hyperintensional concepts, see, for instance [13]. For a summary of λ-calculi, see for instance [1] or [14]. These calculi are characterized as hyperintensional, because individuation of functions is not reduced to set-theoretical mappings. Rather, they operate with Church's functions-in-intensions individuated more finely than mappings. However, these functions-in-intensions (or rules for producing mappings) cannot be displayed as objects on which other procedures operate. Thus, TIL indeed introduces another higher level of abstraction which is the hyperintensional level of displayed procedures.

a given valuation v can be substituted. In an intensional context, equivalent constructions that produce the same function for every valuation v can be substituted. Finally, in a hyperintensional context merely procedural isomorphic constructions (meanings of synonymous terms) can be substituted. Procedurally isomorphic constructions are those that represent the same procedure; they may slightly differ from a formal point of view, yet these differences are negligible from the procedural point of view.[3]

The rest of this paper is organized as follows. In Sect. 2 we briefly introduce the TIL system. Section 3 introduces the rules of substitution that respect a context in which a construction occurs. In Sect. 4 the algorithm implementing these rules in the TIL-Script language is described. Finally, concluding remarks are in Sect. 5.

2 TIL in Brief

A detailed introduction to the TIL system is out of the scope of this paper. Referring to numerous papers and two books (see, e.g., [2,3,12,15]), we just briefly summarize.

The λ-terms of the TIL language denote abstract procedures that produce set-theoretical mappings (functions-in-extension) or lower-order procedures. These procedures are rigorously defined as TIL *constructions*. Being procedural objects, constructions can be executed in order to operate on input objects (of a lower-order type) and produce the object (if any) they are typed to produce, while non-procedural objects, i.e. non-constructions, cannot be executed. There are two atomic constructions that present input objects to be operated on. They are *Trivialization* and *Variables*. The Trivialization presents an object X without the mediation of any other procedures. Using the terminology of programming languages, the Trivialization of X, '^0X' in symbols, is just a *pointer* to X. Variables produce objects dependently on valuations; they v-construct. The execution of a Trivialization or a variable never fails to produce an object. However, the execution of some of the molecular constructions can fail to present an object they are typed to produce. When this happens, we say that a given construction is v-*improper*.

There are two dual molecular constructions which correspond to λ-abstraction and application in λ-calculi, namely *Closure* and *Composition*. λ-Closure, $[\lambda x_1 \ldots x_n X]$, transforms into the very procedure of producing a function by abstracting over the values of the variables $x_1, \ldots, x_n$. The Closure is not v-improper for any valuation v, as it always v-constructs a function. Composition, $[X X_1 \ldots X_n]$, is the very procedure of applying a function produced by the procedure X to the tuple-argument (if any) produced by the procedures $X_1, \ldots, X_n$. While Closure never fails to produce a function, Composition is v-improper if one or more of its constituents $X, X_1, \ldots, X_n$ are v-improper. The main source of improperness is application of a partial function f to an argument

---

[3] For details and definition of procedural isomorphism, see [6].

a such that the function f is not defined at a. Another cause of improperness can be type-theoretical incoherence of the Composition.

Each construction C can occur not only in *execution* mode to produce at most one object, but also be an object on which other (higher-order) constructions operate. In such a case, we say that C is *displayed* and occurs *hyperintensionally*. Constructions are displayed by Trivialization. Yet sometimes we need to cancel the effect of Trivialization and turn the displayed mode of a construction C into the execution mode. To this end, we have the so-called *Double Execution*, ^2C. It literally executes C twice over. Thus, if C v-constructs a construction D that in turn v-constructs an entity E, ^2C v-constructs E. Otherwise ^2C is v-improper.

With constructions of constructions, constructions of functions, functions, and functional values in our stratified ontology, we need to keep track of the traffic between multiple logical strata. The *ramified type hierarchy* does just that. The type of first-order objects includes all objects that are not constructions. Therefore, it includes not only the standard objects of individuals, truth-values, sets, etc., but also functions defined on possible worlds (i.e., intensions germane to possible-world semantics). The type of second-order objects includes constructions of first-order objects and functions that have such constructions in their domain or range. The type of third-order objects includes constructions of first- or second-order objects and functions that have such constructions in their domain or range. And so on, ad infinitum.

Basic types of non-procedural objects are these: o (truth-values **T**, **F**), ι (individuals), τ (times or real numbers), ω (possible worlds). Types of procedural objects are $*_n$, where n is the order of a construction. Functional types are mappings formed by composing these types. Hence, where $\alpha, \beta_1, \ldots, \beta_n$ are types, $(\alpha\beta_1 \ldots \beta_n)$ is a functional type comprising mappings from $(\beta_1 \times \cdots \times \beta_n)$ to α.

3 Valid Substitutions in TIL

In this section we describe the substitution rules for TIL. These rules are an adjusted version of the rules presented in [9]. Based on the context recognition we can validly apply the substitution of an object of the correct type.

3.1 Three Kinds of Contexts

The rigorous formal definition of the three kinds of context is rather complicated, yet the basic ideas are simple.[4] Thus, here we present just those basic ideas.

- *Hyperintensional* context; a subconstruction C of a construction D does not occur in D in executed mode to construct a function. Rather, C itself is *displayed* as an argument of another function presented within D.
- *Intensional* context; a subconstruction C of a construction D occurs in executed mode to construct a *function* but not its value. Moreover, C is not nested within D in a hyperintensional context.

---
[4] For details, see [2, Sect. 2.6].

- *Extensional* context; a subconstruction C of a construction D occurs in executed mode to construct a function and its *value*. Moreover, C is not nested within D in a higher intensional or hyperintensional context.

When C occurs executed, to determine whether C occurs extensionally or intensionally, we must take into account the effect of λ-genericity. Each Closure creates a λ-generic context that turns a construction C occurring with extensional supposition to an intensional occurrence. This is due to the fact that Composition (application of a function to its argument) and Closure (λ-abstraction) are dual operations. Here are a few examples. Where the variables x, y range over τ and the function plus (+) is of type (τττ), the effect of λ-genericity is demonstrated as follows.

$[^0\!+ x\, y]$	non-generic context ($^0\!+$ occurs extensionally)
$\lambda x\, [^0\!+ x\, y]$	τ-generic context ($^0\!+$ occurs intensionally)
$\lambda x\, \lambda y\, [^0\!+ x\, y]$	(ττ)-generic context ($^0\!+$ occurs intensionally)
$[\lambda x\, \lambda y\, [^0\!+ x\, y]\, ^0 4]$	τ-generic context ($^0\!+$ occurs intensionally)
$[[\lambda x\, \lambda y\, [^0\!+ x\, y]\, ^0 4]\, ^0 5]$	non-generic context ($^0\!+$ occurs extensionally)

3.2 v-congruent, Equivalent, and Procedurally Isomorphic Constructions

There are three basic kinds of equivalence between constructions.

v-congruent Constructions. Constructions C, D are *v-congruent* (in symbols '$C \approx_v D$') if for a given valuation v they v-construct the same object or are both v-improper.

Examples: $[^0\!+ x\, ^0 1] \approx_{v(2/x)} [^0\!- \,^0 5\, x]$. Both constructions construct the number 3 for a valuation $v(2/x)$.

$^0 Klaus \approx_v [^0 President\_of_{wt}\, ^0 CR]$. Both constructions v-construct the same individual for a valuation that assigns the actual world to the variable w and any time point from the interval March 7th 2003 and March 7th 2013 to the variable t.

Equivalent Constructions. Constructions C, D are *equivalent* (in symbols '$C \approx D$') if they are v-congruent for every valuation v.

Examples: $^0 Power \approx \lambda x\, [^0\!*\, x\, x]$. Both constructions construct the same function that takes a number to its second power.

$[^0\!*\, x\, ^0 0] \approx [^0\!-\, x\, x]$. Both constructions v-construct the number 0 for every valuation v.

Procedurally Isomorphic Constructions. The problem of procedural isomorphism is the problem of co-hyperintensionality, and in general the problem of the structural isomorphism of meanings, hence synonymy. The individuation of procedures assigned to expressions as their structured meaning cannot be decided in virtue of a universal criterion applicable to every language.[5] Yet, Duží in [6] specifies a set of rigorously defined criteria of fine-grained procedural individuation, partially ordered according to the degree of their being permissive with respect to synonymy. The most suitable criterion for an ordinary, non-professional language is the criterion declaring that procedural isomorphism of TIL constructions obtains whenever the differences between constructions consist just in technical manipulations with λ-bound variables. Thus, in this paper we vote for the safest criterion that includes only α-conversion. Here is a simple example. Where $\approx_*$ is the relation of procedural isomorphism and variables x, y, z range over natural numbers, the following holds:

$$\lambda x\,[^0{+}\,x\,{}^0 1] \approx_* \lambda y\,[^0{+}\,y\,{}^0 1] \approx_* \lambda z\,[^0{+}\,z\,{}^0 1]$$

All the three constructions construct the successor function in a procedurally isomorphic way. Though they are formally distinct, it does not matter which variable is used to supply a number to which the number 1 is added, because the Closure abstracts over the value of the variable.

Remark. Obviously, constructions that are procedurally isomorphic are also equivalent, and thus v-congruent as well, but not vice versa.

3.3 Substitution of v-congruent Constructions

Substitution of v-congruent constructions is valid in an extensional or non-generic context. In other words, we can replace different constructions that happen to v-construct the same value of possibly different functions.

Example:

$$\frac{\lambda w \lambda t\,[^0 President_{wt}\,{}^0 CR]_{wt} \approx_v {}^0 Klaus \qquad [^0 Economist_{wt}\,\lambda w \lambda t\,[^0 President_{wt}\,{}^0 CR]_{wt}]}{[^0 Economist_{wt}\,{}^0 Klaus]}$$

Types: CR/ι; $(Vaclav)Klaus/\iota$; $President/(\iota\iota)_{\tau\omega}$: an attribute, i.e. an empirical function that assigns to an individual another individual dependently on world and time; $Economist/(o\iota)_{\tau\omega}$: a property of individuals. $\lambda w \lambda t [^0 President_{wt}\,{}^0 CR]/*_1 \to \iota_{\tau\omega}$: the individual office.

In both premises the construction $\lambda w \lambda t\,[^0 President_{wt}\,{}^0 CR]$ occurs extensionally, and the constructions $\lambda w \lambda t\,[^0 President_{wt}\,{}^0 CR]_{wt}$ and $^0 Klaus$ are v-congruent. In the first premise, the value of the produced presidential office in

---
[5] Faroldi in [11] makes a similar point. See also [5].

a world w and time t of evaluation is identified with the individual Klaus. The second premise assigns to the same value the property of being an economist. Hence, we can validly substitute ^0Klaus for $\lambda w\lambda t\, [^0President_{wt}\, ^0CR]_{wt}$.

3.4 Substitution of Equivalent Constructions

Substitution of equivalent constructions is valid in an intensional, and thus also in extensional context, of course.

Example:

$$\frac{[^0Wants\_to\_become_{wt}\, ^0Drahos\ \lambda w\lambda t\, [^0President_{wt}\, ^0CR]]\quad \lambda w\lambda t\, [^0President_{wt}\, ^0CR] \approx \lambda w\lambda t\, [^0Highest\_Representative_{wt}\, ^0CR]}{[^0Wants\_to\_become_{wt}\, ^0Drahos\ \lambda w\lambda t\, [^0Highest\_Representative_{wt}\, ^0CR]]}$$

Types: CR/ι; $Drahos/\iota$; $Wants\_to\_become/(o\iota_{\tau\omega})_{\tau\omega}$: a relation between indivdual and individual office; $President/(\iota)_{\tau\omega}$; $Highest\_Representative/(\iota)_{\tau\omega}$; $\lambda w\lambda t\, [^0Highest\_Representative_{wt}\, ^0CR]/*_1 \to \iota_{\tau\omega}$: the individual office.

The substitution is valid, provided both the Closures $\lambda w\lambda t\, [^0President_{wt}\, ^0CR]$ and $\lambda w\lambda t\, [^0Highest\_Representative_{wt}\, ^0CR]$ produce the same office, which we assume in the second premise.

3.5 Substitution of Procedurally Isomorphic Constructions

Substitution of procedurally isomorphic constructions is valid in a hyperintensional and thus also intensional and extensional context, because as mentioned above every procedurally isomorphic pair of constructions is also equivalent and v-congruent.

$$\frac{[^0Solves_{wt}\, ^0Tom\ ^0[\lambda x\, [[^0+ x\, ^05] = {}^010]]\quad ^0[\lambda x\, [[^0+ x\, ^05] = {}^010]] \approx_* {}^0[\lambda y\, [[^0+ y\, ^05] = {}^010]]}{[^0Solves_{wt}\, ^0Tom\ ^0[\lambda y\, [[^0+ y\, ^05] = {}^010]]}$$

Types: Tom/ι; $Solves/(o\iota*_n)_{\tau\omega}$; $x, y \to_v \tau$; $\approx_*/(o*_n*_n)$: the relation of procedural isomorphism.

4 Implementation

In this section we describe the implementation of the above introduced substitution rules in the tool for TIL-Script processing. The TIL-Script language is an operationally isomorphic syntactic variant of TIL. This language is used for computer processing; thus, the TIL language has been slightly syntactically adjusted. For instance, non-ASCII characters have been replaced by ASCII terms, notation for Trivialization of an object C is $'C$ instead of ^0C and instead of Greek λ we use the backslash \.

4.1 Introduction to the Tool for TIL-Script Processing

The substitution rules were implemented as a new functionality in the tool performing TIL-Script analysis, type-checking, and context recognition.[6] The input TIL-Script file consists of a set of constructions together with their typing. First, lexical and syntactic analysis is being performed followed by the decomposition of constructions into their subconstructions. As a result, for each construction we obtain its tree structure to which the depth-first search algorithm is applied. Figure 1 illustrates an example of such a tree. The input root construction is the meaning of the sentence "Tilman is seeking the last decimal of the number π." that in TIL comes down to $(Seek/(o\iota*_n)_{\tau\omega})$.

$$\lambda w \lambda t \,[^0 Seek_{wt}\,{}^0 Tilman\,{}^0[^0 Lastdec\,{}^0\pi]]$$

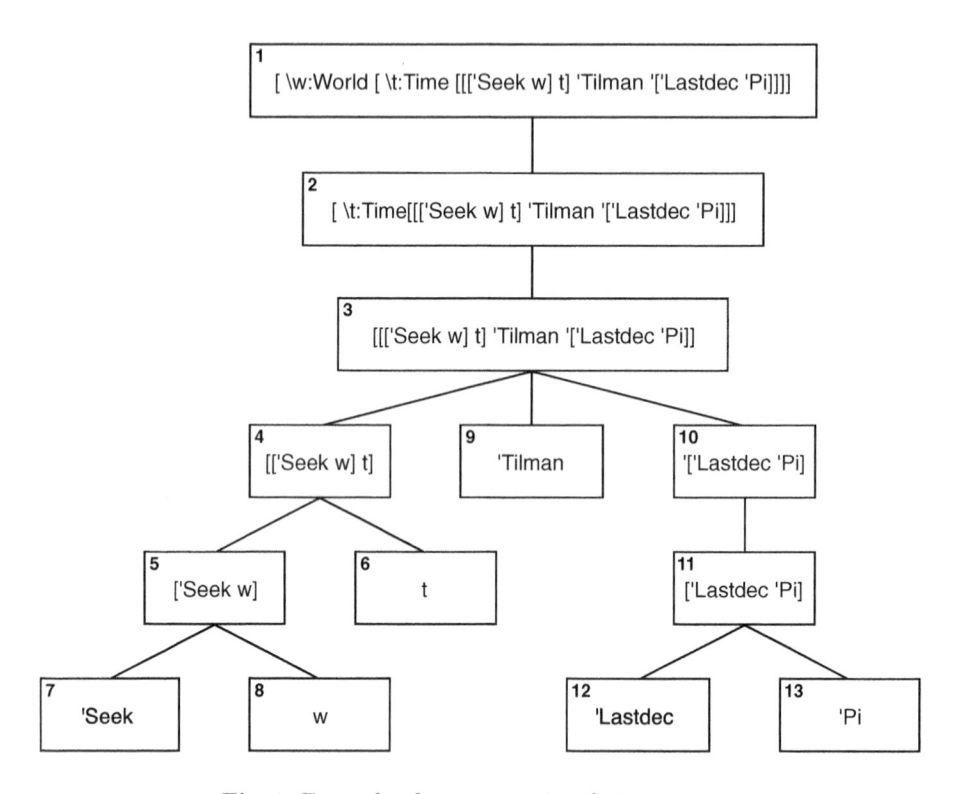

Fig. 1. Example of a construction derivation tree

The results of analysis are exported into XML tree file where each construction is provided by typing and the context of its occurrence. With the new functionality for substitutions, the format of this XML file has been enriched; each

---
[6] For more details on this tool and the algorithm of context recognition see [7].

construction is provided by the 'expl' or 'impl' label that indicates whether the construction has been conveyed by the input TIL-Script file ('expl') or generated by applying a substitution ('impl').

4.2 Equivalence Classes of v-congruent, Equivalent, and Procedurally Isomorphic Constructions

All the three relations on the set of constructions, namely v-congruency, equivalence and procedural isomorphism are obviously reflexive, symmetric, and transitive, hence they are equivalences. Thus, it is useful to create factor sets of equivalence classes of constructions. An effective way to store equivalence classes in a computer program is the disjoint-set (also *union-find*) data structure that tracks a set of elements partitioned into a number of disjoint subsets.

A disjoint-set forest consists of several elements (equivalence classes) each of which has a parent node, and, in efficient algorithms, either a size or a 'rank' value. Each set in the forest is represented by a tree of elements provided with a parent-pointer. If an element's parent pointer points to itself, then the element is the root of a tree and thus the *representative* member of its class. If an element has a parent, the element is part of whatever class is identified by following the chain of parents upwards until a representative element (the one without a parent) is reached at the root of the tree. If two elements have the same representative element then they belong to the same class, which means that they are equivalent.

The algorithm applies three main operations, namely *MakeSet*, *Find* and *Union*. Functioning the *MakeSet* operation is obvious. The $Find(x)$ follows the chain of parent pointers from x up to the representative element. The $Union(x,y)$ uses *Find* to determine the roots of the classes x and y belong to. If the roots are distinct, the classes are combined by attaching the roots to each other. In our algorithm we apply also the 'rank' value that makes the union of two sets more efficient. *Union by rank* always attaches the shorter tree to the root of the taller tree. Thus, the resulting tree is at most one node taller than the originals.

In our algorithm for TIL-Script processing we have implemented three disjoint-set structures, namely VCongruent, Equivalent, ProcIsomorphic.

4.3 Substitution of Constructions

The substitution of constructions has been implemented in the Prolog language. The main procedure Substitute has three parameters that are the identifiers of constructions. They are arguments of the TIL function $Sub/(*_n*_n*_n)$ that operates on constructions in this way: $[^0Sub\,C_1\,C_2\,C_3]$ produces a construction D that is the result of the substitution of D_1 produced by C_1 for D_2 produced by C_2 into D_3 produced by C_3. Thus, the first identifier is that of C_3, the second of C_2 and the third one of C_1. The procedure recursively traverses the structure trees with the depth-first strategy and generates new identifiers for subconstructions that result as a product-tree of the valid substitution (with the label 'impl'). This process is illustrated by Fig. 2.

The result of the process illustrated by Fig. 2 can be found on Fig. 3.

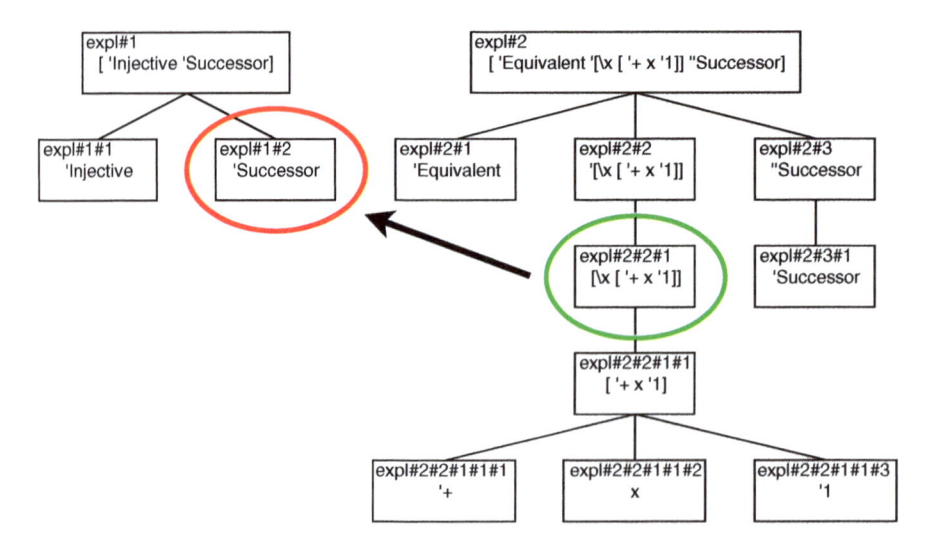

Fig. 2. Example of substitute(expl#1, expl#1#2, expl#2#2#1).

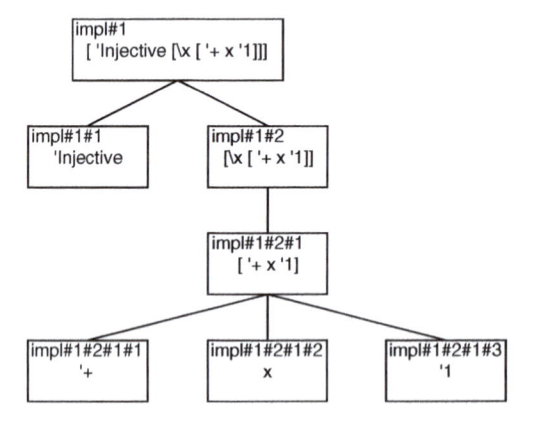

Fig. 3. Result of the substitution.

4.4 Generating Process

With the mechanism of substitution as described above we also need the procedure that goes through the tree-structure of a construction and assigns to a given (sub-) construction the ID of its root predecessor. To this end we have a simple yet important procedure RootID. The following pseudocode illustrates the whole process.

```
For every construction D:
   Determine the context in which D occurs;
   With respect to the context find an equivalence class Eq to
   which D belongs;
       For every construction E belonging to the class Eq
           If E is not identical with D
               IDC = RootID(IDD)
               Substitute(IDC,IDD,IDE)
```

5 Conclusion

In this paper we described the algorithm of substitution with respect to the context in which a given meaning construction occurs. The algorithm has been thoroughly tested on the large number of TIL-Script constructions. All the substitutions have been performed correctly, according to the rules and restrictions as presented in Sect. 3. Even though there have been thousands input constructions, performance of the algorithm turned out to be high so that the substitutions took just hundreds of milliseconds. There is still a space for performance improvements, however; this might be needed in case of larger input-sets of constructions. The presented algorithm is a component of the system of natural-language processing developed within our project. The other components are, inter alia, the algorithm for context recognition and type checking (see [8]). All these components belong to the system of an inference machine that has been developed for the TIL-Script language based on the specification that can be found in [3] or [4]. Future research will concentrate on the implementation of a question-answering system over large textual data as specified in [10] within the TIL-Script inference machine.

Acknowledgments. This research was supported by the Grant Agency of the Czech Republic, project No. GA18-23891S "Hyperintensional Reasoning over Natural Language Texts", by the internal grant agency of VSB-Technical University of Ostrava, project No. SP2018/172, "Application of Formal Methods in Knowledge Modelling and Software Engineering", and also by the Moravian-Silesian region within the program "Support of science and research in Moravian-Silesian region 2017" (RRC/10/2017).

References

1. Barendregt, H.: The Lambda Calculus: Its Syntax and Semantics (Studies in Logic and the Foundations of Mathematics 103), 2nd edn. North-Holland, Amsterdam (1985)
2. Duží, M., Jespersen, B., Materna, P.: Procedural Semantics for Hyperintensional Logic. Foundations and Applications of Trasnsparent Intensional Logic. Springer, Berlin (2010)
3. Duží, M.: Extensional logic of hyperintensions. In: Lecture Notes in Computer Science, vol. 7260, pp. 268–290 (2012)

4. Duží, M.: Towards an extensional calculus of hyperintensions. Organon F **19**(1), 20–45 (2012)
5. Duží, M.: Structural isomorphism of meaning and synonymy. Computacion y Sistemas **18**(3), 439–453 (2014)
6. Duží, M.: If structured propositions are logical procedures then how are procedures individuated? Synthese special issue on the Unity of propositions (2017). https://doi.org/10.1007/s11229-017-1595-5
7. Duží, M., Fait, M.: The algorithm of context recognition in TIL. In: Rychlý, P., Horák, A., Rambousek, A. (eds.) 10th Workshop on Recent Advances in Slavonic Natural Language Processing, RASLAN 2016, pp. 51–62. Tribun, EU s.r.o (2016)
8. Duží, M., Fait, M., Menšík, M.: Context recognition for a hyperintensional inference machine. In: The AIP Proceeding of ICNAAM 2016, International Conference of Numerical Analysis and Applied Mathematics, vol. 1863 (2017). Article No. 330004
9. Duží, M., Materna, P.: Validity and applicability of Leibniz's law of substitution of identicals. In: Arazim, P., Lavička, T. (eds.) The Logica Yearbook 2016, pp. 17–35. College Publications, London (2017)
10. Duží, M., Jespersen, B.: An intelligent question-answer system over natural-language texts. In: These Proceedings (2018)
11. Faroldi, F.L.G.: Co-hyperintensionality. Ratio **30**(3), 270–287 (2017)
12. Jespersen, B., Carrara, M., Duží, M.: Iterated privation and positive predication. J. Appl. Log. **25**, S48–S71 (2017)
13. Nolan, D.: Hyperintensional Metaphysics. Philos. Stud. **171**(1), 149–160 (2014)
14. Revesz, G.E.: Lambda-Calculus, Combinators, and Functional Programming. Cambridge University Press, Cambridge, reprinted 2008 (1988)
15. Tichý, P.: The Foundations of Frege's Logic. Walter de Gruyter, Berlin (1988)

Fuzzy Model Predictive Control for Discrete-Time System with Input Delays

Sofiane Bououden[1(✉)], Ilyes Boulkaibet[2], Mohammed Chadli[3], and Ivan Zelinka[4]

[1] Department of Industrial Engineering, University Abbes Laghrour, Khenchela, Algeria
sofiane.bououden@univ-khenchela.dz
[2] University of Johannesburg, Johannesburg, South Africa
ilyes@aims.ac.za
[3] University of Picardie Jules Verne-MIS, EA 4290 Amiens, France
mohammed.chadli@u-picardie.fr
[4] Faculty of Electrical Engineering and Computer Science, VŠB-TUO, Ostrava, Czech Republic
ivan.zelinka@vsb.cz, zelinkaivan65@gmail.com

Abstract. In this paper, we present a robust fuzzy model predictive control (RFMPC) for a class of discrete-time system with input delays. The system is represented into a Takagi-Sugeno (T-S) discrete fuzzy model. Based on the Lyapunov functions theory, some required sufficient conditions are established in terms of linear-matrix inequalities (LMIs). The provided conditions are obtained through a fuzzy Lyapunov function candidate and a non-PDC control law, which can guarantee that the resulting closed-loop fuzzy system is asymptotically stable. A numerical example is provided to illustrate the effectiveness of the control algorithm.

Keywords: Takagi–Sugeno (T–S) fuzzy system · Time-delays · Model predictive control · Linear-matrix inequalities (LMIs)

1 Introduction

The systems with delays occur extensively in various industrial systems and have attracted many research interests [1–3]. Because the presence of a time-delay often causes serious deterioration of the stability and performance of the system, many researchers have been interested in the control for time delay systems [4–6].

In practice, most industrial plants are nonlinear systems, which lead to additional difficulties for the analysis and design of control systems. Several nonlinear MPC approaches have been presented in order to address the nonlinear MPC stability issues through the Lyapunov stability theorem and Takagi–Sugeno (T-S) fuzzy model [7–9].

On the other hand, as a kind of optimization control methods which can handle constraints, model predictive control (MPC) has been widely investigated and utilized in industrial processes. However, only a few results can be seen on MPC for time-delay systems. In [10], a fuzzy constrained MPC controller is proposed. In [11], a robust MPC for continuous-time T-S systems is established, However, in [11], the delayed T-S system was not considered. In [12], the authors investigated the problem of robust constrained MPC for T-S fuzzy system with PDC and non-PDC law, [13]. In these works, the nonlinear systems with input delay are not addressed. However, the results only apply to linear systems and state and input constraints is not considered [14]. It has not been investigated for nonlinear time-delay systems. For T–S fuzzy-discrete systems, which are subject to PDC law and non-PDC law, the problem of constrained MPC has a lot of room for improvement, which motivates this study. Motivated by this, we investigate the problem of constrained MPC has that a lot of room, for improvement, for T–S fuzzy-discrete systems, which are subject to non-PDC law.

The proposed MPC is based on a Takagi-Sugeno (T-S) fuzzy model, non-parallel distributed compensation (non PDC) fuzzy controller and Lyapunov-Krasovskii function. In the first the T-S fuzzy systems are adopted to approximate nonlinear uncertain discrete-time systems with input delays [15–20]. After defining an optimization problem that minimizes a cost function at each time instant, a state feedback is computed by minimizing the upper bound of the cost function subject to constraints on inputs. Finally, the proposed nonlinear robust FMPC technique is applied to DC-DC converters as a case study.

The remainder of this paper is organized as follows. In Sect. 2, The system dynamics, as well as the robust fuzzy model predictive control (RFMPC) with input delays are introduced. Section 3 presents the main results of this paper, in which the robust MPC approach is obtained using linear matrix inequalities (LMIs). For illustration, a numerical example is provided in Sect. 4. Finally, concluding remarks are presented in Sect. 5.

2 Problem Statement

In this section, a discrete-time nonlinear uncertain system is considered, which can be represented by T-S fuzzy dynamic model in which the ith rule is described as follows:

Plan rule i: If $z_1(k)$ is F_{i1}, and $z_2(k)$ is F_{i2} and .... and $z_b(k)$ is F_{ib} then

$$\begin{cases} x(k+1) = (A_i + \delta A_i)x(k) + (B_i + \delta B_i)u(k-d) \\ y(k) = (C_i + \delta C_i)x(k) \end{cases} \quad (1)$$

Where $z_1(k), z_2(k),\ldots, z_b(k)$ are the premise variable, and F_{ih}, $i = 1, 2, \ldots, r$, $h = 1, 2, \ldots, b$ are fuzzy sets, $x(k) \in R^n$ is the state variable, $u(k) \in R^m$ is the control input, $y(k) \in R^q$ output vectors. Also $d > 0$ is the time-invarying input delay. A_i, B_i and C_i are known system matrices with appropriate dimensions. δA_i, δB_i, δC_i are matrix functions representing parameter uncertainties. r is the number of IF–THEN rules. For the sake of notational convenience, let $\vartheta := 1, 2, \ldots, r$.

Then the equivalent T-S fuzzy model is represented as follows:

$$\begin{cases} x(k+1) = \sum_{i=1}^{r} h_i(z(k))(A_i + \delta A_i)x(k) + (B_i + \delta B_i)u(k-d) \\ y(k) = \sum_{i=1}^{r} h_i(z(k))(C_i + \delta C_i)x(k) \end{cases} \quad (2)$$

Let $F_{ib}(z(k))$ denote the membership function of $z(k)$ in F_{ib}. Then $\mu_i(z(k)) = \prod_{h=1}^{b} F_{ib}(h_h(k))$, and $h_i(z(k)) = \mu_i(z(k))/\sum_{i=1}^{r} \mu_i(z(k))$. Considering $h_i(z(k)) \geq 0$, $i = 1, 2, \ldots, r$, $\sum_{i=1}^{r} \mu_i(z(k)) > 0$. Therefore, we have $h_i(z(k)) \geq 0$, $i = 1, 2, \ldots, r$, and $\sum_{i=1}^{r} h_i(z(k)) = 1$.

Therefore the entire fuzzy converter model corresponds to

$$\begin{cases} x(k+1) = A_v x(k) + B_v u(k) \\ y(k) = C_v x(k) \end{cases} \quad (3)$$

Where

$$A_v = \sum_{i=1}^{r} h_i(z(k))(A_i + \delta A_i),$$

$$B_v = \sum_{i=1}^{r} h_i(z(k))(B_i + \delta B_i)$$

$$C_v = \sum_{i=1}^{r} h_i(z(k))(C_i + \delta C_i),$$

In the following section, the control strategy, which takes into account constraints such as control effort and state variables, is presented.

Lemma 1. The following conditions are equivalent:

There exist a symmetric matrix $P > 0$ and a constant matrix Q such that

$$A^T P A - Q < 0$$

$$\begin{bmatrix} -Q & A^T P \\ PA & -P \end{bmatrix} < 0$$

3 Model Predictive Controller Design

In this section we present the concept and basic procedure of the MPC technique. At each time, the control objective of the MPC problem is to compute the control move $u(k+j/k)$ by minimizing the following performance index:

$$\min_{u(k+i/k),\, i \geq 0} \max J_\infty(k) = \sum_{j=0}^{\infty} \|x(k+j/k)\|_Q^2 + \|u(k+j/k)\|_R^2 \quad (4)$$

$$|uj(k+j/k)| \leq u_{s,\max},\, j \geq 0 \; s = 1, \ldots, m \quad (5)$$

where $Q > 0$, $R > 0$, are two known weighting matrices.

Equations (4), (5) is a constrained min max optimization problem corresponding to a worst-case infinite-horizon MPC with a quadratic objective. Hence, the overall fuzzy control law is represented as

$$u(k+j/k) = \sum_{l=1}^{r} h_n(z(k))F_l x(k+j/k) \qquad (6)$$

Where F_l $(n = 1, 2, \ldots, r)$ are the local control gains.

Applying the fuzzy controller (6) to system (3) yields the closed-loop system can be obtained:

$$x(k+1) = \sum_{l=1}^{r} A_i x(k) - \sum_{i=1}^{r}\sum_{l=1}^{r} B_i F_l x(k-d) \qquad (7)$$

To obtain the stability conditions, Lyapunov-Krasovskii function is defined as:

$$\begin{aligned}V(x(k/k)) &= x^T(k/k)Sx(k/k) \\ &+ \sum_{i=1}^{d} x^T(k-i/k)Wx(k-i) \\ S^T &> 0, W^T > 0 \\ \text{And } V(x(k+j)) &= x^T(k+j)Sx(k+j) \\ &+ \sum_{i=1}^{d} x^T(k+j-i)Wx(k+j-i)\end{aligned} \qquad (8)$$

For any $j \geq 0$, suppose $V(x(k))$ satisfies the following stability constraint:

$$\begin{aligned}&V(x(k+j+1/k)) - V(x(k+j/k)) \\ &\leq -\left[\|x(k+j/k)\|_Q^2 + \|u(k+j/k)\|_R^2\right]\end{aligned} \qquad (9)$$

As it is assumed that summation is up to ∞, i.e., $j \to \infty$, $x(\infty) = 0$. Summing (5nonpdc) from $j = 0$ to ∞ yields

$$J(k) \leq V(x(k/k)) \qquad (10)$$

By defining $V(x(k/k)) \leq \gamma$, an upper bound on the performance index is obtained as $J(k) \leq \gamma$.

Theorem 1. Let us consider the closed-loop time delay systems (6) at each time k and let $x(k/k)$ be the measured state $x(k)$. There exists a state feedback control law, meeting the performance objective (4) if there exist symmetric positive definite matrices $P_v > 0$, N_v and W_v and a positive scalar γ satisfying the following convex optimization: problem:

$$\min_{\gamma, P_v, Y_v, N_v} \gamma \qquad (11)$$

Subject to

$$\begin{bmatrix} -\gamma & X^T(k/k) \\ X(k/k) & -\Gamma \end{bmatrix} < 0 \quad (12)$$

$$\begin{bmatrix} -P_v & P_v A_v^T & P_v & 0 & P_v Q^{1/2} & 0 \\ A_v P_v & -P_v & 0 & 0 & 0 & -X_v^{-T} R^{1/2} \\ P_v & 0 & -M_v & 0 & 0 & 0 \\ 0 & 0 & 0 & -M_v & X_v^{-T} B_v^T & 0 \\ Q^{1/2} P_v & 0 & 0 & B_v X_v^{-1} & -P_v & 0 \\ 0 & -R^{1/2} X_v^{-1} & 0 & 0 & 0 & \gamma I \end{bmatrix} \quad (13)$$

Proof. At sampling time, define the following Lyapunov function:

$$V(x(k+j+1/k)) - V(x(k+j/k)) \leq -\left[\|x(k+j/k)\|_Q^2 + \|u(k+j/k)\|_R^2\right] \quad (14)$$

Using the state-feedback control law

$$u(k+j) = \left(\sum_{l=1}^{r} h_l(z(k)) W_n\right) \left(\sum_{l=1}^{r} h_l(z(k)) X_l\right)^{-1} x(k+j) \quad (15)$$

Let $W_v = \sum_{l=1}^{r} h_l(z(k)) W_l$, $X_v = \sum_{l=1}^{r} h_l(z(k)) X_l$. Then, the following closed-loop system can be obtained:

$$\begin{cases} x(k+1) = A_v x(k) - B_v W_v X_v^{-1} x(k-d) \\ y(k) = C_v x(k) \end{cases} \quad (16)$$

Moreover, we will refer to the Lyapunov-Krasovskii function (7) as

$$V(X(k/k)) = X^T(k/k) \Gamma X(k/k) \quad (17)$$

Where $X(k/k) = [x^T(k/k), x^T(k-k/k), \ldots, x^T(k-d/k)]$ $\Gamma = diag[S_v, W_v, \ldots, W_v]$
With matrix S_v repeated d times
By Schur's complement, the above inequality is equivalent to

$$\begin{bmatrix} -\gamma & X^T(k/k) \\ X(k/k) & -\Gamma \end{bmatrix} < 0$$

Hence, LMI (1) is obtained.

Let $u(k+j/k) = -W_v X_v^{-1} x(k+j/k)$. Then, the following closed-loop system can be obtained:

$$x(k+j+1/k) = A_v x(k+j/k) - B_v W_v X_v^{-1} x(k+j-d/k) \quad (18)$$

According to system (1) we have

$$\begin{aligned}
& [x^T(k+j+1/k) S_v x(k+j+1/k)] - [x^T(k+j/k) S_v x(k+j/k)] \\
& + [x^T(k+j/k) W_v x(k+j/k)] - [x^T(k+j-d/k) W_v x(k+j-d/k)] \\
& < -[(x^T(k+j/k) R x(k+j/k)) + (u^T(k+j/k) Q u(k+j/k))]
\end{aligned} \quad (19)$$

$$\begin{aligned}
& [x^T(k+j/k)\, x^T(k+i-d/k)] \begin{bmatrix} A_v^T \\ -(W_v X_v^{-1})^T B_v^T \end{bmatrix} S_v \\
& [A_v - B_v W_v X_v^{-1}] \begin{bmatrix} x(k+j/k) \\ x(k+i-d/k) \end{bmatrix} \\
& + \\
& [x^T(k+j/k)\, x^T(k+i-d/k)] \begin{bmatrix} -S_v^T + W_v^T & 0 \\ 0 & -W_v^T \end{bmatrix} \\
& \begin{bmatrix} x(k+j/k) \\ x(k+i-d/k) \end{bmatrix}
\end{aligned} \quad (20)$$

$$\leq [x^T(k+j/k)\, x^T(k+i-d/k)]$$

$$\begin{bmatrix} Q^{1/2} & 0 \\ 0 & (-W_v X_v^{-1})^T R^{1/2} \end{bmatrix} I \begin{bmatrix} Q^{1/2} & 0 \\ 0 & R^{1/2}(-W_v X_v^{-1}) \end{bmatrix} \begin{bmatrix} x(k+j/k) \\ x(k+i-d/k) \end{bmatrix}$$

$$\begin{aligned}
& [x^T(k+j/k)\, x^T(k+i-d/k)] \\
& \begin{bmatrix} -S_v + W_v + A_v^T S_v A_v & -A_v^T S_v B_v W_v X_v^{-1} \\ -(W_v X_v^{-1})^T B_v^T S_v A_v & -W_v + (W_v X_v^{-1})^T B_v^T S_v B_v W_v X_v^{-1} \end{bmatrix} \\
& \begin{bmatrix} x(k+j/k) \\ x(k+i-d/k) \end{bmatrix} \\
& \leq [x^T(k+j/k)\, x^T(k+i-d/k)]
\end{aligned} \quad (21)$$

$$\begin{bmatrix} Q^{1/2} & 0 \\ 0 & -(W_v X_v^{-1})^T R^{1/2} \end{bmatrix} I \begin{bmatrix} Q^{1/2} & 0 \\ 0 & -R^{1/2} W_v X_v \end{bmatrix}$$

$$\begin{bmatrix} x(k+j/k) \\ x(k+i-d/k) \end{bmatrix} \quad (22)$$

$$[x^T(k+j/k)\, x^T(k+i-d/k)]$$

$$\begin{bmatrix} -P_v + P_v M_v^{-1} P_v + P_v A_v^T P_v^{-1} A_v P_v & -M_v A_v^T P_v^{-1} B_v X_v^{-1} \\ -X_v^{-T} B_v^T P_v^{-1} A_v M_v & -M_v + X_v^{-T} B_v^T P_v^{-1} B_v X_v^{-1} \end{bmatrix}$$
$$\begin{bmatrix} x(k+j/k) \\ x(k+i-d/k) \end{bmatrix}$$
$$\leq [x^T(k+j/k) x^T(k+i-d/k)]$$
$$\begin{bmatrix} P_v Q^{1/2} & 0 \\ 0 & -X_v^{-T} R^{1/2} \end{bmatrix} \gamma^{-1} I \begin{bmatrix} Q^{1/2} P_v & 0 \\ 0 & -R^{1/2} X_v^{-1} \end{bmatrix}$$
$$\begin{bmatrix} x(k+j/k) \\ x(k+i-d/k) \end{bmatrix}$$

Applying the Schur complement, the above inequality can be rewritten as

$$\begin{bmatrix} -P_v & P_v A_v^T & P_v & 0 & P_v Q^{1/2} & 0 \\ A_v P_v & -P_v & 0 & 0 & 0 & -X_v^{-T} R^{1/2} \\ P_v & 0 & -M_v & 0 & 0 & 0 \\ 0 & 0 & 0 & -M_v & X_v^{-T} B_v^T & 0 \\ Q^{1/2} P_v & 0 & 0 & B_v X_v^{-1} & -P_v & 0 \\ 0 & -R^{1/2} X_v^{-1} & 0 & 0 & 0 & \gamma I \end{bmatrix}$$

When there are constraint bounds on the input $u(k+j/k)$, which can be used to constrain into LMI forms, then the following theorem is obtained.

Theorem 2. If there exists symmetric positive-definite matrices $\{P_i\}_{i\in v}$, $\{\phi_{ii}\}_{i\in v}$, general matrices $\{Y_i\}_{i\in v}$, $\{N_i\}_{i\in v}$ such that the following LMIs are verified:

$$\min_{\gamma, P_i, \phi_i, Y_i, N_i} \gamma \tag{23}$$

Subject to

$$\begin{bmatrix} -\gamma & X^T(k/k) \\ X(k/k) & -\Gamma \end{bmatrix} < 0 \tag{24}$$

$$\Upsilon_{ii} > \phi_{ii}, \quad i \in v \tag{25}$$

$$\Upsilon_{ij} + \Upsilon_{ij}^T > \phi_{ij} + \phi_{ij}^T \quad i \neq j, j \in v \tag{26}$$

$$\Upsilon_{ij} = \begin{bmatrix} -P_j & P_j A_i^T & P_j & 0 & P_j Q^{1/2} & 0 \\ A_i P_j & -P_v & 0 & 0 & 0 & -N_l^T R^{1/2} \\ P_j & 0 & -M_l & 0 & 0 & 0 \\ 0 & 0 & 0 & -M_l & N_l^T B_i^T & 0 \\ Q^{1/2} P_j & 0 & 0 & B_i N_l & -P_j & 0 \\ 0 & -R^{1/2} N_l & 0 & 0 & 0 & \gamma I \end{bmatrix}$$

With $i, j, l \in v$

$$\begin{bmatrix} -u_{max}^2 & -N_l \\ -N_l & -M_l \end{bmatrix} \leq 0 \quad l \in \upsilon \qquad (27)$$

Therefore, the feasible receding horizon state-feedback control law, which is obtained in Theorem 2, asymptotically stabilizes the closed-loop fuzzy system (6).

Proof. The proof can be carried out by a similar as in the proof of Theorem 1 and thus is omitted. This completes the proof.

4 Simulated Example

In this section we present a numerical example to illustrate the proposed RFMPC approach. The MPC is to be fed with a discrete time model which is easily obtained from the continuous model with the *c2d()* routine in MATLAB. A class of fuzzy-discrete system is described as follows.

$$A_1 = \begin{bmatrix} 1 & -1,9 \\ -1 & -0,5 \end{bmatrix}$$

$$A_2 = \begin{bmatrix} 1 & 1,9 \\ -1 & -0,5 \end{bmatrix}$$

$$B_1 = \begin{bmatrix} 6,9 \\ 2,8 \end{bmatrix}$$

$$B_2 = \begin{bmatrix} 3,1 \\ -2,8 \end{bmatrix}$$

$$C_1 = C_2 = \begin{bmatrix} 1 & 0 \end{bmatrix}$$

The number of affine systems used here is $r = 2$.

The weighting matrices Q and R in the cost function are selected as **diag**$\{2,2\}$, 1 respectively.

The simulation results for initial condition $x(-2) = x(-1) = x(0) = [0, 2, 0]T$, $u(-2) = u(-1) = u(0) = 0$.

To improve the response, we can design a fuzzy regulator by adding the constraint on the input, We have $\|u(t)\|_2 = 0, 1$.

For this example, the software LMI control toolbox in MATLAB environment was used to compute the solution of the LMI problem.

The results of the application of the proposed algorithm and its comparison with an MPC algorithm are shown in Figs. 1 and 2. We note that the RFMPC control provides a much better performance than that achieved with the FMPC and we observe that the response of the proposed algorithm is stable and with fewer fluctuations and with faster convergence than the MPC controller. However, the robust MPC method achieves better results, which reduces the fluctuations by 60% compared to the MPC (Fig. 3).

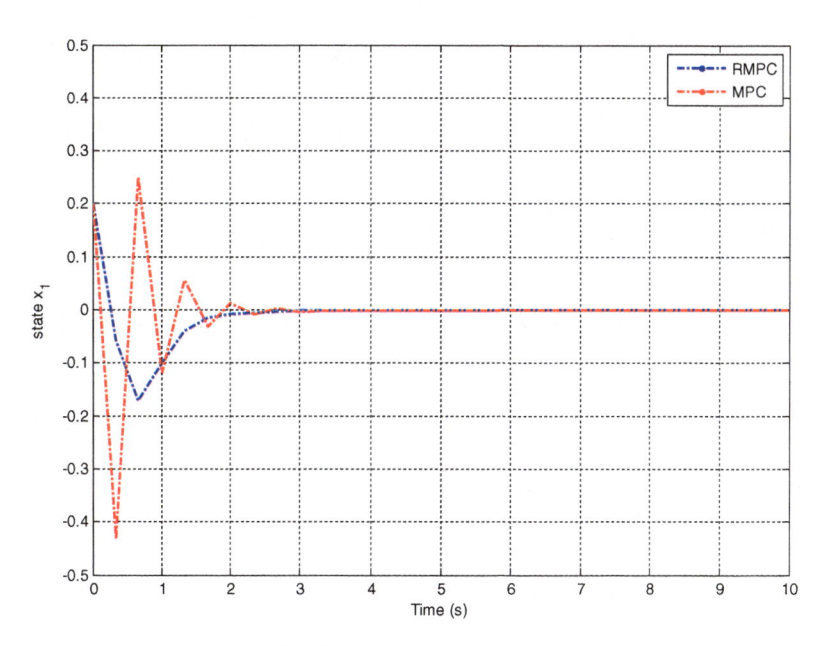

Fig. 1. Time response of the state $x1$.

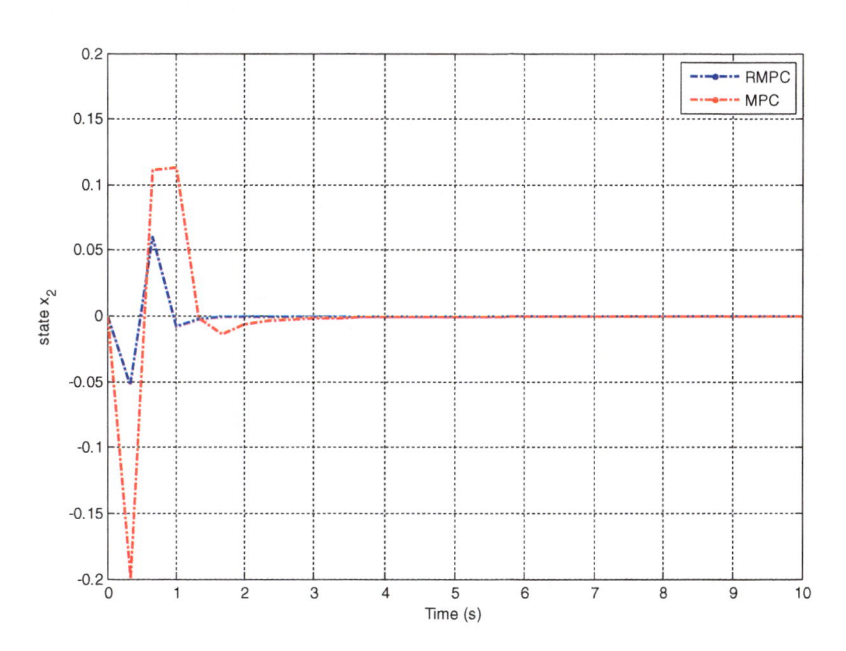

Fig. 2. Time response of the state $x2$.

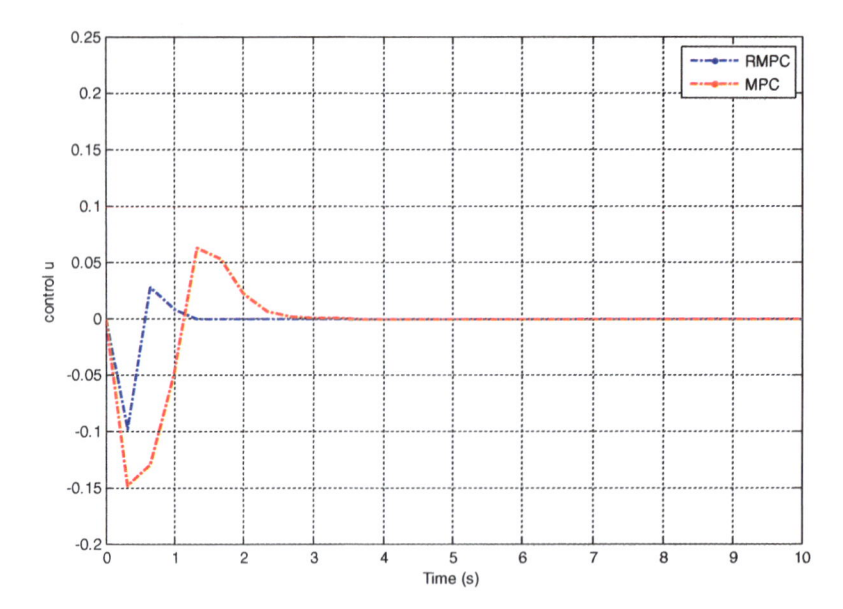

Fig. 3. Control input

5 Conclusion

In this paper, robust model predictive control algorithm for uncertain systems with input delays has been reformulated on optimization LMIs constraints. A nonlinear system considered in this paper can be represented by Takagi-Sugeno fuzzy model. The goal has to design, at each sampling time k, a state feedback control law that minimizes an upper bound of the worst-case objective function, subject to constraints of control inputs. A Lyapunov-Krasovskii function is introduced here for deriving an upper bound on the robust performance index. By minimizing the worst-case objective function, a state feedback controller is designed by using the LMIs, and the sufficient conditions for the existence of the RMPC controller are derived. It is shown that the receding horizon implementation of the feasible solutions guarantees closed-loop fuzzy system is asymptotically stable.

References

1. Liu, L., Yin, S., Zhang, L., Yin, X., Yan, H.: Improved results on asymptotic stabilization for stochastic nonlinear time-delay systems with application to a chemical reactor system. IEEE Trans. Syst. Man Cybern. Syst. **47**(1), 195–204 (2017)
2. Li, J., Chen, Z., Cai, D., Zhen, W., Huang, Q.: Delay-dependent stability control for power system with multiple time-delays. IEEE Trans. Power Syst. **31**(3), 2316–2326 (2016)
3. Qi, T., Zhu, J., Chen, J.: Fundamental limits on uncertain delays: when is a delay system stabilizable by LTI controllers. IEEE Trans. Autom. Control **62**(3), 1314–1328 (2017)

4. Farza, M., M'Saad, M., Ménard, T., Fall, M.L., Gehan, O., Pigeon, E.: Simple cascade observer for a class of nonlinear systems with long output delays. IEEE Trans. Autom. Control **60**(12), 3338–3343 (2015)
5. Bin Zhou; Zhao-Yan Li: Truncated predictor feedback for periodic linear systems with input delays with applications to the elliptical spacecraft rendezvous. IEEE Trans. Control Syst. Technol. **23**(6), 2238–2250 (2015)
6. Yonggang Chen; Shumin Fei; Yongmin Li: Robust stabilization for uncertain saturated time-delay systems: a distributed-delay-dependent polytopic approach. IEEE Trans. Autom. Control **62**(7), 3455–3460 (2017)
7. Bououden, S., Chadli, M., Karimi, H.R.: A robust predictive control design for nonlinear active car suspension systems. Asian J. Control **8**(1), 1–11 (2016)
8. Boulkaibet, I., Belarbi, K., Bououden, S., Marwala, T., Chdali, M.: A new T-S fuzzy model predictive control for nonlinear processes. Expert Syst. Appl. **88**, 132–151 (2017)
9. Bououden, S., Chadli, M., Zhang, L., Yang, T.: Constrained model predictive control for time-varying delays systems: application to an active car suspension. Int. J. Control Autom. Syst. **14**(1), 51–58 (2016)
10. Killian, M., Mayer, B., Schirrer, A., Kozek, M.: Cooperative fuzzy model-predictive control. IEEE Trans. Fuzzy Syst. **24**(2), 471–482 (2016)
11. Khooban, M.H., Vafamand, N., Niknam, T.: T-S fuzzy model predictive speed control of electrical vehicles. ISA Trans. **64**, 231–240 (2016)
12. Lu, Q., Shi, P., Lam, H.K., Zhao, Y.: Interval type-2 fuzzy model predictive control of nonlinear networked control, systems. IEEE Trans. Fuzzy Syst. **23**(6), 2317–2328 (2015)
13. Martins, M.A.F., Yamashita, A.S., Santoro, B.F., Odloak, D.: Robust model predictive control of integrating time delay processes. J. Process Control **23**(7), 917–932 (2013)
14. Bobal, V., Kubalcik, M., Dostal, P., Matejicek, J.: Adaptive predictive control of time-delay systems. Comput. Math. Appl. **66**, 165–176 (2013)
15. Tanaka, K., Wang, H.O.: Fuzzy control systems design and analysis. A Linear Matrix Inequality Approach. Wiley, New York (2001)
16. Ohtake, H., Tanaka, K.: Switching model construction and stability analysis for nonlinear systems. J. Adv. Comput. Intell. Intell. Inform. **10**(1), 3–10 (2006)
17. Ohtake, H., Tanaka, K., Wang, H.O.: Fuzzy modeling via sector nonlinearity concept. Integr. Comput.-Aided Eng. **10**(4), 333–341 (2003)
18. Bououden, S., Chadli, M., Allouani, F., Filali, S.: A new approach for fuzzy predictive adaptive controller design using particle swarm optimization algorithm. Int. J. Innovative Comput. Inf. Control **9**(9), 3741–3758 (2013)
19. Chadli, M., Karimi, H.R.: Robust observer design for unknown inputs Takagi-Sugeno models. IEEE Trans. Fuzzy Syst. **21**(1), 158–164 (2013)
20. Lu, J., Li, D., Xi, Y.: Probability-based constrained MPC for structured uncertain systems with state and random input delays. International Journal of Systems Science **44**(7), 1354–1365 (2013)

An Improvement of Fuzzy-Based Control Strategy for a Series Hydraulic Hybrid Truck

Tri-Vien Vu[1(✉)], Bach H. Dinh[1], Anh-Minh Duc Tran[1], Chih-Keng Chen[2], and Trung-Hieu Vu[3]

[1] Faculty of Electrical and Electronics Engineering, Ton Duc Thang University, 19 Nguyen Huu Tho Rd, Tan Phong, District 7, Ho Chi Minh City, Vietnam
{vutrivien,dinhhoangbach,tranducanhminh}@tdtu.edu.vn
[2] Department of Vehicle Engineering,
National Taipei University of Technology, No. 1, Sec 3,
Zhongxiao East Road Da-an District, Taipei City 106, Taiwan
[3] CT13TIE01, Ho Chi Minh City University of Technology,
Ho Chi Minh City, Vietnam

Abstract. This paper has proposed and investigated a fuzzy-based power management strategy for a light-duty series hydraulic hybrid truck (SHHT). In this approach, a fuzzy-based controller will use the vehicle speed and acceleration commands as the inputs to predict the desired accumulator pressure in order to satisfy the dynamic constraints in efficient manner. The performance of the proposed power management was then evaluated on a high fidelity SHHT model, which is developed in Matlab/Simulink, under typical driving cycles. The simulation results demonstrated that with the proposed supervisory controller, the SHHT system achieves better fuel economy improvement and dynamic performance for both urban and highway driving conditions.

Keywords: Series hydraulic hybrid truck · Multi-input-single-output · Fuzzy logic controller · Dynamic performance

1 Introduction

Nowadays, reducing greenhouse gas (GHG) emissions and energy consumption addressed by automotive transportation is important part of climate policy. According to the International Panel on Climate Change (IPCC) emissions from transport are responsible for 14% of global direct GHG emissions, 0.3% of indirect CO_2 emissions, 26% of global energy use, and 24% of global energy-related GHG emissions. Among of them, light-duty vehicles (LDVs) represent 10% of global energy use and 10% of the world's total GHG [1]. In 2015, the number of passenger cars and commercial vehicles in use worldwide is around 947 million and 335 million units, respectively [2]. Many countries all over the worlds, such as India, the United States of America (USA), the European Union (EU), China, Japan, South Korea, have proposed and established new fuel economy policies and transportation CO_2 emission standard which are revolutionary compared with the former regulations. On November 16, 2017, the United States Environmental Protection Agency and the National Highway Traffic Safety

Administration jointly finalized Phase 2 greenhouse gas emissions and fuel economy standards for certain trailers model years 2018–2027 and semi-trucks, large pickup trucks, vans, and all types and sizes of buses and work trucks for model years 2021–2027. In which, the CO_2 emission is expected to lower by approximately 1.1 billion metric tons and the oil consumption is lower by up to two billion barrels over the lifetime of the vehicles [3]. According to decision no 406/2009/EC of the European Parliament and the Council, the greenhouse gas emissions and energy consumption are expected to reduce at least a 20% by 2020 compared to 1990 [4]. On October 1st 2015, India submitted its Intended Nationally Determined Contribution (INDC), in which the CO_2 emissions is targeted to lower by 33–35%, by 2030 from 2005 level [5]. Some countries are targeted to be non-GHG-emission one by 2020, such as Korea, Iran and Saudi Arabia [5]. In this scenario, electric and hybrid vehicles have great important.

As a short- and mid-term solution, hybrid vehicles have aroused the attention of researchers and manufacturers all over the world. In the recent past, most commercialization efforts have focused on hybrid electric vehicles (HEVs). Recent hydraulic hybrid vehicles (HHVs) have been raised attentions from researchers all over the world. Some of recent published work related to hydraulic hybrid vehicle will be addressed in this paper.

Because of its simplicity on design and implementation, ruled-based control scheme is generally the first choice to study a new configuration of the hybrid vehicle. Most papers regarding rule-based control strategy for hydraulic hybrid vehicle used the State-Of-Charge of the accumulator as the sole state variable for engine power determination [6–10]. The main goal is to move the ICE operating point closer to optimal region of fuel economy, efficiency, and emissions. The rules are designed with the aid of ICE operating maps, power flows within the powertrain, and driving experience. However, how to regulate the engine to its reference power was not presented in these studies.

To realize the control scheme, Kim used a PI controller to control the engine speed [11]. The PI controller uses vehicle velocity error as the input. Engine torque is controlled by adjusting the control signal of the fuel rack. However, the control of the vehicle velocity is not discussed explicitly in this study. Engine speed and vehicle velocity are controlled by individual PID controllers [12]. Controller parameter tuning, input-output interaction and physical constraint are obstacles to this control approach.

In [13], Macor et al. modelled and simulated a hybrid hydro-mechanical transmission for an urban bus in AMESim software. In [14], the authors used power bond graph approach to model the dynamics of series hydraulic hybrid vehicle. In [15], Wu et al. proposed a self-adaptive system that would reduce the control parameters and improve the energy efficiency.

In previous work [16], we developed a fuzzy-based controller that use vehicle speed demand as an input to estimate the desired engine power as an output. Since it is only a single-input-single-output controller, the performance of the system has been limited. The acceleration command was not taken into account. As a consequence, the accumulator is depleted in certain circumstances. In addition, the accumulator is almost fully-charged right before a high decelerated braking event. Hence, braking energy is not captured as much as possible.

To solve above problems, an improvement of the previous fuzzy controller has been proposed in this work. The fuzzy-based power management will be redesigned in which the vehicle acceleration is taken into account as a second input. With the two inputs, the flexibility of the controller will be increased. The control surface will be tuned to guarantee that the SoC of the accumulator is at low level before braking event and high level before acceleration event. In all circumstances, the accumulator is prevented from depletion. To evaluate the effectiveness of the proposed control schemes, Japan 1015 and highway fuel economy test (HWFET) driving cycles are used as the velocity trajectory references. The system fuel economy improvement is also investigated. The simulation results show that with the proposed power management, the SoC of the accumulator is reserved for braking energy recovery before every braking occasion, the high power from the accumulator is available for hard acceleration events, and the depletion of the accumulator is also eliminated. As results, the fuel economy is improved and dynamic performance is guaranteed.

2 System Description and Modeling

A schematic of the series hydraulic hybrid powertrain configuration for a rear-wheel-drive light-duty truck is presented in Fig. 1. A variable displacement hydraulic pump is connected to an engine to convert mechanical power into hydraulic power. The power is stored in the accumulator or converted back to mechanical power by a variable displacement pump/motor. The pump/motor is a reversible energy conversion component. When working as a motor, it converts the hydraulic power into mechanical power to propel the wheels. When the brake pedal is depressed, the pump/motor works as a pump. The momentum of the vehicle is used to pressurize fluid from the reservoir and stores it in the accumulator. In this manner, the mechanical brake power is converted into hydraulic power. The braking energy is captured and stored in the accumulator in the form of hydraulic high pressure. The low-pressure reservoir provides fluid for the hydraulic system.

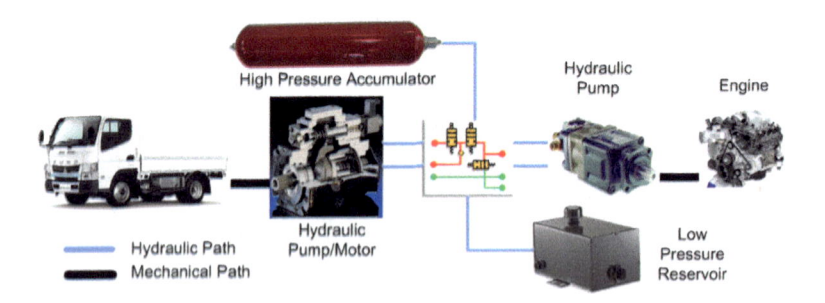

Fig. 1. Schematic and control signal paths of SHHV

The system is modeled in Matlab/Simulink as showed in Fig. 2. The vehicle in this work is equipped with a 2.6 L diesel engine. It reaches its maximal torque at 2500 [rpm]. Its lowest specific fuel consumption is 248.89 [g/kWh] at about 1250 [rpm] when the power output is 20.23 [kW]. The brake-specific-fuel-consumption (BSFC) map is shown in Fig. 3. Details of system description can be found in [17].

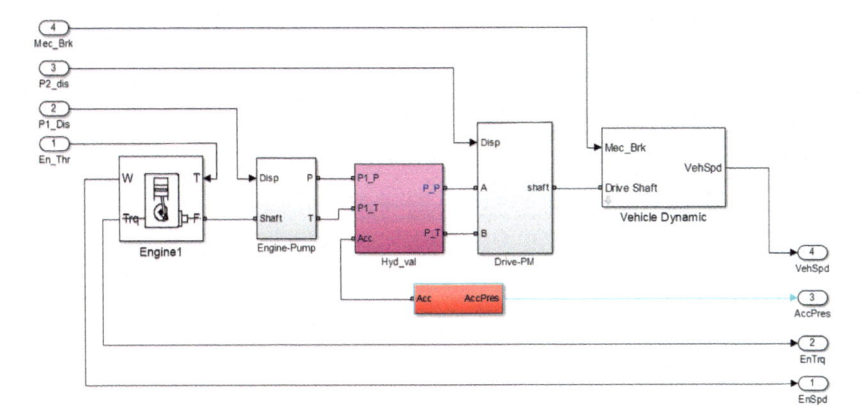

Fig. 2. The simulink model of SHHV

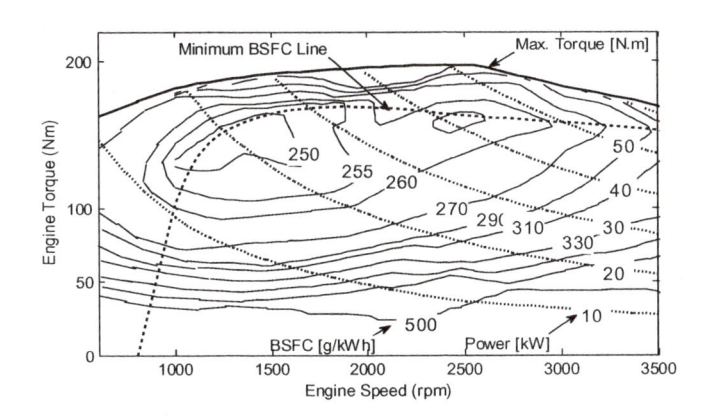

Fig. 3. Experimental engine map

3   Control System Development

In hybrid vehicles, the function of the control system is to determine how to coordinate the power sources to satisfy power demand and the dynamic constraints in the most convenient way. The main objective is to reduce the overall energy consumption. In this work, the power management system is decomposed into two levels as presented in Fig. 4. The first level or supervisory controller determines the desired reference value for engine torque and engine speed to achieve desired vehicle velocity. The second level, known as actuator controllers, calculates the amount of fuel mass injected into cylinder of the engine, the displacements of hydraulic pump and pump/motor to drive the outputs of the SHHV to their corresponding references. In this work, three individual PID controllers are used as the actuator controllers. The aim of this paper is on the development of the supervisory controller using fuzzy logic algorithm.

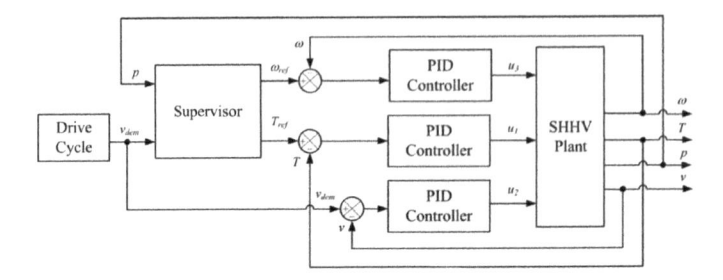

Fig. 4. Hierarchical control system of the SHHV

The diagram of the proposed fuzzy-based supervisory controller is depicted in Fig. 5 below. The fuzzy logic controller (FLC) will use the speed and acceleration demands as the inputs. The output of the FLC is the desired pressure level of the accumulator. The PID controller is used to estimate the desired power of the engine in order to keep the accumulator pressure follow the reference value. An optimal engine speed and engine torque are interpolated from the experimental BSFC map.

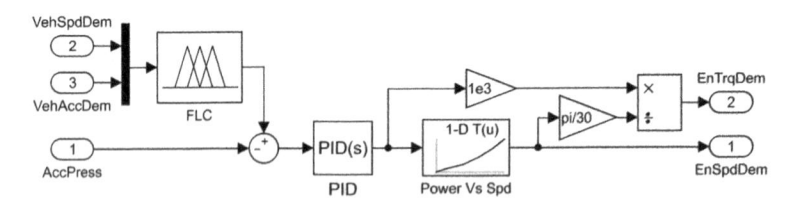

Fig. 5. Fuzzy-based supervisory controller

The structure of the FLC is shown in Fig. 6. The system has two inputs, one output, and 9 rules. The rules are defined as below:

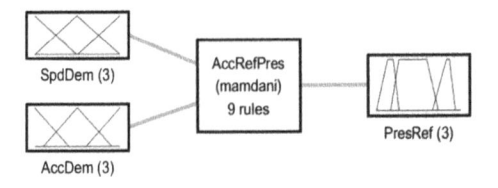

Fig. 6. Fuzzy-based supervisory controller

1. (SpdDem==low) & (AccDem==low) => (PresRef=high) (1)
2. (SpdDem==low) & (AccDem==med) => (PresRef=high) (1)
3. (SpdDem==low) & (AccDem==high) => (PresRef=high) (1)
4. (SpdDem==med) & (AccDem==low) => (PresRef=med) (1)
5. (SpdDem==med) & (AccDem==med) => (PresRef=med) (1)
6. (SpdDem==med) & (AccDem==high) => (PresRef=med) (1)
7. (SpdDem==high) & (AccDem==low) => (PresRef=low) (1)
8. (SpdDem==high) & (AccDem==med) => (PresRef=low) (1)
9. (SpdDem==high) & (AccDem==high) => (PresRef=low) (1)

The control surface of the FLC is shown in Fig. 7. It can be seen that at low-speed demand the pressure of the accumulator is at a high value and when the vehicle speed demand is high the pressure of the accumulator is kept at low level. By this approach, the accumulator supports power for high acceleration rate at low speed and reserves enough space for braking energy recovering.

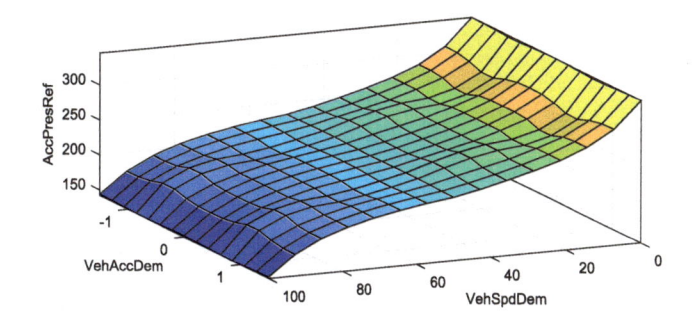

Fig. 7. Control surface of the FLC

The membership function of the inputs and the output are shown in Fig. 8. The speed demand is limited to 100 kmh (60 mph), the acceleration is within ± 1.5 m/s^2, and the pressure is from 150 bar to 350 bar.

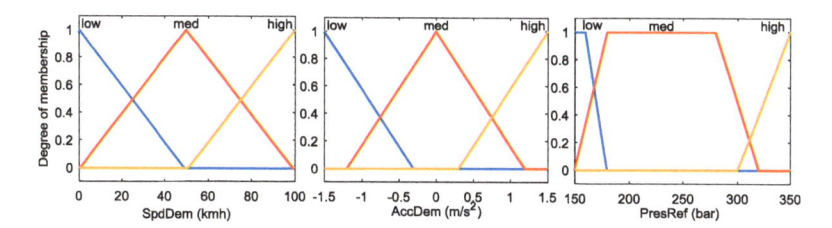

Fig. 8. Inputs and output membership function of the FLC

4 Simulation Results and Discussion

The resulted reference trajectory of the accumulator pressure generated by proposed control strategy under Japan 1015 driving cycle is compared with the speed-based [9] and the SISO-Fuzzy-based [16] as shown in Fig. 9a. Similarly, the reference trajectories for Highway Fuel Economy Test (HWFET) driving cycle is shown in Fig. 9b.

It can be seen that by using two input, the MISO FLC has more degree of freedom, the control surface can be turned to ensure that the stored energy in the accumulator is used as much as possible but depletion prevention.

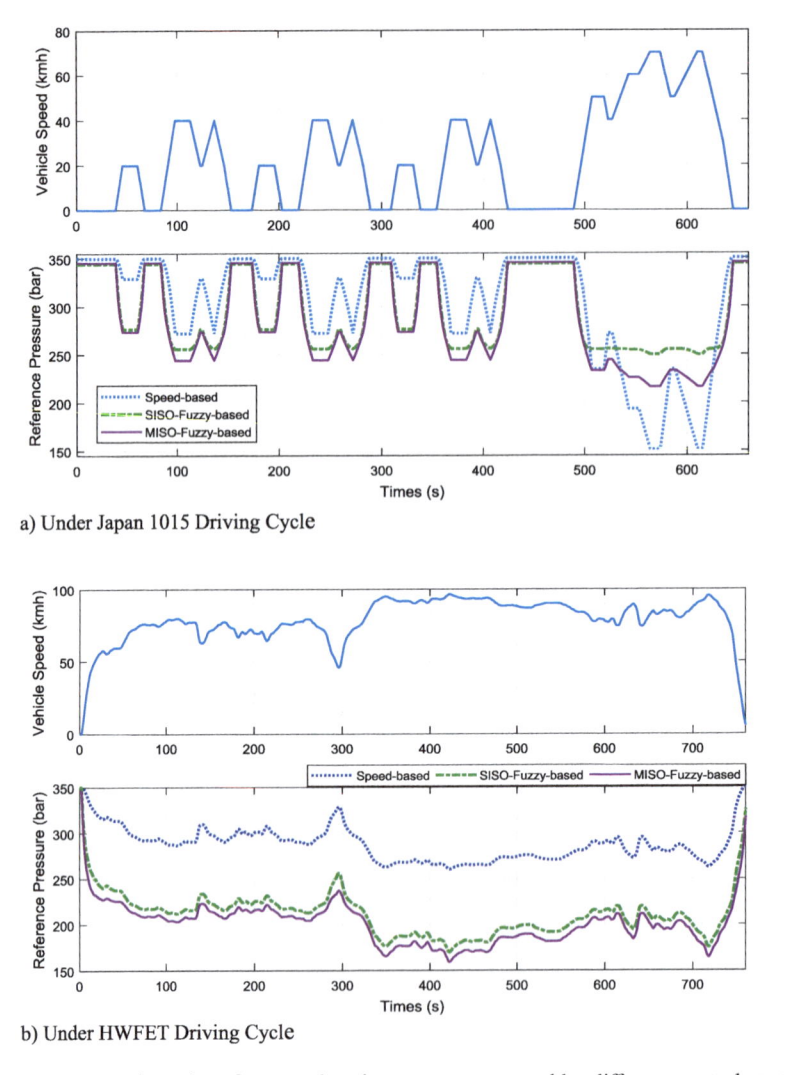

Fig. 9. Reference trajectories of accumulator's pressure generated by different control strategies

The dynamic performance of the system under different driving cycles are shown in Figs. 10a and b respectively. It can be seen that the MISO-Fuzzy-based control strategy in this work achieves better dynamic performance than the SISO-Fuzzy-based one. Details of the performance of the SHHT system under HWFET driving cycles with application of the MISO-Fuzzy-based control strategies can be seen from Fig. 11.

a) Under Japan 1015 Driving Cycle

b) Under HWFET Driving Cycle

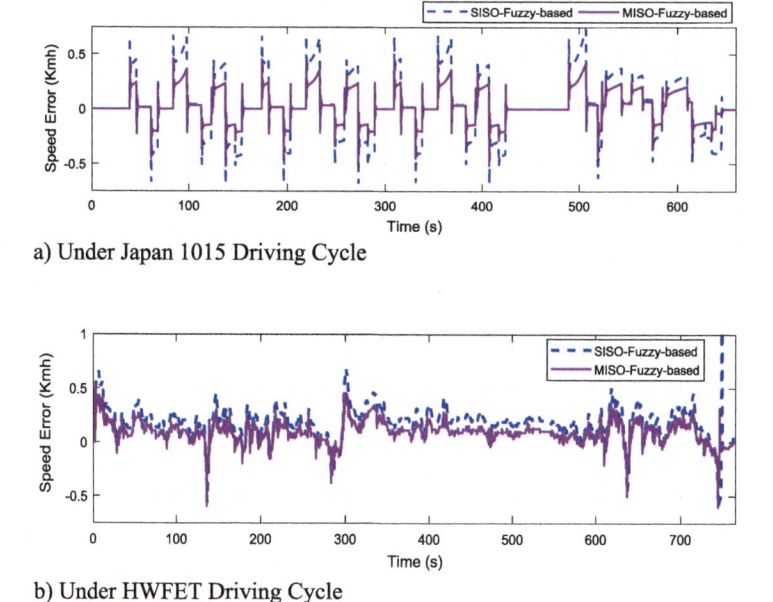

Fig. 10. Dynamic performance of the SHHT with different fuzzy-based control strategies

The effectiveness of the proposed control strategy is also evaluated by using the fuel economy improvement in comparison with conventional vehicle and the SHHT with other control strategies. The conventional fuel economy here is obtained by using the corresponding conventional vehicle model with the same value for similar parameters. The fuel economic of a vehicle is defined as volumetric fuel consumption per distance traveled in the unit of kilometers per liter [KPL]. The results are listed in Table 1.

Table 1. Fuel economy improvement (%) of the SHHT

	Conv. vehicle	Thermostatic	Modulated	Speed-based	SISO-Fuzzy-based	MISO-Fuzzy-based
Japan 1015	–	42.67	34.72	22.46	27.11	31.15
HWFET	–	16.52	25.88	36.56	40.11	42.69

The simulation results indicated that the Thermostatic control strategy is suitable for a city driving cycle, such as Japan 1015. Meanwhile, the Modulate is suitable for urban driving cycle, such as HWFET. By considering the accumulator pressure as a function of the vehicle speed demand, the Speed-based achieve better fuel economy for both typical driving pattern. By applying fuzzy interference system to estimate the desired power of the engine, the SISO-Fuzzy-based help to achieve better fuel economy improvement in comparison with the Speed-based. Finally, by using two input, the MISO-Fuzzy-based has more degree-of-freedom in order to estimate the desired engine power. As a results, the MISO-Fuzzy-based help the SHHT to achieve a slightly improvement in comparison with the SISO-Fuzzy-based.

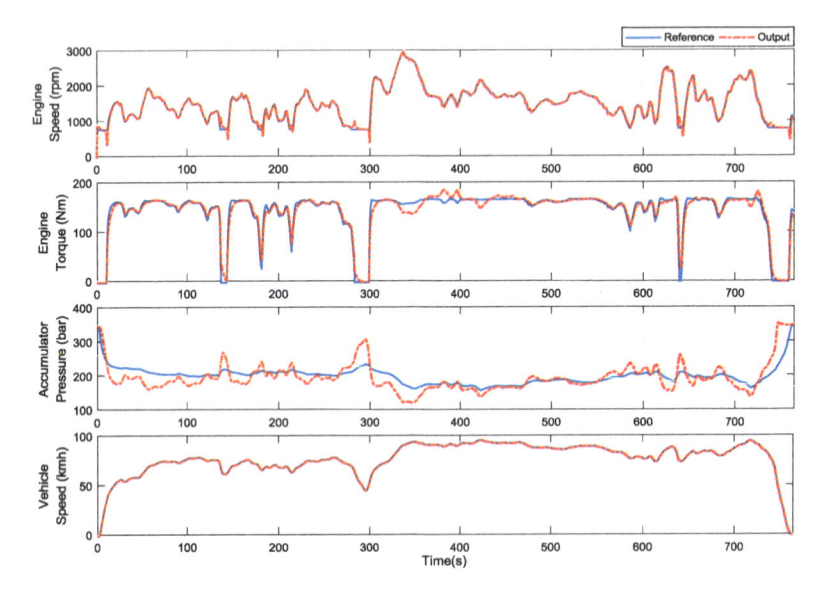

Fig. 11. System performance under HWFET driving cycle

5 Conclusion

This paper presents an improved fuzzy-based approach to estimate the desired power of the engine for a given driving command in a SHHT system. The simulation results indicate that by using vehicle speed and acceleration as the input, the MISO-Fuzzy-based has more flexibility to adjust the desired pressure level of the accumulator for a certain driving command. Hence, the energy stored in the accumulator could be used as much as possible but reserving suitable volume for capturing the braking energy. As a consequence, the MISO-Fuzzy-based help the SHHT to achieve better fuel economy for both typical driving cycles and also dynamic performance.

Currently, the proposed fuzzy logic controller only used to estimate the desired pressure level of the accumulator. An addition PID controller must be used to find out the desired engine power. In addition, the reference value of the engine torque and speed are interpolated from the experiment data, which is only available at steady state condition. In our future work, an adaptive neural network fuzzy interference system can be used to estimate the engine torque and engine speed directly from the vehicle speed demand and the current pressure level of the accumulator. In this manner, the training procedure will be done offline. After that the real-time computing workload could be reduced that will be increase the response of the supervisory controller.

References

1. IPCC: Climate Change 2014: Synthesis Report. Contribution of Working Groups I, II and III to the Fifth Assessment Report of the Intergovernmental Panel on Climate Change (2014)
2. The statistic Portal. https://www.statista.com/statistics/281134/number-of-vehicles-in-use-worldwide/
3. The United State EPA, Regulations for Emissions from Vehicles and Engines. https://www.epa.gov/regulations-emissions-vehicles-and-engines/proposed-rule-repeal-emission-requirements-glider
4. The European Parliament and of the Council, Decision No 406/2009/EC. http://data.euro-pa.eu/eli/dec/2009/406/oj
5. International Energy Agency, CO_2 Emissions from Fuel Combustion. https://www.iea.org/publications/freepublications/publication/CO2EmissionsfromFuelCombustionHighlights-2017.pdf
6. Baer, K.: Simulation-Based Optimization of a Series Hydraulic Hybrid Vehicle, Linköping Studies in Science and Technology. Dissertations, Linköping University Electronic Press, p. 72 (2018)
7. Wu, B., Lin, C.-C., Filipi, Z., Peng, H., Assanis, D.: Optimal power management for a hydraulic hybrid delivery truck. J. Vehicle Syst. Dyn. **42**(1–2), 23–40 (2004)
8. Kim, Y., Filipi, Z.: Series hydraulic hybrid propulsion for a light truck-optimizing the thermostatic power management. SAE Trans. J. Engines **116**, 1597–1609 (2007)
9. Vu, T.V., Vu, T.H.: An improvement of rule-based control strategy for a series hydraulic hybrid vehicle. In: AETA 2016: Recent Advances in Electrical Engineering and Related Sciences. Lecture Notes in Electrical Engineering, vol. 415. Springer, Cham (2016)
10. Tao, L., Jincheng, Z., Shuwen, W., Fangde, G.: Logic threshold based energy control strategy for parallel hydraulic hybrid vehicles. Res. J. Appl. Sci. Eng. Technol. **6**, 2339–2344 (2013)
11. Kim, Y.A.: Integrated Modeling and Hardware-in-the-Loop Study for Systematic Evaluation of Hydraulic Hybrid Propulsion Options. Ph.D. Dissertation, University of Michigan, MI, USA (2008)
12. Shan, M.: Modeling and Control Strategy for Series Hydraulic Hybrid Vehicles. Ph.D. Dissertation, Toledo Univ., OH, USA (2009)
13. Macor, A., Benato, A., Rossetti, A., Bettio, Z.: Study and simulation of a hydraulic hybrid powertrain. Energy Procedia **126**, 1131–1138 (2017)
14. Ramakrishnan, R., Somashekhar, S.H., Singaperumal, M.: Modeling, simulation and design optimization of a series hydraulic hybrid vehicle. Appl. Mech. Mater. **541–542**, 727–731 (2014)
15. Wu, W., Hu, J.B., Yuan, S.H., Di, C.F.: A hydraulic hybrid propulsion method for automobiles with self-adaptive system. Energy **114**, 683–692 (2016)
16. Vu, T.V., Chen, C.K.: A fuzzy-based supervisory controller development for a series hydraulic hybrid vehicle. In: AETA 2017-Recent Advances in Electrical Engineering and Related Sciences: Theory and Application. Lecture Notes in Electrical Engineering, vol. 465. Springer, Cham (2017)
17. Vu, T.V.: Simulation and Design of Hydraulic Hybrid Vehicle, M.S thesis, Da-yeh Univ., Changhua, Taiwan (2011)

A New Approach Newton-Raphson Load Flow Analysis in Power System Networks with STATCOM

Dung Vo Tien[1(✉)], Radomir Gono[1], and Zbigniew Leonowicz[2]

[1] Department of Electrical Power Engineering, FEECS,
VSB - Technical University of Ostrava, Ostrava, Czech Republic
{Dung.vo.tien.st,radomir.gono}@vsb.cz
[2] Faculty of Electrical Engineering,
Wroclaw University of Science and Technology, Wroclaw, Poland
Zbigniew.leonowicz@pwr.edu.pl

Abstract. Load flow analysis is an important tool used by engineers to ensure stable operation of the power system, it is also used in load forecasting, planning, and economic scheduling. In this paper, we propose an improved solution of load flow analysis in the power systems incorporating STATCOM using the Newton-Raphson method. A software based on MATLAB Compiler Runtime (MCR) based on the proposed method is developed to analyze the power system load flow with and without STATCOM. The software has been tested on IEEE 30-bus, and IEEE 57-bus test system. The results have proved the correctness of the proposed method, reliability of the software and high computation speed.

Keywords: Jacobian matrix · Load flow ·
MATLAB Compiler Runtime · Newton-Raphson method ·
Power system analysis · Power quality · STATCOM

1 Introduction

Load flow (or power flow) is a solution for the steady state of the power system network. The studies of load flow provide methods for calculating the magnitudes and phase angles of voltages at each bus, active and reactive power flows through different branches, generators, transformers and loads under steady state conditions.

The load flow computer software was developed in the middle of the 1950s. Since then, a variety of methods has been used in load flow calculation. The most commonly used iterative methods are the Gauss-Seidel, the Newton-Raphson and Fast Decoupled method. The development of these methods is mainly led by the basic requirement of load flow calculation such as convergence properties, computing efficiency, memory requirements, convenience and flexibility of the implementation [13–24]. Among these methods, Newton-Raphson is the most reliable because it converges fast and is most accurate. However, with the expansion of the power system, the dimension of load flow equations grow fast (from several thousands to tens of thousands) [9–24]. Therefore, this situation requires the researchers to continue to seek for methods that can efficiently solve such problems.

To improve the power quality in the electrical power system, STATCOM is one of the most useful devices because it can regulate the voltage quickly, improve transient stability and compensate variable reactive power [1–16]. STATCOM control strategy which presented in several publications, e.g., in [1,5,10,13–16]. Therefore, in this paper, we focus on load flow analysis of power systems incorporating STATCOM. It is an important tool for optimization of system operation. The paper presents Newton-Raphson load flow analysis in power system with and without STATCOM in Sects. 2 and 3. New software has been developed and presented in Sect. 4. The software is tested on IEEE 30-bus and IEEE 57-bus system is Sect. 5. Finally, conclusions are given in Sect. 6.

2 Newton-Raphson Load Flow in Power System Without STATCOM

Taking the node–voltage equation:

$$I = Y_{bus}.V, \tag{1}$$

This equation can be rewritten in form for a n-bus system:

$$I_i = \sum_{j=1}^{n} Y_{ij}.V_j = \sum_{j=1}^{n} |Y_{ij}||V_j|\angle \theta_{ij} + \delta_j, \tag{2}$$

The active and reactive power at bus i is expressed by:

$$P_i - jQ_i = V_i^*.I_i = V_i \angle -\delta_i.(\sum_{j=1}^{n} |Y_{ij}||V_j|\angle \theta_{ij} + \delta_j), \tag{3}$$

Therefore:

$$P_i = Re\{V_i^*.I_i\} = \sum_{j=1}^{n} |V_i||V_j||Y_{ij}|cos(\theta_{ij} - \delta_i + \delta_j), \tag{4}$$

$$Q_i = -Im\{V_i^*.I_i\} = \sum_{j=1}^{n} |V_i||V_j||Y_{ij}|sin(\theta_{ij} - \delta_i + \delta_j). \tag{5}$$

The main equation of Newton-Raphson power flow can be expressed as follows:

$$\begin{bmatrix} \Delta P \\ \Delta Q \end{bmatrix} = \begin{bmatrix} J_{11} & J_{12} \\ J_{21} & J_{22} \end{bmatrix} . \begin{bmatrix} \Delta \delta \\ \Delta |V| \end{bmatrix} \tag{6}$$

The mismatch vectors are defined as follows:

$$\Delta P_i^{(k)} = P_i^{Spec} - P_i^{(k)}, \tag{7}$$

$$\Delta Q_i^{(k)} = Q_i^{Spec} - Q_i^{(k)}. \tag{8}$$

The new estimates for bus voltage are as follows:

$$\delta_i^{(k+1)} = \delta_i^{(k)} + \Delta \delta_i^{(k)}, \tag{9}$$

$$|V_i|^{(k+1)} = |V_i|^{(k)} + \Delta |V_i|^{(k)}. \tag{10}$$

3 Newton-Raphson Load Flow in Power System with STATCOM

In load flow studies, a STATCOM can be represented by an equivalent circuit as shown in Fig. 1 [1–16]. The equation for the ith bus with STATCOM can be rewritten as follows:

$$S_i = (P_{Li} + jQ_{Li}) + (P_{Si} + jQ_{Si}), \tag{11}$$

$$P_i = P_{Li} + P_{Si}, Q_i = Q_{Li} + Q_{Si}, \tag{12}$$

$$I_{Si} = (P_{Si} - jQ_{Si})/V_i^*, \tag{13}$$

$$V_{Si} = V_i - I_{Si}Z_{Si}, \tag{14}$$

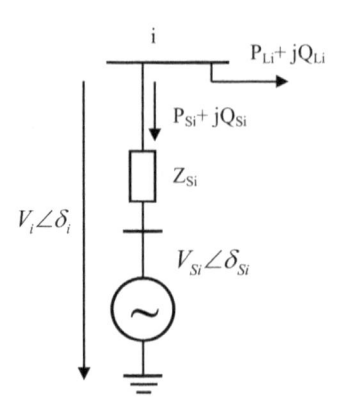

Fig. 1. The equivalent circuit of STATCOM.

According to the equivalent circuit of the STATCOM shown in Fig. 1, the power constrains of STATCOM are:

$$P_{Si} = |V_i|^2 . g_{Si} - |V_i||V_{Si}|.[g_{Si}.cos(\delta_i - \delta_{Si}) + b_{Si}.sin(\delta_i - \delta_{Si})], \qquad (15)$$

$$Q_{Si} = -|V_i|^2 . b_{Si} + |V_i||V_{Si}|.[-g_{Si}.sin(\delta_i - \delta_{Si}) + b_{Si}.cos(\delta_i - \delta_{Si})], \qquad (16)$$

that are the active and reactive power equations obtained from STATCOM at bus i, respectively. The voltage injection and capacity of STATCOM are bounded as follows:

$$V_{Simin} \leq V_{Si} \leq V_{Simax}, \qquad (17)$$

$$Q_{Simin} \leq Q_{Si} \leq Q_{Simax}. \qquad (18)$$

The active and reactive power exchange via the DC-link are described by:

$$P_{STAT,i} - jQ_{STAT,i} = V_{Si}^* . I_{Si}, \qquad (19)$$

Therefore, we have:

$$P_{STAT,i} = |V_{Si}|^2 . g_{Si} - |V_i||V_{Si}|.[g_{Si}.cos(\delta_i - \delta_{Si}) - b_{Si}.sin(\delta_i - \delta_{Si})], \qquad (20)$$

$$Q_{STAT,i} = -|V_{Si}|^2 . b_{Si} + |V_i||V_{Si}|.[g_{Si}.sin(\delta_i - \delta_{Si}) + b_{Si}.cos(\delta_i - \delta_{Si})]. \qquad (21)$$

Many previous publications treated the problem of the load flow in networks incorporating STATCOM [1–16]. Salah Kamel and Francisco Jurado in [16] presented Newton-Raphson load flow based on a current injection model. However, according to [2], it is less effective in term of computation speed and accuracy compared to the power injection model. In [9,11,12], the buses where STATCOM are installed, are assumed to be PV buses. In other publications [13–16], a STATCOM has been represented as independent variables so that the Jacobian matrix needs to be extended. In these publications, the mismatch vectors represented by STATCOM are ΔP_{STAT} and ΔQ_{STAT}. However, the disadvantage of this approach is that of the unspecified reactive power of STATCOM.

In this paper, we propose a new approach by using the parameters of STATCOM as independent variables with the modified Jacobian matrix. The main equations of the power system with STATCOM are given below:

$$\begin{bmatrix} \Delta P \\ \Delta Q \\ \Delta PE \\ \Delta F \end{bmatrix} = \begin{bmatrix} \frac{\partial P}{\partial \delta} & \frac{\partial P}{\partial |V|} & \frac{\partial P}{\partial \delta_S} & \frac{\partial P}{\partial |V_S|} \\ \frac{\partial Q}{\partial \delta} & \frac{\partial Q}{\partial |V|} & \frac{\partial Q}{\partial \delta_S} & \frac{\partial Q}{\partial |V_S|} \\ \frac{\partial PE}{\partial \delta} & \frac{\partial PE}{\partial |V|} & \frac{\partial PE}{\partial \delta_S} & \frac{\partial PE}{\partial |V_S|} \\ \frac{\partial F}{\partial \delta} & \frac{\partial F}{\partial |V|} & \frac{\partial F}{\partial \delta_S} & \frac{\partial F}{\partial |V_S|} \end{bmatrix} \cdot \begin{bmatrix} \Delta \delta \\ \Delta |V| \\ \Delta \delta_S \\ \Delta |V_S| \end{bmatrix} \qquad (22)$$

According to Eq. (22):

- PE is the active power of STATCOM, it is given in Eq. (23).
- For a trial set of variables $\Delta\delta_S$, $\Delta|V_S|$, the mismatch vectors are added, represented by:

$$\Delta PE_i = 0 - P_{STAT,i}, \quad (23)$$

$$\Delta F_i = |V_i^{Spec}| - |V_i|. \quad (24)$$

The flow chart of proposed solution is given in Fig. 2.

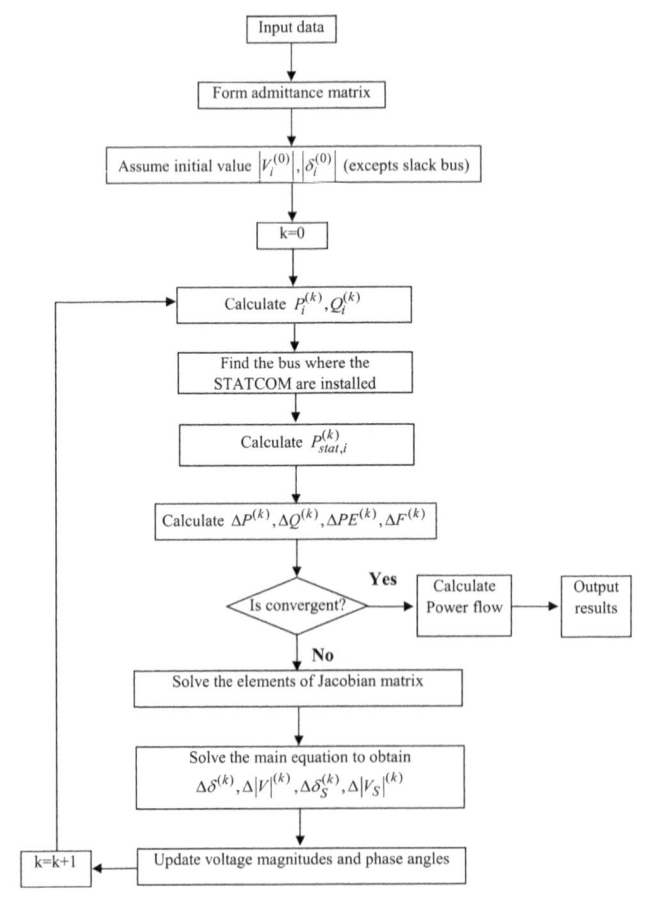

Fig. 2. Flowchart of the Newton-Raphson method for load flow solution of power system network incorporating STATCOM.

4 Design and Development of Software

We designed a stand-alone program for load flow analysis of power systems with and without STATCOM. The MATLAB Compiler Runtime (MCR) provide an environment that allows running MATLAB application as an executable on computers without an installed full version of MATLAB. A software is developed based on MCR Load Flow Analysis (LFA) to analyze the electric power system with and without STATCOM. The screenshot of this software is shown below (Fig. 3).

Fig. 3. Interface of software.

4.1 Technologies and Platform

The software is developed using MCR so that it is required to install MATLAB Compiler Runtime version $9 - 4(2018a)$. The proposed system requirements to use this application are: a 2 GHz Processor or higher, a memory of 4 GB RAM or higher, a hard disk space of 4 GB or more.

4.2 Programming

The software consists of several modules and functions. In addition to two main programs, *Newtonpg* (Newton-Raphson method without STATCOM) and *statpg* (Newton-Raphson method with STATCOM), there are two functions, include *Ybuspg* (form the bus admittance matrix), *displaypg* (prints the power flow solution on the screen). In order to estimate the effectiveness of the method as well as the purpose of comparisons, the function to measure the number of iteration, the time used per iteration and that of the total time to get the solution and the maximum of mismatch vectors were also established. Information of these functions is provided as below:

- *Ybuspg*: This function builds the bus admittance matrix. It reads data from the line data input (import from a text file). The input data contains the transmission line and transformer resistances, reactances and capacitances in per unit, and the tap settings. The function returns the bus admittance matrix.
- *Newtonpg*: This program calculates the load flow solution by the Newton-Raphson method and requires the transmission line data and the bus data (all taken from the text file). The results of load and generation are given in MW and Mvar, bus voltage in per unit and angle in degrees. Constraints of generator reactive power of the PV buses is considered. The violation of reactive power may occur if the specified voltage is either too high or too low. If a limit is reached, the voltage magnitude is adjusted up or down by 0.001 (pu) in order to ensure the reactive power demand within the specified limits.
- *statpg*: This program obtains the load flow solution by the Newton-Raphson method in the power system incorporating STATCOM. It requires the transmission line data, the bus data and the STATCOM data (all taken from the text file). The STATCOM data are arranged as follows: column 1 is bus data where it is installed, column 2 and 3 are initial voltage magnitude and phase angle respectively, column 4 and 5 are minimum and maximum reactive power of STATCOM. A direct algorithm is designed to perform the load flow solution of the power system incorporating STATCOM which is presented in Sect. 3.
- *displaypg*: This function produces the bus output results in a table form. The bus output results include the voltage magnitude (in per unit) and angles (in degrees), active (in MW) and reactive power (in Mvar) of generator and loads. Total generation and total load are also included.

4.3 Data Input and Output

When the user completes the input of data, the program can be started. The output data includes the voltage magnitudes and phase angles of buses, the voltage magnitudes, phase angles and the required capacity of STATCOM, the total losses of the active and reactive power of the system. Furthermore, the maximum of mismatch vectors, the number of iterations, the time needed per iteration and the total time for the solution are included.

4.4 Structure of the Program

The data flow of the program is shown in Fig. 4.

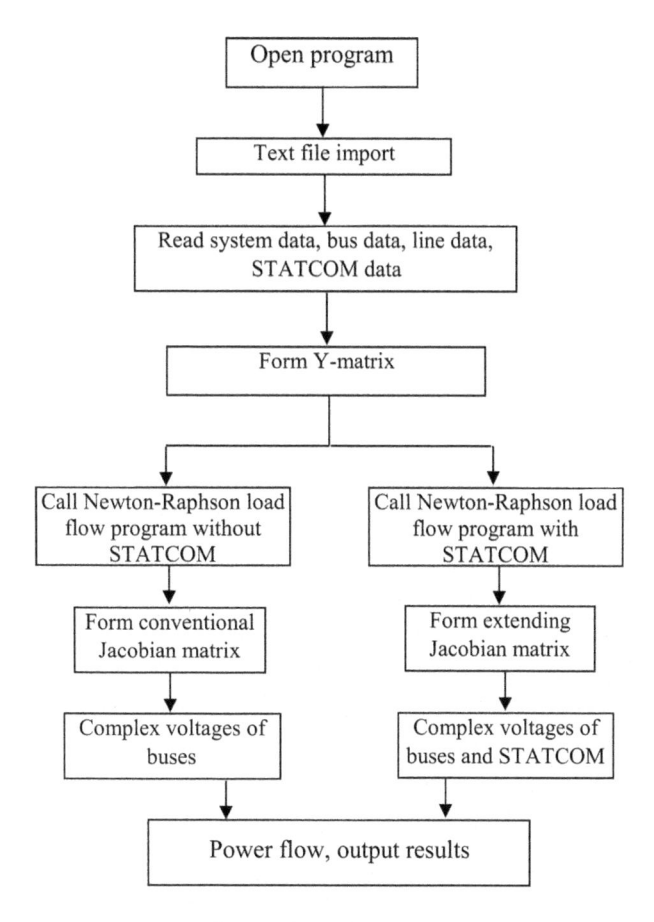

Fig. 4. Data flow diagram.

5 Test Systems and Results

In order to investigate the performance of the method, the software is tested in modified IEEE 30-bus and IEEE 57-bus systems. The STATCOM data is given in Table 1.

5.1 Modified IEEE 30-Bus

The modified IEEE 30 bus is presented in Table 2. Without STATCOM, the lowest bus voltages are 0.8516 pu and 0.9182 pu at bus 30 and 26 respectively. Two STATCOMs have been assumed to be installed at these buses to improve

Table 1. STATCOM parameters.

$R_S(pu)$	$X_S(pu)$	$Q_{Smin}(pu)$	$Q_{Smax}(pu)$
0.01	0.10	-0.5	0.5

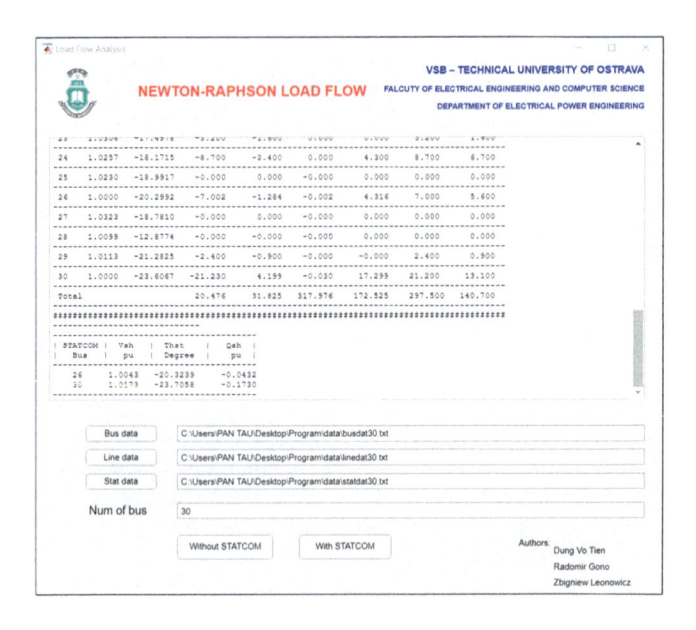

Fig. 5. Display the results in case of IEEE 30-bus.

the voltage at 1.0 pu. The simulation results are illustrated in Fig. 5. The voltage magnitudes of the buses in case of the IEEE-30 bus system with and without STATCOM are shown in Fig. 6. The voltage magnitudes, phases angle, and the injected power of STATCOM are shown in Table 3. In this case, the STATCOMs compensate reactive power, therefore the voltage magnitudes are larger than at bus where the STATCOMs are installed.

Table 2. Modified IEEE 30 bus.

Bus	Initial IEEE-30 bus		Modified IEEE-30 bus	
	P (MW)	Q (Mvar)	P (MW)	Q (Mvar)
26	3.5	2.3	7.0	5.6
30	10.6	1.9	21.2	13.1

Table 3. Comparison of the complex voltage between IEEE-30 bus with and without STATCOM, and parameters of the STATCOM.

Bus no.	Without STATCOM		With STATCOM		STATCOM parameters		
	$|V|(pu)$	$\delta(degree)$	$|V|(pu)$	$\delta(degree)$	$|V|(pu)$	$\delta_S(degree)$	$Q_S(p.u)$
26	0.9106	-19.66	1.0000	-20.30	1.0043	-20.32	-0.0432
30	0.8513	-22.49	1.0000	-23.61	1.0173	-23.71	-0.1730

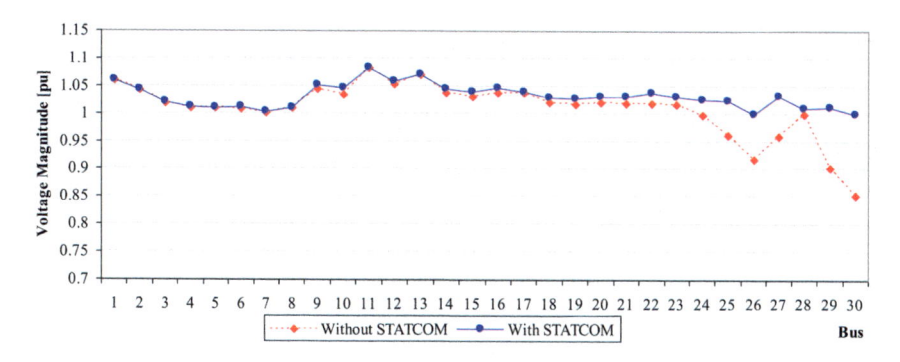

Fig. 6. The graphical representation of the comparison between bus voltages of 30-bus system with and without STATCOM.

5.2 IEEE 57-Bus

Similar tests have also been performed on IEEE-57 bus system. Three STATCOMs are installed at buses 25, 31 and 42. The voltage magnitudes of the buses in case of the IEEE 57 bus system with and without STATCOM are shown in Fig. 7. The voltage magnitudes, phases angles and the injected power of STATCOM are shown in Table 4.

Table 4. Comparison of voltages in IEEE-57 bus with and without STATCOM, and parameters of STATCOM.

Bus no.	Without STATCOM		With STATCOM		STATCOM parameters		
	$|V|(pu)$	$\delta(degree)$	$|V|(pu)$	$\delta(degree)$	$|V|(pu)$	$\delta_S(degree)$	$Q_S(p.u)$
25	0.9377	-18.01	1.0000	-17.92	1.0012	-17.94	-0.0108
31	0.8998	-19.53	1.0000	-20.56	1.0090	-20.61	-0.0889
42	0.9630	-15.56	1.0000	-15.86	1.0064	-15.90	-0.0618

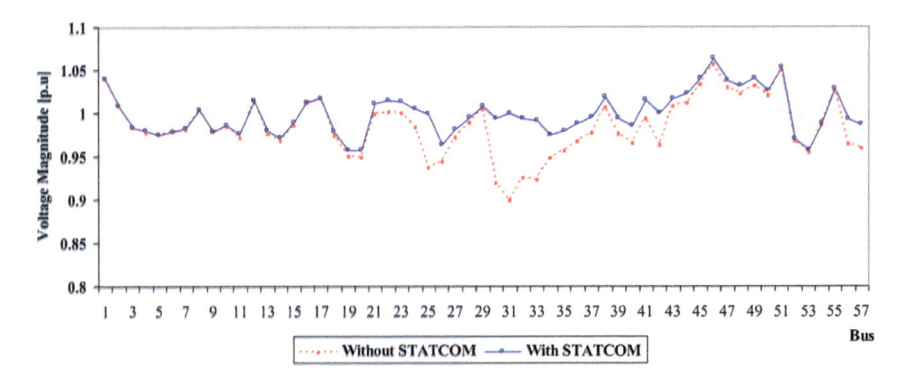

Fig. 7. The graphical representation of the comparison of bus voltages of 57-bus system with and without STATCOM.

5.3 Computation Time and Number of Iterations

The computation time depend on the configuration and processing speed of the computer. The software was tested on computer with processor: $Intel(R)$ $Core(TM)i5 - 6200U$ $CPU@2.30GHz$ $(4CPUs), 2.4GHz$. Table 5 shows the computation time and the iteration number for the load flow solution performed by the program. The number of iterations and the computation time increases with the number of buses and load flow for the system with STATCOM requires more time than without STATCOM because of the extended Jacobian matrix and solution of a larger matrix.

Table 5. The iteration number and the computation time of two case studies by load flow program.

Case study IEEE system	Number of iterations		Computation time (ms)		Maximum mismatch	
	Without STATCOM	With STATCOM	Without STATCOM	With STATCOM	Without STATCOM	With STATCOM
30-bus	4	5	1.582	4.741	1.1738e–05	2.51592e–06
57-bus	22	22	26.265	42.875	8.72853e–10	7.85129e–06

6 Conclusions

In this paper, a new approach for calculating the power system incorporating STATCOM has been presented. A software to analyze the power system with and without STATCOM was designed and tested on IEEE 30-bus and IEEE 57-bus system. From the obtained results, the accuracy of algorithm and the effectiveness of software were confirmed.

This software is a useful tool for teaching and learning purposes and hence provides a solution to help the engineers and operators to calculate the parameters of the power system in order to improve power quality and optimize the operation of the power system.

Acknowledgment. This research was partially supported by the SGS grant from VSB - Technical University of Ostrava (No. SP2018/61) and by the project TUCENET (No. LO1404).

References

1. Canizares, C.A., Pozzi, M., Uzunovic, E.: STATCOM modeling for voltage and angle stability studies. Int. J. Electr. Power Energy Syst. **25**(6), 431–441 (2003)
2. Adepoju, G.A., Komolafe, O.A.: Analysis and modeling of static synchronous compensator (STATCOM)- a comparison of power injection and current injection models in power flow study. Int. J. Adv. Sci. Technol. **36**, 65–76 (2011)
3. Marefatjou, H., Sarvi, M.: Power flow study and performance of STATCOM and TCSC in improvement voltage stability and load ability amplification in power system. Int. J. Appl. Power Eng. (IJAPE) **2**(1), 15–26 (2013)
4. Bisen, P., Shrivastava, A.: Voltage level improvement of power system by the use of STATCOM and UPFC with PSS controller. Int. J. Electr. Electron. Comput. Eng. **2**, 117–126 (2013)
5. Chatterjee, D., Ghosh, A.: Improvement of transient stability of power systems with STATCOM controller using trajectory sensitivity. Int. J. Electr. Power Energy Syst. **33**, 531–539 (2011)
6. Gotham, D.J., Heydt, G.T.: Power flow control and power flow studies for systems with FACTS devices. IEEE Trans. Power Syst. **13**(1), 60–66 (1998)
7. Zhang, Y., Wu, B., Zhou, J.: Power injection model of STATCOM with control and operating limits for power flow and voltage stability analysis. Electr. Power Syst, Res. **76**(12), 1003–1010 (2006)
8. Kamel, S., Abdel-Akher, M., Jurado, F., Ruiz-Rodríguez, F.J.: Modeling and analysis of voltage and power control devices in current injections load flow method. Electr. Power Compon. Syst. **41**, 324–344 (2013)
9. Yang, Z., Shen, C., Crow, M.L., Zhang, L.: An improved STATCOM model for power flow analysis. IEEE Trans. Power Syst. **20**(3), 1121–1126 (2000)
10. Radman, G., Raje, R.S.: Power flow model/calculation for power systems with multiple FACTS controllers. Electr. Power Syst. Res. **77**, 1521–1531 (2007)
11. Bhargava, A., Pant, V., Das, B.: An improved power flow analysis technique with STATCOM, IEEE Trans. Power Syst. (2006)
12. Tien, D.V., Hawliczek, P., Gono, R., Leonowicz, Z.: Analysis and modeling of STATCOM for regulate the voltage in power systems. In: 18th International Scientific Conference on Electric Power Engineering (EPE), pp. 17–20 (2017). https://doi.org/10.1109/EPE.2017.7967358.
13. Zhang, X.-P., Rehtanz, C., Pal, B.: Flexible AC Transmission Systems: Modelling and Control: Modelling and Control. Springer, Berlin (2006). ISBN-13 978-3-540-30606-1
14. Bhowmick, S.: Flexible AC Transmission Systems (FACTS). CRC press, International Standard Book Number-13: 978-1-4987-5620-4

15. Acha, E., Fuerte-Esquivel, C., Ambriz-Pérez, H., Angeles Camacho, C.: FACTS: Modelling and Simulation in Power Networks. Wiley, Chichester (2004). ISBN 0-470-85271-2
16. Shahnia, F., Rajakaruna, S., Ghosh, A.: Static Compensators (STATCOMs) in Power Systems. Springer, Singapore (2015). ISBN 978-981-287-281-4
17. Kothari, I.J., Nagrath, D.P.: Modern Power System Analysis, 3rd edn. Tata McGraw-Hill Education, New York (2007)
18. Glover, J.D., Sarma, M.S.: Power System Analysis and Design, 3rd edn. Brooks/Cole, Pacific Grove (2002)
19. Saatdat, H.: Power System Analysis. McGraw Hill, New York (1999)
20. Milano, F.: Power System Modelling and Scripting. Power Systems. Springer, Heidelberg (2010). ISBN 978-3-642-13668-9
21. Wang, X.-F., Song, Y.S., Malcorm, I.: Modern Power System Analysis. Springer, New York (2008). ISBN 978-0-387-72852-0
22. Afolabi, O.A., Ali, W.H., Cofie, P., Fuller, J., Obiomon, P., Kolawole, E.S.: Analysis of the load flow problem in power system planning studies. Energy Power Eng. **7**, 509–523 (2015)
23. Musti, K.S.S., Ramkhelawan, R.B.: Power System Load Flow Analysis Using Microsoft Excel. University of West Indies (2012)
24. Glavic, M., Alvarado, F.L.: An extension of Newton-Raphson power flow problem. Appl. Math. Comput. **186**, 1192–1204 (2007)

Neural Network for Smart Adjustment of Industrial Camera - Study of Deployed Application

Petr Dolezel[(✉)] and Daniel Honc

University of Pardubice, Pardubice, Czech Republic
{petr.dolezel,daniel.honc}@upce.cz
http://fei.upce.cz

Abstract. Since machine vision is gaining more and more interest lately, it is necessary to deal with correct approaches to visual data acquisition in industry. As a particular part of this complex problematics, a technique for the industrial camera exposure time and image sensor gain tuning is presented in this contribution. In comparison to other approaches, a human expert photographer is used instead of explicitly defined cost function. His knowledge is transformed into an artificial expert system represented by a feedforward neural network. The expert system then provides the suitable exposure time and image sensor gain to gather sharp and balanced images.

Keywords: Industrial camera · Smart camera · Auto-exposure · Artificial neural network

1 Introduction

Machine vision, as an important element in the automation chain, is continuously gaining interest in the industrial area. In order to apply a robust autonomous control system, it is necessary to be able to acquire various state variables and performance characteristics; and the field of machine vision may provide us the suitable tools. As examples, a machine vision based inspection system for printed circuit boards is demonstrated in [14], while in [2], the authors present a robust machine vision apparatus for can inspection. The complex review of machine vision systems, applied in various branches of industry, is available in [8].

A conventional industrial camera with CMOS sensor is one of the most frequently used sources of visual signals for machine vision. The characteristics of the acquired signal are affected by an aperture, an exposure time (shutter speed) and an image sensor gain (qualifies ISO sensitivity). Despite the crucial importance of mentioned camera parameter tuning, dependent on ambient conditions, most of the industrial cameras provide only simple built-in auto-exposure tuning [10,11]. Furthermore, some cameras have a fixed exposure time, gain and aperture level, tuned manually by the user. The mentioned approaches work correctly in an environment with fixed light and motionless conditions. Compared to a fixed manual setting, cited auto-exposure methods tune camera parameters depending on light conditions and, therefore, can be successfully applied to stationary scenes with illumination changes. On the other hand, they can still cause a failure of machine vision algorithms when the scene setup drifts unceasingly or even changes abruptly.

Considering both dynamic light conditions and scene setup motion, the image sensor gain tuning takes on importance in order to get proper exposure time to freeze the motion. However, image noise level might increase with the changes in gain of image sensor. Therefore, not only light balance, but also the blurring level and noise rate should be included into acquired image evaluation.

Apparently, a number of approaches, that exceed the mentioned built-in autoexposure tuning techniques, have been published so far. In [13], the authors use gradient information to tune the camera parameters, while authors in [10] consider the image entropy. With advanced computing technology, more complex and computationally demanding approaches gain bigger attention. As an example, in [12], three images taken at various exposure levels are used for exposure tuning using the method called exposure bracketing. The commonality of all mentioned methods is the effort to set the camera parameters in order to meet the criterion of some explicitly defined cost function. In this paper, conversely, a human expert is used instead of the cost function and his knowledge is transformed into an artificial expert system represented by a feedforward neural network.

In the following sections, the implementation of this approach is demonstrated on a specific application of printed circuit board (PCB) image acquisition (see Fig. 1).

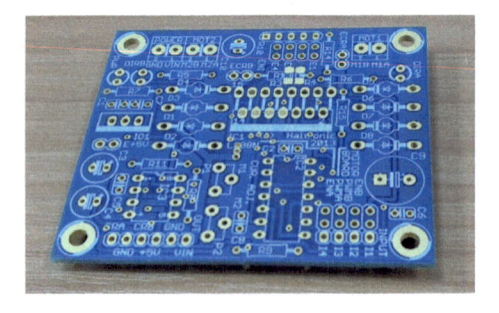

Fig. 1. Printed circuit board for image acquisition system.

2 Problem Formulation

The aim of this paper is to present a neural network based approach to camera exposure time and sensor gain adjustment, in order to obtain the images suitable for further processing by machine vision algorithms. A default aperture level and manual focus are considered, since mainstream industrial cameras rarely offer instrumental actuation of these parameters. The approach should provide well-exposed, sharp images under varying light conditions and dynamic scene setup changes.

Contrary to the most of the methods reviewed in the previous section of the paper, this approach uses a knowledge-base, acquired from an expert photographer rather than any analytical algorithm to determine the camera parameters. In other words, no analytical cost function is defined to specify the quality of camera setting. Instead, a human expert labels the setting as suitable or unsuitable. The knowledge-base is then transformed into an efficient data structure represented by a feedforward neural network.

The whole system should work as shown in Fig. 2.

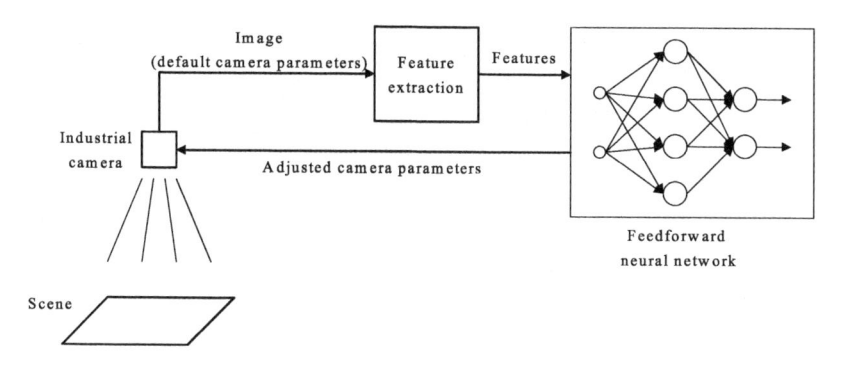

Fig. 2. General diagram of presented approach.

Therefore, the industrial camera takes the shot under the predefined (and fixed) value of gain and exposure time. The acquired image is analyzed and the relevant features are extracted. Then, the extracted features are used as an input vector to a feedforward neural network, which should provide a suitable set of camera parameters.

Apparently, the crucial part of the approach is the neural network, which should be sufficiently trained to provide good results across the whole state space of light conditions and scene setups. Thus, an expert photographer should provide an extensive training set of camera parameters adjusted manually under various experiment conditions and his role is indispensable in this approach.

3 Proposed Solution

As mentioned above, the proposed approach consists of several encapsulated steps; image acquisition by industrial camera, feature extraction from acquired image, parameter adjustment performed by the feedforward neural network, and eventually, image shot under the correctly adjusted camera parameters. The mentioned steps are dealt with in the following paragraphs.

3.1 Image Acquisition by Industrial Camera

For the demonstration system proposed in this paper, a Basler acA1920-40uc is used as an industrial camera together with a TS5014-MP F1.4 f50 mm Basler lens. The default aperture level is set to f/2.8. Apart from the industrial camera, a source of light is included in the system. In this particular case, an AwoX Smart Light Color MESH E27 is considered, since it provides warm white, cool white and combination of 16 million colors, which can be smoothly tuned. Thus, very wide range of light conditions is available. Parallelly, a conveyor is implemented as a last part of the image acquisition system in order to bring the possibility of PCB motion. All the necessary parts are shown in Fig. 3.

Fig. 3. Necessary parts of the image acquisition system (camera, lens, light source and conveyor).

According to the general diagram in Fig. 2, two images are taken during one iteration. The first one is taken under the default parameters (default image). Then, according to the features extracted from the default image, the camera parameters are tuned, and the second image is taken under the "well-adjusted" parameters. The default parameters are set as follows.

Exposure time: 31.25 ms
Image sensor gain: 0

3.2 Feature Extraction

As defined above, the approach should perform correctly under various light conditions as well as for moving objects. Apparently, a correlation needs to exist between features extracted from the default image and mentioned ambient conditions.

Since the images are expected to include only one type of PCB, the light balance of the image can be approximately determined by the average pixel intensity of the image. Pixel intensity of each pixel in the image is formed by the sum of its red, green and blue components [6], as follows.

$$I = 0.2989 I_R + 0.5870 I_G + 0.1140 I_B, \tag{1}$$

where I_R, I_G and I_B are the intensities of the red, green and blue components of the pixel.

However, the weighted sum of intensities may provide insufficient information, especially in the discussed situation. PCB in this case dominantly contains blue color - see Fig. 1, while the weight of the blue component is minor in Eq. (1). In order to prevent this situation, average intensities of all three components (red, green and blue) may be considered as possible features that describe the light balance of the image. Three examples are shown in Fig. 4.

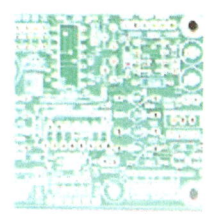

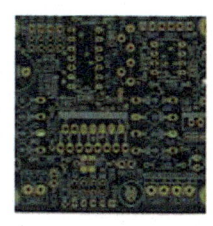

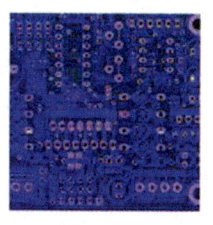

Fig. 4. Average pixel intensities of images (left-hand side: 222, middle: 55, right-hand side: 46) and average pixel intensities of the components (left-hand side - R: 185, G: 242, B: 216, middle - R: 39, G: 66, B: 38, right-hand side - R: 37, G: 33, B: 139).

Another feature should describe the blurring level of the image. As the focus at a constant distance is assumed, blurring of the image indicates motion in the scene. Considering the scene in the presented application (PCB), the steep gradients in the pixel intensity function, with respect to pixel position, are expected in the sharp images. This occurs since there are strong transitions situated in the structures of PCB. The mentioned feature can be represented by many quantities based on variance in the image, high-pass filtering, edge detection, frequency histogram, or entropy. A nice survey about the blurring in the images can be found in [7].

For this particular case, the average value of the local range of an image is applied. It can provide sufficient information, and in addition, it is not computationally demanding. The local range of an image returns the matrix of the same size as the image, where each element contains the range value (the difference between the maximum and minimum value) of the 3 by 3 neighborhood around the corresponding pixel of the original image. The examples of the local range of an image, as well as the average values, are shown in Fig. 5.

Therefore, two possibilities of feature extraction are dealt with in the following paragraphs. The first one considers vector of two components, where the average pixel intensity of the image and the average local range are included. The second possibility considers four components, where the average intensities of red, green and blue color are included, as well as the average local range.

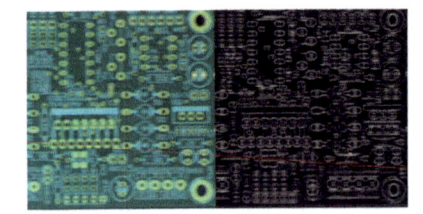

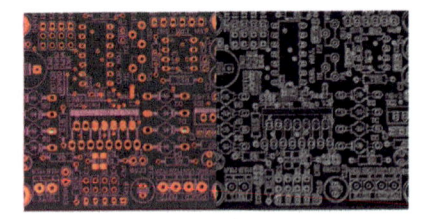

Fig. 5. Local range of of blurred image (left-hand side, average value equal to 18) and sharp image (right-hand side, average value equal to 33).

3.3 Neural Network for Camera Parameters Adjustment

Camera exposure time and image sensor gain need to be adjusted in relation to two, or four, features extracted from the default image. A multilayer feedforward neural network (FFNN) seems to be a perfect tool for this issue, since it provides universal approximation qualities [5]. Thus, a FFNN is expected to transform the extracted features directly to the "optimal" parameters of the industrial camera - see Fig. 6.

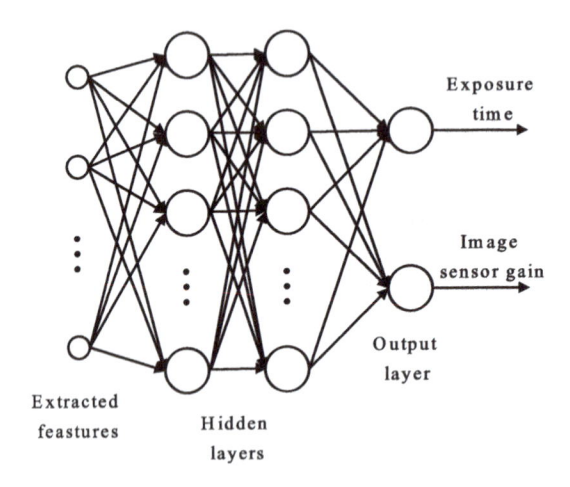

Fig. 6. Feedforward neural network for camera parameters adjustment.

In order to keep the universal approximation qualities, the FFNN has to meet some requirements [4]. In fact, these requests will be met if at least one hidden layer with a sufficient number of neurons is used. Another necessary factor is continuous, bounded and monotonic activation functions, which are applied in hidden neurons. For this particular demonstration, a FFNN with one or two hidden layers is considered. A hyperbolic tangent activation function is used for the neurons in the hidden layer, while the linear identical activation function is applied for the output neurons.

Apparently, the FFNN needs to be trained to provide the expected functionality. The process of training is described in a separate section of the contribution.

4 Design of Feedforward Neural Network

The procedure of FFNN design involves training and validation set acquisition, training, pruning and validating. The essential information related to this procedure is described here. Detailed information about the process, including the discussion to each step of the procedure, can be found in [4].

4.1 Training and Validation Set

As mentioned above, training and validation sets are prepared by an expert photographer using the hardware described in Sect. 3.1. In simple words, the photographer sets the exposure time and the sensor gain under various conditions involving different light intensity and color, and different speed of the photographed object. For each pair of tuned values, a default image is taken using the exposure time and the sensor gain described in Sect. 3.1. The default image is then used for feature extraction. Therefore, the training and validation set contains both input and target data. The first few lines are shown in Table 1, and the corresponding default images and images taken by the expert photographer are shown in Fig. 7. The whole training set is available for the community at [1].

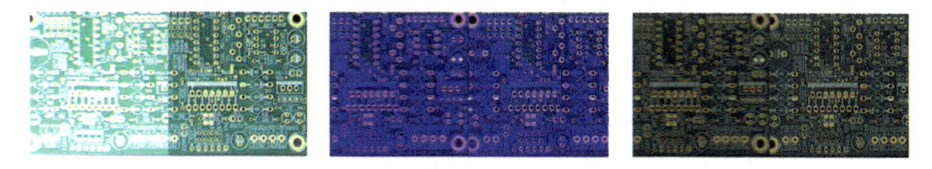

Fig. 7. Three training set examples (Left hand side - default images, right hand side - images shot under the parameters tuned by expert photographer).

4.2 Training and Pruning

While training of a FFNN deals with determining optimal weights and biases, the pruning converts the net into a simpler one, while the performance of the original network is kept. In this contribution, the pruning is performed by training of FFNNs with various topologies. A standard mean square error (E) function computed over the validation set is used as an error function, whereas 100 training experiments for each topology are performed to achieve higher significance of the results. According to the previous authors' experience, the Levenberg-Marquardt algorithm [3] is used for training while the Nguyen-Widrow method [9] is applied for the initialization of the weights. The target values are normalized on a $[-1; 1]$

Table 1. Example of training set values for selected feature extraction techniques (rows correspond to images shown in Fig. 7)

N.	Inputs (pixel intensities)		Inputs (intensities of colors)				Targets	
	PI	LR	RED	GREEN	BLUE	LR	Exp. time, ms	Gain
1	202	27	136	234	212	27	8	0
2	46	21	38	32	144	21	4	188
3	56	13	42	67	38	13	2	267
...	...	...	...	...	...	...	...	...

PI ... Average pixel intensity; LR ... average local range;
RED ... average intensity of red component;
GREEN ... average intensity of green component;
BLUE ... average intensity of blue component

range in order to avoid the unequal influence of individual errors on the training process. All the parameters of the training and pruning procedure are shown below - see Table 2.

The procedure is performed with both discussed possibilities of the feature extraction. Therefore, two sets of resulting data are acquired. In Fig. 8, box graphs with the results obtained from the average pixel intensity and average local range are shown. Conversely, box graphs with the results obtained from the average red, green and blue intensities and average local range are shown in Fig. 9. The central marks of the box graphs are medians, the edges of the boxes are 25th and 75th percentiles and the whiskers extend to the most extreme data points.

Table 2. Parameters of the experiments

Training set ratio	85 %
Validation set ratio	15 %
Training algorithm	Levenberg-Marquardt algorithm
Initialization	Nguyen-Widrow method
Maximum epochs	1000
Stopping criterion	Maximum epochs reached
Adaptive coefficient μ	0.001
Increment μ	10
Decrement μ	0.1

Looking at Figs. 8 and 9, the trends seem to be very similar. Considering one hidden layer, a minimal value of the error function is, in both cases, achieved using the topology with four neurons in the hidden layer. More neurons or another hidden layer do not bring any improvements. Beyond, the resulting value of the error function computed over the validation set is almost identical, using either average pixel intensities or average red, green and blue color component intensities as inputs. Anyway, FFNNs with four neurons (hyperbolic tangent activation function) in one hidden layer are considered for further evaluation. Error histograms for both training and validation sets are shown in Fig. 10.

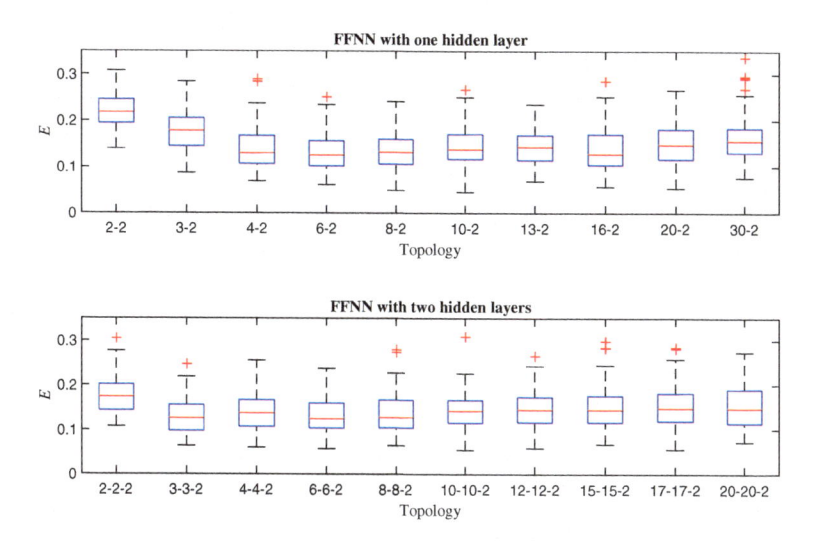

Fig. 8. Neural network pruning with average pixel intensity and average local range as features (X-axis represents network topology to be examined, Y-axis associates resulting values of error function computed over validation set).

5 Evaluation of Proposed Approach

Although the system could be evaluated just on the basis of the results presented in Figs. 8, 9 and 10, there is also available another possible way of evaluation. Accordingly, the system can be used for industrial camera parameters adjustment under the set of expressive ambient conditions. The images taken using the adjusted camera can be then evaluated by a human expert.

Therefore, the machine vision system according to Fig. 2 is prepared. Two FFNNs with four neurons in each hidden layer are used, since both discussed possibilities of feature extraction are still considered. Further, six scene settings are prepared, covering various light conditions and different speeds of the photographed object. The resulting images are shown in Fig. 11.

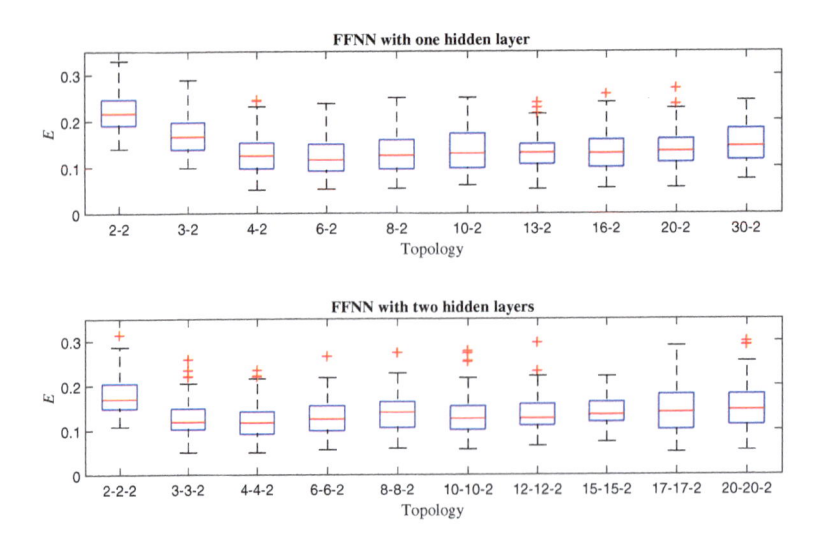

Fig. 9. Neural network pruning with average red, green and blue intensities and average local range as features (X-axis represents network topology to be examined, Y-axis associates resulting values of error function computed over validation set).

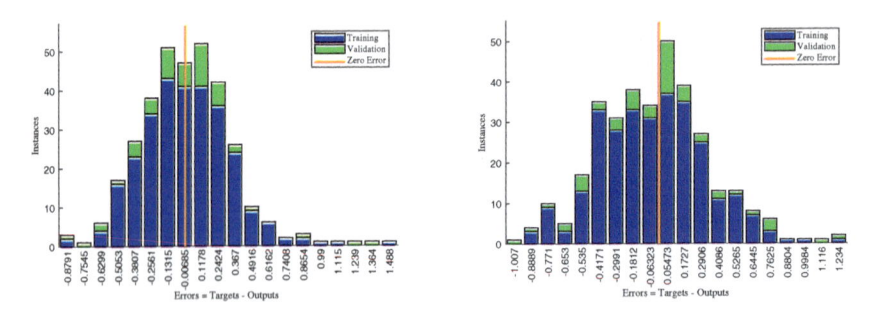

Fig. 10. Error histograms of the most suitable neural networks gained from the procedure presented in Fig. 8 (left hand side) and procedure presented in Fig. 9 (right hand side).

Contrary to the very similar statistical quantities presented in the previous section, the explicit examples shown in Fig. 11 prove decent differences in performance of both discussed approaches.

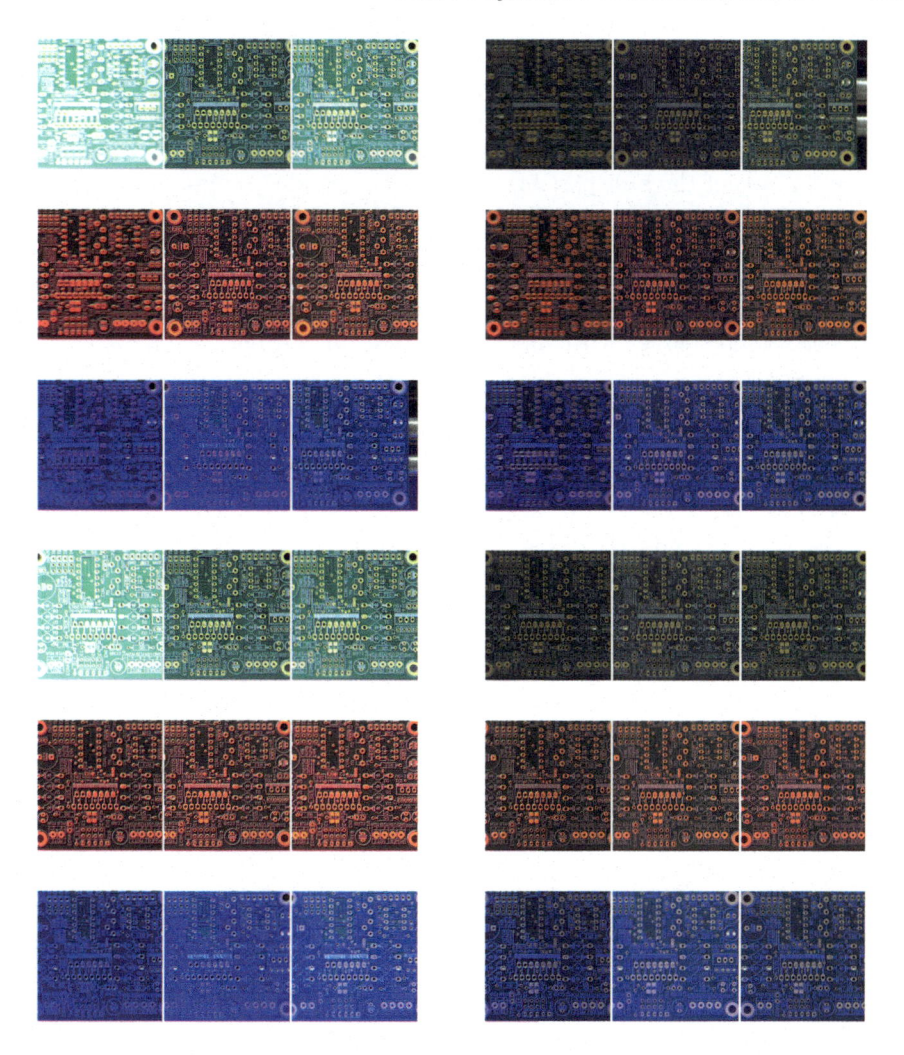

Fig. 11. Evaluation of proposed approach (in each three images, first image is default, second one is obtained using average pixel intensity of default image, while third one is obtained using average red, green and blue color intensities of default image).

6 Conclusion and Future Work

One approach to the industrial camera exposure time and image sensor gain auto-tuning is presented in this paper. Contrary to other approaches, a human expert is used instead of the cost function and his knowledge is transformed into an artificial expert system represented by a feedforward neural network. Rather than a complex analytical solution, the implementation of the introduced approach is demonstrated on a specific application of a printed circuit board image acquisition. Thus, the paper should be understood as the preliminary engineering concept and it deserves to be examined more comprehensively in the future.

In particular, authors are going to deal with the feature extraction techniques as the next step of the research. Many different feature extraction techniques have been published so far and their possible impact to the presented approach should be deeply examined. Then, a systematic methodology for training set acquisition, to get the best possible data for the introduced approach, is considered as a good consequent move. And, eventually, an independent and balanced evaluation procedure for the adjustment technique should be also defined.

Anyway, even as a limited deployed application, the presented approach provides suitable camera parameters, and the acquired images (see Fig. 11) definitely represent a good source for the image processing algorithms.

Acknowledgment. The work was supported from ERDF/ESF "Cooperation in Applied Research between the University of Pardubice and companies, in the Field of Positioning, Detection and Simulation Technology for Transport Systems (PosiTrans)" (No. CZ.02.1.01/0.0/0.0/17_049/0008394).

References

1. Data file with images (2018). https://www.researchgate.net/publication/3265577 28_Images_Basler
2. Chen, T., Wang, Y., Xiao, C., Wu, Q.M.J.: A machine vision apparatus and method for can-end inspection. IEEE Trans. Instrum. Measur. **65**(9), 2055–2066 (2016)
3. Hagan, M., Menhaj, M.: Training feedforward networks with the Marquardt algorithm. IEEE Trans. Neural Netw. **5**(6), 989–993 (1994). https://doi.org/10.1109/72.329697
4. Haykin, S.: Neural Networks: A Comprehensive Foundation. Prentice Hall, New Jersey (1999)
5. Hornik, K., Stinchcombe, M., White, H.: Multilayer feedforward networks are universal approximators. Neural Netw. **2**(5), 359–366 (1989)
6. ITU-R Recommendation BT.601: Studio encoding parameters of digital television for standard 4:3 and wide screen 16:9 aspect ratios (2011)
7. Koik, B.T., Ibrahim, H.: A literature survey on blur detection algorithms for digital imaging. In: 2013 1st International Conference on Artificial Intelligence, Modelling and Simulation, pp. 272–277 (2013). 10.1109/AIMS.2013.50
8. Nandini, V., Vishal, R.D., Prakash, C.A., Aishwarya, S.: A review on applications of machine vision systems in industries. Indian J. Sci. Technol. **9**(48), 1–5 (2016)
9. Nguyen, D., Widrow, B.: Improving the learning speed of 2-layer neural networks by choosing initial values of the adaptive weights. pp. 21–26 (1990)
10. Ning, J., Lu, T., Liu, L., Guo, L., Jin, X.: The optimization and design of the auto-exposure algorithm based on image entropy. In: 2015 8th International Congress on Image and Signal Processing (CISP), pp. 1020–1025 (2015). https://doi.org/10.1109/CISP.2015.7408029
11. Park, S., Kim, G., Jeon, J.: The method of auto exposure control for low-end digital camera. In: 2009 11th International Conference on Advanced Communication Technology, vol. 03, pp. 1712–1714 (2009)
12. Pourreza-Shahri, R., Kehtarnavaz, N.: Exposure bracketing via automatic exposure selection. In: 2015 IEEE International Conference on Image Processing (ICIP), pp. 320–323 (2015). https://doi.org/10.1109/ICIP.2015.7350812

13. Shim, I., Lee, J.Y., Kweon, I.S.: Auto-adjusting camera exposure for outdoor robotics using gradient information. In: 2014 IEEE/RSJ International Conference on Intelligent Robots and Systems, pp. 1011–1017 (2014). https://doi.org/10.1109/IROS.2014.6942682
14. Wang, W.C., Chen, S.L., Chen, L.B., Chang, W.J.: A machine vision based automatic optical inspection system for measuring drilling quality of printed circuit boards. IEEE Access **5**, 10817–10833 (2017)

Risk Assessment Approach to Estimate Security of Cryptographic Keys in Quantum Cryptography

Marcin Niemiec[1(✉)], Miralem Mehic[1,2], and Miroslav Voznak[1]

[1] VSB – Technical University of Ostrava, 17. listopadu 2172/15,
708 33 Ostrava, Czechia
marcin.niemiec@vsb.cz

[2] Department of Telecommunications, Faculty of Electrical Engineering,
University of Sarajevo, Zmaja od Bosne bb, 71000 Sarajevo, Bosnia and Herzegovina

Abstract. Increased interest in quantum cryptography is observed in recent years. Although this technique is characterized by a very high level of security, there are still challenges of quantum key distribution that should be solved. One of the most important problem remains security mechanisms for the key distillation process which can be effectively controlled by end users. This article presents a new idea for security assessment of cryptographic key based on the risk management in quantum cryptography. This proposal assumes the estimation of risk level using two components: likelihood and impact. The likelihood can be defined by the probability of eavesdropping during the quantum bit estimation. The impact is associated with the effect of key reduction in the all steps of key distillation process. Using this novel approach end users of quantum cryptography will be able to control both efficiency and security level of cryptographic keys.

Keywords: Security · Quantum key distribution · Risk assessment · Symmetric cryptography

1 Introduction

Interest in quantum-based mechanisms for communications is growing rapidly in recent years. Solutions such as quantum cryptography or quantum random number generators are not just theories, but are being used in practice. However, these mechanisms are not very widespread, they are likely to form part of the next generation of communications [1].

Quantum cryptography (QC) brings an entirely new way of solving the old problem with cryptographic key distribution. It provides secure key distribution by means of the laws of quantum mechanics. Quantum-based cryptography increases security to a level unachieved by any previous solutions [2]. It seems that even technological problems such as effective quantum repeaters will be solved in the near future. Then, quantum-based solutions will be widely implemented in practical networks.

QC is characterized by a very high level of security, but there are a real risk that eavesdropper can acquire some information about the distributed bits in quantum channel. Although even if eavesdropping modify the quantum information, end users can not uncover and check all distributed bits. Therefore, the security of final cryptographic key is characterized by some risk that eavesdropper has a knowledge about the some bits. Unfortunately, the assessment of this risk is not trivial because of complexity of the key distillation process.

This paper introduces a novel solution to assess risk level of cryptographic keys' security in quantum cryptography. It takes into account the quantum bit error estimation, error correction and privacy amplifications steps. This concept joins the risk management approach and the quantum cryptography technique into the one mechanism. Using proposed solution, end users of quantum cryptography are able to define the security level of cryptographic keys in quantitative way.

The remainder of the paper proceeds as follows. Risk management approach is introduced in the next section. We present basics of quantum cryptography in Sect. 3. The proposed risk-based approach is described in Sect. 4, and the paper closes with a short conclusion.

2 Risk Management

Security is a crucial requirement in contemporary communications systems. Each telecommunication system should offer high guarantees of data confidentiality, integrity or availability. However, it is often difficult to verify the quality of security mechanisms implemented in today's complex IT environments and objectively assess their security in quantitative way. Moreover, it is not possible to consider such systems as fully deterministic because of human creativity, unpredictable incidents, different motives and aims. These influences make impossible to prepare for any type of threats and attacks. This uncertainty is the crucial reason why risk management has been become so useful in nowadays telecommunication systems.

Generally, risk in IT can be described as a threat exploiting a system's weakness. Risk management is a process which supports entities in making decisions related to security issues [3]. It helps prevent malicious individuals or harmful environmental factors from damaging assets. Final decisions are making on the balance between risk, cost and efficiency. Misjudgements in this process may result in severe incidents. That's why the proper assessment of risks is critical part of risk management.

Likelihood of a given threat and magnitude of the impact are the crucial components of risk. They allows to provide risk assessment of particular incident in quantitative or qualitative way [4]. Therefore, risk can be determined as the product of likelihood and impact:

$$Risk = Likelihood \cdot Impact \quad (1)$$

Likelihood is characterized as a measure of how frequently an adverse effect occurs. Usually it is closely related to the malicious individual (called threat agent). If we want to calculate a likelihood, we should consider how likely the adverse event is occurring and how likely the risk is materializing (e.g. the chances that the attack finished successfully).

Likelihood is often assessed in a qualitative way – this is why the term likelihood is used instead of probability. Likelihood estimation is usually subjective to the person performing the assessment. Therefore, specification of likelihood is a challenge during the risk management process. In some cases it is easier to define the likelihood as probability in a mathematical sense – especially if we can make accurate statistical calculations to describe such values (i.e. Mean Time To Failure or estimated downtimes of communication links).

Impact – the second component of risk – describes the degree of harm when the risk will materialize. It provides information on the effect caused by a threat which exploited a given vulnerability. It can be characterized in numerous ways. An example is financial loss providing the monetary equivalent of impact, however it requires a deep understanding of business processes and it is not a proper measure in all cases. Other scale is the reputation caused by risk materialization (i.e. eavesdropping/modification of confidential data). The impact related to the security aspects (confidentiality, integrity, availability or others) is commonly used in the contemporary IT systems. Taking into account these aspects of data security, impact can be measured by the number of compromised records, volume of modified data or number of eavesdropped bits in a cryptographic key. It is worth mentioning that if we want to define the impact – the assets rating is crucial (i.e. loss of confidentiality of cryptographic key for the encrypted session between a user and a server or in contrast, eavesdropping the time of this session).

3 Quantum Cryptography

Secure distribution or agreement of symmetric cryptographic keys are crucial to data confidentiality [5]. Currently, when we use modern ciphers with popular key distribution methods, we are not sure if an intruder is eavesdropping in the communication channel. This means that a hidden intruder can scan the network and obtain sensitive data. Quantum key distribution ensures a very high level of security, because it is not possible to eavesdrop on the communication in a passive way [6].

The rules of quantum mechanics ensure that measurement modifies the state of the transmitted qubit (quantum bit) [7] and it is not possible to clone an unknown quantum state [8]. If an eavesdropper (usually called Eve) reads the distributed key, he/she will change the quantum states of the photons and can be revealed. This modification can be discovered by the sender and the receiver of qubits (usually called Alice and Bob). Therefore, quantum cryptography requires two types of channels: the quantum channel, where qubits with the information about the distributed key are exchanged and the public channel, which is used to check whether the communication through the quantum channel is secure or not. Additionally, the public channel is used for the correction of errors.

In general, quantum cryptography consists of two main steps: the quantum key distribution protocol and the key distillation algorithms [9]. The second step uses the public channel between Alice and Bob to estimate quantum error bit, correct errors and improve security using privacy amplification algorithm.

3.1 Quantum Key Distribution

Quantum key distribution is used to distribute an encryption key for symmetric ciphers but not to transmit any message data between users. As we mentioned, information about a key is transmitted by means of qubits in quantum channel and the security of QKD relies on the foundations of quantum mechanics. In general, we could distinguish two types of QKD protocols: based on single and entangled particles [10].

QKD protocols based on single particles distribute information about the key by means of quantum state of single particles (i.e. polarized photons). Entangled state is a pair of particles which has the following feature: the states of particles are random before the measurement but if we measure the state of one of them, then the state of the second particle is determined.

Popular quantum key distribution protocols, such as BB84 [11], are based on the polarization of single photons, which carry information from a sender (Alice) coded in quantum states (i.e. different polarizations: vertical, horizontal, diagonal). This way, the recipient (Bob) and potential eavesdropper (Eve) do not know which detector should be used to measure the polarization precisely. However, it is not a problem for the intended recipient – when they announce the configuration of the detector which was used during the measurement of a received photon, the sender confirms that the obtained result is correct or asks to delete this bit from the final key because the obtained result is not certain.

Eavesdropping in BB84 protocol can be easily detected. If the eavesdropper chooses the wrong detector to measure the polarization, the polarization of the photon is changed. Then, the sender and the recipient uncover the eavesdropper if they compare a part of the obtained key (wrong bits in the key detect eavesdropper). This way, passive eavesdropping is not possible. If someone tries to eavesdrop some photons and read confidential information, then the quantum states of these photons can change.

3.2 Key Distillation

Quantum key distribution ensures a very high level of security. However, it is only a part of the complete key establishment process. After quantum key distribution process, Alice and Bob must estimate errors in the distributed key by computing quantum bit error rate (QBER). The QBER is defined as the ratio of the number of wrong bits to the total number of bits. It is worth emphasizing that not only Eve can be responsible for errors – the errors may also occur because of disturbance in the quantum channel, optical misalignment, noise in detectors, and others.

In practice, Alice and Bob compare a small portion of a distributed key using public channel and compute the QBER. If QBER exceeds a given threshold, it means that Eve has eavesdropped (or the quantum channel is too noisy to perform a proper key distribution). But if error rate is low enough, Alice and Bob continue further distillation of the key. Unfortunately, the uncovered part of the key must be deleted because the bits are not confidential.

After the quantum bit error estimation, Alice and Bob perform two additional steps:

- error correction – in this step, the sender and recipient must find and correct or delete occurred errors,
- privacy amplification – where the sender and the recipient improve the security of the final key by deleting some of bits.

Currently, two error correction method are popular. The first is BBBSS protocol designed by Bennett and his coworkers [12]. It verifies the parities of block (parts of the raw key). BBBSS uses several passes to correct the errors and a pseudo-random permutation is used after each pass. The second popular method is Cascade algorithm proposed by Brassard and Salvail [13]. Usually, it uses four passes and doubles block length starting from the second pass. This ensures a more efficient error correction process.

The last of the key distillation process is privacy amplification. Because Eve may have gained significant knowledge of the key (mainly eavesdropping in the public channel during the bit error estimation and key reconciliation), Alice and Bob are required to strengthen the security of the final key. They can delete some of the bits and construct the final key usually using universal hash functions.

If Alice and Bob perform all these steps, the final key which can be used to symmetric encryption will be significantly reduced. This reduction of the key length is characteristic for all quantum key distribution protocols [14]. Because each stage reduces the key length, the performance of QKD is also reduced. Sometimes this reduction is significant. The example of this effect was described in [15] where 1024 qubits cause 292 bits of final key. Therefore the key distillation process is crucial to the quantum cryptography implemented in real communication networks because of security and efficiency.

4 Risk-Based Approach in Quantum Cryptography

Security is an abstract concept which causes serious problems if we want to determine it in a quantitative way. Usually, we are able to characterize if a given communication system is secure or not. However, it is not simple to specify the level of security.

In this section we describe a new risk-based approach to specify the security of cryptographic keys in quantum cryptography. We take into account the quantum bit error estimation process and further steps of the key distillation. This novel concept joins the risk management approach and the quantum cryptography technique. Using this solution we are able to define the security level of final cryptographic keys in quantum cryptography systems.

4.1 Probability of Eavesdropping

Let us assume that Alice and Bob finished the communication process via quantum channel and determined the strings of bits (with key length n) using QKD protocol. However, when Bob received qubits from Alice and obtained the bits, he is not sure if this data are really confidential. Therefore, Alice and Bob have to uncover a part of the string to know that nobody was eavesdropped in quantum channel. Uncovered bits allow Alice and Bob to estimate the real error rate and decide if somebody was eavesdropping in quantum channel or not. However, the crucial question is: how many bits they should uncover to know that the encryption key is really secure?

When Alice and Bob uncover and compare a single bit, the acquired information increase the knowledge about the whole string of bits. After that the uncovered bit is not confidential any more and must be removed from the string. When Alice and Bob find too many errors in their strings of bits, they find that qubits were eavesdropped and cryptographic key is not secure. However, if most of uncovered bits are the same in the both strings (the number of errors do not exceed the typical value of QBER in this system), probably the cryptographic key is secure. Unfortunately, it is not certain. There is still a risk that Eve eavesdropped some of the rest bits.

If we take into account the risk-based approach, we can assume that the probability of eavesdropping can be the likelihood (the component of risk defined by Eq. 1). This probability – defined by the function $L(i)$ – depends on the number of uncovered bits (i). We can define two boundary cases:

- if we uncover 0 bits, we know nothing about security of the key (the maximum probability of eavesdropping),
- if we uncover all bits (n bits) we are sure that the key is secure (the minimum probability of eavesdropping).

Moreover, $L(i)$ must be a monotonically decreasing function – it means that if i is growing then $L(i)$ decreases. According to a real scenario: if we compare more bits, the probability that Eve was eavesdropping decreases.

Intuitively, function $L(i)$ should not be linear. Let us consider first uncovered bits: already a few first bits is enough to obtain general information about security of distributed key and check if Eve is eavesdropping continuously. Therefore, these first bits provide more information about security of the key compared to the same number of bits but uncovered later.

The function $L(i)$ which meets these requirements can be defined as the logarithmic function. The inspiration of this choice is information theory – especially the measure of information introduced by Hartley in 1928 [16]. If we assume that a source of information generates messages m_k from the finite alphabet of a source:

$$\varphi = \{m_1, m_2, \ldots, m_n\} \tag{2}$$

and each of the messages occurs with probability:

$$P(M = m_k) = p_k, \quad k = 0, 1, \ldots, n \tag{3}$$

then, the logarithmic function is the most 'natural' measure of information. Therefore Hartley proved that the message m_k carries the following information:

$$I(m_k) = \log\left(\frac{1}{p_k}\right). \tag{4}$$

Taking into account the same criteria, the function $L(i)$ which describes the probability of eavesdropping can be defined as:

$$L(i) = \frac{\log\left(\frac{i}{n}\right)}{\log\left(\frac{1}{n}\right)} = \log_{\frac{1}{n}}\left(\frac{i}{n}\right). \tag{5}$$

In Fig. 1, an example of function $L(i)$ was presented where the key length is $n = 1000$ bits. In Table 1 also the source code of function $L(i)$ in Matlab environment was presented.

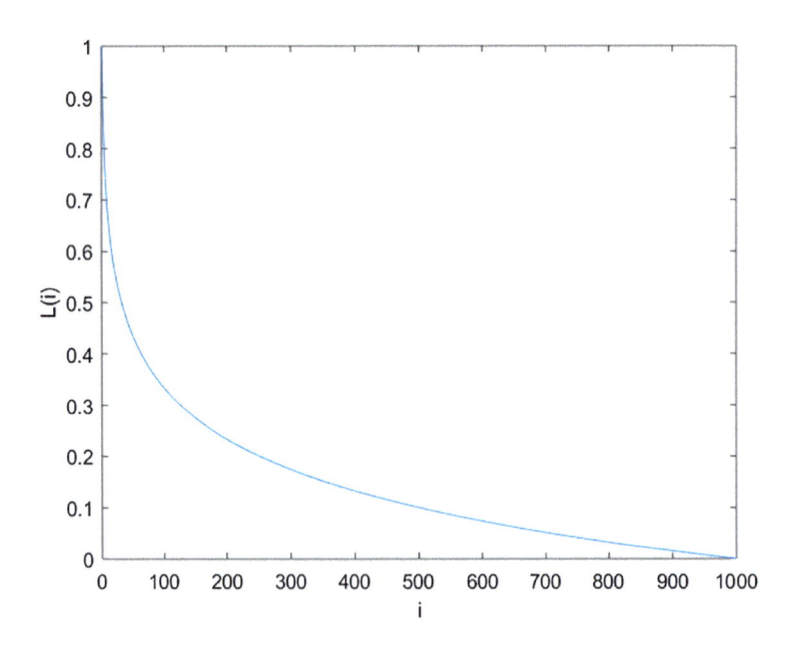

Fig. 1. An example of function L(i) for key length: 1000 bits

We can also define the probability of eavesdropping per single bit in the string of bits after the QBER estimation step. If we assume that Alice and Bob decided to uncovered q bits to estimate the QBER, the probability of eavesdropping per single bit is:

$$\frac{\log_{\frac{1}{n}}\left(\frac{q}{n}\right)}{n-q}. \tag{6}$$

Table 1. Matlab code of function L(i)

Function L(i) (the key contains 1000 bits):
1. i = 1:1000;
2. L = log(i/1000)/(log(1/1000));
3. plot(i,L)
4. xlabel('i');
5. ylabel('$L(i)$');

4.2 Impact to the Final Key

The risk assessment approach takes into account the impact (Eq. 1). This is a proper way also in the case of quantum cryptography because of the key reduction effect during the key reconciliation process. As we mentioned in the previous section, after the QBER estimation, Alice and Bob must delete the bits which were uncovered and compared. Also the next step – errors correction – causes the reduction of the key.

The reduction of bits depends on the used algorithm. Alice and Bob can choose one of several known reconciliation algorithms; however, currently the most popular reconciliation methods are algorithms which are based on a parity check of blocks. The simplest scenario assumes that the key is grouped into blocks of a given size. Alice and Bob compare parities of each block over the public channel. If their parities disagree, the block contains an odd number of errors. This block is cut into two sub-blocks and their parities are compared again. This procedure is continued recursively for all blocks which contain an odd number of errors as long as errors will be corrected. After that, both keys contain an even number of errors or none. Alice and Bob shuffle the positions of bits and repeat the same procedure (sometimes with blocks of bigger size) as long as both keys will be the same.

Unfortunately, each parity control over the public channel discloses a part of the secret key's information. If Eve collects the parities of many blocks, she will be able to calculate parts of the key (let us assume that Eve collected e bits in this way). Therefore, Alice and Bob must reject some bits to reduce the eavesdropper's knowledge about the secret key. Usually, this is realized in the last step of key distillation – the privacy amplification process.

Because Alice and Bob are required to strengthen the security of the key, they can delete some of the bits and construct the final key in a specific way. To achieve this goal, usually the universal hash functions are using in practice. Universal families of hash functions were created by Wegman and Carter [17] and privacy amplification with hash functions was proposed by Bennett, Brassard and Robert [18] in 1988. This algorithm is based on one-way functions which are able to convert a large string of bits into a short binary word. The reduction is defined by the s parameter which ensures that expected amount of information about the key is not bigger than $2^{-S}/\log 2$

Taking into account the reduction of the key, the impact can be defined as:

$$I = n - q - e - s \tag{7}$$

and this value specifies the final length of cryptographic key after the all of key distillation steps. It is worth mentioning that parameter e in Eq. 7 indicates number of deleted bits during the error correction step. However, for some algorithms the value of e is zero because the information leakage during error correction step can be solved by increasing parameter s during the next step (privacy amplification).

Now, we can define in quantitative way the risk R that the final key is not secure, using Eq. 1 and specified likelihood (Eq. 5) and impact (Eq. 7):

$$R = L(q) \cdot I = (n - q - e - s) \log_{\frac{1}{n}} \left(\frac{q}{n}\right). \tag{8}$$

Such defined risk mainly depends on the:

- number of uncovered bits during QBER estimation process (q) and
- s parameter during privacy amplification process.

Alice and Bob can minimize the risk by increasing number of checked bits at the beginning of the key distillation process or/and using one-way function which returns shorter final key in the input.

5 Conclusions

Although the quantum cryptography is characterized by a very high level of security, there are still some problems that limit the widespread use of this solution in practice. One of the biggest challenge remains mechanisms for the security verification and control during the key distillation process. However the passive eavesdropping in quantum channel can be easily detected, the sender and the recipient still must uncover and compare a part of the established key. Unfortunately, the eavesdropped bits may be located in the part of key which was not checked. Therefore, also in the IT systems with quantum cryptography is a risk that cryptographic key is not secure.

In this paper the new solution for security assessment of cryptographic key based on the risk management approach in quantum cryptography was introduced. This idea assumes the estimation of risk using two components – the likelihood that established string of bits is not secure as well as the impact into the final key. The likelihood was defined by the probability of eavesdropping during the quantum bit estimation process. The impact was associated with the effect of key reduction in the QBER estimation, error correction and privacy amplification steps.

Using this novel approach, end users of quantum cryptography are able to control security of cryptographic keys and efficiency of key distribution process. When end users choose the accepted risk level, they will be able to specify the parameters of QBER estimation and privacy amplification steps. Using the proposed risk-based approach, the decision-making process regarding the security of cryptographic key is much simpler. End users of quantum cryptography can easily verify the security of distributed key and choose the proper parameter during privacy amplification process.

Acknowledgment. This work was supported by the ESF in "Science without borders" project, reg. nr. CZ.02.2.69/0.0/0.0/16_027/0008463 within the Operational Programme Research, Development and Education.

References

1. Mulholland, J., Mosca, M., Braun, J.: The day the cryptography dies. IEEE Secur. Priv. **15**(4), 14–21 (2017)
2. Li, J., et al.: A survey on quantum cryptography. Chinese J. Electron. **27**(2), 223–228 (2018)
3. Jekot, M., Niemiec, M.: IT risk assessment and penetration test: comparative analysis of IT controls verification techniques. In: 2016 International Conference on Information and Digital Technologies (IDT), pp. 118–126 (2016)
4. Stoneburner, G., Goguen, A., Feringa, A.: Risk Management Guide for Information Technology Systems. NIST Special Publication 800–30 (2002)
5. Machnik, P.: Comparison of transmission parameters of different IPsec virtual private networks. In: International Conference on Telecommunications and Signal Processing (TSP 2009), Budapest, pp. 132–134 (2009)
6. Assche, G.V.: Quantum Cryptography and Secret-Key Distillation. Cambridge University Press, Cambridge (2006)
7. Benslama, M., Benslama, A., Aris, S.: Quantum Cryptography. Quantum Communications in New Telecommunications Systems. Wiley (2017)
8. Wootters, W.K., Zurek, W.H.: A single quantum cannot be cloned. Nature **299**, 802–803 (1982)
9. Mehic, M., Maurhart, O., Rass, S., Komosny, D., Rezac, F., Voznak, M.: Analysis of the public channel of quantum key distribution link. IEEE J. Quantum Electron. **53**(5), 1–8 (2017)
10. Dusek, M., Lutkenhaus, N., Hendrych, M.: Quantum Cryptography. Progress in Optics, vol. 46. Elsevier (2006)

11. Bennett, C.H., Brassard, G.: Public key distribution and coin tossing. In: IEEE International Conference on Computers, Systems, and Signal Processing, pp. 175–179 (1984)
12. Bennett, C.H., Bessette, F., Brassard, G., Salvail, L., Smolin, J.: Experimental quantum cryptography. J. Cryptol. **5**, 3–28 (1992)
13. Brassard, G., Salvail, L.: Secret-Key Reconciliation by Public Discussion, pp. 410–423. Springer (1994)
14. Mehic, M., Niemiec, M., Voznak, M.: Calculation of the key length for quantum key distribution. Elektronika Ir Elektrotechnika **21**, 81–85 (2015)
15. Niemiec, M., Machnik, P.: Authentication in virtual private networks based on quantum key distribution methods. Multimed. Tools Appl. **75**(17), 10691–10707 (2016)
16. Hartley, R.V.: Transmission of information. Bell Syst. Tech. J. **7**, 535–563 (1928)
17. Carter, L., Wegman, M.N.: Universal classes of hash functions. J. Comput. Syst. Sci. **18**, 143–154 (1979)
18. Bennett, C.H., Brassard, G., Robert, J.M.: Privacy amplifcation by public discussion. SIAM J. Comput. **17**, 210–229 (1988)

Wavelet Transform Decomposition for Fetal Phonocardiogram Extraction from Composite Abdominal Signal

Radana Kahankova[✉] and Radek Martinek

VSB – Technical University of Ostrava, Ostrava, Czech Republic
{radana.kahankova.st,radek.martinek}@vsb.cz

Abstract. This paper deals with the extraction of the fetal and maternal component from the composite abdominal phonocardiogram (PCG). The main method used for this task is the discrete wavelet transform. For the initial tests, we used synthetic PCG data incorporating the maternal heart sound interference. Based on the results, we suggested the suitable wavelet and the level of decomposition for fetal and maternal PCG extraction.

Keywords: fECG · Extraction · DWT · Wavelet decomposition · Fetal monitoring

1 Introduction

The fetal phonocardiography is a promising method used to monitor fetal well-being non-invasively with low initial and operation costs. The recorded signal, fetal phonocardiogram (fPCG), represents the mechanical events (sounds) produced by fetal heart.

Every physiological cardiac cycle includes two major sounds: the first heart sound (S1) and the second heart sound (S2), see Fig. 1. S1 represents the systole starting with the closure of mitral and tricuspid valves, whereas S2 denotes the diastole, it is initiated by the closure of the aortic and pulmonic valves at the end of the systole [1].

The main drawback of the method is that the useful signal's magnitude is low in comparison with the unwanted signals including the ambient noise or the biological artifacts (sounds produced by the maternal body, mainly mPCG). This is a problem especially in case of using a sensitive probe for measuring the fPCG signal [2].

2 State of the Art

For reasons mentioned above, it is necessary to apply advanced signal processing methods obtain the signal with adequate quality [3]. There are many methods that have been used for fPCG signal analysis and extraction, e.g. Blind Source Separation (BSS) techniques [4], Non-adaptive methods [5], and Adaptive methods [2, 3].

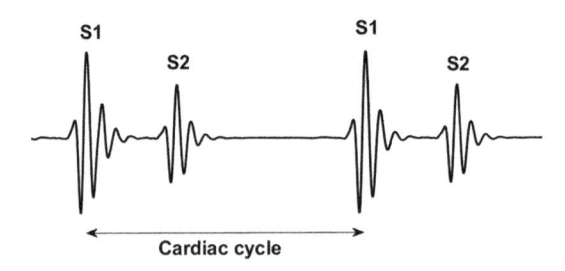

Fig. 1. Illustration of a Phonocardiogram waveform.

One of the methods that seems to be very promising and has been used in several research papers (e.g. [6–8]) is the Discrete Wavelet Transform (DWT). One of the arguments for the use of wavelets for the analysis of acoustic signals mentioned by Daubechies in [9] is that, in contrast with the analysis based on Fourier transform, WT treats frequency in a logarithmic way which is similar to our acoustic perception.

In [10], authors tested different wavelet and concluded that the 4th order Coiflet (coif4) outperforms the others when using the rigorous SURE threshold denoising algorithm with soft thresholding rule. In the mentioned article, the authors used MSE-based assessment, however, in most of the other papers (e.g. [11–13]), the parameters were chosen experimentally.

3 Mathematical Apparatus

In this section, we describe the theoretical background associated with the methods used for fECG extraction, namely the Wavelet Transform.

3.1 Discrete Wavelet Transform

Wavelet transform is defined as function of $\psi \in L_2(\mathbb{R})$. The system of functions $\left\{\psi_{j,k} := 2^{\frac{j}{2}}\psi(2^j x - k)\right\}_{j,k \in \mathbb{Z}}$ create orthonormal basis of $L_2(\mathbb{R})$:

$$\psi \in L_2(\mathbb{R}) \Rightarrow f = \sum_{j \in \mathbb{Z}} \sum_{k \in \mathbb{Z}} \langle f, \psi_{j,k} \rangle \psi_{j,k}. \tag{2.1}$$

The construction of orthonormal wavelet basis is based on multiresolution analysis (MRA). MRA is defined using scale function φ with scaling and wavelet equation as follows:

$$\varphi(x) = \sum_{k} h_k \varphi(2x - k), \tag{2.2}$$

$$\psi(x) = \sum_{k} (-1)^k h_{1-k} \varphi(2x - k), \tag{2.3}$$

$$f \in L_2(\mathbb{R}) \Rightarrow f = \sum_{k\in\mathbb{Z}}\langle f,\psi_{j0,k}\rangle\psi_{j0,k} + \sum_{j=j0}\sum_{k\in\mathbb{Z}}\langle f,\psi_{j,k}\rangle\psi_{j,k}. \quad (2.4)$$

4 Materials and Methods

In this section, we introduce the properties of the extraction system based on DWT as well as the dataset that we used for the initial tests.

4.1 Design of the DWT Based Extraction System

For WT-based signal processing, following parameters must be selected carefully.

Suitable Wavelet Base.
The choice of mother wavelet is very important since it influences the results of filtration. Usually, it is recommended to choose the wavelet base, which is similar to the processed signal – it allows to filter the higher frequencies of the signal.

According [10, 11], the most suitable wavelet family are the coiflets, particularly coif4, which is obvious from the comparison of the coiflets with fPCG waveform (see Fig. 2). Thus, coif 4 was used for the abdominal PCG (aPCG) signal decomposition in this paper.

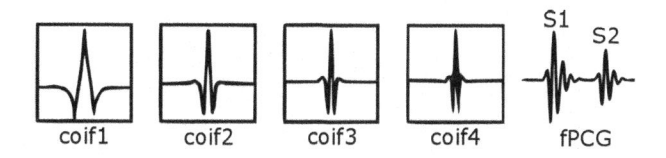

Fig. 2. Coiflet family in comparison with a fPCG waveform

Type of Thresholding.
The most important types of thresholding are the hard thresholding and the soft thresholding defined by Eqs. 3.1 and 3.2, respectively. The *hard* thresholding sets the samples lower than the threshold to zero and the rest of the values remain unchanged whereas for *soft* thresholding, the non-zero coefficients are decreased towards zero.

- Hard

$$x_{j,k}^T = \begin{cases} 0 \text{ for } x_{j,k} < \lambda, \\ x_{j,k} \text{ for } x_{j,k} \geq \lambda, \end{cases} \quad (3.1)$$

- Soft

$$x_{j,k}^M = \begin{cases} 0 \text{ for } x_{j,k} < \lambda, \\ signum(x_{j,k})(x_{j,k} - \lambda) \text{ for } x_{j,k} \geq \lambda, \end{cases} \quad (3.2)$$

where λ is the thresholding constant size defined by the thresholding rules. Four different thresholding rules can be selected in Matlab Wavelet Toolbox:

(a) Rigrsure – based on the principle of Stein's Unbiased Risk Estimate (SURE)
(b) Sqtwolog – Fixed threshold defined as $\lambda = \sigma\sqrt{2 \ln length(x)}$,
(c) Heursure – defined as combination of SURE and fixed threshold.
(d) Minimaxi – based on minimax principle.

Chourasia et al. claim that Coiflet wavelet performs better when using the rigorous SURE threshold denoising algorithm with soft thresholding rule [10].

Level of Decomposition.
One of the stages of the WT-based filtration is the decomposition of the processed signal. The example of wavelet tree decomposition is illustrated by the Fig. 3, which shows the input signal (aPCG) being decomposed at 5 levels.

The suitable level of decomposition depends on the number of the samples (i.e. sampling frequency), type of the signal, and the aim of the wavelet denoising.

In Matlab, the decomposition can be performed using the Wavelet Toolbox or the *wavedec* function, which returns two vectors c and l, where c includes the all of the coefficients and so-called "bookkeeping" vector l includes their lengths (number of coefficients at each level).

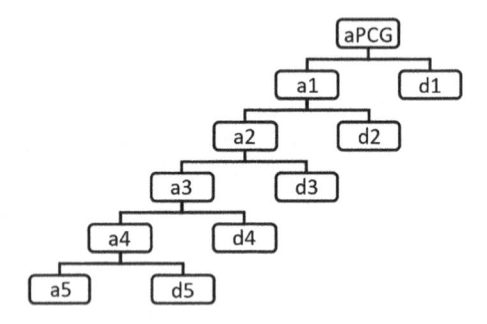

Fig. 3. Wavelet tree decomposition

4.2 Dataset

The data available at the PhysioNet (see [15–17]) include only the Gaussian noise at different noise levels expressed by the value of SNR. The data used in this paper include also maternal heart activity, which in some cases disturbs the desired signal. Since the spectra of mPCG and fPCG overlap, it is not an easy task to separate them.

We generated the signals using a modified version of generator introduced in [20–22]. Besides abdominal PCG signal (aPCG), other signals (reference fPCG, mPCG) are available so the filtration quality can be assessed objectively. Figure 4 shows the examples of the aPCG and mPCG signals.

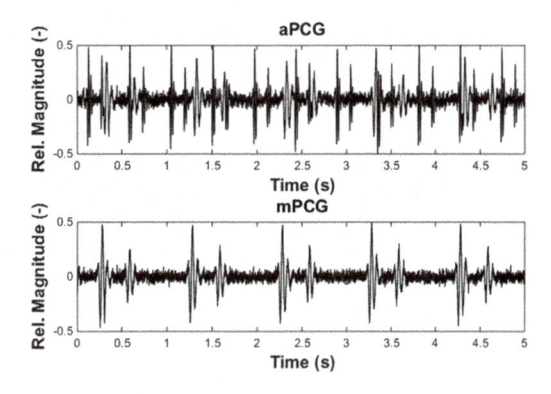

Fig. 4. Examples of the input signals (aPCG and mPCG).

5 Results and Discussion

In this section, we introduce the results of the fPCG extraction using WT. Figure 5 shows the decomposition of the input signal aPCG for levels 1–5 using the wavelet coif4.

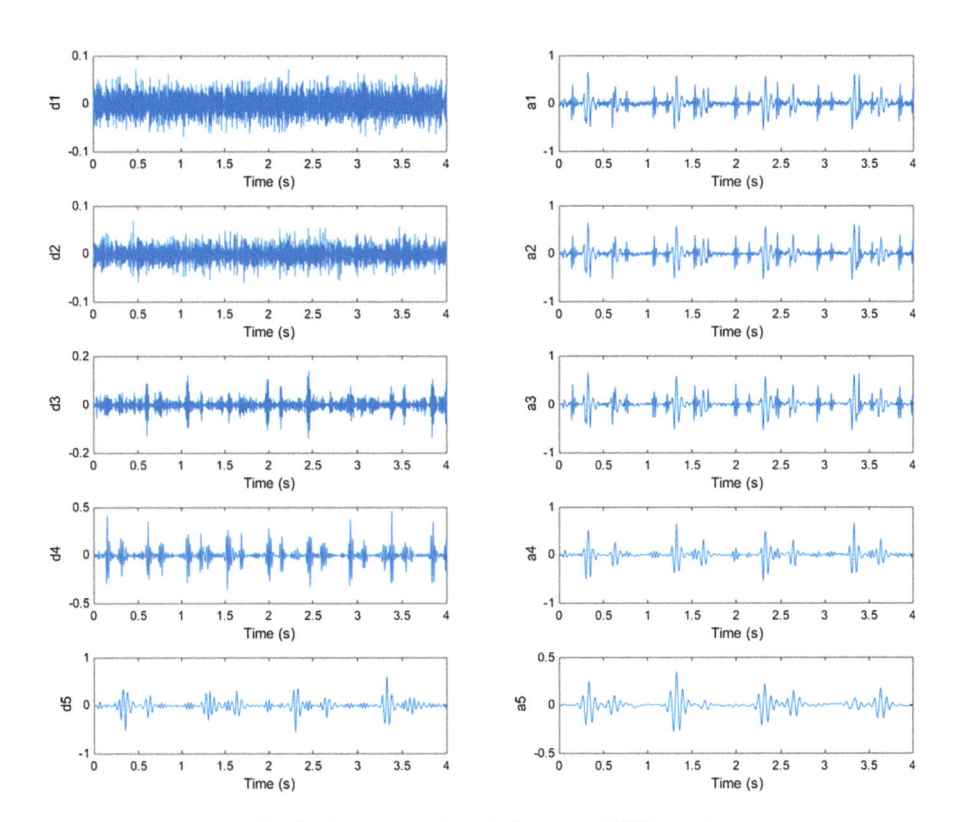

Fig. 5. Decomposition of the input aPCG signal

It can be noticed that the theoretical presumption of the level 4 being the most suitable for fECG extraction was confirmed. Moreover, we found out that the maternal component can be extracted as well as the approximation at level 4. This signal can be used for identification of maternal beats to avoid substitution of the maternal heart signal into the fetal tracing leading to misinterpretation of the fetal health status [18, 19]. Figure 6 shows the comparison of both mECG and fPCG reference signals with the corresponding recovered signals (a4 and d4).

One of the side effects of using wavelet transform is decrease of the output signal magnitude (see comparison with ideal fPCG magnitude in Fig. 7 most significantly marked by ellipse A). Moreover, although the fPCG extraction was quite successful, there was still some maternal component remaining (example B in Fig. 7). This could be a problem in case of detail morphological analysis (e.g. murmur detection, etc.). However, these residua are insignificant in comparison with the useful signal magnitude and thus, for the purpose of the fHR detection.

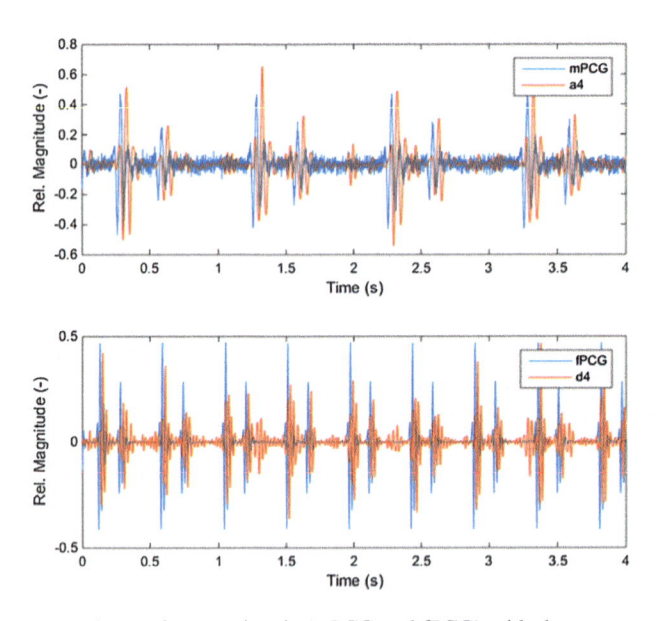

Fig. 6. Comparison of the reference signals (mPCG and fPCG) with the corresponding outputs at level 4 of decomposition (a4 and d4)

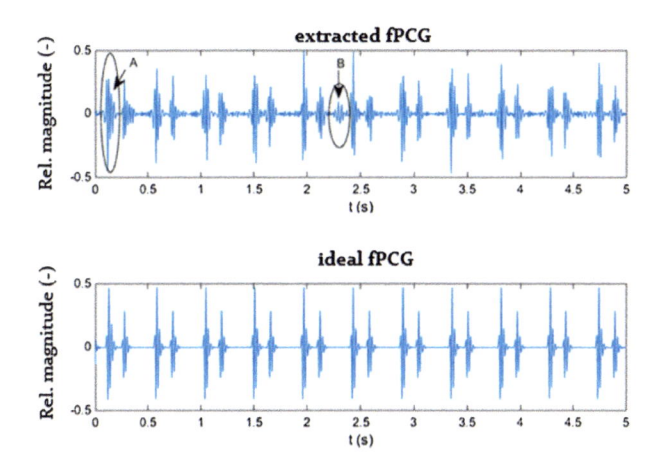

Fig. 7. Examples of the input signals (aPCG and mPCG).

6 Conclusion

In this paper, we introduced DWT as a helpful tool for fPCG extraction and analysis. The results showed that the wavelet base coif4 is suitable for the decomposition of the aPCG signal. The recovered signals using the detailed and approximation coefficients at the level 4 corresponded the most to the fPCG and mPCG signal, respectively. This way, the maternal signal can be easily identified and subsequently used for its identification in order to avoid misinterpretation in the fetal heart rate tracing.

The experiments carried out in this paper used synthetic signals to confirm the theoretical presumption that both maternal and fetal signals can be obtained from the abdominal PCG signal by means of DWT. Currently, our team is developing fPCG measuring system to obtain real signals from various stages of pregnancy to verify our experiments.

Acknowledgment. This article was supported by the Ministry of Education of the Czech Republic (Project No. SP2018/170). This work was supported by the European Regional Development Fund in the Research Centre of Advanced Mechatronic Systems project, project number CZ.02.1.01/0.0/0.0/16_019/0000867 within the Operational Programme Research, Development and Education.

References

1. Bureev, A., Vaganova, E., Zhdanov, D., Zemlyakov, I., Dikman, E.: A mobile full-time daily system for fetal monitoring. In: MATEC Web of Conferences, vol. 79, p. 01026. EDP Sciences (2016)
2. Martinek, R., Nedoma, J., Fajkus, M., Kahankova, R., Konecny, J., Janku, P., Kepak, S., Bilik, P., Nazeran, H.: A phonocardiographic-based fiber-optic sensor and adaptive filtering system for noninvasive continuous fetal heart rate monitoring. Sensors **17**(4), 890 (2017)
3. Martinek, R., Kahankova, R., Nedoma, J., Fajkus, M., Nazeran, H., Nowakova, J.: Adaptive signal processing of fetal PCG recorded by interferometric sensor. In: The Euro-China Conference on Intelligent Data Analysis and Applications, pp. 235–243. Springer, Cham, October 2017
4. Samieinasab, M., Sameni, R.: Fetal phonocardiogram extraction using single channel blind source separation. In: 23rd Iranian Conference on Electrical Engineering (ICEE) 2015, pp. 78–83. IEEE, May 2015
5. Kahankova, R., Jaros, R., Martinek, R., Jezewski, J., He, W., Jezewski, M., Kawala-Janik, A.: Non-adaptive methods of fetal ECG signal processing. Adv. Electr. Electron. Eng. **15**(3), 476 (2017)
6. Misal, A., Sinha, G.R.: Denoising of PCG signal by using wavelet transforms. Adv. Comput. Res. **4**(1), 46–49 (2012)
7. Koutsiana, E., Hadjileontiadis, L.J., Chouvarda, I., Khandoker, A.H.: Fetal heart sounds detection using wavelet transform and fractal dimension. Front. Bioeng. Biotechnol. **5**, 49 (2017)
8. Chourasia, V.S., Mittra, A.K.: Wavelet-based denoising of fetal phonocardiographic signals. Int. J. Med. Eng. Inf. **2**(2), 139–150 (2010)
9. Daubechies, I.: The wavelet transform, time-frequency localization and signal analysis. IEEE Trans. Inf. Theor. **36**(5), 961–1005 (1990)
10. Chourasia, V.S., Mittra, A.K.: Selection of mother wavelet and denoising algorithm for analysis of foetal phonocardiographic signals. J. Med. Eng. Technol. **33**(6), 442–448 (2009)
11. Varady, P.: Wavelet-based adaptive denoising of phonocardiographic records. In: Proceedings of the 23rd Annual International Conference of the IEEE Engineering in Medicine and Biology Society 2001, vol. 2, pp. 1846–1849. IEEE (2001)
12. Jimenez, A., Ortiz, M.R., Pena, M.A., Charleston, S., Aljama, A.T., Gonzalez, R.: The use of wavelet packets to improve the detection of cardiac sounds from the fetal phonocardiogram. In: Computers in Cardiology 1999, pp. 463–466. IEEE (1999)
13. Vaisman, S., Salem, S.Y., Holcberg, G., Geva, A.B.: Passive fetal monitoring by adaptive wavelet denoising method. Comput. Biol. Med. **42**(2), 171–179 (2012)
14. Chourasia, V.S., Mittra, A.K.: Most suitable mother wavelet for fetal phono-cardiographic signal analysis. i-Manager's J. Future Eng. Technol. **4**(3), 23 (2009)
15. Cesarelli, M., Ruffo, M., Romano, M., Bifulco, P.: Simulation of foetal phonocardiographic recordings for testing of FHR extraction algorithms. Comput. Methods Programs Biomed. **107**(3), 513–523 (2012)
16. Ruffo, M., Cesarelli, M., Romano, M., Bifulco, P., Fratini, A.: An algorithm for FHR estimation from foetal phonocardiographic signals. Biomed. Signal Process. Control **5**, 131–141 (2010)

17. Goldberger, A.L., Amaral, L.A.N., Glass, L., Hausdorff, J.M., Ivanov, P.Ch., Mark, R.G., Mietus, J.E., Moody, G.B., Peng, C.-K., Stanley, H.E.: PhysioBank, PhysioToolkit, and PhysioNet: components of a new research resource for complex physiologic signals. Circulation **101**(23), e215–e220 (2000). http://circ.ahajournals.org/cgi/content/full/101/23/e215
18. Emereuwaonu, I.: Fetal heart rate misrepresented by maternal heart rate: a case of signal ambiguity. Am. J. Clin. Med. **9**, 52–57 (2012)
19. Neilson, D.R., Freeman, R.K., Mangan, S.: Signal ambiguity resulting in unexpected outcome with external fetal heart rate monitoring. Am. J. Obstet. Gynecol. **198**(6), 717–724 (2008)
20. Martinek, R., Kelnar, M., Koudelka, P., Vanus, J., Bilik, P., Janku, P., Nazeran, H., Zidek, J.: A novel LabVIEW-based multi-channel non-invasive abdominal maternal-fetal electrocardiogram signal generator. Physiol. Meas. **37**(2), 238 (2016)
21. Martinek, R., Kelnar, M., Koudelka, P., Vanus, J., Bilík, P., Janku, P., Nazeran, H., Zidek, J.: Enhanced processing and analysis of multi-channel non-invasive abdominal foetal ECG signals during labor and delivery. Electron. Lett. **51**(22), 1744–1746 (2015)
22. Martinek, R., Kelnar, M., Vojcinak, P., Koudelka, P., Vanus, J., Bilík, P., Janku, P., Nazeran, H., Zidek, J.: Virtual simulator for the generation of patho-physiological foetal ECGs during the prenatal period. Electron. Lett. **51**(22), 1738–1740 (2015)

A CUDA Approach for Scenario Reduction in Hedging Models

Donald Davendra[1(✉)], Chin-mei Chueh[2], and Emmanuel Hamel[3]

[1] Department of Computer Science, Central Washington University,
400 E University Way, Ellensburg, WA 98926, USA
Donald.Davendra@cwu.edu
[2] Department of Mathematics, Central Washington University,
400 E University Way, Ellensburg, WA 98926, USA
Chin-mei.Chueh@cwu.edu
[3] Autorité des marchés financiers, Québec, QC, Canada
Emmanuel.Hamel@lautorite.qc.ca

Abstract. A CUDA kernel is proposed in this paper for acceleration of the computation of a dynamic hedging model. This is a very useful tool in segregated fund modelling. Current approaches delve on scenario reduction techniques in order to extract meaningful information from a large data set. Parallel programming allows these models to be effectively evaluated within a critical time frame. The GPU execution times shows significant improvement over CPU approaches.

Keywords: CUDA · Scenario reduction · Hedging model

1 Introduction

A segregated fund, as a mutual fund with insurance guarantee, are a valuable estate planning tool to protect from creditors or probate. One of the core components for these funds is the dynamic hedging models.

In practice, real-time calculations for dynamic hedging models for segregated funds are computationally intensive since the latter are nested stochastic scenario calculations. In this paper, we propose a CUDA approach to reduce calculation time by adapting Chueh's scenario reduction algorithm to dynamic hedging models for segregated funds for guaranteed minimum maturity benefit (GMMB) [2] for the univarient scenario. Further details can be found in [3–5].

Graphical processing units (GPU's) have been shown to improve standard financial models with significant time improvement [1,8], and are now becoming more mainstream. However, due to its nature in financial institution modelling, most of the research is still confidential.

The aim of this research is to show that with the application of GPU's [7], the calculation time for the scenario reduction algorithm is significantly improved.

This paper is as follows: Sect. 2 outlines the scenario sampling procedure with Sect. 3 describing the CUDA kernel developed the modified Euclidean distance method. Section 4 presents the experimentation conducted with the research concluded in Sect. 5.

2 Scenario Sampling Procedure

Hedging models have a number of attributes. The core requirements is the sampling process, which is quite large and therefore requires significant computational resources to compute. The Cheuch's scenario is used to extract a suitable sample from the data set, which matches its statistical dynamics. The main components of the hedging model can be given as follows:

Universe: A large number (e.g. 2 million googols) of economic scenarios affecting the model outcome.
Economic Scenario: A high (e.g. 360) dimensional vector of interest rates, stock returns, etc.
Model: Assets and liabilities model which generates asset cash inflows and liability cash outflows. The model is too complex and dynamic to express in closed mathematical form.
Model outcome: Surplus = The Present Value PV (total asset cash flows) minus the Present value PV (total liability cash flows).
Sampling: The process of selecting 1–5% of the economic scenarios from the universe.
Efficient Sampling: The sampled 100 scenarios will be used to project model outcomes so that the model outcomes form a histogram, which is a good representation of the full-run 10,000 scenario projected outcome histogram (probability distribution).
Histogram: A graph showing the possible values and their probabilities.

The main objective is how to effectively select the economic scenarios so that *extreme* or *rare* scenarios, which can cause a big loss, will be selected in a small sample, while not being overweighed or distorting the true probability distribution of the model outcome variable. The following modified weighted Euclidean Distance method approach has been proposed for this task.

2.1 Modified Weighted Euclidean Distance Method

The modified Euclidean Distance method can be given as the following. The distance between a h-period with a scenario path $(i_1, i_2, ..., i_h)$ and a pivot path $(i_1^P, i_2^P, ..., i_h^P)$ is defined as:

$$D = \sqrt{\sum_{i=1}^{h} \left(i_t - i_t^P\right) \cdot V^t} \qquad (1)$$

where:

- i_t, $t = 1, 2, \ldots, h$ is an economic scenario path consisting of short-term (e.g. one-year forward interest rates for 30 years) rates.
- i_t^P, $t = 1, 2, \ldots, h$ is a pivot (i.e. representative) scenario path consisting of short-term (e.g. one-year forward interest rates for 30 years) rates
- V is a weight factor with a value between 0 and 1 that distinguishes the relative importance of interest rate of each projecting year. The author felt that the value is around the one-year discount rate and its value used in this paper is 1/1.06. This is a user input option.

To illustrate how to perform the sampling algorithm, we select 100 representative interest rate paths out of a large number, say N, of interest rate paths by taking the following steps:

Step 1 - Choose an arbitrary interest rate path out of the N simulated ones and call it Pivot #1.
Step 2 - Calculate the distances from Pivot #1 to the remaining $N-1$ interest rate paths.
Step 3 - Name the interest rate path with the largest distance to Pivot #1 as Pivot #2. Randomly decide among ties.
Step 4 - Calculate the distances of the $N-2$ non-pivot interest rate paths to Pivot #1 and Pivot #2.
Step 5 - Assign each of the $N-2$ interest rate paths to the closest of Pivot #1 or Pivot #2, thus forming 2 disjoint sets of interest rate paths. Flip a coin if the distances are equal. Each of the $N-2$ interest rate paths now has a unique distance to its pivot scenario.
Step 6 - Rank these $N-2$ distances in descending order. The interest rate path producing the top distance is called Pivot #3 (Break ties randomly).
Step 7 - Follow the above procedure to select the additional 96 pivot scenarios, Pivot #4, Pivot #5,..., Pivot #100.
Step 8 - If the number of interest rate paths associated to a pivot scenario is Nk, then assign a probability of $Nk/1000$ to this pivot scenario.

3 CUDA Implementation

Compute Unified Device Architecture (CUDA) was developed by NVidia in order to allow General Programming on Graphic Processing Unit (GPGPU) on their proprietary hardware [6]. The main advantages for CUDA is the low level implementation of parallel applications, which does not involve extensive memory management [9].

From the previous section, there is an obvious data dependency, as the next pivot is selected as the maximum distance vector from the current one. Therefore, parallelization of this layer is not possible. However, individual parts of the loop can be evaluated in parallel as:

- Euclidean distance calculation
- calculation of the vector with the maximal distance
- calculation of the next pivot

3.1 Weighted Euclidean Distance Parallelization

Calculation of weighted Euclidean distances is parallelized in a simple way: Each `thread` calculates distance of two vectors and stores the result in the appropriate position of the current row of distance matrix. Total number of `threads` launched is calculated in such way, that there is at least one `thread` per vector. The number of `threads` per `block` was experimentally set to 256, number of `blocks` per `grid` was adaptively calculated from `block` size and the actual number of vectors. This is shown in Algorithm 1.

```
Input:
dist: [in/out] vectorized distance matrix
distPitch: padding size of distance matrix
id: indices of pivots
idMask: flag indicating if vector was used as pivot
vec: vectorized vector matrix vecPitch: pitch of vector matrix
weight: weights for vector elements
n: number of vectors
d: size of vector
di: current distance matrix row

// current thread index
tid=threadIdx.x ;
// determine index of first vector for current thread
vid=blockIdx.x * blockDim.x + tid ;
// determine stride
s=gridDim.x*blockDim.x ;
// load current pivot's id from global memory
pi=id[di] ;
// first thread marks current pivot as used
if (tid==0) then
 |  idMask[pi]=True ;
end
synchronize() ;
// get pointer to current row (di-th) in dist array
distRow=getRowPointer(dist, di, distPitch, sizeof(dist[0])) ;
// get pointer to pivot vector (pi-th) in vector array
pivot=getRowPointer(vec, pi, vecPitch, sizeof(vec[0])) ;
// calculate distances from pivot to other vectors
for j = vid to n by s do
    if idMask[j] then
        // each vector used as pivot in the past has distance set to 0
        distRow[j]=0 ;
    else
        // get pointer to j-th vector
        vector=getRowPointer(vec, j, vecPitch, sizeof(vec[0])) ;
        // calculate weighted euclidean distance between pivot and j-th vector
        distRow[j]=eDistW(pivot, vector, weight, d) ;
    end
end
```

Algorithm 1. Cuda: Weighted Euclidean distance

3.2 Distance Calculation

Finding maximum distance in parallel uses the parallel reduction paradigm, with `block` size set to 256, launching single block, where each `thread` processes (compares) potentially several elements of distance matrix row sequentially, dependent on the number of vectors, until there are fewer than block size competing elements left, the rest of the code performs standard parallel reduction, using `shared` memory as storage of results to speed up the computation. This routine is shown in Algorithm 2.

Input:
maxIx: [in/out] pointer to array to store index of maximal value, of size gridDim.x (each block stores maximal value which it finds in its section of global memory at its index)
maxVal: [in/out] pointer to array to store maximal cost value, same size as maxIx
vals: array containing values to be searched for max.
vs: vals array size
N: number of values to process by one thread

```
ValId=structure(v: value, i: index);
// block size
bs=blockDim.x ;
// alocate shared memory for block: array of bs elements of type ValId
vsh=array(bs, ValId) ;
// current block index
bid=blockIdx.x ;
// current thread index
tid=threadIdx.x ;
// determine index of first value for current thread
vid=bid * bs * N + tid ;
// determine index of last value for current thread
vend=(bid + 1) * bs * N ;
// load data from global memory into shared memory, storing indices
for i = vid to vend by bs do
    if ( vsh[tid].v < vals[i] ) then
    |   vsh[tid].v = vals[i]; vsh[tid].i = i
    end
end
synchronize();
// perform parallel reduction on data in shared memory
for s = bs / 2 to 2 ; s /= 2 do
    // compare values in vsh[tid] and vsh[tid+s]
    if (tid < s) and ( vsh[tid].v < vsh[tid + s].v ) then
        // update vsh[tid] if vsh[tid+s] larger
    |   vsh[tid].v = vsh[tid + s].v; vsh[tid].i = vsh[tid + s].i;
    end
    synchronize();
end
// write result into global memory
if (tid == 0) then
|   maxIx[bid] = vsh[0].i; maxVal[bid] = vsh[0].v;
end
```

Algorithm 2. Cuda: Find Index of Maximal Value In Array

3.3 Pivot Scenario Calculation

Calculation of pivot scenarios, as described in sequential algorithm, is parallelized by-vector, similarly to the Euclidean distances calculation - each `thread` searches the minimal distance pivot for a single vector and updates the counter for such pivot, unless the vector has been already used as pivot (in that case no search or update is performed). `Atomic` operation is used to increment the counter of minimum-distance pivot of each vector (since these might overlap). The same number of `threads` per `block` and `blocks` per `grid` as in the Euclidean distance evaluation was used. The outline is given in Algorithm 3.

Input:
ps: [out] array of pivot scenarios
dist: array containing vectorized distance matrix
distPitch: pitch of dist 2d array
n: number of vectors
di: current distance matrix row (current pivot id)

```
// current thread index
tid=threadIdx.x ;
// determine index of first vector for current thread
vid=blockIdx.x * blockDim.x + tid ;
// determine stride
s=gridDim.x*blockDim.x ;
// initialize auxiliary variables
minVal = 0 ;
minId = 0 ;
valid = False ;
// get pointer to current row (di-th) in dist array
distRow=getRowPointer(dist, di, distPitch, sizeof(dist[0])) ;
// find minimal distance pivot for each vector, which hasn't yet been used as pivot
for j = vid to n by s do
    // vector which has been used as pivot in the past has distance set to 0
    if (distRow[j] ≠ 0) then
        valid = True ;
        minVal = distRow[j] ;
        minId = di ;
        for k = di-1 to 0 do
            // get pointer to (k-th) row in dist array
            distRow2=getRowPointer(dist, k, distPitch, sizeof(dist[0])) ;
            if (minVal > distRow2[j]) then
                | minVal = distRow2[j]; minId = k;
            end
        end
    end
    if (valid) then
        valid = False ;
        atomicAdd(ps[minId],1)
    end
end
```

Algorithm 3. Cuda: Calculate Pivot Scenarios

4 Experimentation and Results

Three separate experimentations were conducted based on differing hardware and software platform to demonstrate the execution time speedups. The experiments were based on the following hardwares and softwares:

- Intel i7-78000X @ 3.5 GHz with Titan Xp GeForce card - 12 GB GDDR5X @ 11 Gbps (dedicated-headless)
- IBM Power8 Cluster with a P100 Tesla card with 3584 core and 4.7 TeraFLOPS
- R code with CUDA-C DLL (heterogeneous code)

Table 1 gives the result of the first experiment. The codes were written in C (CPU) and CUDA-C (GPU) with compiler level optimization. For the CPU's, only a single core was utilized.

The first column is the number of pivots selected, the second column is the CPU time for single core in milliseconds, the third column is the GPU time in milliseconds and the final column is the speedup as given by the following Eq. (2):

$$Speedup = \frac{GPU}{CPU} \qquad (2)$$

Table 1. Results from the Titan Xp experiment

Pivots	CPU (msec)	GPU (msec)	Speedup
100	2794	81.45	0.029
300	12320	215.92	0.017
500	27597	349.85	0.012
700	47305	492.07	0.010
1000	87233	737.17	0.008

GPU: Nvidia Titan Xp
CPU: Intel i7-78000X

The results from the second experiment on the Turing cluster at Central Washington University using a single P100 Tesla card is given in Table 2.

Table 2. Results from the P100 experiment

Pivots	CPU (msec)	GPU (msec)	Speedup
100	4526.24	34.73	0.007
300	15803.69	97.16	0.006
500	29562.57	167.87	0.005
700	46698.59	243.6	0.005
1000	79435.62	369.27	0.004

GPU: Nvidia P100
CPU: IBM Power8

The third experiment is the R code with a CUDA-DLL. The R code is simply a wrapper for the DLL, where the file and data parsing is done in R and the GPU kernel is in CUDA-C. The best approach that could be devised was to implement a dynamic linked library (DLL) of the CUDA kernel. The results for this experiment are given in Table 3.

Table 3. Results from the R-CUDA experiment

Pivots	R (msec)
100	10,580
300	9,270
500	9,560
700	9,630
1000	9,970

GPU: Nvidia Titan Xp
CPU: Intel i7-78000X

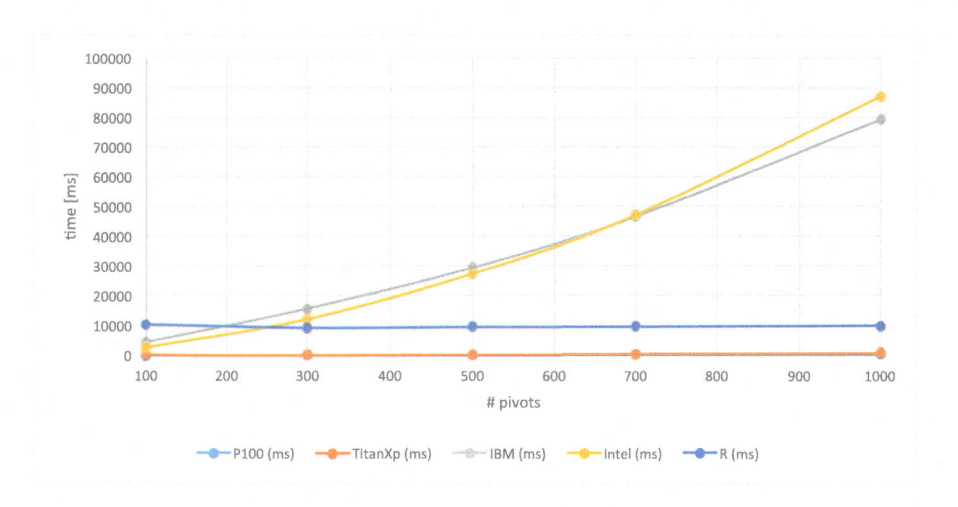

Fig. 1. Performance graph of CPU's and GPU's

Figure 1 gives the comparison for the CPU's and the GPU's for the different platforms. For the CPU's, the Intel architecture is better till 600 pivots from where the Power8 architecture is better performing. The R implementation is basically stagnant at 10,000 ms. From this figure it is obvious that the GPU applications is significantly better than all CPU approaches.

Figure 2 gives the expanded view of the GPU applications. The P100 times is generally half of the Titan Xp.

Figure 3 gives the collated graphs of the GPU and R-CUDA heterogeneous approach. As previously stated, the R time is generally collassing around 10,000 ms for all pivot sizes.

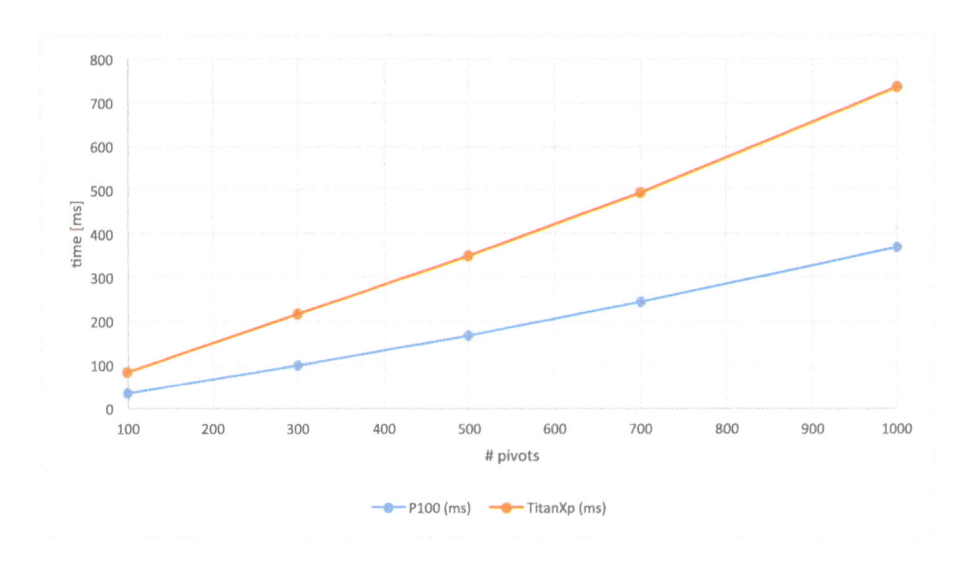

Fig. 2. Performance graph of GPU's and R

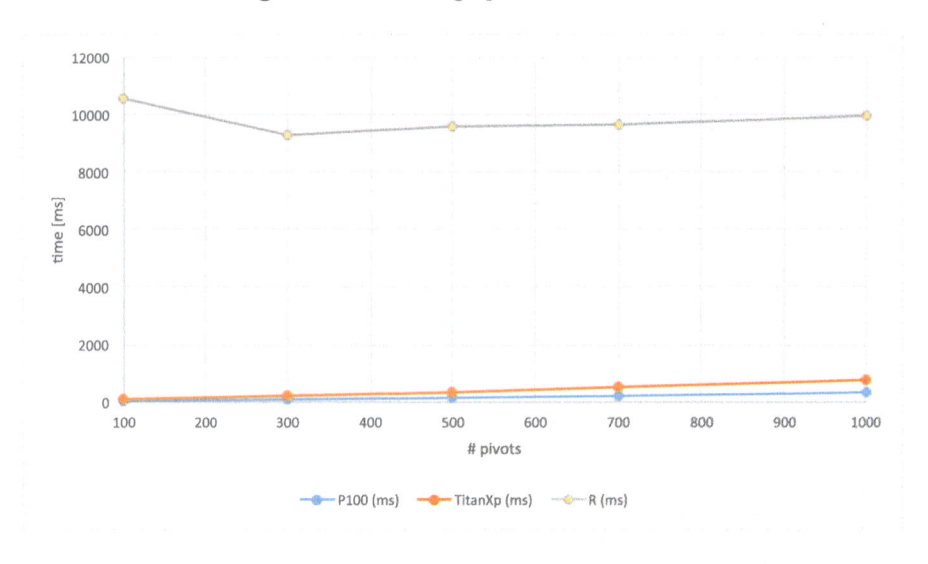

Fig. 3. Performance graph of GPU's

5 Conclusion

A CUDA kernel for modified Euclidean distance computation is described in this paper. The application of this algorithm is in the field of dynamic hedging models for segregated funds. The high computation time complexity of these approaches makes it a prime candidate for a parallization approach.

CUDA is utilized as a low level parallization paradigm to speed up the scenario reduction algorithm. The developed code is tested on two unique CPU and GPU platforms; P100 vs Titan xP and Intel i7 vs IBM Power 8.

Three separate experimentations was conducted, the first on an Intel i7X with Titan xP GPU card, the second on the IBM Power8 with P100 Tesla card and finally a heterogeneous system with R and CUDA-C kernel DLL. For all the experimentations the GPU version was significantly faster as given by the speedup's (the lower the number, the faster the GPU).

The R code was utilized with a DLL in order to create a parser. R language is the primary language used in the field of Actuaries, therefore there is a strong incentive to incorporate R language applications with CUDA.

The next iteration of this research would be to develop native R language packages for the different scenarios with can be then easily disseminated to the community through CRAN repository.

References

1. Castillo, D., Ferreiro, A., García-Rodríguez, J., Vázquez, C.: Numerical methods to solve PDE models for pricing business companies in different regimes and implementation in GPUs. Appl. Math. Comput. **219**(24), 11,233–11,257 (2013)
2. Chueh, Y.: Efficient stochastic modeling for large and consolidated insurance business: interest rate sampling algorithms. N. Am. Actuarial J. **6**(3), 88–103 (2002)
3. Chueh, Y.: Insurance modeling and stochastic cash flow scenario testing: effective sampling algorithms to reduce number of run and SALMS (Stochastic Asset Liability Modeling Sampling). Contingencies, pp. 1–18 (2003)
4. Chueh, Y.: Efficient stochastic modeling: Scenario sampling enhanced by parametric model outcome fitting. Contingencies, pp. 39–43 (2005)
5. Chueh, Y., Johnson, P.J.: CSTEP: a HPC platform for scenario reduction research on efficient stochastic modeling - representative scenario approach. Actuarial Res. Clearing House **1**, 1–12 (2012)
6. Cook, S.: CUDA Programming: A Developer's Guide to Parallel Computing with GPUs, 1st edn. Morgan Kaufmann Publishers Inc., San Francisco (2013)
7. Davendra, D.D., Zelinka, I.: GPU Based Enhanced Differential Evolution Algorithm: A Comparison between CUDA and OpenCL, pp. 845–867. Springer, Heidelberg (2013)
8. Dixon, M.F., Bradley, T., Chong, J., Keutzer, K.: Monte Carlo-based financial market value-at-risk estimation on GPUs. In: Hwu, W.-H. (ed.) GPU Computing Gems Jade Edition, Applications of GPU Computing Series, pp. 337–353. Morgan Kaufmann, Boston (2012)
9. Oliveira, F., Davendra, D., Guimarães, F.: Multi-objective differential evolution on the GPU with C-CUDA, vol. 188, pp. 123–132 (2013)

Using a Strain Gauge Load Cell for Analysis of Round Punch

Dora Lapkova(✉)

Tomas Bata University in Zlín,
Nam. T.G. Masaryka 5555, 760 01 Zlín, Czech Republic
dlapkova@utb.cz

Abstract. This article is focused on the using a strain gauge load cell in a professional defense. In the described experiment, we measured a round punch. The round punch is one of the striking techniques that is used in a professional defense. The professional defense is a very important part of physical security and safety. Nowadays, it is necessary to protect our life and property. We can see many robberies, thefts, murders, vandalism, rapes and many others criminals. The physical security and safety give us effectively possibilities to protect us. In this article, we will describe one of the striking techniques – the round punch (also called a slap). We measured a force during our experiments with the help of a strain gauge load cell L6E-C3-300kg. A total of 194 people took part in our experiment. Then we found dependences on input parameters – a body mass, body height and gender. In previous experiments, we measured a direct punch with the help of the same device. We presented the results in previous articles.

Keywords: Professional defense · Strain gauge load cell · Round punch

1 Introduction

The professional defense [1, 2] is a necessary part of physical security and safety. When we want to protect our life, our health, and our property, we can pay professional defenders. For them, it is important to know how to use defensive techniques and how to use offensive techniques.

We can learn from many martial arts, combat sports, and combat systems. There are many differences among techniques, movement, and quality of instructors. However, many of them have one common part – the striking techniques. This part is the base of all.

In our experiments, we are focused on the basic striking techniques – the direct punch, the round punch, a direct kick and a round kick. The results of the measurement of the direct punch were presented in previous articles [1–7]. In this work, we focus on the round punch. We used the same measuring device – the strain gauge load cell L6E-C3-300kg. For this reason, we present only a brief description of the measurement in this paper. A detailed description of the measurement is provided in the previous work. In this paper, we present a new striking technique, which we analyzed – the round punch.

2 Round Punch

The round punch is one of the striking techniques that is taught during training, especially in the training of women. However, some schools of martial arts [8], combat sports, and combat systems teach this strike also the men. It is a very effective technique. It can seem that the slap is not a professional strike, but it is not true.

The round punch (Figs. 1 and 2) is made with an open hand and in round line to the target [9]. The critical part is a movement of the elbow. The elbow must move along the arc.

Fig. 1. Round punch - view from the right

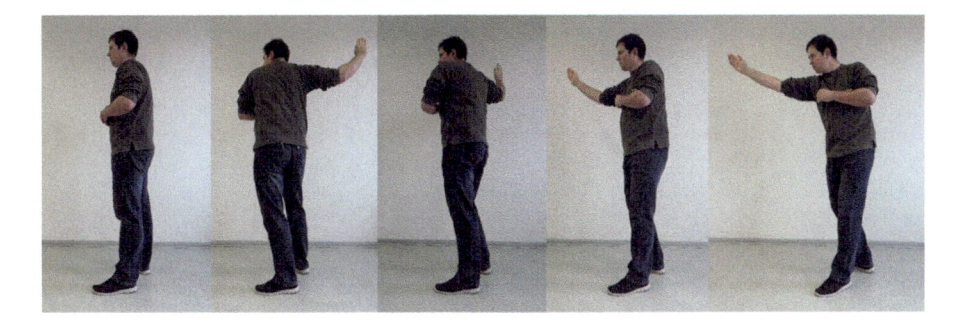

Fig. 2. Round punch - view from the left

During the round punch, there are several important parts. The first is the movement of hips. It is the first part of our body that must move. The motion is much faster with the help of hips. The other problematic part is the height of the hand. For maximum force, it is necessary to have the hand higher than shoulders and in ideal situation higher than the head.

In professional defense, we use the round punch for the destabilization of an attacker. This technique is natural for us, especially for women. The result is a shock to the attacker's body. Another advantage is the possibility to hit the attacker's ears. The results are pain and a longer time period of a shock for the attacker.

3 Measurement of Force

We used the strain gauge load cell L6E-C3-300kg for the measurement of the force. In detail, we described the whole experiment in previous articles [2–4, 7]. Therefore, the measurement is mention here in short.

The measurement station (Fig. 3) consists of a punching bag, a construction of its suspension and from the strain gauge load cell [2]. We used the punching bag for protection of the hand during the punch.

Fig. 3. Process of the slap

We used the strain gauge load cell for measurement of the force of the slap (Fig. 4). The principle is that the sensor changes electrical resistance depending on the deformation of the sensor. The sensor consists of four metal film strain gauges in a Wheatstone bridge configuration that are placed in places with the thinnest wall. For this reason, we can measure the deformation of these several metal film strain gauges very precisely. In the next step, we can calculate the force. This sensor was made in VTS Zlín company, which also calibrated the sensor.

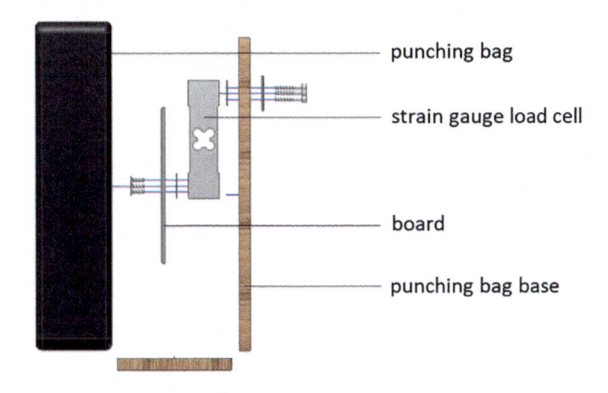

Fig. 4. Measuring station [2, 3]

We have 194 people participating in our measurement (Table 1). They are in age from 19 to 25 years. All of them performed from one to ten punches. It is very important that the target is positioned in such manner that the centre of the strain gauge load cell was in height that is the same as the height of a person's shoulder. The round punch has the maximum force for this position.

On the base of previous knowledge and training, we divided the participants into eight categories. In the beginning, we separated men and women, because we want to compare these two groups. In Table 1, you can see these categories.

Table 1. Participant categories

Categories	Previous experience	Length of training (in years)	Special physical training course	Other combat sports, martial arts or combat systems	Noted	Number of people
Untrained	NO	0	NO	NO	UTM (for men) UTW (for women)	73 11
Mid-trained	YES	0.5–2	YES	NO	MTM (for men) MTW (for women)	55 4
Trained	YES	2 and more	YES	YES	TM (for men) TW (for women)	9 6
Self-trained	YES	0.5–2	NO	YES	STM (for men) STW (for women)	34 2

4 Results

Our goal was the measurement of the force of the round punch. We used the strain gauge load cell. In the other step, we analyzed the data. For this, we used Microsoft Excel and MINITAB. These two pieces of software are analytical tools. We tried to find out the dependences on input parameters – gender, body mass, and height. Limitation of the results is the vibration of the measurement station.

In previous articles [2–4, 7], we presented the same procedure, but we analyzed the direct punch. We can compare these two techniques, and we can evaluate which of them is better for training. This comparison will be published in the following article.

Figure 5 depicts the dependence of average force on time for all categories. The graph is shortened because the ending part is not important for us. All of the dependencies are aligned according to the maximum. This depiction is useful for comparison, especially comparison between men and women. There are significant differences between trained person and other categories.

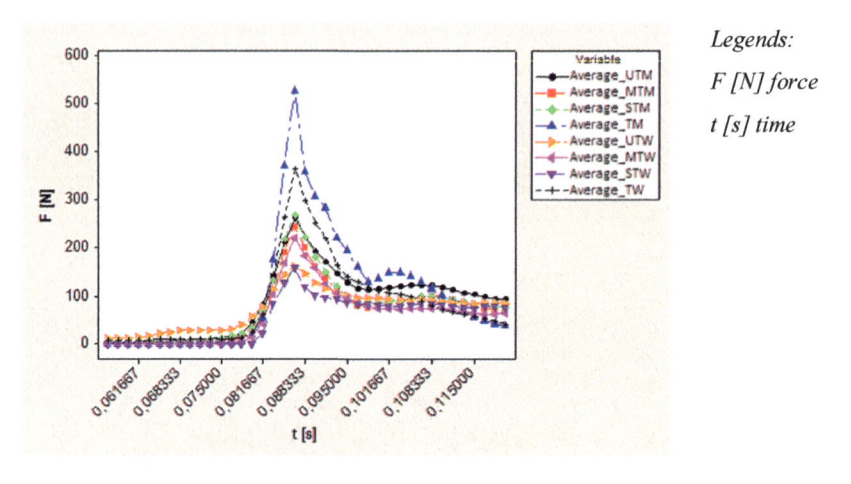

Legends:

F [N] force

t [s] time

Fig. 5. Dependence of average force on time – shortened

Comparison of men and women categories separately is presented in Figs. 7 and 8.

In Fig. 6, the dependence of the average force on time for trained men are significantly higher and sharper than other categories. The interesting point is that categories of the mid-trained men, the self-trained men, and the untrained men are very similar. The differences among them are very small.

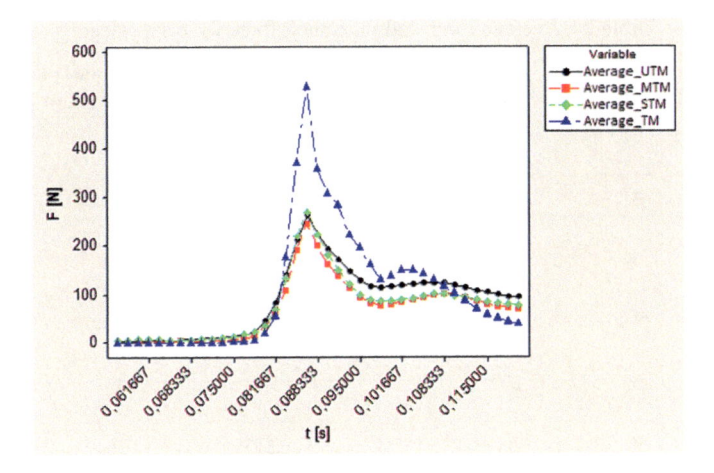

Fig. 6. Dependence of average force on time for men

In Fig. 7, there is the dependence of average force on time for women. The categories of the untrained women and the self-trained women are very similar. The difference is that the untrained women have slower increase and decrease of the force. We can generally say that only few combat sports and combat systems teach the round punch. This is the reason why the category of the self-trained women is similar to the untrained women.

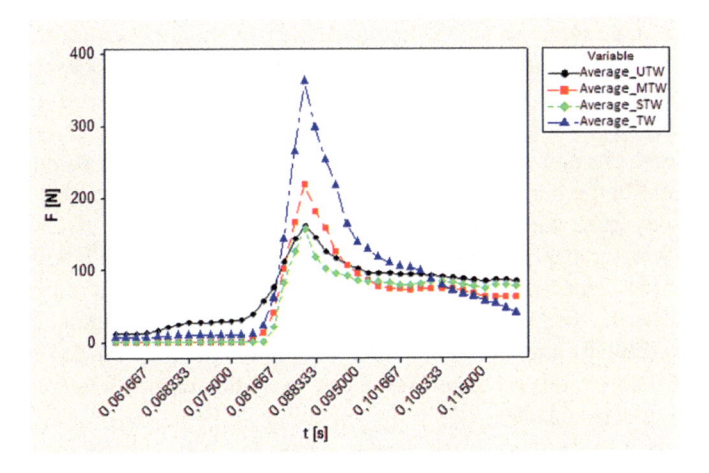

Fig. 7. Dependence of mean force on time for women

We used Microsoft Excel for data analysis. The average force, the standard deviation of average, the maximum and the standard deviation of maximum are important for us.

Table 2. Average force and maximum force for each category

Categories	Average [N]	Standard deviation of average [N]	Maximum [N]	Standard deviation of maximum [N]
Untrained men	70	20	260	90
Mid-trained men	40	20	250	90
Self-trained men	70	30	270	110
Trained men	50	30	530	300
Untrained women	60	20	160	40
Mid-trained women	50	20	220	30
Self-trained women	40	10	160	20
Trained women	60	30	360	210

In Table 2, we can see the average force, the standard deviation of average, the maximum and the standard deviation of the maximum. The highest average values have the categories of the untrained men and the self-trained men. The trained people show the highest maximum force.

The interesting categories are the untrained men, the mid-trained men, and the self-trained men because the maximum force is very similar. In case of women, the same maximum force is in categories of the untrained women and the self-trained women.

In the next step, we analyzed the dependence of maximum force on the body's height and mass. The aim was to find out if the maximum force depends on body height and body mass. In the beginning, we had a premise that men with higher body height and higher body mass will have stronger the round punch.

The following graphs, we created with the help of MINITAB. This software is very useful for graphs, especially for 3D graphs. The graphs with all of three parameters (maximum of force, body height, and body mass) are very important for us. Other software can create 3D graphs, but it is a problem to turn them around for the best view. The MINITAB solves this problem. As we can see in the following figures, the graphs are clear and understandable.

Color is used to distinguish the gender – blue for men and red for women. The darker color means a higher maximum of force.

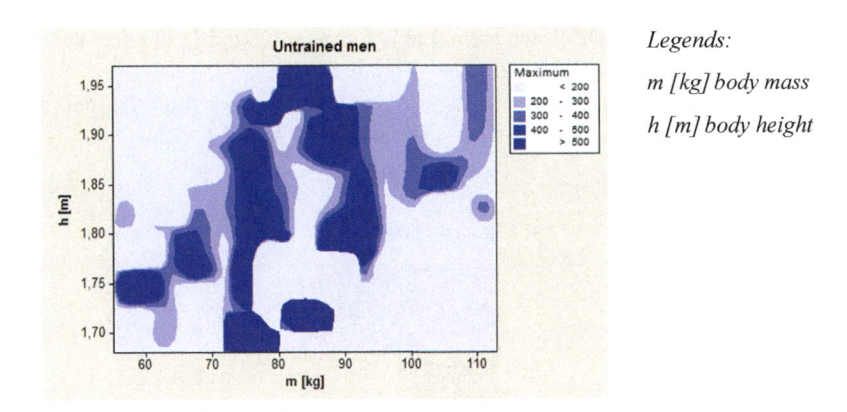

Fig. 8. Dependence of maximum force on body height and mass for untrained men

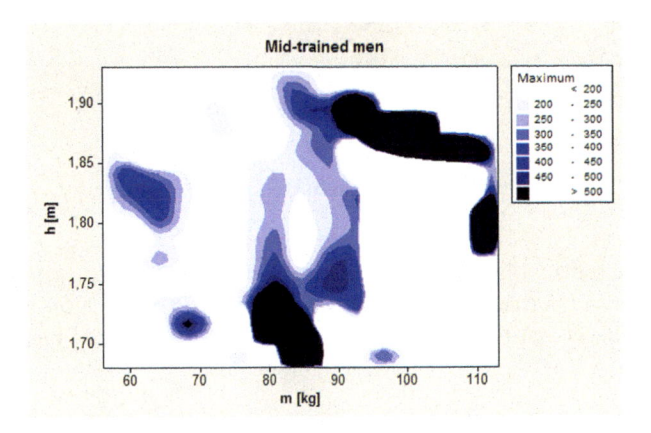

Fig. 9. Dependence of maximum force on body height and mass for mid-trained men

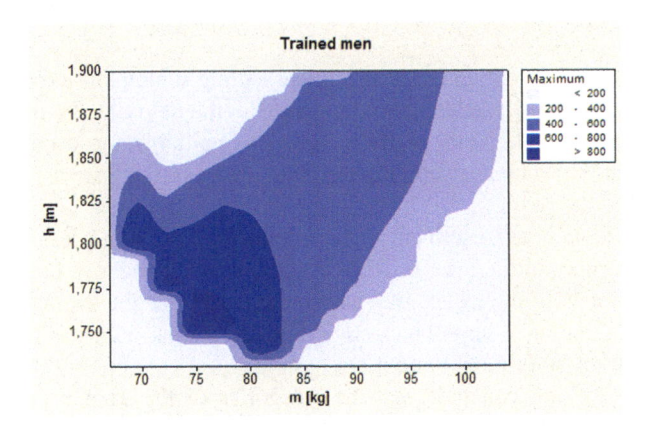

Fig. 10. Dependence of maximum force on body height and mass for trained men

In previously Figs. 8 and 9, we are not able to see any trend in the dependences of maximum force on body height and mass (Fig. 10).

In case of the trained men, we can see the trend that the men with the mass around 75 kg and with height about 1.76 m have the strongest round punch.

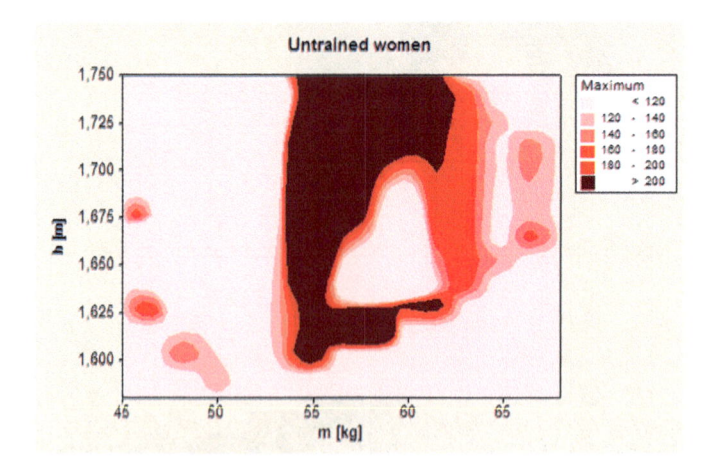

Fig. 11. Dependence of maximum force on body height and mass for untrained women

As we can see in Fig. 11, in the category of the untrained women, is a very interesting trend. Important is the body mass; the ideal is between 54 and 62 kg. The unexpected result is that the maximum force is not related to body height.

5 Conclusion

This article was focused on the round punch that is used in the professional defense. This technique has several advantages, and the instructors of defense should show it to their students.

In this article, we described our goals, the measuring station, the experiment and the results. Our goal was analyzed the round punch with the help of force measurement by the strain gauge load cell. Total of 194 people was participating in our experiment – men and women. For data analysis, we used two pieces of software – Microsoft Excel and MINITAB.

The results have been described in the separate section. We presented the graphs with the dependences of the average force on time for all categories but also separated for men and women. In the tables, we showed the results for the average force and the maximum force. Another important part is the 3D graphs with the dependences of maximum force on body height and mass.

In the end, we can conclude that the categories of the trained people have the highest force. In other categories, the trend is not so significant. In the case of the dependences on inputs parameters - we have demonstrated the dependences on gender and in some categories the dependences on body height and mass (especially in categories of the trained men, the untrained women).

In the future, we will focus on the comparison of the round punch and the direct punch.

Acknowledgment. This work was supported by the Ministry of Education, Youth and Sports of the Czech Republic within the National Sustainability Programme project No. LO1303 (MSMT-7778/2014) and also by the European Regional Development Fund under the project CEBIA-Tech No. CZ.1.05/2.1.00/03.0089 and also by the research project VI20172019073 "Identification and methods of protection of Czech soft targets against violent acts with elaboration of a warning system", supported by the Ministry of the Interior of the Czech Republic in the years 2017–2019 and also by the research project VI20172019054 "An analytical software module for the real-time resilience evaluation from point of the converged security", supported by the Ministry of the Interior of the Czech Republic in the years 2017–2019.

References

1. Lapkova, D., Pospisilik, M., Adamek, M., Malanik, Z.: The utilisation of an impulse of force in self-defence. In: XX IMEKO World Congress: Metrology for Green Growth, Busan, pp. 0–6 (2012)
2. Lapkova, D., Kralik, L., Adamek, M.: Analysis of direct punch in professional defence using multiple methods. In: Tenth International Conference on Emerging Security Information, Systems and Technologies, pp. 34–40 (2016)
3. Lapkova, D., Adamek, M.: Using information technologies in professional defence education – classification of training with help of impulse. In: International Conference on Logistics, Informatics and Service Sciences (LISS), pp. 1–5. IEEE, Kyoto (2017). ISBN 978-1-5386-1047-3
4. Lapkova, D., Adamek, M., Kominkova Oplatkova, Z.: Analysis of direct punch force in professional defence. In: Proceedings 29th European Conference on Modelling and Simulation ECMS 2015, pp. 564–569. Digitaldruck Pirrot GmbH, Germany (2015)
5. Lapkova, D., Adamek, M.: Statistical and mathematical classification of direct punch. In: Proceedings of the 38th International Conference on Telecommunication and Signal Processing (TSP 2015), pp. 486–489. Asszisztencia Szervezo Kft., Prague (2015)
6. Lapkova, D., Adamek, M.: Using strain gauge for measuring of direct punch force. In: IMEKO XXI World Congress, pp. 285–288. Czech Technical University in Prague, Prague (2015)
7. Lapkova, D., Adamek, M.: Using information technologies in professional defence education – classification of training with help of effective punching mass. In: Proceedings of the 12th Iberian Conference on Information Systems and Technologies, pp. 769–774. AISTI, Lisbon, Portugal (2017). ISBN 978-989-98434-7-9
8. Levine, D., Whitman, J.: Complete Krav Maga (2007)
9. Reguli, Z.: Inovation SEBS a ASEBS. Biomechanics of combat sports and martial arts (2011). http://www.fsps.muni.cz/inovace-SEBS-ASEBS/elearning/biomechanika/biomechanika-upolovych-sportu. Accessed 30 July 2018

Geometrical Computational Method to Locate Hypocenter by Signal Readings from a Three Receivers

Alexander D. Krutas[1], Tatyana A. Smaglichenko[2(✉)], Alexander Smaglichenko[3,4], and Maria Sayankina[2]

[1] 18 Vinogradnaya Street, Alushta 298517, Republic of Crimea, Russia
[2] Research Oil and Gas Institute, Russian Academy of Sciences, 3 Gubkina Street, Moscow 119333, Russia
t.a.smaglichenko@gmail.com
[3] V.A. Trapeznikov Institute of Control Sciences, Russian Academy of Sciences, 65 Profsoyuznaya Street, Moscow 117997, Russia
[4] Institute of Seismology and Geodynamics,
V.I. Vernadsky Crimean Federal University, Prospekt Vernadskogo 4, Simferopol 295007, Republic of Crimea, Russia

Abstract. The information about hypocenters is essential for engineering tasks to predict a danger created by earthquakes. A nonlinearity in the earthquake location problem, a one-sided distribution of seismic receivers lead to a search of solutions improving hypocenter location accuracy. The original earthquake location method has been developed by using a physical and geometrical representation of the time difference between first signals from seismic waves registered by three different receivers. Data processing is performed applying algorithms of a computational geometry. The solution error is formulated in terms of an analytical geometry. Thus the method differs from traditional approaches, which are based on a standard statistic error. Taking into account the promising testing results we guess that the geometrical method can be as an additional computational technique or as a tool to get the trial hypocenter, which is an input parameter of standard software for hypocenter determining.

Keywords: Hypocenter location · Calculating geometry · Local earthquake origin time

1 Introduction

There are two groups of methods improving the earthquake location accuracy. The first applies the correlation analysis and operates with identified seismic signals. By using the differences in P- and S- readings for pair of similar earthquakes registered at the same receiver hypocenters are relocated (e.g. [1–5]). The other line of approaches involves 3-D seismic tomography methods, which perform an inversion of the structure model and hypocenter location parameters.

A. D. Krutas—Independent Researcher.

Both parameters can be simultaneously defined in iteration process [6,7] or separately found with every iteration [8–10]. In tomography, the number of P- and S-readings used can reach more than hundreds of thousands. Therefore the arising systems of linear equations are large. Their sparseness is caused by the complexity of the earthquake distribution. The location accuracy of a few thousands of events can be improved owing to the application of algebraic reconstruction techniques. The conventional LSQR technique described in [11] is an algorithm that applies the Lanczos process and bidiagonalization procedure. Other alternative is given in [12] as the projection technique that uses relaxation principles. The next method is a modernization of the Gauss scheme [13]. It was developed to avoid an influence of the data error on the inversion solution [10,13].

Let us make an important note. All relocation methods mentioned above are based on a data of the earthquake absolute locations, which are calculated by real time systems. Amount of data depends on number of regional stations registered signals from an earthquake. The traditional method introduced by Geiger in [14] applies the Taylor expansion and provides inversion scheme, which iteratively derives the hypocenter corrections with respect to the trial hypocenter coordinates by using the time residuals (differences between observed times and calculated from the model). The earthquake location program Hypo71 designed in [15] is the main realization of Geiger's method. Other programs Hypoellipse [16]. Hypoinverse [17], Hypocenter [18] have their distinctive features. However Geiger's method is their common theoretical base. Programs Hypo71, Hypocenter and Hypoinverse are included to SEISAN software [19]. These programs are installed on different seismological centers in a world and as pointed in [20] the problem of common data format exists in order to improve international cooperation. A time for hypocenter parameters calculating varies from a few minutes till a few tens of minutes. It's too long to prevent a seismic hazard if an earthquake is very large. Therefore in Japan an automatic processing system was developed and installed to broadcasting of earthquake alarms (see [21]). Preventive measures of stopping factories, trains, lift and so on can be performed with help of this automatic system. When large earthquake occur the first estimate of hypocenter location is calculated using residuals for P-wave arrival times from two or three stations that detect seismic signal earlier as possible.

Nevertheless, it is known that there is no perfect method to solve the earthquake location nonlinear problem because of insufficient input data as pointed in [22]. We assume that the most difficult case can be when an epicenter is located not under the receiving stations, but on one side away from an epicenter. Then the application of various methods likely will help to get more reliable solution of the problem. In this paper we would like to focus on the method that differs from ordinary way of the residuals minimization and uses principles of calculating geometry.

2 Solution via Tetrahedron

With this article, we describe the earthquake location solution that is based on geometrical representation of seismic signal source and signal receivers as an arbitrary tetrahedron in three-dimensional Euclidean space. Vertices are determined by hypocenter's point and by geographic coordinates of three receivers (stations) with the earliest times of the first signal arrival. Computations are performed for rays passing through simple one-layer homogeneous model of elastic medium.

Let S_1, S_2, and S_3 be points of receivers on a Cartesian coordinate system (Fig. 1). Points have coordinates $(x_1, y_1, 0)$, $(x_2, y_2, 0)$ and $(x_3, y_3, 0)$. Pairs (t_{P1}, t_{S1}), (t_{P2}, t_{S2}), (t_{P3}, t_{S3}) are arrival time data that are P- and S- readings from receivers having indexes $i = 1, 2, 3$. This means that signals of two main types of elastic wave (P- and S-) are arrived from an earthquake at a seismograph of each receiver and times of their arrivals are read on seismograms. Using these data we shall find the coordinate (x_H, y_H, z_H) of the point H (earthquake's hypocenter).

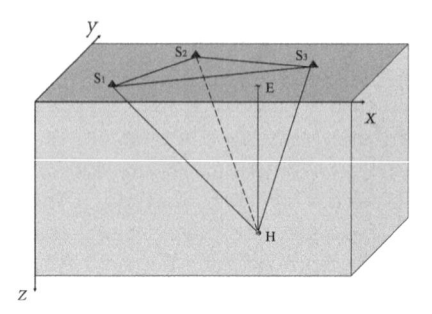

Fig. 1. Tetrahedron. Hypocenter H, epicenter E, receivers S_i are denoted by closed circle, a star and closed triangles, respectively.

2.1 Unfolding of the Faces of a Tetrahedron

Let us consider a three triangular faces having common vertex that is hypocenter's point. Suppose they lie on the same plane (Fig. 2-a). The P-travel time from a hypocenter to the first station S_1 can be found using 4 readings of the P- and S- signals from two receivers S_1 and S_2:

$$T_1 = \frac{t_{S1} - t_{P1}}{(t_{S2} - t_{S1}) - (t_{P2} - t_{P1})}(t_{P2} - t_{P1}) \qquad (1)$$

Notes and Comments. Formula (1) has been derived by revising of the hypocentral distance formula for a single station that given by Savarensky in [23]. However besides of S- and P- arrival difference Savarensky's formula uses P- and S-wave velocities assumed while Eq. (1) involves only observed readings from two receivers.

The P-travel times for the second and third receivers are T_2 and T_3. They are calculated using the following equations:

$$T_2 = T_1 + \Delta t_2 \tag{2}$$

$$T_2 = T_1 + \Delta t_3, \tag{3}$$

where Δt_2 and Δt_3 arrival time differences between the P-wave signals at the second and first receivers and at the third and first receivers. Then sides of the tetrahedron h_i are function of V as follows:

$$h_i = T_i V, \tag{4}$$

where the index i- corresponds to the receiver S_i. Travel times T_i are found using Eqs. (1)–(3). V is a parameter having a sense of average P-wave velocity in the one layer homogeneous model.

In a plane of one triangular face the shortest path from hypocenter's point to the line connecting the receivers pair is determined by the perpendicular to this line that is corresponding side of a tetrahedron. Let P_i be a point, where a perpendicular intersects the line connecting points S_i and S_{i+1} (Fig. 2-b).

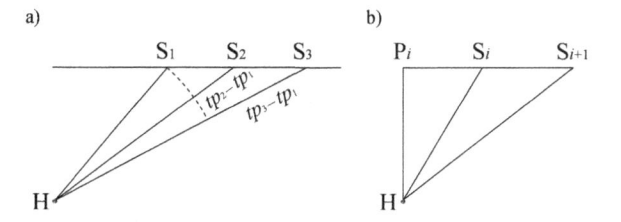

Fig. 2. Triangular faces of a tetrahedron on the same plane: (a) a difference between arrivals of the first signal is proportional to a difference between hypocentral distances; (b) the perpendicular P_i from hypocenter to the line that join receivers S_i and S_{i+1}.

Coordinates of the point P_i (x_{Pi}, y_{Pi}) can be determined by solving the following system:

$$\begin{cases} (x_{Pi} - x_{Si})(y_{Si+1} - y_{Si}) = (x_{Si+1} - x_{Si})(y_{Pi} - y_{Si}) \\ (x_{Pi} - x_{Si})^2 + (y_{Pi} - y_{Si})^2 = d_i^2, \end{cases} \tag{5}$$

where (x_{Si}, y_{Si}) and (x_{Si+1}, y_{Si+1}) are coordinates of two points S_i and S_{i+1} that define the equation of the line through them. The distance d_i between the points P_i and S_i can be calculated applying the Pythagorean Theorem to right triangles having sides are equal to h_i and h_{i+1} according to Eq. (4).

2.2 The Theorem of Three Perpendiculars: The Location Error

Consider the earth's surface plane, to which points P_i and S_i belong. Through each point P_i we will draw perpendiculars to the lines joining the points S_i and S_{i+1}. When the three perpendiculars are intersected on the resulting point E then this point will be orthogonal projection of the hypocenter to the earth's surface. The last statement follows from the Theorem of Three Perpendiculars. By definition the point E is earthquake's epicenter.

Coordinates of the point E are found by solving a system of two linear equations, which are determined from the conditions of orthogonality of vectors $\overrightarrow{S_1 S_2}$ and $\overrightarrow{EP_1}$, and $\overrightarrow{S_2 S_3}$ and $\overrightarrow{EP_2}$:

$$Ag = b, \qquad (6)$$

where A is a matrix of dimension 2×2, whose elements are determined by Euclidean coordinates of vectors $\overrightarrow{S_1 S_2}$ and $\overrightarrow{S_2 S_3}$, g is a unknown vector, whose components correspond to the epicenter coordinates, b is a vector, whose components are determined by coordinates of above mentioned vectors and by coordinates of the points P_1 and P_2 derived from system (5). The depth is easily calculated using the point E coordinates, which are the solution of system (6).

The location error e can be expressed via the scalar product of vectors $f_1 = \overrightarrow{S_3 S_1}$ and $f_2 = \overrightarrow{EP_3}$:

$$e = abs(\langle f_1, f_2 \rangle) \qquad (7)$$

More the value e close to zero more accurate solution is obtained. The value setted in Eq. (7) directly depends on the parameter V value (see Eq. (4)).

Notes and Comments. Equations (1)–(4) as well as all steps of the physical and geometric solution, test outcomes were derived by the first author of this paper. Co-authors developed the algorithms for parameters calculating (Eqs. (5)–(7)) and conducted the control testing by using the designed computer program.

3 Example of a Real Seismic Event

The input data of 6 readings of the signals from the local earthquake that occur in 4 January, 2001 in southern Iceland are presented in Table 1. The data were given in the framework of cooperation on the project RANNIS ID: 152432-051.

Sau, kro and hei are permanent seismic stations of the South Iceland Lowland (SIL) network. We shall compare calculated result of the geometrical method applied to these stations data with the traditional technique result that was obtained by processing P- and S- readings from 9 stations of the SIL network (Table 2).

Taking into a consideration the confidence intervals for the hypocenter solution by the traditional method one can see that the latitude is the same as the latitude found by the geometrical method. The difference between two longitudes is about 1 km, the difference between depths is 1.9 km. Both epicenters are shown in Fig. 3. The image has been constructed using GMT program and ASTER GDEM v.2 data. ASTER GDEM is a product of METI and NASA.

Table 1. Observed P- and S- arrival times from three stations

Name	Signal	Hour	Minute	Second
sau	P	13	25	21.25
sau	S	13	25	25.57
kro	P	13	25	22.96
kro	S	13	25	28.66
hei	P	13	25	23.28
hei	S	13	25	29.15

Table 2. The hypocenter location data by two methods

Traditional method			
Origin time	Latitude (deg.)	Longitude (deg.)	Depth (km)
13 h 25 min 15.5 s	64.279	−20.320	4.4
±0.09 s	±0.013	±0.030	±3.1
Geometrical method			
Origin time	Latitude (deg.)	Longitude (deg.)	Depth (km)
13 h 25 min 15.8 s	64.263	−20.372	9.4

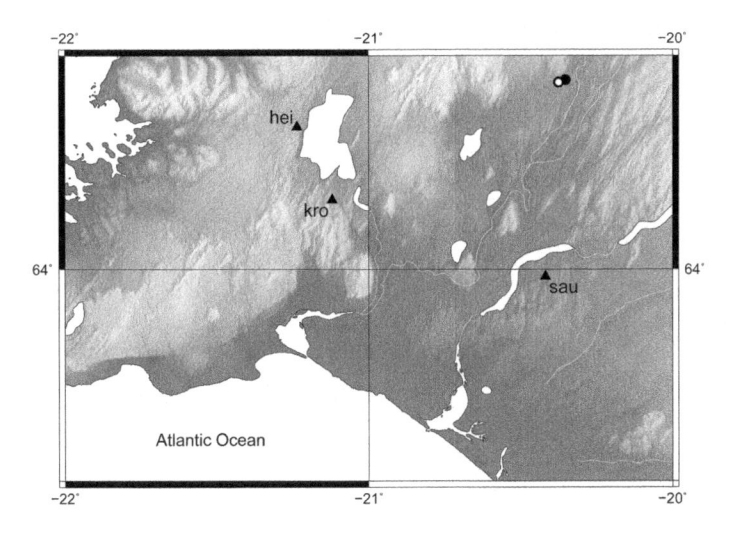

Fig. 3. Permanent seismic stations of the SIL network are denoted by closed triangles. Epicenters calculated by the traditional and geometrical method are denoted by black and white circle.

4 Conclusion

We suppose that the geometrical method can be used to specify the trial hypocenter data when seismic sources are located at sea or ocean. At the same time the proposed method is separately applicable when the model structure is simple. The given method can be further developed for complex models.

We thank the anonymous reviewer for helpful suggestion to verify the geometrical method with data of artificial earthquakes. This can be done for direct waves when seismic signals are produced with the vibrating impact. In the case of explosions, a seismic wave refracts through deep layers. It reaches the maximal dipping point and goes up to surface in accordance with the gradient structure. For such kind of data the geometrical method should be additionally modified. At the same time under some conditions (not heavy dynamite, corresponding geometry of experiment) the explosion data could be useful to apply the method.

Acknowledgement. We thank the staff of the Icelandic Meteorological Office and University of Iceland, Reykjavik for providing us with the data of local earthquakes.

Co-authors will be forever grateful to the main author of this article Alexander D. Krutas, father and grandfather, who tragically died because of car accident on the 24th of November 2015 in Alushta city, Crimea.

References

1. Rognvaldsson, S.T., Slunga, R.: Single and joint fault plane solutions for microearthquakes in south Iceland. Tectonophysics **273**, 73–86 (1994)
2. Got, J.L., Frechet, J., Klein, F.W.: Deep fault plane geometry inferred from multiplet relative relocation beneath the south flank of Kilauea. J. Geophys. Res. **99**, 15375–15386 (1996)
3. Shearer, P.M.: Improving local earthquake locations using the L1 norm and waveform cross-correlation: application to the Whittier, Narrows, aftershock sequence. J. Geophys. Res. **102**, 8269–8283 (1997)
4. Rubin, A.M., Gillard, D., Got, G.L.: Streaks of microearthquakes along creeping faults. Nature **400**, 635–641 (1999)
5. Waldhauser, F., Ellsworth, W.L.: A double-difference earthquake location algorithm: method and application to the Northern Hayward fault, California. Bull. Seismol. Soc. Am. **90**(6), 1353–1368 (2000)
6. Eberhart-Phillips, D.: Local earthquake tomography: earthquake source regions. In: Iyer, H.M., Hirahara, K. (eds.) Seismic Tomography: Theory and Practice, pp. 613–643. Chapman and Hall, New York (1993)
7. Zhang, H., Thurber, C.H.: Double-difference tomography: the method and its application to the Hayward fault, California. Bull. Seismol. Soc. Am. **93**, 1875–1889 (2003)
8. Benz, H.M., Chouet, B.A., Dawson, P.B., Lahr, J.C., Page, R.A., Hole, J.A.: Three-dimensional P and S wave velocity structure of redoubt volcano, Alaska. J. Geophys. Res. **101**, 8111–8128 (1996)
9. Wu, H., Lees, J.M.: Three-dimensional P- and S-wave velocity structures of the Coso geothermal area, California, from microseismic travel time data. J. Geophys. Res. **104**, 13217–13233 (1999)

10. Smaglichenko, T.A., Horiuchi, S., Takai, K.: A differentiated approach to the seismic tomography problem: method, testing and application to the Western Nagano fault area (Japan). Int. J. Appl. Earth Obs. Geoinf. **16**, 27–41 (2012)
11. Paige, C.C., Saunders, M.A.: LSQR: an algorithm for sparse linear equations and sparse least squares. ACM Trans. Math. Softw. **8**(43–71), 195–209 (1982)
12. Smaglichenko, T.A., Nikolaev, A.V., Horiuchi, S., Hasegawa, A.: The method for consecutive subtraction of selected anomalies: the estimated crustal velocity structure in the 1996 Onikobe (M = 5.9) earthquake area, northeastern Japan. Geophys. J. Int. **153**, 627–644 (2003)
13. Smaglichenko, T.A.: Modification of Gaussian elimination for the complex system of seismic observations. Complex Syst. (Journal) **20**(3), 229–241 (2012)
14. Geiger, L.: Herdbestimmung bei Erdbeben aus den Ankunftszeiten. Nachrichten von der Königlichen Gesellschaft der Wissenschaften zu Göttingen. Math. Phys. Klasse **1910**, 331–349 (1910)
15. Lee, W.H.K., Lahr, J.C.: HYPO71 (revised): a computer program for determining hypocenter, magnitude, and first motion pattern of local earthquakes. U.S. Geological Survey Open-File Report (1975)
16. Lahr, J.C.: HYPOELLIPSE: a computer program for determining local earthquake hypocentral parameters, magnitude and first motion pattern. U.S. Geological Survey Open-File Report (1980)
17. Klein, F.W.: User's Guide to Hypoinverse-2000, a Fortran program to solve for earthquake locations and magnitudes. U.S. Geological Survey Open-File Report (2002)
18. Lienert, B.R.E., Havskov, J.: A computer program for locating earthquakes both locally and globally. Seismol. Res. Lett. **66**, 26–36 (1995)
19. Havskov, J., Ottemoller L.: SEISAN: the earthquake analysis software, Version 8.0 preliminary. University of Bergen (2003)
20. Dost, B., Zednik, J., Havskov, J., et al.: Seismic data formats, archival and Exchange. In: Bormann, P. (ed.) New Manual of Seismological Observatory Practice (NMSOP-2). GeoForschungsZentrum, Potsdam (2012). Chap. 10, http://bib.telegrafenberg.de/publizieren/vertrieb/nmsop/
21. Horiuchi, S., Negishi, H., Abe, K., Kamimura, A., Fujinawa, Y.: An automatic processing system for broadcasting earthquake alarms. Bull. Seismol. Soc. Am. **95**(2), 708–718 (2005)
22. Lee, W.H.K., Stewart, S.W.: Principles and Applications of Microearthquake Networks. Academic Press, New York (1981)
23. Savarensky, E.F.: Seismic Waves. Mir Publishers, Moscow (1975)

An Intelligent Question-Answer System over Natural-Language Texts

Marie Duží[1(✉)] and Bjørn Jespersen[1,2]

[1] Department of Computer Science, FEI, VSB-Technical University of Ostrava,
17. listopadu 15, 708 33 Ostrava, Czech Republic
`marie.duzi@vsb.cz, bjornjespersen@gmail.com`
[2] Department of Philosophy and Religious Studies, Utrecht University,
Janskerkhof 13, 3512 BP Utrecht, The Netherlands

Abstract. The success of automated reasoning techniques over large natural-language texts heavily relies on a fine-grained formal analysis of these texts. While there is common agreement that the analysis should be hyperintensional, most automatic reasoning systems are still based on intensional logic. In this paper, we introduce a hyperintensional system of reasoning and answering. We apply Tichy's Transparent Intensional Logic (TIL) which comes with a procedural (as opposed to truth-conditional) semantics. Our goal is to analyse empirical questions that come attached with a presupposition.

Keywords: Hyperintensional logic · Procedural semantics · Presupposition · Question · Answer · Transparent Intensional Logic · TIL

1 Introduction

Questioning and answering plays an important role in our communication and has many logically relevant features. Thus, a formal analysis of interrogative sentences and appropriate answers should not be missing from any formal system dealing with natural language communication. To this end many systems of *erotetic logic* have been developed.[1] In general, these logics specify axioms and rules that are special for questioning and answering.

However, many important features of questions stem from their *presuppositions*. Everybody who is at least partially acquainted with the methods applied in the social sciences has heard of the importance to consider the presuppositions of a question in *questionnaires*. Yet, to our best knowledge, none of the systems of erotetic logic deals with the presuppositions of questions in a satisfactory way. This is unsatisfactory because they fail to consider *properly partial functions*, which lack a value at least one of their arguments. For instance, propositions (in their capacity as truth-bearers) can have *truth-value gaps*.

We have a suitable system at hand, though. It is Tichý's Transparent Intensional Logic (TIL), which comes with a *procedural* semantics that assigns abstract procedures

---
[1] See, for instance, [9, 15–17, 22].

to the terms and expressions of natural language as their meanings. These procedures are rigorously defined as TIL constructions which produce lower-order objects as their products or in well-defined cases fail to produce an object. In case of *empirical* expressions, the produced entity is a *possible-world semantic (PWS) α-intension* viewed as a *partial* function from possible worlds to a function from instants of time to α-typed entities (where α is a placeholder for specific types) such that each world-time pair is taken to *at most one* value of type α.

The goal of this paper is to propose an analysis of questions that come with presuppositions. And since answering is no less important than raising questions, we are also going to propose a method of adequate unambiguous answering of questions that come with presuppositions. The thesis we will be putting forward is this. In case the presupposition of a question is not true (because either false or gappy), there is no unambiguous direct answer, and an adequate complete answer is instead a negated presupposition. We follow Frege and Strawson in treating survival under negation as the most important test for presupposition. But a negative answer to a question is often ambiguous. The ambiguity consists in two kinds of negation, namely narrow-scope (internal) and wide-scope (external) negation. While the former preserves the presupposition, the latter dismisses the presupposition. We show that an adequate unambiguous answer to a question the presupposition of which is not true is the narrow-scope negation of the presupposition.[2]

In terms of practical applications, our theoretical results are being implemented as one of the most important components of an intelligent question-answering system over large corpora of natural language texts. To this end we are making use of the Normal Translation Algorithm (NTA) that has been developed in the Centre for Natural Language Processing at the Faculty of Informatics, MUNI Brno. NTA is a method that integrates logical analysis of sentences with a linguistic approach to semantics. The result of NTA so far is a corpus of 6,272 constructions analysed from newspaper sentences that serve as an input for our inference machine. Furthermore, our procedural approach makes it possible to implement the extensional logic of hyperintensions so as to provide relevant information from a wide range of natural-language resources.

The rest of the paper is organized as follows. Section 2 introduces our background system of Transparent Intensional Logic (TIL). In Sect. 3 we deal with classification of questions and their respective answers. There we also rigorously define the relations of presupposition and mere entailment between propositional constructions, and deal with presuppositions of questions. Section 4 introduces a general analytic schema for questions with presuppositions the application of which is illustrated by examples of a fine-grained (i.e. hyperintensional) analysis. Finally, Sect. 5 contains some concluding remarks.

---

[2] For details on non-equivalence of narrow-scope vs. wide-scope negation with respect to negating propositions that come attached with a presupposition, see [6].

2 TIL in Brief

A detailed description of the TIL system is out of the scope of this paper. Referring to numerous papers and three books (see, e.g., [3–6, 10, 19, 20]), we just provide a brief summary here. TIL is a hyperintensional, typed λ-calculus of partial functions. The λ-terms of the TIL language denote abstract procedures (that could be explicated in terms of Church's 'functions-in-intension') which produce set-theoretical mappings (functions-in-extension) or lower-order procedures. These procedures are rigorously defined as TIL *constructions*. Being procedural objects, constructions can be *executed* so as to operate on input objects (of a lower-order type) and produce at most one object of the type they are typed to produce, while non-procedural objects, i.e. non-constructions, cannot be executed. There are two atomic constructions that present input objects to be operated on by molecular constructions. They are *Trivialization* and *Variables*. A Trivialization presents an object X without the mediation of any other procedures. Using the terminology of programming languages, the Trivialization of X, '^0X' in symbols, is just a *pointer* or *reference* to X. Trivialization can present an object of any type, even another construction C. Hence, if C is a construction, ^0C is said to *display* the construction C, whereby C occurs hyperintensionally, i.e. *non-executed*. Variables produce objects dependently on valuations; they are said to *v-construct*. The execution of a Trivialization or a variable never fails to produce an object. However, the execution of some of the molecular constructions can fail to present an object of the type they are typed to produce. When this happens, we say that a given construction is *v-improper*.

There are two dual molecular constructions which correspond to λ-abstraction and application in the λ-calculi, namely *Closure* and *Composition*. λ-Closure, [λx_1 ... x_n X], transforms into the very procedure of producing a function with the values v-produced by the procedure X, by abstracting over the values of the variables x_1, ..., x_n to provide functional arguments. No Closure is v-improper for any valuation v, as a Closure always v-constructs a function (which may be a degenerate function). *Composition*, [X X_1 ... X_n], is the very procedure of applying a function f produced by X to the tuple argument $\langle a_1, ..., a_n \rangle$ (if any) produced by the procedures X_1, ..., X_n. While a Closure never fails to produce a function, a Composition is v-improper as soon as one or more of its constituents $X, X_1, ..., X_n$ are v-improper, or if the partial function f is not defined at one of its arguments. A third cause of improperness would be type-theoretical incoherence of the Composition, which causes the execution of the procedure to abort at some stage.

Since TIL is a *hyperintensional* system, each construction C can occur not only in *execution* mode to produce at most one object, but also be an object on which other (higher-order) constructions operate. As mentioned above, constructions are displayed by Trivialization. Yet sometimes we need to cancel the effect of Trivialization and turn the displayed mode of a construction C into execution mode. To this end, we have the so-called *Double Execution*, ^2C: it executes C twice over. Thus, if C v-constructs a construction D that in turn v-constructs an entity E then ^2C v-constructs E. Otherwise ^2C is v-improper.

With constructions of constructions, constructions of functions, functions, and functional values in our stratified ontology, we need to keep track of the traffic between multiple logical strata. The *ramified type hierarchy* does just that. The type of first-order objects includes all objects that are not constructions. Therefore, it includes not only the standard objects of individuals and truth-values, but also sets, functions in general and especially functions defined on possible worlds (i.e., the *intensions* germane to possible-world semantics). The type of second-order objects includes constructions of first-order objects and functions that have such constructions in their domain or range. The type of third-order objects includes constructions of first- or second-order objects and functions that have such constructions in their domain or range. And so on, ad infinitum.

The *basic types* of non-procedural objects include the following: o (truth-values **T**, **F**), ι (individuals), τ (times or real numbers), ω (possible worlds). The types of *procedural* objects are $*_n$, where n is the order of a construction. *Functional types* are mappings formed by composing these types. Hence, where $\alpha, \beta_1, \ldots, \beta_n$ are types, $(\alpha \beta_1 \ldots \beta_n)$ is a functional type comprising partial mappings from $(\beta_1 \times \ldots \times \beta_n)$ to α.

Logical objects like *truth-functions* and *quantifiers* are extensional. $\wedge$ (conjunction), $\vee$ (disjunction) and $\supset$ (implication) are of type (ooo), and $\neg$ (negation) of type (oo). The *quantifiers* $\forall^\alpha, \exists^\alpha$ are type-theoretically polymorphous total functions of type (o(oα)), for an arbitrary type α, defined as follows. The *universal quantifier* $\forall^\alpha$ is a function that associates a class A of α-elements with **T** if A contains all elements of the type α, otherwise with **F**. The *existential quantifier* $\exists^\alpha$ is a function that associates a class A of α-elements with **T** if A is a non-empty class, otherwise with **F**. The singularizer I^α is a partial, polymorphic function that assigns to a set S of α-elements the α-element of S if S is a singleton, otherwise (if S is empty or has multiple elements) I^α is undefined.

(PWS-) α-*intensions* are of type (($\alpha\tau$)ω), or '$\alpha_{\tau\omega}$' for short. Typical intensions we frequently deal with are *properties* of individuals of type (oι)$_{\tau\omega}$, individual *offices* or roles of type $\iota_{\tau\omega}$, and *propositions* (empirical truth-conditions) of type o$_{\tau\omega}$.

Below all type indications will be provided outside the formulae in order not to clutter the notation. 'X/α' means that an object X is (a member) of type α. '$X \to \alpha$' means that X is typed to v-construct an object of type α. Throughout, it holds that the variables $w \to \omega$ and $t \to \tau$. If $C \to \alpha_{\tau\omega}$ then the frequently used Composition [[C w] t], which is the intensional descent (a.k.a. extensionalization) of the α-intension v-constructed by C, will be encoded as 'C_{wt}'.

3 Questions and Answers

3.1 Basic Classification of Empirical Questions and Answers

Interrogative empirical sentences can be classified according to many criteria, and various categorizations of questions have been proposed.[3] In this paper we deal only with questions classified into three basic kinds, to wit *Yes-No questions*, *Wh-questions*

---

[3] See, for instance, the questioning toolkit, *The Educational Technology Journal*, vol. 7, no. 3 (1997), http://www.fno.org/nov97/toolkit.html [retrieved on July 27, 2018].

and *exclusive-or questions*. We also work with the notions of *direct* and *complete answer*. An empirical question presents an α-*intension* whose α-value the inquirer desires to know. Thus, a *direct answer* provides directly this α-value. A *complete answer* is the proposition that the α-value of the asked α-intension is an α-object. For instance, the direct answer to the Wh-question "Who is the No. 1 player in the WTA singles ranking"? is 'Simona Halep', while the complete answer is "Simona Halep is the No. 1 player in the WTA singles ranking". Obviously, to each Wh-question there is at most one direct answer, and the respective complete answer, but not vice versa. For instance, the above complete answer could correspond to a direct answer, 'No. 1 player', which would answer another question, namely "Which is the position of Simona Halep in the WTA singles ranking?" For this reason, in our system of intelligent natural-language processing we always pair a given answer with the question that the answer replies.

Yes-No questions like "Did Tilman stop smoking?" and "Did the Pope ever visit Ostrava?" present a proposition (of type $o_{\tau\omega}$) whose actual truth-value the inquirer would like to know. *Wh-questions* like "Who is the Pope?", "When did Tilman stop smoking?", and "Who are the members of the European Union?" present an α-intension of type $\alpha_{\tau\omega}$, where α is the type of a direct answer. It can be an object of any type; an individual, a set of individuals, instants time, location, property, proposition, etc. In case of *exclusive-or questions* like "Are you going by train or by car?" or "Is Tilman an assistant professor or a full professor?" the adequate direct answer does not convey a truth value; instead, it conveys information about which of the alternatives is the case.

3.2 Presupposition of Propositions

Duží [6] defines and analyses presuppositions of propositions denoted by *declarative* sentences. She distinguishes between two kinds of negation, namely wide-scope (or 'external') and narrow-scope ('internal') negation of a proposition. One of the main results is this. While wide-scope negation is presupposition-denying, narrow-scope negation is presupposition-preserving. Furthermore, the distinction between presupposition and mere entailment has been defined. To recapitulate, we first need three properties of propositions, *True*, *False* and *Undef* ('undefined'), all of type $(oo_{\tau\omega})_{\tau\omega}$, that are defined as follows. Let P be a propositional construction ($P/*_n \to o_{\tau\omega}$). Then:

$[^0True_{wt}\ P]$ v-constructs **T** iff P_{wt} v-constructs **T**, otherwise **F**

$[^0False_{wt}\ P]$ v-constructs **T** iff P_{wt} v-constructs **F**, otherwise **F**

$[^0Undefined_{wt}\ P]$ v-constructs **T** iff P_{wt} is v-improper, otherwise **F**

Note that $[^0True_{wt}\ P]$ v-constructs **F** in two cases, namely if P_{wt} v-constructs **F** or if P_{wt} is v-improper. Hence, for instance, if a proposition v-constructed by P is not true at a given $\langle w, t\rangle$-pair, it does not have to be false, because there is the third possibility of being undefined. Formally, we have these relations (=/(ooo)):

$$[^0True_{wt}P] = \neg[^0False_{wt}P] \wedge \neg[^0Undefined_{wt}P]$$
$$[^0False_{wt}P] = \neg[^0True_{wt}P] \wedge \neg[^0Undefined_{wt}P]$$
$$[^0Undefined_{wt}P] = \neg[^0True_{wt}P] \wedge \neg[^0False_{wt}P]$$

Using these properties of propositions, we can rigorously define the difference between presupposition and mere entailment.

Definition 1 (*presupposition and mere entailment*). Let P and S be propositional constructions $(P, S/*_n \to o_{\tau\omega})$. Then:

P is *entailed by* S iff $\forall w \forall t[[^0True_{wt}S] \supset [^0True_{wt}P]]$

P is a *presupposition* of S iff $\forall w \forall t[[[^0True_{wt}S] \vee [^0False_{wt}S]] \supset [^0True_{wt}P]]$. □

Hence, the difference is this. If P is merely entailed by S, then if in a $\langle w, t\rangle$-pair of evaluation P does not take the value **T** the proposition S can take the value **F** or be undefined. While if P is a presupposition of S then if in a $\langle w, t\rangle$-pair of evaluation P does not take the value **T** the proposition S is undefined.

The situation is, however, still a bit more complicated. A given proposition S can have more presuppositions. Consider the sentence "Tom stopped beating his wife". What if Tom never had a wife or, if he did have a wife, never beat her? Then we cannot answer the question whether Tom stopped beating his wife either in the affirmative or the negative. In other words, the proposition that Tom stopped beating his wife is undefined, leaving a truth-value gap, and it has two presuppositions (if not more). The pool of presuppositions of a given proposition S characterizes the necessary conditions for S not to have a truth-value gap.[4]

3.3 Presupposition of Questions

In standard erotetic logic a presupposition of a question is usually characterized by the '*inference from possible answers*' condition like this: a presupposition of a question is entailed by each possible answer to the question.[5] To put this characterization on a more solid footing, we make use of the distinction between a direct and a complete answer. Thus, we define:

Definition 2. A *presupposition of an empirical question* Q is a proposition P that is entailed by each complete answer corresponding to an unambiguous direct answer of the question Q. □

---

[4] We leave aside the question whether the truth of *all* the presuppositions of S a *sufficient* condition for S is having a truth-value. It seems that if S is undefined, we might deduce that it is so *because* there is a presupposition of S that is not true. Yet, we are hesitant to draw this conclusion, since the logic of *because* is much more complicated. For details, see [18, 21].

[5] See, for instance, [7, 11, 12].

To illustrate, consider Levinson's examples (see [13, 178–181]):

(1) John managed [didn't manage] to stop in time
    entailing that John tried to stop in time, and
(2) Martha regrets [doesn't regret] drinking John's home brew
    entailing that Martha drank John's home brew.

If somebody is asking whether John managed to stop in time, a possible answer could be "No, he didn't; he didn't even try". This indicates that the question presupposes that John did try. For instance, Pagin [14] says, "in asking the speaker normally assumes that John tried and is only asking about the success". Indeed, both "John managed to stop in time" and "John didn't manage to stop in time" entail that John tried to stop in time. Only if this is the case can we answer the question "Did John manage to stop in time?" either in the affirmative or the negative. Similarly, when asking (2) in such a situation that the presupposition that Martha drank John's home brew is not true, there is no direct answer Yes/No. The responder should thus provide a complete answer by negating the presupposition thus: "Martha didn't drink John's home brew".

Note that we are applying narrow-scope negation. For instance, the narrow-scope negative form of (2), "Martha doesn't regret drinking John's home beer" implies that Martha did drink John's home beer, unlike the wide-scope negation, "It is not true that Martha regrets drinking John's beer". The latter gets to be true also in case the proposition that Martha regrets drinking John's beer has a truth-value gap, because she didn't drink John's beer. Hence the wide-scope negation of (2) does not entail that Martha drank John's home beer. Therefore, we suggest the following *thesis*:[6]

> *If a presupposition P of a question Q is not true,*
> *then there can be no direct answer*
> *and an adequate answer must be a denial of P*
> *applying narrow-scope negation to P.*

Up to now we have dealt only with Yes/No questions. The situation is a bit more complicated with Wh-questions. Some authors even incline to the opinion that Wh-questions do not come with presuppositions.[7] We are, however, going to show that even Wh-questions have semantic presuppositions, which are frequently existential ones. The decision as to whether there is an existential presupposition depends on an ambiguity that concerns different topic-focus articulations.[8]

For instance, the question "When did the Pope visit Prague?" is ambiguous. If the topic is 'the (current) Pope' (writing in 2018) then 'the Pope' occurs with supposition *de re* (extensionally) and the question has the existential presupposition that the papal office is currently occupied (which in TIL is what it means to say that the Pope exists).

---

[6] The idea of replying by negated presupposition if the presupposition is not true was first proposed by M. Číhalová and presented in [1].

[7] See, for instance, [7].

[8] Hajičová in [8] argues that the analysis of topic-focus articulation is a semantic rather than pragmatic issue. For logical analysis of sentences with topic-focus articulation see [3].

Each positive direct answer completed to a complete answer entails that the Pope exists. The negative answer 'Never' would be ambiguous in case we did not take into account the existential presupposition. It might mean that the Pope exists yet never visited Prague, or that the Pope does not exist. Yet, in the latter case wide-scope negation is applied, which we do not admit as an adequate answer. Hence, in case the papal office goes vacant the correct unambiguous answer is, "The Pope does not exist".

The situation is different if the topic is 'the visit of Prague'. In this case there is no presupposition to the effect that the Pope exists, and the sentence should be better phrased in the passive voice: "When has Prague been visited by the Pope?" Now the answer 'Never' is unambiguous, as it means that the set of dates when Prague was visited by the Pope is empty. 'The Pope' occurs with supposition *de dicto* (intensionally).

There are other presupposition triggers, like *activity* verbs, *factive* verbs, and verbs in *future* or *past tense*. Verbs expressing an activity trigger a presupposition whenever we ask whether the respective activity has ended or is still ongoing. In this case there is a presupposition to the effect that the activity in question has begun. Above we mentioned the example of 'having stopped beating one's wife'. Questions involving attitudes expressed by means of factive verbs like 'to know that', 'to regret that', etc. have the presupposition that the proposition denoted by the embedded clause is true. For instance, the question "Does Tom regret being late?" presupposes that Tom was late. Both positive and negative (complete) answers applying narrow-scope negation entail that Tom was late. As above, if Tom was not late, instead of an answer involving ambiguous wide-scope negation, the adequate answer involves narrow-scope negation, namely "Tom was not late".

Questions phrased in past or future tense and involving a reference time specifying when this or that happened or will happen come with the presupposition that the reference time is in a proper relation to the time of evaluation. For instance, consider the question, "Will Tom show up today at 17:00"? Both the positive and the negative answer entail that the question has been evaluated *before* 17:00. Otherwise, the adequate answer should deny the truth of the presupposition as per "It is now past 17:00". A correct analysis of questions that contain a time of reference is necessary for smooth and consistent communication between agents in a multi-agent system. For instance, a question in future tense might reach the addressee too late due to technical problems (e.g. the agent is not within range of a mobile signal). In such a case the responder informs the inquirer that the message arrived late.[9]

Alternative questions come with a presupposition that exactly one of the two alternatives is the case. For instance, the question "Is the No. 1 player in the singles rankings Serena Williams or Petra Kvitová?" is not a Yes-No question, of course. The inquirer wants to know which one of the two ladies occupies the role of No. 1 in the

---

[9] For a TIL analysis of tenses see [2, 5, 20].

WTA singles rankings, and he/she presupposes that one of the two alternatives is the case. If neither alternative is the case (as per mid-2018), the respondent should convey this information by negating the presupposition thus: "Neither Williams nor Kvitová is the No. 1 player in the WTA singles ranking".[10]

4 Logical Analysis of Questions with Presupposition and Their Answers

4.1 The 'If-then-else' Function and the Analytic Schema

To take presuppositions into account, we apply the *general analytic schema* for sentences with a presupposition that has been introduced in [2, 3] together with the definition of the *If-then-else* function. We define this function in the *strict* way, i.e. as complying with the principle of compositionality. The procedure encoded by "If P ($\rightarrow$ o) then C ($\rightarrow \alpha$), else D ($\rightarrow \beta$)" behaves as follows:

(a) If P v-constructs **T** then execute C (and return the result of type α, provided C is v-proper).
(b) If P v-constructs **F** then execute D (and return the result of type β, provided D is v-proper).
c) If P is v-improper then no result ensues (i.e. the procedure lacks a product).

Hence, *if-then-else* is seen to be a strict function and its definition decomposes into two phases. *First*, select a construction to be executed based on a specific condition P. The choice between C and D comes down to this Composition:

$$[^0\mathrm{I}^* \, \lambda c [[P \wedge [c = {}^0C]] \vee \neg [P \wedge [c = {}^0D]]]]$$

Types: $P \rightarrow$ o v-constructs the condition of the choice between the execution of C or D, $C/*_n, D/*_n \rightarrow \alpha; c \rightarrow *_n; \mathrm{I} * /(*_n(\mathrm{o}*_n))$: the singularizer function on constructions, i.e. the function that associates a singleton of constructions with the construction that is the only element of this singleton, and is otherwise (i.e. if the set is empty or many-valued) undefined.

If P v-constructs **T** then the variable c v-constructs the *construction C*, and if P v-constructs **F** then the variable c v-constructs the *construction D*. In either case, the set produced by

$$\lambda c [[P \wedge [c = {}^0C]] \vee [\neg P \wedge [c = {}^0D]]]$$

is a singleton and the singularizer I* returns as its value either the construction C or the construction D. Note that in this phase C and D are not constituents to be executed; rather they are merely displayed as objects to be selected by the variable c. Without this

---

[10] Alternative questions known from questionnaires are a more general case. They presuppose that at least one of the alternatives is the case. For this reason, there is often the alternative 'other' or 'none of the above', in case none of the alternatives offered applies.

hyperintensional approach, which allows shifting between executed and merely displayed occurrences of procedures, we would not be able to define the *strict* function *if-then-else*.

Second, the selected construction is *executed*; therefore, Double Execution must be applied:

$$^2\left[^0\mathrm{I}^*\lambda c\left[[P \wedge [c = {}^0C]] \vee [\neg P \wedge [c = {}^0D]]\right]\right]$$

To make the schema easier to read, we use the abbreviated notation
'[*If P then C else D*]'.

The schema modified for the analysis of a question $Q \to \alpha_{\tau\omega}$ with a presupposition $P \to o_{\tau\omega}$ is this:

$$\lambda w \lambda t [\textit{If } P_{wt} \textit{ then } Q_{wt} \textit{ else } \neg P_{wt}]$$

Gloss. The schema can be read as an instruction: if the presupposition P is true in a given world w and time t (*If P_{wt}*) then evaluate Q_{wt} to provide the answer of type α, else reply by negating the presupposition P ($\neg P_{wt}$).

4.2 Wh-Questions

The analysis of an empirical interrogative sentence is a construction of an α-intension whose α-value in the actual possible world and time the inquirer wants to know. The type α is given by possible direct answers because these are the values of the relevant α-intension. Consider, for instance, the question

(1) Who are the top ten players in the WTA singles ranking?

Possible direct answers specify a set of individuals, currently {Simona Halep, Caroline Woźniacki, Sloane Stephens, Angelique Kerber, Elina Svitolina, Caroline Garcia, Garbiñe Muguruza, Petra Kvitová, Karolina Plíšková, Julia Görges}, which is an object of type (oι). Hence, the analysis of (1) is a construction of a property of individuals, which is an object of type (oι)$_{\tau\omega}$:

(1*) $\lambda w \lambda t \lambda x [[^0 Order\_WTA_{wt}\ x] \leq {}^0 10] \to (o\iota)_{\tau w}$

Types: $x \to_v \iota$; $10/\tau$; *Order_WTA*/$(\tau\iota)_{\tau\omega}$: an attribute of an individual, that is, an empirical function assigning to individuals their current position (if any) in the WTA singles ranking.

If the set of individuals obtained by evaluating the question (1) in a given world w and at a time t is empty, then the unambiguous direct answer is simply 'No one'. Hence there is no existential presupposition attached to this question.

On the other hand, consider the question:

(2) When did all the top ten WTA players come to Ostrava?

As explained above, if the topic of this question concerns the top ten WTA players then there is the presupposition that the set of top ten players is non-empty. The direct answer 'Never' implies that all the slots in the top ten are occupied. If this is not the case, an adequate answer would be the piece of information that no players are in the top ten.

Applying the above analytic schema to (2) with the topic understood to be *top ten WTA players*, we obtain:

(2*) $\lambda w \lambda t [if [^0\exists \lambda x[[^0Order\_WTA_{wt}\, x] \leq {}^0 10]]$
$then\ \lambda t'[[t' < t] \wedge {}^0\forall \lambda x[[[^0Order\_WTA_{wt}\, x] \leq {}^0 10] \supset [^0Come_{wt'}\, x\, {}^0Ostrava]]]$
$else \neg [^0\exists \lambda x[[^0Order\_WTA_{wt}\, x] \leq {}^0 10]]]$

Additional types: $Come/(o\iota\iota)_{\tau\omega}$: a relation-in-intension between individuals; $Ostrava/\iota$; $x \to \iota$.

Note that the answer is of type $(o\tau)$, that is, the set of past times t', namely those when all the top ten players came to Ostrava. If there were no top ten players, the answer would correspond to the procedure $\neg[^0\exists\, \lambda x\, [[^0Order\_WTA_{wt}\, x] \leq {}^0 10]]$, which is the meaning of "There are no individuals such that their order in the WTA ranking is less than or equal to 10", hence "There are no top ten WTA players".

4.3 Exclusive-or Questions

Exclusive-or questions have only two possible direct answers, to wit one of the two alternatives. For instance, if somebody asks

(3) Is Tom a student or a professor?

they want to know the only property selected from $\{Student/(o\iota)_{\tau\omega},\ Professor/(o\iota)_{\tau\omega}\}$ that Tom has. Thus, we construct an office occupied by individual properties, i.e. an object of type $((o\iota)_{\tau\omega})_{\tau\omega}$:

(3*) $\lambda w \lambda t\, [^0 I\, \lambda p[[p_{wt}\, {}^0Tom] \wedge [[p = {}^0 Professor] \vee [p = {}^0 Student]]]] \to ((o\iota)_{\tau w})_{\tau w}$

Types: $p \to (o\iota)_{\tau\omega}$; Tom/ι; $Professor, Student/(o\iota)_{\tau\omega}$; $I/((o\iota)_{\tau\omega}\ (o(o\iota)_{\tau\omega}))$: the singularizer of properties, i.e. the function that assigns to a singleton M of properties the only element of M and which is otherwise undefined.

Remark. Note that our analyses can be understood as specifications of *instructions* how, in any possible world (λw) at any time (λt), to evaluate the question and obtain an adequate answer, which amounts to the execution of the procedure encoded by the respective interrogative sentence, and thus convey as a direct answer the value (if any) of the so-constructed intension. For instance, the above construction (3*) is a specification of this instruction: In any state-of-affairs ($\lambda w \lambda t$) take the individual Tom (^0Tom), the properties of being a professor and of being a student ($^0Professor$, $^0Student$), assign these properties to the variable p ($[[p = {}^0Professor] \vee [p = {}^0Student]]$), and find out which of them Tom has ($[p_{wt}\ {}^0Tom]$). Finally, produce one of these two properties as the direct answer ($^0I\ \lambda p\ [...]$).

5 Concluding Remarks

In this paper we described a hyperintensional analysis of questions that come with a presupposition and of adequate answers to such questions. We have distinguished direct from complete answers and showed that if a presupposition of a question is not true, then there is no direct answer to such a question. In such a case, an adequate answer is a complete one that negates the presupposition by applying narrow-scope (internal) negation. To formalize these ideas, we applied a general analytic schema for sentences with presuppositions adjusted to interrogative sentences. The results have been illustrated by examples of analysis of Yes/No questions, Wh-questions and exclusive-or alternative questions. These results will be implemented as one of the main components of a system of intelligent natural-language processing and reasoning.

Acknowledgements. This research was supported by the Grant Agency of the Czech Republic, project no. GA18-23891S, *Hyperintensional Reasoning over Natural Language Texts*, and by the internal grant agency of VSB-Technical University of Ostrava, project no. SP2018/172, *Application of Formal Methods in Knowledge Modelling and Software Engineering*.

References

1. Číhalová, M., Duží, M.: Questions, answers and presuppositions. Computación y Sistemas **19**(4), 647–659 (2015)
2. Duží, M.: Tenses and truth-conditions: a plea for if-then-else. In: Peliš, M. (ed.) The Logica Yearbook 2009, pp. 63–80. College Publications, London (2010)
3. Duží M.: Topic-focus articulation from the semantic point of view. In: Computational Linguistics and Intelligent Text Processing. LNCS, vol. 5449, pp. 220–232. Springer (2009)
4. Duží, M., Jespersen, B., Materna, P.: Procedural Semantics for Hyperintensional Logic: Foundations and Applications of Transparent Intensional Logic. Springer, Berlin (2010)
5. Duží, M., Macek, J.: Analysis of time references in natural language by means of Transparent Intensional Logic. Organon F **25**(1), 21–40 (2018)
6. Duží, M.: Presuppositions and two kinds of negation. Logique Anal. **239**, 245–263 (2017). The special issue on How to Say 'Yes' or 'No'
7. Fitzpatrick, J.: The whys and how comes of presupposition and NPI licensing in questions. In: Alderete, J., et al. (eds.) Proceedings of the 24th West Coast Conference on Formal Linguistics, pp. 138–145 (2005)
8. Hajičová, E.: What we are talking about and what we are saying about it. In: Computational Linguistics and Intelligent Text Processing. LNCS, vol. 4919, pp. 241–262. Springer (2008)
9. Harrah, D.: The Logic of questions. In: Handbook of Philosophical Logic, vol. 8, pp. 1–60 (2002)
10. Jespersen, B.: Anatomy of a proposition. Synthese (2017). https://doi.org/10.1007/s11229-017-1512-y
11. Katz, J.J.: Semantic Theory. Harper & Row, New York (1972)
12. Keenan, E.L., Hull, R.D.: The logical presuppositions of questions and answers. In: Petöfi, J. S., Franck, D. (eds.) Präsuppositionen in Philosophie und Linguistik, pp. 441–466. Frankfurt, Athenäum (1973)
13. Levinson, S.C.: Pragmatics. Cambridge University, Cambridge (1983)

14. Pagin, P.: Assertion. In: Zalta, E.N. (ed.) The Stanford Encyclopedia of Philosophy (Spring 2014 Edition). http://plato.stanford.edu/archives/spr2014/entries/assertion/
15. Peliš, M.: Consequence relations in inferential erotetic logic. In: Bílková, M., (ed.) Consequence, Inference, Structure, in the book series Miscellanea Logica VII, Charles University of Prague, pp. 53–88 (2008)
16. Peliš, M., Majer, O.: Logic of questions and public announcements. In: Bezhanishvili, N., Löbner, S., Schwabe, K., Spada, L. (eds.) 8th International Tbilisi Symposium on Logic, Language and Computation 2009. Lecture Notes in Computer Science, pp. 145–157. Springer (2011)
17. Sintonen, M.: On the logic of why-questions. In: PSA: Proceedings of the Biennial Meeting of the Philosophy of Science Association, pp. 168–176. University of Chicago Press (1984)
18. Schnieder, B.: A logic for 'because'. Rev. Symbolic Log. **4**(3), 445–465 (2011)
19. Tichý, P.: Collected papers in logic and philosophy. In: Svoboda, V., Jespersen, B., Cheyne, C. (eds.) Filosofia, Czech Academy of Science, and Dunedin. University of Otago Press, Prague (2004)
20. Tichý, P.: The logic of temporal discourse. In: Linguistics and Philosophy, vol. 3, pp. 343–369 (1980). Reprinted in (Tichý 2004: 373–369)
21. Tsohatzidis, S.L.: A problem for a logic of 'because'. J. Appl. Non-Class. Log. **25**(1), 46–49 (2015)
22. Wiśniewski, A.: The Posing of Questions: Logical Foundations of Erotetic Inferences. Kluwer, Dordrecht (1995)

An Efficient Reduced Basis Construction for Stochastic Galerkin Matrix Equations Using Deflated Conjugate Gradients

Michal Béreš[1,2,3]

[1] FEECS, Department of Applied Mathematics, VŠB - Technical university of Ostrava, 17. listopadu 15, 708 33 Ostrava - Poruba, Czech Republic
michal.beres@vsb.cz
[2] IT4Innovations National Supercomputing Center, VŠB - Technical University of Ostrava, 17. listopadu 15, 708 33 Ostrava - Poruba, Czech Republic
[3] Institute of Geonics of the CAS, Studentská 1768, 708 00 Ostrava - Poruba, Czech Republic

Abstract. In this article, we examine an efficient solution of the stochastic Galerkin (SG) matrix equations coming from the Darcy flow problem with uncertain material parameters on given interfaces. The solution of the SG system of equations, here represented as matrix equations, is usually a very challenging task. A relatively new approach to the solution of the SG matrix equations is the reduced basis (RB) solver, which looks for the low-rank representation of the solution. The construction of the RB is usually done iteratively and consists of multiple solutions of systems of equations. We aim to speed up the process using the deflated conjugate gradients (DCG). Other contributions of this work are a modified specific construction of the RB without the need of Cholesky factor and an adaptive choice of the candidate vectors for the expansion of the RB. The proposed approach allows an efficient parallel implementation.

Keywords: Stochastic Galerkin method · Reduced basis method · Deflated conjugate gradients method · Darcy flow problem

1 Introduction

The motivation for this work is a laboratory experiment on a rock sample. We assume to know the inner structure (separated subdomains with different types of material) of the sample (e.g. from a Computed Tomography scan) and we want to estimate permeabilities of each distinct subdomains in the sample. This is done by a series of tests, where a fluid is pressed to some part of the sample and let out from another part of the sample, the output volume is then measured. We also admit uncertainties of these measurements, this forms a complex inverse problem. This is usually solved using the Bayesian inversion, which relies on many forward problem (Darcy flow) solutions. In this matter, the SG method can be used as a surrogate model substituting the forward solutions.

The SG approach to the solution of PDEs with uncertainties got a lot of attention recently. There is a great recent advancement in the methods for the solution of the SG formulation of problems with uncertainties. Here we present only a small sample of them: Iterative methods with compression [7], reduced basis methods [9], the Generalized spectral decomposition [8] and a tensor approach using higher dimensional tensors [5]. One can also use some of the more classical approaches (not using the SG approach) such are the Monte Carlo methods or stochastic collocation [2].

The work presented in this paper is an extension of the results presented in [9]. Here we focus on alternative approaches to the rational Krylov approximation and expansion vector proposal, see Sect. 3. Main contribution of this work is the use of the DCG method, which is new in this area of application, see Sect. 5.

1.1 Problem Setting

We assume a Darcy flow problem on 2D square $\langle 0,1 \rangle^2$ domain with no forcing term. The area where the fluid is pressed into the rock sample is represented by the part of the Dirichlet boundary with the value equal to 1. The area where the fluid leaves the sample is represented by the part of the Dirichlet boundary with the value equal to 0. The rest of the sample's boundary is sealed, which corresponds to the Neumann boundary condition with the value equal to 0. Pointing out the stochastic nature of the material field, our problem is described by the following equation

$$\begin{cases} -\nabla_x \cdot (k(x; \mathbf{Z}) \nabla_x u(x; \mathbf{Z})) = 0 & \forall x \in \Omega, \mathbf{Z} \in \mathbb{R}^M, \\ u(x; \mathbf{Z}) = g(x) & \forall x \in \Gamma^D, \mathbf{Z} \in \mathbb{R}^M, \\ n(x) \cdot k(x; \mathbf{Z}) \nabla_x u(x; \mathbf{Z}) = 0 & \forall x \in \Gamma^N, \mathbf{Z} \in \mathbb{R}^M. \end{cases} \quad (1)$$

To complete the formulation, we need to specify some probability distribution of the vector $\mathbf{Z} = (Z_1, \ldots, Z_M)$, which reflects the random nature of the material field on given subdomains. We assume a log-normal distribution of permeability on each subdomain, which is natural for the permeability of aquifers, see [4]. Here the random material field takes the form

$$k(x; \mathbf{Z}) = \sum_{i=1}^{M} \chi_{\Omega_i}(x) \exp(\sigma_i Z_i + \mu_i), \quad (2)$$

where $\chi_{\Omega_i}(x)$ is a characteristic function of the subdomain Ω_i and $\exp(\sigma_i Z_i + \mu_i)$ describes the distribution of the permeability on subdomain Ω_i. μ_i, σ_i are the mean value and the standard deviation of the underlying normal distribution (corresponds to the distribution of the porosity). Then the components of $\mathbf{Z}$ are independent standard normal random variables.

2 Stochastic Galerkin Method

The SGM denotes the Galerkin method applied to the PDE with uncertain parameters in the form of functions of a random vector, for a more thorough introduction see [3,6]. The SGM assumes a discretization of both the physical space (functions on the domain) and the stochastic/parametric space (functions of random variables).

The problem (1) can be formulated variationally with bilinear form a in a tensor product of the Sobolev space $H^1(\Omega)$ and $L^2_{dFZ}(\mathbb{R}^M)$, which is the space of square integrable functions on $\mathbb{R}^M$ with respect to the measure/distribution of Z. For the Galerkin discretization we use the finite dimensional space with basis created as a tensor product of basis $\langle \varphi_1(x), \ldots, \varphi_{N_d+N_{Dd}}(x) \rangle$ of standard linear elements on Ω and basis $\langle \psi_1(Z), \ldots, \psi_{N_s}(Z) \rangle$ of orthonormal polynomials (with respect to the distribution of Z) on $\mathbb{R}^M$ (here it means the products of one dimensional Hermite polynomials). Note that we assume that the last basis functions $\varphi_{N_d+1}(x), \ldots, \varphi_{N_d+N_{Dd}}(x)$ correspond to degrees of freedom on the Dirichlet part of the boundary $\partial\Omega$.

Due to the separable nature of the material field, the values of bilinear form on the elements of tensor product basis can be written as

$$a(\varphi_i \psi_j, \varphi_k \psi_l) = \sum_{m=1}^{M} \int_{\mathbb{R}^M} \psi_j \psi_l \exp(a_m Z_m + b_m) \, \mathrm{d}FZ \int_{\Omega_m} \nabla \varphi_i \nabla \varphi_k \mathrm{d}x. \quad (3)$$

This leads to a large system $(N_s \times N_d)$ of linear equations in the form of

$$A \cdot \overline{u}_h = \overline{b}, \quad A = \sum_{m=1}^{M} G_m \otimes K_m, \quad b = \sum_{m=1}^{M} g_m \otimes f_m, \quad (4)$$

where G_m/g_m are matrices/vectors corresponding to the discretization of parameter space and K_m/f_m are matrices/vectors corresponding to the discretization using linear finite elements.

3 Reduced Basis Method

This section extends the work in [9] and deals with technicalities only briefly, for theoretical background see [9]. An advantage of the tensor structure of the matrix $\mathbb{A}$, which we use for the RB solver, is that we can view the system (4) as matrix equations

$$\sum_{m=0}^{M} K_m x G_m^T = \sum_{m=1}^{M} f_m g_m^T. \quad (5)$$

First, we show how to find an approximate solution of the SGM matrix equations with a given RB and then we show the construction of the RB for our problem. In our text, we consider the RB of the matrices K_m, because in usual applications, N_d is significantly larger than N_s. Here the RB of size k denotes a set of k orthonormal vectors (generally needs to be linearly independent) $W_k = [w_1, \ldots, w_k] \in \mathbb{R}^{N_d \times k}$. Then the matrix $x_k = W_k y_k$ is a low-rank approximation of the solution x, where y_k is a reduced solution matrix. The matrix y_k can be obtained from (5) using the Galerkin condition on the residual of x_k

$$R_k := \sum_{m=0}^{M} K_m (W_k y_k) G_m^T - \sum_{m=1}^{M} f_m g_m^T \qquad (6)$$

$$W_k^T R_k = 0 \Rightarrow \sum_{m=0}^{M} W_k^T K_m W_k y_k G_m^T = \sum_{m=1}^{M} W_k^T f_m g_m^T. \qquad (7)$$

The RB approach can be viewed as a standard iterative method, where we can control the relative residual error in each iteration, see Algorithm 1.

Algorithm 1. The reduced basis solver

Require: matrices $\{K_m, G_m\}_{m=1,\ldots,M}$, vectors $\{k_m, g_m\}_{m=1,\ldots,M}$, precision ε
1: $x_0 = 0$; $k = 0$; $W_0 = \emptyset$; $R_0 = \sum_{m=0}^{M} K_m x_0 G_m^T - \sum_{m=1}^{M} f_m g_m^T$
2: **while** $\|R_k\| / \|R_0\| > \varepsilon$
3: $\quad k = k + 1$
4: $\quad$ propose an enhancement of the reduced basis: V_k
5: $\quad$ enrich the reduced basis: $W_k = \text{orth}([W_{k-1}, V_k])$
6: $\quad$ find y_k as a solution of Eq. (7)
7: $\quad$ calculate the residual R_k using Eq. (6)
8: **end**
9: **return** $x_k = W_k y_k$

3.1 Rational Krylov Subspace Methods

We aim to build the RB using an approach from [9] called the rational Krylov subspace approximation. Briefly, the rational Krylov subspace approximation can be performed for a series of symmetric positive definite (SPD) matrices $\{K_m\}_{m=1,\ldots,M}$ and a nonzero vector v. In the first iteration, it generates the basis $\langle K_1^{-1} v, \ldots, K_M^{-1} v \rangle$, in the second the basis $\langle K_1^{-1} K_1^{-1} v, K_1^{-1} K_2^{-1} v, \ldots, K_M^{-1} K_{M-1}^{-1} v, K_M^{-1} K_M^{-1} v \rangle$ and so on, for details see [9]. The reduced basis will be the union of v and all these bases.

In our case the matrices $\{K_m\}_{m=1,\ldots,M}$ are not SPD (they are symmetric, but not positive definite). The approach in [9] proposed the use of the Cholesky factor L of the mean value matrix K_0 (matrix corresponding to the mean value of Z):

$$\sum_{m=0}^{M} L^{-1} K_m L^{-T} L^T x G_m^T = \sum_{m=1}^{M} L^{-1} f_m g_m^T. \qquad (8)$$

This would create the reformulation of the problem with matrices $\tilde{K}_m = L^{-1} K_m L^{-T}$, which have bounded spectra and therefore can be easily shifted to be SPD. This leads to the rational Krylov subspace generated of matrices $\left\{L^{-1}\left(K_m - \alpha K_0\right)L^{-T}\right\}_{m=1,\ldots,M}$ with series of starting vectors $\left\{L^{-1}f_m\right\}_{m=1,\ldots,M}$, which approximates reformulated solution $L^T x$. The α is the shift of the spectrum, for simplicity we assume the same value for all of the matrices.

The aforementioned approach works nicely, but its usefulness is bounded by the need of the Cholesky factor L. Here we propose an alternative approach, which leads to the same approximation space of the solution without the need for the factor L. The use of these matrices $\left\{K_0^{-1}\left(K_m - \alpha K_0\right)\right\}_{m=1,\ldots,M}$ and these starting vectors $\left\{K_0^{-1}f_m\right\}_{m=1,\ldots,M}$ leads directly to a basis approximating the original solution x, while the approximation space is identical to the original approach. The proof of this comes directly from the fact that the rational Krylov subspace generated by the original method equals to the rational Krylov subspace generated by the new method multiplied by L^T.

3.2 Adaptive Selection of Space Expansion

The process of building the rational Krylov subspace is impractical because in each subsequent iteration we need to construct larger bases. The remedy to this is to iteratively select a vector v from the current basis and expand the basis by $\langle K_1^{-1}v, \ldots, K_M^{-1}v \rangle$. Note that the equivalency of approaches in the previous subsection will be broken, if we do not always take the same vector v.

In [9] authors proposed an approach consisting of truncating (e.g. using the singular value decomposition (SVD)) the current expansion $\langle K_1^{-1}v, \ldots, K_M^{-1}v \rangle$ and appending the truncated expansion to the current basis.

Algorithm 2. Orthogonalisation and the choice of the next expansion vector

---

Require: current basis W_{k-1} and its weights β_{k-1} (basis vectors already used for the expansion of the RB have the weights equal the ∞), proposition basis $V_k = \left[V_k^1, \ldots, V_k^M\right]$, tolerance ε

1: $\tilde{W}_0 = W_{k-1}, \tilde{\beta}_0 = \beta_{k-1}$
2: **for** $i = 1, \ldots, M$
3:     $v_i = V_k^i / \left\|V_k^i\right\|$
4:     $w_i = v_i - \tilde{W}_{i-1}\left(v_i^T \tilde{W}_{i-1}\right)$
5:     **if** $\|w_i\| > \varepsilon$
6:        $\tilde{W}_i = \left[\tilde{W}_{i-1}, \frac{w_i}{\|w_i\|}\right]$, $\tilde{\beta}_i = \left[\tilde{\beta}_{i-1}, \|w_i\|\right]$
7:     **else**
8:        $\tilde{W}_i = \tilde{W}_{i-1}$, $\tilde{\beta}_i = \tilde{\beta}_{i-1}$
9:     **end**
10: **end**
11: sort the vectors in $\tilde{W}_M$ and values in $\tilde{\beta}_M$ according to the values of $\tilde{\beta}_M$
12: **return** $W_k = \tilde{W}_M, \beta_k = \tilde{\beta}_M$
     Note: next vector for the expansion is $k+1$ vector of W_k

The candidates for the next expansion of the RB are then picked sequentially. This approach has two drawbacks. First, the truncation using the SVD is computationally expensive and second, it can delay the convergence of the RB solver (e.g. if a good candidate for the expansion is found in i-th iteration, we first need to expand the RB with all of his predecessors).

We propose an approach, which calculates weights for the vectors during the orthogonalisation step (with insignificant additional costs) and then pick the candidates according to these weights. The proposed approach is summarized in Algorithm 2. Compared to the original approach, we save the time spent on the SVD computation. We support the efficiency of this approach with numerical tests in Sect. 5.

4 Deflated Conjugate Gradients

The deflated conjugate gradients (DCG) method is an extension of the standard conjugate gradient (CG or PCG if using a preconditioner) method, see [10]. The DCG method takes an additional parameter in the form of the deflation basis W. The deflation basis W should be able to describe the sought solution reasonably well. Therefore, the DCG method looks for the solution outside of the deflation basis W. The impact on the conjugate gradient implementation is then only in two places. First, the residual of the initial guess should be orthogonal on W (e.g. by using $x_0 = W \left(W^T A W \right)^{-1} W^T b$). Second, in each iteration the DCG projects the preconditioned residual $\tilde{z}_i = P z_i$ using the projector $P = I - W \left(W^T A W \right)^{-1} W^T A$. In our application, we use the current RB as a deflation basis W.

Only significant additional cost compared to the standard CG method is the solution of systems with $Q_i = W_i^T A W_i$ (in i-th iteration of the RB solver). In our application, the size of the matrix Q_i (corresponding to a size of the RB) is reasonably small and we can use e.q. an explicit inversion. Additionally, if we use an explicit inversion, we can exploit the adaptive updates of deflation basis $W_i = [W_{i-1}, V_i]$ and calculate Q_i^{-1} using the Schur complement with known Q_{i-1}^{-1}.

An important note: check the precision of Q^{-1} e.q. $\left\| Q \left(Q^{-1} e \right) - e \right\|$ (e is a unit vector). The DCG precision cannot surpass the precision of Q^{-1}. In our tests, the precision of the solutions using an explicit inversion of Q was approximately 10^{-10}, which was sufficient for our purposes.

5 Numerical Testing

In this section, we present the obtained results. We tested both the convergence of the RB solver (i.e. convergence of the whole SG solution) and the convergence of the DCG with varying size of the deflation basis (i.e. in different iterations of the RB solver). First, we will recall all of the computationally significant steps of the RB solver with corresponding methods:

- solutions of the systems: $\left\langle L^T \left(K_m - \alpha K_0 \right)^{-1} L v \right\rangle_{m=1,\ldots,M}$ respectively $\left\langle \left(K_m - \alpha K_0 \right)^{-1} K_0 v \right\rangle_{m=1,\ldots,M}$, using the DCG method

- we will test three preconditioners: additive Schwarz (with no coarse problem, see [1]), diagonal and ichol (using incomplete Cholesky factors)
– solution of the reduced problem $\sum_{m=0}^{M} W_k^T K_m W_k y_k G_m^T = \sum_{m=1}^{M} W_k^T f_m g_m^T$ using the PCG with the Kronecker type preconditioner, see [11]
  - we use y_{k-1} extended by zeros as an initial guess for y_k and we increase the accuracy of the solver with respect to the current residual of the RB solver
– extending the RB using SVD approach or approach from Algorithm 2

Our model problem consists of $M = 7$ subdomains, with stochastic properties $\mu_i \in (9, 5, 9, 1, 5, 1, 5)$ and $\sigma_i = 0.5$. The used discretizations consist of complete polynomials on 7 variables up to a given degree and an uniform finite element grid on unit square, where grid lvl l equals to the discretization $10l$ on each side.

5.1 Reduced Basis Convergence

Here we compare different settings with 3 different parameters of the RB construction. First, we compare the approach using the Cholesky factor (in figures denoted by "L") with the new approach without it (no special notation in figures). Second, we compare the choice of the expansion candidate vectors using the SVD based approach (in figures denoted by "SVD") with our approach in Algorithm 2 (in figures denoted by "ada"). Third, we compare initial vectors for the RB creation, the two alternatives are: single vector as a sum of f_m (in figures denoted by "sum") and zero iteration of the RB $W_0 = \langle f_m \rangle_{m=1}^{M}$ (in figures denoted by "all").

The results for the tuning of the algorithm can be found in Fig. 1. On the left figure, we compare the impact of the aforementioned parameters on the RB construction. We have found that for our problem it is beneficial to start the RB with the sum of f_m in comparison to $W_0 = \langle f_m \rangle_{m=1}^{M}$. Another observation is that the approach from Algorithm 2 performed better than the SVD based one. In the following tests, we will use exclusively the adaptive approach from Algorithm 2 and the sum of f_m as a starting vector for the RB solver. The results of the impact of the choice of α can be seen in the right figure. In our case, the matrices K_m are positively semi-definite and therefore α can be arbitrarily small. The single values of α do not change the convergence drastically, the best value was $\alpha = 10^{-3}$ and it will be used in the further tests.

Next test takes into account different discretizations of the problem. The results can be seen in Fig. 2. The results show, that the two tested approaches behave very similarly with the slight advantage of our new approach without the use of Cholesky factor.

The number of iterations of the PCG method for the reduced problem using the Kronecker preconditioner was under **20** in all of the RB iterations. This is due to increasing accuracy of the solution together with increasing quality of the initial guess. We start with a bad initial guess, but low desired accuracy in the first few iterations and slowly converge to a good initial guess, but high desired accuracy. This makes the iteration count relatively stable during the RB iterations.

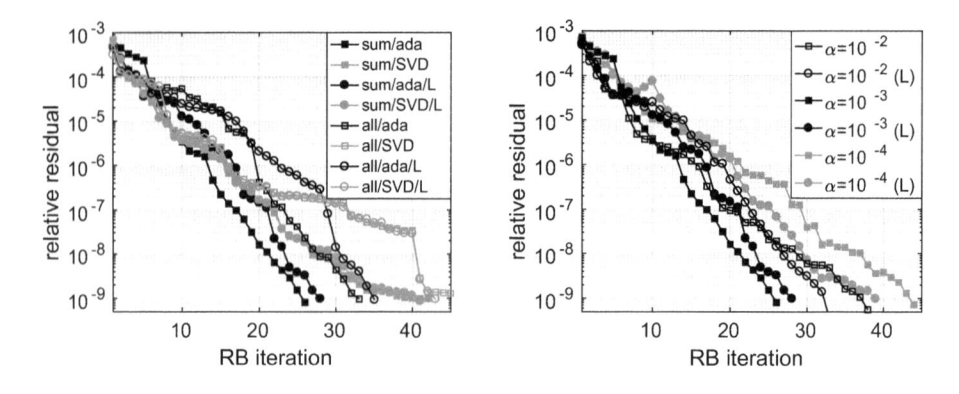

Fig. 1. Comparison of the RB solver settings (degree 4, grid level 10)

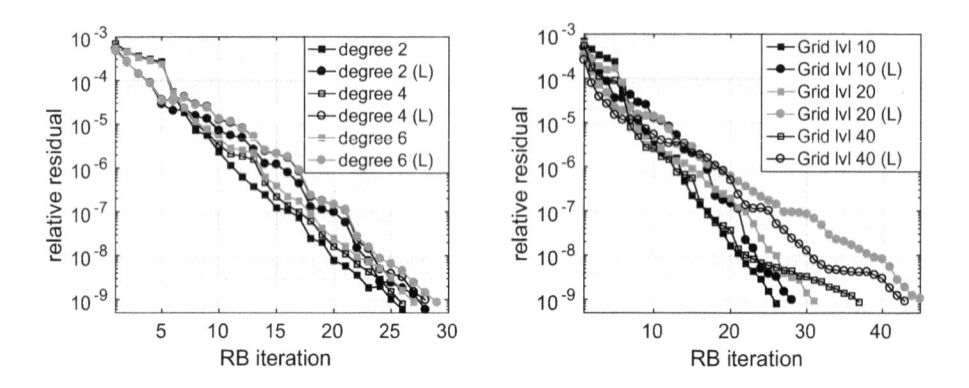

Fig. 2. Comparison of the RB constructions for different problem discretizations

5.2 Deflated CG Convergence

Here we compare the impact of the RB used as a deflation basis together with some preconditioners (Schwarz, diagonal, ichol). The Schwarz preconditioner was set using 30 subdomains = equal column slices of our square domain of the size 1/20. The incomplete Cholesky preconditioner was build with no filling allowed.

The main aim is to reduce the computational cost of the RB construction. We compare the PCG with the DCG working on the current RB. The comparison of the iterations count for both methods is in Fig. 3. In the left figure, we show the mean iterations of DCG/PCG per the RB iteration, in the right figure we show how many iterations were saved by the DCG method (for all 7 systems together). The summary of collected results is in Tab. 1. The RB solver running up to the precision 10^{-9} (higher targeted precision would lead to higher efficiency) saves more than 70% of the time spent on the PCG if the DCG is used.

Finally, in Fig. 4 the convergence of the DCG with snapshots of the RB is shown. The results are for the deterministic alternative of our problem with a random sample of Z.

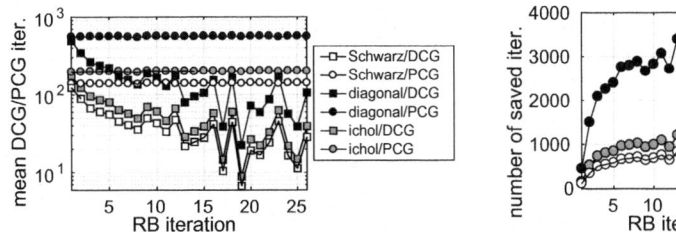

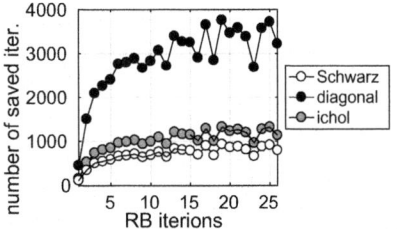

Fig. 3. Num. of the DCG iterations compared to the PCG (inside the RB solver)

Table 1. Computational savings using DCG with the RB as a deflation basis

	Additive Schwarz p.	diagonal p.	ichol (nofill) p.
Sum of saved iterations	18335	75191	26558
Savings in percents	72.32%	73.47%	73.33%

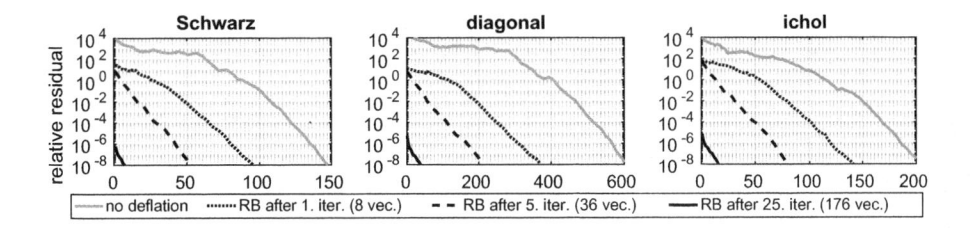

Fig. 4. Convergence of the DCG using different deflation bases

6 Conclusions

We showed that our approach without the use of the Cholesky factor performs similarly (or slightly better) in comparison to the original one from [9], see the results in Fig. 2. Additionally we proposed a cheaper alternative (Algorithm 2) to the RB expansion vector choice using SVD from [9], which performed slightly better than the SVD approach, see the results in Fig. 1.

The main contribution of this paper is the use of the DCG method with the current build of the RB as a deflation basis. With the use of the DCG method, we saved more than 70% of the computational effort during the construction of the RB. The benefit of the DCG was independent of used preconditioner, we tested the additive Schwarz, the diagonal and the incomplete Cholesky preconditioner. The solution of the reduced problem using the PCG with the Kronecker preconditioner was also effective, due to variable accuracy (10 times higher than the last relative residual of the RB solver) and almost precise initial guess based on the solution from the previous RB iteration.

We also aimed at the algorithm, where every step is not bound by memory (like Cholesky decomposition) and can be effectively parallelized. This was fully achieved e.g. with the use of additive Schwarz preconditioner, we can scale up to large supercomputer clusters. Such implementation is the aim of our future work.

Acknowledgments. This work was supported by The Ministry of Education, Youth and Sports from the National Programme of Sustainability (NPS II) project "IT4Innovations excellence in science - LQ1602". The work was also partially supported by Grant of SGS No. SP2018/68 and by Grant of SGS No. SP2018/161, VšB - Technical University of Ostrava, Czech Republic.

References

1. Additive Schwarz Preconditioners. In: Marsden, J.E., Sirovich, L., Antman, S.S. (eds.) The Mathematical Theory of Finite Element Methods, vol. 15, pp. 175–214. Springer, New York (2008). https://doi.org/10.1007/978-0-387-75934-08
2. Babuška, I., Nobile, F., Tempone, R.: A stochastic collocation method for elliptic partial differential equations with random input data. SIAM J. Numer. Anal. **45**(3), 1005–1034 (2007). https://doi.org/10.1137/050645142
3. Béreš, M., Domesová, S.: The stochastic galerkin method for darcy flow problem with log-normal random field coefficients. Adv. Electr. Electron. Eng. **15**(2) (2017). https://doi.org/10.15598/aeee.v15i2.2280
4. Hoeksema, R.J., Kitanidis, P.K.: Analysis of the spatial structure of properties of selected aquifers. Water Resour. Res. **21**(4), 563–572 (1985). https://doi.org/10.1029/WR021i004p00563
5. Khoromskij, B.N., Schwab, C.: Tensor-structured galerkin approximation of parametric and stochastic elliptic PDEs. SIAM J. Sci. Comput. **33**(1), 364–385 (2011). https://doi.org/10.1137/100785715
6. Lord, G.J., Powell, C.E., Shardlow, T.: An Introduction to Computational Stochastic PDEs, 1 edn., No. 50 in Cambridge Texts in Applied Mathematics. Cambridge University Press, New York (2014)
7. Matthies, H.G., Zander, E.: Solving stochastic systems with low-rank tensor compression. Linear Algebra Appl. **436**(10), 3819–3838 (2012). https://doi.org/10.1016/j.laa.2011.04.017
8. Nouy, A.: A generalized spectral decomposition technique to solve a class of linear stochastic partial differential equations. Comput. Methods Appl. Mech. Eng. **196**(45–48), 4521–4537 (2007). https://doi.org/10.1016/j.cma.2007.05.016
9. Powell, C.E., Silvester, D., Simoncini, V.: An efficient reduced basis solver for stochastic galerkin matrix equations. SIAM J. Sci. Comput. **39**(1), A141–A163 (2017). https://doi.org/10.1137/15M1032399
10. Saad, Y., Yeung, M., Erhel, J., Guyomarc'h, F.: A deflated version of the conjugate gradient algorithm. SIAM J. Sci. Comput. **21**(5), 1909–1926 (2000). https://doi.org/10.1137/S1064829598339761
11. Ullmann, E.: A kronecker product preconditioner for stochastic galerkin finite element discretizations. SIAM J. Sci. Comput. **32**(2), 923–946 (2010). https://doi.org/10.1137/080742853

An Investigation on Signal Comparison by Measuring of Numerical Strings Similarity

Alexander Smaglichenko[1,4(✉)], Tatyana A. Smaglichenko[2], Arkady Genkin[1,5], and Boris Melnikov[3]

[1] V.A. Trapeznikov Institute of Control Sciences, Russian Academy of Sciences,
65 Profsoyuznaya Street, Moscow 117997, Russia
losaeylin@gmail.com
[2] Research Oil and Gas Institute, Russian Academy of Sciences,
3 Gubkina Street, Moscow 119333, Russia
[3] Russian State Social University, 4, build.1,
Wilhelm Pieck Street, Moscow 129226, Russia
[4] Institute of Seismology and Geodynamics, V.I. Vernadsky Crimean Federal University, Prospekt Vernadskogo 4, Simferopol 295007, Republic of Crimea, Russia
[5] National University of Science and Technology MISIS,
Leninsky Prospect 4, Moscow 119991, Russia

Abstract. The counter algorithm has been presented to detect pairs of similar numerical strings in order to distinguish between a subset of identical signals and other signals. The pair of similar signals is determined using the matrix of the algorithm. Two elements of the matrix estimate the similarity degree in contrast to the ordinary applied a single value of correlation coefficient. The matching of signal images with the matrix elements has been made on an example of impulse signals. Using this data type we compare the outcomes of two methods: a counter based technique and the correlation method. The difference between the method proposed and the correlation method is discussed.

Keywords: Time series · Similar numerical strings · Signal processing

1 Introduction

Nowadays, science disciplines often related to the study of underground processes on oil and gas production sites [1–3] and to the investigation of natural processes with purpose of environment protection [4,5]. Passive seismic monitoring uses wave traces acquired on downhole and surface arrays [3,6]. Detection of signals from different receiver arrays is most important step for farther estimate various seismic source location. Conventional procedure for signal detecting is its comparison with a known signal template to search similarities [1,3,7]. Cross-correlation analysis is the main technique to pick arrivals of an elastic body waves, denoise and detect an event location (master event template) [8,9].

A recent alternative for adequate event identification is a migration based technique [8,10,11] that locates real events having high signal-to-noise ratio.

An important role is assigned to correlation functions in processing of microseismic noise data. The main mechanism for the formation of primary microseisms was proposed in [12] suggesting that they originate in shallow water. Primary microseisms can be observed both at the Earth surface and at ocean bottom [13,14]. Secondary microseisms have a frequency range that is twice large than the frequency range of primary microseisms. Most of the secondary microseismic noise is presented by the Rayleigh surface wave [15]. Seismic signals detected in microseism records are generated by nature sources arising from volcano activity [16], river flows [17] and other surface processes (e.g. [18]). As in the case of an elastic body waves, techniques of microseismic noise processing involve the surface wave migration algorithms besides of the correlation analysis [5].

Another scientific area, for which algorithms of signal comparison are important is the study of stochastic processes expressed as a time series. A classic approach to predicting the behavior of a process is to use the correlation characteristics of neighboring samples of signals. Based on this one can predict one or more forward steps in time [19]. In this case the stochastic process is estimated averaging the sample of signals [19,20]. However some parts of signal records may knock out from a sample because of the noise arising or measuring device issue. It becomes essential to remove such parts from a signal sequence. Otherwise, further calculation of the stochastic model parameters will be under an influence of distorted measurements.

Notes and Comments. In engineering there are many examples of tasks mentioned above. Stochastic process can be presented by signals transmitted from various transmitters to receivers. In metallurgy the detection of cracks in pipelines is performed analyzing a sequence of ultrasonic signals. The serial probes for excitation of waves and receiving signals after diffraction on the top of the crack were exploited [21].

With this paper, we present a novel approach based on the counter algorithm to compare a pair of signals. We analyze a performance of the method proposed and traditional correlation method under the same conditions of matching experiment.

2 Data Type

Figure 1 shows a several signals, which were radiated by hammer strikes imitating a small seismic energy sources. A description of the field experiment is given in [22]. The signal amplitude was recorded by accelerometers. The signal structure is determinate by conditions of propagations of seismic waves through the ground and proper work of measuring equipment. One can see that some signals are similar and nearly the same while few signals have fluctuation of amplitude values and simultaneously shifted. Selecting identical signals we denoise the structure of signals that is typical for the process.

3   Detection of Pairs of Similar Numerical Strings

In this section we shall describe bases of the counter algorithm for detecting similar numerical strings. It has been developed by the first author in [23] as a greedy algorithm.

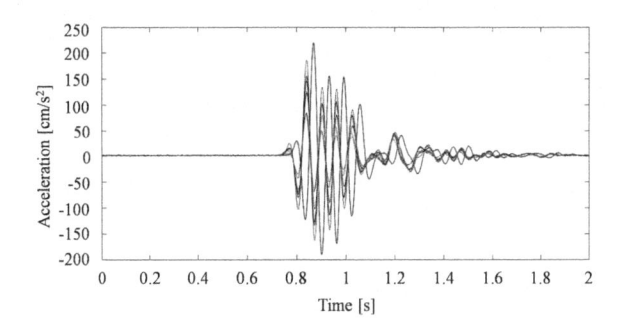

Fig. 1. A sequence of a few impulse seismic signals on the same recording scale.

The distance between numerical strings is estimated by using the matrix of the algorithm that will be defined below. Let us consider a sequence of numerical strings that form a rectangular matrix having size $m \times n$.

$$A = \begin{Bmatrix} a_{11} & a_{12} & \cdots & a_{1n} \\ a_{21} & a_{22} & \cdots & a_{2n} \\ \vdots & \vdots & a_{ij} & \vdots \\ a_{m1} & a_{m2} & \cdots & a_{mn} \end{Bmatrix} \quad (1)$$

$a(i,j)$ is element of matrix A, where i is the row index, whose values vary from 1 to m, j is the column index, whose values vary from 1 to n. We shall find the values of pairs of indices (i_1, i_2), which correspond to pairs of similar rows.

Accordantly to the algorithm each row having index i is subtracted from all rows of matrix A. The result forms matrix C_i, whose elements are expressed as:

$$c_{ij} = |a_{ij} - a_{kj}|, (i = \overline{1,m}; k = \overline{1,m}; j = \overline{1,n}), \quad (2)$$

where $|\ldots|$ denotes the absolute value.

Matrix B_i is formed by the choice of minimal element in each column of matrix C_i. Zero value is setting for other elements of matrix B_i.

The next step of the algorithm is the formation of y that is the Column-Counter (CC) for each row having index i. Values of the CC elements are defined as sums of nonzero elements in corresponding row of matrix B_i.

By using CC obtained for all rows the matrix of the algorithm is being constructed (Table 1).

4 Application of the Counter Algorithm to Identify Similar Signals

Let us consider one impulse seismic signals (Fig. 1) as a row vector, whose components are amplitude values in discrete time $\{0,1,2,\ldots,2000\}$. Consequently, each numerical string has 2000 elements. The data of 10 seismic signals form a rectangular matrix (1) having size 10×2000. By processing these data finally we get the matrix, which consists of 10 CC. An analyze of maximal values presented in Table 2 and their distribution shows that the pair of similar strings correspond to rows having the indexes 1 and 8, 2 and 3, 4 and 10, 5 and 7.

Table 1. The matrix of the algorithm

i	1	...	m
1	0	...	$\sum_{i=1} min_j$
:	:	0	:
m	$\sum_{i=m} min_j$	...	0

Figure 2 demonstrates images of identical signals that present the pair of strings having indexes 4 and 10 (Fig. 2-a); 5 and 7 (Fig. 2-b). For these pairs the elements of the matrix are equal to 170 and 169; 134 and 174.

Table 2. The algorithm matrix obtained for 10 seismic signals

i	1	2	3	4	5	6	7	8	9	10
1	0	6	14	40	20	33	30	221	34	51
2	4	0	133	9	82	10	39	4	18	24
3	5	252	0	10	25	3	24	11	10	45
4	24	9	7	0	15	1	17	15	41	169
5	11	52	18	16	0	6	174	24	21	18
6	202	10	31	26	39	0	13	330	17	87
7	21	33	18	13	134	3	0	21	36	29
8	253	8	4	25	41	48	68	0	20	27
9	12	29	4	52	28	7	62	23	0	63
10	25	14	32	170	16	7	25	8	35	0

Figure 3 demonstrates images of the different signals that present the pair of strings having indexes 1 and 3 (Fig. 3-a); 8 and 10 (Fig. 3-b). For these pairs the elements of the matrix are equal to 5 and 14; 8 and 27 consequently. One can see that small values of the matrix characterize not similar signals with different level of amplitudes in the same discrete time.

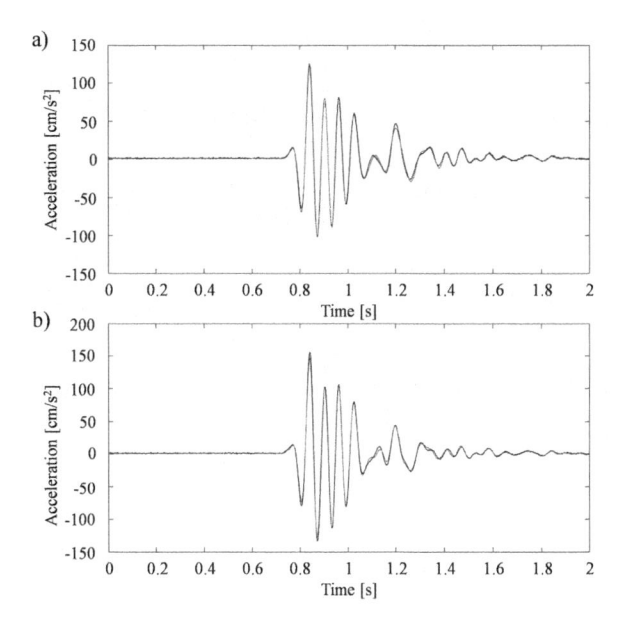

Fig. 2. Identical signals correspond to pair of rows (a) with indexes 4 and 10; (b) 5 and 7.

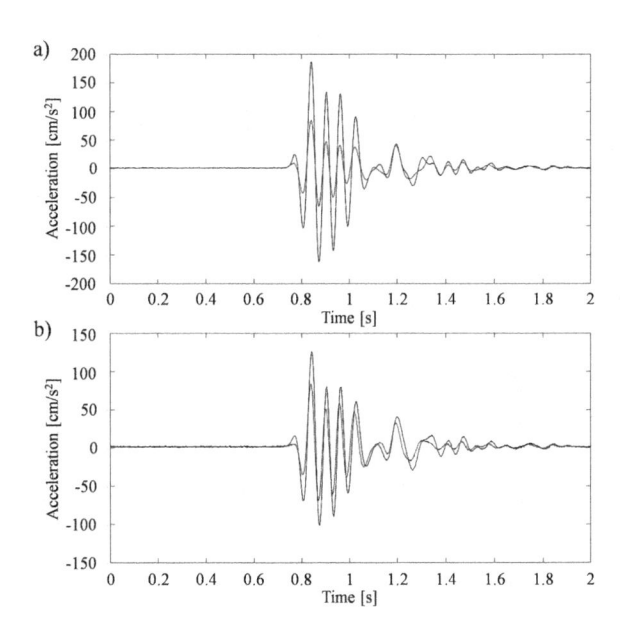

Fig. 3. Not identical signals correspond to pair of rows (a) with indexes 1 and 3; (b) 8 and 10.

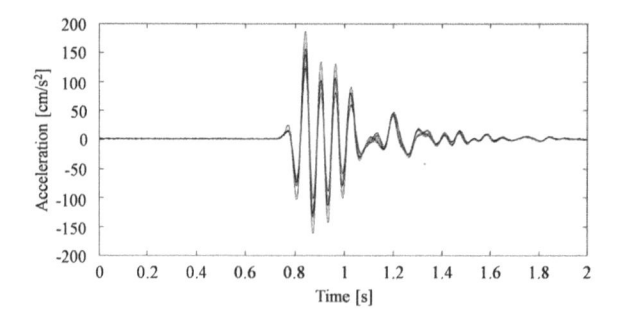

Fig. 4. The structure of the common signal that characterizes the stochastic process.

Figure 4 display the image of a few signals, which were identified by the method as nearly the same. One can see that these signals are similar. Noisy parts from a signal sequence, which is presented in Fig. 1 are absent. Thus we got the signal subsequence that describes the given stochastic process. A model of the structure of the common signal characterizing the stochastic process was improved by identification of stability parts in an initial sequence of signal records.

Notes and Comments. Cross-correlation function is normally used in order to compare two signals [24]. The correlation coefficient is a single value that estimates a similarity. We demonstrated that the matrix of the counter algorithm is able to provide two values, which can be used to characterize a similarity.

5   Comparison of the Proposed Technique with the Correlation Method on the Base the Same Data Set

The experiment conducted in a previous section was extended by comparing elements of the matrix (see Table 2) with the correlation coefficient value conventionally applied in the signal processing. We calculated this coefficient using the standard function "corrcoef" in Matlab [25] (see Table 3).

Table 3. A correlation coefficient value for 10 signals being investigated

i	1	2	3	4	5	6	7	8	9	10
1	1	0.9489	0.9362	0.9599	0.9480	−0.7436	0.9514	0.9404	0.9595	0.9541
2	0.9489	1	0.9966	0.9894	0.9949	−0.8290	0.9941	0.9319	0.9935	0.9876
3	0.9362	0.9966	1	0.9857	0.9940	−0.8376	0.9907	0.9173	0.9922	0.9860
4	0.9599	0.9894	0.9857	1	0.9849	−0.7819	0.9804	0.9103	0.9938	0.9966
5	0.9480	0.9949	0.9940	0.9849	1	−0.8339	0.9968	0.9417	0.9922	0.9819
6	−0.7436	−0.8290	−0.8376	−0.7819	−0.8339	1	−0.8559	−0.8530	−0.8151	−0.7629
7	0.9514	0.9941	0.9907	0.9804	0.9968	−0.8559	1	0.9561	0.9909	0.9743
8	0.9404	0.9319	0.9173	0.9103	0.9417	−0.8530	0.9561	1	0.9232	0.8949
9	0.9595	0.9935	0.9922	0.9938	0.9922	−0.8151	0.9909	0.9232	1	0.9907
10	0.9541	0.9876	0.9860	0.9966	0.9819	−0.7629	0.9743	0.8949	0.9907	1

For each pair of signals a correlation coefficient values has been colated with two values of the matrix of the algorithm proposed. Identical signals found by measuring numerical strings similarity are characterized having maximum correlation coefficient. This confirms an adequacy of the counter algorithm considered. At the same time, it is necessary to note the next. There is a set of signal pairs, for which the correlation coefficient is very high (more than 0.99). Some of these signals were not selected by the counter algorithm because it determined small values of elements of the matrix for them. Figure 5-b demonstrates an example when a correlation coefficient for the pair of signals 4 and 9 is equal to 0.9938 while the elements of the matrix for the same pair are equal to 41 and 52. However visual comparison the similarity shows that the signals correlate not well enough because there are differences in amplitudes for some parts of the signals. Figure 5-a shows the example of good similarity that was detected by both methods.

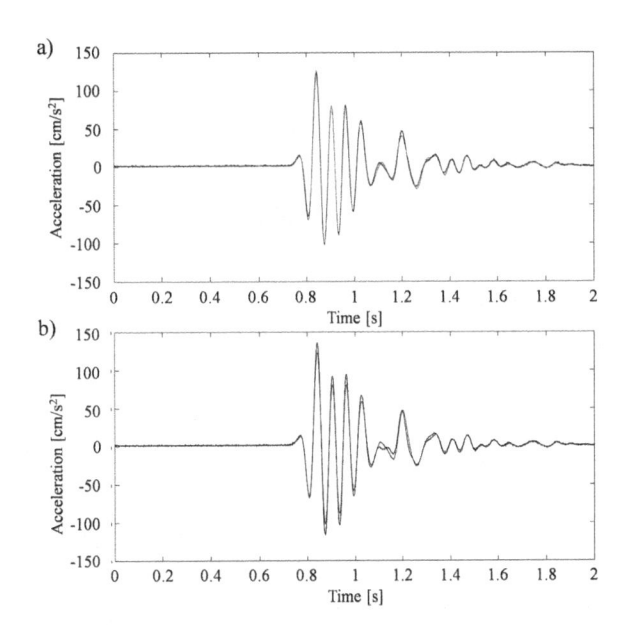

Fig. 5. Signals having the high value of correlation coefficient (a) for indexes 4 and 10; (b) 4 and 9.

Analyzing the results one can see that in general the outcome of the counter algorithm coincides with solution of the correlation method. At the same time, the CC parameters of the similarity of a pair of signals (two elements of the matrix) clearly denote the difference in signals.

6 Conclusion

In this paper we have investigated the approach for comparing of signals by application of the counter algorithm measuring the similarity of numerical strings. Distinctive feature of proposed method is in that it is able to detect the group of pairs of the most similar signals. Comparison with the correlation method shows that low values of the CC parameters indicating the weak similarity of pair of signals can correspond to high values of correlation coefficient. A visual analysis confirms this contradiction. We suggest that the reason for the problem arising from the correlation estimate nonsensitivity is likely in rounding-off error, which is connected with arithmetical operations under calculation of final numerical values of the coefficients. The counter algorithm is characterized by simple arithmetical operations (subtraction and addition) while the formula for the correlation coefficient calculation contains a set of all arithmetical operations with their repetition. At any rate in the future we intend to make theoretical study of this issue. From the other hand we can not claim that the method developed is better than the correlation analyze because there is not enough theoretical proofs. A possibility of a practical application of the method confirmed its efficiency for selecting a group of similar spectra obtained during field observations [23], for separating the noise component of the signal on the example of local earthquakes in Iceland [26].

Acknowledgments. We thank anonymous reviewers for constructive critics that helped to improve the initial version of the paper.

The work was carried out within the framework of the state projects No. 0139-2019-0009, No. 10.331-17, No. 5.6370.2017/BCh.

References

1. Caffagni, E., Eaton, D.W., Jones, J.P., van der Baan, M.: Detection and analysis of microseismic events using a matched filtering algorithm (MFA). Geophys. J. Int. **206**(1), 644–658 (2016)
2. Shtun, S.Y., Golenkin, M.Y., Shtun, A.S., Shabalinskaya, D.D., Cheprasov, A.V., Kuzakov, V.R., Brichikova, M.P., Zolotoi, N.V.: New approach to offshore field development in russia: ultra deep LWD measurements for accurate 3D reservoir model update. Soc. Pet. Eng. (2017). https://doi.org/10.2118/187900-MS
3. Keranen, K.M., Weingarten, M.: Induced seismicity. Annu. Rev. Earth Planet. Sci. **46**, 149–174 (2018). https://doi.org/10.1146/annurev-earth-082517-010054
4. Larose, E., Carrière, S., Voisin, C., Bottelin, P., Baillet, L., Guéguen, P., Walter, F., Jongmans, D., Guillier, B., Garambois, S., Gimbert, F., Massey, C.: Environmental seismology: what can we learn on earth surface processes with ambient noise? J. Appl. Geophys. **116**, 62–74 (2015). https://doi.org/10.1016/j.jappgeo.2015.02.001

5. Dietze, M.: The R package "eseis" - a software toolbox for environmental seismology. Surf. Dynam. **6**, 669–686 (2018). https://doi.org/10.5194/esurf-6-669-2018
6. Eisner, L., Hulsey, B.J., Duncan, P., Jurick, D., Werner, H., Keller, W.: Comparison of surface and borehole locations of induced seismicity. Geophy. Prospect. **58**, 809–820 (2010). https://doi.org/10.1111/j.1365-2478.2010.00867.x
7. Kapetanidis, V., Papadimitriou, P.: Estimation of arrival-times in intense seismic sequences using a Master-Events methodology based on waveform similarity. Geophys. J. Int. **187**, 889–917 (2011). https://doi.org/10.1111/j.1365-246X.2011.05178.x
8. Cieplicki, R., Eisner, L., Mueller, M.: Microseismic event detection: comparing P-wave migration with P- and S-wave crosscorrelation. In: SEG Denver 2014 Annual Meeting, pp. 2168–2172 (2014). https://doi.org/10.1190/segam2014-1614.1
9. Akram, J., Eaton, D.W.: A review and appraisal of arrival-time picking methods for downhole microseismic data. Geophysics **81**(2), 71–91 (2016). https://doi.org/10.1190/GEO2014-0500.1
10. Anikiev, D., Valenta, J., Stanek, F., Eisner, L.: Joint location and source mechanism inversion of microseismic events: benchmarking on seismicity induced by hydraulic fracturing. Geophys. J. Int. **198**, 249–258 (2014). https://doi.org/10.1093/gji/ggu126
11. Stanek, F., Anikiev, D., Valenta, J., Eisner, L.: Semblance for microseismic event detection. Geophys. J. Int. **201**, 1362–1369 (2015). https://doi.org/10.1093/gji/ggv070
12. Hasselman, K.: Statistical analysis of generation of microseisms. Reverend Geophys. **1**(2), 177–210 (1963)
13. Oliver, J.: Worldwide, storm of microseism from the period of about 27 seconds. Bull. Seism. Soc. **52**, 307–517 (1963)
14. Barstow, N., et al.: Particle motion and pressure relationship of the ocean bottom at 3900 m depth: 0.003 to 5 Hz. Geophys. Res. Lett. **16**, 1185–1188 (1989)
15. Shapiro, N.M., et al.: High-resolution surface-wave tomography from ambient seismic noise. Science **307**(5715), 1615–1618 (2005)
16. Schöpa, A., Chao, W.A., Lipovsky, B., Hovius, N., White, R.S., Green, R.G., Turowski, J.M.: Dynamics of the Askja caldera July 2014 landslide, Iceland, from seismic signal analysis: precursor, motion and aftermath. Earth Surf. Dynam. **6**, 467–485 (2018). https://doi.org/10.5194/esurf-6-467-2018
17. Gimbert, F., Tsai, V.C., Lamb, M.P.: A physical model for seismic noise generation by turbulent flow in rivers. J. Geophys. Res. **119**, 2209–2238 (2014). https://doi.org/10.1002/2014JF003201
18. Sens-Schoenfelder, C., Larose, E.: Temporal changes in the lunar soil from correlation of diffuse vibrations. Phys. Rev. E. **78**, 045601 (2008). https://doi.org/10.1103/PhysRevE.78.045601
19. Box, G.E.P., Jenkins, G.M.: Time Series Analysis. Forecasting and Control. Holden-Day, San Francisco (1970)
20. Kashyap, R.L., Rao, A.R.: Dynamic Stochastic Models from Empirical Data. Academic Press. N.Y., San Francisco (1976)
21. Krautkramer, I., Krautkramer, H.: Werkstoffprufunq mit ultraschall. Sprinqer Verlaq, New York (1975)

22. Smaglichenko, A.V., Sayankina, M.K., Smaglichenko, T.A., Volodin, I.A.: Physical experiments and stochastic modeling to clarify the system containing the seismic source and the ground. In: ISCS 2014 International Symposium of Complex Systems, vol. 14, pp. 125–135 (2015)
23. Smaglichenko, A.V., Smaglichenko, T.A., Sayankina, M.K.: An approach to developing greedy algorithms of picking undistorted data in the tasks of seismic exploration. Int. Sci. J. Appl. Discret. Math. Heuristic Algorithms **1**(1), 42–51 (2015). Samara University Press
24. Bendat, J.S., Piersol, A.G.: Engineering applications of correlation and spectral analysis. Wiley-Interscience, New York (1980)
25. Matlab copyright. Version 8.6.0.267246 (R2015b) (2015)
26. Smaglichenko, A.V., Bjarnason, I.Th.: Consecutive Analysis based on the branch and bound method applied to picking P- and S- wave arrival times. Abstract in Materials of "The Science Day of the School of Engineering and Natural Science of the University of Iceland" (2015)

Optimization

A Lightweight SHADE-Based Algorithm for Global Optimization - liteSHADE

Adam Viktorin[✉], Roman Senkerik, Michal Pluhacek, Tomas Kadavy, and Roman Jasek

Faculty of Applied Informatics, Tomas Bata University in Zlin,
T. G. Masaryka 5555, 760 01 Zlin, Czech Republic
{aviktorin,senkerik,pluhacek,kadavy,jasek}@utb.cz

Abstract. In this paper, a novel lightweight version of the Successful-History based Adaptive Differential Evolution (SHADE) is presented as the first step towards a simple, user-friendly, metaheuristic algorithm for global optimization. This simplified algorithm is called liteSHADE and is compared to the original SHADE on the CEC2015 benchmark set in three dimensional settings – $10D$, $30D$ and $50D$. The results support the idea, that simplification may lead to a successful and understandable algorithm with competitive or even better performance.

Keywords: Differential Evolution · SHADE · Simplification · liteSHADE

1 Introduction

The Differential Evolution algorithm (DE) was firstly described by Storn and Price in 1995 [1] and since then, it has been thoroughly studied by a large number of scientists around the world. Selected examples of studies in this field might be found in one of the recent surveys [2–4].

The DE is primarily an algorithm for global optimization in continuous spaces and is based, as all other evolutionary computational techniques, on the Darwinian theory of evolution. It is also very simple and requires a setting of only three control parameters – population size NP, crossover rate CR and scaling factor F. While the number of control parameters is quite small, the influence of their setting to the performance of the algorithm is quite significant and therefore, they require fine-tuning [5, 6]. In order to avoid that task, recent DE variants incorporate self-adaptation of some of these parameters. The self-adaptation is based on the progress of optimisation of a given problem and therefore, it mostly adapts CR and F values, whereas the population size is not affected. Probably a most important recent self-adaptive DE variant is called Success-History based Differential Evolution (SHADE) and it was created by Tanabe and Fukunaga in 2013 [7]. The latest CEC single-objective competition winners were all based on this algorithm – 2014 L-SHADE [8], 2015 SPS-L-SHADE-EIG [9], 2016 LSHADE_EpSin [10] and 2017 jSO [11]. Thus, it is selected as a basis for a simplification in this paper.

According to a recent paper [12], some of the SHADE-based algorithms use unnecessarily complicated mechanisms and should be simplified to made them more clear to their users and might be even improved by such simplification. In this paper, a lightweight SHADE variant (liteSHADE) is presented and it shows some interesting future directions for the simplification of DE-based self-adaptive algorithms. It is shown on the CEC2015 benchmark set, that its performance is competitive with the original SHADE and that it can even provide better results in higher dimensions. The liteSHADE algorithm also uses an updated version of a previously published distance based parameter adaptation [13], which was shown to be quite effective for premature convergence avoidance.

The rest of the paper is structured as follows: The next section describes DE, SHADE and liteSHADE algorithms. Section 3 covers experimental settings, Sect. 4 provides results and their discussion and the whole paper is concluded by Sect. 5.

2 From DE to liteSHADE

In order to describe the liteSHADE, it is important to start from the DE by Storn and Price [1]. The canonical 1995 DE is based on the idea of evolution from a randomly generated set of solutions of the optimization task called population P, which has a preset size of NP. Each individual (solution) in the population consists of a vector x of length D (each vector component corresponds to one attribute of the optimized task) and objective function value $f(x)$, which mirrors the quality of the solution. The number of optimized attributes D is often referred to as the dimensionality of the problem and such generated population P, represent the first generation of solutions.

The individuals in the population are combined in an evolutionary manner to create improved offspring for the next generation. This process is repeated until the stopping criterion is met (either the maximum number of generations, or the maximum number of objective function evaluations, or the population diversity lower limit, or overall computational time), creating a chain of subsequent generations, where each following generation consists of better solutions than those in previous generations – a phenomenon called elitism.

The combination of individuals in the population consists of three main steps: Mutation, crossover and selection.

In the mutation, attribute vectors of selected individuals are combined in simple vector operations to produce a mutated vector v. This operation uses a control parameter – scaling factor F. In the crossover step, a trial vector u is created by selection of attributes either from mutated vector v or the original vector x based on the crossover probability given by a control parameter – crossover rate CR. And finally, in the selection, the quality $f(u)$ of a trial vector is evaluated by an objective function and compared to the quality $f(x)$ of the original vector and the better one is placed into the next generation.

From the basic description of the DE algorithm, it can be seen, that there are three control parameters, which have to be set by the user – population size NP, scaling factor F and crossover rate CR. It was shown in [5, 6], that the setting of these parameters is crucial for the performance of DE. Fine-tuning of the control parameter values is a time-consuming task and therefore, many state-of-the-art DE variants use self-adaptation to avoid this cumbersome task. This is also a case of SHADE algorithm proposed by Tanabe and Fukunaga in 2013 [7] and since it is used in this paper, the algorithm is described in more detail in the next section.

2.1 Shade

As aforementioned, SHADE algorithm was proposed with a self-adaptive mechanism of some of its control parameters to avoid their fine-tuning. Control parameters in question are scaling factor F and crossover rate CR. It is fair to mention, that SHADE algorithm is based on Zhang and Sanderson's JADE [14] and shares a lot of its mechanisms. The main difference is in the historical memories M_F and M_{CR} for successful scaling factor and crossover rate values with their update mechanism.

Following subsections describe individual steps of the SHADE algorithm: Initialization, mutation, crossover, selection and historical memory update.

Initialization

The initial population P is generated randomly and for that matter, a Pseudo-Random Number Generator (PRNG) with uniform distribution is used. Solution vectors x are generated according to the limits of solution space – *lower* and *upper* bounds Eq. (1).

$$x_{j,i} = U[lower_j, upper_j] \text{ for } j = 1,\ldots,D; i = 1,\ldots,NP, \quad (1)$$

where i is the individual index and j is the attribute index. The dimensionality of the problem is represented by D, and NP stands for the population size.

Historical memories are preset to contain only 0.5 values for both, scaling factor and crossover rate parameters Eq. (2).

$$M_{CR,i} = M_{F,i} = 0.5 \text{ for } i = 1,\ldots,H, \quad (2)$$

where H is a user-defined size of historical memories.

Also, the external archive of inferior solutions A has to be initialized. Because of no previous inferior solutions, it is initialized empty, $A = \emptyset$. And index k for historical memory updates is initialized to 1.

The following steps are repeated over the generations until the stopping criterion is met.

Mutation

Mutation strategy "current-to-*p*best/1" was introduced in [14] and it combines four mutually different vectors in creation of the mutated vector v. Therefore, $x_{pbest} \neq x_{r1} \neq x_{r2} \neq x_i$ Eq. (3).

$$v_i = x_i + F_i(x_{pbest} - x_i) + F_i(x_{r1} - x_{r2}), \quad (3)$$

where x_{pbest} is randomly selected individual from the best $NP \times p$ individuals in the current population. The p value is randomly generated for each mutation by PRNG with uniform distribution from the range [p_{min}, 0.2] and $p_{min} = 2/NP$. Vector x_{r1} is randomly selected from the current population P. Vector x_{r2} is randomly selected from the union of the current population P and external archive A. The scaling factor value F_i is given by Eq. (4).

$$F_i = C[M_{F,r}, 0.1], \qquad (4)$$

where $M_{F,r}$ is a randomly selected value (index r is generated by PRNG from the range 1 to H) from M_F memory and C stands for Cauchy distribution. Therefore the F_i value is generated from the Cauchy distribution with location parameter value $M_{F,r}$ and scale parameter value of 0.1. If the generated value F_i higher than 1, it is truncated to 1 and if it is F_i less or equal to 0, it is generated again by Eq. (4).

Crossover

In the crossover step, trial vector u is created from the mutated v and original x vectors. For each vector component, a PRNG with uniform distribution is used to generate a random value. If this random value is less or equal to given crossover rate value CR_i, current vector component will be taken from a trial vector. Otherwise, it will be taken from the original vector Eq. (5). There is also a safety measure, which ensures, that at least one vector component will be taken from the trial vector. This is given by a randomly generated component index j_{rand}.

$$u_{j,i} = \begin{cases} v_{j,i} & \text{if } U[0,1] \leq CR_i \text{ or } j = j_{rand} \\ x_{j,i} & \text{otherwise} \end{cases}. \qquad (5)$$

The crossover rate value CR_i is generated from a Gaussian distribution with a mean parameter value $M_{CR,r}$ selected from the crossover rate historical memory M_{CR} by the same index r as in the scaling factor case and standard deviation value of 0.1 Eq. (6).

$$CR_i = N[M_{CR,r}, 0.1]. \qquad (6)$$

When the generated CR_i value is less than 0, it is replaced by 0 and when it is greater than 1, it is replaced by 1.

Selection

The selection step ensures that the optimization will progress towards better solutions because it allows only individuals of better or at least equal objective function value to proceed into the next generation $G + 1$ Eq. (7).

$$x_{i,G+1} = \begin{cases} u_{i,G} & \text{if } f(u_{i,G}) \leq f(x_{i,G}) \\ x_{i,G} & \text{otherwise} \end{cases}, \qquad (7)$$

where G is the index of the current generation.

Historical Memory Updates

Historical memories M_F and M_{CR} are initialized according to Eq. (2), but their components change during the evolution. These memories serve to hold successful values of F and CR used in mutation and crossover steps. Successful regarding producing trial individual better than the original individual. During every single generation, these successful values are stored in their corresponding arrays S_F and S_{CR}. After each generation, one cell of M_F and M_{CR} memories is updated. This cell is given by the index k, which starts at 1 and increases by 1 after each generation. When it overflows the memory size H, it is reset to 1. The new value of k-th cell for M_F is calculated by Eq. (8) and for M_{CR} by Eq. (9).

$$M_{F,k} = \begin{cases} \text{mean}_{WL}(S_F) & \text{if } S_F \neq \emptyset \\ M_{F,k} & \text{otherwise} \end{cases}, \quad (8)$$

$$M_{CR,k} = \begin{cases} \text{mean}_{WL}(S_{CR}) & \text{if } S_{CR} \neq \emptyset \\ M_{CR,k} & \text{otherwise} \end{cases}, \quad (9)$$

where $\text{mean}_{WL}()$ stands for weighted Lehmer mean Eq. (10).

$$\text{mean}_{WL}(S) = \frac{\sum_{k=1}^{|S|} w_k \cdot S_k^2}{\sum_{k=1}^{|S|} w_k \cdot S_k}, \quad (10)$$

where the weight vector w is given by Eq. (11) and is based on the improvement in objective function value between trial and original individuals in current generation G.

$$w_k = \frac{\text{abs}(f(u_{k,G}) - f(x_{k,G}))}{\sum_{m=1}^{|S_{CR}|} \text{abs}(f(u_{m,G}) - f(x_{m,G}))}. \quad (11)$$

And since both arrays S_F and S_{CR} have the same size, it is arbitrary which size will be used for the upper boundary for m in Eq. (11).

2.2 liteSHADE

The lightweight SHADE variant is based on the previous experiments with the SHADE algorithm and represents a direct approach to some of the unnecessary steps in the original SHADE design. For example, the external archive in the mutation step is not used, and the reason for that came from [15], where it was shown, that there is no direct impact to the performance of the algorithm when the archive is not used. Also, an updated version of the distance based parameter adaptation [13] is used for the update of memory values of F and CR. The list of all changes to the SHADE algorithm is following:

- No archive A is used.
- The p in mutation (Eq. (3)) is no longer generated randomly, but it is set to 10% of the population size, $p = 0.1 * NP$.
- The sizes of historical memories of F and CR values (M_F, M_{CR}) are set to 1, $H = 1$. Therefore, M_F and M_{CR} are no longer vectors, but scalars M_F and M_{CR}
- Historical memories store values of F and CR, which moved the individual furthest (Euclidean distance) in the search space during current generation – updated distance based approach independent of the objective function value improvement.

- These memories (M_F, M_{CR}) are initialized to 0.8 instead of 0.5.
- When the F_i and CR_i values for mutation and crossover (Eqs. (4) and (6)) are generated outside of the predefined range, they are generated again to avoid peaks in boundary values (1 for F and 0 and 1 for CR).

```
Algorithm pseudo-code 1: liteSHADE
1. Set NP and stopping criterion;
2. G = 0, x_best = {}, p_i = 0.1*NP;
3. Randomly initialize (Eq. (1)) population P = (x_1,G,...,x_NP,G);
4. M_F = M_CR = 0.8;
5. P_new = {}, x_best = best from population P;
6. while stopping criterion not met do
7.    for i = 1 to NP do
8.       x_i,G = P[i];
9.       Set F_i by Eq. (4) and CR_i by Eq. (6);
10.         v_i,G by mutation Eq. (3);
11.         u_i,G by crossover Eq. (5);
12.         if f(u_i,G) < f(x_i,G) then
13.            x_i,G+1 = u_i,G;
14.            if distance between u_i,G and x_i,G is the biggest in the
                  current generation then F_i → M_F, CR_i → M_CR;
15.         else
16.            x_i,G+1 = x_i,G;
17.         end
18.         x_i,G+1 → P_new;
19.    end
20.    P = P_new, P_new = {}, x_best = best from population P, G++;
21. end
22. return x_best as the best found solution;
```

3 Experimental Settings

Both algorithms were tested on the CEC 2015 benchmark set of 15 test functions (2 unimodal, 3 simple multimodal, 3 hybrid and 6 composition functions) with accordance to the benchmark requirements. Also the time complexity of both algorithms was measured.

3.1 SHADE and liteSHADE Settings

In order to provide the most comparable results, both algorithms had the same setting of control and other parameters:

- Population size $NP = 100$,
- historical memory size $H = 10$ – SHADE only,

- external archive size $|A| = NP$ – SHADE only,
- dimensionality of problems $D = \{10, 30, 50\}$,
- stopping criterion – maximum number of objective function evaluations $MAXFES = 10{,}000 \times D$,
- number of runs $runs = 51$.

4 Results and Discussion

This section provides the results of both algorithms (SHADE and liteSHADE) on the CEC2015 benchmark set in three different dimensional settings – $10D$ (Table 1), $30D$ (Table 2) and $50D$ (Table 3). These tables provide a basic statistical comparison of the median and mean values over the 51 independent runs on each function from the benchmark set and also a result of the Wilcoxon rank-sum test with the significance level set to 5%. When there is no significant difference in the results between both algorithms on a given function, there is an "=" sign in the last column. When the SHADE algorithm performs better, there is a "–" sign and when the liteSHADE performs better, there is a "+" sign.

Table 1. SHADE vs. liteSHADE on CEC2015 in $10D$.

f	SHADE		liteSHADE		Result
	Median	Mean	Median	Mean	
1	0.00E+00	0.00E+00	0.00E+00	0.00E+00	=
2	0.00E+00	0.00E+00	0.00E+00	0.00E+00	=
3	2.00E+01	1.89E+01	2.01E+01	1.88E+01	–
4	3.07E+00	2.97E+00	3.09E+00	3.13E+00	=
5	2.21E+01	3.42E+01	4.26E+01	5.60E+01	–
6	2.20E–01	2.97E+00	3.36E+00	3.99E+00	–
7	1.67E–01	1.88E–01	1.39E–01	1.70E–01	+
8	8.15E–02	2.69E–01	4.95E–01	4.91E–01	–
9	1.00E+02	1.00E+02	1.00E+02	1.00E+02	=
10	2.17E+02	2.17E+02	2.17E+02	2.17E+02	–
11	3.00E+02	1.66E+02	3.00E+02	2.30E+02	=
12	1.01E+02	1.01E+02	1.01E+02	1.01E+02	–
13	2.78E+01	2.78E+01	2.85E+01	2.84E+01	–
14	2.94E+03	4.28E+03	6.68E+03	4.85E+03	–
15	1.00E+02	1.00E+02	1.00E+02	1.00E+02	=

It can be seen in Table 1, that in lower dimensional setting, the SHADE algorithm performs better over the whole benchmark set (8 wins, 1 lose and 6 draws), which was predictable, since the simplification of the liteSHADE algorithm provides more explorative power than exploitative. This is confirmed in the case of $30D$ and

Table 2. SHADE vs. liteSHADE on CEC2015 in 30D.

f	SHADE		liteSHADE		Result
	Median	Mean	Median	Mean	
1	3.73E+01	2.62E+02	3.85E+02	1.37E+03	−
2	0.00E+00	0.00E+00	0.00E+00	0.00E+00	=
3	2.01E+01	2.01E+01	2.03E+01	2.03E+01	−
4	1.41E+01	1.41E+01	2.11E+01	2.21E+01	−
5	1.55E+03	1.50E+03	2.03E+03	2.03E+03	−
6	5.36E+02	5.73E+02	3.31E+02	3.38E+02	+
7	7.17E+00	7.26E+00	6.78E+00	6.67E+00	+
8	1.26E+02	1.21E+02	8.10E+01	9.52E+01	+
9	1.03E+02	1.03E+02	1.03E+02	1.03E+02	+
10	6.27E+02	6.22E+02	4.33E+02	4.57E+02	+
11	4.53E+02	4.50E+02	4.43E+02	4.28E+02	+
12	1.05E+02	1.05E+02	1.05E+02	1.05E+02	−
13	9.52E+01	9.50E+01	1.01E+02	1.00E+02	−
14	3.21E+04	3.24E+04	3.31E+04	3.25E+04	=
15	1.00E+02	1.00E+02	1.00E+02	1.00E+02	=

Table 3. SHADE vs. liteSHADE on CEC2015 in 50D.

f	SHADE		liteSHADE		Result
	Median	Mean	Median	Mean	
1	1.81E+04	2.14E+04	3.85E+04	4.88E+04	−
2	0.00E+00	0.00E+00	0.00E+00	0.00E+00	=
3	2.01E+01	2.01E+01	2.05E+01	2.05E+01	−
4	3.84E+01	3.92E+01	5.56E+01	5.65E+01	−
5	3.10E+03	3.09E+03	4.38E+03	4.42E+03	−
6	2.87E+03	3.56E+03	2.28E+03	5.90E+03	=
7	4.22E+01	4.25E+01	4.17E+01	4.36E+01	=
8	1.13E+03	1.12E+03	7.35E+02	7.99E+02	+
9	1.06E+02	1.06E+02	1.04E+02	1.04E+02	+
10	1.57E+03	1.59E+03	1.16E+03	1.15E+03	+
11	6.76E+02	6.81E+02	4.77E+02	4.87E+02	+
12	1.08E+02	1.08E+02	1.08E+02	1.08E+02	−
13	1.80E+02	1.80E+02	1.91E+02	1.91E+02	−
14	7.29E+04	6.66E+04	5.92E+04	6.10E+04	+
15	1.00E+02	1.00E+02	1.00E+02	1.00E+02	=

50D problems, where the situation changes and the liteSHADE algorithm can provide better results mostly on the hybrid and composition functions. In 30D the score from the SHADE point of view is 6 wins, 6 loses and 3 draws, and in 50D similarly 6 wins, 5 loses and 4 draws.

These findings support the presumption that the simplified algorithm (liteSHADE) can provide different and competitive results to the original algorithm with higher memory demands (SHADE).

The time complexity measured according to the CEC2015 benchmark set is displayed in Tables 4 and 5, and it can be seen, that the liteSHADE algorithm requires slightly more time to compute. This is most probably caused by the Euclidean distance computation for each individual that improved in the generation. Lowering these time requirements is a subject of future studies in this direction.

Table 4. Time complexity – SHADE.

D	T0	T1	T'2	(T'2−T1)/T0
10	254	422	12753.2	48.5
30		1556	15031.6	53.1
50		2902	18892.4	63.0

Table 5. Time complexity – liteSHADE.

D	T0	T1	T'2	(T'2 − T1)/T0
10	254	422	14275.2	54.5
30		1556	17378.0	62.3
50		2902	20145.6	67.9

5 Conclusion

In this paper, it was shown, that a simplified variant of the SHADE algorithm can provide interesting and competitive results, mostly in higher dimensional settings. The simplified algorithm was coined as liteSHADE. A thorough analysis of its performance and exploration/exploitation abilities will be a next step in the further development.

The goal is to provide a simple, user-friendly, metaheuristic algorithm for global optimization, which would not incorporate complicated mechanisms that introduce new artificial parameters, which should be tuned to the specified problem. The proposed liteSHADE algorithm should be one of the first steps towards reaching this goal.

Acknowledgments. This work was supported by the Ministry of Education, Youth and Sports of the Czech Republic within the National Sustainability Programme Project no. LO1303 (MSMT-7778/2014), further by the European Regional Development Fund under the Project CEBIA-Tech no. CZ.1.05/2.1.00/03.0089 and by Internal Grant Agency of Tomas Bata University under the Projects no. IGA/CebiaTech/2019/002. This work is also based upon support by COST (European Cooperation in Science & Technology) under Action CA15140,

Improving Applicability of Nature-Inspired Optimisation by Joining Theory and Practice (ImAppNIO), and Action IC1406, High-Performance Modelling and Simulation for Big Data Applications (cHiPSet). The work was further supported by resources of A.I.Lab at the Faculty of Applied Informatics, Tomas Bata University in Zlin (ailab.fai.utb.cz).

References

1. Storn, R., Price, K.: Differential Evolution - A Simple and Efficient Adaptive Scheme for Global Optimization Over Continuous Spaces, vol. 3. ICSI, Berkeley (1995)
2. Neri, F., Tirronen, V.: Recent advances in differential evolution: a survey and experimental analysis. Artif. Intell. Rev. **33**(1–2), 61–106 (2010)
3. Das, S., Suganthan, P.N.: Differential evolution: a survey of the state-of-the-art. IEEE Trans. Evol. Comput. **15**(1), 4–31 (2011)
4. Das, S., Mullick, S.S., Suganthan, P.N.: Recent advances in differential evolution–an updated survey. Swarm Evol. Comput. **27**, 1–30 (2016)
5. Gämperle, R., Müller, S.D., Koumoutsakos, P.: A parameter study for differential evolution. Adv. Intell. Syst. Fuzzy Syst. Evol. Comput. **10**, 293–298 (2002)
6. Liu, J., Lampinen, J.: On setting the control parameter of the differential evolution method. In: Proceedings of the 8th International Conference on Soft Computing (MENDEL 2002), pp. 11–18 (2002)
7. Tanabe, R., Fukunaga, A.: Success-history based parameter adaptation for differential evolution. In: IEEE Congress on Evolutionary Computation (CEC), pp. 71–78. IEEE, June 2013
8. Tanabe, R., Fukunaga, A.S.: Improving the search performance of SHADE using linear population size reduction. In: IEEE Congress on Evolutionary Computation (CEC), pp. 1658–1665. IEEE, July 2014
9. Guo, S.M., Tsai, J.S.H., Yang, C.C., Hsu, P.H.: A self-optimization approach for L-SHADE incorporated with eigenvector-based crossover and successful-parent-selecting framework on CEC 2015 benchmark set. In: IEEE Congress on Evolutionary Computation (CEC), pp. 1003–1010. IEEE, May 2015
10. Awad, N.H., Ali, M.Z., Suganthan, P.N., Reynolds, R.G.: An ensemble sinusoidal parameter adaptation incorporated with L-SHADE for solving CEC2014 benchmark problems. In: IEEE Congress on Evolutionary Computation (CEC), pp. 2958–2965. IEEE, July 2016
11. Brest, J., Maučec, M.S., Bošković, B.: Single objective real-parameter optimization: algorithm jSO. In: IEEE Congress on Evolutionary Computation (CEC), pp. 1311–1318. IEEE, June 2017
12. Piotrowski, A.P., Napiorkowski, J.J.: Some metaheuristics should be simplified. Inf. Sci. **427**, 32–62 (2018)
13. Viktorin, A., Senkerik, R., Pluhacek, M., Kadavy, T., Zamuda, A.: Distance based parameter adaptation for differential evolution. In: IEEE Symposium Series on Computational Intelligence (SSCI), pp. 1–7. IEEE, November 2017
14. Zhang, J., Sanderson, A.C.: JADE: adaptive differential evolution with optional external archive. IEEE Trans. Evol. Comput. **13**(5), 945–958 (2009)
15. Viktorin, A., Senkerik, R., Pluhacek, M., Kadavy, T.: Archive analysis in SHADE. In: International Conference on Artificial Intelligence and Soft Computing, pp. 688–699. Springer, Cham (2017)

Pupil Localization Using Self-organizing Migrating Algorithm

Radovan Fusek[✉] and Petr Dobeš

FEECS, Department of Computer Science, Technical University of Ostrava,
17. listopadu 15, 708 33 Ostrava-Poruba, Czech Republic
radovan.fusek@vsb.cz, dobesp.nj@gmail.com

Abstract. In this paper, we propose a new method for pupil localization in images. The main contribution of the proposed method is twofold. Firstly, the method is based on the proposed eye model that takes into account physiological properties of eyes (i.e. reflects the properties of pupil, iris, and sclera). Secondly, the correct shape and the position of the model are determined using an evolutionary algorithm called Self-Organizing Migrating Algorithm (SOMA). Thanks to these ideas, the proposed method is faster than the state-of-the-art methods without reduction of accuracy. We evaluated the algorithms on two publicly available data sets in remote tracking scenarios (namely BioID [7] and GI4E [11]).

Keywords: SOMA · Pupil detection · Evolutionary algorithms · Object detection · Shape analysis

1 Introduction

In the recent years, two main sensor setups are widely used for image-based analysis of eyes (i.e. for iris, pupil, and eyelids localization or eye blink monitoring).

The first setup is represented by the head-mounted systems that are located very close to the human eyes. These systems are capable to capture the eye images in high resolutions. Thanks to this, the image-based approaches can precisely solved many important tasks like eye gaze estimation or eyelid localization.

The second setup is represented by the cameras that are located remotely from the subjects. In these systems, the eye covers only a small region of the image and the extracted images of eyes are usually of a lower quality than in the head-mounted systems in many cases. This fact makes the detection of eye parts, eye tracking and gaze estimation more difficult.

In addition to the detection accuracy, the performance in terms of computational time is equally important. For example, many vehicles are equipped with in-car cameras that monitor the driver fatigue. The fatigue of drivers represents a frequent cause of car accidents. To evaluate and prevent this situation, the approaches (e.g. pupil localization) must work very quickly.

These facts were a motivation for creating a new method for pupil localization that is faster and more precise than the existing state-of-the-art methods and works well in the remote camera systems with low resolution images. In the proposed method, the pupil center is localized by finding the global extrema of the proposed fitness function using an evolutionary algorithm called SOMA (Self-Organizing Migrating Algorithm) [1,13]. Thanks to SOMA and the proposed fitness function that is based on the eye model, the presented method achieved better computational time and recognition performance when compare to the existing methods.

The rest of the paper is organized as follows. The previously presented papers from this area are mentioned in Sect. 2. In Sect. 3, the main ideas of proposed method are described. In Sect. 4, the results of experiments are presented showing the properties of the new method.

2 Related Work

In the recent years, the detection and localization of pupil and iris became very important in many different areas (e.g. medicine, psychology, bio-metric, automotive). In the area of pupil localization, a popular and robust algorithm, named Starburst, was presented in [9]. The corneal reflection is located and removed from the image in the first step. Then the possible pupil edge points are marked using a set of rays (intensity changes along the rays are examined). In the last step, the marked points are used for ellipse fitting using RANSAC. Swirsky et al. [10] proposed the pupil tracking method that uses Haar-like features to estimate the pupil location combined with the k-means clustering to refine the pupil centre. In [3], Exclusive Curve Selector (ExCuSe) for pupil detection was proposed. This method is based on edge filtering and oriented histograms calculated via the angular integral projection function. Another pupil detection method know as SET was proposed in [6]. The SET method consists of image thresholding followed by segmentation to group related pixels. The segments are filtered and the border of each segment is used as an input for ellipse fitting. The next algorithm for pupil detection, named Ellipse Selector (ElSe), was proposed in [4]. The Canny edge filter and morphological operations are used to detect pupil related edges which is followed by ellipse fitting. In the case that no ellipse is found, an advanced blob detection is used to find the pupil center. The method for iris centre localization was proposed in [5]. In the first step, the coarse location of iris center is obtained using a fast convolution based approach. Afterward, the iris center location is refined using boundary tracing and ellipse fitting. In [12], the authors mentioned that the eyelids and eyelashes are noise factors for iris recognition. Therefore, in their paper, the eyelids and eyelashes detection algorithm based on the Canny edge detection technique with the Hough transform is used and the eyelids and eyelashes are removed from the iris images to increase the performance of the iris recognition system. In [8], a pupil localization method based on Hough regression forest was proposed. This method is based on the training process (supervised). A big evaluation of the mentioned pupil detection methods was presented in [2].

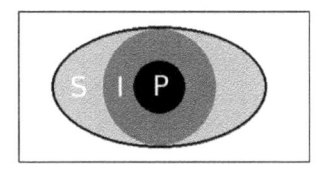

Fig. 1. The eye model for determining the proposed fitness function.

3 Proposed Method

In this section, we describe the main ideas and contributions of the proposed method for pupil localization.

The contribution of the proposed method is twofold. Firstly, the method is based on the proposed eye model (Fig. 1). Thanks to this model, the method gives a lower number of false positive detections compare to the state-of-the-art methods. However, determine the correct location and properties of this model (in real images) represents a challenging optimization problem. Therefore, the second contribution is represented by the use of an evolutionary algorithm SOMA (Self-Organizing Migrating Algorithm). Using SOMA, we are able to find the correct location of the pupil (model fitting) faster than the state-of-the-art methods without reduction of accuracy.

Since our method for pupil detection takes into account physiological properties of eyes, let us consider a following ideal model of eye in Fig. 1. It is important to note that in the flowing text, we suppose that the eye region is obtained beforehand (e.g. using facial landmarks or classical eye or face detectors). In the ideal model, we can assume that the pupil is represented by a dark area (area P in Fig. 1) surrounded by a slightly brighter area of iris (area I in Fig. 1) which is surrounded by a white sclera (area S in Fig. 1). In general, the goal of the pupil (iris) detection method is to find the appropriate location of these parts. As was note in the previous sections, the position of the pupil center is important information for gaze direction recognition, recognition of driver drowsiness, or pupil tracking systems.

As was mentioned, the parts of the eye (pupil, iris, and sclera) have different pixel intensity values in images. For example, we can suppose that the pupil area is the darkest area in the eye region (Fig. 1). However, the darkest area can be detected in the eyebrow. Which is a common problem of many pupil detection approaches. Therefore, in addition to the information of the darkest area of pupil, we use the pixel intensity values of iris (or sclera). It follows that the correct location of the pupil (iris) can be determined as the ratio of mean intensity values between the mentioned areas. Based on this fact, we propose the following function that describes the properties of the presented eye model, and that is designed to maximize the contrast between all three regions (Eq. 1).

$$f = \frac{mean_I}{mean_P} \cdot \frac{mean_S}{mean_P} \qquad (1)$$

Let us consider that the eye model in Fig. 1 represents the image template, and we can perform a template matching procedure. It means that we have to compute the function (Eq. 1) of the template at each location of the input image to find the best match (the location with the maximum value of the proposed function Eq. 1). It is clear that in real images, the eye parts (pupil, iris, and sclera) have different sizes (parameters) and one fixed size template will not be enough for correct detection of the eye parts. It follows, that carry out the template matching procedure (model fitting) for each parameter (different template sizes) represents a time consuming operation.

Therefore, we propose a novelty way for model fitting (localization) of the proposed eye model. Our approach uses an optimization method based on an evolutionary algorithm called Self-Organizing Migrating Algorithm (SOMA [1, 13]). The main idea of this algorithm can be described as individuals cooperating and wandering through the searched space. During the iteration process, the individuals affect each other so they can form the groups migrating together while searching for the global maximum.

Since we focus on the low resolution images (remote tracking scenario), and due to the fact that the computational complexity of pupil detection plays a crucial role (e.g. for the recognition of driver fatigue), the function in Eq. 1 is designed in such a way that it can also be computed very quickly with the use of simplified rectangle eye model. It follows that the proposed eye model in Fig. 1 can be transformed into a square shape model (Fig. 2). The experiments (in the following section) shown that the combination of this model and SOMA allows as to determine the correct location of pupil very quickly, especially in low resolution images, and the achieved results outperform the state-of-the-art methods. Therefore, in the following text, we use this square model.

For the convenience of readers, let us describe the whole SOMA process (with implementation details) in a following example (Fig. 3). In our case, the dimension of searched space is 5 because 5 parameters have to be obtained. The coordinates x and y of the pupil center (center of area P in Fig. 2), and lengths of the sides of pupil, iris, and sclera (lengths of areas P, I, and S in Fig. 2). Each individual is then defined as the vector of size equal to the dimension size of the searched space and represents the position in the searched space.

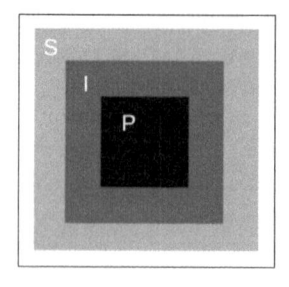

Fig. 2. The square eye model designed for fast computation of the proposed fitness function.

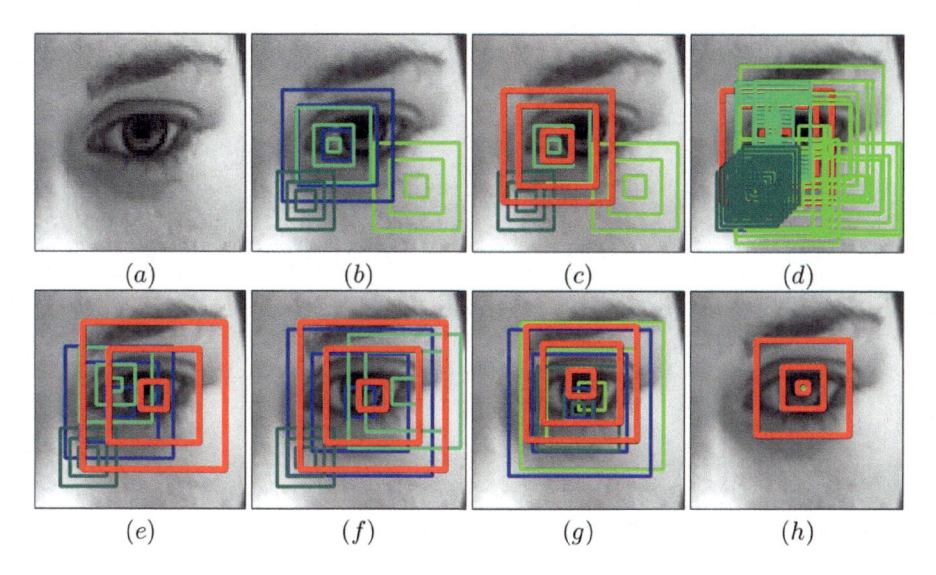

Fig. 3. The detection process of the proposed method. The input image (a). The initial population (the size of population is 4) of individuals (b). The first leader (depicted by the red color) based on the proposed fitness function (c). Examples of jumps of individuals (d). Examples of individuals with the new leaders after three iterations (e–g). The final leader (after 12 iterations) that represents the correct area of pupil with the depicted pupil center (h).

In the first step, an initial population (the size of population is 4) of individuals is generated (Fig. 3(b)). At the beginning of every iteration, so-called *Leader* is found as the individual with the highest fitness function value (Eq. 1) in the population. In Fig. 3(c), the first leader is depicted by the red color.

In the second step, the individuals begin to jump, towards the *Leader*. The jump means, that the individual is moved in every dimension. The jumps are defined by the fixed size step and by the number of jumps. Both parameters are defined in advance. An example of jumps is shown in Fig. 3(d). In every new position (jump) of the individual, the fitness function is evaluated. After evaluating all jumps, the individual is moved to the position with the best fitness function value. After all individuals are processed this way, the algorithm continues with the next iteration by selecting the new *Leader*. Examples of individuals with the new leaders (after each iteration) are shown in Fig. 3(e–g). The final position of the detected area of pupil (final leader after 12 iterations) with the pupil center (after described detection process) is shown in Fig. 3(h).

It is important to note that the individuals begin to jump, towards the *Leader*, according to a perturbation vector. The perturbation vector is generated for every individual at the beginning of iteration. For every dimension of the searched space, a random number (in range $(0,1)$) is generated and compared to a control parameter PRT which is set in advance. If the number is

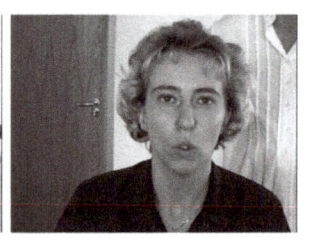

Fig. 4. Examples of BioID dataset images.

Fig. 5. Examples of GI4E dataset images.

greater than the constant PRT, the individual will move in this dimension. We also note that many versions of SOMA exist, in this work, we use a all-to-one version.

4 Experiments

To evaluate the results of the proposed method, we used two public datasets; BioID [7] and GI4E [11]. The BioID dataset contains 1521 gray level images with the resolution of 384×286 pixels. Every image shows one of 23 persons with different illumination conditions in different indoor environments. The dataset was collected for the purpose of testing the face recognition algorithms. The FGNet

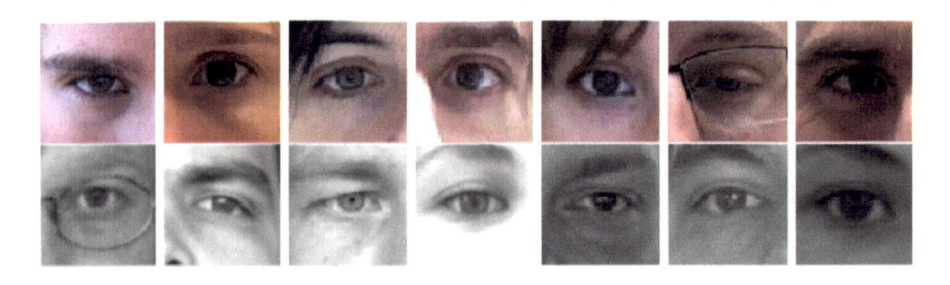

Fig. 6. Examples of used eye regions. The first row: GI4E dataset, the second row: BioID dataset.

project, which focuses on face and gesture recognition provides manually marked up labels for the BioID dataset with several facial landmarks (including the pupil center positions). Figure 4 shows example images from the BioID dataset.

The GI4E database provides 1339 images with the resolution of 800×600 pixels along with the manually labeled ground truth consisting of the iris center position and eye corners positions. In Fig. 5, example images from the GI4E dataset are shown. In the following experiments, the eye regions are selected according to the eye corner positions based on the provided ground truth information. Examples of eye regions are shown in Fig. 6.

To compare the proposed algorithm to state-of-the-art methods, we have chosen three methods, namely ElSe, ExCuSe and Swirski. Even though ExCuSe, ElSe and Swirski were primarily designed to work with high-resolution images acquired by head-mounted cameras, the experiments in [2] show that the methods can be used in the remote images as well. Since the methods that were

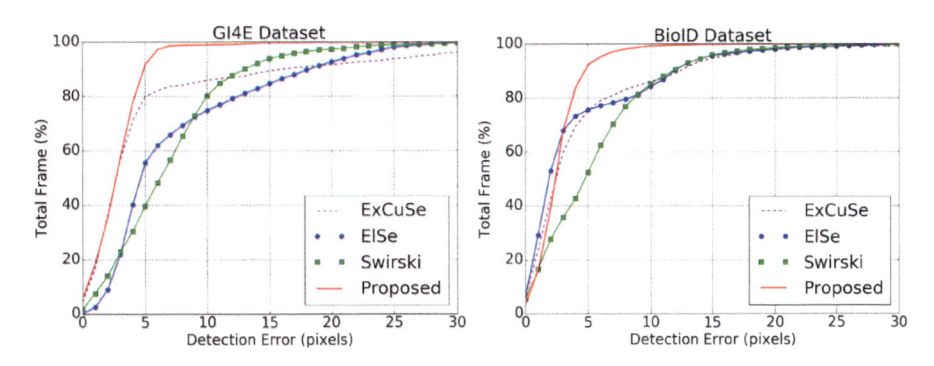

Fig. 7. The cumulative distribution of detection error. The error that is calculated as the Euclidean distance (in pixels) is in the x-axis. The y-axis shows percentage of frames with detection error smaller or equal to a specific error. The names of datasets are placed above the graphs.

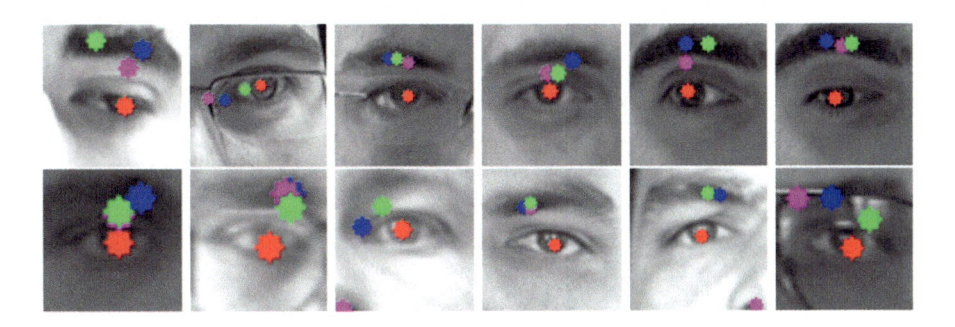

Fig. 8. Examples of images in which the proposed method performs better compared to other tested methods. The results of methods are distinguished by color: proposed method - red, ElSe - blue, ExCuSe - purple, Swirski - green. The first row: GI4E dataset, the second row: BioID dataset.

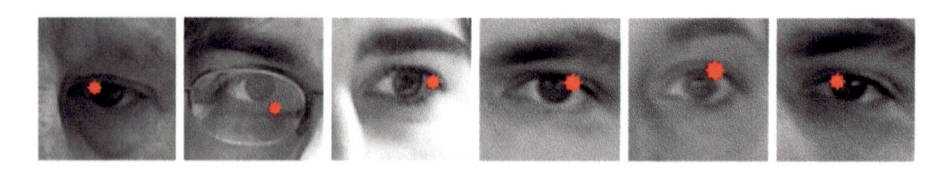

Fig. 9. Examples of images in which the proposed method fails.

Table 1. A comparison of time and errors.

	BioID Mean error (pixels)	GI4E Mean error (pixels)	Time per region (ms)
Proposed method	2.91	2.73	1.0
ElSe	4.17	7.62	2.3
ExCuSe	4.44	6.25	1.5
Swirski	5.58	6.97	7.6

compared require that the values of parameters are properly set, we paid attention to experimenting with their various values. For ElSe, we directly used the setting for remotely acquired images published by the authors of the algorithm.

The resulting plots of experiments are shown in Fig. 7. In the plots, we provide the cumulative distribution of detection errors for each method and dataset (percentage of frames with detection error smaller or equal to a specific error). The error is calculated as the Euclidean distance from the ground truth of the pupil center and the center provided by the particular detection method. In Table 1, we provide the average errors (in pixels) and average times needed for processing one eye region on an Intel core i3 processor (3.7 GHz).

Our results show that the proposed algorithm gives a stable and high detection rate among the evaluated approaches. Especially on the GI4E dataset, our method has a very small average error (2.73 pixels). Based on the results in Fig. 7, it can be observed that the proposed method is able detect approximately 90% of all frames with detection error smaller than 5 pixels. As is shown in Table 1, the proposed method is one of the fastest in the tests (only ExCuSe achieved similar time).

Figure 8 shows several cases in which our method works better compared to other tested methods. Typical cases of errors are caused by the presence of black eyebrow or glasses. Based on the results, we can conclude that the proposed method is better in such cases. However, not all images are successfully detected by proposed method. Figure 9 shows several examples of images in which the proposed method fails.

During the development of proposed method, several parameters were determined experimentally. The number of individuals and jumps was 10, the size step of jump was 2 pixels. The number of iterations of the SOMA algorithm was 12. The higher number of iterations does not lead to perceptible increase the performance. The PRT parameter was set to 0.5. The maximal sizes of the side lengths of pupil, iris, and sclera were determined based on the size of input image.

5 Conclusion

In this paper, we presented a new method for pupil detection in the remote images. The main idea is based on the fact that the pupil can be localized using the eye model that reflects the properties of eyes. Based on this model, we proposed the appropriate fitness function. To find the global extreme of the function, we used Self-Organizing Migrating Algorithm (SOMA). On the basis of the experiments, we can conclude that the newly proposed approach outperforms the state-of-the-art methods (namely ExCuSe, ElSe and Swirski).

In the current version of the presented approach, the rectangular eye model is used due to the fact that it can be computed very quickly. We leave the deeper experiments with another models (e.g. circular, elliptical) that can also be used for pupil detection (e.g. in head-mounted high resolution images) for future work.

Acknowledgments. This work was partially supported by Grant of SGS No. SP2018/42, VŠB - Technical University of Ostrava, Czech Republic.

References

1. Davendra, D., Zelinka, I., et al.: Self-organizing migrating algorithm. New Optimization Techniques in Engineering (2016)
2. Fuhl, W., Geisler, D., Santini, T., Rosenstiel, W., Kasneci, E.: Evaluation of state-of-the-art pupil detection algorithms on remote eye images. In: Proceedings of the 2016 ACM International Joint Conference on Pervasive and Ubiquitous Computing: Adjunct, UbiComp 2016, pp. 1716–1725. ACM, New York (2016). https://doi.org/10.1145/2968219.2968340
3. Fuhl, W., Kübler, T., Sippel, K., Rosenstiel, W., Kasneci, E.: Excuse: Robust pupil detection in real-world scenarios. In: Azzopardi, G., Petkov, N. (eds.) Computer Analysis of Images and Patterns, pp. 39–51. Springer, Cham (2015)
4. Fuhl, W., Santini, T.C., Kübler, T.C., Kasneci, E.: Else: ellipse selection for robust pupil detection in real-world environments. CoRR abs/1511.06575 (2015). http://arxiv.org/abs/1511.06575
5. George, A., Routray, A.: Fast and accurate algorithm for eye localization for gaze tracking in low resolution images. CoRR abs/1605.05272 (2016). http://arxiv.org/abs/1605.05272
6. Javadi, A.H., Hakimi, Z., Barati, M., Walsh, V., Tcheang, L.: Set: a pupil detection method using sinusoidal approximation. Front. Neuroeng. **8**, 4 (2015). https://www.frontiersin.org/article/10.3389/fneng.2015.00004
7. Jesorsky, O., Kirchberg, K.J., Frischholz, R.W.: Robust face detection using the hausdorff distance. In: Bigun, J., Smeraldi, F. (eds.) Audio-and Video-Based Biometric Person Authentication, pp. 90–95. Springer, Heidelberg (2001)

8. Kacete, A., Royan, J., Seguier, R., Collobert, M., Soladie, C.: Real-time eye pupil localization using Hough regression forest. In: 2016 IEEE Winter Conference on Applications of Computer Vision (WACV), pp. 1–8, March 2016
9. Li, D., Winfield, D., Parkhurst, D.J.: Starburst: a hybrid algorithm for video-based eye tracking combining feature-based and model-based approaches. In: 2005 IEEE Computer Society Conference on Computer Vision and Pattern Recognition (CVPR 2005), Workshops, p. 79, June 2005
10. Świrski, L., Bulling, A., Dodgson, N.: Robust real-time pupil tracking in highly off-axis images. In: Proceedings of the Symposium on Eye Tracking Research and Applications, ETRA 2012, pp. 173–176. ACM, New York (2012). https://doi.org/10.1145/2168556.2168585
11. Villanueva, A., Ponz, V., Sesma-Sanchez, L., Ariz, M., Porta, S., Cabeza, R.: Hybrid method based on topography for robust detection of iris center and eye corners. ACM Trans. Multimedia Comput. Commun. Appl. **9**(4), 25:1–25:20 (2013). http://doi.acm.org/10.1145/2501643.2501647
12. Wagh, A.M., Todmal, S.R.: Article: eyelids, eyelashes detection algorithm and hough transform method for noise removal in iris recognition. Int. J. Comput. Appl. **112**(3), 28–31 (2015)
13. Zelinka, I.: SOMA — Self-Organizing Migrating Algorithm. In: New Optimization Techniques in Engineering, pp. 167–217. Springer, Heidelberg (2004). https://doi.org/10.1007/978-3-540-39930-8_7

Differential Evolution Algorithms Used to Optimize Weights of Neural Network Solving Pole-Balancing Problem

Jan Vargovsky[✉] and Lenka Skanderova

Faculty of Electrical Engineering and Computer Science,
VSB – Technical University of Ostrava, 70800 Ostrava, Czech Republic
{jan.vargovsky.st,lenka.skanderova}@vsb.cz

Abstract. Differential evolution (DE) has been successfully used to solve difficult optimization problems. Every year, novel DE algorithms are developed to outperform the previous versions. The JADE is a famous DE algorithm using a mutation strategy *current-to-pbest* and the adaptation of control parameters. The SHADE has been developed to eliminate some bottlenecks of the JADE, especially its tendency to a premature convergence. The performance of these algorithms has been demonstrated on various benchmarks. The goal of this work is to compare the performance of the selected DE algorithms which are used to optimize the weights of the artificial neural network solving the pole-balancing problem.

Keywords: Differential evolution · JADE · SHADE · Artificial neural network · Pole-balancing problem

1 Introduction

Differential evolution (DE) is a popular meta-heuristic introduced by Storn and Price [10]. Its simple principle ensured that it is used in diverse areas of research as well as in real applications. Various DE algorithms outperforming the original DE have been developed. These algorithms can be divided into several groups according to a type of improvement: adaptive and self-adapted algorithms – jDE [1], HSDE [16], DE using combination of mutation strategies – EPSDE [6], CoDE [14], DE algorithms using novel mutation or crossover strategies – 2-Opt based DE [3], IMDE [18], DE with the population size reduction – LSHADE [13] etc.

The JADE algorithm was introduced by Zhang and Sanderson [17] and it became very popular for its robustness and fast convergence. It uses a novel mutation strategy *current-to-pbest* and adaptation of scale factor and crossover probability. The efficiency of this combination has been demonstrated on the well-known benchmark functions [15]. The main bottleneck of the JADE algorithm is its tendency to a premature convergence. Therefore, Tanabe and Fukunaga developed the SHADE algorithm [12]. The SHADE uses the archive of the means of the successful control parameters (scale factor and crossover probability). It has been shown that SHADE outperforms the JADE algorithm. The performance of the SHADE was demonstrated on the benchmarks CEC2013 [5], CEC 2005 [11] as well as on the well-known functions [15].

In [4], Ilonen et al. used the DE to train the feed-forward neural network (NN). Based on the experiments, the authors concluded that the DE does not bring any advantage in terms of learning rate or quality of solution in comparison to the gradient descent method. On the other hand, Slowik and Bialko [8] used the DE algorithm using control parameters adaptation to train an artificial neural network. Their method has been compared with gradient methods as Levenberg-Marquardt and error back-propagation. The results have shown that the DE algorithm can be considered to be an alternative to these training methods.

The evolutionary algorithms, especially genetic algorithms have been successfully used to design an architecture as well as weights of the artificial neural networks (ANN). Stanley and Miikkulainen [9] described the NEAT – NeuroEvolution of Augmenting Topologies, where the genetic algorithm is used to optimize the architecture as well as the weights of the artificial neural network. The NEAT outperformed the best fixed-topology methods on a benchmark reinforcement learning task. Real et al. [7] employed the evolutionary techniques to discover the models of the artificial neural networks for the CIFAR-10 and CIFAR-100 data sets. It has been shown that the ANN designed by the evolutionary techniques achieved the accuracies 94.6% and 77.0%.

In this paper, the original DE, JADE, and SHADE are used to optimize the weights of the NN solving the pole-balancing problem. The goal of the work is to investigate whether there are significant differences between the selected algorithms from the perspective of the convergence rate. The rest of paper is organized as follows: In Sect. 2, the original DE algorithm is described briefly. The pole-balancing problem is mentioned in Sect. 3. In Sect. 4, the experiment settings and results are presented and in Sect. 5, the conclusions are mentioned.

2 Differential Evolution

Differential evolution [10] belongs to a great family of the evolutionary algorithms (EAs) whose principles are inspired by biological evolution. As well as the other EAs, the DE works with a population P_x^G of NP individuals which are represented by D-dimensional vectors $\mathbf{x}_i^G$ of real-valuated parameters such that:

$$\begin{aligned} \mathbf{x}_i^G &= \{x_{i,0}^G, \ldots, x_{i,D-1}^G\}, \\ P_x^G &= \{\mathbf{x}_0^G, \ldots, \mathbf{x}_{NP-1}^G\}, \end{aligned} \quad (1)$$

where G indicates the generation in the range $0, \ldots, G_{max}$. For each j-th parameter in $\mathbf{x}_i^G$ the upper and lower bounds must be specified. These values can be collected into two D-dimensional vectors denoted as $\mathbf{b}_U$ and $\mathbf{b}_L$, where U and L indicate the upper and lower bounds, respectively. The initial population P_x^0

is then composed of the vectors generated randomly in the prescribed range. When a population is initialized, mutation and crossover are used to generate trial vectors. The selection determines which vector will be accepted to the next generation. Unlike genetic algorithms, the DE performs mutation step before crossover operation. For each target vector $\mathbf{x}_i^G$, a mutation vector is generated according to the following equation:

$$\mathbf{v}_i^G = \mathbf{x}_{r_1}^G + F \cdot (\mathbf{x}_{r_2}^G - \mathbf{x}_{r_3}^G), \qquad (2)$$

where $\mathbf{x}_{r_1}^G, \mathbf{x}_{r_2}^G,$ and $\mathbf{x}_{r_3}^G$ ($r_1 \neq r_2 \neq r_3 \neq i$) are solution vectors randomly selected from the actual population and F denotes the scale factor.

In the crossover operation, a trial vector $\mathbf{u}_i^G$ composed of the parameters of a donor vector $\mathbf{v}_i^G$ and a target vector $\mathbf{x}_i^G$. At the beginning of the crossover, a random integer j_{rn} ensuring that at least one parameter will be taken from a donor vector is generated from the interval $[0, D-1]$. A trial vector is then constructed as follows:

$$\mathbf{u}_{i,j}^G = \begin{cases} \mathbf{v}_{i,j}^G & \text{if } rn(j) \leq CR \text{ or } j = j_{rn} \\ \mathbf{x}_{i,j}^G & \text{otherwise.} \end{cases} \qquad (3)$$

Then a trial vector is evaluated. The selection operation is described by the following equation:

$$\mathbf{x}_i^{G+1} = \begin{cases} \mathbf{u}_i^G & \text{if } f(\mathbf{u}_i^G) \leq f(\mathbf{x}_i^G) \\ \mathbf{x}_i^G & \text{otherwise,} \end{cases} \qquad (4)$$

where $f()$ is the function to be minimized. In other words, if an objective function value is not greater than an objective function value of a target vector, a trial vector will replace a corresponding target vector at the next generation; otherwise, a target vector will be preserved.

3 Pole-Balancing Problem

Pole-balancing problem is a problem belonging to the control systems engineering, which deals with controlling behavior of the dynamical systems over time. In the pole-balancing problem, there is a balancing pole connected to a cart by a ball-bearing pivot. The cart can move only to the left or to the right (forward or backward). The pole can move around the pivot (horizontal axis of the pivot). The state of this dynamical system is defined by the four real values: the position of the cart relative to the center of the track x, the velocity of the cart $\dot{x}$, the angle of the pole Θ (in radians), and the angular velocity of the pole $\dot{\theta}$ (in rad/s^2). The output of the system is a direction of the movement of the cart as a fixed force F (in Newtons) [2]. For better illustration, see 1.

The mass of the cart is set to $m_c = 1.0$ kg, the mass of the pole to $m_p = 0.1$ kg, the pole length to $2l = 1.0$ m. The track limit is set to $h = \pm 2.4$. This means that the length of the track is $2k = 4.8$ m. The gravitational acceleration

$g = 9.81$ m/s². Time t is measured in seconds. Pole failure angle is set to $r = 12°$ from $0°$, i.e. approximately ± 0.209 rad. The pole acceleration is denoted as $\ddot{\Theta}$, the cart acceleration as $\ddot{x}$. Time step τ is the discrete time step, $\Delta = 0.02$ s (50 Hz). The force F is a constant, typically $F = \pm 10$ or $F = \pm 1$ Newtons [2].

The dynamical system has some constraints: it is necessary to remain the pole upright (within $\pm r$) and the cart inside the borders (within $\pm h$). The force is always non-zero value. The developed controller must keep the pole balanced within the given time interval. In the beginning, the position of the cart and the angle of the pole are initialized. These values can be selected randomly, or the cart is centered and the pole is balanced [2].

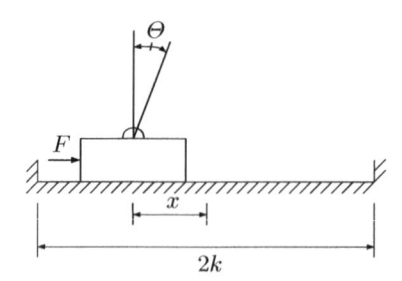

Fig. 1. Cart pole-balancing problem

4 Experiment

As mentioned, the goal of this work was to investigate the performance of the selected DE algorithms used to optimize the weights of the simple NN solving the pole-balancing problem. For this purpose, the environment CartPole-v0 from the toolkit Gym developed by OpenAI (non-profit research company dealing with the artificial intelligence) has been used. We have designed a simple NN as shown in Fig. 2. The inputs were the following values: the position of the cart x, the cart velocity $\dot{x}$, the angle of the pole Θ, and the angular velocity $\dot{Theta}$. The output is the probability to move to the left or to the right (in Fig. 2 denoted as $p(F^-)$ and $p(F^+)$). If the cart moves and the pole remains upright, the value 1 will be added to the score. The simulation ends when the given score (reward) is gained or when the conditions mentioned in Sect. 3 are broken (the cart is outside the borders or the hole fell down). For this experiment four rewards have been selected – 200, 500, 1000, and 2500.

There were 22 weights to be optimized. Therefore, the dimension of the individuals in the DE has been set to $D = 22$. The number of individuals $NP = 50$. The fitness value is a time – how long the hole remains upright. For the fair comparison of the algorithms four static seeds 355, 3244, 6239, and 9844 have been selected. The seeds are used within the Gym, so the deterministic behavior can be achieved. It can be said that the experiments can be then repeated for

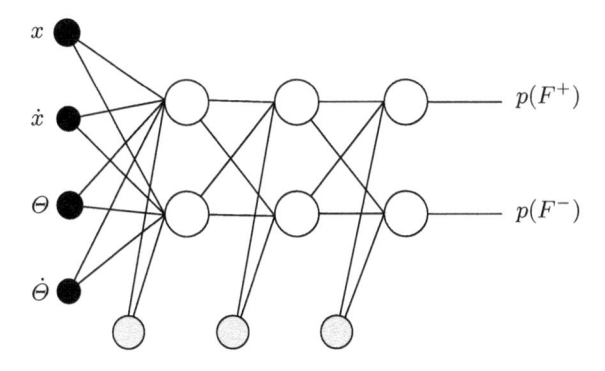

Fig. 2. Architecture of the neural network solving pole-balancing problem

the same problem. For each algorithm, reward, and seed, 50 experiments have been executed.

The neural network as well as the DE algorithms have been implemented in Python 3.5.5 and run on a computer with CPU i5-3570k@3,8 GHz, 8 GB RAM. The number of generations and run-times (in seconds) necessary to achieve the given rewards have been analyzed. To make a fair comparison, Wilcoxon signed-rank test at the significance level $\alpha = 0.05$ has been conducted between each pair of the algorithms. The results of the experiments are mentioned in Tables 1, 2, 3 and 4. If some algorithm achieves better results (smaller number of generations or shorter time) than the others, the results will be highlighted.

4.1 Results

In Table 1, the minimum, maximum, median, mean, and standard deviation of the numbers of generations necessary to achieve the rewards 200, 500, 1000, and 2500 is presented. In Table 2, p-values of the Wilcoxon signed-rank test (related to the results mentioned in Table 1) are mentioned. The statistically significant results are highlighted – in Table 1, the smallest numbers of generations. Based on these results, we have concluded that the JADE needed the smallest number of generations in the case of the seed 3244 and the rewards 500 and 1000, and in the case of the seed 9844 and the reward 1000. When we look at the seed 9844 and the reward 2500 (see Fig. 3), we will see that the smallest numbers of generations were used in JADE and SHADE algorithms. In the rest of the experiments, we can not unambiguously say that some algorithm achieved the given rewards using the smallest number of generations in comparison with the other ones.

In Tables 3 and 4 the minimum, maximum, median, mean, and standard deviations of the times (in seconds) necessary to achieve the rewards 200, 500, 1000, and 2500, and the p-values of the Wilcoxon signed-rank test are presented. The statistically significant results are highlighted – in Table 3, the shortest times. The JADE spent the shortest time to achieve the reward 500 in the case

Table 1. Analysis of the numbers of generations of the original DE (denoted as DE), JADE, and SHADE algorithms to achieve rewards 200, 500, 1000, and 2500

Seed = 355												
Max. rew. Algor.	200			500			1000			2500		
	DE	JADE	SHADE	DE	JADE	SHADE	DE	JADE	SHADE	DE	JADE	SHADE
Min	1	1	1	1	4	2	2	5	1	1	1	1
Max	24	21	21	61	29	35	46	32	54	100	84	48
Median	8.00	7.00	6.00	14.00	12.50	16.00	17.50	16.00	19.00	19.50	13.50	24.00
Mean	8.44	7.26	7.70	15.18	13.60	16.46	18.72	16.64	19.34	21.90	17.02	22.82
Stdev	4.80	3.82	5.22	12.00	6.52	8.83	10.48	6.99	10.85	16.21	12.89	11.83
Seed = 3244												
Max. rew. Algor.	200			500			1000			2500		
	DE	JADE	SHADE	DE	JADE	SHADE	DE	JADE	SHADE	DE	JADE	SHADE
Min	1	1	1	2	1	1	1	1	1	5	1	2
Max	27	24	26	61	**29**	41	71	**32**	48	100	64	97
Median	8.50	9.00	10.00	15.50	**11.00**	14.00	18.00	**15.00**	18.50	24.00	16.00	21.00
Mean	9.24	8.70	10.86	17.18	**11.18**	15.02	21.24	**15.10**	19.56	25.34	19.70	23.08
Stdev	6.06	5.38	6.12	10.74	**5.81**	8.64	14.14	**7.41**	10.07	15.78	13.54	16.73
Seed = 6239												
Max. rew. Algor.	200			500			1000			2500		
	DE	JADE	SHADE	DE	JADE	SHADE	DE	JADE	SHADE	DE	JADE	SHADE
Min	1	1	1	1	1	1	1	1	1	1	1	3
Max	17	17	18	29	25	34	44	27	38	55	46	53
Median	7.00	5.00	8.00	9.50	10.00	12.50	13.00	10.50	13.00	12.00	12.00	15.00
Mean	7.64	6.26	7.96	11.14	10.88	12.60	13.36	11.18	13.56	15.32	14.22	15.78
Stdev	3.53	4.42	4.07	7.06	5.45	7.11	7.71	6.13	8.16	11.37	8.97	8.60
Seed = 9844												
Max. rew. Algor.	200			500			1000			2500		
	DE	JADE	SHADE	DE	JADE	SHADE	DE	JADE	SHADE	DE	JADE	SHADE
Min	1	1	1	1	1	1	1	1	1	3	1	2
Max	21	23	27	36	26	35	52	**41**	56	72	**42**	43
Median	7.00	6.50	7.50	15.50	10.50	12.00	20.00	**13.00**	16.00	20.50	**13.00**	16.50
Mean	7.82	7.90	8.44	14.66	11.88	12.88	19.74	**14.24**	19.10	22.66	**16.28**	17.68
Stdev	4.86	5.20	5.74	8.54	6.25	8.17	10.38	**7.97**	11.52	14.03	**10.03**	9.24

Table 2. P-values of the Wilcoxon signed-rank test related to the Table 1. The original DE algorithm is denoted as DE

	Seed = 355			Seed = 3244		
Max. rew.	DE/JADE	DE/SHADE	JADE/SHADE	DE/JADE	DE/SHADE	JADE/SHADE
200	0.211	0.289	0.920	0.631	0.242	**0.042**
500	0.682	0.298	0.066	**0.003**	0.401	**0.011**
1000	0.529	0.992	0.119	**0.010**	0.603	**0.023**
2500	0.054	0.555	**0.004**	0.077	0.211	0.230
	Seed = 6239			Seed = 9844		
Max. rew.	DE/JADE	DE/SHADE	JADE/SHADE	DE/JADE	DE/SHADE	JADE/SHADE
200	0.064	0.849	**0.036**	0.826	0.682	0.757
500	0.795	0.242	0.222	0.052	0.258	0.772
1000	0.139	0.944	0.091	**0.006**	0.441	**0.008**
2500	0.881	0.453	0.267	**0.015**	**0.038**	0.509

Table 3. Analysis of run-times of the original DE (denoted as DE), JADE, and SHADE algorithms to achieve rewards 200, 500, 1000, and 2500

Seed = 355												
Max. rew. Algor.	200			500			1000			2500		
	DE	JADE	SHADE	DE	JADE	SHADE	DE	JADE	SHADE	DE	JADE	SHADE
Min	0.93	0.96	0.89	1.08	2.84	1.69	2.01	3.32	1.55	2.53	2.51	2.48
Max	17.52	14.15	14.49	133.04	29.44	28.24	89.64	51.88	82.98	901.53	208.37	110.29
Median	4.37	4.03	3.63	11.83	9.13	11.98	18.66	20.03	15.79	24.89	14.14	19.66
Mean	5.13	4.28	4.61	16.42	10.24	12.53	24.80	21.13	19.38	51.52	25.36	25.93
Stdev	3.21	2.34	3.23	21.42	5.65	7.87	21.50	12.99	16.02	128.46	33.66	22.75
Seed = 3244												
Max. rew. Algor.	200			500			1000			2500		
	DE	JADE	SHADE	DE	JADE	SHADE	DE	JADE	SHADE	DE	JADE	SHADE
Min	1.06	1.24	0.93	2.67	**1.97**	2.26	1.56	2.23	1.80	4.98	**3.15**	**3.40**
Max	50.47	27.18	18.09	253.57	**52.44**	66.85	128.50	65.18	75.79	457.51	**281.14**	**612.42**
Median	9.48	9.15	6.15	26.25	**13.45**	17.33	21.49	14.58	16.41	42.26	**22.57**	**22.85**
Mean	11.46	9.24	7.21	33.27	**16.00**	21.15	28.24	17.72	20.22	71.83	**47.03**	**48.84**
Stdev	9.21	5.96	4.41	39.37	**10.64**	14.42	26.20	14.12	16.26	89.23	**65.35**	**103.02**
Seed = 6239												
Max. rew. Algor.	200			500			1000			2500		
	DE	JADE	SHADE	DE	JADE	SHADE	DE	JADE	SHADE	DE	JADE	SHADE
Min	0.91	**0.96**	0.94	1.12	1.39	1.20	1.44	1.47	1.50	2.43	2.44	3.33
Max	11.29	**10.10**	12.44	44.20	29.71	26.82	119.16	31.30	45.96	324.18	59.70	47.13
Median	4.23	**3.20**	4.47	5.91	6.39	8.51	9.69	7.69	8.55	9.99	10.45	12.37
Mean	4.62	**3.74**	4.67	9.03	7.84	9.10	12.72	9.05	11.44	27.59	14.97	14.28
Stdev	2.27	**2.38**	2.36	8.46	5.13	6.22	16.86	6.47	9.50	57.52	12.32	9.80
Seed = 9844												
Max. rew. Algor.	200			500			1000			2500		
	DE	JADE	SHADE	DE	JADE	SHADE	DE	JADE	SHADE	DE	JADE	SHADE
Min	0.88	0.88	0.94	1.07	**1.07**	**1.20**	1.42	**2.09**	**1.59**	3.53	**2.49**	**2.89**
Max	14.69	12.94	16.75	46.08	**29.77**	**27.68**	169.31	**89.10**	**148.86**	547.77	**56.04**	**78.70**
Median	4.01	3.69	4.33	11.75	**7.26**	**7.35**	19.28	**10.89**	**12.60**	27.61	**12.27**	**12.11**
Mean	4.87	4.55	5.46	13.51	**8.25**	**9.23**	27.58	**16.07**	**17.86**	43.33	**17.22**	**17.30**
Stdev	3.38	2.96	3.91	10.89	**5.30**	**6.54**	29.60	**17.79**	**21.84**	79.33	**13.30**	**14.79**

of the seed 3244 and the reward 200 in the case of the seed 6239. The shortest times have been also measured for the JADE and SHADE algorithms and the seed 3244 and the reward 2500 as well as the seed 9844 and the rewards 500, 1000, and 2500 (see Fig. 4). Based on the results of the Wilcoxon signed-rank test, we can not conclude that in the rest of experiments, some algorithm needed significantly shorter time to achieve the given rewards than the others.

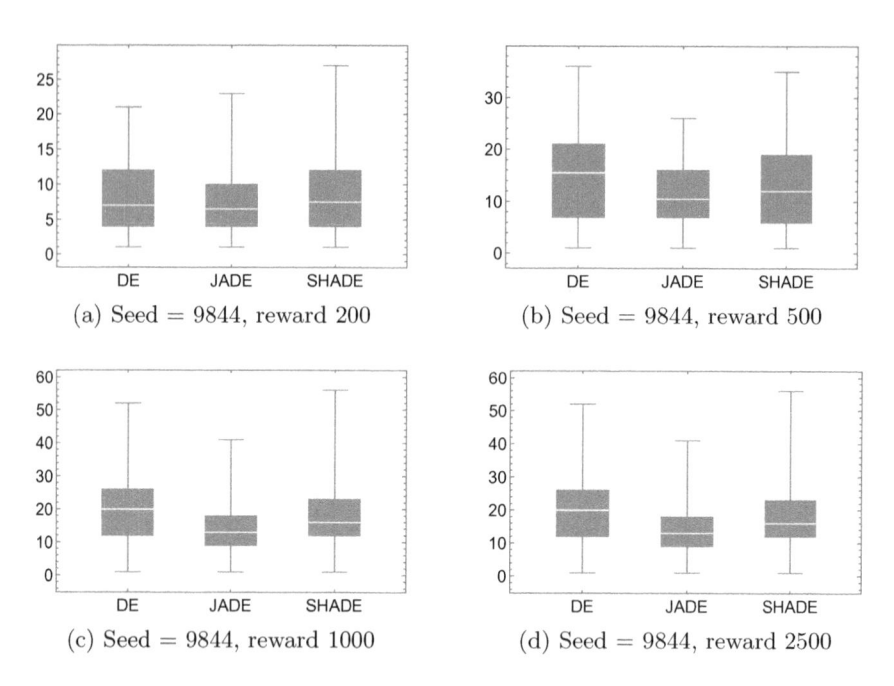

Fig. 3. Numbers of generations used in the original DE, JADE, and SHADE to gain rewards 200, 500, 1000, and 2500

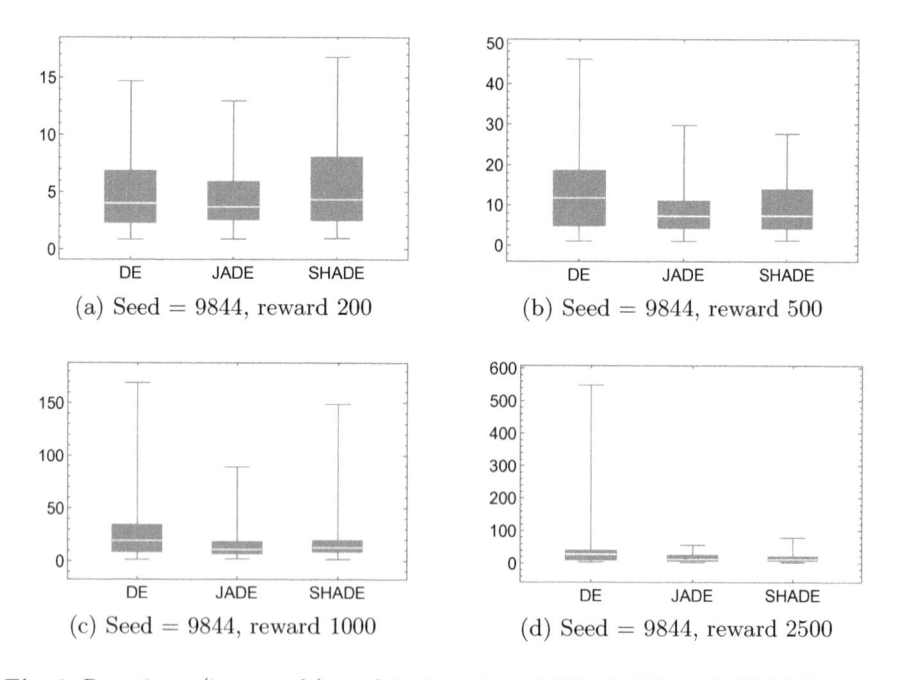

Fig. 4. Run-times (in seconds) used in the original DE, JADE, and SHADE to gain rewards 200, 500, 1000, and 2500

Table 4. P-values of the Wilcoxon signed-rank test related to the Table 3. The original DE algorithm is denoted as DE

	Seed = 355			Seed = 3244		
Max. rew.	DE/JADE	DE/SHADE	JADE/SHADE	DE/JADE	DE/SHADE	JADE/SHADE
200	0.150	0.267	0.968	0.177	**0.005**	0.073
500	0.190	0.889	0.187	**0.001**	0.057	**0.047**
1000	0.834	0.184	0.490	**0.029**	0.082	0.478
2500	**0.035**	0.134	0.289	**0.007**	**0.010**	0.624
	Seed = 6239			Seed = 9844		
Max. rew.	DE/JADE	DE/SHADE	JADE/SHADE	DE/JADE	DE/SHADE	JADE/SHADE
200	**0.027**	0.841	**0.032**	0.992	0.631	0.412
500	0.968	0.459	0.234	**0.004**	**0.038**	0.653
1000	0.131	0.631	0.222	**0.008**	**0.037**	0.368
2500	0.795	0.984	0.920	**0.003**	**0.002**	0.575

5 Conclusion

In this work, the DE algorithms – the original DE, JADE, and SHADE are used to optimize the weights of the NN solving the pole-balancing problem. The goal of the experiments was to investigate whether there is a significant difference between the pairs of the DE algorithms from the perspective of the run-times and numbers of generations necessary to gain the given rewards (in the pole-balancing problem). For the experiment, four rewards have been used: 200, 500, 1000, and 2500. To make a fair comparison, four different seeds of the random generator have been selected. For each DE algorithm, seed and reward, 50 experiments have been done. The simulation ended when the given reward had been gained.

Based on the results, we have concluded that in the most of the cases, there are no differences between run-times and numbers of generations necessary to achieve the rewards 200, 500, 1000, and 2500 in the original DE, JADE, and SHADE. However, the JADE used a significantly smaller number of generations than the original DE and SHADE in 3 from 16 cases. The JADE and SHADE in 1 case. The JADE alone needed the significantly shorter time to gain the given rewards in 2 cases. The JADE and SHADE then in 3 from 16 cases. The original DE algorithm never achieved the better result than the JADE and SHADE algorithm. The results of the experiment have shown an interesting phenomenon: Despite the fact that the JADE and SHADE significantly outperformed the original DE and the SHADE outperformed the JADE algorithm for the selected benchmarks, in the case of the optimization of the weights of the NN used to solve the pole-balancing problem, there are only minimal (statistical) differences between the performances of the aforementioned algorithms. Based on the results we have concluded that the mechanisms of the control parameters adaptation, as well as the mutation strategy used in the JADE and SHADE, have not lead to the higher performance for the given problem. On the other hand, only four seeds have been taken into consideration and it cannot be ruled out that the higher number of the seeds could affect the results of the experiment.

Acknowledgement. The following grants are acknowledged for the financial support provided for this research by Grant of SGS No. 2018/177, VSB - Technical University of Ostrava and under the support of NAVY and MERLIN research lab.

References

1. Brest, J., Greiner, S., Bošković, B., Mernik, M., Zumer, V.: Self-adapting control parameters in differential evolution: a comparative study on numerical benchmark problems. IEEE Trans. Evol. Comput. **10**(6), 646–657 (2006)
2. Brownlee, J., et al.: The pole balancing problem: A Benchmark Control Theory Problem (2005)
3. Chiang, C.-W., Lee, W.-P., Heh, J.-S.: A 2-opt based differential evolution for global optimization. Appl. Soft Comput. **10**(4), 1200–1207 (2010)
4. Ilonen, J., Kamarainen, J.-K., Lampinen, J.: Differential evolution training algorithm for feed-forward neural networks. Neural Process. Lett. **17**(1), 93–105 (2003)
5. Liang, J., Qu, B., Suganthan, P., Hernández-Díaz, A.G.: Problem definitions and evaluation criteria for the CEC 2013 special session on real-parameter optimization. In: Computational Intelligence Laboratory, Zhengzhou University, Zhengzhou, China and Nanyang Technological University, Singapore, Technical Report 201212, 3–18 (2013)
6. Mallipeddi, R., Suganthan, P.N., Pan, Q.-K., Tasgetiren, M.F.: Differential evolution algorithm with ensemble of parameters and mutation strategies. Appl. Soft Comput. **11**(2), 1679–1696 (2011)
7. Real, E., Moore, S., Selle, A., Saxena, S., Suematsu, Y.L., Tan, J., Le, Q., Kurakin, A.: Large-scale evolution of image classifiers. arXiv preprint. arXiv:1703.01041 (2017)
8. Slowik, A., Bialko, M.: Training of artificial neural networks using differential evolution algorithm. In: 2008 Conference on Human System Interactions, pp. 60–65. IEEE (2008)
9. Stanley, K.O., Miikkulainen, R.: Evolving neural networks through augmenting topologies. Evol. Comput. **10**(2), 99–127 (2002)
10. Storn, R., Price, K.: Differential evolution-a simple and efficient adaptive scheme for global optimization over continuous spaces, vol. 3. ICSI, Berkeley (1995)
11. Suganthan, P.N., Hansen, N., Liang, J.J., Deb, K., Chen, Y.-P., Auger, A., Tiwari, S.: Problem definitions and evaluation criteria for the cec 2005 special session on real-parameter optimization. KanGAL report **2005005**, 2005 (2005)
12. Tanabe, R., Fukunaga, A.: Success-history based parameter adaptation for differential evolution. In: 2013 IEEE Congress on Evolutionary Computation (CEC), pp. 71–78. IEEE (2013)
13. Tanabe, R., Fukunaga, A.S.: Improving the search performance of shade using linear population size reduction. In: 2014 IEEE Congress on Evolutionary Computation (CEC), pp. 1658–1665. IEEE (2014)
14. Wang, Y., Cai, Z., Zhang, Q.: Differential evolution with composite trial vector generation strategies and control parameters. IEEE Trans. Evol. Comput. **15**(1), 55–66 (2011)

15. Yao, X., Liu, Y., Lin, G.: Evolutionary programming made faster. IEEE Trans. Evol. Comput. **3**(2), 82–102 (1999)
16. Yi, W., Gao, L., Li, X., Zhou, Y.: A new differential evolution algorithm with a hybrid mutation operator and self-adapting control parameters for global optimization problems. Appl. Intell. **42**(4), 642–660 (2015)
17. Zhang, J., Sanderson, A.C.: JADE: adaptive differential evolution with optional external archive. IEEE Trans. Evol. Comput. **13**(5), 945–958 (2009)
18. Zhou, Y., Li, X., Gao, L.: A differential evolution algorithm with intersect mutation operator. Appl. Soft Comput. **13**(1), 390–401 (2013)

The Use of Radial Basis Function Surrogate Models for Sampling Process Acceleration in Bayesian Inversion

Simona Domesová[1,2,3(✉)]

[1] FEECS, Department of Applied Mathematics, VŠB - Technical University of Ostrava, 17. listopadu 15, 708 33 Ostrava-Poruba, Czech Republic
simona.domesova@vsb.cz
[2] IT4Innovations National Supercomputing Center, VŠB - Technical University of Ostrava, 17. listopadu 15, 708 33 Ostrava-Poruba, Czech Republic
[3] Institute of Geonics of the CAS, Studentska 1768, 708 00 Ostrava-Poruba, Czech Republic

Abstract. The Bayesian approach provides a natural way of solving engineering inverse problems including uncertainties. The objective is to describe unknown parameters of a mathematical model based on noisy measurements. Using the Bayesian approach, the vector of unknown parameters is described by its joint probability distribution, i.e. the posterior distribution. To provide samples, Markov Chain Monte Carlo methods can be used. Their disadvantage lies in the need of repeated evaluations of the mathematical model that are computationally expensive in the case of practical problems.

This paper focuses on the reduction of the number of these evaluations. Specifically, it explores possibilities of the use of radial basis function surrogate models in sampling methods based on the Metropolis-Hastings algorithm. Furthermore, updates of the surrogate model during the sampling process are suggested. The procedure of surrogate model updates and its integration into the sampling algorithm is implemented and supported by numerical experiments.

Keywords: Bayesian inversion · Metropolis-Hastings · Radial basis functions · Surrogate model · Uncertainty quantification

1 Introduction

The topic of this paper is the Bayesian approach to the solution of inverse problems arising in various engineering fields. Discussed procedures aim specifically at problems governed by computationally expensive forward mathematical models, e.g. models of thermal, hydraulic and mechanical processes computed using the finite element method and its variants. An observation that corresponds to a vector of unknown input parameters is given and the aim is to describe these parameters mathematically. In real engineering applications, observed data are almost inevitably corrupted by noise caused e.g. by measurement errors. An advantage of the Bayesian approach compared to deterministic methods is the incorporation of these uncertainties into the solution. Furthermore, information about the unknown parameters available from experiences is taken into account in the form of a prior distribution. The vector of unknown parameters is treated as a random vector and the Bayesian approach provides its joint probability distribution, i.e. the posterior distribution.

Since the resulting posterior distribution depends on the solution of a complex forward model, its probability density function (pdf) cannot be sampled directly. We have to use more advanced sampling procedures such as Markov chain Monte Carlo (MCMC) methods. The principle of these methods basically consists in proposing samples of parameters, which are then accepted or rejected. This way, a chain with limiting distribution equal to the posterior distribution is constructed. MCMC methods usually require high number of evaluations of the forward model. Therefore, much effort has been devoted to improvements of sampling procedures based on MCMC methods and particularly on the Metropolis-Hastings (MH) algorithm, see [12]. An overview of accelerated sampling methods can be found in [10]. From the category of improvements based on approximations of the posterior distribution, the delayed acceptance Metropolis-Hastings (DAMH) algorithm is important; this algorithm is introduced in [2,11] and it is proven that this modification preserves the limiting distribution of the resulting Markov chain. During DAMH, both the exact and the approximated posterior distribution are evaluated, focusing on reducing the number of exact posterior distribution evaluations. DAMH can be understood as an acceleration scheme which has been further developed, e.g. into multilevel variants, see [5,8,9,11].

The use of radial basis functions (RBF) for posterior approximations can be also found in recent literature. In [16], samples are generated from an approximated posterior distribution which is sequentially densified using evaluations of the exact posterior in selected samples of the parameters space. Other sampling methods that use an approximated model, which is iteratively improved can be found in [3,15]; papers [4,13] focus on reusing evaluations of the exact forward model collected during the DAMH sampling. Compared to these approaches, this paper works with a global non-intrusive approximation constructed using RBF.

The paper focuses on the construction and updates of RBF surrogate models and explores the integration of this procedure into the DAMH-based sampling process. It is organized as follows. Section 2 briefly outlines the Bayesian inversion and MH-based algorithms for posterior sampling. Section 3 discusses RBF surrogate models and their construction, updates, and parallel implementation. Numerical experiments can be found in Sect. 4, and Sect. 5 summarizes the results.

2 Bayesian Inversion and Posterior Sampling

This section briefly outlines the Bayesian inversion and illustrates its principle on a two-parametric example. For a more thorough introduction see [12,14].

The objective is to describe posterior distribution of a random vector $\mathbf{u} \in \mathbb{R}^n$ ($n \in \mathbb{N}$) that satisfies

$$\mathbf{y} = G(\mathbf{u}) + \boldsymbol{\eta},$$

where $\mathbf{y} \in \mathbb{R}^m$ ($m \in \mathbb{N}$) is a vector of observations, $G : \mathbb{R}^n \to \mathbb{R}^m$ is a forward mathematical model, and $\boldsymbol{\eta} \in \mathbb{R}^m$ is a random vector of additive noise. The Bayesian approach processes the following known information: vector $\mathbf{y}$, forward model G, joint pdf of the noise f_η, prior pdf π_0. The prior distribution of $\mathbf{u}$ expresses preliminary information about the parameters available from experience. According to the Bayesian theorem, the posterior pdf has the form of

$$\pi(\mathbf{u}|\mathbf{y}) \propto f_\eta(\mathbf{y} - G(\mathbf{u})) \pi_0(\mathbf{u}), \tag{1}$$

where $\propto$ denotes a proportionality.

Figures 1 and 2 illustrate the Bayesian theorem on the Himmelblau's function

$$G(\mathbf{u}) = G((x_1, x_2)) = \left(x_1^2 + x_2 - 11\right)^2 + \left(x_1 + x_2^2 - 7\right)^2. \tag{2}$$

The observation is a scalar value $y = 66.4$. The deterministic solution would be a set of all points in $\mathbb{R}^2$ that satisfy $G(\mathbf{u}) = y$, i.e. points on the corresponding contour line. In the Bayesian approach, we treat this observation as uncertain - corrupted by noise with Gaussian distribution $\mathcal{N}(0; 2)$. According to prior information, $\mathbf{u} = (x_1, x_2)$ has Gaussian distribution $\mathcal{N}((0,0); 1.5)$. Therefore, the pdf of $\mathbf{u}$ conditioned by the observation y is proportional to the multiplication $f_\eta(y - G(\mathbf{u})) \pi_0(\mathbf{u})$, see Fig. 1.

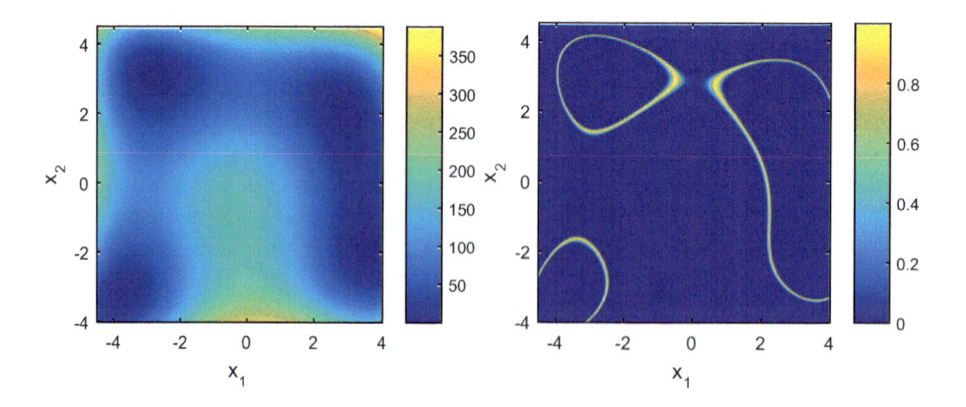

Fig. 1. $G(\mathbf{u})$ (left) and $f_\eta(y - G(\mathbf{u}))$ (right) for $\mathbf{u} = (x_1, x_2) \in \mathbb{R}^2$

A commonly used way to provide samples from $\pi(\mathbf{u}|\mathbf{y})$ is the standard MH algorithm [12]. When G is a complex mathematical model and it is possible to construct its approximation, one should rather and use the DAMH algorithm [2]. This paper builds on previous works [6,7] and suggests a modification of DAMH with surrogate model updates during the sampling process, see Algorithm 1.

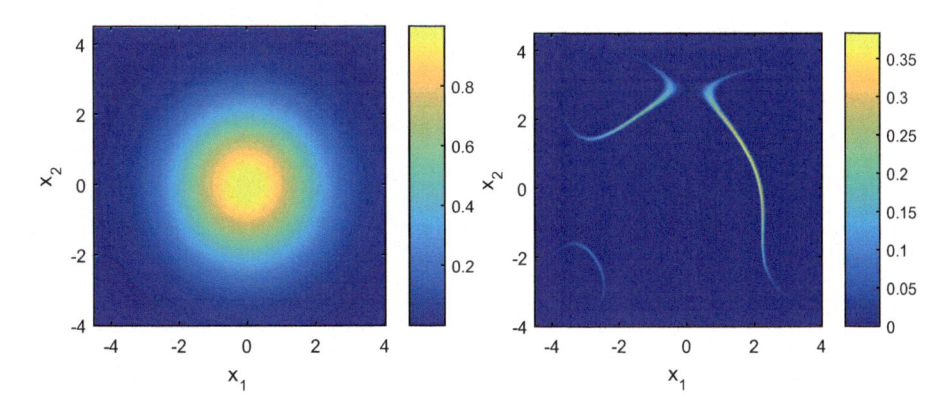

Fig. 2. $\pi_0(\mathbf{u})$ (left) and $f_\eta(y - G(\mathbf{u}))\pi_0(\mathbf{u})$ (right) for $\mathbf{u} = (x_1, x_2) \in \mathbb{R}^2$

The initial approximation $\widetilde{\pi}^{(1)}(\mathbf{u}|\mathbf{y})$ can be obtained using samples from a short preliminary run of the standard MH algorithm, initial sample $\mathbf{u}^{(1)}$ can be chosen e.g. as a random sample from the prior distribution.

Algorithm 1. DAMH with surrogate model updates (DAMH-SMU)

Given: initial sample $\mathbf{u}^{(1)}$, posterior pdf $\pi(\mathbf{u}|\mathbf{y})$ known up to a multiplicative constant, initial approximation $\widetilde{\pi}^{(1)}(\mathbf{u}|\mathbf{y})$, proposal pdf $q(\mathbf{u}|\mathbf{x})$ (for simplicity, a conditional pdf of a symmetric random-walk is considered).
For $t = 1, 2, \ldots, T$:

1. Generate $\mathbf{x}$ from $q\left(\mathbf{x}|\mathbf{u}^{(t)}\right)$, pre-accept $\mathbf{x}$ with probability

$$\widetilde{\alpha}\left(\mathbf{u}^{(t)}, \mathbf{x}\right) = \min\left\{1, \frac{\widetilde{\pi}^{(t)}(\mathbf{x}|\mathbf{y})}{\widetilde{\pi}^{(t)}(\mathbf{u}^{(t)}|\mathbf{y})}\right\}.$$

2. If $\mathbf{x}$ is not pre-accepted, set $\mathbf{u}^{(t+1)} = \mathbf{u}^{(t)}$ and $\widetilde{\pi}^{(t+1)} = \widetilde{\pi}^{(t)}$.
3. If $\mathbf{x}$ is pre-accepted, accept $\mathbf{x}$ with probability

$$\alpha\left(\mathbf{u}^{(t)}, \mathbf{x}\right) = \min\left\{1, \frac{\pi(\mathbf{x}|\mathbf{y})}{\pi(\mathbf{u}^{(t)}|\mathbf{y})} \frac{\widetilde{\pi}^{(t)}(\mathbf{u}^{(t)}|\mathbf{y})}{\widetilde{\pi}^{(t)}(\mathbf{x}|\mathbf{y})}\right\}. \quad (3)$$

   (a) If $\mathbf{x}$ is not accepted, set $\mathbf{u}^{(t+1)} = \mathbf{u}^{(t)}$.
   (b) If $\mathbf{x}$ is accepted, set $\mathbf{u}^{(t+1)} = \mathbf{x}$.
   (c) Update $\widetilde{\pi}^{(t)}$ to $\widetilde{\pi}^{(t+1)}$ using the pair $(\mathbf{x}, \pi(\mathbf{x}|\mathbf{y}))$.

This scheme can be modified by different approaches to surrogate model updates. For example, it is not necessary to update $\widetilde{\pi}^{(t)}$ with each new evaluation of the forward model, it can be updated in batches or using only subsets of available evaluations. According to formula (3), one can see that a more accurate approximation $\widetilde{\pi}^{(t)}$ decreases the probability of the rejection of a pre-accepted sample and therefore increases the quality of the sampling process.

3 Radial Basis Function Surrogate Models

A radial basis function is a multivariate function $\Phi(\mathbf{u}) : \mathbb{R}^n \to \mathbb{R}$ that can be reduced to a function $\phi : \mathbb{R} \to \mathbb{R}$ of the Euclidean norm of the argument $\mathbf{u} \in \mathbb{R}^n$, i.e. $\Phi(\mathbf{u}) = \phi(\|\mathbf{u}\|_2)$. RBF can be used to construct a surrogate model of a multivariate function based on scattered data, see [1]. In the context of the Bayesian approach, we can approximate the function $\pi(\mathbf{u}|\mathbf{y})$ itself or only its part $f_\eta(\mathbf{y} - G(\mathbf{u}))$, or eventually real-valued components of G, etc. In this section, $g : \mathbb{R}^n \to \mathbb{R}$ denotes a function for which a surrogate model is constructed.

If the scattered data $g(\mathbf{u}_1), \ldots, g(\mathbf{u}_K)$ are given, g can be approximated as

$$g(\mathbf{u}) \approx \widetilde{g}(\mathbf{u}) = \sum_{i=1}^{K} \alpha_i \phi(\|\mathbf{u} - \mathbf{u}_i\|_2) + \sum_{k=1}^{Q} \beta_k p_k(\mathbf{u}), \tag{4}$$

where ϕ is a radial basis function, and $p_1, \ldots, p_Q$ for $Q = \begin{pmatrix} q+n \\ n \end{pmatrix}$, is a basis of all n-variate polynomials of degree up to q. In this context, $\mathbf{u}_1, \ldots, \mathbf{u}_K$ are called centers. The coefficients α_i and β_k in (4) are determined by the linear system

$$\sum_{i=1}^{K} \alpha_i \phi(\|\mathbf{u}_j - \mathbf{u}_i\|_2) + \sum_{k=1}^{Q} \beta_k p_k(\mathbf{u}_j) = g(\mathbf{u}_j) \quad \forall j \in \{1, \ldots, K\},$$

$$\sum_{i=1}^{K} \alpha_i p_k(\mathbf{u}_i) = 0 \quad \forall k \in \{1, \ldots, Q\}. \tag{5}$$

In this application, the centers are samples $\mathbf{u}_j$ for which the mathematical model G was evaluated during a MH-based algorithm. The author's implementation constructs several chains in parallel processes (samplers). Pairs $(\mathbf{u}_j, g(\mathbf{u}_j))$ coming from different samplers are used together to construct one shared surrogate model $\widetilde{g}$. This model $\widetilde{g}$ is constructed by another parallel process which receives data from samplers and then constructs $\widetilde{g}$. When the construction finishes and a new batch of data is available, $\widetilde{g}$ is updated. Current matrix from (5) is enlarged and this modified linear system is solved by an appropriate method. With each construction/update, the surrogate model (specifically the vector of the solution of the corresponding linear system) is sent to all samplers. In the following tests, $q = 1$ is considered and several functions ϕ are compared.

4 Numerical Experiments

This section begins with testing of RBF surrogate models construction and updates. This procedure is then integrated into the DAMH-SMU sampling algorithm and tested as a whole on a model two-parametric inverse problem. For comparison, standard MH and DAMH algorithms are also applied.

4.1 Surrogate Model Updates

Consider two artificially chosen test functions: a two-parametric function G_2 defined by (2) and a function $G_5 : \mathbb{R}^5 \to \mathbb{R}$ defined for $\mathbf{u} = (x_1, \ldots, x_5)$ by

$$G_5(\mathbf{u}) = \left(x_1^2 + x_2 - 1\right)^2 + \cdots + \left(x_4^2 + x_5 - 1\right)^2 + \left(x_5^2 + x_1 - 1\right)^2.$$

These functions were used as forward models in the Bayesian approach according to Sect. 2 and scattered data $(\mathbf{u}_j, G_k(\mathbf{u}_j))$, $j = 1, 2, \ldots$ were obtained using 8 separate instances of the standard MH algorithm with different initial samples.

In the first experiment, an initial surrogate model was constructed using 100 centers and then updated using new centers added in batches of 100 up to 4000 centers. Figure 3 shows scattered points used as RBF centers in the case of the two-parametric testing function. Samples accepted and rejected during the sampling process are distinguished by color; however, all of them were utilized.

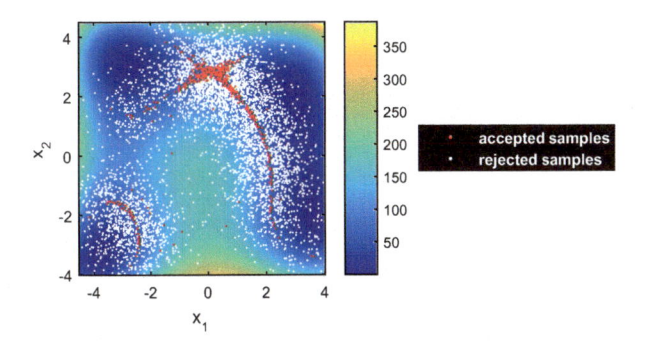

Fig. 3. 4000 samples used as RBF centers for surrogate model construction

Other 4000 scattered points $\mathbf{u}_1, \ldots, \mathbf{u}_{4000}$ were used for accuracy calculation. With each update of surrogate model $\widetilde{G}_k$, approximated posterior pdf $\widetilde{\pi}_k$ was computed (up to a multiplicative constant) according to (1). The error in testing points was then calculated as $|\pi(\mathbf{u}_i|y) - \widetilde{\pi}_k(\mathbf{u}_i|y)|$ for $i \in \{1, \ldots, 4000\}$. Monitored data are mean and 95% quantile of these sets. Figure 4 compares the accuracy achieved using several radial basis functions $\phi(r)$.

According to the first experiment, highest accuracy was achieved using $\phi(r) = r^m$, $m \geq 5$. Low number of centers allowed the use of a direct solver with sufficiently small error. In the second experiment, only G_5 was approximated and centers were added in batches of 1000 up to 8000. Figure 5 shows residual norms of direct solutions and estimated condition numbers of resulting matrices (5). High condition numbers were observed for $\phi(r) = r^5$ and $\phi(r) = r^9$, this may influence an eventual iterative solution negatively, as shown by the next test.

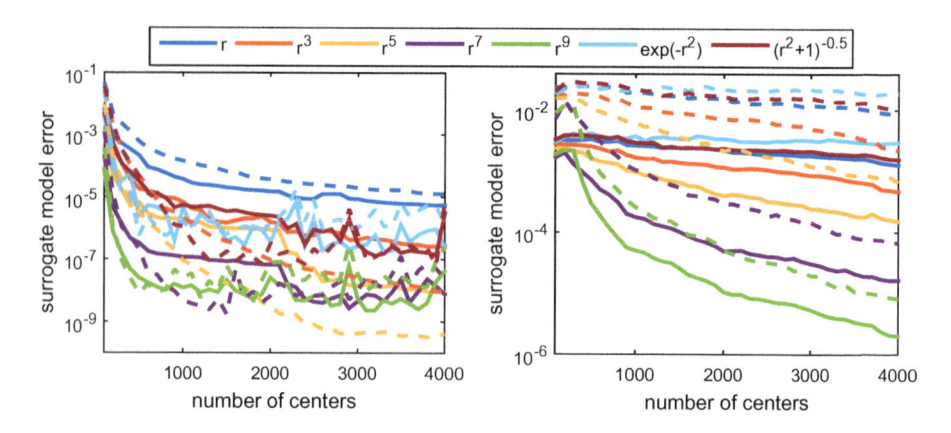

Fig. 4. Comparison of RBF accuracy for G_2 (left) and G_5 (right); mean (solid line) and 95% quantile (dashed line) of posterior error in test samples

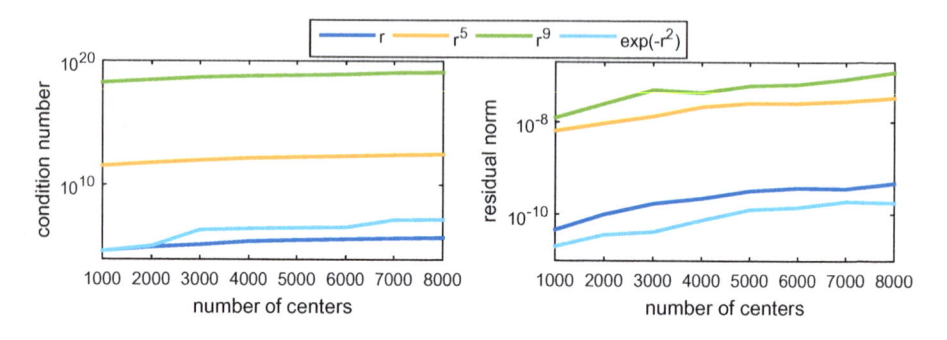

Fig. 5. Estimated condition numbers (left); direct solution residual norms (right)

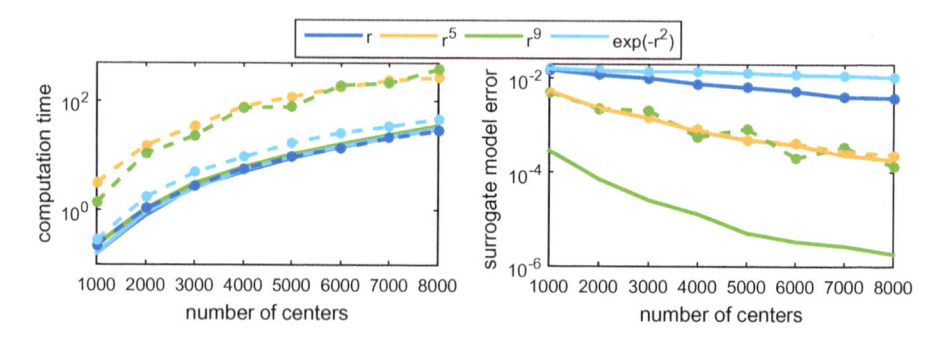

Fig. 6. Computation time of the solution of the linear systems of type (5) in seconds (left) and 95% quantile of posterior error in test samples (right) for direct solver (solid line) and MINRES iterative solver (dashed line with dots)

In the third experiment, the linear systems from experiment 2 were solved using the MINRES method (a Krylov subspace iterative method for symmetric matrices), terminated when the relative residual norm was below $1e-6$. Figure 6 compares the computation time and the accuracy with the direct solver.

4.2 Integration into the Sampling Algorithm

As a model problem for the last experiment, a simple one-dimensional linear elasticity problem with two unknown material parameters was used:

$$\begin{cases} -\left(k\left(x\right)u'\left(x\right)\right)' = -1 & x \in (0,1) \\ u\left(0\right) = 0 \\ ku\left(1\right)' = 0 \end{cases}, \text{ where } k\left(x\right) = \begin{cases} k_1 & x \in \langle 0, 0.5 \rangle \\ k_2 & x \in \langle 0.1, 1 \rangle \end{cases}.$$

This boundary value problem defines a forward model $G : \mathbb{R}^2 \to \mathbb{R}$. The constants k_1 and k_2 are unknown but it is assumed that both of them come from $\mathcal{N}(5; 1.5)$. A noisy measurement $y = -0.01$ of the value $u(1)$ is given and serves as additional information to refine the prior information in the Bayesian way, see Fig. 7. Gaussian noise with distribution $\mathcal{N}\left(0; 2 \cdot 10^{-4}\right)$ is considered.

The resulting posterior was sampled using the DAMH-SMU algorithm with $\phi(k) = k^5$, four chains with different initial samples were generated in parallel and shared one surrogate model. For the histogram of obtained samples see

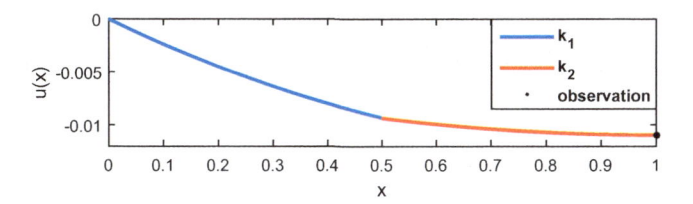

Fig. 7. Illustration of material parameters and measurement

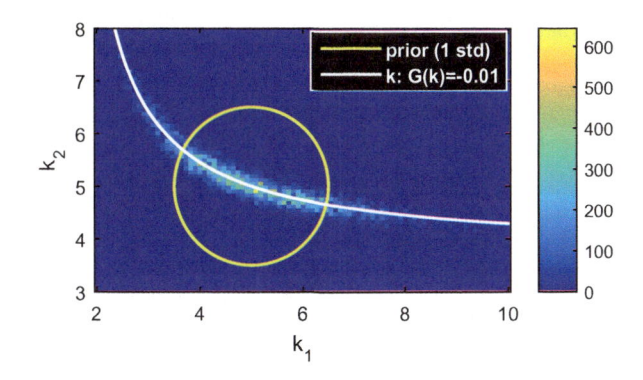

Fig. 8. Two-dimensional histogram of posterior samples; yellow circle illustrates one prior std around the mean $(5,5)$; white curve represents pairs of k_1 and k_2 that would (in the deterministic approach) correspond exactly to the observation $y = -0.01$

Table 1. Comparison of sampling methods; all bounded by 5000 G evaluations

Algorithm	$\phi(r)$	Accepted	Rejected without G evaluation	Rejected after G evaluation	AL	G evaluations per effective sample
MH	-	439	-	4561	53.78	53.78
DAMH	r^5	4757	54329	243	67.56	5.69
DAMH-SMU	r^5	4992	54139	8	53.90	4.56
DAMH-SMU	$\frac{1}{\exp(r^2)}$	4978	52493	22	57.85	5.03
DAMH-SMU	r	4956	54416	44	60.01	5.05

Fig. 8. For comparison, samples were also provided using other radial basis functions and using the DAMH and the MH algorithm (with the same proposal pdf and initial samples). In the case of the DAMH algorithm, a surrogate model was constructed using 200 centers and was not updated. Table 1 summarizes results. The estimation of the autocorrelation length (AL) expresses the distance between two almost uncorrelated (effective) samples the in the resulting chains.

5 Conclusions

This paper explored sampling methods that can be used during the solution of inverse problems governed by complex mathematical models using the Bayesian approach. An acceleration of standard sampling algorithms was proposed and tested on computationally undemanding model problems that allowed to carry out large numerical experiments.

Advantages of sampling methods based on the MH algorithm are mainly their simple implementation and the fact that the sampled pdf can be known only up to a multiplicative constant. However, a disadvantage of the standard MH algorithm is the need of extensive number of evaluations of the forward model. The numerical experiments showed that RBF surrogate models can be successfully used to reduce the number of evaluations of the forward model. Other advantage is that the use of the RBF does not impose strict requirements on the approximated operator G due to its non-intrusive character. A disadvantage of the use of RBF surrogate models can be seen in solving resulting linear systems of type (5) than are often ill-conditioned.

In the case of the model problem, the DAMH-SMU algorithm reduced the number of forward model evaluations needed to obtain one almost uncorrelated sample by 20% in comparison to the DAMH algorithm and by 91% in comparison to the MH algorithm. It the case of computationally intensive forward models, the computation time of surrogate model updates and evaluations is usually almost negligible in comparison to the forward model evaluations. Therefore, the use of RBF can bring significant acceleration to the sampling process.

Author's current work focuses on the development and parallel implementation of a robust framework for efficient sampling of posterior distributions. Future research directions also include combining the current approach with the parallel tempering method.

Acknowledgement. This work was supported by The Ministry of Education, Youth and Sports from the National Programme of Sustainability (NPS II) project "IT4Innovations excellence in science - LQ1602". The work was also partially supported by Grant of SGS No. SP2018/68 and by Grant of SGS No. SP2018/161, VŠB - Technical University of Ostrava, Czech Republic.

References

1. Buhmann, M.D.: Radial Basis Functions: Theory and Implementations. Cambridge University Press, Cambridge (2003). https://doi.org/10.1017/CBO9780511543241
2. Christen, J.A., Fox, C.: Markov chain Monte Carlo using an approximation. J. Comput. Graph. Stat. **14**(4), 795–810 (2005). https://doi.org/10.1198/106186005X76983
3. Cui, T., Fox, C., O'Sullivan, M.J.: Bayesian calibration of a large-scale geothermal reservoir model by a new adaptive delayed acceptance Metropolis Hastings algorithm: adaptive delayed acceptance Metropolis-hastings algorithm. Water Resour. Res. **47**(10) (2011). https://doi.org/10.1029/2010WR010352
4. Cui, T., Marzouk, Y.M., Willcox, K.E.: Data-driven model reduction for the Bayesian solution of inverse problems: Data-driven Model Reduction for Inverse Problems. Int. J. Numer. Methods Eng. **102**(5), 966–990 (2015). https://doi.org/10.1002/nme.4748
5. Dodwell, T.J., Ketelsen, C., Scheichl, R., Teckentrup, A.L.: A hierarchical multi-level Markov Chain Monte Carlo algorithm with applications to uncertainty quantification in subsurface flow. SIAM/ASA J. Uncertain. Quantif. **3**(1), 1075–1108 (2015). https://doi.org/10.1137/130915005
6. Domesová, S., Béreš, M.: A bayesian approach to the identification problem with given material interfaces in the darcy flow. In: High Performance Computing in Science and Engineering, vol. 11087, pp. 203–216. Springer, Cham (2018). https://doi.org/10.1007/978-3-319-97136-0_15
7. Domesova, S., Beres, M.: Inverse problem solution using bayesian approach with application to darcy flow material parameters estimation. Adv. Electr. Electron. Eng. **15**(2), 258–266 (2017). https://doi.org/10.15598/aeee.v15i2.2236
8. Dostert, P., Efendiev, Y., Mohanty, B.: Efficient uncertainty quantification techniques in inverse problems for Richards' equation using coarse-scale simulation models. Adv. Water Resour. **32**(3), 329–339 (2009). https://doi.org/10.1016/j.advwatres.2008.11.009
9. Efendiev, Y., Hou, T., Luo, W.: Preconditioning Markov Chain Monte Carlo simulations using coarse-scale models. SIAM J. Sci. Comput. **28**(2), 776–803 (2006). https://doi.org/10.1137/050628568
10. Kaipio, J.P., Fox, C.: The Bayesian framework for inverse problems in heat transfer. Heat Transfer Eng. **32**(9), 718–753 (2011). https://doi.org/10.1080/01457632.2011.525137
11. Moulton, J.D., Fox, C., Svyatskiy, D.: Multilevel approximations in sample-based inversion from the Dirichlet-to-Neumann map. J. Phys. Conf. Ser. **124**, 012035 (2008). https://doi.org/10.1088/1742-6596/124/1/012035

12. Robert, C.P.: The Bayesian choice: from decision-theoretic foundations to computational implementation, 2nd (edn.) Springer Texts in Statistics. Springer, New York (2007). OCLC: 255965262
13. Sherlock, C., Golightly, A., Henderson, D.A.: Adaptive, delayed-acceptance MCMC for targets with expensive likelihoods. J. Comput. Graph. Stat. **26**(2), 434–444 (2017). https://doi.org/10.1080/10618600.2016.1231064
14. Stuart, A.M.: Inverse problems: a bayesian perspective. Acta Numerica **19**, 451–559 (2010). https://doi.org/10.1017/S0962492910000061
15. Zhang, G., Lu, D., Ye, M., Gunzburger, M., Webster, C.: An adaptive sparse-grid high-order stochastic collocation method for Bayesian inference in groundwater reactive transport modeling: sparse-grid method for bayesian inference. Water Resour. Res. **49**(10), 6871–6892 (2013). https://doi.org/10.1002/wrcr.20467
16. Zhang, W., Liu, J., Cho, C., Han, X.: A fast Bayesian approach using adaptive densifying approximation technique accelerated MCMC. Inverse Prob. Sci. Eng. **24**(2), 247–264 (2016). https://doi.org/10.1080/17415977.2015.1017488

An Optimised Hybrid Group Method in Data Handling (GMDH) Network

Donald Davendra[1(✉)] and Petr Martinek[2]

[1] Department of Computer Science, Central Washington University,
400 E University Way, Ellensburg, WA 98926, USA
Donald.Davendra@cwu.edu
[2] Department of Computer Science,
Faculty of Electrical Engineering and Computer Science, VŠB-Technical University of Ostrava, 17. listopadu 15, 708 33 Ostrava-Poruba, Czech Republic

Abstract. A novel modular optimized hydrid Group Method in Data Handling (GMDH) network is proposed in this paper. A standard GMDH network is optimized using the Discrete Differential Evolution (DDE) algorithm for an optimized network structure, and Singular Value Decomposition (SVD) is further used for coefficient calculations of the network. The developed DE-GMDH algorithm is tested for fitness accuracy, memory usage and maximal error on a manufacturing problem.

Keywords: Group Method in Data Handling · Differential Evolution · Machine learning

1 Introduction

This paper introduces an optimized hybrid Group Method in Data Handling (GMDH) network algorithm. The main attributes is the modularity and expandability of the approach, which uses the Singular Value Decomposition (SVD) as a regression analysis for the calculation of the coefficients of the nodes. As there exists more than one hybrid version of GMDH, this approach of GMDH obtains the whole network structure from an individual of the optimizing algorithm, which implies that any algorithm which works with individuals can be used for the creation of the best network structure. A comprehensive overview of the GMDH application with various evolutionary algorithm is outlined in [5].

The hybridization is done with the Discrete Differential Evolution Algorithm (DDE) [6], to obtain the optimal network structure. This work is a continuation of the Matlab approach of [1], with a focus on better memory optimization of the network structure. This GMDH version also gives more possibilities, when it comes to memory usage optimization and parallelization, because it is composed of two almost independent algorithms (GMDH and DDE) and both can be optimized and parallelized in different ways. Threads have been used for low level parallelization of the algorithm.

The algorithm is tested on a manufacturing engineering problem from [7]. The main attributes tested are the fitness of the developed algorithm, memory usage and the maximal prediction error of the model. This algorithm was created to work efficiently even with larger network models (over 20 layers).

The paper is organized as follows: Sect. 2 describes the GMDH model, SVD and the DDE algorithm. Section 3 outlines the hybrid DE-GMDH approach developed in this paper and Sect. 4 further outlines the optimized network structure. Experimentation and results are given in Sect. 5 followed by the conclusion.

2 Group Method in Data Handling

Group Method of Data Handling (GMDH) was first described by Russian scientist Ivakhnenko [2]. It is a method, which allows to build models of complex systems, without any knowledge about their internal mechanisms. This method creates a model of a system, that was generated based only on input-output relationships and does not contain preconceived ideas of the researcher.

The basic GMDH algorithm constructs a high-order Volterra-Kolmogorov-Gabor (VKG) polynomial of the form (1):

$$y = a_0 + \sum_{i=1}^{n} a_i x_i + \sum_{i=1}^{n}\sum_{j=1}^{n} a_{ij} x_i x_j + \sum_{i=1}^{n}\sum_{j=1}^{n}\sum_{k=1}^{n} a_{ijk} x_i x_j x_k + \ldots \qquad (1)$$

that from n inputs $x_1, x_2, x_3, \ldots, x_n$ generates a single output y.

Ivakhnenko showed that the VKG series can be expressed as a network, that is a cascade of the second order polynomials with only two input variables [2,3].

Each layer in the network uses the nodes in the previous layer as inputs, with the first layer using the input data columns. Nodes that do not contribute to the solution are subsequently removed from the layer.

Because only n layers are needed for the creation of the $n+1$ layer, the whole notion of the network is sometimes omitted and replaced with sets of old and new variables, thereby leading to the implementation of the classical combinatorial GMDH algorithm, which creates new nodes from all the possible combination of the result pairs of the old nodes.

To describe this algorithm, we need the input data set with columns $x_1, x_2, x_3, \ldots, x_n$ (inputs) and column y (result). Data should be split into two subsets; one for training and one for testing. The training set will be the first tsr rows from the total tr rows. Creation of the polynomial (1) can be described as in the following steps:

Construction of the New Variables - To construct new variables $x_1, x_2, x_3, \ldots, x_{\binom{n}{2}}$. We take all $\binom{n}{2}$ combinations of the input pairs u, v from tsr rows of $x_1, x_2, x_3, \ldots, x_n$ and find the best least squares polynomial, which best fits the results.

$$y = C_1 + uC_2 + vC_3 + u^2 C_4 + v^2 C_5 + uv C_6 \qquad (2)$$

Now we use (2) to calculate the y (result) value for every input data row of every u, v input pair and store the results for that pair as z_m, where m is the index of that pair.

Selection of the New Inputs - Calculate least square error (se) between $z_1, \ldots, z_m$ obtained in the previous step and result y, for every input data row as given in (3).

$$se = \sum_{i=1}^{tr}(y - z)^2 \qquad (3)$$

If the result is $se > M$, where M is a predefined threshold, the pair is discarded. Find the smallest se, and if this se is greater than the smallest se obtained during the previous execution of this step, discard every z and go to the next step (unless if this is the first iteration). All remaining results of z are sorted by se and will replace inputs x. The algorithm will iterate back to the 1st step.

Selection of the Result - The first x column will now have the smallest se and this result represents the result of the GMDH polynomial (1). To obtain the coefficient "$a_0, a_i, a_{ij}, a_{ijk}, \ldots$", of the polynomial (2), every input that was used in each iteration must be saved and evaluated.

2.1 Singular Value Decomposition and GMDH Coefficients

It was previously mentioned, that coefficients $C_0, \ldots, C_6$ of the u, v input pair polynomial (2) are obtained by finding the least squares polynomial, that best fits the results y. Singular Value Decomposition (SVD) is a method that can be used for solving most linear least square problems.

The method to obtain the coefficients $C_0, \ldots, C_6$ with SVD, proposed in [4] is described as the following:

1. Create matrix $A = (1, u, v, u^2, v^2, uv)$ from u, v training data columns
2. Perform SVD on A: $A = U\Sigma V^*$
3. Invert every value in Σ
4. Get coefficients by calculating: $C = V^*\Sigma U^T Y$, where Y is array with results.

2.2 Discrete Differential Evolution Algorithm

The optimization algorithm used in this research is the Discrete Differential Evolution Algorithm (DDE) [6] which itself is the modified version of Differential Evolution (DE) algorithm [8]. The DDE algorithm is based on vector perturbation by employing the following steps:

1. Initialization of the population - Population must be initialized randomly of dimension D and population size of NP.

2. Forward transformation - each element in the population is transformed using the following Eq. (4):
$$z = -1 + \frac{z' \times 500}{10^3 - 1} \quad (4)$$
where z is the new real number and z' is the original integer.
3. Mutation - Eq. (5) is used for the mutation of selected objective function parameter of the selected individual:
$$v_{j,i} = z_{j,r_1} + F \cdot (j, z_{r_2} - z_{j,r_3}) \quad (5)$$
where v denotes a value in the trial population and z is the transformed value. $j = [1, D]$ is the index of the objective function parameter, $i = [1, NP]$ is index of the individual in the trial population and $r_1, r_2, r_3 \in [1, NP]$ are randomly selected integers, except $r_1 \neq r_2 \neq r_3 \neq i$, that represents indices of the individuals in the current generation. $F \in (0, 1]$ is a positive scale factor, that is typically less than 1 and remains constant during the execution of the algorithm.
4. Crossover - a crossover probability $CR \in [0, 1]$ is used to select which elements are accepted in the new trial vector.
5. Backward transformation - each value is transformed back to an integer by Eq. (6):
$$z = \frac{(1 + z') \times (10^3 - 1)}{500} \quad (6)$$
6. Selection - if the new solution vector improves on the previous indexed vector, it is included in the population.
7. Stopping Criteria - iteration of the previous steps till a specified number of generations G.

3 Hybrid DE-GMDH

In this paper, the search for the optimal Network Model is done by the optimizer and the entire Network Model is defined by the individuals in the population of the DDE algorithm.

This concept changes the network creation paradigm. In the original GMDH, fitness of every node is evaluated during the network creation and the result of the network creation is the final network. In our algorithm, each individual has its own network, and the network creation and evaluation can be separated and the final network is the network of the individual with the best fitness. The same concept can be used through other evolutionary algorithms [4].

In DE-GMDH, the following **rules** hold true:

1. Nodes are the same as the ones in the standard GMDH (2 inputs, 1 result, coefficients, ...).
   Theoretically, the order of the inputs does not matter, because u, v and u, v pairs should end up with the same coefficient, except C_2 and C_3 being

swapped, however due to floating point errors, this does not hold true and will be addressed in the next section.
2. No nodes will be removed based on $se > M$, therefore M parameter no longer exists.
3. Result of every node must be used.
4. Layer must have at least 1 node.
5. Last network layer will have only 1 node, the result node.
6. Node can use results of the nodes from all previous layer as its inputs.
   This rule was added to expand the number of possible network structures, by creating nodes, that will be used as inputs by the nodes in higher than next layer, which would not be possible in the original GMDH.
7. Structure of the whole network will be presented as individuals in the population.
8. Columns with input data are considered to be results of the nodes of the 0 layer. This rule exists sorely to make the explanation easier.

From these rules, the following **observations** can be made:

1. Number of the nodes in the last layer is 1 as stated in rule 5.
2. Number of nodes in the n layer, where n is not the number of the last layer, is at most two times more than the number of nodes in $n+1$ layer as deduced from rules 1, 6 and 3.
3. Number of nodes in the 1st layer is $2^{(lc-1)}$ where lc is number of positive network layers as deduced from observation 2.
4. Maximal number of nodes in the network is $2^{lc} - 1$ where lc is number of positive network layers as deduced from observation 2 and rule 5.

As stated in rule 6, there is a need to find a way to present the GMDH network as an individual in the population. If rule 6 is changed to having nodes in layer n using the results of the nodes in layer $n-1$ as their input, then observations 3 and 4 will give the exact number of nodes, instead of the maximal. It is assumed, that there are no constraints when in comes to which node uses which input, that is if there is a free choice of inputs of the nodes in the 1st layer and inputs for all the other nodes in the other layer, all possible networks can be obtained.

Figure 1a shows the network with no structural constraints and Fig. 1b shows the network with predefined structure of every layer except the 0 layer. Name of the node is the combination of the names of its inputs, nodes with 1 letter name are not nodes, but input data columns (nodes of the 0 layer).

It is clear, that result nodes of the two networks are the same, therefore the networks themselves must be the same. The only difference is the order in which the nodes were drawn. This demonstrates that the network with a predefined structure in every layer except the 1st one is just a re-ordered version of any other network.

With having this knowledge, the creation of the population that represents such a network is trivial, as such a population is simply the sequence of the nodes in the 0 layer of the network. Because nodes in the 0 layer are input data

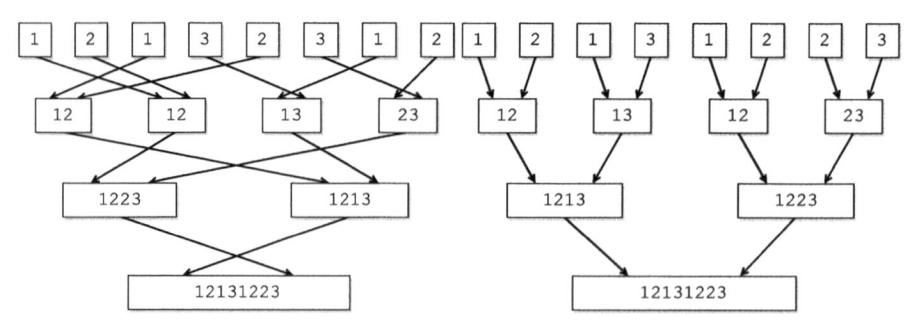

(a) Network of DE GMDH with no structure constraints (b) Network of DE GMDH with structure constraints

Fig. 1. Network structures

columns, we know their range and we also know the number of nodes in the 0 layer is 2^{lc} where lc is the number of positive network layers.

Therefore, the original rule 6 can be reiterated. The assumption remains that the node cannot have the same inputs. Nodes having the same input are permitted, but the nodes will not do any calculation and will simply use one of its inputs as the result. Therefore, such nodes are redundant, however they keep the network structure the same. Figures 2a and b demonstrate this concept.

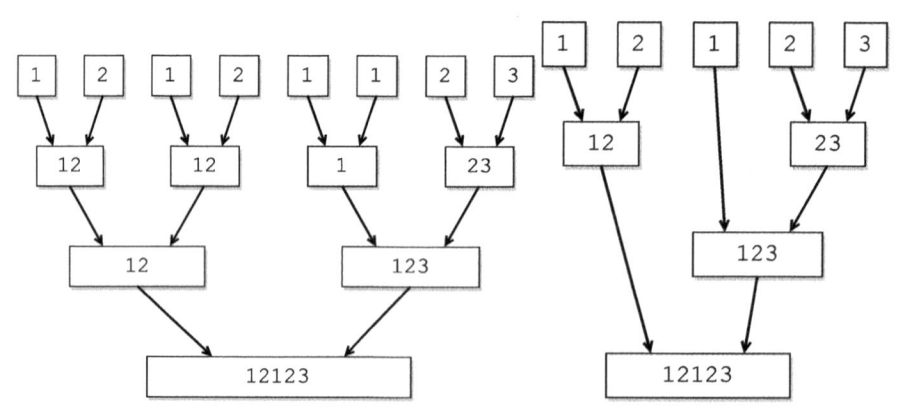

(a) Network of DE GMDH with nodes that do nothing (b) Same network as the one in 2a, but without the useless nodes

Fig. 2. Optimised network structures

The preceding description has described the generation of the network as an individual in the population. The fitness evaluation of the individual is the *square error* of the network. As this point DDE is applied to the population (values in the individuals are indices of the input data columns, therefore integers), until some individual has the sufficient fitness or the population is evaluated.

4 DE-GMDH Network Structure Optimization

From the previous section, it is obvious that the network created from the specification of all the inputs to all the nodes in the 1st layer is not optimized. It may contain nodes with the same inputs having the same result, therefore they can be simply reused. As already mentioned, the order of inputs actually matters.

In fact, most of the Network Model will contain redundant nodes (redundant node is a node that has the same inputs as another node), because the structure of a network with no redundant nodes is severely limited, as the number of nodes in the 1st layer of such network is at most $\binom{idcc}{2} + idcc$ where $idcc$ is the number of input data columns and any next layer would have to have at most half of the nodes that were in the previous layer. Therefore, redundant nodes are in fact a necessary part of any network.

4.1 Network Model

The initial concept is to emphasize that the optimized network is not updated to the individual of the optimizer as it would overwrite the current data. Therefore, a redundant structure must be used to save the current network.

The Network Model will contain the number of layers, number of nodes in each layer and its information, which itself contains index and layer of each input, coefficients and number or nodes and its results.

4.2 Non-redundant Ordered Network

This network contains no redundant nodes and the nodes are ordered. To be able to find redundant nodes, the non-redundant nodes must be stored with a relatively fast addition and search mechanism. Items should also be easily accessible in an ordered form. A Red-Black tree which fulfills these conditions is utilized.

During the optimization, the following are stored; Network Model, **node buffer** - Red-Black tree with non-redundant nodes of the currently processed layer, **reindex 2D array** - translates the index of the node in the **node buffer** to the index of the node in the Network Model, **layer buffer** - holds the current processed unoptimized layer and **layer index** - index of the layer in the Network Model that will be created from the **node buffer**.

The following is the procedure of the processing of the 0th layer:

1. Set **layer index** to 0.
2. For each input pair in the individual, check if the inputs differ.
    - If the inputs are the same, it is assumed that this node does nothing and therefore the **layer buffer**, that corresponds to this pair is set to -1 and index to the value of the pair.

- If the inputs differ, it is ordered and the **layer buffer** of the set layer of the node that corresponds to this pair is set to 0 (1st layer) and index to its index in the **node buffer**. If this node is not in the **node buffer**, then it is added.
3. At this point, the **node buffer** holds all nodes in the 1st layer of the optimized network and **layer buffer** holds the indices and layers of the nodes in the Network Model (after re-indexing) for every node in the unoptimized 1st network layer.
    If **node buffer** contains at least one node:
    - Copy ordered **node buffer** to the 1st layer of the Network Model. and delete the **node buffer**. (in case of binary trees, simply in-order tree traversal and delete)
    - Set **reindex 2D array** for the 1st network layer.
      Get index of the node from the **node buffer** and set the value of the element with that index to the index of that node in the 1st layer of the Network Model.
    - Increase the **layer index**.

Processing the nth layer, where n is positive:

1. For each node pair in the **layer buffer**, check if the inputs differ.
    - If the inputs are the same, then the node, which uses these 2 nodes as inputs does nothing and therefore there is no reason to have it in the optimized network. In the **layer buffer**, set layer and index of that node to the layer and index of one of its input nodes.
    - If the inputs differ, re-index and order them and in the **layer buffer** set layer of the node, which uses these 2 nodes as inputs to **layer index** and index to its index in the **node buffer**. If this node is not in the **node buffer**, then add it.
2. Now the **node buffer** holds all nodes in the nth layer of the optimized network and **layer buffer** holds the indices and layers of the nodes in the Network Model (after re-indexing) for every node in the unoptimized $n + 1$ network layer.
    If **node buffer** contains at least one node:
    - Copy ordered **node buffer** to the **layer index** layer of the Network Model and delete the **node buffer**. (in case of binary tree, simple in-order tree traversal and delete)
    - Set **reindex 2D array** for the **layer index** network layer.
      Get index of the node from the **node buffer** and set the value of the element with that index to the index of that node in the **layer index** layer of the Network Model.
    - Increase **layer index**.

Figure 3 illustrates the entire process. The individual, which is to be optimized has both redundant nodes and wrong input order. Individual and then **layer buffer** are always processed in from 1st to last element order. Ascending order is used during the ordering, with nodes first ordered by input indices,

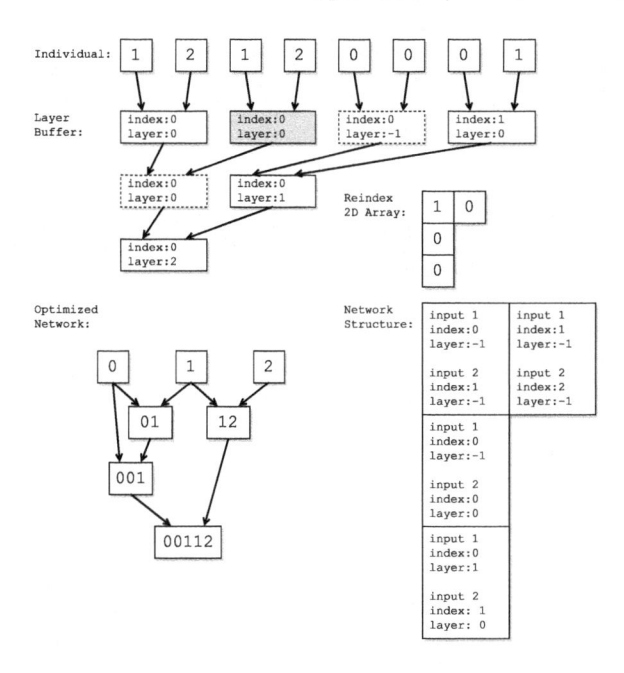

Fig. 3. Example of the network redundancy and order optimization

then by input layers. Elements of the **layer buffer**, that are drawn with dashed line are nodes, which do nothing. Elements with gray background are redundant nodes.

5 Experimentation

The experimentation was conducted on the data sets obtained from [7]. Calculation time, memory usage, fitness of the final network model and maximal

Table 1. DE-GMDH test, data set [7], parameters

Name of the setting	Value
Input data columns	8
Training rows	48
Testing rows	20
Minimal network layers	3
Maximal network layers	5
Population size	200
Number of generations	300
Number of threads	2

Table 2. Results

Exp	T[s]	Mem [kB]	Fitness	Max Error	Exp	T[s]	Mem [kB]	Fitness	Max Error	Exp	T[s]	Mem [kB]	Fitness	Max Error
1	11.05	3312	8.02	0.982515	35	9	3260	8.92	0.998098	69	8.66	3292	11.53	0.892511
2	9.42	3308	9.51	0.792771	36	8.57	3384	10.09	0.826108	70	8.59	5276	8.98	0.834783
3	12.17	3348	9.84	0.790907	37	8.17	3348	9.95	1.00589	71	8.72	3424	10.82	0.855282
4	9.11	3192	9.89	1.21925	38	8.57	3192	10.94	0.829831	72	8.46	3188	10.32	0.85769
5	9.22	3404	10.42	0.741038	39	8.65	3276	10.51	1.07898	73	8.41	3420	9.01	1.29643
6	11.2	3292	11.31	0.723873	40	8.2	3232	11.24	0.703747	74	7.96	3344	8.83	0.944639
7	8.92	3260	9.45	1.06642	41	8.21	3288	8.5	0.830847	75	9.09	3232	10.64	0.820314
8	9.29	3396	11.62	0.874577	42	8.4	3212	9.78	0.798889	76	8.38	3420	10.27	0.887867
9	12.57	3304	11.16	1.05915	43	8.73	3292	8.89	0.890912	77	8.38	3308	9.9	0.888998
10	8.51	3292	10.18	0.866129	44	8.41	3308	9.75	0.928692	78	8.28	3392	10.69	0.934538
11	8.46	3232	9.15	1.03542	45	8.75	3420	8.25	0.869296	79	8.75	3288	8.11	0.920469
12	9.9	3420	8.72	1.05225	46	8.77	3308	11.34	0.745576	80	8.49	3352	10.12	0.922684
13	8.86	3420	8.62	0.770817	47	8.89	3392	10.67	0.921538	81	8.13	3288	10.54	0.750269
14	8.19	3212	8.89	1.06146	48	8.65	3292	9.84	0.818856	82	8.63	3308	9.07	0.796465
15	8.94	3380	11.1	0.776602	49	8.21	3304	9.08	0.950988	83	8.12	3284	9.07	1.25922
16	10.76	3208	9.39	0.909112	50	8.32	3300	7.95	0.811601	84	8.98	3292	9.25	0.893134
17	9.42	3232	9.23	0.876594	51	8.95	3188	9.55	0.900837	85	9.3	3236	9.98	0.984178
18	8.64	3420	9.13	0.907707	52	8.37	3288	11.57	0.848294	86	10.67	3344	9.22	0.899675
19	11.86	3280	9.73	0.762563	53	8.85	3404	9.32	0.737687	87	8.97	3236	9.77	0.87256
20	8.91	3308	8.5	0.776128	54	8.81	3288	9.88	0.826109	88	8.32	3284	8.83	1.14174
21	8.33	3228	9.69	0.826108	55	8.57	3308	11.25	0.962365	89	8.68	3264	9.87	0.940894
22	8.76	3312	11.13	1.04738	56	8.87	3304	13.34	0.679892	90	9.61	3352	8.2	1.05465
23	8.55	3292	10.25	1.06475	57	8.32	3212	9.74	0.883564	91	13.14	3384	9.31	0.938758
24	8.4	3376	8.11	1.04192	58	8.17	3380	9.9	1.01949	92	8.46	3184	8.86	0.857977
25	8.85	3380	8.81	1.09471	59	9.28	3424	9.08	0.817781	93	8.7	3188	11.14	0.791459
26	8.82	3408	12.47	0.855537	60	9.29	3264	9.47	0.926157	94	8.98	3312	9	1.00022
27	8.53	3236	10.23	0.868345	61	8.79	3212	10.31	0.742625	95	8.48	3192	9.87	0.850252
28	8.36	3304	9.75	1.1451	62	8.79	3308	10.94	0.823457	96	9.15	3356	10.06	0.842232
29	8.97	3212	8.7	0.947506	63	8.42	3208	8.92	0.998098	97	8.31	3404	10.31	0.758566
30	8.85	3232	8.72	0.965071	64	8.69	3280	10.03	0.923275	98	8.86	3408	9.95	1.04081
31	9.19	3308	10.09	0.826108	65	8.48	3208	11.47	0.84364	99	7.65	3412	10.7	0.995424
32	8.71	3304	10.91	0.806484	66	8.27	3380	9.06	0.791702	100	8.55	3228	10.78	0.784684
33	9.05	5276	8.33	0.902174	67	8.56	3288	9.9	0.767436					
34	8.58	3304	8.81	0.868797	68	8.79	3408	9.19	0.810517					

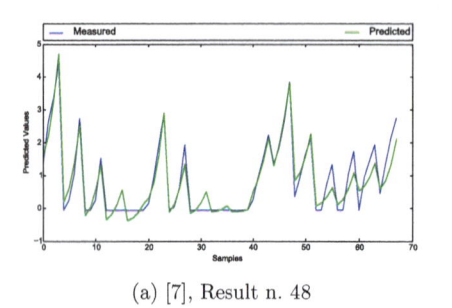

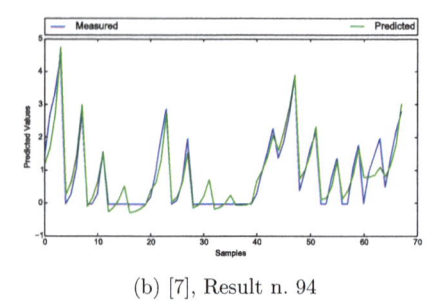

(a) [7], Result n. 48

(b) [7], Result n. 94

Fig. 4. Performance of DE-GMDH on data set

prediction error is given for each evaluation. Graphs with predicted and measured values are shown for two randomly selected results for each dataset.

Fitness of the result fi is calculated by (7):

$$fi = \frac{r}{\sum\limits_{i=1}^{r}(y-z)^2} \qquad (7)$$

where r is number of rows in the input data, y is the measured value of the result and z is the predicted value of the result.

Table 1 shows the parameter setting for the DE-GMDH and Table 2 shows the results obtained. Figures 4a and b show graphs with measured and predicted values of the 48th and 94th experiments respectively.

From the results, the minimal-maximal error in Table 2 is 0.679892 and the maximal error is 1.29643, while the range of the measured results is $[0, 4.5]$, which makes that minimal and maximal error around 15.1% and 28.8% of that range, respectively. Figures 4a and b also show that measured and predicted values never exceed that maximal-maximal error.

6 Conclusion

This paper introduces a unique modular hybrid approach to the GMDH network. The foremost is the application of DDE as the optimizer of the entire network structure. This was the novel approach to removing redundant nodes in the layers and only having valid inputs evaluated. This naturally decreased the execution time and at the same time removes the possibility of more network with the same structure, but different order of inputs of the nodes, to have different results, due to floating point errors and total memory requirements for the network.

SVD was used for the calculation of the coefficients of the nodes in the DE-GMDH network. The implementation of SVD coefficients calculator itself was designed to be easily replaceable, if another method for the coefficients calculation is to be used.

The obtained results show that the developed model is consistently finding good results within an acceptable margin of error, therefore validating the model. Parallelization using threads also improved the execution time.

References

1. Davendra, D., Onwubolu, G., Zelinka, I.: Group method of data handing using discrete differential evolution In: Matlab, pp. 229–260 (2016). https://doi.org/10.1142/9781783266135_0006
2. Ivakhnenko, A.G.: The group method of data handling-a rival of the method of stochastic approximation. Sov. Autom. Control **13**(3), 43–55 (1968)
3. Madala, H., Ivakhnenko, A.: Inductive Learning Algorithms for Complex Systems Modeling. CRC Press Inc., Boca Raton (1994)

4. Nariman-Zadeh, N., Darvizeh, A., Ahmad-Zadeh, G.: Hybrid genetic design of gmdh-type neural networks using singular value decomposition for modelling and prediction of the explosive cutting process. Proc. Inst. Mech. Eng. Part B J. Eng. Manuf. **217**(6), 779–790 (2003)
5. Onwubolu, G.: GMDH-Methodology and Implementation in MATLAB. Imperial College Press, London (2016). https://doi.org/10.1142/p982
6. Onwubolu, G.C., Davendra, D.: Differential Evolution: A Handbook for Global Permutation-Based Combinatorial Optimization, 1st (edn). Springer, Heidelberg (2009)
7. Shi, J., Wang, J.Y., Liu, C.: Modelling white layer thickness based on the cutting parameters of hard machining; data set= 68x8. Proc. Inst. Mech. Eng. Part B J. Eng. Manuf. **220**(2), 119–128 (2006)
8. Storn, R., Price, K.: Differential evolution-a simple and efficient heuristic for global optimization over continuous spaces. J. global Optim. **11**(4), 341–359 (1997)

A Better Indexing Method for Closest Open Location Policy in Forklift Warehouse Operation

Duy Anh Nguyen[✉], Truong Thinh Pham, and Viet Duong Nguyen

Ho Chi Minh City University of Technology, Ho Chi Minh City, Vietnam
duyanhnguyen@hcmut.edu.vn

Abstract. Due to rapid development in logistic industry, there has been growing concern over cold warehouse operation. Four fundamental elements that affect the efficiency in warehouse management include travel, search, pick up, set up. Travel time composes the highest proportion in the group hence recent studies are in favor of promoting time travel in indexing pallet location for COL policy. In fact, because picking also plays an essential role in reducing operating cost, methods optimized both travel and picking are definitely more thorough. This paper aims to point out potential measures by imposing a better indexing method for closest open location policy to address storage location assignment problems, along with the simulation of the forklift moving products in warehouse in order to generate more reliable data. The efficiency of these strategies ensures warehouse can be operated in the most efficient way.

Keywords: COL · Closest open location ·
Storage location assignment problem · Warehouse management problem

1 Introduction

As the ultimate goal of warehouse management is optimize operation cost and time, a generous amount of study has been searched for sufficient measures. According to a well-known research, the figure for order picking is above 50% of the overall cost, away from the other factors. Cutting down the time for this stage can help to save the corporation money easily. Therefore, it is an urgent topic that needs to be taken care of.

Items need to be put into storage location before it can be picked to fill customers' orders. A storage assignment method is a set of rules used to assign items to storage location. Once the order is accepted, the warehouse must produce pick lists to guide the order-picking. These activities are typically accomplished by a warehouse management system, a large software system that co-ordinates the activities of the warehouse.

This paper is design for cold warehouse which is allowed transportation. Major perception holds the view to develop algorithm in order to determinate available storage location in storage process. These acts aim to decrease travel distance and provide storage method to optimized warehouse operation [1–3]. With the COL policy in this article, products will be move to storage place in minimum time. So as to reduce errors, the authors simulate how the forklift moving products in warehouse as well as loading/unloading stocks. The algorithm is run by V- rep software and MATLAB.

2 Methodology

2.1 Assumption Made

Assumptions in this article may not suitable for all type of warehouse and not identical to the condition in reality but it is designed base on the actual size of a warehouse and can be used for some warehouses in real status [1, 4–6].

- The capacity of system are 480 storage locations and ability store of each location is one SKU (stock keeping unit – an inventoried item).
- There are 6 types of frozen shrimp:
    A1: type 100 g/2 shrimp with head intact
    A2: type 100 g/2 shrimp without head
    A3: type 100 g/4 shrimp with head intact
    A4: type 100 g/4 shrimp without head
    A5: type 100 g/6 shrimp with head intact
    A6: type 100 g/6 shrimp without head
- Goods are organized into the pallet. Each pallet is a SKU. This is the smallest item in system. Pallet is placed on single pallet racking and other picker can reach all items in the rack regardless of rack's height.
- Warehouse system use forklift truck to transfer pallets to store and retrieve pallet.
- Because the frozen shrimp are seafood products, the first in first out (FIFO) method is used in the storage and retrieve process. The earliest goods purchased are the first ones removed from the warehouse.

2.2 Layout Design

From assumptions presented in Sect. 2.1, traditional warehouse layout was used in this article with 2 pick aisle and 5 storage aisles. Information about both travel time and picking time are required to apply COL policy. There are 16 forklift – compatible racks. Each rack has 60 storage locations. These lines are named Roman numerals I to VIII for each storage aisle. This design help forklift truck can reach all locations in the pallet racking [4].

Each line is divided into 30 storage locations. Each storage is for one pallet; therefore, the system can store up to 480 pallets. The line in this paper is consist of 5 tiers (A, B, C, D and E) and 6 bays (which are categorized from 1 to 6) (see Fig. 1). Assumptions and mentioned simulated systems are the base for storage algorithms. The storage location is assigned by a coordinate followed by structure Pos- Line- Tier- Bay. For more detail, I/O points are separated into two distinct ones.

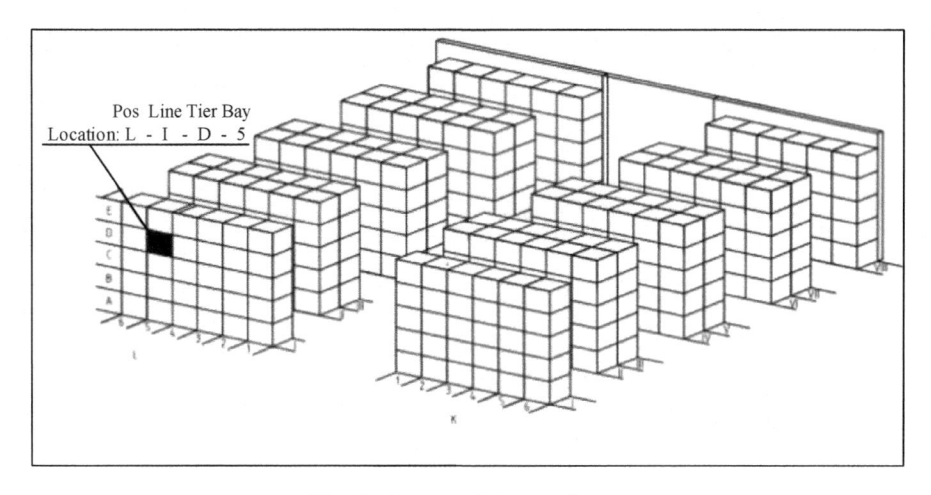

Fig. 1. Layout of the warehouse

2.3 Basic of Closest Open Location – COL Policy

We focus on the COL policy since it is one of basic rules for solving storage location assignment problems. The policy is worked dramatically in case there is no information is available on the characteristics of the arriving items [6] as well as is developed another advanced policy such as Continuous Cluster method which is a combination of dedicated, randomized and COL policy [5]. Under COL policy, SKUs are often allocated to empty location where is closest to the Input/Output (I/O) terminal. "Closest" is always defined by the travel distance from the I/O point to storage location [7] that aims to total transfer time is reduced because travel distance is proportional to warehouse operation time.

In this article, COL policy is based on travel time and picking time of warehouse operation. Closest location is defined by the sum of travel time, loading time and unloading time to reduce the errors between theory and practical, this study also simulates forklift truck base on real forklift truck to run exactly and get its travel time and picking time.

2.4 Time-Based COL Policy

The algorithm was built based on the assumption in Sect. 2.1, COL policy was applied to assign each pallet to a storage location and FIFO policy is used for retrieval process. In this study, the objective of the algorithm is to minimize not only travel time but also picking time and searching time. Closest in this article mean total time are required to move pallet from I/O point to its storage location.

The notation and variable used in this model are listed as follows:
t: Total time
t_R: Total running time
t_l: Total lifting time
d_{es}:   Expected distance

v_r: Running speed in straight line
v_{rc}:   Running speed in curve line
R: Turning radius
n_r: Number of turning radius in actual rout
$v_{l/u}$:   Average loading speed/unloading speed
x_i, y_i: X and y coordinates of the Node
z_i: Height of storage locations (i = 1, 2, 3, 4, 5)
n: Number of picking product
k: Number of Node on forklift truck route

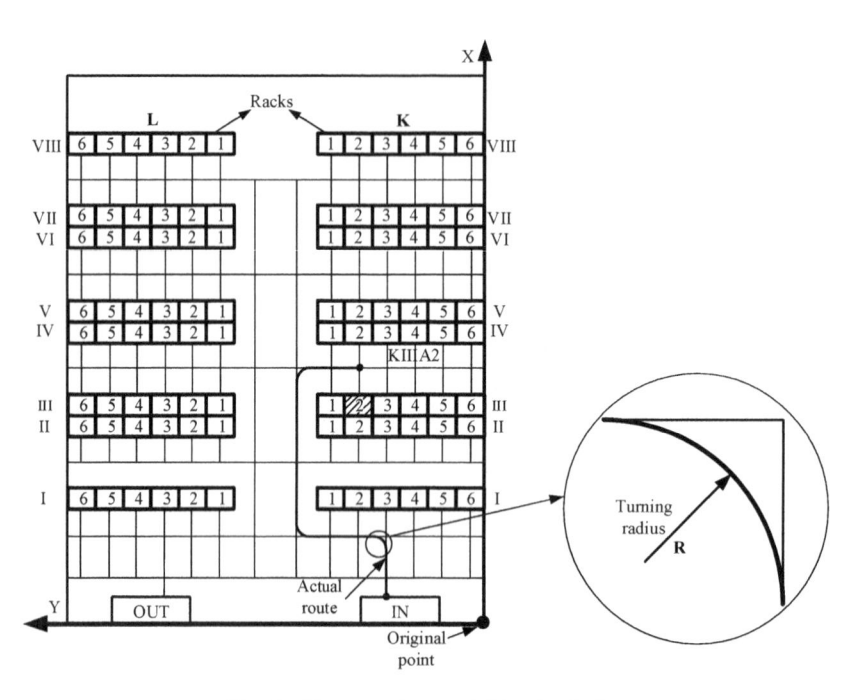

Fig. 2. The mapping model in simulation

Expected distance are determined from the IN terminal to the center of each slot in 2D warehouse layout (see Fig. 2). The expected distance can be estimated by Euclidian distance.

$$d_{es} = \sum_{i=1}^{k-1} \sqrt{(x_{i+1} - x_i)^2 + (y_{i+1} - y_i)^2} \qquad (1)$$

In practice, it is faster that the forklift follows a continuous path rather than making a rigid 90° turn so turning radius is essential. Each node is presented by 3 coordinates (x_i, y_i, z_i) and the turning angle can be calculated:

$$\vec{u} = (x_i - x_{i-1}; y_i - y_{i-1}) \tag{2}$$

$$\vec{v} = (x_{i+1} - x_i; y_{i+1} - y_i) \tag{3}$$

$$cos\theta = \frac{\vec{u}.\vec{v}}{\|\vec{u}.\|\|\vec{v}\|} \tag{4}$$

$$=> |\theta| = \begin{cases} 0, \textit{forklift truck continuod to go straight} \\ 1, \textit{forklift truck begin to turn at Node}(i-1) \end{cases} \tag{5}$$

From (1), (4), and (5)

$$t_r = \frac{d_{es} - 2Rn_r}{v_r} + \frac{\left(\frac{1}{2}n_r\pi R\right)}{v_{rc}} \tag{7}$$

Estimate total loading and unloading time base on the height of storage racks.

$$t_l = \frac{2.z_i}{v_{l/u}} \tag{8}$$

The total time is needed to transfer from I/O point to all storage locations can be calculated. The total time after calculated are stored in 5 × 6 matrix. During storage process, system using this matrix to find which location has lowest time index to sort pallet.

$$t = t_r + t_l = \frac{d_{es} - 2Rn_r}{v_r} + \frac{\left(\frac{1}{2}n_r\pi R\right)}{v_{rc}} + \frac{2z_i}{v_{l/u}} \tag{9}$$

2.5 Distance-Based COL Policy

The distance-based COL policy is based on the value of travel distance are used in this article can be computed by the formulas below [5]:

$$d_{ij} = x_{ij} + y_{ij} + z_{ij} + offset \tag{10}$$

$$x_{ij} = |x_j - x_i| \, y_{ij} = |y_j - y_i| \, z_{ij} = |z_j - z_i| \tag{11}$$

Where, (x_i, y_i, z_i), (x_j, y_j, z_j) are coordinates of I/O point and storage location. Offset is the distance need to be added to suitable with layout in Sect. 2.1.

3 Simulation Software

A simulation environment was built to compare the distance-based COL policy which is defined by travel distance and the time-based COL policy which is defined by operation time. The environment has three separate parts: a V-REP environment including a 3D warehouse with proximity sensor system in each storage location to verify the presentation of the pallets, validate the outcomes of the algorithm and evaluate the impact of COL policy and FIFO policy on warehouse performance, a MATLAB control center is constructed to oversee all operations of warehouse and data about processes is saved in excel file and a MATLAB Simulink to get information about travel time, picking time (Fig. 3).

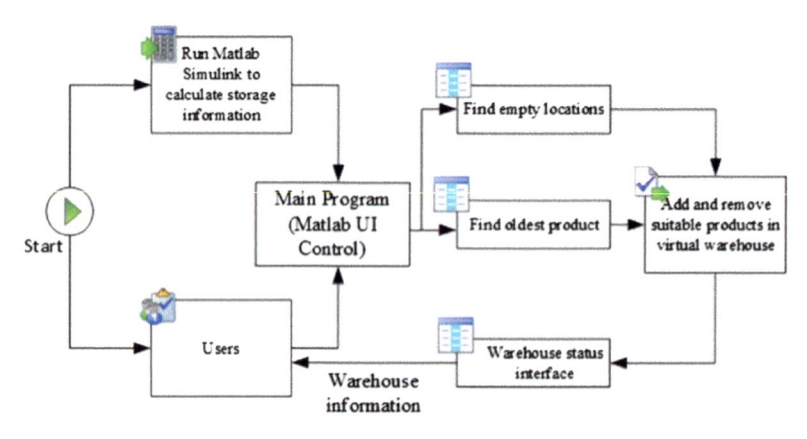

Fig. 3. Simulation software flow

The virtual warehouse in V-REP environment is connected with a MATLAB control UI and MATLAB Simulink. When the user starts the simulation, MATLAB Simulink also starts to get sum of travel time and picking time from Input point to all storage locations. All values are stored in matrices. Based on time values in these matrices, MATLAB UI receives instructions from the user about the number of products and the type of products need to be store/retrieve and applies either the COL policy to assigned storage locations or FIFO policy to determine which pallet is retrieved.

4 Simulation Result

4.1 Rack Stability

The storage locations priority under two difference types of COL policy, the locations (see Figs. 4 and 5) which have darker black color have the higher priority than brighter ones. The pallets are assigned to the empty slot from high priority to low priority.

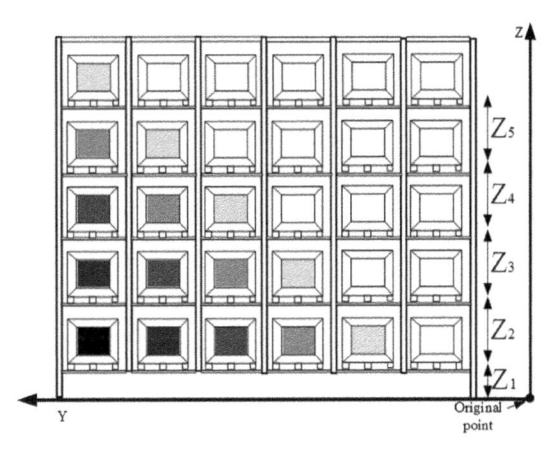

Fig. 4. Priority of each rack location in distance-based method

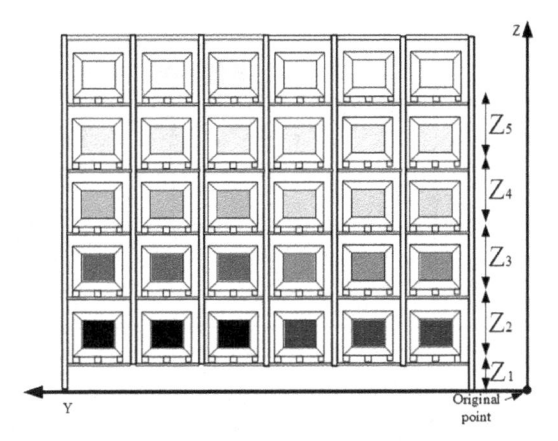

Fig. 5. Priority of each rack location in time-based method

Distance-based method disregards the time required to lift and place the pallet into each location. This affects the distribution of pallets on each rack to tip forward the inner side of the rack.

The time-based method places weight on the loading time hence the higher chance of pallets being placed at the bottom racks and the load is spread more evenly. This information of distribution of pallets can be used in further rack stability analysis.

Based on the stability analysis method of Lewis [8], the pallet racks using distance-based method is less stable than another pallet racks which have products are used COL policy based on the value of travel time (see Fig. 5) and picking time. That is the reason why under COL policy introduced in this article, the racking system is safer and the users can save warehouse operation cost from maintenance process.

4.2 Total Time

See Fig. 6.

Methods \ Fill level	10%	20%	30%	40%	50%	60%	70%	80%	90%
Distance-based (s)	1972	4448	7244	10191	13186	16458	19842	23325	27110
Time-based (s)	1676	3892	6378	9101	12041	15218	18634	22296	26227

Fig. 6. Total operation time of two methods

Overall, at all warehouse fill levels, the COL policy using travel time as a base to index the location requires shorter time to complete order. This proves that the loading time cannot be ignored as assumed in previous studies as there is a visible improvement in cost. The analysis of the operation time is shown in the figures below.

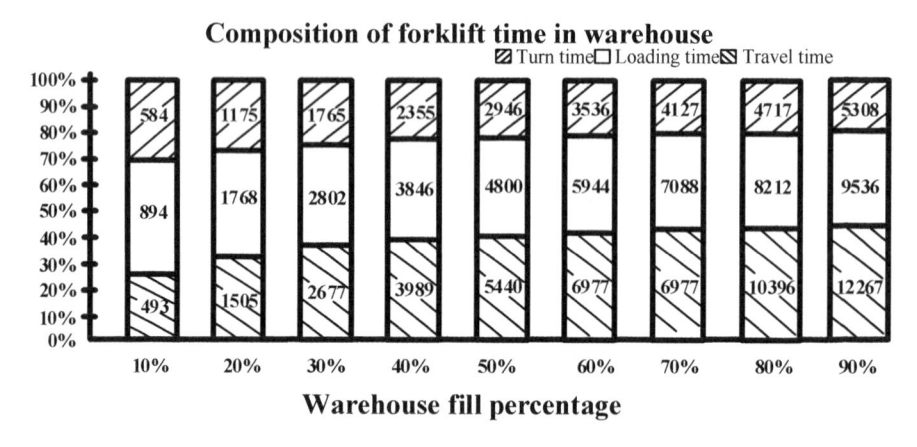

Fig. 7. Composition of the simulation results with distance-based indices

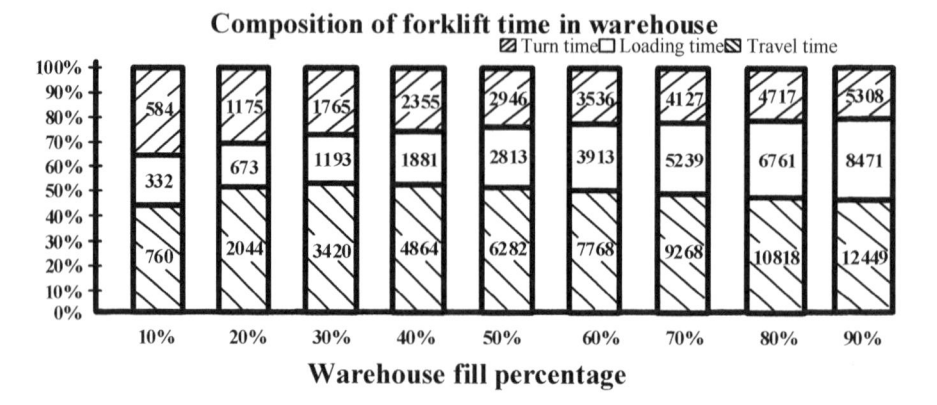

Fig. 8. Composition of the simulation results with time-based indices

Figures 7 and 8 show two results from the simulation with different indexing methods. Forklift warehouse operation can be divided into 3 components: travel time the amount of time the forklift takes to get from point to point, loading time is the time required to lift the pallets to designated positions, and turning time is the cost of getting the forklift into the correct orientation by turning without moving. In this classification of operation time, the lift time and drive time can vary with different indexing method and schedules while assuming that the forklift takes the optimal route every time and required the least amount of orientation correction.

The results in both figures indicate a significant reduction in the loading time at low operating levels. This reduction offsets the increase in forklift travel time and contributes to the savings in the cost of maneuver time shown in Fig. 9. The figures also point out the convergence in travel time and loading time for both methods as the fill level increasing. The finding confirms that at high fill levels both methods provide similar solutions and hence produce similar results. At lower fill levels, when there are more empty lots and the solution space is larger, the time-based indexing method provides better schedule by considering the loading time. The loading time, as shown in Fig. 7, takes up a major portion of the total time.

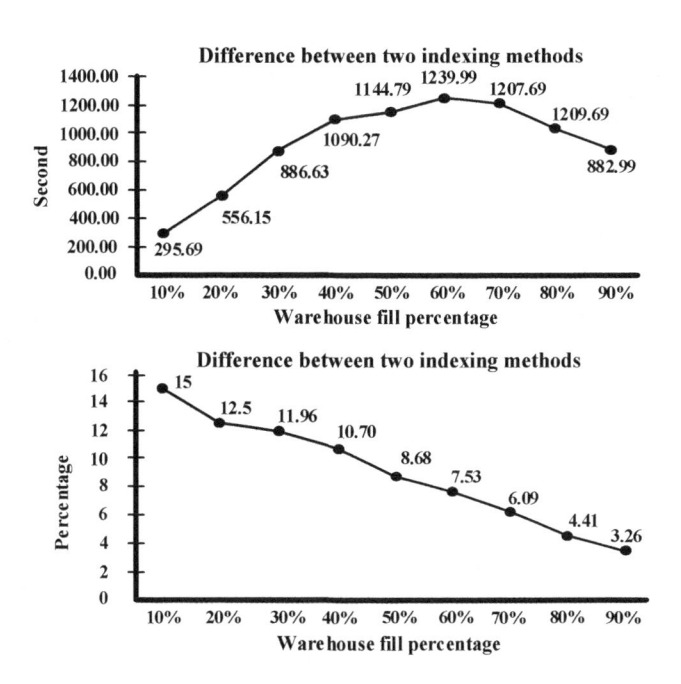

Fig. 9. The saving in cost of maneuvering time between two COL policies

The total time generated by the two methods indicates that the time-based method has moderately improved the total time, especially at low fill levels. This improvement, however, diminishes as the warehouse is filled up. This can be seen in Fig. 8b as the reduction percentage declines almost linearly from a 15% to a 3.26% reduction in operation time. The actual difference in second (see Fig. 9a) peaks at 60% fill level and slowly decreases due to the convergence in solutions of both methods. Due to the nature of warehousing, actual fill level may hover at a 60% to 80% mark. This converts to a saving of roughly 6% in cost of time.

5 Conclusion

The study has shown that the time-based COL policy has achieved better operation time compared to distance-based method. However, the benefit is marginal with high-filled level warehouse. With a normal operation level at 60% to 80%, the saving in cost of time stays within the range of 4 to 8% the total time.

Rack stability wise, the new time-based COL policy distributes the pallets more evenly. In long term, this even distribution prolongs the life span of the rack since there is less concentration of pallets and weight at various rack weak points.

The simulation was done with consideration of various factors including forklift kinematics, orientation, loading and unloading time and actual path taken by the vehicle. This level of simulation provides a more thorough results which helps with the evaluation of the cost of operation.

Extended study is needed to examine the extensive benefit of the algorithm when it is operated in longer period and with more complicated scenarios. With the flow of pallets in and out of the storage, the suggested time-based COL policy can be used in conjunct with proper scheduling and path planning to achieve greater reduction in cost.

References

1. Truong, N.C., Dang, T.G., Nguyen, D.A.: Development and optimization of automated storage and retrieval algorithm warehouse by combining storage location identification and route planning method. In: IEEE International Conference on System Science and Engineering (2017)
2. Wisittipanich, W., Meesuk, P.: Differential evolution algorithm for storage location assignment problem. In: Gen, M., Kim, K., Huang, X., Hiroshi, Y., (eds.) Industrial Engineering, Management Science and Applications (2015)
3. Hsieh, L.F., Tsai, L.: The optimum design of a warehouse system on order picking efficiency. Int. J. Adv. Manuf. Technol. **28**, 626 (2006)
4. Truong, N.C., Dang, T.G., Nguyen, D.A.: Development of automated storage and retrieval algorithm in cold warehouse. In: South East Asian Technical University Consortium Symposium (2017). ISSN 1882-5796
5. Truong, N.C., Dang, T.G., Nguyen, D.A.: Building management algorithms in automated warehouse using continuous cluster analysis method. In: Duy, V., Dao, T., Zelinka, I., Kim, S., Phuong, T., (eds.) AETA (2017)
6. Behnamian, J., Eghtedari, B.: Storage system layout. In: Farahani, R.Z., Hekmatfar, M. (eds.) Facility Location (2009)
7. Fitts, M.O.K.: Warehousing, department of industrial and systems engineering North Carolina State University. In: Kulwiec, R.A. (eds.) Materials Handling Handbook. Wiley, Hoboken (1985)
8. Lewis, G.M.: Stability of rack structures. In: Thin-Walled Structures, vol. 12, pp. 163–174 (1991)

On-Line Efficiency-Optimization Control of Induction Motor Drives Using Particle Swarm Optimization Algorithm

Sang Dang Ho[1,2], Pavel Brandstetter[2], Cuong Tran Dinh[1,2], Thinh Cong Tran[1,2], Minh Chau Huu Nguyen[2], and Bach Hoang Dinh[1(✉)]

[1] Faculty of Electrical and Electronics Engineering, Ton Duc Thang University, 19 Nguyen Huu Tho, Dist. 7, Ho Chi Minh City, Vietnam
{hodangsang,trandinhcuong,trancongthinh, dinhhoangbach}@tdtu.edu.vn
[2] Faculty of Electrical Engineering and Computer Science, VSB-Technical University of Ostrava, 17. Listopadu 15/2172, 708 33 Ostrava-Poruba, Czech Republic
{pavel.brandstetter,huu.chau.minh.nguyen.st}@vsb.cz

Abstract. This paper proposes a method for optimizing the power efficiency of induction motor drives based on the Particle Swarm Optimization algorithm. The power efficiency is improved by adjusting the current magnetization component for a given load torque so that total loss of copper and iron could be minimized. To verify the effectiveness of the proposal method, the simulation in MATLAB/SIMULINK has been implemented and compared with the conventional Rotor Flux Oriented Control. The result shows that the proposed method has improved the power efficiency of the IM drives under the light load regime with a considerable loss reduction.

Keywords: Induction motor (IM) · Rotor flux oriented control (RFOC) · Loss minimization algorithm (LMA) · Particle Swarm Optimization (PSO)

1 Introduction

More than 50% of the electrical energy produced is consumed by motors where induction motors are the most widely used in electrical drives because of their reliability, ruggedness and relatively low cost. Energy saving in induction motor drives aims to minimize the power loss in normal operation which matches the load torque requirement. So far, many approaches have been developed in order to obtain a highly efficient IM drives which have been divided into two categories [1]: online search controllers (SC) and loss-model-based controllers (LMC).

The basic principle of online search methods is to measure the power consumption and iteratively adjust the magnetic flux level so that the minimum value of the power consumption is detected. This approach doesn't need the prior knowledge of motor parameters, but the convergence seems to be slow, and it may cause the chattering of flux and torque during motor's operation.

In the other hand, the idea of LMCs is to establish the power loss function using the mathematic model of IMs and then determine a flux level that minimizes the losses [2–5]. Normally, the LMC methods converse faster than the online search methods but they are more sensitive to noises and/or motor parameter variations. Thus, LMCs are well suited for IM drives controlled by the vector control method requiring motor parameters [4].

There are various optimization search techniques based LMCs to determine an optimal flux level according to each operational point which corresponds a given load torque and machine speed, such as Neural Networks (NN) [5–7], Genetic Algorithm (GA) [8, 9], and Particle Swarm Optimization (PSO) algorithm [10]. Among mentioned approaches, PSO has risen to the most popular method due to its simplicity and accuracy. This algorithm is developed by Kennedy and Eberhart in 1990 [11, 12]. It is characterized as simple in concept, easy to implement, and computationally efficient more than many other heuristic techniques. Especially, PSO has a flexible and well balanced mechanism to enhance the global and local exploitation abilities which possibly obtain a high-quality optimal solution [10].

Because 80% of the total losses is related with copper and iron losses, thus this paper focuses on reducing these electromagnetic losses and neglecting other losses. In this study, an online loss minimization algorithm based on PSO is proposed for RFOC induction motor drives by properly adjusting the magnetic flux which keeps an appropriate balance between copper and iron losses to minimize the electromagnetic losses. To verify the effectiveness of the proposed method, simulations in MATLAB/SIMULINK have been implemented and compared with the result of the classical rotor flux oriented control (RFOC).

2 Induction Motor Loss Model

A mathematical model of IM's losses is developed in [3] by defining the rotor flux as $\Psi_r = L_m i_{mr}$. An iron loss resistor R_{Fe} is connected in parallel to the magnetizing inductance in a reference frame fixed to the rotor magnetizing current where the x-axis has been aligned in the direction of the magnetizing current i_{mr} as shown in Fig. 1. As a result, the steady state equivalent circuits of IMs in x-axis and d-axis are shown in Fig. 2. In the steady state, there is no leakage inductance on the rotor side and the sum of the referred stator current equals the iron current if it is perpendicular to the magnetizing current i_{mr}. As a result, $i_{sx} = i_{mr}$ and it is used to forms the magnetic flux, and $i_{sy} = i_f + i_r$ which is related to the torque control for motors.

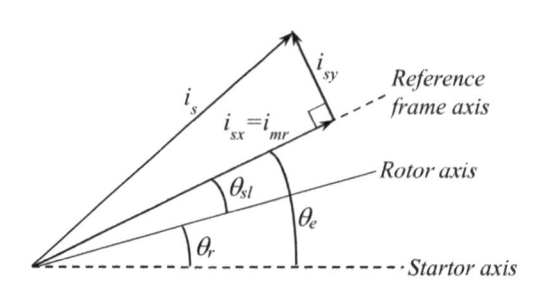

Fig. 1. Phasor diagram of equivalent circuit and rotor field angle definition [3]

The differential equations considering the iron loss [3] are given as following

$$\underline{u}_s = R_s \underline{i}_s + PL'_s \underline{i}_s + j\omega_e L'_s \underline{i}_s + PL'_m \underline{i}_m + j\omega_e L'_m \underline{i}_m \tag{1}$$

$$\begin{aligned}\underline{i}_s &= \underline{i}_m + \underline{i}_f + \underline{i}_r \\ &= \underline{i}_m + (p+j\omega_e)\frac{L'_m}{R_{Fe}}\underline{i}_m + (p+j(\omega_e-\omega_r))\frac{L'_m}{R'_r}\underline{i}_m\end{aligned} \tag{2}$$

where $p \equiv d/dt$ is the differential operator; $\omega_e = d\theta_e/dt$ is the electrical angular speed of the rotor flux; $\omega_r = d\theta_r/dt$ is the electrical rotor speed; R_s, R_r are resistances of the stator and rotor phase windings; L_s, L_r are self-inductances of the stator and rotor windings; L_m is the magnetizing inductance; $\sigma = (1-L_m^2/(L_s L_r))$ is the leakage factor; $R'_r = (L_m/L_r)^2 R_r$ is a referred rotor resistance; $L'_s = \sigma L_s$ is a referred stator inductance.

Replacing $\underline{u}_s = u_{sx} + ju_{sy}$; $\underline{i}_s = i_{sx} + ji_{sy}$; $\underline{i}_m = i_{mx} + ji_{my}$; and $i_{my} = 0$; $i_{mx} = i_{mr} = constant$ to RFOC scheme, we can derive the stator voltages and currents from (1), (2) as follows:

$$u_{sx} = R_s i_{sx} + pL'_s i_{sx} - \omega_e L'_s i_{sx} + pL'_m i_{mr} \tag{3}$$

$$u_{sy} = R_s i_{sy} + pL'_s i_{sy} + \omega_e L'_s i_{sy} + \omega_e L'_m i_{mr} \tag{4}$$

$$i_{sx} = i_{mr} + p\left(\frac{L'_m}{R_{Fe}} + \frac{L'_m}{R'_r}\right) i_{mr} \tag{5}$$

$$i_{sy} = \omega_e \frac{L'_m}{R_{Fe}} i_{mr} + (\omega_e - \omega_r)\frac{L'_m}{R'_r} i_{mr} \tag{6}$$

Let $R_t = R_{Fe} \| R'_r$, from (5) the magnetizing current can be expressed by

$$i_{mr} = \frac{1}{1 + p\left(\frac{L'_m}{R'_r} + \frac{L'_m}{R_{Fe}}\right)} i_{sx} = \frac{1}{1 + p\frac{L'_m}{R_t}} i_{sx} \tag{7}$$

From (6), the slip speed can be derived as

$$\omega_{sl} = \frac{R'_r}{L'_m}\frac{i_{sy}}{i_{mr}} - \omega_e \frac{R'_r}{R_{Fe}} = \frac{R_t}{L'_m}\frac{i_{sy}}{i_{mr}} - \omega_r \frac{R_t}{R_{Fe}} \tag{8}$$

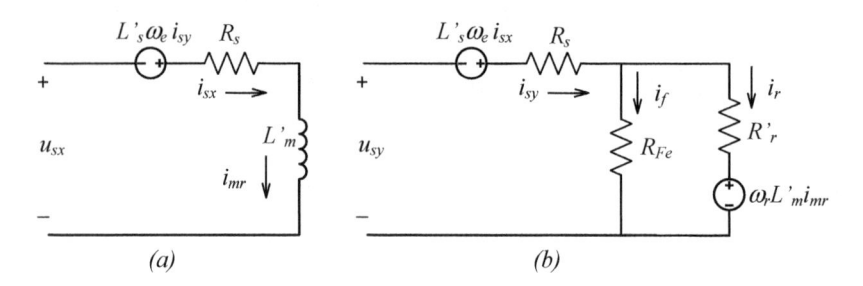

Fig. 2. Steady-state IM equivalent circuit in: (a) x-axis and (b) y-axis.

The power losses of IMs can be classified in five groups: stator copper losses P_{SCL}, rotor copper losses P_{RCL}, magnetic iron losses P_{fe}, mechanical losses P_m and stray losses. Since the electromagnetic losses, such as stator copper, rotor copper and iron losses, dominate the total power loss, thus, the mechanical losses, such as friction, windage and stray losses, are relatively small and can be neglected [2]. From Fig. 2, the total loss is given by

$$P_{loss} = P_{SCL} + P_{RCL} + P_{Fe} \\ = R_s(i_{sx}^2 + i_{sy}^2) + R'_r i_r^2 + R_{Fe}(i_{sy} - i_r)^2 \qquad (9)$$

From Fig. 2(b), the rotor current can be expressed as

$$i_r = i_{sy} - i_f = i_{sy} - \frac{R'_r}{R_{Fe}} - \omega_r \frac{L'_m}{R_{Fe}} i_{sx}$$

then

$$i_r = \frac{R_{Fe}}{R_{Fe} + R'_r} i_{sy} - \omega_r \frac{L'_m}{R_{Fe} + R'_r} i_{sx} \qquad (10)$$

Substituting from (10) into (9) we have

$$P_{loss} = R_d i_{sx}^2 + R_q i_{sy}^2 \qquad (11)$$

where

$$R_d = R_s + \frac{L'^2_m}{R_{Fe} + R'_r} \omega_r^2 ; \qquad R_q = R_s + \frac{R_{Fe} R'_r}{R_{Fe} + R'_r}$$

The developed electrical torque can be expressed as

$$T_e = \tfrac{3}{2} P L'_m i_{mr} i_r \\
= \tfrac{3}{2} P L'_m \left(\frac{R_{Fe}}{R_{Fe}+R'_r}\right) i_{sy} i_{mr} - \tfrac{3}{2} P \frac{(L'_m i_{mr})^2}{R_{Fe}+R'_r} \omega_r \quad (12)$$

Since $R_{Fe} \gg R'_r$ and $(R_{Fe}+R'_r) \gg (L_m i_{mr})^2$, torque (12) can be approximated by

$$T_e \simeq \frac{3}{2} P L'_m i_{sy} i_{mr} = K_t i_{sy} i_{mr} \quad (13)$$

where $K_t = (3/2) P L'_m$

Then in the steady state, we can present the stator current components in relation to the magnetizing current, i_{mr}, as

$$\begin{cases} i_{sy} = \frac{T_e}{K_t \cdot i_{mr}} \\ i_{sx} = i_{mr} \end{cases} \quad (14)$$

Thus, the total loss in the steady state can be written in terms of the magnetizing current as

$$P_{loss} = R_d i_{mr}^2 + R_q \frac{T_e^2}{K_t^2 i_{mr}^2} \quad (15)$$

Finally, the procedure of loss minimization is how to discover the optimal magnetizing current i_{mr} to obtain the minimum power loss in Eq. (15) for a given load torque in the steady state operation.

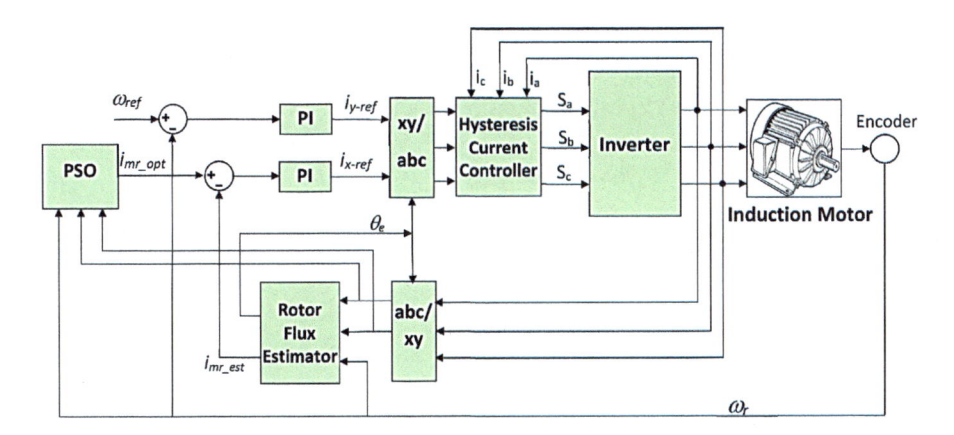

Fig. 3. Block diagram of the proposed optimization control system based PSO

A simplified block diagram of the proposed optimization method is shown in Fig. 3. It is implemented in the base of the conventional RFOC system. In this scheme, the stator currents and rotor speed are measured in order to calculate all parameters used by the PSO block (which were presented in Eq. (15): R_d, R_q, T_e) to minimize the objective function which is the total loss of the motor. The PSO output is a reference value of the magnetizing current which corresponds to minimum loss with a given torque and rotor speed to the RFOC system.

3 Online Optimizing the Efficiency of Induction Motor Drives Based on the Particle Swarm Optimization Algorithm

Particle swarm optimization (PSO) was first proposed by Russell Eberhart and James Kennedy (1995) [11, 12]. PSO is an evolutionary computation technique motivated by the simulation of social behaviors. Among particles "flying" through a multidimensional search space, each particle representing a single intersection of all search dimensions. PSO has two primary operators: updating new solutions via velocity and via position. The advantage of the PSO over the other optimization algorithms is its relative simplicity and stable convergence with a good computational efficiency. The new velocity value for each particle during each iteration is computed based on its current velocity, the distance from previous best position and from global best position. Therefore, the new velocity is used to update the next position of a particle in the search space. This process is iteratively implemented until the minimum error is achieved.

In this paper, the PSO algorithm is proposed to minimize the total power loss by searching the optimum value of the magnetizing current, $i_{mr\text{-}opt}$. The procedure of this algorithm is defined as following:

1. Formulate the initial population and initial velocities randomly of each particle (i_{mr})
2. Calculate the fitness function of each particle using Eq. (15)
3. Find the individual best of each particle.
4. Find the global best in the entire population.
5. Update the velocity and position of each particle [12]:

$$V_i^{(t+1)} = wV_i^{(t)} + C_1 Rand_1 \left(Pbest_i - X_i^{(t)} \right) + C_2 Rand_2 \left(Gbest - X_i^{(t)} \right) \quad (16)$$

$$X_i^{(t+1)} = X_i^{(t)} + V_i^{(t+1)} \quad (17)$$

6. Repeat from step 2 to step 5 until the criterion is satisfied.
7. Apply the optimized magnetizing current i_{mr_opt} to RFOC system.

where: w -inertia weigh; $V_i^{(t)}$ - current velocity of particle i at iteration t; $V_i^{(t+1)}$ - new velocity of particle i at iteration $t + 1$; C_1 - adjustable cognitive acceleration constants (self confidence); C_2 - adjustable social acceleration constant (swarm confidence); $Rand_{1,2}$ - random number between 0 and 1; $X_i^{(t)}$ – current position of particle i at iteration t; $Pbest_i$ – the personal best of particle i; $Gbest$ – the global best of the population; $X_i^{(t+1)}$ - denotes the position of particle i at the next iteration $t + 1$.

4 Simulation Results

In order to verify the effectiveness of the proposed loss minimization scheme, the MATLAB/SIMULINK simulation has been implemented in according to Fig. 3. The simulation is carried out on a three-phase, squirrel cage induction motor, 380 V, 4 kW (5.5HP), 50 Hz, and 4-pole, 1440 rpm, rated load torque $T_n = 26.5$ Nm, rotor flux reference $\Psi_r = 0.954$ Wb. The motor parameters in this experimental tests are similar to the reference parameters in [13]:

$R_s = 1.37\,\Omega$; $R_r = 1.1\,\Omega$; $L_s = 0.1459\,H$; $L_r = 0.149\,H$; $L_m = 0.141\,H$;
$J = 0.1\,kgm^2$.

The equivalent core losses resistance R_{Fe} is also given by [13], which is experimentally analyzed and approximately expressed in terms of frequency as following:

$$R_{Fe}(\Omega) = \begin{cases} 128.92 + 8.242f + 0.0788f^2 & f \leq 50Hz \\ 1841 - 55275/f & f > 50Hz \end{cases} \quad (18)$$

where f is the stator frequency or rotor flux frequency.

Figure 4 shows the speed, torque and stator currents responses before and after the loss minimization algorithm (LMA) based PSO is initiated for the IM drive. The simulation started with the load torque of 5.2Nm (0.2 pu) and the reference speed of 150 rad/s. When the loss minimization scheme is activated at t = 1.5 s, a slight variation in the rotor speed has been observed but it has been recovered at the reference speed very quickly within 0.3 s. When LMA initiated, the magnetizing current as well as two components of the stator currents have been chosen from the searching process in order to achieve the minimum loss while maintaining the required torque. Thus, the calculated total loss has decreased significantly as seen in Fig. 4(e). Moreover, the peak-peak torque ripple has also been reduced as shown in Fig. 4(d).

Figures 5 and 6 present the performance of a variety of simulations where the speeds are 150 rad/s and 75 rad/s, different load torques are set up in a range from 0.1 to 1 pu, and the controller has attached with and without optimization algorithm, respectively.

The results show that a significant efficiency improvement has been achieved in the performance of the proposed LMA based PSO compared to that of the conventional RFOC without optimization algorithm. Moreover, this optimization method can obtain the best results with light loads.

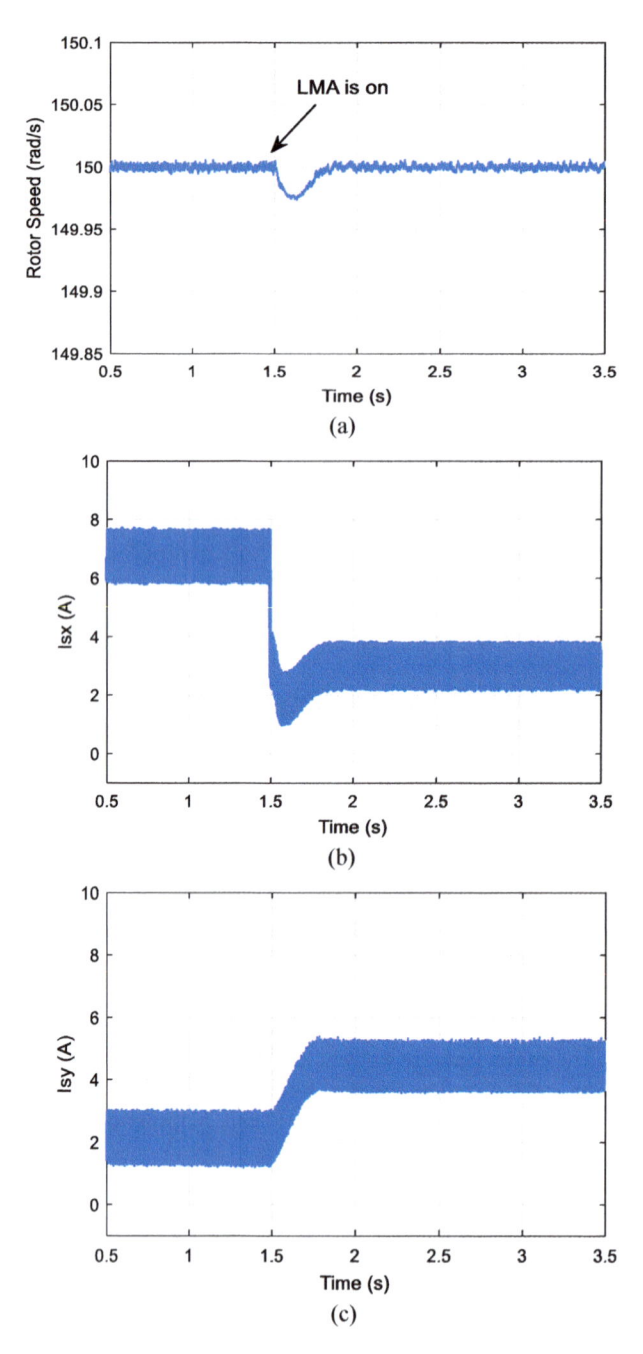

Fig. 4. Simulation results of the proposed drive at a speed of 150 rad/s and load of Nm: (a) rotor speed, (b) stator x-axis current, (c) stator y-axis current, (d) torque and (e) total losses before and after the LMA is on.

(d)

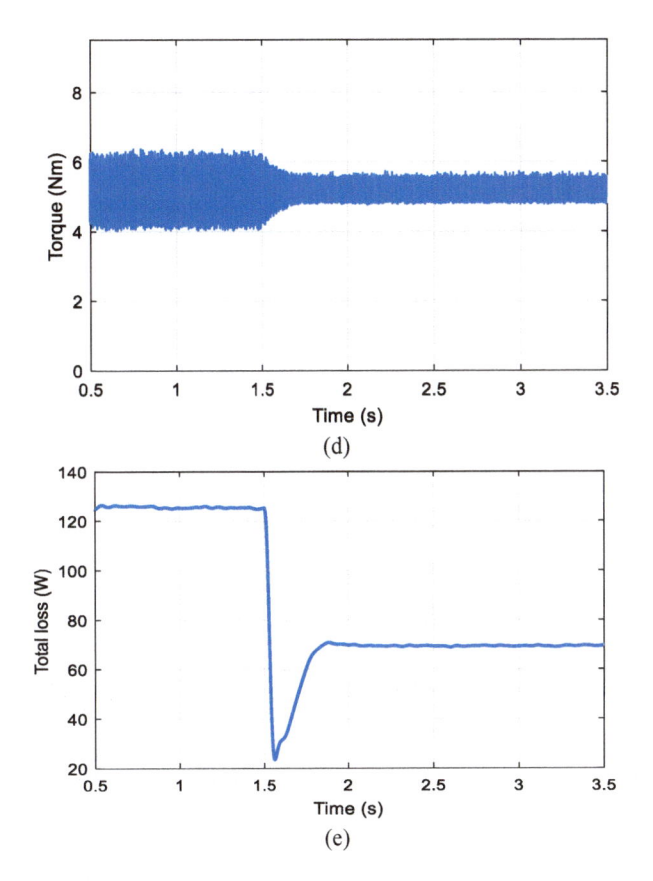

(e)

Fig. 4. (*continued*)

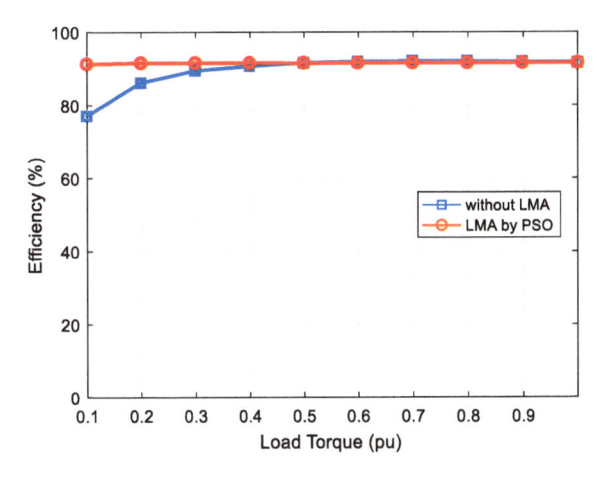

Fig. 5. Motor efficiency for a speed of 150 rad/s

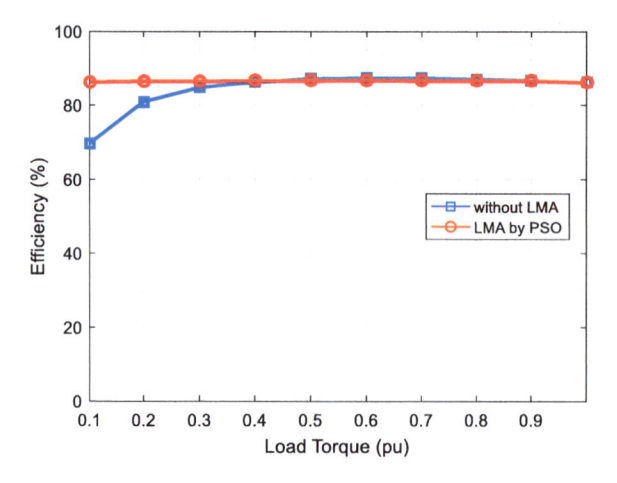

Fig. 6. Motor efficiency for a speed of 75 rad/s

5 Conclusion

In this paper, an online optimizing the efficiency of the induction motor drive based on the Particle Swarm Optimization algorithm has been suggested and applied to improve the IM efficiency. The power loss model of IM drives has been developed by the structure of an iron loss resistor connecting in parallel with a magnetizing inductance. The proposed strategy focuses on decreasing the rotor flux by adjusting the magnetizing current with respect to the torque current component. The obtained results reveal that the efficiency is greatly improved, especially in a range of light loads. Consequently the energy saving is significantly achieved in various operating conditions.

Acknowledgement. The paper was support by the Project reg. no. SP2018/162 – Student Grant Competition of VSB-Technical University of Ostrava, Research and development of advanced control methods of electrical controlled drivers, member of research team, 2018.

References

1. Bazzi, A.M., Krein, P.T.: Review of methods for real-time loss minimization in induction machines. IEEE Trans. Ind. Appl. **46**(6), 2319–2328 (2010)
2. Li, J., Xu, L., Zhang, Z.: A new efficiency optimization method on vector control of induction motors. In: IEEE International Conference on Electric Machines and Drives, pp. 1995–2001 (2005)
3. Uddin, M.N., Nam, S.W.: New online loss-minimization-based control of an induction motor drive. IEEE Trans. Power Electron. **23**(2), 926–933 (2008)
4. Zengcai, Q., Ranta, M., Hinkkanen, M., Luomi, J.: Loss-minimizing flux level control of induction motor drives. IEEE Trans. Ind. Appl. **48**(3), 952–961 (2012)
5. Alexandridis, A.T., Konstantopoulos, G.C., Zhong, Q.-C.: Advanced integrated modeling and analysis for adjustable speed drives of induction motors operating with minimum losses. IEEE Trans. Energy Convers. **30**(3), 1237–1246 (2015)

6. Abdin, E.S., Ghoneem, G.A., Diab, H.M.M., Deraz, S.A.: Efficiency optimization of a vector-controlled induction motor drive using an artificial neural network. In: 29th Annual Conference on IEEE Industrial Electronics Society, IECON 2003, pp. 2543–2548 (2003)
7. Wang, Z., Xie, S., Yang, Y.: A radial basis function neural network based efficiency optimization controller for induction motor with vector control. In: International Conference on Electronic Measurement & Instruments, ICEMI 2009, pp. 866–870 (2009)
8. Pokier, E., Ghribi, M., Kaddouri, A.: Loss minimization control of induction motor drives based on genetic algorithm. In: IEEE International Electric Machines and Drives Conference, IEMDC 2001, pp. 475–478 (2001)
9. Rouabah, Z., Zidani, F., Abdelhadi, B.: Efficiency optimization of induction motor drive using genetic algorithms. In: IEEE Transaction on Energy conservation, pp. 204–208 (2009)
10. Hamid, R.H.A., Amin, A.M.A., Ahmed, R.S., El-Gammal, A.A.A.: New technique for maximum efficiency of induction motors based on particle swarm optimization (PSO). In: IEEE International Symposium on Industrial Electronics, ISIE 2006, pp. 2176–2181 (2006)
11. Kennedy, J., Eberhart, R.: Particle swarm optimization. In: Proceedings of the IEEE International Conference on Neural Networks, pp. 1942–1948 (1995)
12. Shi, Y., Eberhart, R.: A modified particle swarm optimizer. In: Proceedings of the IEEE International Conference on Evolutionary Computation, pp. 69–73 (1998)

Introducing the Run Support Strategy for the Bison Algorithm

Anezka Kazikova[✉], Michal Pluhacek, Tomas Kadavy, and Roman Senkerik

Faculty of Applied Informatics, Tomas Bata University in Zlin,
T. G. Masaryka 5555, 760 01 Zlin, Czech Republic
{kazikova,pluhacek,kadavy,senkerik}@utb.cz

Abstract. Many state-of-the-art optimization algorithms stand against the threat of premature convergence. While some metaheuristics try to avoid it by increasing the diversity in various ways, the Bison Algorithm faces this problem by guaranteeing stable exploitation – exploration ratio throughout the whole optimization process. Still, it is important to ensure, that the newly discovered solutions can affect the overall optimization process. In this paper, we propose a new Run Support Strategy for the Bison Algorithm, that should enhance the utilization of newly discovered solutions, and should be suitable for both continuous and discrete optimization.

Keywords: Bison Algorithm · Run Support Strategy · Exploration optimization

1 Introduction

Sources of inspiration for artificial intelligence applications seem to be limitless. The optimization field employs the bases of evolution [1], chromosomes [2], or even the collective intelligence phenomenon, which created the swarm algorithms [3]. The swarm algorithms are powerful tools for solving both continuous and discrete minimization problems by simulating animal behavior. Though many animal species are social creatures, they do not necessarily need to have a leader to make ultimate decisions, and yet they manage to complete nontrivial optimization tasks, like foraging, hunting, mating, or protecting themselves against predators. Simulating such behavior patterns created a variation of successful optimization techniques like the Particle Swarm Optimization [4], Grey Wolf Optimizer [5], SOMA [6], or Cuckoo Search [7], which were already used to solve some challenging real-life applications [8, 9].

However, many swarm algorithms lean towards premature convergence due to excessive exploitation at the expense of exploration [10]. The so-called *abnormal exploitation* techniques tackle this problem by raising the population diversity [10–12]. One of the latest swarm algorithms, the Bison Algorithm, addresses the same problem differently – by guaranteed exploitation–exploration ratio through the whole optimization process. The Bison Algorithm is a multi-agent system, which divides its population into two groups: the swarming group, and the running group. While the first is exploiting the solutions, the second is exploring the search space.

Since the very first proposal of the Bison Algorithm, the mechanics of the running group evolved rapidly. In [13, 14] the groups were divided solely by the quality of the found solution. Therefore the weaker solutions explored the search space, while the stronger ones managed the exploitation. However, this approach caused a gradual scattering of the running group, as both groups switched their members after each iteration according to their objective function values. Responding to this, [15] proposed a new group arrangement in which the successful exploration solutions were no longer switched, but only copied to the swarming group. The worse swarming solutions were abandoned, and the running group was left intact for further exploration. This redefined the basic Bison Algorithm, as it provided a more logical model.

To enhance the exploration factor even more, [16] proposed a *"run and seek"* variation of the Bison Algorithm, in which the running group temporarily exploited the area of a promising solution on their own and then returned to the exploring behavior. This approach significantly improved the optimization of several functions, yet it was unable to affect the convergence when the swarming group was already stuck in a local optimum quite close to the global one. Also, adding the possibility of switching the exploration and exploitation methods, revoked the guarantee of stable exploration-exploitation ratio.

In this paper, we propose a new Run Support Strategy without changing the exploration behavior of the running group, by redefining the center computation of the swarming movement only. This strategy should provide a robust way of promoting the newly discovered solutions, and should be usable even for large-scale problems, discrete, and real-time optimization.

The paper is structured as follows: Sect. 2 introduces the basic Bison Algorithm, Sect. 3 proposes the new Run Support Strategy and provides an example of the bison movement in 2-dimensional space. Section 4 designs the validation experiments of the new strategy and presents the results. Section 5 discusses the achieved results, and the impact on the future research is evaluated in Sect. 6.

2 Bison Algorithm

The Bison Algorithm is inspired by the exploitation and exploration patterns of bison herds. The former simulates the endangered herd behavior: when attacked, bison create a circle of the strongest individuals to protect the weak. The latter replicates the persistent running behavior [17]. The algorithm divides the population into two groups, each simulating different behavior as outlined in Algorithm 1.

Since many swarm algorithms are based on similar principles [18], we would like to highlight the difference between the Bison Algorithm and other optimization techniques. The main characteristics of the Bison Algorithm is the separation of the exploration and exploitation. There is a unique group of explorers running through the search space with the sole purpose of avoiding the perils of local optima. This exploration mechanism is the main difference between the Bison Algorithm and other swarm algorithm like the PSO, GWO or SOMA (which explores the search space as an integral part of the exploitation movement in a very different manner). The exploitation movement is based on the center of several fittest solutions, while other algorithms usually use only one best solution to move to. The algorithm was compared to other metaheuristics on IEEE CEC 2017 benchmark functions in [14–16].

Algorithm 1: Pseudo code of the basic Bison Algorithm

```
Initialization:
  Objective function: f(x) = (x₁,...,xd)
  Generate: swarming group randomly, running group around
            xbest and run direction vector (Eq. 4)
For every iteration do
  Compute the center of the swarming movement (Eqs. 1, 2)
  For every swarmer do
    Compute new position candidate f(xnew) (Eq. 3)
    if f(xnew) < f(xold) then move to xnew
  End
  Adjust run direction vector (Eq. 5)
  For every runner do
    Move in the run direction vector (Eq. 6)
  End
  Copy successful runners to the swarming group
  Sort the swarming group by f(x) value
End for
```

Swarming Behavior. The swarming behavior computes the center of the strongest solutions (Eqs. 1, 2) and then moves all the solutions from the swarming group closer to the center if it improves their quality (Eq. 3).

$$weight = (10, 20, \ldots, 10 \cdot s) \tag{1}$$

$$center = \sum_{i=1}^{s} \frac{weight_i \cdot x_i}{\sum_{j=1}^{s} weight_j} \tag{2}$$

$$x_{i+1} = x_i + (center - x_i) \cdot rand(0, overstep)_D \tag{3}$$

Where:

- s is the *elite group size* parameter,
- x_i and x_{i+1} represent the current solution and the new solution candidate,
- *rand(from, to)* is a random number in the range of the two given arguments,
- *overstep* defines the maximum length of the swarming movement,
- D represents the dimensionality of the problem.

Running Behavior. Meanwhile, the running group explores the search space by shifting the whole group in the run direction vector (Eq. 6), randomly generated during the initialization (Eq. 4) and slightly altered in each iteration (Eq. 5). When bison outreach the search space boundaries, they appear on the other side of the dimension.

$$run\ direction = rand\left(\frac{ub-lb}{45}, \frac{ub-lb}{15}\right) \quad (4)$$

$$run\ direction = run\ direction \cdot rand(0.9, 1.1)_D \quad (5)$$

$$x_{i+1} = x_i + run\ direction \quad (6)$$

Where:

- *rand(from, to)* is a random number in the range of the two given arguments,
- *ub* and *lb* are the upper and the lower boundaries of the search space,
- *D* represents the dimensionality of the problem,
- x_{i+1} and x_i represent the current solution and its previous state.

Table 1 describes the parameters of the algorithm and their recommended values [18].

Table 1. Parameters of the Bison Algorithm and their recommended values

Parameter	Description	Recommended value
Population		50
Elite group size	No. of best solutions for center computation	20
Swarm group size	No. of bison performing the swarming movement	40
Overstep	The maximum length of the swarming movement 0 – no movement 1 – max to the center	3.5

3 Run Support Strategy

To enhance the impact of the running group, we propose a new strategy for the center computation, applied for successful running solutions. When a runner finds a better solution than a swarmer, the discovered solution should replace the center of the swarming movement for a certain number of iterations, specified by a new parameter called the *run support* and the *overstep* parameter is temporarily changed (Algorithm 2). This enables the swarming group to exploit the area around of the discovered solution. When the iteration limit is met, the center of the swarming movement is computed from the fittest solutions within the swarming group (Eqs. 1, 2).

Algorithm 2: Center computation of the Run Support Strategy

```
if f(x_runner) < f(x_swarmer) do
  for next run support iterations
    center = x_runner
    overstep = rand(0.95, 1.05)
  end for
end if
```

Where:

- x_{runner} and $x_{swarmer}$ are the running and swarming solutions,
- $f(x)$ is the objective function value,
- *run support* is the number of iterations for the planned exploitation of the promising solution,
- *rand(from, to)* is a random number in the range of the two given arguments,
- *center* is the center of the swarming movement,
- *overstep* is the overstep parameter.

3.1 Run Support Strategy Movement Example

Figure 1 shows the application of the Run Support Strategy on population distribution showing the movement on 2-dimensional Schwefel's function. In the first picture, the

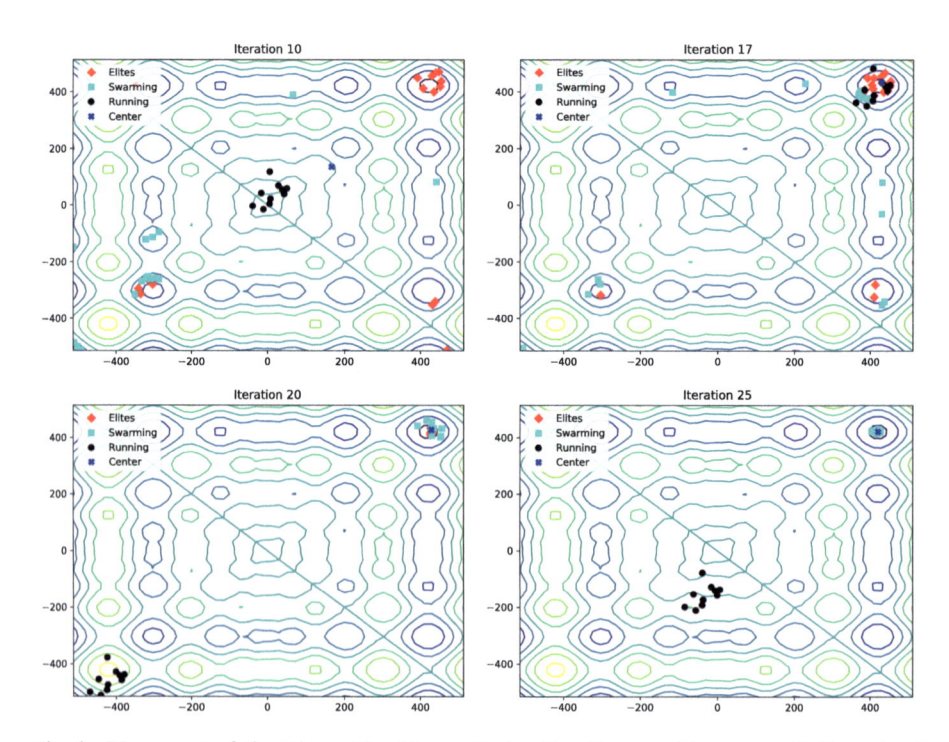

Fig. 1. Movement of the Bison Algorithm with the Run Support Strategy on 2-dimensional Schwefel's function

center is computed from the fittest solutions within the swarming group (elites). When the running group reaches the optimum location area (iteration 17), the center is shifted towards the newly discovered solution. When the run support limit is reached, the center is computed again from the best swarming solutions (iteration 20), and ultimately improves the final solution (iteration 25).

4 Methods and Results

As we were particularly interested in the cases, where the running group found a promising solution and employed the Run Support Strategy, we carried out a success simulation experiment. During this experiment, we placed one member of the running group exactly one run direction vector over the global optimum location (Eq. 7) and generated the rest of the running group around in the original formation. From there the running group explored the search space in the run direction vector, as expected.

$$x_{runner} = x_{opt} + run\, direction \qquad (7)$$

Where:

- x_{runner} represents one solution of the running group,
- x_{opt} is the known optimum location,
- $run\, direction$ is the run direction vector.

All of the experiments were held on 30 independent runs, each consisting of $10000 \cdot dimension$ evaluations of the objective function solving 10, 30 and 50-dimensional problems with well-known locations of global optima (Eqs. 8–11). The parameter configuration used for the experiments are described in Table 2.

Due to the exploration emphasis, the Bison Algorithm was originally intended to solve functions with a particularly narrow decreasing neighborhood around the global optimum. This description notably fits the Easom's function, as can be seen in Fig. 2.

Table 2. Parameter configurations applied for the experiments

Parameter	Basic Bison Algorithm	Success Simulation Run Support Strategy	Tuning of the Run Support Parameter
Population	50	50	50
Elite group size	20	20	20
Swarm group size	40	40	40
Overstep	3.5	3.5	3.5
Run support	0	2	0, 1, 2, 3, 5, 10

Rastrigin's Function

$$f(x) = 10 dim + \sum_{i=1}^{dim} x_i^2 - 10\cos(2\pi x_i) \tag{8}$$

Function minimum for E_n: $(x_1, x_2...x_n) = (0, 0,...,0)$. Value for E_n: $y = 0$.

The 2nd De Jong's Function (Rosenbrock's Valley)

$$f(x) = \sum_{i=1}^{dim-1} 100(x_i^2 - x_{i+1})^2 + (1 - x_i)^2 \tag{9}$$

Function minimum for E_n: $(x_1, x_2...x_n) = (1, 1,...,1)$. Value for E_n: $y = 0$.

Schwefel's Function

$$f(x) = 418.9829 - \sum_{i=1}^{dim} -x_i \sin\left(\sqrt{|x|}\right) \tag{10}$$

Function minimum for E_n: $(x_1, x_2...x_n) = (420.96,..., 420.96)$. Value for E_n: $y = 0$.

Easom's Function

$$f(x) = -\prod_{i=1}^{dim} \cos(x_i) \cdot e^{\sum_{i=1}^{dim} -(x_i - \pi)^2} \tag{11}$$

Function minimum for E_n: $(x_1, x_2...x_n) = (\pi, \pi,..., \pi)$. Value for E_n: $y = -dim + 1$.

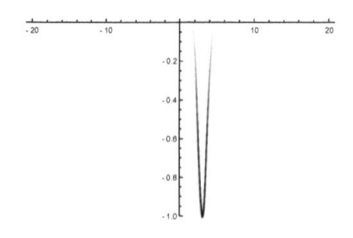

Fig. 2. Easom's function in 2 dimensions

To decide the ideal value of the *run support* parameter, we compared five values of the parameters on all the functions with the Friedman Rank Test ($p < 0.05$). It is worth mentioning that this experiment considered even the basic Bison Algorithm (with *run support* = 0). The results are shown in Table 3.

The results of the success simulation experiments are presented as follows: Table 4 compares the Run Support Strategy and the basic Bison Algorithm with the Wilcoxon rank-sum test ($\alpha = 0.05$). Table 5 shows the statistics of the algorithms. The optimum find rate represents the percentage of all the runs, where the algorithm was able to find a solution with the error $|f(x) - f(x_{global\,optimum})| < E - 8$.

Figure 3 shows a selection of mean convergences of both the success simulation experiments and the standard runs of the algorithms.

Table 3. The Run Support parameter values on functions with simulated success ranked by the Friedman Rank Test (p < 0.05). The best ranks are bold.

Run support parameter	0	1	2	3	5	10	P-Value
10 dimensions	5.00	3.00	3.50	**2.75**	3.25	3.50	0.66
30 dimensions	4.75	4.50	3.00	**1.75**	3.25	3.75	0.22
50 dimensions	5.00	**2.25**	3.25	3.25	4.00	3.25	0.47

Table 4. Winning algorithms on functions with simulated success (Wilcoxon $\alpha = 0.05$)

Dimension	Rastrigin	Rosenbrock	Schwefel	Easom
10 D	–	–	Run Support	–
30 D	–	–	Run Support	Run Support
50 D	–	–	Run Support	Run Support

Table 5. Performance of the Run Support Strategy and the basic Bison Algorithm on functions with simulated success (mean error, standard deviation, and optimum find rate)

	Run Support Strategy			Basic Bison Algorithm		
	avg	std	opt	avg	std	opt
Rastrigin						
10D	3.33	1.96	(7%)	3.97	3.39	(0%)
30D	19.27	6.38	(0%)	22.00	14.69	(0%)
50D	45.24	12.49	(0%)	44.64	10.39	(0%)
Rosenbrock						
10D	1.06	0.80	(0%)	1.23	1.50	(0%)
30D	13.65	5.15	(0%)	13.54	2.94	(0%)
50D	29.84	5.12	(0%)	35.67	16.19	(0%)
Schwefel						
10D	**203.52**	330.81	**(53%)**	741.07	610.45	(27%)
30D	**578.11**	1070.85	**(23%)**	3022.16	1156.57	(3%)
50D	**1091.92**	1785.85	**(10%)**	5129.62	1349.55	(0%)
Easom						
10d	1.84	3.15	(73%)	1.84	2.95	(70%)
30d	**5.11**	9.48	**(73%)**	19.19	10.87	(20%)
50d	**4.37**	11.94	**(80%)**	41.40	9.87	(3%)

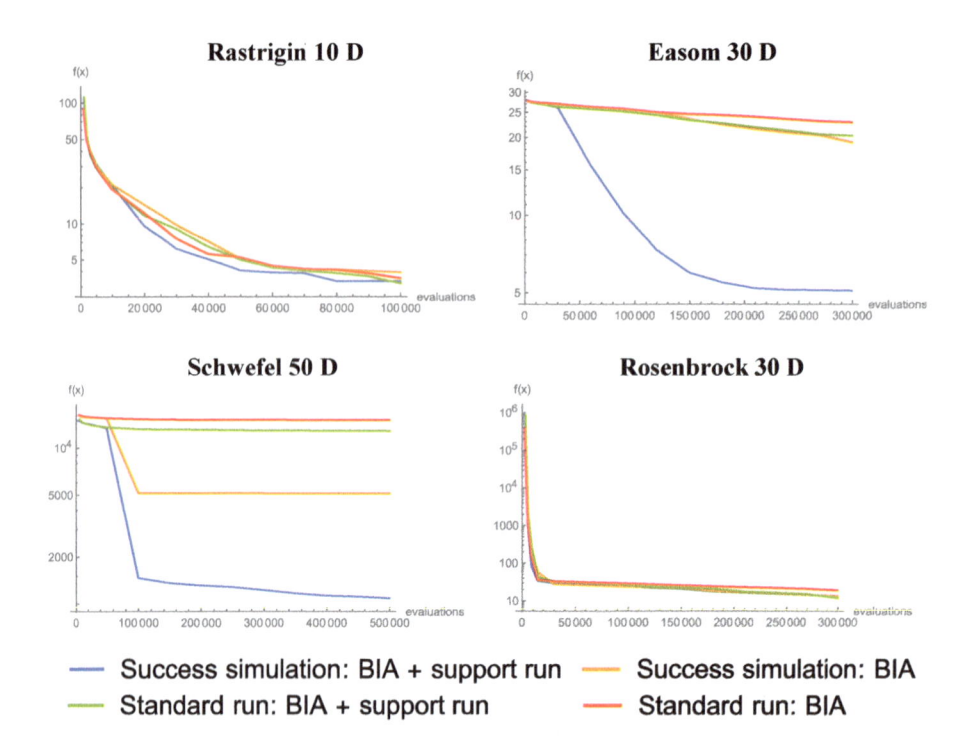

Fig. 3. Mean convergences of the basic Bison Algorithm and the Run Support Strategy both in the success simulation experiment and the standard run (with no bias of the running group)

5 Discussion

The success simulation experiment uncovered the supremacy of the Run Support Strategy over the basic version in all of the tested dimensions of Schwefel's function and high dimensional Easom's function (Wilcoxon $\alpha = 0.05$). At the 50-dimensional Easom's function, the new strategy was ultimately able to find the global optimum in 80% of all the runs. The Run Support Strategy also converged better than the basic algorithm even when comparing the standard run and success simulation alone.

The Friedman Rank Test ($p = 0.05$) experiment compared the *run support* parameter values. Even though that none of the configurations significantly outperformed the other ones, the best results were usually achieved when the *run support* parameter was set to 3 iterations. In contrast, when the *run support* was set to 0 (therefore when the basic algorithm was applied), the results were steadily ranked the worst. However, the results also suggest, the trend might change for higher dimensions.

6 Conclusion

We proposed a new Run Support Strategy for the Bison Algorithm and proved, that it is beneficial for the optimization process when compared to the basic version of the algorithm. The strategy was especially successful when solving the Easom's function, a prototype of a function with the global optimum hidden in a very narrow decreasing neighborhood. These results manifest the asset of the boosted exploration. Accordingly, we highly recommend the employment of the Run Support Strategy.

Furthermore, thanks to the robust design of the Run Support Strategy, the outcomes of this research will certainly be used in future research. We would like to exploit the uncovered benefits in a discrete version of the Bison Algorithm, large-scale, and real-time optimization.

Acknowledgment. This work was supported by the Ministry of Education, Youth and Sports of the Czech Republic within the National Sustainability Programme Project no. LO1303 (MSMT-7778/2014), further by the European Regional Development Fund under the Project CEBIA-Tech no. CZ.1.05/2.1.00/03.0089 and by Internal Grant Agency of Tomas Bata University under the Projects no. IGA/CebiaTech/2019/002. This work is also based upon support by COST (European Cooperation in Science & Technology) under Action CA15140, Improving Applicability of Nature-Inspired Optimisation by Joining Theory and Practice (ImAppNIO), and Action IC1406, High-Performance Modelling, and Simulation for Big Data Applications (cHiPSet). The work was further supported by resources of A.I.Lab at the Faculty of Applied Informatics, Tomas Bata University in Zlin (ailab.fai.utb.cz).

References

1. Back, T.: Evolutionary Algorithms in Theory and Practice: Evolution Strategies, Evolutionary Programming, Genetic Algorithms. Oxford University Press, Oxford (1996)
2. Goldberg, D.E., Holland, J.H.: Genetic algorithms and machine learning. Mach. Learn. **3**(2), 95–99 (1988)
3. Chakraborty, A., Kar, A.K.: Swarm intelligence: a review of algorithms. In: Nature-Inspired Computing and Optimization, pp. 475–494. Springer (2017)
4. Kennedy, J., Eberhart, R.: Particle swarm optimization. In: Proceedings of the IEEE International Conference on Neural Networks, vol. 4 (1995)
5. Mirjalili, S., Mirjalili, S.M., Lewis, A.: Grey wolf optimizer. Adv. Eng. Software. **69**, 46–61 (2014)
6. Zelinka, I.: SOMA—self-organizing migrating algorithm. In: New Optimization Techniques in Engineering, pp. 167–217. Springer, Heidelberg (2004)
7. Yang, X.-S., Deb, S.: Cuckoo search via Levy flights. In: Proceedings of World Congress on Nature & Biologically Inspired Computing (NaBIC 2009), December 2009, India, pp. 210–214. IEEE Publications, USA (2009)
8. Pluhacek, M., Senkerik, R., Viktorin, A., Kadavy, T., Zelinka, I.: A review of real-world applications of particle swarm optimization algorithm. In: Lecture Notes in Electrical Engineering, pp. 115–122. Springer, Cham (2018). ISSN 1876-1100
9. Mohamad, A., Zain, A.M., Bazin, N.E.N., Udin, A.: Cuckoo search algorithm for optimization problems-a literature review. In: Applied Mechanics and Materials, vol. 421, pp. 502–506. Trans Tech Publications (2013)

10. Chen, J., Xin, B., Peng, Z., Dou, L., Zhang, J.: Optimal contraction theorem for exploration–exploitation tradeoff in search and optimization. IEEE Trans. Syst. Man Cybern. Part A Syst. Hum. **39**(3), 680–691 (2009)
11. Pluhacek, M., Senkerik, R., Viktorin, A., Zelinka, I.: Chaos Enhanced Repulsive MC-PSO/DE Hybrid. Springer, Cham (2016)
12. Riget, J., Vesterstrøm, J.S.: A diversity-guided particle swarm optimizer-the ARPSO. Dept. Comput. Sci., Univ. of Aarhus, Denmark, Technical report (2002)
13. Kazikova, A., Pluhacek, M., Viktorin, A., Senkerik, R.: Proposal of a new swarm optimization method inspired in bison behavior. In: Matousek, R. (ed.) Recent Advances in Soft Computing (MENDEL 2017), Advances in Intelligent Systems and Computing. Springer, Cham (2018)
14. Kazikova, A., Pluhacek, M., Senkerik, R.: Performance of the Bison Algorithm on benchmark IEEE CEC 2017. In: Silhavy, R. (ed.) Artificial Intelligence and Algorithms in Intelligent Systems, CSOC2018, Advances in Intelligent Systems and Computing, vol. 764. Springer, Cham (2018)
15. Kazikova, A., Pluhacek, M., Viktorin, A., Senkerik, R.: New Running Technique for the Bison Algorithm. In: Rutkowski, L., Scherer, R., Korytkowski, M., Pedrycz, W., Tadeusiewicz, R., Zurada, J. (eds.) Artificial Intelligence and Soft Computing, ICAISC 2018. Lecture Notes in Computer Science, vol. 10841. Springer, Cham (2018)
16. Kazikova, A., Pluhacek, M., Senkerik, R.: Regarding the behavior of bison runners within the bison algorithm. In: Mendel Journal Series 2018, vol. 24, pp. 63–70 (2018)
17. Berman, R.: American Bison. Nature Watch. Lerner Publications, Minneapolis (2008)
18. Sörensen, K.: Metaheuristics—the metaphor exposed. Int. Trans. Oper. Res. **22**(1), 3–18 (2015)
19. Kazikova, A., Pluhacek, M., Senkerik, R.: Tuning of The Bison Algorithm control parameters. in: 32nd European Conference on Modelling and Simulation, 22nd May–26th May. European Council for Modeling and Simulation (2018)

Optimizing Automated Storage and Retrieval Algorithm in Cold Warehouse by Combining Dynamic Routing and Continuous Cluster Method

Ngoc Cuong Truong[1], Truong Giang Dang[2], and Duy Anh Nguyen[1(✉)]

[1] Ho Chi Minh City University of Technology, Ho Chi Minh City, Viet Nam
duyanhnguyen@hcmut.edu.vn
[2] Ho Chi Minh City University of Transport, Ho Chi Minh City, Viet Nam

Abstract. The effectiveness of a storage and retrieval system in cold warehouse is assessed based on operating cost. In this paper, the cost savings is consider base on the optimization two criteria: optimization of the travel distance and time consuming or in other words, focus on building the advantage algorithm to determining storage location and planning path. A warehouse layout is designed for 480 storage locations on 16 pallet racking, separated by 4 storage aisles and 1 pick aisle. The storage and retrieval of goods is carried out by 2 forklift trucks. The system will be deals with the problem of deadlocks and traffic jams. In order to solve this issue, each vehicle will be assigned different permissions for access to each storage location in the warehouse and the window time algorithm is applied to avoid conflicts between vehicles. On the other hand, Continuous Cluster Method will be used to decrease the travel time. The algorithm is compared to using a single vehicle for a similar storage space to see efficiency. The results show that the algorithm is optimized up to 30% compare with traditional policy.

Keywords: Dynamic routing · Time-window · Collision avoidance · Conflict-free shortest-time · Warehouse management

1 Introduction

Determining storage location - localization is understood as the process of selecting the optimal storage location among different position, so that the travel time is minimization, thus saving the total operating costs of the warehouse. In addition, each storage location should be closely managed based on information such as type of goods, stored time, coordinates.

In storage process, localization will be built based on the following strategy: Random, dedicated, class-based and closest open location (COL) [1–3]. In retrieval process, put-away or expiration day were used to find the storage location, some common strategy are first in first out (FIFO), last in first out (LIFO), first expire first out (FEFO) [1, 2].

Planning path is defined as a process for selecting the most optimal path from all the solutions. The optimal path is determined based on two factors: the distance from I/O point to selected storage location is the shortest and there is no deadlock or traffic jams while vehicles move in the system. Strategy of route planning could be classified into two categories: Static and dynamic routing [4, 5]. For the static routing, there is only 1 storage location and 1 fixed path is choose in advance for each task of the forklift, the selection will not change during the task execution. This strategy is commonly use in the case there is only one vehicle hand on all task in the warehouse. For system with two or more vehicles, collisions could be occur when forklifts move simultaneously such as deadlock and traffic jams, affect drastically the system performance and static algorithm cannot adapt to change [6, 7]. Once the collision occurs, the selected location and path must be changed to avoid accidentally system operating. The process to re-localization and re-routing to avoid accident is carried out throughout once the collision happen is called dynamic routing.

This paper contribute by constructing an algorithm base on dynamic routing strategy to solve the problem. Storage location is determine by continuous cluster method and A-star algorithm, collision is solved through the time windows concept. The paper is organized as following: Warehouse assumption is proposed in Sect. 2, before we introduced auto localization method base on A* algorithm in Sect. 3 and dynamic routing in Sect. 4. In Sect. 5 we present some numerical result and conclusion.

2 Assumption Made and Layout Design

A design of the popular cold warehouse is proposed. The system specifications, storage conditions and import and export scenarios are given to check the efficiency of the algorithm.

2.1 Assumption Made [1, 2]

The system is built based on some special characteristics of cold store for preservation of aquatic products but we can adjust it to suit different types of storage.

- The capacity of system are 480 storage locations, each location contain 1 SKU (stock keeping unit – an inventoried item). System containing 6 types of frozen shrimp which are named A1, A2, A3, A4, A5 and A6.
- Goods are organized into the pallet. Each pallet is a SKU. This is the smallest item in system. Pallet is placed on single pallet racking and other picker can reach all items in the rack regardless of rack's height.
- Pick out time is undefined for all SKUs in system.

For frozen shrimp products, the requirement in storage process is if goods were come first, it will be sorted in pallet racking first (FIFS) and travel distance for each moving cycle is the shortest to prevent damage under wrong temperature. In retrieval process, pallet is removed base on import day, the oldest good in system is the earliest move out. This requirement is necessary to ensure the goods are not in warehouse too long.

2.2 Layout Design

Caron, Marchet, and Perego (2000) found that the layout design greatly affected to order picking distance. According to their study, layouts affect over 60% of the total distance traveled in storage [Dr. Peter]. Therefore, designing layout is an important foundation task before building the management algorithm [7].

From assumptions were presented in Sect. 2.1, warehouse system with 1 single pick aisle and 2 storage aisles is recommended (see Figs. 1 and 2). Warehouse space is divided into 16 pallet rackings (16 lines). The line consider in this paper include 5 tiers (A, B, C, D and E) and 6 bays (are distinguished by the digits from 1 to 8), totally 40 storage locations is located at each line. These lines are named Roman numerals I to XVI. The design help to be easily reach all items in the pallet racking and access to depot by using 2 separate Input and Output points.

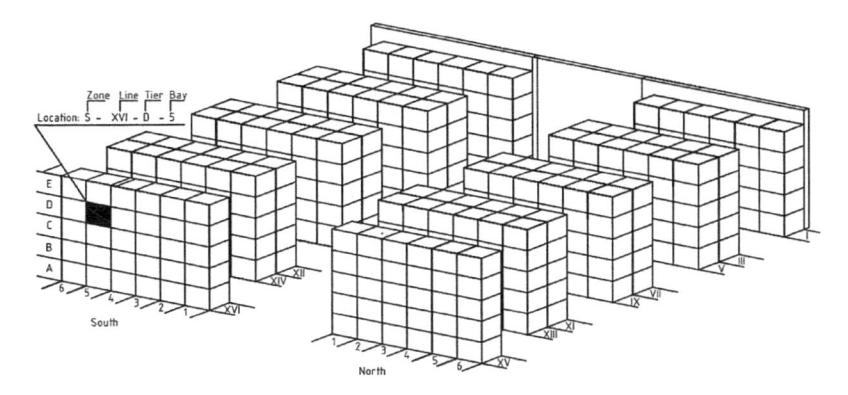

Fig. 1. Warehouse structure

2.3 Vehicle Task Assignment

Since there are two forklifts operate in the warehouse, it is important to separate the tasks for each vehicle to ensure we got the shortest time for each task and to minimize number of collisions. Layout was designed with single pick and storage aisle so that just 1 forklift is present at the same time on each aisle, otherwise deadlock or traffic jams could happen and impacted to performance of operation. For each order, the system will plan the route for each pallet first then task will be assigned to two forklifts.

3 Storage Algorithm Base on Continuous Cluster Method [3]

Continuous cluster method is a combination of ABC storage (class-based storage) and Closest Open Location policy - COL.

In detail, each class of ABC policy is divided into several smaller classes, each of which is called a cluster. In Fig. 2, each cluster is represented by a different color, the cells with the same color have same value d_{ij} (1). The radius of cluster is the travel distance from storage location on racking to I/O point. In storage and retrieval process, SKUs are priority arranged into the cluster which has smallest radius. By this method, the storage location is found always has travel distance less than the ABC class method. If the number of clusters in each class is greater, the algorithm is more efficient. To see the effectiveness of the algorithm, a warehouse system is simulated and analyzed in the following section [5].

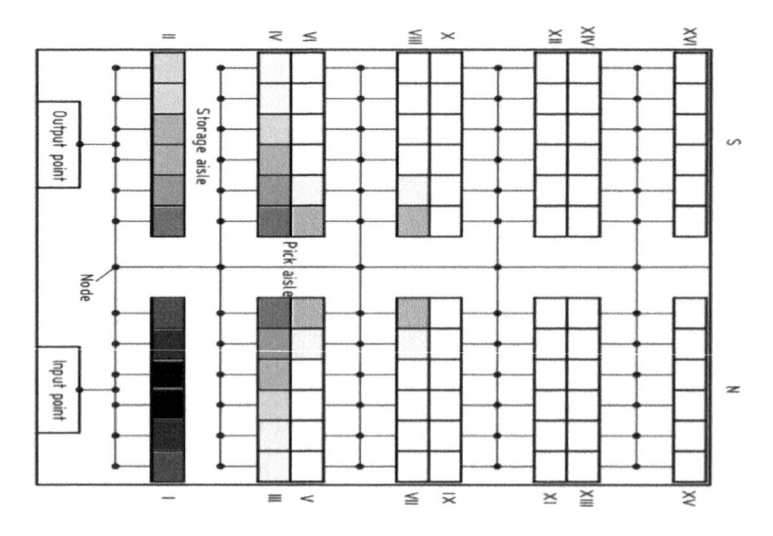

Fig. 2. Coordinate of storage location

In storage process, pallets are transferred from I/O point to storage location in turn by 3 axes in space as Line (x axis), Bay (y axis) and Tier (z axis) [7].

The travel distance of pallet in each circle can be computed as follows:

$$d_{ij} = x_{ij} + y_{ij} + z_{ij} \quad (1)$$

$$\text{Where, } x_{ij} = |x_j - x_i| \quad y_{ij} = |y_j - y_i| \quad z_{ij} = |z_j - z_i| \quad (2)$$

(x_i, y_i, z_i), (x_j, y_j, z_j) are coordinates of I/O point and storage location.

$$\text{Total travel distance for an order of n SKUs}: D = \sum_{k=1}^{n} d_{ij} \quad (3)$$

4 Auto – Localization

4.1 Storage and Retrieval Strategy

For frozen shrimp products, the shorter duration of sorting, the less risk of failure of goods, so pallet should be entered into inventory under FIFO and continuous cluster method. For FIFO policy, if goods are shipped to warehouses before, it will be sort in pallet racking before. With continuous cluster method, each pallet was added into appropriate storage location whose time to I/O point is the shortest and it's represented by distance index is d_{ij}.

To apply those policies, A* algorithm was propose to find the optimal storage location.

4.2 Determining Storage Location by A* Algorithm

First published in 1968 by Peter Hart, Nils Nilsson and Bertram Raphael, A* is an informed search algorithm, meaning that it solves problems by searching among all possible paths to the solution (goal).

Evaluation function:

$$f(n) = g(n) + h(n) \qquad (4)$$

- Operating cost function, $g(n)$ – Actual operating cost having been already traversed
- Heuristic function, $h(n)$ – Information used to find the promising node to traverse, the heuristic function must be admissible.

Figure 3 demonstrates how to determine an optimal storage location using the star algorithm:

The $g(n_{best})$ (choose from d_{ij}) in this flow chart represents the exact travel distance of the path from the starting point to any vertex n_{best} – which is defined as a shortest node in each step of the loop and $h(n_{best})$ represents the heuristic estimated distance from vertex n_{best} to the selected storage location x and $h(n)$ value is calculated using the Euclidean distance formula. Each time through the main loop, it examines the vertex n that has the lowest (1) with each:

$$g(n_{best}) + h(n_{best}, x) < g(x) \qquad (5)$$

One more node in the shortest path is found. The main loop repeat until latest node is determined – which represent selected storage location.

The auto-localization algorithm base on A-star approach is clear. It is easy to implement and allows very fast route computations since this method only cares about the start and end of each row and ignore the time dependent between forklifts. However, when system was performed by 2 forklifts, various drawbacks are caused by deadlock and traffic jam have a deteriorating effect on the system performance (see Fig. 3).

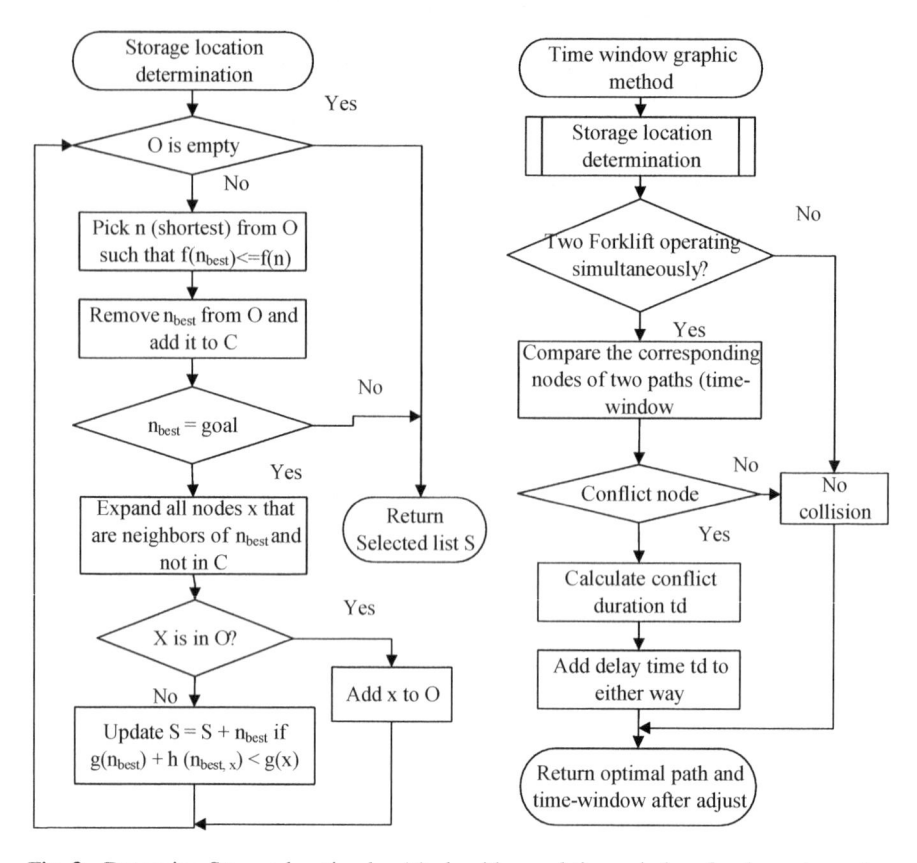

Fig. 3. Determine Storage location by A* algorithm and time window for dynamic routing

5 Dynamic Routing Method by Time Window

To deal with the problems of the model given in previous Section, a different approach that computes shortest (traveling time) and conflict-free routes simultaneously is propose which time-dependent between vehicles is considered [4–6].

The idea of the algorithm is that find a conflict-free shortest-time route in the case there is collision potential in the aisle (see Fig. 4). According to the approach, after the shortest path to the storage location is found by the A* algorithm, a time-dependent histogram is established. Based on distance and velocity data, the position of each vehicle at each time on the map is determined and then a free-conflict path is formed by using the waiting time for a vehicle.

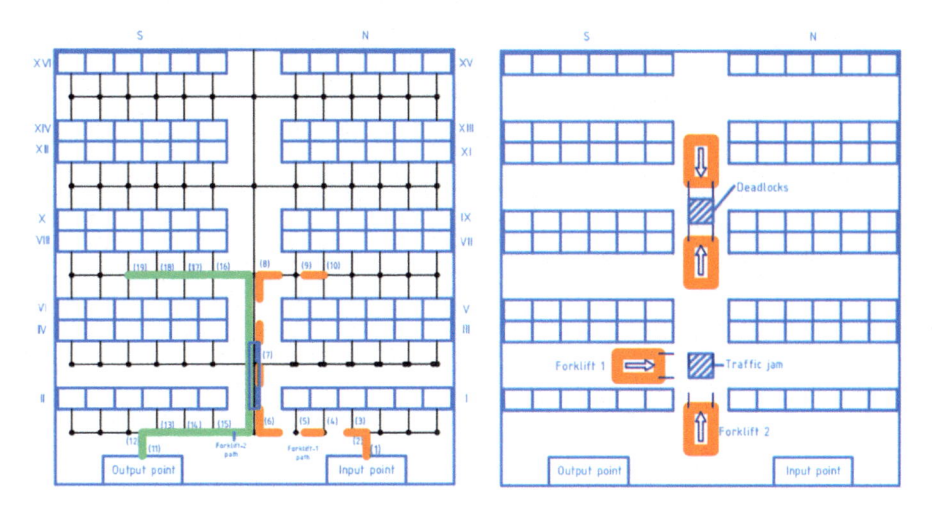

Fig. 4. Deadlock and traffic jams

The algorithm is explained through a scenario as follows: There is 2 tasks were assigned to Forklift 1 and 2, one to store pallet from I/O point to storage location have coordinate N-VII-A-5 and other one to retrieval pallet from selected location S-VIII–A-4. The A* algorithm shows the shortest static path for the two tasks of FL1 and FL2 as follows:

- Forklift1 path (storage): [1 -> 2 -> 3 -> 4 -> 5 -> 6 -> 7 -> 8 -> 9 -> 10]
- Forklift2 path (retrieval): [19 -> 18 -> 17 -> 16 -> 8 -> 7 -> 6 -> 15 -> 14 -> 13 -> 12 -> 11]

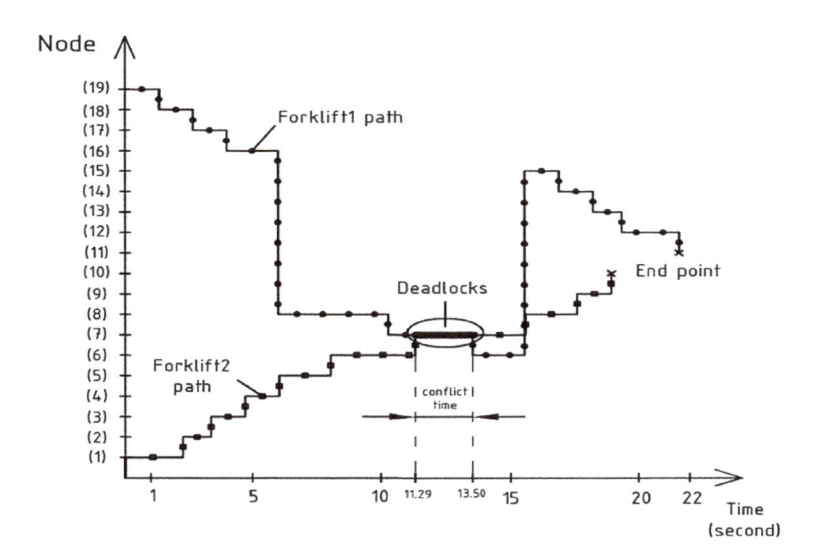

Fig. 5. Time window with deadlock between 2 paths

Time dependence of two vehicles is shown in Fig. 5. The graph shown that there are several nodes overlap between two paths, they will collide at second 11.29^{th} to 13.50^{th} on nodes [6, 7].

To avoid collisions between two vehicles, a delay time is added for Forklift1, which means that the vehicle will paused between node 5 and node 6 a time $\Delta = 2.21$ s. During this time, Forklift2 can pass nodes 8, 7, 6 without Folkift1 on those node (see Fig. 6).

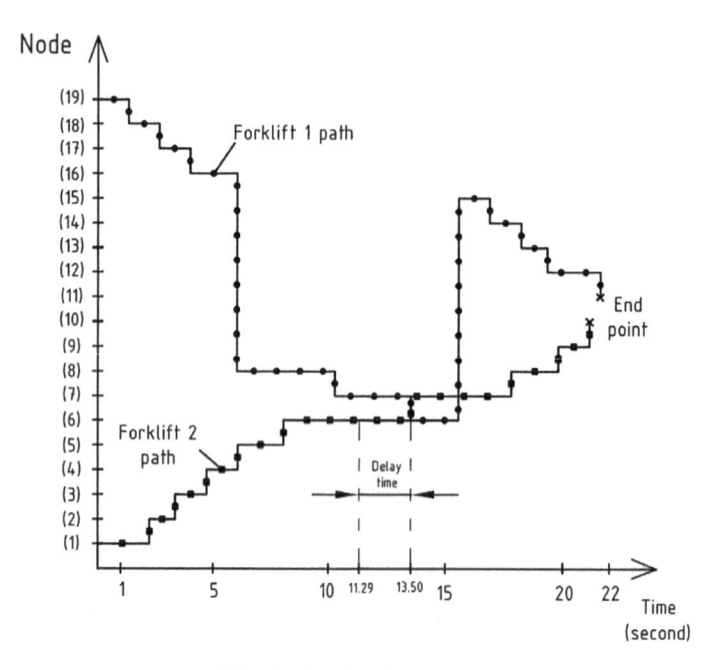

Fig. 6. Conflict-free routes

6 Simulation and Result

Goods management is done through a UI interface on MATLAB as shown in Fig. 7, the warehouse space is simulated on V-REP (see Fig. 8). The constructed algorithms will be tested by two comparisons: the time efficiency between the COL strategy and random algorithms; dynamic routing by time window with conflict-free routing by using double pick aisle.

Fig. 7. Software interface

Fig. 8. Warehouse layout simulation

6.1 Comparison About Travel Distance of 3 Storage Algorithms

Table 1 is a comparison among solution groups: the combination of dynamic routing - Random, dynamic routing - ABC storage and dynamic routing - Continuous cluster with volume are 35%, 60% and 85% warehouse capacity.

Under the continuous cluster method with 35% capacity of warehouse is used, travel distance is lower than 54.49% and 23.74% compare with algorithm base on Random and ABC policy respectively.

This efficiency is maintained at 43.22% and 28.41% when the total stock is filled up 60%. The indicator dropped to 13.91% and 7.12% when volume is reached 85% warehouse capacity. Base on this data, the smaller of volume is filled up, the more effective of dynamic routing combine continuous cluster method.

Table 1. Travel distance of 3 algorithms

Dynamic routing combine with…	Fill up 35% volume	Fill up 60% volume	Fill up 85% volume
Random (mm)	4622100	7725300	10920000
ABC (mm)	2758200	6127800	10121400
Continuous cluster method (mm)	2103300	4386700	9400700

6.2 The Efficiency of the Dynamic and Static Routing Algorithm Through the Time Consumption Comparison

Table 2 is the total time consumption on tasks A, B and C, respectively, through two static and dynamic routing algorithms. With dynamic algorithms, two forklift are used in parallel to perform all tasks while in static routing algorithm, the picking up and retrieval process is done by a single vehicle (the travel distance of both cases are the same). Based on data, the dynamic management algorithm is always faster than traditional algorithms, which are 21.25%, 31.74% and 27.97%, respectively.

Table 2. Time consumption of static and dynamic routing

	Fill up from 0% volume to 35% (A)	Fill it up to 60% from 35% and then drain it down to 45% level (B)	Fill up to 85% from 45% and bring it down to 50% (C)
Dynamic routing (s)	2150	4121	6217
Static routing (s)	2730	6237	8631

7 Conclusion

The article presents a combination of dynamic routing and determinate storage location for simultaneous 2 forklifts, combining the first in first out and the continuous cluster method to help reduce the time and travel distance consumption. This is also a positive aspect in the reduction of warehouse operating costs - a top priority in cold storage management. The 2 comparisons point out the approach is more effective when the volume is lower than warehouse capacity.

Future work will include more complex comparisons such as the quantity of goods are delivered must be greater like the real environment. The frequency of storage and retrieval tasks need to higher than so that could validate the stable of the designed system. Moreover, mechanical system design to connect with software need to be implement, this is next step to completely build an automated storage and retrieval system in warehouse.

References

1. Truong, N.C., Dang, T.G., Nguyen, D.A.: Development of automated storage and retrieval algorithm in cold warehouse. In: South East Asian Technical University Consortium Symposium (2017). ISSN: 1882-5796
2. Truong, N.C., Dang, T.G., Nguyen, D.A.: Development and optimization of automated storage and retrieval algorithm warehouse by combining storage location identification and route planning method. In: IEEE International Conference on System Science and Engineering (2017)

3. Truong, N.C., Dang, T.G., Nguyen, D.A.: Building management algorithms in automated warehouse using continuous cluster analysis method. In: AETA (2017)
4. Vivaldini, K.C.T., Becker, M., Caurin, G.A.P.: Automatic routing of forklift robots in warehouse applications. In: ABCM Symposium Series in Mechatronics (2010)
5. Vivaldini, K.T., Galdames, J.P.M., Pasqual, T.B., Sobral, R.M., Araújo, R.C.: Automatic Routing System for Intelligent Warehouses. IEEE (2010)
6. Maza, S., Castagna, P.: Conflict-free AGV Routing in Bi-directional Network. IEEE (2001)
7. Mohring, R.H., Kohler, E., Gawrilow, E., Stenzel, B.: Dynamic Routing of Automated Guided Vehicles in Real-time. Springer, Heidelberg (2008)

Dependency of GPA-ES Algorithm Efficiency on ES Parameters Optimization Strength

Tomas Brandejsky[✉]

University of Pardubice, 532 10 Pardubice, Czech Republic
tomas.brandejsky@upce.cz

Abstract. In this work, the relation between number of ES iterations and convergence of the whole GPA-ES hybrid algorithm will be studied due to increasing needs to analyze and model large data sets. Evolutionary algorithms are applicable in the areas which are not covered by neural networks and deep learning like search of algebraic model of data. The difference between time and algorithmic complexity will be also mentioned as well as the problems of multitasking implementation of GPA, where external influences complicate increasing of GPA efficiency via Pseudo Random Number Generator (PRNG) choice optimization.

Hybrid evolutionary algorithms like GPA-ES uses GPA for solution structure development and Evolutionary Strategy (ES) for parameters identification are controlled by many parameters. The most significant are sizes of GPA population and sizes of ES populations related to each particular individual in GPA population. There is also limit of ES algorithm evolutionary cycles. This limit plays two contradictory roles. On one side bigger number of ES iterations means less chance to omit good solution for wrongly identified parameters, on the opposite side large number of ES iterations significantly increases computational time and thus limits application domain of GPA-ES algorithm.

Keywords: Genetic Programming Algorithm · Evolutionary Strategy · Hybrid Evolutionary System · Algorithm efficiency · Optimization

1 Introduction

Increasing amount of data to be processed forces needs for improving efficiency of existing algorithms. While artificial neural networks and deep learning technology are rather data representation or interpolation, recall and approximation tool, evolutionary algorithms are capable to transform data into models which can be understand by humans, search for optima, and solve many next tasks on data. In the area of model development by evolutionary algorithms called symbolic regression Genetic Programming Algorithms (GPA) and related techniques are used.

Hybrid evolutionary algorithms like GPA-ES [1] uses GPA for solution structure development and Evolutionary Strategy (ES) for parameters identification are controlled by many parameters. The most significant are sizes of GPA population and sizes of ES populations related to each particular individual in GPA population. There is also limit of ES algorithm evolutionary cycles. This limit plays two contradictory roles. On one side bigger number of ES iterations means less chance to omit good solution for wrongly identified parameters [2] and it was the main idea of GPA-ES hybrid algorithm development, on the opposite side large number of ES iterations significantly increases computational time and thus limits application domain of GPA-ES algorithm.

In this study the relation between number of ES iterations and convergence of the whole GPA-ES hybrid algorithm will be studied. The difference between time and algorithmic complexity will be also mentioned as well as the problems of multitasking implementation of GPA, where external influences complicates increasing of GPA efficiency via Pseudo Random Number Generator (PRNG) choice optimization.

2 Hybrid GPA-ES Algorithm

GPA-ES hybrid algorithm combines GPA algorithm for solution structure development and ES algorithm for optimization of parameters of each individual in GPA population. Such design of evolutionary algorithm prevents situations when good structure solution (e.g. well composed equation) but with wrongly estimated constants (coefficient) is eliminated from population and replaced by individual of worse structure but better constants, which has worse evolutionary potential.

Algorithm 1. Studied GPA algorithm.

```
1) FOR ALL individuals DO Initialize() END FOR;
2) FOR ALL individuals DO Evaluate()=>fitness END FOR;
3) Sort(individuals);
4) IF Terminal_condition() THEN STOP END IF;
5) FOR ALL individuals DO
     SELECT Rand() OF
       CASE a DO Mutate()=> new_individuals;
       CASE b DO Symmetric_crossover() => new_individuals;
       CASE c DO One_point_crossover() => new_individuals;
       CASE d DO Re-gerating() => new_individuals;
     END SELECT;
   END FOR;
6) FOR ALL individuals DO
     New ES_algorithm_object
     FOR ALL ES_individuals DO Initialize()END;
     evaluate() => ES_fitness
     FOR ALL ES_cycles DO
       FOR ALL ES_individuals DO
```

```
            Evaluate => ES_fitness
          END FOR;
          FOR ALL ES_individuals DO
            intelligent_crossover() => new_ES_individuals
            Evaluate => new_ES_fitness
          END FOR;
          FOR ALL ES_individuals DO
              IF new_ES_fitness>ES_fitness THEN
                  ES_individual = new_ES_individual;
                  fitness = new_fitness;
              END IF;
          Sort(ES_individuals, ES_fitness);
          END FOR;
        END FOR;
        new_individual=ES_individual[0]; new_fitness = ES_fitness[0];
    7) FOR ALL individuals DO IF new_fitness<fitness THEN
          individual = new_individual;
          fitness = new_fitness;
        END IF;
    8) GOTO 3);
```

Size of GPA population and sizes of ES populations related to each particular individual in GPA population are the most significant parameters of GPA-ES algorithm. Influence of population sizes was studied in many publications, see [1]. Small populations forces evolutionary pressure and in the case of specific conditions it might speed up evolution. On the opposite side, in the small populations there is increased risk of movement into local optima on the place of global one. Very large populations brings problems with small speed an efficiency of evolution frequently. Problem is that precise meaning of terms small or big population depends on fitness function landscape. Extremely small populations also bring big dispersion of needed evolutionary cycles and thus also of computational time.

It is possible to accept result of experiments concluding that large populations might be replaced by bigger number of generations and vice versa. But such reasons are about ability to find solution. When the number of fitness function evaluations (computational complexity) or computational time is evaluated, dependencies between them and population sizes are not the same as it will be presented in the next sections.

There is also parameter representing number of ES algorithm evolutionary cycles. This number plays two contradictory roles. The bigger number of ES iterations means less chance to omit good solution for wrongly identified parameters [2–8] and it was the main idea of GPA-ES hybrid algorithm development but the large number of ES iterations significantly increases computational time and thus limits application domain of GPA-ES algorithm. Standard set-up of this algorithm is 40 populations of ES

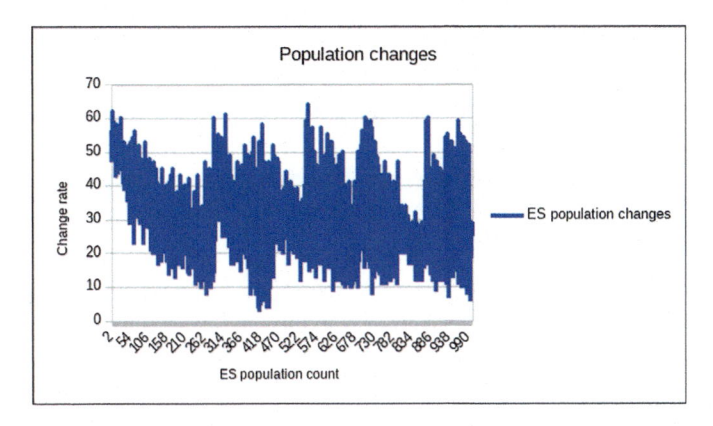

Fig. 1. Count of the ES population improvements for whole GPA population evaluated from the best to the worst individual.

algorithm for each GPA individual. Equivalent of pure GPA is none ES population. The experiments published in the next section describe influence of this parameter.

Detail dynamics of GPA-ES system is complicated and influenced by many control parameters. E.g. Fig. 1 illustrates decrease of population activity in the time. We can identify periods of 100 given by size of GPA population because individuals are evaluated sequentially. Figure 1 also confirms that after first three evolutionary cycles when only mutation is applied evolutionary activity decreased. In the rest part of data influence of crossover operations which are not used to whole populations but only randomly. They cause periodic boosting of activity. Periods of 100 individuals of decreasing activity also illustrates that activity of individuals with better fitness function is bigger than activity of worse ones non looking to the fact that in each period to each individual one evolutionary operator is applied but this complicated microdynamics study is not the main subject of this work.

Small number of individuals in ES population represents analogy of small populations in GPA part of algorithm. Also there they can bring faster convergence in average (from the viewpoint of time), but because the number of generations is fixed, it cannot correspond to larger number of generations. Thus the quality of results (resulting fitness function magnitudes) must be worse for small ES populations due to ill-identified parameters. In the superior GPA it might cause worse recognition between good and wrong structures against original ideas of GPA-ES design. As the conclusion, if the number of ES population decreases, superior GPA will need more population to achieve the comparable quality results. The significant question is if the faster is to decrease of ES populations or to increase of GPA ones.

Paper [9] was focused to relations between GPA and ES population sizes of composed GPA-ES algorithm. Now the influence of ES population size and ES population number limit is studied applying modified methodology of experiments. Very low limit of generation number for constant optimizing embedded ES part of the algorithm was not studied yet. This modification focuses to elimination influences affecting PRNG used in evolutionary algorithm as

- influence of other tasks running on the same node of computing cluster
- PRNG number series stationarity
- thread switching (if the used number generator is hared between threads or if threads mutually communicate)
- task allocating on cluster nodes.

The first point is caused by the fact that other tasks in the operating system influence in the next point described task switching and also function of (P)RNGs if they are not implemented as thread safe. Problem is, that each operating system also runs different, it is possible to say "service", tasks like network communication support, cluster operation support etc. These tasks cannot be stopped during experiments and they can cause changes of parallely executed sub-task run order during experiments. There plays its role also task scheduler of used operating system. Linux systems frequently offers normal, batch, round-robin and FIFO CPU schedulers influencing order of tasks and threads. On clusters there is also the last mentioned source of non-determinism, allocation of tasks on computational nodes which also might not be deterministic.

Presented experiments eliminate these sources of non-determinism by the simplest way – by elimination of multitask and multithread execution Each task was running as singlethread one without any communication with others.

3 Experiments and Obtained Data

Lorenz attractor system served as test case for experiments with symbolic regression of differential equations describing this dynamic system on the base of pre-computed data set. Lorenz attractor system is described by (1) and (2).

$$\begin{aligned} x'(t) &= \sigma(y(t) - x(t)), \\ y'(t) &= x(t)(\rho - z(t)) - y(t), \\ z'(t) &= x(t)\,y(t) - \beta\,z(t) \end{aligned} \quad (1)$$

$$\begin{aligned} \sigma &= 16 \\ \beta &= 4 \\ \rho &= 45.91 \end{aligned} \quad (2)$$

The limit of GPA algorithm iteration number was set to 10000 cycles, the termination condition was set to sum of error squares less than 10^{-8} for applied 599 samples of training data, while the number of ES algorithm iterations has varying between magnitudes 1, 10 and 100, as well as the sizes of ES populations was varying magnitudes 10, 40 and 100. Size of GPA population was 64 individuals. Experiments were repeated 10000 times for different seeds of used PRNGs.

In the presented study, we use two variable parameters, number of ES algorithm iterations and size of ES populations. They determine number of evaluations of fitness function both in ES and composed GPA-ES algorithms. To analyze computational efficiency of GPA-ES algorithm, both number of needed GPA cycle iterations and computational time are measured (computational time can be replaced by another HW independent measure, number of fitness function iterations).

Results of experiments are presented on the following Figs. 2, 3, 4, 5, 6 and 7.

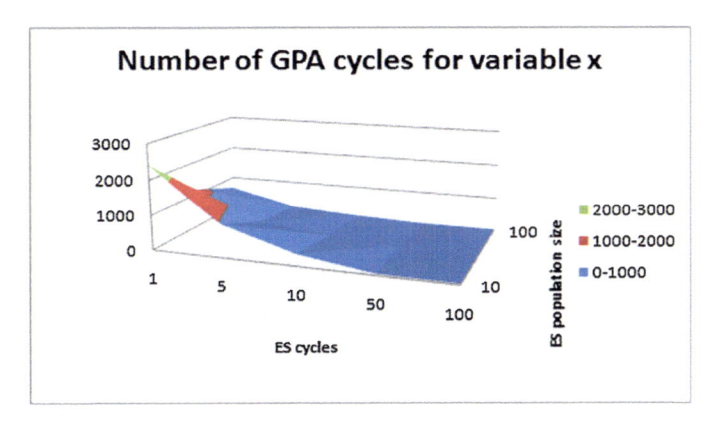

Fig. 2. Number of iterations of encapsulating GPA for x variable depending on ES cycle limit and ES population size.

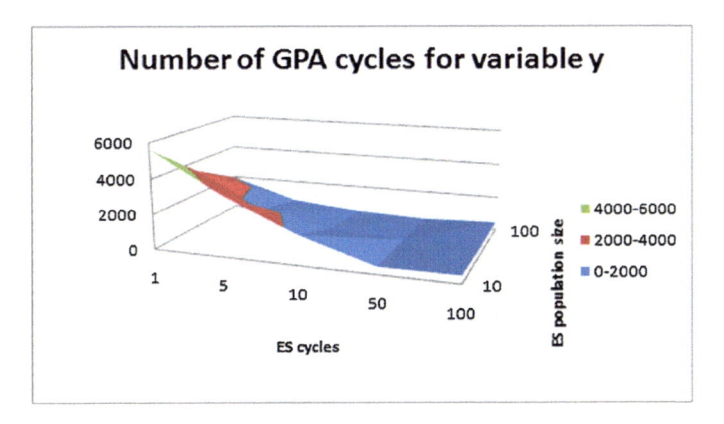

Fig. 3. Number of iterations of encapsulating GPA for y variable depending on ES cycle limit and ES population size.

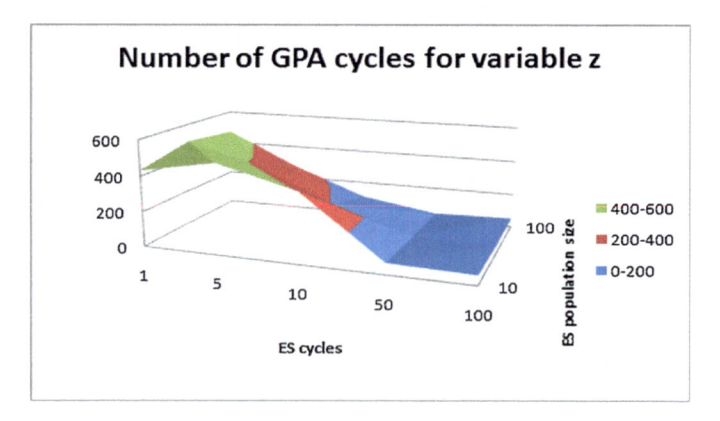

Fig. 4. Number of iterations of encapsulating GPA for x variable depending on ES cycle limit and ES population size.

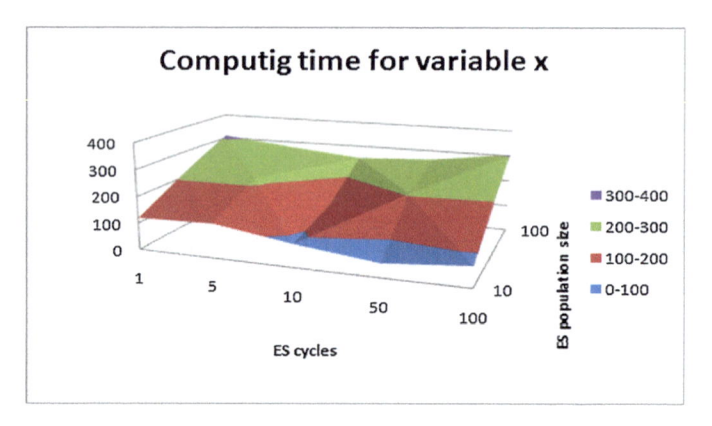

Fig. 5. Computation time of whole GPA-ES for x variable depending on ES cycle limit and ES population size.

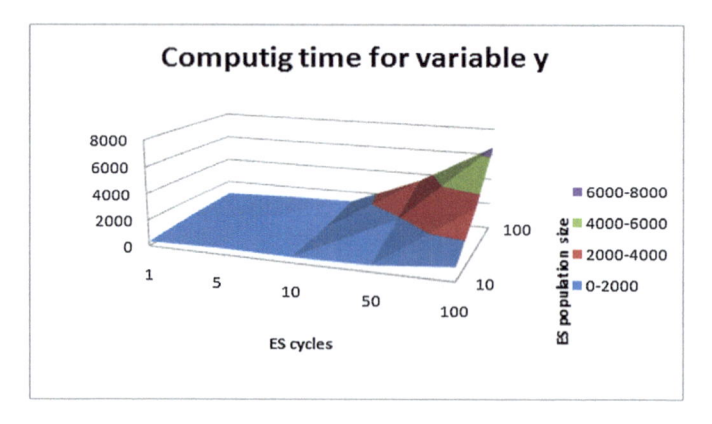

Fig. 6. Computation time of whole GPA-ES for y variable depending on ES cycle limit and ES population size.

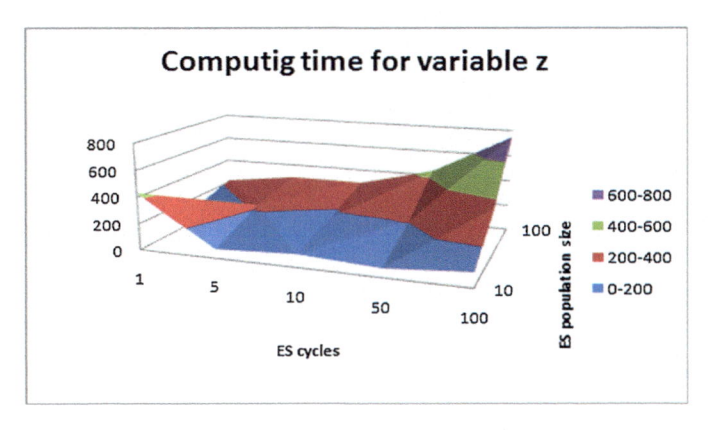

Fig. 7. Computation time of whole GPA-ES for z variable depending on ES cycle limit and ES population size.

4 Conclusions

Above presented data are depending on two parameters - number of ES algorithm iterations and size of ES populations. While for simplest equation describing x variable in Fig. 2 was for smallest ES population of 10 individuals, more complex equations describing dynamics of y and z variables required larger ES population of 100 individuals.

Computing time was different. There was smaller variance of results and the optima were between 5 and 50 populations and smallest populations of 10 individuals. Such results partially defers to above described expectations and they are caused by computational complexity of ES algorithm which depends quadratically on the number of individuals but the improvement of larger population to convergence is of the lower order.

Presented data points that GPA-ES algorithm behavior corresponds to expectations in computational complexity. Time complexity of algorithms corresponds less and it is given by used hardware properties.

Non looking that Fig. 1 presents decrease of GPA efficiency, small populations and higher numbers of iterations are interesting way of efficiency increasing except extremely low magnitudes. Moreover, with decreasing of population members number and with increasing of count of evolutionary cycles there increases dispersion of needed iterations and thus even if the average number is small, some runs can be extremely long.

Presented experiments also points that computational time is controversial measure of algorithm efficiency – from the algorithm complexity viewpoint numbers of individuals can be replaced by amount of iterations but GPA and ES algorithms has incomparable complexity and their implementation might have different efficiency. ES works faster, thus in time space comparison the results displayed on Figs. 5, 6 and 7 are not significant and do not copy results from Figs. 2, 3 and 4 representing numbers of iterations (evolutionary cycles).

Presented study uses relatively simple problem as case study. In the future, when more computational time will be obtained, there will be need to repeat herein presented experiments on more complex problems where the area of highest efficiency is expected far from the lowest population sizes and iteration numbers of ES.

Acknowledgements. Access to computing and storage facilities owned by parties and projects contributing to the National Grid Infrastructure MetaCentrum provided under the programme "Projects of Large Research, Development, and Innovations Infrastructures" (CESNET LM2015042), is greatly appreciated.

References

1. Brandejsky, T.: Evolutionary system to model structure and parameters regression. Neural Netw. World **12**(2), 181–194 (2012). ISSN 1210-0552
2. Brandejsky, T.: The use of local models optimized by genetic programming algorithm in biomedical-signal analysis. In: Zelinka, I., Snasel, V., Abraham, A. (eds.) Handbook of Optimization from Classical to Modern Approach, pp. 697–716 (2012). ISSN 1868-4394, ISBN 978-3-642-30503-0
3. Alander, T.: On optimal population size of genetic algorithms. In: Proceedings of the IEEE Computer Systems and Software Engineering, pp. 65–69 (1992)
4. Eiben, A.E., Hinterding, R., Michalewic, Z.: Parameter control in evolutionary algorithms. Trans. Evol. Comput. **3**(2), 124–141 (1999). https://doi.org/10.1109/4235.771166
5. Koumousis, K., Katsaras, C.P.: A saw-tooth genetic algorithm combining the effects of variable population size and reinitialization to enhance performance. IEEE Trans. Evol. Comput. **10**(1), 19–28 (2006). https://doi.org/10.1109/TEVC.2005.860765
6. Lobo, G., Lima, C.F., Michalewicz, Z. (eds.): Parameter Setting in Evolutionary Algorithms. Studies in Computational Intelligence, vol. 54. Springer, Heidelberg (2007). ISBN 978-3-540-69431-1
7. Reeves, C.R.: Using genetic algorithms with small populations. In: Proceedings of the Fifth International Conference on Genetic Algorithms, San Mateo, pp. 92–99 (1993). ISBN 1-55860-299-2
8. Piszcz, A., Soul, T.: Genetic programming: optimal population sizes for varying complexity problems. In: Proceedings of the Genetic and Evolutionary Computation Conference GECCO, Seattle, pp. 953–954 (2006). https://doi.org/10.1145/1143997.1144166
9. Brandejsky, T.: Small populations in GPA-ES algorithm. In: Matousek, R., (ed.) 19th International Conference on Soft Computing, MENDEL 2013, Brno, pp. 31–36 (2013). ISSN 1803-3814, ISBN 978-80-214-4755-4

A Modified Bat Algorithm to Improve the Search Performance Applying for the Optimal Combined Heat and Power Generations

Bach H. Dinh[1(✉)], An H. Ngo[2], and Thang T. Nguyen[1]

[1] Faculty of Electrical and Electronics Engineering,
Ton Duc Thang University, Ho Chi Minh City, Vietnam
{dinhhoangbach, nguyentrungthang}@tdtu.edu.vn
[2] Faculty of Electrical and Electronic Technology, HCM City University
of Food Industry, Ho Chi Minh City, Vietnam
annh@cntp.edu.vn

Abstract. Cogeneration technology known as combined heat and power generations can achieve much more energy-efficient than separately generating electricity and useful heat for electric/heat demand loads. Thus, the economic dispatch of cogeneration systems is very complex optimization problems in power systems because many complicated constrains for combined demands of heat and power loads as well as operating zone of cogeneration units have to take into consideration. In this paper, a Modified Bat Algorithm (MBA) with three improvements has been proposed to solve the optimal operation of combined heat and power generations (OOCHPG). To evaluate the effectiveness of the proposed ideas, both Conventional Bat Algorithm (CBA) and MBA have been applied for a test case of 7 generations and the results have proved that the proposed MBA is persuasively superior to CBA and other methods reported in the literature in terms of optimal value quality, low fitness evaluations and fast convergence. Consequently, the proposed MBA is an efficient method for solving OOCHPG problem.

Keywords: Bat algorithm · Combined heat and power generations · Non-convex cost function

1 Introduction

Cogeneration or Combined Heat and Power Generations (CHPG) is an interesting issue in modern power systems because it can improve the efficiency in using fuels due to the heat, occurred by the electrical generation process which can be uselessly discharged into the air, reutilized for industrial zones or manufacturers [1]. In the Optimal Operation of CHPG (OOCHPG), a set of suppliers consisting of pure power and pure heat units as well as cogeneration units are produced both electricity and heat to electrical loads and heat loads located far away. In fact, the generation characteristic of pure units is a function of the fuel cost with respect to either power or heat only, whereas the fuel cost function of cogeneration units contains both power and heat outputs making the complexity of operating zone of cogeneration units. Therefore, the OOCHPG problem makes a huge challenge for any optimization algorithms in searching a global optimal solution.

During over several decades, authors in Refs. [2–5] have introduced differential methods to solve the OOCHPG. The Newton [2] and Lagrange relaxation (LR) [3] methods are two methods of those that first applied to solve the OOCHPG. However, they have a common main disadvantage to be limited when dealing with a large-scale system. In order to overcome this disadvantage, authors in Ref. [5] have proposed the combined method between the augmented Lagrange and Hopfield network (ALHN). A number of studies [6–9] used the meta-heuristic algorithms based on random search with considering the non-convex objective issue for the OOCHPG. Moreover, the OOCHPG considering the valve-point loading effects on pure power units has been sufficiently solved by the Bee Colony Optimization (BCO) in [6]. In recent years, some of the promised methods have proposed for the OOCHPG as the Group Search Optimization [10], Oppositional Teaching Learning Based Optimization [11], Opposition-Based Group Search Optimization [12], and eight versions of PSO [13], Grey Wolf Optimization [14], Machine Learning Algorithm [15], Cuckoo Search Algorithm [16] and Artificial Immune System [17].

This paper proposes an alternative method for solving the OOCHPG on the basis of the Bat algorithm (BA) [18], considering three modifications that are optimal range of updated velocity, use of fixed loudness, and new formula for producing new solutions of the second new solution generation to improve the searching performance of conventional BA. To examine our modified approach, several simulation cases of the proposed method and conventional BA are analyzed based on a test system which consists of 7 generation units, considering transmission power losses and non-differential objective function. The remainder of this paper is organized as follow: Sect. 2 introduces the methodologies for the OOCHPG problem while Sect. 3 gives the proposed algorithm. Simulation results and discussion are presented in Sect. 4. Finally, conclusions are reported in Sect. 5.

2 Formulation of Optimal Operation of Combined Heat and Power Generation Problem

Suppose there is a complex CHPG system including N_{pp} pure power units, N_{ph} pure heat units and N_c cogeneration units and the task of the optimal operation of this CHPG problem is how to determine the proper values of heat and/or power outputs at all sources (units) so that the total energy cost of both heat and power generations is minimized while the heat supply – demand balance, the power supply - demand balance and the capacity limits of all units must be satisfied.

The objective of the problem is to minimize the total fuel cost for heat and power production [1–3]. It is formulated as

$$\text{Min} \left\{ \sum_{i=1}^{N_{pp}} F_{pi}(P_{pi}) + \sum_{j=1}^{N_c} F_{cj}(P_{cj}, H_{cj}) + \sum_{k=1}^{N_{ph}} F_{hk}(H_{hk}) \right\} \quad (1)$$

where
- $F_{pi}(P_{pi})$ — cost function of the pure power unit i, and
- $F_{hk}(H_{hk})$ — cost function of the pure heat unit k, and
- $F_{cj}(P_{cj}, H_{cj})$ — cost function of the cogeneration unit j.
- P_{pi} — power output of the pure power unit i,
- P_{cj} — power output of the cogeneration unit j,
- H_{cj} — heat output of the cogeneration unit j,
- H_{hk} — heat output of the pure heat unit k

Furthermore, the cost function of a pure power unit [1–3] can be presented by

$$F_{pi}(P_{pi}) = a_{pi} + b_{pi}P_{pi} + c_{pi}P_{pi}^2 + |e_{pi} \times \sin(f_{pi} \times (P_{pi,\min} - P_{pi}))| \quad (2)$$

where $a_{pi}, b_{pi}, c_{pi}, e_{pi}$ and f_{pi} are cost function coefficients of pure power unit i.

The cost function of a pure heat unit [1–3] is obtained by

$$F_{hk}(H_{hk}) = a_{hk} + b_{hk}H_{hk} + c_{hk}H_{hk}^2 \quad (3)$$

where a_{hk}, b_{hk} and c_{hk} are cost function coefficients of pure heat unit k.

The cost function of a cogeneration unit [1–3] depending on both power and heat as shown by

$$F_{cj}(P_{cj}, H_{cj}) = a_{cj} + b_{cj}P_{cj} + c_{cj}P_{cj}^2 + k_{cj}H_{cj} + l_{cj}H_{cj}^2 + m_{cj}H_{cj}P_{cj} \quad (4)$$

where $a_{cj}, b_{cj}, c_{cj}, k_{cj}, l_{cj}$ and m_{cj} are cost function coefficients of cogeneration unit j.

In practice, the operation of power systems is constrained by the balance of power, the balance of heat and physical capacities of electrical and mechanical equipment. Therefore, there are three constraints representing the three such manners as follows [1–3].

(a) *Power balance constraint*:

The total power supply generated by pure power units and cogeneration units must satisfy the power demand of power systems as below

$$P_D + P_L - \sum_{i=1}^{N_{pp}} P_{pi} - \sum_{j=1}^{N_c} P_{cj} = 0 \quad (5)$$

where P_D and P_L are power demand and transmission power loss, respectively. Moreover, the power loss in transmission lines can be calculated by

$$P_L = \sum_{i=1}^{Npp+Nc} \sum_{j=1}^{Npp+Nc} P_i B_{ij} P_j + \sum_{i=1}^{Npp+Nc} B_{0i} P_i + B_{00} \qquad (6)$$

(b) *Heat balance constraint:*

The total heat produced by pure heat and cogeneration units must satisfy the heat demand neglecting heat loss as

$$H_D - \sum_{j=1}^{N_c} H_{cj} - \sum_{k=1}^{N_{ph}} H_{hk} = 0 \qquad (7)$$

where H_D is heat demand.

(c) *Power and heat unbalanced constraints (physical limit conditions):*

Each unit must operate within their upper and lower bounds as shown below:

$$\begin{cases} P_{pi}^{\min} \leq P_{pi} \leq P_{pi}^{\max} \\ P_{cj}^{\min}(H_{cj}) \leq P_{cj} \leq P_{cj}^{\max}(H_{cj}) \\ H_{cj}^{\min}(P_{cj}) \leq H_{cj} \leq H_{cj}^{\max}(P_{cj}) \\ H_{hk}^{\min} \leq H_{hk} \leq H_{hk}^{\max} \end{cases} \qquad (8)$$

where

$\left[P_{pi}^{\min}, P_{pi}^{\max}\right]; \left[P_{cj}^{\min}, P_{cj}^{\max}\right]$ – minimum and maximum power of pure power unit i and cogeneration unit j, respectively, and

$\left[H_{cj}^{\min}, H_{cj}^{\max}\right]; \left[H_{hk}^{\min}, H_{hk}^{\max}\right]$ – minimum and maximum heat of pure heat unit k and cogeneration unit j, respectively.

Furthermore, the operation zone of dependent power and heat outputs of a cogeneration unit must be limited by:

$$\begin{cases} P_{cj}^{\max}(H_{cj}) = \min\left\{P_{cj}(H_{cj})|_{AB}, P_{cj}(H_{cj})|_{BC}\right\} \\ P_{cj}^{\min}(H_{cj}) = \max\left\{P_{cj}(H_{cj})|_{CD}, P_{cj}(H_{cj})|_{DE}, P_{cj}(H_{cj})|_{EF}\right\} \\ H_{cj}^{\max}(P_{cj}) = \min\left\{H_{cj}(P_{cj})|_{BC}, H_{cj}(P_{cj})|_{CD}\right\} \\ H_{cj}^{\min}(P_{cj}) = 0 \end{cases} \qquad (9)$$

3 Modified Bat Algorithms

Bats are species very strong in positioning strategy. Using the supersonic radar a bat navigates the distance and direction to a predator in order to decide an optimal strategy to approach it. In Bat optimization algorithm [18], each Bat's position is the indices of a solution of optimization problems, which will be gradually enhance while the number of iterations is increased, and the velocity is used to determine the following positions emblematizing the possibility of changing positions of a Bat. Increasing the number of iterations, normally, the searching process could generate a new position which would be closer the predator than the previous one. Thus, the quality of a new solution tends to be improved during the searching. Similar to other meta-heuristics approaches, to generate new positions, two random walks including the global and local search and a ranking step based on the competitive mechanism, will be implemented.

In order to improve the searching performance of the CBA for solving the OOCHPG, in this paper, we propose three modifications, including an optimal range of updated velocity, a fixed loudness value, and a new way to generate new solutions in the second random walk. Velocity is one main factor to update new position for each bat. If the velocity is high, the bats' positions, i.e. solutions, could move outside the feasibility area whereas the small velocity could limit the discovering areas. Many studies to improve PSO algorithm in [13] have mentioned that the optimal range of the velocity can be selected from ten to fifteen percent of the difference between the boundary positions (maximum and minimum positions). Thus, the first modification recommends that updated velocities of the bats should be narrowed in a range of $[V_d^{min}, V_d^{max}]$ where

$$V_d^{max} = 0.15 * (X_d^{max} - X_d^{min}), \tag{10}$$

$$V_d^{min} = -V_d^{max}, \text{ and} \tag{11}$$

$[X_d^{min}, X_d^{max}]$ is limitation of the searching area.

Moreover, in CBA, a new solution is retained only if satisfying both conditions, its fitness function smaller than that of the old solution (of the same Bat) and its assigned random value lower than a current loudness varying at each iteration [18]. In our second modification, the loudness is fixed at 1 in order to keep all better solutions without considering any random decision. Thus, the current loudness is set to

$$A = 1 \tag{12}$$

For local search in CBA [18], each Bat has a chance to generate a new position which is around the best location, can be described by

$$X_d^{new} = G_{best} + rand * alpha \tag{13}$$

where G_{best} is the current best solution; alpha is the positive scaling factor.

In our third modification, the new generation of a current position (solution), X_d, should relate to the distance between that current position and the current best solution as shown in the equation below

$$X_d^{new} = G_{best} + rand * (G_{best} - X_d) \tag{14}$$

4 Numerical Results

A case study with 7 generation units, including four pure power units, two cogeneration units, and one pure heat unit, is examined to verify the effectiveness and robust of the proposed method. The study case has considered the transmission power losses and the non-convex fuel cost function as well. The data of this system has been detailed in [6]. The obtained results with different pulse rate values through CBA and MBA are listed in Tables 1 and 2, respectively. It is pointed out that the best minimum cost from CBA is obtained at pulse rate of 0.6 whilst that from MBA is at pulse rate of 0.9. Clearly, the proposed method can obtain higher efficiency than CBA in terms of optimal solutions and the stable quality of solutions. The successful rate (SR) reflecting the ability of handling all constraints is also reported. It shows that the proposed method (SR = 100%) can handle all constraints more effectively than CBA (SR = 70% to 81%). The whole search process by using CBA and MBA at the best run with the lowest fuel cost has been plotted in Fig. 1. From the 18th iteration onwards, it indicates that the proposed method can find better optimal solution than CBA at each iteration and still improves to the 80th iteration and not change afterward. Table 3 presents the minimum cost, execution time, and associated information with the comparison of computational time (CPU time) by using the proposed method and other methods. As shown in this table, the fuel cost comparison points out that MBA find a better optimal solution than most related methods except GCPSO [13] and GWO [14] with small differences of $ 33.54 and $ 66.09, respectively. Despite of this disadvantage point, MBA is still a the strongest method since its CPU time is the fastest among the compared methods, especially only needs 0.1 s for MBA but need 1.01 and 5.26 s for GCPSO [13] and GWO [14], respectively. Moreover, as indicated on columns of FE_{max} and SCT (s), MBA has used 4,000 fitness evaluations and spent the runtime of 0.1 s, which are lower than all other methods, e.g., GCPSO [13] of 20,000 and 0.76 s as well as GWO [14] with no population size reported and 5.04 s. For the SCTs (PU), MBA is much faster than other methods with 50.41, 64.45 and 80.90 times corresponding to GWO [14], BCO [6] and RCGA [6], respectively. Based on three evaluated criteria, we conclude that the proposed method is a very efficient method for solving the OOCHPG problem to the system with 7 units. The optimal solutions obtained by CBA and MBA are clearly shown in Table 4 for validating purposes.

Table 1. Result obtained by conventional BA with different pulse rates.

r_0	Min. cost ($/h)	Average cost ($/h)	Max. cost ($/h)	Std. dev. ($/h)	CPU time (s)	Successful rate
0.1	10863.57	12958.54	16329.95	1089.64	0.16	70%
0.2	12419.69	12719.72	16999.38	1136.131	0.15	74%
0.3	10775.64	12654.37	14982.6	882.6889	0.15	80%
0.4	10778.97	12644.21	16476.53	1057.704	0.15	74%
0.5	10769.19	12683.31	15214.04	899.8228	0.16	81%
0.6	**10411.58**	**12826.65**	**15088.96**	**1024.726**	**0.15**	**74%**

(*continued*)

Table 1. (*continued*)

r_0	Min. cost ($/h)	Average cost ($/h)	Max. cost ($/h)	Std. dev. ($/h)	CPU time (s)	Successful rate
0.7	10748.25	12887.54	16515.85	1308.058	0.15	79%
0.8	10956.46	12784.38	14871.65	971.5433	0.15	73%
0.9	10771.13	12730.56	15182.12	959.9498	0.15	77%

Table 2. Result obtained by proposed method with different pulse rates.

r_0	Min. cost ($/h)	Average cost ($/h)	Max. cost ($/h)	Std. dev. ($/h)	CPU time (s)	SR
0.1	10294.19	12654.84	17827.34	1672.36	0.099	100%
0.2	10234.11	12068.74	17408.49	1139.03	0.098	100%
0.3	10260.77	11762.08	17302.79	983.35	0.1	100%
0.4	10192.99	11584.49	15810.23	824.24	0.107	100%
0.5	10258.59	11520.55	14800.12	755.94	0.097	100%
0.6	10355.84	11622.42	16331.10	826.10	0.097	100%
0.7	10223.35	11551.46	14338.14	794.46	0.099	100%
0.8	10197.48	11724.59	15898.86	964.50	0.099	100%
0.9	**10168.16**	**11054.44**	**13173.13**	**690.30**	**0.099**	**100%**

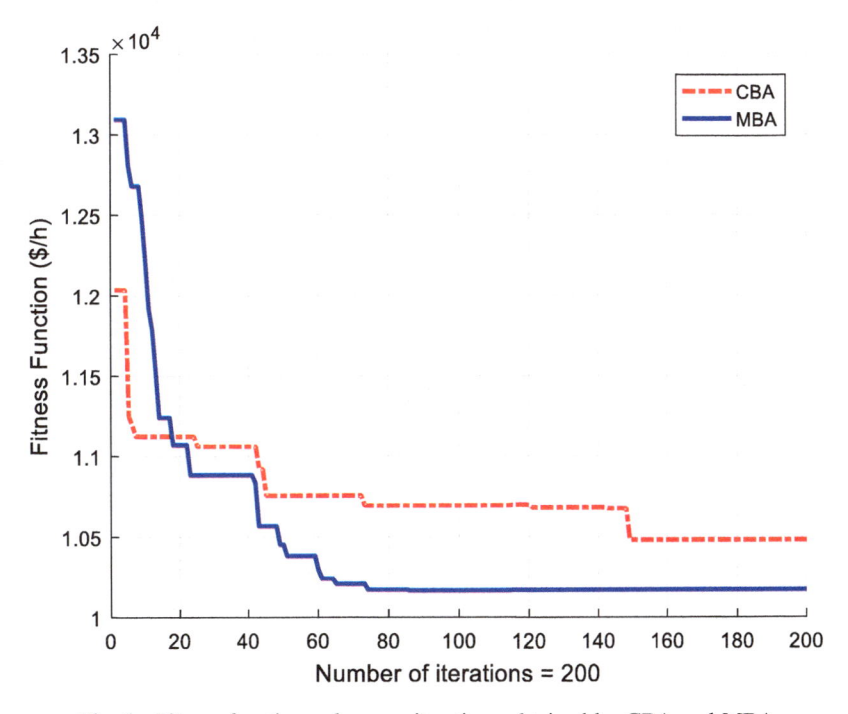

Fig. 1. Fitness function values vs. iterations obtained by CBA and MBA

Table 3. Result comparison among methods.

Method	Min cost ($/h)	N_P/G_{max}	FE_{max}	CPU (s)	Processor	SCT (s)	SCT (pu)
BCO [6]	10,317	50/100	5,000	5.1563	3	6.45	64.45
RCGA [6]	10,667	100/100	10,000	6.4723	3	8.09	80.90
AIS [17]	10,355	50/100	5,000	5.2956	3	6.62	66.20
EP [17]	10,390	100/100	10,000	5.3274	3	6.66	66.59
PSO [17]	10,613	100/100	10,000	5.3944	3	6.74	67.43
LCPSO [13]	10199.54	10/2000	20,000	1.02	1.8	0.77	7.65
LCPSO-CD [13]	10279.10	10/2000	20,000	1.02	1.8	0.77	7.65
LWPSO [13]	10194.021	10/2000	20,000	1.03	1.8	0.77	7.73
LWPSO-CD [13]	10265.67	10/2000	20,000	1.02	1.8	0.77	7.65
GCPSO [13]	10143.78	10/2000	20,000	1.01	1.8	0.76	7.58
GCPSO-CD [13]	10294.23	10/2000	20,000	1.02	1.8	0.77	7.65
GWPSO [13]	10243.02	10/2000	20,000	1.03	1.8	0.77	7.73
GWPSO-CD [13]	10281.68	10/2000	20,000	1.03	1.8	0.77	7.73
CSA [16]	10168.16	10/1500	30,000	1.2	1.8	0.90	9.00
GWO [14]	10111.24	-/200	–	5.26	2.3	5.04	50.41
CBA	10376.4676	40/200	8,000	0.154	2.4	0.15	1.54
MBA	**10168.15**	**20/200**	**4,000**	**0.1**	**2.4**	**0.1**	**1**

Table 4. The optimal solutions obtained by BA and MBA for the considered test system

Variable	BA	MBA	Variable	BA	MBA
P_{p1} (MW)	19.3747	46.0437	P_{c2} (MW)	41.5676	40.0423
P_{p2} (MW)	98.8901	98.4748	H_{c1} (MWth)	25.8386	5.9938
P_{p3} (MW)	167.926	114.8225	H_{c2} (MWth)	72.104	74.3784
P_{p4} (MW)	185.0687	210.8631	H_{k1} (MWth)	52.0573	69.6279
P_{c1} (MW)	95.2766	97.7825			

5 Conclusion

This paper has proposed the modifications of Bat Algorithm to deal with the complexity of the combined heat and power generations considering valve-point loading effects on pure power thermal units. The MBA has enhanced the advantages of CBA and its effectiveness has been pointed out by comparing to other methods based on three criteria such as the quality of optimal solution, maximum number of fitness evaluations and convergence speed. Especially, the convergence speed of MBA is always superior to all others. Furthermore, MBA also outperforms CBA in charge of improving solution quality and avoid converging to a local optimum. Therefore, MBA is a promising method for yielding optimal solution for the problem of optimal operation of combined heat and power generations.

References

1. Vasebi, A., Fesanghary, M., Bathaee, S.M.T.: Combined heat and power economic dispatch by harmony search algorithm. Electr. Power Energy Syst. **29**, 713–719 (2007). https://doi.org/10.1016/j.ijepes.2007.06.006
2. Rooijers, F.J., Van Amerongen, R.A.M.: Static economic dispatch for co-generation systems. IEEE Trans. Power Syst. **3**(9), 1392–1398 (1994). https://doi.org/10.1109/59.336125
3. Tao, G., Henwood, M.I., Van, O.M.: An algorithm for heat and power dispatch. IEEE Trans. Power Syst. **11**(4), 1778–1784 (1996). https://doi.org/10.1109/59.544642
4. Chapa, G., Galaz, V.: An economic dispatch algorithm for cogeneration systems. In: Proceedings of IEEE Power Engineering Society General Meeting 2004, vol. 1, pp. 989–994 (2004). https://doi.org/10.1109/PES.2004.1372985
5. Dieu, V.N., Ongsakul, W.: Augmented Lagrange Hopfield network for economic load dispatch with combined heat and power. Electr. Pow. Compo. Syst. **37**(12), 1289–1304 (2009). https://doi.org/10.1080/15325000903054969
6. Basu, M.: Bee colony optimization for combined heat and power economic dispatch. Expert Syst. Appl. **38**, 13527–13531 (2011). https://doi.org/10.1016/j.eswa.2011.03.067
7. Esmaile, K., Majid, J.: Harmony search algorithm for solving combined heat and power economic dispatch problems. Energy Convers. Manage. **52**, 1550–1554 (2011). https://doi.org/10.1016/j.enconman.2010.10.017
8. Behnam, M.I., Mohammad, M.D., Abbas, R.: Combined heat and power economic dispatch problem solution using particle swarm optimization with time varying acceleration coefficients. Electr. Pow. Syst. Res. **95**, 9–18 (2013). https://doi.org/10.1016/j.epsr.2012.08.005
9. Mehrdad, T.H., Saeed, T., Manijeh, A., Parinaz, A.: Improved group search optimization method for solving CHPG in large scale power systems. Energy Convers. Manage. **80**, 446–456 (2014). https://doi.org/10.1016/j.enconman.2014.01.051
10. Basu, M.: Group search optimization for combined heat and power economic dispatch. Electr. Power Energy Syst. **78**, 138–147 (2016). https://doi.org/10.1016/j.ijepes.2016.02.051
11. Provas, K.R., Chandan, P., Sneha, S.: Oppositional teaching learning based optimization approach for combined heat and power dispatch. Electr. Power Energy Syst. **57**, 392–403 (2014). https://doi.org/10.1016/j.ijepes.2013.12.006
12. Basu, M.: Combined heat and power economic dispatch using opposition-based group search optimization. Electr. Power Energy Syst. **73**, 819–829 (2015). https://doi.org/10.1016/j.ijepes.2015.06.023
13. Nguyen, T.T., Vo, D.N.: Improved particle swarm optimization for combined heat and power economic dispatch. Sci. Iran. D **23**(3), 1318–1334 (2016)
14. Jayakumar, N., Subramanian, S., Ganesan, S., Elanchezhian, E.B.: Grey wolf optimization for combined heat and power dispatch with cogeneration systems. Electr. Power Energy Syst. **74**, 252–264 (2016). https://doi.org/10.1155/2015/120975
15. Tüfekci, P.: Prediction of full load electrical power output of a base load operated combined cycle power plant using machine learning methods. Electr. Power Energy Syst. **60**, 126–140 (2014). https://doi.org/10.1016/j.ijepes.2014.02.027

16. Nguyen, T.T., Vo, D.N., Dinh, B.H.: Cuckoo search algorithm for combined heat and power economic dispatch. Electr. Power Energy Syst. **81**, 204–214 (2016). https://doi.org/10.1016/j.ijepes.2016.02.026
17. Basu, M.: Artificial immune system for combined heat and power economic dispatch. Electr. Power Energy Syst. **43**, 1–5 (2012). https://doi.org/10.1016/j.ijepes.2012.05.016
18. Yang, X.S.: A new meta-heuristic bat-inspired algorithm. In: Nature Inspired Cooperative Strategies for Optimization (NICSO 2010), pp. 65–74 (2010). https://doi.org/10.1007/978-3-642-12538-6_6

Prediction of Hourly Vehicle Flows by Optimized Evolutionary Fuzzy Rules

Pavel Krömer[1(✉)], Jana Nowaková[1], Martin Hasal[2], and Jan Platoš[1]

[1] Department of Computer Science, VŠB Technical University of Ostrava, Ostrava, Czech Republic
{pavel.kromer,jana.nowakova,jan.platos}@vsb.cz
[2] IT4Innovations, VŠB Technical University of Ostrava, Ostrava, Czech Republic
martin.hasal@vsb.cz

Abstract. The prediction of traffic situation at different time periods is essential for intelligent management of transportation systems and represents a key concept of smart cognitive environments. Road traffic is a complex dynamic system with many stochastic elements and many internal and external dependencies. Real–world traffic patterns in large cities are very complicated to model and simulate analytically. Road traffic monitoring, on the other hand, can be easily achieved by inexpensive sensing and monitoring systems and is often readily available. It can be even obtained as a by–product of other transportation services, for example, toll collection. In this work, we use a modified version of a recent machine–learning method, evolutionary fuzzy rules, to learn location–specific estimators of hourly traffic flow at specific locations.

1 Introduction

The growing number of vehicles and the increasing intensity of road traffic represent a major challenge for large urban areas. The estimation of traffic flows at different seasons of the year, days of the week, and times of day can be used for road network design and optimization, construction and repair work planning, and many other real–world applications in the domain of intelligent transportation systems [14]. However, road traffic is a complex phenomenon affected by a large number of internal and external factors. Traffic patterns exhibit both stochastic (randomized), periodic, and social properties [11]. Analytical models, considering multiple of these influences at the same time, are hard to design and usually computationally expensive [10]. Moreover, deterministic models are often not able to embrace the stochastic and social component of road traffic.

On the other hand, the collection of traffic data with high spatiotemporal resolution commonly happens in today's road and transportation system. Traffic data with sufficient resolution can be used as an input for various machine-learning algorithm that can build accurate site–specific classification, prediction, and regression models of various traffic phenomena (e.g. flow, congestion, accidents). Nature–inspired approaches (e.g. swarm and evolutionary computation) and soft computing methods (e.g. fuzzy systems) enable efficient stochastic learning of complex phenomena from data. Fuzzy sets and fuzzy logic provide methods and tools for accurate and sensitive data analysis and processing [2]. Fuzzy algorithms are especially useful in situations where modelling of vagueness, imprecision, and stochasticity is needed. Besides their modelling capabilities and accuracy, fuzzy models are popular due to their linguistic character that allows their comprehension.

In this work, a recent multi–paradigm data mining method named evolutionary fuzzy rules (FR) [6,7] is used to learn site–specific models of hourly traffic flow from data. The rest of this paper is organized as follows: Sect. 2 summarizes related work, Sect. 3 details the FR algorithm and its proposed modification. Section 4 describes its application to hourly vehicle flow prediction and provides experimental evaluation of the accuracy of the approach. Finally, major conclusions are drawn and future work is outlined in Sect. 5.

2 Related Work

The processing and analysis of traffic flow data is an open problem and many techniques have been developed. One of the powerful group of used techniques is the non-conventional or bio-inspired approach. Application of neural networks to traffic flow modelling is proposed by Ledoux [8] already in 1997, which is then integrated into a real time adaptive urban traffic control system. It is a conjunction of networks, where local networks are used for modelling of the traffic flow on a signalized link, which communicate between themselves to model a wide traffic flow. Time-lag recurrent neural network for short-term traffic forecasting up to 15 m into the future is suggested for usage in Australia [4].

A combination of bio-inspired techniques – neural networks and fuzzy logic in fuzzy-neural model for prediction of the traffic flow, which is responsible for traffic control system is described in [17]. It consists of two parts – a gate network and an expert network. Vlahogianni et al. [16] states that neural networks are one of the best way for traffic parameters modelling and prediction. So they present genetic approach for multilayer structural optimization for easy representation of traffic flow data and to determine structure of the neural network.

The combination of neural and Bayesian network is suggested by Zhen et al. [20] for short-term traffic flow prediction by two neural network predictors with a Bayesian combination approach. Fuzzy rule-based system for urban traffic flow modelling and forecasting combines genetic algorithms for offline and also online setting of membership function of the proposed fuzzy rule-based system [5]. This system could be complemented by an online adaptive Kalman filter [15]. An unsupervised machine learning method – clustering is used for vessel traffic flow on Yangzi river in [19]. It is not proposed any new method, it is only a application of well-known clustering method using Weka. Online support vector machine for regression, which is a supervised statistical learning method is used for prediction of traffic flow under typical and atypical conditions in [3]. Experimental comparison of supervised learning is presented in [9].

3 Evolutionary Fuzzy Rules

Evolutionary fuzzy rules [6] are simple yet successful classification and regression instruments based on the merger of fuzzy information retrieval and genetic programming. The fuzzy set theory provides the theoretical background behind fuzzy information retrieval and evolutionary fuzzy rules. The concepts of information retrieval are employed to interpret processed data and to define the classification and regression model. Genetic programming (GP) [1] is used as a generic, problem–independent meta–heuristic machine learning algorithm for the evolution of symbolic rules.

Essentially, evolutionary fuzzy rules are soft classifiers heavily inspired by the area of information retrieval (IR). In the IR, extended Boolean IR model utilizes fuzzy set theory and fuzzy logic for flexible and accurate search [12]. It uses extended Boolean queries that consist of search terms, operators, and weights, and evaluates them against an internal representation (index) of a collection of documents. Evolutionary fuzzy rules use similar basic concepts, data structures, and operations, and employ them for general data processing tasks such as classification, prediction, and so forth.

The data, processed by a fuzzy rule, is a real–valued matrix. Each row of the matrix corresponds to a single data record which is interpreted as a fuzzy set of features. A general data matrix, D, with m rows (records) and n columns (attributes, features) can be mapped to an IR index that describes a collection of objects. Each fuzzy rule is a symbolic expression that can be parsed into a tree structure. The tree structure consists of nodes and leafs (terminal nodes). In the fuzzy rule, three types of terminal nodes are recognized. A *feature* node represents the name of an input feature (variable). A *past feature* node defines a requirement on certain feature in a previous data record. The index of the previous data record (current - 1, current - 2 etc.) is a parameter of the node. A *past output* node puts a requirement on a previous output of the predictor. The index of the previous output (current - 1, current - 2) is a parameter of the node. An example of a fuzzy rule written down using an infix notation is given below:

$$f1 : 0.5 \; and : 0.4 \; (f2[1] : 0.3 \; or : 0.1 \; ([1] : 0.1 \; and : 0.2 \; [2] : 0.3))$$

where *f1:0.5* is a feature node, *f2[1]:0.3* is a past feature node, and *[1]:0.5* is a past output node. A graphical illustration of the FR is provided in Fig. 1

Different node types can be used when dealing with different data sets. For example, past feature past output nodes are useful for time series analysis and to process data sets where the ordering of records matters. However, they are infeasible for models of regular data sets. The feature node is the basic building block of all FR–based classifiers and value estimators for arbitrary data sets.

The operator nodes, supported within the framework of evolutionary fuzzy rules, are *and*, *or*, *not*, *prod*, and *sum* nodes. However, more general and domain specific operators can be used as well. Both nodes and leafs are weighted to soften the criteria they represent. The operators *and*, *or*, *not*, *prod*, and *sum*

```
           and:0.4
          /      \
      f1:0.5    or:0.1
                /    \
           f2[1]:0.9  and:0.2
                      /    \
                  [1]:0.1  [2]:0.3
```

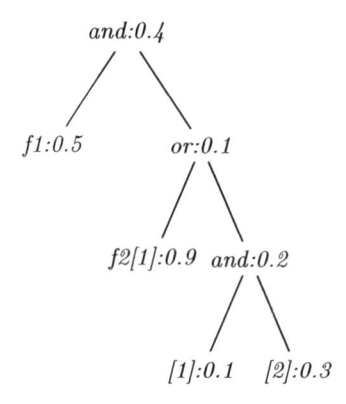

Fig. 1. Example of a fuzzy rule.

are evaluated using fuzzy set operations, extensions of crisp set operations to fuzzy sets. They are defined using the characteristic functions of combined fuzzy sets [18]. The standard t-norm (1) and t-conorm (2) are used to implement *and* and *or* operators, respectively. Fuzzy complement is used to evaluate of the *not* operator (3). Product t-norm (4) is employed to evaluate the *prod* operator, and its dual product t-conorm (5) is employed to evaluate the *sum* operator.

$$t(x,y) = \min(x,y) \quad (1)$$
$$s(x,y) = \max(x,y) \quad (2)$$
$$c(x) = 1 - x \quad (3)$$
$$t_{\text{prod}}(x,y) = xy \quad (4)$$
$$s_{\text{prod}}(x,y) = a + b - ab \quad (5)$$

The FRs have been successfully used for classification and regression [6]. Although machine–generated, they retain the understandable structure and ease of interpretation inherited from extended Boolean search expressions and allow a soft classification/regression without the complexity and computational costs of full–featured fuzzy rule–based systems [6]. Although originally a single–output predictor, they have been recently generalized into an evolutionary fuzzy rule forest (FRF) for the modelling and prediction of multiple target variables at the same time [7].

3.1 EFR Internals

The FRs are learned by genetic programming. The GP is an evolutionary algorithm that browses a search space of encoded problem solutions and seeks globally or at least locally optimum ones. A FR is represented by a linear chromosome composed of a series of instructions in reverse polish notation. They encode the operations of the rule and are interpreted by a virtual stack machine. The instruction set contains instructions corresponding to the operations of FR nodes and, in the case of the FRF, one special instruction, SI. SI is a reserved system instruction that represents the end of the code of one rule in the FRF. Using this approach, an entire FR, composed of n instructions, can be stored in a single linear chromosome, as illustrated in Fig. 2.

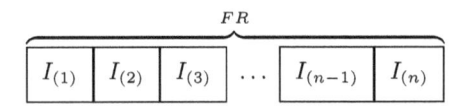

$$\underbrace{\boxed{I_{(1)}\mid I_{(2)}\mid I_{(3)}\mid \cdots \mid I_{(n-1)} \mid I_{(n)}}}_{FR}$$

Fig. 2. An example of a FR chromosome. $I_{(i)}$ stands for an executable FR instruction.

The length of the chromosomes depends on the number and size of the encoded FRs and may vary between the individuals in the population. GP operators were implemented with respect to the semantics of the FR and are applied directly to the chromosomes. They were designed so that when applied to correctly formed FRs (valid instruction sequences), they guarantee that the modified chromosome will contain only correctly formed FRs as well.

Mutation is implemented by a stochastic application of the following operations: (i) removal of a randomly selected subtree, (ii) replacement of a randomly selected node by a new randomly generated subtree, (iii) replacement of a randomly selected node by another compatible node, and (iv) a combination of the above. The linear structure of FR chromosomes allows the use of traditional genetic recombination strategies with only minor modifications. A modified one–point crossover, implemented in a way that maintains the number and correctness of the FRs in the forest, was employed. The crossover operator selects a random gene, x, from a FR located in the first parent chromosome. Then, a random gene compatible with x is selected from the corresponding FR in the second parent chromosome. Finally, the marked parts of parent chromosomes (single genes or complete subtrees) are exchanged to form two new offspring chromosomes. The actions, required to implement both proposed genetic operators, can be executed by the stack machine very efficiently.

The fitness function employed by FRs, can be an arbitrary similarity, error, or goodness–of–fit measure. In this work, a fuzzy version of the generalized F–score IR measure (F_β), used in the original FR design [6], was employed. It is a simple aggregation method that combines the precision (specificity) and recall (sensitivity) of the estimator into a single quality score. In the case of the compound FRF, the fitness function is a weighted linear combination of the fitness values of all FRs in the forest [7].

3.2 Optimized FR

Although the GP is able to find good FRs and FRFs [6,7], the nature of the learned models reveals potential for further tuning and optimization. The GP has to deal with a vast search space due to the fact that it learns the structure and the parameters of the models at the same time. When the structure of a FR is fixed, it can be seen as a model with a number of real–valued parameters and a real–parameter optimization methods can be employed to fine–tune the FR discovered by the GP. In this work, an efficient nature–inspired real–parameter optimization method, the differential evolution (DE), is adopted to optimize learned FRs and to constitute a 2–stage nature–inspired learning and optimization pipeline that produces accurate models of the underlying data.

The DE is a population–based evolutionary optimization algorithm [13]. It starts with an initial population of N real-valued vectors. The vectors are initialized with real values either randomly or so, that they are evenly spread over the problem space. The latter initialization leads to better results of the optimization [13]. During the optimization, the DE generates new vectors that are scaled perturbations of existing population vectors. The algorithm perturbs selected base vectors with the scaled difference of two (or more) other population vectors in order to produce the trial vectors. The trial vectors compete with members of the current population with the same index called the target vectors. If a trial vector represents a better solution than the corresponding target vector, it takes its place in the population [13].

The two most significant parameters of the DE are scaling factor and mutation probability [13]. The scaling factor $F \in [0, \infty]$ controls the rate at which the population evolves and the crossover probability $C \in [0, 1]$ determines the ratio of elements that are transferred to the trial vector from its opponent. The size of the population and the choice of operators are other important parameters of the optimization process.

The basic operations of the classic DE can be summarized using the following formulae [13]: the random initialization of the ith vector with N parameters is defined by

$$x_j^i = \mathrm{rand}(b_j^\mathrm{L}, b_j^\mathrm{U}), j \in \{1, \ldots, N\}, \tag{6}$$

where b_j^L is the lower bound of j-th parameter, b_j^U is the upper bound of j-th parameter, and $\mathrm{rand}(a,b)$ is a function generating a random number from the range $[a, b]$. A simple form of the standard differential mutation is given by

$$\mathbf{v}^i = \mathbf{v}^{r1} + F(\mathbf{v}^{r2} - \mathbf{v}^{r3}), \tag{7}$$

where F is the scaling factor and $\mathbf{v}^{r1}$, $\mathbf{v}^{r2}$, and $\mathbf{v}^{r3}$ are three random vectors from the population. The vector $\mathbf{v}^{r1}$ is the base vector, $\mathbf{v}^{r2}$ and $\mathbf{v}^{r3}$ are the difference vectors, and $\mathbf{v}^i$ is the trial vector. It is required that $i \neq r1 \neq r2 \neq r3$.

The uniform (binomial) crossover that combines the target vector, $\mathbf{x}^i$, with the trial vector, $\mathbf{v}^i$, is given by

$$v_j^i = \begin{cases} v_j^i & \text{if } (\mathrm{rand}(0,1) < C) \text{ or } j = j_\mathrm{rand} \\ x_j^i, & \text{otherwise} \end{cases} \tag{8}$$

for each $j \in \{1, \ldots, N\}$. The random index j_rand is in the above selected randomly as $j_\mathrm{rand} = \mathrm{rand}(1, N)$. The uniform crossover replaces the parameters in $\mathbf{v}^i$ by the parameters from the target vector $\mathbf{x}^i$ with probability $1 - C$.

The DE is a successful evolutionary algorithm designed for continuous parameter optimization driven by the idea of scaled vector differentials. In this work, it is employed to optimize the parameters of fuzzy rules learned by genetic programming in an attempt to improve the machine–learning procedure.

4 Experiments

In this study, the FR algorithm is employed as a location–specific predictor of hourly vehicle flows. The predictors are machine–learned from real–world vehicle flow data collected at specific locations. The data comes from the Metropolitan Transportation Authority (MTA) of the New York City and is freely available on the Web[1]. It contains the number of vehicles (cars, buses, trucks and motorcycles) passing through toll plazas at each of the bridges and tunnels operated by the MTA every hour of the day. The data is updated weekly and extracted directly from the EZ-Pass system database. MTA Bridges and Tunnels serve more than 800,000 vehicles every day of the week and nearly 290 million vehicles every year. The dataset, used in this work, was last updated on April 17, 2018. For each plaza, it contained the information about the number of incoming vehicles (the target variable), the number of incoming vehicles from the previous hour, and information about current date and time. However, not all toll plazas were observed every year and the results, reported in this paper, are related to locations that have been monitored in the course of the last couple of years (i.e. 2015–2017).

A high–level view of the proposed approach is outlined on a block diagram shown in Fig. 3. A FR is learned by the GP for each MTA plaza for every month of a year and its parameters are further optimized by the DE using the

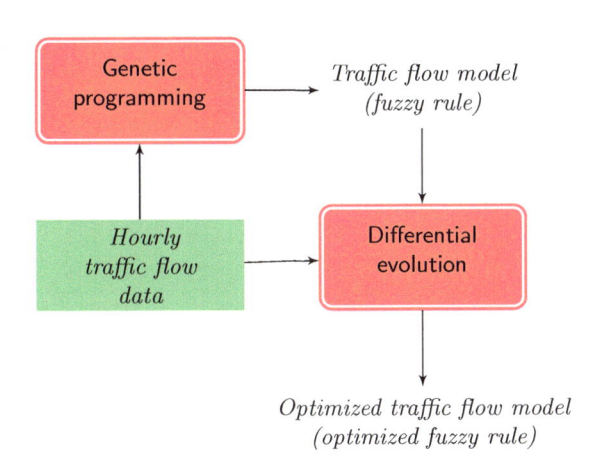

Fig. 3. Block diagram of FR evolution and optimization.

---

[1] https://data.ny.gov/Transportation/Hourly-Traffic-on-Metropolitan-Transportation-Auth/qzve-kjga/data.

same training dataset. A separate model is learned for each month to cope with seasonal changes and trend differences. For each plaza and each month, data from 2015 was taken as the training dataset and data from 2016 and 207 as the first and the second test dataset.

The 2–stage algorithm takes the following fixed parameters, determined on the basis of best–practices, previous experience, and extensive initial trial–and–error runs: the GP was a steady–state version of the algorithm with generation gap 2, population of 100 candidate fuzzy rules, mutation probability, $m = 0.2$, crossover probability $c = 0.9$, a limit of 1,000 generations, and the F-score measure, F_β, as the fitness function. The DE was the traditional $/DE/rand/1$ version of the algorithm with population size 100, scaling factor, F, and crossover probability, C, equal to 0.9, and the maximum number of DE generations was set to 1,000. In order to assess the ability of the stochastic algorithm to find FRs modelling location–specific traffic flows with a good accuracy, the evolution was repeated for each location in the data set 31 times independently and average results are discussed in this section.

Table 1. Fitness (F-score) of the best found FR on the train and test data sets.

Plaza id/month	Train	Test	
	(2015)	(2016)	(2017)
01/01	0.954559	0.950119	0.953673
02/01	0.946422	0.947057	0.947352
03/01	0.948988	0.951161	0.947703
05/01	0.926275	0.925213	0.920952
06/01	0.944300	0.943465	0.938518
07/01	0.944554	0.945582	0.935459
08/01	0.932167	0.935708	0.907091
09/01	0.945929	0.948238	0.946890
11/01	0.955433	0.954460	0.952403
01/(01–12)	0.952949	0.951231	0.945832
02/(01–12)	0.944806	0.942598	0.938638
03/(01–12)	0.953836	0.949489	0.946732
05/(01–12)	0.931501	0.92921	0.922281
06/(01–12)	0.941717	0.93917	0.933913
07/(01–12)	0.948487	0.943912	0.935459
08/(01–12)	0.938011	0.933582	0.907091
09/(01–12)	0.950512	0.949846	0.947273
11/(01–12)	0.955381	0.953195	0.951127

The ability of FRs to learn and predict location–specific hourly vehicle flows is summarized in Table 1. The top half of the table shows the fitness of the best found FRs for each plaza for the month of January. The bottom half of the table shows for each plaza the average fitness of the best found FRs through the year. The table clearly shows that the algorithm is able to learn the hourly vehicle flow patterns and to use them to predict the flows.

The average fitness of the best found FR models on the training set was between 0.932 and 0.955 (note that the maximum F-score fitness is 1). The average fitness of the learned models (i.e. prediction accuracy), applied to previously unknown test data sets (2016, 2017), is similar to that achieved on the training data set and was between 0.929 and 0.953 on 2016 data and 0.9071 and 0.9458 on 2017 data. This shows that the learned models are able to generalize well and can be used to estimate the hourly vehicle flows at different time periods of the year.

Nevertheless, the learning and prediction did not achieve the same level of accuracy for every day of the year. The examples of a typical day with good prediction and a typical day with bad prediction are shown in Fig. 4.

Table 2. Improvement in average final fitness achieved by FR optimization.

Plaza id/month	Fitness (F-score)		Improvement [%]
	FR	Optimized FR	
01/01	0.925136	0.927584	0.263877
02/01	0.911004	0.913266	0.247671
03/01	0.915607	0.918125	0.274257
05/01	0.897604	0.900790	0.353655
06/01	0.893823	0.900584	0.750648
07/01	0.908422	0.912129	0.406413
08/01	0.889509	0.894569	0.565636
09/01	0.914312	0.917837	0.384083
11/01	0.919446	0.922126	0.290593
01/(01–12)	0.925345	0.928358	0.324208
02/(01–12)	0.915145	0.917181	0.2221
03/(01–12)	0.920011	0.92312	0.336644
05/(01–12)	0.899547	0.901686	0.236708
06/(01–12)	0.897228	0.901363	0.459328
07/(01–12)	0.914409	0.917438	0.32996
08/(01–12)	0.89786	0.902839	0.551639
09/(01–12)	0.918663	0.921574	0.315517
11/(01–12)	0.923397	0.925431	0.219822

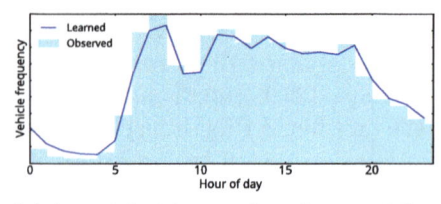

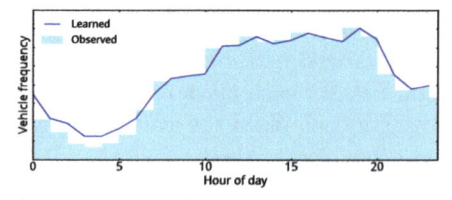

(a) A model with a good prediction ability.

(b) A model with worse prediction ability (overestimation of morning traffic).

Fig. 4. A visualisation of traffic prediction at typical days with good (left) and worse (right) predictionaccuracy.

The effect of FR optimization is summarized in Table 2. The table shows the average final fitness of the FRs found for each plaza by the GP, its change after the optimization by the DE, and summarizes the improvement. The average fitness improvement is between 0.22 and 0.75% which documents that the optimization process is useful and improves the accuracy of the machine–learned predictors.

5 Conclusions

A recent multi–paradigm data mining algorithm, evolutionary fuzzy rules, was used to model and predict hourly vehicle traffic at specific locations. The method, previously successfully used in a number of real-world applications, was modified and extended with the ability to optimize the parameters of the data–driven models by an arbitrary real–parameter optimization method. In this work, a nature–inspired optimization algorithm called the differential evolution was used for this task.

The ability of FRs to estimate hourly vehicle traffic on the basis of date, time, and past traffic intensity information was evaluated on a real–world dataset. The conducted experiments revealed a good potential of the method and suggest that it is able to learn location–specific traffic patterns. Nevertheless, further experiments with different data sets are needed thoroughly evaluate and fine–tune the method for such a specific task as hourly vehicle traffic prediction is. The advantage of FRs is their ability to learn complex patterns from data and the ease of their evaluation at the online classification/prediction stage. That has potential for the use in embedded and low–power devices such as wireless sensor nodes and traffic monitoring devices.

Acknowledgement. This work was supported by the European Regional Development Fund under the project AI&Reasoning (reg. no. CZ.02.1.01/0.0/0.0/15_003/0000466), by the Czech Science Foundation under the grant no. GJ16-25694Y, and by the project SP2018/126 of the Student Grant System, VŠB-Technical University of Ostrava.

References

1. Affenzeller, M., Winkler, S., Wagner, S., Beham, A.: Genetic Algorithms and Genetic Programming: Modern Concepts and Practical Applications. Chapman & Hall/CRC, New York (2009)
2. Bezdek, J.C., Keller, J., Krisnapuram, R., Pal, N.R.: Fuzzy Models and Algorithms for Pattern Recognition and Image Processing (The Handbooks of Fuzzy Sets). Springer, New York (2005)
3. Castro-Neto, M., Jeong, Y.S., Jeong, M.K., Han, L.D.: Online-SVR for short-term traffic flow prediction under typical and atypical traffic conditions. Expert Syst. Appl. **36**(3), 6164–6173 (2009)
4. Dia, H.: An object-oriented neural network approach to short-term traffic forecasting. Eur. J. Oper. Res. **131**(2), 253–261 (2001)
5. Dimitriou, L., Tsekeris, T., Stathopoulos, A.: Adaptive hybrid fuzzy rule-based system approach for modeling and predicting urban traffic flow. Transp. Res. Part C Emerg. Technol. **16**(5), 554–573 (2008)
6. Krömer, P., Owais, S.S.J., Platos, J., Snásel, V.: Towards new directions of data mining by evolutionary fuzzy rules and symbolic regression. Comput. Math. Appl. **66**(2), 190–200 (2013)
7. Krömer, P., Platos, J.: Simultaneous prediction of wind speed and direction by evolutionary fuzzy rule forest. In: International Conference on Computational Science, ICCS 2017, 12-14 June 2017, Zurich, Switzerland, pp. 295–304 (2017)
8. Ledoux, C.: An urban traffic flow model integrating neural networks. Transp. Res. Part C Emerg. Technol. **5**(5), 287–300 (1997)
9. Lippi, M., Bertini, M., Frasconi, P.: Short-term traffic flow forecasting: an experimental comparison of time-series analysis and supervised learning. IEEE Trans. Intell. Transp. Syst. **14**(2), 871–882 (2013)
10. Osorio, C., Selvam, K.K.: Solving large-scale urban transportation problems by combining the use of multiple traffic simulation models. Transp. Res. Procedia **6**, 272–284 (2015). 4th International Symposium of Transport Simulation (ISTS 2014) Selected Proceedings, Ajaccio, France, 1-4 June 2014
11. Pan, B., Demiryurek, U., Shahabi, C.: Utilizing real-world transportation data for accurate traffic prediction. In: IEEE 12th International Conference on Data Mining, pp. 595–604 (2012)
12. Pasi, G.: Fuzzy sets in information retrieval: state of the art and research trends. In: Bustince, H., Herrera, F., Montero, J. (eds.) Fuzzy Sets and Their Extensions: Representation, Aggregation and Models, Studies in Fuzziness and Soft Computing, vol. 220, pp. 517–535. Springer, Heidelberg (2008)
13. Price, K.V., Storn, R.M., Lampinen, J.A.: Differential Evolution A Practical Approach to Global Optimization. Natural Computing Series. Springer, Heidelberg (2005)
14. Smith, B.L., Demetsky, M.J.: Traffic flow forecasting: comparison of modeling approaches. J. Transp. Eng. **123**(4), 261–266 (1997)
15. Stathopoulos, A., Dimitriou, L., Tsekeris, T.: Fuzzy modeling approach for combined forecasting of urban traffic flow. Comput. Aided Civ. Inf. Eng. **23**(7), 521–535 (2008)
16. Vlahogianni, E.I., Karlaftis, M.G., Golias, J.C.: Optimized and meta-optimized neural networks for short-term traffic flow prediction: a genetic approach. Transp. Res. Part C Emerg. Technol. **13**(3), 211–234 (2005)

17. Yin, H., Wong, S., Xu, J., Wong, C.: Urban traffic flow prediction using a fuzzy-neural approach. Transp. Res. Part C Emerg. Technol. **10**(2), 85–98 (2002)
18. Zadeh, L.A.: Fuzzy sets. Inf. Control **8**, 338–353 (1965)
19. Zheng, B., Chen, J., Xia, S., Jin, Y.: Data analysis of vessel traffic flow using clustering algorithms. In: International Conference on Intelligent Computation Technology and Automation (ICICTA) 2008, vol. 2, pp. 243–246. IEEE (2008)
20. Zheng, W., Lee, D.H., Shi, Q.: Short-term freeway traffic flow prediction: Bayesian combined neural network approach. J. Transp. Eng. ASCE **132**(2), 114–121 (2006)

A New Simple, Fast and Robust Total Least Square Error Computation in E2: Experimental Comparison

Michal Smolik[(✉)], Vaclav Skala, and Zuzana Majdisova

Faculty of Applied Sciences, University of West Bohemia, Plzen, Czech Republic
{smolik,skala,majdisz}@kiv.zcu.cz

Abstract. Many problems, not only in signal processing, image processing, digital imaging, computer vision and visualization, lead to the Least Square Error (LSE) problem or Total (Orthogonal) Least Square Error (TLSE) problem computation. Usually the standard least square error approximation method is used due to its simplicity, but it is not an optimal solution, as it does not optimize the orthogonal distances, but only the vertical distances. There are many problems for which the LSE is not convenient and the TLSE is to be used. Unfortunately, the TLSE is computationally much more expensive. This paper presents a new, simple, robust and fast algorithm for the total least square error computation in E^2.

Keywords: Least squares · Total least squares · Orthogonal distance · Approximation

1 Introduction

The fitting of geometric features to given $2D$ or $3D$ points is desired in various fields of science and engineering. The least squares error approximation method (LSE) [6,7] is one of the best known, and most often applied, mathematical tools in various disciplines of science and engineering, e.g. signal processing [4], optics [18], surface modeling [12], regression modelling [8], nonlinear systems [14]. In the past, fitting problems have usually been solved through the LSE method with respect to the effective implementation and acceptable computing cost. However the natural and best choice of the error distance is the shortest distance between the given point and the model feature. This error definition is used in total (orthogonal) least squares error approximation method (TLSE), which fits the model more accurately than the standard LSE method [1,2].

2 Least Square Error Approximation

In the vast majority the Least Square Error (LSE) methods measuring vertical distances are used. This approach is acceptable in the case of explicit functions

$f(x, y) = h$, resp. $f(x, y, z) = h$. However, it should be noted that a user should keep in a mind, that smaller differences than 1.0, will have significantly smaller weight than higher differences than 1.0 as the differences are taken in a square resulting to dependence in scaling of the approximated data, i.e. the result will depend on physical units used, etc. The main advantage of the LSE method is its simplicity for fitting polynomial curves and it is easy to implement. The standard LSE method leads to an over determined system of linear equations. This approach is also known as polynomial regression.

Let us consider a data set $\Omega = \{\langle x_i, y_i, f_i \rangle\}_{i=1}^{N}$, i.e. data set containing for x_i and y_i measured functional value f_i and we want to find parameters $\boldsymbol{a} = [a, b, c, d]^T$ for optimal fitting function, as an example:

$$f(x, y, \boldsymbol{a}) = a + bx + cy + dxy \tag{1}$$

by minimizing the vertical squared distance D, i.e.:

$$D = \min_{a,b,c,d} \sum_{i=1}^{N} (f_i - f(x_i, y_i, \boldsymbol{a}))^2. \tag{2}$$

Conditions for an extreme are given as a vector equation:

$$\frac{\partial D}{\partial \boldsymbol{a}} = \sum_{i=1}^{N} (f_i - (a + bx_i + cy_i + dx_i y_i)) \frac{\partial f(x, y, \boldsymbol{a})}{\partial \boldsymbol{a}} = \boldsymbol{0}. \tag{3}$$

All those conditions can be rewritten in a matrix form as $\boldsymbol{Ax} = \boldsymbol{b}$. The selection of bilinear form was used to show the LSE method application to a non-linear case, if the case of linear function, i.e. $f(x, y, \boldsymbol{a}) = a + bx + cy$, the 4th row and column of the matrix $\boldsymbol{A}$ is to be removed.

Several methods for LSE have been derived [5,11], however those methods are sensitive to the vector $\boldsymbol{a}$ orientation and not robust in general as a value of $\sum_{i=1}^{N} x_i^2 y_i^2$ might be too high in comparison with the value N which has an influence to robustness of numerical solution. Also the LSE methods are sensitive to a rotation as they measure vertical distances. Rotational and translation invariance is fundamental requirement not only in geometrically oriented applications.

3 Total Least Square Error Approximation

The Total (Orthogonal) Least Square (TLSE) method takes another approach as it measures distances orthogonally and approximation by a line or plane is used nearly exclusively [19,22]. One significant property of the TLSE method is its rotational and translational invariance [20,21]. This approach leads to an approximation by an implicit function $F(x, y) = 0$ in the E^2 case, resp. $F(x, y, z) = 0$, in the E^3 case, i.e. dependence expressible as an implicit function.

There are several approaches how to solve TLSE problem and comprehensive analysis is given in [9]. Many algorithms are based on Singular Value Decomposition (SVD) or on a "simple" solution based on the explicit line representation [13] in the form $y = bx + a$. This formulation leads to a simple formula for calculation of the a, b coefficients. However, it is not robust and it is sensitive to a rotation. If we rotate the input points and then compute the LSE approximation, the result is different than when we first compute the LSE approximation and then rotate the line. Although when using the TLSE approximation, there is no such problem with rotation and the results are consistent. Also when a line is close to a vertical one, there is a high numerical imprecision for the LSE approximation and an overflow can appear as well, etc. If TLSE method is to be used many times, it is reasonable to consider robust and fast method specialized for the E^2 case. In the E^2 case and the linear case a linear function $F(\boldsymbol{x}) = ax + by + c = 0$ is used, the orthogonal distance d of the given point $\boldsymbol{x}$ and the line p is determined as:

$$d = \frac{|ax + by + c|}{\sqrt{a^2 + b^2}}, \tag{4}$$

where $\boldsymbol{x} = [x, y]^T$ is the given point and a line p is given as $ax + by + c = 0$. The computational problem is determination of coefficients a, b, c of a line $p \in E^2$.

There is a difference in approximation results for the LSE and the TLSE method as each of them use a different minimization criteria. The difference can be high for higher angles between x axis and the line, see Fig. 1 for an example.

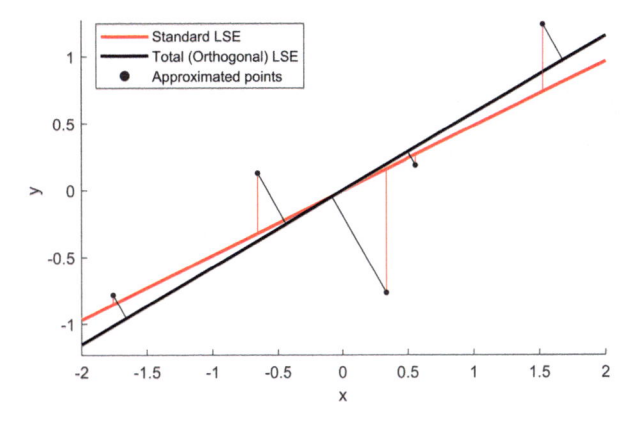

Fig. 1. Difference between standard least square error method and total (orthogonal) least square method.

In image processing, signal processing, digital imaging and computer graphics specialized algorithms should be used in the E^2 case. Such a solution for the E^2 case was published in [3] which is based on a line representation in the polar coordinates. Some specialized algorithms for a circle, resp. ellipse fitting were developed recently as well. The algorithm fully described in [3] is based on polar representation and leads to a formula which is stable. The derivation of the algorithm is not simple and uses goniometric functions, i.e. $sin(\theta)$ and $cos(\theta)$. A special case for perfectly circular data is to be solved. The algorithm [3] is not extensible to the E^3 case.

3.1 Total Least Square Error - Goniometric Functions

The approach proposed by Alciatore [3] defines an implicit line as

$$x\sin(\theta) + y\cos(\theta) + \rho = 0, \qquad (5)$$

where θ is chosen to be line orientation with respect to the x axis. The LSE method is computed as the sum of the squares of the related perpendicular distances is minimal. That is, minimize the value

$$\sum_{i=1}^{N} r_i^2(\theta, \rho), \qquad (6)$$

where N is the number of points and r_i are perpendicular distances.

After lengthy derivation, the authors end up with the following formulas

$$\begin{aligned} A &= 2b' \\ B &= -\left(a' + \sqrt{(a')^2 + 4(b')^2}\right) \\ C &= A\bar{x} + B\bar{y}, \end{aligned} \qquad (7)$$

where A, B and C are parameters of implicit line $Ax + By + C = 0$, $[\bar{x}, \bar{y}]^T$ is the centroid of the dataset $\{(x_i, y_i)\}$. Values of a' and b' are computed using

$$a' = \sum_{i=1}^{N}(x_i - \bar{x})^2 - \sum_{i=1}^{N}(y_i - \bar{y})^2 \qquad b' = \sum_{i=1}^{N}(x_i - \bar{x})(y_i - \bar{y}). \qquad (8)$$

3.2 Total Least Square Error - Parametric Form

The approach proposed by Skala [16] defines a parametric line as

$$y = x_T + ts, \qquad (9)$$

where x_T is the centroid of the dataset $\{(x_i, y_i)\}$, s is a directional vector and t a parameter $t \in \Re$. The TLSE method is computed as the sum of the squares of the related perpendicular distances is minimal. That is, minimize the value

$$\sum_{i=1}^{N} r_i^2 = \sum_{i=1}^{N} \frac{((x_i - x_T) \times s)^T ((x_i - x_T) \times s)}{s^T s}, \qquad (10)$$

where N is the number of points, r_i are perpendicular distances and vector s is to be calculated. After derivation, the author end up with the following formula

$$\left(\Omega - (s^T \Omega s)\, I\right) s = 0, \qquad (11)$$

where Ω is

$$\Omega = \sum_{i=1}^{N} x_i \otimes x_i^T, \qquad (12)$$

where $\otimes$ means a tensor product (result is a matrix).

4 Proposed Approach

In the following a new approach to TLSE computation will be described with experimental verification of the proposed method.

Fundamental requirement for any algorithm is its robustness. It should be fast and simple to implement as well. The proposed TLSE algorithm [17] is based on a squared orthogonal distance computation. As the TSLE method has to be translationally and rotationally invariant, the centroid of the given point set is to be $x_0 = 0$, this was shown also in [3]. As it is not a general case, the first step is a data set transformation:

$$x_i = x_i - x_0 \qquad x_0 = \frac{1}{2}\sum_{i=1}^{N} x_i, \qquad (13)$$

where N is the number of the given points, $x_i = [x, y_i]^T$ are the given points, $i = 1, \ldots, N$. This step has two consequences, the line p $(ax + by + c = 0)$ passes the origin of the coordinate system and therefore the coefficient c of the line p is set $c = 0$ by definition, now.

There is a simple formulation of the TLSE problem using optimization and Lagrange multipliers, i.e.

$$\min_{a,b,\lambda} D(a,b,\lambda) = \min_{a,b,\lambda}\left((ax_i + by_i)^2 + \lambda g(a,b)\right) \quad \& \quad g(a,b) = 0, \qquad (14)$$

where

$$g(a,b) = a^2 + b^2 - 1. \qquad (15)$$

Unfortunately, this approach does not lead to a simple solution.

The proposed algorithm is based on direct minimization of a distance given as:

$$D(a,b) = \sum_{i=1}^{N} d_i^2 = \sum_{i=1}^{N} \frac{(ax_i + by_i)^2}{a^2 + b^2}. \qquad (16)$$

For a minimum the following conditions must be fulfilled

$$\frac{\partial D(a,b)}{\partial a} = 0 \quad \& \quad \frac{\partial D(a,b)}{\partial b} = 0. \qquad (17)$$

Using (16) and two conditions (17) we get two conditions for an extreme

$$\begin{aligned}\frac{\partial D(a,b)}{\partial a} &= ab\sum_{i=1}^{N}\left(x_i^2 - y_i^2\right) + (b^2 - a^2)\sum_{i=1}^{N} x_i y_i = 0 \\ \frac{\partial D(a,b)}{\partial b} &= ab\sum_{i=1}^{N}\left(x_i^2 - y_i^2\right) + (b^2 - a^2)\sum_{i=1}^{N} x_i y_i = 0.\end{aligned} \qquad (18)$$

It can be seen that both equations above are equivalent and actually we have got just one equation

$$ab\sum_{i=1}^{N}\left(x_i^2 - y_i^2\right) + (b^2 - a^2)\sum_{i=1}^{N} x_i y_i = 0, \qquad (19)$$

using substitutions

$$\xi = \sum_{i=1}^{N}(x_i^2 - y_i^2) \quad \& \quad \eta = \sum_{i=1}^{N} x_i y_i, \tag{20}$$

(19) can be rewritten as

$$ab\xi + (b^2 - a^2)\eta = 0, \tag{21}$$

Now, we need to determine values a, b. As we keep the normalization condition for coefficients a, b during the extreme conditions, if

$$\sum_{i=1}^{N} x_i^2 \geq \sum_{i=1}^{N} y_i^2 \tag{22}$$

we can select the value a, e.g. $a = 1$, and solve the equation for b or vice versa. This leads to a quadratic equation

$$b\xi + (b^2 - 1)\eta = 0 \quad i.e. \quad \eta b^2 + \xi b - \eta = 0 \tag{23}$$

and therefore

$$b_1 = \frac{-\xi + \sqrt{\xi^2 + 4\eta^2}}{2\eta} \quad i.e. \quad b_2 = \frac{-\xi - \sqrt{\xi^2 + 4\eta^2}}{2\eta}. \tag{24}$$

The minimum distance is given by the b_2 value and

$$b = b_2 \quad i.e. \quad a = 1 \tag{25}$$

and the a, b values are of general values

Now, the computed line $p : ax + by = 0$, which is represented by the vector $\boldsymbol{p} = [a, b : 0]^T$, passes the origin of the coordinated system is to be "moved" back to the original coordinate system of the original data set using the standard geometric transformation represented by a matrix $\boldsymbol{T}$ [15], i.e.

$$\begin{bmatrix} a' \\ b' \\ c' \end{bmatrix} = \begin{bmatrix} 1 & 0 & 0 \\ 0 & 1 & 0 \\ -x_0 & -y_0 & 1 \end{bmatrix} \begin{bmatrix} a \\ b \\ 0 \end{bmatrix} \tag{26}$$

(26) can be rewritten as

$$\boldsymbol{p}' = \boldsymbol{T}\boldsymbol{p}. \tag{27}$$

Now, the line $\boldsymbol{p}'$ represents the line which optimally fits data in the sense of the TLSE method.

The formula is simple, easy and robust. However, it should be noted, that the proposed method above is not directly extensible to the E^3 case as a line in E^3 cannot be represented by an implicit formula.

5 Experimental Results

In this section we will compare our proposed approach for total least squares approximation with other approaches. The firs approach we will compare with, is the method presented in [3]. This method computes total orthogonal least squares approximation like our proposed method. It uses trigonometrical functions to derive the final formula for approximation. However the result of [3] is the same as result of our proposed approach, the derivation of final approximation formula in [3] is much more complicated. Derivation of our proposed approach is much more simpler. Next we will compare with the traditional least squares approximation method. This method minimizes the sum of square differences in y axis, i.e. approximated line formula is $y = f(x)$.

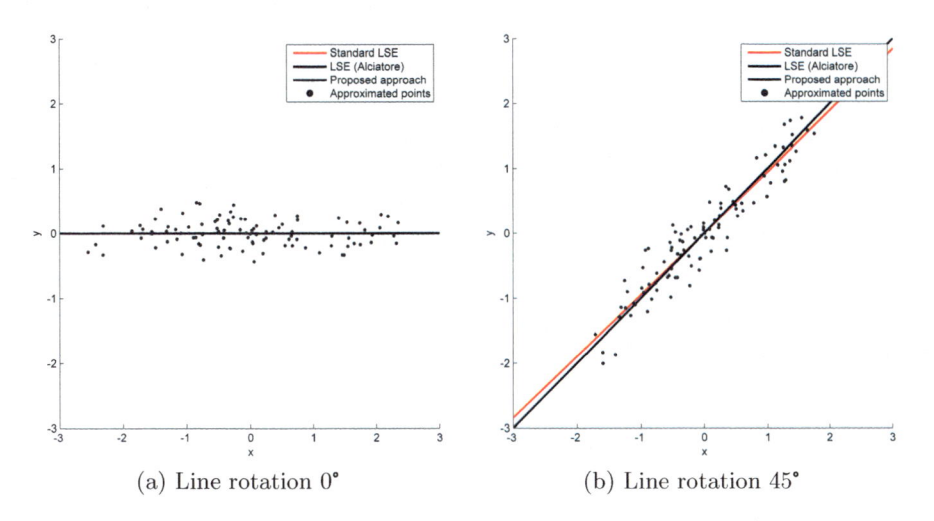

Fig. 2. Results of line approximation for different line rotation angles. Rotation angle is with respect to the x axis.

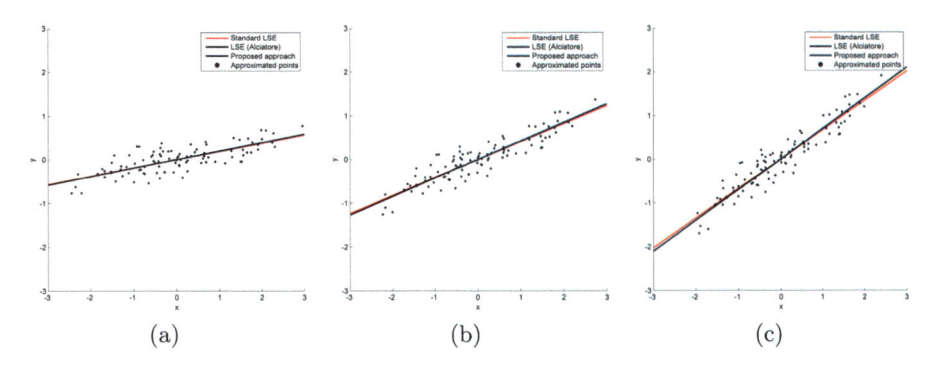

Fig. 3. Results of line approximation for different line rotation angles. Line rotation 11° (a), 23° (b) and 35° (c)

We used 100 uniformly generated points for testing purposes. The visualization of generated input points is in Fig. 2a. The generated input points were rotated by an angle $\varphi = \{0°, 1°, \ldots, 45°\}$.

We computed all the three tested least squares approximation methods and visualize the results in Figs. 2 and 3. It can be seen that the proposed approach for total least squares approximation has the same results as Alciatore's method presented in [3], i.e. the both approximated lines have the same mathematical formula. However the proposed approach is 21% faster than the Alciatore's method.

The results of traditional least squares approximation method differs from the results of the proposed approach. The reason for this is a different criteria for minimization, as the traditional method minimizes the sum of squared distances in y axis and the proposed approach minimizes the sum of squared orthogonal distances. For line that is parallel with the x axis both methods give the same result as the orthogonal distances are identical to the distances in y axis. The more the angle of line with x axis increases, the more differs orthogonal distances from the distances in y axis and thus both approximated lines differs more and more, see Figs. 2 and 3. The additional comparison of TLSE approximation method with other approximation methods is in [10] and the TLSE method gives the most accurate results.

The more the angle of line with x axis increases, the more differs the standard approximation and the proposed approach. To compute how much the standard approximation differs from our proposed method for total least squares approximation we need to define the distance measurement

$$dist\left([a,b,c],\boldsymbol{x}\right) = \frac{\sum_{i=1}^{N} |ax_i + by_i + c|}{\sqrt{a^2 + b^2}}, \tag{28}$$

where $[a, b, c]$ are coefficients of approximated line $ax + by + c = 0$ and $\boldsymbol{x} = \{[x_1, y_1], \ldots, [x_N, y_N]\}$ are approximated points. To compute how much the standard LSE approximation differs from our proposed method, which is the correct one, we define the following formula

$$error = \frac{(dist\left([a,b,c]_{StandardLSE},\boldsymbol{x}\right) - dist\left([a,b,c]_{ProposedApproach},\boldsymbol{x}\right))}{dist\left([a,b,c]_{ProposedApproach},\boldsymbol{x}\right)} \tag{29}$$

Using (29) we can compute the approximation line distance error, see Fig. 4. It can be seen that for line with angle 0° the approximation error is 0%. The error of the standard LSE increases with increasing the line angle. We performed the error computation for line angle up to 45°, as for angle from 45° to 90° we should use $x = g(y)$ for the standard LSE approximation, otherwise the approximation error will be too high for the standard LSE method ($y = f(x)$).

It should be noted that the TLSE approximation method by Alciatore [3] and the proposed TLSE method give the identical results.

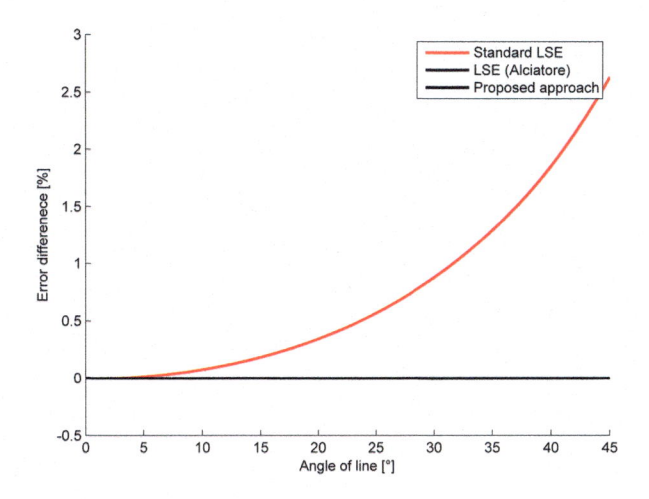

Fig. 4. Approximation line distance error in %. The error is computed using (29).

6 Conclusion

We presented a new approach for total least squares error approximation. The algorithm proved its simplicity and robustness and was compared with a LSE and a TLSE approaches. The proposed TLSE method has similar computational costs as standard LSE method and moreover, the approximation results are more correct.

In the future we plan to extend this approach for implicit plane approximation in E^3 as the extension is not straightforward due to non-linearity.

Acknowledgment. The authors would like to thank their colleagues at the University of West Bohemia, Plzen, for their discussions and suggestions, and anonymous reviewers for their valuable comments and hints provided. The research was supported by projects Czech Science Foundation (GACR) No. 17-05534S and SGS 2016-013.

References

1. Ahn, S.J.: Least Squares Orthogonal Distance Fitting of Curves and Surfaces in Space, vol. 3151. Springer Science & Business Media, Heidelberg (2004)
2. Ahn, S.J., Rauh, W., Warnecke, H.-J.: Least-squares orthogonal distances fitting of circle, sphere, ellipse, hyperbola, and parabola. Pattern Recogn. **34**(12), 2283–2303 (2001)
3. Alciatore, D., Miranda, R.: The best least-squares line fit. In: Paeth, A.W. (ed.) Graphics Gems V, pp. 91–97. Academic Press, Boston (1995)
4. Cadzow, J.A.: Signal processing via least squares error modeling. IEEE ASSP Mag. **7**(4), 12–31 (1990)
5. Chapra, S.C., Canale, R.P.: Numerical Methods for Engineers, vol. 2. McGraw-Hill, New York (1998)

6. Chernov, N.: Circular and Linear Regression: Fitting Circles and Lines by Least Squares. CRC Press, Boca Raton (2010)
7. Chernov, N., Lesort, C.: Least squares fitting of circles. J. Math. Imaging Vis. **23**(3), 239–252 (2005)
8. Duong, T.B.A., Tsuchida, J., Yadohisa, H.: Multivariate multiple orthogonal linear regression. In: International Conference on Intelligent Decision Technologies, pp. 44–53. Springer (2018)
9. Golub, G.H., Van Loan, C.F.: An analysis of the total least squares problem. SIAM J. Numer. Anal. **17**(6), 883–893 (1980)
10. Gutta, S., Bhatt, M., Kalva, S.K., Pramanik, M., Yalavarthy, P.K.: Modeling errors compensation with total least squares for limited data photoacoustic tomography. IEEE J. Sele. Top. Quantum Electron. **25**(1), 1–14 (2019)
11. Kreyszig, E.: Advanced Engineering Mathematics. Wiley, Hoboken (2010)
12. Lancaster, P., Salkauskas, K.: Surfaces generated by moving least squares methods. Math. Comput. **37**(155), 141–158 (1981)
13. Lee, S.L.: A note on the total least square fit to coplanar points. Technical report, ORNL-TM-12852, Oak Ridge National Laboratory (1994)
14. Rusnak, I., Peled-Eitan, L.: Least squares error criterion based estimator of non-linear systems. In: 2017 11th Asian Control Conference (ASCC), pp. 2522–2527. IEEE (2017)
15. Skala, V.: Projective geometry and duality for graphics, games and visualization. In: SIGGRAPH Asia 2012 Courses, SA 2012, pp. 10:1–10:47. ACM, New York (2012)
16. Skala, V.: A new formulation for total least square error method in d-dimensional space with mapping to a parametric line. In: AIP Conference Proceedings, vol. 1738, p. 480106. AIP Publishing (2016)
17. Skala, V.: Total least square error computation in E2: a new simple, fast and robust algorithm. In: Proceedings of the 33rd Computer Graphics International, pp. 1–4. ACM (2016)
18. Stutz, J., Platt, U.: Numerical analysis and estimation of the statistical error of differential optical absorption spectroscopy measurements with least-squares methods. Appl. Opt. **35**(30), 6041–6053 (1996)
19. Van Huffel, S., Cheng, C.-L., Mastronardi, N., Paige, C., Kukush, A.: Total least squares and errors-in-variables modeling (2007)
20. Van Huffel, S., Lemmerling, P.: Total Least Squares and Errors-in-Variables Modeling: Analysis. Algorithms and Applications. Springer Science & Business Media, Dordrecht (2013)
21. Van Huffel, S., Vandewalle, J.: The Total Least Squares Problem: Computational Aspects and Analysis, vol. 9. SIAM, Philadelphia (1991)
22. Xu, P., Liu, J., Shi, C.: Total least squares adjustment in partial errors-in-variables models: algorithm and statistical analysis. J. Geodesy **86**(8), 661–675 (2012)

On the Self-organizing Migrating Algorithm Comparison by Means of Centrality Measures

Lukas Tomaszek[(✉)], Patrik Lycka, and Ivan Zelinka

VSB-TU Ostrava, 17. listopadu, Ostrava, Czech Republic
{lukas.tomaszek,ivan.zelinka}@vsb.cz, zelinkaivan65@gmail.com

Abstract. In this article we continue in our research which combines three different areas - swarm and evolutionary algorithms, networks and coupled map lattices control. Main aim of this article is to compare networks obtained from best and worst self-organizing migrating algorithm runs. All experiments were done on well known CEC 2014 benchmark functions. For each selected function we picked 30 best and 30 worst runs, converted each run into a network, counted selected properties and compared the results. All obtained results are reported in this article.

Keywords: Self-organizing migrating algorithm · Networks · Centrality measure

1 Introduction

Networks provide us a powerful tool for a data analysis. We can use them for analyzing world wide web [2], brain connectivity [18], social interactions [19,24] and for many other datasets.

In our research, we are attempting to capture a run of swarm or evolutionary algorithm (SEA), into a network. Network vertices represent individuals (particles, fireflies), and edges capture interactions between them. As an interaction example, we can mention leader-individuals migration in the self-organizing migrating algorithm (SOMA) or crossing between individuals in the differential evolution. More about the conversions, you can find in [27].

Given networks may be analyzed by a well-developed theory about networks [4,5,13,15]. For example, we can calculate global properties like diameter, average path length or clustering coefficient to determine which network, and consequently which algorithm run, is better. Alternatively, we can count metrics like degree centrality, closeness centrality or eigenvector centrality to differentiate between good and bad vertices, and if there is a correlation between good and bad individuals.

With the information about the network, we can make algorithm better and faster. E.g., we can improve the algorithm by replacing or removing the individuals from the population, or by changing the algorithm parameters. Some improvements have been presented in [11,21].

Also, we can convert the network into a coupled map lattices (CML) [13]. Each row of the CML is understood as a time development of a vertex from a network. We can analyze and study given CMLs for a different kind of behavior (deterministic and chaotic regimes, intermittency) or we may attempt to control it [25,28].

2 Motivation

Until now, we have been able to demonstrate, that it is possible to control the SEAs. We showed that it is possible to change the algorithm dynamics [22], but we do not know, what is the optimal state, we want to reach. What parameters the network must have to get an optimal run. In this article, we want to compare the network properties of SOMA for better and worse runs, and find out what is the state we want to reach during the control.

3 Experiment Design

In the following parts, we firstly describe the SOMA and basic network properties we use. Also we show, how we can convert the SOMA run into a network. After, we will propose the experiment for better and worse SOMA runs comparison, and at the end, we will present some results.

3.1 Self-organizing Migrating Algorithm

SOMA is a swarm optimization algorithm based on the social behavior of competitive-cooperating individuals [7,26]. Before the start of the algorithm, we have to set up the parameters, and we have to define a cost function. We can see all required parameters, also with recommended range, in Table 1.

Table 1. SOMA parameters

Parameter name	Recommended range	Remark
$PathLength$	[1.1, 5]	Controlling parameter
$Step$	[0.11, $PathLength$]	Controlling parameter
PRT	[0, 1]	Controlling parameter
$Dimemsion$	Given by problem	Number of arguments in cost function
$PopSize$	[10, up to user]	Controlling parameter
$Migrations$	[10, up to user]	Stopping parameter

The run of the algorithm starts with creating an initial population. This population is randomly generated and distributed over searching space.

After that, we start migration loops. In each migration loop, firstly we select a leader. The leader is the best individual (the individual with the best cost value). After the leader is selected, each individual jumps towards him according to (1).

$$x_{i,j}^{ML+1} = x_{i,j,start}^{ML} + (x_{L,j}^{ML} - x_{i,j,start}^{ML})\, t\, PRTVector_j \qquad (1)$$

$x_{i,j}^{ML+1}$ The value of i-th individual's j-th parameter in $Step\ t$ and migration loop $ML + 1$.
$x_{i,j,start}^{ML}$ The value of i-th individual's j-th parameter staring position in actual migration loop.
$x_{L,j}^{ML}$ The value of leader's j-th parameter in actual migration loop.
t A step from 0 by $Step$ to $PathLength$.
$PRTVector$ A random vector of ones and zeros depended on PRT. If a randomly generated number is less than PRT, then in $PRTVector$ will be 1 otherwise 0.

Each individual remembers all positions and values of the cost function reached on these positions. After all jumps, individual returns to the best position. Migration loops repeat until we reach a maximum number of migration loops.

Above described SOMA is called SOMA with AllToOne strategy. This is not the only SOMA strategy. As an example of other strategies we can name:

- AllToOneAdaptive - In this strategy, the leader is replaced immediately, when we improve him, so it is replaced during the migration loop and not after it.
- AllToOneRand - In this strategy, the leader is not the best individual, but for each migration and each jumping individual, the leader is generated randomly.
- AllToAll - In this last mentioned strategy, each individual jumps to all other individuals in each migration loop.

For more details about the algorithm, please look at [7, 26].

3.2 Networks

Networks [5, 13] are a very powerful tool for a data analysis. A network, or a graph $G = (V, E)$, is a collection of n vertices V (also called nodes or actors) joined by m edges E (also called links, ties or arcs). According to an edge type, we can distinguish several types of networks:

- **Undirected.** In these networks, we can not determine from which and to which vertex the edge flows.
- **Directed.** These networks are the opposite of undirected networks. We can distinguish from and to which vertex the edge flows.
- **Weighted.** In weighted networks, each edge is assigned a real number w, which denotes the strength of the bonding between connected vertices.

3.3 Network Model for the SOMA

For our experiments, we proposed a network model [29], which is based on an ant behavioral [9]. Ants possess unique pheromones and they release them during the traveling for food or materials for an anthill. The pheromones form paths which cross, and the paths create a network. In the network, crossroads are represented by the vertices, and the edges connect the vertices (crossroads), which are joined by a road. When many ants travel on the same road, this road will contain significantly more pheromones according to other roads, so each road is as big (as strong) as how many ants travel on it. Also, the pheromones are vaporized by the wind, so when ants stop walking on some road, this road disappears. The strength of the road, we represent as an edge weight. This model, we can use for capturing the dynamic of the SOMA, but also for other SEA [23].

In the network which captures the dynamic of the SOMA, the vertices are individuals in the population, the edges represent the interactions between these individuals, and the edge weights captures the strength of these interactions. If we want to create a network, from this algorithm, we have to follow two simple rules:

1. **Movement of ants, improvement.** If during the migration loop, the jumping individual x_i improve itself, we raise the edge weight from vertex V_i to vertex V_l by one. The vertex V_l corresponds to the leader x_l.
2. **Vaporization of the pheromones.** After each generation loop, we reduce all edges weights by $N\%$.

Given network will be dynamical, weighted and directed. The network will change during the run of the algorithm. The obtained networks can be analyzed [3,15].

3.4 Centrality Measures

In this paper, we will concentrate on the centrality measures [10,17]. Centralities inform us how important the vertex is according to the other vertices. Since our networks are directed and weighted, we will focus on incoming strength, outcoming strength, closeness centrality, and betweenness centrality.

Vertex Strength. The strength arises from the degree property, but instead of counting the vertex neighbors, it takes into account the edges weights. In directed networks, we distinguish between incoming and outcoming strength. Formally, we count incoming strength SC_i^{in} of vertex i according to equation

$$SC_i^{in} = \sum_j w_{j,i}, \tag{2}$$

and outcoming strength SC_i^{out} of vertex i like this:

$$SC_i^{out} = \sum_j w_{i,j}. \tag{3}$$

$w_{i,j}$ is the edge weight connecting the vertex i with the vertex j.

Closeness and Betweenness Centralities. The closeness and the betweenness centralities are based on the shortest path. In many networks, the weights represent a cost, a travel time, or a distance. E.g., the cost of transmitting in the Internet network [8], the distance between two points on a map [20], or the travel time between airports [1]. In other words, lower weights are better, than higher ones. In such networks, the shortest path $d_{i,j}$ between vertex i and vertex j is given by equation

$$d_{i,j} = min\,(w_{i,k},\ldots,w_{k,j}), \tag{4}$$

where k is the vertex we go through on the road from i to j.

In our networks, the weights do not determine the cost or the distance, but they represent the tie strength. They have an opposite meaning. Higher weights are better than lower ones. For the analysis of such networks, in [6,14], authors propose to invert the weights. They count shortest path according to equation

$$d_{i,j} = min\left(\frac{1}{w_{i,k}},\ldots,\frac{1}{w_{k,j}}\right). \tag{5}$$

After we define the proper shortest path calculation, we can count the closeness centrality CC_i of vertex i according to equation

$$CC_i = \sum_j \frac{1}{d_{i,j}} \tag{6}$$

and the betweenness centrality BC_i of vertex i by using

$$BC_i = \sum_{j,k} \frac{g^i_{j,k}}{g_{j,k}}, \tag{7}$$

where $g_{j,k}$ is the number of shortest paths between vertex j and vertex k, and $g^i_{j,k}$ is the number of such paths going throw the vertex i.

3.5 Experiment Design

In the experiment, we want to compare the networks created from better and worse runs, so we run SOMA algorithm for each tested function 500 times. After the simulation were done, we ordered the runs according to the found optimal solutions, and we picked 30 best, and 30 worst runs. After, each of the run were converted into 3 networks. First network captured the run after 50 migration loops, second after 200 loops, and third at the end of the run, so after 500 loops. For each network we counted incoming and outcoming strength, closeness and betweenness of all vertices and compared them.

Parameters. For the experiment, we set up parameters like this: $Dimension = 10, pathLength = 3.1, Step = 0.3, PRT = 0.3, PopSize = 30, Migrations = 500$.

Test Functions. The experiments were done on 10 selected multimodal CEC 2014 Benchmark functions [12]. The selected functions are marked by numbers 4 to 13.

Used Software. The selected network properties were counted by JUNG framework [16]. Note that the this framework count the betweenness just as a number of shorted paths going through the selected vertex.

4 Results

In the first part, all counted properties (incoming and outcoming strength, closeness and betweenness) were visualized as a distribution plot, but the shape of the distribution were similar within the given property. In other words, the number of high quality vertices and low quality vertices in the best runs is similar as in worst runs. Incoming strength and betweenness have most vertices with value 0, so we do not use them for next experiments.

In the second part, we counted the average vertex property value. The counted values for best runs are visualized in Table 2 and for worst runs in Table 3. Also the higher values are bold.

Table 2. Average vertex property after 50, 200 and 500 migration loops for 10 selected functions and best runs

Function no.	Avg. vertex strength			Avg. vertex closeness		
	50	200	500	50	200	500
4	34.239	**70.318**	**57.437**	1.243	**1.729**	**1.109**
5	**5.477**	**8.679**	**8.442**	**0.760**	0.332	0.053
6	27.414	**42.823**	8.475	1.248	**1.075**	0.217
7	24.553	16.466	1.273	1.094	0.343	0.017
8	28.647	**21.578**	**7.249**	1.213	0.306	0.015
9	12.902	**20.273**	**5.115**	0.961	0.399	0.024
10	15.218	**26.415**	**10.208**	0.829	**1.175**	**0.457**
11	**6.410**	**4.792**	**2.790**	**0.475**	**0.317**	**0.190**
12	5.613	**2.906**	0.978	0.437	**1.199**	0.061
13	**8.962**	**3.126**	0.631	**0.562**	0.189	0.027

As we can see the average strength and closeness is higher for the best runs according to the worse runs. In some cases, the average strength for best runs is five times higher according to worst runs.

In the last part, we check the results by statistical test (Mann-Whitney U Test). The comparison you can see in Table 4. The + symbol tells us that the

Table 3. Average vertex property after 50, 200 and 500 migration loops for 10 selected functions and worst runs

Function no.	Avg. vertex strength			Avg. vertex closeness		
	50	200	500	50	200	500
4	**35.633**	24.248	1.270	**1.267**	0.620	0.031
5	4.991	2.221	0.386	0.720	0.212	0.012
6	22.742	38.858	3.626	**1.391**	0.953	0.064
7	21.876	13.798	**2.217**	0.994	0.298	0.016
8	26.128	17.326	2.389	1.195	0.292	0.014
9	**14.000**	15.883	2.581	**1.059**	0.364	0.020
10	**17.674**	5.633	0.363	**1.134**	0.267	0.013
11	6.227	4.378	2.312	0.466	0.291	0.166
12	**5.644**	2.868	0.928	**0.441**	1.197	0.054
13	8.645	3.034	0.600	0.547	0.183	0.024

Table 4. Statistical property comparison

Function no.	Outcoming strength			Closeness		
	50	200	500	50	200	500
4	− −	+ + +	+ + +	− − −	+ + +	+ + +
5	=	+ + +	+ + +	+ + +	+ + +	+ + +
6	+ + +	=	+ + +	+ + +	=	=
7	=	+ + +	=	=	+ + +	+ + +
8	+ + +	+ + +	+ + +	+ + +	+ + +	+ + +
9	=	+ + +	+ + +	− −	+ + +	+ + +
10	+ + +	+ + +	+ + +	+ + +	+ + +	+ + +
11	=	+	+ +	=	+ + +	+ + +
12	=	=	=	− −	=	+ + +
13	+ + +	+ +	+	+ + +	+ +	+ + +

best run has statistically higher property value than worst run. The number of symbols represents the difference according to p-value (0.05, 0.01, 0.001).

As we can see, the best runs have statistically higher outcoming strength and also statistically higher closeness than worst runs for most of the test functions and also for each picked migration loop, where we counted the selected properties.

5 Conclusion

In this article we mainly want to check, if there is a difference between the bet and worst SOMA runs. As it showed the best runs have statistically higher average strength and also statistically higher closeness. We also checked the distributions, but as it showed, best and worst runs have similar number of important individuals and also similar numbers of unimportant individuals.

In the past we showed that it is possible to control the SOMA. We were able to change the dynamic of this algorithm. Now, we also know what state we want to reach, so in the future we can make the algorithm better and make it faster, by controlling to the higher closeness and outcoming strength.

Acknowledgement. The following grants are acknowledged for the financial support provided for this research by Grant of SGS No. 2018/177, VSB-Technical University of Ostrava and under the support of NAVY and MERLIN research lab.

References

1. Bagler, G.: Analysis of the airport network of India as a complex weighted network. Phys. A: Stat. Mech. Appl. **387**(12), 2972–2980 (2008)
2. Barabási, A.L., Albert, R., Jeong, H.: Scale-free characteristics of random networks: the topology of the world-wide web. Phys. A: Stat. Mech. Appl. **281**(1–4), 69–77 (2000)
3. Barrat, A., Barthelemy, M., Pastor-Satorras, R., Vespignani, A.: The architecture of complex weighted networks. Proc. Nat. Acad. Sci. U.S.A. **101**(11), 3747–3752 (2004)
4. Barrat, A., Barthelemy, M., Vespignani, A.: The architecture of complex weighted networks: measurements and models. In: Large Scale Structure and Dynamics of Complex Networks: From Information Technology to Finance and Natural Science, pp. 67–92. World Scientific (2007)
5. Boccaletti, S., Latora, V., Moreno, Y., Chavez, M., Hwang, D.U.: Complex networks: structure and dynamics. Phys. Rep. **424**(4–5), 175–308 (2006)
6. Brandes, U.: A faster algorithm for betweenness centrality. J. Math. Sociol. **25**(2), 163–177 (2001)
7. Davendra, D., Zelinka, I., et al.: Self-organizing migrating algorithm. In: New Optimization Techniques in Engineering (2016)
8. Dijkstra, E.W.: A note on two problems in connexion with graphs. Numer. Math. **1**(1), 269–271 (1959)
9. Dorigo, M., Birattari, M., Stutzle, T.: Ant colony optimization. IEEE Comput. Intell. Mag. **1**(4), 28–39 (2006)
10. Freeman, L.C.: Centrality in social networks conceptual clarification. Soc. Netw. **1**(3), 215–239 (1978)
11. Krömer, P., Kudělka, M., Senkerik, R., Pluhacek, M.: Differential evolution with preferential interaction network. In: 2017 IEEE Congress on Evolutionary Computation (CEC), pp. 1916–1923. IEEE (2017)
12. Liang, J., Qu, B., Suganthan, P.: Problem definitions and evaluation criteria for the CEC 2014 special session and competition on single objective real-parameter numerical optimization. Computational Intelligence Laboratory, Zhengzhou University, Zhengzhou China and Technical Report, Nanyang Technological University, Singapore (2013)
13. Newman, M.: Networks. Oxford University Press, Oxford (2018)
14. Newman, M.E.: Scientific collaboration networks. II. Shortest paths, weighted networks, and centrality. Phys. Rev. E **64**(1), 016132 (2001)

15. Newman, M.E.: Analysis of weighted networks. Phys. Rev. E **70**(5), 056131 (2004)
16. O'Madadhain, J., Fisher, D., Smyth, P., White, S., Boey, Y.B.: Analysis and visualization of network data using jung. J. Stat. Softw. **10**(2), 1–35 (2005)
17. Opsahl, T., Agneessens, F., Skvoretz, J.: Node centrality in weighted networks: generalizing degree and shortest paths. Soc. Netw. **32**(3), 245–251 (2010)
18. Rubinov, M., Sporns, O.: Complex network measures of brain connectivity: uses and interpretations. Neuroimage **52**(3), 1059–1069 (2010)
19. Scott, J.: Social Network Analysis. Sage, Thousand Oaks (2017)
20. Soh, H., Lim, S., Zhang, T., Fu, X., Lee, G.K.K., Hung, T.G.G., Di, P., Prakasam, S., Wong, L.: Weighted complex network analysis of travel routes on the singapore public transportation system. Phys. A: Stat. Mech. Appl. **389**(24), 5852–5863 (2010)
21. Tomaszek, L., Zelinka, I.: On performance improvement of the soma swarm based algorithm and its complex network duality. In: 2016 IEEE Congress on Evolutionary Computation (CEC), pp. 4494–4500. IEEE (2016)
22. Tomaszek, L., Zelinka, I.: On static control of swarm systems. In: 2017 IEEE Symposium Series on Computational Intelligence (SSCI), pp. 1–7. IEEE (2017)
23. Tomaszek, L., Zelinka, I.: Conversion of soma algorithm into complex networks. In: Evolutionary Algorithms, Swarm Dynamics and Complex Networks, pp. 101–114. Springer (2018)
24. Wasserman, S., Faust, K.: Social Network Analysis: Methods and Applications, vol. 8. Cambridge University Press, Cambridge (1994)
25. Zelinka, I.: Investigation on evolutionary deterministic chaos control-extended study. Heuristica **1000**, 2 (2005)
26. Zelinka, I.: SOMA–self-organizing migrating algorithm. In: Self-Organizing Migrating Algorithm, pp. 3–49. Springer (2016)
27. Zelinka, I.: On mutual relations amongst evolutionary algorithm dynamics and its hidden complex network structures: an overview and recent advances. In: Nature-Inspired Computing: Concepts, Methodologies, Tools, and Applications, pp. 215–239. IGI Global (2017)
28. Zelinka, I., Senkerik, R., Navratil, E.: Investigation on evolutionary optimization of chaos control. Chaos Solitons Fractals **40**(1), 111–129 (2009)
29. Zelinka, I., Tomaszek, L., Kojecky, L.: On evolutionary dynamics modeled by ant algorithm. In: 2016 International Conference on Intelligent Networking and Collaborative Systems (INCoS), pp. 193–198. IEEE (2016)

A Brief Overview of the Synergy Between Metaheuristics and Unconventional Dynamics

Roman Senkerik[✉]

Faculty of Applied Informatics, Tomas Bata University in Zlin,
Nam T.G. Masaryka 5555, 760 01 Zlin, Czech Republic
senkerik@utb.cz

Abstract. This brief review paper focuses on the modern and original hybridization of the unconventional dynamics and the metaheuristic optimization algorithms. It discusses the concept of chaos-based optimization in general, i.e. the influence of chaotic sequences on the population diversity as well as at the metaheuristics performance. Further, the non-random processes used in evolutionary algorithms, and finally also the examples of the evolving complex network dynamics as the unconventional tool for the visualization and analysis of the population in popular optimization metaheuristics. This work should inspire the researchers for applying such methods and take advantage of possible performance improvements for the optimization tasks.

Keywords: Optimization · Metaheuristics · Evolutionary algorithms · Complex networks · Chaotic systems

1 Introduction

This brief survey explores the unconventional synergy of several different research fields belonging to the computational intelligence paradigm, which are the stochastics processes, complex chaotic dynamics, complex networks (CN), and metaheuristics algorithms, specifically evolutionary computation techniques (ECT's). The algorithms of the interest here are Differential Evolution (DE) [1], Particle Swarm Optimization (PSO) [2], Self Organizing Migrating Algorithm (SOMA) [3], Firefly Algorithm (FA) [4], and Firework algorithm (FWA) [5].

The motivation behind this survey is quite simple. In recent decades, the metaheuristic techniques became well-established and frequently used tools for solving engineering and research optimization tasks with a various level of complexity in both real and discrete domains. Despite the fact, that the ongoing research has brought many powerful and robust metaheuristic algorithms, the researchers have to deal with a well-known phenomenon so-called *no free lunch theorem* [6] forcing them to test various methods, techniques, adaptations and parameter settings leading to the acceptable results. The importance of finding a well-performing algorithm is growing together with the increase of the dimensionality and number of complex objectives in current optimization tasks. Moreover, it is necessary to emphasize the fact that, like most of these methods, they are inspired by natural evolution, and their development can be considered as a form of evolution. Such a fact is mentioned in the article [7] that even incremental steps in algorithm development, including failures, may be the inspiration for the development of robust and powerful metaheuristics. Also, it is always advisable to focus on simplifying algorithms, as stated in [8].

Thus this survey introduces several simple, yet successful, modifications and unconventional approaches applied for metaheuristic algorithms. This paper represents a follow-up and summarization of previous research [9–12]. The organization is the following: firstly, hybridization of chaotic dynamics and optimization algorithms is introduced, followed by the short chapter discussing the utilization of non-random processes. Finally, the original concept of the mutual connection between CN and evolutionary algorithms (EAs) is described.

2 Chaos Driven Metaheuristics

The key operation in metaheuristic algorithms is the randomness. Together with the persistent development of metaheuristic algorithms, the popularity of hybridizing them with deterministic chaos is growing every year, due to its properties like ergodicity, stochasticity, self-similarity, and density of periodic orbits. Also, vice-versa, the metaheuristic approach in chaos control/synchronization is more popular in recent years.

Recent research in chaotic approach for metaheuristics mostly uses straightforwardly various chaotic maps in the place of pseudo-random number generators (PRNG).

The original chaos-based approach is tightly connected with the importance of randomization within heuristics as compensation of a limited amount of search moves. This idea has been carried out in several papers describing different techniques to modify the randomization process [13, 14], as well as the influence of randomization operations to parameter adaptation was profoundly experimentally tested in [15].

The original concept of embedding chaotic dynamics into the evolutionary/swarm algorithms as chaotic pseudo-random number generator (CPRNG) is given in [16]. Firstly, the PSO algorithm with elements of chaos was introduced as CPSO [17], followed by the initial testing of chaos embedded DE [18], DE with chaotic mutation factor (SACDE) [19], and with the deterministic chaos for the initialization (CIDE algorithm) [20]. Original inertia weight based PSO strategy driven by CPRNGs was also profoundly investigated [21–23]. Besides the continuous space domain, chaos-driven metaheuristic proved to be successful also in the discrete domain [24, 25].

Recently the chaos driven heuristic concept has been utilized in several swarm-based algorithms [26–30], as well as many applications with DE [31, 32].

2.1 Chaos as the CPRNG

The general idea of CPRNG is to replace the default PRNG with the chaotic system (either discrete map or discretized time-continuous flow). Following nine well known and frequently studied discrete dissipative chaotic maps were used as the CPRNGs for various metaheuristics. Systems of the interest were: *Arnold Cat Map, Burgers Map, Delayed Logistic Map, Dissipative Standard Map, Henon Map, Ikeda Map, Lozi Map, Sinai Map,* and *Tinkerbell Map*. With the typical settings and definitions as in [33], systems exhibit typical chaotic behavior. Please refer to the Table 1 and Eqs. (1)–(9)

for the definition of the maps. Also, Fig. 1 shows the short chaotic sequences for all nine above mentioned maps. These plots support the claims that due to the presence of self-similar chaotic sequences, the heuristic is forced to neighborhood-based selection (or alternative communication in swarms).

Table 1. Definition of popular chaotic maps

Chaotic system	Notation	Parameters
Arnold Cat map	$X_{n+1} = X_n + Y_n \pmod 1$ $Y_{n+1} = X_n + kY_n \pmod 1$ (1)	$k = 2.0$
Burgers map	$X_{n+1} = aX_n - Y_n^2$ $Y_{n+1} = bY_n + X_n Y_n$ (2)	$a = 0.75$ and $b = 1.75$
Delayed Logistic	$X_{n+1} = AX_n(1 - Y_n)$ $Y_{n+1} = X_n$ (3)	$A = 2.27$
Dissipative Standard map	$X_{n+1} = X_n + Y_{n+1} \pmod{2\pi}$ $Y_{n+1} = bY_n + k \sin X_n \pmod{2\pi}$ (4)	$b = 0.1$ and $k = 8.8$
Hénon map	$x_{n+1} = a - x_n^2 + by_n$ $y_{n+1} = x_n$ (5)	$a = 1.4$ and $b = 0.3$
Ikeda map	$X_{n+1} = \gamma + \mu(X_n \cos \phi + Y_n \sin \phi)$ $Y_{n+1} = \mu(X_n \sin \phi + Y_n \cos \phi)$ (6) $\phi = \beta - \alpha/(1 + X_n^2 + Y_n^2)$	$\alpha = 6, \beta = 0.4, \gamma = 1$ and $\mu = 0.9$
Lozi Map	$X_{n+1} = 1 - a\|X_n\| + bY_n$ $Y_{n+1} = X_n$ (7)	$a = 1.7$ and $b = 0.5$
Sinai map	$X_{n+1} = X_n + Y_n + \delta \cos 2\pi Y_n \pmod 1$ $Y_{n+1} = X_n + 2Y_n \pmod 1$ (8)	$\delta = 0.1$
Tinkerbell map	$X_{n+1} = X_n^2 - Y_n^2 + aX_n + bY_n$ $Y_{n+1} = 2X_n Y_n + cX_n + dY_n$ (9)	$a = 0.9, b = -0.6, c = 2$ and $d = 0.5$

Obtaining the CPRNG output is quite simple to implement. As the chaotic system is a set of equations with a static start position (See Table 1), we created random start positions of the chaotic systems, to have different start position for different experiments. Thus we are utilizing the typical feature of chaotic systems, which is extreme sensitivity to the initial conditions, popularly known as "butterfly effect," as the random seed. This random position is initialized with the default implementation PRNG. Once the start position of the chaotic system has been obtained, the system generates the next sequence using its current position. Getting the normalized pseudo-random value from the typical range of 0–1 can be done in several ways:

- The abs value of current output iteration of the chaotic map (x-axis), is divided by the maximum value from generated chaotic series. This simple scaling approach is causing so-called folding of the attractor around y-axis.

- The whole generated chaotic series is shifted to the positive real number region and then normalized to the range of 0–1.

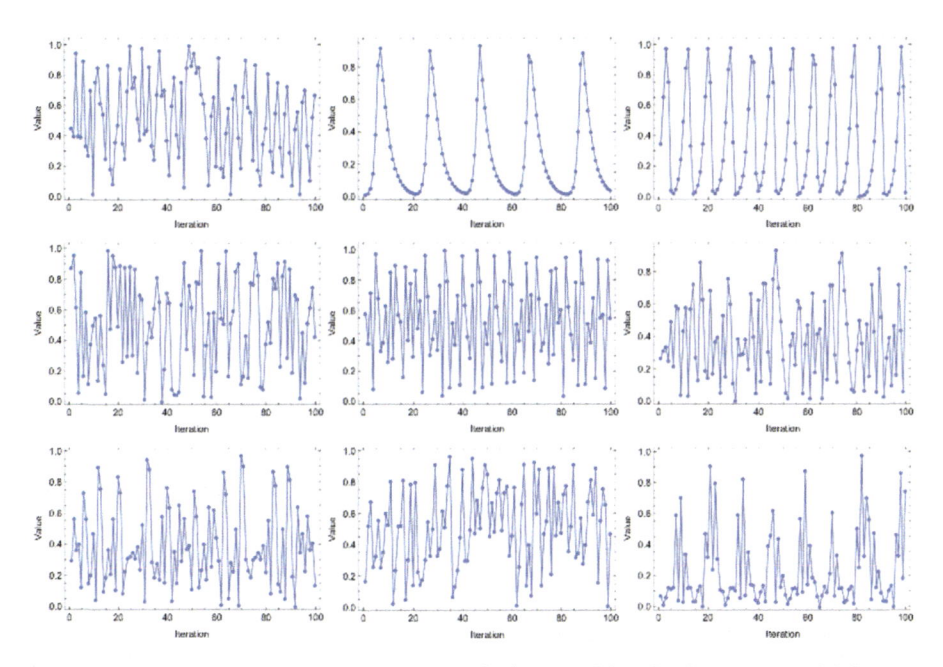

Fig. 1. Chaotic sequences normalized to the typical range of 0–1 for CPRNG; Arnold Cat map (upper left), Burgers Map (upper middle), and so on for the other maps (Delayed Logistic, Dissipative, Henon, Ikeda, Lozi, Sinai, Tinkerbell).

Also, the chaotic flows and oscillators have been widely studied, as well as other physical or chemical phenomenon showing chaos [9].

Nevertheless, most of the referred studies have used the chaotic dynamics to improve the properties of algorithms in a particular application, and unfortunately, deeper insights or theoretical explanations are not present in those research papers. The questions remain unanswered:

- Why does it work? Why may it be beneficial to use the chaotic sequences for pseudo-random numbers driving the selection, mutation, crossover or other processes in particular heuristics?
- Are there any chaotic features in the evolving population?

The first question was experimentally investigated in [34], and [35], where different sampling rates applied to the chaotic sequences were resulting in either keeping, or partially/fully removing of traces of chaos. These works show that not the distribution of the CPRNG used, but the unique sequencing given by the chaotic attractor hidden dynamics seems to be the key feature that may improve the heuristic performance.

In some instances, the population dynamics (which can also be considered as a chaotic or at the edge of chaos) seems to self-synchronize with the chaotic attractor dynamics and sequencing. This was not fully examined at all.

Based on the growing popularity of the complex/ensemble adaptation approaches within metaheuristics, the multi-chaotic CPRNGs were introduced in papers [21], and [36].

Moreover, as can be seen in Table 1, chaotic systems have easy-accessible parameters, which can be tuned resulting in different CPRNG distributions, sequencing of chaotic series, thus different influence on the metaheuristics being driven, as studied in [37]. The findings from referred research papers can be summarized as follows:

- Above referred research papers confirmed that used optimization metaheuristic is sensitive to the chaotic dynamics driving the selection, mutation, or communication process inside swarms through CPRNG. At the same time, it is clear that (selection of) the best CPRNGs are problem-dependent. The observed performance of enhanced optimizer is (significantly) different: either better or worse against other compared versions. Such a worse performance was repeatedly observed for three chaotic maps: Delayed logistic, Burgers, and Tinkerbell. On the other hand, these maps usually secured very fast progress towards function extreme (local) followed by premature population stagnation phase, thus repeatedly secured the finding of minimum values [12].
- The multi-chaotic generators [38] or ensemble systems could be beneficial since the used metaheuristic algorithm can profit from the combined/selective population diversity (i.e., exploration/exploitation) tendencies, sequencing-based either stronger or moderate progress towards the function extreme, all given by the smart combination of multi-randomization schemes.
- Swarm algorithms seem to benefit more from chaotic dynamics than "classical" EAs like Genetic Algorithms (GA) and DE.
- The population diversity analysis presented in [12] supports the theory, that unique features of the chaos transformed into the sequencing of CPRNG values may create the subpopulations or inner neighborhood selection schemes. Thus the metaheuristic can benefit from the searching within those sub-populations and quasi-periodic exchanges of information between individuals.
- There are many unexplored aspects and theories, like an auto-parameter adaptation for attractors driving metaheuristics, synchronization with optimized (dynamical) systems, and many more.
- The complex sequencing given by the chaotic system used seems to be the essence, not just the different distribution of CPRNG.
- Although intensive benchmarking (CEC benchmark suites) showed mixed results, in real-life optimization problems, the chaos driven heuristics is performing very well [39, 40], especially for some instances in the discrete domain [24, 25] – with the exception mentioned below.
- Comparison of PRNGs and CPRNGs is given in [41], whereas a critical review is in [42].

2.2 Other Unconventional Approaches with Chaos

Besides the direct utilization of CPRNGs, there exist two more approaches for interconnection between chaos and metaheuristics.

- The complexity of chaotic systems and its movement in the space is used for dynamical mapping of the search space mostly within the local search techniques [43].
- Finally, the hybridization of searching/optimization process and chaotic systems is represented by chaos based random walk technique [44].

3 Non-random Processes and Evolutionary Algorithms

As stated in the introduction, an inherent part of EAs, are random processes. Interesting study [10] is discussing whether random processes are needed EAs. Simple experiments revealed the fact that random number generators can be replaced by deterministic processes with short periodicity. Authors are claiming, that an advantage of the deterministic processes utilization is the possibility to repeat the experiment, analysis of algorithm behavior, mapping of its full path on the searched fitness landscape, and finally the possibility of easier construction of any kind of (mathematical) proofs for the used class of the EAs.

Also, different examples of non-random processes can be found in [45], where the sinusoidal function is used within the control parameters adaptation process.

4 Metaheuristics and Complex Networks

In this chapter, that represents a follow-up of more detailed studies [11, 46, 47], a fusion of two different attractive areas of research: (complex) networks and evolutionary computation, is described. Interactions in a swarm/evolutionary algorithms during the optimization process can be considered like user interactions in social networks or just people in society. It has been observed that networks generated by evolutionary dynamics show properties of CN in certain time frames and conditions [48]. The CN approach is utilized to show the linkage between different individuals in the population. Each individual in the population can be taken as a node in the CN graph, where its links specify the successful exchange of information in the population.

The population is visualized as an evolving CN that exhibits non-trivial features – e.g., degree distribution, clustering, and centralities. These features are important markers for a population used in evolutionary/swarm-based algorithms and can be utilized for the adaptive population control as well as parameter control during the metaheuristic run. The initial studies [49, 50] describing the possibilities of transforming population dynamics into CN were followed by the successful adaptation and control of the metaheuristic algorithm during the run through the given CN frameworks [51–55].

This paper briefly reviews the CN frameworks for DE, PSO, FA, and FWA. All referred works have shared a common motivation:

- To show the different approaches in building CN to capture the dynamics either of evolutionary or swarm-based algorithms.
- To investigate the time development of the influence of either individual selection for the mutation/crossover process or communication inside a swarm transferred into the CN.
- To show the usability of CN attributes, that can be extracted from graph visualizations, for adaptive population and parameter control during the metaheuristic run.

Since the internal principles are different for "classical" evolutionary-based (DE, GA), and swarm-based algorithms (PSO, FA, and FWA), several different approaches for capturing the population dynamics have been developed and tested:

- Capturing of the evolution process, i.e., the contribution of individuals from the population (DE, GA).
- Capturing of the communication in the swarm algorithms (PSO, FA)
- Capturing of the history of contributions for particles in randomized local search engines (FWA).

In the case of the classical EAs, mostly an Adjacency Graph was used. In each generation, the node is only active for the successful transfer of information, i.e., if the individual is successful in generating a new better individual who is accepted for the next generation of the population, one establishes the connections between the newly created individual and the (for the DE – several) sources; otherwise, no connections are recorded in the Adjacency Matrix. An illustrative example is given in Fig. 2 containing Adjacency Graph for the short snapshot (10 iterations at the beginning of the optimization process), and also the example of the corresponding community plot. The Degree Centrality value is highlighted by the size of the node (red color).

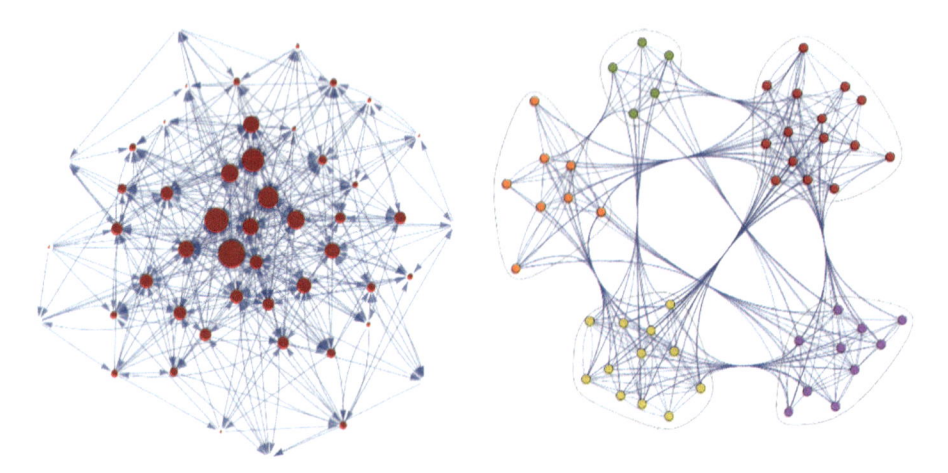

Fig. 2. Adjacency graph for a short time snapshot of DE algorithm (left), and corresponding community plot (right).

Analysis of CNs from DE algorithm can be found in [51, 56–59]; and also in a comprehensive study discussing the usability of network types [60].

The transition between evolutionary based and swarm-based algorithms can be seen in the FA. Although it is a swarm type, the situation here is very similar to the classical EAs. Every firefly is depicted as a node. The connection between nodes is plotted for every successful interaction when firefly flies towards another and improves own brightness. Details are discussed in [61].

For the PSO algorithm, the main interest is in the communications that lead to population quality improvement. Therefore, only communication leading to improvement of the particles personal best results (noted in original PSO as *pBest*) was tracked [55]. Another study was dealing with capturing of the density of communication [62], that can reveal the relations between the density of communication and convergence speed of the PSO.

Alternatively, it is possible to construct an Adjacency Graph and to benefit from its statistical features - as with the DE/GA/FA case. The direct link is created between the particle that has improved the global shared best solution and the particle that has been improved later based on such update. More investigation aimed at PSO and CN framework are in [63]. The very interesting approach of fitness landscape classifying based on the CN features and PSO is in [64].

The last studied algorithm (FWA) is the original representative of the random search/local search engine type algorithm. The paper [65] has shown, that even for this type, it is possible to develop a scheme for capturing the communication in the form of a graph. An interesting phenomenon has been discovered. The network seems to have a lack of any other usable information, besides the ability to identify the surface type of optimized function.

Besides the above-presented approaches, more have been explored for a wider portfolio of algorithms [66–68]. Overall, findings from the all referred papers can be summarized in the following way:

- *The building of the Network*: Since there is a direct link between parent solutions and offspring in the classical EAs, this information is used to build a CN. In the case of swarm algorithms, it depends on the inner swarm mechanisms, but mostly, it is possible to capture the communications within the swarm based on the points of attraction driven information updating.
- *Any original approach* leads to the different graph visualizations and possible subsequent analyses.
- *Complex Network Features*: Centralities and clustering/community analyses may contain direct information about the selection of individuals and their success; therefore, many network features can be used for controlling a population during an EA run. It is possible to use such information either for the injection or replacement of individuals or to modify/alternate the evolutionary strategy [57, 58]. In the case of swarm algorithms, the communication dynamics are captured - thus the level of particle performance (usefulness) can be calculated, or some sub-clusters and centralities of such communication can also be identified [63].

- *Fitness Landscape*: The capturing of communications (swarm dynamics) is sensitive to the fitness landscape. Thus, network features can be used for the raw estimation of a fitness landscape [64].
- *Dimensional independence* – since the size of the network is given only by the number of nodes (individuals from the populations), the resulting features analyses are not directly connected to the dimensionality of the search space.
- The advantage is that the CN framework can be used almost on any metaheuristic.

5 Conclusions

The primary aim of this original work is to provide more in-depth insights into the synergy between popular optimization metaheuristic algorithms and unconventional dynamics. The research of randomization issues and insights into the inner dynamic of metaheuristic algorithms was many times addressed as essential and beneficial. The influence of chaotic sequences on the metaheuristics performance, further the non-random processes used in EAs, and finally, the original concept of the evolving CN analysis for the better understanding of the population dynamics in popular optimization metaheuristics is briefly reviewed here. Important conclusions and findings are summarized at the end of each chapter.

Acknowledgments. This work was supported by the Ministry of Education, Youth and Sports of the Czech Republic within the National Sustainability Programme Project no. LO1303 (MSMT-7778/2014), further by the European Regional Development Fund under the Project CEBIA-Tech no. CZ.1.05/2.1.00/03.0089. This work is also based upon support by COST Action CA15140 (ImAppNIO), and COST Action IC1406 (cHiPSet).

References

1. Das, S., Mullick, S.S., Suganthan, P.N.: Recent advances in differential evolution–an updated survey. Swarm Evol. Comput. **27**, 1–30 (2016)
2. Engelbrecht, A.P.: Heterogeneous particle swarm optimization. In: International Conference on Swarm Intelligence, pp. 191–202. Springer, Heidelberg, September 2010
3. Zelinka, I.: SOMA—self-organizing migrating algorithm. In: Self-Organizing Migrating Algorithm, pp. 3–49. Springer, Cham (2016)
4. Fister, I., Fister Jr., I., Yang, X.S., Brest, J.: A comprehensive review of firefly algorithms. Swarm Evol. Comput. **13**, 34–46 (2013)
5. Tan, Y., Zhu, Y.: Fireworks algorithm for optimization. In: International Conference in Swarm Intelligence, pp. 355–364. Springer, Heidelberg, June 2010
6. Droste, S., Jansen, T., Wegener, I.: Perhaps not a free lunch but at least a free appetizer. In: Proceedings of the 1st Annual Conference on Genetic and Evolutionary Computation, vol. 1, pp. 833–839. Morgan Kaufmann Publishers Inc., July 1999
7. Piotrowski, A.P., Napiorkowski, J.J.: Step-by-step improvement of JADE and SHADE-based algorithms: success or failure? Swarm Evol. Comput. **43**, 88–108 (2018)
8. Piotrowski, A.P., Napiorkowski, J.J.: Some metaheuristics should be simplified. Inf. Sci. **427**, 32–62 (2018)

9. Senkerik, R., Zelinka, I., Pluhacek, M.: Chaos-based optimization-a review. J. Adv. Eng. Comput. **1**(1), 68–79 (2017)
10. Zelinka, I., Lampinen, J., Senkerik, R., Pluhacek, M.: Investigation on evolutionary algorithms powered by nonrandom processes. Soft. Comput. **22**(6), 1791–1801 (2018)
11. Senkerik, R., Zelinka, I., Pluhacek, M., Viktorin, A.: Study on the development of complex network for evolutionary and swarm based algorithms. In: Mexican International Conference on Artificial Intelligence, pp. 151–161. Springer, Cham, October 2016
12. Senkerik, R., Viktorin, A., Pluhacek, M., Kadavy, T.: Population diversity analysis for the chaotic based selection of individuals in differential evolution. In: International Conference on Bioinspired Methods and Their Applications, pp. 283–294. Springer, Cham, May 2018
13. Weber, M., Neri, F., Tirronen, V.: A study on scale factor in distributed differential evolution. Inf. Sci. **181**(12), 2488–2511 (2011)
14. Neri, F., Iacca, G., Mininno, E.: Disturbed exploitation compact differential evolution for limited memory optimization problems. Inf. Sci. **181**(12), 2469–2487 (2011)
15. Zamuda, A., Brest, J.: Self-adaptive control parameters' randomization frequency and propagations in differential evolution. Swarm Evol. Comput. **25**, 72–99 (2015)
16. Caponetto, R., Fortuna, L., Fazzino, S., Xibilia, M.G.: Chaotic sequences to improve the performance of evolutionary algorithms. IEEE Trans. Evol. Comput. **7**(3), 289–304 (2003)
17. Coelho, L.d.S, Mariani, V.C.: A novel chaotic particle swarm optimization approach using Hénon map and implicit filtering local search for economic load dispatch. Chaos, Solitons Fractals **39**(2), 510–518 (2009)
18. Davendra, D., Zelinka, I., Senkerik, R.: Chaos driven evolutionary algorithms for the task of PID control. Comput. Math Appl. **60**(4), 1088–1104 (2010)
19. Zhenyu, G., Bo, C., Min, Y., Binggang, C.: Self-adaptive chaos differential evolution. In: International Conference on Natural Computation, pp. 972–975. Springer, Heidelberg, September 2006
20. Ozer, A.B.: CIDE: chaotically initialized differential evolution. Expert Syst. Appl. **37**(6), 4632–4641 (2010)
21. Pluhacek, M., Senkerik, R., Davendra, D.: Chaos particle swarm optimization with Eensemble of chaotic systems. Swarm Evol. Comput. **25**, 29–35 (2015)
22. Pluhacek, M., Senkerik, R., Davendra, D., Oplatkova, Z.K., Zelinka, I.: On the behavior and performance of chaos driven PSO algorithm with inertia weight. Comput. Math Appl. **66**(2), 122–134 (2013)
23. Pluhacek, M., Senkerik, R., Viktorin, A., Kadavy, T.: Chaos-enhanced multiple-choice strategy for particle swarm optimisation. Int. J. Parallel Emergent Distrib. Syst. 1–14 (2018)
24. Metlicka, M., Davendra, D.: Chaos driven discrete artificial bee algorithm for location and assignment optimisation problems. Swarm Evol. Comput. **25**, 15–28 (2015)
25. Davendra, D., Bialic-Davendra, M., Senkerik, R.: Scheduling the lot-streaming flowshop scheduling problem with setup time with the chaos-induced enhanced differential evolution. In: 2013 IEEE Symposium on Differential Evolution (SDE), pp. 119–126. IEEE, April 2013
26. Gandomi, A.H., Yang, X.S., Talatahari, S., Alavi, A.H.: Firefly algorithm with chaos. Commun. Nonlinear Sci. Numer. Simul. **18**(1), 89–98 (2013)
27. Wang, G.G., Guo, L., Gandomi, A.H., Hao, G.S., Wang, H.: Chaotic krill herd algorithm. Inf. Sci. **274**, 17–34 (2014)
28. Zhang, C., Cui, G., Peng, F.: A novel hybrid chaotic ant swarm algorithm for heat exchanger networks synthesis. Appl. Thermal Eng. **104**, 707–719 (2016)
29. Jordehi, A.R.: Chaotic bat swarm optimisation (CBSO). Appl. Soft Comput. **26**, 523–530 (2015)
30. Wang, G.G., Deb, S., Gandomi, A.H., Zhang, Z., Alavi, A.H.: Chaotic cuckoo search. Soft. Comput. **20**(9), 3349–3362 (2016)

31. Coelho, L.d.S., Ayala, H.V.H., Mariani, V.C.: A self-adaptive chaotic differential evolution algorithm using gamma distribution for unconstrained global optimization. Appl. Math. Comput. **234**(0), 452–459 (2014)
32. Coelho, L.d.S., Pessôa, M.W.: A tuning strategy for multivariable PI and PID controllers using differential evolution combined with chaotic Zaslavskii map. Expert Syst. Appl. **38**(11), 13694–13701 (2011)
33. Sprott, J.C.: Chaos and Time-Series Analysis. Oxford University Press, Oxford (2003)
34. Senkerik, R., Pluhacek, M., Zelinka, I., Davendra, D., Janostik, J.: Preliminary study on the randomization and sequencing for the chaos embedded heuristic. In: Proceedings of the Second International Afro-European Conference for Industrial Advancement AECIA 2015, pp. 591–601. Springer, Cham (2016)
35. Senkerik, R., Pluhacek, M., Viktorin, A., Kadavy, T., Oplatkova, Z.K.: Randomization of individuals selection in differential evolution. In: 23rd International Conference on Soft Computing, pp. 180–191. Springer, Cham, June 2017
36. Senkerik, R., Pluhacek, M., Zelinka, I., Viktorin, A., Oplatkova, Z.K.: Hybridization of multi-chaotic dynamics and adaptive control parameter adjusting jDE strategy. In: International Conference on Soft Computing-MENDEL, pp. 77–87. Springer, Heidelberg, June 2016
37. Senkerik, R., Pluhacek, M., Oplatkova, Z.K., Davendra, D.: On the parameter settings for the chaotic dynamics embedded differential evolution. In: 2015 IEEE Congress on Evolutionary Computation (CEC), pp. 1410–1417. IEEE, May 2015
38. Viktorin, A., Pluhacek, M., Senkerik, R.: Success-history based adaptive differential evolution algorithm with multi-chaotic framework for parent selection performance on CEC2014 benchmark set. In: 2016 IEEE Congress on Evolutionary Computation (CEC), pp. 4797–4803. IEEE, July 2016
39. Senkerik, R., Pluhacek, M., Oplatkova, Z.K., Davendra, D., Zelinka, I.: Investigation on the differential evolution driven by selected six chaotic systems in the task of reactor geometry optimization. In: 2013 IEEE Congress on Evolutionary Computation (CEC), pp. 3087–3094. IEEE, June 2013
40. Senkerik, R., Zelinka, I., Pluhacek, M., Davendra, D., Oplatková Kominkova, Z.: Chaos enhanced differential evolution in the task of evolutionary control of selected set of discrete chaotic systems. Sci. World J. (2014)
41. Skanderova, L., Řehoř, A.: Comparison of pseudorandom numbers generators and chaotic numbers generators used in differential evolution. In: Nostradamus 2014: Prediction, Modeling and Analysis of Complex Systems, pp. 111–121. Springer, Cham (2014)
42. Krömer, P., Zelinka, I., Snášel, V.: Can deterministic chaos improve differential evolution for the linear ordering problem? In: 2014 IEEE Congress on Evolutionary Computation (CEC), pp. 1443–1448. IEEE, July 2014
43. Hamaizia, T., Lozi, R.: Improving chaotic optimization algorithm using a new global locally averaged strategy. In: Emergent Properties in Natural and Artificial Complex Systems, pp. pp-17, September 2011
44. Viktorin, A., Senkerik, R., Pluhacek, M., Kadavy, T.: Modified progressive random walk with chaotic PRNG. Int. J. Parallel Emergent Distrib. Syst. 1–10 (2017)
45. Awad, N.H., Ali, M.Z., Suganthan, P.N.: Ensemble of parameters in a sinusoidal differential evolution with niching-based population reduction. Swarm Evol. Comput. **39**, 141–156 (2018)
46. Chen, G., Zelinka, I.: Evolutionary Algorithms, Swarm Dynamics and Complex Networks (2018)

47. Senkerik, R., Pluhacek, M., Viktorin, A., Kadavy, T., Janostik, J., Oplatková, Z.K.: A review on the simulation of social networks inside heuristic algorithms. In: ECMS, pp. 176–182 (2018)
48. Skanderova, L., Fabian, T., Zelinka, I.: Small-world hidden in differential evolution. In: 2016 IEEE Congress on Evolutionary Computation (CEC), pp. 3354–3361. IEEE, July 2016
49. Zelinka, I., Davendra, D., Lampinen, J., Senkerik, R., Pluhacek, M.: Evolutionary algorithms dynamics and its hidden complex network structures. In: 2014 IEEE Congress on Evolutionary Computation (CEC), pp. 3246–3251. IEEE, July 2014
50. Davendra, D., Zelinka, I., Metlicka, M., Senkerik, R., Pluhacek, M.: Complex network analysis of differential evolution algorithm applied to flowshop with no-wait problem. In: 2014 IEEE Symposium on Differential Evolution (SDE), pp. 1–8. IEEE, December 2014
51. Skanderova, L., Fabian, T.: Differential evolution dynamics analysis by complex networks. Soft. Comput. **21**(7), 1817–1831 (2017)
52. Metlicka, M., Davendra, D.: Ensemble centralities based adaptive Artificial Bee algorithm. In: 2015 IEEE Congress on Evolutionary Computation (CEC), pp. 3370–3376. IEEE, May 2015
53. Gajdos, P., Kromer, P., Zelinka, I.: Network visualization of population dynamics in the differential evolution. In: 2015 IEEE Symposium Series on Computational Intelligence, pp. 1522–1528. IEEE, December 2015
54. Janostik, J., Pluhacek, M., Senkerik, R., Zelinka, I., Spacek, F.: Capturing inner dynamics of firefly algorithm in complex network—initial study. In: Proceedings of the Second International Afro-European Conference for Industrial Advancement AECIA 2015, pp. 571–577. Springer, Cham (2016)
55. Pluhacek, M., Janostik, J., Senkerik, R., Zelinka, I., Davendra, D.: PSO as complex network—capturing the inner dynamics—initial study. In: Proceedings of the Second International Afro-European Conference for Industrial Advancement AECIA 2015, pp. 551–559. Springer, Cham (2016)
56. Skanderova, L., Fabian, T., Zelinka, I.: Differential evolution dynamics modeled by longitudinal social network. J. Intell. Syst. **26**(3), 523–529 (2017)
57. Viktorin, A., Senkerik, R., Pluhacek, M., Kadavy, T.: Towards better population sizing for differential evolution through active population analysis with complex network. In: Conference on Complex, Intelligent, and Software Intensive Systems, pp. 225–235. Springer, Cham, July 2017
58. Viktorin, A., Pluhacek, M., Senkerik, R.: Network based linear population size reduction in SHADE. In: 2016 International Conference on Intelligent Networking and Collaborative Systems (INCoS), pp. 86–93. IEEE, September 2016
59. Senkerik, R., Viktorin, A., Pluhacek, M., Janostik, J., Davendra, D.: On the influence of different randomization and complex network analysis for differential evolution. In: 2016 IEEE Congress on Evolutionary Computation (CEC), pp. 3346–3353. IEEE, July 2016
60. Skanderova, L., Fabian, T., Zelinka, I.: Analysis of causality-driven changes of diffusion speed in non-Markovian temporal networks generated on the basis of differential evolution dynamics. Swarm Evol. Comput. **44**, 212–227 (2018)
61. Janostik, J., Pluhacek, M., Senkerik, R., Zelinka, I.: Particle swarm optimizer with diversity measure based on swarm representation in complex network. In: Proceedings of the Second International Afro-European Conference for Industrial Advancement AECIA 2015, pp. 561–569. Springer, Cham (2016)
62. Pluhacek, M., Senkerik, R., Viktorin, A., Kadavy, T.: Uncovering communication density in PSO using complex network (2017)

63. Pluhacek, M., Viktorin, A., Senkerik, R., Kadavy, T., Zelinka, I.: PSO with partial population restart based on complex network analysis. In: International Conference on Hybrid Artificial Intelligence Systems, pp. 183–192. Springer, Cham, June 2017
64. Pluhacek, M., Senkerik, R., Janostik, A.V.J., Davendra, D.: Complex network analysis in PSO as an fitness landscape classifier. In: 2016 IEEE Congress on Evolutionary Computation (CEC), pp. 3332–3337. IEEE, July 2016
65. Kadavý, T., Pluháček, M., Viktorin, A., Šenkeřík, R.: Firework algorithm dynamics simulated and analyzed with the aid of complex network. In: Proceedings-31st European Conference on Modelling and Simulation, ECMS 2017. European Council for Modelling and Simulation (2017)
66. Tomaszek, L., Zelinka, I.: On performance improvement of the SOMA swarm based algorithm and its complex network duality. In: 2016 IEEE Congress on Evolutionary Computation (CEC), pp. 4494–4500. IEEE, July 2016
67. Krömer, P., Gajdo, P., Zelinka, I.: Towards a network interpretation of agent interaction in ant colony optimization. In: 2015 IEEE Symposium Series on Computational Intelligence, pp. 1126–1132. IEEE, December 2015
68. Skanderova, L., Zelinka, I., Saloun, P.: Complex network construction based on SOMA: vertices in-degree reliance on fitness value evolution. In: ISCS 2013: Interdisciplinary Symposium on Complex Systems, pp. 291–297. Springer, Heidelberg (2014)

Telecommunications

An Examination of Outage Performance for Selected Relay and Fixed Relay in Cognitive Radio-Aided NOMA

Tam Nguyen Kieu[1,2](✉), Hong Nhu Nguyen[1], Long Nguyen Ngoc[1], Tu-Trinh Thi Nguyen[3], Jaroslav Zdralek[1], and Miroslav Voznak[1]

[1] VSB-Technical University of Ostrava, 17. listopadu 15, Ostrava, Czech Republic
nhu.nh@sgu.edu.vn, nguyenngoclong@tdt.edu.vn,
{Jaroslav.Zdralek,miroslav.voznak}@vsb.cz
[2] Ton Duc Thang University,
19 Nguyen Huu Tho Street, 7th District, Ho Chi Minh City, Vietnam
nguyenkieutam@tdtu.edu.vn
[3] Faculty of Electronics Technology, Industrial University of Ho Chi Minh City,
12 Nguyen Van Bao Street, Go Vap District, Ho Chi Minh City, Vietnam
tutrinhamber@gmail.com

Abstract. In this paper, we consider impacts of relay selection scheme on primary network performance in cognitive radio (CR) assisted non-orthogonal multiple access (NOMA). We suggest relay selection model to improve the outage performance at primary network. In this model, the fixed power allocation in NOMA is deployed to achieve considered simulation results. To completely study the profits of the NOMA scheme, we derive closed-form formula for the outage probability. We find that the CR-NOMA enhances its performance as suitable selection of relay intending to forward signal to NOMA users. It is shown several performance comparison in such CR-NOMA as varying the target rates at relay and destination. Numerical results are extensively studied to confirm that our proposed CR-NOMA outperforms the other designs in terms of outage probability.

Keywords: Non-orthogonal multiple access (NOMA) · Decode-and-forward (DF) · Primary network · Outage probability

1 Introduction

NOMA has drawn a lot of interests in recent times since it can remarkable promote the network spectral effect as in [1]. Various from the normal orthogonal multiple access approaches such as frequency division multiple access, time division orthogonal multiple access, and code division multiple access, NOMA permits multiple users to share the same source block (i.e., time/frequency/code), in that successive interference cancellation (SIC) has to be carried at one receiver to reduce the noise breed by other users' data. In [2], cooperation NOMA with optimum rate associating was investigated to enlarge the spatial diversity.

As in [3], various relay selective plans were suggested and studied in cooperation NOMA networks. Besides, in [4], outage probability and ergodic sum capacity were studied in a NOMA network with coordinated running and relay transmittance. On the other hand, in [5], two resource-destination user pairs sharing a public half-duplex relay was studied in NOMA cooperation network. It is paid attention to cooperation NOMA taken in [2–5], all approve half-duplex cooperative relaying model. Moreover, in [6] maximised the sum rate of users in an uplink NOMA network. In practice, the authors in [7] examined the effect of user pairing on the sum rate in a flat-power allocation NOMA network and a cognition radio inspired NOMA network.

Especially, authors in [8] decreased effect power allocation models to sure the quality of service in both down link and uplink NOMA networks. In addition, [9] examined instantaneous/average channel state information at the transmitter and set up power allocation models to optimize the minimum achievement rate in the NOMA networks. While, CR has been largely known as a hoped remedy for the spectrum scarcity of wireless applications as in [10–12]. The fresh technic permits secondary users (SUs) to opportunistically use the licensed spectrum taken by primary users (PUs). Lately, results in [13–16] shown energy harvesting network model can be combined in the NOMA network and supplied the green NOMA information systems.

Motivated by above analysis and novel results presented in [17], we consider such analysis in CR-NOMA network. The main contributions of this paper can be sum up as follows. First, we suggest a dynamic cooperation CR-NOMA model, whereby the multi cast NOMA users are served by potential relays for both the primary system and the secondary system, will also support better the trust of the such transfers. Next, we compare outage operation of two suggested models.

The remainder of this paper is organized as follows. Section 2 describes the system model. In Sect. 3, DF relaying in CR-NOMA are analyzed. The simulation results are presented in Sect. 4. Finally, the conclusion is drawn in Sect. 5.

2 System Model

We examine a CR-NOMA network as in Fig. 1 with one base station as primary transmitter (namely PS), two relays (PR, SR) located in primary and secondary network and two representative users in each network of such CR-NOMA system. Suppose that all nodes are set up with a single antenna, and between the base station and the end users there is no direct link. It worth noting that the base station with the users can be able to communicate via a selected DF relay from the two available relays. Suppose that all links are quasi-static Rayleigh fading. Recall that the quality of channels from the relays to users in CR-NOMA are a coefficient to order users as in [10,12].

Not the same these available papers, we suppose that channel conditions between the relays and users are statistically the same, i.e., the relay-user channel

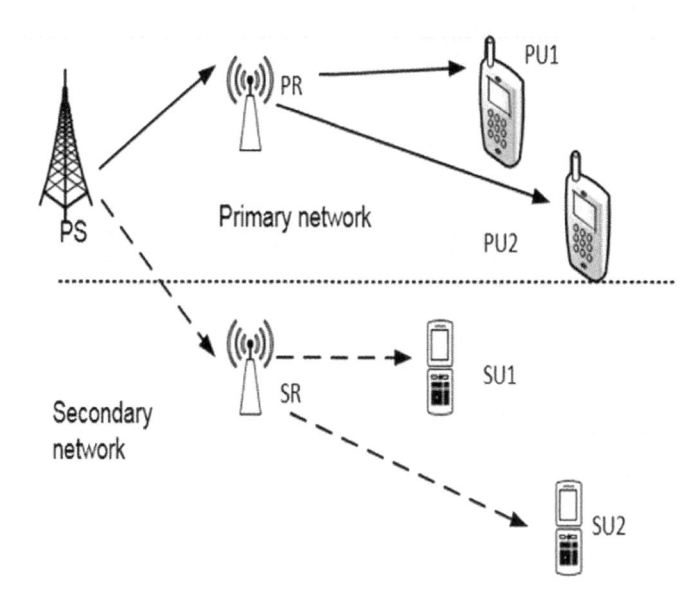

Fig. 1. System model.

fading gains are supposed to be independent and identically allocated, and the users are not set up relied on their channel constrains. The users are set up following to their various QoS constrains instead in [14,15], e.g., the users are realized not by their QoS requirements, not their channel conditions.

Without loss of generality, we suppose that the user PU_1 or SU_1 would like to be attend to quickly services but has a low achieved rate, such as an Internet of Things sensor or a health care equipment. Besides, the user PU_2 or SU_2 would like to download a movie with high resolution. The two-phase signal transmittance models for DF relaying are introduced in the two sections below.

3 Performance of CR-NOMA with DF Relaying

First, the base station broadcasts a mixture signal, $x_s = \sum_{i=1}^{2} \sqrt{\gamma_i} x_i$, to two relays, where x_i is the signal for $xU_i, x = P, S$, γ_i defines the power allocation coefficient for message x_i with $\sum_{i=1}^{2} \gamma_i = 1$ and $\gamma_i > 0$. Accordingly, the gotten signal at relay n, with $n = PR, SR$, is taken by:

$$y_n^r = \sqrt{P\gamma_1} g_n x_1 + \sqrt{P\gamma_2} g_n x_2 + \omega_n^r. \tag{1}$$

where P and ω_n^r define the transmittance power at the resource and the additive white Gaussian noise (AWGN) with zero mean and variance σ^2, and g_n is the channel factor from the base station to relay n. Because of the xU_2 needs QoS demands to be much more than that of xU_1, we suppose that xU_1 is always decoded before xU_2. Hence, to decode the signals x_1 and x_2, the achievable rates for relay n are taken by [14]:

$$R_1^n = \frac{1}{2}\log_2\left(1 + \frac{\gamma_1 \rho |g_n|^2}{1 + \gamma_2 \rho |g_n|^2}\right). \tag{2}$$

and

$$R_2^n = \frac{1}{2}\log_2\left(1 + \gamma_2 \rho |g_n|^2\right). \tag{3}$$

In turn, where the transmittance SNR is defined by $\rho = \frac{P}{\sigma^2}$.

Remark that the R_2^n in Eq. (3) is achievable under the constrain that $R_1^n \geq R_1$, where R_1 defines the target information rate of U_1. i.e., the signal x_1 is limited at relay utilising the success interference cancellation (SIC). If in the first phase, both the signals x_1 and x_2 can be decoded successfully by relay PR or SR, and then the selected relay can utilize NOMA to transfer a mixture signal to the users in the second phase.

Suppose the mixture signal is $x_n^r = \sum_{i=1}^{2} \sqrt{\alpha_i} x_i$, where α_i is the power allocation factor for PU_i with $\alpha_1 + \alpha_2 = 1$ and $\alpha_i > 0$. So, the signals at PU_k, $k = 1, 2$ can be taken by:

$$y_k^d = \sqrt{P\alpha_1} h_{nk} x_1 + \sqrt{P\alpha_2} h_{nk} x_2 + w_k^d. \tag{4}$$

From relay PR or SR to PU_k, we have the channel factor h_{nk}, the transfer power at relay n is deputized by P, and w_k^d defines the AWGN at PU_k, i.e., $w_k^d \sim CN(0, \sigma^2)$. In accordance with Eq. (4), the instantaneous rate for PU_1 to find its signal x_1 is taken by:

$$R_1^d = \frac{1}{2}\log_2\left(1 + \frac{\alpha_1 \rho |h_{n1}|^2}{1 + \alpha_2 \rho |h_{n1}|^2}\right). \tag{5}$$

The signal for PU_1 will be replaced when the SIC is carried out at PU_2, so the instantaneous rate for PU_2 to find the signal x_1 is taken by:

$$R_{1\to 2}^d = \frac{1}{2}\log_2\left(1 + \frac{\alpha_1 \rho |h_{n1}|^2}{1 + \alpha_2 \rho |h_{n1}|^2}\right). \tag{6}$$

In the constrain that $R_{1\to 2}^d \geq R_1$, the view at PU_2 in Eq. (6) can replace the signal x_1 before finding its own message. So, the achievement rate for PU_2 to decode the message x_2 is taken by:

$$R_2^d = \frac{1}{2}\log_2\left(1 + \alpha_2 \rho |h_{n2}|^2\right). \tag{7}$$

Remark that the decoding order of most existing papers about NOMA, such as [5] and [10], bases on utilising the users' channel conditions. Nevertheless, in this paper, the user will be set up by the their QoS demands, which means NOMA can be implemented even if users have statistically the same channel conditions.

3.1 Model 1: Selected Relay to Improve Performance in Primary Network

In this part, a two-phase DF relay selective model is studied, then we will investigate its outage performance. For a taken relay PR or SR, the messages x_1 and x_2 can be decoded completely only if the instantaneous rates for PU_1 and PU_2 are bigger or equal to their achievement communication rates R_1 and R_2, in turn.

$$A_1 = \frac{1}{2}\log_2\left(1 + \frac{\gamma_1|g_n|^2}{\frac{1}{\rho} + \gamma_2|g_n|^2}\right) \geq R_1, \quad A_2 = \frac{1}{2}\log_2\left(1 + \gamma_2\rho|g_n|^2\right) \geq R_2. \quad (8)$$

if $A_1 = \frac{1}{2}\log_2\left(1 + \frac{\gamma_1|g_n|^2}{\frac{1}{\rho} + \gamma_2|g_n|^2}\right) \geq R_1$ and then

$$|g_n|^2 \geq \frac{\varepsilon_1}{\rho(\gamma_1 - \varepsilon_1\gamma_2)}. \quad (9)$$

and $A_2 = \frac{1}{2}\log_2\left(1 + \gamma_2\rho|g_n|^2\right) \geq R_2$ and then

$$|g_n|^2 \geq \frac{\varepsilon_2}{\gamma_2\rho}. \quad (10)$$

Relied on the rate demands of the users, we can select various values for R_1 and R_2 depending on the rate demands of the users, and we will illustrate in the numerical result part how the R_1 and R_2 impact on the outage performance. Relied on Eq. (8), we can realize that:

$$|g_n| > \frac{\eta}{\rho}. \quad (11)$$

where $\eta = \max\left\{\frac{\varepsilon_1}{\gamma_1 - \lambda_2\varepsilon_1}, \frac{\varepsilon_2}{\gamma_2}\right\}, \varepsilon_1 = 2^{2R_1} - 1, \varepsilon_2 = 2^{2R_2} - 1$.

Remark that the relay n can exactly decode both x_1 and x_2 if we utilize Eq. (11) as the constraint. Recall that PU_1 is always attended to first, i.e., the achieve rate demands of PU_1 need to be respond as follows:

$$\frac{1}{2}\log_2\left(1 + \frac{\alpha_1\rho|h_{n1}|^2}{1 + \alpha_2\rho|h_{n1}|^2}\right) \geq R_1. \quad (12)$$

Relied on Eq. (12), we can know that the optimal power allocation coefficient α_2 can be shown as follows:

$$\alpha_2 = \max\left\{0, \frac{\rho|h_{n1}|^2 - \varepsilon_1}{\rho|h_{n1}|^2(1 + \varepsilon_1)}\right\}. \quad (13)$$

if $\rho|h_{n1}|^2 - \varepsilon_1 > 0 \Leftrightarrow |h_{n1}|^2 > \frac{\varepsilon_1}{\rho}$.

Remark that the condition in Eq. (12) is always possible if $|h_{n1}|^2 > \frac{\varepsilon_1}{\rho}$. It means, under the condition $|h_{n1}|^2 > \frac{\varepsilon_1}{\rho}$, PU_1 can always successfully find its own message. In accordance with Eqs. (11), (12), (13), a set of active relays adapting the QoS requirements of PU_1 can be shown as follows:

$$S_n = \left\{ n : |g_n|^2 > \frac{\eta}{\rho}, |h_{n1}|^2 \geq \frac{\varepsilon_1}{\rho} \right\}. \tag{14}$$

In Eq. (14), just relay with the optimal quality is chosen to attend to PU_2, which can be shown as follows:

$$n^* = \arg\max \left\{ |h_{n2}|^2 \right\}, \quad \forall n \in S_n. \tag{15}$$

Furthermore, to guarantee PU_2 can get a high data rate, the SIC must to be successfully carry out at PU_2, i.e., $\frac{1}{2}\log_2\left(1 + \frac{\alpha_1 \rho |h_{n*2}|^2}{1+\alpha_2 \rho |h_{n*2}|^2}\right) \geq R_1$. In addition, remark that the affect of this constraint will be given into examination for the outage probability. Then the optimal power allocation coefficient α_2 can be rewritten as follows:

$$\alpha_2 = \max\left\{ 0, \frac{\rho |h_{n*2}|^2 - \varepsilon_1}{\rho |h_{n*2}|^2 (1+\varepsilon_1)} \right\}. \tag{16}$$

Relied on Eqs. (13), (14), (15) and (16), PU_2 has the power allocation coefficient α_2, which can be expressed as follows:

$$\alpha_2 = \min\left\{ \frac{\rho |h_{n*1}|^2 - \varepsilon_1}{\rho |h_{n*1}|^2 (1+\varepsilon_1)}, \max\left\{ 0, \frac{\rho |h_{n*2}|^2 - \varepsilon_1}{\rho |h_{n*2}|^2 (1+\varepsilon_1)} \right\} \right\}. \tag{17}$$

If $\rho |h_{n*2}|^2 - \varepsilon_1 > 0 \Leftrightarrow |h_{n*2}|^2 > \frac{\varepsilon_1}{\rho}$, we can see from Eq. (17) that ε_1 and the channel fading gains from the selected relay to the two users will specify the power allocation factor. α_2.

Besides, remark that if $|h_{n*2}|^2 < \frac{\varepsilon_1}{\rho}$, since the achieve rate of the PU_1 is high or the channel condition between the optimal relay and PU_2 is evil, the relay will distribute all the power to attend to PU_1. In accordance with the above discussion of the two-phase DF relay selective model, this model supplies us an correct expression for the outage probability.

Theorem: The outage probability for PU_2 in cooperation NOMA with the examined two-phase DF relay model is taken by:

$$P_2 = B_1 + B_2. \tag{18}$$

Where:

$B_1 = e^{-\frac{a_1}{\rho}}\left(1 - e^{-\frac{a}{\rho}}\right)^2 + \frac{a_1}{\rho}\left(1 - e^{-\frac{a}{\rho}}\right)^2 q_0(k) + e^{-\frac{a_1}{\rho}} e^{-\frac{a}{\rho}}\left(1 - e^{-\frac{a}{\rho}}\right)\left(1 - e^{-\frac{a_2}{\rho}}\right).$

$B_2 = \frac{a_1}{\rho} e^{-\frac{a}{\rho}}\left(1 - e^{-\frac{a}{\rho}}\right) q_1(k) + e^{-\frac{a_1}{\rho}} e^{-\frac{2a}{\rho}}\left(1 - e^{-\frac{a_2}{\rho}}\right)^2 + \frac{a_1}{\rho} e^{-\frac{2a}{\rho}} q_2(k).$

$a_2 = a_1 + \varepsilon_1, a = \eta + \varepsilon_1.$

$q_0(k) = \int_0^1 e^{-\frac{a_1 t}{\rho}} dt.$

$q_1(k) = \int_0^1 e^{-\frac{a_1 t}{\rho}}\left(1 - e^{-\frac{1}{\rho}\left(\frac{\varepsilon_1}{t} + a_1\right)}\right) dt.$

$q_2(k) = \int_0^1 e^{-\frac{a_1 t}{\rho}}\left(1 - e^{-\frac{1}{\rho}\left(\frac{\varepsilon_1}{t} + a_1\right)}\right)^2 dt.$

Proof: Relied on Eqs. (7), (14), (15) and (25), the outage probability of PU_2 can be gotten as follows:

$$\overline{P_2} = \underbrace{\Pr\left\{\alpha_2 \geq \frac{\varepsilon_2}{\rho |h_{n*2}|^2} \Big| |S_n| = 1\right\} \Pr\{|S_n| = 1\}}_{Q_1}$$

$$+ \underbrace{\Pr\left\{\alpha_2 \geq \frac{\varepsilon_2}{\rho |h_{n*2}|^2} \Big| |S_n| = 2\right\} \Pr\{|S_n| = 2\}}_{Q_2}. \tag{19}$$

For a given $k = 1$, the above probability Q_1 can be assessed by using Eq. (27) as follows:

$$Q_1 = \Pr\left\{|h_{12}|^2 > \frac{\varepsilon_1}{\rho}, \frac{|h_{11}|^2 - \frac{\varepsilon_1}{\rho}}{|h_{11}|^2 (1 + \varepsilon_1)} \geq \frac{\varepsilon_2}{\rho |h_{12}|^2}, \frac{|h_{12}|^2 - \frac{\varepsilon_1}{\rho}}{|h_{12}|^2 (1 + \varepsilon_1)} \geq \frac{\varepsilon_2}{\rho |h_{12}|^2} \Big| n^* = 1\right\}. \tag{20}$$

The above Q_1 can be assessed as follows:

$$Q_1 = \Pr\left\{\begin{array}{c} |h_{12}|^2 > \frac{\varepsilon_1}{\rho}, \dfrac{|h_{11}|^2 - \frac{\varepsilon_1}{\rho}}{|h_{11}|^2 (1 + \varepsilon_1)} \geq \dfrac{\varepsilon_2}{\rho |h_{12}|^2}, \\ \dfrac{|h_{12}|^2 - \frac{\varepsilon_1}{\rho}}{|h_{12}|^2 (1 + \varepsilon_1)} \geq \dfrac{\varepsilon_2}{\rho |h_{12}|^2} \Big| n^* = 1 \end{array}\right\}. \tag{21}$$

The above Q_1 can be assessed as follows:

$$Q_1 = 1 - e^{-\frac{a_1}{\rho}}\left(1 - e^{-\frac{a_2}{\rho}}\right) - \frac{a_1}{\rho}\int_0^1 e^{-\frac{a_1 t}{\rho}}\left(1 - e^{-\frac{1}{\rho}\left(\frac{\varepsilon_1}{t} + a_1\right)}\right) dt. \tag{22}$$

For a given $k = 2$, the above probability Q_2 can be assessed by using Eq. (27) as follows:

$$Q_2 = \Pr\left\{\min\left\{\dfrac{\rho|h_{n*1}|^2 - \varepsilon_1}{\rho|h_{n*1}|^2 (1 + \varepsilon_1)}, \max\left\{0, \dfrac{\rho|h_{n*2}|^2 - \varepsilon_1}{\rho|h_{n*2}|^2 (1 + \varepsilon_1)}\right\}\right\} \geq \dfrac{\varepsilon_2}{\rho|h_{n*2}|^2} \Big| n^* = 2\right\}.$$

and then
$$Q_2 = \Pr \begin{cases} |h_{22}|^2 > \frac{\varepsilon_1}{\rho}, \frac{|h_{21}|^2 - \frac{\varepsilon_1}{\rho}}{|h_{21}|^2(1+\varepsilon_1)} \geq \frac{\varepsilon_2}{\rho|h_{22}|^2}, \\ \frac{|h_{22}|^2 - \frac{\varepsilon_1}{\rho}}{|h_{22}|^2(1+\varepsilon_1)} \geq \frac{\varepsilon_2}{\rho|h_{22}|^2} |n^* = 2 \end{cases}.$$

The above Q_2 can be taken as follows:
$$Q_2 = 1 - e^{-\frac{a_1}{\rho}}\left(1 - e^{-\frac{a_2}{\rho}}\right)^2 - \frac{a_1}{\rho}\int_0^1 e^{-\frac{a_1 t}{\rho}}\left(1 - e^{-\frac{1}{\rho}\left(\frac{\varepsilon_1}{t}+a_1\right)}\right)^2 dt. \quad (23)$$

Setting $a = \eta + \varepsilon_1$ and $\lambda_h = 1$, we get:
$$\Pr\left\{|g_n|^2 \geq \frac{\eta}{\rho}, |h_{n1}|^2 \geq \frac{\varepsilon_1}{\rho}\right\}.$$
$$\Leftrightarrow \Pr\left\{|g_n|^2 \geq \frac{\eta}{\rho}\right\} \Pr\left\{|h_{n1}|^2 \geq \frac{\varepsilon_1}{\rho}\right\}. \quad (24)$$

Using the formulas on PDF, we get:
$$\begin{cases} \Pr\{|g_n|^2 \geq \frac{\eta}{\rho}\} = e^{\frac{\eta}{\rho\lambda_n}}. \\ \Pr\{|h_{n1}|^2 \geq \frac{\varepsilon_1}{\rho}\} = e^{\frac{\varepsilon_1}{\rho\lambda_{n1}}}. \end{cases}$$

where $\lambda_n = \lambda_{n1} = 1$, we have:
$$\Pr\left\{|g_n|^2 \geq \frac{\eta}{\rho}, |h_{n1}|^2 \geq \frac{\varepsilon_1}{\rho}\right\} = e^{\frac{\eta}{\rho}} \times e^{\frac{\varepsilon_1}{\rho}} = e^{\frac{\eta}{\rho}+\frac{\varepsilon_1}{\rho}} = e^{\frac{\eta+\varepsilon_1}{\rho}}.$$

Setting $a = \eta + \varepsilon_1$, we get:
$$\Pr\left\{|g_n|^2 \geq \frac{\eta}{\rho}, |h_{n1}|^2 \geq \frac{\varepsilon_1}{\rho}\right\} = e^{-\frac{a}{\rho}}.$$

Then, the probability for the happening that there are k action relays in S_n can be assessed as follows:
$$\Pr\{|S_n| = k\} = \binom{M}{k} e^{-\frac{a}{\rho}k}\left(1 - e^{-\frac{a}{\rho}}\right)^{M-k}.$$
$$\Pr\{|S_n| = 0\} = \left(1 - e^{-\frac{a}{\rho}}\right)^2.$$
$$\Pr\{|S_n| = 1\} = \binom{2}{1} e^{-\frac{a}{\rho}}\left(1 - e^{-\frac{a}{\rho}}\right)^{2-1} = e^{-\frac{a}{\rho}}\left(1 - e^{-\frac{a}{\rho}}\right).$$
$$\Pr\{|S_n| = 2\} = \binom{2}{1} e^{-\frac{2a}{\rho}}\left(1 - e^{-\frac{a}{\rho}}\right)^0.$$

where $\binom{M}{k}$ is the association of k active relays in M relays, $e^{-\frac{a}{\rho}k}$ is the probability of action relay, is the probability of the relay which is not active. Utilizing the fact that the outage probability $\rho_2 = 1 - \overline{\rho_2}$. After some manipulations, we get the expected thing.

3.2 Model 2: Fixed Relay in Primary Network

We study outage performance on primary network in such CR-NOMA system:

$$\overline{P_2} = \Pr\left\{|g_n|^2 \geq \frac{\eta}{\rho}, |h_{12}|^2 \geq \frac{\varepsilon_2}{\rho\alpha_2}\right\}. \tag{25}$$

or $\overline{P_2} = \Pr\left\{|g_n|^2 \geq \frac{\eta}{\rho}\right\} \times \Pr\left\{|h_{12}|^2 \geq \frac{\varepsilon_2}{\rho\alpha_2}\right\}$.
or $\overline{P_2} = \exp\left(-\frac{\eta\alpha_2\lambda_h + \varepsilon_2\lambda_g}{\rho\alpha_2\lambda_g\lambda_h}\right)$.
with

$$P_2 = 1 - \overline{P_2}. \tag{26}$$

The probability of the happening that one casual relay is chosen as an action relaying in S_n is taken by:

$$\alpha_2 = \min\left(\frac{\rho|h_{11}|^2 - \varepsilon_1}{\rho|h_{11}|^2(1+\varepsilon_1)}, \max\left(0, \frac{\rho|h_{12}|^2 - \varepsilon_1}{\rho|h_{12}|^2(1+\varepsilon_1)}\right)\right). \tag{27}$$

where $a_1 = \varepsilon_2(\varepsilon_1 + 1)$, $a_2 = a_1 + \varepsilon_1$, $\varepsilon_1 = 2^{2R_1} - 1$, $\varepsilon_2 = 2^{2R_2} - 1$.

4 Numerical Results

This part supplies computer simulations to assess the exactness of the outage performance of cooperation NOMA with the two-phase DF relay model. Assume that the channels between two nodes are independent and identically distributed complex Gaussian distributed, i.e., CN(0,1).

Figure 2 plots the outage probability for cooperative NOMA as comparing the proposed schemes with max-min scheme using two-stage DF relaying.

Observing the Fig. 2, one can conclude that compared to cooperative NOMA with two-stage max-min relaying, the proposed CR NOMA in model 1 can realize better outage performance compared with model 2. At high SNR, outage performance of two schemes will be improved remarkably. Such observation proved that selecting proper relay contribute to primary network with enhanced performance.

Figures 3 and 4 confirmed impact of target rate to outage performance of such NOMA. As clear observation, higher target rate leads to worse outage event. We also see strict matching lines between simulation and analytical results.

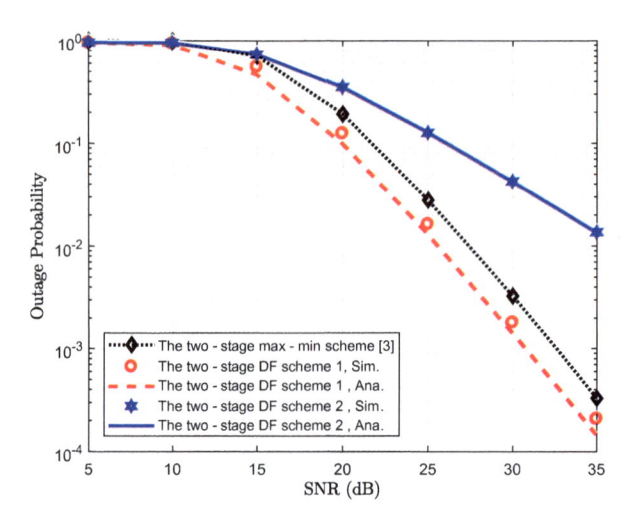

Fig. 2. The outage performance for cooperative NOMA with different DF relaying schemes, where R1 = 0.5 bits/s/Hz, R2 = 1.5 bits/s/Hz, the power allocation factors of two-stage max-min relay selection for cooperative NOMA are $\gamma_1 = \alpha_1 = 3/4$, while the power allocation coefficient of two-stage DF relaying networks is $\gamma_1 = 3/4$

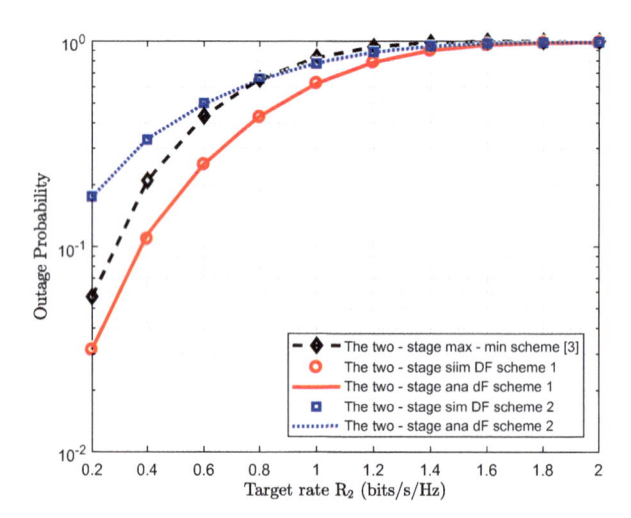

Fig. 3. Outage cooperative NOMA with DF relaying, where R2 = 0.1 bits/s/Hz, SNR = 10 dB, the power allocation coefficients of two-stage DF relaying is $\gamma_1 = 3/4$. The power allocation coefficients of two-stage max-min are $\gamma_1 = \alpha_1 = 3/4$.

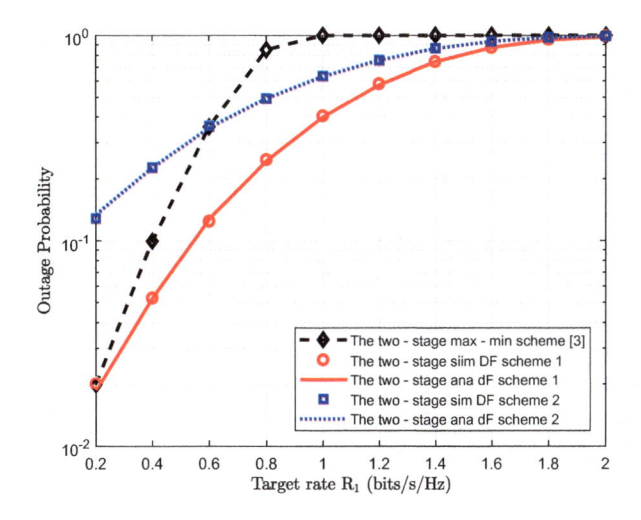

Fig. 4. Outage probability comparison between cooperative NOMA with DF where R1 = 0.1 bits/s/Hz, SNR = 10 dB, the power allocation coefficients of two-stage DF relaying is $\gamma_1 = 3/4$. The power allocation coefficients of two-stage max-min are $\gamma_1 = \alpha_1 = 3/4$.

5 Conclusion

In this paper, we investigate and compare the outage performance in primary network of dual-hop relaying system adopting DF strategy in CR-NOMA. The closed-form expressions for outage probability is extracted, relied on which, a tighter upper bound for outage probability in CR-NOMA is introduced. Moreover, relied on the operation analysis, we compare fixed forwarding model and relay selection strategy choice applied in primary network of CR-NOMA under fixed user power allocation to evaluate performance in the context of different network metrics.

Acknowledgment. This research received funding from the grant No. SP2018/59 conducted by VSB-Technical University of Ostrava, Czech Republic.

References

1. Ding, Z., Liu, Y., Choi, J., Sun, Q., Elkashlan, M., Poor, H.V.: Application of non-orthogonal multiple access in LTE and 5G networks. IEEE Commun. Mag. **55**(2), 185–191 (2017)
2. Ding, Z., Peng, M., Poor, H.V.: Cooperative non-orthogonal multiple access in 5G systems. IEEE Commun. Lett. **19**(8), 1462–1465 (2015)
3. Ding, Z., Dai, H., Poor, H.V.: Relay selection for cooperative NOMA. IEEE Commun. Lett. **5**(4), 416–419 (2016)
4. Kim, J.B., Lee, I.H.: Capacity analysis of cooperative relaying systems using non-orthogonal multiple access. IEEE Commun. Lett. **19**(11), 1949–1952 (2015)

5. Kader, M., Shahab, M., Shin, S.: Exploiting non-orthogonal multiple access in cooperative relay sharing. IEEE Commun. Lett
6. Al-Imari, M., Xiao, P., Imran, M.A., Tafazolli, R.: Uplink nonorthogonal multiple access for 5G wireless networks. In: Proceedings of 11th International Symposium Wireless Communications Systems (ISWCS), pp. 781-785, August 2014
7. Ding, Z., Fan, P., Poor, V.C.: Impact of user pairing on 5G nonorthogonal multiple-access downlink transmissions. IEEE Trans. Veh. Technol. **65**(8), 6010–6023 (2016)
8. Yang, Z., Ding, Z., Fan, P., Al-Dhair, N.: A general power allocation scheme to guarantee quality of service in downlink and uplink NOMA system. IEEE Trans. Wirel. Commun. **15**(11), 7244–7257 (2016)
9. Timotheou, S., Krikidis, I.: Fairness for non-orthogonal multiple access in 5G systems. IEEE Sig. Process. Lett. **22**(10), 1647–1651 (2015)
10. Nguyen, T.T., Pham, M.-N., Do, D.-T.: Wireless powered underlay cognitive radio network with multiple primary transceivers: energy constraint, node arrangement and energy harvesting policies. Int. J. Commun. Syst. (Wiley) **30**(18), e3372 (2017)
11. Pham, M.-N., Do, D.-T., Nguyen, T.-T., Phu, T.-T.: Energy harvesting assisted cognitive radio: random location-based transceivers scheme and performance analysis. Telecommun. Syst. **67**(1), 123–132 (2018)
12. Zheng, G., Ho, Z., Jorswieck, E.A., Ottersten, B.: Information and energy cooperation in cognitive radio networks. IEEE Trans. Sig. Process. **69**(2), 2290–2303 (2014)
13. Nguyen, T.-L., Do, D.-T.: A new look at AF two-way relaying networks: energy harvesting architecture and impact of co-channel interference. Ann. Telecommun. **72**(11), 669–678 (2017)
14. Do, D.-T., Nguyen, H.-S.: A tractable approach to analyze the energy-aware two-way relaying networks in presence of co-channel interference. EURASIP J. Wirel. Commun. Netw. **2016**, 271 (2016)
15. Nguyen, H.-S., Bui, A.-H., Do, D.-T., Voznak, M.: Imperfect channel state information of AF and DF energy harvesting cooperative networks. China Commun. **13**(10), 11–19 (2016)
16. Nguyen, K.T., Do, D.-T., Nguyen, X.X., Nguyen, N.T., Ha, D.H.: Wireless information and power transfer for full duplex relaying networks: performance analysis. In: Proceeding of Recent Advances in Electrical Engineering and Related Sciences (AETA 2015), HCMC, Vietnam, pp. 53–62 (2015)
17. Zheng, Y., et al.: Novel relay selection strategies for cooperative NOMA. IEEE Trans. Veh. Technol. **66**(11), 10114–10123 (2017)

Throughput Analysis of Power Beacon-Aided Multi-hop Relaying Networks Employing Non-orthogonal Multiple Access with Hardware Impairments

Phu Tran Tin[1,2], Pham Minh Nam[2,3], Tran Trung Duy[4], Phuong T. Tran[5(✉)], Tam Nguyen Kieu[1], and Miroslav Voznak[1]

[1] VSB - Technical University of Ostrava,
17. Listopadu 15/2172, 708 33 Ostrava - Poruba, Czech Republic
phutrantin@iuh.edu.vn, nguyenkieutam@tdtu.edu.vn,
miroslav.voznak@vsb.cz

[2] Faculty of Electronics Technology, Industrial University of Ho Chi Minh City,
Ho Chi Minh City, Vietnam
1727002@student.hcmute.edu.vn

[3] Faculty of Electrical and Electronics Engineering,
HCMC University of Technology and Education, Ho Chi Minh City, Vietnam

[4] Department of Telecommunications, Posts and Telecommunications Institute
of Technology, Ho Chi Minh City, Vietnam
trantrungduy@ptithcm.edu.vn

[5] Wireless Communications Research Group,
Faculty of Electrical and Electronics Engineering, Ton Duc Thang University,
Ho Chi Minh City, Vietnam
tranthanhphuong@tdtu.edu.vn

Abstract. In this paper, we evaluate throughput of a power beacon-aided multi-hop relaying networks employing non-orthogonal multiple access (NOMA). In the proposed protocol, the source data are sent to the destination via the multi-hop transmission model. In addition, the source and relay nodes have to harvest energy the radio frequency (RF) signals generated by a power beacon. For performance evaluation, we derive an exact closed-form expression of throughput for the proposed scheme over Rayleigh fading channel and under impact of imperfect transceiver hardware. We finally perform simulation results to verify the theoretical results.

Keywords: Multi-hop relaying networks · Beacon-aided energy harvesting · NOMA · Throughput

1 Introduction

Multi-hop relaying technique [1–5] is an efficient solution for the source-destination communication when the source is far the destination. With the data transmission at short distances, the multi-hop relaying technique can reduce outage probability (OP) and error rates (i.e., bit error rate (BER), symbol error rate (SER)), enhance ergodic capacity, extend network coverage and obtain energy efficiency as compared with the direct transmission between the source and the destination. However, the disadvantages of this technique are high time delay and low throughput since the data transmission is realized via multiple orthogonal time slots [1–5].

Recently, Non-Orthogonal Multiple Access (NOMA) [6–10] has gained much attention of researchers because it can significantly enhance throughput for wireless communication systems. In NOMA, a transmitter can simultaneously transmit different data to intended receivers by linearly combining these data with different transmit power levels. At the receivers, a Successive Interference Cancellation (SIC) method is employed to extract the data. In [10], the authors used the NOMA technique to enhance the end-to-end throughput for the multi-hop relaying technique under the attack of an active eavesdropper.

Power beacon-aided energy harvesting (PB-EH) [11–15] is an efficient method to solve the energy-constrained problem for wireless communication networks. In PB-EH, PBs are deployed to support energy for the wireless devices. In [11–13], the performance of the PB-EH underlay cognitive radio networks was evaluated. In [14, 15], the authors proposed multi-path multi-hop PB-EH secured communication protocols, where the transmitters must adjust their transmit power to avoid the eavesdroppers intercept the source data.

In this paper, we propose a PB-EH multi-hop relaying networks employing NOMA. In the proposed scheme, the source and relay nodes harvest the energy from the PB to transmit the source data to the destination. The main motivation and contributions can be summarized as follows:

- Similar to [16, 17], we consider the impact of hardware impairments on the performance of the proposed scheme. However, the references [16, 17] only considered NOMA-based one-hop and dual-hop systems, respectively.
- Different with the previous work [10], the source and relay nodes in our scheme combine multiple data to enhance the system throughput.
- We derive an exact closed-form expression of throughput for the proposed scheme over Rayleigh fading channel.
- Computer simulations using Monte Carlo method are performed to verify the theoretical results.

The rest of this paper is organized as follows. The system model of the proposed protocol is described in Sect. 2. In Sect. 3, we derive expression of throughput for the proposed scheme. The simulation results are presented in Sect. 4. Finally, Sect. 5 concludes the paper.

2 Network Model

As illustrated in Fig. 1, the source T_0 wants to send its data to the destination T_K with the help of $K - 1$ relay nodes denoted as $T_1, T_2, \ldots, T_{K-1}$. The transmitter T_k has to harvest energy from the power beacon (B) to use for the data transmission, where

$k = 0, 1, \ldots, K-1$. Assume that all the nodes are equipped with a single-antenna and operate on the half-duplex mode. As a result, the data transmission is split into K orthogonal time slots.

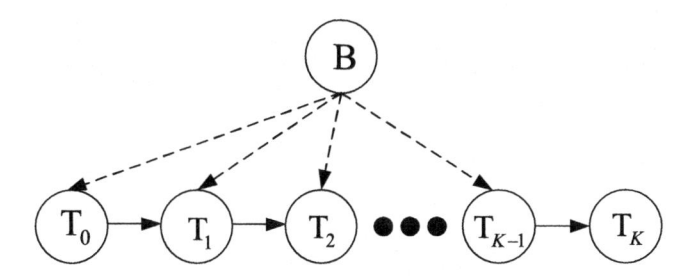

Fig. 1. System model of the proposed protocol.

Let us denote Q as the total transmission time between the source and the destination. Hence, the transmission time allocated for each time slot is given as $\tau = Q/K$. Moreover, at each time slot, the transmitter T_k spends a duration of $\alpha\tau$ for harvesting the energy from the radio frequency (RF) signals generated by B, and the remaining time $(1-\alpha)\tau$ is used for forwarding the source data to the next hop, where $0 < \alpha < 1$. Then, the energy that T_k can harvest is formulated by

$$E_k = \eta\alpha\tau P \gamma_{B,k}, \tag{1}$$

where η $(0 \leq \eta \leq 1)$ is energy conversion efficiency, P is transmit power of B, and $\gamma_{B,k}$ is channel gain between B and T_k.

From (1), the transmit power of T_k is calculated by

$$P_k = \frac{E_k}{(1-\alpha)\tau} = \mu P \gamma_{B,k}, \tag{2}$$

where

$$\mu = \frac{\eta\alpha}{1-\alpha}. \tag{3}$$

Comment 1. We assume that the frequencies used for the EH phase are different with those used for the data transmission so that there is no interference in the signals received at the receivers.

Let us consider the data transmission at the k-th time slot, where the transmitter T_{k-1} sends the source data to the receiver T_k, where $k = 1, \ldots, K$. To enhance the system throughput, the node T_{k-1} combines N signals to create a superimposed data as

$$x_c = \sum_{n=1}^{N} \sqrt{a_n P_{k-1}} x_n, \tag{4}$$

where a_n are power allocation coefficients, x_n is the transmitted signal, $n = 1, 2, \ldots, N$, $\sum_{n=1}^{N} a_n = 1$, and $a_1 > a_2 > \ldots > a_N$.

Comment 2. Conventionally, K-hop relaying protocol using the orthogonal multiple access (OMA) technique only obtains a data rate of $1/K$. Hence, by simultaneously transmitting N signals at the same time, code and frequency, our proposed scheme can obtain a data rate of N/K.

Assume the SIC process is perfect [6–10, 17], the instantaneous SNR obtained at T_k used for decoding the signal x_n under the impact of the hardware impairments can be formulated as

$$\psi_k^n = \begin{cases} \dfrac{a_n P_{k-1} \gamma_{D,k}}{\kappa^2 P_{k-1} \gamma_{D,k} + \sum_{i=n+1}^{N} a_i P_{k-1} \gamma_{D,k} + \sigma^2}, & \text{if } n < N \\ \dfrac{a_N P_{k-1} \gamma_{D,k}}{\kappa^2 P_k \gamma_{D,k} + \sigma^2}, & \text{if } n = N \end{cases} \tag{5}$$

where $\gamma_{D,k}$ is channel gain between T_{k-1} and T_k, κ^2 is total hardware impairment level on all of the data links [18–20], and σ^2 is variance of Gaussian noises at all of the receivers.

Substituting (2) into (5), which yields

$$\psi_k^n = \begin{cases} \dfrac{\mu a_n \Delta \gamma_{B,k-1} \gamma_{D,k}}{\left(\kappa^2 + \sum_{i=n+1}^{N} a_i\right) \mu \Delta \gamma_{B,k-1} \gamma_{D,k} + 1}, & \text{if } n < N \\ \dfrac{\mu a_N \Delta \gamma_{B,k-1} \gamma_{D,k}}{\kappa^2 \mu \Delta \gamma_{B,k-1} \gamma_{D,k} + 1}, & \text{if } n = N \end{cases} \tag{6}$$

where $\Delta = P/\sigma^2$ is transmit SNR.

Moreover, the instantaneous channel capacity of the signal x_n is calculated as

$$C_k^n = (1 - \alpha)\tau \log_2\left(1 + \psi_k^n\right). \tag{7}$$

Using decode-and-forward (DF) relaying technique, the end-to-end channel capacity of the signal x_n can be formulated by

$$C_{e2e}^n = \min_{k=1,2,\ldots,K} \left(C_k^n\right). \tag{8}$$

Finally, the throughput of the proposed scheme can be defined, similar to [10]:

$$TP_{NOMA} = (1 - \alpha)\tau C_{th} \sum_{n=1}^{N} \Pr\left(C_{e2e}^n \geq C_{th}\right), \tag{9}$$

where C_{th} is the target rate.

For baseline comparison, this paper also considers the PB-EH multi-hop relaying scheme without using NOMA (named OMA). In this method, T_{k-1} only sends one signal to T_k using the transmit power P_{k-1}. The throughput of this scheme is defined as

$$\text{TP}_{\text{OMA}} = (1-\alpha)\tau C_{\text{th}} \Pr\left(C_{\text{e2e}}^{\text{wo}} \geq C_{\text{th}}\right), \tag{10}$$

where

$$C_{\text{e2e}}^{\text{wo}} = \min_{k=1,2,\ldots,K}\left((1-\alpha)\tau \log_2\left(1 + \frac{\mu\Delta\gamma_{\text{B},k-1}\gamma_{\text{D},k}}{\kappa^2\mu\Delta\gamma_{\text{B},k-1}\gamma_{\text{D},k}+1}\right)\right). \tag{11}$$

3 Throughput Evaluation

3.1 Channel Model

Assume that all of channels are Rayleigh fading; the channel gains $\gamma_{\text{B},k}$ and $\gamma_{\text{D},k}$ are exponential random variables (RVs). Let us denote $\lambda_{\text{B},k}$ and $\lambda_{\text{D},k}$ as channel parameters of the RVs $\gamma_{\text{B},k}$ and $\gamma_{\text{D},k}$, respectively. Therefore, cumulative distribution functions (CDFs) of RVs $\gamma_{\text{B},k}$ and $\gamma_{\text{D},k}$ can be expressed, respectively by

$$\begin{aligned} F_{\gamma_{\text{B},k}}(x) &= 1 - \exp(-\lambda_{\text{B},k}x), \\ F_{\gamma_{\text{D},k}}(x) &= 1 - \exp(-\lambda_{\text{D},k}x). \end{aligned} \tag{12}$$

Then, the corresponding probability density functions (PDF) of $\gamma_{\text{B},k}$ and $\gamma_{\text{D},k}$ are given, respectively as

$$\begin{aligned} f_{\gamma_{\text{B},k}}(x) &= \lambda_{\text{B},k}\exp(-\lambda_{\text{B},k}x), \\ f_{\gamma_{\text{D},k}}(x) &= \lambda_{\text{D},k}\exp(-\lambda_{\text{D},k}x). \end{aligned} \tag{13}$$

To take path loss into account, the channel parameters $\lambda_{\text{B},k}$ and $\lambda_{\text{D},k}$ can be modeled as in [21–23]:

$$\lambda_{\text{B},k} = (d_{\text{B},k})^\beta, \lambda_{\text{D},k} = (d_{\text{D},k})^\beta, \tag{14}$$

where $d_{\text{B},k}$ and $d_{\text{D},k}$ are distances of the B $\to T_k$ and $T_k \to T_{k+1}$, respectively, and β is path-loss exponent.

3.2 Throughput Analysis

At first, our objective is to calculate the probability $\Pr(C_{e2e}^n \geq C_{th})$. Considering the case of $n < N$; combining (6)–(8), we have

$$\Pr(C_{e2e}^n \geq C_{th}) = \prod_{k=1}^{K} \Pr(C_k^n \geq C_{th})$$

$$= \prod_{k=1}^{K} \Pr\left(\frac{\mu a_n \Delta \gamma_{B,k-1} \gamma_{D,k}}{\left(\kappa^2 + \sum_{i=n+1}^{N} a_i\right) \mu \Delta \gamma_{B,k-1} \gamma_{D,k} + 1} \geq \theta \right), \quad (15)$$

where

$$\theta = 2^{\frac{C_{th}}{(1-\alpha)\tau}} - 1.$$

It is obvious from (15) that if $a_n - \theta\left(\kappa^2 + \sum_{i=n+1}^{N} a_i\right) \leq 0$ then $\Pr(C_{e2e}^n \geq C_{th}) = 0$, and if $a_n - \theta\left(\kappa^2 + \sum_{i=n+1}^{N} a_i\right) > 0$, we can rewrite (15) as

$$\Pr(C_{e2e}^n \geq C_{th}) = \prod_{k=1}^{K} \Pr(\gamma_{B,k-1} \gamma_{D,k} \geq \rho_n), \quad (16)$$

where

$$\rho_n = \frac{\theta}{\left(a_n - \left(\kappa^2 + \sum_{i=n+1}^{N} a_i\right)\theta\right)\mu\Delta}.$$

Comment 3. The fractions of transmit power (a_n) need to be designed carefully so that the conditions, i.e., $a_n - \theta\left(\kappa^2 + \sum_{i=n+1}^{N} a_i\right) > 0$, are satisfied.

Now, considering the probability $\Pr(\gamma_{B,k-1} \gamma_{D,k} \geq \rho_n)$ which can be formulated by

$$\Pr(\gamma_{B,k-1} \gamma_{D,k} \geq \rho_n) = \int_0^{+\infty} \left(1 - F_{\gamma_{B,k-1}}\left(\frac{\rho_n}{x}\right)\right) f_{\gamma_{D,k}}(x) dx. \quad (17)$$

Substituting (12) and (13) into (17), and then using [24, Eq. (3.324.1)], we obtain

$$\Pr(C_{e2e}^n \geq C_{th}) = \prod_{k=1}^{K} \left[2\sqrt{\lambda_{B,k-1} \lambda_{D,k} \rho_n} K_1\left(2\sqrt{\lambda_{B,k-1} \lambda_{D,k} \rho_n}\right) \right], \quad (18)$$

where $K_1(.)$ is modified Bessel function of the second kind [24].

When $n = N$, with the same manner, we have

$$\Pr\left(C_{\text{e2e}}^N \geq C_{\text{th}}\right) = \prod_{k=1}^{K} \left[2\sqrt{\lambda_{\text{B},k-1}\lambda_{\text{D},k}\rho_N} K_1\left(2\sqrt{\lambda_{\text{B},k-1}\lambda_{\text{D},k}\rho_N}\right)\right], \tag{19}$$

where

$$\rho_N = \frac{\theta}{(a_N - \kappa^2\theta)\mu\Delta}.$$

Plugging (9), (18) and (19) together, an exact closed-form formula of TP_{NOMA} can be obtained as

$$\text{TP}_{\text{NOMA}} = (1-\alpha)\tau C_{\text{th}} \left\{ \sum_{n=1}^{N} \prod_{k=1}^{K} \left[2\sqrt{\lambda_{\text{B},k-1}\lambda_{\text{D},k}\rho_N} K_1\left(2\sqrt{\lambda_{\text{B},k-1}\lambda_{\text{D},k}\rho_N}\right)\right] \right\}. \tag{20}$$

For the multi-hop relaying protocol without using NOMA, similarly, we have

$$\Pr\left(C_{\text{e2e}}^{\text{wo}} \geq C_{\text{th}}\right) = \prod_{k=1}^{K} \Pr\left(\gamma_{\text{B},k-1}\gamma_{\text{D},k} \geq \omega\right)$$

$$= \prod_{k=1}^{K} \left[2\sqrt{\lambda_{\text{B},k-1}\lambda_{\text{D},k}\omega} K_1\left(2\sqrt{\lambda_{\text{B},k-1}\lambda_{\text{D},k}\omega}\right)\right], \tag{21}$$

where

$$\omega = \frac{\theta}{(1-\kappa^2\theta)\mu\Delta}.$$

From (21), the throughput TP_{OMA} is expressed as

$$\text{TP}_{\text{OMA}} = (1-\alpha)\tau C_{\text{th}} \prod_{k=1}^{K} \left[2\sqrt{\lambda_{\text{B},k-1}\lambda_{\text{D},k}\omega} K_1\left(2\sqrt{\lambda_{\text{B},k-1}\lambda_{\text{D},k}\omega}\right)\right]. \tag{22}$$

4 Simulation Results

In this section, Monte Carlo simulations are presented to verify our derivations. In simulation environment, a two-dimensional Oxy plane is employed, where we place the source at the origin (0, 0) and the destination at (1, 0) so that the distance between the source and the destination is fixed by 1. Then, the coordinates of the relay T_k and the power beacon B are $(k/K, 0)$ and $(0.5, 0.5)$, respectively, where $k = 1, 2, \ldots, K-1$. In all of the simulations, we fix the values of path-loss exponent (β), the total transmission time

(Q), the target rate C_{th} and the energy conversion efficiency (η) by 3, 1, 0.1 and 1, respectively. For only illustration purpose, we consider three NOMA schemes as follows. In scheme 1, we set $N = 2$, $a_1 = 0.85$ and $a_2 = 0.15$; in scheme 2, we have $N = 3$, $a_1 = 0.85$, $a_2 = 0.12$ and $a_3 = 0.15$; and the parameters in scheme 3 are $N = 4$ and $a_n \in \{0.85, 0.12, 0.025, 0.005\}$, where $n = 1, 2, 3, 4$.

In Fig. 2, we present the throughput of the considered protocols as a function of the transmit SNR Δ in dB. We can see that the throughput of the proposed scheme at very low Δ regions is same with that of the OMA scheme. However, at medium and high Δ regimes, our scheme obtains much higher performance than OMA. It is also seen that the performance of the NOMA scheme at high Δ values is better with the increasing of the number of signals N.

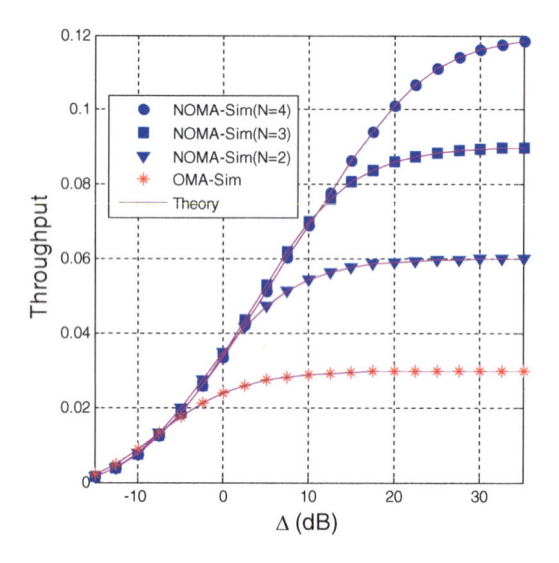

Fig. 2. Throughput as a function of Δ in dB when $K = 3$, $\alpha = 0.1$ and $\kappa^2 = 0.01$.

In Fig. 3, the throughput is presented as a function of the number of hops (K). Similar to the results of Fig. 2, Fig. 3 shows that the proposed scheme obtains higher throughput than the OMA scheme. Moreover, we can observe that the performance of all of the schemes is worse when the number of hops increases.

In Fig. 4, we investigate the impact of α on the throughput. As seen, designing this value appropriately can provide higher value of throughput. For example, with $N = 3$, the system can set $\alpha = 0.125$ to obtain the highest value of throughput.

From Figs. 2, 3 and 4, it is worth noting that the simulation results match very well with the theoretical ones, which validates our derivations.

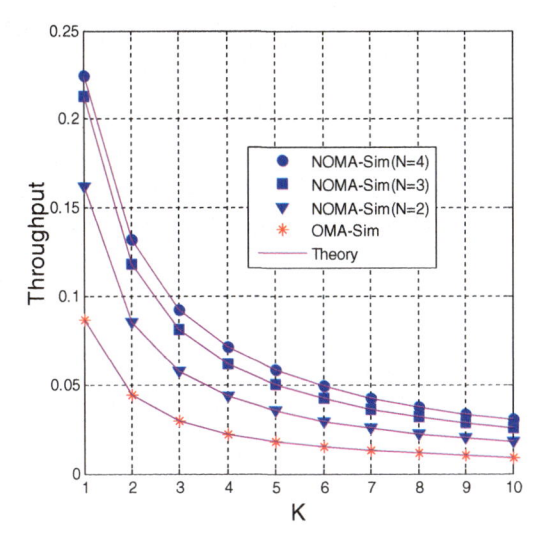

Fig. 3. Throughput as a function of K when $\Delta = 15\,\text{dB}$, $\alpha = 0.1$ and $\kappa^2 = 0$.

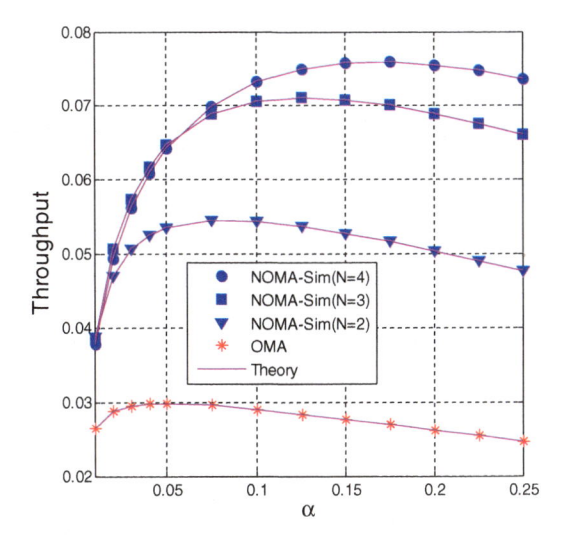

Fig. 4. Throughput as a function of α when $\Delta = 10\,\text{dB}$, $K = 3$ and $\kappa^2 = 0$.

5 Conclusion

In this paper, we proposed of the power beacon-aided multi-hop relaying networks employing non-orthogonal multiple access. The performance of the proposed protocol was evaluated via both simulations and analyses. The results showed that the proposed method outperforms the traditional orthogonal multiple access method, in terms of the system throughput. Moreover, the performance of the proposed scheme can be enhanced by increasing the number of signals, reducing the number of hops and designing the energy harvesting time appropriately.

Acknowledgment. This work was supported by the VSB-Technical University of Ostrava, Czech Republic - Networks and Telecommunications Technologies for Smart Cities under SGS Grant SP2018/59.

References

1. Hasna, M.O., Alouini, M.S.: Outage probability of multihop transmission over Nakagami fading channels. IEEE Commun. Lett. **7**(5), 216–218 (2003)
2. Conne, C., Kim, I.M.: Outage probability of multi-hop amplify-and-forward relay systems. IEEE Trans. Wirel. Commun. **9**(2), 1139–1149 (2010)
3. Farhadi, G., Beaulieu, N.: Fixed relaying versus selective relaying in multi-hop diversity transmission systems. IEEE Trans. Wirel. Commun. **58**(3), 956–965 (2010)
4. Tin, P.T., Hung, D.T., Duy, T.T., Voznak, M.: Analysis of probability of non-zero secrecy capacity for multi-hop networks in presence of hardware impairments over Nakagami-m fading Channels. RadioEngineering **25**(4), 774–782 (2016)
5. Tin, P.T., Nam, P.M., Duy, T.T., Voznak, M.: Security-reliability analysis for a cognitive multi-hop protocol in cluster networks with hardware imperfections. IEIE Trans. Smart Process. Comput. **6**(3), 200–209 (2017)
6. Ding, Z., Yang, Z., Fan, P., Poor, H.V.: On the performance of non-orthogonal multiple access in 5G systems with randomly deployed users. IEEE Sig. Process. Lett. **21**(12), 1501–1505 (2014)
7. Ding, Z., Peng, M., Poor, H.V.: Cooperative non-orthogonal multiple access in 5G systems. IEEE Commun. Lett. **19**(8), 1462–1465 (2015)
8. Ding, Z., Dai, H., Poor, H.V.: Relay selection for cooperative NOMA. IEEE Wirel. Commun. Lett. **5**(4), 419–426 (2016)
9. Liang, X., Wu, Y., Kwan, D.W., Zuo, Y., Jin, S., Zhu, H.: Outage performance for cooperative NOMA transmission with an AF relay. IEEE Commun. Lett. **21**(11), 2428–2431 (2017)
10. Tin, P.T., Hung, D.T., Duy, T.T., Voznak, M.: Security-reliability analysis of NOMA – based multi-hop relay networks in presence of an active eavesdropper with imperfect eavesdropping CSI. Adv. Electr. Electron. Eng. **15**(4), 591–597 (2017)
11. Xu, C., Zheng, M., Liang, W., Yu, H., Liang, Y.C.: Outage performance of underlay multihop cognitive relay networks with energy harvesting. IEEE Commun. Lett. **20**(6), 1148–1151 (2016)
12. Xu, C., Zheng, M., Liang, W., Yu, H., Liang, Y.C.: End-to-end throughput maximization for underlay multi-hop cognitive radio networks with RF energy harvesting. IEEE Trans. Wirel. Commun. **16**(6), 3561–3572 (2017)
13. Hieu, T.D., Duy, T.T., Dung, L.T., Choi, S.G.: Performance evaluation of relay selection schemes in beacon-assisted dual-hop cognitive radio wireless sensor networks under impact of hardware noises. Sensors **18**(6), 1–24 (2018)
14. Hieu, T.D., Duy, T.T., Choi, S.G.: Performance enhancement for harvest-to-transmit cognitive multi-hop networks with best path selection method under presence of eavesdropper. In: The 20th IEEE International Conference on Advanced Communications Technology (ICACT 2018), Gangwondo, Korea, pp. 323–328 (2018)

15. Hieu, T.D., Duy, T.T., Kim, B.-S.: Performance enhancement for multi-hop harvest-to-transmit WSNs with path-selection methods in presence of eavesdroppers and hardware noises. IEEE Sens. J. **18**(12), 5173–5186 (2018)
16. Selim, B., Muhaidat, S., Sofotasios, P.C., Sharif, B.S., Stouraitis, T., Karagiannidis, G.K., Al-Dhahir, N.: Performance analysis of non-orthogonal multiple access under I/Q imbalance. IEEE Access **6**, 18453–18468 (2018)
17. Ding, F., Wang, H., Zhang, S., Dai, M.: Impact of residual hardware impairments on non-orthogonal multiple access based amplify-and-forward relaying networks. IEEE Access **6**, 15117–15131 (2018)
18. Matthaiou, M., Papadogiannis, A.: Two-way relaying under the presence of relay transceiver hardware impairments. IEEE Commun. Lett. **17**(6), 1136–1139 (2013)
19. Bjornson, E., Matthaiou, M., Debbah, M.: A new look at dual-hop relaying: performance limits with hardware impairments. IEEE Trans. Commun. **61**(11), 4512–4525 (2013)
20. Duy, T.T., Son, P.N.: A novel adaptive spectrum access protocol in cognitive radio with primary multicast network, secondary user selection and hardware impairments. Telecommun. Syst. **65**(3), 525–538 (2017)
21. Herhold, P., Zimmermann, E., Fettweis, G.: A simple cooperative extension to wireless relaying. In: 2004 International Zurich Seminar on Communications, Zurich, Switzerland, February 2004
22. Laneman, J.N., Tse, D., Wornell, G.: Cooperative diversity in wireless networks: efficient protocols and outage behavior. IEEE Trans. Inf. Theory **50**(12), 3062–3080 (2004)
23. Duy, T.T., Son, P.N.: Secrecy performances of multicast underlay cognitive protocols with partial relay selection and without eavesdropper's information. KSII Trans. Internet Inf. Syst. **9**(11), 4623–4643 (2015)
24. Gradshteyn, I.S., Ryzhik, I.M.: Table of Integrals, Series, and Products, 6th edn. Academic Press, San Diego (2000)

Optimum Selection of the Reference Signal for Correlation Receiver Applied to Marker Localization

Martin Vestenický[1(✉)] and Peter Vestenický[2]

[1] Faculty of Electrical Engineering and Information Technology,
Department of Multimedia and Information-Communication Technologies,
University of Žilina, Univerzitná 8215/1, 010 26 Žilina, Slovakia
martin.vestenicky@fel.uniza.sk
[2] Faculty of Electrical Engineering,
Department of Control and Information Systems,
University of Žilina, Univerzitná 8215/1, 010 26 Žilina, Slovakia
peter.vestenicky@fel.uniza.sk

Abstract. The localization of RFID transponders is very popular task today in industrial RFID applications. One of the special applications of RFID is the localization and identification of the underground facility networks by special RFID transponders called markers. This paper describes an optimum selection of reference signal for the correlation receiver applied to RSSI based localization of the inductive coupled RFID markers. The reference responses of the marker are generated by using the model of localization device created in Matlab – Simulink software package. The analysis takes into account the quality factor of the marker resonant circuit. The results show that the correlations coefficients between any two marker responses are not significantly dependent on the marker parameters.

Keywords: Localization · RFID · Correlation receiver · Marker · Response

1 Introduction

The underground facilities (i.e. electrical and telecommunication cables, water and gas pipes, etc.) can be identified and localized by multiple techniques, for example by connecting a signal generator to the continual metal conductor whose signal is sensed from the terrain surface. But modern underground facilities do not use metallic, but plastic material which does not conduct electrical current so the localization can be performed only by RFID (Radiofrequency Identification) transponders (for this purpose these transponders are called markers) which are buried under surface near the underground facility.

The marker is, essentially, a single bit RFID transponder i.e. tuned LC circuit [1], or it can contain an identification chip with serial number for identification of the important underground facility points to fulfill the requirements of the Industry 4.0 [2]. The markers work in low frequency band and they are tuned on frequency in range from 77 kHz up to 170 kHz depending on the type of underground facility, for example gas pipes are marked by 83 kHz marker, telecommunication cables are marked by 101.4 kHz markers etc. Because the markers are inductively coupled with the localization device, the signal returned from the marker into the locator antenna is strongly dependent on the distance between them. The analysis in [3] shows that the amplitude of signal is inversely proportional on the sixth power of distance between the locator antenna and the marker i.e. if the marker has to be localized under terrain surface in unknown depth, the receiver must be able to process the signal with level near the level of noise. The localization is based on the RSSI (Received Signal Strength Indicator) method [4].

Theory of the optimum receiver in [5] shows that for AWGN (Additive White Gaussian Noise) channel the correlation receiver is an appropriate selection. Therefore this paper is focused on the mathematical and simulation analyses of the correlation receiver applied on the signal from the marker to be localized. Initial experiments with this approach were performed and described in [3].

2 Related Works

The correlation receiver or its special form called matched filter [5] is widely being used in the field of RFID transponder data signal demodulation or position estimation.

The paper [6] describes a special RFID transponder equipped with chirp signal generator which transmits the chirp signal into the base stations where the matched filters are used to obtain chirp signal autocorrelation function. The position of RFID transponder is then estimated by ToA (Time of Arrival) method. The matched filter is also applied for baseband data demodulation in UHF (Ultra High Frequencies) RFID reader working with EPCglobal Class-1 Gen-2 RFID transponders [7]. Such robust data demodulator allows compensation of signal distortion and frequency deviation. In [8] the novel digital matched filter without adders is described and applied to demodulation of signals spread by the Barker code which is being used in IEEE 802.11 compatible RFID transponders. Another application of correlation receiver is proposed in [9] where the digital correlation demodulator for UHF RFID reader can work with low SNR (Signal-to-Noise Ratio). Similar digital receiver architecture for demodulation of low SNR signals from UHF RFID transponders is proposed in [10]. The paper [11] demonstrates lower bit error ratio in compare to the traditional RFID decoder when the matched filter in the decoder is used.

3 Model of the Locator – Marker System

The model in Fig. 1 was created in Matlab – Simulink environment. It is based on the two inductively coupled lossy resonant circuits – serial resonant circuit of the locator and parallel resonant circuit of the marker. The coils of the resonant circuits are modeled by LS_ANT_L_MKR component, which has three parameters – inductances of the coils (L_1, L_2) and the coupling factor k. The marker excitation signal is obtained as a product of harmonic carrier with frequency $f_C = 125$ kHz and amplitude $A_C = 12$ V and two state modulation signal. The resulting signal controls an ideal voltage source through auxiliary block S→PS#1 and this signal powers the serial resonant circuit of the locator. The information about the current I_ant in the serial resonant circuit is exported into the Matlab workspace by the auxiliary S→PS#2 and To Workspace block. The obtained time course of the current in the serial resonant circuit can be divided into three time intervals (Fig. 2):

- Marker excitation, when the current amplitude graduates exponentially and approaches asymptotically the value of amplitude in steady state. In this time interval (from 0 ms up to 0.75 ms) the voltage is delivered from the harmonic carrier source. Note that the excitation signal is cut off in Fig. 2 due to small amplitude scale.
- Damping of the locator resonant circuit self-oscillations (from 0.75 ms up to 0.79 ms) by using the controlled resistor R whose value is critical for the given resonant circuit i.e. the resonant circuit has the quality factor equal to ½. In this time interval the voltage source has zero value.
- Sensing of the marker response – this time interval is the longest (from 0.79 up to 2 ms), the I_ant current course is almost exclusively given by position and properties of the marker. The voltage source has zero value, too. The course of I_ant current in this time interval is interesting for further signal processing.

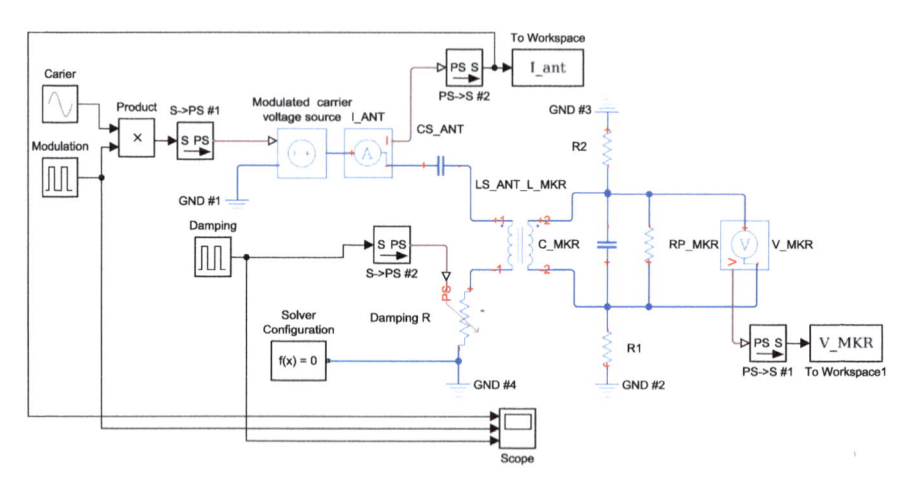

Fig. 1. Simulink model of the locator – marker

The inductances of coils L_1, L_2 were calculated taking their real geometrical parameters (radiuses $r_r = r_m = 0.1$ m, wire radius $r_w = 0.5$ mm) and numbers of turns ($N_r = N_m = 37$) into account by Eqs. (1) and (2).

$$L_1 = \mu_0 \cdot N_r^2 \cdot r_r \cdot \ln\left(\frac{8 \cdot r_r}{r_w} - 2\right) \tag{1}$$

$$L_2 = \mu_0 \cdot N_m^2 \cdot r_m \cdot \ln\left(\frac{8 \cdot r_m}{r_w} - 2\right) \tag{2}$$

The coupling factor k was calculated by Eq. (3), M is mutual inductance calculated by approximate Eq. (4), where x is distance between centers of the coils L_1 and L_2 and Θ is the angle between their planes. The capacities CS_ANT and C_MKR were calculated from the well-known Thomson's equation that both resonant frequencies to be 125 kHz. The lossy resistor RP_MKR in marker parallel resonant circuit was calculated by Eq. (5) for various quality factors Q_m. Special attention is paid to the damping resistor R which is switched between two values according to the Eq. (6). In the damping time interval this resistors has to quickly damp the self-oscillations of the locator antenna serial resonant circuit.

$$k = \frac{M}{\sqrt{L_1 \cdot L_2}} \tag{3}$$

$$M = \frac{\pi \cdot \mu_0 \cdot N_r \cdot N_m \cdot r_r^2 \cdot r_m^2 \cos(\Theta)}{2 \cdot \sqrt{(r_r^2 + x^2)^3}} \tag{4}$$

$$RP\_MKR = \sqrt{\frac{L_2}{C\_MKR}} \cdot Q_m \tag{5}$$

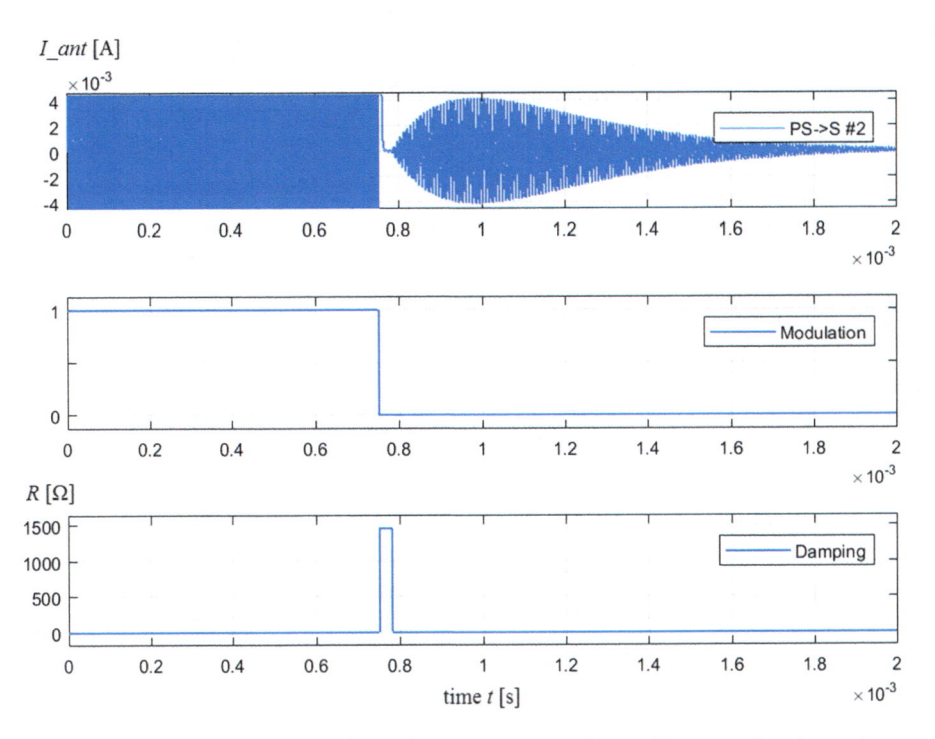

Fig. 2. Scope view of the complete reference response with auxiliary signal and quantity

$$R = \begin{Bmatrix} 10 & t < 0.75 \text{ ms}, t > 0.79 \text{ ms} \\ 2 \cdot \sqrt{\dfrac{L_1}{CS\_ANT}} & t \in (0.75, 0.79) \text{ ms} \end{Bmatrix} \quad (6)$$

4 Selection of the Optimum Reference Signal for the Correlation Receiver

In the correlation receiver focused on the marker localization only the signal in the third time interval (response from the marker) is interesting to be processed because the signals in the first two time intervals are not significantly influenced by the marker presence, but the marker response signal in the third interval is explicitly affected by the marker properties and its position.

The correlation receiver calculates the correlation value from the actual signal being received and the reference signal which is stored in memory. To select the optimum reference signal the model described in previous chapter was used to generate a set of marker responses with variable parameters i.e. the marker quality factor Q_m and distance x between the locator antenna coil and the marker coil. The distance x was recalculated into the coupling factor k by Eqs. (3) and (4). Then the all combinations of actual response and the reference response were correlated by the well-known Pearson formula. The result is a symmetrical matrix of correlation coefficients in Table 1 which is calculated from the marker responses for $Q_m = 90$.

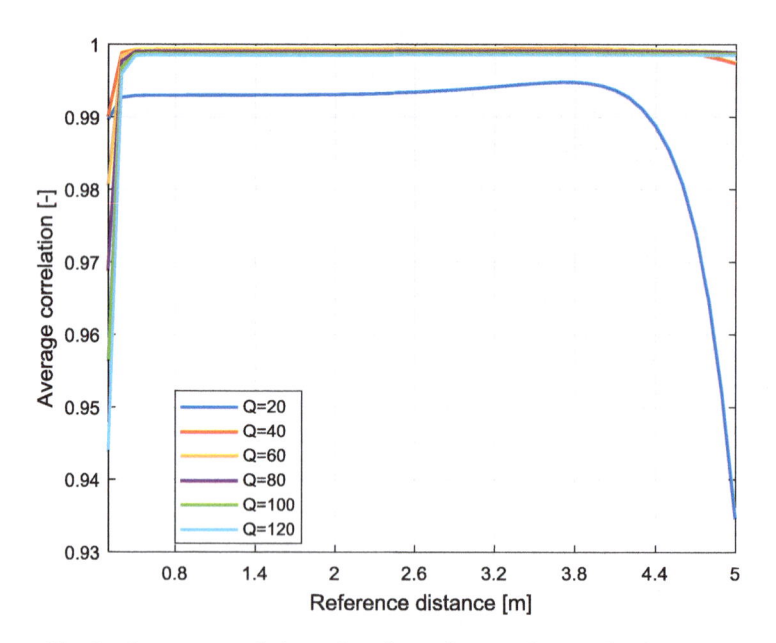

Fig. 3. Average correlation values for various marker quality factor Q_m

Table 1. Correlations between the marker response and the reference response for $Q_m = 90$

		Distance of reference response [m]								
		0.3	0.8	1.3	1.8	2.3	2.8	3.3	3.8	4.3	4.8
Distance of response [m]	0.3	1.0000	0.9621	0.9618	0.9618	0.9618	0.9618	0.9617	0.9615	0.9612	0.9606
	0.8	0.9621	1.0000	1.0000	1.0000	1.0000	1.0000	1.0000	1.0000	1.0000	0.9999
	1.3	0.9618	1.0000	1.0000	1.0000	1.0000	1.0000	1.0000	1.0000	1.0000	0.9999
	1.8	0.9618	1.0000	1.0000	1.0000	1.0000	1.0000	1.0000	1.0000	1.0000	0.9999
	2.3	0.9618	1.0000	1.0000	1.0000	1.0000	1.0000	1.0000	1.0000	1.0000	0.9999
	2.8	0.9618	1.0000	1.0000	1.0000	1.0000	1.0000	1.0000	1.0000	1.0000	0.9999
	3.3	0.9617	1.0000	1.0000	1.0000	1.0000	1.0000	1.0000	1.0000	1.0000	0.9999
	3.8	0.9615	1.0000	1.0000	1.0000	1.0000	1.0000	1.0000	1.0000	1.0000	0.9999
	4.3	0.9612	1.0000	1.0000	1.0000	1.0000	1.0000	1.0000	1.0000	1.0000	1.0000
	4.8	0.9606	0.9999	0.9999	0.9999	0.9999	0.9999	0.9999	0.9999	1.0000	1.0000
Average correlation		0.9654	0.9962	0.9962	0.9962	0.9962	0.9962	0.9962	0.9961	0.9961	0.9960

The optimum distance for calculation of the reference response was selected from the average correlation coefficients (last row in the Table 1) by finding their maximum. Because the maximum average correlation lies in wide range of distances (from 0.8 up to 3.3 m) the good selection is to generate the reference response for distance $x = 1$ m. The values of average correlations were calculated not only for single value of the marker quality factor Q_m, as it is shown in the Table 1, but for the range from 20 up to 120 with step 20. The results are shown in Fig. 3 and as 3D graph in Fig. 4.

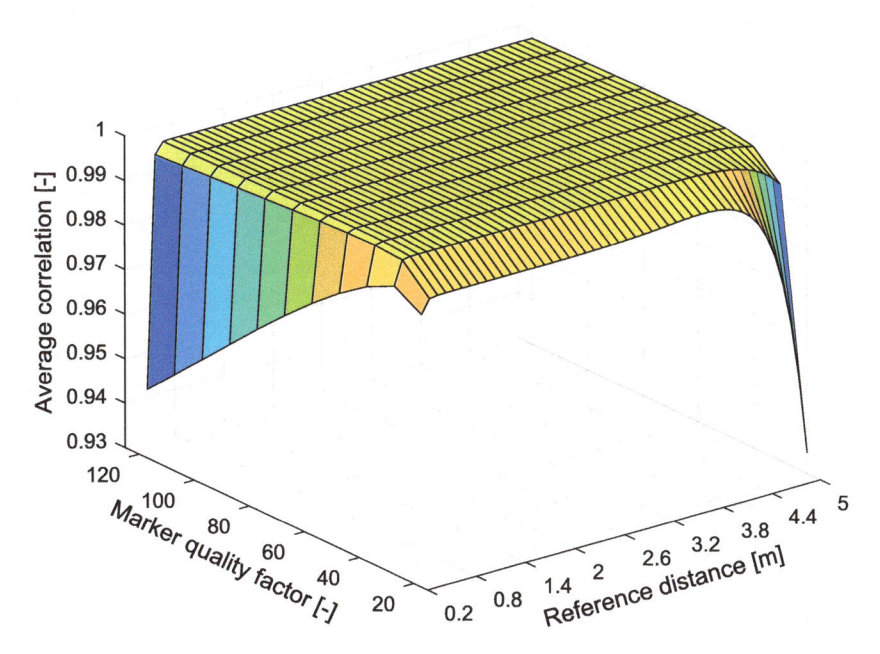

Fig. 4. Average correlation values for various marker quality factor Q_m in 3D view

5 Conclusion

The simulation and mathematic analyses gives answer on the question how to select the appropriate reference response for the correlation receiver for marker localization. The results given in the Table 1 and the Figs. 3, 4 show that the selection of proper reference response is not critical. The correlation coefficients are very close to the one i.e. the responses are linearly interdependent without significant influence of the distance between the locator and the marker so the reference response can be calculated from the model described in the Sect. 3 for distance $x = 1$ m with very good accuracy. Small decrease of correlation is visible only for small distances $x < 0.5$ m or small marker quality factor $Q_m < 20$ when the parasitic resistance RP_MKR negatively influences the serial resonant circuit of the locator through inductive coupling between the locator and marker coils.

On the other side the maximum amplitude of the response depends inversely proportionally on the sixth power of the distance x (see analysis in [3]) and this fact does not allow to use the simpler implementation of the correlation receiver i.e. the matched filter in the localization device because the matched filter does not calculates relative correlation in the range from −1 up to 1, but a covariance whose value would be strongly dependent on the signal levels.

The future works will be focused on the analyses how the markers on the adjacent frequency channels influence the correlation receiver. Also the influence of random noise and deterministic disturbance signals (for example telemetric transmitters) must be researched to consider if the correlation receiver is appropriate technology for the marker localization.

Acknowledgment. This work has been supported by the grant of Cultural and Educational Grant Agency of the Slovak Republic (KEGA) No. 038ŽU-4/2017: "Laboratory education methods of automatic identification and localization using radiofrequency identification technology."

References

1. Finkenzeller, K.: RFID Handbook: Fundamentals and Applications in Contactless Smart Cards, Radio Frequency Identification and Near-Field Communication, 3rd edn. Wiley, Chichester (2010). ISBN 978-0-470-69506-7
2. Peniak, P., Franeková, M.: Open communication protocols for integration of embedded systems within Industry 4. In: 20th International Conference on Applied Electronics, September 8th–9th, pp. 181–184. IEEE, Pilsen (2015). ISBN 978-8-0261-0385-1, ISSN 1803-7232
3. Mravec, T., Vestenický, P., Hruboš, M.: Application of correlation receiver on the RFID marker localization signals. In: 11th International Conference ELEKTRO 2016, May 16th–18th, pp. 440–444. IEEE, Štrbské Pleso (2016). ISBN 978-1-4673-8698-2
4. Brída, P., Machaj, J., Benikovský, J., Dúha, J.: An experimental evaluation of AGA algorithm for RSS positioning in GSM networks. Electron. Electr. Eng. **104**(8), 113–118 (2010). ISSN 1392–1215

5. Proakis, J.G., Salehi, M.: Digital Communication, 5th edn. McGraw-Hill, New York (2008). ISBN 978-0-07-295716-7
6. Brandl, M., Posnicek, T., Kellner, K.: Position estimation of RFID-based sensors using SAW compressive receivers. Sens. Actuators A: Phys. **244**, 277–284 (2016). ISSN 0924-4247
7. Jin, C., Cho, S.H.: A robust baseband demodulator for ISO 18000-6C RFID reader systems. Int. J. Distrib. Sens. Netw. **8**(9), 1–12 (2012). ISSN 1550-1477
8. Amin, M.S., Reaz, M.B.I., Jalil, J.: Design of a novel adder-less Barker matched filter for RFID. Int. J. Circuit Theory Appl. **42**(3), 321–329 (2014). ISSN 1097-007X
9. Liu, Y., Huang, C., Min, H., Li, G., Han, Y.: Digital correlation demodulator design for RFID reader receiver. In: Wireless Communications and Networking Conference, March 11th–15th, pp. 1666–1670. IEEE, Kowloon (2007). ISBN 1-4244-0658-7, ISSN 1525-3511
10. Angerer, C.: A digital receiver architecture for RFID readers. In: International Symposium on Industrial Embedded Systems, June 11th–13th, pp. 97–102. IEEE, Le Grande Motte (2008). ISBN 978-1-4244-1994-4, ISSN 2150-3109
11. Xi, L., Cho, S.H.: A RFID decoder using a matched filter for compensation of the frequency variation. In: 5th International Conference on Wireless Communications, Networking and Mobile Computing, Beijing, China, 24th–26th September, pp. 1–5 (2009). ISBN 978-1-4244-3693-4, ISSN 2161-9654

Comparing of Transfer Process Data in PLC and MCU Based on IoT

Antonin Gavlas[1(✉)], Jiri Koziorek[1], and Robert Rakay[2]

[1] Department of Cybernetics and Biomedical Engineering, Faculty of Electrical Engineering and Computer Science, VSB-Technical University of Ostrava, 17. listopadu 15, 70833 Ostrava, Czech Republic
{antonin.gavlas,jiri.koziorek}@vsb.cz
[2] Department of Automation, Control and Human Machine Interactions, Faculty of Mechanical Engineering, Technical University of Košice, Park Komenského 8, 042 00 Košice, Slovak Republic
robert.rakay@tuke.sk

Abstract. An article is based on comparing of data processing within PLC and MCU with using Cloud platforms. The last advances in computation and communication technologies are taking shape in the form of IoT (Internet of Things), M2M (Machine to Machine) technology, Industry 4.0 and Cyber Physical Systems. Internet of Things connects an enormous number of devices. Devices within IoT need small number of sensors and control data, but with numerous messages. Today Cloud services work with the volume of these data and support large scale IoT systems. In this paper we would like to represent a comparison of solutions Industrial and Commercial IoT. The proposed systems use one of the newest communication protocols MQTT.

Keywords: Industrial IoT · Commercial IoT · PLC · MCU · Cloud

1 Introduction

The world around us develops very quickly. Gradually it increases demands of people and companies on the technologies that surround us every day. The industry is increasingly focusing on more efficient use of resources, increasing productivity, reducing operating costs or safety of workers. All of this makes it possible to use technology Internet of Things. At the beginning it is advisable to explain what exactly it is. Internet of Things is a network of interconnected objects (things) that are uniquely addressable with the fact that this network is based on standardized communications protocols allowing the exchange and sharing of data and information. Everything happens to achieve higher added value. IoT is therefore a concept for physical and virtual objects (things) that can share data over the Internet. It is important to achieve higher goals (new features, processing complex tasks, using predictive maintenance algorithms, etc.). All used "things" can be linked to IoT within different contexts [1, 2].

1.1 Division and Development of IoT

In terms of the development of IoT was gradually formed two main directions, but they do not compete with each other. These directions are focused on their unique area of use (see Fig. 1). The first segment (Industrial Internet of Things) is based on M2M (Machine to Machine) and it adds the possibility of analyzing data through Cloud platform [3].

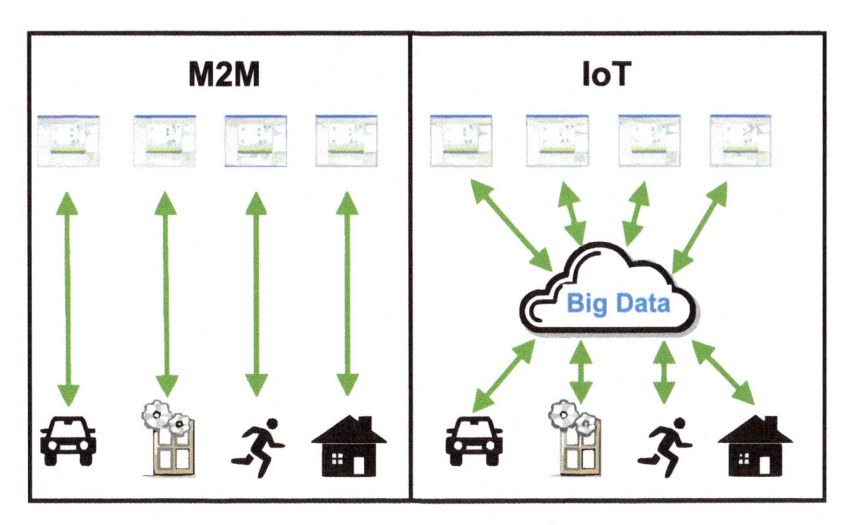

Fig. 1. Elementary difference between M2M and IoT [4]

Communication Machine to Machine means two machines, which are able to communicate or exchange data without human interactions. This communication includes serial and powerline connection and also different forms of wireless communication. M2M works on cellular communication of embedded devices. M2M applications can be divided into 4 main categories – Manufacturing, Home automation, Healthcare devices and Smart grid systems. Communication between devices within IoT is built on a different principle. Existing M2M solution generally works on one-time and pre-programmed communication. IoT contains "things", which communicate with each other randomly and all the time. It is used, for example, in industrial automation, transport or energetic industry [5, 6].

Within this segment it is possible to create a definition for Industrial Internet of Things (IIoT), which represents application of the IoT to the manufacturing industry. It improves manufacturing by allowing accessibility of far greater amounts of data and far more efficiently than before. Companies look for benefits from lower consumption as a result of predictive maintenance, improved safety and more operational efficiency. Leaders of companies can use IIoT data to get a complete and accurate idea of how their business works, which will help them to make better decisions in different situations. Interoperability and safety are probably the two biggest challenges within implementation of IIoT [7–9].

The second segment focuses on consumers - consumer products, appliances, IT, telecommunication equipment and more (see Fig. 2). Electronic devices are used to simplify and improve everyday life. It is possible to find few examples in home automation, smart devices (washing machines, televisions, lighting and more) or portable electronics. The main focus of this segment is to increase user experience [3].

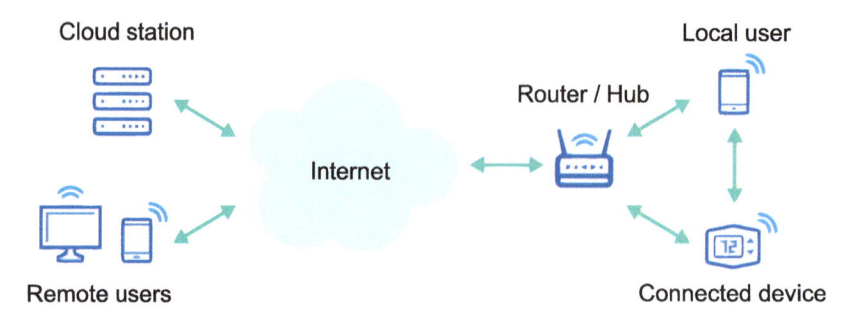

Fig. 2. Communication between devices based on Commercial IoT [7]

2 Possibilities of Data Processing

From the technological point of view, there are several possible ways to communicate within different devices. In industry, however, it is very often better to use a cable connection (e.g. industrial ethernet or profibus), which leads to increased reliability of the solution. It is also possible to use wireless communication and not only for the Commercial IoT segment. It mostly includes Wi-Fi, Bluetooth, NFC, ZigBee, Z-Wave and more [10, 11].

There are also many different communication protocols for IoT solutions. Nowadays the most usable IoT and IIoT protocols of application layer are:

- HTTP/HTTPS & WebSockets
  It represents the foundation of the client-server form of communication. For IoT solutions it is safer to implement the client part directly to the device. WebSockets are defined as a two-way communication between the servers and clients, which means both parties communicate and exchange information at the same time. The main benefits are real time communication possibilities and similar schema as HTTP [12].
- CoAP, XMPP
  The most of current application protocols are energy consuming for the IoT devices. The CoAP was designed for use with low-power networks. Within CoAP the client-server connection is controlled by the requests of the client to the server. The server then answers to the requests, where the clients can use PUT/GET commands [13].
- MQTT or MQTT-SN (for sensor networks)
  In this study is MQTT the most important protocol because it uses in a practical demonstration of reached results. MQTT works on principle of transmissioning messages between clients who are connected to server MQTT Broker (see Fig. 3).

Each client can be either Subscriber or Publisher. In MQTT every message is a small piece of data. Each message is published to an exact address, named topic. One client can use multiple topics, and when subscribed to a topic every message is received by the subscriber. The protocol also defines and determines what happens after the clients will be offline [14–16].

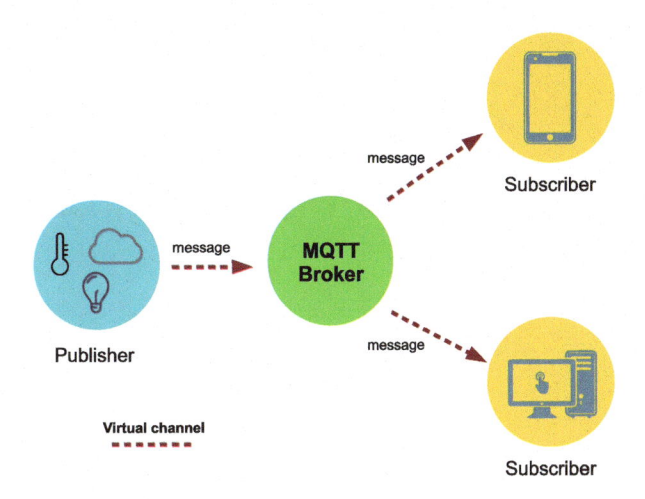

Fig. 3. Communication between publishers and subscribers based on MQTT [17]

3 Design of Solution

At the beginning it is important to determine difference between traditional control (programming in available software interface from manufacturer) and offered control. It is necessary to compile wrote program and upload it to PLC or MCU. During testing these steps have to be often repeated and results are mostly immediately visible.

3.1 Industrial IoT Solution

Industrial solution represents cooperation between PLC, IoT Gateway and Cloud platforms (see Fig. 4). It allows the device will be remotely controlled through transferring of process data between the PLC and IoT Gateway (it have to be connected to Internet).

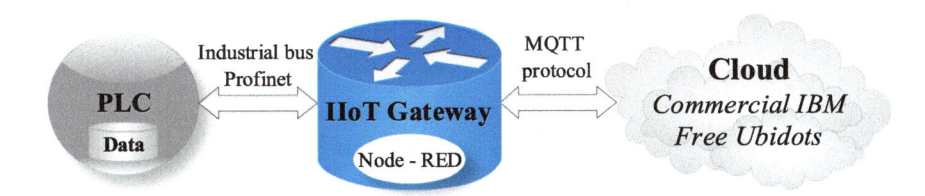

Fig. 4. Data processing between PLC and Cloud platforms

PLC controls mechatronic tasks - production of pucks. Pucks are step by step modified through six worked tables for the potential customer who will receive the finished product. Last worked table used to store finished products - stock of pucks (see Fig. 5).

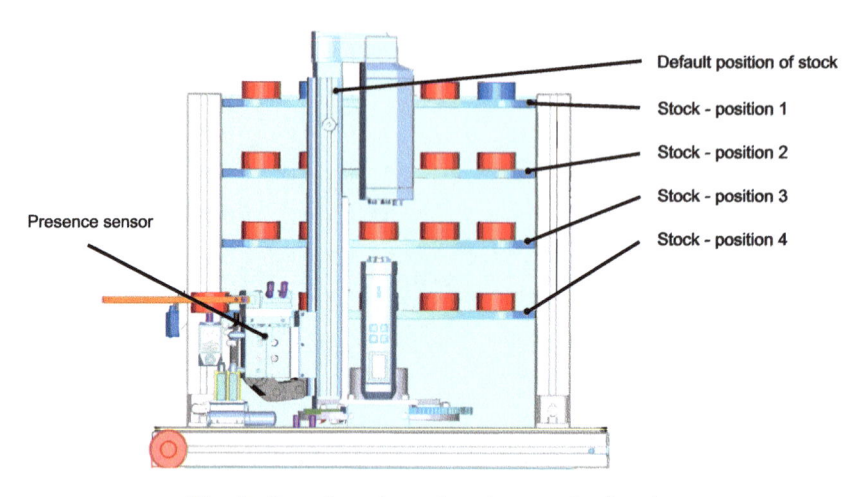

Fig. 5. Part of mechatronic tasks – stock of pucks

Stock of pucks uses conveyor belt, which is powered by a DC motor. At the end of the belt there is a handling unit for transferring the pucks from loading position to color testing position. After that a puck is transferred to a certain position of the store, which contains 4 levels (positions) based on loaded color. Elementary visualization, which counts pucks and also it alerts on full of stock, is available on Fig. 6.

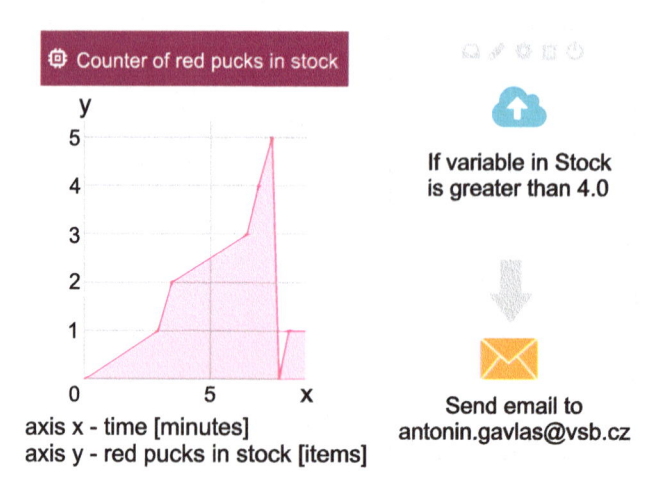

Fig. 6. Visualization of red pucks in stock

Within Industrial IoT it is necessary to use PLC and device, which is able to send process data to Cloud platforms via internet. In this study it used an intelligent industrial gateway. The IIoT gateway is used to store process data from PLC to Cloud platforms and then it controls the real devices remotely through the web application. Siemens IIoT Gateway (Simatic IoT 2040) was selected for the collection, processing and transmission of process data. It's a reliable gateway that can be used as an interface between Cloud platforms and production. The Simatic IoT2040 is an economical and safe alternative for control based on PLC. This device supports several programming languages, such as C++, Java or Python. It is possible to use various communication protocols (MQTT, SQLITE3 and more). The IOT2040 is also equipped with an IP20 protection. The shield is made of durable industrial components. It allows continuous operations even in demanding conditions [18].

3.2 Commercial IoT Solution

Commercial solution uses MCU and Cloud platform. It allows the device will be remotely controlled and received relevant information for tracking current status. For control of whole process, it used MCU ESP8266, which manufactured by the company Espressif. The main part of this system is a Tensilica l06, 32-bit microcontroller unit and a Wi-Fi transceiver. The device can be programmed like an Arduino or similar microcontroller. On the NodeMCU we can set the function of the pins. After that it is possible to work with inputs, outputs or communication ports. For testing our IoT system we used several NodeMCU modules. The main characteristics of NodeMCU ESP32 are available on Table 1. It is also possible to use integrated Wi-Fi communication to connect to a network, create a web server with system data or sending information with other devices [19].

Table 1. NodeMCU features [19]

Characteristics	Value
GPIO	11
USB to serial	It supports
Flash	4 MB
ADC range	(0–0.3) V

The programming is possible in different programming environments as Arduino IDE, Scratch, Processing and others. The device supports programming through many languages, for example C, Python. For purpose of this paper we used the Arduino IDE and "USB to Serial" communication for programming and checking the status of our system. There are also different communication interfaces to cooperate with other devices - Serial, I2C and SPI.

ESP32 is a microcontroller (MCU) combining Wi-Fi and Bluetooth interface with a Xtensa microprocessor. In general, it's a powerful version of the NodeMCU, with more communication and interaction options. This device was designed for mobile, wearable

Table 2. EPS32 features [20]

Characteristics	Value
GPIO	34
SPI	4x
I2C	2x
UART	3x

and IoT applications. The main characteristics of MCU ESP32 is shown on following Table 2 [19, 20].

The ESP32 unit was connected through MQTT protocol to receive information about active alarm states. Each alarm situation was connected to one of LED on the ESP32. Example of receiving alarm messages is shown on following Fig. 7. For testing the MQTT communication we used buttons on the Cloud dashboard. Each button was controlled manually. It is also possible to create this signal through alarm. Each alarm state includes a logical statement of received process data, which controls an output, or generates a message. Distant monitoring system was designed for experimental testing of MQTT protocol. Temperature and humidity were monitored. When triggered alarm events were used to represent MQTT messages. Our experimental system consists of described NodeMCU and ESP32. The data acquisition was performed with NodeMCU units with using HTTP protocol. Each unit measured temperature and humidity in the place of deployment. For this measurement were used digital temperature and humidity sensors.

Fig. 7. Using of MQTT communication through ESP32 unit

3.3 Comparing of Industrial and Commercial IoT Solutions

The average latency of Ubidots Cloud (free platform) was evaluated on MCU and PLC. The latency was measured by timer and it was calculated average latency of every 10 messages. The latency was probably affected by the quality of service of MQTT broker. The latency of the MQTT communication between the Ubidots broker and the subscribed ESP32 is shown on following Fig. 8.

Graph on Fig. 9 shows comparison of different Cloud platforms. First Cloud Ubidots (free platform) was evaluated by MCU. Second one, commercial platform

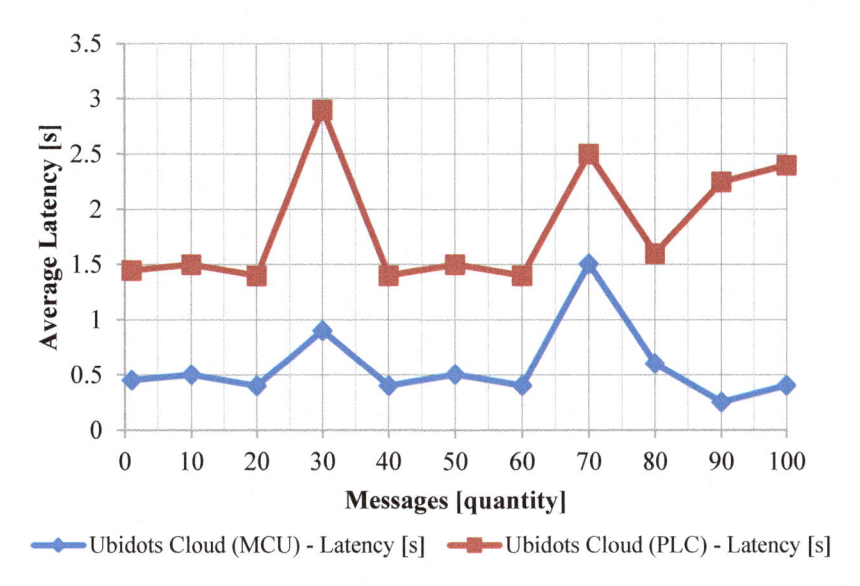

Fig. 8. Comparing of average latency of MQTT communication (Ubidots Cloud)

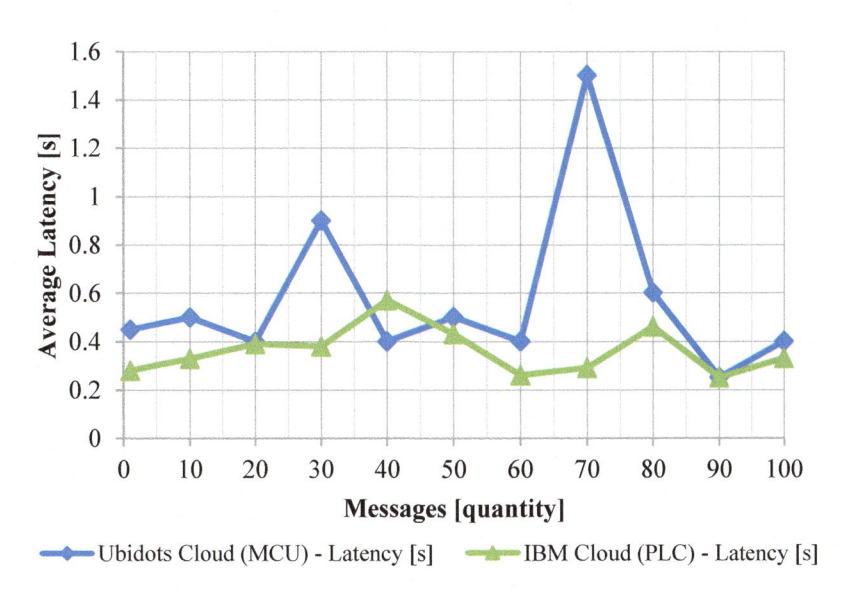

Fig. 9. Comparing of average latency of MQTT communication (Ubidots and IBM Cloud)

IBM Cloud, was evaluated by PLC. Commercial solution from company IBM, IBM Cloud, reacts with maximum time, which is equals about 0.6 s. The second Cloud, free platform Ubidots shows (when we compared it with MCU) worse results. Delay through comparing with commercial IBM Cloud was up to almost 1.5 s.

4 Conclusion

It is not existing best solution, which someone will be used within IoT. Using of Industrial and Commercial IoT depends on the use case. As they are very vast fields covering everything from home to industrial automation, it is definitely possible to use some kinds of very diversified protocols. In this study was used protocol MQTT, which worked very well, but only in some situations. In terms of random moments was changed a latency, which increased up to 4 s per new unique value. Commercial part of our solution, Free Cloud platform Ubidots, was mostly at least about a second slower than industrial part of similar solution. It can be dangerous in time critical tasks. Possibilities of security are available for example in form "Last Will and Testament". It works as a fuse in the event of faults and a failure of communications or device. Commercial IBM Cloud platform was fast enough with a maximum average latency about 0.6 s.

In terms of other implementation algorithm, we would like to focus on another big company Siemens, which offers very sophisticated platform MindSphere. For other results it will be probably interesting comparing between platforms within commercial IoT solution.

Acknowledgement. This work was supported by the European Regional Development Fund in the Research Centre of Advanced Mechatronic Systems project, project number CZ.02.1.01/0.0/0.0/16_019/0000867 within the Operational Programme Research, Development and Education.

This work was also supported by the project SP2018/160, "Development of algorithms and systems for control, measurement and safety applications IV" of Student Grant System, VSB-TU Ostrava.

References

1. Greengard, S.: The Internet of Things. MIT Press, Cambridge (2015). ISBN 978-0-262-52773-6
2. Bloom, G., Alsulami, B., Nwafor, E., Bertolotti, I.C.: Design patterns for the industrial Internet of Things. In: 14th IEEE International Workshop on Factory Communication Systems (WFCS), Imperia, Italy, pp. 1–10 (2018). https://doi.org/10.1109/wfcs.2018.8402353
3. Internet of Things. http://i2ot.eu/internet-of-things/. Accessed 18 July 2018
4. M2M vs IoT. http://gyantemple.com/full.php?ID=196. Accessed 18 July 2018
5. What is M2M. https://www.link-labs.com/blog/what-is-m2m. Accessed 18 July 2018
6. Meng, Z., Wu, Z., Muvianto, C., Gray, J.: A data-oriented M2M messaging mechanism for industrial IoT applications. IEEE Internet Things J. **4**(1), 236–246 (2017)
7. IoT Security. https://www.whitehatsec.com/blog/a-model-for-successful-iot-security-assessment/. Accessed 19 July 2018

8. What is IoT. https://inductiveautomation.com/what-is-iiot. Accessed 19 July 2018
9. Ruppert, T., Abonyi, J.: Industrial Internet of Things based cycle time control of assembly lines. In: IEEE International Conference on Future IoT Technologies (Future IoT), Eger, pp. 1–4 (2018). https://doi.org/10.1109/fiot.2018.8325590
10. Petrenko, A.S., Petrenko, S.A., Makoveichuk, K.A., Chetyrbok, P.V.: The IIoT/IoT device control model based on narrow-band IoT (NB-IoT). In: IEEE Conference of Russian Young Researchers in Electrical and Electronic Engineering (EIConRus), Moscow, pp. 950–953 (2018). https://doi.org/10.1109/eiconrus.2018.8317246
11. Shinde, K.S., Bhagat, P.H.: Industrial process monitoring using IoT. In: International Conference on I-SMAC (IoT in Social, Mobile, Analytics and Cloud) (I-SMAC), Palladam, pp. 38–42 (2017). https://doi.org/10.1109/i-smac.2017.8058374
12. WebSockets overview. https://www.tutorialspoint.com/websockets/websockets_overview.htm Accessed 20 July 2018
13. COAP technology. http://coap.technology/. Accessed 20 July 2018
14. MQTT. https://www.mqtt.org. Accessed 20 July 2018
15. Katsikeas, S., et al.: Lightweight & secure industrial IoT communications via the MQ telemetry transport protocol. In: IEEE Symposium on Computers and Communications (ISCC), Heraklion, pp. 1193–1200 (2017). https://doi.org/10.1109/iscc.2017.8024687
16. Ferrari, P., Sisinni, E., Brandão, D., Rocha, M.: Evaluation of communication latency in industrial IoT applications. In: IEEE International Workshop on Measurement and Networking (M&N), Naples, pp. 1–6 (2017). https://doi.org/10.1109/iwmn.2017.8078359
17. IoT MQTT prakticky v automatizaci. https://automatizace.hw.cz/iot-mqtt-prakticky-vautomatizaci-1dil-uvod.html. Accessed 20 July 2018
18. Industrial IoT Gateway. https://w3.siemens.com/mcms/pc-based-automation/en/industrial-iot/pages/default.aspx. Accessed 22 July 2018
19. ESP 8266 Datasheeet. https://www.espressif.com/sites/default/files/documentation/0a-esp8266ex_datasheet_en.pdf. Accessed 22 July 2018
20. ESP32 Datasheeet. https://everythingesp.com/. Accessed 22 July 2018

Protecting Gateway from ABP Replay Attack on LoRaWAN

Erik Gresak[✉] and Miroslav Voznak

Faculty of Electrical Engineering and Computer Science,
VSB - Technical University of Ostrava, 17. listopadu 15,
708 00 Ostrava-Poruba, Czech Republic
{erik.gresak,miroslav.voznak}@vsb.cz
https://www.fei.vsb.cz/en

Abstract. This paper discusses the problem of replay attacks with the ABP (Activation By Personalisation) authentication method on the LoRaWAN infrastructure and proposes effective gateway protection. To solve the problem, an experiment is replicated that simulates the attacker and is embedded in a real infrastructure environment. Subsequently, a detector is proposed and implemented based on knowledge of attacker's steps. The paper brings a proposed and verified detection algorithm that is implemented directly on the gate with an attack incident report. The aim of this approach is to prevent server-side spoofing and dosing attack on the end-device.

Keywords: LoRaWAN · LoRA · Replay attack · ABP · Detection · Internet of Things · IoT · LPWAN

1 Introduction

The concept of IoT (Internet of Things) includes countless devices that are already working in practice today [1–5]. There are many industries where IoT facilities have found employment, logistics, health, energy, transport, meteorology and more. An ever-expanding trend is for many manufacturers of common electronics to upgrade and modernize their devices using IoT. For ordinary users, the device delivers new communications, low power, and remote management capabilities.

In this paper, we focused on the LoRa (Long Range) network and its standard LoRaWAN (Long Range Wide Area Network), where we dealt with a replay attack on communication with the ABP (Authentication By Personalization) authentication method. Standard belong to the low-power wireless network protocols designed for low-energy and secure two-way communication in IoT [11]. Based on the replay attack analysis, we designed and created a detection method to capture and eliminate the attacker on the gateway side of the LoRaWAN infrastructure.

The current state of the art is described in articles describing the use of the Join Procedure in OTAA (Over the Air Activation) authentication method that has a security breach. They describe the attack (Information Gathering Phase - Pattern Analysis Phase - Attack Phase) and propose countermeasures [6]. The attack scenario assumes an environment in which the end device is on and off at regular intervals. As a countermeasure against this scenario proposes a value DevNonce (random number to prevent replay attack) in the "Join request" message to XOR with the network key from the previous join. If the attacker captures the Join request message, DevNonce values do not know. Here the authors also dealt with the vulnerability of Join Procedure and the possible replay attack [7]. They determined the probabilities of dosing devices based on the repeated DevNonce. In the next step, they investigated the random value generators used for DevNonce. The articles are focused on the Join Procedures themselves and their abuse in the OTAA authentication method. Some of them describe the combination using the DevNonce identification method and the physical properties of the RSSI (Received Signal Strength Indication) and a new technique called Proprietary Hand-Shaking [8]. However, none of them resolves the of the replay attack in the authentication method of the ABP method, which occurs when the message counter is reset.

The paper is organized into the following sections. In the first part, we are focusing on a test environment where we describe the elements of the infrastructure deployed in real traffic over which we performed an experimental attack to verify the proposed detection method. In the next section, which focuses on authentication methods, we describe their principle and types. In Sect. 3, we describe the replay attack and its implementation in the LoRaWAN infrastructure. Section 4 explains how to of designed and applied detector next part shows his testing in traffic. Finally, in Sect. 5, we discuss the conclusion.

2 Testbed Infrastructure

As a test environment, we used the LoRaWAN infrastructure, which is part of the campus of the Mining University - the Technical University of Ostrava where it is in working for testing, data processing, sensor development, and gateways. This infrastructure currently covers the entire city of Ostrava and the city of Opava, located in the Czech Republic. The main element of the infrastructure is the server by which we are able to collect and then analyze the data. The application server is equipped with a web administration interface that allows you to register additional network elements, both endpoints, and gateways, into the testbed. An integral part of the network is the gateway that we made on a shield equipped with the iC880A radio concentrator, whereas an MCU unit was Raspberry Pi. The iC880A Concentrator is able to receive packets from different devices and send with a different spreadability factor of up to eight channels in parallel [13]. This allows multiple packets to be received at one time. Concentrator can be connected to any MCU with SPI interface. The received packets store in a FIFO queue and are then reading over the SPI interface to MCU. The integrated SX1301 digital board module provides 10 programmable reception options. The methods are divided into levels ranging from IF0 to IF9, which allow us different uses. Demodulation LoRa channels IF0 to IF7 have a fixed bandwidth of the 125 kHz channel that can not be changed or configured. For each individual IF channel, the frequency can be set arbitrarily. Individual channels allow multiple packets to be received using different baud rates and spreading factors. Channels are designed to serve tens of thousands of endpoints from star topology. Channel IF8 is designed for LoRa and is linked together with SX1257. Demodulation bandwidth can be configured at 125, 250, or 500 kHz and is compatible with the SX1272 and SX1276 family transmitter [12]. An omnidirectional antenna with 10dBi gain is used to receive the signal and is connected to the lightning arrester. Power is fed into the gate by Power over Ethernet (PoE). There are 5 gateways so constructed that have sufficient network coverage for the entire territory.

3 Authentication Methods

The LoRaWAN protocol provides packet encryption using symmetric keys known for network or application networks and end devices. Distribution is done in one of two authentication methods, depending on how the end node connects to the network [15].

Authentication method OTTA (Over The Air Activation) Authentication method uses key and signature. Each node is equipped with a unique 128-bit application key (AppKey), this key is used for Join-Request. The request is not encrypted but is signed with (AppKey). The content of the connection request is also unique values AppEUI, DevEUI and DevNonce. DevNonce is a randomly generated double-byte value, AppEUI is unique to device owners and DevEUI is a globally unique device identifier. AppKey is used to create a Message Integrity Code (MIC) message integrity code, server then checks the MIC with AppKey, if the key is valid, the server generates two new 128-bit keys. For this method, you need to confirm the message using the downlink.

Authentication method ABP (Authentication By Personalization) The ABP method differs from OTAA in hard-defined unique keys (NwkSKey), (AppSKey) and device addresses (DevAddr) that are stored on the end-device. These keys are usually entered directly in the factory, and the end-devices can directly begin

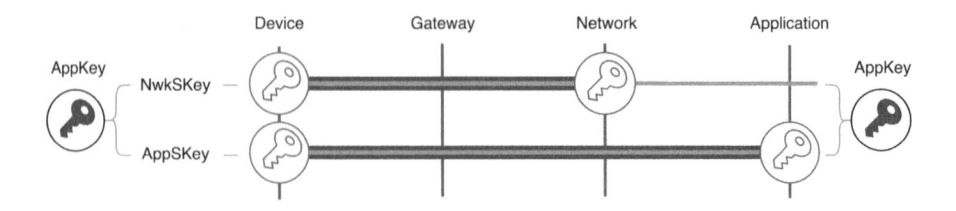

Fig. 1. Keys visualization.

communication with the server. In this method, the device may not send a (Join-Request) message and does not need confirmation via the downlink.

LoRaWAN standard specifies the three AES-128 keys shown in the Fig. 1.

- Network Session Key (NwkSKey), which is used when exchanging between the end device and the network. Ensures device authentication by computing and validating the MIC message from the embedded encrypted header code and payload.
- Application Session Key (AppSKey), a specific session key for a specific device, is used to encrypt and decrypt the payload.
- Application Key (AppKey), application key only knows the application and the end device, and which makes it possible to derive the two previous keys.

4 Experimental Attack

The replay attack is, from the point of view of definition, the kind of attack on network data transmission. Valid data is repeated or fraudulent to the end user for fraudulent purposes. The attacker can capture the data during transmission and later replicate or alter it.

The principle of attacking LoRaWAN infrastructure is described and defined through ABP authentication methods where replay attack is performed [10]. If an attacker receives a message with the highest counter value for the end device, it can periodically repeat this message to permanently block the end device. The device will have to change its session key so that its messages can be retrieved. Device restart occurs when the counter overflows, but also when a power failure or hardware restart of the end-device. In networks with fewer end-devices, the attacker probably waiting longer for the device that overflows the message counter. With an increasing number of terminal devices, it is easier to find the device that has overflowed. This attack is extremely damaging to the device and is quite simple to do.

Describing an experiment is based on the creation of a test environment where any LoRa module for transmitting and receiving wireless messages can be used to

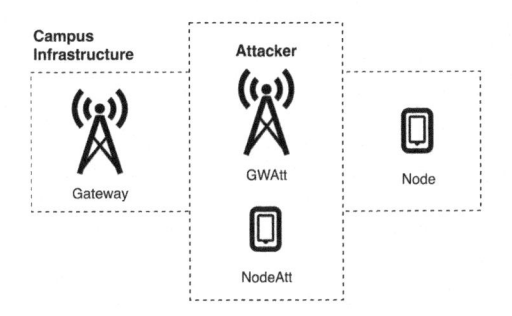

Fig. 2. Elements for LoRaWAN replay attack.

reach the attack. To replicate the experiment, we used the existing LoRaWAN VSB campus infrastructure, where we used one of the gates as an attacker, receiving messages through this gateway. We used the USB LoRa dongle with the RN2483 chip connected to the gateway to send the harmful messages. The attack is depicted in Fig. 2, showing on one side a harmful host with a receiver (GWAtt) and end-device (NodeAtt) for replicating a message. On the other said, the campus network is composed of the network server, application server, backend, and gateway that contains the devices we intend to attack (Node). The replicated experiment procedure from this points.

- The end device to be attack communicates with the network server via the real gateway.
- The attacker tunes the gate frequency according to the rules for the area. In our case this is for Europe 868 MHz or 468 MHz.
- For capturing messages, the attacker uses the harmful gateway GWAtt to record all messages from the Node. From the collected messages, the DevNonce message counter information and the DevAddr device address can be retrieved from the open text header. We store all these messages and information in the database.
- The end-device restarts and therefore sets the message counter to 0. Fixed-defined keys will not change and therefore remain the same.
- If a GWAtt gateway detects a reboot, searches the database of stored messages for that device by DevAddr. An attacker finds the message that has the largest message counter value. Restart is defined so that the message counter value for the same device is less than the last one received.
- After finding a message, an attacker using the DevAtt attacker sends a physical payload.
- If we received a message from DevAtt on the backend, the replay attack is successful.

The result of the experiment is its successful replication and commissioning on the LoRaWAN campus network. Thanks to which we were able to observe the attacker's behavior under realistic conditions.

5 ABP Detector for Protect Gateway

The detector works based on the stored data in the form of a list (DeviceList). The spreadsheet contains the received message information that is retrieved from PHYPayload by parsing and reversing octets. Each line in the DeviceList list contains a device with a message counter (FCnt) of the received message, a message checksum (CSum), a message size, and RESTART. If the device is restarted, the RESTART value is set to 1, differently is the value 0. If the device does not restart, the data in the list is updated with the latest message. Device address (DevAddr) is used as a line index. The attacker is recognized if his last message is the same as the message after the device restarts. An attacker's ID is based on a comparison of FCnt, CSum, and Size for that device. The CSum checksum is counted as the sum of decimal values, where the function f_x converts the hexadecimal to decimal. The sums (1) of messages are compared where n is the size of the received message, p is the size of the previous message, x is the set of octets of the received message, and y is the set of octets from the previous message. Subsequently, this sum is stored as a parameter in DeviceList.

$$\sum_{i=x}^{n} f(x_i) = \sum_{i=y}^{p} f(y_i) \qquad (1)$$

The values are compared if they are not equal, if all values meet the equality conditions, the attacker is identified and the message is discarded. Otherwise, the message is forwarded to the server. The incident is reported to the DevAddr administrator and the user if the device belongs between the registered in the infrastructure.

6 Description of the Algorithm

In the first part of the algorithm depicted in the diagram in Fig. 3, a packet is inserted into the input, which is then the input parameter for the (Packet decoder). The decoder will parse and reverse octets to retrieve packet information. The information we obtained from the decoder is then the input parameters for the next algorithm processing. The GetDeviceObject (DevAddr) command obtains an object from the worksheet list if the object is not found to return NULL. Under the following condition, the object is checked whether it is empty, ie NULL. If the object was not found and the condition was FALSE, the next step is to define the variables for the new object and insert object by using the Insert statement to DeviceList. This will bring the algorithm to an end.

Otherwise, when the object is not empty (TRUE), it follows the compound condition (RESTART == 1) AND (LAST_FCNT == FCnt) AND (FCnt! = 0), if this condition is true, it is probably the attacker. The next step is followed by computing the sum of the previous payload LAST_CSUM and the current f (PHYPayload). If the sums are the same, the ATTACK DETECTION output command is issued. Subsequently, the reset command is called RESTART = 0 and also if compound conditions are not followed by the exit command. The algorithm continues to check the size of the current packet message (FCnt > 0), if the FCnt is greater than zero, then the values of the found object (Device) are set where the current values for LAST_FCNT = FCnt and LAST_CSUM = CSum. Otherwise, the original values of the object remain. The last check is to determine whether this is a restart using the condition FCnt == 0. If this is a restart, which means that the message counter has been reset and set to 0, the RESTART indicator will be set to state 1, using RESTART = 1. This is the end of the algorithm, it is called again when the packet arrives from the network again.

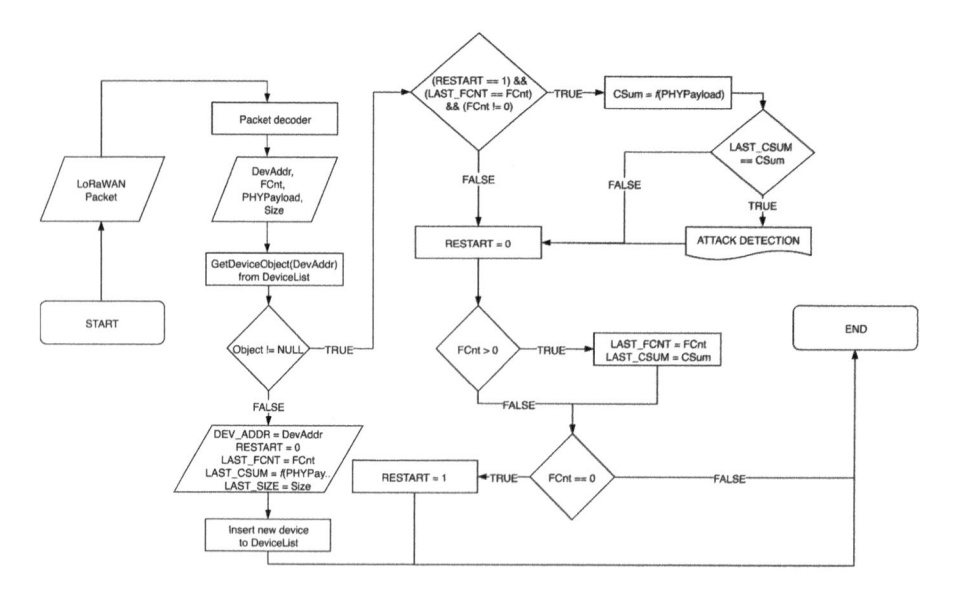

Fig. 3. Flow diagram.

7 Test Process

For the first step of testing, we created data sets where we modified data including FCnt to simulate a network attack. The sets were created in the Comma-separated values (CSV) format, which was an input to the algorithm and contained an incoming set of packets. Above these data, we launched an algorithm that checked the individual packets and tested the attacker using the detector. If it is an attacker, the incidents have been listed on the console, so we have verified the proper functioning of the detector.

Live event logs

Time	Device	Counter	Payload
10:13:58	260121EE	25	74303A2032352E36322074313A2032342E3530
10:12:51	260121EE	24	74303A2032352E36362074313A2032342E3633
10:11:46	260121EE	23	74303A2032352E36352074313A2032342E3536
09:54:56	260121EE	22	74303A2032352E37322074313A2032342E3434
09:54:48	260121EE	0	74303A2032352E36332074313A2032342E3530
09:53:40	260121EE	22	74303A2032352E37322074313A2032342E3434
09:52:32	260121EE	21	74303A2032352E35382074313A2032342E3633
09:51:24	260121EE	20	74303A2032352E36372074313A2032352E3133
09:50:16	260121EE	19	74303A2032352E37302074313A2032352E3934
09:59:08	260121EE	18	74303A2032352E35392074313A2032362E3139
09:58:00	260121EE	17	74303A2032352E35392074313A2032362E3030
09:46:52	260121EE	16	74303A2032352E36382074313A2032362E3530

Live event logs

Time	Device	Counter	Payload
12:05:17	260121EE	4	74303A2032352E37352074313A2032362E3838
12:04:12	260121EE	3	74303A2032352E36382074313A2032372E3036
12:03:09	260121EE	2	74303A2032352E36302074313A2032362E3831
12:02:01	260121EE	1	74303A2032352E35362074313A2032362E3639
12:00:58	260121EE	0	74303A2032352E34342074313A2032362E3633
12:00:38	260121EE	14	74303A2032352E36392074313A2032362E3831
11:59:29	260121EE	13	74303A2032352E36372074313A2032362E3633
11:58:21	260121EE	12	74303A2032352E35312074313A2032362E3735
11:57:18	260121EE	11	74303A2032352E36342074313A2032362E3530
11:56:06	260121EE	10	74303A2032352E38312074313A2032362E3934
11:55:02	260121EE	9	74303A2032352E37382074313A2032362E3838
11:53:51	260121EE	8	74303A2032352E36312074313A2032372E3030

Fig. 4. Live events log.

We introduced the detection method to all 5 gateways in our campus LoRaWAN infrastructure and for each gate we implemented a detection module that was linked to the liblotagw from SEMTECH using which gateway receives the packets. We then launched a simulated attack using a replicated experiment and observed the results on the application server through the web interface. Using a simulated attack, we verified the predicted behavior of the detector. The prerequisite was to eliminate the attacking message and respond to the resulting threat by administrator and user notifications. We observed the behavior of the device from received messages on the application server in the web interface. On the left part Fig. 4, you can see communication without a detector, where after a hardware reboot the attacker perform a successful replay attack with the message counter (FCnt = 22). The corrupted device tries to send messages successively, but the server rejects them because it waits for the following message with (FCnt = 23). The end device is the dosing until it sends the request message, in this case it is 16 min. The right part of the Fig. 4 shows the result with the applied detector where it is obvious that the attack did not occur and the end device was successfully restarted.

8 Conslusion

In this article, we analyzed the area of attacks of end devices activated by the ABP authentication method and found countermeasures to replay attack by detection. We assumed the attacker's behavior, according to a simulated experimental attack and summed up this attack to points that we then replicated over the campus network. We then described how to prevent such an attack by applying this method to the algorithm and describing it in the Fig. 3. In the implementation of the detector to the gateway, we performed real-time testing the data set. From the results, we described the behavior of the received data on the application server web server with a detector applied and without a detector. The results from the article [10] of the experiments show the management vulnerabilities and static keys in the LoRaWAN protocol. Our replicated experiment results also confirm this vulnerability with our simpler model, where the dongle is used by the intruder to connect directly to the gateway and not as a separate attacker sensor. Thanks to the experiment, we managed to come to the principle of attack detection. Extending this work helps resolve this vulnerability, as can be seen from the results in the table in Fig. 4, where the correct data from the gateway comes to the server. By using our designed detector, it is expected that users who have received a replay attack can be protected while maintaining an existing connection.

The gateway has the potential to prevent and detect the first attacks originating from the terminal equipment. We want to expand the paper on additional attack detection capabilities that can be analyzed and detected directly on the gateway on the basis of the payload. The design of this work is verified in practice and the extension of the security of the gate is kept as a future work.

Acknowledgment. This work was supported by the Secure Gateway for Internet of Things (SIoT) project No. VI20172020079 funded by the Ministry of the Interior of the Czech Republic and partially by the project Networks and Telecommunications Technologies for Smart Cities under SGS Grant SP2018/59 conducted by the VSB-Technical University of Ostrava, Czech Republic.

References

1. Gubbi, J., Buyya, R., Marusic, S., Palaniswami, M.: Internet of Things (IoT): a vision, architectural elements, and future directions. Future Gener. Comput. Syst. **29**(7), 1645–1660 (2013)
2. Miorandi, D., Sicari, S., De Pellegrini, F., Chlamtac, I.: Internet of Things: vision, applications and research challenges. Ad Hoc Netw. **10**(7), 1497–1516 (2012)
3. Al-Fuqaha, A., Guizani, M., Mohammadi, M., Aledhari, M., Ayyash, M.: Internet of Things: a survey on enabling technologies, protocols, and applications. IEEE Commun. Surv. Tutor. **17**(4), 2347–2376 (2015)
4. Stankovic, J.A.: Research directions for the Internet of Things. IEEE Internet Things J. **1**(1), 3–9 (2014). Article no. 6774858
5. Bandyopadhyay, D., Sen, J.: Internet of Things: applications and challenges in technology and standardization. Wirel. Pers. Commun. **58**(1), 49–69 (2011)
6. Na, S., Hwang, D., Shin, W., Kim, K.-H.: Scenario and countermeasure for replay attack using join request messages in LoRaWAN. In: 2017 International Conference on Information Networking (ICOIN), pp. 718–720. IEEE (2017). https://doi.org/10.1109/ICOIN.2017.7899580, http://ieeexplore.ieee.org/document/7899580/. Accessed 31 July 2018. ISBN 978-1-5090-5124-3
7. Tomasin, S., Zulian, S., Vangelista, L.: Security analysis of LoRaWAN join procedure for Internet of Things networks. In: 2017 IEEE Wireless Communications and Networking Conference Workshops (WCNCW), pp. 1–6. IEEE (2017). https://doi.org/10.1109/WCNCW.2017.7919091, http://ieeexplore.ieee.org/document/7919091/. Accessed 31 July 2018. ISBN 978-1-5090-5908-9
8. Sung, W.-J., Ahn, H.-G., Kim, J.-B., Choi, S.-G.: Protecting end-device from replay attack on LoRaWAN. In: 2018 20th International Conference on Advanced Communication Technology (ICACT), pp. 167–171. IEEE (2018). https://doi.org/10.23919/ICACT.2018.8323684, https://ieeexplore.ieee.org/document/8323684/. ISBN 979-11-88428-01-4
9. Miller, R.: LoRa Security - Building a Secure LoRa Solution, MWR Labs. https://labs.mwrinfosecurity.com/assets/BlogFiles/mwri-LoRa-securityguide-1.2-2016-03-22.pdf
10. Yang, X.: LoRaWAN: Vulnerability Analysis and Practical Exploitation (2017). https://repository.tudelft.nl/islandora/object/uuid%3A87730790-6166-4424-9d82-8fe815733f1e
11. LoRa Alliance: A technical overview of LoRa and LoRaWAN (2015). https://www.lora-alliance.org/what-is-lora
12. Semtech Sx1272/73: Datasheet (2015). http://www.semtech.com/images/datasheet/sx1272.pdf. Accessed 12 May 2015
13. IMST GmbH Germany: WiMOD iC880A datasheet (2015). https://wireless-solutions.de/products/radiomodules/ic880a.html
14. LoRa App Server – open-source LoRaWAN application-server. https://docs.loraserver.io/lora-app-server/. Accessed 05 July 2017
15. The LoRa Alliance: LoRaWAN 1.1 Specification, October 2017

Development of a Distributed VoIP Honeypot System with Advanced Malicious Traffic Detection

Ladislav Behan[(✉)], Lukas Sevcik, and Miroslav Voznak

Faculty of Electrical Engineering and Computer Science, VSB-Technical University
of Ostrava, 17. listopadu 15/2172, 708 33 Ostrava, Czech Republic
ladislav.behan@vsb.cz

Abstract. The number of active users using Voice over IP (VoIP) services has an increasing tendency. With an expanding number of users, there is also a rapid increase in the number of hackers interested in attacking the VoIP communication system. This paper aims at detecting malicious SIP traffic and also deals with the security of the VoIP architecture issue. It is not a trivial matter to secure the VoIP system because exploiting the vulnerabilities of IP based telecommunication systems have increased. It is crucial to develop a tool that would be able to detect these attacks, analyse collected data, monitor attackers progress and to prepare an effective way of how to defend against VoIP attackers. That was the primary motivation why we have decided to develop our honeypot solution which can detect attacks on VoIP infrastructure, and it is adapted to the new security threats and which is designed according to the needs of the telecommunications market. Our VoIP honeypot is implemented purely in JAVA programming language and is capable of capturing and processing various types of attacks. The whole project is based on a Linux distribution, ready for the easiest deployment because it is prepared as a virtual machine image.

Keywords: VoIP · Honeypot · Flood · DoS · Spit · Attacks

1 Introduction

Security is often underestimated in VoIP telephony, so unsecured voice transmission over the Internet is easily deductible. In addition to end-to-end devices, communications servers are frequently attacked when attackers attempt to access their user accounts and subsequently abuse them. For sufficient protection of VoIP technology, it is necessary to keep up with hackers and to continue to improve security measures [1,2]. Currently, a large number of corporations use internal IP telephony services [3,4].

This growing interest in VoIP services attracts increasing attention in hacking circles. While we are continually discovering new defence mechanisms and trying to improve security systems to protect VoIP infrastructure regularly, we sometimes fail to keep pace with attackers who are mostly one step ahead of us in discovering system weaknesses. For this reason, a VoIP Phoneypot has been developed, the main feature of this tool is to detect insufficient system security and to gather information about the course of the attack. A precisely implemented honeypot could lead to increased protection of the telephone system and also would be able to predict further attacks [5,6].

2 State of the Art

In the area of network security, the honeypot tool includes a range of services, tools, and implementations that are deliberately exposed to the risk of attack. As the name suggests, it is a kind of attraction for network intruders. These are then used to identify new attack methods and to prevent incidents. Honeypots can also be expediently exhibited to attacks to sidetrack the attention from significant system resources [1,5,6].

Currently, most of the VoIP attack detection projects have already ended their development process. The most popular VoIP honeypots are Artemisa and Dionaea. The Artemisa project released the last stable version in 2011, since this year, apparently the development has not progressed and has not released any new honeypot update. Artemisa also does not simulate the PBX but instead acts as an active end-to-end device [9,10]. Artemisa is not the silver bullet solution for discovering all security threats and cannot handle large amounts of SIP messages. Another honeypot project is Dionaea. However, the project only deals marginally with VoIP services and support for this type of technology is not processed in a sophisticated way [1,7,8].

This paper builds on the "Development and implementation of VoIP honeypots with a wide range of analysis", which was presented at the SPIE 2018 conference that took place in Orlando. In this article, we introduced our initial concept of VoIP honeypot, which was developed as part of the project under the CESNET association [1].

3 Honeypot Components and Functions

Our honeypot solution is prepared to be deployed in any VoIP infrastructure that uses the SIP (Session Initiation Protocol) protocol. Figure 1 shows an architecture in which it performs the role of a regular SIP PBX and a SIP phone. The honeypot can be deployed on both the physical and the virtual machine. Current trends in communications technology have an increasing tendency to use virtualisation constraints. Honeypots are developed in the VMware ESXi virtualisation environment, enabling simple deployment and portability of the system. They can be deployed in various network infrastructures of our partners and also across multiple virtualisation environments such as VMware Workstation, XEN server, Virtualbox.

The honeypot can be developed and customised by different individuals or organisations because it is primarily implemented as an open-source project with appropriate documentation. As mentioned above, we are also prepared to provide pre-built virtual systems.

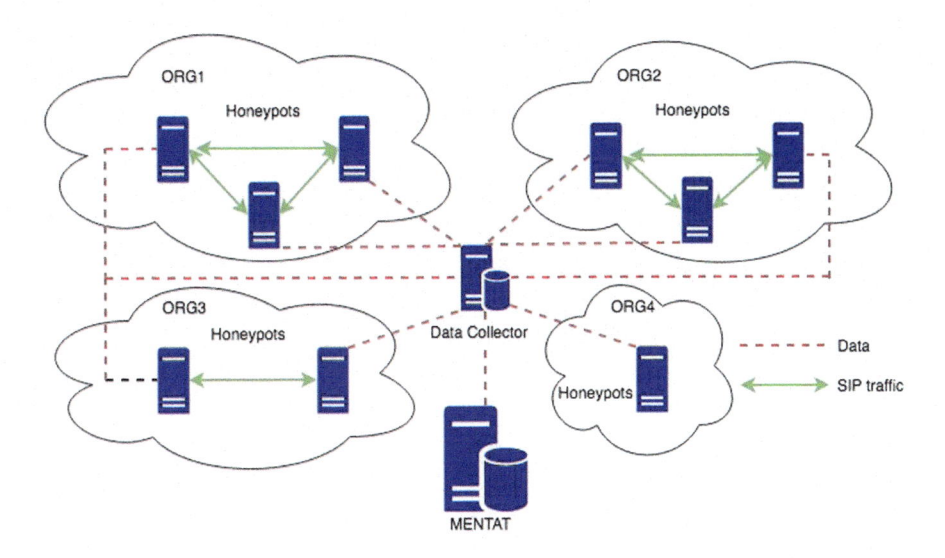

Fig. 1. Honeypot vs regular SIP traffic

Our VoIP honeypot is capable of both threat detection and it allows proactive responses on the attacks coming to VoIP infrastructure. These responses are performed by the other network security tools such as IPS systems or information gathered by honeypots could be used to protect the real network infrastructure (configuration for firewalls etc.). The VoIP honeypot solution offers following features:

- INVITE, OPTIONS, REGISTER flood attack detection - In this type of attacks, the attackers are generating thousands of messages and transmitting them to the victims. SIP PBX system can barely process all these messages. Flood attacks are representing typical attacks that can deplete available sources of VoIP servers and they can also affect speech quality or overall performance of infrastructure.
- SPIT detection mechanism - SPIT (Spam over Internet Telephony) belongs to one of the important threats in VOIP infrastructure. Actually, SPIT is bit of similar to email spams, it represents unsolicited bulk of calls sent via VoIP networks.
- Audio message recording - All data comes from attackers are the recorded in honeypot storage.
- Automatic search for sources of attacks by using WhoIS service - Various information about attackers is stored in database.
- Data exchange among VoIP honeypots - Information of source and type of attacks are passed among honeypots.

- VoIP Honeypots are creating fake VoIP traffic among themselves.
- Logged data are stored in the structured JSON format.
- Integration of MENTAT data.

A very important feature of all honeypots is the export and sharing of all captured attacks and possibility of the subsequent analysis of these attacks by the standard tools. We have integrated the export into standard format and also into the CESNET MENTAT system.

Mentat system is an open-source distributed modular SIEM (Security Information and Event Management System) which was developed to monitor networks of different sizes. Mentat's architecture is capable of reception, storage, analysis, processing and response to a huge amount of security incidents originating from various sources, such as honeypots, network probes, log analysers, etc. [13].

4 Design and Implementation of the VoIP Honeypot

The first honeypot solution consisted of three main parts. The honeypot concentrator that collects data from all honeypots, the honeypot master which analyzes SIP traffic a has the Asterisk PBX integrated in it and honeypot client which acted like as VoIP SIP client, and was generating SIP INVITE messages. Since the honeypot clients were never under attack, our new honeypot solution performs both honeypot master and honeypot client functions. The main advantage is the placement of multiple honeypots with the integrated Asterisk PBX, where, in parallel with the increasing number of attackers, the number of data collected is also increased. As it was mentioned before the new version of VoIP honeypot is implemented in JAVA programming language and consists of four parallel running threads.

The first thread captures and filters all incoming packets on a defined interface by using jnetpcap library. In applications configuration file you have to define the interface on which the application has to listen to and set up a pcap filter to capture only certain packets. We have essentially focused on the UDP

Table 1. AttackInfo object variables.

Variable name	Data type	Description
addressSource	String	IP address of attacker
addressDestination	String	IP address of the target of attack
attackType	String	INVITE, REGISTER or OPTIONS flood
requestUrl	String	Request URL string
packetCount	Integer	Current number of incoming packets
updateTime	Date	Timestamp of last incoming packet
timestamp	Date	Timestamp of first incoming packet

protocol filtration, which is used by Asterisk PBX to provide the SIP communication. If one of the INVITE, REGISTER or OPTIONS messages appears at the interface, the processAttackFunction() first checks if the object with the same IP address and type of attacks already exists. If it does not exist it will create a new object, if the function finds out that the object was already created it will increase the number of packet counter and these messages are further processed. In Table 1 you can see the list of object variable parsed out of SIP message.

The second thread feature is to detect SPIT attacks. If there is an incoming call, honeypot will automatically answer it after a random time of ringing. If the total call duration is less than for example 15 s, the program will classify this call as a potential SPIT attack. All the caller information is stored in the database. If the same inbound calls were marked as a suspicious more than five times, the program would consider it as a SPIT attack. The Whois function will determinate its origin, and other honeypots are informed about this attack. This SPIT detection system is not so sophisticated right now, and we are currently working on a neural-based VoIP spam detection.

Third thread creates fake VoIP traffic. For its functionality, it uses properties from the javax.sip library. Based on the values in the configuration file, honeypot is registered to the Asterisk PBX as a SIP phone. In the case of successful registration, it subsequently send INVITE messages too other honeypots and they will establish a telephone connection.

```
{"id":"5a98f3ccb057e121e9c37253"},
"destIP":"195.113.113.152",
"duration":"00:01:29",
"connCount":1,
"sourceIP":"145.239.82.47",
"whoIS":{"RestRipeData":{"country":"PL",
        "address":"Ul. Szkocka 5 lok. 1, Wroclaw,Poland"},
        "FreegeoIP":  {"country_code":"PL"
        "latitude":52.2394,longitude":21.0362
                "country_name":"Poland",
                "time_zone":"Europe/Warsaw",}},
"attack":"REGISTER",
"from": 1007 sip:1007@195.113.113.152;
"timestamp":"2018.03.02 07:30:42","port=5172"}
```

Listing 1.1. JSON summary of attack

The last thread has been designed for inter-honeypots communication. Unlike the previous version of honeypot, we modified the classification of VoIP DoS attacks. In the new version, we integrated data aggregation feature. Honeypot does not classify VoIP attacks by achieving a certain number of equal packets per unit of time. As already mentioned above, the honeypot is storing attacks information in separate objects where they are aggregating themselves during ongoing attacks. The application is then storing these objects in a list. The thread has an implemented function that periodically passes through a list of objects, and examines whether the time between the last incoming SIP packet and current timestamp is higher than 15 min. If so, an attack report is created, which is then sent to the data collector, which then sends it to all honeypots in the network. If the calculated time is less than 15 min, we consider the attack to be still ongoing. All messages exchanged among honeypots are in JSON format (see Listing 1.1).

5 Interim Results

During the two-month honeypot operation, we managed to collect a fair amount of data. Gathered data coming from various attacks are processed by another application implemented in the JAVA programming language and the results of this application are presented below.

5.1 Attacks Analysis

In Table 1, you can see an overview of the attacks that were made on our honeypots in two months. Honeypots were deployed in two universities within the Czech Republic and Slovakia. Overall, over this period, we were committed to capturing 954 attacks on VoIP services. Each attack involves approximately 400 SIP messages, which means that in total, the attackers generated and sent over 381600 SIP messages. The most common attack was REGISTER flood, which

Table 2. Floods by country.

	Invite	Register	Options	Total	Duration	Avg. duration
Canada	1	1	0	2	10.8	5.4
Netherlands	3	8	1	12	344	28
Latvia	0	1	0	1	4	4
United States	95	207	5	307	118338	385
Czechia	1	7	30	38	530	13
Poland	0	81	1	82	19361	236
United Kingdom	1	15	0	16	9719	607
Malaysia	1	0	0	1	8	8
France	22	113	3	138	52562	380
Germany	5	194	1	200	63942	319
Russia	26	125	1	152	60137	395
Palestina	1	2	2	5	658	131
Total	156	754	44	954	325614	341

Table 3. Hourly statistics by country.

	0<x<6	6<x<9	9<x<12	12<x<15	15<x<18	18<x<21	21<x<0
Canada	0	1	1	0	0	0	0
Netherlands	1	1	2	0	0	4	4
Latvia	0	1	0	0	0	0	0
United States	89	35	31	22	27	32	71
Czechia	6	5	5	5	5	5	7
Poland	17	10	9	13	13	10	10
United Kingdom	12	0	3	0	0	1	0
Malaysia	0	0	1	0	0	0	0
France	52	14	4	3	7	17	41
Germany	42	24	24	28	28	21	33
Russia	38	19	16	9	21	24	25
Palestina	0	0	1	1	0	0	3
Total	257	110	97	81	101	114	194

attackers used in 79% of all attacks. At least times the attackers flooded our honeypots by using the OPTIONS flood attack, which was detected only in 5 % of cases. In 45 % of cases, attackers used INVITE flood.

Attackers from the United States most often targeted our honeypots. In overall, they hit honeypots 307 times (32%) over the course of two months. On the contrary, the least active attackers are from Latvia and Malaysia, who presented themselves with a single attack. In overall, the total duration of all attacks is 325 614 s, which means that honeypots were 3 days 18 h 26 min and 54 s under attack. The most time devoted to generating a high number of SIP messages and targeting VoIP services was spent by the Americans with a total time of 1 day 8 h 52 min 18 s with an average attack time of 6 min 25 s. Another interesting statistic is the processing of collected data based on the day and time of their execution. Table 2 shows the frequency of attacks divided by time. From the gathered data, we can notice that with the total number of 257 attacks, honeypots are most likely tend to be under attack between midnight and six o'clock in the morning. Honeypots gathered the least data between 12 am and 3 pm. We can also notice that attackers from the USA, Germany, Russia are attacking honeypots permanently throughout the day, while the attacks from other countries can be considered sporadic.

In Table 3, we can see the frequency of attacks divided by the date of their execution. The least busy days are Thursdays, Fridays, Saturdays and Sundays. Most attacks take place on Mondays, Tuesday and Wednesdays.

By looking at the frequency chart in Fig. 2, we can see that honeypots are the most loaded by DoS attacks on Tuesday specifically between 6 pm and 11:59 pm. On the contrary, the least burdened time interval took place on Saturdays between 12 am and 12 pm.

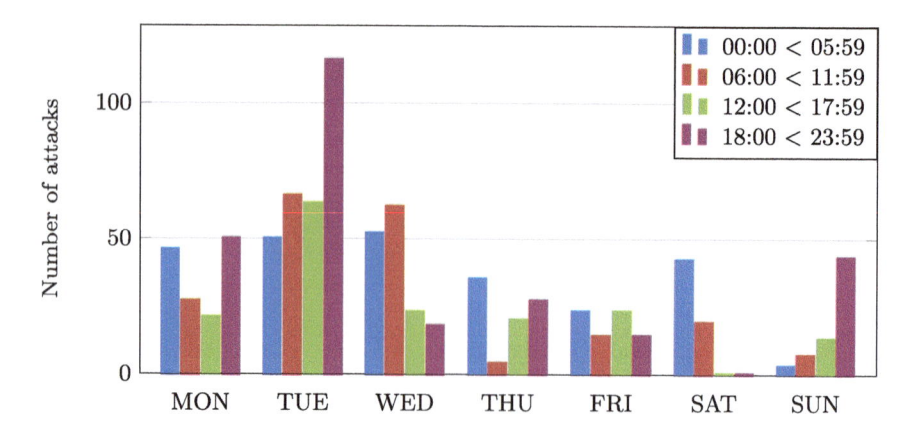

Fig. 2. Daily and hourly frequency chart

5.2 Whois Providers Comparison

As previously mentioned, the honeypot is using Whois IP address based queries to search for various information about the attacker. In the previous version of honeypot, we implemented a connector on the WHOIS Rest API provider available at rest.db.ripe.net. This web service can be used for both queries and updates. When honeypot detects some VoIP DoS attack, it will query Whois service that will return an answer containing various information about the attacker. Since in many cases the aforementioned service was unable to provide the correct information, we decided to extend the honeypot to another connector. We implemented

Table 4. Daily statistics by country.

	Mon	Tue	Wed	Thu	Fri	Sat	Sun
Canada	2	0	0	0	0	0	0
Netherlands	5	2	3	0	0	1	1
Latvia	0	1	0	0	0	0	0
United States	31	83	118	52	6	7	10
Czechia	6	9	9	2	1	6	5
Poland	0	65	0	3	7	7	0
United Kingdom	0	0	0	0	15	2	1
Malaysia	0	0	1	0	0	0	0
France	46	17	4	7	11	23	30
Germany	43	43	12	0	37	57	8
Russia	12	79	11	26	1	10	13
Palestina	3	0	1	0	0	0	1
Total	148	299	159	90	78	113	69

the connector to the FreeGeoIP web service. Based on our data, we decided to compare these two mentioned web services. The results of our analysis are processed in Tables 4 and 5.

Table 5. WhoIS providers comparison.

	Failures	Success	Percentage(%)
Rest Ripe Api	311	643	67
FreeGeoIP	0	954	100

Out of a total of 954 attacks, REST RIPE API was unable to provide information about the attacker in 311 (33%) cases. The FreeGeoIP service was 100% successful, so it was able to determine the origin of the attacker in all cases. Only 495 (49%) of the attackers found their services in 465 (49%) cases, and 178 (19%) did not agree. The results obtained indicate that web service FreeGeoIP by far dominates to providers of Whois services, and we can consider it as one of the most extensive and most widely used APIs for IP to location services worldwide. The old FreeGeoIP API is now deprecated and is discontinued from July 1st, 2018 and a wholly re-designed API is now available at http://api.ipstack.com.

6 Conclusion

In this article, we introduced a new version of VoIP honeypot. Compared to the original version, we can find a few differences. For ease of implementation, the new honeypot solution was written in the JAVA programming language. This version also calculates a new parameter, and that is a total attack time. It also allows detecting INVITE, OPTIONS, and REGISTER flood attacks. The honeypot is also capable of detecting a SPIT attack, and it has a connector on the FreeGeoIP web service implemented that can determine the origin of the attacker and provides other essential data that we can work with in the future. In the previous version, the honeypot was connected to the REST RIPE API web service, but during the test period, it did not prove to be reliable. We are currently working intensively on SPIT attack detection improvement. Based on the data from a real VoIP service provider, we want to employ neural networks, which will be able to distinguish between the correct VoIP traffic and the SIP traffic flow that is generated by the SPIT attacker.

Individual honeypots exchange information about attacks and create records in their databases. Throughout the architecture, there is one central database that performs the data collector role, so all attack records are stored here.

Honeypots create fake SIP dialogues and transactions among themselves. This functionality will be redesigned to a more sophisticated look soon. Based on the processed data from a real VoIP PBX and the use of Markov chains, we will be able to simulate SIP traffic, and making the honeypot indistinguishable from the real full-featured SIP PBX. This should increase the interest of attackers and possibility of the attack. For the statistical analyses contained in the article, we implemented a custom program that processes attack data stored in the MongoDB database. The new version of honeypots are currently in the test phase, and they run on servers in different locations like Ostrava, Prague, Opava and Slovakian Zilina. The article contains processed data that we were collecting for two months of running since launching the new JAVA version of honeypot.

We are also planning to deploy honeypots to other countries like Italy or Colombia. Using these honeypots for analysis is not a primary goal. We plan to work with some VoIP providers who will use the results of our research to make our network more secure. Honeypot VoIP management will be done through VoIP Honeypot admin in the near future. VoIP Honeypot admin is basic interface for the honeypot management and soon we will implement Web management.

Acknowledgment. This work was supported by the VSB-Technical University of Ostrava, Czech Republic - Networks and Telecommunications Technologies for Smart Cities under SGS Grant SP2018/59.

References

1. Behan, L., Kapicak, L., Jalowiczor, J.: Development and implementation of VoIP honeypots with wide range of analysis. In: Proceedings of SPIE 10630, Cyber Sensing 2018, vol. 106300S, 3 May 2018. https://doi.org/10.1117/12.2304602
2. Voznak, M., Kapicak, L., Zdralek, J., Nevlud, P., Plucar, J.: Multimedia services in asterisk based on voiceXML. Int. J. Math. Models Methods Appl. Sci. **5**(5), 857–865 (2011)
3. Voznak, M., Rezac, F.: Threats to voice over IP communications systems. WSEAS Trans. Comput. **9**(11), 1348–1358 (2010)
4. Nevlud, P., Bures, M., Kapicak, L., Zdralek, J.: Anomaly-based network intrusion detection methods. Adv. Electr. Electron. Eng. **11**(6), 468–474 (2013)
5. Sisalem, D., Floroiu, J., Kuthan, J., Abend, U., Schulzrinne, H.: SIP Security. Wiley Blackwell, Hoboken (2009)
6. Rezac, F., Voznak, M., Tomala, K., Rozhon, J., Vychodil, J.: Security analysis system to detect threats on a SIP VoIP infrasctructure elements. Adv. Electr. Electron. Eng. **9**(5), 225–232 (2011)
7. Safarik, J., Partila, P., Rezac, F., Macura, L., Voznak, M.: Automatic classification of attacks on IP telephony. Adv. Electr. Electron. Eng. **11**(6), 481–486 (2013)
8. Voznak, M., Safarik, J., Rezac, F.: Threat prevention and intrusion detection in VoIP infrastructures. Int. J. Math. Comput. Simul. **7**(1), 69–76 (2013)
9. Voznak, M., Rozhon, J.: SIP infrastructure performance testing. In: 9th WSEAS International Conference on Telecommunications and Informatics, TELE-INFO 2010 , pp. 153–158 (2010)
10. Rozhon, J., Voznak, M.: SIP registration burst load test. In: Communications in Computer and Information Science, vol. 189. CCIS(PART 2), pp. 329–336 (2011)
11. Vennila, G., Manikandan, M., Suresh, M.: Detection and prevention of spam over internet telephony in voice over internet protocol networks using Markov chain with incremental SVM. Int. J. Commun. Syst. **30**(11) (2017)

12. Voznak, M., Rezac, F.: The implementation of SPAM over Internet telephony and a defence against this attack. In: TSP 2009: 32nd International Conference on Telecommunications and Signal Processing, pp. 200–203 (2009)
13. Open-source project. Mentat - distributed modular Security Information and Event Management System. Cesnet, 25 August 2017. https://mentat.cesnet.cz/en/index. Accessed 17 Mar 2018

Proposal and Implementation of Probe for Sigfox Technology

Jakub Jalowiczor[(✉)] and Miroslav Voznak

Faculty of Electrical Engineering and Computer Science, VSB-Technical University of Ostrava, 17. listopadu 15/2172, 708 33 Ostrava, Czech Republic
jakub.jalowiczor@vsb.cz

Abstract. The paper deals with designing and implementation of a probe that measures the quality of Sigfox radio network coverage. This probe consists of the mobile application for Android operation system, Sigfox RF module, and Bluetooth module. Finding the coverage quality is solved by wireless communication between mobile device and Sigfox module, which sends Sigfox message. The developed solution mainly offers a possibility to display base stations on a map and show a list of all measurements. The probe represents a tool, which is intended primarily to find areas with weak or non-existent Sigfox network coverage and as the authors know, the presented solution is unique so far.

Keywords: Internet of Things · Sigfox · Measuring · Probe

1 Introduction

The Internet of Things is a very hot topic, and many people expect considerable changes in the living of their lives, increasing efficiency and cost reduction. Moreover, the Internet of Things accelerates many different and sophisticated applications across various fields [1]. It is estimated, that the number of connected devices around the world could increase dramatically, with expectations ranging from 25 billion to 50 billion devices in 2025 [2]. Even Cisco estimated 50 billion connected devices, but this contains a greater range of equipment, including computers, then we include in the Internet of Things [3]. Because of this significant expansion and unique requirements of devices, new technologies have been created exclusively for this purpose to provide wireless connectivity for the Internet of Things devices (the most common are various kinds of sensors). Among these technologies, we can include Sigfox technology.

The aim of this paper is proposal and implementation of a probe for Sigfox technology with the necessary mobile application for Android operation system. Together with the connected Sigfox RF module, this mobile application can measure the qualitative parameters of a radio signal and based on this information it can determine the quality of coverage in the given measured area.

The rest of this paper is structured as follows: Sect. 2 provides analysis of the current work in LPWAN with the focus on Sigfox technology and measuring the essential parameters of these technologies, Sect. 3 describes overall concept of the solution and used technology. Section 4 includes details about implementation and its results. Finally, Sect. 5 summarizes our conclusions.

2 State of the Art

In the following, we briefly review a few works that are relevant to our work. Authors in many papers discuss the overall overview of various LPWAN technologies including Sigfox, that is provided for example in [4]. The authors describe the techniques LPWAN technologies use to achieve their goals and they highlight and compare proprietary technologies as Sigfox and LoRa. The main conclusion is that each of the technology has its pros and cons, leading to a fragmented market.

The authors in paper [5] discuss the technical differences between Sigfox, LoRa and NB-IoT, their advantages in different use cases and major issues.

The simulation work in [6] compares GPRS, NB-IoT, LoRa and Sigfox technologies in a scenario with an area of 7800 km^2, and the target is to determine which of the technologies provides the best coverage for the Internet of Things. Their solution is based on calibrated Matlab simulator using the 3GPP Rural Macro non-line-of-sight (NLOS) model for the rural areas and the 3GPP Urban Macro NLOS model for the urban areas.

Although we can find professional Sigfox field test devices as for example Adeunis ARF8121AA on the market, there is not any low price solution, that could be sent to the end user to test coverage and parameters from all available base stations in a designated area and forward results back to the network provider. Adeunis is equipped with an LCD screen, where you can see various parameters relating to how the network is operating (Uplink, Downlink, PER, etc.) [7]. A solution with more advanced graphical user interface that could serve also for presentations is not available. The foregoing facts have led to the design and development of the probe.

3 Overall Concept and Used Technology

First of all, important is to describe the overall concept of our proposal, or in other words what the probe with the mobile application should do. To fulfill its function, the probe must always be used with the mobile application, without that it is only usual Sigfox RF module connected with Bluetooth module, therefore further when we talk about the probe, it means the probe itself with the corresponding mobile application.

The primary function of the probe is to define a quality of coverage in the desired area. This area is determined by the actual position of the probe. Measurement of the coverage can be done by wireless communication between a mobile device with the application installed and Sigfox RF module TD1208. Mentioned wireless communication is realized by using Bluetooth technology. The application is developed for Android operation system, and it is intended primarily for devices with a larger screen - tablets. The main part of this application is a map on which base stations that receive the sent message are graphically marked. Information, as values of RSSI and SNR parameters of the actual received message, is tied up with every marked base station. This information is visible when you click on given base station marker on the map. Marker color depends on the actual RSSI or SNR parameter value. History from all of the measurements carried out on a given device is saved into the local database, and it is available through the application menu. Figure 1 shows a sequence diagram of the measurement process.

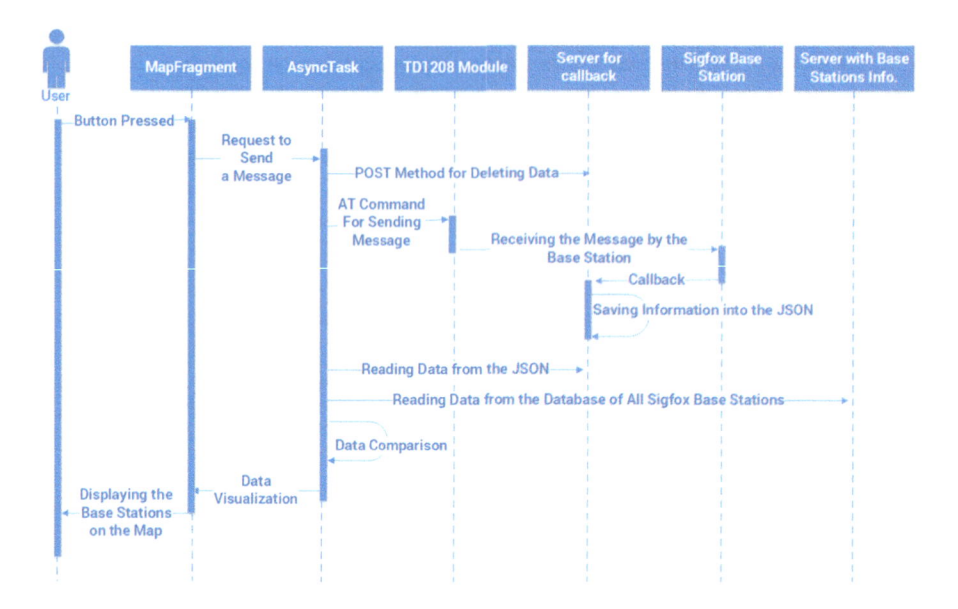

Fig. 1. Sequence diagram of the measurement process

As for the used technologies to realize this concept, we used Sigfox RF module TD1208, Bluetooth module HC-05, YwRobot MB102 breadboard power supply module and a server receiving the callback from Sigfox base stations.

3.1 Sigfox Radio Network Technology

Sigfox [8] is radio technology designed especially for the Internet of Things area, and it belongs to Low Power Wide Area Networks (LPWANs). It uses Ultra Narrow Band (UNB) modulation technique to provide a scalable high-capacity and low energy consumption for end devices, with Differential Binary Phase-Shift Keying at 100 bps (DBPSK). Devices connected to the Sigfox network should least several years powered from a battery, in many cases up to ten years. The advantage is the ability to design simple, small-size antennas at the end of the network infrastructure and cheap, easy-to-match antennas on the device side. Using lower data transfer rates leads to a narrower bandwidth, which reduces interference level and increases sensitivity for the receiver, and this all makes it possible to cover large geographic locations with a smaller number of base stations [5]. The technology operates in ISM (Industrial, Scientific, and Medical) band. This band is an unlicensed frequency band, which was originally intended for industrial, scientific and medical applications. In Europe, it is the frequency band in the area of 868 MHz [9]. Table 1 lists the different frequency bands for other areas.

Table 1. Sigfox technology frequency bands for individual countries

Area	Frequency
Europe, South Africa (Radio zone 1)	868 MHz
USA, Mexico, Brazil (Radio zone 2)	902 MHz
Japan (Radio zone 3)	923 MHz
Argentina, Colombia, New Zealand, Hong Kong, Singapore (Radio zone 4)	920 MHz

Unlicensed frequency band brings the lower price for transferred data, but on the other hand, there is higher interference than in licensed frequency band. Devices communicating in an unlicensed band must be designed to be able to coexist together with other technologies on these frequencies without risk of collision.

For device management or data collection, there is Sigfox Cloud, which provides a web application interface. Once device sent a message to Sigfox network, one of the three main ways to get data and work with them can be used:

- Graphical user interface - via web page backend.sigfox.com
- REST API - this interface can be used to reach data and related information
- Callbacks - a mechanism for forwarding each new message to the user's application server. Callbacks are fully configurable HTTPS requests, for example, you can set method, headers, type of content and structure of content within the message body.

An advantage of this technology is global coverage of a large number of countries, including Czech Republic. This advantage is especially useful there, where the subscribers do not want to build a private network for their application and solve all of the other things which are related to it.

Sigfox technology is appropriate for more straightforward applications, where the device sends an uplink message only a few times per day because the number of uplink messages for one device is limited to value 144 and the transmit power is limited to 25 mW in this frequency band. Sigfox uses the duty cycle method as a mechanism for sharing the spectrum in ISM band. The duty cycle is 1% in

Europe, and that is why a Sigfox device can only transmit 36 s per hour. With the time on air 6 s per package, the maximum is 6 messages per hour with a payload of 4, 8 or 12 bytes [10]. Radio messages handled by the Sigfox network are small (12 bytes payload in uplink, 8 bytes in downlink) thanks to lightweight protocol. From the above implies, this technology is not appropriate there, where it is necessary to communicate bidirectionally and many times per day. Sigfox is closed technology, and it does not have many detailed technical specifications freely available. Subscriber depends on the network provider, and data storages for all received messages are under the provider's management.

3.2 Sigfox Module Telecom Design 1208

Module TD1208 is Sigfox gateway with high performance and low energy consumption. The wide operating voltage range of 2.3–3.3 V and low current consumption make the TD1208 an ideal solution for the battery-powered applications. TD1208 Evaluation Board provides access to the different module interfaces, USB connectivity using USB cable with FTDI LVTTL RS232 3.3 V interface and development facility [11].

3.3 Bluetooth Module HC-05

Module HC-05 is easy to use Bluetooth version 2.0 Serial Port Protocol module designed for transparent wireless connection setting [12]. SPP Bluetooth profile defines the requirements for setting up emulated serial connections between Bluetooth devices [13].

3.4 Power Supply YwRobot MB102

YwRobot MB102 is breadboard power supply module which provides dual 5 V and 3.3 V power rails with jumpers, DC barrel connector and female USB socket for power connection.

4 Implementation and Results

Mobile device with application installed uses Bluetooth technology to connect wirelessly to the HC-05 module. Module HC-05 allows wireless communication by emulating the serial cable connection. Sigfox module TD1208 is connected with Bluetooth module HC-05. In this manner, a mobile device can send AT commands via Bluetooth technology, communicate with the TD1208 module and control its functions, e.g. message sending.

The coverage quality measurement is implemented as follows. First of all, a user who wants to realize a measurement must push an appropriate button in the application. As a response, the mobile device sends AT command via Bluetooth, intended for sending messages, to TD1208 module. Then, the mobile device sends HTTP POST method to clear the JSON file, where is the information from the previous measurement and the module sends Sigfox message based on received AT command into the Sigfox network. At this moment, Sigfox callback is used. This callback ensures forwarding of every message received by Sigfox base station to the own application server. Together with the received message, callback forwards also other information as base station ID number and values of SNR and RSSI parameters. You can set up callback format on Sigfox web interface. In this case, the callback is sent by the HTTP POST method, and in its body, there is a JSON object with the information of the given base station about the received message. The content of this JSON object can be seen below. The variable in brackets is replaced with an actual value in a real situation.

```
{
    "device" : "{deviceID}",
    "station" : "{baseStationID}",
    "rssi" : "{rssiValue}",
    "snr" : "{snrValue}",
    "avgSnr" : "{averageSnrValue}",
    "time" : "{time}",
    "data":"{messageData}",
    "seqNumber" : "{seqenceNumber}",
    "lat" : "{latitude}",
    "lng" : "{longitude}"
}
```

There is PHP script on the application server, which ensures saving of obtained information into the JSON file. Then, the mobile application accesses this file and carries out a comparison of the base station ID number with a database of all Sigfox base stations, in order to get more specific information. In the database of all Sigfox base stations, there are GPS coordinates of given base station among others. This database is under Sigfox network provider management. After equal base station ID number is found, the base station information is saved into the mobile device's local database, and the base station is marked on the map according to GPS coordinates.

We measured the quality of Sigfox radio network coverage with the probe in many places in Czech Republic. In a bigger cities, there is not problem with coverage and the probe detected multiple base stations. When we tested coverage in city district Kralovo Pole in Brno, seven base stations reached the sent Sigfox message. For tests performed in Ostrava inside of the new building of FEI VSB-TU Ostrava, eight base stations received the sent Sigfox message. That is a very good result, bearing in mind, of what materials the building is constructed. Figure 2 shows results from the measurement realized in Ostrava, where we can find out that Sigfox message was received also by base stations located outside of Ostrava city, namely in Cesky Tesin, Havirov, and Frydek Mistek.

In Table 2, you can see some information from measurements realized with the probe at various locations in Czech Republic. As you can see, in smaller cities as for example in Krnov or Opava, only one base station received the Sigfox message, but that should be enough because utilization of this base station is

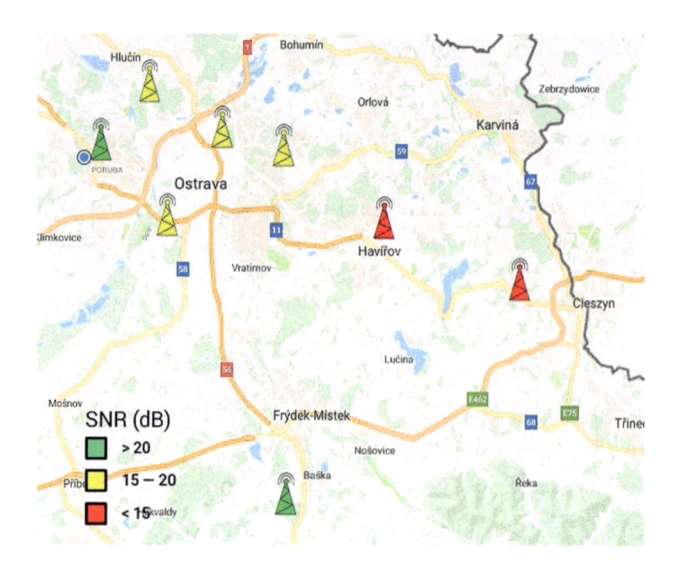

Fig. 2. Map of coverage measured from inside of the FEI VSB-TU building in Ostrava

Table 2. Measurement details from different places

Place	Number of measurements	Average number of base stations	Average SNR (dB)	Average RSSI (dBm)
Ostrava - FEI build.	35	8	14.17	−130.21
Ostrava - N build.	35	8	9.62	−129.71
Ostrava - city center	35	9	24.48	−120.19
Brno - Kralovo Pole	25	7	21.57	−118.88
Prague - Zlicin	25	12	19.30	−128.00
Frydek Mistek	25	8	18.76	−125.63
Cesky Tesin	25	9	20.12	−121.26
Opava	25	1	18.42	−125.25
Krnov	25	1	21.15	−111.10

not planned to be high and also interference is smaller in these locations. The first two locations in the table have worse results (lower values for the average SNR and RSSI parameters), as the measurements were realized from the inside of the buildings.

Initially, we think of about possible extension of the application to perform drive tests, when the probe would be placed inside a car and after turning on this function, the continuous measurement would be realized. In the end, we decided to skip this function. The reason is the limited number of messages, which can be sent between device and base station per one day. This limitation is 144 messages. Another reason is a long delay between the moment of message sending and the moment when all of the involved base stations answer with information that they received the message.

5 Conclusion

The mobile application allows wireless communication between a mobile device (tablet, cell phone) and Sigfox RF module via Bluetooth and well-arranged representation of all base stations that received the Sigfox message. These messages are saved in the mobile device database.

The developed probe serves as a measuring tool with a purpose of defining the quality of radio signal coverage in the determined area. This measurement helps Sigfox network provider to solve problems with inappropriate or non-existing coverage at the customer side.

As a future work, we intend to replace the native Android mobile application with a web application, which can work on any device that has a web browser.

Acknowledgment. This work was supported by the VSB-Technical University of Ostrava, Czech Republic - Networks and Telecommunications Technologies for Smart Cities under SGS Grant SP2018/59.

References

1. Krivtsova, I., et al.: Implementing a broadcast storm attack on a mission-critical wireless sensor network. In: Mamatas, L., Matta, I., Papadimitriou, P., Koucheryavy Y. (eds) Wired/Wireless Internet Communications. WWIC 2016. LNCS, vol 9674. Springer, Cham (2016)
2. Manyika, J., Chui, M., Bisson, P., Woetzel, J., Dobbs, R., Bughin, J., Aharon, D.: The Internet of things: mapping the value beyond the hype. McKinsey Global Institute (2015)
3. Bradley, J., Barbier, J., Handler, D.: Embracing the Internet of everything to capture your share of $14.4 trillion. Cisco (2013)
4. Raza, U., Kulkarni, P., Sooriyabandara, M.: Low power wide area networks: an overview. IEEE Commun. Surveys Tuts. **19**(2), 855–873 (2017)
5. Mekki, K., Bajic, E., Chaxel, F., Meyer, F.: A comparative study of LPWAN technologies for large-scale IoT deployment. ICT Express (2018)
6. Lauridsen, M., Nguyen, H., Vejlgaard, B., Kovacs, I., Mogensen, P., Sørensen, M.: Coverage comparison of GPRS, NB-IoT, LoRa, and SigFox in a 7800 km^2 area. In: VTC Spring. Accepted, June 2017
7. Adeunis. https://www.adeunis.com/
8. SigFox. https://www.sigfox.com/
9. ETSI EN 300 220-1 V2.4.1. Electromagnetic compatibility and Radio spectrum Matters; Short Range Devices; Radio equipment to be used in the 25 MHz to 1 000 MHz frequency range with power levels ranging up to 500 mW; Part 1, 1 (2012)
10. Vejlgaard, B., Lauridsen, M., Nguyen, H., Kovacs, I., Mogensen, P., Sorensen, M.: Coverage and capacity analysis of Sigfox LoRa GPRS and NB-IoT. In: IEEE 85th Vehicular Technology Conference, June 2017

11. TD Next RF modules. http://rfmodules.td-next.com/
12. Cotta, A., Devidas, N.T., Ekoskar, V.K.N.: Wireless communication using HC-05 bluetooth module interfaced with arduino. Int. J. Sci., Eng. and Tech. Res. (IJSETR) **5**(4), 1–4 (2016)
13. Serial Port Profile. https://www.bluetooth.org/

IoT Approach to Street Lighting Control Using MQTT Protocol

Radim Kuncicky[1(✉)], Jakub Kolarik[2], Lukas Soustek[2], Lumir Kuncicky[3], and Radek Martinek[2]

[1] Department of Computer Science, Faculty of Electrical Engineering and Computer Science, VSB–Technical University of Ostrava, Ostrava, Czech Republic
radim.kuncicky@vsb.cz

[2] Department of Cybernetics and Biomedical Engineering, Faculty of Electrical Engineering and Computer Science, VSB–Technical University of Ostrava, Ostrava, Czech Republic
{jakub.kolarik,lukas.soustek,radek.martinek}@vsb.cz

[3] Department of Electrical Power Engineering, Faculty of Electrical Engineering and Computer Science, VSB–Technical University of Ostrava, Ostrava, Czech Republic
lumir.kuncicky@vsb.cz

Abstract. This article shows modern approach to street light control based on Ethernet network with Message Queuing Telemetry Transport (MQTT) protocol with ideas of Internet of Things (IoT). The real implementation of such a luminaire is shown, with following testing on BroadbandLIGHT polygon. We demonstrates advantages of our solution, which meet requirements of future cities, today many times called as a smart city. It obvious that this approach to connection and controlling of public light system allows fast expansion of Smart technologies and related services. In experimental part we primary discus requirements and influence of technology to network bandwidth. The original benefit of the study is to verify the functionality of IoT approach to street lighting control using real-time MQTT protocol. The realized experiments clearly confirmed the usability of the real-world public lighting infrastructure to cover the intravilan city with SMART technologies.

Keywords: sMQTT protocol · IoT · Street lighting control · DALI

1 Introduction

Undoubtedly, everyone has already wondered what it would be like to pass in time into the future to a place where the smart technology solutions are used to improve people's lives with the ecologically efficient energy management involved. The smart city like this should have been covered by data connection and advanced security features which would notified the competent authorities in a case of traffic accident for example, and make contact with wounded or pedestrians and begin to solve the situation before the rescue teams arrive and divert the traffic elsewhere too. Similarly, the city could have a network of sensors to monitor the weather and the environment to improve the standard of living of the population. Today, some of the parking lots can already count the availability and, moreover, navigate the vehicle directly to the free spot. Street lighting lamps may not be only boring poles, but they could for example enable charging of electrical appliances or provide wirelessly important information such as a timetables, shop opening hours or show times for free. They could be also used to play music, imitate the singing birds or the sound of the river and bring a little bit nature to the city. At the same time, the city should smartly dispose of energy and use it where it is really needed. For example, it is not necessary to illuminate the entire street with the maximum power of the luminaire when there is no pedestrian or moving car.

All this would not be possible without sufficiently robust data communication between streetlights. On such technologies we are already working at the VSB - technical university of Ostrava. We are striving to exploit the potential of street lighting on our parking lot by deploying the latest technology. In the case of half-empty parking lot, only the occupied spots can be light on. The public lightning could be set up in storm and bad weather to a low luminous flux. All this contributes to improving the lighting conditions of the city and reducing light smog.

2 Street Light Control Solutions

Proper implementation of whole control system can have a significant impact to electricity savings [1, 2]. Many efforts have been made to found appropriate communication solution [3] in the past. A lot of these systems was built on extension of current systems [4] (Digital Addressable Lighting Interface - DALI - over Ethernet, ArtNet [5], ...) or they tried to use new way of communication [6]. As today available solution we can mention Petra System Small Cell Network [7], Street Light Control (SLC) [8], Philips CityTouch [9]. All these systems supports enhanced lighting control and management. These solutions are example of bottom to up approach, where the solution is tied to technologies like general purpose radio or powerline communication and secondary services have to run over and many times have to be supported by manufacturer. The extensibility of such a system is questionable and can easily lead to vendor lock problem. We suggest layered approach as better, where each part like network connectivity, data transfer, control and monitoring system can be independent. Our solution is an attempt to transfer this IoT approaches to world of street light control using broadband connection and integration of secondary services.

2.1 Luminaires

From basic point of view we can divide luminaires to two groups – with limited regulation and fully dimmable. In limited regulation group we can found especially discharge lamps, fluorescent lamps, sodium-vapor or metal halide lamps where for example metal halide lamps can be regulated between 60 and 100% of rated output. Fully dimmable are today especially LED luminaires which can be regulated in range from 0 to 100% of rated power. The largest part of installed luminaires in Czech Republic is still from the first group, and thanks to higher purchase price the transition to LED lights is quiet slow.

2.2 Control Systems

The most basic system and widespread solution is switching on or off the electricity from commons substation for whole blocks of streets based on presets clocks. Usage of dedicated communication interface, which enables power regulation and backward telemetry of luminaire, is better solution. In this case we are limited just with the abilities of lamp driver or supporting devices.

2.3 Advantages and Disadvantages of Common Control Solutions

The electricity switching solution is suitable only with one control element and just a few measurement elements. And obvious advantage is that we know for sure that the all lamps are off when electricity is disconnected. The changeover to new device is easy, because even fully controlled devices are backward compatible with this system, and it is not necessary to evaluate the compatibility of control system.

The electrical installation for this control system is cheaper by using simple (3 or 5 core) power cables. In this case, the possibilities of telemetry are very limited. We cannot find whether all luminaires are working properly. The system status can be just supposed based on electric current measurements, which have a large variance depending on the age of luminaires. We are not able to determine which one is faulty without visual control.

More modern light are equipped with control logic and communication interface. The most widespread interface today is Digital Addressable Lighting Interface (DALI). DALI is an industrial standard developed to luminaire control. It is defined in IEC 62386 and IEC60929. It was developed as replacement of 0–10 V system that is inappropriate to long distance regulation due to its sensitivity to noise.

DALI system is limited with number of devices (64) working on one branch and for more devices is required to buy a new master device and split the branch. All DALI model layers are protocol specific (line levels, codding, messages) and it is almost impossible to modify the system. The bandwidth is limited to 1200 bps what cases propagation delay and sometimes slow response of the system. Advantage of system is ability to define scenes, groups and preset timing, and a basic telemetry is available too. Such a light can share permanently powered line with other city systems.

The Ethernet based systems are still from future, our research group together with company L2Led has developed luminaire that can work on Ethernet network. Usage of TCP/IP stack has a lot of advantages. For example it is easy apply encryption, we are not limited with number of device, it is simple to add sub service, it is easy to implement an open protocol etc. All these possibilities are enabled thanks to universal layered communication model.

3 Street Light as IoT

From Road Act [10] point of view the streets lights are the accessories of road. Most often the operator is town government or holder of the land. In the municipalities, public lighting is abundantly widespread and in most cases covers more than 80% of the built-up area. The condition of public light systems in Czech Republic Is very well know and can be easily found [11], especially thanks to Strategy One's survey from 2015 that was based on the Freedom Information Act requests that were send to all municipalities with more than 2000 people living in. These documents can be found on municipalities web sites [12].

Public lighting is commonly seen as a public service without any added value or usable feedback. Added value of public lighting can be pylon usage for installation of surveillance systems, warning systems etc. The idea of integration of these systems to public lighting arises with expansion of modern technology. A luminaire with abilities in video, data, information and extended telemetry services can be called as a Smart device.

With this changes we have to change the point of view to primary function of public lighting. Extended sensor networks can be used to create fully autonomous system or even autonomous luminaire that can work without superior control system. This point of view is the exactly one of the ideas of the IoT world.

4 Protocol MQTT

For IoT purposes there are available many commonly used protocols like Constrained Application protocol (CoAP) [13], Message Queuing Telemetry Transport (MQTT) [14], Streaming MQTT (MQTT-SN) [14], Extensible Messaging and Presence Protocol (XMPP) [15], Advanced Message Queuing Protocol (AMQP) and Simple Sensor Interface (SSI). Selection of proper solution is not simple and have to consider different point requirements as a throughput [16], network overhead [17], propagation delay and jitter [18, 19], security [20] and finally platform support and development speed.

We selected MQTT as a reliable and stable protocol with good support over all required platforms (Linux, Windows, microcotrolers) and short prototype-to-product cycle.

MQTT is machine to machine protocol build on Transmission Control Protocol and designed for IoT appliances. Protocol is low bandwidth with low overhead [21]. MQTT is implementation of design pattern publisher/subscriber, where message distribution is provide by server software called broker which clients are connected to. Messages belong to topics which are very similar to url address. Every client can subscribe to any topic and the new messages are then send to him.

MQTT can solve many problems from IoT world. One of them is quality of service (QOS) which is available in three levels 0, 1, 2, where 0 stands for at most once deliver, 1 stands for at last once delivery and 2 for exactly one delivery.

MQTT supports last will, or sometimes called testament. The last will is message that is published when connection is lost.

Individual pats of topic name are separated by slashes. It is possible to use specific symbols + and # when subscribing topic. Plus sign substitute one level (Fig. 1a) and sharp substitute multiple (Fig. 1b) levels but placing at the end is required. Design of such a structure is nontrivial. Appropriate design can minimize complexity of following higher layers – control systems.

```
a) building/room/+/door
b) building/room/1/#
```

Fig. 1. Topic examples (a) Topic for receiving all information about doors (b) Topic for receiving information about everything in room number 1

5   L2Led L2LCM Luminaire

The test light was developed in cooperation with L2Led. The high-performance "smart" luminaires series L2LCM are designed for use on public lighting columns in so-called "Smart City" systems, offering something extra beyond perfect and highly economical lighting. The luminaires are equipped with high-efficiency LED chips with definable shades of white light (chromaticity temperature). The luminaire body is made of a special alloy of aluminum, magnesium and silicon with an electrically passivated surface and arbitrary color shade of the outdoor protective layer. The design of the luminaire, out of mechanical resistance and perfect cooling, ensures sufficient shielding to prevent glare.

The luminaires are equipped with a system of highly efficient power supply that allows remotely control the luminosity. The remote control is based on direct connectivity with the high speed Ethernet network for which the luminaires series L2LCM can be equipped with an optical or metallic connection module, or eventually with module for a Wi-Fi or other wireless connection. If the existing power line cannot be complemented by other data line wiring, the luminaires series L2LCM can be equipped with power line data transmission – PowerLine system.

Ethernet connections may not only be used to control the light, but can also be used for other functions. Patented power supplies allow high speed data to be transferred directly to LED chips and spread further into light beam (Li-Fi technology, BroadbandLight, etc.). For example, the illuminated street with these lights can be covered by high-speed internet.

6   Testing Polygon BroadbandLIGHT

Whole concept was tested in real environment at Broadband Light testing polygon Fig. 2. Newly build probationary street light polygon is part of Faculty of Electrical Engineering and Computer Science's parking lot. Consist of totally ten pylons that are equipped with different luminaires Fig. 3. Whole facility is designed for testing of the cooperation between street light and other smart technologies including video surveillance systems, sensor etc.

Polygon is equipped with power quality measurements corresponding to the requirements of ČSN EN 50160 [22].

Fig. 2. Actually mounted luminaires - Thorn R2L2 (blue), Schreder Teceo (green), Boss Naica Large (red) a L2Led (yellow). (Color figure online)

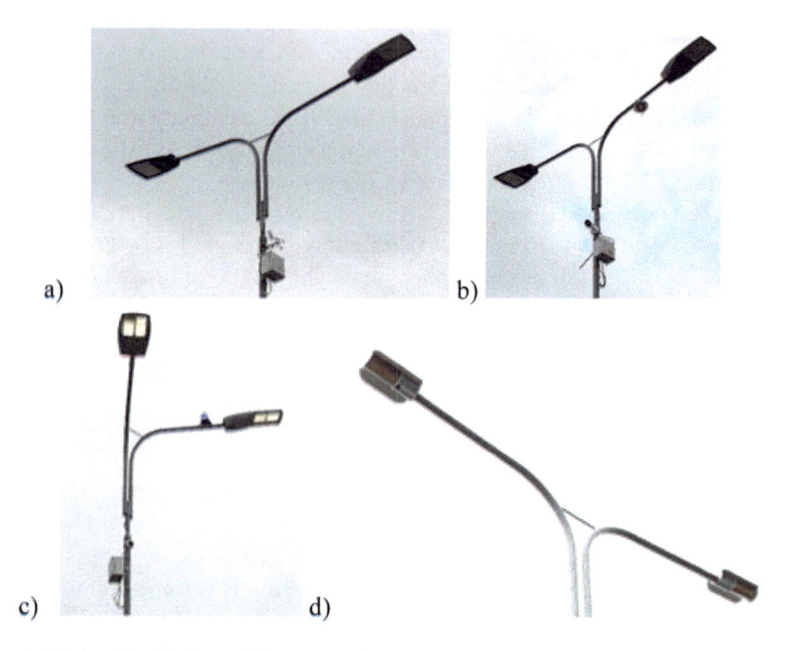

Fig. 3. (a) Pylon No. 5 (Boss, Thorn, Weather station), Pylon No. 7 (Thorn, Schreder, AXIS video camera, megaphone), Pylon No. 8 (Schreder, Thorn, beacon), Pylon No. 7 (2x L2Led L2LCM).

7 Implementation

The L2LCM prototype luminaire was implemented on ST Microelectronics STM32 platform using STM32F103 ARM based microcontroller unit and WizNet W5500 Ethernet communication interface. For MQTT processing the Paho library was used. We found the topics structure design as most complicated. We considered multiple approaches. For example we will mention two of them. The first one is topological where the structure is based on physical or electrical location (Fig. 4.). This solution is more difficult to implement and configure because the structure has to be changed in the time of luminaire installation.

```
/ulice/slabihoudka-st/line/2/column/5/lamp/1/power
/parkinglot/1/phase/3/lamp/37/power
```

Fig. 4. Example of topic structure for topological approach

The second approach is systematic, where the topics are broken down by device types and following grouping of fixtures is control system based. We decided to implement this variant though it is obvious that this design is higher bandwidth requirements, especially when controlling of multiple fixtures is required. Used structure is shown in Fig. 5.

```
device/identifier/{power|telemetry|settings|}/
           propertie/action
```

Fig. 5. Proposed topic structure

We used the device MAC address as identifier. All are actually delivered with QOS 1. Because of MQTT payload agnostic nature, we had to design the payload format too. In our system the payload is always coded as ANSI symbols, where numbers have decimal format with dot and two decimal places.

As control system we developed web based application called SmartLight Manager that is based on Django framework. SmartLight Manager uses Asynchronous Server Gateway Interface (ASGI) with MQTT to ASGI bridge, where ASGI together with websocket is devoted for information redistribution to all web browser instances. This flexible layered structure enables multiple control system connection just through MQTT network.

For easier configuration we implemented designated discovery topic, where the all required data about luminaire are published when the power is turned on or discovery request is send.

8 Testing and Results

The L2Led lights are located on the Testing Polygon of the VSB-TUO, where they make the regular service. At the same time, telemetry data is collected from the sensors and the availability of the device is monitored both by means of the telemetry of the given luminaire (uptime information) as well as by automatic availability check (PING). During the existence of the luminaire it shows good stability and protocol MQTT reports high stability without losing the connection. Initially, there were blackouts caused by an error in the polygon distribution system and consequent loss of connection between the control center and the polygon. When the network connection was restored, the communication between and control center was re-established. Nowadays, this problem is solved by the multi-connection of a polygon (power line, optics, Wi-Fi) to the control room. In the future, the plan is to deploy a distributed MQTT broker to enable this system to be backed up.

During the operation, the demands on data network load was tested. Most of the used messages did not exceed a total of 150 Bps, which at 100 Mbit permits theoretical throughput in thousands of messages per second - of course, in reality it will be limited by the power of the processor in the luminaire and the capabilities of the broker. MQTT delivers a negligible load in this application even when the QOS 1 ACK message is counted, which increases the operation by 56B per message in the direction from the broker to the end device. The size of each message is shown in the Table 1.

Table 1. Messages, values and lengths of MQTT packets

Command	QOS	Topic	Value	Total (B) *	MQTT (B)	Topic (B)	Message (B)
Get power	1	Lights/device/deadbeef0001/segment/1/power	0.00	106	52	42	4
Set power	1	Lights/device/deadbeef0001/segment/0/power/set	100.00	110	56	46	6
Telemetry	1	Lights/device/deadbeef0001/telemetry/humidity	45.63	110	56	45	5
Uptime	1	Lights/device/deadbeef0001/telemetry/uptime	11549	108	54	43	5

* With all headers (TCP, IP, Ethernet frame)

9 Conclusion

The primary aim of this research was to realistically verify the possibility of controlling public lighting on the basis of Ethernet using the MQTT Protocol with the IoT ideology. The implementation itself took place on the BroadbandLIGHT test polygon. This nowadays unused approach to the management of public lighting has many advantages like possibility of secondary services (video, sensors, active traffic signs, Wi-Fi), easy expansion through payload agnostic protocol, good operation results in high latency networks, possibility of integrating to existing systems (DALI over MQTT or DALI/MQTT Bridging).

Our solution is not the optimal IoT approach, but we consider it as a good enough hybrid between current street light control and IoT world. Testing polygon is actually still central controlled, the reason for this solution is required integration of luminaires without smart possibilities and ability to evaluate controlled operation of modern luminaires before autonomous system will be applied. This evaluation is important especially to meet the legal requirements.

For whole testing period the telemetry data was collected – error messages, temperature, humidity, illuminance uptime and current power of each luminaire. This data was received and stored every minute and our database has more than one million measurements. Together with another polygon systems (video, meteorological stations) this data will be used to automatic control systems development. As we mentioned in cap. 6, we can compare our measurements with measured power quality. Our plan is to include power quality and electric load forecast.

In future we would like to work on overall system cooperation. For example we suggest maintaining constant luminosity based on measured data and brightness calibrated video system or dynamic luminosity control based on parking lot usage. Thanks to out Ethernet anywhere approach and layered control system integration of these features can be easily added as another service in network.

Acknowledgment. This article was supported by the Ministry of Education of the Czech Republic (Project No. SP2018/170). This work was supported by the European Regional Development Fund in the Research Centre of Advanced Mechatronic Systems project, project number CZ.02.1.01/0.0/0.0/16_019/0000867 within the Operational Programme Research, Development and Education and by Grant of SGS No. 2018/177, VSB-Technical University of Ostrava and under the support of NAVY and MERLIN research lab.

References

1. Wojnicki, I., Kotulski, L.: Empirical study of how traffic intensity detector parameters influence dynamic street lighting energy consumption: a case study in Krakow, Poland. Sustainability **10**(4), 1221 (2018). https://doi.org/10.3390/su10041221. ISSN 2071-1050. Accessed 25 July 2018
2. Wojnicki, I., Ernst, S., Kotulski, L.: Economic impact of intelligent dynamic control in urban outdoor lighting. Energies **9**(5), 314 (2016). https://doi.org/10.3390/en9050314. ISSN 1996-1073. Accessed 25 July 2018
3. Muhendra, R., Arzi, Y.H.: Development of street lights controller using wifi mesh network. In: Proceeding of 2017 International Conference on Smart Cities, Automation and Intelligent Computing System. IEEE (2017). ISBN 978-1-5090-6280-5
4. Bellido-Outeiriño, F., Quiles-Latorre, F., Moreno-Moreno, C., Flores-Arias, J., Moreno-García, I., Ortiz-López, M.: Streetlight control system based on wireless communication over DALI protocol. Sensors, **16**(5), 597 (2016). https://doi.org/10.3390/s16050597. ISSN 1424-8220. http://www.mdpi.com/1424-8220/16/5/597. Accessed 30 July 2018
5. Artistic Licence. https://artisticlicence.com/. Accessed 22 Aug 2018
6. Tseng, K.H., Hsieh, CL.: A Solution for intelligent street lamp monitoring and energy management. In: Proceeding of 2016 IEEE 11th Conference on Industrial Electronics and Application. IEEE, Hefei, pp. 843–847 (2016)
7. Petra Energy Solution Brochure. Petra Systems. http://www.petrasystems.com/pdf/Petra_Energy_Solutions_Brochure.pdf. Accessed 22 Aug 2018
8. Street Light Control Brochure: Osram. https://www.osram.com/media/resource/HIRES/341262/6195320/street-light-control-innovative-light-control.pdf. Accessed 22 Aug 2018

9. CityTouch Brochure: Philips. http://images.philips.com/is/content/PhilipsConsumer/PDFDownloads/Global/ODLI20160905_001-UPD-en_AA-CityTouch.pdf. Accessed 22 Aug 2018
10. ACT No. 13/1997 Sb, Zákon o pozemních komunikacích (In Czech)
11. Tesar, J.: Veřejné osvětlení a jeho současný stav v České republice. Srvo.cz (2008). (In Czech). http://www.srvo.cz/wp-content/uploads/2017/12/VO_a_jeho_soucasny_stav.pdf. Accessed 24 July 2018
12. Žádost o informaci podle zákona č. 106/1999 Sb. Chvaletice: Chvaletice (2015). (In Czech). http://www.chvaletice.cz/modules/file_storage/download.php?file=3a4f50b5%7C131. Accessed 30 July 2018
13. Constrained Application protocol. http://coap.technology/
14. MQTT - Message Queuing Telemetry Transport. http://mqtt.org
15. Extensible Messaging and Presence Protocol. https://xmpp.org/
16. Happ, D., Karowski, N., Menzel, T., Handziski, V., Wolisz, A.: Meeting IoT platform requirements with open pub/sub solutions. Ann. Telecommun. **72**(1–2), 41–52 (2017). https://doi.org/10.1007/s12243-016-0537-4. ISSN 0003-4347. Accessed 24 Aug 2018
17. Năstase, L., Sandu, I.E., Popescu, N.: An experimental evaluation of application layer protocols for the internet of things. Stud. Inf. Control **26**(4), 403–412 (2017). https://doi.org/10.24846/v26i4y201704. ISSN 12201766. Accessed 24 Aug 2018
18. Crespi, N., Manzalini, A., Secci, S.: Proceedings of the 2017 20th International Conference on Innovations in Clouds, Internet and Networks (ICIN), 7–9 March 2017 in Paris, France. IEEE, Piscataway (2017)
19. Safaei, B., Monazzah, A.M.H., Bafroei, M.B., Ejlali, A.: 2017 2nd International Conference on System Reliability and Safety, ICSRS 2017, 20–22 December 2017, Milan, Italy. IEEE, Piscataway (2017)
20. Swamy, S.N., Jadhav, D., Kulkarni, N.: Proceedings of the International Conference on IoT in Social, Mobile, Analytics and Cloud (I-SMAC 2017), 10–11 February 2017. IEEE, Piscataway (2017)
21. Yokotani, T., SASAKI, Y.: Comparison with HTTP and MQTT on required network resources for IoT. In: 2016 International Conference on Control, Electronics, Renewable Energy and Communications (ICCEREC), pp. 1–6. IEEE (2016). ISBN 978-1-5090-0744-8. https://doi.org/10.1109/iccerec.2016.7814989. Accessed 24 July 2018
22. ČSN EN 50160 ED. 3. Charakteristiky napětí elektrické energie dodávané z veřejných distribučních sítí. (2011). (In Czech)

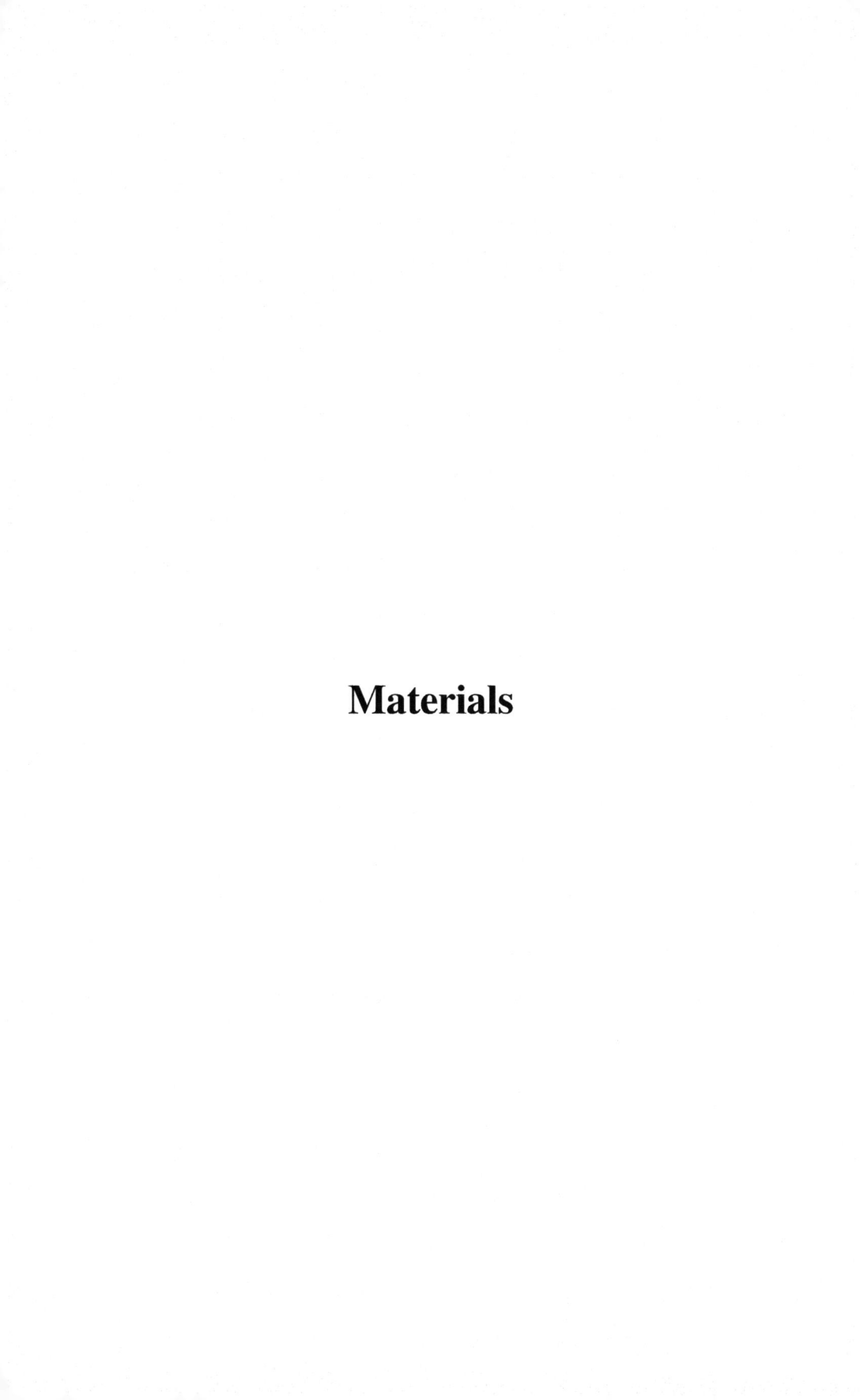

Materials

Temperature Dependence of Microstructure in Liquid Aluminosilicate

Mai Van Dung[2,4], Le The Vinh[1(✉)], Vo Hoang Duy[1],
Nguyen Kieu Tam[1], Tran Thanh Nam[1], Nguyen Manh Tuan[2],
Truong Duc Quynh[3], and Nguyen Van Yen[5]

[1] Faculty of Electrical and Electronics Engineering, Ton Duc Thang University,
No. 19 Nguyen Huu Tho Street, Tan Phong Ward, District 7,
Ho Chi Minh City, Vietnam
{lethevinh,vohoangduy,nguyenkieutam,
tranthanhnam}@tdtu.edu.vn
[2] Institute of Applied Materials Science, Vietnam Academy of Science
and Technology, No. 1A TL29 Street, Thanh Loc Ward, District 12,
Ho Chi Minh City, Vietnam
[3] Ho Chi Minh of University Transport, No. 2 Vo Oanh Street, Ward 25,
Binh Thanh District, Ho Chi Minh City, Vietnam
[4] Thu Dau Mot University, No. 6, Tran Van on Street, Phu Hoa Ward,
Thu Dau Mot City, Binh Duong Province, Vietnam
[5] Institute of Research and Development,
Duy Tan University, Da Nang 550000, Vietnam

Abstract. The structure of liquid $Al_2O_3.2SiO_2$ (AS2) have been investigated by means molecular dynamics simulation with the Born-Mayer potential at different temperatures. The structural characteristics are analyzed via the partial radial distribution functions, coordination number, bond angle and bond length distributions. The results show that, the structure of the liquid aluminosilicate consist the basic structural units TO_x (T = Al, Si; x = 3, 4, 5). The fraction of TO_x units have a small change, in which the shape and size of the basic structural units are identical and do not depended on temperature. Calculations also show that calculated data agree well with the experimental ones.

Keywords: Structure · Materials · Temperature · Molecular dynamics · Spatial distribution

1 Introduction

Silica (SiO_2) is the main component and most important in the glass and ceramics industry. The results of molecular dynamics (MD) simulations show that the network structure of SiO_2 almost is basic structural unit SiO_4 at low pressure. As pressure increases, the network structure of SiO_2 consists SiO_4, SiO_5 and SiO_6 units. It means there is a gradual transformation from tetrahedral- to octahedral-network structure and SiO_5 unit thought to be structural defects. Therefore the characteristics of network forming liquids depend on both the fraction and the spatial distributions of SiO_x (x = 4, 5, 6) units. The difference of fraction of SiO_x units and network structure leads to the polymorphisms and the dynamical heterogeneity in liquid SiO_2 [1–10]. For alumina (Al_2O_3), the results of MD simulations reveal that the network structure of Al_2O_3 mainly consists AlO_4 and AlO_5 units and it also comprises a small fraction of AlO_3 and AlO_6 units at low pressure, as pressure increases the network structure of Al_2O_3 main comprises AlO_5 and AlO_6 units. This means as pressure increases there is a gradual transformation from tetrahedral-to octahedral-network structure [11–14].

Recently, there are many studies for silicate materials which are its mixtures of SiO_2 with various oxides (Al_2O_3, MgO, Na_2O, etc.). In which $Al_2O_3.2SiO_2$ materials as well as other multi-component oxides play an important role in many fields of the new materials [15–17]. Therefore, a lot of studies experimental and simulated focus on the structural units TO_x (T = Al or Si; x = 4, 5, 6). The results show that the structure of aluminosilicate consist basic structural units TO_x, in which basic structural units TO_4 are dominant. The fraction of this structure units significant changes as the pressure, Al_2O_3 content and temperature increase. However, the size and shape of basic structural units TO_x are identical and do not depended on the composition, pressure and temperature [18–37].

However, up to now information about the structure of aluminosilicate materials under influence of the temperature still many the debate. In this paper, we will clarify the structure of aluminosilicate materials under influence of temperature by means molecular dynamics simulation and analytical structure.

2 Calculation Method

The model $Al_2O_3.2SiO_2$ (abbreviated as AS2) consist of 2090 atoms (380 Si, 380 Al and 1330 O atoms) was constructed by means molecular dynamics simulation with boundary condition and the potential interaction Born-Mayer. Details about this potential can be found in reference [11],

$$U_{ij} = \frac{q_i q_j}{r_{ij}} + A_{ij} \exp(-B_{ij} r_{ij}) \qquad (1)$$

with i, j = Al, Si, O, r is interaction distance. The initial configuration is obtained by randomly placing all atoms in a simulation box. This sample is equilibrated the temperature of 7000 K for 50.000 MD steps. After that, this sample was cooled down to the temperature of 2100 K. A consequent long relaxation was performed in the NPT ensemble to obtain a sample at ambient pressure. The model at different temperature were constructed by heating the model the temperature of 2100 K to different temperature and the relaxed for a long time to reach the equilibrium. In order to improve the statistics of the measure, the parameters characteristics are computed by averaging over 1000 configurations separated by 5 MD steps.

3 Results and Discussion

To assure the reliability, the structural and characteristics are calculated and compared to experimental data as well as other simulation results. These characteristics of liquid AS2 show in Table 1.

Table 1. Comparison the simulation results with the experimental data.

r_{ij}(Å)	This paper	Simulation [22]	Experiment [18]
Al-O	1.64	1.66	1.69
Si-O	1.58	1.60	1.61

Table 2. The bond length of atomic pairs at different temperatures

r_{ij}(Å)						
T(K)	Si-Si	Si-O	O-O	Si-Al	O-Al	Al-Al
2100	3.14	1.58	2.62	3.18	1.64	3.16
2300	3.14	1.60	2.64	3.16	1.64	3.16
2500	3.18	1.60	2.62	3.16	1.66	3.14
2700	3.16	1.60	2.62	3.16	1.64	3.14
2900	3.16	1.60	2.62	3.16	1.64	3.10
3100	3.18	1.58	2.66	3.18	1.66	3.16
3300	3.16	1.58	2.64	3.16	1.66	3.16
3500	3.16	1.58	2.64	3.16	1.64	3.14

The pair radial distribution function of all atomic pairs of the liquid AS2 at 2100, 2700, 3100 and 3500 K show in Fig. 1. In general, the height of the first peaks of all atomic pairs decrease as the temperature increases and becomes broader, except Al-Al. This means the degree of structural order decreases as the temperature increases. The atomic pairs Al-Al, Si-O, Si-Al, Si-Si and Al-O have the height of the first peaks decrease significantly, in while the atomic pair O-O decreases not significantly and the position of the first peaks of all atomic pairs also has a small change as the temperature increases.

The bond length of all atomic pairs of liquid AS2 display in Table 2. The result shows that the bond length of all atomic pairs change not significantly with temperature, this means the position of the first peaks of all atomic pairs are almost unchanged as the temperature increases. This result is consistent with the result in Fig. 1. The bond length of the atomic pairs Al-O and Si-O are 1.64 and 1.58 Å respectively at 3500 K, as well as the bond length of the atomic pairs remaining in the 2100–3500 K temperature range are in good agreement with the experimental and simulated results in references [18, 21–26]. The Si-Si, Al-Al, Si-Al and O-O bond lengths relate to short range order (SRO) and the Si-O and Al-O bond lengths relate to intermediate range order (IRO), this means that the SRO and IRO change not significantly under influence of the temperature.

Figure 2a shows the distribution of the SiO_x basic structural units at the different temperature. In which, at 2100 K most of the Si atoms are surrounded by four O atoms (94%) forming the SiO_4 coordination unit, this fraction slightly increases as the temperature increases and at 3500 K is about 97%. In while the fraction of the SiO_3 and SiO_6 units are very small, the SiO_5 unit has the fraction at about 6.17% at 2100 K and these fractions tend to decrease as the temperature increases. This means in the model the SiO_4 unit is dominant. These results are in good agreement with the results in references [21, 22]. So, the fraction of the SiO_4 and SiO_5 units slightly change with the increasing temperature. This can cause lead to the self-diffusion coefficient of Si and O atoms are small and form the immobile regions. In Fig. 2b displays the distribution of AlO_x units. At 2100 K the Al atoms are surrounded by three, four and five O atoms forming AlO_3, AlO_4 and AlO_5 coordination units. The fraction of AlO_3, AlO_4 and AlO_5 units are 12%, 75% and 12% respectively. As the temperature increases, the fraction of AlO_4 and AlO_5 units decreases, meanwhile the fraction of AlO_3 unit increases. At 3500 K the fractions of AlO_3, AlO_4 and AlO_5 units are 28%, 63% and 8% respectively, these values are in agreement with the result in reference [21]. This means that as the temperature increases the fraction of high coordination units decreases, in contrast the fraction of low coordination units increases. So, the fraction of AlO_x units have a significant change with the temperature, this can also lead to the self-diffusion coefficient of Al and O atoms in AlO_x unit are bigger the self-diffusion coefficient of Si and O atoms in SiO_x and form the mobile regions. This result is similar with prediction in reference [22].

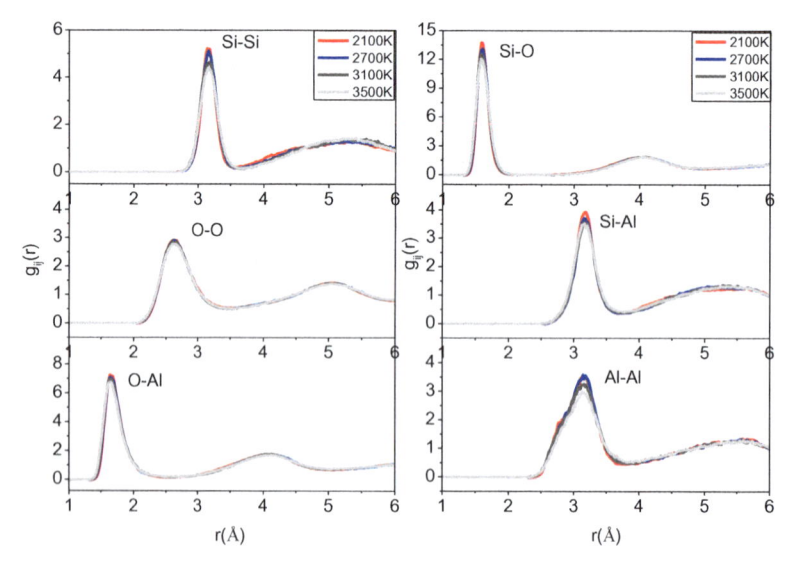

Fig. 1. Radial distribution functions of $Al_2O_3.2SiO_2$ pairs at different temperatures.

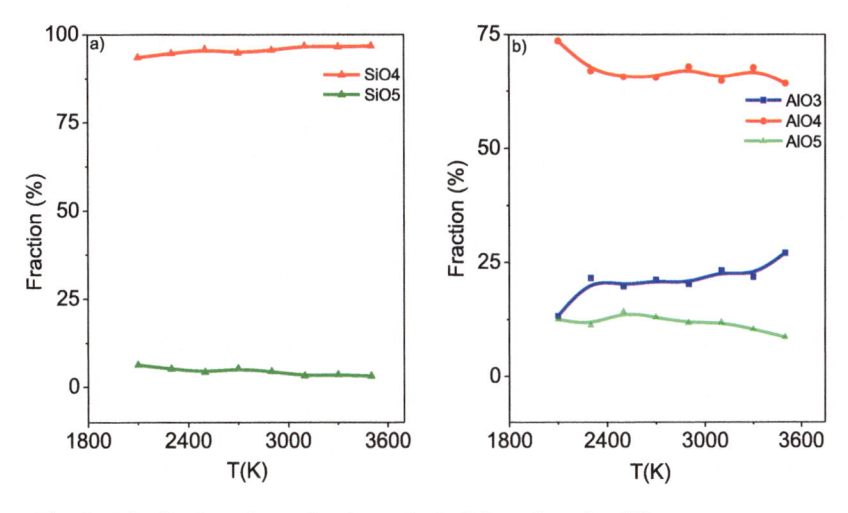

Fig. 2. Distribution of coordination units in TO_x unit at the different temperatures

Figure 3 displays the O-Si-O bond angle distribution in SiO_x units at the different temperature. The result shows that the O-Si-O bond angle distribution in SiO_4 unit has a peak at about 105°. This result is agreement with the experimental and simulated results in references [21, 35–37]. Besides, the height of peak of the O-Si-O bond angle distribution has a small change as the temperature increases. For SiO_5 unit at 2100 K, the O-Si-O bond angle distribution has a first peak at about 92° and a shoulder at about 156°, this result similar with the simulated result in reference [21]. The first peak shifts from the location of 92° to the one of 90° as the temperature increases. This means SiO_5 unit is slightly distorted and the position of the peak has a small change, in which the height of the peak significant change as the temperature increases. This cause is due to the fraction of SiO_5 unit has a significant change with the increasing temperature.

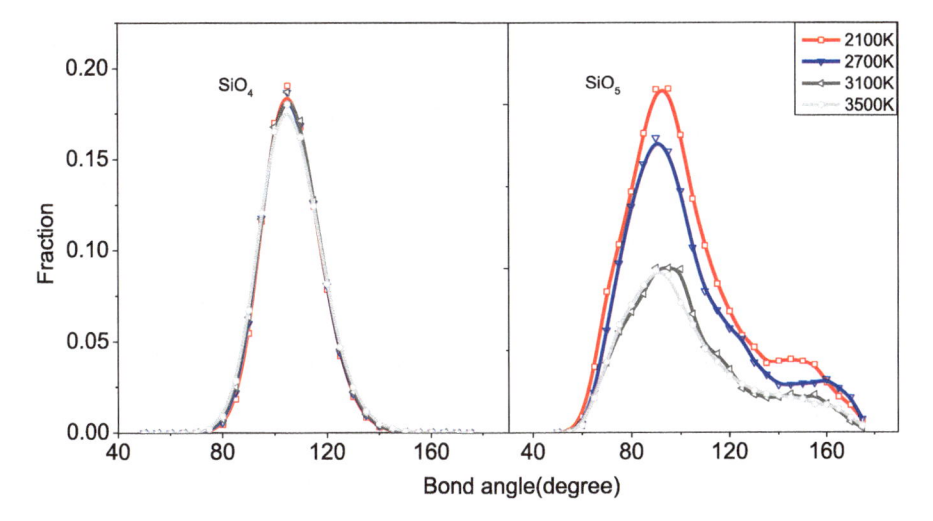

Fig. 3. Distribution of the bond angle in SiO_x unit at the different temperatures.

The Si-O bond length distribution in SiO_x units shows in Fig. 4. In which, the Si-O bond length in SiO_4 and SiO_5 units are 1.60 and 1.62 Å respectively at 2100 K. This reveals that the Coulomb repulsive force between anion and anion in SiO_5 unit is much stronger than in SiO_4 unit, this is cause of the increase of Si-O bond length in SiO_5 unit. As the temperature increases the Si-O bond length distribution in SiO_4 unit has the height of peak slightly decreases and the position of peak is almost unchanged, meanwhile the position of peak in SO_5 sifts from at 1.62 Å to at 1.66 Å. This means as the temperature increases the Si-O bond length increases, this can cause lead to significant change the diffusion of atoms. The height of peak of the Si-O bond length distribution in SiO_5 unit has a significantly change, this is due to the fraction of SiO_5 unit has a significantly change as the temperature increases.

Figure 5 displays the distribution of O-Al-O bond angle in AlO_x units at the different temperatures. The result shows that the O-Al-O bond angle in AlO_x units is strongly change as the temperature increases, in which the position of peak of O-Al-O bond angle in AlO_3 and AlO_4 units have the peaks at 115° and 105° respectively and almost unchanged as the temperature increases. For AlO_5 unit has a peak at 100° and the position of peak sifts to the position of peak at 95° as temperature increases, this means AlO_5 unit is distorted as the temperature increases. As the temperature increases the height of peak of the O-Al-O bond angle in AlO_4 and AlO_5 units decrease, in while the height of peak of O-Al-O bond angle in AlO_3 unit increases. This is due to the fraction of AlO_3, AlO_4 and AlO_5 units significant change with the increasing temperature. This result is similar with the result in reference [21].

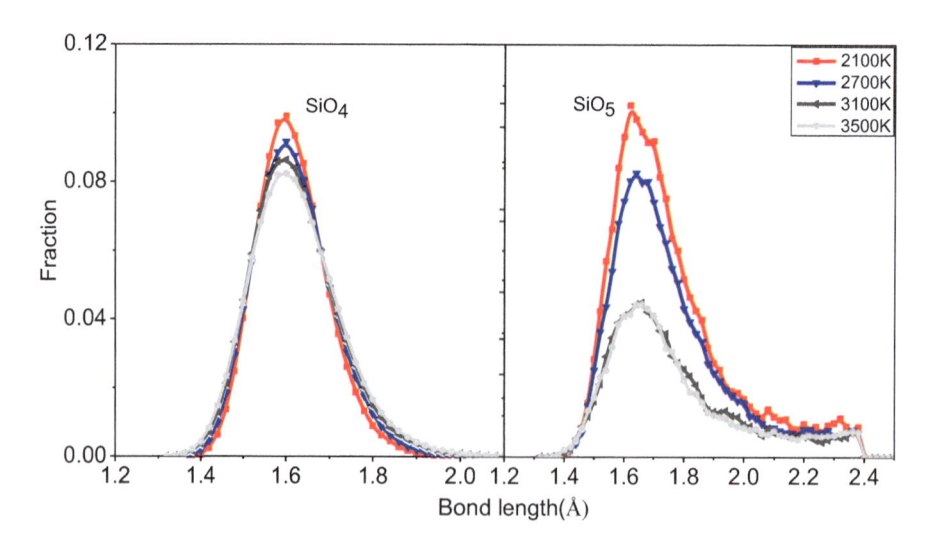

Fig. 4. Distribution of the bond length in SiO_x unit at the different temperatures.

Figure 6 shows the Al-O bond length in AlO_x units at the different temperatures. The position of peaks of the Al-O bond length distribution in AlO_3, AlO_4 and AlO_5 units have the values about 1.62, 1.68 and 1.70 Å respectively. This means the average

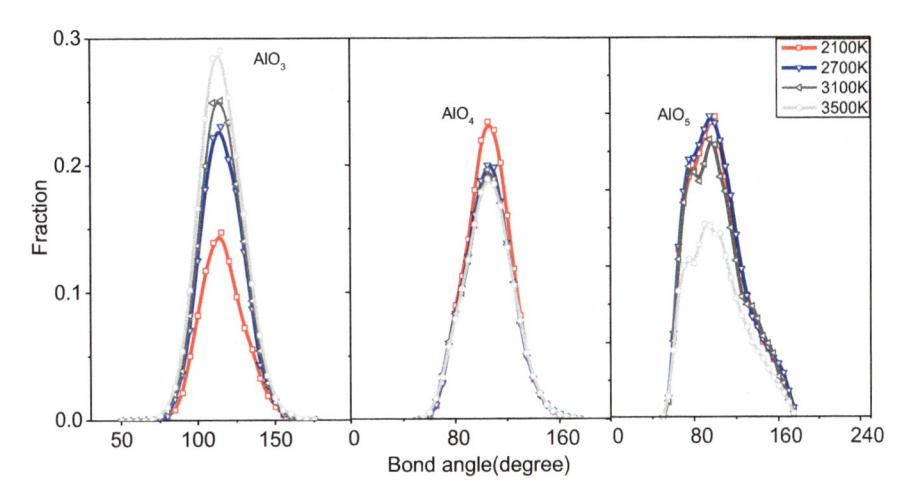

Fig. 5. Distribution of bond angle in AlO_x unit at the different temperatures.

distances between Al and O atoms are about 1.62, 1.68 and 1.70 Å respectively, these values show that the Coulomb repulsive force between anion and anion in AlO_4 and AlO_5 units is stronger than in AlO_3 unit and this is one of the cause leads to the Al-O bond length in AlO_4 and AlO_5 units is bigger than in AlO_3 unit. As the temperature increases the position of peaks of the Al-O bond length distribution in AlO_3 and AlO_4 units almost unchanged, in which the position of peak of the Al-O bond length distribution in AlO_5 unit increases. The height of peaks of the Al-O bond length distribution in AlO_4, AlO_5 units decrease and in AlO_3 unit increases. This is due to the fraction of AlO_3, AlO_4 and AlO_5 units significant change as the temperature increases.

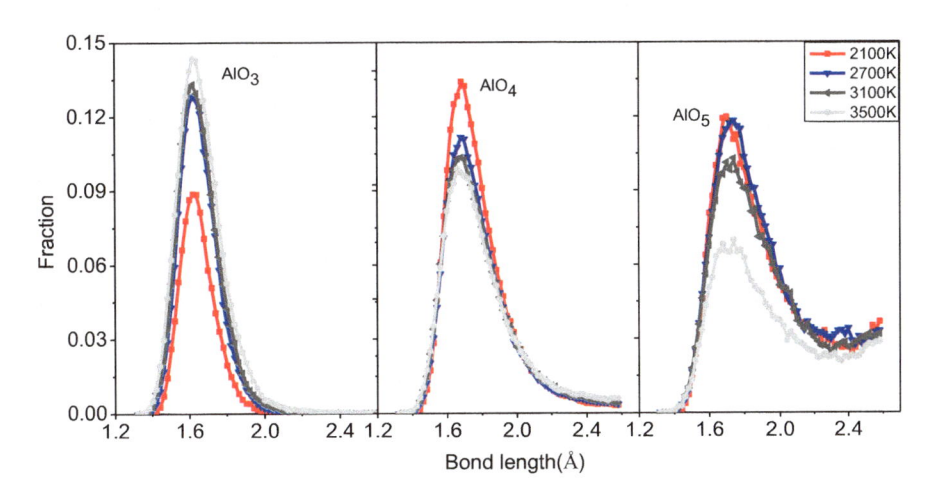

Fig. 6. Distribution of bond length in AlO_x unit at the different temperatures.

4 Conclusion

The models are constructed in the 2100–3500 K temperature range. The network structure of liquid AS2 consist the basic structure units of SiO_x (x = 4, 5) and AlO_y (y = 3, 4, 5) units. The distribution of TO_x units are not uniform at different temperatures.

The results also show the shape and size in AlO_3, AlO_4 and SiO_4 units are almost unchanged under temperatures, in contrast the shape and size of SiO_5 and AlO_5 units are slightly distorted as the temperature increases. The simulated results are in good agreement with previously the experimental and simulation data.

Acknowledgment. This research is funded by Vietnam National Foundation for Science and Technology Development (NAFOSTED) under grant number 103.05-2017.345.

References

1. Horbach, J., Kob, W.: Static and dynamic properties of a viscous silica melt. Phys. Rev. B **60**, 3169–3181 (1999)
2. Oligschleger, C.: Dynamics of SiO_2 glasses. Phys. Rev. B **60**, 3182–3193 (1999)
3. Vollmayr-Lee, K., Zippelius, A.: Temperature-dependent defect dynamics in the network glass SiO_2. Phys. Rev. E **88**, 052145 (2013)
4. Koziatek, P., Barrat, J.L., Rodney, D.: Short- and medium-range orders in as-quenched and deformed SiO_2 glasses: an atomistic study. J. Non-Cryst. Solids **414**, 7–15 (2015)
5. Jin, W., Kalia, R.K., Vashishta, P., et al.: Structural transformation in densified silica glass: a molecular-dynamics study. Phys. Rev. B **50**, 118–131 (1994)
6. Sato, T., Funamori, N.: High-pressure structural transformation of SiO_2 glass up to 100 Gpa. Phys. Rev. B **82**, 184102 (2010)
7. Trachenko, K., Dove, M.T.: Densification of silica glass under pressure. J. Phys.: Condens. Matter **14**, 7449–7459 (2002)
8. Inamura, Y., Arai, M., Nakamura, M., et al.: Intermediate range structure and lowenergy dynamics of densified vitreous silica. J. Non-Cryst. Solids **293–295**, 389–393 (2002)
9. Liang, Y., Miranda, C.R., Scandolo, S.: Mechanical strength and coordination defects in compressed silica glass: molecular dynamics simulations. Phys. Rev. B **75**, 024205 (2007)
10. Trachenko, K., Dove, M.T.: Compressibility, kinetics, and phase transition in pressurized amorphous silica. Phys. Rev. B **67**, 064107 (2003)
11. Gutierrez, G., Belonoshko, A.B., Ahuja, R., et al.: Structural properties of liquid Al_2O_3: a molecular dynamics study. Phys. Rev. E **61**(3), 2723–2729 (2000)
12. Hoang, V.V.: About an order of liquid–liquid phase transition in simulated liquid Al_2O_3. Phys. Lett. A **335**, 439–443 (2005)
13. Hemmati, M.: Structure of liquid Al_2O_3 from a computer simulation model. J. Phys. Chem. B **103**, 4023–4028 (1999)
14. Vashishta, P., Kalia, R.K., Nakano, A., et al.: Interaction potentials for alumina and molecular dynamics simulations of amorphous and liquid alumina. J. Appl. Phys. **103**, 083504 (2008)
15. Kushiro, I.: Changes in viscosity and structure of melt of $NaAlSiO_6$ composition at high pressures. J. Geophys. Res. **81**, 6347 (1976)

16. Watson, E.B.: Calcium diffusion in a simple silicate melt to 30 kbar. Geochim. Cosmochim. Acta **43**, 313 (1979)
17. Watson, E.B.: Diffusion in magmas at depth in the earth: the effects of pressure and dissolved H_2O. Earth Planet. Sci. Lett. **52**, 291 (1981)
18. Morikawa, H., Miwa, S.I., Miyake, M., Marumo, F.: Structural analysis of SiO_2-Al_2O_3. J. Am. Ceram. Soc. **65**, 78 (1982)
19. Okuno, M., Zotov, N., Schmucker, M., Schneider, H.: Structure of SiO_2–Al_2O_3 glasses: combined X-ray diffraction, IR and Raman studies. J. Non-Cryst. Solids **351**, 1032 (2005)
20. Hong, N.V., Yen, N.V., Lan, M.T., Hung, P.K.: Coordination and polyamorphism of aluminium silicate under high pressure: insight from analysis and visualization of molecular dynamics data. Can. J. Phys. **92**, 1573–1580 (2014)
21. Mai, L.T., Yen, N.V., Hong, N.V., Hung, P.K.: Visualisation based analysis of structure and dynamics of liquid aluminosilicate under compression. Phys. Chem. Liq. **55**(1), 62–84 (2017)
22. Winkler, A., Horbach, J., Kob, W., et al.: Structure and diffusion in amorphous aluminum silicate: a molecular dynamics computer simulation. J. Chem. Phys. **120**, 384–393 (2004)
23. Hoang, V.V., Linh, N.N., Hung, N.H.: Structure and dynamics of liquid and amorphous Al_2O_3.$2SiO_2$. Eur. Phys. J. Appl. Phys. **37**, 111–118 (2007)
24. Linh, N.N., Hoang, V.V.: Evolution of structure of liquid and amorphous Al_2O_3.$2SiO_2$ nanoparticles upon cooling from the melts. World Sci. **2**(4), 227–232 (2007)
25. Hoang, V.V.: Dynamical heterogeneity and diffusion in high-density Al_2O_3.$2SiO_2$ melts. Physica B **400**, 278–286 (2007)
26. Hoang, V.V., Hung, N.H., Linh, N.N.: Liquid–liquid phase transition in simulated liquid Al_2O_3·$2SiO_2$. Phys. Scr. **74**, 697–701 (2006)
27. Narayanan, B., Reimanis, I.E., Ciobanu, C.V.: Atomic-scale mechanism for pressure-induced amorphization of β-eucryptite. J. Appl. Phys. **114**, 083520 (2013)
28. Grandi, S., Costa, L.: Lanthanide-doped SiO_2–Al_2O_3 aerogels and densified glasses. J. Non-Crystall. Solids **225**, 141–145 (1998)
29. Yang, Y., Takahashi, M., Abe, H., Kawazoe, Y.: Structural, electronic and optical properties of the Al_2O_3 doped SiO_2: first principles calculations. Mater. Trans. **49**(11), 2474–2479 (2008)
30. Boe, P.T., Mcmillan, P.F.: Al and Si coordination in SiO_2-Al_2O_3 glasses and liquids: a study by NMR and IR spectroscopy and MD simulations. Chem. Geol. **96**, 333–349 (1992)
31. Binder, K., Horbach, J., Winkler, A., Kob, W.: Modeling glass materials. Ceram. Int. **31**, 713–717 (2005)
32. Shimoda, K., Saito, K.: Detailed structure elucidation of the blast furnace slag by molecular dynamics simulation. ISIJ Int. **47**, 1275–1279 (2007)
33. Zheng, K., Zhang, Z., Yang, F., Sridhar, S.: Molecular dynamics study of the structural properties of calcium aluminosilicate slags with varying Al_2O_3/SiO_2 ratios. ISIJ Int. **52**(3), 342–349 (2012)
34. Takei, T., Kameshima, Y., Yasumori, A., Okada, K.: Crystallization kinetics of mullite from Al_2O_3– SiO_2 glasses under non-isothermal conditions. J. Mater. Res. **15**(1) (2000)
35. Pfleiderer, P., Horbach, J., Binder, K.: Structure and transport properties of amorphous aluminium silicates: computer simulation studies. Chem. Geol. **229**, 186–197 (2006)
36. Bauchy, M.: Structural, vibrational, and elastic properties of a calcium aluminosilicate glass from molecular dynamics simulations: the role of the potential. J. Chem. Phys. **141**(2), 024507 (2014)
37. Tossell, J.A., Cohen, R.E.: Calculation of the electric field gradients at tricluster-like O atoms in the polymorphs of Al_2SiO_5 and in aluminosilicate molecules: models for tricluster O atoms in glasses. J. Non-Cryst. Solids **286**, 187–199 (2001)

Study on Effect of Parameters on Friction Stir Welding Process of 6061 Aluminum Alloy Tubes

Van Vu Nguyen[1], Hoang Linh Nguyen[1], Tan Tien Nguyen[1], Thien Phuc Tran[1(✉)], and Sang Bong Kim[2]

[1] Ho Chi Minh City University of Technology, Ho Chi Minh City, Vietnam
ttphuc.rectie@hcmut.edu.vn
[2] Pukyong National University, Busan, Korea
kimsb@pknu.ac.kr

Abstract. This paper presents the experimental results on friction stir welding (FSW) with 6061 aluminum alloy cylinder. This operation is implemented on a universal milling machine with specialized holder and the stepless spindle speed which can be adjusted by using frequency converter.

Two pieces of aluminum tube with the same dimensions ($\phi 100$ mm × 5 mm × 40 mm) are facing machined and located in mandrel by screw clamp in order to be concentric and stable. The mandrel is installed on a continuously variable transmission with various speeds. The model was designed to find out the best value of each technical parameters such as tool speed, traverse speed, welding shoulder radius. This paper also presents the effect of technical parameters to the tensile strength, normal force and the welding force of FSW.

Keywords: Friction stir welding · 6061 aluminum tube · Spindle speed · Traverse speed · Radius of tool shoulder · Tensional stress

1 Introduction

The friction stir welding (FSW) method is used in aeronautics and petrochemical industry. FSW is known as a green technology with a high efficiency and environmental protection due to the fact that this operation does not emit arc and welding fumes, and do not require mask. FSW is a non-heated welding technology which produces pre-eminent welding links, especially in the case of poor weldability metals or alloys such as aluminum alloy, copper alloy, etc. (Fig. 1).

Moreover, FSW do not require electrode to fill the welding links, which makes the welding links less deformation and fracture (Fig. 2).

Nowadays, this method is used commonly for aluminum alloys because of the superiority compared to fusion welding method. Take aeronautics industry as an example, plasma arc welding method has a high defect rate while FSW method reduce dramatically this defect rate to nearly zero [2].

In previous study, FSW method is applied for plate sheet materials. This study focuses on FSW used for 6061 aluminum alloys tube with small diameter.

Fig. 1. Friction stir welding of aluminum tube

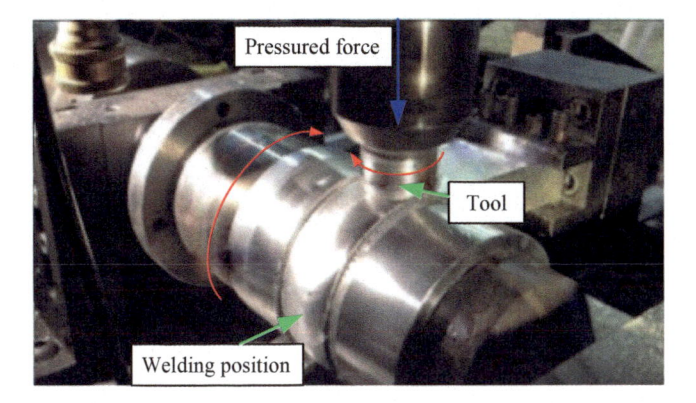

Fig. 2. Design for friction stir welding

2 Mathematical Model

2.1 Mathematical Thermal Equilibrium Model

Mathematical Thermal Model of Stir Tool. The contact between welding tool and materials produces heat in FSW process. Heat can be created at different positions on welding tool, divided into three field (see Fig. 3) [2].
Total heat created by the welding tool

$$Q_{total} = Q_1 + Q_2 + Q_3 \qquad (1)$$

To determine the value of each term

$$dQ = \omega dM = \omega r dF \qquad (2)$$

where, ω: rotation speed, dF: force exerting on the surface at the position r from the center of tool.

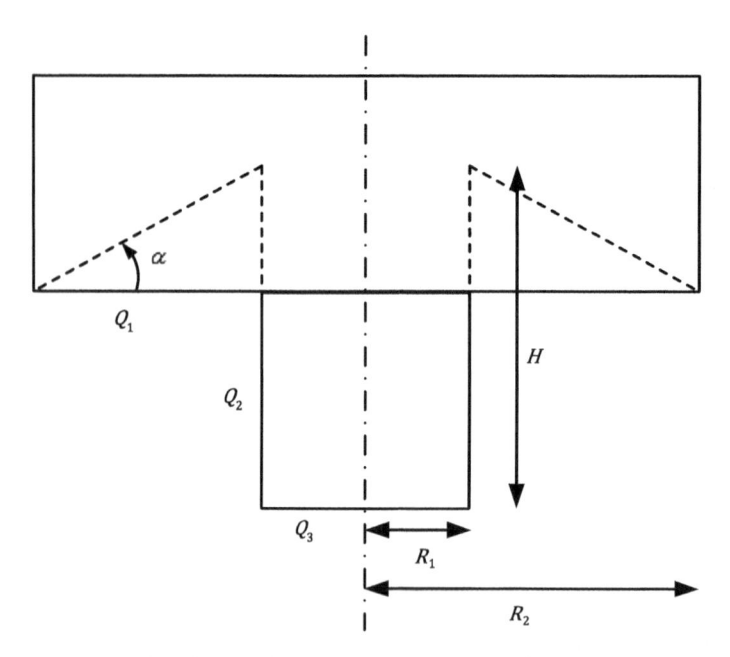

Q_1: heat created under the shoulder (W) Q_2: head created at side of tool (W)
Q_3: head created at end of tool (W) R_1: radius of tool
R_2: radius of shoulder

Fig. 3. Heat distribution on welding tool surface

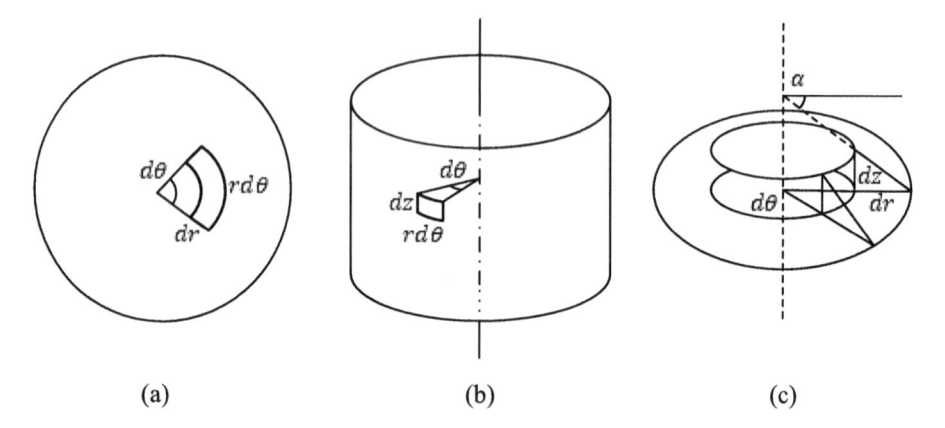

(a)   (b)   (c)

Fig. 4. Geometry of tool and shoulder

Vertical differential surface of tool are presented in Fig. 4(a). $dA = r.d\theta.dr$ is exerted by the vertical force or moment, which produces a differential heat.

$$dQ_3 = \omega r dF_{vertical} = \omega r^2 \tau_{contact} d\theta dr \tag{3}$$

So that

$$Q_3 = \int_{r=0}^{R_1} \int_{\theta=0}^{2\pi} \omega r^2 \tau_{contact} d\theta dr = \frac{2}{3}\pi \tau_{contact} \omega R_1^3 \tag{4}$$

Figure 4(b) the differential area of the pin $dA = rd\theta dz$ is acted by the horizontal force or moment, create

$$dQ_2 = \omega R_1^2 \tau_{contact} d\theta dz \tag{5}$$

$$Q_2 = \int_{\theta=0}^{2\pi} \int_{z=0}^{H} \omega R_1^2 \tau_{contact} d\theta dz = 2\pi \tau_{contact} \omega R_1^2 \tag{6}$$

The conical surface in Fig. 4(c) with differential area

$$dA = rd\theta(dz+dr) = rd\theta(dr + dr tan\alpha) = rd\theta dr(1+tan\alpha) \tag{7}$$

This area is exerted by the vertical and horizontal force, produces the heat

$$dQ_1 = \omega r^2 \tau_{contact}(1+\tan\alpha)drd\theta \tag{8}$$

Integrate the differential equation above

$$\begin{aligned}Q_1 &= \int_{r=R_1}^{R_2} \int_{\theta=0}^{2\pi} \omega R_1^2 \tau_{contact}(1+tan\alpha)drd\theta \\ &= 2\pi(1+tan\alpha)\tau_{contact}(R_2^3 - R_1^3)\end{aligned} \tag{9}$$

The total heat

$$\begin{aligned}Q_{total} &= Q_1 + Q_2 + Q_3 \\ &= \frac{2}{3}\pi \tau_{contact} \omega ((R_2^3 - R_1^3)(1+tan\alpha) + R_1^3 + 3R_1^2 H)\end{aligned} \tag{10}$$

If the material is cooled lower than the critical temperature where the deformation by yield stress is higher than the shear stress by friction, the interaction between tool and workpiece could be changed from deformation to friction. If the shear phase happen between welding tool and workpiece, the temperature will be reduced. The roster of boundary condition at contact section could result in the thermal disequilibrium and the oscillation between stickiness and sliding.

- Sticky condition: the net sticks on the tool if the friction shear stress higher than yield shear stress. In this case, one part of the net will increase along to the tool's surface until that the equilibrium between tangential shear stress and internal shear stress.

- Sliding condition: if the friction shear stress lower than yield shear stress, a net volume will experience a deformation where shear stress equal to tangential dynamics shear stress.
- Simultaneous sticky and sliding condition: the velocity of one part of net is lower than the velocity of tool's surface. The equilibrium is established while tangential dynamics shear stress equal to yield shear stress.

Temperature at sticky condition

$$\tau_{contact} = \tau_{yield} = \frac{\sigma_{yield}}{\sqrt{3}} \quad (11)$$

So that:

$$Q_{total,sticky} = \frac{2}{3}\pi \frac{\sigma_{yield}}{\sqrt{3}} \omega\left((R_2^3 - R_1^3)(1+\tan\alpha) + R_1^3 + R_1^3 H\right) \quad (12)$$

Temperature at sliding condition:

$$\tau_{contact} = \tau_{friction} = \mu.P \quad (13)$$

So that

$$Q_{total\ sliding} = \frac{2}{3}\pi\mu P\omega\left((R_2^3 - R_1^3)(1+\tan\alpha) + R_1^3 + R_1^3 H\right) \quad (14)$$

where, μ: friction coefficient and P: acting force (N).

Mathematical Thermal Model of the End of Stir Tools. With welding tube, the shoulder of tool do not contact completely with the workpiece [3]. Total input heat is determined by using Schmidt's equation, basing on the sliding contact condition on surface (Schimidt, Hattel and West) [5].

$$Q_A = \tau_{contact}\omega AD \quad (15)$$

where,
- $\tau_{contact}$: sliding contact stress of workpiece, $\tau_{contact} = \mu P$
- Q_A: heat created at area A
- μ: friction coefficient
- P: acting force
- ω: spindle speed

For a simple one head stir head cylinder tools, the equation becomes

$$Q_{total} = \frac{2}{3}\pi\tau_{contact}\omega\left(R_2^3 + 3R_1^2 H\right) \quad (16)$$

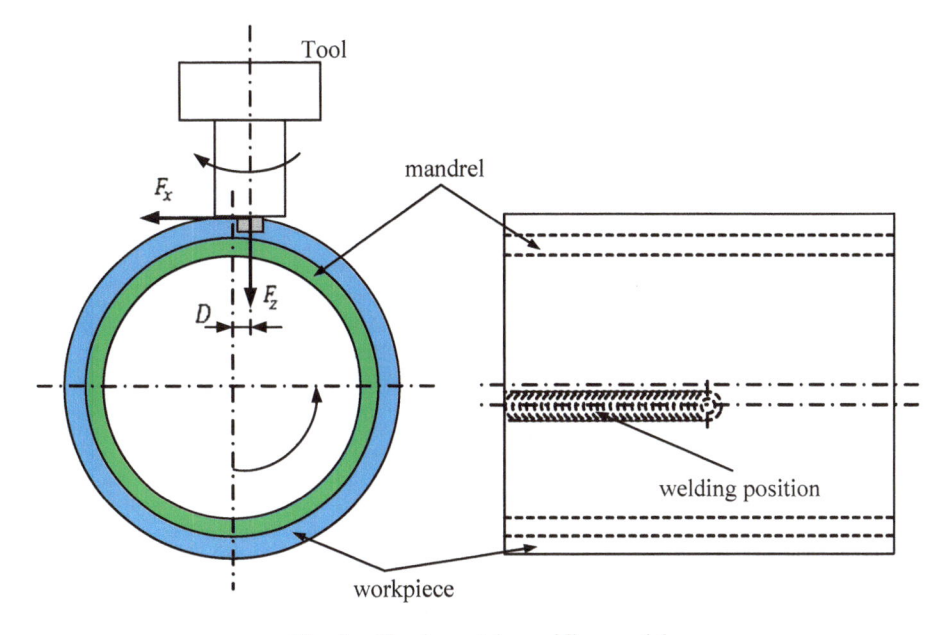

Fig. 5. Aluminum tube welding model

Where R_2 is radius of tool shoulder, R_1 is radius of stir head and H is height of stir head. This area is determined by θ (Fig. 5).

The contactless area can be calculated

$$A = \frac{1}{2}R_2^2(\theta - sin\theta) \qquad (17)$$

Distance between center of this area and workpiece's rotary axis

$$D = \frac{4R sin\left(\frac{\theta}{2}\right)^3}{3(\theta - \sin(\theta))} \qquad (18)$$

The acting force P can be calculated by divide the normal force of tools

$$P = \frac{F_z}{A_x} = \frac{F_z}{\pi R^2 - A} \qquad (19)$$

For the experiment condition, Schmidt's equation become

$$Q_{tool} = \mu P \omega \left[\frac{2}{3}\pi\left(R^3 + 3r^2 h\right) - AD\right] \qquad (20)$$

According to the mathematical thermal equilibrium model, the effect of input parameters on welding link can be observed: (1) Geometry parameters: size and profile of friction welding tool, radius of tool shoulder, length of friction welding tool; (2) Kinematic and dynamic parameters: spindle speed, traverse speed, holding force, eccentricity; and (3) Welding material parameter: size of workpiece, friction coefficient.

In this research, three selected parameters affect to welding temperature is: spindle speed, radius of tool shoulder, and traverse speed.

2.2 Experiment

Two pieces of aluminum tube must have the same $\phi 100$ mm diameter, 5 mm thickness 40 mm length, face-milled are fixed by nut. It is fixed with a stepless-speed driven shaft, which can change to various speeds [1] (Fig. 6).

Fig. 6. Machine, holder and workpiece

2.3 Design Option for Experiment Parameters Value Field

Spindle Speed of Tool n(rpm). The temperature is not high enough at $n < 985$(rpm). At $n > 2200$(rpm), the welding links overheated. So, spindle can be chosen from 1200 to 1600 rpm.

Traverse Speed (mm/min). For a $\phi 100$ mm aluminum tube, these parameters can be chosen from 25 mm/min to 125 mm/min.

Radius of Tool Shoulder (mm). 7 – 15 mm, and a fixed eccentricity 6 mm. Therefore, these parameters are determined and shown in Table 1.

Table 1. Parameters value field for tube-welding

Spindle speed n (rpm)	1000 rpm–2200 rpm
Traverse speed v (mm/min)	25 mm/min–125 mm/min
Radius of tool shoulder R_2 (mm)	7–15 mm

Table 2. Limited value field

Level	Low	Standard	High
Spindle speed n(rpm)	1200	1400	1600
Traverse speed v(mm/min)	75	87,5	100
Radius of tool shoulder R_2(mm)	9	10	11

2.4 Experiment

- **Step 1** (Table 1). Investigate value of parameter for tube welding.
- **Step 2** (Table 2). From the result of harsh experimental results, limit the parameter value.
- **Step 3.** Choose the optimal y_1 (welding force F_x), y_2 (normal force F_z), y_3 (tensile stress σ_k), and three element affect to optimal parameter radius of tool shoulder $R_2(x_1)$, spindle speed $n(x_2)$ and traverse speed $v(x_3)$.
- **Step 4.** Choose second order regression equation model for each component.

2.5 Tensile Strength Testing for Welding Link

The welded aluminum tube will be cut into four equal part as Fig. 7. The area of doughnut $= \frac{\pi r^2 \alpha}{360}$, tensile stress $\sigma = \frac{F}{S}$.

2.6 Normal Force F_z and Welding Force F_x Testing

Holders are positioned on a system include two loadcell to measure the normal force F_z, one loadcell on side to measure the force F_x along the welding path (Fig. 8).

The loadcell's signal is display on the monitor and connect with a computer to collect the data.

2.7 Analyze the Result

In this experiment, we choose the optimal parameters y_1 (welding force F_x), y_2 (normal force F_z), y_3 (tensile stress σ_k) and three elements affect to the optimal parameters radius of tool shoulder $R_2(x_1)$, spindle speed $n(x_2)$ and traverse speed $v(x_3)$. Choose second order regression equation model, 3^k type, as

$$\bar{y} = b_0 + b_1 x_1 + b_2 x_2 + b_3 x_3 + b_{12} x_1 x_2 + b_{13} x_1 x_3 \\ + b_{23} x_2 x_3 + b_{11} x_1^2 + b_{22} x_2^2 + b_{33} x_1 x_2 x_3 \tag{21}$$

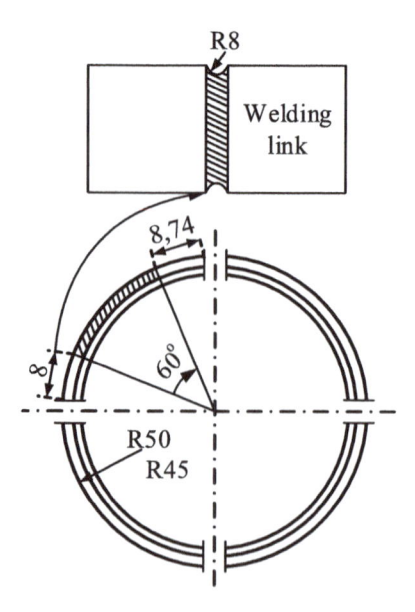

Fig. 7. Specifications of the sample

Fig. 8. Force measuring system

With 3_k type method, we must to do $3^3 = 27$ experiments. To calculate and evaluate the significance level of the equation's coefficient, this process is repeated twice (Figs. 9, 10 and 11).

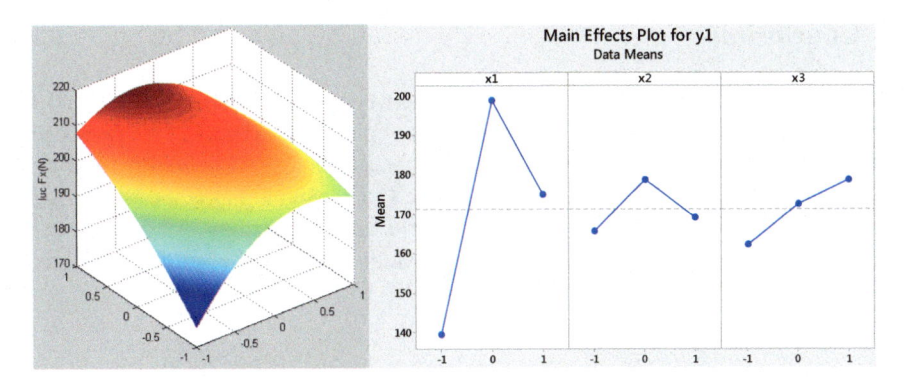

Fig. 9. The effect of technical parameter on welding force F_x

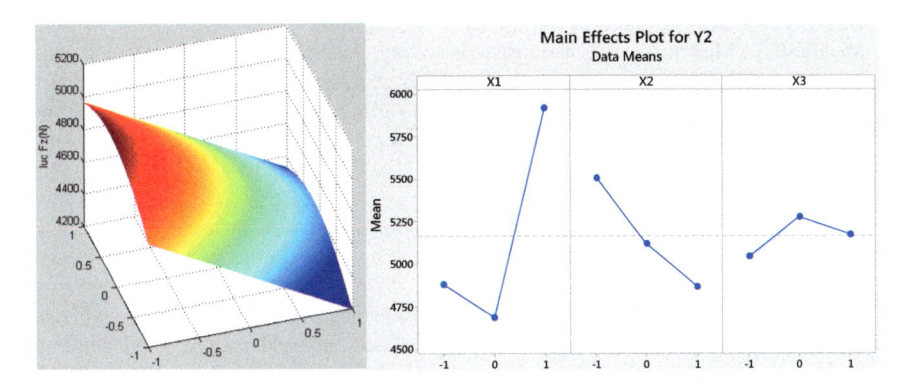

Fig. 10. The effect of technical parameter on normal force F_z

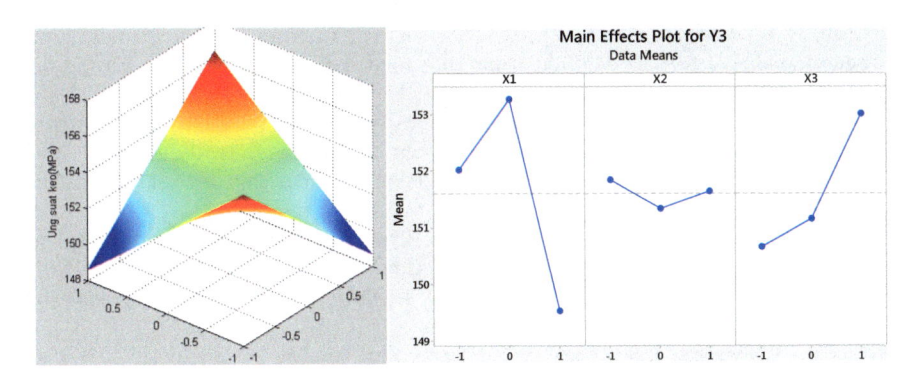

Fig. 11. The effect of technical parameters on tensile stress σ_k

3 Conclusion

The paper has determined: (1) The relation between practical tensile stress and the technical parameters radius of tool's shoulder R_2, spindle speed n and traverse speed v. The results show that the welding links tensile stress achieve 58% – 70% comparing to the original tubes; (2) The practical normal force F_z and welding force F_x for aluminum alloy tube. Moreover, the quality of welding links are evaluated by using radiographic method with three samples. The experimental study provides the set of optimal technical parameters for FSW of aluminum tube.

References

1. Thomas, W.M., et al.: Friction stir butt welding, International Patent Application PCT/GB92, Patent Application GB9125978.8, 6 December 1991
2. Mishra, R.S.: Minerals, metals and materials society, minerals, metals and materials society. In: Friction Stir Welding and Processing VI: Proceedings of a Symposium Sponsored by the Shaping and Forming Committee of the Materials Processing & Manufacturing Division of TMS (The Minerals, Metals & Materials Society): Held During the TMS 2011 Annual Meeting & Exhibition, San Diego, California, USA, 27 February–3 March 2011. Wiley, Hoboken (2011)
3. Yuan, S.J., Hu, Z.L., Wang, X.S.: Evaluation of formability and material characteristics of aluminum alloy friction stir welded tube produced by a novel process. Mater. Sci. Eng., A **543**, 210–216 (2012)
4. Lindner, K., Khandkar, Z., Khan, J., Tang, W., Reynolds, A.P.: Rationalization of hardness distributions in alloy 7050 friction stir welds based on weld energy, weld power and time/temperature history. In: 4th International Symposium on Friction Stir Welding (4ISFSW), Park City, Utah, USA, 14–16 May 2003
5. Schmidt, H., Hattel, J., Wert, J.: An analytical model for the heat generation in friction stir welding. Model. Simul. Mater. Sci. Eng. **12**(1), 143 (2004)
6. Chen, H.-B., Yan, K., Lin, T., Chen, S.-B., Jiang, C.-Y., Zhao, Y.: The investigation of typical welding defects for 5456 aluminum alloy friction stir welds. Mater. Sci. Eng., A **433**, 64–69 (2006)
7. Smith, C.B., Mishra, R.S.: Friction Stir Processing for Enhanced Low Temperature Formability. Butterworth-Heinemann is an imprint of Elsevier, Amsterdam (2014)
8. Mishra, R.S., Mahoney, M.W.: Friction Stir Welding and Processing. In: ASM International (2007)
9. Chen, B., Chen, K., Hao, W., Liang, Z., Yao, J., Zhang, L., Shan, A.: Friction stir welding of small-dimension Al3003 and pure Cu pipes. J. Mater. Process. Technol. **223**, 48–57 (2015)
10. Rai, R., De, A., Bhadeshia, H.K.D.H., DebRoy, T.: Review: friction stir welding tools. Sci. Technol. Weld. Join. **16**, 325–342 (2011)
11. Kumar, N., Mishra, R.S., Yuan, W.: Friction Stir Welding of Dissimilar Alloys and Materials. Elsevier, Amsterdam (2015)
12. Norrish, J.: Advanced Welding Processes: Technologies and Process Control. Woodhead Publ., Cambridge (2006)

Convergence Study of Different Approaches of Solving the Hartree-Fock Equation on the Potential Curve of the Hydrogen Fluoride

Martin Mrovec[1,2(✉)]

[1] FEECS, Department of Applied Mathematics, VŠB - Technical University of Ostrava, 17. listopadu 15, 708 33 Ostrava - Poruba, Czech Republic
martin.mrovec@vsb.cz

[2] IT4Innovations National Supercomputing Center, VŠB - Technical University of Ostrava, 17. listopadu 15, 708 33 Ostrava - Poruba, Czech Republic

Abstract. The aim of the paper is to compare the convergence of chosen numerical methods, namely the Direct Inversion of the Iterative Subspace and the Inexact Restoration Method, for solving the nonlinear eigenvalue problem occurring in the electronic structure calculations. We have selected the Hartree-Fock approximation where the behavior of the energy functional is known. The numerical experiments are performed on the modeling of the potential curve of the Hydrogen fluoride molecule. The results will be used as a clue for the development of optimization methods in the area of the Density Functional Theory.

Keywords: Electronic structure calculations · Direct Inversion of the Iterative Subspace · Inexact Restoration Method · Hartree-Fock approximation

1 Introduction

Study of electronic structures of materials has a significant influence to many engineering fields. As an example we should mention semiconductor components as an integral part of modern electronic devices. Their physical properties are well described by principles of the Quantum Mechanics. Understanding the behavior of electrons is also necessary for studying mechanical properties of materials [1]. In Electronic Structure Calculations (ESC) we encounter many mathematical challenges. ESC can be categorized using several criteria. Depending on the structure of the particle system, ESC can be divided to calculations of periodic systems (crystalline solids) and calculations of finite systems (molecules). Furthermore, there exist many approximation approaches, such as Hartree-Fock (HF) method, post Hartree-Fock methods or the Density Functional Theory (DFT) that are used in both fields [2]. Our research is focused on the development of mathematical methods in the ESC of finite systems where we have primarily used the HF method and currently we are getting first results from the DFT.

Generally, each ESC is structured as follows. First we have to choose a finite dimensional basis (or a direct spatial discretization). This choice determines how matrices of given operators are precomputed. First problems occur when we try to choose different functions than Gaussian-type orbitals (GTO) [2], because we have to calculate those matrices numerically. Such calculations are computationally very demanding. We solve this problem by using the Tensor Numerical Methods (TNM) combined with a parallelization. The whole precomputing process is out of the scope of this paper. For more details see e.g. [3–5].

When all the matrices are precomputed we obtain a nonlinear eigenvalue problem that has to be solved. Currently, the most commonly used approach is the iteration of the Self Consistent Field (SCF) accelerated by the Direct Inversion of the Iterative Subspace (DIIS) [6]. Since its introduction, the DIIS approach has been extensively developed. As a reference to current algorithms we should mention [7–9]. Despite significant improvements we may still encounter convergence issues related to DIIS (even in current prominent computational codes, such as Molpro [10]). Thus, we should consider an alternative approach, e.g. the optimization one, where (in the case of the HF method) the problem is transformed to a Quadratically Constrained Quadratic Program (QCQP). There are known methods, such as Trust Region Methods [11], for solving such kind of problem. We use methods inspired by the Inexact Restoration Method (IRM) published in [12].

Each of the mentioned approaches has its pros and cons. The aim of this study is to localize problematic areas of the convergence in the case of the HF method. The results will serve as an important clue for numerical experiments with optimization methods in the area of the DFT which we are currently working on. The idea of using the optimization methods in the DFT calculations already exists. We should mention [13] as one of the state of the art methods based on the Conjugate Gradient Method (CGM). Our goal is to modify the IRM algorithm to be usable within DFT calculations as well.

As a testing suite we have chosen a modeling of the potential curve of the Hydrogen fluoride molecule. It means that we do the ESC for a couple of distances between the nuclei and watch the convergence of the solution of the nonlinear eigenproblem and of the optimization methods. The choice of such a system is justified by its simplicity and thus computational cheapness. Simultaneously, such system is fully sufficient for determining problematic areas of the convergence.

The text is structured as follows. At first, in Sect. 2, we briefly describe the eigenproblems and the QCQP that come from the HF approximation. In Sect. 3 we describe methods of their solution. In Sect. 4 we demonstrate results of our calculations and compare mentioned approaches. Finally the conclusions for the application of the optimization method in the area of DFT are made.

2 Problem Setting

As we have mentioned in the introduction, we are going to describe the second part of the ESC process. That means we assume that for a given electronic system the precomputing phase has been done so we are interested in the structure of the following numerical problem. This section focuses on a brief description of the standard eigenproblem approach and the alternative optimization one.

2.1 Standard Eigenproblem Approach

The discretized form of the HF equations using a standard Galerkin approach with N_b basis functions can be written as

$$\begin{cases} \mathbf{F}(\mathbf{C})\,\mathbf{C} = \mathbf{SC}\boldsymbol{\Lambda}, \\ \mathbf{C}^T\mathbf{SC} = \mathbf{I} \end{cases} \quad (1)$$

with coefficient matrix (representing the solution) $\mathbf{C} \in \mathbb{R}^{N_b \times N}$, $(\mathbf{C})_{ij} = c_{ij}$, overlap matrix $\mathbf{S} \in \mathbb{R}^{N_b \times N_b}$, diagonal matrix $\boldsymbol{\Lambda} = \text{diag}(\lambda_1, \ldots, \lambda_N)$ with eigenvalues and Fock matrix $\mathbf{F} \in \mathbb{R}^{N_b \times N_b}$,

$$\mathbf{F}(\mathbf{C}) = \mathbf{H} + \mathbf{J}(\mathbf{C}) + \mathbf{K}(\mathbf{C}) \quad (2)$$

where submatrices $\mathbf{H}, \mathbf{J}, \mathbf{K}$ represent discretized operators of the HF equation. All the data required for their assembly are obtained during the precomputing phase. We should notice that those matrices are assembled using matrix multiplication operations (for more details see [3]). The second line of Eq. (1) represents an orthogonality condition which naturally comes from the first line, thus it can be omitted. We have mentioned it because of its relation to the constraints occurring in the optimization approach described below. Equation (1) is in principle a nonlinear generalized eigenproblem where we want to find N lowest eigenvalues and corresponding eigenvectors. The nonlinearity consists in the fact that the Fock matrix depends on coefficient matrix $\mathbf{C}$ itself. Such a problem has to be solved iteratively. We chose matrix $\mathbf{C_0}$ as an arbitrary initial guess and perform an eigenvalue calculation with fixed matrix $\mathbf{F}(\mathbf{C_0})$. As a result, we obtain a coefficient matrix which in some way has to be used in the next iteration. This process is described in Sect. 3.1.

We would like to notice that the Kohn-Sham equation that comes from the DFT can be represented by the similar eigenproblem. The only difference is the way how matrix $\mathbf{F}$ is constructed from the coefficient matrix $\mathbf{C}$

$$\mathbf{F}(\mathbf{C}) = \mathbf{H} + \mathbf{J}(\mathbf{C}) + V_{xc}(\rho(\mathbf{C})). \quad (3)$$

In the case of the DFT a construction of matrix $\mathbf{F}$ includes an additional exchange-correlation term $V_{xc}(\rho(\mathbf{C}))$ [2]. In contrast with the HF approximation, it cannot be evaluated as a simple matrix multiplication. It requires an additional numerical integration which makes it computationally more demanding.

2.2 Optimization Approach

An alternative to the standard approach is the optimization approach based on a direct minimization of the functional representing the enegry of the system.

Using the same notation as in previous subsection, the Hartree-Fock energy functional can be written as

$$E(\mathbf{Z}) = \text{Trace}\left[2\mathbf{HZ} + (\mathbf{J}(\mathbf{Z}) + \mathbf{K}(\mathbf{Z}))\mathbf{Z}\right] \quad (4)$$

where matrix $\mathbf{Z} \in \mathbb{R}^{N_b \times N_b}$ is defined as $\mathbf{Z} := \mathbf{CC}^T$. It should be noticed that matrices which are dependent on the density matrix $\mathbf{Z}$ are calculated in the same way as in the case of the eigenproblem approach. We use a different notation, because the knowledge of matrix $\mathbf{C}$ is not needed in the optimization approach. For the purposes of the optimization we should mention the gradient of the functional in Eq. (4) with respect to matrix $\mathbf{Z}$:

$$\nabla_{\mathbf{Z}} E(\mathbf{Z}) = 2(\mathbf{H} + \mathbf{J}(\mathbf{Z}) + \mathbf{K}(\mathbf{Z})) = 2\mathbf{F}(\mathbf{Z}). \quad (5)$$

The solution of the eigenproblem in Eq. (1) is equivalent to finding a minimum of the functional in Eq. (4) with respect to equality constraints

$$\mathbf{Z} = \mathbf{Z}^T, \quad (6)$$

$$\mathbf{ZSZ} = \mathbf{Z}, \quad (7)$$

$$\text{Trace}(\mathbf{ZS}) = N. \quad (8)$$

The structure of the problem can be simplified using substitution $\mathbf{X} = \mathbf{S}^{\frac{1}{2}}\mathbf{Z}\mathbf{S}^{\frac{1}{2}}$. Then we can simply write the optimization problem as

$$\text{find } \min f(\mathbf{X}) = E\left(\mathbf{S}^{-\frac{1}{2}}\mathbf{X}\mathbf{S}^{-\frac{1}{2}}\right) \quad (9)$$

subject to

$$\mathbf{X} = \mathbf{X}^T, \quad (10)$$

$$\mathbf{XX} = \mathbf{X}, \quad (11)$$

$$\text{Trace}(\mathbf{X}) = N, \quad (12)$$

which fulfills conditions of a QCQP.

Similarly to the previous subsection, we would like to mention the differences in the optimization approach if the DFT with the Kohn-Sham equation is used. The energy functional contains the exchange-correlation term which corresponds to the above mentioned one, thus $f(\mathbf{X})$ is no longer quadratic. However, the equality constraints are preserved so we can use similar optimization methods with just little modifications.

3 Methods of Solution

Each of the discussed approaches requires using of efficient numerical methods. In this section we focus on the DIIS algorithm which is the most popular method used for the eigenproblem approach. As an optimization method we use a modification of the IRM which we plan to develop outside the scope of the HF method, e.g. in DFT.

3.1 Solution of the Eigenproblem

As a first idea of solving problem in Eq. (1) one may consider a naive iteration of the SCF, which can be described by scheme:

$$\mathbf{F}\left(\mathbf{C}_{k-1}\right)\mathbf{C}_k = \mathbf{SC}_k \Lambda \tag{13}$$

where $k \in \mathbb{N}$ is a number of the iteration. Unfortunately, such algorithm usually does not converge. As a fix we may consider a simple mixing scheme

$$\mathbf{C}_{k+1} = (1-\alpha) \cdot \mathbf{C}^* + \alpha \cdot \mathbf{C}_k \tag{14}$$

where $\mathbf{C}_k$ has been used to assemble the Fock matrix, $\mathbf{C}^*$ is the solution of the current eigenproblem, $\alpha \in (0,1)$ is a parameter. The next iteration Fock matrix is assembled using matrix $\mathbf{C}_{k+1}$. The choice of parameter α significantly affects the convergence. In some cases we may encounter an oscillating behavior of iterates or a very slow convergence. Unfortunately we have no recipes for setting the right value of α before starting the iterative process. Current solvers use a more sophisticated mixing approach DIIS introduced by Pulay in 1980 [6]. The algorithm is based on the idea of representing the next iterate density matrix $\mathbf{Z}_{k+1}$ as a linear combination of previous iterates $\mathbf{Z}_i$, $i = 1, \ldots, k$. Let us denote $\mathbf{e}_i$ the error matrix of the i-th iterate $\mathbf{Z}_i$ (difference between $\mathbf{Z}_i$ and the exact solution). We would like to find coefficients α_i of the linear combination of the previous error matrices which approximates the zero matrix in the least-square sense and the sum of the coefficients is equal to 1 (see [6]). Such requirements lead to the system of linear equations

$$\begin{bmatrix} B_{11} & B_{12} & \cdots & B_{1k} & -1 \\ B_{21} & B_{22} & \cdots & B_{2k} & -1 \\ \vdots & \vdots & \ddots & \vdots & \vdots \\ B_{k1} & B_{k2} & \cdots & B_{kk} & -1 \\ -1 & -1 & \cdots & -1 & 0 \end{bmatrix} \begin{bmatrix} \alpha_1 \\ \alpha_2 \\ \vdots \\ \alpha_k \\ \lambda \end{bmatrix} = \begin{bmatrix} 0 \\ 0 \\ \vdots \\ 0 \\ -1 \end{bmatrix} \tag{15}$$

where $B_{ij} = \langle \mathbf{e}_i, \mathbf{e}_j \rangle$ is a dot product of error matrices (e.g. Frobenian dot product). As we do not know the exact error formula, we have to approximate it. For example, we can substitute it by residual

$$\mathbf{e}_i = \mathbf{F}(\mathbf{Z}_i)\mathbf{Z}_i\mathbf{S} - \mathbf{SZ}_i\mathbf{F}(\mathbf{Z}_i) \tag{16}$$

which is equal to zero matrix if $\mathbf{Z}_i$ represents the ground state density [3]. As a stopping criterion for the iterative process we may for example choose the norm of the residue $\mathbf{e}_{k+1}$.

The DIIS method is very popular and belongs to the state-of-the-art algorithms for solving system in Eq. (1). However, it does not give us a certainty of the convergence. Furthermore, we cannot be sure that the solution is correct. The reason is, that excited states also fulfill necessary conditions for the solution. This phenomenon is demonstrated in Sect. 4.

3.2 Solution of the Optimization Problem

In Sect. 1 we have mentioned the state of the art algorithm based on the CGM published in [13]. The optimization method we use is based on the IRM that has been published sooner in [12]. These two algorithms are based on similar ideas (Grassmann Manifold, tangent space, linesearch). The reason why we have chosen the IRM is that it comes with ideas that are not mentioned in the current CGM-based algorithm, such as projection to the Grassmann Manifold without calculating the matrix inverse or using of the Lagrangian for finding the descent direction.

First we are going to describe the fundamental structure of the method. The algorithm is based on two alternating steps. Given an iterate $\mathbf{X}_i$ that is not necessarily in the feasible set (set of points, that fulfill equality constraints), the first step (restoration phase) finds a point $\mathbf{Y}_i$, that is closer to the feasible set or that directly belongs to the feasible set (as the projection of $\mathbf{X}_i$ onto the feasible set in the ideal case). Next, in the optimality phase, an appropriate descent direction $\mathbf{E}_i$ is found within the tangent set which can be characterized as

$$S(\mathbf{Y}) = \{\mathbf{E} \in \mathbb{R}^{N_b \times N_b} | \mathbf{E} = \mathbf{E}^T \wedge \mathbf{YE} + \mathbf{EY} - \mathbf{E} = \mathbf{O}\}. \tag{17}$$

The next iterate $\mathbf{X}_{i+1}$ is searched in this direction and thus leaves the feasible set. These two phases are repeated until the convergence is reached.

Unlike the eigenproblem approach we do not need any eigenvalue calculation (with one exception). The restoration phase is performed using a simple iterative scheme

$$\mathbf{Y}_i^{j+1} = 3\left(\mathbf{Y}_i^j\right)^2 - 2\left(\mathbf{Y}_i^j\right)^3. \tag{18}$$

Before the iterative process we have to make a spectral radius estimation of matrix $\mathbf{X}_i$ which is much more cheap operation than calculation of N lowest eigenvalues. Within the optimization phase we have to find a descent direction in the tangent set. This can be done by projecting the gradient descent direction onto that using the following formula:

$$\mathbf{E}_i = \mathbf{Y}_i\left(-\nabla f(\mathbf{Y}_i)\right) + \left(-\nabla f(\mathbf{Y}_i)\right)\mathbf{Y}_i - 2\mathbf{Y}_i\left(-\nabla f(\mathbf{Y}_i)\right)\mathbf{Y}_i. \tag{19}$$

Finally the new iterate $\mathbf{X}_{i+1}$ is found using a simple linesearch procedure, such as the bisection method. As a stopping criterion we may choose a norm of the projection of the normalized gradient descent direction onto the tangent set.

4 Numerical Experiments

For the numerical experiments we have chosen the calculation of the potential curve of the Hydrogen fluoride. It means that we want to calculate the dependence of the ground state energy of the system on the distance between nuclei. The interval of distances we are interested is between 0.173 a_0[1] to 17.329 a_0.

---

[1] a_0 is Bohr radius (1 a_0 = $5.2917721067 \times 10^{-11}$ m). As an energy unit we use the Hartree energy (1 E_h = $4.359744650 \times 10^{-18}$ J). These units are commonly used in the quantum chemistry.

Potential curve of the Hydrogen fluoride

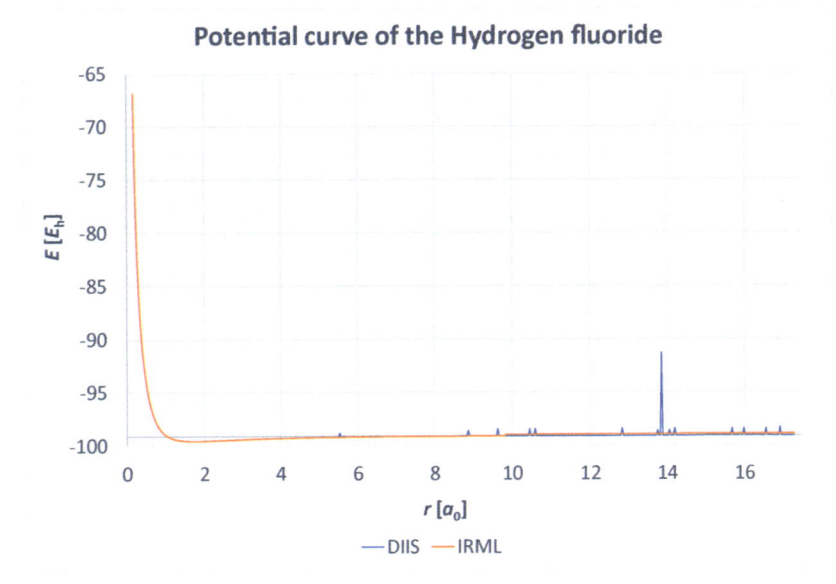

Fig. 1. Potential curve of the Hydrogen fluoride calculated using DIIS and IRM.

The interval has been divided into 500 points. For each distance we have performed an electronic structure calculation for obtaining the ground state energy. We have focused on the correctness of the result and on the number of the iterations required to get the same accuracy. The stopping criterion threshold has been set to 10^{-5}. As a basis set we have chosen 6–31G basis [14,15], which consisted of $N_b = 11$ contracted Gaussian-type orbital basis functions. Number of electrons is equal to 10, i.e. $N = 5$ eigenvectors had to be computed. In all cases the initial approximation has been chosen as the zero matrix. On Fig. 1 we can see the results of the calculation. The whole computational interval can be divided into several areas:

1. descending part $(0.173\ a_0, 1.789\ a_0)$,
2. ascending part $(1.789\ a_0, 4.000\ a_0)$,
3. ascending part $(4.000\ a_0, 9.000\ a_0)$,
4. ascending part $(9.000\ a_0, 17.329\ a_0)$.

We have observed that in first two subintervals the calculation resulted in the same values. However, in the second interval we can observe some differences. While the optimization approach (IRM) is stable, the eigenproblem approach encounter convergence issues. The detail of the third interval is shown on Fig. 2.

We may observe that in the case of DIIS algorithm the calculation reached an excited states at some points. This behavior can be undesirable especially if we are interested only in the ground state energy.

Conversely we can observe convergence issues in the optimization approach within the last interval. The detail of the third interval is shown on Fig. 3.

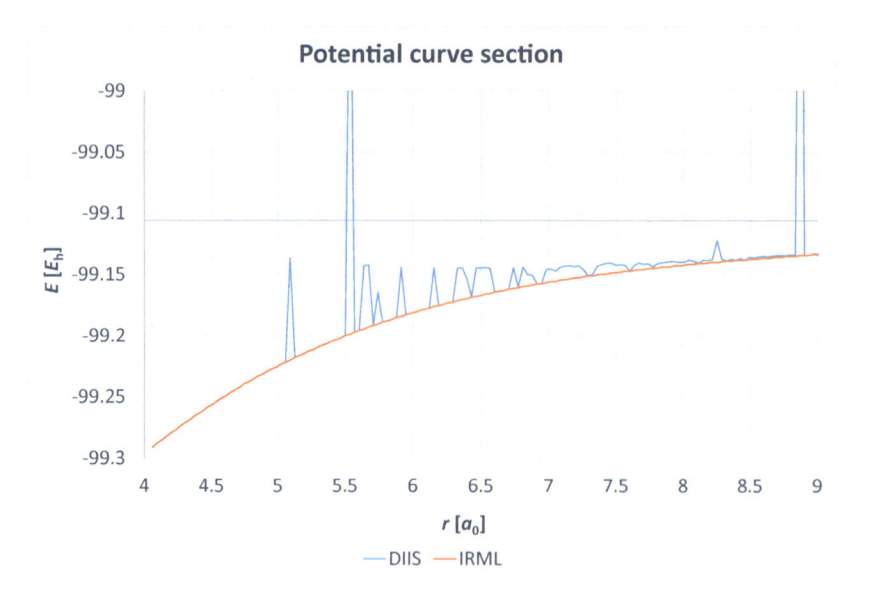

Fig. 2. Section (interval $(4.000\ a_0, 9.000\ a_0)$) of the potential curve of the Hydrogen fluoride calculated using DIIS and IRM.

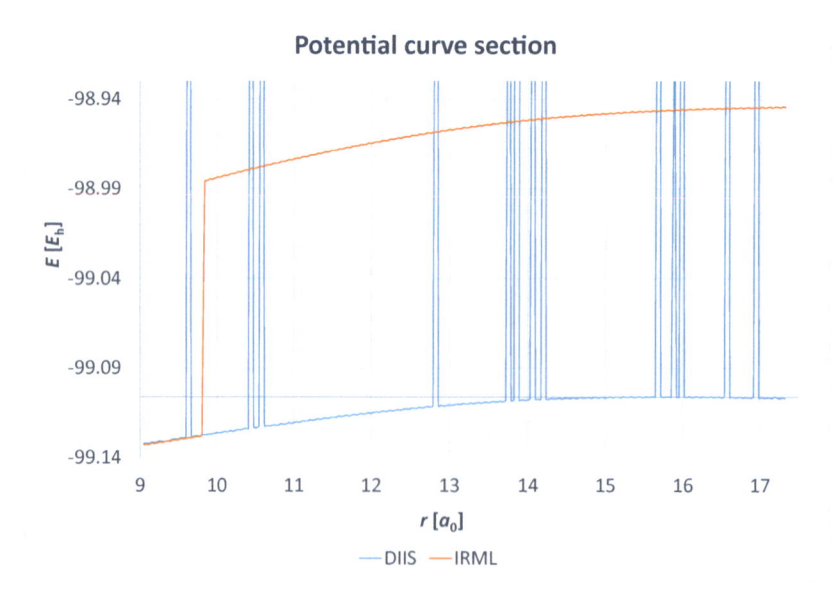

Fig. 3. Section (interval $(9.000\ a_0, 17.329\ a_0)$) of the potential curve of the Hydrogen fluoride calculated using DIIS and IRM.

Comparison of the number of iterations

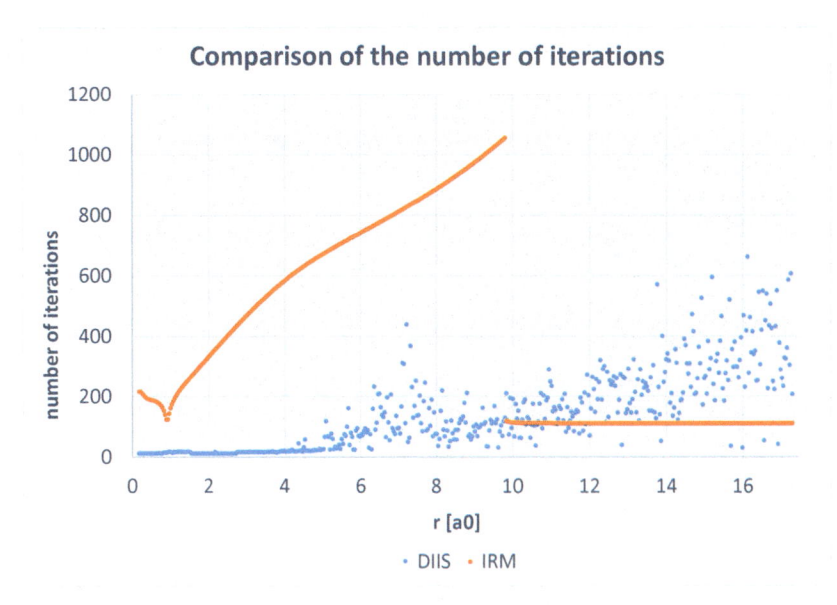

Fig. 4. Comparison of the number of iterations required to find a solution.

It is evident that the optimization approach is not able to reach the ground state energy for large distances. The DIIS method is much more efficient in this area. Looking at the graph of the number of iterations (Fig. 4), we can observe that the problematic area of the optimization approach is related to a significantly lower number of iterations required to converge. It suggests that there exists an excited state which is represented by a local minimum or a stationary point with a significant basin of attraction. Furthermore this basin becomes dominant within the feasible set as the internuclear distance grows. Fortunately, the last interval is not as physically interesting because it represents a dissociated system in principle. The last thing we can observe on the graph of the iterations is the lower number of iterations for the DIIS at first three intervals. However we have to take in account that the DIIS method requires performing of eigenvalue calculations which may become a computationally dominant part for larger systems of electrons. Especially at the third interval this fact is crucial.

5 Conclusion

The results of the study can be advantageously used for preparing the tests of mentioned algorithms in the area of the DFT calculations. As a starting point we should choose the area around the minimum of the potential well. There we can expect good convergence properties for both approaches. Furthermore the number of iterations is quite low. Then should follow experiments within the third interval, when we can expect more stability in a favor of the optimization approach. Finally there can be performed tests for large distances. However we expect the failure of the optimization approach.

Thanks to this study we does not have to perform the DFT calculations in such many points. We can instead concentrate on chosen intervals. If the results of the optimization approach in the area of the DFT is positive, we may concentrate its further development.

Acknowledgements. This work was supported by The Ministry of Education, Youth and Sports from the National Programme of Sustainability (NPS II) project IT4Innovations excellence in science - LQ1602, by Grant of SGS No. SP2018/165 and SP2018/178, VŠB - Technical University of Ostrava, Czech Republic. An important support has been provided by project OPEN-10-35 within the Open Access Call announced by IT4Innovations which enabled us to use the supercomputing services.

References

1. Pan, Y., Lin, Y.: Influence of re concentration on the mechanical properties of tungsten borides from first-principles calculations. JOM **69**(10), 2009–2013 (2017). https://doi.org/10.1007/s11837-017-2483-7. ISSN 1047-4838
2. Saad, Y., Chelikowsky, J.R., Shontz, S.M.: Numerical methods for electronic structure calculations of materials. SIAM Rev. (2010). https://doi.org/10.1137/060651653. ISSN 0036-1445, Zbl 1185.82004
3. Khoromskaia, V., Khoromskij, B.N., Schneider, R.: Tensor-structured factorized calculation of two-electron integrals in a general basis. SIAM J. Sci. Comput. **35**(2), A987–A1010 (2013). https://doi.org/10.1137/120884067. ISSN 1064-8275, Zbl 1266.65069
4. Mrovec, M.: Tensor approximation of slater-type orbital basis functions. Adv. Electr. Electron. Eng. **15**(2) (2017). https://doi.org/10.15598/aeee.v15i2.2235, http://advances.utc.sk/index.php/AEEE/article/view/2235. ISSN 1804-3119
5. Mrovec, M.: Low-rank tensor representation of Slater-type and Hydrogen-like orbitals. Appl. Math. **62**(6), 679–698 (2017). https://doi.org/10.21136/AM.2017.0177-17, http://articles.math.cas.cz/10.21136/AM.2017.0177-17
6. Pulay, P.: Convergence acceleration of iterative sequences. The case of SCF iteration. Chem. Phys. Lett. **73**(2), 393–398 (1980)
7. Hu, W., Lin, L., Yang, C.: Projected commutator DIIS method for accelerating hybrid functional electronic structure calculations. J. Chem. Theory Comput. **13**(11), 5458–5467 (2017). https://doi.org/10.1021/acs.jctc.7b00892. ISSN 1549-9618
8. Pratapa, P.P., Suryanarayana, P.: Restarted Pulay mixing for efficient and robust acceleration of fixed-point iterations. Chem. Phys. Lett. **635**, 69–74 (2015). https://doi.org/10.1016/j.cplett.2015.06.029. ISSN 00092614
9. Kudin, K.N., Scuseria, G.E., Cances, E.: A black-box self-consistent field convergence algorithm: one step closer. J. Chem. Phys. **116**(19), 8255 (2002). https://doi.org/10.1063/1.1470195. ISSN 00219606
10. https://www.molpro.net
11. Francisco, J.B., Martínez, J.M., Martínez, L.: Globally convergent trust-region methods for self-consistent field electronic structure calculations. J. Chem. Phys. **121**(22), 10863 (2004). https://doi.org/10.1063/1.1814935. ISSN 00219606

12. Martínez, J.M., Martínez, L., Pisnitchenko, F.: Inexact restoration method for minimization problems arising in electronic structure calculations. Comput. Optim. Appl. **50**(3), 555–590 (2011)
13. Dai, X., Liu, Z., Zhang, L., Zhou, A.: A conjugate gradient method for electronic structure calculations. SIAM J. Sci. Comput. **39**(6), A2702–A2740 (2017). https://doi.org/10.1137/16M1072929. ISSN 1064-8275
14. Hehre, W.J., Ditchfield, R., Pople, J.A.: Self-consistent molecular orbital methods. XII. Further extensions of gaussian-type basis sets for use in molecular orbital studies of organic molecules. J. Chem. Physi. **56**(5), 2257–2261 (1972). https://doi.org/10.1063/1.1677527. ISSN 0021-9606
15. Dill, J.D., Pople, J.A.: Self-consistent molecular orbital methods. XV. Extended Gaussian-type basis sets for lithium, beryllium, and boron. J. Chem. Phys. **62**(7), 2921–2923 (1975). https://doi.org/10.1063/1.430801. ISSN 0021-9606

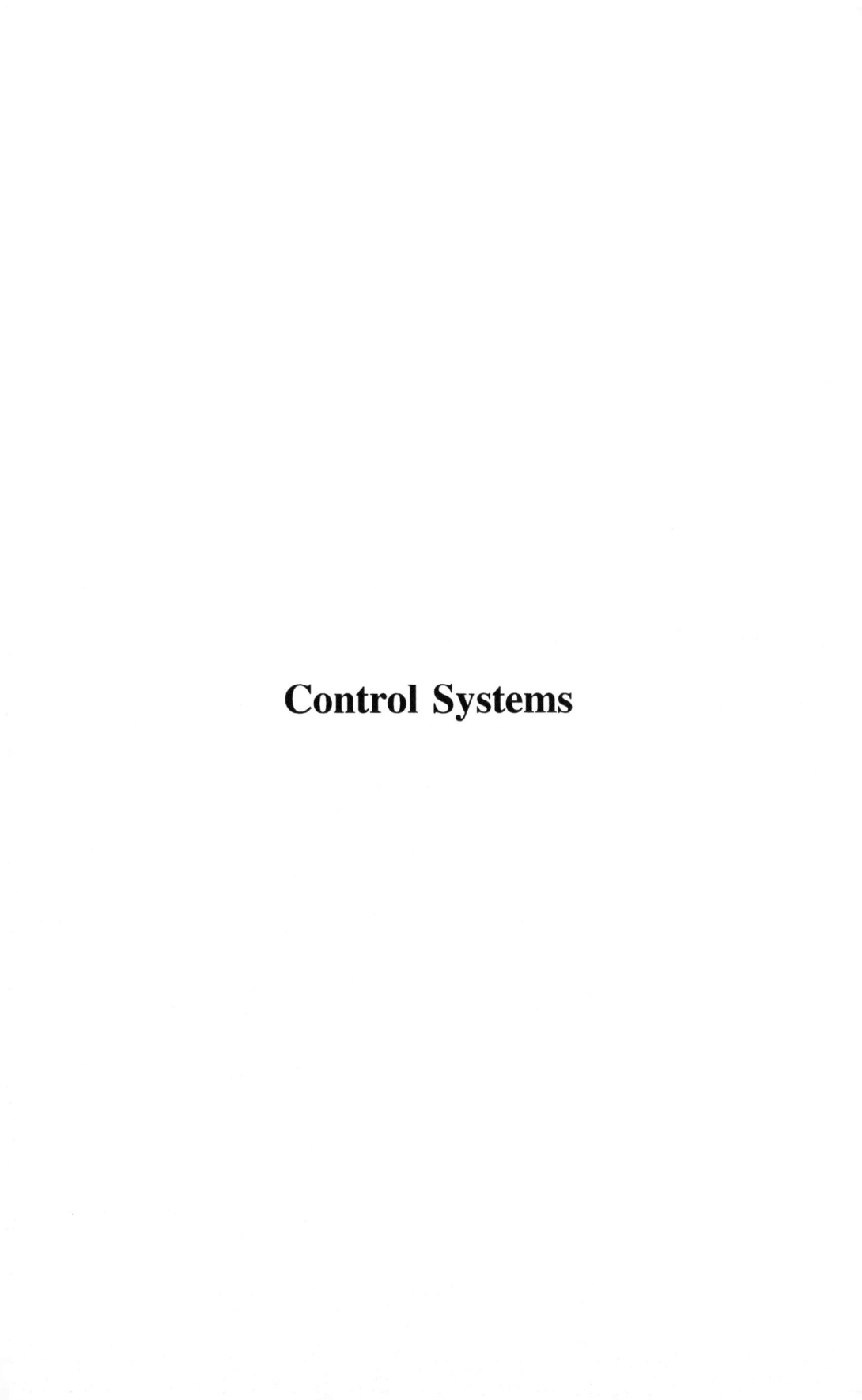

Control Systems

Network Traffic Anomaly Detection in Railway Intelligent Control Systems Using Nonlinear Dynamics Approach

Maria A. Butakova[1], Andrey V. Chernov[1], Sergey M. Kovalev[1], Andrey V. Sukhanov[1,2(✉)], and Stanislav Zajaczek[2]

[1] Rostov State Transport University, Rostov-on-Don, Russia
drewnia@rambler.ru
[2] VSB - Technical University of Ostrava, Ostrava, Czech Republic

Abstract. The work presents an approach for anomaly detection in network traffic based on nonlinear dynamics techniques. The main attention is paid to nonlinear-dynamical models of telecommunication traffic in the distributed network subsystems of railway intelligent control systems. In the considered system, telecommunication traffic is presented in time series form. The time series is used as the basis for reconstructed nonlinear dynamic system with chaotic behavior. The calculation algorithms for embedding dimension, correlation dimension and spectrum of Lyapunov exponents are given. The computational implementations for assessment of dynamical characteristics reconstructed from noisy time series of network traffic are presented. Anomaly detection algorithm based on Lyapunov exponents calculation is presented for nonlinear system generating network traffic.

Keywords: Anomaly detection · Network traffic ·
Nonlinear dynamics · Lyapunov exponents ·
Railway intelligent control system

1 Introduction

Railway intelligent control system (RICS) [1] is the distributed complex of information control networks and automated workstations intended for efficient control over transportation processes on railways. The main tool of data transmission between RICS is network telecommunication traffic. Modeling of information exchanges in distributed systems allows to estimate capacity of telecommunication networks and give recommendations regarding optimization of network processing. It is logical that increasing of information exchanges and traffic growth are caused by increased number of queries from users and signals from various input devices. However, in practice, it can be caused by non-natural reasons. The known examples are Denial-of-Service attacks and Distributed-Denial-of-Service.

Anomaly detection in discrete sequences [2] takes a specific place in the traffic modeling. Anomalies are usually referred to the observations, which do not conform to the expected behavior of a process. Traffic anomaly detection usually suggests searching for faults, large deviations and frequent errors in transmitted data. Anomalies in telecommunication traffic lead to various incidents. Because of this, it is relevant task in area of RICS security [1,3,4].

The present work proposes the approach for anomaly detection in traffic. The approach is based on nonlinear dynamics reconstruction using discrete time series. The main feature of the proposed approach is usability of proposed algorithms for noisy time series. The work is organized as follows. Section 2 shows the state-of-art in the considered area. Section 3 presents algorithms of dynamics reconstruction and algorithm of anomaly detection in noisy time series. Section 4 provides the experimental part of the presented work. Conclusions and future work are presented in Sect. 5.

2 Related Works

Common teletraffic-based approaches connected with traffic modelling assume it in form of data flows described by queuing theory [5] and in form of stochastic processes [6]. As described there, the main modelled properties of teletraffic are its self-similarity and stable correlation dependency. It was shown that these models are linear, but recent researches prove the possibility of information exchange modelling from the point of determined chaos view [7,8]. Determined chaos and fractals closely connected because dynamic attractors described by nonlinear equations are elements of fractal sets. Therefore, the following study of fractal properties of teletraffic can be done utilizing nonlinear dynamics theory, for example, by implementing phase space reconstruction for telecommunication systems.

Chaotic process is something intermediate between determined process and stochastic one. To define its affinity to one or another there are many properties, which computation deals with phase space reconstruction. These properties are [9]: time delay τ, embedded dimension m, correlation dimension d_c and Lyapunov spectrum λ. Usually, time series is one-parameter observation, when it is obtained during teletraffic measuring. It can be mapped into m-dimensioned space, which characterizes properties of generating dynamic system [10]. In such case, phase space reconstruction is performed by time series embedding $f = \{x(t_i)\}_{i=1}^{N}$ into Euclid space $\mathbb{R}^m$ with time delay τ:

$$x(t + i\tau) = f\Big(x\big(t + (i-1)\tau\big), x\big(t + (i-2)\tau\big), \ldots, x\big(t + (i-m)\tau\big)\Big),$$

where $i \in N$ is the observation id (N is the number of observations), m is the embedded dimension.

The value of correlation dimension d_c is also used in reconstructed phase space. It denotes the number of variables, which are required to correctly reconstruct the system dynamics. In [10], the relation between m and d_c is shown in form:
$$m \geq 2[d_c] + 1,$$
where $[d_c]$ is the integer part of d_c.

To describe the sense of d_c, the following descriptions are useful. Let $d \in Z^+$ be integer dimension of dynamic system. Then, $V(\varepsilon) \propto \varepsilon^d$, where V is the volume measure, ε is the corresponding scale measure. Therefore, d is defined by $d \propto \dfrac{\log(V(\varepsilon))}{\varepsilon}$. Analogically d_c has the following property

$$d_c = \lim_{\varepsilon \to 0} \lim_{m \to \infty} \frac{\partial C_m(\varepsilon)}{\partial \log(\varepsilon)}. \qquad (1)$$

It is followed from (1) that $C_m(\varepsilon) \propto \varepsilon^{d_c}$, where $C_m(\varepsilon)$ is called as correlation function. This function is defined as probability that distance between points belonging to reconstructed attractor is less than $\varepsilon > 0$, i.e. $C_m(\varepsilon) = P\Big(||x-y|| < \varepsilon | x, y \in A\Big)$.

As a rule, correlation dimension depends on m and τ. The reconstructed phase space will be isomorphic to the initial phase space of the dynamic system if the parameters are correctly chosen. There are two branches of assumptions regarding these parameters: one considers m and τ to be independent from each other and another considers them to be correlated because of noises in real time series. The changing of τ has a simple geometric meaning: if τ is chosen rather small, than two neighbouring points are very close and correlate with each other. In opposite case, two neighbouring points are very far and attractor is curved.

Correlation dimension can be estimated by correlation integral represented by cumulative function of distance distribution between vectors in phase space, i.e.

$$C_m(\varepsilon) = \int \int K(\frac{r}{\varepsilon}) \mu_m dx \mu_m dy, \qquad (2)$$

where μ_m is independent probability measure, $K(x)$ is so-called kernel function, ε is kernel scale and $r = ||x - y||$ is normalized distance between vectors $x = x_i$ and $y = y_i$.

Heaviside step function can be used as a kernel and maximal distance between points can be used as a normalized distance, i.e.:

$$C_m^H(\varepsilon) = \int \int H(\varepsilon - ||x - y||) \mu_m dx \mu_m dy, \qquad (3)$$

where

$$H(x) = \begin{cases} 1, & x < 0 \\ 1, & x \geq 0 \end{cases},$$

$$||x - y|| = \sup_i (x_i - y_i).$$

In this area, the key task is reconstruction in noisy conditions. The following section proposes algorithms applicable in such cases together with specification for anomaly detection.

3 Proposed Approach

In this section, we propose several algorithms which are used to find the primary parameters of the reconstructed nonlinear system and then the anomaly detection algorithm. First, we propose the three-step approximation of the $\log C_m^H(N,\varepsilon)$ function. We note, that correlation function is estimated by the probability, thus $0 \le C_m^H(N,\varepsilon) \le 1$. In the numerical calculations, we restrict some approximation precision, i.e. if $\varepsilon = 0.0001$ then we must take it to account. Therefore, we can define the interval $\varepsilon_{\min} \le \varepsilon \le \varepsilon_{\max}$ and find the linear part of a plot through this interval, as it shown in Fig. 1.

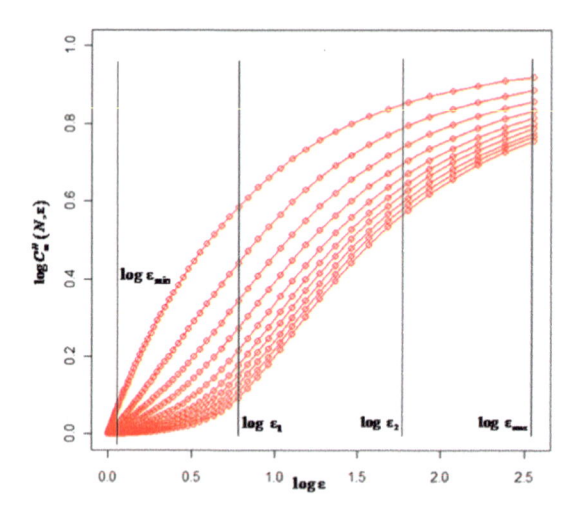

Fig. 1. Finding the linear part of correlation function plot

The ε value can be determined in one of three parts: (1) $\varepsilon_{\min} \le \varepsilon \le \varepsilon_1$; (2) $\varepsilon_1 \le \varepsilon \le \varepsilon_2$; (3) $\varepsilon_2 \le \varepsilon \le \varepsilon_{\max}$. For this purpose we propose Algorithm 1. One can substitute a Gaussian kernel $\mathrm{Ker}(x) = e^{-\frac{x^2}{4}}$ into (1) and Euclidean distance $\|\mathbf{x} - \mathbf{y}\|^2 = \sum_{i=1}^{m}(x_i - y_i)^2$ into (2), thus

$$C_m^G(\varepsilon) = \int\int e^{-\frac{\|\mathbf{x}-\mathbf{y}\|^2}{4\varepsilon^2}} \mu_m d\mathbf{x}\, \mu_m d\mathbf{y}. \tag{4}$$

From (3) it follows the estimation of correlation dimension

$$C_m^G(\varepsilon) \sim \left(\frac{\varepsilon}{\sqrt{m}}\right)^{d_c} e^{(-Km\tau)} \tag{5}$$

where K is a metric Kolmogorov-Sinai entropy.

With Gaussian noise Eq. (4) has the following form

$$C_m^G(\varepsilon) = \left(\frac{\varepsilon^2}{\varepsilon^2 + \sigma^2}\right)^{\frac{m}{2}} \int\int e^{-\frac{\|\mathbf{x}-\mathbf{y}\|^2}{4(\varepsilon^2+\sigma^2)}} \mu_m d\mathbf{x}\ \mu_m d\mathbf{y} \tag{6}$$

The estimation of correlation dimension can be evaluated as

$$C_m^G(\varepsilon) \sim \psi \cdot \left(\frac{\varepsilon^2}{\varepsilon^2 + \sigma^2}\right)^{\frac{m}{2}} e^{(-Km\tau)} \left(\frac{\varepsilon^2 + \sigma^2}{m}\right)^{\frac{d_c}{2}}, \tag{7}$$

where ψ is a normalizing constant.

Algorithm 1. Three-step approximation of the correlation dimension estimation

---
1: **Input parameters:** RICS traffic time series $\{x(t_i)\}$, $i = 1,...N$; $\varepsilon, \varepsilon_1, \varepsilon_2, \varepsilon_{\min}, \varepsilon_{\max}$
2: Creating m-dimensional arrays $V_m(i)$,
   $V_m(i) = (x(i), x(i+\tau), ..., x(i+(m-1)\tau))$, $i = 1, ..., L$,
   L is a number of the reconstructed vectors, $\tau > 0$ time delay
3: **for** 1 to m **do**
4:   Compute sums $C_m^H(N, \varepsilon) = \frac{2}{(N-m+1)(N-m)} \sum_{i=2}^{N-m+1} \sum_{j=1}^{i-1} H(\varepsilon - \|V_m(j) - V_m(i)\|)$,
     $\|V_m(j) - V_m(i)\| = \sup\limits_{i=1,...,m}(V_m(j) - V_m(i))$, ε is precision.
5: **end for**
6: **if** $\varepsilon_{\min} \leq \varepsilon \leq \varepsilon_1$ **then**
7:   Approximate by $v = \varphi_1(u)$, $\delta_1 = \sum\limits_{\varepsilon=\varepsilon_{\min}}^{\varepsilon_1} (\varphi_1(\log(\varepsilon)) - \log(C_m^H(N,\varepsilon)))^2$.
8: **end if**
9: **if** $\varepsilon_1 \leq \varepsilon \leq \varepsilon_2$ **then**
10: Approximate by $v = au + b$, $\delta_2 = \sum\limits_{\varepsilon=\varepsilon_1}^{\varepsilon_2} (a\log(\varepsilon) + b - \log(C_m^H(N,\varepsilon)))^2$.
11: **end if**
12: **if** $\varepsilon_2 \leq \varepsilon \leq \varepsilon_{\max}$ **then**
13: Approximate by $v = \varphi_2(u)$, $\delta_3 = \sum\limits_{\varepsilon=\varepsilon_2}^{\varepsilon_{\max}} (\varphi_2(\log(\varepsilon)) - \log(C_m^H(N,\varepsilon)))^2$
14: **end if**
15: Compute $\delta = \delta_1 + \delta_2 + \delta_3$.
16: Change ε, repeat 6-15 to find $min(\delta)$
17: **Output:** $\varepsilon_1, \varepsilon_2$

On the basis of Algorithm 1, we propose a modified Algorithm 2 for numerical calculation of the embedded dimension of the reconstructed attractor of a nonlinear dynamical system.

Next, an Algorithm 3 is proposed that is suitable for calculating the correlation dimension under Gaussian noise conditions for traffic time-series data.

After reconstruction the phase space from traffic time series we use Algorithm 4 for the maximal Lyapunov exponent calculation. Finally, we apply Algorithm 5 to determine abnormal activity in the traffic time series.

Algorithm 2. Numerical calculation of the embedded dimension

1: **Input parameters:** RICS traffic time series $\{x(t_i)\}$, $i = 1,...N$; m
2: Compute $C_m^H(N, \varepsilon)$
3: Run Algorithm 1 for $\varepsilon^m, \varepsilon_1^m, \varepsilon_2^m, \varepsilon_{min}^m, \varepsilon_{max}^m$
4: **if** $[\varepsilon_1^m, \varepsilon_2^m] = [\varepsilon_1^{m-1}, \varepsilon_2^{m-1}]$ **then** Goto 7
5: **else** $m = m + 1$, Goto 2
6: **end if**
7: Approximate function $\log C_m^H(N, \varepsilon)$ for $[\varepsilon_1^m, \varepsilon_2^m]$ and get the correspondent line y equation.
8: Compute $s = \text{tg } y$.
9: **Output:** $d_c = s$, m

Algorithm 3. Numerical calculation of the embedded dimension under Gaussian noise

1: **Input parameters:** RICS traffic time series $\{x(t_i)\}$, $i = 1,...N$; $L \geq 0$; ε_k; σ
2: **for** 1 to N **do**
3:    Normalize values $\left\{ z(t_i) = \frac{x(t_i) - \bar{x}}{\sigma} \right\}$
4: **end for**
5: **for** 1 to L **do**
6:    Run Algorithm 2
7: **end for**
8: Implement nonlinear estimation of (7).
9: **Output:** d_c, K

Algorithm 4. Numerical calculation of the maximal Lyapunov exponent

1: **Input parameters:** RICS traffic time series $\{x(t_i)\}$, $i = 1,...N$
2: Run Algorithm 3
3: Choose the initial coordinate $x_0(t_0)$ from traffic time series.
4: **for** $j = 1$ to m **do**
5:    Find the nearest to $x_{j0}(t_0)$ coordinate $x_j(t_0)$
6:    $d_{0j}(t_0) = \|x_{0j}(t_0) - x_j(t_0)\|$
7:    Input ε_j
8:    $d_{1j}(t_1) = d_{j0} + \|x(t_1) - x(t_0)\|$.
9:    **while** $i \leq N, d_j(t_i) > \varepsilon_j$ **do**
10:       Find the nearest to $x_j(t_i)$ coordinate $x_j(t_{i-1})$.
11:       $d_{1j}(t_i) = \|x_j(t_i) - x_j(t_{i-1})\|$
12:    **end while**
13: **end for**
14: Calculate the maximal Lyapunov exponent $\lambda_{\max} = \frac{1}{t_N - t_0} \sum_{i=1}^{N} \log \frac{d_1(t_i)}{d_0(t_{i-1})}$
15: **Output:** $\lambda_{\max}$

Algorithm 5. Anomaly detection by indicators

1: **Input parameters:** RICS traffic time series $\{x(t_i)\}$, $i = 1,...N$; $h > 0$ is a threshold; $D_k = 0$ is indicator value, k is integer.
2: Run Algorithm 4 i-times.
3: Form the $\left\{\lambda_{\max}^{(i)}\right\}$.
4: Calculate $\Delta\lambda_{\max}^{(i)} = \left|\log \lambda_{\max}^{(i)} - \log \lambda_{\max}^{(i-1)}\right|$.
5: **if** $\Delta\lambda_{\max}^{(i)} \geq h$ **then** $D_k = 1$; $k = k+1$;
6: **end if**
7: Form the array of anomaly indicators $\{D_k\}_{i=1}^{M}$.

4 Computational Experiment

The computation experiments were performed on well-known chaos Lorenz dataset:

$$\begin{cases} \dfrac{dx}{dt} = \sigma(y-x) \\ \dfrac{dy}{dt} = rx - y - xz \\ \dfrac{dz}{dt} = xy - bz \end{cases}, \qquad (8)$$

where σ, r and b are the model parameters, known as Prandtl number, Rayleigh number and the geometric factor, respectively [9]. The time series generated by the Lorenz system with the commonly used values of the parameters ($\sigma = 10$, $r = 28$ and $b = 8/3$) is shown in Fig. 2.

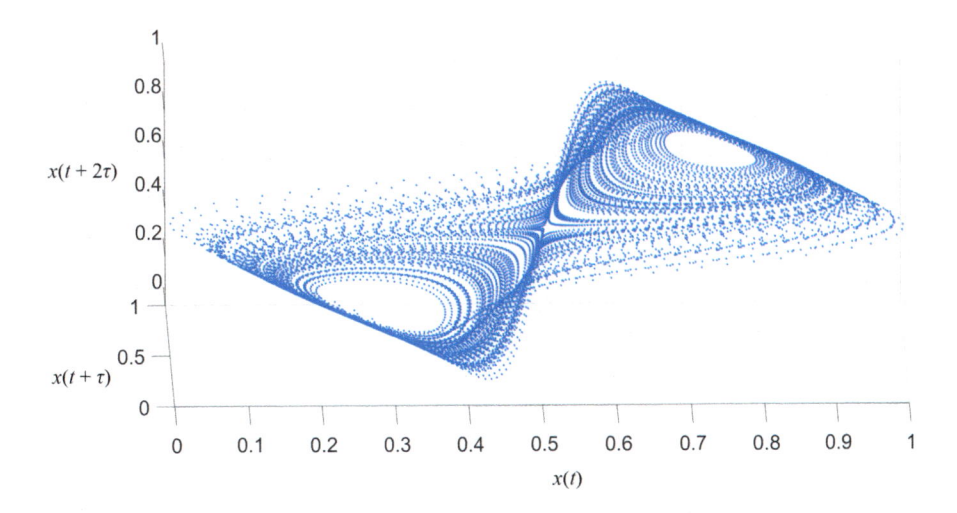

Fig. 2. Initial dataset used in the experiments

Initial phase space was calculated using Algorithms 1 and 2. As a result, $\tau = 0.12\,s$ and $m = 3$ are obtained.

The anomalous data was added into parts of the initial dataset using random variables. The number of windows equals to 20. Anomalous behavior considered to be started at window no. 10 and finished at window no. 18. Figure 3 shows that the Lyapunov exponent rises, when system becomes more stochastic and anomalous behavior can be detected if threshold equals to 2.

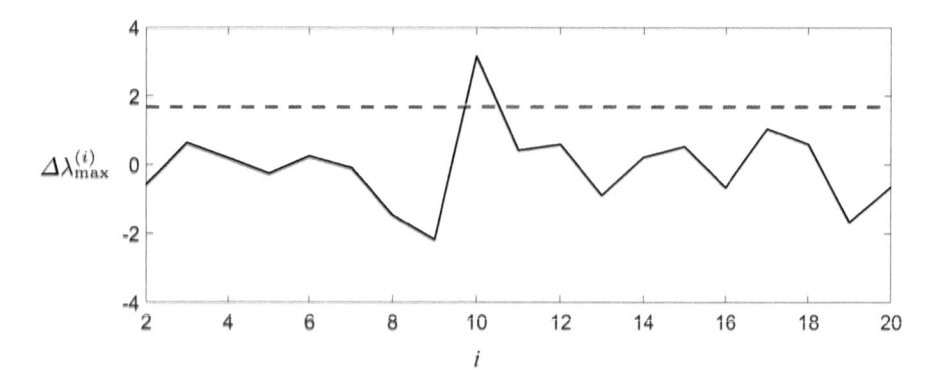

Fig. 3. Anomaly detection by threshold and indicators

5 Conclusions and Future Work

In the presented work, an approach for anomalous detection in network traffic is proposed. The network traffic is presented in form of discrete time series described by determined chaos. The reconstructed phase space is utilized to map the time series into convenient form for processing and Lyapunov exponent is proposed as a criterion for the classification. It is proposed that anomalous behaviour rises if the Lyapunov coefficient is greater than user-defined threshold. The presented benchmarking experiments show that the proposed approach allows to detect anomalous observations in the time series. However, it is preliminary research. The following experiments will be provided by computation of the presented technique with conventional ones on the real traffic data. As well, automatic-defined threshold will be performed.

Acknowledgement. The work was supported Grant No. SP2018/163 "Diagnostics, reliability and efficiency of electrical machines and devices, problems of antenna systems" and by Russian Foundation for Basic Research (Grants No. 16-07-00888-a, 18-08-00549-a, 17-20-01040 ofi_m_RZD, No. 16-07-00032-a and No. 16-07-00086-a).

References

1. Chernov, A.V., Butakova, M.A., Karpenko, E.V.: Security incident detection technique for multilevel intelligent control systems on railway transport in Russia. In: 2015 23rd Telecommunications Forum Telfor (TELFOR), pp. 1–4. IEEE (2015)
2. Chandola, V., Banerjee, A., Kumar, V.: Anomaly detection for discrete sequences: a survey. IEEE Trans. Knowl. Data Eng. **24**(5), 823–839 (2012)
3. Chemov, A.V., Butakova, M.A., Karpenko, E.V., Kartashov, O.O.: Improving security incidents detection for networked multilevel intelligent control systems in railway transport. Telfor J. **8**(1), 14–19 (2016)
4. Butakova, M.A., Chernov, A.V., Shevchuk, P.S., Vereskun, V.D.: Complex event processing for network anomaly detection in digital railway communication services. In: 2017 25th Telecommunication Forum (TELFOR), pp. 1–4. IEEE (2017)
5. Akimaru, H., Kawashima, K.: Teletraffic: Theory and Applications. Springer Science & Business Media, New York (2012)
6. Sheluhin, O., Smolskiy, S., Osin, A.: Self-similar processes in telecommunications. John Wiley & Sons, New York (2007)
7. Fu, C., Jiang, H.Y.: On the chaotic dynamics analysis of internet traffic. In: 2008 International Conference on Intelligent Computation Technology and Automation, pp. 840–844. IEEE (2008)
8. Guo, X., Vogel, D., Zhou, Z., Zhang, X., Chen, H.: Chaos theory as a model for interpreting weblog traffic. In: Proceedings of the 41st Annual Hawaii International Conference on System Sciences, p. 289. IEEE (2008)
9. Kantz, H., Schreiber, T.: Nonlinear time series analysis, vol. 7. Cambridge University Press, New York (2004)
10. Takens, F.: Detecting strange attractors in turbulence. In: Dynamical systems and turbulence, Warwick 1980, pp. 366–381. Springer (1981)

Advanced Methods of Detection of the Steganography Content

Jakub Hendrych[✉] and Lačezar Ličev

VŠB-TU Ostrava, 708 33 Ostrava-Poruba, Czech Republic
{jakub.hendrych,lacezar.licev}@vsb.cz

Abstract. In this paper, we deal with the classification of the steganography content. Some illegal activities can perform steganography for stealing the secret information from a company internal network. Therefore, we must be prepared to protect our data. To detect steganography content, we have counter-technique known as steganalysis. There are different types of steganalysis, based on the existence of the original artifact (cover work) or if we know which algorithm embed a secret message. For practical use, most important are methods of blind steganalysis, that can be applied to the most compact and ordinary cover work - JPEG image files. This paper describes the methodology to the issues of JPEG image steganalysis. It is crucial to understand the behavior of the targeted steganography algorithm. Then we can use it is weaknesses to increase the detection capability and success of classification. We are primarily focusing on breaking the DCT steganography algorithm OutGuess2.0 and secondary on breaking the F5 algorithm. We are analyzing the ability of the detector, which utilizes the calibration process, blockiness calculation, and shallow neural network to identify the presence of steganography message in the suspected image. This approach is an improvement over our previous researches. Contribution and new results are discussed.

Keywords: Steganography · Steganalysis · Shallow neural network · ANN · JPEG · DCT · Calibration · Blockiness · OutGuess2.0 · F5

1 Introduction

The term Steganography refers to the art of secret communications. By using this art, it is possible for Alice to send a secret message to Bob in such a way that the third party does not know that the message even exists. Steganography embedding process manipulates a cover works properties by the specific way. The output is known as a stegogramme that is very similar to cover work, but it also carries the hidden message. If Alice sends this stegogramme to Bob, someone who intercepts this communication will obtain only a stegogramme. It is a difficult task to tell that the stegogramme is not innocent. Moreover, this is the main advantage of the steganography, to create an illusion of an innocent communication [1].

The development of the information technologies has brought a new opportunity for the steganography methods. In modern terms, steganography is usually implemented computationally, where a cover work can be anything from a text file to a video file.

Then a secret message is embedded within those files. Imagine the scenario with an employee and a secret company data. The data can be easily embedded by a steganography algorithm into a prepared JPEG image file (cover work) and send by email to a competitor. Then a recipient uses a steganography decoding algorithm on an image and retrieves a company data. Of course, there are many other scenarios with a social network, internet forums, and not to mention that terrorists also use steganography [2].

The very idea of the steganography does not necessarily mean an equivalent to some illegal activity. It can also be used, like digital watermarking, to protect our data, as well as an additional data layer (alternative to metadata). However, there is always a chance that the steganography will be misusage and that is why we must be ready for it. The counter technique of the steganography is the steganalysis, which serves us for detection of a stegogrammes. The primary goal of the steganalysis is to identify a presence of a secret message, but not a successful extraction of a message. The extraction is a non-trivial task, because of an involved cryptography method on a secret message. There are different types of the steganalysis. It depends on the fact if a cover work is known or not (known-cover attack steganalysis). Also, it depends on the fact if we know which steganography algorithm was used (known-stego attack steganalysis).

2 State of the Art

Many types of research, for example [3–5], were tested and developed on the low-resolution grayscale JPEG images. This type of data is sufficient for the research and testing. However, we want to test our methodology on the data that reflect the realistic scenarios - high-resolution and colorful images.

Authors of [4], presents a pattern recognition system to detect anomalies in JPEG images, especially steganography content. The system consists of the feature generation, the feature ranking and selection, the feature extraction, and pattern classification. This lead for capture image characteristics, reduce the dimensionality problem, elimination of the noise inferences between the features, and further improves the classification accuracies on a clear and a steganography JPEG images. This approach provides a classification capability of 97.4% for a 1000 B secret message created by OutGuess algorithm [6]. However, when the size of a secret message falls to 200 B, the classification capability also falls to 67.9%. Authors performed all testing on the 512×512 resolution images. In the case of the F5 algorithm [7], under the same conditions, the classification capability was in the range 91.6%–52.2%.

Authors of [8], proposed the steganalysis method for steganography at any domain (spatial or transform domain). The particular interest authors placed on the steganalysis of Highly Undetectable Steganography. The proposed method extracts the features via applying a function to the image, constructing the k-variate probability density function, and downsampling it by a suitable algorithm. The extracted feature vectors are then further optimized in order to increase the detection performance and reduce the computational time. Finally, using a supervised classification algorithm such as Support Vector Machine (SVM), the steganalysis is performed. The proposed method is capable of detecting the BOSSRank image set with an accuracy of 85%.

In research [9], the authors proposed a generic hybrid deep-learning framework for JPEG steganalysis incorporating the domain knowledge behind rich steganalysis models. The proposed framework involves two main stages. The first stage is handcrafted, corresponding to the convolution phase and the quantization and truncation phase of the rich models. The second stage is a complex deep neural network containing multiple deep subnets, where training procedure is learning the model parameters. Authors performed the experiments on large-scale datasets extracted from the ImageNet.

Next research [10], providing the analysis of the neighboring joint density of the Discrete Cosine Transform (DCT) coefficients and reveal the difference between a clear image and a modified version. By exploring the self-calibration, under different shift recompressions, authors propose the calibrated neighboring joint density-based on approaches with a simple feature set to distinguish a stegogramme and clear images. Authors claim that this approach has multiple promising applications in image forensics. If we compare this steganalysis detector, it delivers better or comparable detection performances. We see a significant contribution in the universality that this methodology can adequately classify input images into the class - the used embedded algorithm. Clear images were classified with 82.2%, F5 stegogrammes were classified with 95.6%, and OutGuess stegogrammes were classified with 89.6% success rate.

3 Proposed Method

As we mentioned, this proposed method is primarily focused on the breaking the steganography algorithm OutGuess2.0 and secondary on the algorithm F5. The encoding process of the OutGuess2.0, developed by authors [11], is a combination of the randomized Hide & Seek [11] and the JSteg [12] algorithm. The main principle is an embedding of a secret message data into the DCT coefficients used by JPEG compression. OutGuess2.0 provides corrections to the coefficients to make them appear similar to a clear image, regarding frequencies of the values. This operation causes a new compression, and the stegogramme is not changed from the statistical analysis as was described in [13]. The decoding process is irrelevant by steganalysis. At this moment a question arises - Where exactly OutGuess2.0 embeds a secret message data? In other words, which of YCbCr components are modified? There is a possibility that the embedding process excludes the Y component (luminance) because human vision is more sensitive to the luminance change and therefore the modification could be visible. On the other hand, the DCT coefficients of the Y component have mostly larger values than the coefficients of the other two components. Therefore, these values provide a better solution for the LSB technique. At our previous researches [14, 15], we were not distinguished YCbCr as separate independent values. Therefore, this was leading to the distortion of the correct stegogramme and the clear image classification. Our new proposed method approaches these three components independently, during the entire classification process. We cannot confidently say that the OutGuess2.0 always modified all three components in the same way. However, by using the separate approach for YCbCr, we are increasing the likelihood of the correct classification. In other words, we do not need to care about which component is modified. The main principle of F5 encoding algorithm is calculating the embedding potential of encoding message within a cover work. Hamming coding is then applied to embed a potentially more than one bit per value by making no more than one modification to the DCT coefficient.

The following Fig. 1 represents the workflow of the classification process:

```
Suspected image      Calibration      Calibrated image
(Colorful JPEG)  →    process     →   (Colorful JPEG)
       ↓                                      ↓
   Blockiness                             Blockiness
   calculation                            calculation
       ↓                                      ↓
 $B_{SY}$ $B_{SCb}$ $B_{SCr}$          $B_{CY}$ $B_{CCb}$ $B_{CCr}$
   (3x integer)                         (3x integer)
       └──────────→  Learned ANN  ←──────────┘
                          ↓
                   1 – Stegogramme
                   0 – Clear image
```

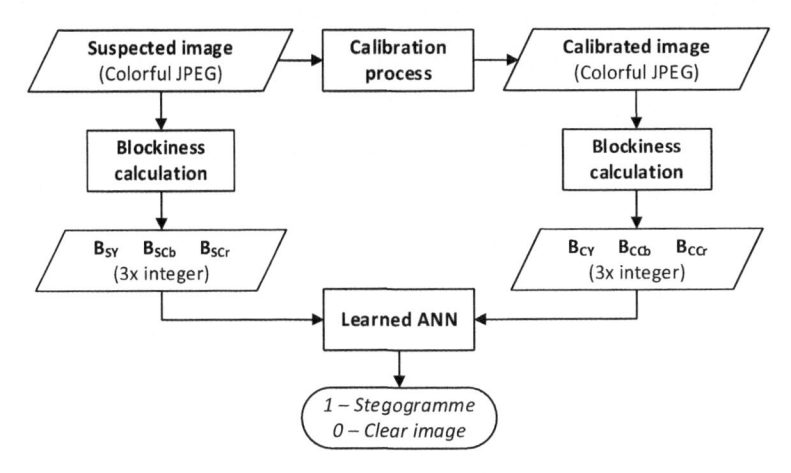

Fig. 1. Workflow of the classification process

For the calibration process, we adopted the method of steganalysis [3]. The input of this method is a suspected image, and the output is a calibrated image. The calibrated image should represent the state of the image after the secret message embedding. We adopt the calculation of cover work quality factor Qc, where Qc is now computed for every component separately, and then it is sum together. Also, we deploy the macroblock filtering feature - we exclude (based on the standard deviation of every macroblock) the low-frequency domains of the suspected image because these areas distort the result. This feature improves the sensitivity of the test.

For the blockiness calculation, we modified research [16]. One step of the calibration process was cropping the suspected image by four pixels from every side. This modification ensures removing the entire block structure. The 8x8 macroblocks are shifted from both directions, and thus a more accurate estimation of the calibrated image is derived. The blockiness calculation takes advantage of the fact that JPEG steganography embeds the secret message in the same 8x8 macroblocks that are used for compression. Again, we must adopt this part of the process for a new YCbCr approach. Therefore, the blockiness values are calculated for all components separately, and the output is six integer values - three for the suspected image and other three for the calibrated image. We can get the YCbCr values of the pixel by calculation via the RGB pixel values.

To take advantage of the separate approach, we need to modify the topology of the Artificial Neural Network (ANN). We add new input neurons. This modification involved recreation of the training sets. Now, to the input, we sent the six values representing the six results of the blockiness calculation. We also try to test other topologies, for example, we include some hidden layers with a different number of neurons. However, for this topology (see Fig. 2) we get still the best results.

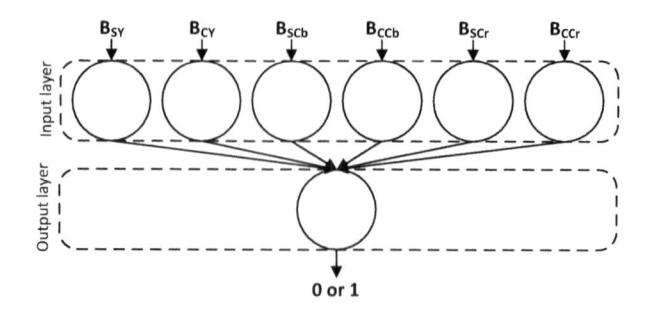

Fig. 2. Modified topology of the artificial neural network

The classification border between the stegogramme and the clear image will be defined as the poly-plane in a six-dimensional space. The output then generates the value of 1 or 0. The value 1 indicates stegogramme and the value 0 indicates a clear image without any secret message. We must emphasize, that we use our simple ANN for finding the best possible border between the classification of the stegogramme and the clear image. For the excitation of the neuron, the potential P (defined by the Eq. 1) must be larger than w_0, where w_0 is the threshold value that must be overcome.

$$P = w_1x_1 + w_2x_2 + w_3x_3 + w_4x_4 + w_5x_5 + w_6x_6, \qquad (1)$$

where w_n indicates the weights of the neurons that are set by a learning process, x_1 indicates the blockiness value B_{SY} of the Y component of the suspected image, x_2 indicates the blockiness value B_{CY} of the Y component of the calibrated image, x_3 indicates the blockiness value B_{SCb} of the Cb component of the suspected image, x_4 indicates the blockiness value B_{CCb} of the Cb component of the calibrated image, x_5 indicates the blockiness value B_{SCr} of the Cr component of the suspected image and x_6 indicates the blockiness value B_{CCr} of the Cr component of the calibrated image.

4 Results and Discussion

For development and testing purposes, we create a database of color JPEG images. The database contains the clear images and the stegogrammes created by the steganography algorithm OutGuess2.0 and F5. Individual images are available in five different resolutions – 800 × 449, 1024 × 575, 1440 × 809, 2560 × 1438 and 4200 × 2358. For every resolution, we prepared 320 images with a message length of 10 B and the same amount of the images for message lengths 50 B, 200 B, 500 B, 800 B and 1000 B. For every resolution we also have 320 images without an embedded secret message.

4.1 Classification Success Rate

On the Fig. 3 we can see the chart of the classification success rates of our proposed steganography detector compared to our previous results. We can confidently say that we improve the classification success rate for the low-resolution images, but for the cost of small reduction of high-resolution images success rate.

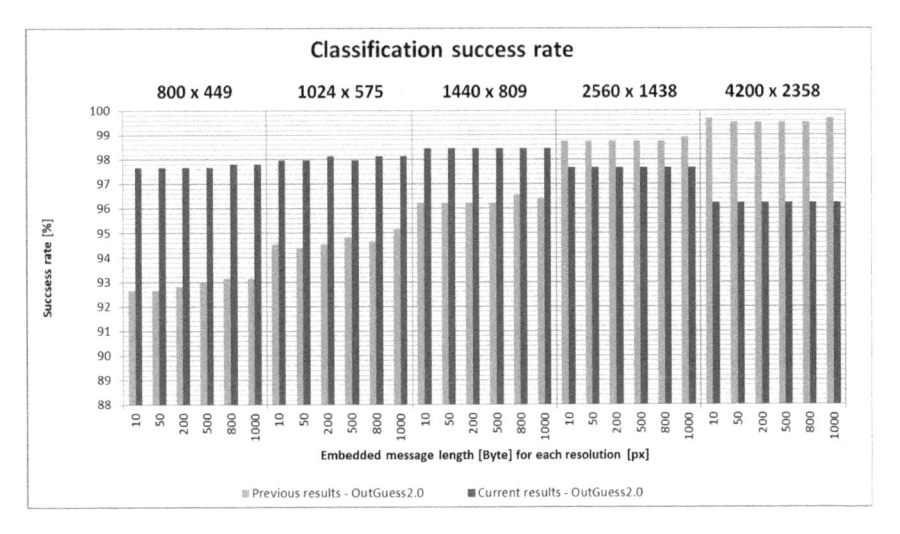

Fig. 3. Chart of the classification success rate of the proposed method with the comparison of previous results [14, 15]

The following results Table 1 for OutGuess2.0 and Table 2 for F5 also includes the statistical values for the sensitivity and the specificity of the test. Especially the high sensitivity of the test is essential for us (more in Sect. 4.3). The sensitivity is defined as the probability that a test result will be positive when the stegogramme is present. We must emphasize that the following tables do not include separate results for a different secret message length for each resolution. We also performed an analysis of variance (ANOVA) statistical hypothesis test for the invariance of the sensitivity against a secret message length. As we have found, there is no significant statistical difference. This small difference is also visible in Fig. 3. Therefore, there is no need to distinguish the results for the individual message payloads, and all result values are presented as average for each tested resolution.

It is clear from the previous Table 1 that for the correct classification of the stegogramme created by the OutGuess2.0 algorithm and the clear image, we achieved an accuracy of **96.25%–98.44%** (with an error of 3.75% - 1.56%). However, the emphasized sensitivity of the test that was **97.19%–99.53%**. From the Table 2, we can see the achieved correct classification of the stegogramme created by F5 algorithm and the clear image with an accuracy of **89.95%–95.23%** (with an error of 10.05%–4.77%). The sensitivity of the test was **82.40%–94.53%**. For the last resolution we

Table 1. Results for OutGuess2.0 algorithm

Resolution [px]	800 × 449	1024 × 575	1440 × 809	2560 × 1438	4200 × 2358
Sensitivity [%]	99.48	99.53	99.38	98.44	97.19
Specificity [%]	95.94	96.56	97.50	96.88	95.31
Accuracy [%]	97.71	98.05	98.44	97.66	96.25
Error [%]	2.29	1.95	1.56	2.34	3.75

Table 2. Results for F5 algorithm

Resolution [px]	800 × 449	1024 × 575	1440 × 809	2560 × 1438	4200 × 2358
Sensitivity [%]	94.53	90.83	82.40	89.38	–
Specificity [%]	95.94	96.95	97.50	96.288	–
Accuracy [%]	95.23	93.70	89.95	93.13	–
Error [%]	4.77	6.30	10.05	6.88	–

cannot create the stegogrammes - with this resolution, F5 create an empty stegogrammes with file size of 0 B.

4.2 Results of the Macroblock Filtering

In this chapter, we will present the differences in the classification with and without the use of the macroblock filtering. The following Table 3, for OutGuess2.0 and Table 4 for F5, shows the results representing only the difference values between tests with and without the use of macroblock filtering feature.

Table 3. Difference between the use of macroblock filtering and without – OutGuess2.0

Resolution [px]	800 × 449	1024 × 575	1440 × 809	2560 × 1438	4200 × 2358
Sensitivity dif. [%]	3.65	3.02	4.11	1.30	18.49
Specificity dif. [%]	0.00	0.00	−0.63	−1.25	−4.69
Accuracy dif. [%]	1.82	1.51	1.74	0.03	6.90
Error dif. [%]	−1.82	−1.51	−1.74	−0.03	−6.90

Table 4. Difference between the use of macroblock filtering and without – F5

Resolution [px]	800 × 449	1024 × 575	1440 × 809	2560 × 1438	4200 × 2358
Sensitivity dif. [%]	8.48	10.31	21.46	8.54	–
Specificity dif. [%]	0.00	0.00	−0.63	−1.25	–
Accuracy dif. [%]	4.24	5.16	10.42	3.65	–
Error dif. [%]	−4.24	−5.16	−10.42	−3.65	–

From the Table 3, we can see using of the macroblock filtering function has a positive effect on the sensitivity of the test, in each of the tests performed. The improvements in sensitivity were in the range from **1.30%–18.49%**. This feature also improved the overall test accuracy up to **6.90%**. However, in some cases, the specificity of the test was reduced to **−4.69%**. However, priority is given to the sensitivity of the test (more in Sect. 4.3). Same as before, we also summarize the results presented in Table 4. Again, this feature had a positive effect on the sensitivity of the test. Sensitivity was improved in the range of **8.48%–21.46%**. This feature also improved the accuracy by up to **10.42%**. Same as OutGuess2.0 algorithm, there was a small reduction in the specificity of the test down to **−1.25%**.

If we compare the results of both algorithms, the application of the macroblock filtering function was successful, especially for the F5 algorithm.

4.3 Overall Contribution of the Proposed Method

In our effort, we cannot concentrate on breaking all the steganography algorithms and the related issues. Instead, we contributed to the area of detection of the stegogrammes created by steganography techniques based on the transform domain modification of JPEG images. Now, we would like to summarize our contributions that bring our proposed method:

- High classification success rate - with regard to the sensitivity of the test.
- Invariance of the sensitivity of the test against the secret message length - this is critical. The secret message can contain anything from short passwords to the long description of the customers. Therefore, it is essential to correctly classified stegogrammes with the approximately same success rate with no relation to a message payload. Our proposed methodology provides this invariance.
- Image resolution and color depth - development and testing on several resolutions including high-resolution, colorful JPEG images (up to 10 Mpx). High-resolution and color depth provide better solutions for DCT steganography and also, this kind of images are commonly used, so they are not suspicious.
- Application of the ANN (we are using the specific type called as shallow neural network) - our methodology implements the ANN for the classification between a stegogramme and clear image. With the ANN, we can replace the trivial process of classification such as comparison of two values with a more complex solution.
- JPEG macroblock filtering - filtering of non-modified macroblocks. By the observation of the OutGuess2.0, we apply a new feature to the classification process. If we remove macroblocks that are not used for embedding of the secret message, we can remove the distortion from the blockiness calculation and therefore, increase the classification success rate.

5 Conclusion and Future Work

In this paper, we present our research about the detection of the steganography content. We propose a new methodology for the stegogrammes detection, primary created by the DCT steganography algorithm OutGuess2.0 and then by DCT algorithm F5. Such research can be used in the corporate sector to secure communication in the internal network and protect the company data. These algorithms are readily available for download. Also, they do not require a lot of IT knowledge. Therefore, there is a great possibility for using in the corporate sector.

The proposed methodology with a separate approach to YCbCr components had promising results. These results were presented in the previous chapter. In the first part of this chapter, we compare current results with our previous research. Our methodology achieves very good results against stegogrammes created by OutGuess2.0 algorithm. Mainly the emphasized sensitivity of the test was up to 99.53%. For the higher resolution images, we observe the small drop in the classification capability. This drop may be caused due to the character of the ANN training set. The training set was learned only on resolutions up to 2560 × 1438. Therefore, the lack of "experience" could lead to this drop in the classification capabilities for the higher resolutions. Next reason could be a feature of OutGuess2.0 where a user is allowed to insert a message to a specific length only, based on the composition of the cover image. This feature was most prominent in the high-resolution images (e.g., 4200 × 2358), especially for 1000 B message length. The number of such "reduced" stegogrammes was in the range from about ten images (for the small resolutions) to the half of the stegogramme series (for the high resolution). This fact can confuse the ANN and subsequent determination of a clear image and stegogramme, based on the individual blockiness values. This feature also demonstrates how interesting the OutGuess2.0 algorithm is. We also provide the results of classification capability on the F5 algorithm, where we achieved a satisfying accuracy with the sensitivity of the test 94.53%. Even though these results are not as good as the classification of OutGuess2.0 stegogrammes, they are still decent. It is important to mention that this methodology is not primarily intended for F5 classification. In the second part of this chapter, we present the positive effect of the macroblock filtering function that we also tested.

Finally, we want to mention the possible ideas for the future work. Very interesting would be to design a complete framework that would be able to classify the stegogrammes from the different steganography algorithms. We already made and published some research [14]. Since the terms of the steganography and the steganalysis are relatively unknown, the most of companies have no idea how quickly they can lose their secret internal data. Another exciting direction that development could take would be the calculation of the length of an embedded secret message. This feature could be the first step in the extraction of the contents of a secret message itself. However, this is out of the scope of the steganalysis. On top of that, as we mentioned before, the secret message itself is protected by encryption.

Acknowledgements. The following grants are acknowledged for the financial support provided for this research by Grant of SGS No. 2018/177, VSB-Technical University of Ostrava and under the support of NAVY and MERLIN research lab.

References

1. Johnson, N.F., Jajodia, S.: Exploring steganography: seeing the unseen. Computer **31**(2), 26–34 (1998)
2. Conway, M.: Code wars: steganography, signals intelligence, and terrorism. Knowl. Technol. Policy **16**(2), 45–62 (2003)
3. Fridrich, J., Goljan, M., Hogea, D.: Attacking the outguess. In: Proceedings of the ACM Workshop on Multimedia and Security. Juan-les-Pins, France (2002)
4. Chen, C.L.P., et al.: A pattern recognition system for JPEG steganography detection. Opt. Commun. **285**(21), 4252–4261 (2012)
5. Liu, Q., et al.: An improved approach to steganalysis of JPEG images. Inf. Sci. **180**(9), 1643–1655 (2010)
6. Provos, N.: Outguess 0.2 (2001). http://www.outguess.org/S
7. Westfeld, A.: F5—a steganographic algorithm. In: Information Hiding, pp. 289–302. Springer, Heidelberg (2001)
8. Gul, G., Kurugollu, F.: A new methodology in steganalysis: breaking highly undetectable steganograpy (HUGO). In: Information Hiding, pp. 71–84. Springer, Heidelberg (2011)
9. Zeng, J., et al.: Large-scale JPEG steganalysis using hybrid deep-learning framework. arXiv preprint arXiv:1611.03233 (2016)
10. Liu, Q., Chen, Z.: Z.: Improved approaches with calibrated neighboring joint density to steganalysis and seam-carved forgery detection in JPEG images. ACM Trans. Intell. Syst. Technol. (TIST) **5(4)**, 63 (2015)
11. Provos, N., Honeyman, P.: Hide and seek: an introduction to steganography. IEEE Secur. Priv. **99**(3), 32–44 (2003)
12. Upham, D.: Steganographic algorithm JSteg. Software (1993). http://zooid.org/~paul/crypto/jsteg
13. Provos, N.: Defending Against Statistical Steganalysis. In: Usenix Security Symposium, pp. 323–336 (2001)
14. Hendrych, J., Kunčický, R., Ličev, L.: New approach to steganography detection via steganalysis framework. In: International Conference on Intelligent Information Technologies for Industry, pp. 496–503. Springer, Cham (2017)
15. Hendrych, J., Ličev, L., Kunčický, R.: Detector of the steganography images with the application of artificial neural network. In: 17th International Multidisciplinary Scientific GeoConference SGEM2017, Informatics, Geoinformatics and Remote Sensing (2017). https://doi.org/10.5593/sgem2017/21/s07.033
16. Fu, D., et al.: JPEG steganalysis using empirical transition matrix in block DCT domain. In: 2006 IEEE 8th Workshop on Multimedia Signal Processing, pp. 310–313. IEEE (2006)

Robust Servo Controller Design Based on Linear Shift Invariant Differential Operator

Dae Hwan Kim[✉] and Sang Bong Kim

Department of Mechanical Design Engineering,
Pukyong National University, Busan 608-739, South Korea
kimdh2599@pknu.ac.kr

Abstract. This paper proposes a robust servo controller deign for MIMO systems based on a linear shift invariant differential (LSID) operator and its inverse operator on Schwartz space using the internal model principle (IMP). To do this task, the followings are done. First, the basic concept idea of LSID operator and its invertible operator are described. Furthermore, the IMP is rearranged based on LSID operator and the IMP is modified by the properties of the LSID operator. Second, an extended system operated by the LSID operator to a given MIMO system with the given reference is obtained. Third, the controllability checking of extended systems is done. Fourth, a state feedback law is obtained by solving the pole assignment problem with all the poles of the extended system. Fifth, The proposed controller is applied for controlling the sideslip angle and yaw rate of a 4-wheel steering vehicle as a MIMO system with 2 inputs of steering angles of front and rear wheels and two outputs of the sideslip angle and yaw rate. Finally, simulation results are shown to verify the effectiveness of the proposed controller.

Keywords: Robust servo controller · MIMO systems ·
Linear shift invariant differential (LSID) operator · Four wheel steering vehicle ·
Internal model principle (IMP)

1 Introduction

The multivariable robust servo system controller design problem is one of the most interested problems in control engineering field. The servo controller design problem is usually desired to find a controller for a plant to solve the robust servomechanism problem such that the stability of the closed loop system and asymptotic regulation occur, and also other desirable properties of the controlled system are satisfied under existences of disturbance and system parameter perturbation.

To solve the servomechanism problem, several researchers have studied several design concepts. Davison and Smith [1] considered the servo controller design problem for the special case that disturbance term is a constant un-measurable case. Davison [2] considered the output control problem of multivariable systems as being a multivariable generalization of the classical single-input, single-output servomechanism problem. The method taken to develop the result relies extensively on properties of the asymptotic solution of a stable linear constant system subject to a specified class of forcing function inputs. It was initially shown that the closed loop system with the proposed controller is controllable if and only if the original system is controllable. Kim et al. [3] introduced a servo control method with disturbance rejection and reference signal tracking by adopting the uses of the internal model principle and bilinear transformation method. However, there was not shown for the explicit condition and its proof for the controller existence of the extended system to achieve the robust servo control object.

In this paper, the main result on robust controller is obtained by modifying the well known concept of the internal model principle shown in Kim et al. [3], introducing a polynomial differential operator for the state space model and error signal, and getting its extended system. The servo controller can be easily designed by using several kinds of regulator design methods. In order to get the robust controller for the extended system, an existence condition is shown in the view how to get the controllability for the design of the time-invariant feedback control system.

2 Preliminaries

A real plant can be expressed by the linear time invariant model as follows:

$$\frac{dx}{dt} = Ax + Bu + \omega \tag{1}$$

$$y = Cx \tag{2}$$

where $A \in R^{n \times n}$ is the system matrix, $B \in R^{n \times m}$ is the input matrix, $C \in R^{p \times n}$ is the out matrix, $x = [x_1 \ x_2 \ \cdots \ x_n]^T \in R^n$ is the system state vector, $u = [u_1 \ u_2 \ \cdots \ u_m] \in R^m$ is the control input vector, $y = [y_1 \ y_2 \ \cdots \ y_p]^T \in R^p$ is the system output vector, $\omega = [\omega_1 \ \omega_2 \ \cdots \ \omega_n]^T \in R^n$ is the unmeasurable disturbance vector, and $m \geq p$.

The output error vector is defined by

$$e = y_r - y \tag{3}$$

where $y_r = [y_{r1} \ y_{r2} \ \cdots \ y_{rp}]^T \in R^p$ is the reference output vector, and $e = [e_1 \ e_2 \ \cdots \ e_p]^T \in R^p$ is the output error vector.

The following homogeneous differential equations for the ith disturbance $\omega_i(t)$ and the ith reference output vector y_{ri} are assumed to be described respectively as:

$$L_r(D)y_{ri}(t) = 0 \ for \ i = 1 \sim p \ and \ L_\omega(D)\omega_i(t) = 0 \ for \ i = 1 \sim n \tag{4}$$

where $L_\omega(D)$ and $L_r(D)$ are assumed to be described as the following differential polynomial operators with constant coefficients.

$$L_r(D) = D^\sigma + \rho_{\sigma-1}D^{\sigma-1} + \cdots + \rho_0 \text{ and } L_\omega(D) = D^l + \mu_{l-1}D^{l-1} + \cdots + \mu_0 \quad (5)$$

where $D = d/dt$ is the differential operator, ρ_i, μ_i are constant coefficients, σ, l are orders of differential polynomials. $L_\omega(D)$ and $L_r(D)$ can be expressed as

$$L_r(D) = R(D)U(D) \text{ and } L_\omega(D) = R(D)V(D) \quad (6)$$

where $R(D)$ is the greatest common divisor of $L_r(D)$ and $L_\omega(D)$, and $U(D), V(D)$ are factors of $L_r(D)$ and $L_\omega(D)$, respectively.

$L(D)$ is defined as the least common multiple of $L_r(D)$ and $L_\omega(D)$ and can be obtained from Eqs. (5)–(6) as the polynomial differential operator as follows:

$$L(D) = \frac{L_\omega(D)L_r(D)}{R(D)} = U(D)R(D)V(D) = V(D)L_r(D) \text{ or } U(D)L_\omega(D) \quad (7)$$
$$= D^{(q)} + \alpha_{q-1}D^{(q-1)} + \cdots + \alpha_0$$

where $\dim\{V(D)\} = q - \sigma$, $\dim\{U(D)\} = q - l$, $\dim\{R(D)\} = l + \sigma - q$, and $\dim\{L(D)\} = q \geq \dim\{L_r(D)\}$ or $\dim\{L_w(D)\}$.

For simplicity, let us consider a single-input single-output (SISO) system case with disturbance in Eq. (1). Then the block diagram of the closed loop control system can be shown in Fig. 1.

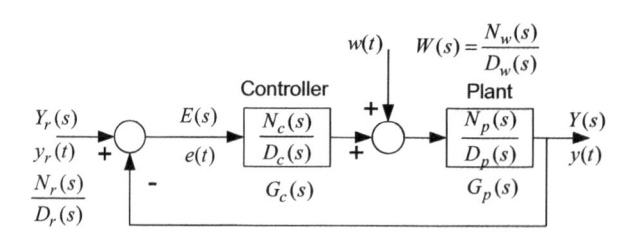

Fig. 1. Block diagram of a closed-loop control system

From Fig. 1, the output error after Laplace transform for the reference signal $Y_r(s)$ with disturbance $W(s)$ is obtained as:

$$E(s) = \frac{1}{1 + G_c(s)G_p(s)} Y_r(s) - \frac{G_p(s)}{1 + G_c(s)G_p(s)} W(s) = E_r(s) - E_w(s) \quad (8)$$

where

$$\begin{cases} E_r(s) = \frac{D_p(s)D_c(s)}{D_p(s)D_c(s) + N_p(s)N_c(s)} \frac{N_r(s)}{D_r(s)} \\ E_w(s) = \frac{D_c(s)N_p(s)}{D_p(s)D_c(s) + N_p(s)N_c(s)} \frac{N_w(s)}{D_w(s)} \end{cases} \quad (9)$$

The following polynomial equation obtained from Eq. (9) is the closed-loop characteristic polynomial equation of the closed-loop system depicted in Fig. 1 consisting of the polynomial poles and polynomial zeros of transfer functions of plant and controller and its roots are the closed-loop poles:

$$D_p(s)D_c(s) + N_p(s)N_c(s) = 0 \tag{10}$$

Theorem 1 (Internal model principle based on least common multiple model). Let us assume that the given system of Eq. (1) and the controller of $G_c(s)$ has no transmission zeros at the origin point and the closed loop poles of Fig. 1 are located in open left half plane under disturbance condition. Under assumptions of Eq. (4) for the disturbance and reference signals, the output error $e(t)$ of Eq. (3) becomes zero: $\lim_{t \to \infty} e(t) = 0$ if and only if the least common multiple polynomial for disturbance and reference signals is a factor of $D_c(s)$.

3 Robust Tracking Controller Design Method for MIMO System

Kim et al. [3] applied the tracking controller concept based on the previous stated Theorem 1 to the single input and single output system(SISO), but in the case of multi-input and multi-output(MIMO) system, it is not directly applicable and also it is very difficult to incorporate the design concept. Therefore, we should consider a different type of the robust tracking controller design concept.

In order to adopt the IMP of Sect. 2 to the robust MIMO tracking control system, we will introduce the polynomial differential operator stated in the previous section and get an extended system.

Operating $L(D)$ for ω_i and y_{ri} of Eq. (4), the following are obtained

$$\begin{cases} L(D)y_{ri} = U(D)R(D)V(D)y_{ri} = V(D)L_r(D)y_{ri} = 0 \\ L(D)\omega_i = U(D)R(D)V(D)\omega_i = U(D)L_w(D)\omega_i = 0 \end{cases} \tag{11}$$

where the dimension of q holds $q \geq l$ or $q \geq \sigma$.

Firstly, to eliminate the effect of disturbance in Eq. (1), operating the polynomial differential operator of $L(D)$ to both sides of Eq. (1) by using Eq. (11), Eq. (1) can be written as

$$\frac{d}{dt}\{L(D)x\} = AL(D)x + BL(D)u \tag{12}$$

The ith output error of Eq. (3) can be written as

$$e_i(t) = y_i(t) - y_{ri}(t) \text{ for } i = 1 \sim p \tag{13}$$

Secondly operating $L(D)$ to Eq. (13) and using the property of Eq. (11), the followings can be obtained.

$$L(D)e_i(t) = D^q e_i + \alpha_{q-1} D^{q-1} e_i + \cdots + \alpha_1 D e_i + \alpha_0 e_i$$
$$= L(D)y_i - L(D)y_{ri} = c_i L(D)x_i \quad \text{for} \quad i = 1, \cdots, p \tag{14}$$

From Eq. (14), the following can be obtained as the matrix form:

$$\dot{z}_i = M_i L(D)x + N z_i = \begin{bmatrix} 0 \\ 0 \\ \vdots \\ c_i^T \end{bmatrix} L(D)x + N z_i \tag{15}$$

where

$$N = \begin{bmatrix} 0 & 1 & 0 & 0 & \cdots & 0 \\ 0 & 0 & 1 & 0 & \cdots & 0 \\ 0 & 0 & 0 & 1 & \cdots & 0 \\ \vdots & \vdots & \vdots & \vdots & \ddots & \vdots \\ 0 & 0 & 0 & 0 & \cdots & 1 \\ -\alpha_0 & -\alpha_1 & -\alpha_2 & -\alpha_3 & \cdots & -\alpha_{q-1} \end{bmatrix} \in R^{q \times q},$$

$$M_i = \begin{bmatrix} 0 \\ \vdots \\ c_i^T \end{bmatrix} = \begin{bmatrix} 0 \\ 0 \\ \vdots \\ 1 \end{bmatrix} c_i^T \in R^{q \times n}, \quad c_i \in R^n$$

$C^T = \begin{bmatrix} c_1 & c_2 & \cdots & c_p \end{bmatrix} \in R^{n \times p}$, and $z_i = \begin{bmatrix} e_i & De_i & \cdots & D^{q-1} e_i \end{bmatrix}^T \in R^q$

By combining the operated system, Eqs. (12) and (15), an extended system can be obtained as follows:

$$\dot{x}_e = A_e x_e + B_e v \tag{16}$$

where

$$A_e = \begin{bmatrix} A & 0 & \cdots & 0 & 0 \\ \begin{bmatrix} 0 \\ c_1^T \end{bmatrix} & N & 0 & \vdots & 0 \\ \begin{bmatrix} 0 \\ c_2^T \end{bmatrix} & 0 & N & 0 & \vdots \\ \vdots & \vdots & \ddots & \ddots & 0 \\ \begin{bmatrix} 0 \\ c_2^T \end{bmatrix} & 0 & \cdots & 0 & N \end{bmatrix} \in R^{(n+pq) \times (n+pq)}, \quad B_e = \begin{bmatrix} B \\ 0 \\ \vdots \\ 0 \end{bmatrix} \in R^{(n+pq) \times m},$$

$$x_e = \begin{bmatrix} L(D)x \\ z_1 \\ z_2 \\ \vdots \\ z_p \end{bmatrix} \in R^{n+pq}$$

$x_e = \begin{bmatrix} L(D)x^T & z^T \end{bmatrix}^T$ is an extended system state variable vector, $v = L(D)u \in R^m$ is a new control law for the extended system, and $z = \begin{bmatrix} z_1^T & z_2^T & \cdots & z_p^T \end{bmatrix}^T \in R^{pq}$ is an error variable vector for the extended system.

A new control law for the extended system is defined by the following form:

$$v = L(D)u = -Fx_e \in R^m \qquad (17)$$

where $F = \begin{bmatrix} F_x & F_z \end{bmatrix} \in R^{m \times (n+pq)}$ is a feedback control gain matrix, and $F_x \in R^{m \times n}$ and $F_z \in R^{m \times pq}$ are feedback control gain matrices for $L(D)x$ and z, respectively.

A new error variable vector for the extended system can be defined as

$$\zeta = L^{-1}(D)z \qquad (18)$$

where $\zeta = \begin{bmatrix} \zeta_1^T & \zeta_2^T & \cdots & \zeta_p^T \end{bmatrix}^T \in R^{pq}$, and $\zeta_i \in R^q$ for $i = 1, \cdots, p$.

Using Eqs. (17) and (18), the control law of Eq. (1) can be obtained as follows:

$$u = -Fx_\zeta = -\begin{bmatrix} F_x & F_z \end{bmatrix} \begin{bmatrix} x \\ \zeta \end{bmatrix} \qquad (19)$$

where $x_\zeta \in R^{n+pq}$ is a new extended system variable vector.

The configuration of the proposed servo control system can be described as shown in Fig. 2.

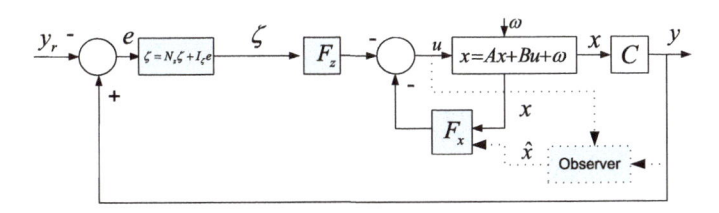

Fig. 2. Configuration of the proposed servo control system

4 Simulation Results

To verify the effectiveness of the proposed controller design method, our controller is compared with PI controller for a four steering wheel vehicle as a MIMO system with two inputs and two outputs as given in [4]. By defining the state vector $x = \begin{bmatrix} \kappa & \gamma \end{bmatrix}^T$, input vector $u = \begin{bmatrix} \delta_f & \delta_r \end{bmatrix}^T$ and output vector $y = \begin{bmatrix} \kappa & \gamma \end{bmatrix}^T$, the system matrices of the four steering wheel vehicle model in [4] can be written in the state-space form as follows:

$$A = \begin{bmatrix} -\frac{2(C_f+C_r)}{mV} & -\frac{2(l_fC_f-l_rC_r)}{mV^2}-1 \\ -\frac{2(l_fC_f-l_rC_r)}{J_z} & -\frac{2(l_f^2C_f-l_r^2C_r)}{J_zV} \end{bmatrix} = \begin{bmatrix} -3.41 & -0.9045 \\ 46.5451 & 3.173 \end{bmatrix}$$

$$B = \begin{bmatrix} \frac{2C_f}{mV} & \frac{2C_r}{mV} \\ \frac{2C_f}{J_z} & -\frac{2C_r}{J_z} \end{bmatrix} = \begin{bmatrix} 1000 & 2069 \\ 18.046 & -37.3367 \end{bmatrix}, \quad C = \begin{bmatrix} 1 & 0 \\ 0 & 1 \end{bmatrix}$$

where κ and γ denote the sideslip angle and yaw rate of vehicle at the CG; m is the vehicle mass; J_z is the yaw moment of inertia about its mass centre z – axis; V denotes the velocity of vehicle; l_f and l_r are the distances from the CG (centre of gravity) to the front and rear axles; δ_f and δ_r denote the steering angles of front and rear tires; C_f and C_r denote the lateral stiffness of the front and rear tires, respectively.

4.1 PI-MIMO Controller

The tunable gain matrices of K_P and K_I have eight tunable variables. Initial values for the controller are generated as shown in [4]. The gain value for PI-MIMO controller in this simulation was chosen as (Fig. 3):

$$K_P = \begin{bmatrix} 0.5 & 2 \\ 2 & 0.5 \end{bmatrix} \quad K_I = \begin{bmatrix} 0.5 & 0.1 \\ 0.1 & 0.5 \end{bmatrix}$$

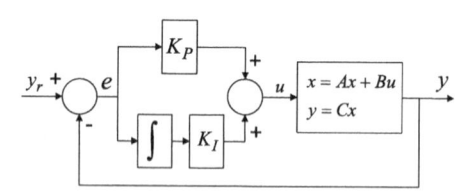

Fig. 3. PI-MIMO controller

4.2 The Proposed Servo Control System

The proposed servo control system is shown in Fig. 2. The parameters, controller gains and servo compensators for 3 types of reference signals are obtained by using the proposed robust servo controller design method as shown in Table 1.

4.3 Step Reference

Simulation results using step reference for the proposed method and the PI-MIMO controller are shown in Figs. 4, 5 and 6. Figure 4 shows the control inputs using for the proposed method and the PI-MIMO. Figure 5 shows that the outputs of the controllers used for both methods track the reference values and stabilize the plant after finite time. As we can see, the outputs of controller using the proposed method can track the

Table 1. Parameters and gains for proposed method

Design parameter	Reference types		
	Step	Ramp	Parabolic
Reference model	$y_{r1}=0.1, y_{r2}=0.01$	$\dot{y}_{r1}=0.1, \dot{y}_{r2}=0.01$	$\ddot{y}_{r1}=0.1, \ddot{y}_{r2}=0.01$
Matrix N	$N=[0]$	$N=\begin{bmatrix}0 & 1\\ 0 & 0\end{bmatrix}$	$N=\begin{bmatrix}0 & 1 & 0\\ 0 & 0 & 1\\ 0 & 0 & 0\end{bmatrix}$
Assigned poles	$\{-10,-11,-12,-13\}$	$\{-10,-11,-12,-13,-14,-15\}$	$\{-10,-11,-12,-13,-14,-15,-16,-17\}$
Gain matrix F_x	$\begin{bmatrix}1.3004 & 0.6693\\ -0.6181 & -0.3239\end{bmatrix}$	$\begin{bmatrix}1.3449 & 1.1334\\ -0.6337 & -0.5475\end{bmatrix}$	$\begin{bmatrix}1.2630 & 1.5923\\ -0.5862 & -0.7711\end{bmatrix}$
Gain matrix F_z	$\begin{bmatrix}0.0780 & 3.0478\\ 0.0377 & -1.4731\end{bmatrix}$	$\begin{bmatrix}6.848 & -2.4031\\ 1.1859 & -0.3509\\ 54.1180 & -26.0385\\ 13.0744 & -6.3002\end{bmatrix}^T$	$\begin{bmatrix}-100.7003 & 66.9356\\ -23.1968 & 15.7626\\ -1.5543 & 1.2687\\ 899.5088 & -437.1666\\ 271.1928 & -131.6198\\ 30.4197 & -14.7434\end{bmatrix}^T$
Compensator	$\dot{\zeta}=\begin{bmatrix}0 & 0\\ 0 & 0\end{bmatrix}\zeta+\begin{bmatrix}1 & 0\\ 0 & 1\end{bmatrix}e$	$\dot{\zeta}=\begin{bmatrix}0 & 1 & 0 & 0\\ 0 & 0 & 0 & 0\\ 0 & 0 & 0 & 1\\ 0 & 0 & 0 & 0\end{bmatrix}\zeta+\begin{bmatrix}0 & 0\\ 0 & 1\\ 0 & 0\\ 0 & 1\end{bmatrix}e$	$\dot{\zeta}=\begin{bmatrix}0 & 1 & 0 & 0 & 0 & 0\\ 0 & 0 & 1 & 0 & 0 & 0\\ 0 & 0 & 0 & 0 & 0 & 0\\ 0 & 0 & 0 & 0 & 1 & 0\\ 0 & 0 & 0 & 0 & 0 & 1\\ 0 & 0 & 0 & 0 & 0 & 0\end{bmatrix}\zeta+\begin{bmatrix}0 & 0\\ 0 & 0\\ 1 & 0\\ 0 & 0\\ 0 & 0\\ 0 & 1\end{bmatrix}e$

reference signal faster than the output of the controller using the PI-MIMO. Figure 6 shows that the output errors of the controllers using both methods converge to zero and stabilize the plant after finite time. However, the output errors of the closed loop system using the proposed method can become zero faster than the output errors of the controller using the PI-MIMO.

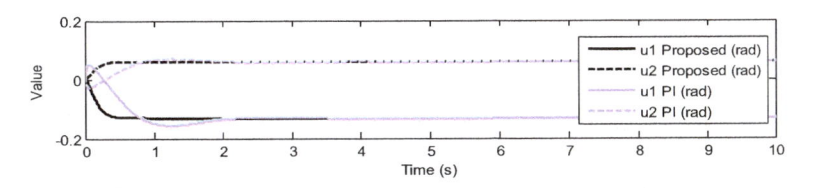

Fig. 4. Control inputs of PI and proposed methods for step reference

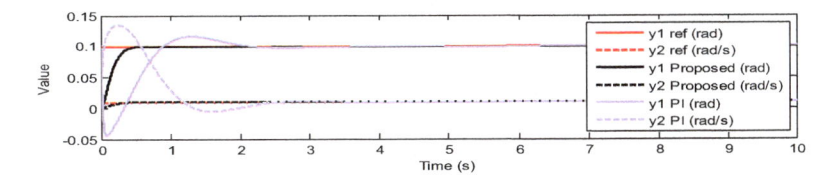

Fig. 5. Outputs of PI and proposed methods for step reference

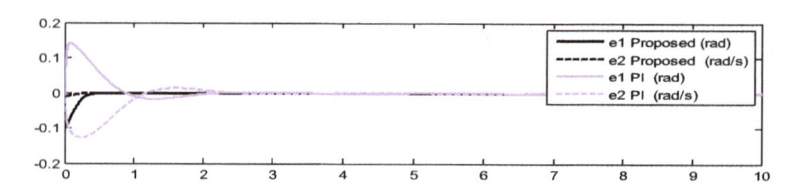

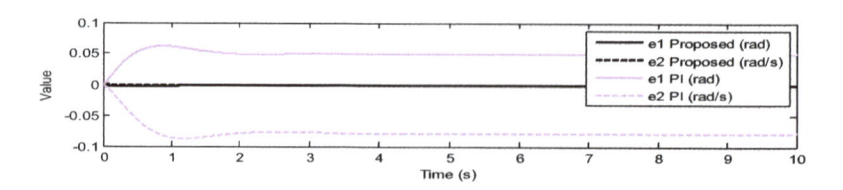

Fig. 6. Output errors of PI and proposed methods for step reference

4.4 Ramp Input

Simulation results using ramp input for the proposed method and the PI-MIMO controller are shown in Figs. 7, 8 and 9. Figure 7 shows the control inputs using both methods. Figure 8 shows that only the outputs of the controller using the proposed method can track the reference signals and stabilize the plant after finite time. The outputs of the controller using the PI-MIMO cannot track the reference signal. Furthermore, Fig. 9 shows that the output errors of the controller using the proposed method converge to zero and stabilize the plant after finite time. On the other hand, in the controller using the PI-MIMO, the steady state errors exist.

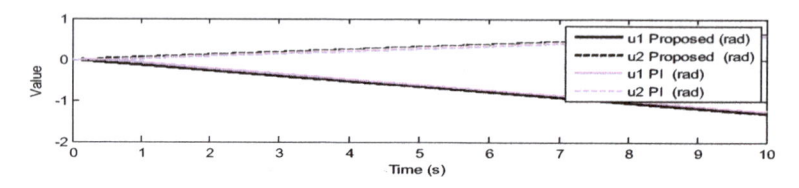

Fig. 7. Control inputs of PI and proposed methods for ramp reference

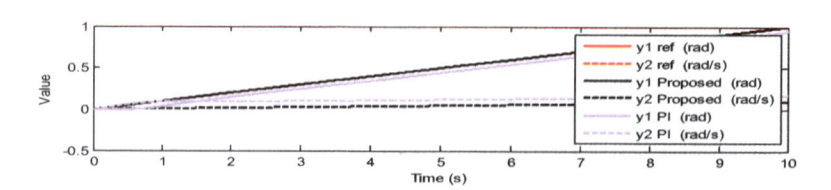

Fig. 8. Outputs of PI and proposed methods for ramp reference

Fig. 9. Output errors of PI and proposed methods for step reference

4.5 Parabolic Input

Simulation results using parabolic input for the proposed method and the PI-MIMO controller are shown in Figs. 9, 10 and 11. Figure 9 shows the control inputs using both methods. Figures 10 and 11 show the same results as shown in the case of ramp for the parabolic type of reference signal, that is, PI controller has steady error for the given parabolic signals (Fig. 12).

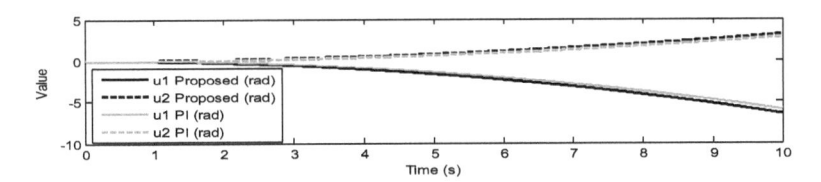

Fig. 10. Control inputs of PI and proposed methods for parabolic reference

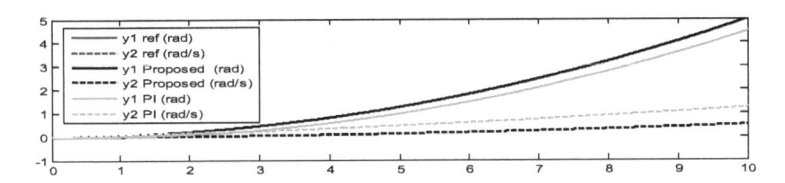

Fig. 11. Outputs of PI and proposed methods for parabolic reference

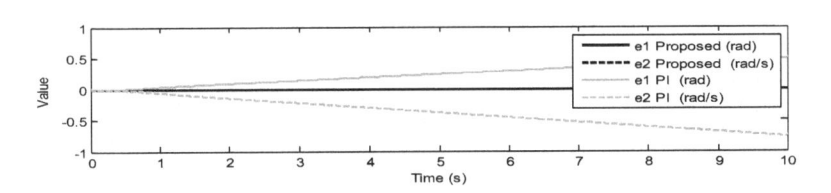

Fig. 12. Outputs of PI and proposed methods for parabolic reference

5 Conclusions

This paper proposed a robust servo controller deign for MIMO systems based on a linear shift invariant differential (LSID) operator and its inverse operator on Schwartz space using the internal model principle (IMP). To do this task, the followings were done. First, the basic concept idea of LSID operator and its invertible operator were described clearly. Furthermore, the IMP was rearranged based on LSID operator and the IMP was modified by the properties of the LSID operator. Second, an extended system operated by the LSID operator to a given MIMO system with the given reference and disturbance was obtained. Third, the controllability checking of extended systems was done. Fourth, a state feedback law was obtained by solving the pole assignment problem with all the poles of the extended system. The simulation results were shown to verify the effectiveness of the proposed controller for a 4-wheel steering vehicle. It showed that the proposed method could overcome the difficult tracking problem for more high order types of reference signals under disturbances. That is, the proposed controller could track the step reference for the system within desired finite errors under the step disturbance, the ramp reference for the system under the ramp disturbance, and the parabolic reference for the system under the parabolic disturbance.

Acknowledgement. This research was conducted under the Pukyong National University Research Park (PKURP) for Industry-Academic Convergence R&D support program, which is funded by the Busan Metropolitan City, Korea.

References

1. Davison, E.J., Smith, H.W.: Pole assignment in linear time-invariant multivariable systems with constant disturbances. Automatics **7**, 489–498 (1971)
2. Davison, E.J.: The output control of linear time-invariant multivariable systems with unmeasurable arbitrary disturbances. IEEE Trans. Autom. Control **17**(5) (1972)
3. Kim, S.B., Oh, S.J., Jung, Y.G., Kim, H.S.: Application of bilinear transformation method to servo system design and position control for a cart system. Trans. KIEE **40**(3) (1991)
4. Lv, H., Liu, S.: Closed-loop handling stability of 4WS vehicle with yaw rate control. J. Mech. Eng. **59**(10), 595–603 (2013)
5. Furuta, K., Sano, A., Atherton, D.: State Variable Methods in Automatic Control. Wiley, Chichester (1988)
6. Furuta, K., Kim, S.B.: Pole assignment in a specified disk. IEEE Trans. Auto. Control **32**(5), 423–427 (1987)
7. Kim, S.B., Furuta, K.: Regulator design with poles on a specified region. Int. J. Control **47**(1), 143–160 (1988)
8. Kim, D.H: Servo Controller Design Using Polynomial Differential Operator Method and Its Applications, Doctor Thesis, Mechanical Design Engineering Department, Pukyong National University (2015)
9. Kim, D.-H., Nguyen, T.H., Pratama, P.S., Gulakari, A.V., Kim, H.-K., Kim, S.-B.: Servo system design for speed control of AC induction motors using polynomial differential operator. Int. J. Control Autom. Syst. **15**(3), 1207–1216 (2017)

Servo Controller Design and Fault Detection Algorithm for Speed Control of a Conveyor System

Trong Hai Nguyen[✉], Nguyen Thanh Phuong, and Hung Nguyen

Hutech Institute of Engineering, Ho Chi Minh City University
of Technology (HUTECH), 475A Dien Bien Phu Street, Ward 25,
Binh Thanh District, Ho Chi Minh City, Vietnam
{nt.hai75,nt.phuong,n.hung}@hutech.edu.vn

Abstract. This paper proposes a servo controller design and fault detection algorithm for speed control of a conveyor system. Firstly, modeling for a conveyor system is described. Secondly, the robust servo controller based on polynomial differential operator is applied to track the trapezoidal velocity profile reference input. Thirdly, a fault detection algorithm based on Extended Kalman Filter (EKF) is proposed. From the EKF, the estimated angular velocity indicates the encoder failure. The estimated friction indicates the mechanical failure. Fault isolation is obtained by the friction bound. Finally, the simulation and experimental results are shown to verify the effectiveness of the proposed algorithm.

Keywords: Internal model principle · Servo system · Speed control · Fault detection · Kalman Filter

1 Introduction

Conveyors are material handling devices not only to transport loads/units but also to sort or store them temporarily. Most AC motor drives in conveyor systems have used PI control [1] but difficult for applying to obtain sufficiently high performance in the tracking application. To satisfy the tracking requirements, huge amount of algorithms were proposed by researchers [2–5]. However, the transient performances are always unsatisfactory. To solve this problem, Kim et al. [6, 7] proposed a new concept of servo controller design method by introducing a polynomial differential operator. In this paper, a robust servo controller proposed by Kim et al. [7] is applied for speed control of a conveyor system. The experimental results of the proposed servo controller are compared to the conventional PI controller and MRAC controller.

Since the conveyor system work automatically, their safety is considered. In order to maintain the conveyor system, several features such as energy consumption, output, temperature, vibrations, belt speed etc. might be monitor. The fault detection algorithm helps user to detect faults and prevents serious damage in the conveyor. Therefore, many fault detection algorithms were proposed to increase the safety and reliability of conveyor system [8–10]. Li et al. [8] proposed the fault detection algorithm named

Modified Regular Bands based on X-ray image to monitor the status of steel cord conveyor belt. Jiang et al. [9] proposed a belt conveyor roller fault audio detection based on the wavelet neural network. Kanmani [10] proposed faults such as belt tear up faults, oil level reduction fault, fire occurrence fault by using PLC and SCADA. However, this method had high computational cost. Therefore, to deal with these problems, this paper proposes an improved fault detection algorithm for the conveyor system with known mathematical model and low computational cost. In this paper, a fault detection algorithm based on EKF is proposed. From the EKF, the estimated angular velocity indicates the encoder failure. The estimated friction indicates the mechanical failure Fault isolation is obtained by the friction bound.

2 Servo Controller Design

2.1 Modeling of a Belt Driven Transmission Section

Figure 1 shows the schematic of a BDTS where ω_m, θ_m, R_a, J_m, b_m, τ_m denote the angular velocity, angular speed, driving pulley radius, moment of inertia, viscous damping coefficient, and electromagnetic torque of induction motor, respectively. ω_L, θ_L, R_b, J_L, b_L, τ_L, T denote the angular velocity, angular speed, driven pulley radius, moment of inertia, viscous damping coefficient, and torque of driven pulley, belt tension, respectively.

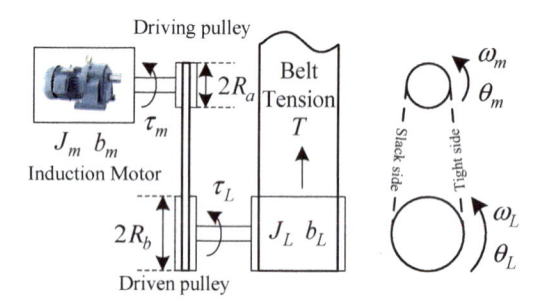

Fig. 1. Schematic of a conveyor system.

The state dynamic equation of a given system with disturbance ε can be expressed as follows:

$$\begin{cases} \dot{x} = Ax + Bu + \varepsilon \\ y = v_L = Cx \end{cases} \quad (1)$$

where $A = \begin{bmatrix} 0 & 1 \\ -a_2 & -a_1 \end{bmatrix}$, $B = \begin{bmatrix} 0 \\ b_1 \end{bmatrix}$, $C = [R_b \ 0]$, $a_1 = (b_L/J_L)$, $a_2 = (R_b^2 K/J_L)$, $b_1 = (R_a R_b K/J_L)$, $x = [x_1 \ x_2]^T = [\omega_L \ \dot{\omega}_L]^T$, $u = \omega_m$, $v_L = R_b \omega_L$

where v_L is the belt linear velocity, K is the spring constant of belt and ε is related to load, tension, etc. It is assumed that (A,B) is controllable and (A,C) is observable.

2.2 Controller Design

The output error is defined by

$$e = y - y_r \tag{2}$$

The servo control system design is attempted by 3 steps as follows:

[Step 1] eliminate the effect of disturbance in Eq. (1) by operating the polynomial differential operator of $L(D) = D^q - \alpha_{q-1}D^{q-1} + \cdots + \alpha_0$ to both sides of Eqs. (1) and (2):

$$\begin{cases} \frac{d}{dt}\{L(D)x\} = AL(D)x + BL(D)u \\ L(D)e = CL(D)x \end{cases} \tag{3}$$

where $D = d/dt$, $L(D)y_r = 0$ and $L(D)\varepsilon = 0$.

[Step 2] an extended system is obtained by using the operated system through the step 1 as follows:

$$\dot{x}_e = A_e x_e + B_e v \tag{4}$$

where $A_e = \begin{bmatrix} A & \mathbf{0} \\ M & N \end{bmatrix}$, $B_e = \begin{bmatrix} B \\ \mathbf{0} \end{bmatrix}$, $M = \begin{bmatrix} \mathbf{0} \\ \mathbf{c}_1^T \end{bmatrix}$,

$$x_e = \begin{bmatrix} L(D)x \\ e \\ e^{(1)} \\ \vdots \\ e^{(q-1)} \end{bmatrix}, \quad N = \begin{bmatrix} 0 & 1 & 0 & \cdots & 0 \\ 0 & 0 & 1 & \cdots & 0 \\ \vdots & \vdots & \vdots & \ddots & \vdots \\ 0 & 0 & 0 & \cdots & 1 \\ -\alpha_0 & -\alpha_1 & -\alpha_2 & \cdots & -\alpha_{q-1} \end{bmatrix}$$

[Step 3] by defining a new control law, v, and a new error variable vector, z, as Eq. (5), the extended system of Eq. (4) can be written by Eq. (6) as follows:

$$\begin{cases} v = L(D)u = -Fx_e \\ \zeta = L^{-1}(D)z \end{cases} \tag{5}$$

$$\begin{cases} \frac{d}{dt}x = [A - BF_x]x - BF_z\zeta + \varepsilon = Ax + Bu + \varepsilon \\ \frac{d}{dt}\zeta = N\zeta + I_\zeta e \end{cases} \tag{6}$$

where $\zeta = \begin{bmatrix} \zeta_1 & \zeta_2 & \cdots & \zeta_q \end{bmatrix}^T$, $I_\zeta = \begin{bmatrix} 0 & 0 & \cdots & 1 \end{bmatrix}^T$, $L(D)\varepsilon = 0$, $F = \begin{bmatrix} F_x & F_z \end{bmatrix}$ is a feedback control gain matrix, and the second term of Eq. (6) is a servo compensator for Eq. (1).

The configuration of the proposed servo control system can be described as shown in Fig. 2.

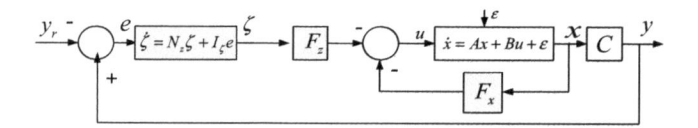

Fig. 2. Proposed servo control system

3 Fault Detection Algorithm

3.1 Extended Kalman Filter

In the EKF [11], the state transition model and the observation model are not linear functions of the state but may be differentiable functions as follows:

$$\begin{cases} \mathbf{x}_k = f(\mathbf{x}_{k-1}, \mathbf{u}_{k-1}) + \mathbf{w}_{k-1} \\ \mathbf{z}_k = h(\mathbf{x}_k) + \mathbf{v}_k \end{cases} \quad (7)$$

where $\mathbf{x}_k$ is the state vector at time k, $\mathbf{w}_{k-1}$ is the process noise at previous time $k-1$ and $\mathbf{v}_k$ is the observation noise at time k which are assumed to be zero mean multivariate Gaussian noises with covariance $\mathbf{Q}_k$ and $\mathbf{R}_k$ at time k, respectively. $f(\cdot)$ is the process nonlinear vector function, $h(\cdot)$ is the observation nonlinear vector function, and $\mathbf{z}_k$ is the output vector current time k.

To drive the conveyor system, the inverter with DC voltage input controls the induction motor to create sufficient torque can be expressed as follow:

$$\tau_m = J\dot{\omega}_m + b\omega_m \quad (8)$$

where $J = J_m + J_L$, $b = b_m + b_L$ and τ_m is a sufficient torque and:

$$\tau_m = K_m u_m \quad (9)$$

where K_m is a torque constant, u_m is a control input

To estimate the friction b using EKF, introduce an auxiliary state of friction coefficient as follow:

$$\dot{b} = 0 \quad (10)$$

The model state, $\mathbf{x} = [\omega \ b]^T$, and measurement, $\mathbf{z} = [\omega \ \dot{\omega}]^T$, equations are:

$$\dot{\mathbf{x}} = \begin{bmatrix} \dot{\omega}_m \\ \dot{b} \end{bmatrix} = \begin{bmatrix} \frac{K_m u_m - b\omega_m}{J} \\ 0 \end{bmatrix} \quad (11)$$

$$\mathbf{z} = \begin{bmatrix} \omega \\ \frac{K_m u_m - b\omega_m}{J} \end{bmatrix} \quad (12)$$

or in discrete type

$$\mathbf{x}_k = \begin{bmatrix} \omega_{mk} \\ b_k \end{bmatrix} = \begin{bmatrix} \omega_{m(k-1)} \frac{K_{m(k-1)} u_{m(k-1)} - b_{(k-1)} \omega_{m(k-1)}}{J} \\ b_{(k-1)} \end{bmatrix} + \mathbf{w}_{k-1} \quad (13)$$

$$\mathbf{z}_k = \begin{bmatrix} \omega_k \\ \frac{K_{mk} u_{mk} - b_k \omega_{mk}}{J} \end{bmatrix} + \mathbf{v}_k \quad (14)$$

The EKF consists of two steps as follows:
Prediction step:

$$\begin{cases} \hat{\mathbf{x}}_{k|k-1} = f(\hat{\mathbf{x}}_{k-1|k-1}, \mathbf{u}_{k-1}) \\ \mathbf{P}_{k|k-1} = \mathbf{F}_{k-1} \mathbf{P}_{k-1|k-1} \mathbf{F}_{k-1}^T + \mathbf{W}_k \mathbf{Q}_{k-1} \mathbf{W}_k^T \\ \mathbf{F}_{k-1} = \frac{\partial f}{\partial \mathbf{x}} \Big|_{\hat{\mathbf{x}}_{k-1|k-1}, \mathbf{u}_{k-1}} \\ \mathbf{W}_k = \frac{\partial f}{\partial \mathbf{u}} \Big|_{\hat{\mathbf{x}}_{k-1|k-1}, \mathbf{u}_{k-1}} \end{cases} \quad (15)$$

where $\hat{\mathbf{x}}_{k|k-1}$ is the predicted state estimate at time k, and $\mathbf{P}_{k|k-1}$ is the predicted covariance matrix estimate at time k.

Update step:

$$\begin{cases} \tilde{\mathbf{y}}_k = \mathbf{z}_k - h(\hat{\mathbf{x}}_{k|k-1}) \\ \mathbf{S}_k = \mathbf{H}_k \mathbf{P}_{k|k-1} \mathbf{H}_k^T + \mathbf{R}_k \\ \mathbf{K}_k = \mathbf{P}_{k|k-1} \mathbf{H}_k^T (\mathbf{S}_k)^{-1} \\ \hat{\mathbf{x}}_{k|k} = \hat{\mathbf{x}}_{k|k-1} + \mathbf{K}_k \tilde{\mathbf{y}}_k \quad \tilde{\mathbf{y}}_k = \mathbf{z}_k - h(\hat{\mathbf{x}}_{k|k-1}) \\ \mathbf{P}_{k|k} = (\mathbf{I} - \mathbf{K}_k \mathbf{H}_k) \mathbf{P}_{k|k-1} \\ \mathbf{H}_{ik} = \frac{\partial h_i}{\partial \mathbf{x}_i} \Big|_{\mathbf{x}_{ik|k-1}}, \end{cases} \quad (16)$$

where $\tilde{\mathbf{y}}_k$ is the measurement innovation at time k, $\mathbf{z}_k$ is the output vector from estimation at time k, $h(\hat{\mathbf{x}}_{k|k-1})$ is the measurement result, $\mathbf{S}_k$ is the innovation covariance at time k, $\mathbf{R}_k$ is the measurement noise covariance at time k, $\mathbf{K}_k$ is the Kalman gain at time k, $\hat{\mathbf{x}}_{k|k}$ is the updated state estimation at time k, and $\mathbf{P}_{k|k}$ is the updated covariance matrix estimation at time k.

3.2 Fault Detection and Isolation Using EKF

From the EKF, the estimated angular velocity, $\tilde{\omega}$, and the estimated friction, $\tilde{b}$, can be obtained. The block diagram of the proposed algorithm is shown in Fig. 3. The estimated angular velocity indicate the encoder failure. The estimated friction indicate the mechanical failure. The fault condition is as follow:

$$\text{Fault condition} \begin{cases} \text{normal 0 if } \tilde{\omega} < Th_1 \text{ and } \tilde{b} < Th_2 \\ \text{fault 1 if } \tilde{\omega} > Th_1 \text{ or } \tilde{b} > Th_2 \end{cases} \quad (17)$$

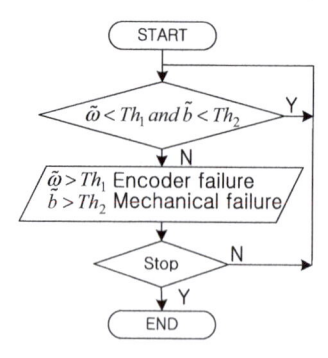

Fig. 3. The block diagram of the proposed algorithm

In this paper, the threshold value Th_1 and Th_2 is predetermined and constant based on experiment data.

4 Experimental Results

Experiments are done to verify the effectiveness of the designed controller and fault detection algorithm. The inverter is a custom-built inverter (controller board and power driver) shown in Fig. 4. Table 1 shows the parameters of conveyor.

Fig. 4. Controller board and power driver of inverter

Table 1. Parameters values of conveyor system.

Symbol	Value	Symbol	Value
J_L	2.10^{-5} Kgm²	$b_m = b_L$	0.5 Nms
K	4716 N/m	$R_a = R_b$	0.055 m

Table 2 shows the parameters, controller gains and servo compensators for 2 types of reference signals are obtained by using the proposed servo controller design method.

Table 2. Parameters and gains for the proposed method.

Reference types (m/s)	Reference model	Matrix N	Assigned poles	F_x	F_z	Servo compensator
Step $y_r = 4$	$\dot{y}_r = 0$ $y_r(0) = 4$	$[0]$	$\{-200, -100, -90\}$	$\begin{bmatrix} -0.9341 \\ 0.0005 \end{bmatrix}^T$	$[45.8818]$	$\dot{\zeta} = e$
Ramp $y_r(t) = t$	$\ddot{y}_r(t) = 0$ $\dot{y}_r(0) = 1$ $y_r(0) = 0$	$\begin{bmatrix} 0 & 1 \\ 0 & 0 \end{bmatrix}$	$\{-200, -100, -20, -21\}$	$\begin{bmatrix} -0.9541 \\ -0.0005 \end{bmatrix}^T$	$\begin{bmatrix} 214.1152 \\ 24.1134 \end{bmatrix}^T$	$\dot{\zeta} = \begin{bmatrix} 0 & 1 \\ 0 & 0 \end{bmatrix} \zeta + \begin{bmatrix} 0 \\ 1 \end{bmatrix} e$

It is assumed that all the initial state variable values of system are zero and the disturbance is given as step type model.

Figure 5 shows the profile using trapezoidal velocity reference input of a conveyor system used in this paper.

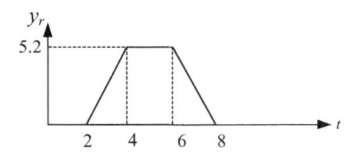

Fig. 5. Trapezoidal velocity profile reference input

4.1 Experimental Results of Servo Controller Design

Experimental results are shown in Fig. 6. The outputs of the PI controller have steady state errors in cases of ramp reference input. The outputs of MRAC controller can track the reference input more slowly than those of the proposed method. Only the outputs of the controller using the proposed method can track this type of reference input and the plant can be stabilized after finite time.

4.2 Experimental Result of Fault Detection Algorithm

The experimental results for fault detection are shown in Fig. 7. Figure 7(a) shows the experimental results for normal condition. Figure 7(b) shows that the estimated friction values increased over the threshold value $Th_2 = 2$ Nms at $t = 5$ s. It indicate the mechanical fault. Figure 7(c) shows that the estimated angular velocity increased over the threshold value $Th_1 = 300$ rpm at $t = 4$ s. It indicate the encoder fault.

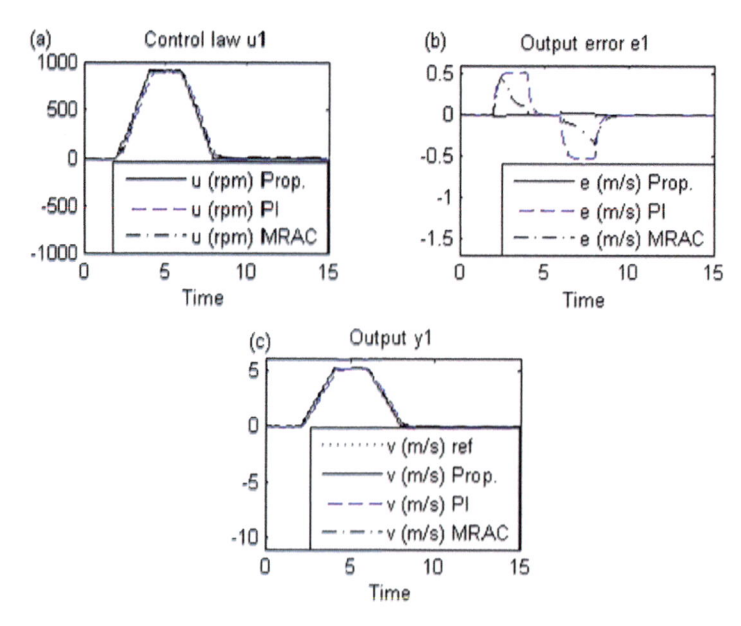

Fig. 6. Control law u, output error e, output y for trapezoidal velocity profile y_r

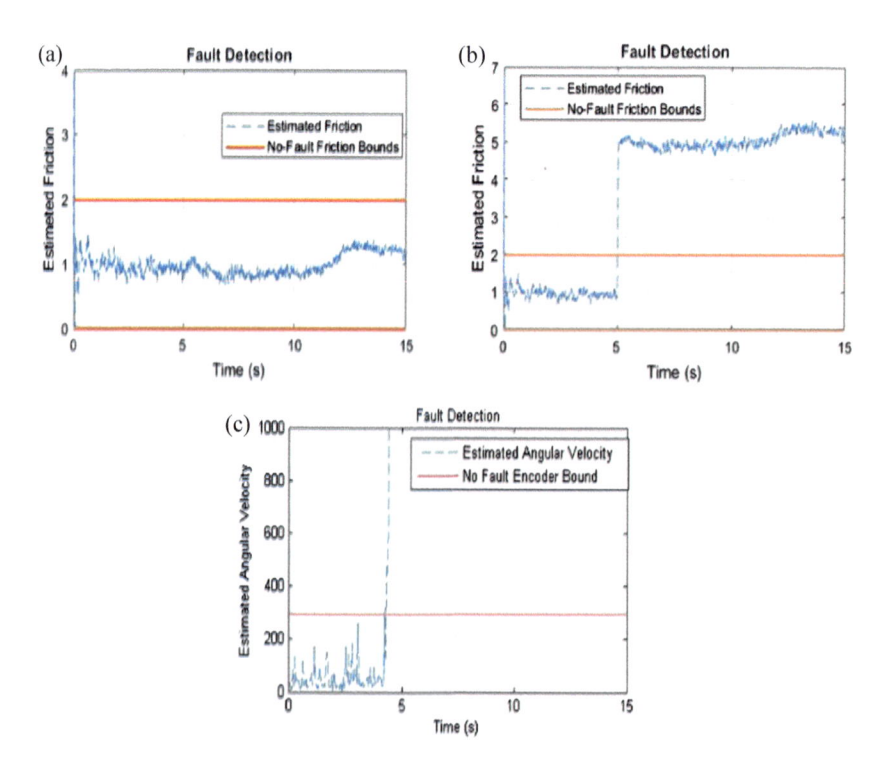

Fig. 7. (a) Normal condition, (b) Mechanical fault detection, (c) Encoder fault detection

5 Conclusion

This paper proposed a servo controller design and fault detection algorithm for conveyor system. The experimental results of the servo controller design showed that the performance of the proposed method satisfied the requirement at a speed of 5.2 m/s. The experimental results of fault detection showed that the proposed algorithm successfully detected the mechanical fault conditions and encoder fault detection. The algorithm calculated the estimated based on EKF. In the future, this system can be improved so that it can detect more than one fault at a time.

Acknowledgement. This research was supported by a Research Grant of Ho Chi Minh City University of Technology (HUTECH) (2018 year).

References

1. Lin, C.T., Hung, C.W., Liu, C.W.: Fuzzy PI controller for BLDC motors considering variable sampling effect. In: 33rd Annual Conference of the IEEE Industrial Electronics Society, IECON 2007, pp. 5–8 (2007)
2. Tseng, C.S., Chen, B.S., Uang, H.J.: Fuzzy tracking control design for nonlinear dynamic systems via T-S fuzzy model. IEEE Trans. Fuzzy Syst. **9**(3), 381–392 (2001)
3. Marino, R., Peresada, S., Tomei, P.: Adaptive output feedback control of induction motor with incertain rotor resistance. IEEE Trans. Autom. Control **44**, 967–983 (1983)
4. Fu, T.J., Xie, W.F.: A novel sliding-mode control of induction motor using space vector modulation technique. ISA Trans. **44**(4), 481–490 (2005)
5. Kumamoto, A., Tada, S., Hirane, Y.: Speed regulation of an induction motor using model reference adaptive control. IEEE Control Syst. Mag. **6**(5), 25–29 (2003)
6. Nguyen, T.H., Lee, J.W., Kim, H.K., Tran, T.P., Choe, Y.W., Dinh, V.T., Kim, S.B.: Servo controller design for speed control of BLDC motors using polynomial differential operator method. In: Recent Advances in Electrical Engineering and Related Sciences, LNEE, vol. 371, pp. 353–365 (2015)
7. Kim, D.H., Nguyen, T.H., Pratama, P.S., Kim, H.K., Jung, Y.S., Kim, S.B.: Servo system design for speed control of AC induction motors using polynomial differential operator. Int. J. Control Autom. Syst. **15**, 1207–1216 (2017)
8. Li, X.G., Miao, C.Y., Wang, W., Zhang, Y.: Fault automatic detection method for steel cord conveyor belt based on the regularity analysis. Int. J. Dig. Content Technol. Appl. **6**(1), 226–234 (2012)
9. Jiang, X.P., Cao, G.Q.: Belt conveyor roller fault audio detection based on the wavelet neural network. In: 11th International Conference on Natural Computation, pp. 954–958 (2015)
10. Kanmani, M., Nivedha, J., Sundar, G.: Belt conveyor monitoring and fault detecting using PLC and SCADA. Int. J. Adv. Res. Electr. Electron. Instrum. Eng. **3**, 243–248 (2014)
11. Krener, A.J.: The convergence of the extended Kalman filter. In: LNCIS, vol. 286, pp. 173–182 (2003)

A Control System for Power Electronics with an NXP Kinetis Series Microcontroller

Daniel Kouřil[✉], Martin Sobek, and Petr Chamrád

Department of Electronics, FEECS, VŠB-Technical University of Ostrava,
17. listopadu 15, 708 33 Ostrava-Poruba, Czech Republic
{daniel.kouril,martin.sobek,petr.chamrad}@vsb.cz

Abstract. This article describes the design and implementation of a Power Electronics Control System, which is based on the selected NXP Kinetis series microcontroller, the KV58 family. The introduction describes the specific requirements of the Power Electronics for the Control System and units. The properties of the selected microcontroller, topology design and realization are described in the following sections. The experimental results and application of the proposed Control System are shown in the final part of this article.

Keywords: Control system for power electronics · Hardware design · Microcomputer electronics · Microcontroller · Industrial electronics

1 Introduction

Power and Control Electronics use semiconductor components connected in specific topologies to provide efficient power conversion, operation and control accuracy. Large emphasis is placed on operation reliability and cost savings with regards to the purchase and maintenance. Thus, it deals with fields of power engineering, automation, electric vehicles, transport and especially applications with induction motor drives, where their massive expansion to almost every industrial sector leads to the development of efficient safe modern inverters and control systems with small dimensions and extensive functions that provide compact modular solutions which communicate through industrial bus interfaces.

The main task for the Power Electronics Control System is to implement a created switching algorithm and control signals toward the driver's gate drive. Thus, the Pulse-width Modulation (PWM) modulator periphery with a sufficient number of output channels and functions is suitable. For effective regulation and diagnostics, it is necessary to implement a feedback loop, so the system is enhanced with components for signal processing from real quantity sensors (voltage, current, speed, position). It is equipped with peripherals of industrial communication standards) for contact with the nearby devices.

Modern Power Electronic controllers are built on Digital Signal Processors (DSPs) and Microcontroller Units (MCUs) that can provide industry-leading computing power, special peripherals, relatively low cost and speed. This has enabled real implementation of adaptive and robust control methods.

The described Control Unit is designed modularly to provide the ability for easy system reconfiguration even on a hardware level, which may result in the maximum possible utilisation of the computing power for a concrete application [2–4].

2 The Control System Based on the NXP Kinetis MKV58F Microcontroller

The MKV58F1M0VLQ24 MCU allows top-of-the-line peripherals for real-time work. This MCU is specifically designed for high-processing performance applications with industrial motor drives and advanced motor control, industrial automation, multi-motor drives, photovoltaic systems, an industrial robot controllers based on ARM and FPGA [6], UPSs and more.

The entire family of the Kinetis KV5x series is built on a single-core, 32-bit floating point architecture ARM Cortex-M7, which is the latest and most powerful cortex-M core available nowadays, due to this fact, the MKV58F1M0VLQ24 is one of the most powerful industrial motor drive MCUs at this time. The core includes a DSP and a Floating-Point Unit (FPU), which enables the rapid energy-saving processing of algorithms and many more features. A list of the most important peripherals for Power Electronics control unit follows. For more information, see reference [1].

2.1 eFlex PWM Modulators

As mentioned in the introduction section, this is the most important periphery, the main purpose of which is to create switching combinations for power semiconductor devices. The MCU contains two PWM modules with a so-called *"nano-edge"* function that allows you to set a duty cycle with a decimal, floating resolution. Both PWM modules contain 4 submodules, each submodule having 3 PWM outputs, 12 PWM outputs in total that can be driven independently or as a complementary output pair. They can also be configured as inputs for detecting the polarity of current flowing through the complementary pair of the so-called "dead time effect" correction function. The modulator is able to create *a Center aligned PWM, Edge Aligned PWM, Phase Shifted PWM, Double Switching PWM* and *PWM output trigger by Analogue to Digital Converter (ADC)* [1].

2.2 ADC and HSADC Converter

The MCU is equipped with one linearly approximated 16-bit ADC converter module. It is possible to create 4 pairs of differential inputs and 24 single-acting analogue inputs. The module supports several output modes with adjustable resolution (16, 13, 11 and 9 bits).

The MCU includes two High Speed Analogue to Digital Converters (HSADC), a 12-bit, 19-channel, fast converter with a maximum sampling rate of 5Ms/s at 12-bit resolution and 10Ms/s for 6-bit resolution. Each HSADC converter module contains two internal A and B converters. The modules allow for the internal synchronisation with other peripherals in the microcontroller. The maximum conversion time is 200 ns at 75 MHz. The internal multiplexer selects one of the 19 channels, allowing parallel operation to record and store up to 8 measurements, with serial operation of up to 16 measurements on both HSADCs [1].

2.3 DAC Converter

A Digital to Analogue (DAC) converter periphery, an external integrated circuit AD5624, a 12-bit serial "nanoDAC" converter with low power consumption was selected. It allows conversion up to 50 MHz. The MCU is connected via the Serial Peripheral Interface (SPI) to this external periphery. The converter has 4 output channels [1].

2.4 Rotary and Position Sensor Interface (ENC)

The design of the control unit includes the interface circuitry in order to connect the rotary position and speed sensors. A typical quadrature encoder configuration is with three outputs. The signals PHASE_A, PHASE_B and INDEX. By evaluating the phase shift between PHASE_A and PHASE_B, it is possible to determine the direction of motor drive rotation, by integrating the phase shift values, it is possible to determine the position and the speed information by derivation. The INDEX resets the position counter [1].

2.5 Serial Communication Interfaces (SPI, I2C, CAN, UART)

These are serial communication peripherals that the MCU uses to communicate with other devices, external peripherals, and possibly a superior control system. The proposed control system has two Universal Asynchronous Receiver-Transmitter (UART) peripherals running on RS422 and RS485 industrial standards. Both industrial interfaces are galvanically isolated [1, 5, 7].

3 Control System Topology Design

Figure 1 shows the topology of the proposed system unit. The control system is designed modularly, physically it is multiple printed circuit boards (PCBs), and it makes the design very versatile for further development. **The MCU's main-board** is a PCB consisting of the fed circuits, the MCU, an ENC interface, and added peripheral communication circuits. **Expansion PCBs for analogue inputs and outputs** generally adapt analogue signals using OPAMP to ADC and HSAD levels, and digital signals to analogue DACs, respectively. The PCB can be connected to the motherboard [1, 8].

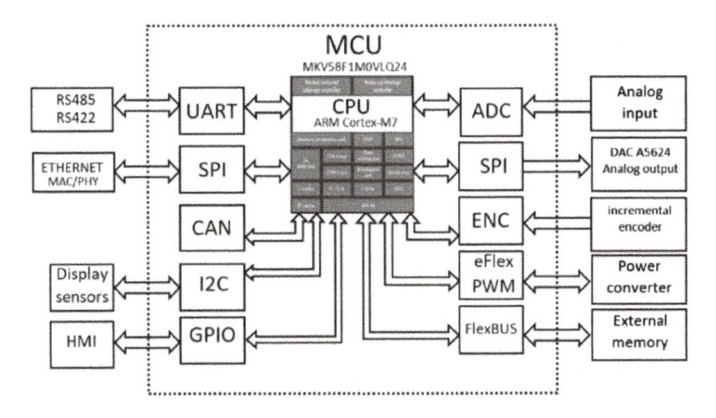

Fig. 1. The control unit topology block diagram

4 PCB Design and 3D Modelling of the Control Unit

Based on the topology, the hardware part of the Control Unit was designed in EAGLE software. The designs of the individual PCBs were exported from EAGLE to Sketch up. By inserting all the necessary components, the 3D models of all the PCBs, **the mainboard, the analogue input board, the analogue output board, the PWM driver board** were created. The illustration in Fig. 2 shows examples of the mainboard design of the control system in the programme and EAGLE and the export of the 3D model.

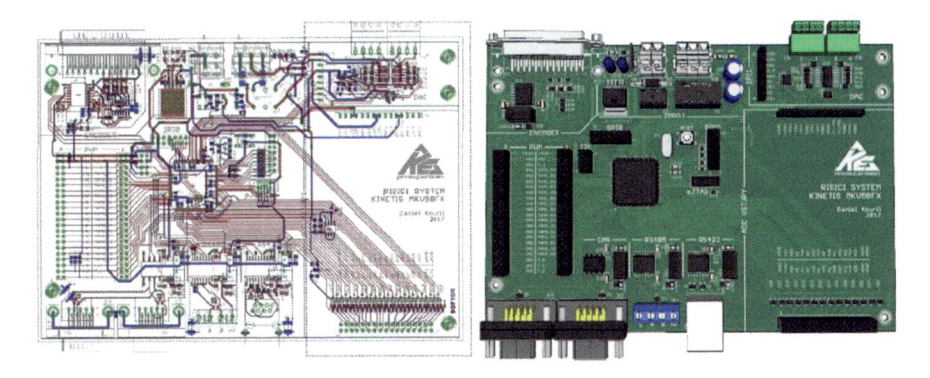

Fig. 2. The design of a control unit MCU main-board PCB and 3D model

5 Software Development Platform and Support

5.1 MCUXpresso IDE

NXP has developed the MCUXpresso IDE for the new MCU series Kinetis KV5 and LPC. The Processor Expert is replaced by the MCUXpresso configuration module, which can be run as an online or offline version of the application. In the tool, you can assign signals from the peripherals to the microcontroller's outlets using the "Pins Tool|", and set the system clocks with the "Clocks Tool". You can then download and import the configuration into the project. The exported SDK archive also includes sample algorithms.

5.2 FreeMASTER

FreeMASTER is a graphical tool for reading or writing memory to a destination, either by using a serial COM port on the host computer or using a dedicated communication protocol (JTAG, CAN…). The designed unit is connected via the RS422 communication standard. FreeMASTER uses the binary application file and debugging information for the location of the global variable in the target memory and is able to display it in real-time [7]. It does not affect code execution. This makes it a great tool for visualising or editing the programme variable.

Figure 3 below shows the described software development options and a connection of the Control Unit for the Power Electronics.

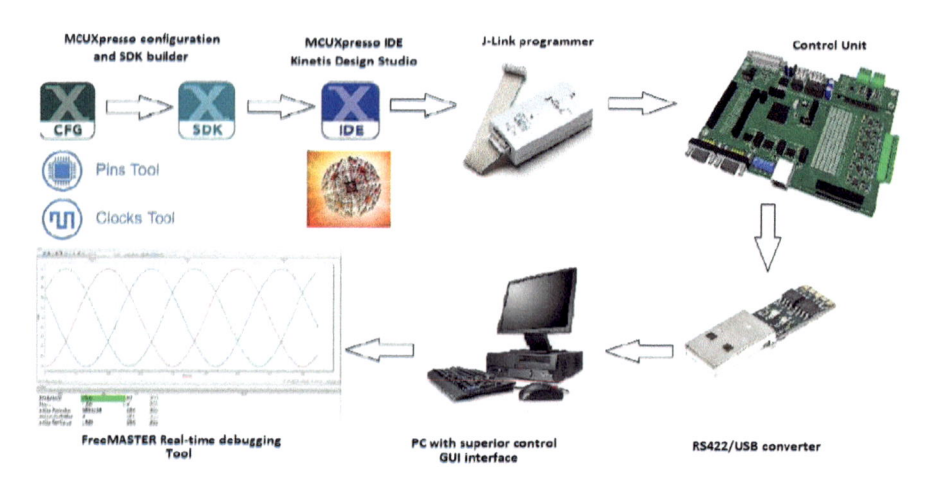

Fig. 3. The software development platform and real-time debugging tools connection

6 Experimental Results

For application, the system was connected to a motor drive consisting of a three-phase indirect frequency converter with a DC link and an induction motor. The output voltage from the inverter was controlled by PWM signals with a modulation depth of 1 and basic harmonics of 25 Hz.

The following time courses in Fig. 4 shows the measured three phase stator voltage waveforms.

The waveforms in Fig. 5 show the voltage of one phase (yellow waveform) and the current (green waveform) during the motor start up process.

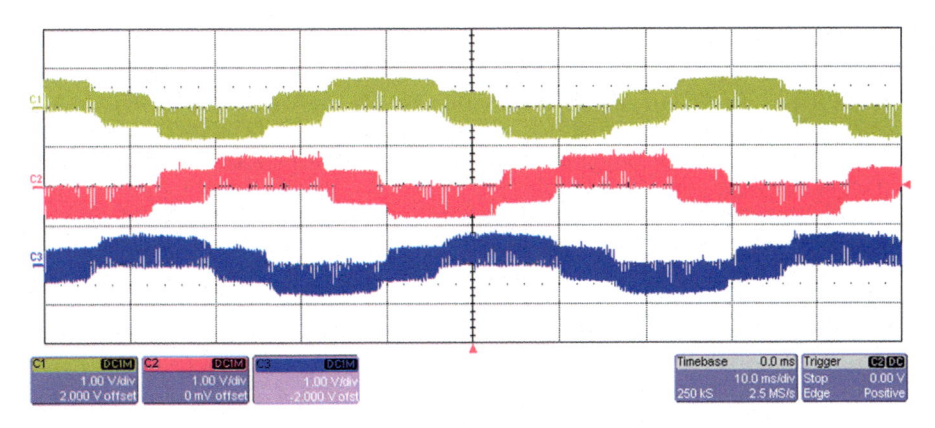

Fig. 4. Width control of 3 phase indirect voltage frequency converter, the stator voltage waveforms of an induction motor

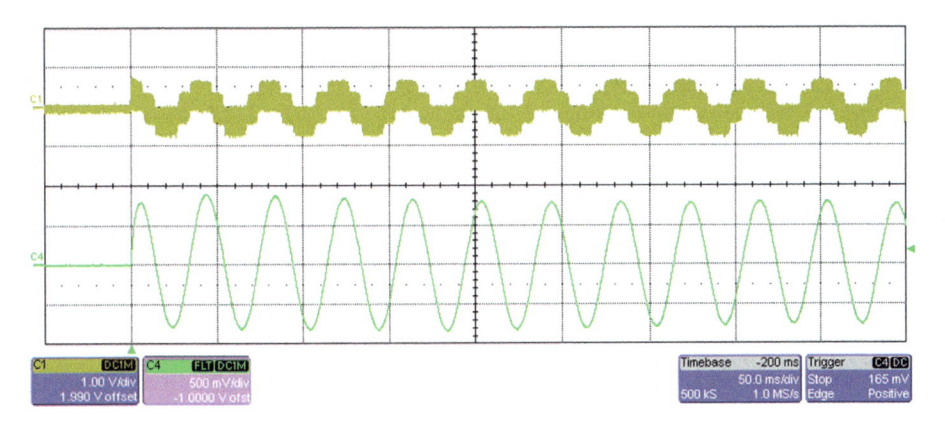

Fig. 5. Voltage and current waveforms during process of motor start up

7 Conclusion

The paper introduces the design and implementation of a Control System with an NXP microcontroller, Kinetis series KV58F1M0VLQ24. The paper describes the specific requirements of the Control System for the Power Electronics and peripherals used in general. Another part of the paper is focused on the description of the selected microcontroller and its used periphery. Furthermore, the paper describes the design and practical implementation of the whole control system. The last part of the paper is devoted to the description of the basic software tools and to an application on induction motor drive.

Acknowledgement. This paper was supported by the project Reg. No. SP2018/162 funded by the Student Grant Competition of VSB – Technical University of Ostrava.

References

1. KV5x Sub-Family Reference Manual: Supports:MKV58F1M0Vxx24, MKV56F1M0Vxx24, MKV58F512Vxx24, MKV56F512Vxx24. In: Reference Manuals. NXP Semiconductors (2016). http://www.nxp.com/assets/documents/data/en/reference-manuals/KV5XP144M240 RM.pdf. Accessed 20 July 2018
2. Kommaraju, N.V.: Design trends in PLC-based control systems. In: Control Engineering, September 2011. http://www.controleng.com/search/search-single-display/design-trends-in-plc-based-control-systems/6d130ad861.html
3. Bychkov, M., Fedorenko, A.: Microcontroller with ARM kernel and real time operating system. In: 2016 (IX) International Conference on Power Drives Systems (ICPDS) (2016). https://doi.org/10.1109/icpds.2016.7756684
4. Šetka, V., Tolar, D.: Motor controller designed for robotics based on microcontroller with integrated EtherCAT. In: 19th International Carpathian Control Conference (ICCC) (2018). https://doi.org/10.1109/carpathiancc.2018.8399643
5. Kommu, A., Rao Kanchi, R.: Designing a learning platform for the implementation of serial standards using ARM microcontroller LPC2148. In: 1st International Conference on Recent Advances and Innovations in Engineering (2014). https://doi.org/10.1109/icraie.2014.6909185
6. He, X., Wang, Z., Fang, H.: An embedded robot controller based on ARM and FPGA. In: 4th IEEE International Conference on Information Science and Technology (2014). https://doi.org/10.1109/icist.2014.6920574
7. Slivka, D., Palacký, P., Havel, A.: Electric vehicle control units communication. Adv. Electr. Electron. Eng. **10**(1), 17–21 (2012). https://doi.org/10.15598/aeee.v10i1.557
8. Rao Kanchi, R., Varadarajula, R.P.G., Kommu, A.: Design and development of a low-cost student experiments for teaching ARM based embedded system laboratory. In: IEEE International Conference on Teaching, Assessment and Learning for Engineering (TALE) (2013). https://doi.org/10.1109/tale.2013.6654491

A MIMO Robust Servo Controller Design Method for Omnidirectional Automated Guided Vehicles Using Polynomial Differential Operator

Van Lanh Nguyen[1], Sung Won Kim[1], Choong Hwan Lee[2], Dae Hwan Kim[1], Hak Kyeong Kim[1], and Sang Bong Kim[1(✉)]

[1] Department of Mechanical Design Engineering, Pukyong National University, Busan 608-739, Republic of Korea
kimsb@pknu.ac.kr
[2] Department of Machine System Engineering, Dongwon Institute of Science and Technology, Yangsan 50598, Republic of Korea

Abstract. This paper proposes a MIMO robust servo controller design method for a three wheeled Omnidirectional Automated Guided Vehicles (OAGVs) with a disturbance to track desired references using a polynomial differential operator. The process for designing the proposed controller can be described as follows: Firstly, modeling of the MIMO three-wheeled OAGV are presented. Secondly, a new extended system is obtained by applying the polynomial differential operator to the state space model and the output velocity error vector. Thirdly, the proposed controller for the given plant is designed by using the pole assignment method. By applying an inverse polynomial differential operator, a servo compensator is obtained. Finally, in order to verify the effectiveness of the proposed controller, the numerical simulation results are shown. The simulation results show that the proposed controller has good tracking performance under a step type of disturbance and the complicated higher order reference signals such as ramp and parabola. These simulation results are compared with those of the adaptive controller proposed by Bui, T. L. in 2013. The proposed controller shows the better tracking performance than the adaptive controller.

Keywords: MIMO · OAGV · Three-wheeled · Polynomial differential operator

1 Introduction

For recent decades, the automation technologies replace workers in the factories, nurses in the hospitals, and farmers in the agriculture. Among many types of the automation machines, Automatic Guided Vehicles (AGVs) are used for many different applications in the last few years. Especially, the Omnidirectional Automatic Guided Vehicles (OAGVs) have three degrees of freedom in the motion plane with three omnidirectional wheels arranged 120° apart. Thus, it can easily move any direction and any desired orientation in the work environment. However, the control design for OAGVs is a challenging mission.

To control the motion of OAGVs, many control algorithms have been proposed. Watanabe, K. et al. proposed the feedback control method for an autonomous platform with a three lateral orthogonal-wheel assemblies in [1]. Kim et al. proposed a fuzzy azimuth estimator for tracking control of a three-wheeled omnidirectional mobile robot in [2]. These control algorithms are easy to be applied to OAGVs system but are not robust. Liu et al. proposed a nonlinear controller for a three-wheeled omnidirectional mobile robot based on trajectory linearization to improve the path following performance in [3]. However, this the trajectory tracking control algorithm is rather complex. Ranjbar et al. proposed a novel robust decentralized adaptive fuzzy control in [4]. The main advantage of this adaptive control algorithm is insensitive to robot dynamic uncertainties, external disturbances, and input nonlinearities. However, Xu et al. proposed a robust neural network-based sliding mode control approach for trajectory tracking control of omnidirectional three-wheeled mobile manipulators in [5]. In this paper, integration of sliding mode control with neural network was concluded in robust formation control design for dynamic agents. Hung et al. proposed an integral sliding mode controller (ISMC) for a three-wheeled omnidirectional mobile platform in [6]. The advantage of this control method was strong robustness. Nevertheless, response time was slow and the control signal showed chattering.

To solve the robust servo system design problem, Kim et al. [8, 9, 12] introduced a new design concept for the robust servo control system by using the polynomial differential operator based on internal model principle. Kim et al. [10, 11, 13] suggested a speed controller for AC induction motors using the polynomial differential operator. The servo speed controller is able to track high order types of references. Those recommended controllers are designed for a single input single output (SISO) system. However, these were not applied to three-wheeled OAGVs.

To track the linear output velocity of the three-wheeled OAGV with the external disturbances to the desired linear output velocity perfectly, this paper proposes a MIMO robust servo controller design method for three-wheeled OAGVs by using a polynomial differential operator (PDO). The main advantages of PDO are fast response and strong robustness against external disturbances. In this paper, the proposed MIMO robust servo controller design process for a three-wheeled OAGV is described as follows: Firstly, modeling of the MIMO three-wheeled OAGV are presented. Secondly, a new extended system is obtained by applying the polynomial differential operator to the state space model and the output velocity error vector. Thirdly, the proposed controller for the given plant is designed by using the pole assignment method [14, 15]. Fourthly, by applying an inverse polynomial differential operator, a servo compensator is obtained. Finally, in order to verify the effectiveness of the proposed controller, the numerical simulation results are shown. The simulation results show that the proposed controller has good tracking performance under the step type of disturbance and the complicated higher order reference signals such as ramp and parabola. These simulation results of the proposed controller are compared with those of the adaptive controller in [7]. The adaptive controller is designed based on Lyapunov stability theory and an adaptive backstepping control theory to track a desired sharp edge trajectory. The proposed controller shows the better tracking performance than the adaptive controller.

2 System Modeling

The geometric description of the three-wheeled OAGV is described and its dynamic equation in [7] can be represented as the state space equation by defining $x = \dot{q}_C = [\dot{x}_C \; \dot{y}_C \; \dot{\Phi}_C]^T$:

$$\begin{cases} \dot{x} = Ax + Bu + \varepsilon \\ y = Cx \end{cases} \quad (1)$$

$$A = \begin{bmatrix} -\frac{1.5\rho_2}{m} & 0 & 0 \\ 0 & -\frac{1.5\rho_2}{m} & 0 \\ 0 & 0 & -\frac{3\rho_2 L^2}{I} \end{bmatrix}, \quad \begin{aligned} C &= diag([1 \; 1 \; 1]) \\ &= [c_1^T \; c_2^T \; c_3^T]^T \end{aligned} \quad (2)$$

$$B = \rho_1 \begin{bmatrix} \frac{-\sin \Phi_C}{m} & \frac{-\sin(\frac{\pi}{3}-\Phi_C)}{m} & \frac{\sin(\frac{\pi}{3}+\Phi_C)}{m} \\ \frac{\cos \Phi_C}{m} & \frac{-\cos(\frac{\pi}{3}-\Phi_C)}{m} & \frac{-\cos(\frac{\pi}{3}+\Phi_C)}{m} \\ \frac{L}{I} & \frac{L}{I} & \frac{L}{I} \end{bmatrix} \quad \varepsilon = \begin{bmatrix} -\frac{f_{1d}}{m} \\ -\frac{f_{2d}}{m} \\ -\frac{f_{3d}}{I} \end{bmatrix} \quad (3)$$

where A, B and C are parameter matrices for the three-wheeled OAGV, $x = [x_1 \; x_2 \; x_3]^T \in R^3$ is the system state vector, $y = [y_1 \; y_2 \; y_3]^T \in R^3$ is the velocity output vector of the three-wheeled OAGV, $u = [u_1 \; u_2 \; u_3] \in R^3$ is the control input vector as the voltage applied to the DC motors, $\varepsilon = [\varepsilon_1 \; \varepsilon_2 \; \varepsilon_3]^T \in R^3$ is the disturbance vector, $f_d = [f_{1d} \; f_{2d} \; f_{3d}]^T \in R^{3 \times 1}$ is a friction and slip force disturbance vector related to $f_{Ai}, f_{Mi}(i=1,2,3)$, ρ_1 and ρ_2 are motor characteristic coefficients of DC motor, m is the mass of the three-wheeled OAGV and I is the moment of inertia for the three-wheeled OAGV.

3 MIMO Robust Servo Controller Design

An output error vector is defined as follows:

$$e = [e_1 \; e_2 \; e_3]^T = y - y_r. \quad (4)$$

where $y_r = [y_{r1} \; y_{r2} \; y_{r3}]^T \in R^3$ is a reference input vector.

A MIMO robust servo controller design method is implemented by 3 steps [9]:

Firstly, to eliminate the effect of the disturbance in Eq. (1), operating a differential polynomial operator $L(D)$ to both sides of Eq. (1) can be written as

$$\frac{d}{dt}\{L(D)x\} = AL(D)x + BL(D)u + L(D)\varepsilon \quad (5)$$

$$L(D) = D^{(q)} + \alpha_{q-1}D^{(q-1)} + \cdots + \alpha_0 \quad (6)$$

where $D = d/dt$ is the differential operator with $dim\{L(D)\} = q$.

It is supposed that the following differential equation forms:

$$L(D)y_r(t) = 0 \text{ and } L(D)\varepsilon(t) = 0 \qquad (7)$$

Secondly, by operating $L(D)$ to Eqs. (4) and (5), the followings can be obtained:

$$D^{(q)}e_i = -\alpha_{q-1}D^{(q-1)}e_i - \cdots - \alpha_1 D^{(1)}e_i - \alpha_0 e_i + c_i L(D)x \quad \text{for} \quad i = 1,2,3 \qquad (8)$$

Equation (8) can be described into the matrix form as follows:

$$\dot{z}_i = M_i L(D)x + N z_i \text{ for } i = 1,2,3 \qquad (9)$$

where

$$N = \begin{bmatrix} 0 & 1 & 0 & 0 & \cdots & 0 \\ 0 & 0 & 1 & 0 & \cdots & 0 \\ 0 & 0 & 0 & 1 & \cdots & 0 \\ \vdots & \vdots & \vdots & \vdots & \ddots & \vdots \\ 0 & 0 & 0 & 0 & \cdots & 1 \\ -\alpha_0 & -\alpha_1 & -\alpha_2 & -\alpha_3 & \cdots & -\alpha_{q-1} \end{bmatrix} \in R^{q \times q}, \; M_i = \begin{bmatrix} 0 \\ \vdots \\ c_i^T \end{bmatrix}$$

$$= \begin{bmatrix} 0 & 0 & 0 \\ \vdots & \vdots & \vdots \\ c_{i1} & c_{i2} & c_{i3} \end{bmatrix} \in R^{q \times 3}$$

and $z_i = \begin{bmatrix} e_i & e_i^{(1)} & \cdots & e_i^{(q-1)} \end{bmatrix}^T \in R^q$ for $i = 1, 2, 3$.

By combining the operated system of Eqs. (5) and (9), an extended system can be obtained as follows:

$$\dot{x}_e = A_e x_e + B_e v \qquad (10)$$

$$A_e = \begin{bmatrix} A & 0 & 0 & 0 \\ \begin{bmatrix} 0 \\ c_1^T \end{bmatrix} & N & 0 & 0 \\ \begin{bmatrix} 0 \\ c_2^T \end{bmatrix} & 0 & N & 0 \\ \begin{bmatrix} 0 \\ c_3^T \end{bmatrix} & 0 & 0 & N \end{bmatrix} \in R^{(3+3q) \times (3+3q)}, B_e = \begin{bmatrix} B \\ 0 \\ \vdots \\ 0 \end{bmatrix} \in R^{(3+3q) \times 3}, x_e$$

$$= \begin{bmatrix} L(D)x \\ z_1 \\ z_2 \\ z_3 \end{bmatrix} \in R^{3+3q}$$

where x_e is an extended system state variable vector, $v \in R^3$ is a new control law for the extended system, and $z = \begin{bmatrix} z_1^T & z_2^T & z_3^T \end{bmatrix}^T \in R^{3q}$ is an error variable vector for the extended system.

A new control law for the extended system is defined by the following form:

$$v = L(D)u = -Fx_e \in R^3 \tag{11}$$

Where $F = [F_x \ F_z] \in R^{3 \times (3+3q)}$ is a feedback control gain matrix, $F_x \in R^{3 \times 3}$ and $F_z \in R^{3 \times 3q}$ are feedback control gain matrices for $L(D)x$ and z, respectively.

A new error variable vector for the extended system can be defined as

$$\zeta = L^{-1}(D)z \ for \ \zeta = [\zeta_1^T \ \zeta_2^T \ \zeta_3^T]^T \in R^{3q} \tag{12}$$

Using Eqs. (11) and (12), the control law of Eq. (1) can be obtained as follows:

$$u = -Fx_\zeta = -[F_x \ F_z]\begin{bmatrix} x \\ \zeta \end{bmatrix} \tag{13}$$

where $x_\zeta = L^{-1}(D)x_e = [x^T \ \zeta^T]^T \in R^{3+3q}$ is a new extended system variable vector.
Controllability of the extended system can be checked in [9].
From Eqs. (10) and (11), the closed loop system is obtained as

$$\dot{x}_e = (A_e - B_e F)x_e \tag{14}$$

Thirdly, consider the system of Eq. (1) and assume that controllability holds, then there exists gain matrix $F = [F_x \ F_z]$ so that the closed loop control system obtained by applying the feedback control law of Eq. (11) to the extended system of Eq. (10) is asymptotically stable, i.e. There exists a gain matrix $F = [F_x \ F_z]$ so that the following matrix is asymptotically stable.

$$A_F = A_e - B_e F = A_e - B_e[F_x \ F_z] \tag{15}$$

When the feedback control law of the extended system of Eq. (10) is designed based on controllability of the extended system and regulator design problem, the output error vector of Eq. (4) becomes $e(t) \to 0$ as $t \to \infty$.

$x_e \to 0$ by the regulator design result, that is, $x_e \to 0$ means $[Lx^T \ z^T]^T \to 0$ as follows:

$$x_e = [L(D)x^T \ z_1^T \ z_2^T \ z_3^T]^T$$
$$= \left[L(D)x^T \ \left[e_1 \ e_1^{(1)} \ \cdots \ e_1^{(q-1)}\right]^T \ \cdots \ \left[e_3 \ e_3^{(1)} \ \cdots \ e_3^{(q-1)}\right]^T \right]^T \tag{16}$$

As the result, the error $e(t) = [e_1(t) \ e_2(t) \ e_3(t)]^T \to 0$ as $t \to \infty$.

Using the new input vector v, the extended system of Eq. (10) can be written by

$$\frac{d}{dt}\{L(D)x\} = [A - BF_x]L(D)x - BF_z z + L(D)\varepsilon \tag{17}$$

$$\frac{d}{dt}z = N_z z + I_\zeta L(D)e \quad for\ L(D)\varepsilon = 0 \tag{18}$$

where

$$N_z = \begin{bmatrix} N & 0 & \cdots & 0 \\ 0 & N & \cdots & 0 \\ \vdots & \vdots & \ddots & \vdots \\ 0 & 0 & \cdots & N \end{bmatrix} \in R^{3q \times 3q}, \ I_\zeta = \begin{bmatrix} \lambda & 0 & \cdots & 0 \\ 0 & \lambda & \cdots & 0 \\ \vdots & \vdots & \ddots & \vdots \\ 0 & 0 & \cdots & \lambda \end{bmatrix} \in R^{3q \times 3} \quad and$$

$$\lambda = [\dot{0} \ \vdots \ \dot{0} \cdot 1]^T \in R^q.$$

By operating an inverse polynomial differential operator $L^{-1}(D)$ for Eqs. (17) and (18), the following equations can be obtained:

$$\frac{d}{dt}x = [A - BF_x]x - BF_z \zeta + \varepsilon \tag{19}$$

$$\frac{d\zeta}{dt} = N_z \zeta + I_\zeta e \quad for\ L(D)\varepsilon = 0 \tag{20}$$

The servo compensator of Eq. (20) includes the model of reference and disturbance signals since the matrix N_z is composed of the least common multiple model of two signals.

The configuration of the proposed MIMO robust servo controller under the disturbance can be described as shown in Fig. 1.

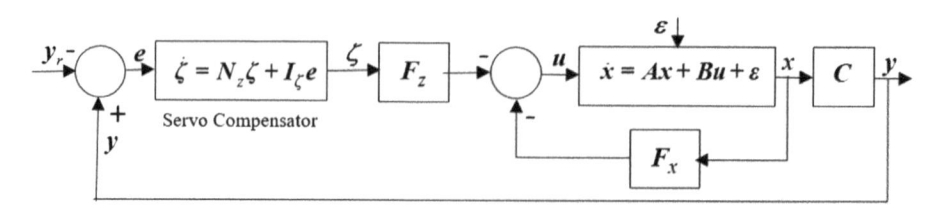

Fig. 1. Configuration of the proposed MIMO robust servo controller

4 Simulation Results

Simulations are done to verify the effectiveness of the proposed controller. This controller is compared with adaptive controller in [7] for the three-wheeled OAGV. The OAGV is a MIMO system with three inputs $\boldsymbol{u} = \begin{bmatrix} u_1 & u_2 & u_3 \end{bmatrix}^T$ and three outputs $\boldsymbol{y} = \begin{bmatrix} y_1 & y_2 & y_3 \end{bmatrix}^T$. It is assumed that the OAGV does not self-rotation and the disturbance is given as a step type model of the following $\boldsymbol{\varepsilon} = \begin{bmatrix} 0.1 & 0.05 & 0.05 \end{bmatrix}^T$. The reference inputs are chosen as follows: $\boldsymbol{y}_r = \begin{bmatrix} 0.01t \text{ m/s} & 0.005t \text{ m/s} & 0 \text{ rad/s} \end{bmatrix}^T$ for the ramp reference input and $\boldsymbol{y}_r = \begin{bmatrix} 0.001t^2 \text{ m/s} & 0.0005t^2 \text{ m/s} & 0 \text{ rad/s} \end{bmatrix}^T$ for the parabolic reference input.

Tables 1, 2 and 3 show parameters of the OAGV system, matrices values of MIMO system and initial values for the proposed controller.

Table 1. Parameters of the OAGV system

Symbols	Descriptions	Values	Units
r	Radius of wheels	0.0625	[m]
L	Radius of OAGV	0.275	[m]
m	Weight of OAGV	45	[kg]
I	Moment of inertia	12	[kg.m^2]
ρ_1	Motor characteristic coefficients	1.5	[N/v]
ρ_2	Motor characteristic coefficients	1.924	[Ns/m]

Table 2. Matrices values of MIMO system

Parameters	Values
Matrix A	$A = \begin{bmatrix} -0.0641 & 0 & 0 \\ 0 & -0.0641 & 0 \\ 0 & 0 & -0.0364 \end{bmatrix}$
Matrix B	$B = \begin{bmatrix} 0 & -0.0289 & 0.0289 \\ 0.0333 & -0.0167 & -0.0167 \\ 0.0344 & 0.0344 & 0.0344 \end{bmatrix}$
Matrix C	$C = \begin{bmatrix} 1 & 0 & 0 \\ 0 & 1 & 0 \\ 0 & 0 & 1 \end{bmatrix}$

Table 3. Initial values for the proposed controller

Initial values $x(0)$	Ramp reference $\zeta(0)$	Parabola reference $\zeta(0)$
Values $\begin{bmatrix} 0 & 0 & 0 \end{bmatrix}^T$	$\begin{bmatrix} 0 & 0 & 0 & 0 & 0 \end{bmatrix}^T$	$\begin{bmatrix} 0 & 0 & 0 & 0 & 0 & 0 & 0 & 0 \end{bmatrix}^T$

4.1 Proposed Controller

The controller gains and servo compensators for two types of reference signals are obtained as shown in Tables 4 and 5.

Table 4. Parameter values for ramp reference input in the proposed controller

Parameters	Values		
Gain matrix F_x, F_z	$F_x = \begin{bmatrix} 35 & 846 & 382 \\ -715 & -476 & 434 \\ 675 & 403 & 431 \end{bmatrix}$	$F_z = \begin{bmatrix} 6200 & 943 & 54212 & 11823 & 20444 & 4905 \\ -43279 & -9715 & -37150 & -7387 & 31635 & 6445 \\ 35771 & 8616 & -23594 & -5367 & 30142 & 6293 \end{bmatrix}$	
Matrices of compensator	$N_z = \begin{bmatrix} 0 & 1 & 0 & 0 & 0 & 0 \\ 0 & 0 & 0 & 0 & 0 & 0 \\ 0 & 0 & 0 & 1 & 0 & 0 \\ 0 & 0 & 0 & 0 & 0 & 0 \\ 0 & 0 & 0 & 0 & 0 & 1 \\ 0 & 0 & 0 & 0 & 0 & 0 \end{bmatrix}$,	$I_\zeta = \begin{bmatrix} 0 & 0 & 0 \\ 1 & 0 & 0 \\ 0 & 0 & 0 \\ 0 & 1 & 0 \\ 0 & 0 & 0 \\ 0 & 0 & 1 \end{bmatrix}$	

Table 5. Parameter values for parabolic reference input in the proposed controller

Parameters	Values
Gain matrix F_x	$F_x = \begin{bmatrix} 0 & 1200 & 600 \\ -1100 & -600 & 600 \\ 1100 & -600 & 700 \end{bmatrix}$
Gain matrix F_z	$F_z = \begin{bmatrix} 48400 & 9100 & 500 & 763200 & 225400 & 24400 & 546700 & 145100 & 14200 \\ -907700 & -246200 & -24600 & -475700 & -131200 & -13400 & 505600 & 136900 & 13700 \\ 1023500 & 269500 & 26100 & -446400 & -125500 & -13000 & 784200 & 191000 & 17100 \end{bmatrix}$
Matrices of compensator	$N_z = \begin{bmatrix} 0 & 1 & 0 & 0 & 0 & 0 & 0 & 0 & 0 \\ 0 & 0 & 1 & 0 & 0 & 0 & 0 & 0 & 0 \\ 0 & 0 & 0 & 0 & 0 & 0 & 0 & 0 & 0 \\ 0 & 0 & 0 & 0 & 1 & 0 & 0 & 0 & 0 \\ 0 & 0 & 0 & 0 & 0 & 1 & 0 & 0 & 0 \\ 0 & 0 & 0 & 0 & 0 & 0 & 0 & 0 & 0 \\ 0 & 0 & 0 & 0 & 0 & 0 & 1 & 0 & 0 \\ 0 & 0 & 0 & 0 & 0 & 0 & 0 & 1 & 0 \\ 0 & 0 & 0 & 0 & 0 & 0 & 0 & 0 & 0 \end{bmatrix}, \quad I_\zeta = \begin{bmatrix} 0 & 0 & 0 \\ 0 & 0 & 0 \\ 1 & 0 & 0 \\ 0 & 0 & 0 \\ 0 & 0 & 0 \\ 0 & 1 & 0 \\ 0 & 0 & 0 \\ 0 & 0 & 0 \\ 0 & 0 & 1 \end{bmatrix}$

4.2 Adaptive Controller

The adaptive controller design of OAGV is described in [7]. The parameters and gain values of the adaptive controller are shown in Table 6.

Table 6. Parameters and gain values of Adaptive controller

Parameters	Descriptions	Values	Units
λ_m	Positive numbers	25.5	[kg s^2/m^2]
λ_k	Positive numbers	0.01	[kg/s^2]
K_p	Diagonal positive definite matrix	diag ([22 22 0])	diag ([1/s 1/s 1/s])
K_b	Diagonal positive definite matrix	diag ([40 40 0])	diag ([kg/s kg/s kg m^2/s])

4.3 Ramp Reference Input

Simulation results using the ramp reference input vector $y_r = \begin{bmatrix} 0.01t & 0.005t & 0 \end{bmatrix}^T$ for the proposed controller and the adaptive controller are shown in Figs. 2, 3, 4 and 5. Figure 2 shows the control laws using both controllers. The control laws u_1 and u_3 increase approximately linearly after 0.8 s, but the control laws u_2 decreases approximately linearly after 0.8 s. Figure 3 shows that the output vector y of both controllers. The output y of the proposed controller tracks the ramp reference input vector y_r well after about **0.8 s**, while the output vector y of the adaptive controller tracks the ramp reference input vector of y_r with the **steady state error**. Figure 4 shows that the output error vector e of both controllers. The output error vector the proposed controller converges to zero after about **0.8 s**, while the output error vector of the adaptive controller has **steady state error** of $\begin{bmatrix} 0.022 \text{ m/s} & 0.011 \text{ m/s} & 0 \text{ rad/s} \end{bmatrix}$ after about **5 s**.

Therefore, the proposed controller tracks the ramp reference input faster and better than the adaptive controller. Figure 5 shows the servo compensator output vector $\zeta = [\zeta_1 \ \zeta_2 \ \zeta_3]^T$ for the ramp reference input vector of the proposed controller. The servo compensator output ζ_2 converges to 2×10^{-4} m/s after about 0.8 s but the servo compensator outputs ζ_1 and ζ_3 decrease approximately linearly after about 0.8 s.

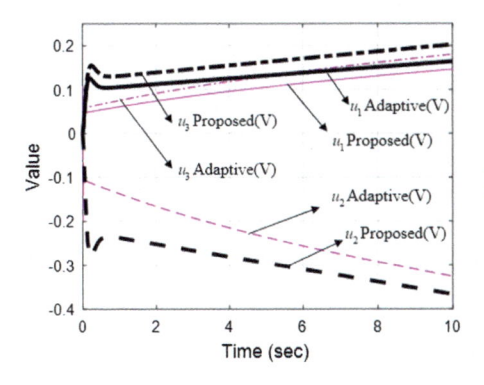

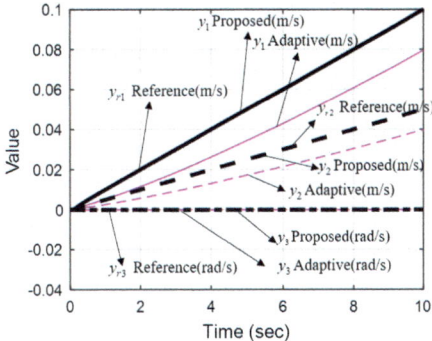

Fig. 2. Control law u for ramp reference input

Fig. 3. Output y for ramp reference input

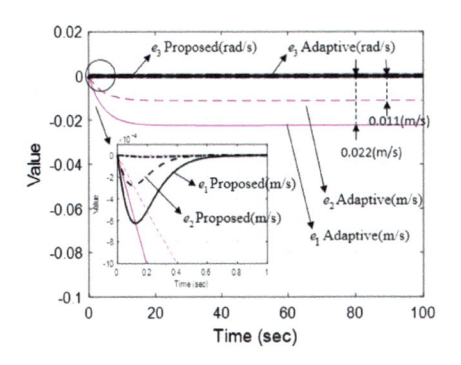

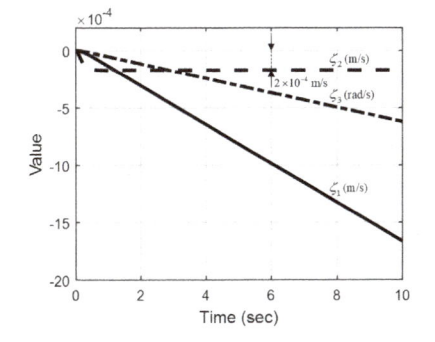

Fig. 4. Output e for ramp reference input

Fig. 5. Servo compensator output ζ

4.4 Parabolic Reference Input

Simulation results using the parabolic reference input vector $y_r = [0.001t^2 \ 0.0005t^2 \ 0]^T$ for the proposed controller and the adaptive controller are shown in Figs. 6, 7, 8 and 9. Figure 6 shows the control laws using both controllers. The control laws u_1 and u_3 increase parabolically after about 0.8 s, but the control laws u_2 decreases parabolically after about 0.8 s. Figure 7 shows that the output vector y of both controllers. The output vector y of the proposed controller tracks the parabolic reference input vector y_r well after about **0.8 s**, while the output vector y of the adaptive controller doesn't track the parabolic reference input of y_r. Figure 8 shows that the output error vector e of the both controllers. The output error vector of the proposed controller converges to zero after about **0.8 s**, while the output error vector of the adaptive controller doesn't converge to zero or constant. Therefore, the proposed controller using the parabolic reference input vector is stable, while the adaptive controller using the parabolic reference input is unstable. Figure 9 shows servo compensator output vector ζ of the proposed controller using the parabolic reference input vector. The servo compensator output ζ_3 converges to 2×10^{-6} rad/s after about 0.8 s but the servo compensator outputs ζ_1 and ζ_2 decrease parabolically after about 0.8 s.

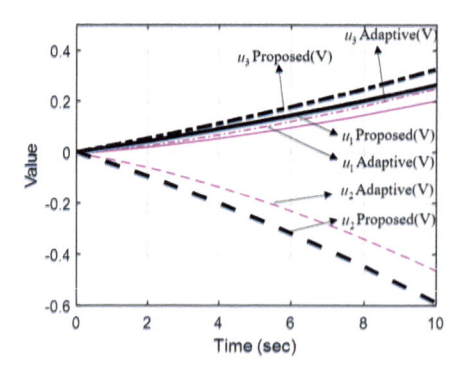

Fig. 6. Control law u for parabolic reference input

Fig. 7. Output y for parabolic reference input

Fig. 8. Output error e for parabolic reference input

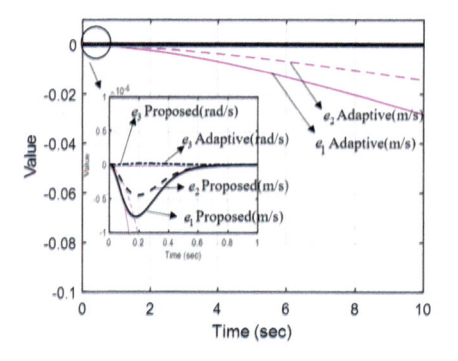

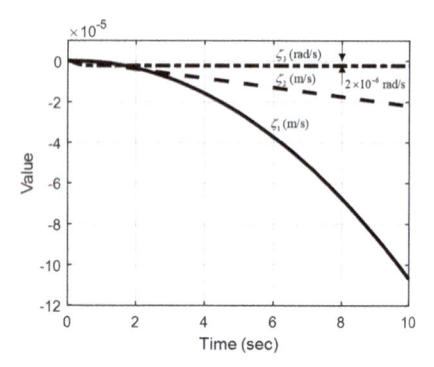

Fig. 9. Servo compensator output ζ

5 Conclusions

This paper proposed a MIMO robust servo controller design method for a three-wheeled OAGV with a step type of disturbance to track desired reference inputs using a polynomial differential operator. The process for the proposed controller design was implemented as follows: Firstly, to eliminate the effect of the step type of disturbance, operating the differential polynomial operator $L(D)$ to the state space model. Secondly, by operating the polynomial differential operator to the state space model and the output velocity error vector, a new extended system was obtained. Thirdly, by applying an inverse polynomial differential operator $L^{-1}(D)$, a servo compensator of the proposed controller was obtained. Fourthly, the control laws was designed by using the pole assignment method. The simulation results showed that the proposed controller had good tracking performance for ramp reference input and parabolic reference input under the step type of disturbance. For simulation results using ramp reference, the output of the proposed controller tracked the ramp reference input well after about 0.8 s, while the output the adaptive controller tracked the ramp reference input with the steady state error. For simulation results using parabolic reference, the output of the proposed controller tracked the parabolic reference input well after about 0.8 s, while the adaptive controller didn't track the parabolic reference input and was unstable. Therefore, the proposed controller showed the faster and better tracking performance than the adaptive controller.

Acknowledgement. This work (Grants No. S2608022) was supported by Business for Cooperative R&D between Industry, Academy, and Research Institute funded Korea Small and Medium Business Administration in 20.

References

1. Watanabe, K., Shiraishi, Y., Spyros, G.: Feedback control of an omnidirectional autonomous platform for mobile service robots. Intell. Rob. Syst. **22**, 315–330 (1998)
2. Kim, S.D., Hyun, C.H., Cho, Y.W., Kim, S.W.: Tracking control of 3-wheels omnidirectional mobile robot using fuzzy azimuth estimator. In: 10th WSEAS International Conference on Robotics Control and Manufacturing Technology, vol. 11, no. 10, pp. 3873–3879 (2010)
3. Liu, Y., Zhu, J.J., Robert, L., Williams, I.I., Wu, J.: Omni-directional mobile robot controller based on trajectory linearization. Robot. Auton. Syst. **56**, 461–479 (2008)
4. Ranjbar, S.B., Shabaninia, F., Nemati, A., Stan, S.: A novel robust decentralized adaptive fuzzy control for swarm formation of multiagent systems. Ind. Electron. **59**(8), 3124–3134 (2012)
5. Xu, D., Zhao, D., Yi, J., Tan, X., Chen, Z.: Trajectory tracking control of omnidirectional wheeled mobile manipulators: robust neural network-based sliding mode approach. Robot. Autom. **39**(3), 1653–1658 (2008)
6. Hung, N., Tuan, D.V., Sung, J.K., Kim, H.K., Kim, S.B.: Motion control of an omnidirectional mobile platform for trajectory tracking using an integral sliding mode controller. Autom. Syst. **8**(6), 1221–1231 (2010)

7. Bui, T.L., Doan, P.T., Kim, H.K., Nguyen, V.G., Kim, S.B.: Adaptive motion controller design for an omnidirectional AGV based on laser sensor. In: Lecture Notes in Electrical Engineering, vol. 282, pp. 509–523. Springer, Heidelberg (2013)
8. Kim, S.B., Kim, D.H., Pratama, P.S., Kim, J.W., Kim, H.K., Oh, S.J., Jung, Y.S.: MIMO robust servo controller design based on internal model principle using polynomial differential operator. In: Lecture Notes in Electric Engineering, vol. 371, pp. 469–484. Springer, Cham (2015)
9. Kim, S.B., Nguyen, H.H., Kim, D.H., Kim, H.K.: Robust servo controller design for MIMO systems based on linear shift invariant differential operator. J. Inst. Control Robot. Syst. **24**(6), 501–511 (2018)
10. Kim, D.H., Pratama, P.S., Doan, P.T., Oh, S.J., Min, J.H., Jung, Y.S., Duong, V.T., Kim, S.B.: Application of servo controller design for speed control of AC Induction Motors using polynomial differential operator. In: Lecture Notes in Electric Engineering, vol. 371, pp. 337–352. Springer, Cham (2015)
11. Kim, D.H., Nguyen, T.H., Pratama, P.S., Gulakari, A.V., Kim, H.K., Kim, S.B.: Servo system design for speed control of AC induction motors using polynomial differential operator. Int. J. Control Autom. Syst. **15**(3), 1207–1216 (2017)
12. Nguyen, T.H., Lee, J.W., Kim, H.K., Tran, T.P., Choe, Y.W., Dinh, V.T., Kim, S.B.: Servo controller design for speed control of BLDC motors using polynomial differential operator method. In: Lecture Notes in Electric Engineering, vol. 371, pp. 353–365. Springer, Cham (2015)
13. Nguyen, T.H., Kim, D.H., Oh, S.J., Kim, H.K., Kim, S.B.: Controller design for MIMO servo system using polynomial differential operator–A solution for increasing speed of an induction conveyor system. In: Lecture Notes in Electric Engineering, vol. 415, pp. 529–542. Springer, Cham (2016)
14. Furuta, K., Kim, S.B.: Pole assignment in a specified disk. IEEE Trans. Autom. Control **32**(5), 423–427 (1987)
15. Kim, S.B., Furuta, K.: Regulator design with poles on a specified region. Int. J. Control **47**(1), 143–160 (1988)

Model Reference Adaptive Control Strategy for Application to Robot Manipulators

Manh Son Tran[1,2], Suk Ho Jung[1], Nhat Binh Le[1,3],
Huy Hung Nguyen[4], Dac Chi Dang[5], Anh Minh Duc Tran[6],
and Young Bok Kim[1(✉)]

[1] Department of Mechanical System Engineering, Pukyong National University,
Busan 48547, Republic of Korea
kpjiwoo@pknu.ac.kr
[2] Faculty of Electrical and Electronics Engineering,
HCMC University of Technology and Education, Ho Chi Minh City, Vietnam
[3] Faculty of Aviation Electronics and Telecommunication,
Vietnam Aviation Academy, Ho Chi Minh City, Vietnam
[4] Faculty of Electronics and Telecommunications,
Saigon University, Ho Chi Minh City, Vietnam
[5] Faculty of Electrical and Refrigeration Engineering,
Cao Thang Technical College, Ho Chi Minh City, Vietnam
[6] Faculty of Electrical and Electronics Engineering,
Ton Duc Thang University, Ho Chi Minh City, Vietnam

Abstract. The geometric nonlinearities, strong couplings, and the dependence on the inertia payload in the system dynamics of the robot manipulators lead to the difficulty in achieving good control performance. Conventional control methods cannot compensate for the payload variation effect. On the other hand, the mathematical model of the robot systems is extremely complicated and consumes an excessive amount of time in computing the robot dynamics. Moreover, deriving an exact mathematical model of the manipulator is very difficult. To handle the above issues, the model reference adaptive controller for motion control applied to robot manipulators is presented in this paper. The control law is based on the decentralized linear joint control strategy. In this approach, the control law does not require the exact model of the joint. Experiments are conducted on the 4-DOF robot manipulator to demonstrate the practicality and feasibility of the proposed control scheme, and the results are compared to those of the Ziegler-Nichols method-based PID controller and those of the model-independent controller based on time-delay estimation technique. The comparison results show that the control performance of the proposed scheme is better than that of the other controllers.

Keywords: Motion control · Decentralized control · Model reference adaptive control (MRAC) · Time-delay estimation (TDE)

1 Introduction

The user-oriented demand of robot manipulators used in small industries such as painting and welding work is increasingly remarkable in the last few decades. The basic requirement of motion control for this kind of applications is usually described as follows, given the prescribed trajectory, a well-known mathematical model of the robot system and its interactions with surroundings, design a control law that generate the control signals to the actuators to reconstruct accurately the motion. However, control system design for such robot systems there are some obstacles such as uncertainties in parameter identification, natural nonlinearities in the system dynamics, variations of the system parameters due to the posture change of the robot, and the effect of external disturbances. There has been much effort in developing robust advanced control techniques to obtain acceptable control performance. These robust advanced control techniques can be listed as the conventional computed-torque control [1], sliding mode control [2, 3], and adaptive control [4]. However, the aforementioned controllers are the model-based control methods which require the complicated calculations of the high nonlinearity terms of the robot dynamics equations. Henceforth, it is not easy to apply these conventional model-based control techniques for practical implementation. To realize the simplicity in estimating system dynamics parameters, fuzzy logic control [5] and neural network [6] are incorporated to approximate unknown functions. By introducing these intelligent control techniques, neither offline identification nor prior knowledge of the perfect model of the manipulator is required. However, such implementation of intelligent control methods is hindered because they introduce many design parameters and complicated rules which may heavily affect the overall control performance. Therefore, intelligent-based control techniques may be not a good solution.

As a simple alternative choice to the aforementioned techniques, the model reference adaptive control method was proposed for control of robot manipulators in the 1980s. The first idea to apply model reference adaptive control to manipulators was introduced by Dubowsky and DesFores [7]. In [8], the MRAC method for a dynamically uncertain hydraulic robot for position tracking was implemented. The authors employed a recursive parameter estimator for the parameter estimation. Although the tracking performance obtained from MRAC method was better than that of a PID control system, there was an issue was that a small fluctuation exists at the beginning of the movement. The authors [9] illustrated the application of the MRAC method for tracking the endpoint of a flexible joint manipulator which was used in space exploration. Some numerical simulations pointed out that the adaptive control strategy could maintain the stability and good tracking performance in spite of large uncertainties in the joint stiffness coefficients. A nonlinear model reference adaptive controller working together with impedance control approach was proposed in [10] for asymptotic tracking control of the robot end-effector in physical human-robot interaction issue. A novel enhanced human-robot interaction system based on model reference adaptive control was presented in [11]. In this paper, the control system was included two control loops, the one is a robot-specific inner integrated the neural network, it was utilized to learn the robot dynamics online and make the robot behaves as the prescribed impedance model, while the other one is the task-specific loop which took the human dynamics into account and updated the desired impedance model in order to obtain the desired characteristics for the human-robot task performance. A model-free model reference adaptive fuzzy sliding mode control was proposed to control a 5-DOF robot manipulator in [12]. The proposed method coped with the implementation problem in a multivariable robotic system. The shortcoming idea comes from the paper is that the controller drove the system state variables to reach a user-prescribed sliding surface and then slid along it to approach the desired reference model. Experimental results showed that the controller has good control performance, stability, and robustness. The authors in [13] presented an adaptive global asymptotic stable control method for compensating the friction and disturbance effects on the manipulators. The control system combined MRAC and exact linearization. In order to control the system to track the desired reference model response, the disturbance compensation scheme was employed to counteract the modeling error, disturbance, and noise. Simulation results were shown to verify the effectiveness of the proposed method. The MRAC PID controller for controlling a compliant flexures-based XY micro-positioning stage was presented in [14], in which the control parameters were systematically tuned based on intuitive desired performance and robustness. Experimental results indicated that the designed controller performed the expected performance.

However, most of the above-mentioned papers either just gave the simulation results, or extreme difficulty to carry out the control system due to the heavy calculations of the nonlinear terms of the robot dynamic equations. In this paper, an independent joint controller using the direct model reference adaptive control method is proposed for the position tracking control problem applied to robot manipulators. To do this task, the process is described as follows: Firstly, in order to cope with the difficulty in deriving the nonlinear mathematical model of the robot system, the second-order, linear time-invariant model is employed as the reference model for each joint of the manipulator. Secondly, the adaptive control algorithm based on the reference model is developed to assure the stability of the system and obtain good control performance. Finally, the efficiency and the performance of the proposed control method are validated through experiments on a 4-DOF robot arm, and the results are compared with those of the Ziegler-Nichols method-based PID controller and the ones of the model-independent controller based on time-delay estimation technique.

The rest of the paper is structured as follows: The system modeling and problem statement are presented in Sect. 2. Controller design is derived in Sect. 3. Section 4 shows experimental results and discussion. The conclusion and future work are summarized in Sect. 5.

2 System Modeling

2.1 Problem Statement

Consider the standard form of the dynamic equations of an n-DOF robot manipulator:

$$\mathbf{M}(\mathbf{q})\ddot{\mathbf{q}} + \mathbf{C}(\mathbf{q},\dot{\mathbf{q}}) + \mathbf{G}(\mathbf{q}) + \mathbf{F}(\mathbf{q},\dot{\mathbf{q}}) + \tau_d = \tau \quad (1)$$

where $\mathbf{q}, \dot{\mathbf{q}}, \ddot{\mathbf{q}} \in R^n$ stand for the joint position, velocity, and acceleration, respectively; $\mathbf{M}(\mathbf{q}) \in R^{n \times n}$ is the positive definite inertia matrix; $\mathbf{C}(\mathbf{q},\dot{\mathbf{q}}) \in R^{n \times 1}$ stands for the Coriolis (centrifugal) vector; $\mathbf{G}(\mathbf{q}) \in R^n$ the gravitational vector; $\mathbf{F}(\mathbf{q},\dot{\mathbf{q}}) \in R^n$ denotes the friction forces; $\tau_d \in R^n$ is the disturbance torques; and $\tau \in R^n$ is the joint torque. Basically, this equation is a set of strong coupling and high nonlinearity differential equations which lead to the control design and practical implementation based on Eq. (1) are extremely complicated. In this study, we present a simple method for mathematical modeling of joints on a robot system. Our studied system is a 4-DOF robot system (RRRR) in which each joint is driven by a permanent magnet DC motor with a reduction gear, and an incremental encoder to sense angular displacement is attached to each joint. The following section will precisely describe the method.

2.2 Mathematical Modeling

As well known, the robot dynamics is highly nonlinear and strongly coupled due to the nonlinear terms and the disturbance terms. A more practical controller design solution is to decide the control inputs to the actuators that drive the corresponding joints. Nowadays, most industrial manipulators utilize DC motors with high gear ratio as the joint.

The dynamic equation of a joint of such a robot manipulator can be written as:

$$J_m \ddot{\theta} + B_m \dot{\theta} = K_a v_a - K_g \tau_L \quad (2)$$

where J_m the total inertia of rotor and gear, B_m viscous friction coefficient, v_a applied voltage, K_a voltage constant, K_g gear ratio, τ_L is the load torque caused by the joint of the robot arm, θ joint angular position. In the structure of an n-DOF manipulator, τ_{Li} at the ith joint can be expressed as in Eq. (1) as follows:

$$\tau_{Li} = \sum_{j=1}^{n} m_{ij}(\mathbf{q})\ddot{q}_j + c_i(\mathbf{q},\dot{\mathbf{q}}) + g_i(\mathbf{q}) + f_i(\dot{\mathbf{q}}) + \tau_{di} \quad (3)$$

It should be noted that the typical values of gear ratios range from 30:1 to 300:1. Subsequently, the nonlinear terms in the dynamic equations and coupling interactions among the joints are noticeably reduced. This fact allows designers to design controllers for joints individually based on Eq. (2) with external interactions can be considered as a disturbance. The studied system structure is illustrated schematically in Fig. 1(a) in which the positive direction is indicated by the arrows, whereas the system for the experiments is shown in Fig. 1(b).

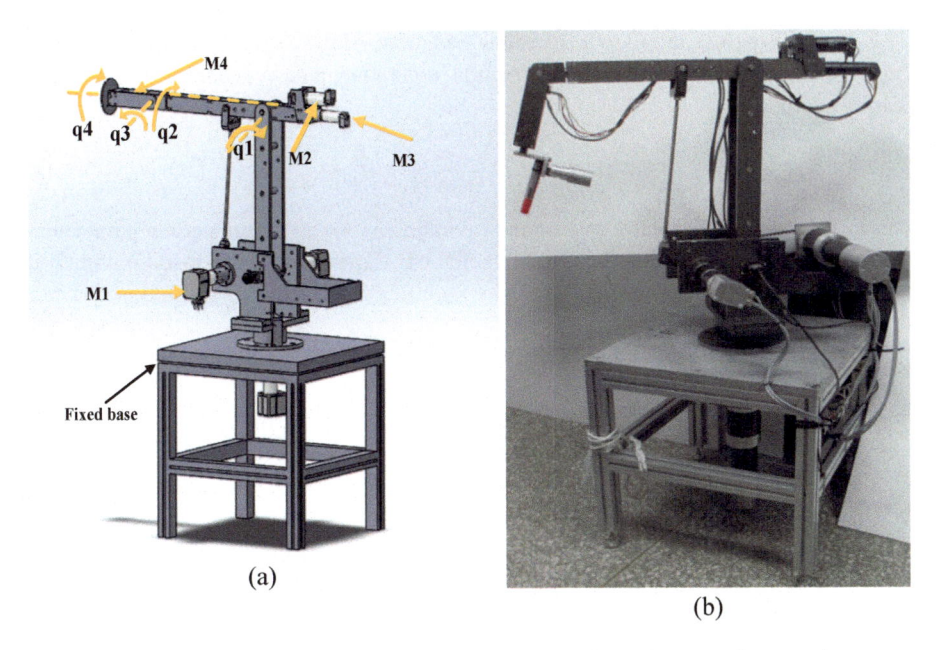

Fig. 1. System demonstration: (a) schematic structure, (b) real system for experiments

3 Model Reference Adaptive Controller Design

3.1 Overview of Adaptive Control

Since the fact that the modeling of the joint in a robot system has some unknown parameters such as friction, moment of inertia, couplings, etc. To identify the system parameters, it takes a great deal of time, some special hardware and software equipment, and experimental studies to analyze the dynamic characteristics of the joint model. On the other hand, there are large variations in the system dynamics of the robot due to the posture change during the operation, carrying different payloads, the effect of Coulomb friction, etc. Adaptive control, since has been widely deployed to handle the above issues. Among many adaptive control methods, the direct MRAC method based on the Lyapunov stability theory is one of the main schemes widely utilized. The interesting thing in this method is that the control scheme directly adjusts the control parameters in the controller under the condition that the plant parameters are poorly known, uncertain, or even unknown.

3.2 Model Reference Adaptive Controller Design

The control objective is to determine the control inputs for the model reference adaptive control system such that the plant position output tracks a reference model position output. In addition, a reference model is chosen for the reference outputs to track the desired reference inputs.

Consider Eqs. (2) and (3), the dynamic equations of the ith joint of the robot system can be formulated as second-order differential equations, and in general, can be formed as follows:

$$\alpha_{0i}\ddot{\theta}_i + \alpha_{1i}\dot{\theta}_i + \alpha_{2i}\theta_i = \beta_{0i}u_i \tag{4}$$

where $i = 1 \div 4$ indicates the corresponding joint; $\alpha_{0i}, \alpha_{1i}, \alpha_{2i}$ are uncertain parameters; u_i applied voltage, β_{0i} voltage constant coefficient. Equation (4) can also be written in matrix form as:

$$\begin{cases} \dot{\mathbf{x}}_i = \mathbf{A}_i \mathbf{x}_i + \mathbf{b}_i u_i \\ y_i = \mathbf{c}_i \mathbf{x}_i \end{cases} \tag{5}$$

where $\mathbf{x}_i = [\theta_i \; \dot{\theta}_i]^T$ state variable vector of the ith joint, θ_i, $\dot{\theta}_i$ angular position and angular velocity output, u_i the control input, $\mathbf{c}_i = [1 \; 0]$, and unknown constant matrices $\mathbf{A}_i \in R^{2 \times 2}$, $\mathbf{b}_i \in R^{2 \times 1}$ are given as follows:

$$\mathbf{A}_i = \begin{bmatrix} 0 & 1 \\ -\alpha_{2i}/-\alpha_{0i} & -\alpha_{1i}/-\alpha_{0i} \end{bmatrix}, \; \mathbf{b}_i = \begin{bmatrix} 0 \\ \beta_{0i}/\alpha_{0i} \end{bmatrix}, \text{ and } y_i = \theta_i \text{ is the position output.}$$

As mentioned earlier, in this brief, the control strategy of the system is developed based on the decentralized control technique. For the purpose of position tracking control, a linear time-invariant, second-order reference model for the ith joint is given as:

$$\begin{cases} \dot{\mathbf{x}}_{mi} = \mathbf{A}_{mi} \mathbf{x}_{mi} + \mathbf{b}_{mi} r_i \\ y_{mi} = \mathbf{c}_{mi} \mathbf{x}_{mi} \end{cases} \tag{6}$$

where $\mathbf{x}_{mi} = [\theta_{mi} \; \dot{\theta}_{mi}]$ is the state vector of the reference model of the ith joint, θ_{mi} and $\dot{\theta}_{mi}$ are the reference model angular position and angular velocity output, respectively; r_i is the reference input; $y_{mi} = \theta_{mi}$ is the angular position output; $\mathbf{A}_{mi}$ and $\mathbf{b}_{mi}$ are the reference model system matrix and control distribution vector which can be easily determined through the minimum set of parameters comprising of the undamped natural frequency ω_{ni}, and the damping ratio ξ_i.

The model reference adaptive controller for position tracking control is developed based on the following assumptions:

- Given a known Hurwitz matrix $\mathbf{A}_{mi} \in R^{2 \times 2}$ and known vector $\mathbf{b}_{mi}$, there exists an unknown ideal gain vector $\mathbf{k}_i \in R^{1 \times 2}$ and an unknown ideal gain constant ϕ_i such that the following equations are held

$$\begin{cases} \mathbf{A}_i + \mathbf{b}_i \mathbf{k}_i = \mathbf{A}_{mi} \\ \mathbf{b}_i \phi_i = \mathbf{b}_{mi} \end{cases} \tag{7}$$

- There exists a symmetric positive definite matrix $\mathbf{P}_i$ such that the following equation is satisfied:

$$\mathbf{A}_{mi}^T \mathbf{P}_i + \mathbf{P}_i \mathbf{A}_{mi} = -\mathbf{Q}_i \tag{8}$$

where $\mathbf{Q}_i$ is a symmetric positive definite matrix, and p_{ij} is the $ij-th$ element of the matrix $\mathbf{P}_i$ with $i,j = 1 \div 2$.

The control input can be designed as follows:

$$u_i = \hat{\mathbf{k}}_i \mathbf{x}_i + \hat{\phi}_i r_i \tag{9}$$

where $\hat{\mathbf{k}}_i, \hat{\phi}_i$ are the estimated parameters of the unknown ideal gains $\mathbf{k}_i$ and ϕ_i. These estimated parameters are updated online according to adaptation laws in Theorem 1 in the following.

Define the tracking error as

$$\mathbf{e}_i = \mathbf{x}_i - \mathbf{x}_{mi} \tag{10}$$

then, subtracting (6) from (5), combining with (7) lead to the error dynamics as

$$\begin{aligned}\dot{\mathbf{e}}_i &= \dot{\mathbf{x}}_i - \dot{\mathbf{x}}_{mi} = \mathbf{A}_i \mathbf{x}_i + \mathbf{b}_i u_i - \mathbf{A}_{mi} \mathbf{x}_{mi} - \mathbf{b}_{mi} r_i \\ &= \mathbf{A}_{mi} \mathbf{e}_i + \mathbf{b}_i (\tilde{\mathbf{k}}_i \mathbf{x}_i + \tilde{\phi}_i r_i)\end{aligned} \tag{11}$$

where $\tilde{\mathbf{k}}_i = \hat{\mathbf{k}}_i - \mathbf{k}_i, \tilde{\phi}_i = \hat{\phi}_i - \phi_i$

Theorem 1. The MRAC system defined by (4)–(6) is stable provided that the system control input in (9) and adaptation laws are given as

$$\dot{\hat{\mathbf{k}}}_i^T = -\gamma_{1i} \mathbf{e}_i^T \mathbf{P}_i \mathbf{b}_i \mathbf{x}_i, \; \dot{\hat{\phi}}_i = -\gamma_{2i} \mathbf{e}_i^T \mathbf{P}_i \mathbf{b}_i r_i \tag{12}$$

where $\hat{\mathbf{k}}_i, \hat{\phi}_i$ are estimated parameters and γ_{1i}, γ_{2i} are positive gain constants.

Proof of Theorem 1. To demonstrate the stability of the controller, a Lyapunov function candidate is defined as follows:

$$V_i = \mathbf{e}_i^T \mathbf{P}_i \mathbf{e}_i + \frac{1}{\gamma_{1i}} \tilde{\mathbf{k}}_i \tilde{\mathbf{k}}_i^T + \frac{1}{\gamma_{2i}} \tilde{\phi}_i^2 \geq 0 \tag{13}$$

Consider Eqs. (8), (11)–(13), taking the time derivative of V_i, yields:

$$\dot{V}_i = -\mathbf{e}_i^T \mathbf{Q}_i \mathbf{e}_i + 2\tilde{\mathbf{k}}_i (\frac{\dot{\hat{\mathbf{k}}}_i^T}{\gamma_{1i}} + \mathbf{e}_i^T \mathbf{P}_i \mathbf{b}_i \mathbf{x}_i) + 2\tilde{\phi}_i (\frac{\dot{\hat{\phi}}_i}{\gamma_{2i}} + \mathbf{e}_i^T \mathbf{P}_i \mathbf{b}_i r_i) \tag{14}$$

Next, inserting the adaptive laws (12) into (14), the time derivative becomes

$$\dot{V}_i = -\mathbf{e}_i^T Q_i \mathbf{e}_i \leq 0 \tag{15}$$

This implies that $\mathbf{e}_i$, $\hat{\mathbf{k}}_i$, $\hat{\boldsymbol{\phi}}_i$ are bounded from (13) and (15). Thus, the stability condition of the closed-loop system is satisfied.

3.3 Reference Motion Reconstruction Scheme

Seeing the control law (9), it has two inputs, namely the desired control voltage signal r_i and angular displacement θ_i; the output u_i that will be utilized to force the corresponding joint actuator to track the reference model output. However, instead of using the predefined voltage as the reference input, the desired motion trajectory is preferred for practical applications. Thus, from this point of view, once the reference–motion is given, the corresponding voltage should be generated. For generating the desired voltage, a simple mapping method is implemented via the inverse model of the reference model (6), which is expressed by the following transfer function:

$$G_{mi}^{-1}(s) = \frac{R_i(s)}{\theta_{di}(s)} = \frac{1}{\omega_{ni}^2}(s^2 + 2\xi_i \omega_{ni} s + \omega_{ni}^2) \tag{16}$$

where R_i and θ_{di} stand for the Laplace form of r_i and θ_{di}, respectively. For easy understanding, Fig. 2 illustrates the structure of the control system.

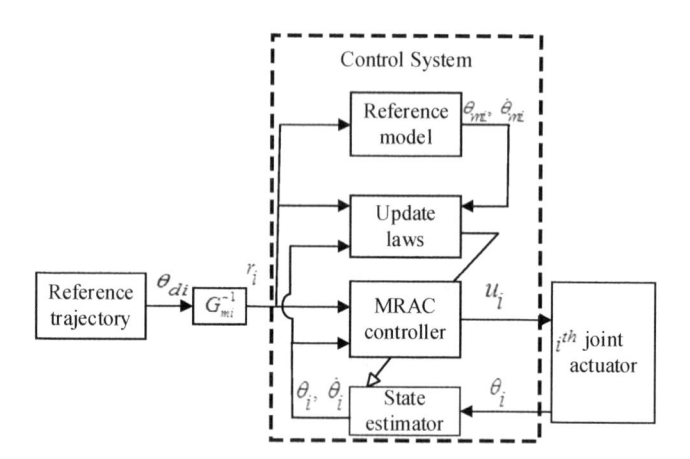

Fig. 2. The structure diagram of the control scheme.

4 Experimental Studies

In order to verify the effectiveness of the proposed control scheme in practical situations, some experiments on the 4-DOF robot manipulator developed in our laboratory as shown in Fig. 1(b) have been conducted. In the experiments, the robot arm is commanded to follow a low-speed sinusoidal trajectory, and afterwards a high-speed sinusoidal trajectory. Besides, two model-free based controllers namely PID and time-delay control (TDC) are utilized for comparing control performances. All parameters of the three controllers are summarized in Table 1.

4.1 Implementation of the Proposed Controller

In this paper, the reference model parameters for all joints are $\omega_{ni} = 16.5$ and $\xi_i = 0.9103$.

Based on the fact that the dynamic equation of the joint is a second-order differential equation, through simulation and an extensive of trial and error times, the plant parameters and the controller gains for all joints are determined.

4.2 Controllers for Comparison

PID Controller
In the PID control system, the controller has the general form as:

$$G_{PID}(s) = \left(K_P + \frac{K_I}{s} + K_D s\right) \tag{17}$$

In this paper, the controller gains are determined through the second experiment method proposed by Zeigler-Nichols.

TDE-Based Controller
In recent years, the control method based on time-delay estimation, especially applied to the robotic field, has attracted a great deal of attention from the robotic community due to its simplicity, efficiency, model-free, and robustness. Herein, the TDE-based nonsingular terminal sliding mode controller (TDE-NTSM) in [15] is applied. Basically, the control method does not require any complex numerical computation of the robot model, nor does it require any online estimation of robot dynamic parameters. The structure of the controller can be briefly expressed as follows:

$$\begin{aligned}v_{ai} =\,& v_{ai}(t-T) - M_i \ddot{\theta}_i(t-T) \\&+ M_i \left[\ddot{\theta}_{di} + \frac{b_i}{a_i} c_i^{-1} \dot{e}_i^{2-a_i/b_i} + k_{swi}\,\text{sgn}(s_i)\right]\end{aligned} \tag{18}$$

where v_{ai} is the current control input, $v_{ai}(t-T)$ the time-delayed control input, $\ddot{\theta}_i(t-T)$ the time-delayed value of acceleration, T the sampling time, M_i represents the equivalent inertia moment of the ith joint, $\text{sgn}(s_i)$ is the sign function then replaced by the signum-like function. More details about the controller can be found in [15].

4.3 Experimental Setup

All joints of the manipulator are driven by Maxon DC servo motors comes with drivers through a planetary gear. The maximum continuous torque are 0.192, 0.106, 0.0897, and 0.0279 Nm for joints 1, 2, 3, and 4, respectively. The reduction gear ratio of joints 1, 2, 3, and 4 are 53:1, 53:1, 51:1, and 33:1, respectively. The resolution of the angular sensors at joints 1, 2, 3, 4 are 0.03°, 0.0035°, 0.0035°, and 0.0055°, respectively. In addition, the National Instrument PXIe-8115 controller combined with PXIe-1078 chassis is adopted to implement the control schemes. The chassis is equipped with the data acquisition card PXIe-6363 and PXI-6221. The control algorithms are programmed using the NI LabVIEW 2009 software with the sampling time 0.01 s. The control signals sent to the driver boards of the motors are limited to ±5V.

4.4 Experimental Results

All joints of the manipulator are forced to track the following sinusoidal trajectory

$$q_i(t) = 30\sin(2\pi t/T) \text{ [degree]} \quad (19)$$

where T denotes the period in second. Two different periods of 20 s and 10 s which denotes the slow speed and high speed are used. The experimental results are arranged in Figs. 3 and 4. Moreover, to make the comparison more precise, the root-mean-square errors (RMS) of the experimental data are listed in Tables 2 and 3.

Table 1. Plant model and controllers' parameters of the three control algorithms.

Control	Joint			
	1	2	3	4
PID	$K_P = 0.552$, $K_I = 0$, $K_D = 0.006$	$K_P = 0.17$, $K_I = 0$, $K_D = 0.004$	$K_P = 0.15$, $K_I = 0$, $K_D = 0.0039$	$K_P = 0.122$, $K_I = 0$, $K_D = 0.0015$
TDE-NTSM	$M = 4.2 \times 10^{-4}$, $a = 19$, $b = 15$, $c = 0.1$, $k_{sw} = 180$, $\delta = 5$	$M = 9.1 \times 10^{-5}$, $a = 21$, $b = 19$, $c = 0.08$, $k_{sw} = 180$, $\delta = 4$	$M = 9.7 \times 10^{-5}$, $a = 21$, $b = 19$, $c = 0.08$, $k_{sw} = 180$, $\delta = 2$	$M = 2.23 \times 10^{-5}$, $a = 9$, $b = 7$, $c = 0.95$, $k_{sw} = 180$, $\delta = 3$
MRAC	$\gamma_1 = 5 \times 10^{-3}$, $\gamma_2 = 1 \times 10^{-3}$, $p_{12} = 2 \times 10^{-2}$, $p_{22} = 5 \times 10^{-3}$	$\gamma_1 = 1 \times 10^{-3}$, $\gamma_2 = 5 \times 10^{-4}$, $p_{12} = 6 \times 10^{-3}$, $p_{22} = 8 \times 10^{-4}$	$\gamma_1 = 5 \times 10^{-3}$, $\gamma_2 = 1 \times 10^{-3}$, $p_{12} = 5 \times 10^{-3}$, $p_{22} = 8 \times 10^{-4}$	$\gamma_1 = 2 \times 10^{-3}$, $\gamma_2 = 5 \times 10^{-4}$, $p_{12} = 2 \times 10^{-3}$, $p_{22} = 1 \times 10^{-3}$

Table 2. RMS errors under low-speed commands [deg].

Control	Joint 1	Joint 2	Joint 3	Joint 4
PID	0.6125	0.4219	0.5584	0.5470
TDE-NTSM	0.3408	0.1695	0.2592	0.5023
MRAC	0.1960	0.1458	0.1701	0.1687

As shown in Figs. 3 and 4, all three controllers can drive the manipulator to accurately track the desired trajectory. In the model reference adaptive control method and the control method based on time-delay estimation, due to the control method do not use the robot dynamic model, they take a short period of time to update the controller gains, after a few seconds the control schemes can drive the manipulator to follow the trajectory very well. Moreover, especially in the MRAC scheme, if the controller's parameters are optimally tuned, it will absolutely take less time to adaptively update the control gains. Particularly, as illustrated in Tables 2 and 3, the proposed control shows the smallest RMS errors among the three controllers of joints under two different periods.

Table 3. RMS errors under high-speed commands [deg].

Control	Joint 1	Joint 2	Joint 3	Joint 4
PID	1.2380	0.8374	1.1138	0.9630
TDE-NTSM	0.9515	0.4391	0.5598	0.9557
MRAC	0.3156	0.1448	0.2724	0.2321

Furthermore, as shown in Fig. 3, when the desired trajectory become more quickly, the control performance has a degradation. Especially, the results obtained from the TDE-NTSM are significantly degraded by a high ratio as shown in Table 3 as compared with those in Table 2. This fact comes as a result because the time-delay estimation–based control techniques presume that the unknown nonlinear dynamics do not change much for a relatively small period of time.

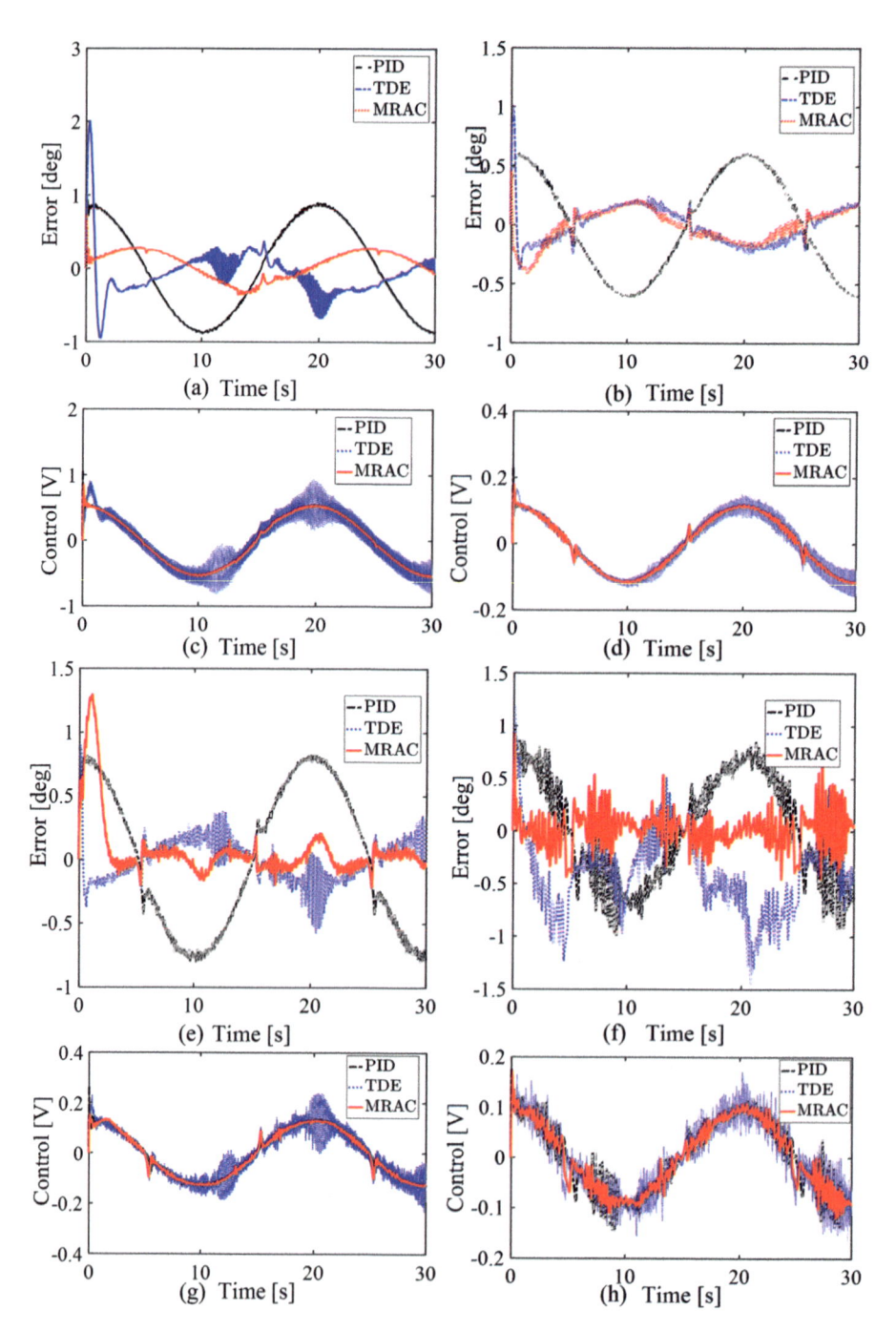

Fig. 3. Experimental results under low speed sinusoidal desired trajectory. (a), (b), (e), and (f) Tracking errors of joints 1, 2, 3, and 4, respectively. (c), (d), (g), and (h) Control inputs of joints 1, 2, 3, and 4, respectively.

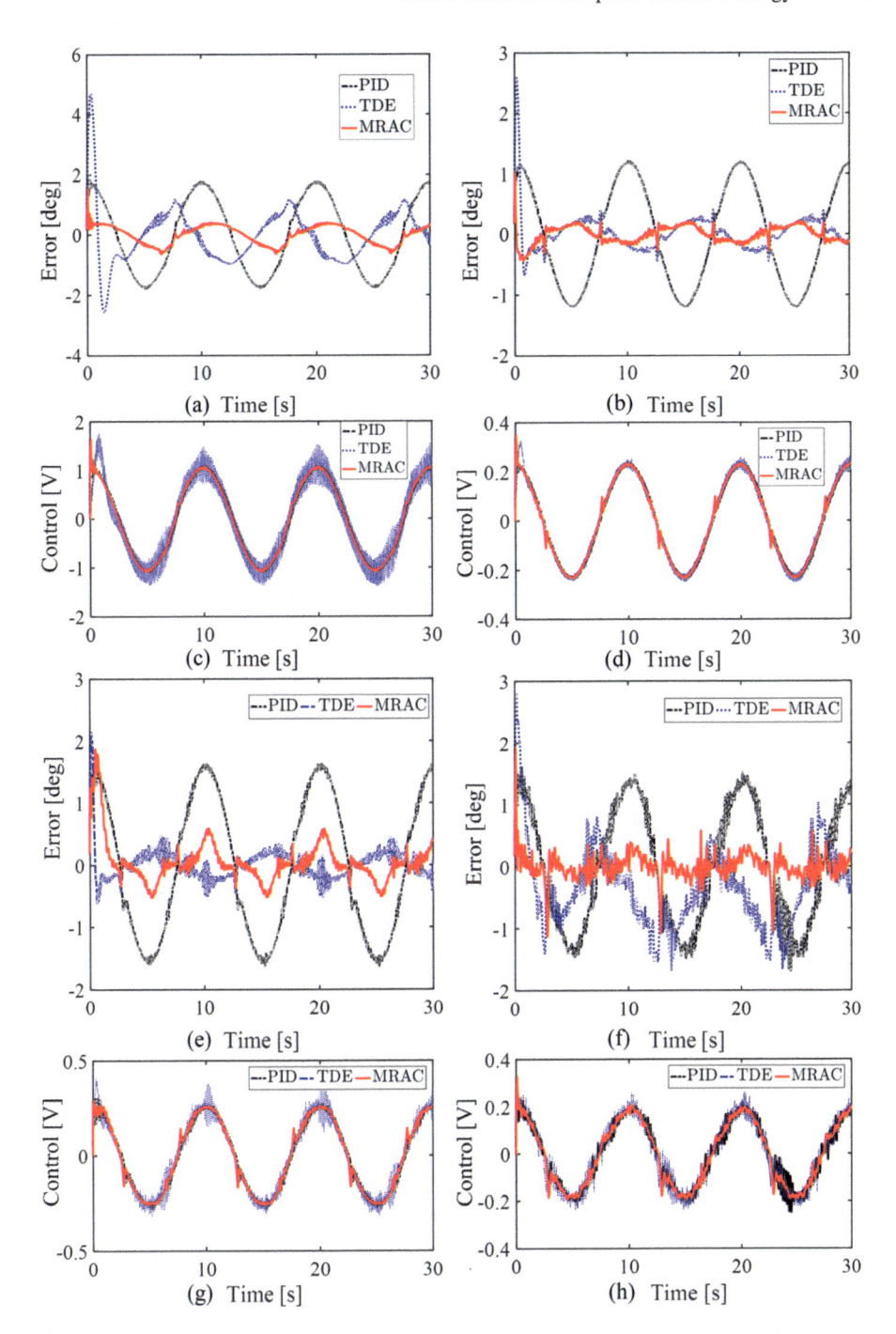

Fig. 4. Experimental results under high speed sinusoidal desired trajectory. (a), (b), (e), and (f) Tracking errors of joints 1, 2, 3, and 4, respectively. (c), (d), (g), and (h) Control inputs of joints 1, 2, 3, and 4, respectively.

5 Conclusion

In this study, we have proposed the decentralized adaptive control strategy using the MRAC method for motion control of robot manipulators. Due to high nonlinearities, strong couplings, and uncertainties in the robot dynamics, it is not easy to obtain the mathematical model of manipulators, thus, three control methods which do not require the exact model have been deployed. Especially, in the TDE–NTSM control method and the MRAC method, the stability of the closed-loop system is guaranteed based on the Lyapunov stability theory. It was verified through experiments that the proposed MRAC scheme can ensure better control performance than that of the conventional PID control and that of TDE-NTSM control despite the presence of unknown parameters and nonlinear uncertainties in the robot dynamics. For further work, in order to enhance the control performance as well as counteract the parameter variations during operation, adaptive control method combining with disturbance rejection techniques should be deployed.

Acknowledgment. This work was supported by the National Research Foundation of Korea (NRF) grant funded by the Korea Government (Ministry of Education) (No.NRF-2015R1D1A1A09056885). This work was also supported by the INNOPOLIS Foundation of Korea (INNOPOLIS BUSAN) (Project Name: Development of a Practical Technology of Mobile Fender System, No. 17BSI1008).

References

1. Jazar, R.N.: Theory of Applied Robotics, 2nd edn. Springer, New York (2010)
2. Soltanpour, M.R., Khooban, M.H., Soltani, M.: Robust fuzzy sliding mode control for tracking the robot manipulator in joint space and in the presence of uncertainties. Robotica **32**(3), 433–446 (2014)
3. Geng, J., Sheng, Y., Liu, X.: Time-varying nonsingular terminal sliding mode control for robot manipulators. Trans Inst. Measur. Control **36**(5), 604–617 (2013)
4. Zhou, Q., Li, H., Shi, P.: Decentralized adaptive fuzzy tracking control for robot finger dynamics. IEEE Trans. Fuzzy Syst. **23**(3), 501–509 (2015)
5. Huang, S.J., Lee, J.S.: A stable self-organizing fuzzy controller for robotic motion control. IEEE Trans. Industr. Electron. **47**(2), 421–428 (2000)
6. Han, S.I., Lee, J.M.: Fuzzy echo state neural networks and funnel dynamic surface control for prescribed performance of a nonlinear dynamic system. IEEE Trans. Industr. Electron. **61**(2), 1099–1112 (2014)
7. Dubowsky, S., DesForges, D.T.: The application of model-referenced adaptive control to robotic manipulators. J. Dyn. Syst. Measur. Control **101**, 193–200 (1979)
8. Kirecci, A., Topalbekiroglu, M., Eker, I.: Experimental evaluation of a model reference adaptive control for a hydraulic robot. Robotica **21**(1), 433–446 (2003)
9. Ulrich, S., Sasiadek, J.Z.: Direct model reference adaptive control of a flexible joint robot. In: Proceedings of the AIAA Guidance, Navigation, and Control Conference, pp. 7844–7853. The American Institute of Aeronautics and Astronautics, Toronto, Ontario Canada, 2–5 August 2010 (2010)

10. Sharifi, M., Behzadipour, S., Vossoughi, G.R.: Model reference adaptive impedance control in cartesian coordinates for physical human-robot interaction. Adv. Robot. **28**(19), 1277–1290 (2014)
11. Alqaudi, B., Modares, H., Ranatunga, I., Tousif, S.M., Lewis, F.L., Popa, D.O.: Model reference adaptive impedance control for physical human-robot interaction. Control Theor. Technol. **14**(1), 68–82 (2016)
12. Chiou, K.C., Huang, S.J.: An adaptive fuzzy controller for robot manipulators. Mechatronics **15**, 151–177 (2004)
13. Le, H.T., Lee, S.R., Lee, C.Y.: Integration model reference adaptive control and exact linearization with disturbance rejection for control of robot manipulators. Int. J. Innov. Comput. Inf. Control **7**(6), 3255–3267 (2011)
14. Xiao, S., Li, Y., Liu, J.: A model reference adaptive PID control for electromagnetic actuated micro-positioning stage. In: Proceedings of the 8th IEEE International Conference on Automation Science and Engineering, pp. 20–22, Seoul, Korea (2010)
15. Jin, M., Lee, J., Chang, P.H., Choi, C.: Practical nonsingular terminal sliding-mode control of robot manipulators for high-accuracy tracking control. IEEE Trans. Industr. Electron. **46**(9), 3593–3601 (2009)

Stabilization of Time-Varying Systems Subject to Actuator Saturation: A Takagi-Sugeno Approach

Sabrina Aouaouda[1(✉)] and Mohammed Chadli[2]

[1] Faculty of Sciences and Technologies, Univ. Souk Ahras-LEER,
BP 1553, 41000 Souk-Ahras, Algeria
`Sabrina.aouaouda@univ-soukahras.dz`
[2] MIS (E.A.4290), University of Picardie Jules Verne,
33, Rue Saint-Leu, 80039 Amiens, France
`mchadli@u-picardie.fr`

Abstract. This paper investigates the problem of control design for a class of time-varying parameter systems subject to input and state constraints. The aim is to synthesize a control strategy by combining a descriptor approach and a dynamic state feedback control laws. This ensures the closed-loop system stability with respect to the given saturation constraints on the control input. The optimization problem is formulated in terms of linear matrix inequality constraints (LMIs). Nonlinear vehicle model is used to highlight the effectiveness of the proposed approach.

Keywords: Time-varying systems · Actuator saturation ·
Descriptor approach · Stability · $\mathcal{L}_2$-gain performance ·
Linear Matrix Inequality (LMI)

1 Introduction

Recently, due to the increasing demand of higher control performance of complex modern systems, new design strategies in the fuzzy control field has been developed to deal with system nonlinearities and environment constraints. Particularly, control design based on Takagi-Sugeno (T-S) fuzzy models has received considerable interest [1]. In fact nonlinear systems can be accurately approximated with T-S models by the nonlinear sector decomposition approach [2]. Besides, stability analysis of T-S systems with Lyapunov functions is of practical importance with the so-called parallel distributed compensation (PDC) structure [3, 4], which provides to a systematic design of optimization problems with linear matrix inequalities (LMIs) constraints. Moreover, control design of dynamical processes with time varying behaviour is considered now as a challenging problem with a higher level of complexity since the time-varying parameters must be taken into account during the system modeling phase. Hence, different solutions for this subject have been proposed in the literature for various engineering areas (see [5, 6] and the references therein). Most of the proposed approaches consist in representing the varying parameters and the nonlinearities in a polytopic form and resulting in T-S model representation.

Plan uncertainty and actuator saturation and/or sensor saturation are almost encountered in all real physical applications. Consequently, an immense deal of attention has been focused on the control design of T-S models with input saturation constraint [7–9]. Among the most popular works dealing with saturated input constraints, the convexity based approach to the saturation function (see [9, 10] and the references therein). The main idea is to consider a bounded ellipsoidal symmetric region of stability by solving a set of LMIs. In addition, descriptor design technique has been recently investigated in [11] to deal with problem of input saturated T-S systems using a polytopic representation of the saturation function. Control design of saturated LPV (linear parameter varying) systems has been also investigated, see for instance [12, 13]. Recall that most existing design control approaches are based on state feedback or dynamic output feedback with anti-windup (AW) mechanisms.

In this paper, a robust control law reconfiguration to guarantee the stability of disturbed T-S systems subject to input saturation and time varying parameters is studied. A descriptor approach is used with a modified dynamic state feedback control law. The derived conditions of asymptotic stability based on the $\mathcal{L}_2$-gain performance, are established using Lyapunov function and solved by means of LMI optimization.

The rest of this paper is structured as follows: Sect. 2 describes the considered class of systems together with the control problem statement. The control problem is then defined in Sect. 3 and solved in Sect. 4 where LMI-based design conditions are proposed. In Sect. 4 simulation results of lateral vehicle dynamic model are reported, and finally some conclusions are drawn in Sect. 6.

Notations. $\mathbb{N}_r$ denotes the set $\{1, 2, \cdots, r\}$, I denotes the identity matrix. $\mathcal{H}(A)$ denotes the Hermitian of the matrix A, i.e. $\mathcal{H}(A) = A + A^T$. Given a matrix $P > 0$, and positive scalar $\mathfrak{S}(P, \gamma)$ denotes the ellipsoidal set defined by $\mathfrak{S}(P) = \{x \in \mathbb{R}^n : x^T p^{-1} x \leq \gamma\}$ and $\mathfrak{S}(P) \equiv \mathfrak{S}(P, 1)$.

2 Problem Statement

2.1 Preliminaries: Time-Varying Parameters Polytopic Modeling

Consider the following time-varying linear system

$$\begin{cases} \dot{x}(t) = A(\rho(t))x(t) + B(\rho(t))u(t) \\ y(t) = Cx(t) \end{cases} \quad (1)$$

where $x(t) \in \mathbb{R}^n$ is the state vector, $u(t) \in \mathbb{R}^{n_u}$ is the control input and $y(t) \in \mathbb{R}^p$ is the measured output vector. $A(\rho(t)) \in \mathbb{R}^{n \times n}$ is the parameter varying state matrix, $B(\rho(t)) \in \mathbb{R}^{n \times n_u}$ is the input matrix, and $C \in \mathbb{R}^{p \times n}$ is the output matrix. The time varying linear system represented by Eq. (1) can be rewritten using the polytopic modeling representation. Accordingly, the parameter varying state and input matrices $A(\rho(t))$, $B(\rho(t))$ are defined by:

$$A(\rho(t)) = A_0 + \rho(t)\bar{A} \\ B(\rho(t)) = B_0 + \rho(t)\bar{B} \tag{2}$$

with $\rho(t) \in [\rho^{min}, \rho^{max}]$ is a bounded non measurable time-varying parameter. A_0, $\bar{A}$, B_0, and $\bar{B}$ are known matrices with appropriate dimension. The bounded parameter $\theta(t)$ can be modeled as a polytopic representation by means of the so-called sector non-linearity approach. The obtained structure is given by:

$$\rho(t) = \eta_1(\rho(t))\rho^{min} + \eta_2(\rho(t)\rho^{max}) \tag{3}$$

with

$$\eta_1(\rho(t)) = \frac{\rho^{max} - \rho(t)}{\rho^{max} - \rho^{min}}, \eta_2(\rho(t)) = \frac{\rho(t) - \rho^{min}}{\rho^{max} - \rho^{min}} \tag{4}$$

The weighting functions $\eta_1(\rho(t))$ and $\eta_2(\rho(t))$ verifies the convex sum property

$$\begin{cases} \sum_{i=1}^{2} \eta_i(\rho(t)) = 1 \\ 0 \leq \eta_i(\rho(t)) \leq 1, \quad i = 1, 2 \end{cases} \tag{5}$$

Now, using Eqs. (3) and (4), the initial time-varying parameter linear system is written as a polytopic linear model (PLM) structure, such that:

$$\begin{cases} \dot{x}(t) = \sum_{i=1}^{2} \eta_i(\rho(t))(A_i x(t) + B_i u(t)) \\ y(t) = Cx(t) \end{cases} \tag{6}$$

with

$$\begin{cases} A_1 = A_0 + \rho^{min}\bar{A} \\ A_2 = A_0 + \rho^{max}\bar{A} \end{cases} \text{and} \begin{cases} B_1 = B_0 + \rho^{min}\bar{B} \\ B_2 = B_0 + \rho^{max}\bar{B} \end{cases} \tag{7}$$

Remark 1. Practically, changes appearing in system parameter are directly related to malfunctions in the system itself or variations in their environment setting. The system performance and specifically system stability is directly influenced by this changes. In this paper, only the case of system affected by one time-varying parameter is considered. However, the proposed approach can be generalized for multiple parameters $\rho_j(t)$ affecting the system matrices and a compact T-S form is derived.

2.2 Saturated Control Problem Statement

A parameter time-varying nonlinear system under actuator saturation and may be subject to unknown inputs is represented by the following PLM structure:

$$\dot{x}(t) = \sum_{i=1}^{r}\sum_{j=1}^{2} \eta_i(\xi(t))\eta_j(\rho(t))\left(A_{ij}x(t) + B_{ij}sat(u(t))\right) + B_w w(t) \quad (8)$$

where, the saturation function $sat : \mathbb{R}^{n_u} \to \mathbb{R}^{n_u}$ is defined as:

$$\begin{cases} sat(u) = \left[sat_1(u_1) \cdots sat_j(u_j) \cdots sat_{n_u}(u_{n_u})\right]^T \\ sat_j(u_j) = sign(u_j) min\left(|u_j|, u_j^{max}\right) \end{cases} \quad (9)$$

with $u_j^{max} > 0$ denote the saturation level. The disturbance $w(t)$ in (8) belongs to the set of functions

$$\mathcal{Q}_\delta = \left\{ w(t) : \mathbb{R}^+ \to \mathbb{R}^{n_d + p}, \int_0^\infty w(\tau)^T w(\tau) d\tau \leq \delta \right\} \quad (10)$$

In closed loop control system, the nonlinear behavior of the input saturation has been investigated as a convex combination of 2^m linear models in [7, 8], and as an alternative polytopic structure using Takagi-Sugeno modeling in [13]. In this work only adjustment on the controller input will be considered as a reminiscent of the classical anti-windup configuration.

3 Saturated State Feedback Control Law

3.1 Control Law

In this section, the objective is to design a state-feedback control ensuring the desired control performance and guarantying the stability of the closed loop system, despite the presence of input control saturation. Accordingly, the proposed dynamic state-feedback controller law (DSFC) is designed as follows:

$$\begin{cases} \dot{x}_c(t) = \sum_{i=1}^{r} \eta_i(\xi(t))\left(A_i^c x_c(t) + B_i^c \psi(u)\right), x_c(0) = 0 \\ u(t) = \sum_{j=1}^{r} \eta_j(\xi(t))\left(K_j x(t) + C_j^c x_c(t)\right) \\ \psi(u) = u(t) - sat(u(t)) \end{cases} \quad (11)$$

where, $x_c(t) \in \mathbb{R}^n$ is the state variable of the proposed controller, and A_i^c, B_i^c, C_i^c, K_i are matrices to be determined. By introducing the augmented state variable $\bar{x} = [x^T x_c^T u^T]^T$ the closed loop system is described by the following augmented descriptor form:

$$\mathbb{E}\dot{\bar{x}}(t) = \sum_{i=1}^{r}\sum_{j=1}^{2} \eta_i(\xi(t))\eta_j(\rho(t))\left(\bar{A}_{ij}\bar{x}(t) + \bar{B}_{ij}\psi(u)\right) + \bar{B}_w w(t) \qquad (12)$$

where

$$\mathbb{E} = \begin{bmatrix} I & 0 & 0 \\ 0 & I & 0 \\ 0 & 0 & 0 \end{bmatrix}, \bar{A}_{ij} = \begin{bmatrix} A_{ij} + B_{ij}K_j & 0 & 0 \\ 0 & A_i^c & 0 \\ K_j & C_j^c & -I \end{bmatrix}, \bar{B}_{ij} = \begin{bmatrix} -B_{ij} \\ B_i^c \\ 0 \end{bmatrix}, \bar{B}_w = \begin{bmatrix} B_w \\ 0 \\ 0 \end{bmatrix} \qquad (13)$$

Remark 2. For control design, the augmented state vector $\bar{x}(t)$ in (12) can be limited to $\bar{x}(t) = [x^T x_c^T]^T$ which lead to a corresponding set of design conditions different to those obtained later. Here, the defined descriptor form (12) is impulse-free. This conveniently allows deriving of LMI condition by avoiding products of system and controller gains, and thus obtaining more tractable and less restrictive LMI conditions.

3.2 Control Problem Definition

The aim of this paper is to propose a new LMI-based technique for the design of a DSF controller (11) to preserve the closed-loop stability of the constrained polytopic model (8). Specifically, the following properties will have to be satisfied:

State Constraints: Given vectors $N_k \in \mathbb{R}^n, k \in \mathbb{N}_q$ for any $x(0) \in \mathfrak{S}(P,\gamma)$ and $w(t) \in \mathfrak{Q}_\delta$ the closed loop system remains inside the admissible polyhedral set $\mathcal{P}_x$ defined by:

$$\mathcal{P}_x = \{x \in \mathbb{R}^n;\ N_k^T x \leq 1,\ k \in \mathbb{N}_q\} \qquad (14)$$

Local Stability Convergence: There exists an ellipsoidal set $\mathfrak{S}(p) \subset \mathcal{P}_x$ such that any closed-loop trajectory of the undisturbed system (*i.e.* w(t) = 0) with initial state in this ellipsoid converges exponentially to the origin with a decay rate less than a given positive real number α.

$\mathcal{L}_2$-*gain Performance:* The input signal $w(t)$ belong to $\mathfrak{Q}_\delta$ set. In fact, the control problem is designed by minimizing the $\mathcal{L}_2$-gain from the disturbance signal to the solution of the closed loop system. That is, for $x(0) = 0$ and all allowable values $w(t) \in \mathfrak{Q}_\delta$, one has $\|x(t)\|_2 < \sqrt{\lambda}\|w(t)\|_2$.

4 Main Results

The objective of this section is to design a stabilizing DSF controller for the time-varying parameter T-S system (8) even in the presence of control input saturation. As previously discussed, the solution is obtained by representing the time varying parameter as a TS system and by solving an optimization problem under LMI constraints. The following lemmas are needed for the theoretical development:

Lemma 1: *Given two matrices K_i and $S_i \in \mathbb{R}^{m \times n}, i \in \mathbb{N}_r$ and let $\mathcal{P}_u$ be the polyhedral set related with this matrices and defined by:*

$$\mathcal{P}_u = \bigcap_{i=1}^{r} \mathfrak{S}(K_i - S_i)$$

where

$$\mathfrak{S}(K_i - S_i) = \left\{ x \in \mathbb{R}^n : \left| (K_{i(l)} - S_{i(l)}) x \right| \leq u_{max(l)} \right\}, l \in \mathbb{N}_m$$

Then if $x \in \mathcal{P}_u$, the inequality

$$\psi(u)^T \left(\sum_{i=1}^{r} \eta_i \mathcal{Q}_i \right)^{-1} \left(\psi(u) - \sum_{i=1}^{r} \eta_i S_i x \right) \leq 0$$

is verified, for any positive diagonal matrix $\mathcal{Q}_i$ and any family of functions $\eta_i, i \in \mathbb{N}_r$ satisfying the convex sum propriety.

The results given by Lemma 1 represent an extension of the sector bound condition form [10] to deal with the control input nonlinearity.

Lemma 2 [14]: *Consider symmetric matrices $\mathfrak{T}_{ij}, i,j \in \mathbb{N}_r$ of appropriate dimensions and a family of scalar functions $\eta_1, \cdots, \eta_r$ satisfying the convex sum property. The following inequality $\sum_{i=1}^{r} \sum_{j=1}^{r} \mathfrak{T}_{ij} < 0$ is verified if*

$$\begin{cases} \mathfrak{T}_{ii} < 0 \\ \frac{2}{r-1} \mathfrak{T}_{ii} + \mathfrak{T}_{ij} + \mathfrak{T}_{ji} \end{cases} \qquad i,j \in \mathbb{N}_r, \ i \neq j$$

The next theorem provides LMI conditions for the synthesis of a DSF controller satisfying the design problem properties expressed in the previous section.

Theorem 1: *Given positive scalar α, assume there exist positive definite matrices X_{1i}, X_{5i}, positive diagonal matrices $\mathcal{Q}_i$, matrices X_{7i}, X_{8i}, X_{9i}, $\mathcal{M}_i$, $\mathcal{N}_i$, $\mathcal{A}_i^c$, $\mathcal{B}_i^c$ $\mathcal{C}_i^c$ and positive scalars λ such that*

$$\begin{bmatrix} X_{1i} & * \\ \mathcal{M}_{i(l)} - \mathcal{N}_{i(l)} & u_{max(l)}^2 \end{bmatrix} \geq 0, \quad i \in \mathbb{N}_r, \quad l \in \mathbb{N}_m \tag{15}$$

$$\begin{bmatrix} X_{1i} & * \\ N_k X_{1i} & 1 \end{bmatrix} \geq 0, \quad k \in \mathbb{N}_q \tag{16}$$

$$Y_{ii} < 0, \quad i \in \mathbb{N}_r \tag{17}$$

$$\frac{2}{r-1} Y_{ii} + Y_{ij} + Y_{ji} < 0, \quad i, j \in \mathbb{N}_r, i \neq j \tag{18}$$

where Y_{ij} is defined by:

$$Y_{ij} = \begin{bmatrix} Y_{ij}^{11} & * & * & * & * & * \\ 0 & Y_{ij}^{22} & * & * & * & * \\ \mathcal{M}_j - X_{7i} & \mathcal{C}_j^c - X_{8i} & \mathcal{H}(X_{9i}) & * & * & * \\ X_{1i} & 0 & 0 & -I & * & * \\ -\mathcal{Q}_i^T B_{ij}^T + \mathcal{N}_i & \mathcal{B}_i^c & 0 & 0 & -2\mathcal{Q}_i & * \\ B_w^T & 0 & 0 & 0 & 0 & -\lambda I \end{bmatrix} \tag{19}$$

with $Y_{ij}^{11} = \mathcal{H}(A_{ij} X_{1i} + B_{ij} \mathcal{M}_j) + 2\alpha X_{1i}^T$, $Y_{ij}^{22} = \mathcal{H}(\mathcal{A}_i^c) + 2\alpha X_{5i}^T$. Then the DSFC (11) with the feedback gains given by $K_j = \mathcal{M}_j X_{1i}^{-1}$, $A_i^c = \mathcal{A}_i^c X_{5i}^{-1}$, $C_j^c = \mathcal{C}_j^c X_{5i}^{-1}$, $B_i^c = \mathcal{Q}_i^{-1} \mathcal{B}_i^c$ solves the design problem stated in Sect. 3 for the closed loop polytopic LPV system (8).

Proof. From the inequality (15) (respectively (16)) we can prove the inclusion $\mathfrak{S}(P) \subseteq \mathcal{p}_u$ (respectively $\mathfrak{S}(P) \subseteq \mathcal{p}_x$). Now, consider the following candidate Lyapunov function V defined as:

$$V(\bar{x}(t)) = \bar{x}(t)^T \mathbb{E} \mathbb{P}_i \bar{x}(t) \tag{20}$$

with

$$\mathbb{E} \mathbb{P}_i = \mathbb{P}_i^T \mathbb{E} \geq 0 \tag{21}$$

Let $\dot{V}(\bar{x}(t))$ be the derivative of $V(\bar{x}(t))$ along the trajectories of (8). For $x(t) \in \mathcal{P}_{x}$, the closed loop system converge asymptotically to zero if:

$$\dot{V}(\bar{x}) + 2\alpha V(\bar{x}) + \bar{x}^T \mathcal{Q}_w \bar{x} - \lambda w^T w - 2\psi(u)^T \mathcal{Q}_i^{-1}(\psi(u) - \mathcal{S}_i x) < 0 \quad (22)$$

With (12) and (20), the inequality (22) is fulfilled if:

$$\begin{bmatrix} \mathcal{H}(\bar{A}_{ij}\mathbb{P}_i) + 2\alpha\mathbb{E}\mathbb{P}_i + \mathcal{Q}_w & * & * \\ \bar{B}_{ij}^T \mathbb{P}_i + \mathcal{Q}_i^{-1}\mathcal{S}_i^* & -2\mathcal{Q}_i^{-1} & * \\ \bar{B}_w^T \mathbb{P}_i & 0 & -\lambda I \end{bmatrix} < 0 \quad (23)$$

where $\mathcal{S}_i^* = [\mathcal{S}_i \ 0 \ 0]$. Let $\mathbb{X}_i = \mathbb{P}_i^{-1}$. Pre- and post-multiplying (23) with $diag(\mathbb{X}_i, \mathcal{Q}_i, I)$ and its transpose yields:

$$\begin{bmatrix} \mathcal{H}(\bar{A}_{ij}\mathbb{X}_i) + 2\alpha\mathbb{X}_i^T\mathbb{E} + \mathbb{X}_i^T\mathcal{Q}_w\mathbb{X}_i & * & * \\ \mathcal{Q}_i^T \bar{B}_{ij}^T + \mathcal{S}_i^*\mathbb{X}_i & -2\mathcal{Q}_i & * \\ \bar{B}_w^T & 0 & -\lambda I \end{bmatrix} < 0 \quad (24)$$

In order to derive easily LMI conditions, the structure of the matrix $\mathbb{X}_i$ is considered as follows:

$$\mathbb{X}_i = \begin{bmatrix} X_{1i} & 0 & 0 \\ 0 & X_{5i} & 0 \\ X_{7i} & X_{8i} & X_{9i} \end{bmatrix} \quad (25)$$

According to (21), it follows that $X_{1i} = X_{1i}^T$, $X_{5i} = X_{5i}^T$, and X_{7i}, X_{8i}, X_{9i} are free slack matrices. Let $\mathcal{N}_i = \mathcal{S}_i X_{1i}$. Using Shur complement lemma and system matrices defined by (13) we can prove that the condition (24) is developed to:

$$\begin{bmatrix} \phi_{ij} & * & * & * & * & * \\ 0 & \mathcal{H}(A_i^c X_{5i}) + 2\alpha X_{5i}^T & * & * & * & * \\ K_j X_{1i} - X_{7i} & C_j^c X_{5i} - X_{8i} & \mathcal{H}(X_{9i}) & * & * & * \\ X_{1i} & 0 & 0 & -I & * & * \\ -\mathcal{Q}_i^T \bar{B}_{ij}^T + \mathcal{N}_i & \mathcal{Q}_i^T B_i^c & 0 & 0 & -2\mathcal{Q}_i & * \\ B_w^T & 0 & 0 & 0 & 0 & -\lambda I \end{bmatrix} < 0 \quad (26)$$

with $\phi_{ij} = \mathcal{H}(A_{ij}X_{1i} + B_{ij}K_j X_{1i}) + 2\alpha X_{1i}^T$. This inequality is equivalent to (19). Accordingly, the proof of Theorem 1 can be now concluded.

5 Numerical Example

In this section, the proposed approach is applied to a vehicle model. A reduced form of lateral dynamics of the so-called bicycle model (see the Fig. 1) is considered with F_f and F_r are the lumped lateral tire force of the front and rear tires, respectively. Starting from the nonlinear system equations, a T-S representation is derived with only the dynamics of the yaw rate $r(t)$ and the vehicle sideslip angle $\beta(t)$ as follows [15]:

$$\begin{cases} \dot{x}(t) = \sum_{i=1}^{2} \eta_i(|\alpha_f|)\left(A_i(v)x(t) + B_{fi}(v)\delta_f(t) + B_i M_z(t)\right) \\ y(t) = Cx(t) \end{cases} \quad (27)$$

where

$$A_i = \begin{bmatrix} -2\frac{S_{fi}+S_{ri}}{mv} & -2\frac{S_{fi}a_f - S_{ri}a_r}{mv^2} - 1 \\ -2\frac{S_{fi}a_f - S_{ri}a_r}{J} & -2\frac{S_{fi}a_f^2 + S_{ri}a_r^2}{Jv} \end{bmatrix}, \; B_{fi} = \begin{bmatrix} \frac{2S_{fi}}{mv} \\ \frac{2a_f S_{fi}}{J} \end{bmatrix}, \; B_i = B = \begin{bmatrix} 0 \\ \frac{1}{J} \end{bmatrix},$$

$C = \begin{bmatrix} 0 & 1 \end{bmatrix}$. With $x(t) = \begin{bmatrix} \beta^T(t) r^T(t) \end{bmatrix}^T$ is the state vector, $y(t)$ is the measured output. The control input to be designed is the front steering angle $u(t) = \delta_f(t)$ while the disturbance input is the yaw moment $M_z(t)$. The membership functions $\mu_i, i \in \{1,2\}$ satisfy the convex sum property are defined as follows:

$$\eta_i(|\alpha_f|) = \frac{\vartheta_i(|\alpha_f|)}{\sum_{i=1}^{2} \vartheta_i(|\alpha_f|)}, \; \vartheta_i(|\alpha_f|) = \frac{1}{\left[1 + \left|\left(\frac{|\alpha_f| - c_i}{a_i}\right)\right|\right]^{2b_i}} \quad (28)$$

with the following parameter values obtained for a dry road, [15]: $a_1 = 0.5077, a_2 = 0.4748, b_1 = 3.1893, b_2 = 5.3907, c_1 = -0.4356, c_2 = 0.5622$.

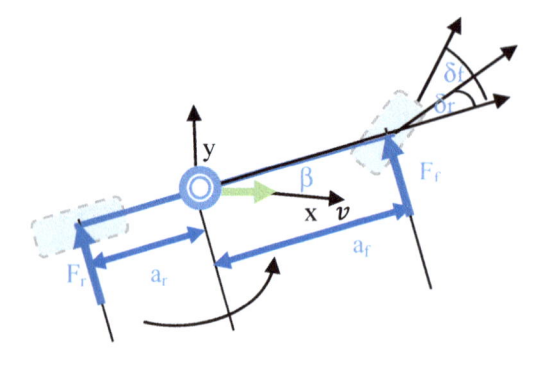

Fig. 1. Vehicle lateral dynamics modeling

Table 1. Vehicle parameters

Symbol	Value
J	3000 kg . m²
m	1500 kg
a_f	1.3 m
a_r	1.2 m
S_{f10}	60712 N/rad
S_{f20}	4812 N/rad
S_{r10}	60088 N/rad
S_{r20}	3455 N/rad

The vehicle parameters are given in Table 1 [15]. Due to the physical limitation of the steering system, the control input $\delta_f(t)$ of the vehicle system (27) is subject to actuator saturation: $\delta_f(t) = sat(u(t) = sign(u(t))min(|u(t)|, \delta_{fmax})$ where δ_{fmax} is the limitation of the steering angle. The dynamics of the system (27) depend nonlinearly on the vehicle speed which is measured and bounded $v^{min} \leq v \leq v^{max}$, $v^{min} = 5$ [m/s], $=30$ [m/s]. Using the first order Taylor approximation $1/v$ and $1/v^2$ are written as:

$$\frac{1}{v} = \frac{1}{v_0} + \frac{1}{v_1}\rho(t), \quad \frac{1}{v^2} = \frac{1}{v_0^2} + 2\frac{1}{v_0 v_1}\rho(t) \qquad \rho(t) \in [-1 \; 1] \qquad (29)$$

Where the two constants v_0 and v_1 in (29) are given by $v_0 = \frac{2v^{min}v^{max}}{v^{min}+v^{max}}$, $v_1 = \frac{2v^{min}v^{max}}{v^{min}-v^{max}}$. Using the approach presented in Sect. 2, the vehicle model (27) can be equivalently represented in the following polytopic form:

$$\begin{cases} \dot{x}(t) = \sum_{j=1}^{2}\sum_{i=1}^{2} \eta_i(|\alpha_f|)\eta_i(\rho)\left(A_{ij}x(t) + B_{fij}\delta_f(t) + B_i M_z(t)\right) \\ y(t) = Cx(t) \end{cases} \qquad (30)$$

where

$$A_{i1} = \begin{bmatrix} -2\frac{S_{fi}+S_{ri}}{m}\left(\frac{1}{v_0}+\frac{1}{v_1}\rho^{max}\right) & -2\frac{S_{fi}a_f - S_{ri}a_r}{m}\left(\frac{1}{v_0^2}+2\frac{1}{v_0 v_1}\rho^{max}\right)-1 \\ -2\frac{S_{fi}a_f - S_{ri}a_r}{J} & -2\frac{S_{fi}a_f^2 + S_{ri}a_r^2}{J}\left(\frac{1}{v_0}+\frac{1}{v_1}\rho^{max}\right) \end{bmatrix}$$

$$A_{i2} = \begin{bmatrix} -2\frac{S_{fi}+S_{ri}}{m}\left(\frac{1}{v_0}+\frac{1}{v_1}\rho^{min}\right) & -2\frac{S_{fi}a_f - S_{ri}a_r}{m}\left(\frac{1}{v_0^2}+2\frac{1}{v_0 v_1}\rho^{min}\right)-1 \\ -2\frac{S_{fi}a_f - S_{ri}a_r}{J} & -2\frac{S_{fi}a_f^2 + S_{ri}a_r^2}{J}\left(\frac{1}{v_0}+\frac{1}{v_1}\rho^{min}\right) \end{bmatrix}$$

$$B_{fi1} = \begin{bmatrix} \frac{2S_{fi}}{m}\left(\frac{1}{v_0}+\frac{1}{v_1}\rho^{max}\right) \\ \frac{2a_f S_{fi}}{J} \end{bmatrix}, \quad B_{fi2} = \begin{bmatrix} \frac{2S_{fi}}{m}\left(\frac{1}{v_0}+\frac{1}{v_1}\rho^{min}\right) \\ \frac{2a_f S_{fi}}{J} \end{bmatrix}$$

and

$$\eta_1(\rho) = \frac{1-\rho(t)}{2}, \eta_2(\rho) = 1 - \eta_1(\rho)$$

Now the bicycle lateral dynamics polytopic model (30) is employed to simulate the performance of the designed DSFcontroller. The control gains K_j, A_j^c, C_j^c, and B_i^c can be easily calculated from the optimization conditions in Theorem 1. The details on the solution data are not given here for brevity. In order to illustrate the influence of the time varying parameters corresponding to the velocity, Fig. 2 represents the vehicle state responses $x_n(t)$ of the nominal system (i.e. without varying parameters) and the time varying system states $x_v(t)$ given by (30). Figure 3 shows the different control inputs applied to the proposed system where $u(t)$ represent the derived DSFcontroller input. As depicted in Fig. 2, the designed DSF controller allows the vehicle to achieve a good path performance despite an important level of steering saturation at the beginning. Accordingly, the obtained results illustrate the effectiveness of the designed approach for the vehicle dynamic systems.

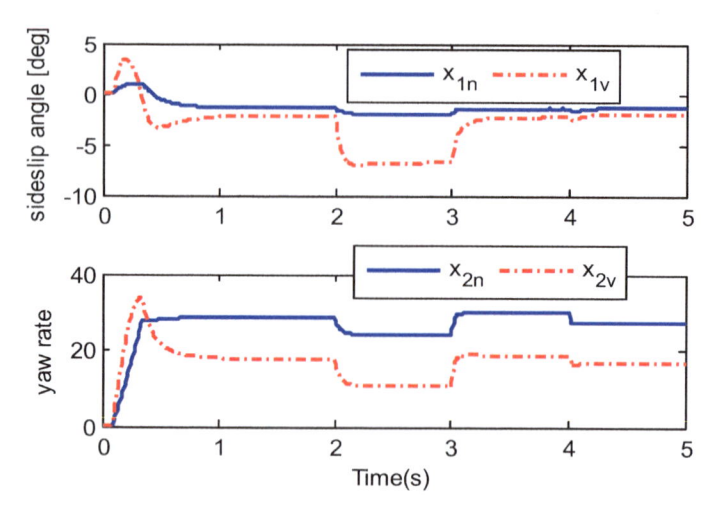

Fig. 2. Vehicle responses in the presence of actuator saturation- with and without $\rho(t)$

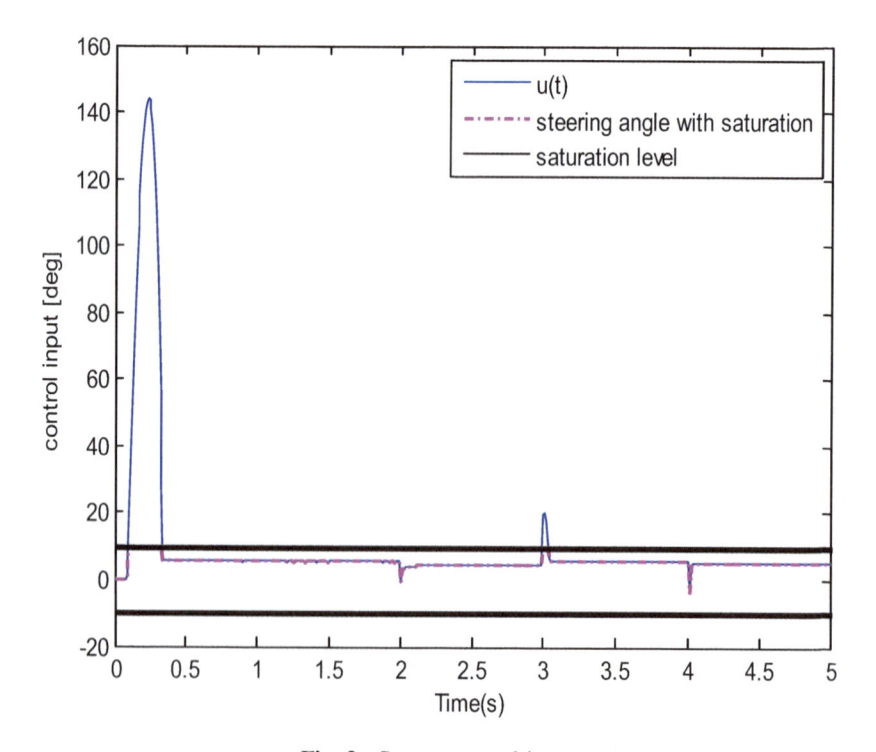

Fig. 3. System control input

6 Conclusion

A new LMI-based control solution has been established for nonlinear system represented by T-S model with time varying parameters and subject to input and state constraints. The main advantage of the proposed approach is to synthesize the control law by considering the saturation limits. The controller gains are then obtained by solving an optimization problem under LMI constraints. A vehicle lateral dynamics is used to illustrate the effectiveness of the proposed approach. First, the vehicle system is transformed to a polytopic LPV model to deal with the speed variation. Then, the steering actuator saturation is considered in the control design procedure. As future works, the proposed approach may be generalized to T-S systems with unmeasurable decision variable and/or subject to sensor saturations.

References

1. Takagi, T., Sugeno, M.: Fuzzy identification of systems and its applications to modeling and control. IEEE Trans. Syst. Men Cybern. **SMC-15**(1), 116–132 (1985)
2. Tanaka, K., Wang, H.: Fuzzy Control Systems Design and Analysis: A Linear Matrix Inequality Approach. Wiley-Interscience, New York (2001)
3. Fang, C.H., Liu, Y.S., Kau, S.W., Hong, L., Lee, C.H.: A new LMI-based approach to relaxed quadratic stabilization of T-S fuzzy control systems. IEEE Trans. Fuzzy Syst. **14**(3), 386–394 (2006)
4. Sala, A., Arino, C.: Relaxed stability and performance conditions for Takagi-Sugeno fuzzy systems with knowledge on membership function overlap. IEEE Trans. Syst. Men Cybern. **37**(3), 727–732 (2007)
5. Hoffmann, C., Werner, H.: A survey of linear parameter-varying control applications validated by experiments or high-fidelity simulations. IEEE Trans. Control Syst. Technol. **23**(2), 416–433 (2015)
6. Fergani, S., Menhour, L., Sename, O., Dugard, L., D'Andrea-Novel, B.: Integrated vehicle control through the coordination of longitudinal/lateral and vertical dynamics controllers: flatness and LPV/H∞-based design. Int. J. Robust Nonlinear Control **27**(18), 4992–5007 (2017)
7. Tarbouriech, S., Garcia, G., da Silva, J.G., Queinnec, I.: Stability and Stabilization of Linear Systems with Saturating Actuators. Springer, London (2011)
8. Saifia, D., Chadli, M., Labiod, S., Guerra, T.M.: Robust H∞ static output feedback stabilization of T-S fuzzy systems subject to actuator saturation. Int. J. Control Autom. Syst. **10**(3), 613–622 (2012)
9. Nguyen, A.-T., Dambrine, M., Lauber, J.: Simultaneous design parallel distributed output feedback and anti-windup compensators for constrained Takagi-Sugeno fuzzy systems. Asian J. Control **18**(5), 1641–1654 (2016)
10. Dang, Q.V., Vermeiren, L., Dequidt, A., Dambrine, M.: Robust stabilizing controller design for Takagi-Sugeno fuzzy descriptor systems under state constraints and actuator saturation. Fuzzy Sets Syst. **329**, 77–90 (2017)

11. Bezzaoucha, S., Marks, B., Maquin, D., Ragot, J.: Stabilization of nonlinear systems subject to uncertainties and actuator saturation. In: Proceeding of the American Control Conference, pp. 2403–2408 (2013)
12. Gao, Y.-Y., Lin, Z., Shamash, Y.: Set invariance analysis and gain-scheduling control for LPV systems subject to actuator saturation. Syst. Control Lett. **46**(2), 137–151 (2002)
13. Nguyen, A.-T., Chevrel, P., Claveau, F.: Gain-scheduled static output feedback control for saturated LPV systems with bounded parameter variations. Automatica **89**, 420–424 (2018)
14. Tuan, H., Apkarian, P., Narikiyo, T., Yamamoto, Y.: Parameterized linear matrix inequality techniques in fuzzy control design. IEEE Trans. Fuzzy Syst. **9**, 324–332 (2001)
15. Aouaouda, S., Chadli, M., Boukhnifer, M., Karimi, H.k.: Robust fault tracking controller design for vehicle dynamics: A descriptor approach. Mechatronics **30**, 316–326 (2015)

Observer Based Control for Systems with Mismatched Uncertainties in Output Matrix

Van Van Huynh[1(✉)], Tran Thanh Phong[2], and Bach Hoang Dinh[2]

[1] Modeling Evolutionary Algorithms Simulation and Artificial Intelligence, Faculty of Electrical and Electronics Engineering, Ton Duc Thang University, Ho Chi Minh City, Vietnam
huynhvanvan@tdtu.edu.vn
[2] Faculty of Electrical and Electronics Engineering, Ton Duc Thang University, Ho Chi Minh City, Vietnam

Abstract. This paper presents a new analysis method to design an observer-based control for a class of mismatched uncertain time-delay system with mismatched uncertainties in the output matrix. One of the contributions is to estimate the current true value of the system state variables, avoiding the effect of the delayed and noised measurement output. Linear matrix inequality (LMI) approach is used to design the observer-based control. The control and observer gains matrices are characterized using the solution of the LMI existence condition.

Keywords: Mismatched uncertain time-delay systems · Linear Matrix Inequalities (LMI) · Observer-based control

1 Introduction

The observer-based control design problem for both deterministic and stochastic linear systems is well characterized in [1–6]. In [1], a state observer for mismatched uncertain systems has been proposed. Conditions for exponential observation error decay are given. The same control method has been applied for a class of uncertain systems with disturbance input in [2]. The author of [3] has presented a new disturbance observer based control techniques for nonlinear systems under disturbances. In [4] observer-based nonlinear control approaches have been developed for classes of nonlinear systems. An LMI-based sliding-mode observer design method is proposed in [5, 6]. In order to reduce conservative, the authors of [11] have presented a new observer-based stabilization for mismatched uncertain systems. The problem of design of observers and controllers for a class of uncertain systems against signal quantization occurring in both measurement output and control input channels were proposed in [16]. However, time delays are not included in the above approaches [1, 2, 4, 11, 16]. Time delays are frequently encountered in control systems. Time delay commonly leads to a degradation and/or instability in system performance [14, 15]. That is why the observer-based control design for time delay systems has attracted so much attention. For example, the authors of [12] proposed a delay-dependent design method of an observer-based H∞ control. That paper provides a linear-matrix-inequality-based algorithm by adopting the idea of the cone-complementarity problem. In order to improve the estimation error dynamics, a robust proportional integral observer is proposed for time-delay systems [7]. For nominal fuzzy systems, a delay-independent design of observer-based H_∞ control is developed, and the stabilization conditions are proposed in LMIs terms [8]. In [9], the problem of adaptive robust state observer design is proposed for a class of uncertain time-delay systems with the unknown upper bounds of the non-linear delayed state perturbations. In [13], the unknown input observer was designed for the systems involving both delay in the state, and delay appears in the input and output. The other observers for some classes of time-delay systems can be found in [12] and references therein.

However, most of the existing works on state observer are focused on time-delay systems whose outputs are not affected by unknown inputs and mismatched parameter uncertainties in the state matrix and in the delayed state matrix are not considered. However, this situation might exist in many practical applications because the mathematical model always contains some uncertain elements; these uncertainties may be due to additive unknown internal or external noise, environmental influence, poor plant knowledge, or slowly varying parameters. This motivates the work of this paper. Compared to the existing results in the literature, this paper deals with the observer-based control design problem for a class of mismatched uncertain time-delay systems where mismatched parameter uncertainties are involved in the state matrix and the delayed state matrix, as well as in the output matrix. In addition, the control and observers gain could be found from the LMI formulation.

Notation: The notation used throughout this paper is fairly standard. X^T denotes the transpose of matrix X. $I_{n \times m}$ and $0_{n \times m}$ are used to denote the $n \times m$ identity matrix and the $n \times m$ zero matrix, respectively. The subscripts n and $n \times m$ are omitted where the dimension is irrelevant or can be determined from the context. $\|x\|$ stands for the Euclidean norm of vector x and $\|A\|$ stands for the matrix induced norm of the matrix A. The expression $A > 0$ means that A is symmetric positive definite. R^n denotes the n-dimensional Euclidean space. For the sake of simplicity, sometimes function $x(t)$ is denoted by x.

2 Problem Formulation

Consider the following mismatched uncertain system with time varying delay described by

$$\begin{aligned}\dot{x}(t) &= [A + \Delta A]x(t) + [A_d + \Delta A_d]x(t - d) + Bu(t) \\ y(t) &= (C + \Delta C)x(t)\end{aligned} \quad (1)$$

where $x(t) \in R^n$ is the state vector, $u(t) \in R^m$ is the control input vector, $y(t) \in R^p$ is the output measurement. The symbols $x(t - d(t))$ and $y(t - d(t))$ represent delayed states and delayed outputs, respectively. The known function $d = d(t)$ is the time-varying delay which is assumed to be continuous, non-negative and bounded in $\Re^+$, that is, $\bar{d} := \sup_{t \in \Re^+} \{d(t)\} < \infty$. The initial conditions for the system is given by $x(t) = \phi(t)$ ($t \in [-\bar{d}, 0]$) where $\phi(t)$ are continuous in $[-\bar{d}, 0]$. The matrices A, B, C and A_d are constant matrices with appropriate dimensions. First, we consider the following assumptions:

Assumption 1. The matrix pair (A, C) is observable.

Assumption 2. There exist matrices M_i, N_i and $F_i(t)$, $i = 1, 2, 3$ of appropriate dimensions so that

$$\Delta A = M_1 F_1(t) N_1, \Delta A_d = M_2 F_2(t) N_2, \Delta C = M_3 F_3(t) N_3 \tag{2}$$

where the unknown matrices $F_i(t)$ satisfy the condition $F_i^T(t) F_i(t) \leq I$ for $i = 1, 2, 3$.

Now, we consider the following full-order observer

$$\begin{aligned} \dot{\hat{x}}(t) &= A\hat{x}(t) + A_d \hat{x}(t-d) + Bu(t) + L[y(t) - \hat{y}(t)] \\ \hat{y}(t) &= C\hat{x}(t) \end{aligned} \tag{3}$$

and the observer-based state feedback controller

$$u(t) = -K\hat{x}(t) \tag{4}$$

where $\hat{x}(t) \in R^n$ is the observer state, $\hat{y}(t) \in R^p$ is the plant output estimate. The matrices A, B, C and A_d are the same as that described in (1). $L \in R^{n \times p}$ and $K \in R^{m \times n}$ are the determined observer and controller gain matrices, respectively.

Let the error state be $e(t) = x(t) - \hat{x}(t)$. Combining (1) and (3) yields the error dynamic system described as

$$\dot{e}(t) = (A - LC)e(t) + A_d e(t-d) + (\Delta A - L\Delta C)x(t) + \Delta A_d x(t-d) \tag{5}$$

According to Eqs. (1) and (5) we achieve

$$\begin{bmatrix} \dot{x}(t) \\ \dot{e}(t) \end{bmatrix} = \begin{bmatrix} A - BK + \Delta A & BK \\ \Delta A - L\Delta C & A - LC \end{bmatrix} \begin{bmatrix} x(t) \\ e(t) \end{bmatrix} + \begin{bmatrix} A_d + \Delta A_d & 0 \\ \Delta A_d & A_d \end{bmatrix} \begin{bmatrix} x(t-d) \\ e(t-d) \end{bmatrix} \tag{6}$$

Before ending this section, we would like to introduce the following lemmas, which will be used in the development of our main results.

Lemma 1 [1]: Let X, Y and F are real matrices of suitable dimension with $F^T F \leq I$ then, for any scalar $\varphi > 0$, the following matrix inequality holds:

$$XFY + Y^T F^T X^T \leq \varphi^{-1} XX^T + \varphi Y^T Y.$$

Lemma 2 [6]: The following matrix inequality:

$$\begin{bmatrix} Q(x) & \Pi(x) \\ \Pi(x)^T & R(x) \end{bmatrix} > 0$$

where $Q(x) = Q(x)^T$, $R(x) = R(x)^T$ and $\Pi(x)$ depend affinity on x, is equivalent to $R(x) > 0$, $Q(x) - \Pi(x) R(x)^{-1} \Pi(x)^T > 0$.

3 Main Theoretical Results

A simultaneous stabilization controller and observer gains computation approach is presented in the following theorem.

Theorem 1: Consider the augmented system (6) obtained by estimation dynamics (5) and the closed-loop system (1). Then the augmented system (6) is asymptotically stable if there exist symmetric matrices $R \in R^{n\times n}$, $P \in R^{n\times n}$, $Q_1 \in R^{n\times n}$, $Q_2 \in R^{n\times n}$ validating $R > 0$, $P > 0$, $Q_1 > 0$, $Q_2 > 0$ and positive scalars $\varepsilon > 0$, $\varepsilon_1 > 0$, $\varepsilon_2 > 0$. $\varepsilon_3 > 0$, $\varepsilon_4 > 0$ and $\varepsilon_5 > 0$ such that the matrix inequalities

$$\begin{bmatrix} \begin{bmatrix} A_{11} & PBK \\ (BK)^T P & A_{22} \end{bmatrix} & \begin{bmatrix} PA_d & 0 \\ 0 & RA_d \end{bmatrix} & \begin{bmatrix} 0 \\ RM_1 \end{bmatrix} & \begin{bmatrix} 0 \\ -RLM_3 \end{bmatrix} & \begin{bmatrix} PM_1 \\ 0 \end{bmatrix} & \begin{bmatrix} PM_2 \\ 0 \end{bmatrix} & \begin{bmatrix} 0 \\ RM_2 \end{bmatrix} \\ \begin{bmatrix} A_d^T P & 0 \\ 0 & A_d^T R \end{bmatrix} - \begin{bmatrix} A_{33} & 0 \\ 0 & Q_2 \end{bmatrix} & 0 & 0 & 0 & 0 & 0 \\ \begin{bmatrix} 0 & M_1^T R \end{bmatrix} & 0 & -\varepsilon_1^{-1} I & 0 & 0 & 0 & 0 \\ \begin{bmatrix} 0 & -M_3^T L^T R \end{bmatrix} & 0 & 0 & -\varepsilon_3^{-1} I & 0 & 0 & 0 \\ \begin{bmatrix} M_1^T P & 0 \end{bmatrix} & 0 & 0 & 0 & -\varepsilon_2^{-1} I & 0 & 0 \\ \begin{bmatrix} M_2^T P & 0 \end{bmatrix} & 0 & 0 & 0 & 0 & -\varepsilon_4^{-1} I & 0 \\ \begin{bmatrix} 0 & M_2^T R \end{bmatrix} & 0 & 0 & 0 & 0 & 0 & -\varepsilon_5^{-1} I \end{bmatrix} < 0 \quad (7)$$

where $A_{22} = R(A - LC) + (A - LC)^T R + Q_2$, $A_{11} = P(A - BK) + (A - BK)^T P + \varepsilon N_1^T N_1 + \frac{1}{\varepsilon_3} N_3^T N_3 + Q_1$ and $A_{33} = Q_1 + \left(\frac{1}{\varepsilon_4} + \frac{1}{\varepsilon_5}\right) N_2^T N_2$.

Proof. Defining an Lyapunov functional as

$$V(x(t), e(t)) = \begin{bmatrix} x(t) \\ e(t) \end{bmatrix}^T \begin{bmatrix} P & 0 \\ 0 & R \end{bmatrix} \begin{bmatrix} x(t) \\ e(t) \end{bmatrix} + \int_{t-\tau}^{t} \begin{bmatrix} x(s) \\ e(s) \end{bmatrix}^T \begin{bmatrix} Q_1 & 0 \\ 0 & Q_2 \end{bmatrix} \begin{bmatrix} x(s) \\ e(s) \end{bmatrix} ds \quad (8)$$

Taking the time-derivative of (8) is given by

$$\dot{V}(x(t), e(t)) = \begin{bmatrix} x(t) \\ e(t) \end{bmatrix}^T \begin{bmatrix} P & 0 \\ 0 & R \end{bmatrix} \begin{bmatrix} \dot{x}(t) \\ \dot{e}(t) \end{bmatrix} + \begin{bmatrix} \dot{x}(t) \\ \dot{e}(t) \end{bmatrix}^T \begin{bmatrix} P & 0 \\ 0 & R \end{bmatrix} \begin{bmatrix} x(t) \\ e(t) \end{bmatrix} + \begin{bmatrix} x(t) \\ e(t) \end{bmatrix}^T \begin{bmatrix} Q_1 & 0 \\ 0 & Q_2 \end{bmatrix} \begin{bmatrix} x(t) \\ e(t) \end{bmatrix} - \begin{bmatrix} x(t-d) \\ e(t-d) \end{bmatrix}^T \begin{bmatrix} Q_1 & 0 \\ 0 & Q_2 \end{bmatrix} \begin{bmatrix} x(t-d) \\ e(t-d) \end{bmatrix} \quad (9)$$

Substituting Eq. (6) into Eq. (9), we can obtain that

$$\begin{aligned}
\dot{V}(x(t),e(t)) &= \begin{bmatrix} x(t) \\ e(t) \end{bmatrix}^T \begin{bmatrix} P & 0 \\ 0 & R \end{bmatrix} \begin{bmatrix} A-BK+\Delta A & BK \\ \Delta A - L\Delta C & A-LC \end{bmatrix} \begin{bmatrix} x(t) \\ e(t) \end{bmatrix} \\
&+ \begin{bmatrix} x(t) \\ e(t) \end{bmatrix}^T \begin{bmatrix} A-BK+\Delta A & BK \\ \Delta A - L\Delta C & A-LC \end{bmatrix}^T \begin{bmatrix} P & 0 \\ 0 & R \end{bmatrix} \begin{bmatrix} x(t) \\ e(t) \end{bmatrix} \\
&+ \begin{bmatrix} x(t) \\ e(t) \end{bmatrix}^T \begin{bmatrix} P & 0 \\ 0 & R \end{bmatrix} \begin{bmatrix} A_d+\Delta A_d & 0 \\ \Delta A_d & A_d \end{bmatrix} \begin{bmatrix} x(t-d) \\ e(t-d) \end{bmatrix} \\
&+ \begin{bmatrix} x(t-d) \\ e(t-d) \end{bmatrix}^T \begin{bmatrix} A_d+\Delta A_d & 0 \\ \Delta A_d & A_d \end{bmatrix}^T \begin{bmatrix} P & 0 \\ 0 & R \end{bmatrix} \begin{bmatrix} x(t) \\ e(t) \end{bmatrix} \\
&+ \begin{bmatrix} x(t) \\ e(t) \end{bmatrix}^T \begin{bmatrix} Q_1 & 0 \\ 0 & Q_2 \end{bmatrix} \begin{bmatrix} x(t) \\ e(t) \end{bmatrix} - \begin{bmatrix} x(t-d) \\ e(t-d) \end{bmatrix}^T \begin{bmatrix} Q_1 & 0 \\ 0 & Q_2 \end{bmatrix} \begin{bmatrix} x(t-d) \\ e(t-d) \end{bmatrix} \\
&= \begin{bmatrix} x(t) \\ e(t) \end{bmatrix}^T \left\{ \begin{bmatrix} P(A-BK)+(A-BK)^T P & PBK \\ (BK)^T P & R(A-LC)+(A-LC)^T R \end{bmatrix} \right. \\
&+ \begin{bmatrix} 0 \\ RM_1 \end{bmatrix} F_1(t)[N_1 \;\; 0] + \begin{bmatrix} N_1^T \\ 0 \end{bmatrix} F_1^T(t)[0 \;\; M_1^T R] + \begin{bmatrix} PM_1 \\ 0 \end{bmatrix} F_1(t)[N_1 \;\; 0] \\
&+ \begin{bmatrix} N_1^T \\ 0 \end{bmatrix} F_1^T(t)[M_1^T P \;\; 0] + \begin{bmatrix} 0 \\ -RLM_3 \end{bmatrix} F_3(t)[N_3 \;\; 0] \\
&+ \begin{bmatrix} N_3^T \\ 0 \end{bmatrix} F_3^T(t)[0 \;\; -M_3^T L^T R] \left. \right\} \begin{bmatrix} x(t) \\ e(t) \end{bmatrix} \\
&+ \begin{bmatrix} x(t) \\ e(t) \end{bmatrix}^T \left\{ \begin{bmatrix} PA_d & 0 \\ 0 & RA_d \end{bmatrix} + \begin{bmatrix} PM_2 \\ 0 \end{bmatrix} F_2(t)[N_2 \;\; 0] \right. \\
&+ \begin{bmatrix} 0 \\ RM_2 \end{bmatrix} F_2(t)[N_2 \;\; 0] \left. \right\} \begin{bmatrix} x(t-d) \\ e(t-d) \end{bmatrix} \\
&+ \begin{bmatrix} x(t-d) \\ e(t-d) \end{bmatrix}^T \left\{ \begin{bmatrix} A_d^T P & 0 \\ 0 & A_d^T R \end{bmatrix} + \begin{bmatrix} N_2^T \\ 0 \end{bmatrix} F_2^T(t)[M_2^T P \;\; 0] \right. \\
&+ \begin{bmatrix} N_2^T \\ 0 \end{bmatrix} F_2^T(t)[0 \;\; M_2^T R] \left. \right\} \begin{bmatrix} x(t) \\ e(t) \end{bmatrix} \\
&+ \begin{bmatrix} x(t) \\ e(t) \end{bmatrix}^T \begin{bmatrix} Q_1 & 0 \\ 0 & Q_2 \end{bmatrix} \begin{bmatrix} x(t) \\ e(t) \end{bmatrix} - \begin{bmatrix} x(t-d) \\ e(t-d) \end{bmatrix}^T \begin{bmatrix} Q_1 & 0 \\ 0 & Q_2 \end{bmatrix} \begin{bmatrix} x(t-d) \\ e(t-d) \end{bmatrix}
\end{aligned} \quad (10)$$

Applying Lemma 1 to Eq. (10), we achieve

$$\dot{V}(x(t),e(t)) \leq \begin{bmatrix} x(t) \\ e(t) \end{bmatrix}^T \{ \begin{bmatrix} P(A-BK)+(A-BK)^TP & PBK \\ (BK)^TP & R(A-LC)+(A-LC)^TR \end{bmatrix}$$

$$+ \varepsilon_1 \begin{bmatrix} 0 \\ RM_1 \end{bmatrix} \begin{bmatrix} 0 & M_1^TR \end{bmatrix} + \frac{1}{\varepsilon_1}\begin{bmatrix} N_1^T \\ 0 \end{bmatrix}\begin{bmatrix} N_1 & 0 \end{bmatrix}$$

$$+ \varepsilon_2 \begin{bmatrix} PM_1 \\ 0 \end{bmatrix} \begin{bmatrix} M_1^TP & 0 \end{bmatrix} + \frac{1}{\varepsilon_2}\begin{bmatrix} N_1^T \\ 0 \end{bmatrix}\begin{bmatrix} N_1 & 0 \end{bmatrix}$$

$$+ \varepsilon_3 \begin{bmatrix} 0 \\ -RLM_3 \end{bmatrix} \begin{bmatrix} 0 & -M_3^TL^TR \end{bmatrix} + \frac{1}{\varepsilon_3}\begin{bmatrix} N_3^T \\ 0 \end{bmatrix}\begin{bmatrix} N_3 & 0 \end{bmatrix} \} \begin{bmatrix} x(t) \\ e(t) \end{bmatrix}$$

$$+ \begin{bmatrix} x(t) \\ e(t) \end{bmatrix}^T \begin{bmatrix} PA_d & 0 \\ 0 & RA_d \end{bmatrix} \begin{bmatrix} x(t-d) \\ e(t-d) \end{bmatrix} \quad (11)$$

$$+ \begin{bmatrix} x(t-d) \\ e(t-d) \end{bmatrix}^T \begin{bmatrix} A_d^TP & 0 \\ 0 & A_d^TR \end{bmatrix} \begin{bmatrix} x(t) \\ e(t) \end{bmatrix}$$

$$+ \begin{bmatrix} x(t) \\ e(t) \end{bmatrix}^T \{ \varepsilon_4 \begin{bmatrix} PM_2 \\ 0 \end{bmatrix} \begin{bmatrix} M_2^TP & 0 \end{bmatrix} + \varepsilon_5 \begin{bmatrix} 0 \\ RM_2 \end{bmatrix} \begin{bmatrix} 0 & M_2^TR \end{bmatrix} \} \begin{bmatrix} x(t) \\ e(t) \end{bmatrix}$$

$$+ \begin{bmatrix} x(t-d) \\ e(t-d) \end{bmatrix}^T \{ \frac{1}{\varepsilon_4}\begin{bmatrix} N_2^T \\ 0 \end{bmatrix}\begin{bmatrix} N_2 & 0 \end{bmatrix} + \frac{1}{\varepsilon_5}\begin{bmatrix} N_2^T \\ 0 \end{bmatrix}\begin{bmatrix} N_2 & 0 \end{bmatrix} \} \begin{bmatrix} x(t-d) \\ e(t-d) \end{bmatrix}$$

$$+ \begin{bmatrix} x(t) \\ e(t) \end{bmatrix}^T \begin{bmatrix} Q_1 & 0 \\ 0 & Q_2 \end{bmatrix}\begin{bmatrix} x(t) \\ e(t) \end{bmatrix} - \begin{bmatrix} x(t-d) \\ e(t-d) \end{bmatrix}^T \begin{bmatrix} Q_1 & 0 \\ 0 & Q_2 \end{bmatrix}\begin{bmatrix} x(t-d) \\ e(t-d) \end{bmatrix}$$

Define the augmented vector

$$\xi(t) = \begin{bmatrix} x^T(t) & e^T(t) & x^T(t-d) & e^T(t-d) \end{bmatrix}^T \quad (12)$$

Then, we can get the following equation

$$\dot{V}(x(t),e(t)) \leq \xi^T(t)\Omega\xi(t) \quad (13)$$

where

$$\Omega = \begin{bmatrix} \begin{bmatrix} A_{11} & PBK \\ (BK)^TP & A_{22} \end{bmatrix} + \varepsilon_1 \begin{bmatrix} 0 \\ RM_1 \end{bmatrix}\begin{bmatrix} 0 & M_1^TR \end{bmatrix} & \begin{bmatrix} PA_d & 0 \\ 0 & RA_d \end{bmatrix} \\ \begin{bmatrix} A_d^TP & 0 \\ 0 & A_d^TR \end{bmatrix} & -\begin{bmatrix} A_{33} & 0 \\ 0 & Q_2 \end{bmatrix} \end{bmatrix} + \Xi \quad (14)$$

in which $A_{11} = P(A-BK)+(A-BK)^TP + \varepsilon N_1^TN_1 + \frac{1}{\varepsilon_3}N_3^TN_3 + Q_1$ $A_{33} = Q_1 + (\frac{1}{\varepsilon_4} + \frac{1}{\varepsilon_5})N_2^TN_2$,

$$\Xi = \varepsilon_2 \begin{bmatrix} PM_1 \\ 0 \end{bmatrix} \begin{bmatrix} M_1^T P & 0 \end{bmatrix} + \varepsilon_3 \begin{bmatrix} 0 \\ -RLM_3 \end{bmatrix} \begin{bmatrix} 0 & -M_3^T L^T R \end{bmatrix} + \varepsilon_4 \begin{bmatrix} PM_2 \\ 0 \end{bmatrix} \begin{bmatrix} M_2^T P & 0 \end{bmatrix}$$
$$+ \varepsilon_5 \begin{bmatrix} 0 \\ RM_2 \end{bmatrix} \begin{bmatrix} 0 & M_2^T R \end{bmatrix}$$

and $A_{22} = R(A - LC) + (A - LC)^T R + Q_2$.

It follows from Lemma 2 that LMI (7) is equivalent to

$$\Omega = \left[\begin{bmatrix} A_{11} & PBK \\ (BK)^T P & A_{22} \\ A_d^T P & 0 \\ 0 & A_d^T R \end{bmatrix} + \varepsilon_1 \begin{bmatrix} 0 \\ RM_1 \end{bmatrix} \begin{bmatrix} 0 & M_1^T R \end{bmatrix} \begin{bmatrix} PA_d & 0 \\ 0 & RA_d \\ -\begin{bmatrix} A_{33} & 0 \\ 0 & Q_2 \end{bmatrix} \end{bmatrix} \right] + \Xi < 0 \quad (15)$$

Therefore Eq. (15) can guarantee that $\dot{V}(x(t), e(t)) < 0$. Then the stability of system (6) is satisfied. Thus, this completes the proof.

The condition in Theorem 1 can be used to select controller and observer gain matrices $K \in R^{m \times n}$ and $L \in R^{n \times p}$ such that convergence of both the state of system (1) and the estimation error is guaranteed.

Remark 1: The controller (4) uses only the output measurements, which is very beneficial and more feasible since it can be easily carried out in many practice systems. One of the contributions is to estimate the current true value of the system state variables to guarantee the feedback-controlled system is asymptotically stabilizable, avoiding the effect of the delayed and noised measurement output. Linear matrix inequality (LMI) approach is used to design the observer-based control. The control and observer gains matrices are characterized using the solution of the LMI existence condition.

4 Conclusion

In this paper, the observer-based control design problem has been proposed for a class of mismatched uncertain time-delay systems where mismatched parameter uncertainties are involved in the state matrix and the delayed state matrix, as well as in the output matrix. Thus the mismatched uncertain time-delay systems investigated in this paper is more general structure than is considered in [2, 3, 10]. One of the contributions is to estimate the current true value of the system state variables to guarantee the feedback-controlled system is asymptotically stabilizable, avoiding the effect of the delayed and noised measurement output. Moreover, the control and observer gains matrices are characterized using the solution of the LMI existence condition. The proposed observer-based control design can be applied to a wider class of mismatched uncertain time-delay systems.

References

1. Lien, C.H.: Robust observer-based control of systems with state perturbations via LMI approach. IEEE Trans. Autom. Control **49**(8), 1365–1370 (2004)
2. Lien, C.H.: An efficient method to design robust observer-based control of uncertain linear systems. Appl. Math. Comput. **158**, 29–44 (2004)
3. Chen, W.H.: Disturbance observer based control for nonlinear systems. IEEE/ASME Trans. Mechatron. **9**(4), 706–710 (2004)
4. Song, B., Hedrick, J.K.: Observer-based dynamic surface control for a class of nonlinear systems: an LMI approach. IEEE Trans. Autom. Control **49**(11), 1995–2001 (2004)
5. Yan, X.G., Sarah, K.S., Edwards, C.: Dynamic sliding mode control for a class of systems with mismatched uncertaint. Eur. J. Control **11**, 1–20 (2005)
6. Choi, H.H., Ro, K.S.: LMI-based sliding-mode observer design method. In: IEE Proceedings on Control Theory and Applications, vol. 152, no. 1 (2005)
7. Yao, Y.X., Radun, A.V.: Proportional integral observer design for linear systems with time delay. IET Control Theor. Appl. **1**(4), 887–892 (2007)
8. Gassara, H., Hajjaji, A.E., Chaabane, M.: Observer-based robust H∞ reliable control for uncertain t–s fuzzy systems with state time delay. IEEE Trans. Fuzzy Syst. **18**(6), 1027–1040 (2010)
9. Wu, H.: A class of adaptive robust state observers with simpler structure for uncertain nonlinear systems with time-varying delays. IET Control Theor. Appl. **7**(2), 218–227 (2013)
10. Ghanes, M., Leon, J.D., Barbot, J.-P.: Observer design for nonlinear systems under unknown time-varying delays. IEEE Trans. Autom. Control **58**(6), 1529–1534 (2012)
11. Wang, S., Jiang, Y.: On LMI conditions to design observer-based controllers for linear systems with parameter uncertainties. Automatica **49**(12), 3700–3704 (2013)
12. Wang, H.P., Tian, Y., Vasseur, V.: Piecewise continuous hybrid systems based observer design for linear systems with variable sampling periods and delay output. Signal Process. **114**, 75–84 (2015)
13. Zheng, G., Bejarano, F.J., Peruquetti, W.: Unknown input observer for linear time-delay systems. Automatica **61**, 35–43 (2015)
14. Xia, Y., Jia, Y.: Robust sliding-mode control for uncertain time-delay systems: an LMI approach. IEEE Trans. Autom. Control **48**(6), 1086–1092 (2003)
15. Yan, X.G., Spurgeon, S.K., Edwards, C.: Static output feedback sliding mode control for time-varying delay systems with time-delayed nonlinear disturbances. Int. J. Robust Nonlinear Control **20**(7), 777–788 (2010)
16. Zhang, L., Ning, Z., Zheng, W.X.: Observer-based control for piecewise-affine systems with both input and output quantization. IEEE Trans. Autom. Control **62**(11), 5858–5865 (2017)

Nonlinear Disturbance Observer with Recurrent Neural Network Compensator

Shihono Yamada and Jun Ishikawa[✉]

Tokyo Denki University, Tokyo, Japan
ishikawa@fr.dendai.ac.jp

Abstract. This paper proposes a nonlinear disturbance observer with recurrent neural network compensator applicable to nonlinear systems that is velocity controlled. In the proposed method, the recurrent neural network is trained to generate nonlinear velocity defined as the difference between the measured velocity of joints and a nominal linear velocity calculated via a linear model from the reference input to the velocity control system. The training results is evaluated based on the leave-one-out method, and it is shown that the RNN can be well-trained to estimate the nonlinear velocity. Then, a common DOB can be estimate disturbance from the nominal linear velocity restored by the trained RNN output and the velocity measurement. The validity of the proposed method is evaluated by simulation for a vertically-articulated two-link manipulator, comparing a conventional disturbance observer. The simulation result showed that the proposed disturbance observer can be comparable to a conventional DOB that works in an ideal condition where all the parameters of the manipulator are known, while the data needed to construct the proposed DOB is only the velocity measurement and its reference.

Keywords: Recurrent neural network · Disturbance observer · Nonlinear system

1 Introduction

Disturbance observers (DOBs) as a method to compensate unknown disturbances and its applications to robust control have been widely used in industries [1–3]. The DOB estimates the disturbance based the difference between outputs of a nominal model of the controlled plant and the actual measurement [3–5] .Research of the DOB started in late 1980's by Ohnishi et al. [3] and it has been applied for many control systems such as process control, damping control system, flight control system [6–11]. It will be expected to establish methodologies to design DOBs for nonlinear system, but frequency domain approaches are said to be difficult to extend it to nonlinear systems [12]. Thus, it is a challenge to develop design method of DOB applicable to nonlinear systems. For example, Li et al., proposed a neural network disturbance observer that can deal with both minimum and non-minimum phase systems with nonlinearity [13]. Ma et al., a developed disturbance observer based on the radial basic function for systems with model uncertainties [14].

This article proposes a nonlinear disturbance observer with recurrent neural network compensator (nonlinear DOB with RNN compensator) for manipulator systems. In the proposed method, the recurrent neural network (RNN) is trained to generate nonlinear angular velocity that is defined as the difference between the measured angular velocity of joints and an estimate by using a linear model from the desired inputs to the angular velocity control system of the manipulator. By using the trained RNN, nominal linear angular velocity, which is used to estimate disturbances by using a common DOB, can be calculated.

In the following parts of the paper, after providing a brief explanation of RNN that estimates nonlinearities, the proposed nonlinear DOB with RNN compensator will be explained in Sect. 2. In Sect. 3, manipulator dynamics and its control system of joint angular velocity are described, and a conventional DOB to be compared with the proposed method is explained. Section 4 gives simulation results to show the effectiveness of the proposed method, and Sect. 5 concludes the paper with the achievements and the future works.

2 Disturbance Observer (DOB) for Nonlinear Systems Using Recurrent Neural Network

2.1 Recurrent Neural Network (RNN)

As shown in Fig. 1, a recurrent neural network (RNN) is composed of an input layer, a hidden layer, and an output layer and is said to be a neural network suitable for learning time series data. An activation function of the hidden layer of the RNN used here is a hyperbolic tangent sigmoid function defined as

$$S_t(u) = \frac{2}{1+e^{-2u}} - 1, \qquad (1)$$

and that of the output layer is a linear function with slope 1. For these activation functions, the RNN with the Nth tapped delay line is given by

$$h(k) = S_t\left\{b_h + W_x \cdot x(k) + W_h\left[h(k-1)^T, h(k-2)^T, \cdots, h(k-N)^T\right]^T\right\}, \qquad (2)$$

$$y(k) = b_o + W_o \cdot h(k) \qquad (3)$$

where $x(k)$ is the input (feature vector) to the RNN, $h(k)$ is the output of the hidden layer, and $y(k)$ is the output from the RNN. W_x, W_h, and W_o are connecting weights respectively for the input, the delayed output of the hidden layer and the current output of the hidden layer. b_n and b_o are biases. The training algorithm used in this research is a back propagation method based on Bayesian regularization [15].

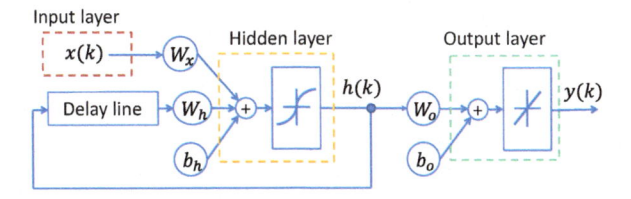

Fig. 1. Structure of recurrent neural network (RNN).

2.2 Nonlinear Disturbance Observer with Recurrent Neural Network Compensator (Nonlinear DOB with RNN Compensator)

In this section, a nonlinear disturbance observer with recurrent neural network compensator (hereinafter referred to as "nonlinear DOB with RNN compensator") is proposed. As shown in the upper half of Fig. 2, the proposed observer a combination of the conventional linear DOB and the RNN. The nonlinear system targeted by the proposed system is a velocity-controlled system. Thus, the system input is the desired velocity $\dot{\theta}_d$ and the output is the actual joint angular velocity measurement $\dot{\theta}$. As shown in the lower half of Fig. 2, the RNN has $\dot{\theta}_d$ as the input and is trained so as to generate nonlinear angular velocity $\Delta\dot{\theta}_n$ that is defined as the difference between the measured angular velocity $\dot{\theta}$ and the linear estimate $\dot{\theta}_l$ by using a linear velocity controlled system given by

$$\frac{d}{dt}\begin{bmatrix}\theta_l \\ \dot{\theta}_l\end{bmatrix} = \begin{bmatrix}0 & 1 \\ 0 & -K_b\end{bmatrix}\begin{bmatrix}\theta_l \\ \dot{\theta}_l\end{bmatrix} + \begin{bmatrix}0 \\ K_b\end{bmatrix}\dot{\theta}_d$$
$$y = \begin{bmatrix}0 & 1\end{bmatrix}\begin{bmatrix}\theta_l \\ \dot{\theta}_l\end{bmatrix} \quad (4)$$

where K_b is the feedback gain. Note that since the proposed system is implemented as a discrete time system, the velocity-controlled system (4) is used after discretized.

After the training finished, as shown in the upper half of Fig. 2 again, the RNN is used to generate $\Delta\dot{\theta}_n$ from $\dot{\theta}_d$, and by subtracting $\Delta\dot{\theta}_n$ from $\dot{\theta}$, the linear velocity $\dot{\theta}_l$ is restored. The augmented system of Eq. (4) so as to include the disturbance d in the dimension of angular acceleration is given by

$$\frac{d}{dt}\begin{bmatrix} \theta_l \\ \dot{\theta}_l \\ d \end{bmatrix} = \begin{bmatrix} 0 & 1 & 0 \\ 0 & -K_b & 1 \\ 0 & 0 & 0 \end{bmatrix} \begin{bmatrix} \theta_l \\ \dot{\theta}_l \\ d \end{bmatrix} + \begin{bmatrix} 0 \\ K_b \\ 0 \end{bmatrix} \dot{\theta}_d$$

$$:= A_L \cdot x_L + B_L \cdot u_L$$

$$y_L = \begin{bmatrix} 0 & 1 & 0 \end{bmatrix} \begin{bmatrix} \theta_l \\ \dot{\theta}_l \\ d \end{bmatrix}$$

$$:= C_L \cdot x_L.$$

(5)

For the augmented system (5), a DOB is designed as follows:

$$\frac{d}{dt}\begin{bmatrix} \hat{\theta}_l \\ \hat{\dot{\theta}}_l \\ \hat{d} \end{bmatrix} = (A_L - K_L \cdot C_L) \begin{bmatrix} \hat{\theta}_l \\ \hat{\dot{\theta}}_l \\ \hat{d} \end{bmatrix} + \begin{bmatrix} B_L & K_L \end{bmatrix} \begin{bmatrix} u_L \\ y_L \end{bmatrix} \qquad (6)$$

where the observer gain K_L is designed according to a framework of the steady-state Kalman filter. It is, however, used as a gain tuning method by adjusting weights that are not based on actual covariance of the system noise and sensor noise.

Thus, by providing $u_L = \dot{\theta}_d$ and $y_L = \dot{\theta}_l$ to the DOB (6), the DOB can estimate the disturbance $\hat{d}$ in the dimension of angular acceleration as well as the linear angular velocity $\hat{\dot{\theta}}_l$. The role of the RNN is to compensate the nonlinear components in the angular velocity and to make the linear DOB functioning.

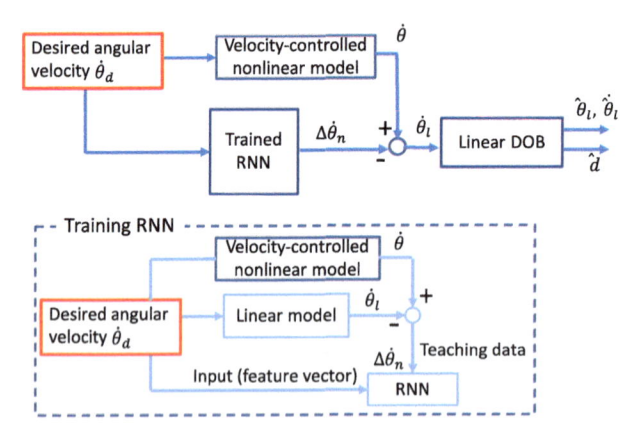

Fig. 2. Concept of nonlinear disturbance observer with recurrent neural network compensator: The upper half shows an implementation configuration, and the lower half shows a configuration in training RNN.

3 Vertically-Articulated Two-Link Manipulator System for Simulation Study

3.1 Modeling

For simulation study to show the validity of the proposed method, a two-link manipulator, which is vertically-articulated and selective compliance assembly robot arm (SCARA) type, is selected as a nonlinear system (Fig. 3). The equation of motion of the two-link manipulator is given by

$$\begin{bmatrix} J_1 + J_2 + 2r\cos\theta_2 & J_2 + r\cos\theta_2 \\ J_2 + r\cos\theta_2 & J_2 \end{bmatrix} \begin{bmatrix} \ddot{\theta}_1 \\ \ddot{\theta}_2 \end{bmatrix} + \begin{bmatrix} -r\sin\theta_2 \cdot \dot{\theta}_2 \left(2\dot{\theta}_1 + \dot{\theta}_2\right) \\ r\sin\theta_2 \cdot \dot{\theta}_1^2 \end{bmatrix}$$
$$+ \begin{bmatrix} m_1 g a_1 \cos\theta_1 + m_2 g (l_1 \cos\theta_1 + a_2 \cos(\theta_1 + \theta_2)) \\ m_2 g a_2 \cos(\theta_1 + \theta_2) \end{bmatrix} + \begin{bmatrix} b_1 & 0 \\ 0 & b_2 \end{bmatrix} \begin{bmatrix} \dot{\theta}_1 \\ \dot{\theta}_2 \end{bmatrix} = \begin{bmatrix} \tau_1 \\ \tau_2 \end{bmatrix}$$
(7)

where

$$J_1 := I_1 + m_1 a_1^2 + m_2 l_1^2, \tag{8}$$

$$J_2 := I_2 + m_2 a_2^2, \tag{9}$$

$$r := m_2 a_2 l_1. \tag{10}$$

Subscripts 1 and 2 show the link number, the joint angle θ, the joint torque τ, the link length l, the distance between the joint axis and the center of gravity a, the moment of inertia around the center of gravity I, the link mass m, and the viscous friction coefficient b. Simulation study in Sect. 4 discuss to estimate an external force $f = [f_x \ f_y]^T$ applied to the manipulator hand.

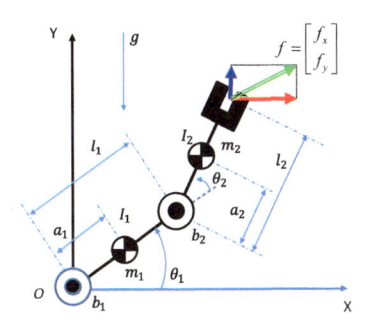

Fig. 3. Vertically-articulated SCARA type two-link manipulator affected by external force in the absolute x-y coordinate system.

3.2 Angular Velocity Control System

The equation of motion (5) is rewritten as

$$\tau = M(\theta) \cdot \ddot{\theta} + h(\theta, \dot{\theta}) + g(\theta) + D \cdot \dot{\theta} \tag{11}$$

where $M(\theta)$ is the inertia tensor, $h(\theta, \dot{\theta})$ is the term representing centrifugal force or Coriolis force, $g(\theta)$ is the gravity term, and D is the coefficient matrix of viscous friction. Then, by choosing u_θ as the new system input and generating the joint torque as

$$\tau = M(\theta) \cdot u_\theta + h(\theta, \dot{\theta}) + g(\theta) + D \cdot \dot{\theta}, \tag{12}$$

the dynamics of the two-link manipulator from u_θ to $\dot{\theta}$ surrounding by the dotted line can be linearized as

$$\dot{\theta} = \frac{1}{s} u_\theta. \tag{13}$$

For the linearized system, a simple angular velocity controlled system can be designed by linear control theory as shown in Fig. 4.

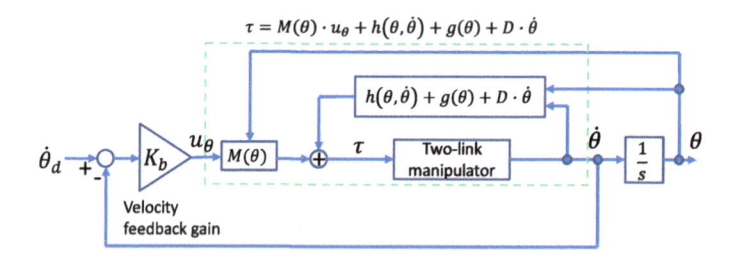

Fig. 4. Velocity-controlled system for linearized two-link manipulator dynamics by nonlinear feedback.

3.3 Conventional Disturbance Observer to Be Compared

In this section, a conventional DOB, which is used as a benchmark to evaluate the performance of the proposed DOB with RNN compensator, is briefly introduced. The linearized two-link manipulator dynamics is given by

$$\frac{d}{dt}\begin{bmatrix} \theta \\ \dot{\theta} \end{bmatrix} = \begin{bmatrix} 0 & 1 \\ 0 & 0 \end{bmatrix}\begin{bmatrix} \theta \\ \dot{\theta} \end{bmatrix} + \begin{bmatrix} 0 \\ 1 \end{bmatrix} u_\theta$$

$$y_\theta = \begin{bmatrix} 1 & 0 \end{bmatrix}\begin{bmatrix} \theta \\ \dot{\theta} \end{bmatrix} \tag{14}$$

For the system, a type-1 DOB can be implemented as follows:

$$\frac{d}{dt}\begin{bmatrix}\hat{\theta}\\\hat{\dot{\theta}}\\\hat{d}\end{bmatrix} = (A - K \cdot C)\begin{bmatrix}\hat{\theta}\\\hat{\dot{\theta}}\\\hat{d}\end{bmatrix} + [B \quad K]\begin{bmatrix}u_\theta\\y_\theta\end{bmatrix} \quad (15)$$

where K is the observer gain and

$$A = \begin{bmatrix}0 & 1 & 0\\0 & 0 & 1\\0 & 0 & 0\end{bmatrix}, B = \begin{bmatrix}0\\1\\0\end{bmatrix}, C = [1 \quad 0 \quad 0]. \quad (16)$$

Note that all the nonlinear dynamic parameters must be known in order to implement the DOB (15) because the two-link manipulator system must be linearized by the nonlinear feedback (12). On the other hand, the proposed DOB with RNN compensator needs only $\dot{\theta}$ and $\dot{\theta}_d$.

4 Simulation Study

4.1 RNN Training

Before a simulation study to show the validity of the proposed method is conducted, the RNN is trained first using 10 sets of random trajectories generated from M series. The controlled plant is simulated so as to have full nonlinear dynamics and is actuated according to the control system block diagram in Fig. 4. The training results is evaluated based on the leave-one-out method. The leave-one-out method is a cross-validation where only one case is extracted from the sample group and is set as a test case, and all the other cases used as training examples. The verification is repeated so that all the cases will be tested once. Thus, 9 trajectory sets are used for training data to make the RNN generate the nonlinear angular velocity $\Delta\dot{\theta}_n$ ($= \dot{\theta} - \dot{\theta}_l$) from the desired velocity $\dot{\theta}_d$. Then, the standard deviation of the estimation error for $\Delta\dot{\theta}_n$ when the one data set left out (an unknown data set for RNN) is input to the system is evaluated.

As an example, Fig. 5 shows how the two-link manipulator behaves when $\dot{\theta}_d$ of Data set 1 is given for each link as the desired input. Figure 6(a) shows the actual time response of $\Delta\dot{\theta}_n$ and its estimate generated from the RNN trained by using Data sets 2 to 10. Figure 6(b) shows the estimation error. Table 1 lists the standard deviation of the estimation error of each link for all the 10 data sets. As shown in Fig. 6 and Table 1, the RNN can be well-trained to estimate the nonlinear angular velocity $\Delta\dot{\theta}_n$.

Table 1. Evaluation results of leave-one-out cross-validation: Standard deviation of estimation error for 10 data sets is listed.

	Data set #	Standard deviation [deg/s]
1	Link 1	0.0016
	Link 2	0.0016
2	Link 1	0.0010
	Link 2	0.0011
3	Link 1	0.0008
	Link 2	0.0006
4	Link 1	0.0019
	Link 2	0.0022
5	Link 1	0.0164
	Link 2	0.0172
6	Link 1	0.0142
	Link 2	0.0191
7	Link 1	0.0021
	Link 2	0.0017
8	Link 1	0.0026
	Link 2	0.0009
9	Link 1	0.0069
	Link 2	0.0022
10	Link 1	0.0024
	Link 2	0.0031

(a) Desired angular velocity $\dot{\theta}_d$

(b) Joint angular velocity $\dot{\theta}$ and $\dot{\theta}_l$

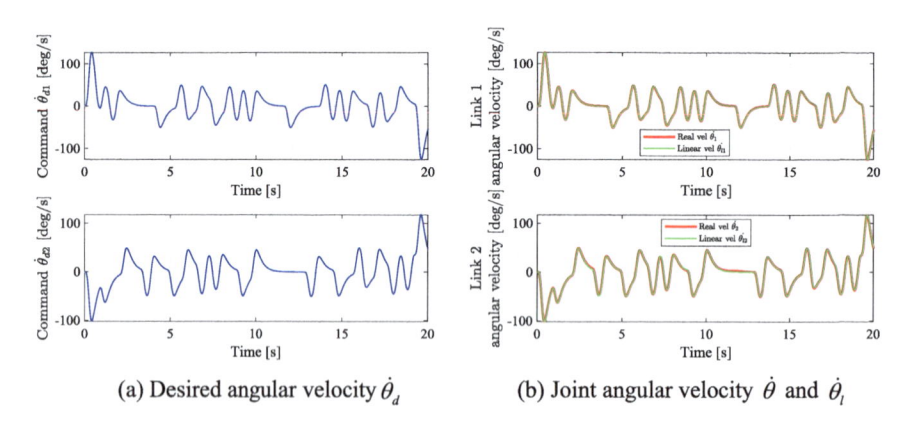

Fig. 5. Example (Data 1) to show how the manipulator behaves. Note that $\dot{\theta}$ and $\dot{\theta}_l$ are NOT identical to each other as shown in Fig. 6(a).

(a) Actual $\Delta\dot{\theta}_n \ (= \dot{\theta} - \dot{\theta}_t)$ and its estimate   (b) Estimation error

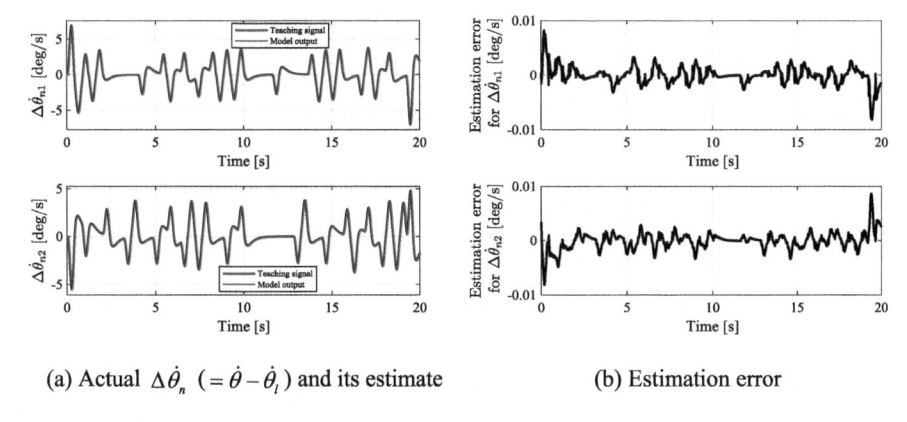

Fig. 6. Example of evaluation result of trained RNN for Data 1.

4.2 Comparison with Conventional DOB

Finally, to show the validity of the proposed DOB, simulation to estimate an external force $f = [f_x \ f_y]^T$ applied to the hand of the two-link manipulator is conducted. The external force f_x in the x direction of the absolute coordinate system is a sine wave with a frequency of 0.1 Hz and an amplitude 2 N, and that in the y direction is with a frequency of 0.1 Hz and an amplitude 1 N. The nonlinear DOB with RNN compensator is generate the estimate of the disturbance $\hat{d}$ in the dimension of angular acceleration while the two-link manipulator is controlled to track the desired angular velocity $\dot{\theta}_d$ shown in Fig. 5(a). Blue solid lines in Fig. 7 show the time responses of the estimated disturbance $\hat{d}$ for each joint. Red solid lines in Fig. 7 show those estimated by the conventional DOB (15). Both the results agree with each other, and this means that the proposed DOB can be comparable to the conventional DOB that works in an ideal condition where all the parameters of the manipulator are known. Note that data needed to construct the proposed DOB is only $\dot{\theta}$ and $\dot{\theta}_d$.

The external force in the absolute coordinate system can be calculated from $\hat{d}$ as follows:

$$\hat{f} = (J^T)^{-1} \cdot M(\theta) \cdot \hat{d} \qquad (17)$$

where J is Jacobian of the two-link manipulator. Figure 8 shows the calculation results for both the conventional and proposed DOBs, which are accompanied with the true values depicted by the black dotted line. Although the estimate by the proposed method has small deviation from the true values due to the small differences observed in Fig. 7, the performance can be thought to be acceptable for several applications.

(a) Disturbance estimate for joint 1

(b) Disturbance estimate for joint 2

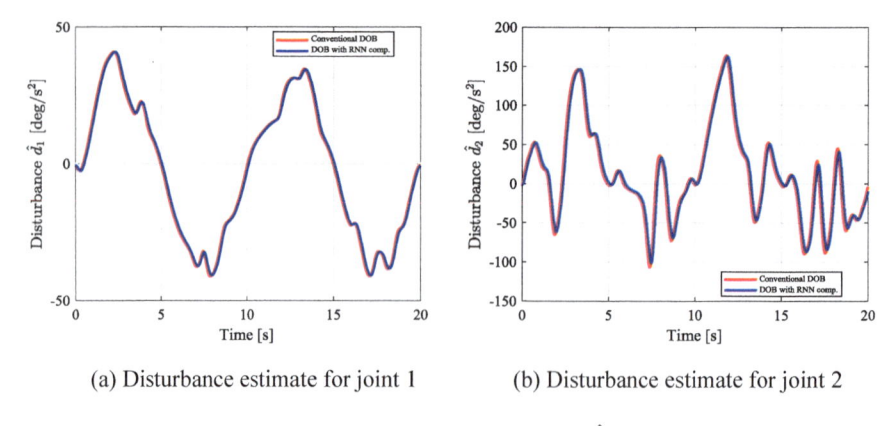

Fig. 7. Comparison of time responses of disturbance estimate $\hat{d}$ for conventional and proposed DOBs while the manipulator is controlled according to the desired angular trajectories in Fig. 5 (a) under the influence of the external force f.

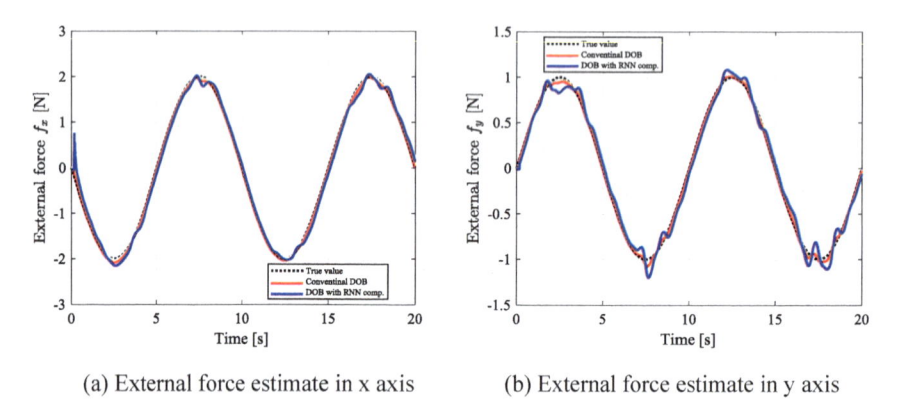

(a) External force estimate in x axis

(b) External force estimate in y axis

Fig. 8. Calculated external force f in the absolute coordinate system from f in the joint coordinate system.

5 Conclusion

This paper proposed a disturbance observer with recurrent neural network compensator applicable to nonlinear systems that is velocity controlled. The validity of the proposed method is evaluated by simulation for a two-link manipulator, comparing a conventional disturbance observer. The simulation result showed that the proposed disturbance observer can be well-worked for the given two-link manipulator system.

References

1. Zhang, X.H., Liu, X.H., Ding, H.S., Li, H.S.: PMSM speed-adjusting system based on disturbance observer and finite-time control. Control Decis. **24**(7), 1028–1032 (2009)
2. Ma, J., Yang, T., Hou, G.Z., Tan, M.: Adaptive neural network controller of a Stewart platform with unknown dynamics for active vibration isolation. In: Proceedings of the IEEE International Conference on Robotics and Biomimetics, pp. 1631–1636 (2008)
3. Nakao, M., Ohnishi, K., Miyachi, K.: A robust decentralize joint control based on interference estimation. In: Proceedings of the IEEE International Conference on Robotics and Automation, pp. 326–331 (1987)
4. Yang, J., Li, H.S., Chen, X., Li, Q.: Disturbance rejection of ball mill grinding circuits using DOB and MP C. Powder Technol. **198**(2), 219–228 (2010)
5. Choi, Y., Yang, K., Chung, K.W., Kim, H.R., Suh, H.I.: On the robustness and performance of disturbance observer for second-order systems. Autom. Control **48**(2), 315–320 (2003)
6. Zhou, P., Xiang, B., Chai, Y.T.: Improved disturbance observer (DOB) based advanced feedback control for optimal operation of a mineral grinding process. Chin. J. Chem. Eng. **20**(6), 1206–1212 (2012)
7. Li, Q.S., Qiu, H.J., Ji, H.L., Zhu, J.K., Li, J.: Piezoelectric vibration control for all-clamped panel using DOB-based optimal control. Mechatronics **21**(7), 1212–1221 (2011)
8. Li, Q.S., Qiu, H.J., Li, J., Ji, H.L., Zhu, J.K.: Multi-modal vibration control using amended disturbance observer compensation. IET Control Theor. Appl. **6**(1), 72–83 (2012)
9. Li, H.S., Yang, J.: Robust autopilot design for bank-to-turn missiles using disturbance observers. IEEE Trans. Aerosp. Electron. Syst. **49**(1), 558–579 (2013)
10. Guo, L., Chen, H.W.: Disturbance attenuation and rejection for systems with nonlinearity via DOBC approach. Int. J. Robust Nonlinear Control **15**(3), 109–125 (2005)
11. Li, S., Liu, Z.: Adaptive speed control for permanent magnet synchronous motor system with variations of load inertia. IEEE Trans. Ind. Electron. **56**(8), 3050–3059 (2009)
12. Chen, W.: Disturbance observer based control for nonlinear systems. Mechatronics **9**(4), 706–710 (2004)
13. Li, J., Jun, Y., Shihua, L., Xisong, C.: Design of neural network disturbance observer using RBFN for complex nonlinear systems. In: Proceedings of the IEEE International Conference on Chinese Control Conference, pp. 6187–6192 (2011)
14. Ma, J., Yang, T., Hou, Z.G., Tan, M.: Neural network disturbance observer based controller of an electrically driven stewart platform using backstepping for active vibration isolation. In: Proceedings of the IEEE International Conference on International Joint Conference on Neural Networks, pp. 1939–1944 (2009)
15. Beale, H.M., Hagan, T.M., Demuth, B.H.: MATLAB Neural Network Toolbox Getting Started Guide (R2016b). Math Works Inc. (2016)

Parameters Estimation for Sensorless Control of Induction Motor Drive Using Modify GA and CSA Algorithm

Thinh Cong Tran[1,2](✉), Pavel Brandstetter[2], Cuong Dinh Tran[1,2], Sang Dang Ho[1,2], Minh Chau Huu Nguyen[2], and Pham Nhat Phuong[1,2]

[1] Faculty of Electrical and Electronics Engineering, Ton Duc Thang University, 19 Nguyen Huu Tho Street, Tan Phong Ward, District 7, Ho Chi Minh City, Vietnam
{trancongthinh, trandinhcuong, hodangsang, phamnhatphuong}@tdt.edu.vn
[2] Faculty of Electrical Engineering and Computer Science, VSB-Technical University of Ostrava, 17. listopadu 15/2172, 708 33 Ostrava-Poruba, Czech Republic
{pavel.brandstetter, huu.chau.minh.nguyen.st}@vsb.cz

Abstract. This paper presents methods for estimating CB-MRAS model parameters such as $K_{1(CB)}$, $K_{2(CB)}$, $K_{3(CB)}$, $T_{i(CB)}$, K_{Lm}, K_{Tr} by binary Genetic Algorithm (GA), real number GA, modify GA, and CucKoo Search Algorithm (CSA). The first part of the paper is the vector model of the induction motor and the CB-MRAS model for estimating parameters by the above algorithms; the second part is the detailed way to implement the algorithms; the third part is simulation and as a result of the simulation, the results show that it is possible to estimate the parameters of this model by the modify GA or CSA algorithm.

Keywords: A.C. drive · Induction motor · Vector control · Modify · Genetic Algorithm · Estimating parameters · Model · CSA · Optimization

1 Introduction

Nowadays, the sensorless control of the AC machines whose basic advantages such as reduction of hardware complexity and cost, increasing mechanical robustness, higher reliability, lower maintenance demands and cost, working in hostile environments, etc., so the development of different sensorless methods for rotor position and mechanical speed estimation of electrical drives with the induction motor. The sensorless control method can be classified as follows [2, 5, 9]:

1-Methods with machine model: Model reference adaptive system (MRAS), Observers: Kalman filter, Luenberger observer, Sliding mode observer [5]. 2-Methods without machine model: Estimators with injection methods, estimators using artificial intelligence [6, 7, 11, 15].

The speed estimation base on CB-MRAS is a sample, have high reliability, etc. It is easy to apply in the practice. However, the accuracy of the speed estimation parameter depends heavily on the parameters of the model. This paper presents the CB-MRAS model and the Genetic Algorithm (GA) such as the Binary GA, the real number GA, the modify GA and the CucKoo search algorithm (CSA) to estimate these parameters.

Recently, the GA has been used for motor parameter identification [1, 4, 12, 14, 15]. GA is a stochastic search technique borrowing concepts from biological evolutionary theory. Unlike other conventional techniques, the GA enables the acquisition of the best set of parameter values by minimizing errors between the model evaluated and the plant. GA does not require derivative of the cost function. Therefore, the estimated parameters can cover all of the values in their research space. The genetic procedure consists in minimizing a cost function. The cost function is computed as a weighted sum either of square or absolute differences of the output variables, generally acquired experimentally, or those of the model computed by simulation.

The Cuckoo Search algorithm (CSA), developed in 2009 by Xin-She Yang Suash Deb, is an algorithm inspired by Cuckoo lay their eggs in the nest of another nest of birds [3, 8, 10]. This algorithm is very positive in solving the optimal problem.

The simulation results show that the prospect of the algorithms is proposed.

2 The Current Based Model Reference Adaptive System (CB-MRAS) Model to Estimate the Speed of the IM

2.1 The Vector Controlled Model of IM with Sensorless Speed Based on CB-MRAS Method

We built a model with sensorless speed based on the CB-MRAS method as the following figure [2, 5, 7]:

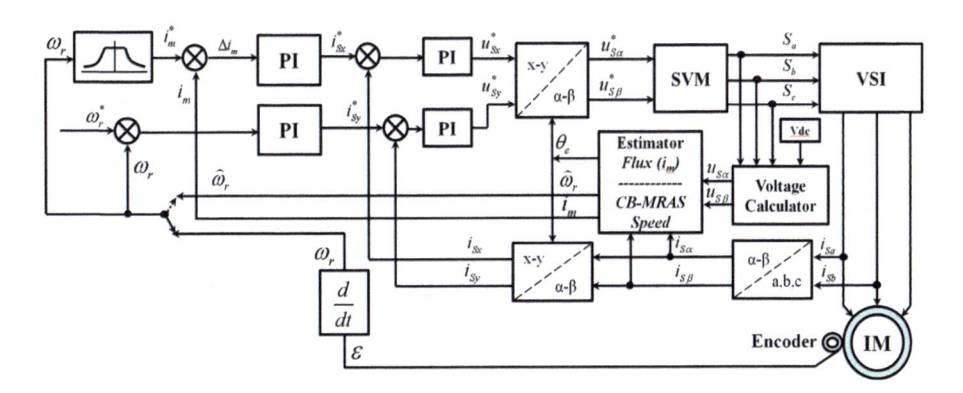

Fig. 1. The block scheme of CB-MRAS method for sensorless control of IM drive.

2.2 The CB-MRAS Model to Estimate the Speed of the Induction Motor

We have:

$$\hat{\psi}_R^S = \psi_R^S \ [5,7] \quad (1)$$

$$\int \left[\left(j\hat{\omega}_R - \frac{1}{T_R} \right) \hat{\psi}_{R(AM)}^S + \frac{1}{T_R} L_m i_s^s \right] dt = \frac{L_R}{L_m} \left[\int (u_s^s - R_s i_s^s) dt - \frac{L_S L_R - L_m^2}{L_R} i_s^s \right] \quad (2)$$

Then resulting equation of the stator current estimator has the following form:

$$i_{s(CB)}^s + T_{i(CB)} \frac{di_{s(CB)}^s}{dt} = K_{1(CB)} u_s^s + K_{2(CB)} \psi_R^S + j\omega_R K_{3(CB)} \psi_R^S \quad (3)$$

We have the coefficients of the equation:

$$K_{1(CB)}, K_{2(CB)}, K_{3(CB)} = f(L_R, L_m, R_S, T_R) \quad (4)$$

$$T_{i(CB)} = f(L_S, L_R, L_m, R_S, T_R) \quad (5)$$

The adaptation algorithm is presented with the following equations:

$$\xi = (i_s^s - i_{s(CB)}^s) \hat{\psi}_R^S = (i_{s\alpha} - i_{s\alpha(CB)}) \hat{\psi}_{R\beta} - (i_{s\beta} - i_{s\beta(CB)}) \hat{\psi}_{R\alpha}, \ \hat{\omega}_R = K_P \xi + K_I \int_0^t \xi dt \quad (6)$$

Where:
Components of the rotor flux vector and its estimated vector are outputs of the reference model and the adaptive model. These outputs are calculated according to equations:

$$\hat{\psi}_{R\alpha} = \int \left(\frac{L_m}{T_r} i_{s\alpha} - \frac{1}{T_r} \hat{\psi}_{R\alpha} - \hat{\omega}_R \hat{\psi}_{R\beta} \right) dt, \ \hat{\psi}_{R\beta} = \int \left(\frac{L_m}{T_r} i_{s\beta} - \frac{1}{T_r} \hat{\psi}_{R\beta} + \hat{\omega}_R \hat{\psi}_{R\alpha} \right) dt \quad (7)$$

The current model in this scheme and the adaptive model, the reference model in Fig. 2 are identical respectively. The input-output relationships of the current estimator are described by the Eqs. (8).

$$\begin{aligned} i_{s\alpha(CB)} &= \frac{1}{T_i} \int \left(K_{1(CB)} u_{s\alpha} + K_{2(CB)} \hat{\psi}_{R\alpha} + K_{3(CB)} \hat{\omega}_R \hat{\psi}_{R\beta} - i_{s\alpha} \right) dt, \\ i_{s\beta(CB)} &= \frac{1}{T_i} \int \left(K_{1(CB)} u_{s\beta} + K_{2(CB)} \hat{\psi}_{R\beta} - K_{3(CB)} \hat{\omega}_R \hat{\psi}_{R\alpha} - i_{s\beta} \right) dt, \end{aligned} \quad (8)$$

The model coefficients depend on the induction motor parameters as follows [5]:

$$K_{1(CB)} = \frac{L_r}{L_m} \times \frac{1}{\frac{L_r R_S}{L_m} + \frac{L_m}{T_r}}; K_{2(CB)} = \frac{L_m}{(L_r R_s T_r + L_m^2)}; K_{3(CB)} = \frac{1}{\frac{L_r R_S}{L_m} + \frac{L_m}{T_r}}; T_{i(CB)} = \frac{L_s L_r - L_m^2}{L_r R_S + \frac{L_m^2}{T}} \quad (9)$$

In addition, the Eq. (7) we see that it has two parameters which we don't know are L_m, T_R; we need to define two coefficients are K_{Lm}, K_{Tr}.

Because the parameters of the motor which we do not know. The inference is the six parameters of CB-MRAS model are also not determined $(K_{1(CB)}, K_{2(CB)}, K_{3(CB)}, T_{i(CB)}, K_{Lm}, K_{Tr})$.

From the system of the above equations, we establish the CB-MRAS estimated model of speed. It is described in detail as shown below [5, 13].

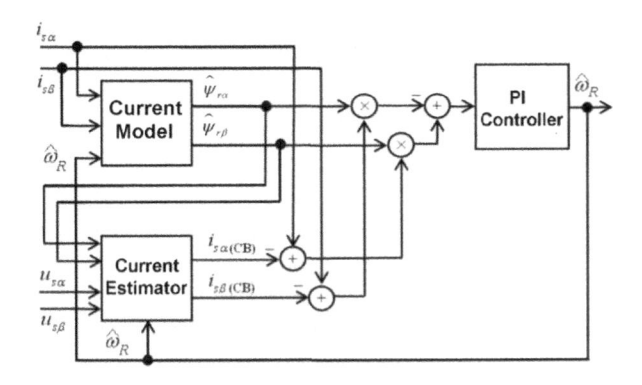

Fig. 2. The CB-MRAS observer

3 Estimate Factors of CB-MRAS Equations Using GA and CucKoo Search Algorithm

3.1 How to Determine Optimal Parameters for CB-MRAS Model

The CB-MRAS model is one of very good speed estimate model, but we must specify its parameters. We have six coefficients $K_{1(CB)}, K_{2(CB)}, K_{3(CB)}, T_{i(CB)}, K_{Lm}, K_{Tr}$ which we need to determine. If we do not know six above parameters exactly then the estimated speed of the induction motor is incorrect.

In this part, we show how to estimate six those parameters of this model. We will use the GA algorithm and CucKoo search algorithm with a structure as Fig. 3.

How to implement this algorithm in the Simulink of Matlab as follows [9]:
First, we determine the cost function of this topic is:

$$Cost\_function = (\omega_{actual} - \hat{\omega}_{CB-MRAS})^2 = (\omega_{actual} - \hat{\omega}_R)^2 = \sum_{i=1}^{n}(\omega_{actual\_i} - \hat{\omega}_{R\_i})^2 \quad (10)$$

With: $n = \frac{time\ of\ the\ simulation}{time\ step\ of\ the\ run} = \frac{1\ (S)}{0.000001\ (S)} = 10^6$

This is the Method of Least Squares is used in our paper. Next, we need to find the values of the parameters: $K_{1(CB)}, K_{2(CB)}, K_{3(CB)}, T_{i(CB)}, K_{Lm}, K_{Tr}$ will be introduced into CB-MRAS model. The values that make the cost function approximated by zero are the optimal values that we need to find. This value will make the speed response of the CB-MRAS model approximate to the actual value.

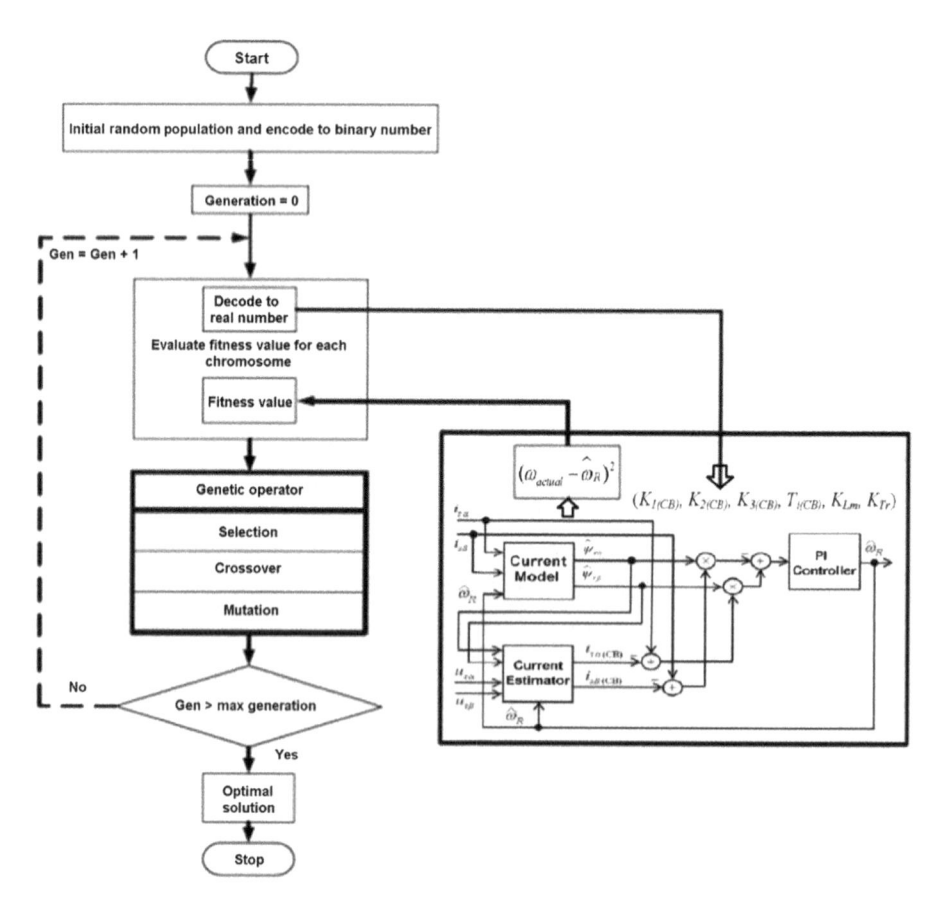

Fig. 3. The block scheme of using GA algorithm for CB-MRAS method.

To implement this algorithm on Matlab and Simulink we need to follow the steps in below diagram Fig. 4 to find the optimal parameters of the CB-MRAS model.

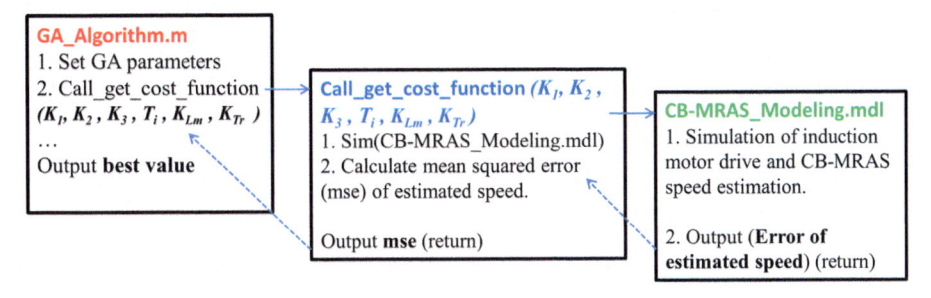

```
GA_Algorithm.m
1. Set GA parameters
2. Call_get_cost_function
($K_1, K_2, K_3, T_i, K_{Lm}, K_{Tr}$)
...
Output best value
```
→
```
Call_get_cost_function ($K_1, K_2,$
$K_3, T_i, K_{Lm}, K_{Tr}$)
1. Sim(CB-MRAS_Modeling.mdl)
2. Calculate mean squared error
(mse) of estimated speed.

Output mse (return)
```
→
```
CB-MRAS_Modeling.mdl
1. Simulation of induction
motor drive and CB-MRAS
speed estimation.

2. Output (Error of
estimated speed) (return)
```

Fig. 4. The steps for performing the optimal parameter search on Matlab using GA algorithm.

3.2 The Difference Methods to Estimate Parameters of CB-MRAS Model

a. The binary GA algorithm [12]:

Each parameter that we use 12 bits binary to encode; so six parameters, we have 72 bits binary and how to calculate these parameters as follows:

101010101010	100110011001	...	110110110110
Parameter $K_{1(CB)}$	Parameter $K_{2(CB)}$	...	Parameter K_{Tr}

$$\text{Each step for the parameter } K_{1(CB)} : step\_K_{1(CB)} = \frac{K_{1(CB)\max} - K_{1(CB)\min}}{2^{12} - 1} \quad (11)$$

We transfer the binary numbers to the decimal as follows:

```
value_K₁(j,1)=0;
For j=1:M
    For i=1:N
        value_K₁(j,1)= pop_K₁(j,i)*(2^(N-i)) + value_K₁(j,1);
    End
End
```

$$Real\_value\_of\_parameter = step\_K_{1(CB)} * value + K_{1(CB)\min} \quad (12)$$

These values will be transferred to CB-MRAS modeling to evaluate them.

b. **The real number GA algorithm [9]:**

How to match two random chromosomes to form two new chromosomes. Suppose we have two parent chromosomes as following:

$$Parent_1 = [p_{m1}\ p_{m2}...\ p_{m\alpha}...\ p_{mN}]$$
$$parent_2 = [p_{d1}\ p_{d2}...\ p_{d\alpha}...\ p_{dN}]$$

The α is a random number between 1 and N: α = round(rand * N). Where the m and d subscripts discriminate between the mom and the dad. Then the selected variables are combined to form new variables that will appear in the children:

$$\begin{aligned} p_{new1} &= p_{m\alpha} - \beta(p_{m\alpha} - p_{d\alpha}) \\ p_{new2} &= p_{d\alpha} + \beta(p_{m\alpha} - p_{d\alpha}) \end{aligned} \quad (13)$$

Where β is also a random value between 0 and 1. Finally, we create two new chromosomes:

$$offspring_1 = [p_{m1}\ p_{m2}...\ p_{d\gamma}...\ p_{new1}...\ p_{mN}]$$
$$offspring_2 = [p_{d1}\ p_{d2}...\ p_{m\gamma}...\ p_{new2}...\ p_{dN}]$$

The γ is also a random value between 1 and N but different from α.

c. **The modify real number GA algorithm:**

After initializing, we calculate the cost function for all chromosomes, then arrange them in order from best to worst. We choose the retention rate as k% and match them together. When hybridizing a pair of chromosomes together we make a change in all the components in each chromosome as the following equation (Fig. 5):

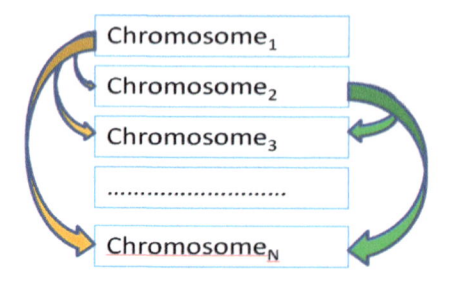

Fig. 5. The steps for performing the improved GA algorithm.

$$\begin{aligned}
&for\ i = 2 : M\\
&\beta = rand;\\
&for\ \alpha = 1 : N\\
&temp(1,\alpha) = kept(1,\alpha) + \beta * (kept(1,\alpha) - \text{kept}(i,\alpha));\\
&temp(2,\alpha) = kept(i,\alpha) - \beta * (kept(1,\alpha) - \text{kept}(i,\alpha));\\
&\text{End of } \alpha\\
&\text{End of } i
\end{aligned} \qquad (14)$$

In addition, we created N = 6 chromosomes as shown below:

$$\begin{aligned}
Pop_{new\_k1} &= [kept(1,1) + \Delta K_1\ kept(1,2)\ldots kept(1,6)]\\
Pop_{new\_k2} &= [kept(1,1)\ kept(1,2) + \Delta K_2 \ldots kept(1,6)]\\
Pop_{new\_k3} &= [kept(1,1)\ldots kept(1,3) + \Delta K_3 \ldots kept(1,6)]\\
Pop_{new\_Ti} &= [kept(1,1)\ldots kept(1,4) + \Delta T_i \ldots kept(1,6)]\\
Pop_{new\_K_{Lm}} &= [kept(1,1)\ldots kept(1,5) + \Delta K_{Lm} kept(1,6)]\\
Pop_{new\_K_{Tr}} &= [kept(1,1)\ldots\ kept(1,5)\ kept(1,6) + \Delta K_{Tr}]
\end{aligned} \qquad (15)$$

Where $\Delta K_i = rand * (K_j - K_m)$ with j, m are the random number between 1 and M. The total number of produced chromosomes are now (M + N), then we will use the cost function to re-evaluate.

d. **Cuckoo Search Algorithm (CSA)** [3, 10]:

The main steps for CSA are described as follows:

1. **Initialization:** A population of N host nests is representing by $X = [X, X, \ldots, X]^T$, where each nest $X_d = [P_{d1}, \ldots, P_{dN}]$ representing for the parameters: $(K_{1(CB)}, K_{2(CB)}, K_{3(CB)}, T_{i(CB)}, K_{Lm}, K_{Tr})$ these parameters are initialized by:

$$X_{di} = p_{imin} + rand_1 * (p_{imax} - p_{imin}) \qquad (16)$$

Where rand is a distributed random number in [0, 1] for each population of host nests. The P_{imin}, P_{imax} are the maximum and minimum of the parameters.

With cost function of CB−MRAS model is $\cos t\_function = (\omega_{actual} - \widehat{\omega}_R)^2$ (17)

The initial population of the host nests have best value of each nest $Xbest_d(d = 1, \ldots, N_d)$ and the nest corresponding to the best cost function and have also the best nest Gbest among all nests in the population.

2. **Generation of New Solution via Lévy Flights:** The new solution by each nest is created as follows [10]:

$$X_d^{new} = Xbest_d + \alpha * rand_2 * \Delta X_d^{new} \tag{18}$$

The cost function (17) will be re-evaluated for the new solution to find out the newly best value of each nest $Xbest_d$ and the best nest of all nests Gbest by comparing the stored cost value and the newly created ones.

3. **Alien Egg Discovery and Randomization [10]:** The discovery of an alien egg in the nest of a host bird with the probability of P_a. This event also creates a new solution for the similar to the Lévy Flights. The new solution is created as follows:

$$X_d^{dis} = Xbest_d + K * \Delta X_d^{dis} \tag{19}$$

Where K is the updated and the determined factor based on the probability of the host bird to discover an alien egg in their nest.

$$if\ rand_3 < p_a\ then\ K = 1\ else\ K = 0$$

$$\Delta X_d^{dis} = rand_4 * (randp_1(Xbest_d) - randp_3(Xbest_d)) \tag{20}$$

Similar to the solution of Lévy flights, this new solution is also determined again for each nest $Xbest_d$ and the best value of all nests $Gbest$ are set based on cost value obtained from (17).

4. **Stopping Criteria:** The CSA is terminated when the current iteration is equal or greater the maximum number of iteration.

4 Simulation Results

The basic parameters of the induction motor: $P = 1.500\ kW$, $U_{dc} = 300\ V$, $p = 2$. The remaining parameters will be estimated $(K_{1(CB)}, K_{2(CB)}, K_{3(CB)}, T_{i(CB)}, K_{Lm}, K_{Tr})$.

4.1 The Used Parameters for Simulation, Graphs and Table of Data

We follow these algorithms [a], [b], [c], [d]. Because the model in Fig. 1 has a relatively small run time step = 10^{-6} s so long runtime. If we give the algorithms to run with a population of 20 chromosomes, each chromosome has 6 parameters and the number of generations is 40, it takes about 48 h.

Table 1. The parameters are selected for the algorithms.

Parameter	Value
Probability of mutation	0.30
Probability of crossover	0.30
Population size	20
Generation	40

In this topic, we use the parameters in Table 1 to simulate. After the implementation we obtain the following graphs (Figs. 6, 7, 8 and 9):

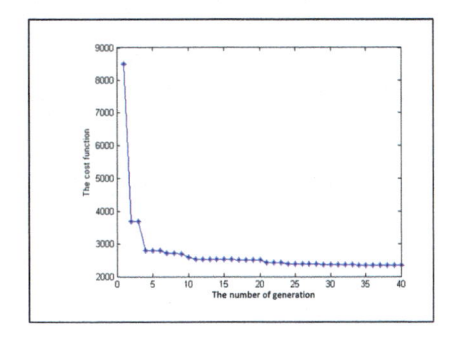

Fig. 6. The Binary GA convergence scheme

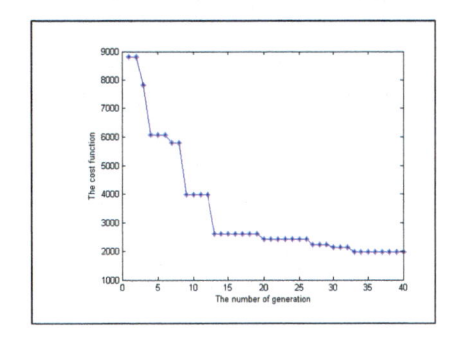

Fig. 7. The real number GA convergence scheme

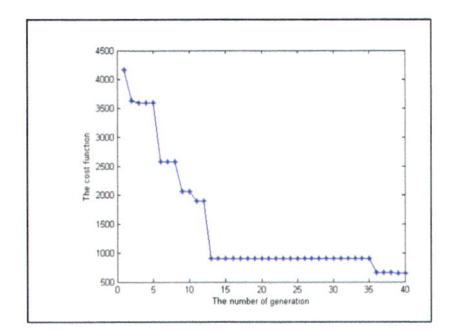

Fig. 9. The CSA convergence scheme

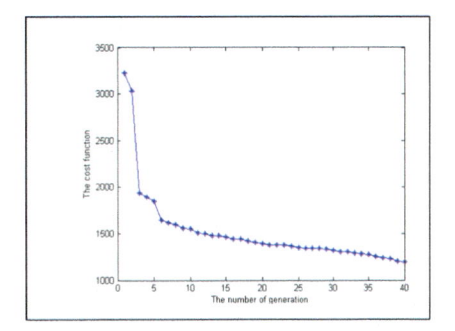

Fig. 8. The modify real number GA convergence scheme

Table 2. The result table of the Simulink for above cases.

The methods	K_1	K_2	K_3	T_i	K_{Lm}	K_{Tr}	The best value
[a]binary	1.9916	0.1000	1.7506	0.0336	0.1722	0.0489	2346.7
[b]org GA	1.3614	0.9626	1.9936	0.0252	0.1251	0.0545	1959.3
[c]modify	1.1975	0.1440	3.7997	0.0167	0.0523	0.0054	1193.7
[d]CucKoo	0.2763	5.0000	2.3968	0.0038	0.0431	0.2990	642.75
Best value	0.2311	3.988	0.2176	0.0035	0.1290	0.0545	10.000

4.2 The Speed Response of CB-MRAS Model with the Algorithms

We obtain the parameters of the four methods, as shown in the Table 2 above and supply these parameters to the CB-MRAS model. We collect the following responses (Fig. 10):

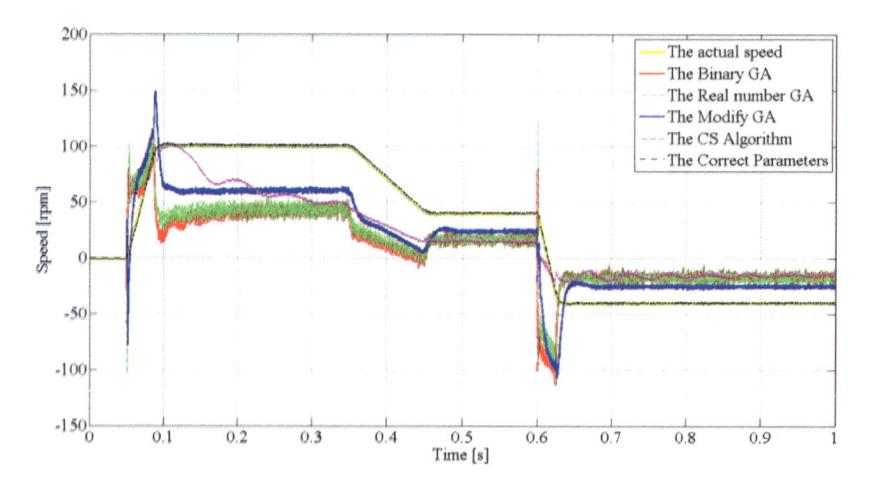

Fig. 10. The speed response of four above algorithms

We see that with the binary GA algorithm or real number GA algorithm, the results are similar, the improved GA and CSA algorithms for better results, the improved GA algorithm with good slope, the CSA algorithm converges faster. This problem can have many solutions, which make the cost function of the smallest value, do the response speed closer to the actual value [13]. For improved GA or CSA algorithms, if we give a larger number of loops (number of generations), better results will be obtained.

5 Conclusion

The CB-MRAS speed estimation model is one of the most powerful models, but it is highly dependent on the parameters of the induction motor model, which are $K_{1(CB)}, K_{2(CB)}, K_{3(CB)}, T_{i(CB)}, K_{Lm}, K_{Tr}$. The paper presents four algorithms for estimating; these parameters and simulation results as well as responding to the speed of induction motors for each case; especially the improved GA algorithm and CucKoo search algorithm, these two algorithms have very good convergence results and can be used to find optimal parameters for this model or similar models if the number of loops or the population size increases when running this algorithm.

Acknowledgement. This paper was supported by the projects: Centre for Intelligent Drives and Advance Machine control (CIDAM) project, Reg. No. TE02000103 funded by the Technology Agency of the Czech Republic and Project Reg. No. SP2018/162 funded by the Student Grant Competition of VSB – Technical University of Ostrava.

References

1. Haupt, R.L., Haupt, S.E.: Practical Genetic Algorithms, 2nd edn. A Wiley-Interscience, New York (2004)
2. Brandstetter, P., Dobrovsky, M., Kuchar, M.: Implementation of genetic algorithm in control structure of induction motor A.C. drive. In: Advances in Electrical and Computer Engineering, vol. 14, no. 4 (2014)
3. Thao, N.T.P., Thang, N.T.: Environmental economic load dispatch with quadratic fuel cost function using CucKoo search algorithm. Int. J. u-and e-Serv. Sci. Technol. **7**(2), 199–210 (2014)
4. Datta, M., Rafiq, M.A., Ghosh, B.C.: Genetic algorithm based fast speed response induction motor drive without speed encoder. In: POWERENG 2007, Setubal, Portugal (2007)
5. Brandstetter, P.: Sensorless control of induction motor using modified MRAS. Int. Rev. Electr. Eng. IREE **7**(3), 4404–4411 (2012)
6. Saghafinia, A., Ping, H.W., Rahman, M.A.: High performance induction motor drive using hybrid fuzzy-PI and PI controllers: a review. Int. Rev. Electr. Eng. IREE **5**(5), 2000–2012 (2010)
7. Girovsky, P., Timko, J., Zilkova, J.: Shaft sensor-less FOC control of an induction motor using neural estimators. Acta Polytechnica Hungarica **9**(4), 31–45 (2012)
8. Rajasekhar, A., Abraham, A., Jatoth, R.K.: Controller tuning using a Cauchy mutated artificial bee colony algorithm. In: Advances in Intelligent and Soft Computing, vol. 87, pp. 11–18. Springer, Berlin (2011)
9. Tran, T.C., Brandstetter, P., Duy, V.H., Dong, C., Tran, C.D., Ho, S.D.: Estimate parameters of induction motor using ANN and GA algorithm. In: AETA: Recent Advances in Electrical Engineering and Related Sciences (2017)
10. Yang, X.-S., Deb, S.: CucKoo search via Lévy flights. In: Proceedings of World Congress on Nature & Biologically Inspired Computing (NaBic 2009), India, pp. 210–214 (2009)
11. Ben Regaya, C., Zaafouri, A., Chaari, A.: Electric drive control with rotor resistance and rotor speed observers based on fuzzy logic. Math. Prob. Eng. **2014**, 9 (2014)
12. Megherbi, A.C., Megherbi, H., Benmahamed, K., Aissaoui, A.G., Tahour, A.: Parameter identification of induction motors using variable-weighted cost function of genetic algorithms. J. Electr. Eng. Technol. **5**(4), 597–605 (2010)
13. Timer, J., Adžic, E., Porobic, V., Marcetic, D.: Influence of rotor time constant error on IFOC control structure. Electronics **13**(1) (2009)
14. Eissa, M.M., Virk, G.S., AbdelGhany, A.M., Ghith, E.S.: Optimum induction motor speed control technique using genetic algorithm. Am. J. Intell. Syst. **3**(1), 1–12 (2013)
15. Chacko, S., Bhende, C.N., Jain, S., Nema, R.K.: Rotor resistance estimation of vector controlled induction motor drive using GA/PSO tuned fuzzy controller. Int. J. Electr. Eng. Inf. **8**(1), 218 (2016)

# Study on Algorithms and Path-Optimization for USV's Obstacle Avoidance

Ngoc-Huy Tran[1(✉)] and Nguyen Nhut-Thanh Pham[2]

[1] Ho Chi Minh City University of Technology,
VNUHCM, Ho Chi Minh City, Vietnam
tnhuy@hcmut.edu.vn
[2] National Key Laboratory of Digital Control and System Engineering,
VNUHCM, Ho Chi Minh City, Vietnam

Abstract. An unmanned surface vessel indispensable tasks to avoid obstacles when it moves in real environment. In order to solve this problem, it has been classified stationary obstacles and moving obstacles as well as offering many different algorithms each type. Because the structure of the unmanned surface vessel is indispensable in two main components: Guidance and Control, in which Guidance will receive waypoints for creating the desired trajectory and then combine the current location of ship to calculate and provide input data for Control to drive the ship following the desired trajectory. Therefore an obstacle avoidance algorithms are integrated in the block Guidance to provide the most suitable input data for the controller is essential. In addition, some cases because the priority of avoiding the collision lead to the path will be long and inefficient. This paper will use the Line of Sight (LOS) algorithm with the lookahead distance to design the Guidance as well as propose an obstacle avoidance algorithm to integrate with it. Besides proposing a method to optimize the way while avoiding obstacles. The results as well as the effectiveness of the proposed method will be shown in the MATLAB/SIMULINK simulation.

Keywords: Avoid obstacles · Path-following · Line of Sight (LOS)

1 Introduction

Human habitat is shrinking steadily because of the rapid growth of population. Besides the development of society has caused depletion of resources and serious environmental pollution. Faced with this problem, many robotic technologies has been studied as an effective technological solutions to survey environment in a large area and high accuracy as well as search for new resources.

Outstanding is autonomous vehicles were used in the task of environmental surveys including surface [1–3] and underwater [4–6]. These tasks can be understood in physical bottom morphology characterization (bathymetry, obstacles, bottom sediment accumulation, geological properties), in water volume characterization (water parameters, biological and ecosystem information, plume identification,...) and above water environment characterization, either in the air or air-water layer or in land border modeling. Unmanned surface vessel (USV) is very common in the surface vehicles for surveying environment.

To plan and control the motion of an USV, a guidance, navigation and control (GNC) system is required [7]. Normally, surface vessels generally only have actuators in surge (thruster force) and yaw (rudder angle or azimuth thrusters), while they do not have any actuators in the sideways direction (sway). Therefore, path following tasks will usually be resolved by a guidance system consists of guidance laws for heading and surge velocity that ensure convergence to the desired path if the control system can satisfy it. A commonly used approach for this guidance system is Light of Sight (LOS) method. LOS is very famous because it can be used for both straight line paths and curve paths as well as expanded to counteract environmental disturbances and the presence of ocean currents.

However, this approach often ignores obstructions during movement of the vessel. So this paper considers a guidance and control system that allows an USV to avoid stationary obstacles while following a given path. Also consider special cases and provide a method for optimizing the way while avoiding obstacles.

2 Mathematical Model of a Ship

Define the three DOF $\eta = [x, y, \psi]^T$ indicate position (x, y) and heading (ψ) of the ship in an earth-fixed inertial frame $\{e\}$, and $\upsilon = [u, v, r]^T$ be the corresponding linear velocities called surge (u), sway (v) and angular rate (r) called yaw in the body-fixed frame $\{b\}$ as Fig. 1. According to [7] the dynamic model of the ship is

$$\begin{cases} \dot{\eta} = R(\psi)\upsilon \\ M\dot{\upsilon} + C(\upsilon)\upsilon + D(\upsilon)\upsilon = \tau \end{cases} \quad (1)$$

where $R(.)$ is the three DOF rotation matrix with

$$R(\psi) = \begin{pmatrix} cos(\psi) & -sin(\psi) & 0 \\ sin(\psi) & cos(\psi) & 0 \\ 0 & 0 & 1 \end{pmatrix} \quad (2)$$

The system inertia matrix $M = M^T > 0$ is

$$M = \begin{pmatrix} m_{11} & 0 & 0 \\ 0 & m_{22} & m_{23} \\ 0 & m_{32} & m_{33} \end{pmatrix} = \begin{pmatrix} m - X_{\dot{u}} & 0 & 0 \\ 0 & m - Y_{\dot{v}} & mx_G - Y_{\dot{r}} \\ 0 & mx_G - N_{\dot{v}} & I_z - N_{\dot{r}} \end{pmatrix} \quad (3)$$

A skew-symmetric matrix of Coriolis and centripetal terms $C(\upsilon)$ is

$$C(\upsilon) = \begin{pmatrix} 0 & 0 & c_{13} \\ 0 & 0 & c_{23} \\ -c_{13} & -c_{23} & 0 \end{pmatrix} = \begin{pmatrix} 0 & 0 & -(m - Y_{\dot{v}})v - (mx_G - Y_{\dot{r}})r \\ 0 & 0 & (m - X_{\dot{u}})u \\ (m - Y_{\dot{v}})v + (mx_G - Y_{\dot{r}})r & -(m - X_{\dot{u}})u & 0 \end{pmatrix} \quad (4)$$

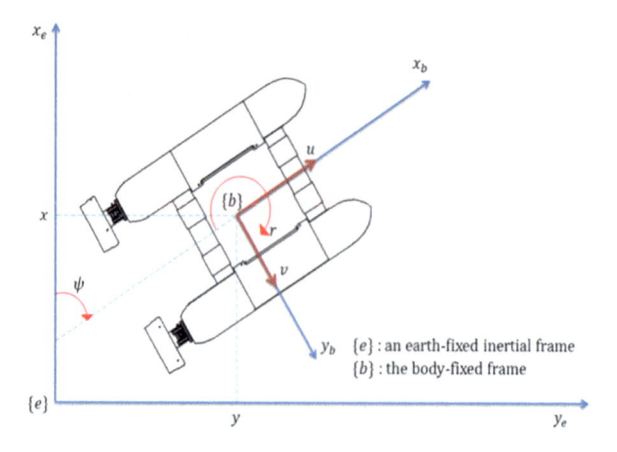

Fig. 1. Reference frame

The damping matrix $D(v)$ is

$$D(v) = \begin{pmatrix} d_{11} & 0 & 0 \\ 0 & d_{22} & d_{23} \\ 0 & d_{32} & d_{33} \end{pmatrix}$$
$$= \begin{pmatrix} -X_u - X_{|u|u}|u| & 0 & 0 \\ 0 & -Y_v - Y_{|v|v}|v| - Y_{|r|v}|r| & -Y_r - Y_{|v|r}|v| - Y_{|r|r}|r| \\ 0 & -N_v - N_{|v|v}|v| - N_{|r|v}|r| & -N_r - N_{|v|r}|v| - N_{|r|r}|r| \end{pmatrix} \quad (5)$$

where x_G is the distance from the center of gravity of ship to the origin of the body-fixed frame {b}. The coefficients $\{X_{(.)}, Y_{(.)}, N_{(.)}\}$ are hydrodynamic parameters according to the notation in [8] and $\tau = [\tau_1, \tau_2, \tau_3]^T$ is the control input. The arrangement of the thrusters in the USV is shown in Fig. 2 and the thruster forces and torque are related to the force vector τ through the equation:

$$\tau = \begin{bmatrix} \tau_1 \\ \tau_2 \\ \tau_3 \end{bmatrix} = \begin{bmatrix} 1 & 1 & 0 & 0 \\ 0 & 0 & 1 & 1 \\ L_{y1} & -L_{y2} & -L_{x1} & L_{x2} \end{bmatrix} \begin{bmatrix} F_1 \\ F_2 \\ F_3 \\ F_4 \end{bmatrix} \quad (6)$$

From (6) we can choose the force $F_3 = -F_4$ so

$$\tau = [\tau_1, 0, \tau_3]^T \quad (7)$$

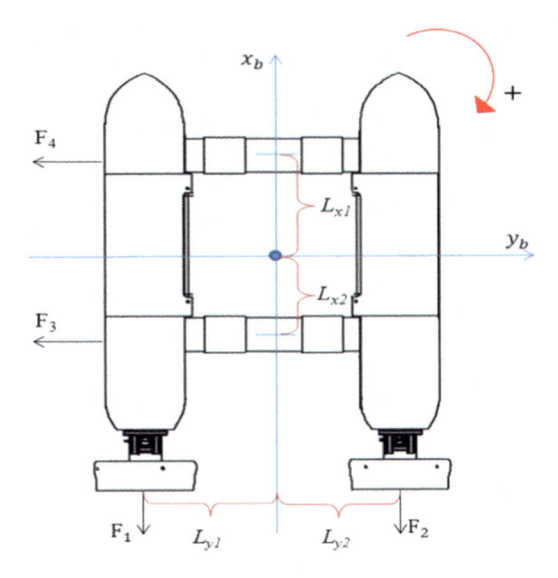

Fig. 2. Thruster configuration

3 Guidance and Control Design

3.1 Guidance Design

This paper focus on "guidance for path following" to serve the environmental survey, especially square or zig-zac path. Numerous methods have been proposed for path following, where popular methods are Line of Sight (LOS), Pure Pursuit (PP), Constant Bearing (CB). For marine craft, LOS has proved very effective because of the way it works similar to the helmsman, which will typically steer the vessel towards a point lying a constant distance, called the look-ahead distance, ahead of the vessel, along the desired path. Beside, many experiments as well as practical applications have proven LOS guidance not only simple but also effective. LOS guidance algorithms allow the vehicle at any initial position outside the desired path to converge on the path and stay on the path. In addition, LOS will calculate the desired heading so is usually associated with the heading controller or autopilots.

A. Cross-track Error

Assuming that USV needs to be convergence on the path that are connected by two way-points wp(k) and wp($k + 1$), the path reference frame is rotated an angle:

$$\alpha_p = \text{atan2}(y_{k+1} - y_k, x_{k+1} - x_k) \tag{8}$$

For the USV located at (x, y), the along-track (x_e) and cross-track (y_e) are defined by:

$$\begin{bmatrix} x_e \\ y_e \end{bmatrix} = R^T(\alpha_p) \begin{bmatrix} x - x_k \\ y - y_k \end{bmatrix} \tag{9}$$

where (x_k, y_k) is the coordinates of wp(k) in an earth-fixed inertial frame ($k = 1 \ldots N$) and $R^T(\alpha_p)$ is the rotation matrix which denotes the path reference frame is rotated an angle α_p with

$$R^T(\alpha_p) = \begin{bmatrix} \cos(\alpha_P) & -\sin(\alpha_P) \\ \sin(\alpha_P) & \cos(\alpha_P) \end{bmatrix}^T \in SO(2) \tag{10}$$

Expressing (9) we get:

$$\begin{aligned} x_e &= (x - x_k)\cos(\alpha_P) + (y - y_k)\sin(\alpha_P) \\ y_e &= -(x - x_k)\sin(\alpha_P) + (y - y_k)\cos(\alpha_P) \end{aligned} \tag{11}$$

The goal is making the cross-track converge to zero or in other words:

$$\lim_{t \to \infty} y_e(t) = 0$$

B. LOS guidance

Depending on the purpose of application, vehicle used or method of guidance, the LOS vector can be defined in different ways. Specifically in the application path following for the surface vehicle, the LOS vector is considered a vector with tail at the origin of body-fixed frame and head is located at a point (x_{los}, y_{los}) on the tangent line connecting two way-points wp(k) and wp(k + 1). The point (x_{los}, y_{los}) will satisfy the distance between it and projection of vehicle on the tangent line by lookahead distance $\Delta > 0$ as illustrated in Fig. 3. The lookahead distance is selected to fit in the experiment.

The desired heading can determine by formula:

$$\psi_d = \alpha_p + \arctan\left(\frac{-y_e}{\Delta}\right) \tag{12}$$

C. Avoid Obstacles

Because the LOS will provide the desired heading ψ_d, the idea is proposed an algorithm to avoid obstacles by calculating and provide the avoid angle ψ_{ao} so that it can be traversed between path following mode and avoid obstacles mode.

For an obstacle with any profile, we define "the circle of obstacle" is the circle can cover this obstacle with the minimum radius. Then considering the new obstacle is "the circle of obstacle" with the location in the NED is (x_c, y_c) and radius is r_c to replace for this obstacle. Next drawing a concentric circle with radius R_o which defined as "the safety circle".

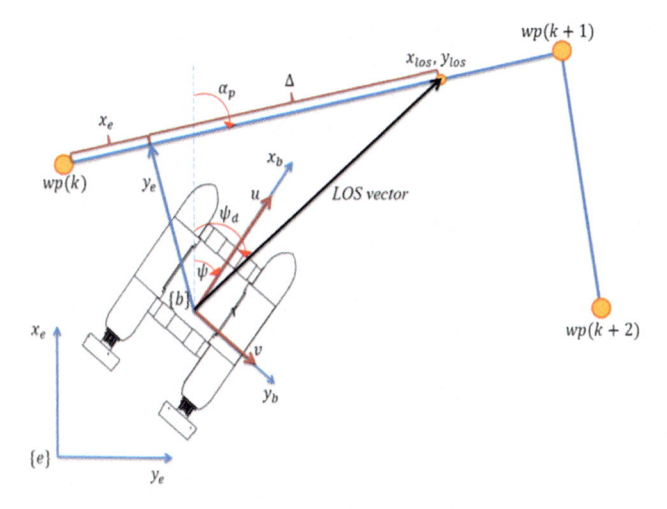

Fig. 3. LOS guidance geometry

Denote

$$\phi = arctan\left(\frac{y_c - y}{x_c - x}\right) \quad (13)$$

$$d_c = \sqrt{(x - x_c)^2 + (y - y_c)^2} \quad (14)$$

$$e = d_c - R_o \quad (15)$$

where ϕ and e are illustrated in Fig. 4.

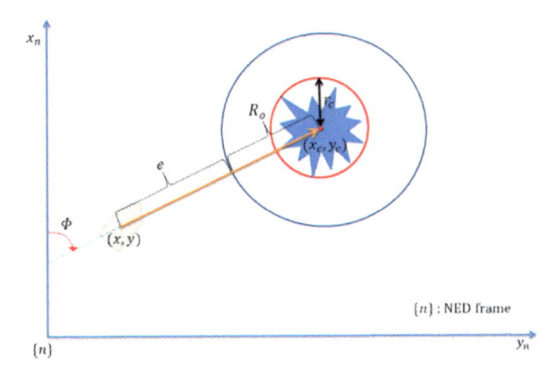

Fig. 4. Illustration of parameters used for avoiding obstacles

The desire is that when the desired path cut "the safety circle" and the USV moved into this circle, USV must move on the boundary of the circle until returning to the desired path. Based on the idea of LOS we will consider a vector has magnitude Δ_{ao} is the tangent vector with "the safety circle" as Fig. 5. The direction of Δ_{ao} is denoted λ where $\lambda = 1$ corresponds to clockwise motion and $\lambda = -1$ to counter-clockwise motion. In this paper λ will be selected so that USV can switch between the two modes smoothly. When λ can be determined by the following formula:

$$\lambda = \begin{bmatrix} 1 & |\psi_{ao,c} - \psi| < |\psi_{ao,cc} - \psi| \\ -1 & |\psi_{ao,c} - \psi| \geq |\psi_{ao,cc} - \psi| \end{bmatrix} \tag{16}$$

where $\psi_{ao,cc}$ and $\psi_{ao,c}$ are illustrated in Fig. 5. From the above definition, we can determine the general formula for the avoid angle ψ_{ao} is as follows:

$$\psi_{ao} = \Phi + \lambda \left(\frac{pi}{2} - atan\left(\frac{e}{\Delta}\right) \right) \tag{17}$$

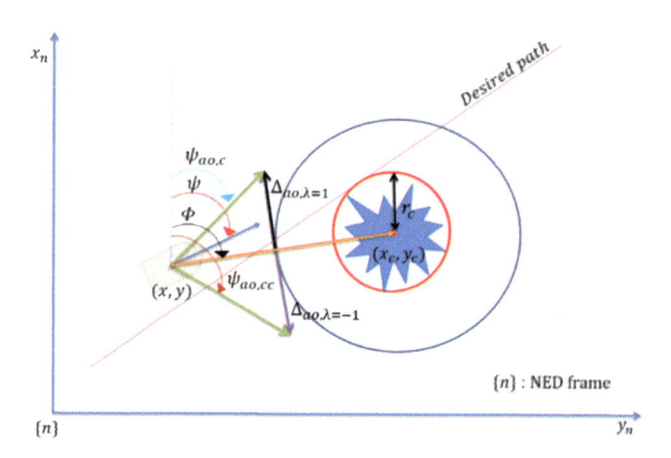

Fig. 5. Illustration of vector used for avoiding obstacles

With selecting the λ as above we give priority to work smoothly when switching between the two modes, so it appears the case the path is not optimal or ineffective. This can be seen in the simulation Fig. 8. Note that these cases only appear when the waypoint is inside "the safety circle". We can accept that the USV will not achieve this waypoint so we can ignore to continue the path and consider a new waypoint. This paper proposes a method for solving these cases by considering "the optimal circle" which will be tangent to "the safety circle" and the desired path as Fig. 6. Then the ship will move on "the optimal circle" instead of "the safety circle".

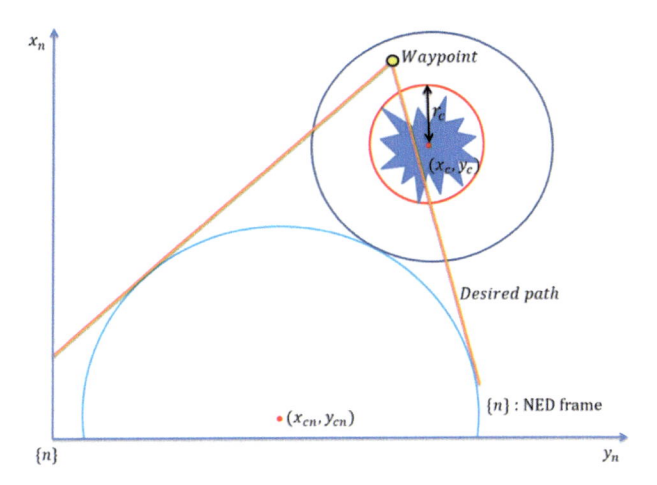

Fig. 6. The optimal circle (x_{cn}, y_{cn}) of obstacle (x_c, y_c)

3.2 Controller Design

Because LOS guidance algorithms providing the desired heading angle, we can divide the problem into two control problems include:

- Heading control: Control the heading angle tracking the desired heading angle.

$$\lim_{t \to \infty} |\psi(t) - \psi_d(t)| = 0$$

- Speed assignment: Speed control according to the desired value.

$$\lim_{t \to \infty} |u(t) - u_d(t)| = 0$$

So we can use two Sliding mode controllers to solve these two problems. We have a block diagram of the control model in Fig. 7.

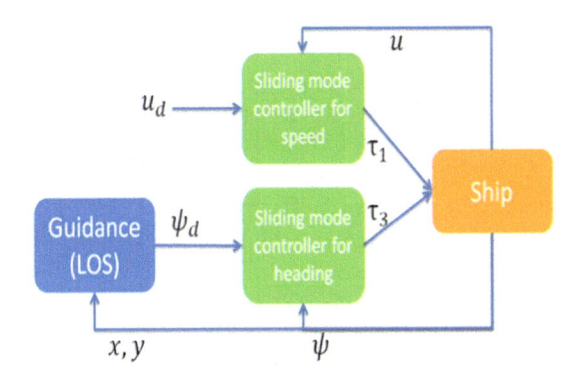

Fig. 7. Block diagram of the control model

Expressing (1) for second equation, we have

$$\begin{cases} m_{11}\dot{u} = \tau_1 - c_{13}r - d_{11}u \\ m_{22}\dot{v} + m_{23}\dot{r} = -c_{23}r - d_{22}v - d_{23}r \\ m_{32}\dot{v} + m_{33}\dot{r} = \tau_3 + c_{13}u + c_{23}v - d_{32}v - d_{33}r \end{cases} \quad (18)$$

D. Speed Controller

Because property of control input is $\tau_y = 0$, we can approximate $U = \sqrt{u^2 + v^2} \approx u$. Suppose the desired velocity is u_d, define the speed error $e_u = u - u_d$ so

$$\dot{e}_u = \dot{u} - \dot{u}_d = (\tau_1 - c_{13}r - d_{11}u)/m_{11} - \dot{u}_d$$

Select the sliding surface $s_u = e_u$ and define the control Lyapunov function $V_u = s_u^2/2 > 0$ whose time derivative is

$$\dot{V}_u = s_u \dot{s}_u = s_u[(\tau_1 - c_{13}r - d_{11}u)/m_{11} - \dot{u}_d]$$

Select the control law

$$\tau_1 = c_{13}r + d_{11}u + m_{11}(\dot{u}_d - K_u sat(s_u)) \quad (19)$$

where K_u is positive constant, we get $\dot{V}_u = -K_u sat(s_u) s_u < 0$ so $s_u \to 0$ or $e_u \to 0$.

E. Heading Controller

From (18) we have $\dot{r} = f_{\dot{r}}(u, v, r) + g_{\dot{r}}\tau_3$ where

$$f_{\dot{r}}(u, v, r) = \frac{1}{m_{22}m_{33} - m_{32}m_{23}}[m_{22}c_{13}u + c_{23}(m_{22}v + m_{32}r) \\ + v(d_{22}m_{32} - d_{32}m_{22}) + r(d_{23}m_{32} - d_{33}m_{22})] \quad (20)$$

$$g_{\dot{r}} = \frac{m_{22}\tau_3}{m_{22}m_{33} - m_{32}m_{23}} \quad (21)$$

Define the heading error $e_\psi = \psi - \psi_d$. The first and second derivative are

$$\dot{e}_\psi = \dot{\psi} - \dot{\psi}_d = r - \dot{\psi}_d$$
$$\ddot{e}_\psi = \dot{r} - \ddot{\psi}_d = f_{\dot{r}}(u, v, r) + g_{\dot{r}}\tau_3 + d - \ddot{\psi}_d$$

Select the sliding surface $s_\psi = \dot{e}_\psi + ce_\psi$ with c is positive constant. Define the control Lyapunov function $V_\psi = s_\psi^2/2 > 0$ whose time derivative is

$$\dot{V}_\psi = s_\psi \dot{s}_\psi = s_\psi[f_{\dot{r}}(u, v, r) + g_{\dot{r}}\tau_3 + d - \ddot{\psi}_d + c\dot{e}_\psi]$$

Select the control law for normal sliding mode controller

$$\tau_3 = \left(-f_{\dot{r}}(u,v,r) + \ddot{\psi}_d - c\dot{e}_\psi - K_\psi sat(s_\psi)\right)/g_{\dot{r}} \tag{22}$$

where K_ψ is positive constant, we get $\dot{V}_\psi = -K_\psi sat(s_\psi)s_\psi < 0$ so $s_\psi \to 0$ or $e_\psi \to 0$.

4 Simulation Results

This section presents the main results of the proposed method in two case: normal and optimize the path. The desired surge velocity is chosen to be 1 m/s both during path following and obstacle avoidance. The lookahead distance Δ is chosen as $\Delta = 5$ m and $\Delta_{ao} = 10$ m. The controller gains are chosen as $K_u = 10$, $c = 4$, $K_\psi = 7$. The distance between "the safety circle" and "the circle of obstacle" is usually chosen according to the length of the ship, in the simulation $R_o - r_c = 2.5$ m

The result in Fig. 8 shows that the combination of controller and guidance is very effective which helps the USV converge on the path, stay on it and avoid obstacles. Figure 9 illustrate that the heading angel is tracking the desired heading very well. Figure 10 show the Coriolis and centripetal terms caused an affect on the speed control process, which reduced the quality of the control but overall the surge velocity achieved the desired value. However there are some cases of USV have to detour to avoid obstacles, this causes loss of energy and less effective.

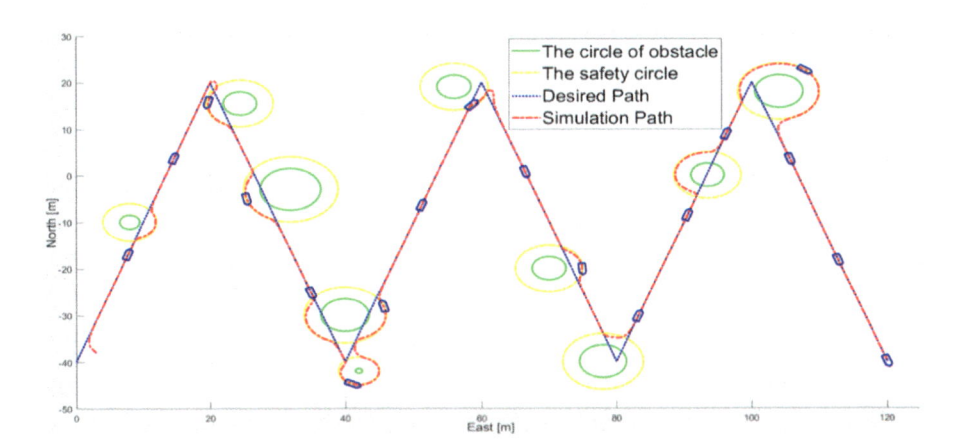

Fig. 8. Desired path and simulation path of USV

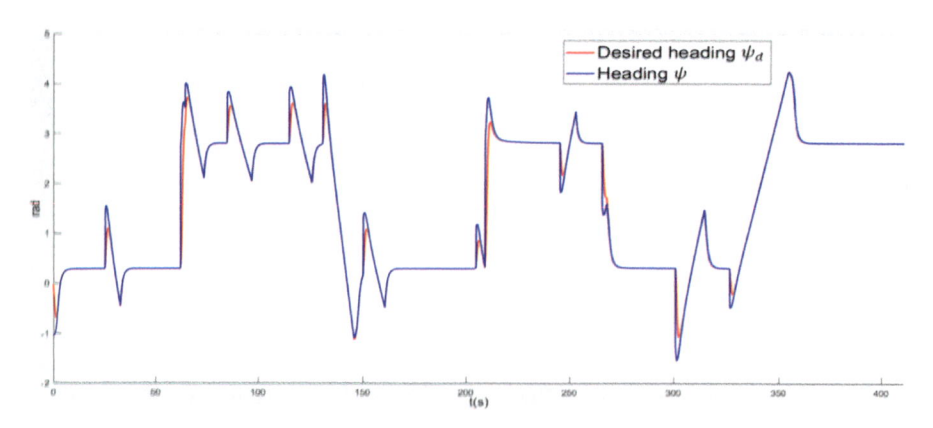

Fig. 9. The ship heading (ψ) closely follow the desired heading (ψ_d)

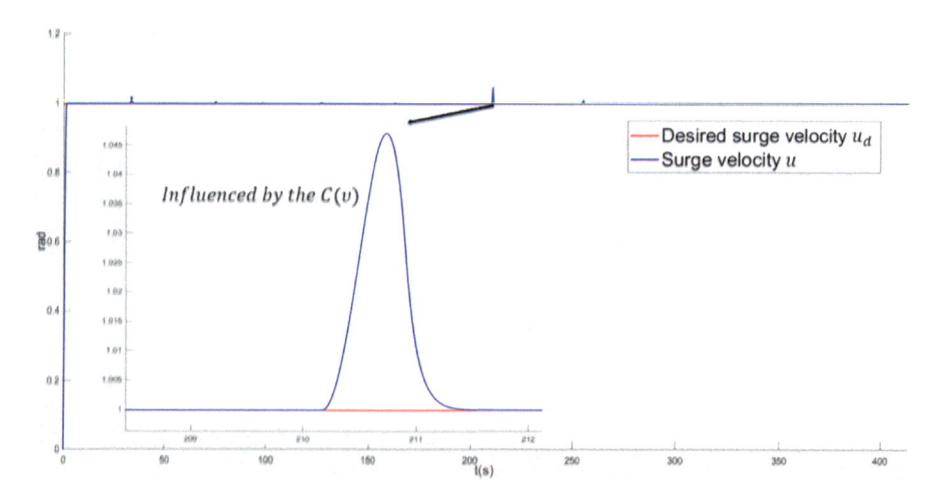

Fig. 10. The surge velocity converges to the desired velocity provided by the guidance system

Figure 11 show USV trajectory after optimization. Obviously this helps to make the path shorter and more efficient. Note that the obstacle cover the third waypoint which the trajectory after the optimization is not tangential to it. This is because "the optimal circle" of the obstacle has radius too small for the USV can maneuvering. So we have to add the condition of increasing the radius of "the optimal circle" to conformity with the ship.

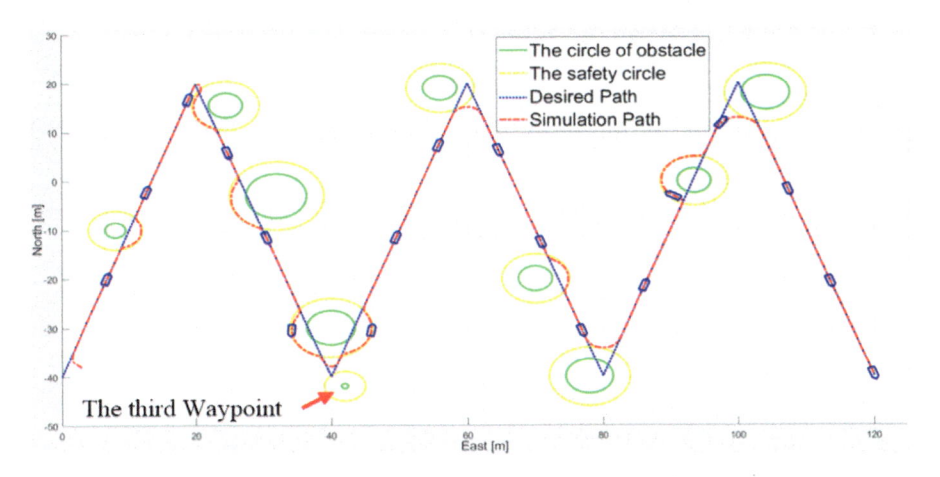

Fig. 11. USV's trajectory when optimized

5 Conclusion

This paper presents about the LOS algorithm for guidance as well as the design of controller to combine with the guidance. Besides that an obstacle avoidance algorithm based on the idea of LOS is proposed to integrate in the Guidance. The goal is to solve the problem of avoiding collision when ship moving in practice. This obstacle avoidance algorithm has suggested a specific guidance law for the collision avoidance that will, if satisfied, make the ship track a circle with a constant safe radius about the moving obstacle center. In addition, a method for optimal path is proposed in some special cases. The control object is used here as USV and stationary obstacles. The target of this USV is apply to environmental surveys on a river, lake or sea. The further work is extending the algorithm for moving obstacles. The simulation results show the effectiveness of the obstacle avoidance algorithm and the optimal method.

Acknowledgement. This research is supported by National Key Lab. of Digital Control and System Engineering (DCSELAB), HCMUT and funded by Vietnam National University Ho Chi Minh city (VNU-HCM) under grant number C2018-20b-02.

References

1. Ferreira, H., Almeida, C., Martins, A., Almeida, J., Dias, N., Dias, A., Silva, E.: Autonomous bathymetry for risk assessment with ROAZ robotic surface vehicle. In: OCEANS 2009-EUROPE, pp. 1–6 (2009)
2. Caccia, M., Bibuli, M., Bono, R., Bruzzone, G., Bruzzone, G., Spirandelli, E.: Aluminum hull USV for coastal water and seafloor monitoring. In: OCEANS 2009-EUROPE, pp. 1–5 (2009)
3. Tokekar, P., Bhadauria, D., Studenski, A., Isler, V.: A robotic system for monitoring carp in Minnesota lakes. J. Field Robot. **27**(6), 779–789 (2010)

4. Ramos, P., Cruz, N., Matos, A., Neves, M.V., Pereira, F.L.: Monitoring an ocean outfall using an AUV. In: Conference Proceedings of MTS/IEEE Oceans 2001, An Ocean Odyssey (IEEE Cat. No. 01CH37295), vol. 3 (2009–2014)
5. Fiorelli, E., Leonard, N.E., Bhatta, P., Paley, D.A., Bachmayer, R., Fratantoni, D.M.: Multi-AUV control and adaptive sampling in Monterey Bay. IEEE J. Oceanic Eng. **31**(4), 935–948 (2006)
6. Kim, A., Eustice, R.M.: Toward AUV survey design for optimal coverage and localization using the Cramer Rao lower bound. IEEE, 26 October 2009
7. Fossen, T.I.: Handbook of Marine Craft Hydrodynamics and Motion Control. Wiley, New York (2011)
8. The Society of Naval Architects and Marine Engineers: Nomenclature for treating the motion of a submerged body through a fluid. Technical and Research Bulletin No. 1–5
9. Fossen, T.I.: Marine Control Systems Guidance, Navigation, and Control of Ships, Rigs and Underwater Vehicles. Marine Cybernetics AS (2002)
10. Fossen, T.I., Breivik, M., RSkjetne, R.: Line-of-sight path following of underactuated marine craft. In: Proceedings of 6th IFAC Conference on Manoeuvering and Control of Marine Craft, pp. 244–249 (2003)
11. Zereik, E., Sorbara, A., Bibuli, M., Bruzzone, G., Caccia, M.: Priority task approach for USV's path following missions with obstacle avoidance and speed regulation. In: Proceedings of 10th Conference of Manoeuvering and Control of Marine Craft (2015)
12. Antonelli, G., Moe, S., Pettersen, K.Y.: Incorporating Set-based control within the singularity-robust multiple task-priority inverse kinematics. In: Proceedings of 23rd Mediterranean Conference on Control and Automation (2015)

Visual Servoing Controller Design Based on Barrier Lyapunov Function for a Picking System

Jong Min Oh[1], Jotje Rantung[1], Sung Rak Kim[1], Sang Kwun Jeong[2], Hak Kyeong Kim[1], Sea June Oh[3], and Sang Bong Kim[1(✉)]

[1] Department of Mechanical Design Engineering,
Pukyong National University, Busan, Republic of Korea
kimsb@pknu.ac.kr
[2] Department of Automation System, Korea Polytechnics,
Jinju 660-996, Republic of Korea
[3] Korea Maritime and Ocean University, Busan, Republic of Korea

Abstract. This paper proposes a visual servoing controller design based on Barrier Lyapunov function for a picking system. Visual servoing uses feedback data provided by the camera to control the movement of a picking system in a closed loop system. Visual servoing requires an object in the field of view of the camera in order to control the picking system. To improve the visual servoing controller, the image-based visual servoing and the position-based visual servoing are presented. To apply this method an offline trajectory is developed to perform the image-based visual servoing and the position-based visual servoing tasks for the picking system. Two different control approaches i.e. the visual servoing controller with the limit orientation using the Barrier Lyapunov function and the visual servoing controller with a quadratic Lyapunov function are presented. The proof of asymptotic stability is presented and simulation results from two visual servoing controllers are presented to verify the effectiveness of the proposed controller.

Keywords: Image–based visual servoing (IBVS) · Position–based visual servoing (PBVS) · Quadratic Lyapunov function (QLF) · Barrier Lyapunov function (BLF)

1 Introduction

Position-based visual servoing (PBVS) and image-based visual servoing (IBVS) were introduced in [1]. The positioning error will be minimized with feedback features by using Position Based Visual Servoing (PBVS), and Image-Based Visual Servoing (IBVS). The main problems of implementing visual servoing were presented in [2], i.e. instability of system in long distance tasks, interaction matrix singularity, local minima, pure rotation about center of camera, having no control on the speed of the robot during the visual servoing task, unknown path of the robot prior to the tasks, and features leaving the Field of View (FOV).

Various potential field methods [3, 4], have been presented in literature to overcome the shortage of visual servoing problems. An image based dynamic visual feedback controller was proposed [5]. The controller was designed by considering the robot's dynamics. The stability proof of the control system using Lyapunov analysis method was described. However, in this method, the controller separated the visual servoing task from the robot controller and could be implemented on a robot with a pre-designed computed torque controller.

The stability of the IBVS was analyzed in the presence of camera calibration errors [6]. In this work, the IBVS interaction matrix had a complex structure because of a large number of variables and their nonlinear formation in this matrix. Thus, no analytical solution is yet available for the pseudo-inverse of the interaction matrix. In PBVS, the error between the initial and the desired poses in the 3-D work space were computed and used for a control law for the positioning task of a robotic manipulator [7]. However, there is apparently no direct control over the feature points on the image plane. Consequently, it is possible that the feature points get out of the Vield of View (FOV). This causes the stability analysis problem to prove the stability of the IBVS and PBVS.

In this paper, a new visual servoing controller design method of the control system based on Barrier Lyapunov function (BLF) is proposed. To apply this method, an offline trajectory is developed to perform image-based visual servoing (IBVS) and position-based visual servoing (PBVS) tasks for a picking system. Two different control approaches i.e. the visual servoing controller with Barrier Lyapunov function (BLF) and the visual servoing controller with quadratic Lyapunov function (QLF) are presented. To do this task, the followings are done. Firstly, obtain the interaction matrix or image Jacobian by adopting a pinhole camera model. Secondly, obtain the camera velocity vector of an IBVS and the camera velocity vector of a PBVS. Thirdly, prove the asymptotic stability of proposed controller using Barrier Lyapunov function (BLF). Fourthly, compute through simulations the velocity vectors of the classical IBVS and PBVS controllers and proposed IBVS and PBVS controllers.

2 Basic Elements of Image Motion

The pinhole camera, Fig. 1, as the simplest camera model is used to build the relationship between the camera frame which moves through the scene and object frame. The transformation from the world frames to the camera frame is represented by $\mathbf{R}_c^w$ and $\mathbf{t}_c^w$. The $\mathbf{R}_c^w$ is a rotation matrix and $\mathbf{t}_c^w$ is a translation vector from the world frame to the camera frame, respectively. The coordinates (x, y) can be calculated as follows:

$$\begin{bmatrix} x \\ y \end{bmatrix} = \frac{f}{Z} \begin{bmatrix} X \\ Y \end{bmatrix} \quad (1)$$

where f denotes the focal length of the camera.

The first time derivative of Eq. (1) is obtained as follows:

$$\begin{bmatrix} \dot{x} \\ \dot{y} \end{bmatrix} = f\left[\left(\frac{\dot{X}}{Z} - \frac{X\dot{Z}}{Z^2}\right), \left(\frac{\dot{Y}}{Z} - \frac{Y\dot{Z}}{Z^2}\right)\right]^T \qquad (2)$$

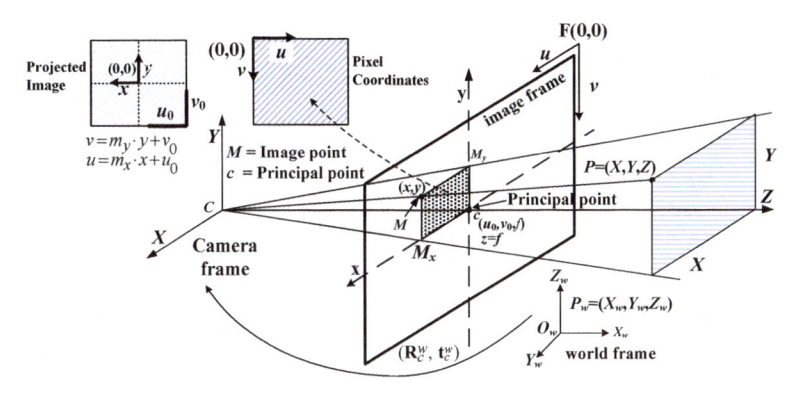

Fig. 1. Pinhole camera model

The velocity of the point with respect to the camera frame can be written as follows:

$$\dot{P} = -\omega \times P - v \qquad (3)$$

where $P = [X, Y, Z]^T$ is a pose vector of an object with respect to the camera frame, the symbol $\times$ is the cross product, $v = [v_x, v_y, v_z]^T$ and $\omega = [\omega_x, \omega_y, \omega_z]^T$ are the linear velocity vector and the angular velocity vector of the origin of the camera frame, respectively.

From Eq. (2), $x = X/Z$ and $y = Y/Z$, and for normalized coordinates (for $f = 1$), the linear velocity vector of a feature in the image frame can be written in the matrix form as follows:

$$\dot{m}_i = \begin{bmatrix} -\frac{1}{Z_i} & 0 & \frac{x_i}{Z_i} & x_i y_i & -(1+x_i^2) & y_i \\ 0 & -\frac{1}{Z_i} & \frac{y_i}{Z_i} & (1+y_i^2) & -x_i y_i & -x_i \end{bmatrix} v_c = L_i v_c \qquad (4)$$

where i is the number of a target feature point, Z_i is the depth of the ith target point frame, $m_i = (x_i, y_i)$ is a non-homogenous coordinate vector of the ith target feature point, $v_c = [v_x, v_x, v_x, \omega_x, \omega_x, \omega_x]^T \in \mathbb{R}^6$ is a velocity vector of the camera or the camera twist-screw. L_i is called *Interaction Matrix* or *Image Jacobian* of the ith target feature point and is the function of the state system vector (visual features) and of the ith 3D camera pose (X_i, Y_i, Z_i).

For n target points, the visual feature space vector can be built as follows:

$$\mathbf{s} = [\mathbf{m}_1^T \quad \mathbf{m}_2^T \quad \cdots \quad \mathbf{m}_n^T]^T \tag{5}$$

From Eq. (5), the dynamics of the system to be controlled is obtained as follows:

$$\dot{\mathbf{s}} = \mathbf{L}\mathbf{v}_c \tag{6}$$

where $\mathbf{v}_c$ is the camera velocity vector as the input vector to the picking system controller.

2.1 Image-Based Visual Servoing (IBVS)

In the image-based visual servoing, the control problem can be expressed in terms of image coordinates, and the error signal is defined directly in terms of image feature parameters. In the present application, the target feature is a square on the top surface of an image plane.

The aim of all vision-based control schemes is to compute a camera velocity vector $\mathbf{v}_c$ to minimize the following feature error vector

$$\mathbf{e}(t) = (\mathbf{s} - \mathbf{s}^*) \tag{7}$$

where $\mathbf{s}$ is the current feature vector, and $\mathbf{s}^*$ is a desired fixed feature vector of the image plane.

By using Eqs. (6) and (7), the relationship between camera velocity and the time derivative of the error is obtained as follows:

$$\dot{\mathbf{e}} = \dot{\mathbf{s}} - \dot{\mathbf{s}}^* = \mathbf{L}\mathbf{v}_c \tag{8}$$

To decrease exponential decoupling of the error ($\dot{\mathbf{e}} = -\lambda\mathbf{e}$) with $\lambda > 0$ as a proportional coefficient from Eq. (8), the camera velocity is obtained as follows:

$$\mathbf{v}_c = -\lambda \mathbf{L}^\dagger \mathbf{e} \tag{9}$$

where $\mathbf{L}^\dagger = (\mathbf{L}^T\mathbf{L})^{-1}\mathbf{L}^T \in \mathbb{R}^{6 \times k}$ is chosen as the pseudo-inverse of $\mathbf{L}$.

2.2 Position-Based Visual Servoing (PBVS)

In a PBVS system, the pose of the target with respect to the camera frame $\mathcal{P}_c = (\mathbf{t}_c^o, \mathbf{R}_c^o)$ is estimated. The relationship between the camera pose and target pose is shown in Fig. 2. Frame {C} is the current camera pose and frame {C*} is the frame representing the desired camera pose, and frame {C^o} is the frame representing the object camera pose. To determine the motion required to move the camera from its initial pose {C} to the desired pose {C*}, the desired relative pose with respect to the target $\mathcal{P}_c^* = (\mathbf{t}_{c^*}^o, \mathbf{R}_{c^*}^o)$ is specified.

By considering $\{\mathbf{C^o}\}$ as an object frame, $\{\mathbf{C}\}$ as a camera frame at the current position, and $\{\mathbf{C*}\}$ as a camera frame in the desired position, the feature error can be calculated as a function of these relative positions. A current camera pose with respect to the object frame is denoted by $\mathcal{P}_c = (\mathbf{t}_c^o, \mathbf{R}_c^o)$, where $\mathbf{t}_c^o$ is the position vector and $\mathbf{R}_c^o$ is the orientation expressed with the triplet of RPY Euler angles: $\mathbf{R}_c^o = [\varphi \ \theta \ \psi]^T$. The desired camera pose with respect to the object frame is denoted as $\mathcal{P}_c^* = (\mathbf{t}_{c^*}^o, \mathbf{R}_{c^*}^o)$.

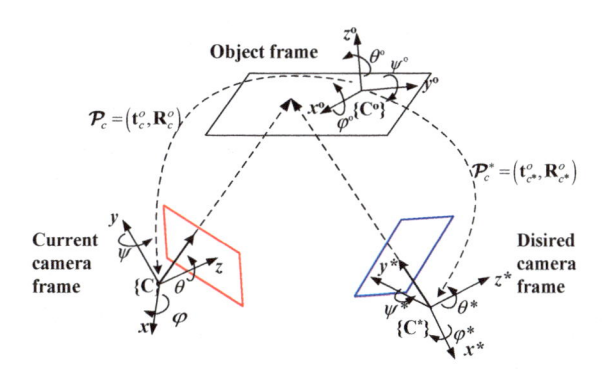

Fig. 2. Frames used in the modeling of PBVS.

If the visual feature space vector is defined with respect to the object frame $\{\mathbf{C^o}\}$, the error feature vector is given as

$$\mathbf{e} = \mathbf{s} - \mathbf{s}^* = \left[(\mathbf{t}_c^o - \mathbf{t}_{c^*}^o)^T, (\theta \mathbf{u}_c^o)^T\right] = \left[\mathbf{e}_v^T \ \mathbf{e}_\omega^T\right]^T \tag{10}$$

where $\theta \mathbf{u}_c^o$ is the angle/axis parameterization for the rotation, $\mathbf{e}_v$ and $\mathbf{e}_\omega$ are the position error vector and the rotation error vector, respectively.

By assuming exponential decrease of the feature error $\dot{\mathbf{e}}_v = -\lambda \mathbf{e}_\omega$, the linear camera velocity v_c^o and angular camera velocity ω_c^o with respect to the object frame are written as follows:

$$\begin{cases} v_c^o = -\lambda[(\mathbf{t}_{c^*}^o - \mathbf{t}_c^o) + (\lambda \theta \mathbf{u}_c^o) \times \mathbf{t}_c^o] \\ \omega_c^o = -\lambda \theta \mathbf{u}_c^o \end{cases} \tag{11}$$

If desired camera pose is defined with respect to the object frame, a feature error vector between the current feature vector and the desired feature vector is written as $\mathbf{e}(t) = \mathbf{s}(t) - \mathbf{s}^* = [(\mathbf{t}_{c^*}^o)^T, (\theta \mathbf{u}_{c^*}^o)]^T$.

The relation between the first time derivative of the feature error vector $\mathbf{e}$ and the camera velocity vector is given by:

$$\dot{\mathbf{e}} = \mathbf{L} v_{c^*}^o \tag{12}$$

3 Lyapunov Function

To assess the stability of the closed-loop visual servo systems, a clasical Lyapunov function is used as follows:

$$V = \frac{1}{2}\|e(t)\|^2 = \frac{1}{2}e^T e \geq 0 \tag{13}$$

By using Eq. (7)–Eq. (9), the time derivative of Eq. (13) can be obtained as follows:

$$\dot{V} = e^T \dot{e} = -\lambda e^T L L^\dagger e \tag{14}$$

Notice that the controlled system is locally stable if the following condition is satisfied:

$$\dot{V} \leq 0 \tag{15}$$

Thus, the dynamic system $-\lambda e^T L\hat{L}^\dagger e$ in Eq. (14) is globally stable since $L\hat{L}^\dagger \geq 0$. However, the matrix $L\hat{L}^\dagger$ (a $2k \times 2k$ matrix) is at most of rank 6, since the matrix L is a $2k \times 6$ matrix with $k > 3$. Therefore, the necessary condition $L\hat{L}^\dagger > 0$ for global asymptotic stability does not hold. Practically, there does not exist the exact value of L since an estimation of the depth information corresponding to the object is used. Therefore, it is assumed that a full-rank approximation $\hat{L}$ of the interaction matrix or an approximation $\hat{L}^\dagger$ of the matrix $L^\dagger$ is provided.

As shown earlier, even under the perfect condition of having the exact interaction matrix, global asymptotic stability cannot be established. However, the local asymptotic stability can be considered as follows:

A time derivative of a new error is given as follows:

$$\dot{e} = \hat{L}^\dagger \dot{e} + \frac{d\hat{L}^\dagger}{dt} e \tag{16}$$

From [7], it can be shown that

$$\frac{d\hat{L}^\dagger}{dt} e = O(e)v \tag{17}$$

From Eqs. (10), (12), and (17), the following can be obtained as $O(e) \to 0$ as $e \to 0$.

$$\dot{e} = \hat{L}^\dagger \dot{e} = -\lambda \hat{L}^\dagger L \hat{e} + O(e)v = -\lambda \hat{L}^\dagger L \hat{e} - \lambda O(e) \hat{L}^\dagger e \tag{18}$$

This system is known to be locally asymptotically stable in a small neighborhood of $e = 0$ if $\hat{L}^\dagger L > 0$.

4 Controller Design Using Barrier Lyapunov Function

The aim of this section is to develop a controller design method based on Barrier Lyapunov function (BLF) of the IBVS and the PBVS tasks for a picking system. The design method of the controller based on Barrier Lyapunov function (BLF) is applied to both IBVS and PBVS. The PBVS scheme is applied based on current camera pose with respect to the object frame and desired camera pose with respect to the object frame. Lyapunov analysis is used for classical IBVS and PBVS.

To determine the condition of the state variable for not violating the state variable constraint, a Barrier Lyapunov function is defined. The Barrier Lyapunov function is designed for every state of interval Ω_i with respect to all state variables error $\mathbf{e}_i$ Eq. (7) in domain $(-\bar{\mathbf{e}}_i, \bar{\mathbf{e}}_i)$ where $\bar{\mathbf{e}}_i$ is a boundary constant that satisfies $|\mathbf{e}_i| \leq \bar{\mathbf{e}}_i$.

$$\Omega_i(x_i(t)) : (-\bar{\mathbf{e}}_i, \bar{\mathbf{e}}_i) \to R^+ \, ; i = 1, \cdots, n \tag{19}$$

The Barrier Lyapunov function [8] candidate is defined as follows:

$$\mathbf{V} = \frac{1}{2} \sum_{i=1}^{n} \ln \left(\frac{\bar{\mathbf{e}}_i^2}{\bar{\mathbf{e}}_i^2 - \mathbf{e}_i^2} \right) \geq 0 \tag{20}$$

The first time derivative of $\mathbf{V}$ is satisfied as follows:

$$\dot{\mathbf{V}} = \sum_{i=1}^{n} \frac{\mathbf{e}_i \dot{\mathbf{e}}_i}{\bar{\mathbf{e}}_i^2 - \mathbf{e}_i^2} < 0 \tag{21}$$

The proof of Eq. (21) is shown in Appendix.

5 Simulation Result

The simulation results of the proposed schemes are shown in Figs. 3, 4 and 5. Figure 3 shows the classical and proposed IBVS controllers, Figs. 4 and 5 show the classical and proposed IBVS controllers, respectively. The start point is rectangle (□), and the goal is to position the camera to observe a sign (**x**) in the image. To do the simulation, four feature points in the coplanar case are chosen, and $\lambda = 0.5$. The image feature is defined from the four points forming the square as the desired pose as shown in Figs. 3 (a), 4(a), and 5(a). The initial camera pose is selected far away from the desired pose, particularly with regard to the rotational motions, which are known to be the most problematic for visual servoing. As inputs, a random set of target features and initial and desired final camera positions with respect to the target are chosen. For these comparisons, simulation runs for the 45 s with time gap of 0.1 s. The camera rotates 45 degrees around the camera's principal axis. The camera moves in translational and rotational. The camera is assumed to be fully controllable (i.e., 6 DOF). Figure 3 shows the simulation results for classical IBVS and proposed IBVS trajectories and errors generated. Figure 3 shows that the feature parameter error and the velocity are satisfied at 30 s and 10 s, respectively. Figure 4 shows the simulation results for classical PBVS and proposed PBVS trajectories of controller with respect to desired frame. Figure 4 shows that the feature parameter error and the velocity are satisfied at 40 s and 10 s, respectively. Figure 5 shows the simulation results for classical PBVS and proposed

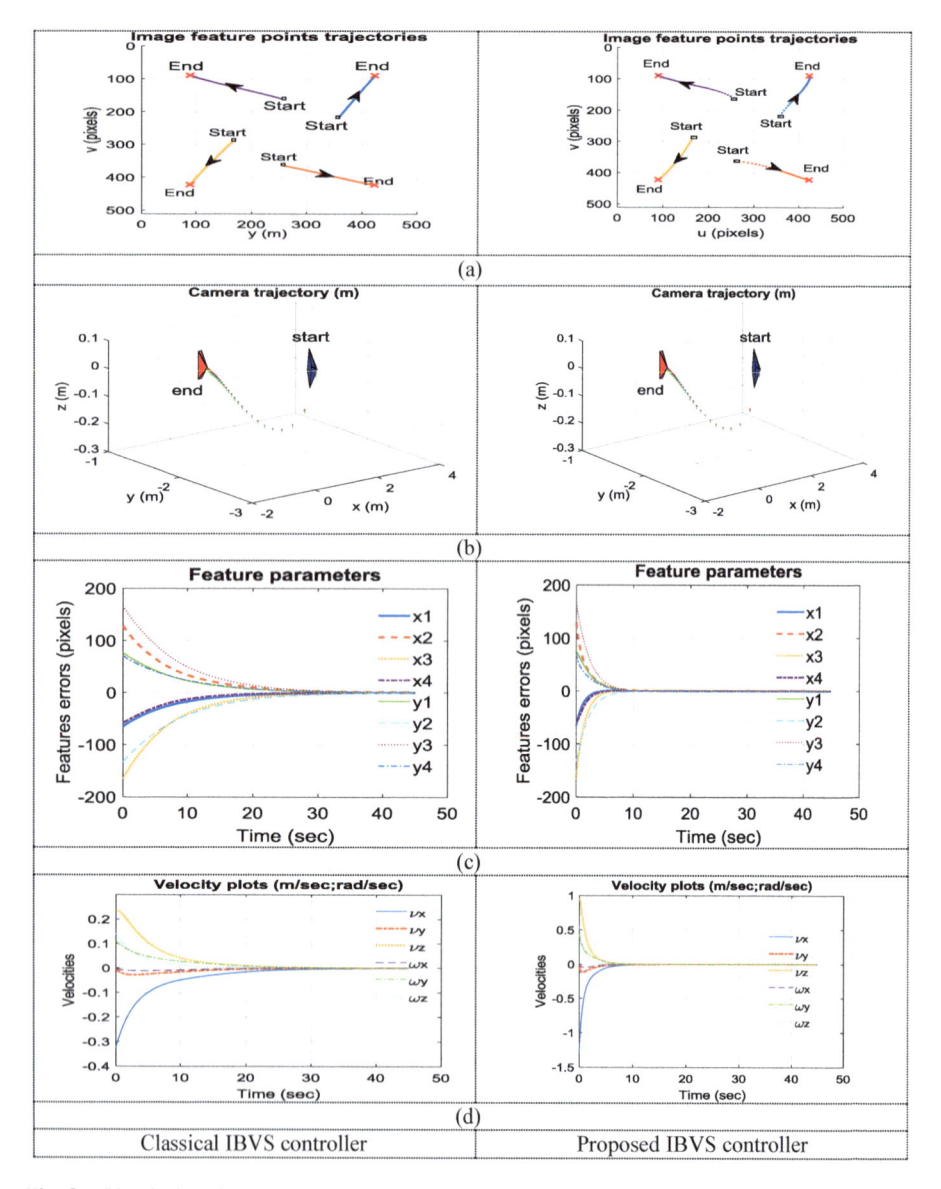

Fig. 3. Classical and proposed IBVS controller: (a) Feature trajectories; (b) Camera trajectory; (c) Feature parameter vector; (d) Camera pose vector.

PBVS trajectories of controller with respect to object frame. Figure 5 shows that the feature parameter error and the velocity are satisfied at 45 s and 10 s, respectively. Therefore, the proposed IBVS and PBVS controllers is better in camera position and velocity tracking performances than the classical IBVS and PBVS controllers.

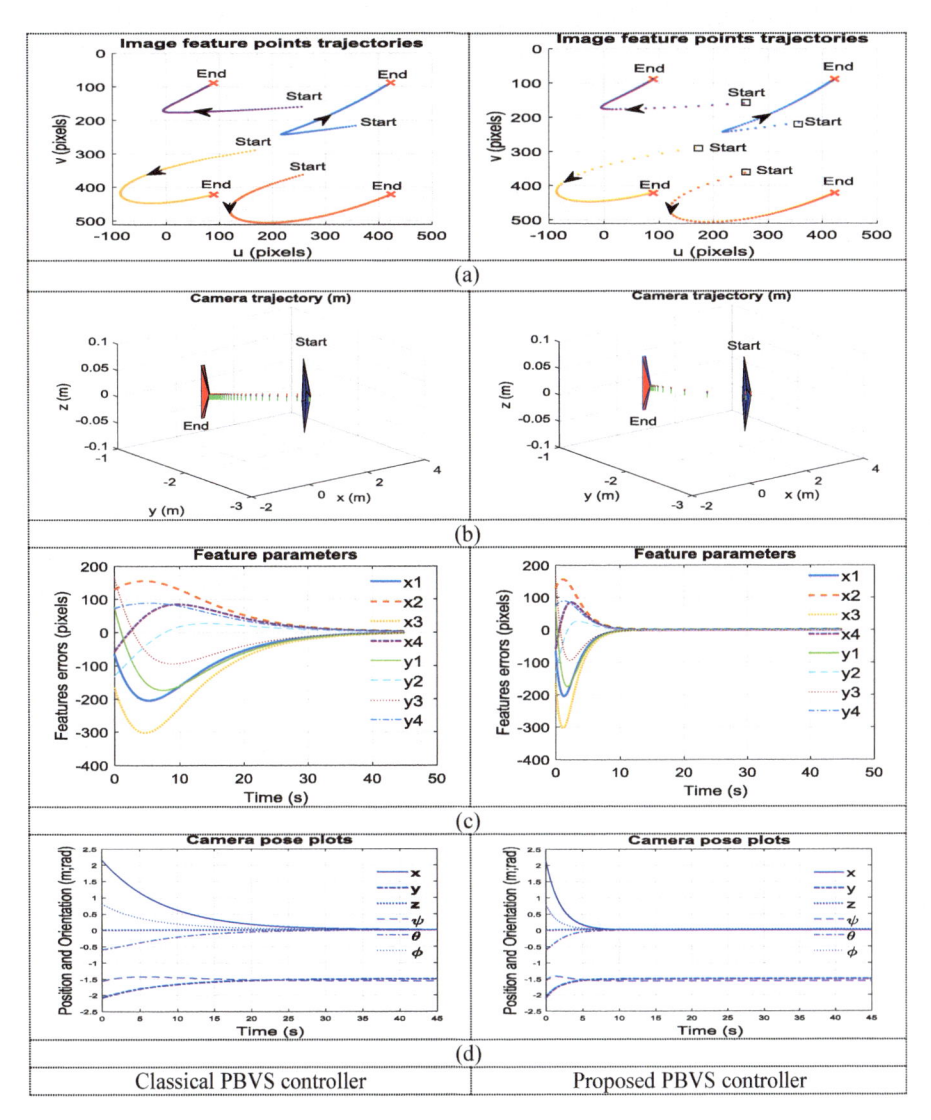

Fig. 4. Classical and proposed PBVS controller with respect to desired frame: (a) Feature trajectories; (b) Camera trajectory; (c) Feature parameter vector; (d) Camera pose vector.

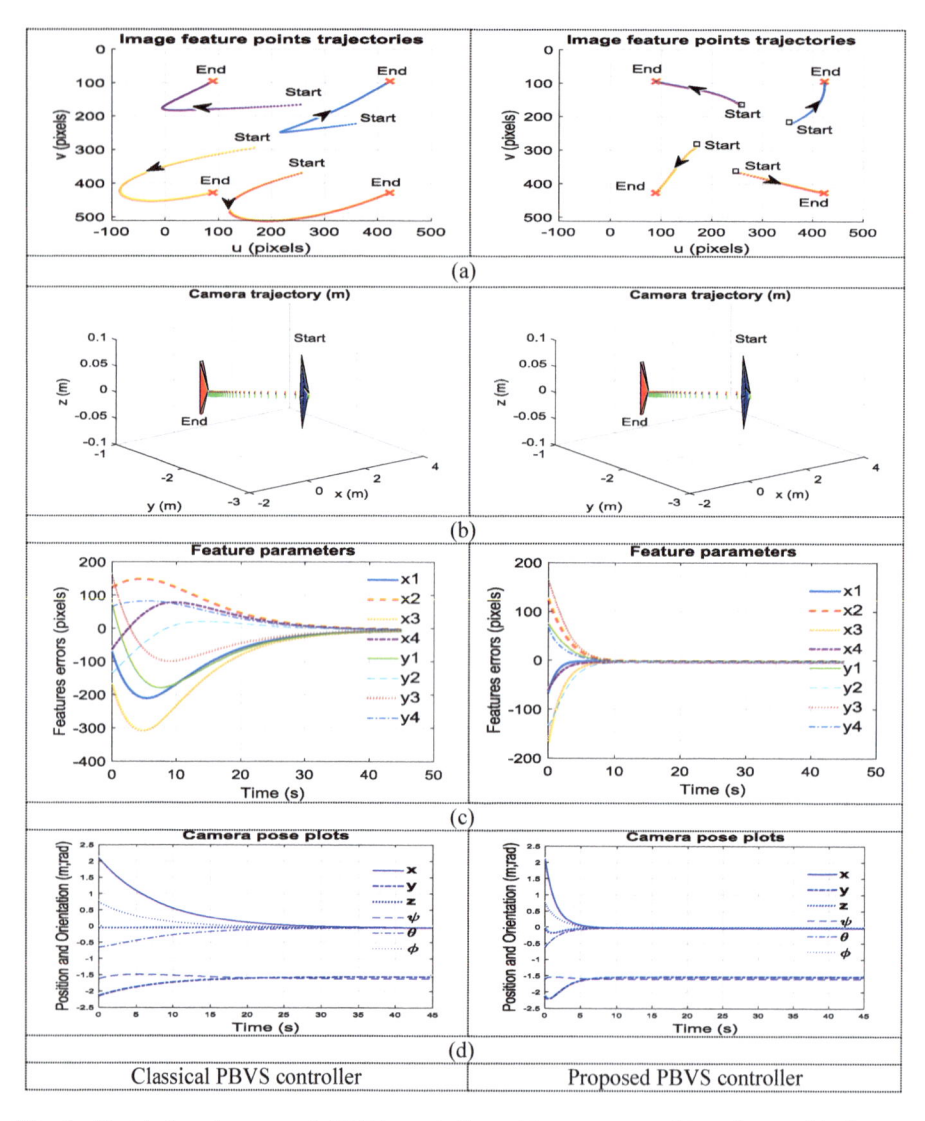

Fig. 5. Classical and proposed PBVS controller with respect to object frame: (a) Feature trajectories; (b) Camera trajectory; (c) Feature parameter vector; (d) Camera pose vector

6 Conclusion

This paper presented a visual servoing controller based on the Barrier Lyapunov function (BLF) schemes for IBVS and PBVS applied to a picking system. The asymptotic stability of the proposed controller was proven using Barrier Lyapunov function (BLF). Further, based on derived model i.e. the interaction matrix or image Jacobian, the camera velocity vector of an IBVS and the camera velocity vector of a PBVS, and the Lyapunov stability theory of the proposed controller using Barrier

Lyapunov function (BLF), the system could achieve the asymptotic stability. The superiority and potentials of the proposed method were shown through some simulation results. The reliability of the proposed method was verified compared to classical IBVS and PBVS techniques. The simulation results showed that the proposed method using Barrier Lyapunov function was better than the classical method using quadratic Lyapunov function.

Acknowledgment. This work was supported by the Materials and Components Technology Development Program of MOTIE/KEIT. [10063273, Development of Picking Tool for Logistic Robots to Automate Picking Process of Atypical Parcels]

Appendix

The Barrier Lyapunov function is designed for every state of interval Ω_i with respect to all state variables error $\mathbf{e}_i$ Eq. (7) in domain $(-\bar{\mathbf{e}}_i, \bar{\mathbf{e}}_i)$.

The Barrier Lyapunov function candidate is defined as follows:

$$\mathbf{V} = \frac{1}{2}\sum_{i=1}^{n}\mathbf{V}_i = \frac{1}{2}\sum_{i=1}^{n}\ln\left(\frac{\bar{\mathbf{e}}_i^2}{\bar{\mathbf{e}}_i^2 - \mathbf{e}_i^2}\right) \tag{A1}$$

where $\bar{\mathbf{e}}_i$ is a boundary constant that satisfies $|\mathbf{e}_i| \leq \bar{\mathbf{e}}_i$.

Since the number of target feature point is one, the first time derivative of $\mathbf{V}$ is given by

$$\dot{\mathbf{V}} = \sum_{i=1}^{n}\frac{1}{2}\left(\frac{\bar{\mathbf{e}}_i^2 - \mathbf{e}_i^2}{\bar{\mathbf{e}}^2}\right)\frac{-\bar{\mathbf{e}}_i^2(-2\mathbf{e}_i\dot{\mathbf{e}}_i)}{(\bar{\mathbf{e}}_i^2 - \mathbf{e}_i^2)^2} = \sum_{i=1}^{n}\frac{\mathbf{e}_i\dot{\mathbf{e}}_i}{\bar{\mathbf{e}}_i^2 - \mathbf{e}_i^2} \tag{A2}$$

Eq. (A2) can also be written in a compact form as

$$\dot{\mathbf{V}} = (\mathbf{B}_L \mathbf{E})^T \dot{\mathbf{E}} \tag{A3}$$

where $\mathbf{E} = [\mathbf{e}_1, \mathbf{e}_2, \cdots, \mathbf{e}_n]^T$ is an error vector and $\mathbf{B}_L \in \mathbb{R}^{n \times n}$ is diagonal matrix, defined as

$$\mathbf{B}_L = \text{diag}\left(\frac{1}{\bar{\mathbf{e}}_1^2 - \mathbf{e}_1^2}, \frac{1}{\bar{\mathbf{e}}_2^2 - \mathbf{e}_2^2}, \cdots, \frac{1}{\bar{\mathbf{e}}_n^2 - \mathbf{e}_n^2}\right) \tag{A4}$$

To fulfill the requirement of Lyapunov stability theory, the following is chosen as

$$\dot{\mathbf{e}}_i = -\frac{k_i \mathbf{e}_i}{(\bar{\mathbf{e}}_i^2 - \mathbf{e}_i^2)^{\lambda_i - 1}}, \quad i = 1, 2, \cdots, n. \tag{A5}$$

From Eq. (A5), Eq. (A3) becomes

$$\dot{V} = -E^T K E < 0 \tag{A6}$$

where $K \in \mathbb{R}^{n \times n}$ is positive definite diagonal matrix expressed as

$$K = \text{diag}\left(\frac{k_1}{(\bar{e}_1^2 - e_1^2)^{\lambda_1}}, \frac{k_2}{(\bar{e}_2^2 - e_2^2)^{\lambda_2}}, \cdots, \frac{k_n}{(\bar{e}_n^2 - e_n^2)^{\lambda_n}} \right) \tag{A7}$$

where k_i for $\forall i$ are positive controller gain and λ_i for $\forall i$ are positive even integer.

By using Eqs. (A3) and (A5), since $\dot{V} < 0$, $V(t) = \sum_{i=1}^{n} V_i < V_0, \forall t \geq 0$, where V_0 is the value of Lyapunov function Eq. (A1) at time $t = 0$. Since each individual element of $V(t)$ defined in Eq. (A1) is positive, the following relation holds

$$V = \frac{1}{2} \sum_{i=1}^{n} \ln\left(\frac{\bar{e}_i^2}{\bar{e}_i^2 - e_i^2} \right) < V_0 \text{ for } \forall i \tag{A8}$$

The ith Barrier Lyapunov function candidate is satisfied as follows:

$$V_i = \frac{1}{2} \ln\left(\frac{\bar{e}_i^2}{\bar{e}_i^2 - e_i^2} \right) \leq \frac{1}{2} \sum_{i=1}^{n} \ln\left(\frac{\bar{e}_i^2}{\bar{e}_i^2 - e_i^2} \right) < V_0 \text{ for } \forall i \tag{A9}$$

which can be simplified as

$$\bar{e}_i^2 e^{-2V_0} < (\bar{e}_i^2 - e_i^2) \tag{A10}$$

where e represents exponential.

Eq. (A10) implies that

$$|e_i| < \sqrt{\bar{e}_i^2 (1 - e^{2V_0})} \leq \sqrt{\bar{e}_i^2} = \bar{e}_i \tag{A11}$$

Since $V_0 \in [0, \infty)$, the following can be concluded

$$|e_i| < \bar{e}_i \text{ for } \forall i \tag{A12}$$

References

1. Chaumette, F., Hutchinson, S.: Visual servo control–part I: basic approaches. IEEE Robot. Autom. Mag. **13**, 82–90 (2006)
2. Chaumette, F.: Potential problems of stability and convergence in image-based and position-based visual servoing. In: The Confluence of Vision and Control, LNCIS Series, vol. 237, pp. 66–78. Springer, London (1998)

3. Pandya, H., Krishna, K.M., Jawahar, C.V.: Discriminative learning based visual servoing across object instances. In: IEEE International Conference on Robotics and Automation, Stockholm, Sweden, pp. 447–454 (2016)
4. Deng, L., Janabi, S.F., Wilson, W.: Hybrid motion control and planning strategies for visual servoing. IEEE Trans. Industr. Electron. **52**, 1024–1040 (2005)
5. Kawai, H., Murao, T., Fujita, M.: Image-based dynamic visual feedback control via passivity approach. In: IEEE International Symposium on Intelligent Control, Munich, pp. 740–745 (2006)
6. Malis, E.: Visual servoing invariant to changes in camera-intrinsic parameters. IEEE Trans. Robot. Autom. **20**(1), 72–81 (2004)
7. Wilson, W., Hulls, C., Bell, G.: Relative end-effector control using Cartesian position-based visual servoing. IEEE Trans. Robot. Autom. **12**(5), 684–696 (1996)
8. Tee, K.P., Ge, S.S., Tay, E.H.: Barrier Lyapunov functions for the control of output-constrained nonlinear systems. Automatica **45**(4), 918–927 (2009)

Designing a PID Controller for Ship Autopilot System

Dinh Due Vo[1], Viet Anh Pham[2], Phung Hung Nguyen[2], and Duy Anh Nguyen[1(✉)]

[1] Ho Chi Minh University of Technology, Ho Chi Minh City, Vietnam
duyanhnguyen@hcmut.edu.vn
[2] Ho Chi Minh University of Transport, Ho Chi Minh City, Vietnam

Abstract. Many means of transportation in these days have autopilot system – a modern system made from the ultimate growth of technology that can handle a partial or a complete control and help the drivers get relaxed and restored for a long drive, especially in aeronautics and maritime in which a trip usually lasts for plenty of hours and a normal man cannot control the whole system for such long duration, due to mental and physical fitness issues. Thus, in this paper, we are going to present a popular controller in autopilot system – named "PID controller" – including its fundamentals, its mathematical models and its performance for both theoretical and practical result. In experimentation, we used the controller to navigate a ship model along a predefined tour and tested at a pool in order to compare and evaluate, as well as promote solutions to deal with the drawbacks.

Keywords: Ship autopilot system · PID controller · Adaptive controller · Nomoto model

1 Introduction

Major factors that make it difficult to control ships comprise the highly nonlinear, time-varying dynamic behavior of the ship, uncertainties in hydrodynamic coefficients and disturbances by ocean currents. Therefore, it is highly desirable to have an autopilot system that could update and learn itself when the control performance is degraded during operation due to changes in the dynamics of the AUV and its environment.

The main motivation of this research is to remove the necessity of mathematical ship model by using an "intelligent" control technique. In the control systems where the controlled plants are highly nonlinear and external disturbances are highly nonlinear uncertainties, intelligent control techniques are useful. Especially, the NN (Neural Network) control, one of the artificial intelligence branches, has grown rapidly in recent years. Many NN control systems of different structures have been proposed and widely applied in a range of technical practices.

In this research, a hybrid neural adaptive control scheme which can perform the heading and depth control of a ship. The aim is to take advantage of the learning ability of NN and to derive a NN-based control algorithm which is independent of the exact mathematical model of the ship. Furthermore, it is not necessary to estimate the bias term representing slowly-varying external environmental forces and moments. A conventional PD-controller for nonlinear model [2] is modified and combined with the ANNAI controller to design a hybrid neural adaptive controller. In this proposed controller, PD-controller provides an approximate control and ANNAI controller with online training ability is introduced to improve the system performance.

2 The Kinematic and Kinetic Equations

In order to construct kinematic and kinetic equations for the ship, we can assume the ship in space as a 6-degree independent (6 DOF) robot, and assume the ship as a solid. In addition, the rotation of the Earth does not affect the acceleration of the central of mass.

2.1 Reference Frames [1]

According to SNAME, 1950, the coordinate axes used to represent the movement of vehicles on the ship include the North-East-Down (NED) coordinate and body-fixed reference frame as shown in Table 1 and Fig. 1.

ECI: The Earth-centered inertial (ECI) frame $\{i\} = (xi, yi, zi)$ is an inertial frame for terrestrial navigation, that is a nonaccelerating reference frame in which Newton's laws of motion apply. This includes inertial navigation systems. The origin of $\{i\}$ is located at the centroid of the Earth with axes as shown in Fig. 1.

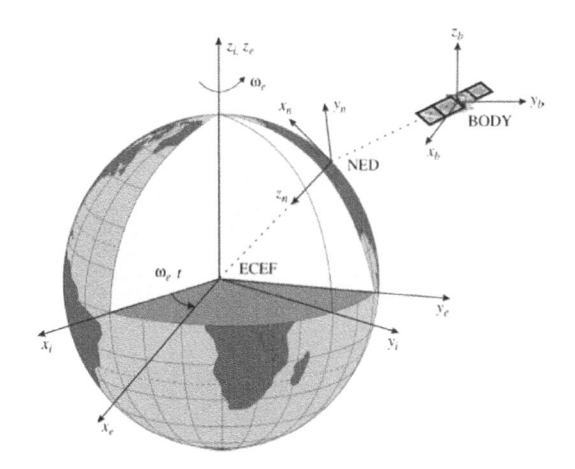

Fig. 1. Reference frames

ECEF: The Earth-centered Earth-fixed (ECEF) reference frame $\{e\} = (xe, ye, ze)$ has its origin Oe fixed to the center of the Earth but the axes rotate relative to the inertial frame ECI, which is fixed in space. The angular rate of rotation is $\omega_e = 7.2921 \times 10^{-5}$ rad/s. For marine craft moving at relatively low speed, the Earth rotation can be neglected and hence $\{e\}$ can be considered to be inertial. Drifting ships, however, should not neglect the Earth rotation. The coordinate system {e} is usually used for global guidance, navigation and control, for instance to describe the motion and location of ships in transit between different continents.

NED: The North-East-Down (NED) coordinate system $\{n\} = (xn, yn, zn)$ with origin On is defined relative to the Earth's reference. This is the coordinate system we refer to in our everyday life. It is usually defined as the tangent plane on the surface of the Earth moving with the craft, but with axes pointing in different directions than the body-fixed axes of the craft. For this system the x axis points towards true North, the y axis points towards East while the z axis points downwards normal to the Earth's surface. The location of {n} relative to {e} is determined by using two angles l and μ denoting the longitude and latitude, respectively.

BODY: The body-fixed reference frame $\{b\} = (xb, yb, zb)$ with origin Ob is a moving coordinate frame that is fixed to the craft. The position and orientation of the craft are described relative to the inertial reference frame (approximated by {e} or {n} for marine craft) while the linear and angular velocities of the craft should be expressed in the body-fixed coordinate system (Fig. 2).

Table 1. Notation of SNAME (1950) for ship

DOF	Motion	Forces and moments	Linear and angular velocities	Positions and Euler angles
1	Motions in the x direction (surge)	X	u	x
2	Motions in the y direction (sway)	Y	v	y
3	Motions in the z direction (heave)	Z	w	z
4	Rotation about the x axis (roll)	K	p	ϕ
5	Rotation about the y axis (pitch)	M	q	θ
6	Rotation about the z axis (yaw)	N	r	ψ

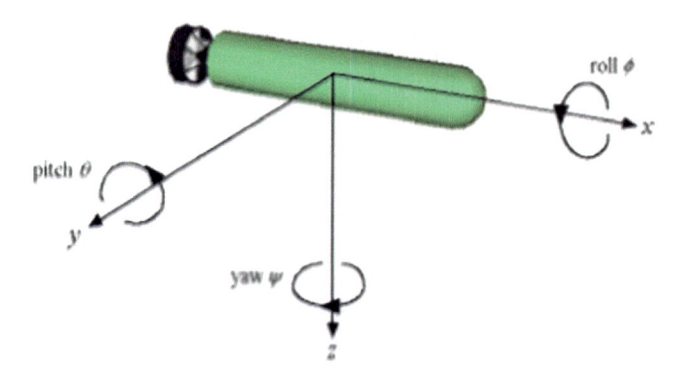

Fig. 2. Six motion of a ship

For ship the following notation will be adopted for vectors in the coordinate systems {b}, {e} and {n}:

$v_{b/n}^e$: Linear velocity of the point O_b with respect to {n} expressed in {e}.
$\omega_{n/e}^b$: Angular velocity of {n} with respect to {e} expressed in {b}.
f_b^n: Force with line of action through the point O_b expressed in {n}.
m_b^n: Moment about the point O_b expressed in {n}.
Θ_{nb}: Euler angles between {n} and {b}.

Hence, the general motion of a ship in 6 DOF with O_b as coordinate origin is described by the following vectors:

$$\eta = \begin{bmatrix} p_{b/n}^n \\ \Theta_{nb} \end{bmatrix}; v = \begin{bmatrix} v_{b/n}^n \\ \omega_{b/n}^b \end{bmatrix}; \tau = \begin{bmatrix} f_b^b \\ m_b^b \end{bmatrix} \quad (1)$$

2.2 Ship Kinematic Equations

The direction of the frame {b} is determined in relation to the reference frame {n} by three angles roll (ϕ), pitch (θ) and yaw (ψ).

$$\begin{bmatrix} \dot{p}_{b/n}^n \\ \dot{\Theta}_{nb} \end{bmatrix} = \begin{bmatrix} R_b^n(\Theta_{nb}) & 0_{3\times 3} \\ 0_{3\times 3} & T_\Theta(\Theta_{nb}) \end{bmatrix} \begin{bmatrix} v_{b/n}^n \\ \omega_{b/n}^b \end{bmatrix} \quad (2)$$

The kinetic equations for ship can be represented by Fossen (1994) [1]:

$$M_{RB}\dot{v} + C_{RB}(v)v = \tau_{RB} \quad (3)$$

Where:

$v = \begin{bmatrix} v_{b/n}^b & \omega_{b/n}^b \end{bmatrix} = [u, v, w, p, q, r]^T$ is the generalized velocity vector expressed in {b}. $\tau_{RB} = \begin{bmatrix} f_b^b & m_b^b \end{bmatrix} = [X, Y, Z, K, M, N]^T$ is a generalized vector of external forces and moments expressed in {b}. It can be analyzed into: $\tau_{RB} = \tau_H + \tau_{wind,wave} + \tau$.

M_{RB} is the rigid-body mass matrix.
$C_{RB}(v)$ is the rigid-body Coriolis and centripetal matrix.

3 PID Controller

3.1 Design of PID Controller with Acceleration Feedback

To design a PID controller with acceleration feedback to control ship's heading angle, we will first use the second order mass-damper-spring system as the control object with the differential equation as follows:

$$m\ddot{x} + d\dot{x} + kx = \tau + w \tag{4}$$

where:
- τ: The control signal needs to be determined.
- w: Environment disturbances.

We calculate K_p and K_d as below:

$$K_P = (m + K_m)\omega_n^2 - k \tag{5}$$

$$K_D = 2\zeta\omega_n(m + K_m) - d \tag{6}$$

According to [1], we can calculate K_i as follow:

$$K_I = \frac{\omega_n}{10}K_P = \frac{\omega_n}{10}\left[(m + K_m)\omega_n^2 - k\right] \tag{7}$$

According to [1], the natural frequency ω_n can be calculated based on system bandwidth ω_n as below:

$$\omega_b = \omega_n\sqrt{1 - 2\zeta^2 + \sqrt{4\zeta^4 - 4\zeta^2 + 2}} \tag{8}$$

3.2 PID Controller with Acceleration Feedback for the Ship Autopilot

On the first-order Nomoto model (1957) we obtain the characteristic equation of the system as follows:

$$T\ddot{\psi} + \dot{\psi} = K\delta \tag{9}$$

where ψ is heading angle, δ is rudder angle.

$$\begin{aligned} K_m &\geq 0 \\ K_P &= \frac{T + KK_m}{K}\omega_n^2 > 0 \\ K_D &= \frac{T + KK_m}{K}2\zeta\omega_n - \frac{1}{K} > 0 \\ K_I &= \frac{T + KK_m}{10K}\omega_n^3 \end{aligned} \tag{10}$$

The coefficients of this PID controller will be applied to ship autopilot simulation in simulation section to test the response of our system.

4 Simulation of the Ship Autopilot

4.1 Simulation Conditions

Figure 3 shows the actual trajectory of the ship with the automatic steering system using a PID controller.

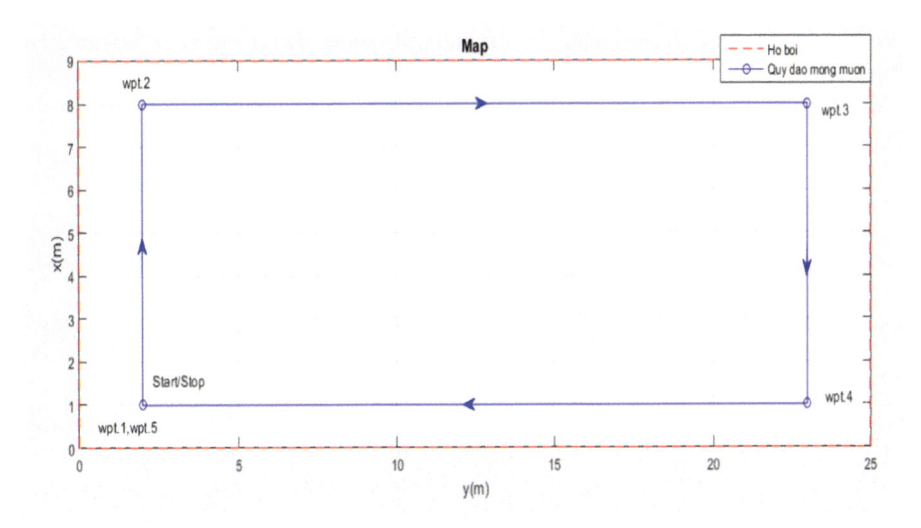

Fig. 3. Ship's desired trajectory

Fig. 4. Actual experiment conditions of the ship

Figure 4 illustrates ship's desired trajectory which represent for test location (a swimming pool 9×25 m (Fig. 5) at Ho Chi Minh City University of Transport). So, to ensure consistency between simulation and experiment, the ship is controlled running to track the profile of the swimming pool as depicted in Fig. 5. The desired trajectory of the ship is described in five waypoints and the ship will turn three times with the steer angle values as 90° - 90° - 90°. At t = 0 s the ship is placed along the width of the swimming pool corresponding to the initial ship's heading angle of 10°.

During the simulation, to check the response of the controller, three types of disturbance are added as follows. First, wave disturbance described by PM spectrum with the dominant frequency of $\omega_0 = 0.60625$, relative damping factor $\xi = 0.3$, and $K_\omega = 0.1979$. Next, the noise of ocean current would be constant and the ship's heading angle would be deflected 10°. Finally, the third disturbance is measurement error which is a random number. The rudder angle is bounded from $-20°$ to $20°$. The rudder velocity is also bounded from $-5°/s$ to $5°/s$. The sampling time is 0.1 s and ship velocity are constant 0.4 m/s.

4.2 Simulation of the Ship Autopilot Using the PID Controller

From Fig. 5, when going through waypoint 2, the ship tends to be deflected out of the desired trajectory. In other words, the ship's cross tracking error is very large when going through waypoint 2 with the largest error is -0.3395 m at point B in Figs. 7 and 8. After going through point B, the ship tends to oscillate around the desired trajectory and the cross-tracking error also decreases, but it takes a large amount of time to correct the cross-tracking error toward zero. It can be explained as the response of the PID controller is too slow. Similarly, when going through waypoint 3, the ship also tends to be deflected out of desired trajectory. On the other hand, when going through waypoint 3, the ship was not able to follow the desired trajectory created by waypoint 3 and waypoint 4, then it got into the next limit circle at point D as shown in Fig. 5. Therefore, the have to start turning to follow the new trajectory created by waypoint 4 and waypoint 5.

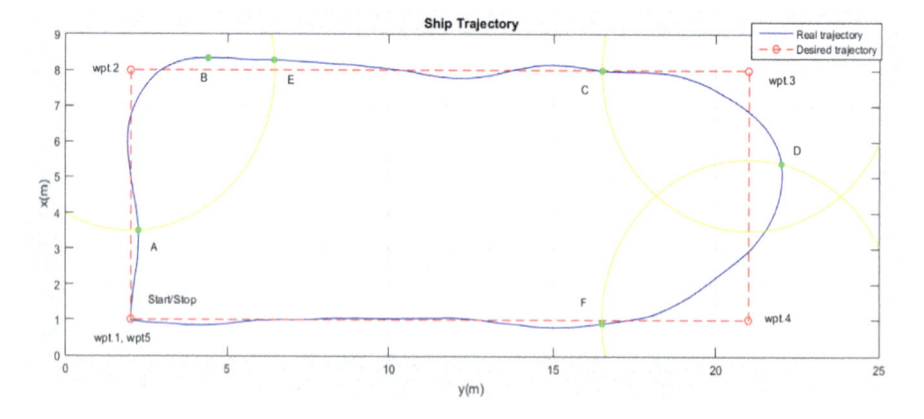

Fig. 5. Actual ship's trajectory with the PID controller

At time t = 0 s, the ship starts from waypoint 1, and the ship goes to point A at time t = 6.3 s, as shown in Figs. 5 and 6. At this time, the cross-tracking error is 0.2308 m. After that, the cross-tracking error jumps to 4.495 m. It can be explained that at that time the ship enters the limitation circle centered at waypoint 2, whose radius equal to three times the ship length ($R = 3L_{ship} = 4.5$ m). Then the Waypoint Selection system will detect this and instruct the Navigation system to guide the ship to follow the next part of desired trajectory built by waypoints 2 and waypoint 3. Thus, ship's cross tracking error will be determined as the distance from the ship to the new part of desired trajectory. As a result, the cross-tracking error increases dramatically at time t = 6.3 s. The jump on cross tracking error at t = 52.6 s, 69.12 s will also be interpreted similarly.

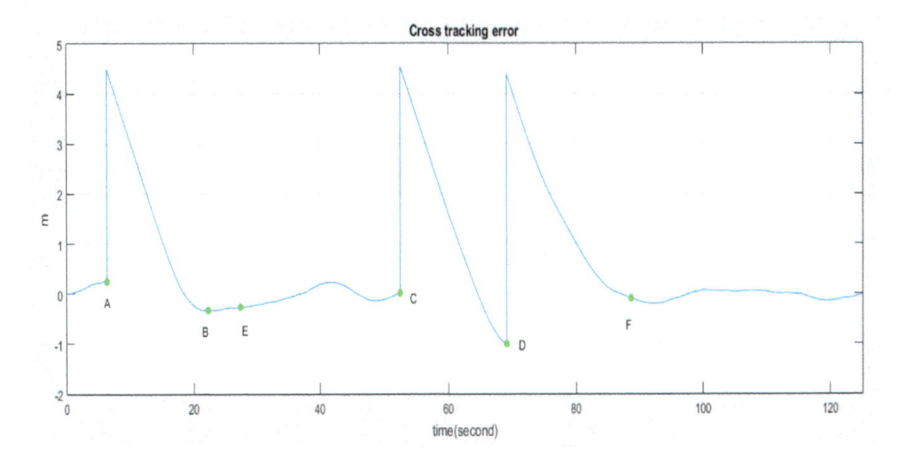

Fig. 6. Cross tracking error with the PID controller

As shown in Fig. 7, we find that at times t = 6.3 s, 52.6 s, 69.12 s corresponding to points A, C, D. This is also explained in by same way because at these times the ship enters the limited circle, so the ship must divert to follow the new straight trajectory. The angle between the bow angle and the angle of the new orbital line so that the angle of the bow will have big jump.

In addition, it can be seen from Fig. 7 that after the ship leaves the center circle at the waypoint 2 point, the bow angle error in the stable region from point E to point C is shown in Figs. 5 and 7 does not move at 0°, the angle of the bow is fluctuated due to interference from the environment and the measurement error of the compass so the angle of the bow cannot be 0°, the maximum angular error in the segment this is 13°. Similarly, when the ship leaves the center circle at the waypoint point 4, the bow angle error in the stable region from point F to waypoint 5 as on Figs. 5 and 7 cannot be 0°, invalid number corner to the maximum ship in this section is 7.5°.

Figure 8 shows the steering angle of the ship over time. From Fig. 8, the rudder angle will gradually advance to 0 after the boat passes waypoint 2 and waypoint 4. However, the steering wheel angle is still slightly oscillated after passing waypoint 2

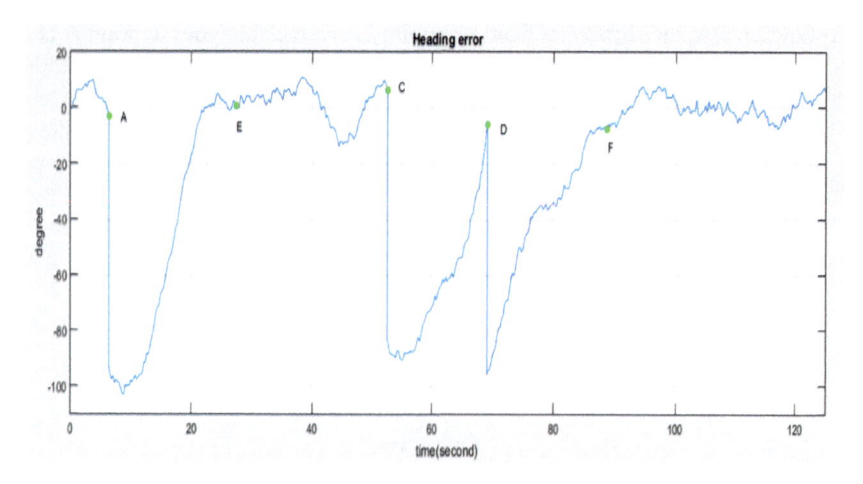

Fig. 7. Heading error with the PID controller

and waypoint 4. Due to the effects of environmental noise, the angle of the rudder will be shifted at an angle to compensate for the noise so that the vessel can follow the desired trajectory and help maintain the bow angle.

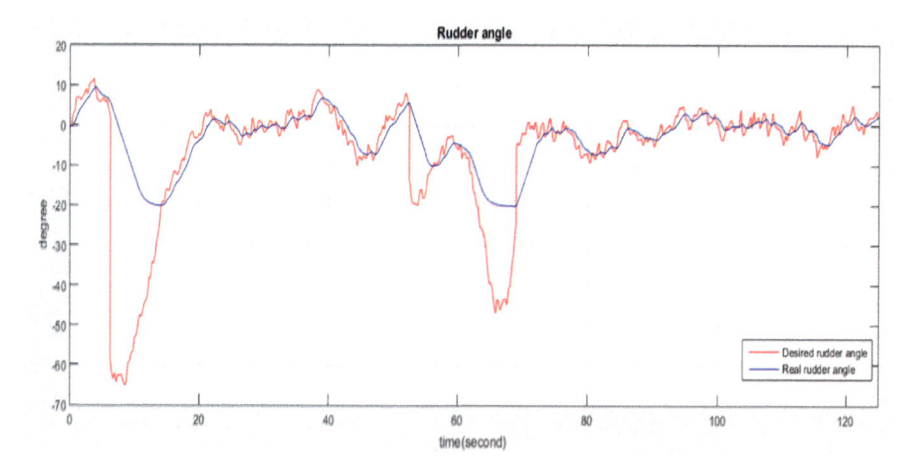

Fig. 8. Rudder angle with the PID controller

Figure 9 displays the bow angle before and after passing through the Kalman filter over time. Also, this Figure displays the bow angle after passing through the Kalman filter compared to the angle of the bow returned from the HMC 5883L compass sensor, the underside was partially removed. The smoothness of the bow angle signal after passing through the Kalman filter is still not high and is still undulating. This can be explained by the fact that the selection of two values of Q and R of the Kalman filter has not yet reached an optimal value.

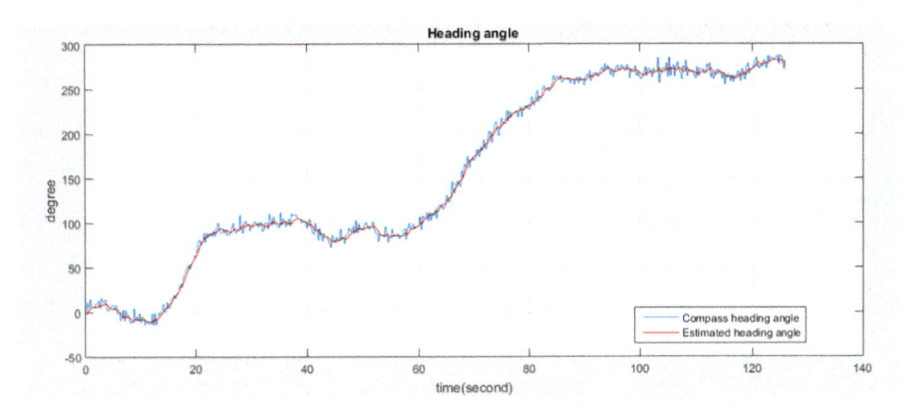

Fig. 9. Ship heading angle with the PID controller

5 Conclusion

To sum up, in this paper, we have successfully applied first-order Nomoto model and PID controller to build a mathematical model for ship and to experiment in real conditions. Obviously, we hardly expect that the theoretical and practical result would be the same, however the drawback of PID controller is that the experimental trajectory is still so far different from the computed one, as shown in Fig. 6. It is explained that the response of PID controller is not quick enough for a smooth turn while moving fast, since the gradient of velocity changes dramatically.

In future, we suggest using another controller with faster response, such as adaptive controller, as it could adapt its working parameter continuously and reduce the rate of changing states, in comparison with PID controller. Moreover, we attempt to apply and test the controller on some different means of vehicle, for instance AUV (Autonomous Underwater Vehicle).

References

1. Fossen, T.I.: Handbook of Marine Craft Hydrodynamics and Motion Control (2011)
2. Velagic, J., Vnkic, Z., Omerdic, E.: Adaptive fuzzy ship autopilot for track-keeping. Control Eng. Pract. **11**, 433–443 (2003)
3. Lee, S.-D., Tzeng, C.-Y., Chen, B.-J.: Design and experiment of a fuzzy PID track-keeping ship autopilot. Ocean Engineering (2014)
4. Fossen, T.I.: Marine control systems. Norwegian University of Science and Technology Trondheim, Norway (2002)
5. Fossen, T.I.: Handbook of marine craft hydrodynamics and motion control. Norwegian University of Science and Technology Trondheim, Norway (2011)
6. Omerdic, E., Roberts, G.N., Vukic, Z.: A fuzzy track-keeping autopilot for ship steering. J. Mar. Eng. Technol. **2**(1), 23–35 (2003)
7. Yunsheng, F., Xiaojie, S., Guofeng, W., Chen, G.: On fuzzy self-adaptive PID control for USV course. In: Proceedings of the 34th Chinese Control Conference, pp. 8472–8478 (2015)

The Rotor Initial Position Determination of the Hi-Speed Switch-Reluctance Electrical Generator for the Steam-Microturbine

Pavel G. Kolpakhchyan[1(✉)], Vladimir I. Parshukov[2], Boris N. Lobov[3], Nikolay N. Efimov[3], and Vadim V. Kopitza[2]

[1] Rostov State Transport University, Rostov-on-Don, Russian Federation
kolpahchyan@mail.ru
[2] ETC "DonEnergoMash" Ltd., Rostov-on-Don, Russian Federation
[3] Platov South-Russian State Polytechnic University (NPI), Novocherkassk, Russian Federation

Abstract. The article focuses on the control of the hi-speed switch-reluctance electrical machine, which works in conjunction with a steam microturbine. The use of sensorless control is efficient in the case under consideration. We described the use of test pulses in the electric generator phases to determine the rotor position. We study the matters of rotor position determination, the method accuracy and its implementation opportunity. The results of mathematical modeling of the initial rotor position determination are given.

Keywords: Electric generator · Switch-reluctance electrical machine · Sensorless control · Steam-turbine power system

1 Introduction

Increasing the efficiency of renewable energy sources is possible when using the distributed energy systems. The combined power units allows providing thermal and electrical energy to consumers. The renewable sources energy (solar thermal panels, geothermal sources), the energy from organic fuel (biogas) or elimination of waste is mostly the thermal energy. Therefore, the creation of an equipment for converting thermal energy into electrical energy is an important task. The equipment with the power from 1–5 kW to 50–100 kW is the most highly sought [1–3]. In this case, they use the microturbines operating on a saturated vapor. Such systems they also use to utilize excess heat energy and convert it to electric power at enterprises or in the urban economy.

The efficiency of a microturbine is possible at high rotational speeds. It is one of the problems of implementing these power-generating systems.

The use of a traditional electrical machine in conjunction with a microturbine requires the use of a reducer. Therefore, the use of the gearless generators directly connected to the turbine is useful [4]. The creation of a high-speed electric generator with a power of up to 100 kW for the use in the distributed and autonomous power systems is actual task.

In the power range under consideration in power systems based on steam microturbines they use the steam with the following parameters: the temperature is from 180 °C to 300 °C, the pressure is 3 atm to 15 atm. With these parameters of the steam, turbine can effectively operate at a speed of 12,000 rpm and higher. At such speeds of rotation, brushless electrical machines are used as electric generators. The use of the electric machines with the permanent magnets on the rotor is effective and allows us to obtain the high specific parameters. The high cost of permanent magnets, the complex design of the rotor and the need to cool the magnets. The reactive reluctance-flux electrical machine has the simple construction and the high manufacturability and can be used as an alternative to the high-speed generators with the permanent magnets on the rotor.

The authors participated in the development of a reactive reluctance-flux electrical generator working with a steam microturbine that is described in the article [5]. The need of the using the rotor position sensorless control is one of the challenge when developing such an electrical machine [6–12]. The article [5] describes one of the variant of the sensorless control adapted for the use in the high-speed generator. In this article we describe the development of the sensorless control method.

2 Problem Formulation

The most common method of the switch-reluctance electrical machine control is the feeding of its phases by unipolar current pulses [6–8,12,13]. The mutual position of the rotor and stator determines moment of the current pulse and its duration. The deviation from the calculated position of the rotor when the current pulse is applied leads to a significant deterioration in performance of the electric machine or to its nonoperability.

Therefore, to ensure the work of the switch-reluctance electrical machine we need to know the rotor position relative to the stator. The use of rotor position sensors for this purpose for high-speed switched-reluctance electrical machine is difficult. Therefor for such electric machine, they usually determine the rotor position by indirect methods [5,11,12].

The test pulse method does not give good results for a high-speed power generator. We considered the method for rotor position determining based on the phase current shape analysis, in the article [5]. The application of this method requires information about the rotor initial position. The start of the high-speed power generator working with steam microturbine has its features. The start of the turbine is carried out without a load; the generation process begins after the rotor is already rotating at a speed close to the working one. The capacitor in the DC link of the converter is charged from a third-party source to a voltage of 20–25% of the nominal value to start the generator. The energy reserve in the capacitors is minimal and sufficient to create one or two current pulses in the phases of an electric generator sufficient to start the generation process with the correct timing of the pulses. A significant error in the rotor initial position determining can lead to a generation failure. Therefore, a method for the initial position determining of the hi-speed switch-reluctance electrical machine rotor working in conjunction with a steam microturbine must be implemented.

3 Method for the Rotor Initial Position Determination

The use of short test voltage pulses supplied to the stator windings is the most widespread method of the reactive reluctance-flux electrical machine rotor position determining [14,15]. The phase inductance of the switch-reluctance electrical machine uniquely depends on the position of the rotor relative to the stator. The phase inductance has a maximum in a aligned position; it has a minimum in the unaligned position. The amplitude of the test pulse current has also a minimum and a maximum, respectively. They fix the moments of the rotor passage through the coordinated or mismatched position, tracking the current pulses amplitude extremes.

The application of the described method is difficult in case of the start of the hi-speed switch-reluctance electrical generator without a power supply with a reduced voltage charge of the capacitor in the DC link of the electric power converter. The reduced voltage and limited energy supply in the condenser complicate the current pulses maxima and minima recording. The control system requires information about the state of the rotor and its rotational frequency when starting an electric generator with a rotating rotor. To determine the rotational speed with acceptable accuracy we must track the rotor position for five to six rotor passes through the aligned position.

To prevent a significant voltage reduction on the capacitor, the test pulses should have the minimum possible duration. In this case, the current pulses amplitude may not be sufficient to determine the maxima and minima and the moments of rotor passes through the aligned and unaligned position. Therefore, the use of the discussed principle of rotors position determining is difficult for the start of the hi-speed reactive reluctance-flux electrical generator with a rotating rotor and a reduced voltage on the capacitor.

To determine the starting position and rotor speed when starting the hi-speed switch-reluctance electrical generator we propose the using of test pulses method with some changes. The method reside in the fact that when the rotor passes a coordinated position with one of the phases of three-phases reluctance-flux electrical machine, two other phases are in the same magnetic conditions, their inductances will be equal and, accordingly, the test pulses amplitudes in these phases will be equal. If the test pulses are simultaneously fed into all phases of the switch-reluctance electrical machine, the fixing the equality of the test pulses amplitudes in other phases is sufficient to determine the coordinated position of one of the phases. The advantage of this method is the fact that is no need for determining the maxima or minima of the current pulses amplitude when the rotor passes through the aligned or unaligned position. When the rotor is in an intermediate position relative to the phase, the inductance of this phase changes rapidly as the rotor moves. It makes much easier to record the change in the current pulses amplitude with short test voltage pulses.

The phase inductance change of switch-reluctance electrical machine can also occur because of saturation of the magnetic system. Therefore, to increase the accuracy of the proposed method, the current pulses must not saturate the magnetic system.

4 An Example of Using the Proposed Method of the Hi-Speed Switch-Reluctance Electrical Generator Initial Position Determination

The hi-speed switch-reluctance electrical generator for the steam microturbine (power and rotational speed in the rated mode are 35 kW and 12,000 rpm, respectively) was considered in the article [5] as an example of the use of the proposed method. The Fig. 1 shows the sketch of the active part of this high-speed electric generator. It has a performance of 6/4 (6 teeth on the stator, 4 on the rotor, 3 phases), its air gap diameter is 100 mm, the active length of magnetic core is 200 mm, the air gap is 0, 2 mm.

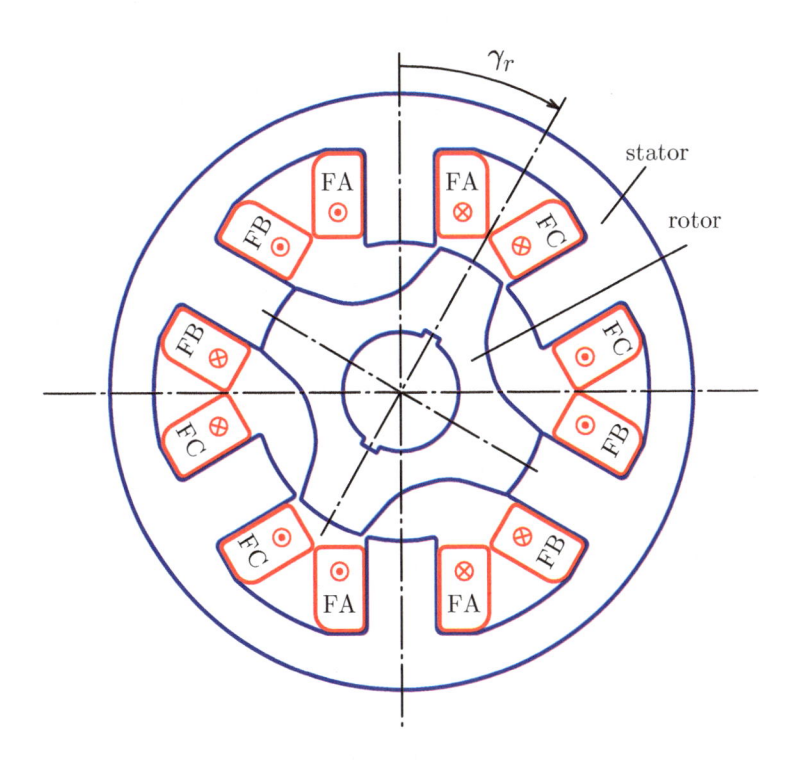

Fig. 1. Magnetic system of the high-speed switch-reluctance electrical generator

The switch-reluctance electric generator is powered by an electrical power converter consisting of three phase modules - half-bridge self-commutate voltage inverters with a common DC link. Figure 2 shows a schematic diagram of an energy converter for powering the switch-reluctance electric generator.

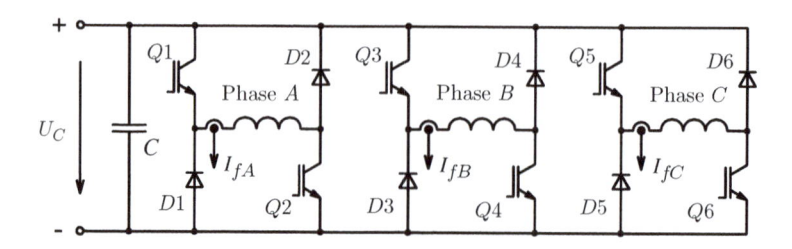

Fig. 2. Schematic diagram of an energy converter for powering the switch-reluctance electric generator

We used the field theory methods to calculate the flux-current characteristics of the stator phase at different mutual positions of the stator and the rotor of the high-speed switch-reluctance electrical generator under consideration. For this purpose, we solved a series of magnetostatic problems in a plane-parallel formulation using a FEMM (Finite Element Method Magnetic © David Meeker) software package. Figure 3 shows the stator phases fluxlinkage dependences on current at different mutual positions of the stator and the rotor.

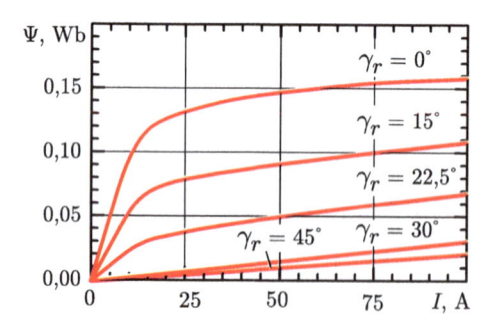

Fig. 3. The stator phases fluxlinkage dependences on current at various rotor rotational angles

Based on flux-current characteristics showed on the Fig. 3, we have constructed the dependences of the stator differential inductance in the absence of saturation (Fig. 4)

The analysis of changes in the obtained dependencies shows that when one of the phases is in an aligned position, the inductances of the other phases are

⊣⊢ - phase aligned position   ⊥⊤ - phase unaligned position

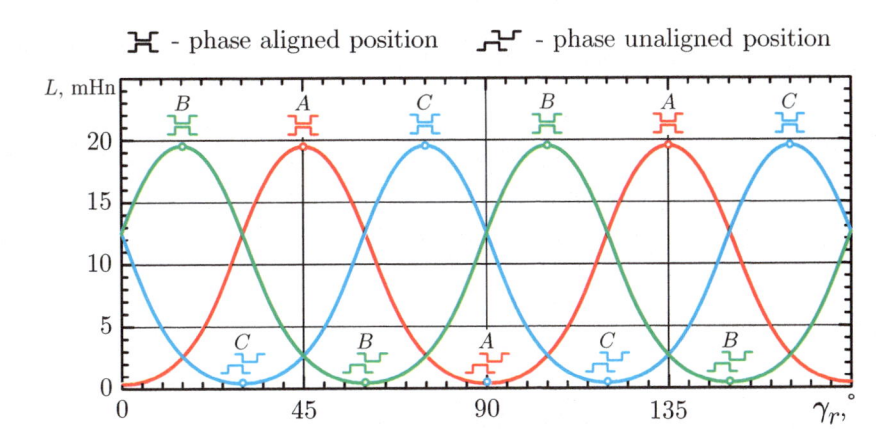

Fig. 4. Stator differential inductance of the hi-speed switch-reluctance electrical generator

equal and are about 20% of the maximum value. The amplitude of the current pulses in these phases will be sufficient to determine the moment of their equality. Also, the inductance of these phases varies significantly when the rotor position is changed. It facilitates the determination of the equality moment of the current pulses amplitude.

5   Simulation Results and Analysis

We have designed the mathematical model of the electric machine, electric power converter and control system to assess the correctness of the developed principle of control of the high-speed switch-reluctance electrical generator. When compiling the model, the following assumptions were adopted, taking into account the problem particularities [5,10,15–17]. The main particularity is that the phases of the electrical machine have magnetic circuits not connected magnetically, therefore the processes in them are considered independently. With the assumptions made, the following equation describes the electromagnetic processes in the electrical machine [5,10,15–17]:

$$\frac{d\Psi_f}{dt} = U_f - I_f\left(\Psi_f, \gamma_r\right), \qquad (1)$$

where Ψ_f, U_f and R_f – fluxlinkage, voltage and phase resistance, respectively; $I_f\left(\Psi_f, \gamma_r\right)$ – phase current, which depends on the phase linkage magni-tude and the rotor position.

The formation of the voltage applied to the phases must ensure the formation of a current pulse of a given magnitude and duration. The following conditions determine the magnitude of the voltage applied to the phase:

$$U_f = \begin{cases} U_C & \text{if } \left(0 \leq I_f < I_m - \frac{\Delta I}{2}\right) \vee \left(I_m - \frac{\Delta I}{2} \leq I_m \leq I_m + \frac{\Delta I}{2}\right) \wedge \frac{dI_f}{dt} > 0; \\ 0 & \text{if } \left(I_m - \frac{\Delta I}{2} \leq I_m \leq I_m + \frac{\Delta I}{2}\right) \wedge \frac{dI_f}{dt} < 0; \\ -U_C & \text{if } (I_m = 0) \wedge (I_f > 0), \end{cases} \quad (2)$$

where I_m – the amplitude of the phase current pulse;
ΔI – the current corridor width;
U_C – DC-link voltage.

The initial and final angles $\alpha 0$ and $\alpha 1$, respectively, counted from the coordinated phase position determine the duration and current pulse phase of rectangular shape with a specified amplitude.

Preliminary analysis of the processes in the simulated system showed that during the determination of the initial position of the rotor, the capacitor in the DC link is practically not discharged. The voltage on the capacitor at the beginning and end of the initial rotor position determination process is different by less than 1 V. Therefore, the processes in the DC link of the electric power converter are not modeled. It is believed that the phase modules are powered by voltage sources.

The Eq. (1) taking into account expression (2) constitute mathematical model, describing the processes in reluctance-flux electrical machine. It is used to modeling the process in the electrical machine to determine the rotor initial position. The capacitor voltage in the DC-link (U_C) is 160 V.

We simulated the process in the high-speed switch-reluctance electrical generator to determine the rotor initial position by using the described mathematic model. The following conditions have been met. The rotor speed is 12000 rpm. The voltage pulses duration is 10 µs and the frequency is 30 kHz at each phase of electrical machine. Figure 5 shows the simulation results. It shows the phase currents values and the rotor position.

The obtained result shows that the current pulses amplitude does not exceed 15 A the magnetic system is not saturated at the voltage pulses specified above. It is one of the conditions of the proposed method applicability. As the Fig. 4 shows, when the phase A is in the coordinated position, the current pulses amplitudes are above 10 A in phases B and C and they differ from the amplitude of the preceding and subsequent pulses by 2 to 3 A. It allows us to identify the moment of equality of the current pulses amplitude in the phases B and C when using the current sensors in the control system of a high-speed switch-reluctance electrical generator.

The accuracy of the rotor position determining depends on the rotor speed and the pulse repetition rate. The error in determining the moment when the rotor passes the coordinated position with one of the phases is 33.3 µs for the 30 kHz frequency. With a rotor speed of 12,000 rpm, it corresponds to a rotor angle of 2.4°. The results on the Fig. 5 confirm the accuracy of the method.

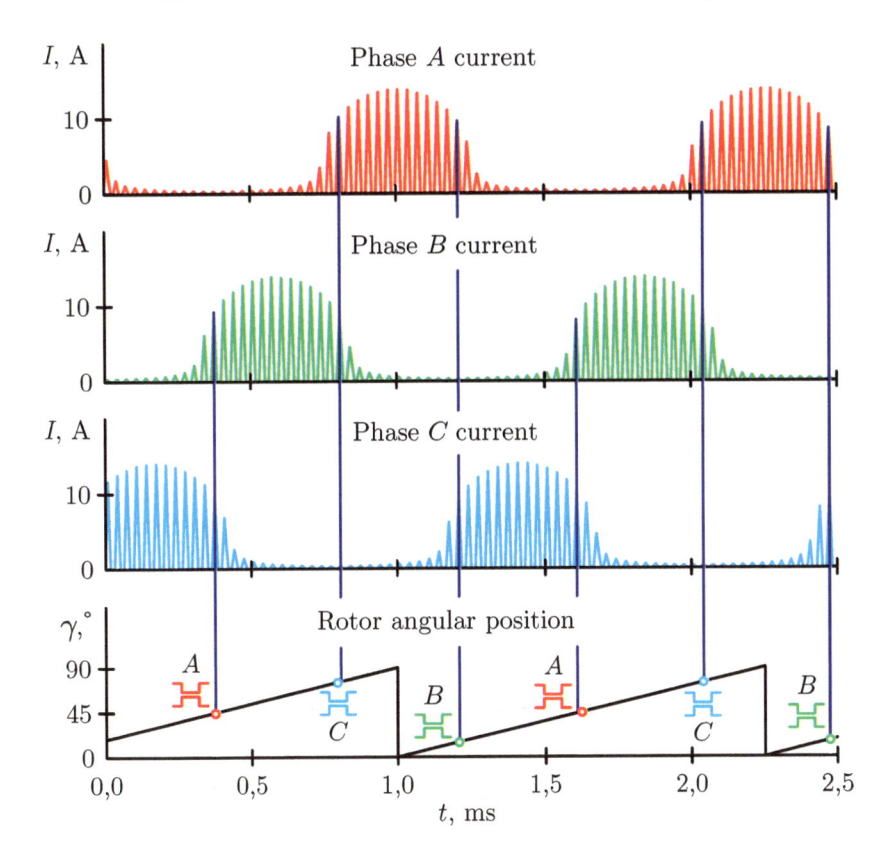

Fig. 5. The determination of the rotor initial position

The application of this method is possible only when electrical machine rotates without load for the reasons indicated in the article [5]. To determine the position of the rotor in the operating modes of the high-speed generator, the article proposes the use of another method based on the analysis of the phase current shape. The use of this method let effectively control the electrical machine in the deviation of the estimated position of the rotor from the real one by 3 to 5°.

Therefore, the application of the method proposed in this article allows us to determine with sufficient accuracy the rotor initial position and its rotation speed at start.

In the known methods of sensorless position determination of the rotor of switch-reluctance electrical machine [6–8, 12, 13], an inductive measurement of each of their stator phases is performed indirectly. When the phase is in a aligned or unaligned position, its inductance varies little when the rotor is rotated. In this case, the registration of the extremes of inductance is difficult due to the presence of measurement errors, especially in the case of using short test pulses. The difference between the proposed methods of determining the position of the rotor from the known is that instead of an indirect measurement of the inductance of a separate phase of the switch-reluctance electrical machine, the inductances of all phases are compared. This makes it possible to improve the accuracy of determining the position of the rotor, which is especially true under conditions where the test pulses should have a short duration and a low voltage. The application of this method is possible when electrical machine is operating without current in all phases, therefore it can be used, for example, to determine the initial position of the generator's rotor without load.

6 Conclusion and Perspectives

The obtained results show that the sensorless control of the hi-speed switch-reluctance electrical machine is efficient. The rotor initial position and rotor speed determination when starting an electric generator with a rotating rotor is one of the most difficult challenges. The generator start is carried out at a reduced voltage of the capacitor in the DC-link, without external power, that is why the methods based on the indirect measurement of the stator phase inductance is impossible. To solve this problem we proposed the modified method that includes the simultaneous feeding in all phases of the electrical machine the series of the test pulses of the minimum possible duration, the subsequent determination of the coordinated position of the phase from the equality of the current pulses amplitudes in the other two phases. The rotor position determination accuracy is 2 to 3°. This accuracy is sufficient for the efficient operation of the reluctance-flux electric machine control system under load, using the phase current shape to determine the rotor position.

The combined application of the proposed method for determining the initial position of the rotor and the method of sensorless control of the switch-reluctance electrical machine based on the analysis of the phase current shape considered in [5] allows solving the problem of controlling a high-speed electric generator at start-up with a rotating rotor.

Acknowledgements. The article was published with financial support by the Ministry of Education and Science of the Russian Federation within the framework of the Federal Target Program "Research and development in the priority directions of the scientific-technological complex of Russia for 2014–2020" (No. 14.577.21.0260 Agreement on "Development of an autonomous mobile micro-energy complex functioning on the basis of technologies of processing of the industrial, municipal and agricultural wastes with power supply with the trigeneration mode". The unique identifier of the applied scientific research and experimental developments (of the project) is RFMEFI57717X0260)

References

1. Monti, A., Pesch, D., Ellis, K., Mancarella, P.: Energy Positive Neighborhoods and Smart Energy Districts: Methods, Tools, and Experiences from the Field. Elsevier Science (2016). https://books.google.ru/books?id=KHm0CwAAQBAJ
2. Boicea, A.V., Chicco, G., Mancarella, P.: Optimal operation of a microturbine cluster with partial-load efficiency and emission characterization. In: 2009 IEEE Bucharest PowerTech, pp. 1–8 (2009)
3. Sioshansi, F.: Smart Grid: Integrating Renewable, Distributed & Efficient Energy. Academic Press (2012). https://books.google.ru/books?id=MQMrLNPjZVcC
4. Gerada, D., Mebarki, A., Brown, N.L., Gerada, C., Cavagnino, A., Boglietti, A.: High-speed electrical machines: Technologies, trends, and developments. IEEE Trans. Ind. Electron. **61**(6), 2946–2959 (2014)
5. Kolpakhchyan, P.G., Shaikhiev, A.R., Kochin, A.E.: Sensorless control of thehigh-speed switched-reluctance generator for the steam turbine. In: Abraham, A., Kovalev, S., Tarassov, V., Snasel, V., Vasileva, M., Sukhanov, A. (eds.) Advances in Intelligent Systems and Computing, pp. 349–358. Springer International Publishing (2017). https://doi.org/10.1007/978-3-319-68324-9_38
6. Ye, J., Bilgin, B., Emadi, A.: An offline torque sharing function for torque ripple reduction in switched reluctance motor drives. IEEE Trans. Energy Convers. **30**(2), 726–735 (2015). https://doi.org/10.1109/tec.2014.2383991
7. Torrey, D.: Switched reluctance generators and their control. IEEE Trans. Ind. Electron. **49**(1), 3–14 (2002). https://doi.org/10.1109/41.982243
8. Shao, B., Emadi, A.: A digital PWM control for switched reluctance motor drives. In: 2010 IEEE Vehicle Power and Propulsion Conference. IEEE (2010). https://doi.org/10.1109/vppc.2010.5729103
9. Uygun, D., Bal, G., Sefa, I.: Linear model of a novel 5-phase segment type switched reluctance motor. Elektronika ir Elektrotechnika **20**(1), 3–7 (2014)
10. Do, V., Ta, M.C.: Modeling, simulation and control of reluctance motor drives for high speed operation. In: 2009 IEEE Energy Conversion Congress and Exposition, pp. 1–6. IEEE (2009). https://doi.org/10.1109/ecce.2009.5316067
11. Cai, J., Deng, Z.: A position sensorless control of switched reluctance motors based on phase inductance slope. J. Power Electron. **13**(2), 264–274 (2013). https://doi.org/10.6113/JPE.2013.13.2.264. http://www.koreascience.or.kr/article/ArticleFullRecord.jsp?cn=E1PWAX2013v13n2264
12. Borges, T., de Andrade, D., de Azevedo, H., Luciano, M.: Switched reluctance motor drive at high speeds, with control of current. In: 1997 IEEE International Electric Machines and Drives Conference Record, pp. TB1/12.1–TB1/12.3. IEEE (1997). https://doi.org/10.1109/iemdc.1997.604216. https://ieeexplore.ieee.org/document/604216/
13. Ye, J., Bilgin, B., Emadi, A.: An extended-speed low-ripple torque control of switched reluctance motor drives. IEEE Trans. Power Electron. **30**(3), 1457–1470 (2015). https://doi.org/10.1109/tpel.2014.2316272. https://ieeexplore.ieee.org/document/6797980/
14. Krna, P.: Sensorless control of the SRM using nonlinear observer. Ph.D. thesis, VSB Technical University of Ostrava (2013)
15. Kolpakhchyan, P., Shcherbakov, V., Kochin, A., Shaikhiev, A.: Sensorless control of a linear reciprocating switched-reluctance electric machine. Russ. Electr. Eng. **88**(6), 366–371 (2017). https://doi.org/10.3103/S1068371217060086

16. Chi, H.P., Lin, R.L., Chen, J.F.: Simplified flux-linkage model for switched-reluctance motors. IEE Proc. Electric Power Appl. **152**(3), 577 (2005). https://doi.org/10.1049/ip-epa:20045207. https://ieeexplore.ieee.org/document/1436181/
17. Fahimi, B., Suresh, G., Mahdavi, J., Ehsami, M.: A new approach to model switched reluctance motor drive application to dynamic performance prediction, control and design. In: PESC 98 Record. 29th Annual IEEE Power Electronics Specialists Conference (Cat. No. 98CH36196), vol. 2, pp. 1061–1067. IEEE (1998). https://doi.org/10.1109/pesc.1998.703469. https://ieeexplore.ieee.org/document/703469/

Stability and Chaotic Attractors of Memristor-Based Circuit with a Line of Equilibria

N. V. Kuznetsov[1,2,3(✉)], T. N. Mokaev[1], E. V. Kudryashova[1],
O. A. Kuznetsova[1], R. N. Mokaev[1,2], M. V. Yuldashev[1], and R. V. Yuldashev[1]

[1] St. Petersburg State University, Peterhof, St. Petersburg, Russia
nkuznetsov239@gmail.com
[2] University of Jyväskylä, Jyväskylä, Finland
[3] Institute of Problems of Mechanical Engineering RAS, St. Petersburg, Russia

Abstract. This report investigates the stability problem of memristive systems with a line of equilibria on the example of SBT memristor-based Wien-bridge circuit. For the considered system, conditions of local and global partial stability are obtained, and chaotic dynamics is studied.

Keywords: Partial stability · Memristor · Chaos · Hidden attractors

1 Introduction

In 1971, Leon Chua suggested the concept of a memristor [1] as an electrical component that regulates the flow of electrical current in a circuit and remembers the amount of charge that has previously flowed through it. Nowadays, various types of memristors are developing for the realization of memory, computations and many other applications (see e.g. [2–4]).

Consider the dynamical model of SBT memristor-based Wien-bridge circuit [5]:

$$\begin{cases} \dot{x} = \frac{1}{C_1}\left(\frac{1}{R_5}(y-x) - (A+B+g(\varphi)+G)x\right), \\ \dot{y} = \frac{1}{C_2}\left(\frac{1}{R_2}\left(\frac{R_4}{R_3}y - z\right) - \frac{1}{R_1}y - \frac{1}{R_5}(y-x)\right), \\ \dot{z} = \frac{1}{C_3}\left(\frac{1}{R_2}\left(\frac{R_4}{R_3}y - z\right)\right), \\ \dot{\varphi} = x, \end{cases} \quad (1)$$

where $g(\varphi) = |\varphi|$, the parameters $C_{1,2,3}$, $R_{1,2,3,4,5}$, A, B are positive, and G is negative. Using the notation $\alpha_i = \frac{1}{C_i}$, $(i=1,2,3)$, $\beta_1 = \frac{1}{R_1}$, $\beta_2 = \frac{1}{R_2}$, $\beta_3 = \frac{R_4}{R_3}$, $\beta_4 = \frac{1}{R_5}$ we rewrite system (1) as follows:

$$\begin{cases} \dot{x} = f_1(x,y,z,\varphi) = \alpha_1\big(\beta_4(y-x) - (A+B+g(\varphi)+G)x\big), \\ \dot{y} = f_2(x,y,z,\varphi) = \alpha_2\big(\beta_2(\beta_3 y - z) - \beta_1 y - \beta_4(y-x)\big), \\ \dot{z} = f_3(x,y,z,\varphi) = \alpha_3\big(\beta_2(\beta_3 y - z)\big), \\ \dot{\varphi} = f_4(x,y,z,\varphi) = x. \end{cases} \quad (2)$$

Equating the right-hand of system (2) to zero we obtain a line of equilibria:

$$E = \{(x, y, z, \varphi) \,|\, x = y = z = 0, \varphi \in \mathbb{R}\}. \tag{3}$$

2 Local Stability Analysis

Let us analyze the local stability of the equilibrium points on the line of equilibria E. Here for simplicity, we approximate continuous function $g(\varphi) = |\varphi|$ by a smooth function $g(\varphi) = \varphi \tanh(\rho\varphi) \geq 0$, where $\rho \gg 1$. Since for an arbitrary equilibrium $(0, 0, 0, \varphi_0) \in E$ we have

$$\left.\frac{\partial f_1(x,y,z,\varphi)}{\partial \varphi}\right|_{(0,0,0,\varphi_0)} = \lim_{h\to 0} \frac{f_1(x,y,z,\varphi+h)|_{(0,0,0,\varphi_0)} - f_1(x,y,z,\varphi)|_{(0,0,0,\varphi_0)}}{h} =$$

$$= \lim_{h\to 0} \frac{(\alpha_1 x(g(\varphi+h) - g(\varphi)))|_{(0,0,0,\varphi_0)}}{h} = 0,$$

the Jacobi matrix at $(0, 0, 0, \varphi_0)$ can be expressed as:

$$J = \begin{pmatrix} -\alpha_1(A + Bg(\varphi_0) + G + \beta_4) & \alpha_1\beta_4 & 0 & 0 \\ \alpha_2\beta_4 & \alpha_2(\beta_2\beta_3 - \beta_1 - \beta_4) & -\alpha_2\beta_2 & 0 \\ 0 & \alpha_3\beta_2\beta_3 & -\alpha_3\beta_2 & 0 \\ 1 & 0 & 0 & 0 \end{pmatrix}. \tag{4}$$

The characteristic polynomial for the Jacobi matrix J is as follows:

$$\det(\lambda I - J) = \lambda(\lambda^3 + P_2\lambda^2 + P_1\lambda + P_0), \tag{5}$$

where

$$P_2 = \alpha_1 (A + Bg(\varphi_0) + G + \beta_4) + \alpha_2 (\beta_1 - \beta_2\beta_3 + \beta_4) + \alpha_3\beta_2,$$
$$P_1 = \alpha_1\alpha_2 \left((A + Bg(\varphi_0) + G + \beta_4)(\beta_1 - \beta_2\beta_3 + \beta_4) - \beta_4^2\right) +$$
$$\qquad + \alpha_1\alpha_3 \left((A + Bg(\varphi_0) + G + \beta_4)\beta_2\right) + \alpha_2\alpha_3\beta_2 (\beta_1 + \beta_4),$$
$$P_0 = \alpha_1\alpha_2\alpha_3\beta_2 \left((A + Bg(\varphi_0) + G + \beta_4)(\beta_1 + \beta_4) - \beta_4^2\right). \tag{6}$$

Expression (5) indicates that the characteristic equation of Jacobi matrix J has a zero eigenvalue with corresponding eigenvector $(0, 0, 0, 1)^*$, and three non-zero eigenvalues. Since the central manifold of each equilibrium $p \in E$ is placed on the line of equilibria E, local dynamics of the nonlinear system near p is described by the local dynamics of linearized system (see, e.g. *Shoshitaishvili reduction principle* [6] and related results). Thus, if p has three eigenvalues with negative real parts, then it is locally stable.

According to the Routh-Hurwitz criterion of stability, all the non-zero eigenvalues of (5) have negative real parts, iff $P_2 > 0$, $P_0 > 0$ and $P_2P_1 - P_0 > 0$. Left-hand side of the latter inequality has the form of quadratic equation:

$$P_2 P_1 - P_0 = Q_2 \nu^2 + Q_1 \nu + Q_0 \tag{7}$$

with respect to $\nu = \alpha_1 B g(\varphi_0)$, where

$$\begin{aligned}
Q_2 &= \alpha_2\left(\beta_1 - \beta_2\beta_3 + \beta_4\right) + \alpha_3\beta_2, \\
Q_1 &= Q_2^2 + \alpha_1\left(2\left(A + G + \beta_4\right)Q_2 - \alpha_2\beta_4^2\right), \\
Q_0 &= \alpha_1\left(\left(A + G + \beta_4\right)Q_2 - \alpha_2\beta_4^2\right)\left(\left(A + G + \beta_4\right)\alpha_1 + Q_2\right) \\
&\quad + \alpha_2\alpha_3\beta_2\left(\alpha_1\beta_4^2 + \left(\beta_1 + \beta_4\right)Q_2\right).
\end{aligned} \tag{8}$$

The discriminant of (7) has the following form:

$$\mathcal{D} = \left(Q_2^2 + \alpha_1\alpha_2\beta_4^2\right)^2 - 4\alpha_2\alpha_3\beta_2 Q_2\left(\left(\beta_1 + \beta_4\right)Q_2 + \alpha_1\beta_4^4\right). \tag{9}$$

For stability of all points on the line of equilibria (3), the branches of parabola Eq. (7) has to be directed upwards, i.e. the inequality $Q_2 > 0$ is needed. Since $\nu = \alpha_1 B g(\varphi_0) \geq 0$ for all $\varphi_0 \in \mathbb{R}$, to satisfy the inequality $P_2 P_1 - P_0 > 0$ it is necessary and sufficient to have either no real roots (i.e $\mathcal{D} < 0$), or all negative roots of the Eq. (7). The latter condition is satisfied, iff $\mathcal{D} \geq 0$, $Q_1 > 0$, $Q_0 > 0$. Inequalities $P_2 > 0$, $P_0 > 0$ are satisfied for all $\varphi_0 \in \mathbb{R}$, iff

$$\begin{aligned}
(P_2 \geq)\quad & \underbrace{\alpha_1\left(A + G + \beta_4\right) + \alpha_2\left(\beta_1 - \beta_2\beta_3 + \beta_4\right) + \alpha_3\beta_2}_{=\kappa_1} > 0, \\
(P_0 \geq)\quad & \underbrace{\alpha_1\alpha_2\alpha_3\beta_2\left(\left(A + G + \beta_4\right)\left(\beta_1 + \beta_4\right) - \beta_4^2\right)}_{=\kappa_2} > 0.
\end{aligned} \tag{10}$$

Thus, it is possible to formulate the following statement

Lemma 1. *If the values of parameters $\alpha_{1,2,3}$, $\beta_{1,2,3,4}$, A, B, G are such that the conditions $\kappa_1 > 0$, $\kappa_2 > 0$, $Q_2 > 0$, and*

$$\mathcal{D} \geq 0, \quad Q_1 > 0, \quad Q_0 > 0, \quad \text{or} \quad \mathcal{D} < 0 \tag{11}$$

hold, then each point at the line of equilibria E is locally Lyapunov stable[1].

3 Global Stability Analysis

In order to study the global stability of the line (3) let us consider the following Lyapunov function:

$$V = \tfrac{1}{2}\left(\tfrac{x^2}{\alpha_1} + \tfrac{y^2}{\alpha_2} + \tfrac{z^2}{\alpha_3}\right), \tag{12}$$

which has the following derivative along the solutions of system (2):

$$\begin{aligned}
\dot{V} &= -(\beta_4 + A + B g(\varphi) + G)x^2 + 2\beta_4 xy - (\beta_1 - \beta_2\beta_3 + \beta_4)y^2 + (\beta_2(\beta_3 - 1))yz - \beta_2 z^2 \\
&= -\gamma_1\left(x - \tfrac{\beta_4 y}{\gamma_1}\right)^2 - \gamma_2\left(y - \tfrac{1}{2}\tfrac{\beta_2(\beta_3-1)z}{\gamma_2}\right)^2 - \gamma_3 z^2,
\end{aligned} \tag{13}$$

---

[1] For any $\varepsilon > 0$ there exists $\delta > 0$, such that, if $|u(0) - u_{eq}| < \delta$, then $|u(t) - u_{eq}| < \varepsilon$ is valid for all $t > 0$. Recall that *local asymptotic stability* of u_{eq} means that u_{eq} is locally Lyapunov stable and also there exists $\delta > 0$, such that if $|u(0) - u_{eq}| < \delta$, then $\lim_{t\to\infty}|u(t) - u_{eq}| = 0$. Thus, due to the noise, the state of the physical model may drift along the line of equilibria.

where

$$\gamma_1 = \beta_4 + A + Bg(\varphi) + G, \qquad \gamma_2 = \beta_1 - \beta_2\beta_3 + \beta_4\left(1 - \tfrac{\beta_4}{\gamma_1}\right), \qquad \gamma_3 = \beta_2\left(1 - \tfrac{\beta_2(\beta_3-1)^2}{4\gamma_2}\right). \tag{14}$$

Since $B > 0$, we have

$$\gamma_1 \geq \underbrace{\beta_4 + A + G}_{=\mu_1}, \qquad \gamma_2 \geq \underbrace{\beta_1 - \beta_2\beta_3 + \beta_4\left(1 - \tfrac{\beta_4}{\mu_1}\right)}_{=\mu_2}, \qquad \gamma_3 \geq \underbrace{\beta_2\left(1 - \tfrac{\beta_2(\beta_3-1)^2}{4\mu_2}\right)}_{=\mu_3}. \tag{15}$$

Thus, it is possible to formulate the following statement

Lemma 2. *If the values of parameters $\alpha_{1,2,3}$, $\beta_{1,2,3,4}$, A, B, G are such that the conditions $\mu_1 > 0$, $\mu_2 > 0$, $\mu_3 > 0$ hold, then the line of equilibria E is partially globally stable, i.e. for any ellipsoidal cylinder $\varepsilon = V(x,y,z)$ defined by (12) with a sufficiently small radius ε and for any trajectory from outside it there exists a moment of time T after which the trajectory enters the cylinder and remain there (see, e.g. [7]).*

4 Chaotic Attractors

For parameters $\alpha_1 = 10^8$, $\alpha_2 = \alpha_3 = 5 \cdot 10^7$, $\beta_1 = \beta_2 = 4 \cdot 10^{-5}$, $\beta_3 = 2.5$, $\beta_4 = 2.22 \cdot 10^{-5}$, $A = 0.0676$, $B = 0.3682$, $G = -0.0677$ chaotic attractors [5] can be found in system (2) (see Fig. 1). These attractors are self-excited ones with respect to some unstable points on the line of equilibria E according to the definition from [8–13]. However since there is a continuum of unstable equilibria on E, the unstable manifold of which may form attractors, the revealing of all co-existing attractors is a challenging task, and, thus, attractors in such systems sometimes are also called "hidden". For their search one can use, e.g., various evolutionary algorithms [14,15]. The search of all co-existing attractors and determination of their mutual disposition in dynamical systems can be regarded as a generalization [16] of the second part of Hilbert's 16th problem on the number and mutual disposition of limit cycles in two-dimensional polynomial systems. Remark, that since there is an unbounded line of equilibria E in system (1), one has to consider cylindrical absorbing sets and unbounded attractors.

One can see that the region of parameters given by the conditions of Lemma 1 does not coincide with the region of parameters corresponding to the conditions of Lemma 2. When all equilibria are locally stable, the following cases are of interest:

(a) system (1) can be partially globally stable when all trajectories tend to the line of equilibria E;
(b) system (1) may have hidden attractors with respect to E;
(c) system (1) can be dichotomic (some trajectories can tend to infinity in the (x,y,z) subspace).

(a) Initial point $u_0 = (10^{-2}, 0, 0, 0)$.

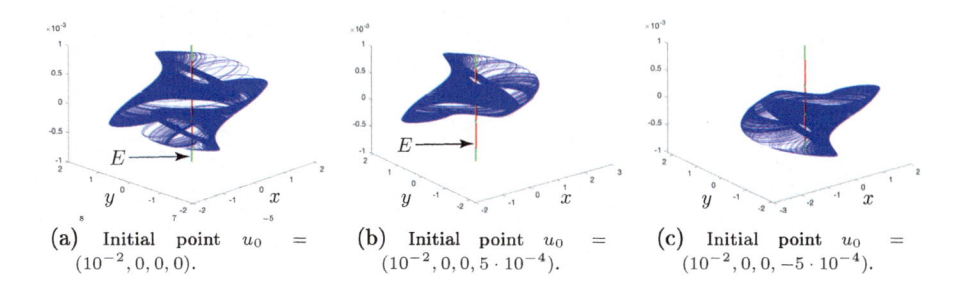

(b) Initial point $u_0 = (10^{-2}, 0, 0, 5 \cdot 10^{-4})$.

(c) Initial point $u_0 = (10^{-2}, 0, 0, -5 \cdot 10^{-4})$.

Fig. 1. Self-excited chaotic attractors (blue) in system (2) for parameters $\alpha_1 = 10^8$, $\alpha_2 = \alpha_3 = 5 \cdot 10^7$, $\beta_1 = \beta_2 = 4 \cdot 10^{-5}$, $\beta_3 = 2.5$, $\beta_4 = 2.22 \cdot 10^{-5}$, $A = 0.0676$, $B = 0.3682$, $G = -0.0677$, which are visualized by trajectories with initial data in vicinity of the line of equilibria E (stable equilibria are green, unstable – red).

When some of the equilibria on E are unstable, the following cases are of interest:

(a) system (1) can be gradient-like (i.e. when all trajectories except unstable equilibria tend to the stable equilibria on E);
(b) system (1) can be partially dissipative (all trajectories do not leave an absorbing cylinder; in this case system (1) can have self-excited attractors);
(c) system (1) can be dichotomic.

5 Conclusion

In this report we discussed some basic ideas of the stability theory for memristive systems with a line of equilibria. For the SBT memristor-based Wien-bridge chaotic circuit the conditions of local and global partial stability are obtained. Using [17,18], various other memristive circuits can be studied similarly. More detailed studies and results will be included in the forthcoming survey *"Theory of stability for memristive systems with a line of equilibria"* [19].

Acknowledgements. The authors wish to thank Prof. Leon Chua (University of California, Berkeley, USA) for the fruitful discussions and valuable comments on memristive systems. This work was supported by Russian Scientific Foundation project 19-41-02002 (Sects. 2–4) the Leading Scientific Schools of Russia grant NSh-2858.2018.1 (Sect. 1).

References

1. Chua, L.: Memristor-the missing circuit element. IEEE Trans. Circuit Theory **18**(5), 507–519 (1971)
2. Tetzlaff, R.: Memristors and memristive systems. Springer, Heidelberg (2013)
3. Adamatzky, A., Chen, G.: Chaos, CNN, Memristors and Beyond: A Festschrift for Leon Chua. World Scientific, Singapore (2013)
4. Vaidyanathan, S., Volos, C.: Advances in Memristors, Memristive Devices and Systems, vol. 701. Springer (2017)

5. Guo, M., Gao, Z., Xue, Y., Dou, G., Li, Y.: Dynamics of a physical SBT memristor-based Wien-bridge circuit. Nonlinear Dyn. **93**(3), 1681–1693 (2018)
6. Shoshitaishvili, A.N.: Bifurcations of topological type of a vector field near a singular point. Trudy Semin. Im. IG Petrovskogo **1**, 279–309 (1975)
7. Rumyantsev, V.V., Oziraner, A.S.: Stability and Partial Motion Stabilization. Nauka, Moscow (1987). (in Russian)
8. Leonov, G., Kuznetsov, N.: Hidden attractors in dynamical systems. From hidden oscillations in Hilbert-Kolmogorov, Aizerman, and Kalman problems to hidden chaotic attractors in Chua circuits. International Journal of Bifurcation and Chaos 23(1) (2013), art. no. 1330002
9. Leonov, G., Kuznetsov, N., Mokaev, T.: Homoclinic orbits, and self-excited and hidden attractors in a Lorenz-like system describing convective fluid motion. Eur. Phys. J. Spec. Top. **224**(8), 1421–1458 (2015)
10. Dudkowski, D., Jafari, S., Kapitaniak, T., Kuznetsov, N., Leonov, G., Prasad, A.: Hidden attractors in dynamical systems. Phys. Rep. **637**, 1–50 (2016)
11. Stankevich, N., Kuznetsov, N., Leonov, G., Chua, L.: Scenario of the birth of hidden attractors in the Chua circuit. International Journal of Bifurcation and Chaos 27(12) (2017), art. num. 1730038
12. Chen, G., Kuznetsov, N., Leonov, G., Mokaev, T.: Hidden attractors on one path: Glukhovsky-Dolzhansky, Lorenz, and Rabinovich systems. International Journal of Bifurcation and Chaos 27(8) (2017), art. num. 1750115
13. Kuznetsov, N., Leonov, G., Mokaev, T., Prasad, A., Shrimali, M.: Finite-time Lyapunov dimension and hidden attractor of the Rabinovich system. Nonlinear Dyn. **92**(2), 267–285 (2018)
14. Zelinka, I.: A survey on evolutionary algorithms dynamics and its complexity - mutual relations, past, present and future. Swarm Evol. Comput. **25**, 2–14 (2015)
15. Zelinka, I.: Evolutionary identification of hidden chaotic attractors. Engineering ApplicatiEng. Appl. Artif. Intell. **50**, 159–167 (2016)
16. Leonov, G., Kuznetsov, N.: On differences and similarities in the analysis of Lorenz, Chen, and Lu systems. Appl. Math. Comput. **256**, 334–343 (2015)
17. Corinto, F., Ascoli, A., Gilli, M.: Nonlinear dynamics of memristor oscillators. IEEE Trans. Circuits Syst. I: Regul. Pap. **58**(6), 1323–1336 (2011)
18. Corinto, F., Forti, M.: Memristor circuits: flux-charge analysis method. IEEE Trans. Circuits Syst. **63**(11), 1997–2009 (2016)
19. Kuznetsov, N.V., Kuznetsova, O.A., Kudryashova, E.V., Mokaev, T.N., Mokaev, R.N., Yuldashev, M.V., Yuldashev, R.V., Chua, L.: Theory of stability for memristive systems with a line of equilibria (2019, in prepartion)

Mechanical Engineering

Behavior of Five-Pad Tilting–Pad Journal Bearings with Different Pivot Stiffness

Phuoc Vinh Dang[1(✉)], Steven Chatterton[2], and Paolo Pennacchi[2]

[1] Department of Mechanical Engineering,
The University of Danang - University of Science and Technology,
54 Nguyen Luong Bang Street, Danang, Vietnam
dpvinh@dut.udn.vn
[2] Department of Mechanical Engineering, Politecnico di Milano,
Via G. La Masa 1, 20145 Milan, Italy
{steven.chatterton,paolo.pennacchi}@polimi.it

Abstract. In this paper, two different five-shoe tilting-pad journal bearings, namely rocker-backed pivot and spherical pivot with different pivot stiffness have been characterized with different working conditions. For the spherical pivot bearing, elastic shims and displacement restriction component are introduced to recognize the role of the variable stiffness of pivot. The copper is used for the material of the shim. An analysis of the dynamic behavior of the two bearings using a thermo-elasto-hydrodynamic model is presented first. This model considers the flexibility for both pad as well as pivot and a simple thermal model only for the fluid film temperature to accurately calculate the performance of bearings. The model also accounts for pivot stiffness of pads. The predicted dynamic coefficients of the two bearings were compared with the experimentally measured ones. The results show that the pivot stiffness or the pivot flexibility plays an important role in the dynamic coefficients estimation.

Keywords: Tilting-pad journal bearings · Pivot stiffness · Dynamic characteristics · Thermo-elasto-hydrodynamic model · Experimental tests

1 Introduction

Over the years, the topic of pivot flexibility of tilting-pad journal bearings (TPJBs) has become more and more important, especially for the bearings operating with high applied static load and high rotational speed. Studies including flexibility of the pad and the pivot on the modeling in order to obtain accurate results of TPJBs are well documented.

Chen et al. [1] studied numerically the behavior of the tilting-pad journal gas bearings with variable pivot stiffness. Based on their finding, they concluded that compared with the fixed pads, the minimum oil-film thickness increases if flexibility of the pad and the pivot is considered.

In 2013, San Andres and Tao [2] investigated the role of pivot flexibility in TPJBs on the dynamic coefficients. It was found that if pivot stiffness is one order of magnitude larger than the oil film stiffness, the effect of pivot flexibility on the TPJB dynamic coefficients is negligible. Wilkes et al. [3] studied the effects of the pad and pivot flexibility in forecasting the dynamic coefficients for the TPJB. They considered the variations of the bearing clearance with the working temperature and determined the differences between predicted and measured dynamic coefficients as functions of dynamic excitation. The study on the influence of flexibility of the pad and pivot on the dynamic behaviors of the TPJBs was also carried out [4–6].

In this paper, the dynamic characteristics of two different five-shoe TPJB, namely rocker-backed pivots and spherical pivots with different pivot stiffness have been studied by means of a TEHD model. The model accounts for pivot stiffness of pads which are determined by a model based on the Hertz contact theory and the experimental measurements. The pivot stiffness calculation using the experimental measurement is presented. The experimental tests were performed for these two bearings using a suitable test-rig in order to validate the model. The predicted dynamic coefficients of the two bearings were compared with the measured ones.

2 TEHD Bearing Model

Figure 1 represents the geometry of a single pad from the five-pad tilting-pad journal bearing, where O_b and O_j is the centre of the bearing and the journal, respectively.

A thorough description of the TEHD model to estimate the static and dynamic behaviors of a five-pad TPJB is provided in [7–12].

The hydrodynamic model is based on the well-known Reynolds equation:

$$\frac{\partial}{\partial x}\left(\frac{\rho h^3}{\mu}\frac{\partial p}{\partial x}\right) + \frac{\partial}{\partial z}\left(\frac{\rho h^3}{\mu}\frac{\partial p}{\partial z}\right)$$
$$= 6\left[(U_1 - U_2)\frac{\partial h}{\partial x} + h\frac{\partial}{\partial x}(U_1 + U_2)\right.$$
$$\left. + (W_1 - W_2)\frac{\partial h}{\partial z} + h\frac{\partial}{\partial z}(W_1 + W_2) + 2(V_1 + V_2)\right] \quad (1)$$

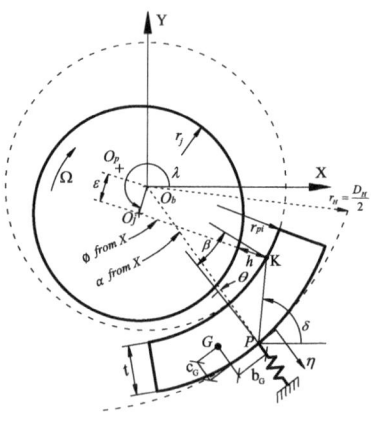

Fig. 1. Sketch of a rocker-backed TPJB [7].

where p is the pressure in the oil-film, h is the oil-film thickness, μ is the dynamic viscosity, ρ is the density of oil, z is the axial direction, x is the tangential direction. The velocity vector component of the shaft and the pads are defined by U_1, V_1, W_1 and U_2, V_2, W_2, respectively.

The effect of temperature on the dynamic viscosity $\mu(T) = \mu_{40°C} \exp[\kappa(T_{40°C} - T)]$ and oil density $\rho(T) = \rho_{40°C}[1 + \alpha_v(T_{40°C} - T)]$ is considered using a simple two-dimensional thermal model, governed, at steady state, by the energy equation:

$$\rho c_p \left(u \frac{\partial T}{\partial x} + w \frac{\partial T}{\partial z} \right) = k_{OIL} \left(\frac{\partial^2 T}{\partial x^2} + \frac{\partial^2 T}{\partial z^2} \right) + \mu \left[\left(\frac{\partial u}{\partial y} \right)^2 + \left(\frac{\partial w}{\partial y} \right)^2 \right] \quad (2)$$

where κ is the viscosity index and α_v is the thermal expansion coefficient of the oil, c_p and k_L are the heat capacity and the conductivity of the lubricant respectively. Equation (2) has been integrated using a finite difference method, where adiabatic conditions at the shaft, pad surfaces and constant oil temperature in the oil film thickness direction are considered.

3 Test-Rig and Bearing Under Test

A detail description of the test-rig and a five-pad rocker-backed TPJB is provided in [7] and [12]. In the test rig, the rotor axis can describe orbits similar to the real configuration of a rotating machine in which two identical five-pad TPJBs (in the condition of nominal dimensions) which are labeled as **1** and **2** in Fig. 2 support the shaft. The shaft is driven using a 6.0 kW inverter-driven electric motor by means of a flexible coupling up to the maximum rotational speed of 3000 rpm.

Fig. 2. Test rig for experimental tests [7]

The load is applied in the middle of the shaft by means of two hydraulic actuators arranged in an orthogonal configuration at ±45° with respect to the load cells. The actuators are connected to the shaft by means of two deep-groove high-precision ball bearings. Due to this configuration, a static load, as well as a dynamic load, can be applied in any direction.

During the tests, the oil inlet was kept at approximately 40 ± 0.5 °C using a closed-loop PI. The bearing under test, shown in Fig. 3a, is a five shoe rocker-backed TPJB. The bearing is installed in its housing in a standard LOP configuration with a nominal diameter of 100 mm and a length-to-diameter ratio (L/D) of 0.7.

For the spherical pivot bearing, elastic shims and displacement restriction component are introduced to recognize the role of the variable stiffness of pivot. The material of the shim can be changed to offer different stiffness in the radial direction (see Fig. 3b). In this paper, copper has been used for the shims. The pivot system (including pivot, cap, and shim) is installed on each pad (see Fig. 4).

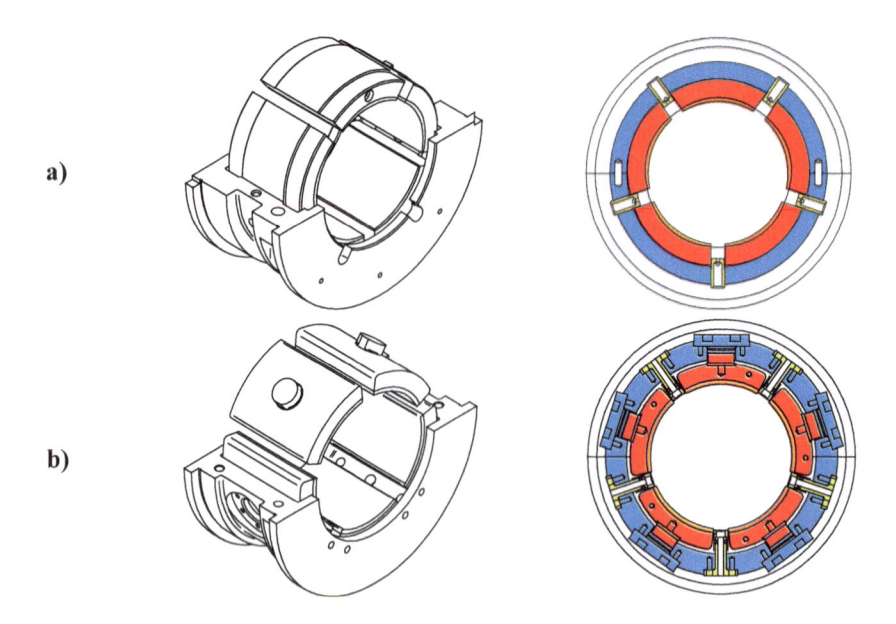

Fig. 3. Five-shoe TPJB under test (a) rocker-backed and (b) spherical pivot

The rocker-backed bearing and the spherical pivot bearing specifications and operating conditions are listed in Table 1.

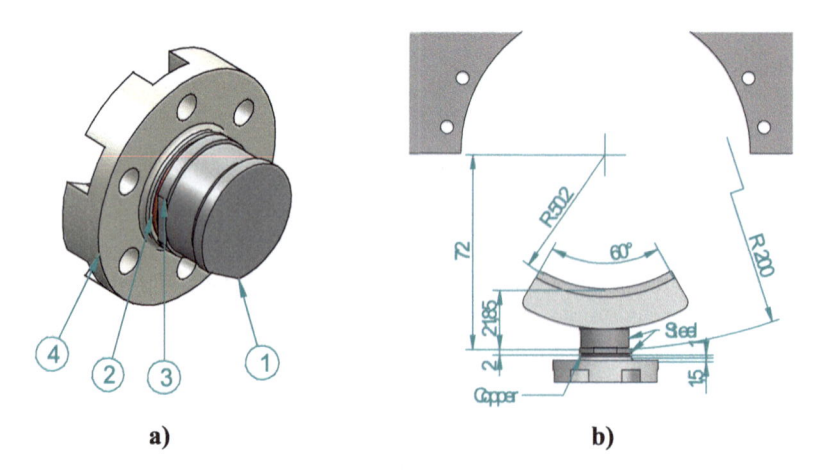

Fig. 4. (a) 3D drawing of the pivot and (b) drawing of the pad and pivot (nominal dimension). 1: pivot (pad side); 2: copper liner (shim); 3: pivot (housing side); 4: pads cap.

Table 1. Specifications and operating conditions of rocker-backed and spherical pivot bearing.

Item	Unit	Value/Span	
		Rocker back	Spherical pivot
Number of pads		5	5
Configuration w.r.t bearing housing		LOP	LOP
Shaft diameter	mm	99.86	99.86
Bearing length	mm	69.6	75
Housing radius	mm	66	–
Pad outer radius	mm	59.8	–
Radius of pad inner surface	mm	50.06	50.15
Radius of spherical pivot	mm	–	200
Nominal pad thickness	mm	16	–
Nominal pad thickness (pad + pivot)	mm	–	21.85
Angular amplitude of pads	degree	63.5	60
Oil inlet temperature	°C	40	40
Pad mass	kg	0.540	0.823
Pad mass moment of inertia w.r.t pivot	kg m^2	0.256E−3	1.380E−3

4 Pivot Stiffness Calculation

Assuming the same material properties for the pad pivot and its contact housing, the pivot stiffness equation considering the contact Hertz theory [13] are given by:

$$K_{\text{rockerbacked}} = \frac{\pi E L}{2(1-v^2)} \left(\frac{2}{3} + \ln \frac{\pi E L (D_H - D_P)}{4W(1-v^2)} \right)$$

$$K_{\text{spherical}} = \frac{W}{\sqrt[3]{\frac{9W^2(1-v^2)^2}{2E^2 D_s}}} \qquad (3)$$

where E and v are the Young's modulus and the Poisson coefficient respectively, D_H and D_P are the diameter of the bearing housing and pivot respectively, D_s is the diameter of spherical pivot, and W is the load along the radial direction of the pivot. For the rocker-backed pad a contact between an outer (pad) and an inner (bearing housing) cylinders is considered, where the length of the contact is equal to $L_P = 39.5$ mm. For the spherical pivot, the contact between a sphere and a flat surface is considered.

However, in order to validate the model, also the pivot stiffness for each pad of TPJB needs to be identified by experimental activities.

In the initial step for the test, a sufficiently high static load is applied in the middle of the shaft in the vertical direction to make the shaft in contact with the pad. The static load and the corresponding position of the shaft and the pad can be considered as a reference one. In order to obtain more accurate results, the tests are performed in a dry condition to avoid any possible effect of the oil.

The static load is then increased from the reference load up to 3.5 kN, in steps of 0.5 kN. The actual static load applied on the rotor bearing system and the relative displacement of the shaft with respect to the bearing is recorded using two load cells and two proximity probes. As mentioned before, owing to the installation configuration of the hydraulic actuators, the static load can be applied in any directions. The same procedure is applied for all the pads by varying the load direction (see Fig. 5): −90° (LOP pad #1), −18° (LOP pad #2), 54° (LOP pad #3), 126° (LOP pad #4) and 198° (LOP pad #5).

Fig. 5. Varying load configurations.

The local pivot stiffness is the slope of the load versus the displacement fitting curve and can be calculated by:

$$k_{pivot}^{experiment} = \frac{\partial F}{\partial X} \tag{4}$$

where ∂F is the increment of applied static load and ∂X is the pivot radial deflection.

This procedure is repeated to calculate the pivot stiffness for the spherical pivot TPJBs. These pivot stiffness values are then introduced in the numerical model for the bearing static as well as dynamic characteristics prediction.

For easy of visualization, Fig. 6 shows the pivot stiffness obtained using the Hertz contact theory and only the identified pivot stiffness of pad #1 of two bearings using the experimental tests.

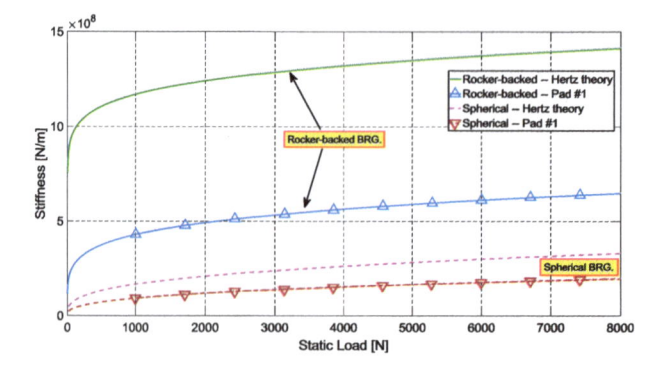

Fig. 6. Pivot stiffness of pad #1 of rocker-backed and spherical pivot TPJB obtained by using the Hertz contact theory and experimental data.

5 Results and Discussions

The dynamic stiffness and damping coefficients as functions of applied static load of the rocker-backed and spherical pivot bearing are shown in Figs. 7 and 8, respectively. All the tests were carried out with the rotational speed of 22 Hz and the force excitation frequency of 25 Hz. In this manner, a quasi-synchronous excitation was supposed. The static load applied in the bearing increased from 2.5 kN to 8.75 kN. The amplitude of dynamic load was chosen approximately 10% of static load in order to avoid nonlinearities of the rotor bearing system.

From Figs. 7 and 8, it is clear that the direct stiffness coefficients of two bearings increase with the increase of the static load, in which the coefficients in the loaded direction k_{yy} increase at a rate much greater than those in the unloaded direction, k_{xx}. Note that the k_{yy} of two bearings increase significantly approximately two times when the applied static load increases from 2.5 kN to 8.75 kN. Cross-coupled stiffness coefficients are much smaller than the direct coefficients and are of the same sign (positive sign) for k_{yx} and opposite sign (negative sign) for k_{xy}. However, these coefficients are quite stable with excitation frequency and their values can be considered as zero.

In contrast, the measured damping coefficients of two bearings are found to be quite independent with the applied static load, except for the damping coefficient in the loaded direction of the spherical pivot bearing, c_{yy}. It decreases dramatically approximately 200%, from about 3.8×10^5 N-s/m to about 0.2×10^5 N-s/m in the considered range of applied static load.

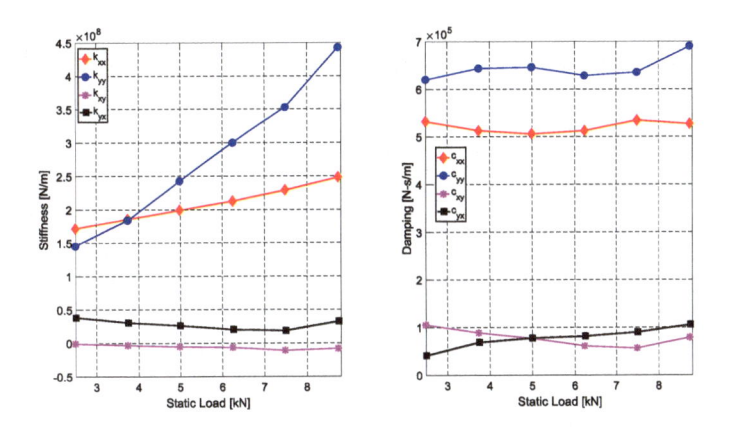

Fig. 7. Experimental dynamic coefficients vs. static load (rocker-backed bearing)

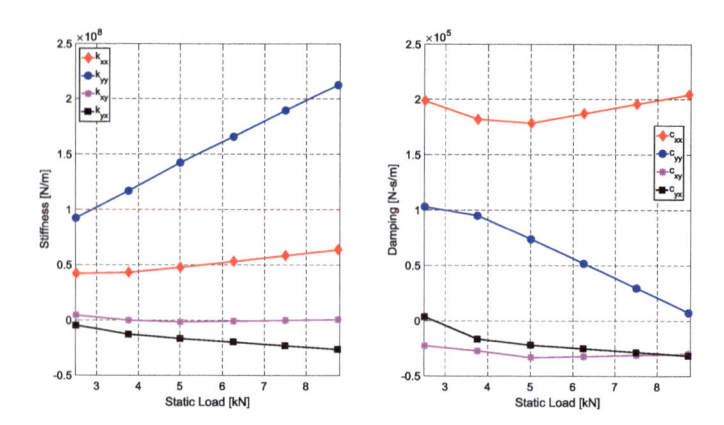

Fig. 8. Experimental dynamic coefficients vs. static load (spherical pivot bearing)

The predicted (solid line) and measured (dashed line) stiffness and damping coefficients of the rocker-backed TPJB are shown in Fig. 9a and b, respectively. The predicted results using the Hertz contact theory and experiment measurements for the pivot stiffness are plotted as well.

The model based on the Hertz contact theory correctly predicts the trend of increasing stiffness coefficients in the unloaded direction, k_{xx}, with the increase of static load. However, this model overestimates k_{xx} approximately 80% that obtained by experiments. Conversely, by using the experimental pivot stiffness, the predicted coefficient k_{xx} shows very good agreement with measured result.

For the stiffness coefficient in the loaded direction, k_{yy}, the predicted result shows a better agreement with the measured one if the model uses the Hertz theory for the pivot stiffness evaluation, especially at low applied static load. Nevertheless, the code always underestimates the value of k_{yy} approximately 10% and 50% at the lowest and largest applied static load, respectively.

In Fig. 9b, the code over predicts the damping coefficients around 300% for the c_{xx} and about 200% for the c_{yy} in the case of using the Hertz theory for the pivot stiffness. The better predicted results can be obtained if the pivot stiffness is identified from experiments. The discrepancy between the prediction and the measurement in this case reduces to about 20%. It is worthy of notice that the pivot stiffness or the pivot flexibility plays a key role in the dynamic coefficients estimation, particularly the damping coefficients.

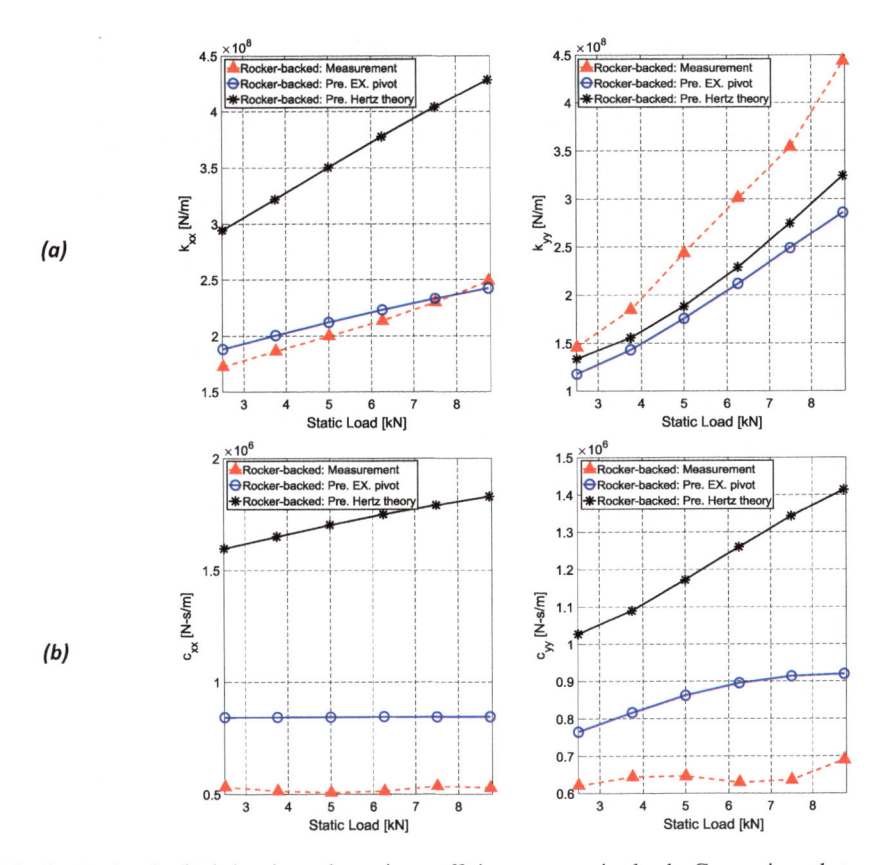

Fig. 9. Rocker-backed bearing: dynamic coefficient vs. static load. Comparison between measurement and prediction.

Figure 10 represents the comparison between the predicted and measured dynamic coefficients of the spherical pivot bearing as a function of applied static load. It is interesting to note that the results for this bearing are similar to the rocker-backed one. For instance, the predicted coefficient k_{xx} using the experimental pivot stiffness shows very good agreement with measured one. Besides, by using the Hertz contact theory for the pivot stiffness estimation the model correctly predicts the stiffness coefficient in the loaded direction, k_{yy}, especially at low applied static load.

(a)

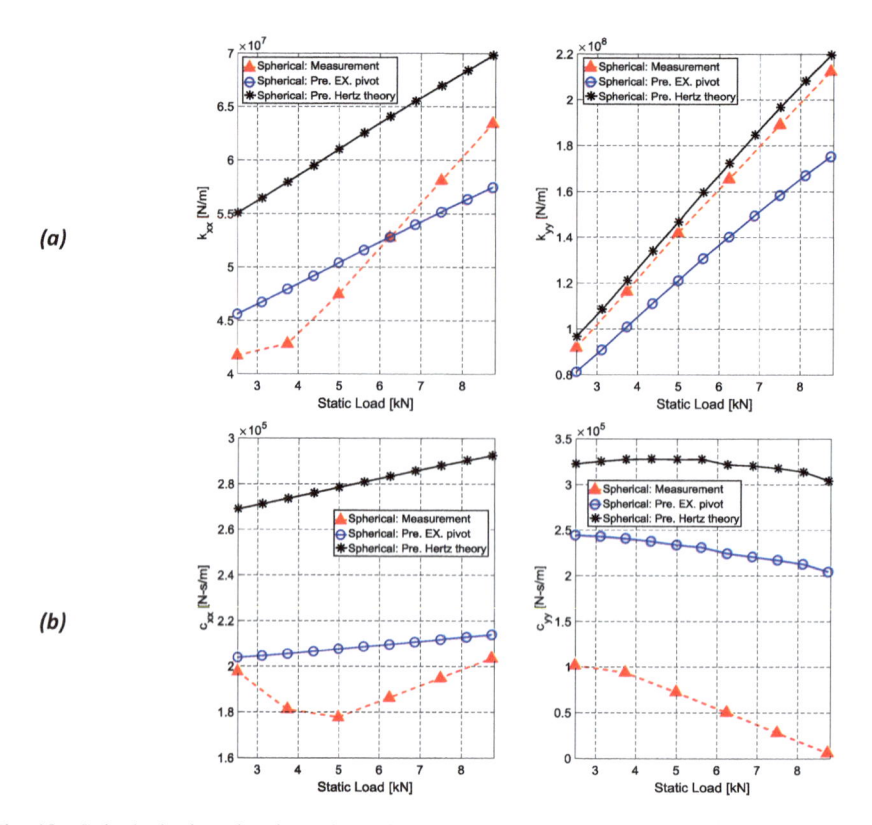

(b)

Fig. 10. Spherical pivot bearing: dynamic coefficient vs. static load. Comparison between measurement and prediction

6 Conclusions

In this study, the dynamic characteristics of two different five-shoe TPJBs, namely rocker-backed pivots and spherical pivots with different pivot stiffness were studied using a TEHD model. The experimental tests were performed for these two bearings using a suitable test-rig. The predicted dynamic coefficients of the two tilting-pad bearings were compared with the measured ones. In particular the following conclusions can be drawn:

1. The pivot stiffness values of the spherical pivot pad bearing are approximately one order of magnitude lower than that of the rocker-backed one. In addition, the Hertz contact theory overestimates the pivot stiffness, particularly the rocker-backed bearing.
2. For both bearings, the predicted coefficients in the unloaded direction k_{xx} using the experimental pivot stiffness shows very good agreement with measured ones.
3. It is worthy of notice that the pivot stiffness or the pivot flexibility plays a key role in the estimation of the dynamic coefficients. For both bearings, the predicted coefficients in the unloaded direction k_{xx} using the experimental pivot stiffness

shows very good agreement with measured ones. However, by using the Hertz contact theory for the pivot stiffness estimation the model correctly predicts the stiffness coefficient in the loaded direction, k_{yx}, especially at low applied static load.
4. Regarding the excitation, the more flexible the pad pivot, the lower the dynamic stiffness and damping coefficients, especially the damping coefficients.

References

1. Chen, S., Yang, S., Zhou, Q., Hou, Y., Lai, T.: Numerical study on tilting pad journal gas bearing with variable stiffness springs. J. Mech. Sci. Technol. **29**(8), 3059–3067 (2015)
2. San Andres, L., Tao, Y.: The role of pivot stiffness on the dynamic force coefficients of tilting pad journal bearings. J. Eng. Gas Turbines Power **135**(11), 112505 (2013)
3. Wilkes, J.C., Childs, D.W.: Tilting pad journal bearing - a discussion on stability calculation, frequency dependence, and pad and pivot. J. Eng. Gas Turbines Power **134**(12), 122508 (2015)
4. Feng, K., Liu, W., Zhang, Z., Zhang, T.: Theoretical model of flexure pivot tilting pad gas bearings with metal mesh dampers in parallel. Tribol. Int. **94**, 26–38 (2016)
5. Cha, M., Glavatskih, S.: Nonlinear dynamic behaviour of vertical and horizontal rotors in compliant liner tilting pad journal bearings: some design considerations. Tribol. Int. **82**, 142–152 (2015)
6. Kuznetsov, E., Glavatskih, S.: Dynamic characteristics of compliant journal bearings considering thermal effects. Tribol. Int. **94**, 288–305 (2016)
7. Dang, P.V., Chatterton, S., Pennacchi, P., Vania, A.: Effect of the load direction on non-nominal five-pad tilting-pad journal bearings. Tribol. Int. **98**, 197–211 (2016)
8. Dang, P.V., Chatterton, S., Pennacchi, P., Vania, A.: Numerical investigation of the effect of manufacturing errors in pads on the behaviour of tilting-pad journal bearings. Proc. Inst. Mech. Eng. Part J: J. Eng. Tribol. **233**(4), 480–500 (2018)
9. Dang, P.V., Chatterton, S., Pennacchi, P., Vania, A., Cangioli, F.: Behavior of a tilting–pad journal bearing with different load directions. In: ASME Paper No. DETC2015-46598, 8 (2015). https://doi.org/10.1115/detc2015-46598
10. Dang, P.V., Chatterton, S., Pennacchi, P., Vania, A., Cangioli, F.: Behavior of tilting-pad journal bearings with large machining error on pads. In: ASME Turbo Expo, Paper No. GT GT2016-56674 (2016)
11. Chatterton, S., Dang, P.V., Pennacchi, P., Luca, A.D., Flumian, F.: Experimental evidence of a two-axial groove hydrodynamic journal bearing under severe operation conditions. Tribol. Int. **109**, 416–427 (2017)
12. Chatterton, S., Dang, P.V., Pennacchi, P., Vania, A.: A test rig for evaluating tilting–pad journal bearing characteristics. In: 9th International Conference on Rotor Dynamics IFToMM ICORD, Milan, Italy, pp. 921–930 (2014)
13. Kirk, R.G., Reedy, S.W.: Evaluation of pivot stiffness for typical tilting-pad journal bearing designs. J. Vib. Acoust. **10**, 165–171 (1988)

Dynamic Characteristics of a Non-symmetric Tilting Pad Journal Bearing

Phuoc Vinh Dang[1(✉)], Steven Chatterton[2], and Paolo Pennacchi[2]

[1] Department of Mechanical Engineering,
The University of Danang-University of Science and Technology,
54 Nguyen Luong Bang Street, Danang, Vietnam
`dpvinh@dut.udn.vn`
[2] Department of Mechanical Engineering, Politecnico di Milano,
Via G. La Masa 1, 20145 Milan, Italy
`{steven.chatterton,paolo.pennacchi}@polimi.it`

Abstract. Because tilting-pad journal bearings are more stable and efficient than conventional bearings, they have been commonly applied to many rotating machinery applications. Most of the studies about steady state and dynamic characteristics of tilting-pad journal bearings are usually evaluated by means of thermo hydrodynamic models assuming nominal dimensions for the bearing. However machining errors could lead to actual bearing geometry and dimensions different from the nominal ones. In particular for tilting-pad journal bearing the asymmetry of the bearing geometry is the principal cause of unexpected behavior. In this paper a theoretical analysis on dynamic characteristics of a five-pad tilting-pad journal bearing is investigated with non-nominal geometry, that is, different thickness for each pad. The dynamic coefficients of a five-pad tilting-pad journal bearing with a nominal diameter of 100 mm, length-to-diameter ratio (L/D) of 0.7 are evaluated versus rotor rotational speed, load direction and static load. Then, the analytical results of the non-nominal bearing are compared to those of a bearing having nominal (i.e. ideal) geometry.

Keywords: Tilting-pad journal bearing · Analytical model · Non-nominal bearing · Dynamic coefficients · Five pads

1 Introduction

Because tilting-pad journal bearings (TPJBs) are more stable and efficient than conventional bearings, they have been commonly applied to many rotating machinery applications. The chief feature of tilting-pads is that they modify their configuration to adapt to every operating condition, creating several convergent-divergent gaps around the circumference and thus making the system highly stable. Since Lund [1] developed a numerical method for calculating dynamic coefficients for tilting-pad journal bearings, extensive theoretical and experimental studies on dynamic and stability analysis have been conducted. In the course of the development of journal bearings, many effective methods have been applied, such as Newton-Raphson method, pad assembly technique, finite elements method, and Genetic Algorithm to calculate static as well as dynamic characteristics of journal bearing [2, 3].

The majority of the papers on TPJB in the literature consider the tilting pad bearings with nominal dimensions [4–10]. Jones et al [11] studied theoretical the effects of bearing clearance and pad clearance on the steady-state and dynamic behavior in tilting pad journal bearings. They concluded that the pad clearance has little effect on the K_{yy} term at small values of bearing clearance ratio and the bearing clearance has more effect than pad clearance.

Strzelecki [12] has studied the dynamic characteristics of tilting-pad bearings with asymmetric pad support using Reynolds equation and an adiabatic model for the oil film. It was applied to a five pad bearing, and then the pad relative clearance has influenced the stiffness and damping coefficients, while the length to the diameter ratio has just affected the direct stiffness coefficients.

However, most papers studied steady-state and dynamic characteristics of tilting-pad journal bearing by means of thermo hydrodynamic (THD) models assuming nominal dimensions for the bearing. It means that the physical dimensions of all pads are identical or the bearing is uniform configuration. In this paper, effects of non-uniform clearance on TPJBs' performances such as dynamic coefficients, clearance profile, shaft locus, power loss, pad tilt angle, and pressure distribution are evaluated using an analytical model. A five-pad tilting-pad journal bearing with different thickness in one pad is analytically modeled. Its steady state and dynamic behaviors are compared with a symmetric tilting-pad journal bearing. Experimental measurement of the clearance profile for a bearing with intentional machining error is also compared to results obtained from the analytical model.

2 Bearing Description

The sketch of the five-pads TPJB considered in this paper is shown in Fig. 1. The bearing is installed in its housing as a standard LOP configuration, with nominal diameter of 100 mm and length–to–diameter ratio of 0.7. The geometric characteristics of the bearing and the operating conditions are listed in Table 1. O_b and O_j which are located at the origin of the $X - Y$ coordinate system denote for the center of the bearing and journal, respectively.

The main advantage of TPJB consists in the pads capability to follow the displace-

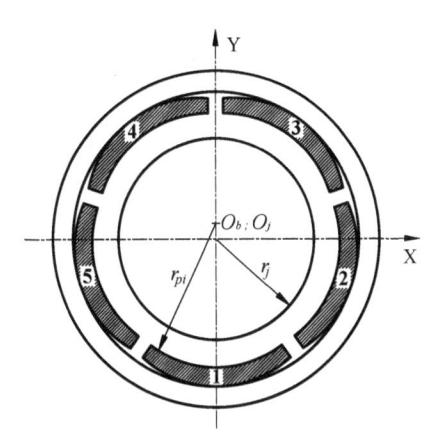

Fig. 1. Sketch of the five-pads TPJB

ments of the rotor. During operation, each pad rotates so that the resultant of the fluid-film forces passes through the pivot. Thus, the pivot location influences the pad rotation and the magnitude of the hydrodynamic pressure distribution.

Table 1. Bearing geometric characteristics and operating conditions.

Item	Unit	Value/Span
Number of pad	–	5
Configuration w.r.t bearing housing	–	LOP
Bearing diameter (D)	mm	100
Machined clearance (C_p)	mm	0.125
Bearing length (L)	mm	70
Angular amplitude of pads	Degree (°)	60
Lubricant	–	ISO VG46
Oil inlet temperature	°C	38–40
Rotational speed	rpm	1200
Static load (on each bearing)	kN	5

In this paper, a non-ideal TPJB, in which the thicknesses of the five pads are different from each other due to machining errors, is considered. The effect is a different assembled clearance or preload factor $(m = 1 - C_b/C_p)$ for each pad (see in Table 2).

Table 2. Specifications of the real tilting-pad journal bearing.

Item	Measurement					Nominal dimension
	Pad #1	Pad #2	Pad #3	Pad #4	Pad #5	
Thickness (mm)	15.9942	16.0146	15.9995	15.9812	16.0179	16.0
Assembled clearance C_b (mm)	0.0658	0.0454	0.0605	0.0788	0.0421	0.07
Preload factor m	0.4739	0.6365	0.5157	0.3698	0.6630	0.44

3 TEHD Bearing Model

A thorough description of the TEHD model to estimate the static and dynamic behaviors of a five-pad TPJB is provided in [5–8] (Fig. 2).

The hydrodynamic model is based on the well-known Reynolds equation:

$$\frac{\partial}{\partial x}\left(\frac{\rho h^3}{\mu}\frac{\partial p}{\partial x}\right) + \frac{\partial}{\partial z}\left(\frac{\rho h^3}{\mu}\frac{\partial p}{\partial z}\right)$$
$$= 6\left[(U_1 - U_2)\frac{\partial h}{\partial x} + h\frac{\partial}{\partial x}(U_1 + U_2) \right.$$
$$\left. + (W_1 - W_2)\frac{\partial h}{\partial z} + h\frac{\partial}{\partial z}(W_1 + W_2) + 2(V_1 + V_2)\right] \quad (1)$$

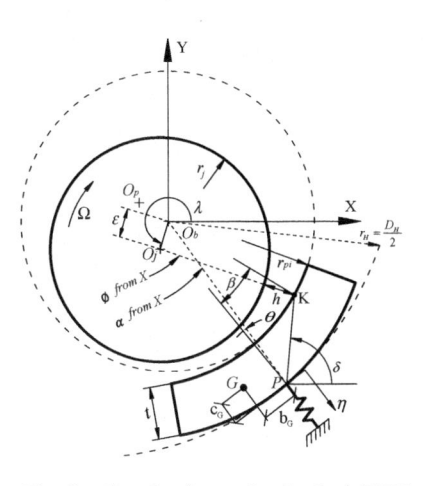

Fig. 2. Sketch of a rocker-backed TPJB [5].

where p is the pressure in the oil-film, h is the oil-film thickness, μ is the dynamic viscosity, ρ is the density of oil, z is the axial direction, x is the tangential direction. The velocity vector component of the shaft and the pads are defined by U_1, V_1, W_1 and U_2, V_2, W_2, respectively.

The effect of temperature on the dynamic viscosity $\mu(T) = \mu_{40°C} \exp[\kappa(T_{40°C} - T)]$ and oil density $\rho(T) = \rho_{40°C}[1 + \alpha_v(T_{40°C} - T)]$ is considered using a simple two-dimensional thermal model, governed, at steady state, by the energy equation:

$$\rho c_p \left(u\frac{\partial T}{\partial x} + w\frac{\partial T}{\partial z}\right) = k_{OIL}\left(\frac{\partial^2 T}{\partial x^2} + \frac{\partial^2 T}{\partial z^2}\right) + \mu\left[\left(\frac{\partial u}{\partial y}\right)^2 + \left(\frac{\partial w}{\partial y}\right)^2\right] \quad (2)$$

where κ is the viscosity index and α_v is the thermal expansion coefficient of the oil, c_p and k_L are the heat capacity and the conductivity of the lubricant respectively. Equation (2) has been integrated using a finite difference method, where adiabatic conditions at the shaft, pad surfaces and constant oil temperature in the oil film thickness direction are considered.

The dynamic coefficients of the bearing because of the lateral motion of the shaft (represented by four impedance coefficients of the impedance matrix $[\mathbf{Z}_{BRG}]$) in Eq. (3) are obtained in Eq. (4) by reducing the complete set of impedance coefficients of the pads (impedance matrix $[\mathbf{Z}^k]$) in Eq. (5), assuming a harmonic motion of the system at frequency ω [5]:

$$\begin{bmatrix} \Delta F_{x,oil} \\ \Delta F_{y,oil} \end{bmatrix} = -[\mathbf{Z}_{BRG}]\begin{bmatrix} \Delta X \\ \Delta Y \end{bmatrix} = -\begin{bmatrix} Z_{xx} & Z_{xy} \\ Z_{yx} & Z_{yy} \end{bmatrix}\begin{bmatrix} \Delta X \\ \Delta Y \end{bmatrix} \quad (3)$$

$$[\mathbf{Z}_{BRG}] = -\sum_k \left(\begin{bmatrix} Z_{xx} & Z_{xy} \\ Z_{yx} & Z_{yy} \end{bmatrix}^k - \begin{bmatrix} Z_{x\theta} & Z_{x\eta} \\ Z_{y\theta} & Z_{y\eta} \end{bmatrix}^k \left(\begin{bmatrix} Z_{\theta\theta} & Z_{\theta\eta} \\ Z_{\eta\theta} & Z_{\eta\eta} \end{bmatrix}^k - \omega^2 [M_{pad}]^k + i\omega [C_{pad}]^k + [K_{pad}]^k \right)^{-1} \begin{bmatrix} Z_{\theta x} & Z_{\theta y} \\ Z_{\eta x} & Z_{\eta y} \end{bmatrix}^k \right)$$

(4)

$$[\mathbf{Z}]^k = [\mathbf{K}]^k + i\omega [\mathbf{C}]^k = \begin{bmatrix} Z_{xx} & Z_{xy} & Z_{x\theta} & Z_{x\eta} \\ Z_{yx} & Z_{yy} & Z_{y\theta} & Z_{y\eta} \\ Z_{\theta x} & Z_{\theta y} & Z_{\theta\theta} & Z_{\theta\eta} \\ Z_{\eta x} & Z_{\eta y} & Z_{\eta\theta} & Z_{\eta\eta} \end{bmatrix}^k$$

(5)

where $[\mathbf{K}]^k$ and $[\mathbf{C}]^k$ are the linear stiffness and damping coefficient matrices, respectively, which are calculated for the k-th pad, and:

$$[M_{pad}]^k = \begin{bmatrix} J_P & m\, b_G \\ m\, b_G & m \end{bmatrix}^k \quad [C_{pad}]^k = \begin{bmatrix} c_\theta & 0 \\ 0 & c_\eta \end{bmatrix}^k \quad [K_{pad}]^k = \begin{bmatrix} k_\theta & 0 \\ 0 & k_\eta \end{bmatrix}^k$$

(6)

where J_p is the mass moment of inertia of the pad, m is the mass, and b_G is the position of the barycenter. The pivot stiffness of the pivot along the direction η, which depends on the applied static load W, is obtained using the contact Hertz theory.

4 Results and Discussion

4.1 Dynamic Coefficients Versus Rotor Rotational Speed

Figure 3 shows the stiffness coefficients of two bearings as a function of rotational speed in LOP configuration when a static load of 5 kN is applied on the bearing in the vertical direction. The direct stiffness coefficients k_{xx} and k_{yy} increase more or less linearly with increasing of rotational speed from 1000 rpm to 3000 rpm. The direct stiffness coefficient of the non-nominal bearing in unloaded direction (k_{xx}) is always larger than (about 50%) that of in the nominal bearing. On the other hand, the stiffness coefficients in the orthogonal loaded direction (k_{yy}) show an opposite trend. These results are consistent with the level of orthotropy expected for a LOP configuration, as presented in [13].

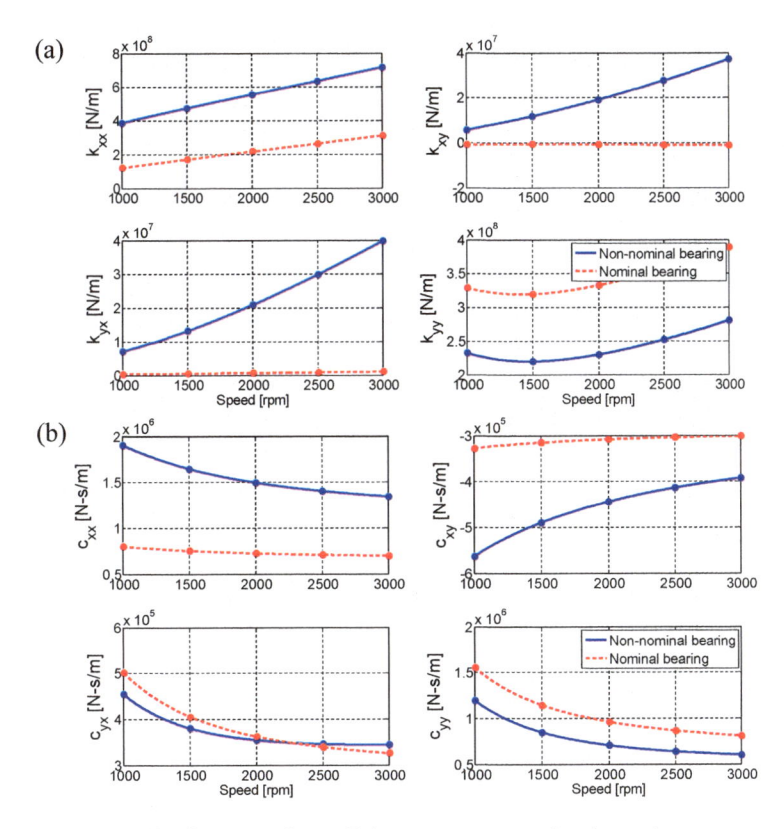

Fig. 3. Dynamic coefficients vs. rotor rotational speed

The non-nominal dimension has strong impact on cross-coupled stiffness coefficients (k_{xy} and k_{yx}). While the k_{xy} and k_{yx} of the nominal bearing are almost zero with rotational speed, these coefficients of the non-nominal bearing strongly depend on rotor speed, increase linearly with increasing of rotational speed. It could be explained that despite the applied load is vertical; the center locus of the non-nominal bearing is not vertical (see Fig. 4). Hence, there is a cross-coupling effect, so it is necessary to take into account also the cross-coupling dynamic coefficients k_{xy} and k_{yx} when the dynamic characteristics of the non-nominal bearing are evaluated.

Contrary results are shown for the direct dynamic damping coefficients. They decrease with increasing of rotational speed. For the bearing with non-nominal dimension, the direct damping in loaded direction c_{yy} is less than in case of the nominal bearing, nearly about 10%.

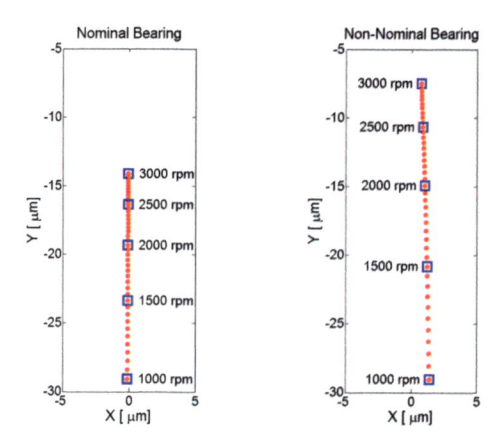

Fig. 4. Predicted shaft loci vs. rotor rotational speed

4.2 Dynamic Coefficients Versus Static Load

Figure 5 shows the effect of increasing static load on the dynamic stiffness and damping coefficients for rotor speed of 1200 rpm (20 Hz) with LOP configuration.

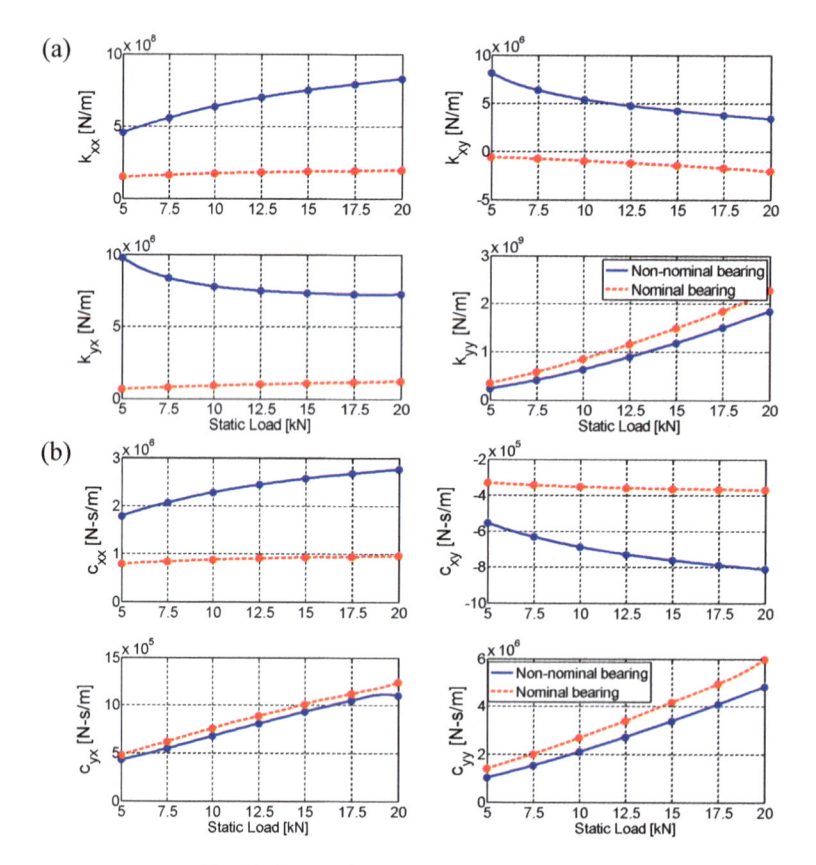

Fig. 5. Dynamic coefficients vs. static load

While it should be noted that the direct coefficients in the unloaded direction (k_{xx}) of the nominal bearing is quite stable with static load, this coefficient in the non-nominal bearing increases linearly with increasing of static load, from 4.6×10^8 N/m at 5 kN to 8.35×10^8 N/m at 20 kN.

On the other hand, the dynamic direct stiffness coefficients in the loaded direction (k_{yy}) on both kinds of bearing strongly depend on the static load and the stiffness coefficient of the nominal bearing is always greater (about 10%) than that of the non-nominal one. This coefficient increases significantly, more than 600%, from 3.57×10^8 N/m to 2.28×10^9 N/m for the nominal bearing and from 2.48×10^8 N/m to 1.86×10^9 N/m for the non-nominal bearing corresponding static load from 5 kN to 20 kN. It can be seen that the cross-coupled stiffness coefficients of non-nominal bearing (k_{xy} and k_{yx}) are much larger than those of the nominal one. However these coefficients are very small in comparison with the direct coefficients on both bearings.

The similar results are shown for the dynamic damping coefficients as shown in Fig. 5 except for the cross-coupled damping coefficients (c_{xy} and c_{yx}) in which the cross coefficients of the nominal bearing are always higher than those of the non-nominal one.

4.3 Dynamic Coefficients Versus Load Direction

Generally, a TPJB is loaded symmetrically, i.e. load-on-pad or load-between-pad configuration, the former being the more common situation. However in practical the bearing may be loaded in any direction, not just in these two special directions. For instance, high bearing loads out of the vertical direction may occur in industrial rotating machines, like in turbo-generators, owing to bad alignment conditions of the shaft-line or in gearboxes. For the operating conditions that load direction change rapidly, it needs to perform the analysis of the bearing in its special load forms to enhance the precision and efficiency of bearing design.

Two reference systems are introduced in this section to evaluate dynamic coefficients, namely the absolute reference and the load reference systems. For the first one, a varying load direction is considered, as shown in Fig. 6. In this case, the static load is rotated in a full revolution (360°), with steps of 18°. It means that, all the five pads will be loaded under LOP and LBP configuration.

Fig. 6. Varying load configuration

Figure 7 shows the influence of the load direction on the calculated stiffness and damping coefficients of the nominal and non-nominal bearing. Obviously, the load direction has a strong effect on both of bearings' dynamic coefficients.

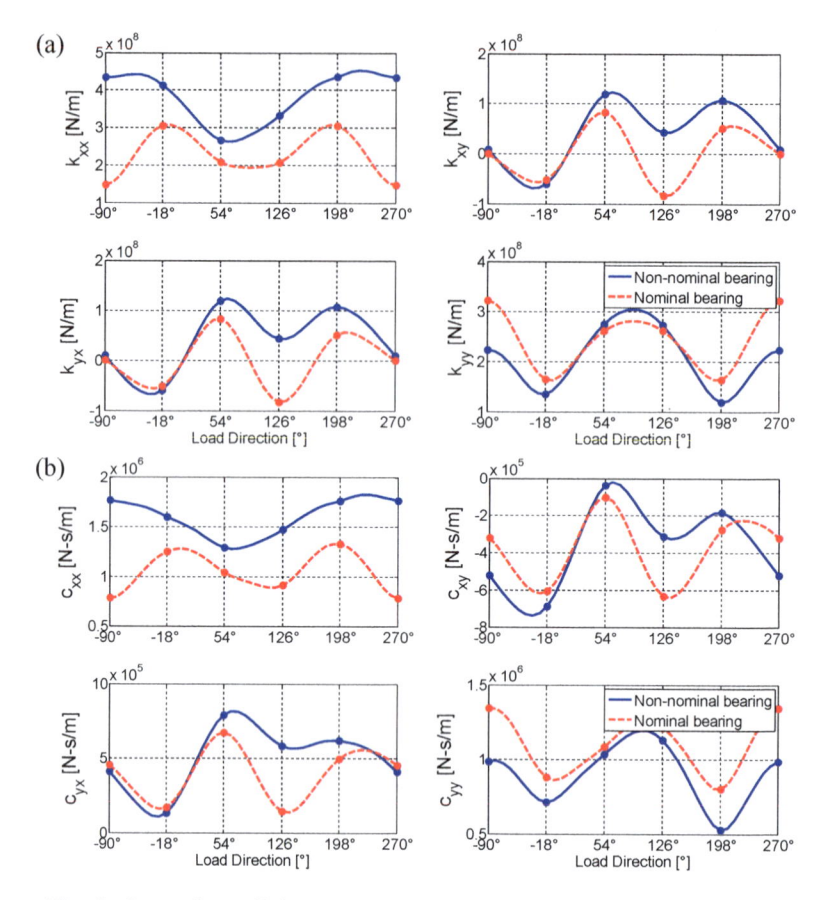

Fig. 7. Dynamic coefficients vs. load directions in absolute reference system

The dynamic stiffness coefficients of the nominal bearing show a general good agreement with the calculated values of the non-nominal one apart from the direct term in horizontal direction (k_{xx}). The k_{xx} of the non-nominal bearing is always larger than that of the nominal one, especially when the static load is applied in the vertical (load direction is −90°) and pad 1 is loaded. At −90°, the direct stiffness coefficient in the horizontal direction k_{xx} of the non-nominal bearing is about 4.3×10^8 N/m while this value for the nominal one is nearly 1.5×10^8 N/m. The effects of load directions on the damping coefficients (Fig. 7b) are very similar to those on the stiffness terms.

In the case of load reference system, the load direction is kept fixed in the vertical direction and the bearing is rotated, from −90° to 270°, with steps of 18° with respect to the load direction, as shown in Fig. 8.

Fig. 8. Varying load configuration

The influence of load directions on the dynamic coefficients in the load reference is shown in Fig. 9.

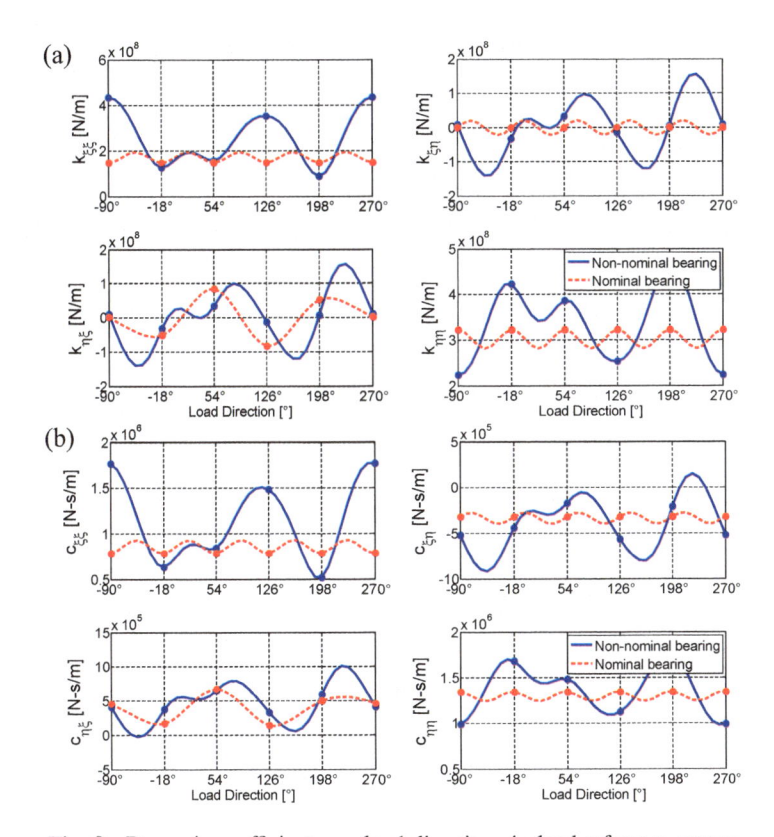

Fig. 9. Dynamic coefficients vs. load directions in load reference system

It also possible to observe that the calculated dynamic coefficients of the nominal bearing maintain a certain symmetry as a function of the direction of the load with a small variation when the static load rotates. This is less evident for the non-nominal bearing, due to asymmetric geometry. The dynamic coefficients of the non-nominal bearing show a large fluctuation with varying load directions. Similar to the previous varying load configuration, also in this case, direct term of dynamic stiffness and damping coefficients in unloaded direction ($k_{\xi\xi}$ and $c_{\xi\xi}$) show the worse agreement between the nominal and the non-nominal one, among the dynamic coefficients.

5 Conclusions

This paper presents the dynamic characteristics of a five-pads TPJB, which is characterized by not constant assembled clearance for all the pads distribution, due to pad machining and assembling. A comparison is performed with a bearing having nominal (i.e. ideal) geometry. The following conclusions can be drawn:

1. It is necessary to take into account the cross-coupling dynamic coefficients k_{xy} and k_{yx} when the dynamic characteristics of non-ideal TPJB are evaluated.
2. The effect of static load on direct dynamic coefficients in the loaded direction (k_{yy} and c_{yy}) is stronger than those calculated in the unloaded direction (k_{xx} and c_{xx}).
3. The reference system can be considered in absolute reference or in load reference when evaluating effect of load directions.
4. The load direction has a strong effect on the characteristics of TPJBs. This influence can be increased or decreased by a non-nominal geometry of the bearing, for which big differences can be identified between the experimental performance of a real bearing and the predicted behavior obtained using a model of a nominal bearing.

References

1. Lund, J.W.: Spring and damping coefficients for tilting pad journal bearing. ASLE Trans. **7**, 342–352 (1964)
2. Zheng, T.S., Hasebe, N.: Calculation of equilibrium position and dynamic coefficients of a journal bearing using free boundary theory. Trans. ASME J. Tribol. **122**, 616–621 (2000)
3. Feng, K., Liu, W., Zhang, Z., Zhang, T.: Theoretical model of flexure pivot tilting pad gas bearings with metal mesh dampers in parallel. Tribol. Int. **94**, 26–38 (2016)
4. Cha, M., Glavatskih, S.: Nonlinear dynamic behaviour of vertical and horizontal rotors in compliant liner tilting pad journal bearings: some design considerations. Tribol. Int. **82**, 142–152 (2015)
5. Dang, P.V., Chatterton, S., Pennacchi, P., Vania, A.: Effect of the load direction on non-nominal five-pad tilting-pad journal bearings. Tribol. Int. **98**, 197–211 (2016)
6. Dang, P.V., Chatterton, S., Pennacchi, P., Vania, A., Cangioli, F.: Behavior of a tilting–pad journal bearing with different load directions. ASME Paper No. DETC2015-46598, vol. 8 (2015). https://doi.org/10.1115/detc2015-46598
7. Dang, P.V., Chatterton, S., Pennacchi, P., Vania, A.: Numerical investigation of the effect of manufacturing errors in pads on the behaviour of tilting-pad journal bearings. Proc. Inst. Mech. Eng. Part J.: J. Eng. Tribol. **233**(4), 480–500 (2018)
8. Dang, P.V., Chatterton, S., Pennacchi, P., Vania, A., Cangioli, F.: Behavior of tilting-pad journal bearings with large machining error on pads. ASME Turbo Expo, Paper No. GT GT2016-56674 (2016)
9. Chatterton, S., Dang, P.V., Pennacchi, P., Luca, A.D., Flumian, F.: Experimental evidence of a two-axial groove hydrodynamic journal bearing under severe operation conditions. Tribol. Int. **109**, 416–427 (2017)

10. Chatterton, S., Dang, P.V., Pennacchi, P., Vania, A.: A test rig for evaluating tilting–pad journal bearing characteristics. In: 9th International Conference on Rotor Dynamics IFToMM ICORD, Milan, Italy, pp. 921–930 (2014)
11. Jones, G.J., Martin, F.A.: Geometry effects in tilting-pad journal bearings. In: The 33rd Annual Meeting in Dearborn, Michigan (1978)
12. Strzelecki, S.: Dynamic characteristics of tilting 5-pads journal bearing with asymmetric support of pads. In: Sixth International Conference on Rotor Dynamics (IFToMM), University of New South Wales, Sydney, Australia, pp. 807–814 (2002)
13. Delgado, A., Vannini, G., Ertas, B.M., Drexel, M., Naldi, L.: Identification and prediction of force coefficients in a five-pad and four-pad tilting pad bearing for load-on-pad and load-between-pad configurations. J. Eng. Gas Turbines Power **133**(9), 092503–092503-9 (2011)

Energy

DCM Boost Converter in CPM Operation for Tuning Piezoelectric Energy Harvesters

Andrés Gomez-Casseres[1(✉)], David Florez[2], and Darío Cortes[3]

[1] Corporación Unificada Nacional de Educación Superior - CUN, Bogotá, Colombia
andres_gomezcasseres@cun.edu.co
[2] Departamento de Ingeniería Electrónica, Pontificia Universidad Javeriana, Bogotá, Colombia
[3] Corporación Unificada Nacional de Educación Superior - CUN, Bogotá, Colombia

Abstract. The power extraction from piezoelectric energy harvesters is considered an important alternative to the employment of batteries when powering ultra-low power circuits. However, the amount of extracted power and the frequency range where extraction is possible remain as key challenges for practical implementations. In this paper, a boost rectifier in *Current Programmed Mode* (CPM) able to emulate a complex load at its input terminals is presented. This circuit is validated through circuit simulation using PSIM9. From the results, the circuit is capable of extracting the maximum available power from a piezoelectric harvester, modeled by an electric equivalent circuit, at its first resonant frequency. This is achieved by the emulation of an *RC* network at the harvester's terminals by controlling the peak current through its inductance.

Keywords: Boost rectifier · Current programed mode · Energy harvesting · Piezoelectric

1 Introduction

Energy harvesting (EH) has received a lot of attention in the last decade due to the possible reduction, or even elimination, of battery employment in low power applications. This is achievable by the harvesting of power present in the surroundings. Several applications, such as Wireless Sensor Networks (WSN), Internet of Things (IoT), biomedical wearables, among others, have been benefited from the increased portability, access to hazardous environments and reduction in maintenance costs feasible with EH.

Although EH offers important benefits when implementing low power electronics in different areas, there are some challenges that limit its impact and application range. According to Todaro *et al.* [9], there are two main challenges that remain unsolved: low output power and limited bandwidth of vibrational EH. Commonly, the approach to attain higher power outputs have been related to the mechanical design of the harvester [5,9,10], while the solutions for higher power bandwidth also require an electrical approach. Nevertheless, some attempts to improve the output power within the electrical domain are present in the literature, regarding the consumption and efficiency of the power and control circuitry [4,6–8].

In many real applications, the harvesters used are high Q systems with a highly peaked frequency response. Although this narrow-band characteristic improves the output power at resonance, any variation in the input frequency of the system will decrease the output power, even preventing system operation. This problem is stressed when the mechanical system is excited in a intermittent way [11]. Therefore, in order to maintain the requested output power, the bandwidth of the EH must allow to harvest energy from such nondeterministic sources.

One attempt to increase the bandwidth of an energy harvester, presented in [2], proposes the emulation (synthesis) of a complex impedance by using a digital platform. This system enables the synthesizing of a series RC network, presented as the load of an electromagnetic kinetic harvester, thus, implementing maximum power transfer (MPT). However, the employment of a battery as the converter's load reduces the usage of this system in real applications.

In this work, the design and simulation of a digitally controlled harvesting circuit is presented. The circuit, composed by a boost rectifier, is able to synthesize a complex RC load to attain MPT from a piezoelectric energy harvester. In order to simulate the behavior of a piezoelectric energy harvester, the model developed by Yang et al. [12] is used.

This paper is organized as follows: in Sect. 2 the proposed architecture, along with its related requirements, is presented. In Sect. 3 the design of the harvester subsystems and component selection is performed. Section 4 presents the simulated results and discusses the benefits and disadvantages of the proposed system. Finally, in Sect. 5, conclusions are drawn and further work is discussed.

2 Architecture Description

In Fig. 1 the architecture of the EH system is presented. It is conformed by a piezoelectric energy harvester, a rectifier circuit, a DC-DC power converter, a storage capacitor and the load. As stated previously, the piezoelectric energy harvester is modeled by the equivalent circuit presented in Fig. 2. Furthermore, the load is composed by a 3.3 kΩ resistor in series with a switching MOSFET. Such MOSFET is turned on when the voltage on the storage capacitor C_{str} reaches 3.8 V, connecting the 3.3 kΩ resistor to the storage capacitor, and, when the capacitor voltage falls below 3.3 V, the MOSFET is turned off. Should be pointed that, in order to decrease the burden imposed by the ancillary circuits, the system will not implement any protection.

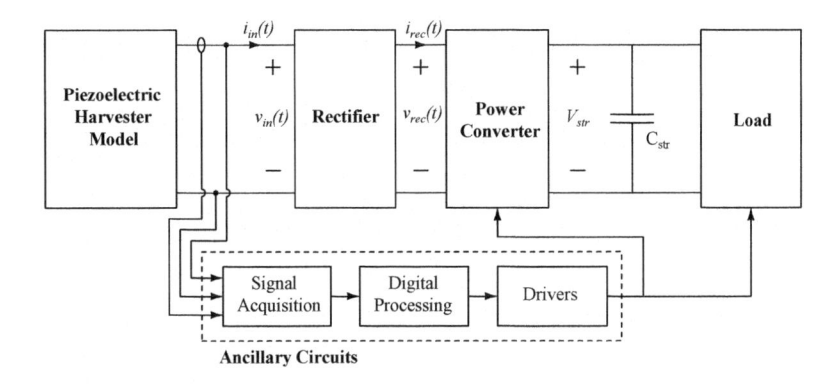

Fig. 1. Proposed architecture

2.1 Model Transfer Function

In order to accomplish MPT from the piezoelectric harvester, a complex load impedance must be synthesized at the piezoelectric terminals [1]. Such impedance must present a capacitive behavior to extract the maximum power at the first vibrational mode, given the electric equivalent model of the piezoelectric harvester, shown in Fig. 2. A parallel configuration of an *RC* network is used as a load, where the resistance and capacitance values that attain MPT are 31.781 kΩ and 32.479 nF, respectively. These values were obtained through the computation of the piezoelectric admittance, presented in Eq. 1, and the employment of the MPT theorem. The magnitude and phase of the harvester's admittance are presented in Fig. 3. Both, the harvester parameters f_r, C_r, L_r, R_r and N_r (where r and f_r are the vibrational mode and the frequency where such mode appears, respectively), and the obtained values for R_{opt} and L_{opt} are presented in Table 1 for each resonant frequency.

$$Y_s = j\omega \left(C_e + \sum_{r=1}^{\infty} \frac{N_r^2}{j\omega R_r - \omega^2 L_r + \frac{1}{C_r}} \right) \quad (1)$$

From Fig. 3 is clear that the harvester has a very peaked response. This limits the power extraction to the frequencies where the admittance phase differs from 90°, which are the regions in close proximity to the resonant frequencies. It should also be pointed out that, around the first resonant frequency, the harvester presents and inductive behavior characterized by a negative phase shift. This leads to a capacitive optimum load where the first vibrational mode is present. However, this is not valid for the other resonant frequencies, where an inductive optimum load is required.

Given that the optimum admittance at the first resonant frequency is an *RC* network, the difference equation that computes the piezoelectric current from its output voltage is presented in Eq. 2. This equation was obtained using the backward Euler method on the admittance of an *RC* network. The backward

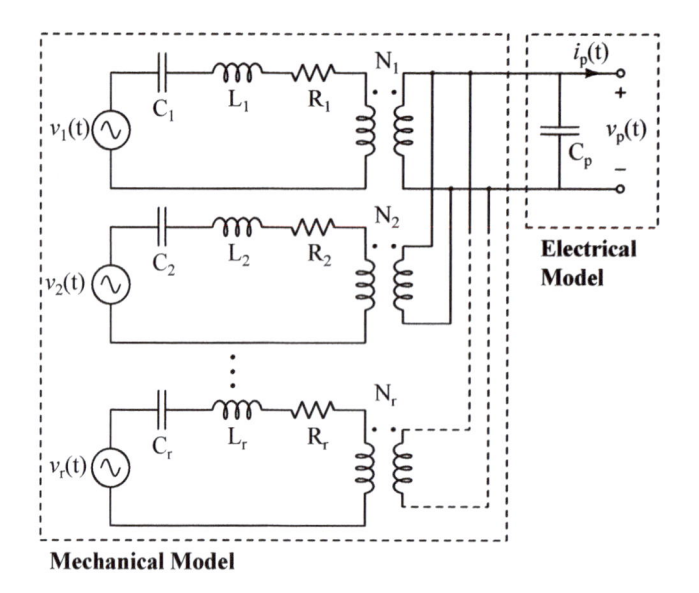

Fig. 2. Piezoelectric circuit model obtained from FEA.

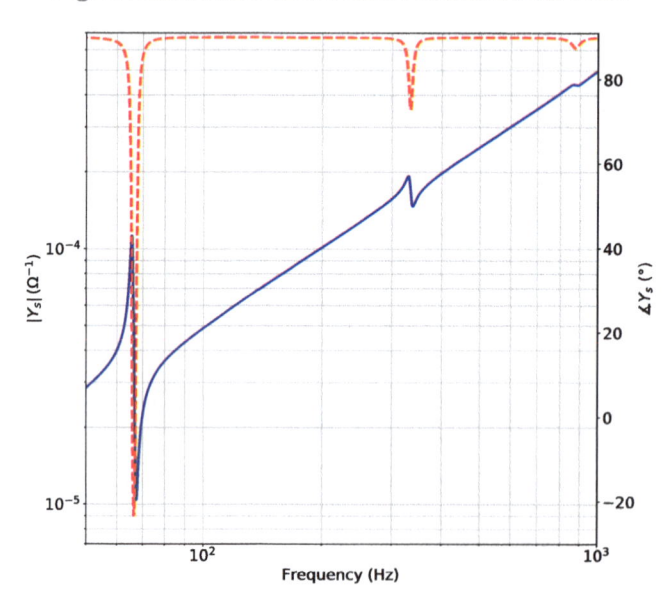

Fig. 3. Frequency response of the admittance presented by the piezoelectric material.

Euler method was employed given that this model will be implemented using a digital processor.

$$I_p[k] = \left(\frac{1}{R_{opt}} + \frac{C_{opt}}{T_s}\right)V_p[k] - \frac{C_{opt}}{T_s}V_p[k-1] \qquad (2)$$

Table 1. Piezoelectric model parameters and optimal load values

Parameter	1st Resonant frequency	2nd Resonant frequency	3rd Resonant frequency
f_r	66.505 Hz	334.409 Hz	876.822 Hz
C_r	5.874 µF	227.4 nF	32.96 nF
L_r	1 H	1 H	1 H
R_r	6.9366 Ω	57.868 Ω	343.56 Ω
N_r	0.0267198	0.0541673	0.0817594
L_{opt}	32.4787 nF*	2.9493 H	0.4122 H
R_{opt}	31.781 kΩ	20.116 kΩ	51.377 kΩ

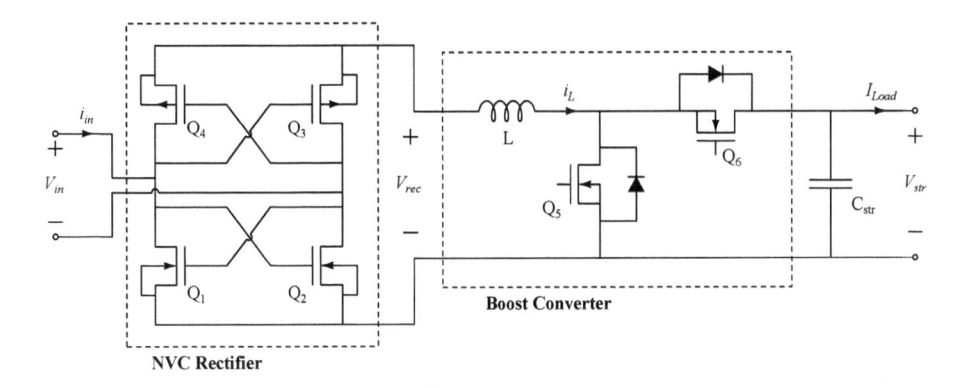

Fig. 4. Power stage of the energy harvester

2.2 Circuit Description and Operation

In Fig. 4, the boost rectifier is presented, where the rectifier stage is implemented with a Negative Voltage Converter (NVC) whilst the converter stage is conformed by a bidirectional boost converter. The NVC was selected due to its low losses, compared with a diode bridge rectifier, and its ability to operate without additional circuitry [3]. Also, in order to obtain high efficiency operation, a bidirectional boost, operated in Discontinuous Conduction Mode (DCM), is employed as a DC-DC converter. The improvement in power efficiency is achieved by the Zero Current Switching (ZCS), intrinsic in the DCM operation [6].

The bidirectional power flow through the converter is achieved by the use of the body diodes in MOSFETs Q_5 and Q_6. When a power sink operation is needed, transistor Q_6 remains off while Q_5 turn on and off at a fixed switching frequency. This leads to the flow of current from the harvester to the load and the circuit operating as a boost converter with the body diode of Q_6 conducting current when Q_5 is turned off. On the other hand, when a sourcing operation is required, transistor Q_5 remains off while Q_6 is switched, hence, the converter operates as a buck converter, where its input is connected to the reservoir capacitor while the output is the piezoelectric harvester. It should be pointed out that, under the source operation, the body diode in Q_5 behaves as a freewheeling diode that provides a current path for the discharge of the converter's inductance (L).

3 Digital Implementation

In Fig. 5 a block diagram of the modules used in a TMS320F28069 microcontroller to obtain CPM operation of the boost rectifier is presented. In this work, the COMP1, COMP2, COMP3, CLA and PWM modules are employed. The COMP2 and COMP3 modules verify the polarity of the voltages $v_{in,a}$ and $v_{i_{in},a}$, which are the output voltages of the signal acquisition block and are proportional to the output voltage and current of the piezoelectric harvester. This is performed with the aim to establish the direction of the power flow within the converter. After the COMP2 and COMP3 modules, the CLA (Control Law Accelerator) computed the current from the piezoelectric harvester to obtain MPT. Finally, the COMP1 and PWM modules implement a CPM operation, feeding the signals to the Gate's drivers.

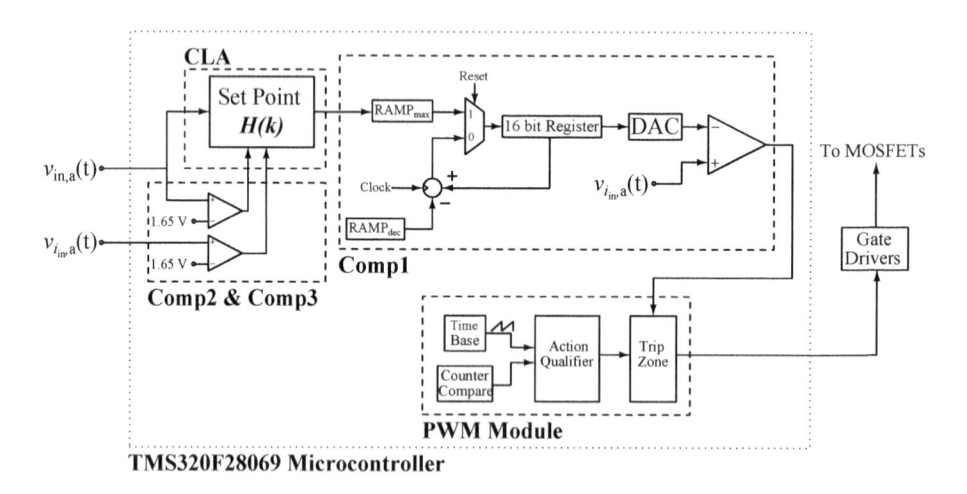

Fig. 5. Block diagram of the system implemented in the TMS320F28069 microcontroller

The flow chart that implements the set point function $H(k)$ is presented in Fig. 6. First, the computation of the piezoelectric input voltage is obtained from an ADC result. Given that the TMS320F28069 microcontroller has a 12 bit ADC with a reference voltage equal to 3.3 V, the expression to obtain the piezoelectric input voltage is shown in Eq. 3, where N, V_{ref} and A_{cond} are the bit number, voltage reference and the acquisition gain.

$$\frac{2^N}{V_{ref} A_{cond}} (ADC_{result} - 2^{N-1}) \qquad (3)$$

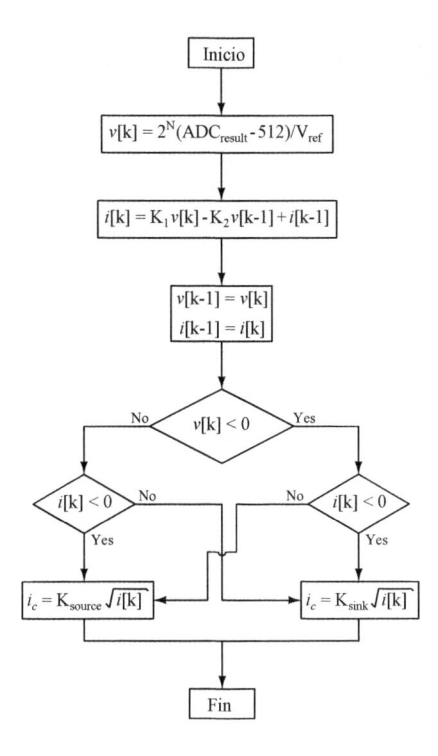

Fig. 6. Bidirectional boost converter

Followed by the computation of the piezoelectric input voltage, the calculation of the piezoelectric current when an optimum complex load is connected at its terminals is carried out. Such optimum complex load is the complex conjugate of the Eq. 1. After this step, the computed voltage and current are storaged to be taken as the previous value in subsequent calculations. Finally, the direction of the power flow is determined by the sign of the current and voltage waveforms, which is made by the result from the Comp2 and Comp3 modules and determines the gain to compute i_c.

4 Validation

The boost rectifier, along with the algorithm presented in Fig. 6, were simulated in PSIM version 9. In Figs. 7 and 8 the simulation results are shown. In Fig. 7 the input voltage, divided by 1×10^6, current and reference current are presented. From the figure, it is clear that the input current follows the reference and a phase shift of approximate 20° between the current reference and the input voltage is established. Furthermore, in Fig. 8, the DCM operation of the boost rectifier is validated. From this figure, it is possible to visualize a distortion when the input voltage crosses zero. This arises from the low input voltage that reduces the current slope.

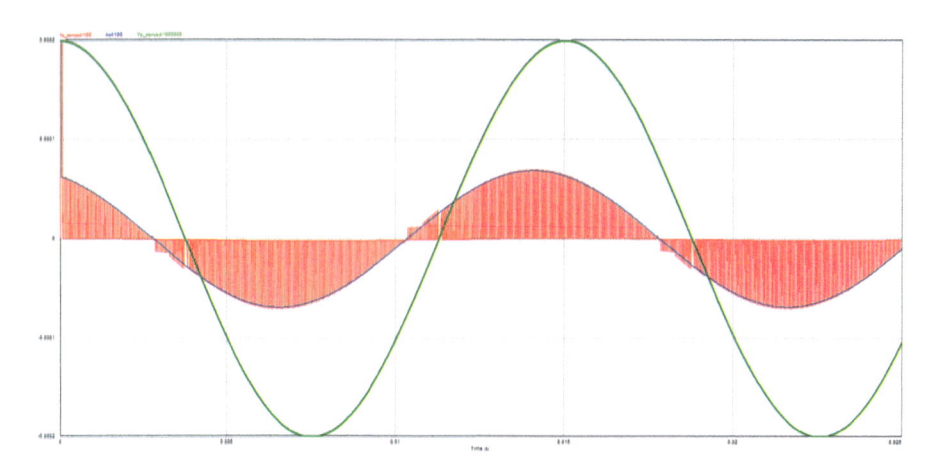

Fig. 7. Voltage (green), measured current (red) and reference current(blue)

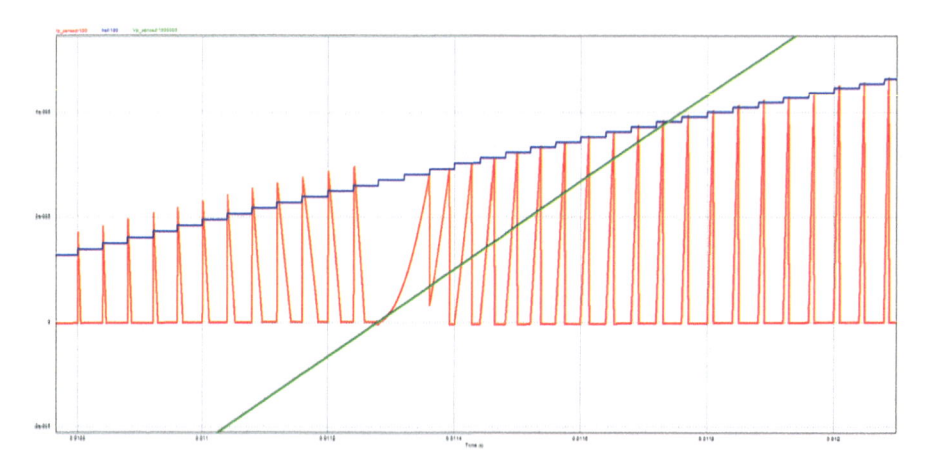

Fig. 8. Zoom of Fig. 7.

Futhermore, in Fig. 9, the Fast Fourier Transform of the measured and reference current obtained from simulation are presented. The peak values of such signals, which appear at a frequency equal to 66 Hz, are 1.3754 mA and 6.8655 mA, respectively. It is clear that the control strategy must take into account the relation between the average and the peak current in CPM operation to reduce such a difference. Furthermore, the relation between the piezoelectric voltage and current must be considered at the acquisition stage. This in order to avoid unexpected behavior when connecting the system to the piezoelectric harvester.

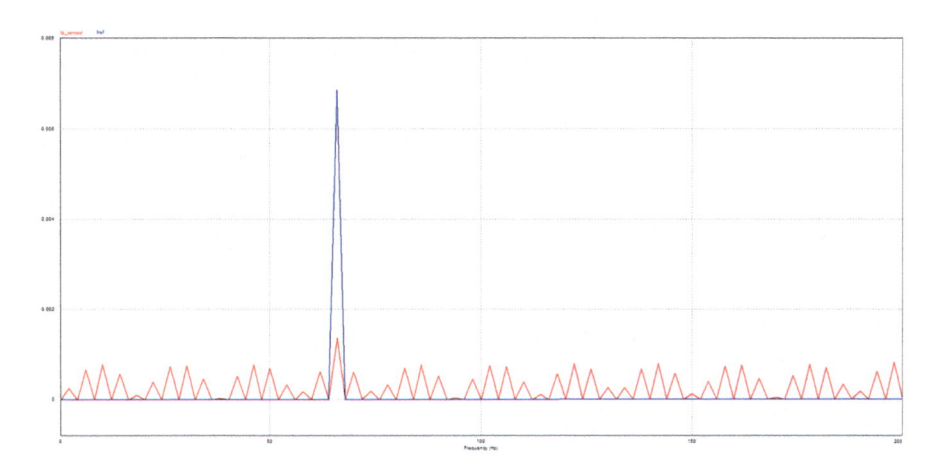

Fig. 9. FFT of the reference current (blue) and the measured current signal (red)

5 Conclusion

In this article, the operation of a DCM boost rectifier, composed by a NVC and a boost converter, as a harvesting circuit is validated. This is achieved by the emulation of a complex RC impedance at the input port of the boost rectifier through a digitally-implemented control strategy. This impedance emulation assures the maximum power extraction from the piezoelectric harvester at the first system resonance. In order to achieve a better performance, the control strategy might be improved by considering the variations of the input and output voltages and the differences between the peak and the averaged current waveforms. Furthermore, it is necessary that the voltage measurement is considered when the input port of the rectifier is connected to the piezoelectric harvester in order to avoid system unexpected behavior.

References

1. Abdelmoula, H., Abdelkefi, A.: Ultra-wide bandwidth improvement of piezoelectric energy harvesters through electrical inductance coupling. Eur. Phys. J. Spec. Top. **224**(14–15), 2733–2753 (2015). https://doi.org/10.1140/epjst/e2015-02586-4
2. Bowden, J.A., Burrow, S.G., Cammarano, A., Clare, L.R., Mitcheson, P.D.: Switched-mode load impedance synthesis to parametrically tune electromagnetic vibration energy harvesters. IEEE/ASME Trans. Mechatron. **20**(2), 603–610 (2015). https://doi.org/10.1109/TMECH.2014.2325825
3. Gomez-Casseres, E.A., Arbulu, S.M., Franco, R.J., Contreras, R., Martinez, J.: Comparison of passive rectifier circuits for energy harvesting applications. In: 2016 IEEE Canadian Conference on Electrical and Computer Engineering (CCECE), pp. 1–6 (2016). https://doi.org/10.1109/CCECE.2016.7726840
4. Le, T.T., Han, J., von Jouanne, A., Mayaram, K., Fiez, T.S.: Piezoelectric micropower generation interface circuits. IEEE J. Solid-State Circuits **41**(6), 1411–1420 (2006). https://doi.org/10.1109/JSSC.2006.874286

5. Liu, H., Qian, Y., Lee, C.: A multi-frequency vibration-based MEMS electromagnetic energy harvesting device. Sens. Actuators A: Phys. **204**, 37–43 (2013). https://doi.org/10.1016/j.sna.2013.09.015
6. Sankman, J., Ma, D.: A 12-µW to 1.1-mW AIM piezoelectric energy harvester for time-varying vibrations with 450-nA IQ. IEEE Trans. Power Electron. **30**(2), 632–643 (2015). https://doi.org/10.1109/TPEL.2014.2313738
7. Szarka, G.D., Burrow, S.G., Stark, B.H.: Ultralow power, fully autonomous boost rectifier for electromagnetic energy harvesters. IEEE Trans. Power Electron. **28**(7), 3353–3362 (2013). https://doi.org/10.1109/TPEL.2012.2219594
8. Szarka, G.D., Stark, B.H., Burrow, S.G.: Review of power conditioning for kinetic energy harvesting systems. IEEE Trans. Power Electron. **27**(2), 803–815 (2012). https://doi.org/10.1109/TPEL.2011.2161675
9. Todaro, M.T., Guido, F., Mastronardi, V., Desmaele, D., Epifani, G., Algieri, L., De Vittorio, M.: Piezoelectric MEMS vibrational energy harvesters: advances and outlook. Microelectron. Eng. (2017). https://doi.org/10.1016/j.mee.2017.10.005. http://www.sciencedirect.com/science/article/pii/S0167931717303349
10. Toprak, A., Tigli, O.: Piezoelectric energy harvesting: state-of-the-art and challenges. Appl. Phys. Rev. **1**(3), 031104 (2014). https://doi.org/10.1063/1.4896166
11. Yang, G., Stark, B.H., Hollis, S.J., Burrow, S.G.: Challenges for energy harvesting systems under intermittent excitation. IEEE J. Emerg. Sel. Top. Circuits Syst. **4**(3), 364–374 (2014)
12. Yang, Y., Tang, L.: Equivalent circuit modeling of piezoelectric energy harvesters. J. Intell. Mater. Syst. Struct. **20**(18), 2223–2235 (2009). https://doi.org/10.1177/1045389X09351757

Effect of Weighting Coefficients on Behavior of the DTC Method with Direct Calculation of Voltage Vector

Jakub Baca[✉] [iD], Martin Kuchar [iD], and Petr Palacky [iD]

VSB – Technical University of Ostrava, 17. listopadu 15,
708 33 Ostrava – Poruba, Czechia
{jakub.baca,martin.kuchar,petr.palacky}@vsb.cz

Abstract. Direct torque control and vector control are modern ways of controlling AC machines that allow an electric drive to achieve a very fast response. This paper deals with the method of direct torque control of an induction motor, which is called Direct Torque Control with Direct Calculation of Voltage Vector (DVC-DTC). The method is dependent on the setting of the coefficients referred to as k_1 and k_2, so the aim of this contribution is to show how their setting affects the behavior of a drive. A DSC-based implementation of this method has been developed using a modern control system with the TMS320F28335 digital signal controller from Texas Instruments. The first part of this paper includes a theoretical description of the method. The second part describes the courses of the most important quantities, which were measured for different settings of the weighting coefficients. The experimental results were obtained by measuring a real drive in laboratory conditions.

Keywords: Direct torque control · DTC · Induction motor · Weighting coefficients

1 Introduction

Developments in the field of digital computing technologies and power electronics have made wider use of AC machines possible. There are currently three basic ways of controlling AC machines: scalar control, vector control and direct torque control. The scalar control is used only for dynamically non-problematic drives that work mainly in the steady state. For more demanding applications, vector control or direct torque control is used. These two methods are characterized by very high dynamics and the accuracy of the quantities in both steady states and transient states.

The aim of vector control and direct torque control is to achieve AC drive characteristics comparable to DC drives. The basic idea of these two methods is, therefore, to allow the independent control of the torque and excitation (magnetic flux) of a machine. For the vector control, these two variables are indirectly controlled by two mutually perpendicular components of the stator current vector, the so-called flux-generating and torque-generating current. The direct torque control of an induction motor is based on the control of the electromagnetic torque and amplitude of the stator magnetic flux in the required tolerance band; the endpoint of stator magnetic flux vector moves along the specified curve (ideally a circle in a steady state). The torque is given by the rotation speed of the stator flux vector and the magnetization of an induction motor is given by the modulus of the vector [1, 2].

Many direct torque control methods and their modifications have been developed, e.g. [3–7]. The method of direct torque control with direct calculation of voltage vector was proposed in the 1990s at the Department of Electronics, VSB – Technical University of Ostrava [8, 9], it was originally called the New Method [10]. The weighting coefficients k_1 and k_2 are used in the calculation of a voltage vector that is to appear at the inverter output. It has been found that these coefficients have a significant effect on the quality of the quantities of an induction motor and on the switching frequency of an inverter.

2 Description of the Method

The algorithm of the DVC-DTC method computes directly the inverter output voltage vector which to be switched. Unlike other methods, it is not necessary to divide the α-β plane into sectors and create a switching table; the voltage vector is determined by a calculation from the torque and flux error values where coefficients k_1 and k_2 are used. This section briefly describes the principle of the method; a detailed description can be found in [8] or [9].

In Fig. 1, we can see vectors $\mathbf{g}_1$ and $\mathbf{g}_2$ plotted in the stator coordinate system [α, β]. The vector $\mathbf{g}_1$ always has the same direction as the stator flux vector, while $\mathbf{g}_2$ is always perpendicular to the flux vector. By the vector sum of these two vectors, the resultant vector $\mathbf{g}$ is given. If it is true that:

$$\Delta \Psi_1 = \Psi_{\text{ref}} - |\Psi_1^S| \tag{1}$$

$$\Delta T = T_{\text{ref}} - T \tag{2}$$

$$|\mathbf{g}_1| = k_1 \Delta \Psi_1 \tag{3}$$

$$|\mathbf{g}_2| = k_2 \Delta T \tag{4}$$

then the vector $\mathbf{g}$ determines the direction of the voltage vector that should be switched. The constants k_1 and k_2 are the weighting coefficients. If a voltage vector whose direction is identical to the vector $\mathbf{g}$ is applied at the inverter output, then the size of vector $\mathbf{g}_1$ (3) determines the degree of the motor excitation change and the size of vector $\mathbf{g}_2$ (4) determines the degree of the rotation speed of the stator flux vector and thereby torque size. Since two-level voltage inverter has only eight voltage vectors – see Fig. 1, two of which are zero, it is necessary to select the vector that is in the direction closest to the desired vector $\mathbf{g}$. This can simply be done by determining in which sector from I to VI the vector $\mathbf{g}$ is found and by assignment of the voltage vector which lies in the same sector (e.g. III. $\Rightarrow$ $\mathbf{u}_2$).

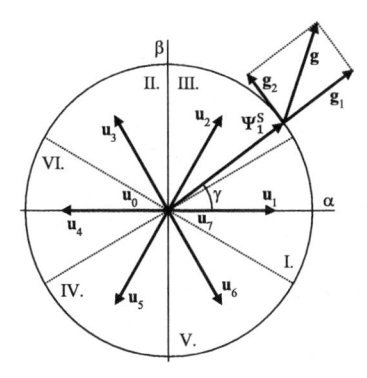

Fig. 1. Trajectory of the endpoint of the stator magnetic flux vector

Figure 2 shows a block diagram of the control structure. The phase currents i_{1a} and i_{1b} and the DC link voltage u_d are measured. The voltage at the motor terminals (stator voltage) is estimated from the u_d value and the switching combination. The actual electromagnetic torque and magnetic flux values are obtained by calculation using the values of the measured quantities and a mathematical model of an induction motor.

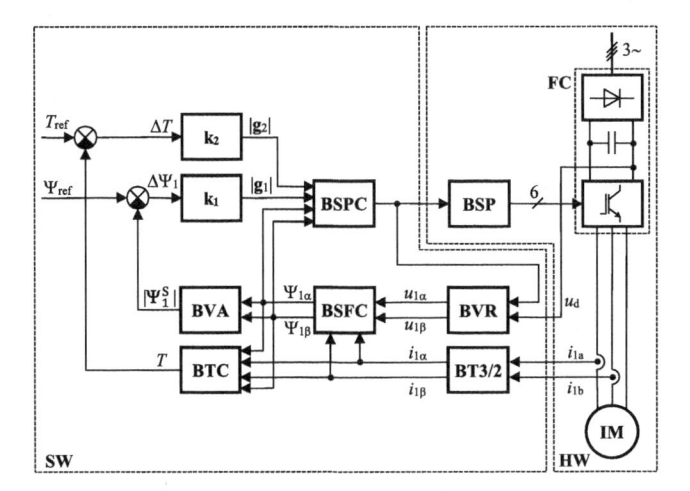

Fig. 2. Control structure of the DVC-DTC method (*Meaning of blocks:* **SW** – *software;* **HW** – *hardware;* **IM** – *induction motor;* **BT3/2** – *3/2 coordinate transformation (Clarke's transformation);* **BVR** – *voltage reconstruction;* **BSFC** – *stator flux calculation (voltage model);* **BVA** – *vector analyzer (modulus calculation);* **BTC** – *torque calculation (equation for electromagnetic torque);* **BSPC** – *switching pulses calculation;* **BSP** – *switching pulses creation.*)

The key part of the control structure is the BSPC block. The output of this block is the switching combination of the voltage vector to be applied at the inverter output. In the block, the coordinates of the vector **g** expressed in the stator coordinate system [α, β] are calculated first, they are marked as variables dx and dy in the algorithm. Then another calculations are performed to select a voltage vector ($\mathbf{u}_1$ to $\mathbf{u}_6$). In the case where a torque reduction is requested[1], and a reversal is not being performed, a zero-voltage vector ($\mathbf{u}_0$ or $\mathbf{u}_7$) is preferred before the active voltage vector. Using the zero vectors to decrease the motor torque, the torque ripple, stator voltage distortion, harmonic losses and switching losses are reduced [9]. The zero vector should be selected according to the previous active vector to minimize the switching losses. The switching combination of such zero vector differs from the previous only in one phase of the inverter.

It is possible to adapt the algorithm for use of a vector modulator, which then accurately approximates the vector **g**. The weighting coefficients k_1 and k_2 are replaced by PI controllers in this case. Application of the vector modulator provides a constant switching frequency and a minimum ripple of the stator flux and torque is achieved [8].

3 Specification of the Measurement Conditions

A software that implements the DVC-DTC algorithm was created for laboratory purposes. The control system used was developed at the Department of Electronics, VSB – TUO and is adapted for the control applications from the field of electric drives and power electronics [11, 12]. The key element of this control system is the TMS320F28335 Digital Signal Controller (DSC) from Texas Instruments, which belongs to the C2000 family of microcontrollers. TMS320F28335 has been designed to control embedded real-time systems, including electric drives [13, 14].

Measurements were carried out on a laboratory drive; a block diagram is in Fig. 3. The laboratory drive consisted of an induction motor, a loading mechanism (separately excited DC motor), a frequency converter with a DC link and a control system. The internal variables were being written to the analogue output of the control system (4-channel D/A converter) and displayed on the *LeCroy Wave Surfer 424* oscilloscope. The DC link voltage was set to $U_{DC} = 200$ V. Parameters of the induction motor summarizes the Table 1. The Table 2 includes the values that were constant during all measurements.

Table 1. Rated parameters of the induction motor used

Parameter name	Value	Unit
Power	2,7	kW
Rotor speed	1360	$\min^{-1}$
Torque	19,0	Nm
Stator flux (vector modulus)	0,877	Wb
Stator voltage	380Y/220Δ	V
Stator current	7,51	A
Frequency	50	Hz

---

[1] I.e.: $\mathbf{g}_2$ has the opposite direction to the direction of the rotation of the stator flux vector Ψ_1^S.

Table 2. Configuration used in the experiment

Parameter name	Symbol	Value	Unit
DC link voltage	U_{DC}	200	V
Sampling frequency	f_{sa}	10	kHz
Stator flux reference	Ψ_{ref}	0.6	Wb
Torque reference	T_{ref}	7	Nm
Rotor speed	n	400	$\min^{-1}$

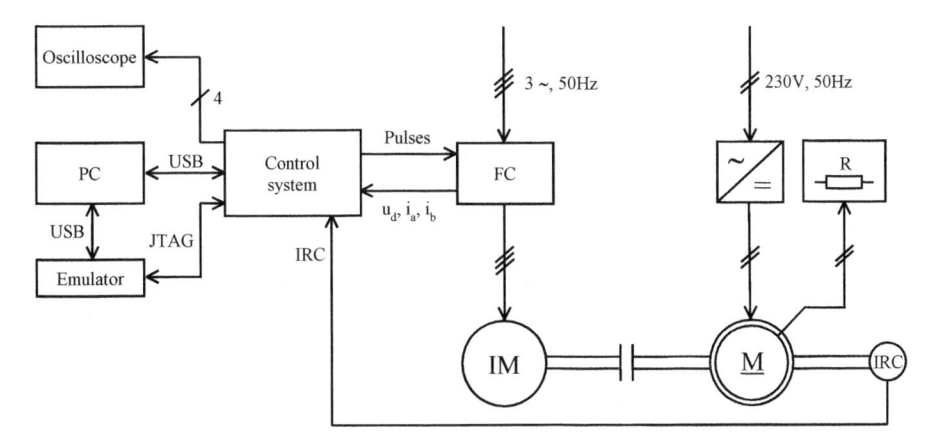

Fig. 3. Block diagram of the laboratory stand

4 Experimental Results

With k_1 and k_2, it is possible to modify the ratio between the magnetic flux (1) and torque (2) deviations, in other words, ratio of the size of vector $\mathbf{g}_1$ (3) to the size of $\mathbf{g}_2$ (4). The target is to obtain an optimal ratio between the priority of the control loops of torque and magnetic flux. There are not any formulas to calculate the setting of these coefficients; they have to be set experimentally. The following results should clarify what criteria to consider when adjusting the weighting coefficients.

Figures 5 and 6 show screenshots for three different settings of the weighting coefficients. In Fig. 5, there are the courses of the flux, voltage, current and torque of the induction motor; Fig. 6 shows the corresponding trajectories of the endpoint of stator flux vector.

The switching frequency of the frequency converter was also monitored; the measured values are included in the screenshot captions. In addition, a graph of the switching frequency as a function of the k_1/k_2 ratio is displayed in Fig. 4. The switching frequency was measured as an average number of voltage vector changes per second.

It was found that the ratio of k_1 and k_2 has a decisive influence, so k_2 was set to a constant value of 0.1 and k_1 was being changed. It is important to note that in the algorithm the sizes of the components $|\mathbf{g}_1|$ and $|\mathbf{g}_2|$ are limited. For this reason, it was necessary to suppress the torque deviation.

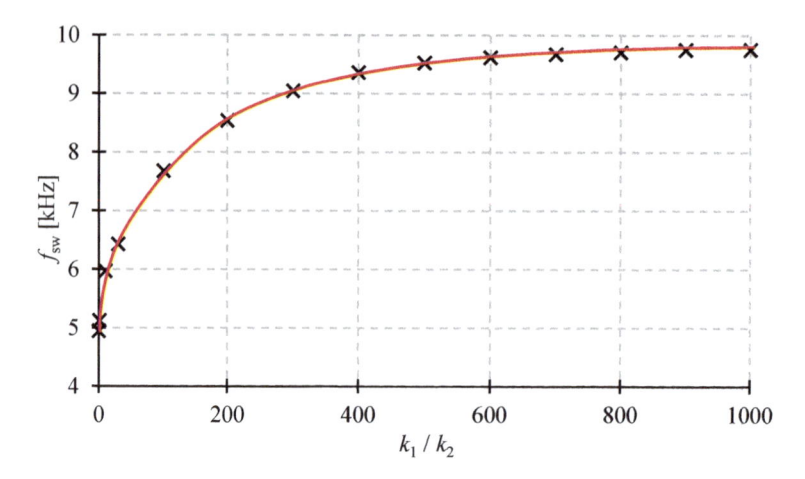

Fig. 4. Dependence of the switching frequency on the ratio of the weighting coefficients

In Figs. 5(b) and 6(b), the time courses at the value of $k_1/k_2 = 30$ are shown. This setting seemed to be optimal. However, it depends on the actual operating state of a drive. The most significant is the influence of the degree of motor excitation because it influences the torque ripple, but this is beyond the scope of this paper.

In Figs. 5(a) and 6(a), the case for $k_1/k_2 = 1$ is shown. This setting caused that the torque deviation had a greater weight than in the previous case, so the voltage vectors that acted more in the direction of the rotation of the stator flux vector were being preferred. In other words, the torque control loop had a higher priority than the control loop of magnetic flux. We can see a strong distortion in the stator current and flux courses, and the trajectory of the stator flux has a hexagonal shape. By further decreasing of k_1/k_2, the magnetic flux, voltage and current courses approaches the courses of the Depenbrock's DTC method [10]. In addition to the magnetic flux deformation, its amplitude (motor excitation) decreases which results in a growth of the frequency of rotating magnetic field and amplitude of the current.

Figures 5(c) and 6(c) shows the state when the torque deviation was significantly suppressed by setting the value of $k_1/k_2 = 200$. In this case, the control loop of stator flux had a higher priority, so the active voltage vectors that had a small angle to the stator flux vector, were being preferred. It led to a higher frequency of vectors changes. This can be explained by the fact that the active vector that has a small angle to the stator flux vector causes a lower torque deviation, and the shorter the deviation is, the shorter the time it takes to compensate this deviation by a zero vector. The change is the most visible in the time course of stator voltage, but we can see, also, that the torque ripple increased, and the torque mean value decreased slightly.

Some interesting behaviour can be observed when k_1 or k_2 is set to zero. When $k_2 = 0$, the torque control loop is disabled. The rotating magnetic field stops and only the excitation of a motor is controlled by switching between two opposite active vectors; the switching frequency equals the sampling frequency. By contrast, setting $k_1 = 0$ gives the highest priority to the torque control loop, however, the motor stops.

a) $k_1 = 0.1$
$k_2 = 0.1$
$f_{sw} = 5.1$ kHz

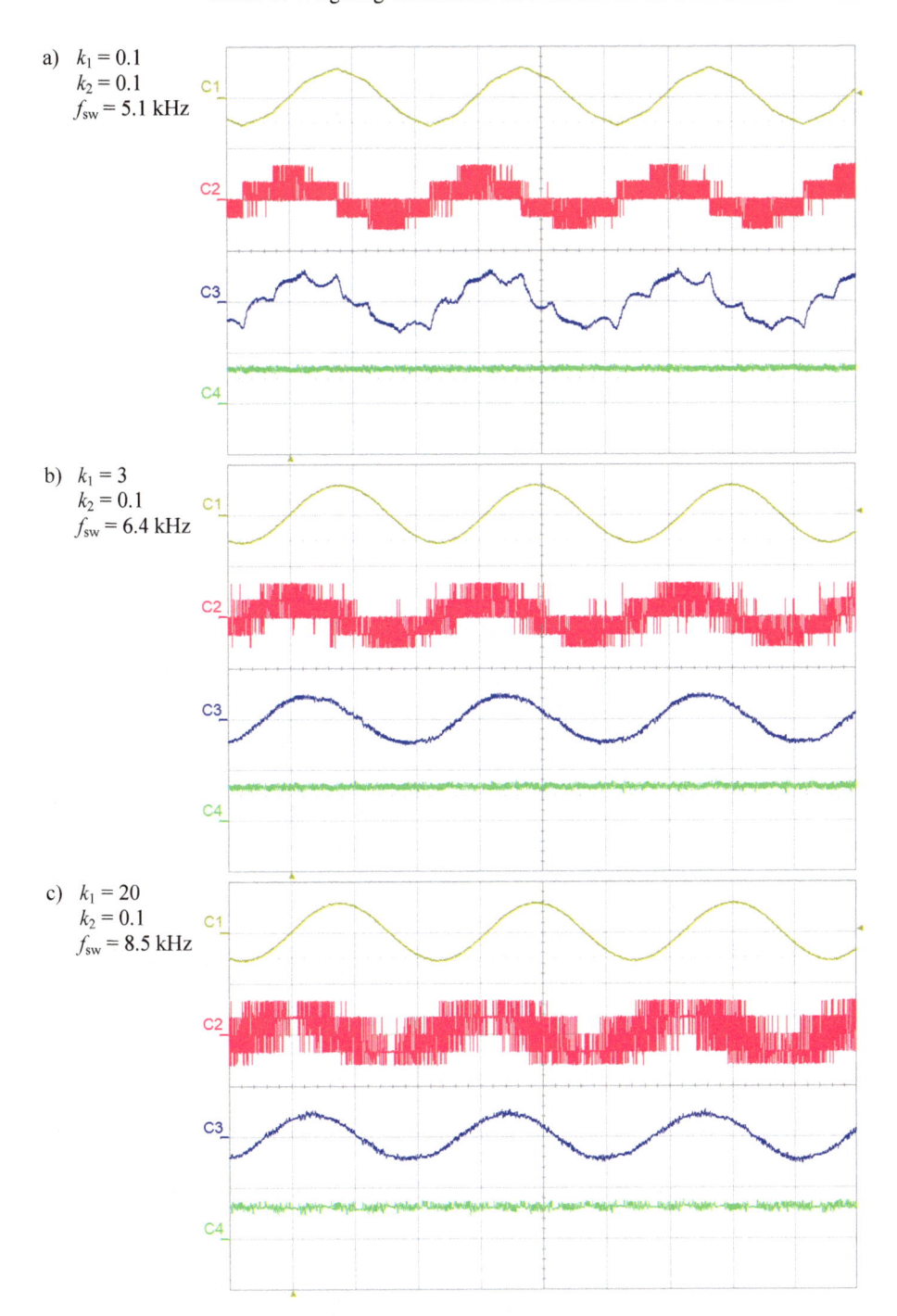

b) $k_1 = 3$
$k_2 = 0.1$
$f_{sw} = 6.4$ kHz

c) $k_1 = 20$
$k_2 = 0.1$
$f_{sw} = 8.5$ kHz

Fig. 5. Time courses of the quantities in a steady state. (**t**: 20 ms/d; **C1**: $\Psi_{1\alpha} = f(t)$, 1 Wb/d; **C2**: $u_{1\alpha} = f(t)$, 200 V/d; **C3**: $i_{1\alpha} = f(t)$, 10 A/d; **C4**: $T = f(t)$, 10 Nm/d)

a) $k_1 = 0.1$
$k_2 = 0.1$
$f_{sw} = 5.1$ kHz

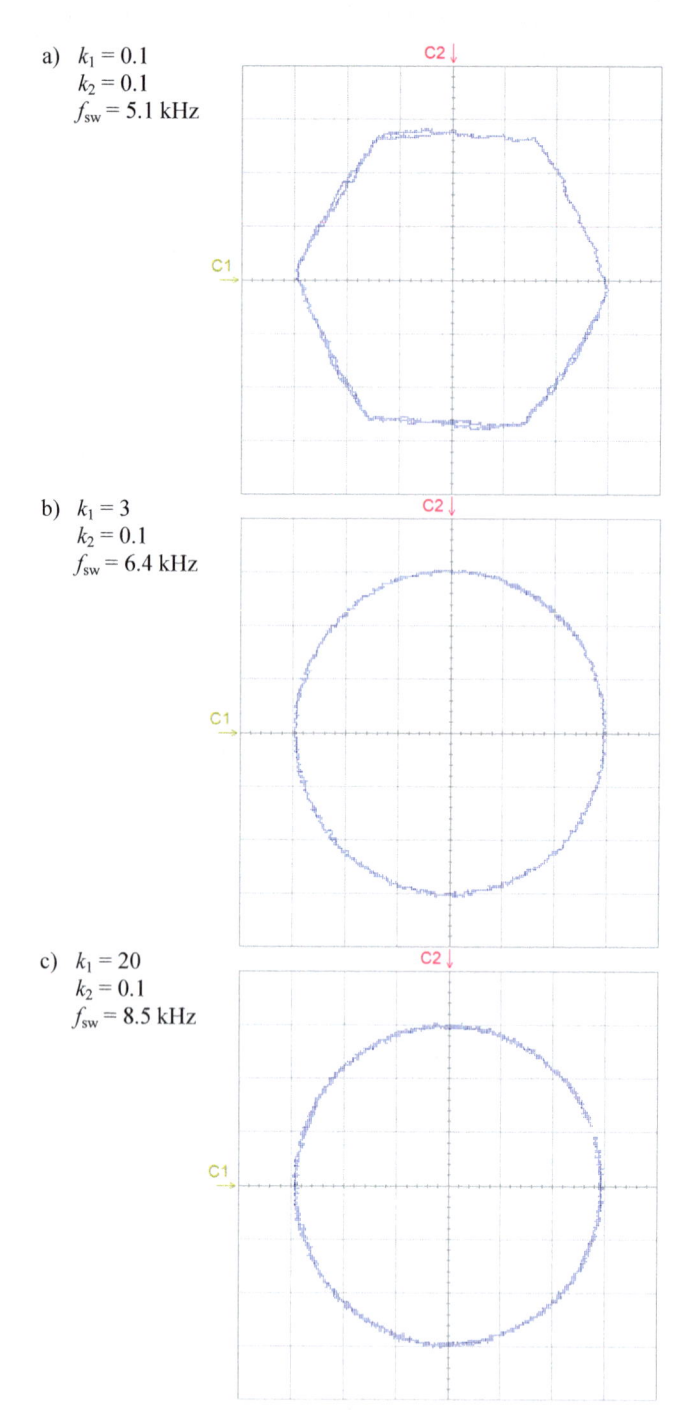

b) $k_1 = 3$
$k_2 = 0.1$
$f_{sw} = 6.4$ kHz

c) $k_1 = 20$
$k_2 = 0.1$
$f_{sw} = 8.5$ kHz

Fig. 6. Trajectory of the endpoint of stator flux vector in a steady state. (**t**: 10 ms/d; **C1**: $\Psi_{1\alpha}$ = f(*t*), 0.2 Wb/d; **C2**: $\Psi_{1\beta}$ = f(*t*), 0.2 Wb/d)

The frequency of rotating magnetic field increases to such a high value that the motor stops due to the breakdown torque decrease caused by the field suppression of the motor.

5 Conclusion

An assessment of the influence of the weighting coefficients has been accomplished. The presented findings can be useful when realizing a practical application of the DVC-DTC method. The weighting coefficients have a significant impact on the operation of the DVC-DTC algorithm. With these coefficients, it is possible to adjust the ratio between the priority of torque control and stator flux control. Actual values of the coefficients do not matter so much; the ratio between the values is essential. The ratio affects switching frequency and thus the switching losses of a frequency converter. It has been confirmed that the higher the priority of the torque control is, the lower the switching frequency and the higher the level of harmonic distortion of the stator current and magnetic flux, and vice versa.

The ratio $k_1/k_2 = 30$ seemed to be an optimal setting for the given configuration of the drive (see Table 2), the related courses of quantities are in Figs. 5(b) and 6(b). The following criteria should be considered when performing an experimental setting of the coefficients: the quality of the stator current and stator flux time courses (sinusoidal time courses and a circular trajectory of the flux vector), switching frequency and torque ripple.

Acknowledgement. This paper was supported by the projects: Center for Intelligent Drives and Advanced Machine Control (CIDAM) project, Reg. No. TE02000103, funded by the Technology Agency of the Czech Republic and Project Reg. No. SP2018/162 funded by the Student Grant Competition of VSB – Technical University of Ostrava.

Appendix – Meaning of Symbols

Symbol	Unit	Description
T_{ref}	Nm	Torque reference
T	Nm	Torque (calculated)
$i_{1\alpha}$	A	Stator current (phase a)
$u_{1\alpha}$	V	Stator voltage (phase a)
Ψ_{ref}	Wb	Reference of the modulus of stator flux vector (the same as $\left|\Psi_1^S\right|_{ref}$)
$\Psi_{1\alpha}$	Wb	α-component of stator flux vector (calculated)
$\Psi_{1\beta}$	Wb	β-component of stator flux vector (calculated)
f_{sa}	Hz	Sampling frequency (frequency of control loop execution)
f_{sw}	Hz	Switching frequency

References

1. Vas, P.: Sensorless Vector and Direct Torque Control. Oxford University Press, New York (1998). ISBN 0198564651
2. Leonhard, W.: Control of Electrical Drives, 3rd edn. Springer, New York (2001). ISBN 3-540-41820-2
3. Takahashi, I., Noguchi, T.: A new quick-response and high-efficiency control strategy of an induction motor. IEEE Trans. Ind. Appl. **IA-22**(5), 820–827 (1986). https://doi.org/10.1109/tia.1986.4504799. ISSN 0093-9994
4. Baader, U., Depenbrock, M., Gierse, G.: Direct self control (DSC) of inverter-fed induction machine: a basis for speed control without speed measurement. IEEE Trans. Ind. Appl. **28**(3), 581–588. https://doi.org/10.1109/28.137442. ISSN 00939994
5. Stando, D., Kazmierkowski, M.P.: Novel speed sensorless DTC-SVM scheme for induction motor drives. In: 2013 International Conference-Workshop Compatibility and Power Electronics, pp. 225–230. IEEE (2013). https://doi.org/10.1109/cpe.2013.6601159. ISBN 978-1-4673-4913-0
6. Sikorski, A., Korzeniewski, M.: Improved algorithms of Direct Torque Control method. Automatika **54**(2), 188–198 (2017). https://doi.org/10.7305/automatika.54-2.173. ISSN 0005-1144
7. Metidji, B., Taib, N., Baghli, L., Rekioua, T., Bacha, S.: Low-cost Direct Torque Control algorithm for induction motor without AC phase current sensors. IEEE Trans. Power Electron. **27**(9), 4132–4139 (2012). https://doi.org/10.1109/tpel.2012.2190101. ISSN 0885-8993
8. Brandstetter, P., Chlebis, P., Palacky, P.: Direct Torque Control of induction motor with direct calculation of voltage vector. Adv. Electr. Comput. Eng. **10**(4), 17–22 (2010). https://doi.org/10.4316/aece.2010.04003. ISSN 1582-7445
9. Hrdina, L.: Direct Torque Control methods of induction motors. Ph.D. thesis. VSB - Technical University of Ostrava, Ostrava (2008)
10. Brandstetter, P., Kusyn, Q.: Induction motor drive with DSP-based control system. In: Proceedings of IEEE International Symposium on Industrial Electronics, pp. 163–167. IEEE (1996). https://doi.org/10.1109/isie.1996.548412. ISBN 0-7803-3334-9
11. Havel, A., Sobek, M., Chamrad, P.: Control methods of modern systems utilizing accumulation of electrical energy. In: 2017 18th International Scientific Conference on Electric Power Engineering (EPE), pp. 1–6. IEEE (2017). https://doi.org/10.1109/epe.2017.7967338. ISBN 978-1-5090-6406-9
12. Havel, A., Sobek, M., Palacky, P.: The measuring and control center for accumulation systems of electrical energy. Electr. Eng. **99**(4), 1295–1303 (2017). https://doi.org/10.1007/s00202-017-0619-y. ISSN 0948-7921
13. Elrajoubi, A., Ang, S.S., Abushaiba, A.: TMS320F28335 DSP programming using MATLAB Simulink embedded coder: techniques and advancements. In: 2017 IEEE 18th Workshop on Control and Modeling for Power Electronics (COMPEL), pp. 1–7. IEEE (2017). https://doi.org/10.1109/compel.2017.8013418. ISBN 978-1-5090-5326-1
14. Talanov, M.V., Karasev, A.V., Talanov, V.M.: Implementation of extended Kalman filtering algorithm with improved flux estimator on TMS320F28335 processor for induction sensorless drive. In: 2014 6th European Embedded Design in Education and Research Conference (EDERC), pp. 119–123. IEEE (2014). https://doi.org/10.1109/ederc.2014.6924371. ISBN 978-1-4799-6843-5

A New Protocol for Energy Harvesting Decode-and-Forward Relaying Networks

Duy-Hung Ha[1,3(✉)], Dac-Binh Ha[2], Jaroslav Zdralek[1], Miroslav Voznak[1], and Tan N. Nguyen[1]

[1] VSB-Technical University of Ostrava, 17. listopadu 15, Ostrava, Czech Republic
{jaroslav.zdralek,miroslav.voznak,tan.nhat.nguyen.st}@vsb.cz
[2] Faculty of Electrical and Electronics Engineering, Duy Tan University, Danang, Vietnam
hadacbinh@duytan.edu.vn
[3] Ton Duc Thang University, 19 Nguyen Huu Tho Street, 7th District, Ho Chi Minh City, Vietnam
haduyhung@tdtu.edu.vn

Abstract. This paper investigates radio frequency energy harvesting decode-and-forward (DF) multi-relay networks with hybrid power transfer architecture, i.e. time switching (TS) combines with power splitting (PS). Specifically, this system consists of one source, one destination and multiple energy constraint relay nodes which help to the source transfer information to the destination over Nakagami fading channels. In order to improve the performance and reduce the load of this considered network, an efficient protocol for this system is proposed based on adaptive power splitting ratio adjustment and best relay selection. We aim to evaluate the performance of this considered system by deriving the closed-form expressions for the outage probability (OP) based on the statistical characteristics of signal-to-noise ratio (SNR). In addition, the results show that this scheme outperforms the random relay selection scheme, including transmit SNR, number of relays, active energy and fading severity factor. Our analysis is also verified by Monte Carlo simulation.

Keywords: Energy harvesting · Relay selection · Power splitting · Decode-and-forward · Relaying network

1 Introduction

The next generation networks, i.e., 5G, has been attracting a lot of attention from the academia and industry in very recent years. The radio frequency (RF) energy harvesting (EH) relaying technique that can prolong the lifetime of energy-constrained networks, spread out coverage and improve the performance as well as maintain network connectivity of wireless networks is introduced for 5G [1–7]. Two information and power transfer architectures, namely time switching (TS) and power splitting (PS), have been proposed to be used to extract the energy and information from RF signals. With the TS architecture, the total received RF signal is used for both information processing and energy harvesting but in different time blocks; whereas in the PS architecture, the received signal is split into two parts: one for information detection and the other for energy harvesting [8]. The work in [7] studied energy harvesting multi-relay-assisted relaying system, where multiple relay nodes simultaneously assist the transmission from source to destination using the concept of distributed space-time coding. The optimization problems of PS ratios at the relays are formulated for both decode-and-forward (DF) and amplify-and-forward (AF) relaying protocols with PS scheme. The efficient algorithms were also proposed to find the optimal solutions. The authors in [8] studied the hybrid TS-PS SWIPT architecture in full-duplex massive multiple-input multiple-output (MIMO) system without relaying. The evaluated results have shown that the system achievable sum rate performance can be maximized by using multiple antennas. The AF relay network was studied in the work [9], where the relay nodes need to harvest energy from the source's RF signal to forward information to the destination. To improve the performance of information transmission, two distributed relay selection protocols, i.e., Maximum Harvested Energy (MHE) protocol and Maximum Signal-to-Noise Ratio (MSNR) protocol were proposed. The research results proved that the proposed selection protocols can improve this system performance and the MSNR protocol outperforms the MHE protocol. In order to achieve optimal throughput, the hybrid TS-based and PS-based AF relaying (HTPSR) protocol was proposed in the work [10]. In this protocol, both TS-based and PS-based coefficients and the impact of outdated channel state information over Rayleigh fading channels were studied. The hybrid TS and adaptive PS (TSAPS) protocol for EH DF relay networks was proposed in [11]. The closed-form expressions for its outage probability and effective transmission rate under DF scheme were derived to analyze the performance of this considered system by using the discrete level energy battery model. The results have shown that the proposed TS-APS approach with optimized TS ratio can achieve more effective transmission rate then the previous works. The work in [12] investigated the outage performance of an energy harvesting relay-aided cooperative network, where the source node transmits information to its destination node with the help of multiple energy harvesting AF cooperative relay nodes. The authors derived the closed-form expression of outage probability over Nakagami-m fading channels, where the on-off relay-aided cooperative protocol with the help of the Markov property of energy buffer status was considered for this system. The investigated results have shown that the system performance can be improved by using the multiple source EH relays.

Motivated from [11,12], in this paper we study the energy-harvesting-aware DF multi-relay networks in Nakagami-m fading with hybrid TS and adaptive PS EH architecture in the case of continuous battery model. More specifically, the main contributions of this paper can be summarized as follows

- First, we propose a protocol for energy-harvesting-aware DF multi-relay networks based on hybrid fixed time switching and adaptive power splitting EH architecture and best relay selection scheme, terms FTSAPS-BRS.
- Next, the closed-form expressions of outage probability are derived based on continuous battery model and the statistical characteristics of signal-to-noise ratio (SNR).
- Then, we compare the performance of our proposed schemes to the fixed TS - fixed PS scheme and random relay selection scheme, namely FTSFPS-RRS, to verify the effectiveness of our proposed protocol.
- Finally, the numerical results are also provided to understand the behaviors of our considered system based on the key parameters, i.e, transmit power, number of relays, active energy, and fading severity factor.

The rest of this paper is organized as follows. Section 2 describes network and channel models. Section 3 presents outage probability analysis. Section 4 discusses some numerical results. Finally, the conclusions are given in Sect. 5.

2 Network and Channel Models

The dual-hop DF relay network with energy harvesting considered in this paper is illustrated in Fig. 1. This system consists of one power transfer/information source, denoted by $\mathbb{S}$, one destination, denoted by $\mathbb{D}$, and N energy-constrained relay nodes, denoted by $\mathbb{R}_i, 1 \leq i \leq N$. Supposing that all nodes are equipped with single antenna and half-duplex constraint. We assume that source $\mathbb{S}$ has a fixed power supply, i.e., P_0, where there is no fixed energy supply for the relay nodes and thus needs to harvest energy from RF signals with continuous battery model [2], while $\mathbb{D}$ is not an EH node and has continuous supply of energy. Due to poor transmission condition, the direct link between source $\mathbb{S}$ and destination $\mathbb{D}$ is unavailable and communication from $\mathbb{S}$ to $\mathbb{D}$ is performed by the help of relay $\mathbb{R}$. The $\mathbb{S} - \mathbb{R}_i$ and $\mathbb{R}_i - \mathbb{D}$ frequency-flat fading channels, described by random variables g_i and h_i, respectively, are both assumed to follow Nakagami-m distributions. This distribution is convenient to work with because it can include the well-known Rayleigh and Rice distributions by properly selecting parameter m. It is also assumed that in each block time T, these channels remain constant, whereas they are independently and identically distributed (i.i.d) over different block times (i.e., quasi-static block fading). Moreover, we assume that the additive white Gaussian noises (AWGN) at all nodes are independent circular symmetric complex Gaussian random variables, each having zero mean and variance N_0, i.e., $\mathcal{CN}(0, N_0)$.

The whole process can be divided into two phases, as Fig. 2. In the first phase with the duration time αT ($0 \leq \alpha \leq 1$), the source broadcasts its information/energy signal to EH relays and EH relays dynamically split the received signal into two parts with power splitting ratio ε and $(1 - \varepsilon)$ for information decoding and energy harvesting, respectively. PS ratio ε can be adjusted by relay itself to ensure successful detection at the relay for a given target data rate R. In the second phase with $(1 - \alpha)T$, the best relay node selected from

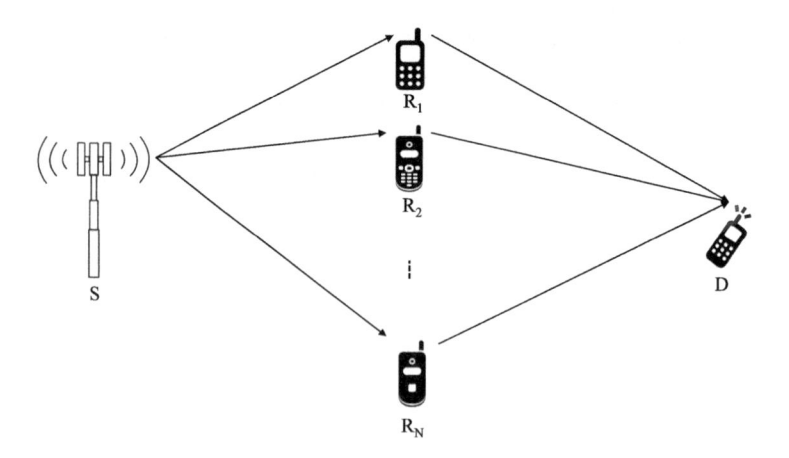

Fig. 1. System model for energy constraint DF relaying network.

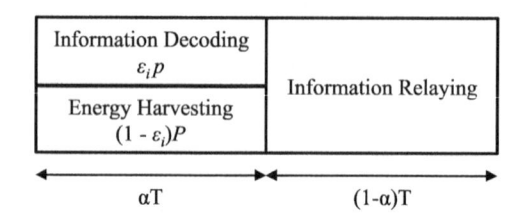

Fig. 2. Hybrid TS-adaptive PS energy harvesting architecture.

N relay nodes assists the source by applying the DF scheme to forward source's information to the destination.

2.1 Random Relay Selection (RRS) Scheme

Without loss of generality, hereafter we assume that the ith relay is randomly selected for relaying. Specifically, in the first phase the signal received at the relay can be written as

$$y_i(t) = \frac{\sqrt{\varepsilon_i P_0} g_i}{\sqrt{d_{1i}^{\theta_{1i}}}} x(t) + n_{R_i}, \tag{1}$$

where d_{1i} is the distance from $\mathbb{S}$ to $\mathbb{R}_i$, θ_{1i} is the path loss exponent, and n_R represents AWGN. The instantaneous SNR γ_{1i} at $\mathbb{R}_i$ is given as

$$\gamma_{1i} = \frac{\varepsilon_i P_0 |g_i|^2}{N_0 d_{1i}^{\theta_{1i}}} = \varepsilon_i \gamma_0 X_{1i}, \tag{2}$$

where $\gamma_0 = \frac{P_0}{N_0}$, namely transmit SNR, and $X_{1i} = \frac{|g_i|^2}{d_{1i}^{\theta_{1i}}}$. The data rate supported by $\mathbb{S} - \mathbb{R}_i$ link can be written as

$$C_{1i} = \alpha T \log_2(1 + \gamma_{1i}). \tag{3}$$

The PS ratio ε_i required to ensure successful detection at the relay for a given target data rate R is given by [11].

$$\varepsilon_i = \min\left\{1, \frac{2^{\frac{R}{\alpha T}} - 1}{\gamma_0 X_{1i}}\right\}. \tag{4}$$

The energy harvested by EH relay $\mathbb{R}_i$ over the time αT can be written as

$$E_{h_i} = \frac{\eta(1-\varepsilon_i)P_0|g_i|^2 \alpha T}{d_{1i}^{\theta_{1i}}} = \eta(1-\varepsilon_i)\alpha T P_0 X_{1i}, \tag{5}$$

where $0 \leq \eta \leq 1$ is the energy conversion efficiency of the energy receiver [1,2].

Assuming that the relay node uses harvested energy for information decoding, re-encoding, and forwarding, thus the transmit power of the ith relay is given as

$$P_{R_i} = \frac{E_{h_i} - E_0}{(1-\alpha)T} = \frac{\eta(1-\varepsilon_i)\alpha P_0 X_{1i}}{(1-\alpha)} - \frac{E_0}{(1-\alpha)T}, \tag{6}$$

where E_0 is the minimum required energy to activate relay [12].

The received signal at the destination can be expressed as

$$z_i(t) = \sqrt{\frac{P_{R_i}}{d_{2i}^{\theta_{2i}}}} h_i x(t) + n_{D_i}, \tag{7}$$

where d_{2i} and θ_{2i}, are the distance and path-loss exponent, respectively. n_{D_i} is denoted by AWGN. According to the Eqs. (6) and (7), the instantaneous SNR γ_{2i} at $\mathbb{D}$ is given as

$$\gamma_{2i} = (a_i \gamma_0 X_{1i} - \beta) X_{2i}, \tag{8}$$

where $a_i = \frac{\eta(1-\varepsilon_i)\alpha}{(1-\alpha)}$, $\beta = \frac{E_0}{(1-\alpha)TN_0}$, and $X_{2i} = \frac{|h_i|^2}{d_{2i}^{\theta_{2i}}}$. The channel capacity of $\mathbb{R}_i$-$\mathbb{D}$ link is written as

$$C_{2i} = (1-\alpha)T \log_2(1+\gamma_{2i}). \tag{9}$$

2.2 Best Relay Selection (BRS) Scheme

In this scheme, a best relay is selected among N relays based on the criteria of best system capacity. The implementation based on the algorithm of channel estimation in request-to-send (RTS)/clear-to-send (CTS) transmission from the source and destination allows relays estimate the instantaneous channel gain [11]. As soon as the relay receives the RTS packet from the source, it starts a timer with setting time as

$$T_i = \frac{\Delta}{\min\{C_{1i}, C_{2i}\}}, \tag{10}$$

where Δ has a unit of time [11]. The relay whose timeout expires first broadcasts its flag signal to other relays who listen to the flag should keep silence of relaying.

Here, we assume that all relays can hear to each other. The best relay here is defined with the minimum waiting time, and the corresponding waiting time can be given as

$$T_{best} = \min\{T_i\}, \quad 1 \leq i \leq N. \tag{11}$$

For system performance analysis carried out in the next section, the following expressions of the probability density function (PDF) and cumulative density function (CDF) of γ_k, $k = 1, 2$ are useful

$$f_{\gamma_k}(x) = \frac{m_k^{m_k} x^{m_k-1}}{(m_k-1)!\lambda_k^{m_k}} e^{-\frac{m_k x}{\lambda_k}}, \tag{12}$$

$$F_{\gamma_k}(x) = \int_0^x f_{\gamma_k}(t)dt = 1 - \sum_{i=0}^{m_k-1} \frac{m_k^i}{i!\lambda_k^i} x^i e^{-\frac{m_k x}{\lambda_k}}, \tag{13}$$

where $\lambda_k = \mathbb{E}\{X_k\}$ is interpreted as the average power gain of the source-relay or relay-destination channel, $m_k \geq 1/2$ is the fading severity factor. Note that the case of $m_k = 1$ corresponds to Rayleigh fading, whereas the case $m_k = (K_k+1)^2/(2K_k+1)$ approximates Rician fading with parameter K_k. In order to simplify the notation in our subsequent analyses, the statistical properties of the path gains are assumed to be identical, i.e., $m_{S,R_i} = m_1$, $m_{R_i,D} = m_2$, $\lambda_{S,R_i} = \lambda_1$ and $\lambda_{R_i,D} = \lambda_2$ for all i.

3 Performance Analysis

This section analyses performance of the considered relay network in terms of outage probability. The outage probability is defined as the probability that the instantaneous end-to-end, C, falls below a predetermined threshold R (a given target data rate R). It means that this metric is for evaluating the outage probability at a fixed source transmission rate – R bps/Hz.

3.1 Random Relay Selection Scheme

In RRS scheme, for the selected relay $\mathbb{R}_i$ the outage event at the destination occur when the link of $\mathbb{S}$-$\mathbb{R}_i$ is in outage (i.e., $E_{hi} \leq E_0$) or the link of $\mathbb{S}$-$\mathbb{R}_i$ is not in outage (i.e., $E_{hi} > E_0$) but the link of $\mathbb{R}_i$-$\mathbb{D}$ is in outage. Therefore, the overall outage event can be obtained as

$$P_{out} = 1 - \Pr[(E_{hi} > E_0) \cap (C_{2i} \geq R)], \tag{14}$$

Proposition 1. *Under Nakagami fading, the outage probability of the considered system for RRS scheme can be derived as*

$$P_{out} = 1 - 2 \sum_{k=0}^{m_2-1} \sum_{j=0}^{m_1-1} \frac{m_2^k \Omega^k b^{m_1-j-1} e^{-\frac{m_1 b}{\lambda_1 \phi}}}{k! j! (m_1-j-1)! \lambda_2^k} \left(\frac{m_1}{\lambda_1 \phi}\right)^{m_1+k-j-1} u^{j-k+1} \mathcal{K}_{j-k+1}(2u), \tag{15}$$

where $\Omega = 2^{\frac{R}{(1-\alpha)T}} - 1$, $\phi = \frac{\eta\alpha\gamma_0}{1-\alpha}$, $b = \beta + \frac{(2^{\frac{R}{\alpha T}}-1)\phi}{\gamma_0}$, $u = \sqrt{\frac{m_1 m_2 \Omega}{\lambda_1 \lambda_2 \phi}}$, and $\mathcal{K}_\nu(.)$ is the modified Bessel function of the second kind and ν^{th} order.

Proof. See Appendix A.

3.2 Best Relay Selection Scheme

In BRS scheme, the overall outage event can be expressed as

$$P^*_{out} = 1 - \Pr[C^* \geq R)], \quad (16)$$

where $C^* = \max\{\min[C_{1i}, C_{2i}]\}$, $1 \leq i \leq N$. We obtain the following proposition:

Proposition 2. *Under Nakagami fading, the outage probability of the considered system for BRS scheme can be derived as*

$$P^*_{out} = \left[1 - 2 \sum_{k=0}^{m_2-1} \sum_{j=0}^{m_1-1} \frac{m_2^k \Omega^k b^{m_1-j-1} e^{-\frac{m_1 b}{\lambda_1 \phi}}}{k! j! (m_1-j-1)! \lambda_2^k} \left(\frac{m_1}{\lambda_1 \phi} \right)^{m_1+k-j-1} \right.$$
$$\left. \times u^{j-k+1} \mathcal{K}_{j-k+1}(2u) \right]^N. \quad (17)$$

Proof. In BRS scheme, this system is said to be in outage state when the best selected relay fails to decode the information transmitted by source or the destination cannot decode successfully the signal transmitted by the best relay. Therefore, the overall outage event can be obtained as (17). This concludes the proof.

4 Numerical Results and Discussion

This section presents simulation and analytical results in terms of the outage probability $Pout$ to verify the effectiveness of our proposed policy and investigate the impact of key system parameters, such as transmit power, number of relays, minimum active energy, and fading severity factor (m).

4.1 Verification of the Effectiveness of Proposed Policy

Figure 3 depicts the theoretical and simulated results in terms of the outage probabilities for FTSAPS scheme and FTSFPS scheme versus the transmit SNR. From Fig. 3, we can observe that the $Pout$ curve of FTSAPS is below the one of FTSFPS when transmit SNR increases from around 15 dB to 30 dB. This result confirms that the proposed FTSAPS protocol outperforms to the FTSFPS. This is explained that the relay uses the adaptive PS ratio varied according to the variation of channel coefficient to split just the right amount of power to ensure successful detection at the relay for a given target data rate and the remain part is left for relaying.

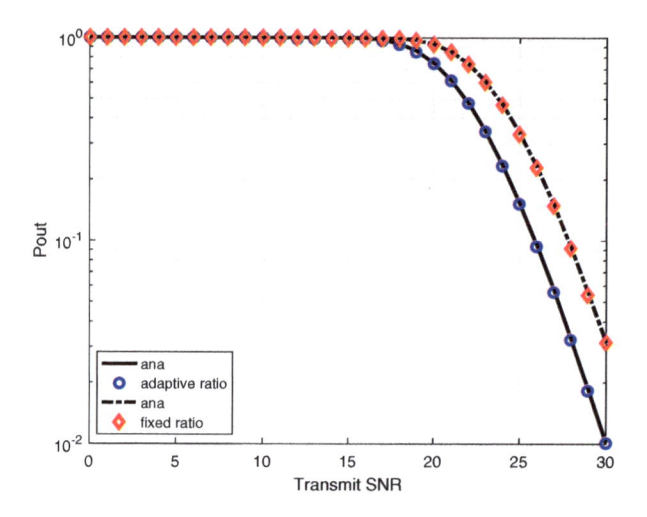

Fig. 3. P_{out} versus transmit SNR for FTSAPS and FTSFPS with $N = 3$, $\alpha = 0.5$, $m_1 = 1$, $m_2 = 2$, $R = 3$ bps/Hz, $d_1 = d_2 = 1$, $\theta_1 = \theta_2 = 2$.

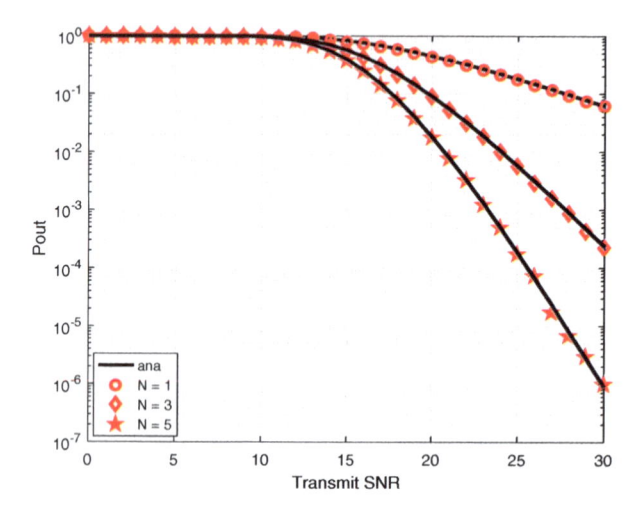

Fig. 4. P_{out} versus transmit SNR with number of relays and $E_0 = 1$ dB, $\alpha = 0.5$, $m_1 = 1$, $m_2 = 2$, $R = 3$ bps/Hz, $d_1 = d_2 = 1$, $\theta_1 = \theta_2 = 2$.

4.2 Effect of Number of Relays (N)

The effects of number of relays on OP are shown in Fig. 4. From this figure we can see that, the outage probability decreases when the number of relays increases. The reason is that, when the number of relays increase we have more chances to select the relay with better capacity. In other words, the BRS scheme is better than RRS scheme in terms of outage probability. From Figs. 3 and 4, we conclude that the FTSAPS-BRS policy outperforms to the FTSFPS-RRS policy.

4.3 Effect of Minimum Active Energy (E_0)

Figure 5 plots *Pout* versus transmit SNR with different minimum active energy. We can observe that the more active energy the worse performance, this is because the more power is consumed on the decoding and encoding processing and the less power for retransmission. In order to improve the performance of this system, the active energy should be reduced.

Figures 3, 4 and 5 show that the performance of this considered system can be improved by increasing the transmit power.

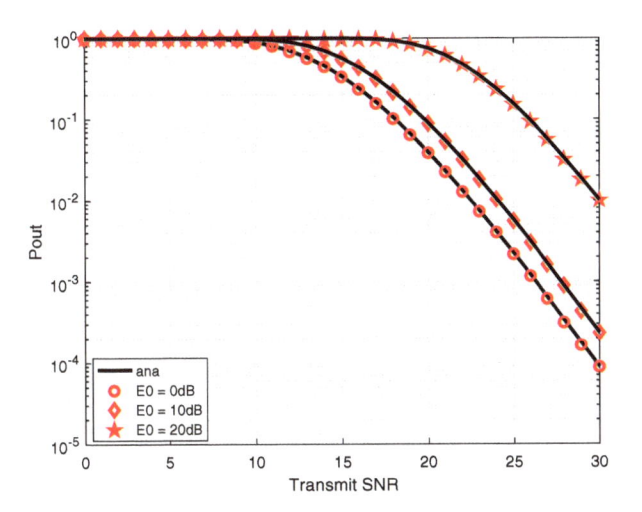

Fig. 5. *Pout* versus transmit SNR and minimum active energy with $N = 3$, $\alpha = 0.5$, $m_1 = 1$, $m_2 = 2$, $R = 3\,\text{bps/Hz}$, $d_1 = d_2 = 1$, $\theta_1 = \theta_2 = 2$.

4.4 Effect of Fading Severity Parameters (m_1 and m_2)

Figure 6 depicts the theoretical and simulated results in terms of the outage probabilities versus fading severity parameters m_1 and m_2. From this figure, we can know that the performance is better when m_1 or m_2 increases because the channel is better.

From Figs. 3, 4, 5 and 6, it is clear to conclude that the theoretical and simulated results are very nice matching. In other words, the correctness of our analysis is verified.

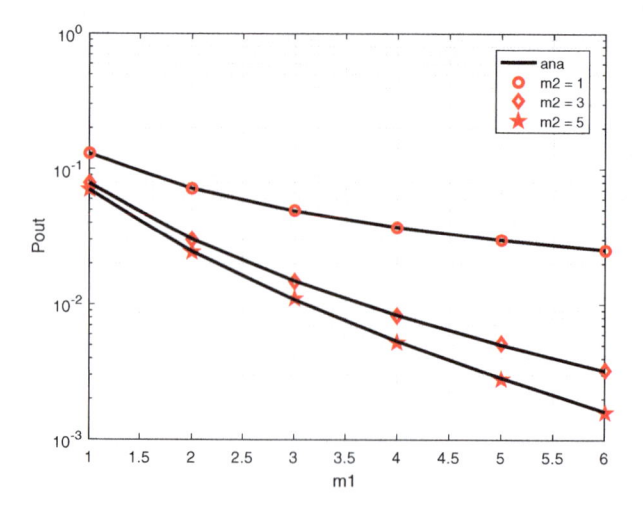

Fig. 6. P_{out} versus fading severity parameters with $E_0 = 20\,\text{dB}$, $\alpha = 0.5$, $\gamma_0 = 20\,\text{dB}$, $R = 3\,\text{bps/Hz}$, $N = 3$, $d_1 = d_2 = 1$, $\theta_1 = \theta_2 = 2$.

5 Conclusions

In order to improve the performance of EH DF relaying networks, we have proposed an effective protocol based on FTSAPS-BRS scheme. Moreover, the exact closed-form expressions of outage probability for RRS and BRS have been derived. Finally, the theoretical and simulated results have been provided to confirm the correctness of our analysis and to understand the behaviour of our system. Our results have also shown that the performance of this considered system can be improved by increasing the number of relays or transmit power or decreasing the active energy.

Acknowledgment. This work was supported by the VSB-Technical University of Ostrava, Czech Republic - Networks and Telecommunications Technologies for Smart Cities under SGS Grant SP2018/59.

Appendix A: Proof of Proposition 1

From (4), (5) and (8), the Eq. (10) can be rewritten as

$$P_{out} = 1 - \Pr[\eta(1-\varepsilon_i)\alpha T P_0 X_{1i} > E_0), (a_i \gamma_0 X_{1i} - \beta)X_{2i} \geq (2^{\frac{R}{(1-\alpha)T}} - 1)]$$

$$= 1 - \Pr\left[\phi_1\left(1 - \frac{\Omega_1}{X_{1i}}\right)X_{1i} > E_0, \left(\phi_2\left(1 - \frac{\Omega_1}{X_{1i}}\right)X_{1i} - \beta\right)X_{2i} \geq \Omega_2\right]$$

$$= 1 - \Pr\left[\phi_1 X_{1i} > E_0 + \phi_1 \Omega_1, \left(\phi_2 X_{1i} - \beta - \phi_2 \Omega_1\right)X_{2i} \geq \Omega_2\right]$$

$$= 1 - \Pr\left[X_{1i} > b_1, X_{2i} \geq \frac{\Omega_2}{(\phi_2 X_{1i} - b_2)}\right]$$

$$= 1 - \int_{b_1}^{\infty} \left[1 - F_{X_{2i}}\left(\frac{\Omega_2}{\phi_2 x - b_2}\right)\right] f_{X_{1i}}(x)\, dx$$

$$P_{out} = 1 - \sum_{k=0}^{m_2-1} \frac{m_1^{m_1} m_2^k}{k!(m_1-1)!\lambda_1^{m_1}\lambda_2^k} \int_{b_1}^{\infty} \left(\frac{\Omega_2}{\phi_2 x - b_2}\right)^k x^{m_1-1} e^{-\frac{m_2}{\lambda_2}\frac{\Omega_2}{\phi_2 x - b_2} - \frac{m_1 x}{\lambda_1}}\, dx$$

$$= 1 - \sum_{k=0}^{m_2-1} \frac{m_1^{m_1} m_2^k \Omega_2^k e^{-\frac{m_1 b_2}{\lambda_1 \phi_2}}}{k!(m_1-1)!\lambda_1^{m_1}\lambda_2^k \phi_2^{m_1}} \int_0^{\infty} y^{-k}(y+b_2)^{m_1-1} e^{-\frac{m_2 \Omega_2}{\lambda_2 y} - \frac{m_1 y}{\lambda_1 \phi_2}}\, dy$$

$$= 1 - \sum_{k=0}^{m_2-1}\sum_{j=0}^{m_1-1} \frac{C_{m_1-1}^j m_1^{m_1} m_2^k \Omega_2^k b_2^{m_1-j-1} e^{-\frac{m_1 b_2}{\lambda_1 \phi_2}}}{k!(m_1-1)!\lambda_1^{m_1}\lambda_2^k \phi_2^{m_1}} \int_0^{\infty} y^{j-k} e^{-\frac{m_2 \Omega_2}{\lambda_2 y} - \frac{m_1 y}{\lambda_1 \phi_2}}\, dy$$

$$= 1 - 2\sum_{k=0}^{m_2-1}\sum_{j=0}^{m_1-1} \frac{m_2^k \Omega_2^k b_2^{m_1-j-1} e^{-\frac{m_1 b_2}{\lambda_1 \phi_2}}}{k!j!(m_1-j-1)!\lambda_2^k}\left(\frac{m_1}{\lambda_1 \phi_2}\right)^{m_1+k-j-1} u^{j-k+1}\mathcal{K}_{j-k+1}(2u), (18)$$

where $\Omega_1 = \frac{2^{\frac{R}{\alpha T}}-1}{\gamma_0}$, $\Omega_2 = 2^{\frac{R}{(1-\alpha)T}}-1$, $\phi_1 = \eta\alpha T P_0$, $\phi_2 = \frac{\eta\alpha\gamma_0}{1-\alpha}$, $b_1 = \Omega_1 + \frac{E_0}{\phi_1}$, $b_2 = \beta + \phi_2\Omega_1$, $u = \sqrt{\frac{m_1 m_2 \Omega_2}{\lambda_1 \lambda_2 \phi_2}}$, and $\mathcal{K}_\nu(.)$ is the modified Bessel function of the second kind and ν^{th} order [13]. Substituting $\Omega_2 = \Omega$, $b_2 = b$ and $\phi_2 = \phi$ into (18), the Proposition 1 is completely proved.

References

1. Ulukus, S., Yener, A., Erkip, E., Simeone, O., Zorzi, M., Grover, P., Huang, K.: Energy harvesting wireless communications: a review of recent advances. IEEE J. Sel. Areas Commun. **33**(3), 360–381 (2015)
2. Ha, D.-B., Tran, D.-D., Tran-Ha, V., Hong, E.-K.: Performance of amplify-and-forward relaying with wireless power transfer over dissimilar channels. Elektronika ir Elektrotechnika J. **21**(5), 90–95 (2015)
3. De Rango, F., Gerla, M., Marano, S.: A scalable routing scheme with group motion support in large and dense wireless ad hoc networks. Comput. Electr. Eng. **32**(1–3), 224–240 (2006)

4. Tran, D.D., Tran-Ha, V., Ha, D.B., Tran, H., Kaddoum, G.: Performance analysis of two-way relaying system with radio frequency energy harvesting and multiple antennas. In: 2016 IEEE 84th Vehicular Technology Conference, VTC 2016-Fall, Montréal, Canada, 18–21 September 2016 (2016)
5. Zhou, B., Lee, Y.-Z., Gerla, M., De Rango, F.: Geo-LANMAR: a scalable routing protocol for ad hoc networks with group motion. Wirel. Commun. Mob. Comput. **6**(7), 989–1002 (2006)
6. Ha, D.B., Nguyen, Q.S.: Outage performance of energy harvesting DF relaying NOMA networks. Mob. Netw. Appl. (2017)
7. Liu, Y.: Wireless information and power transfer for multirelay-assisted cooperative communication. IEEE Commun. Lett. **20**(4), 784–787 (2016)
8. Xu, K., Shen, Z., Wang, Y., Xia, X.: Beam-domain hybrid times witching and power splitting SWIPT in full-duplex massive MIMO system. EURASIP J. Wirel. Commun. Network. (2018)
9. Yan, J., Zhang, C., Gao, Z.: Distributed relay selection protocols for simultaneous wireless information and power transfer. In: 2014 IEEE 25th International Symposium on Personal, Indoor and Mobile Radio Communications, Montreal, QC, Canada, 08–13 October (2017)
10. Nguyen, H.-S., Do, D.-T., Nguyen, T.-S., Voznak, M.: Exploiting hybrid time switching-based and power splitting-based relaying protocol in wireless powered communication networks with outdated channel state information. J. Control Meas. Electron. Comput. Commun. **58**(1), 111–118 (2017)
11. Singh, V., Ochiai, H.: A efficient time switching protocol with adaptive power splitting for wireless energy harvesting relay networks. In: IEEE 85th Vehicular Technology Conference (VTC Spring) (2017)
12. Zhong, S., Huang, H., Li, R.: Performance analysis of energy-harvesting-aware multi-relay networks in Nakagami-m fading. EURASIP J. Wirel. Commun. Netw. (2018)
13. Gradshteyn, I.S., Ryzhik, I.M.: Table of Integrals, Series and Products, 7th edn. Academic Press, Cambridge (2007)

Average Bit Error Probability Analysis for Cooperative DF Relaying in Wireless Energy Harvesting Networks

Hoang-Sy Nguyen[1,2], Thanh-Sang Nguyen[1,2], Tan N. Nguyen[3(✉)], and Miroslav Voznak[1]

[1] VSB-Technical University of Ostrava, Ostrava, Czech Republic
{nhsy,ntsang}@bdu.edu.vn, miroslav.voznak@vsb.cz
[2] Institute of Artificial Intelligence, Binh Duong University, Thu Dau Mot City, Binh Duong Province, Vietnam
[3] Wireless Communications Research Group, Faculty of Electrical and Electronics Engineering, Ton Duc Thang University, Ho Chi Minh City, Vietnam
nguyennhattan@tdt.edu.vn

Abstract. Thanks to the benefits of energy harvesting (EH) in cooperative decode-and-forward (DF) relaying networks, we decided to consider a CRN deploying time-switching based relaying protocol (TSR) to study EH. To clearly evaluate the system performance, we derive the expressions for outage probability at high end-to-end signal-to-noise ratio (SNR), ergodic capacity, and the average bit error probability (ABEP). After finishing the performance analysis, we provide Monte-Carlo simulations to prove the performance and the correctness of the obtained numerical results.

Keywords: Energy harvesting · Cooperative relaying networks · Time-switching · Outage probability · Ergodic capacity · Bit error probability

1 Introduction

Recently, a lot of attention has been paid to the prolongation of wireless networks' lifetime since it is often is undesirable to replace or recharge batteries. Therefore, it is more beneficial to make wireless networks become power-independent by harvesting energy from the ambient environment, i.e., electromagnetic waves, which is able to simultaneously transfer both information and energy, is refereed to simultaneous wireless information and power transfer (SWIPT) [1–5]. In fact, most studies on SWIPT concentrate on two relaying transmission schemes, i.e., amplify-and-forward (AF) and decode-and-forward (DF) with energy harvesting (EH) relays [6–8]. Besides that, EH receiver architectures include two prime relaying models, such as power splitting relaying (PSR) and time switching relaying (TSR) which were discussed comprehensively in [9,10]. In particular, PSR protocol slits the received signal power into two parts, one for information processing and the other for EH, whereas TSR switches between receiving information and EH [11–13]. In [11], a point-to-point wireless link over the flat-fading channel considering a dynamic PS scheme was studied. The authors in [12] took a small cell cognitive relay network into consideration, where two TS policies to maximize the transmit power at source and relay, namely Optimal Time for Transmit Power at Source and Optimal Time for Transmit Power at Relay were proposed. Besides that, to optimize the achievable throughput, the instantaneous channel state information was carefully studied by formulating two problems with respect to PS ratio and TS ratio [13].

In addition, a number of investigations have been carried out on the performance metrics of cooperative relaying networks (CRNs) [14–20]. In particular, the authors in [14] focused on proposing wireless power supply policies, such as separated power (SP) and harvested power (HP) to come up with a flexible architecture in full-duplex (FD) DF relaying networks deploying TSR to achieve optimal time. When applying SP and HP, the proposed optimal power constraints can achieve better power consumption at the relay node. In [15], thanks to the proposed relay selection scheme, they evaluated the outage performance of CRN with cognitive radio. The authors in [16] evaluated relay selection (RS) schemes in multi-relay networks, in which the exact and approximate closed-form expressions for outage probability in three proposed optimal RS schemes, namely HD deploying maximal ratio combine (HDMRC) and FD deploying joint decoding (FDJD) and hybrid FD/HD relaying transmission scheme (HTS) were studied. In [17], a two-way CRN in the presence of hardware impairments was considered, where the authors obtained analytical expressions for ergodic capacity and the achievable sum rate for AF and DF protocols. In addition, the work in [19] obtained a closed-form average expression for bit-error rate (BER) for an AF CRN with SWIPT.

Our paper is organized as follows: In Sect. 2, the system model is presented. We analyze the system performance by obtaining expressions for outage probability, the ergodic capacity, and the average bit error probability (ABEP) in Sect. 3. In Sect. 4, the numerical results are provided. Meanwhile, Sect. 5 concludes the paper.

Notation: The SNR of any specific links is denoted by γ, and we denote $\bar{\gamma}$ as the average γ. The cumulative distribution function (CDF) and the probability density function (PDF) of the variable are denoted by $F_X(.)$ and $f_X(.)$, respectively. $\Pr(.)$ is the outage probability function. The expectation operation is $E\{.\}$. $K_1(.)$ is the first order modified Bessel function of the second kind.

2 System Model

As illustrated in Fig. 1, a two-hop wireless relaying network (CRN) is considered, in which it consists of a source (S), a energy harvesting (EH) relay, and a destination (D). It is noted that due to far distance between S and D, R is deployed to assist the direct communication between the two nodes and the coverage extension. In such a network, S can be deployed as a base station (BS),

R is often an in-building BS while and D is normally referred to a mobile user. S and D are assumed to be equipped with a single-antenna nodes while two antennas, including one receive and one transmit antenna are equipped at R.

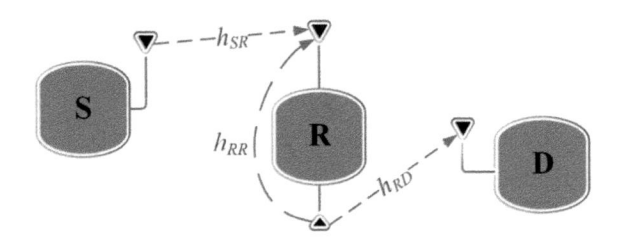

Fig. 1. The system model for the EH CRN considering TSR

In this paper, we deploy time switching-based relaying (TSR) protocol to investigate EH [9]. In particular, αT is denoted as time for EH, where T is the whole time slot for each signal frame.

Regarding the system model, it comprises three links, including S-R, the residual loop interference (LR) and RD. We respectively denote the communication channels of these links as $h_{SR}, h_{RR}, and h_{RD}$, and they are modeled as frequency-flat and quasi-static. It is worth noting that these aforementioned channels are independent and affected by Rayleigh fading, where they vary independently from one slot to another. The transmit powers at S and R are respectively represented by E_S and E_R. Besides that, the channel power gains are exponentially distributed random variables (RVs).

We parametrize the system with the signal-to-noise ratios (SNRs) expressed by $\gamma_{SR} = |h_{SR}|^2/N_0^2$, $\gamma_{RD} = |h_{RD}|^2/N_0^2$, and $\gamma_{RR} = |h_{RR}|^2/N_0^2$. In particular, $|.|^2$ is denoted as the channel gain, and N_0^2 is the variances of the additive white Gaussian noise (AWGNs) at all node. In principle, relay nodes deploying FD are able to receive and transmit data at the same time with the same frequency band while the residual self-interference (SI) denoted as γ_{RR} always has an impact on the system despite applying SI cancellation methods at R.

The received signal at R and D in conventional wireless networks can be written respectively as

$$y_R = \sqrt{E_S} h_{SR} x_S + \sqrt{E_R} h_{RR} x_R + N_0, \quad (1)$$

and

$$y_D = \sqrt{E_R} h_{RD} x_S + N_0, \quad (2)$$

In the down-link transmission, we compute the SNR at R and D respectively as

$$\gamma_R = \frac{E_S}{E_R} \times \frac{\gamma_{SR}}{\gamma_{RR}}, \quad (3)$$

and

$$\gamma_D = E_R \times \gamma_{RD}. \quad (4)$$

Considering decode-and-forward (DF) transmission scheme, the equivalent end-to-end (e2e) SNR of the network can be expressed as

$$\gamma_{eq} = \min\{\gamma_R, \gamma_D\}. \tag{5}$$

In practice, EH from the received signals from S is only considered at R. Therefore, we express the received signal at R as

$$E_R = \frac{E_h}{(1-\alpha)T/2} = 2\eta\alpha(1-\alpha)^{-1}E_S\gamma_{SR}, \tag{6}$$

where the energy conversion efficiency is $0 < \eta < 1$. Furthermore, the harvested energy can be generally defined as $E_h = \eta\alpha E_S\gamma_{SR}T$.

Following that, the instantaneous rate of the system is given by

$$R = B\log_2(1+\gamma_{eq}), \tag{7}$$

where the signal bandwidth is denoted as B, and we set $B = 1$.

We assume that the optimal TS ratio is α^*. Thus, the optimal instantaneous transmission rate considering TS defined in (7) can be calculated by

$$(\alpha^*) = \mathrm{argmax} R(\alpha), \tag{8}$$

where it is subject to $0 < \alpha < 1$.

Following from (5), the below expression has to be solved to optimize the e2e SNR as

$$\left(\frac{1-\alpha^*}{\alpha^*}\right)^2 = 4\eta^2 E_S \overline{\gamma}_{RR} \overline{\gamma}_{SR} \overline{\gamma}_{RD}. \tag{9}$$

The optimal TS ratio can be obtained after several algebraic manipulations as follows

$$\alpha^* = \frac{1}{1+2\eta\sqrt{\overline{\gamma}_{RR}\overline{\gamma}_{SR}\overline{\gamma}_{RD}}}. \tag{10}$$

3 Performance Analysis

In this section, to carefully evaluate the system performance, we take the outage probability, ergodic capacity and average bit error probability (ABEP) into consideration. In particular, expressions for these performance metrics at high SNR are going to be obtained. Let us first start with the outage performance analysis.

3.1 Outage Performance with the End-to-end SNR

In principle, the outage probability (OP) related to the e2e SNR CRNs can be defined as

$$OP_{\gamma_{eq}}(x) = 1 - Q_1(x) + Q_2(x), \tag{11}$$

where $\Psi(x) = \sqrt{\frac{2x(1-\alpha)}{(\eta\alpha E_s \bar{\gamma}_{SR} \bar{\gamma}_{RD})}}$, $Q_1(x) = \Psi(x) K_1(\Psi(x))$, and

$$Q_2(x) = exp\left(-\frac{(1-\alpha)}{2\eta\alpha x \bar{\gamma}_{RR}}\right) \Psi(x) K_1(\Psi(x)).$$

Proof: It is noted that the PDF and CDF in the first hop based on (3) and (6) are expressed by

$$f_{\gamma_{SR}}(x) = \frac{1}{\bar{\gamma}_{SR}} exp\left(-\frac{x}{\bar{\gamma}_{SR}}\right), x \geq 0 \tag{12}$$

and

$$F_{\gamma_{SR}}(x) = 1 - exp\left(-\frac{x}{\bar{\gamma}_{SR}}\right). \tag{13}$$

Following that, the approximated CDF for SI at R at high SNR is given as

$$\begin{aligned} F_{\gamma_R}(x) &= \Pr\{\gamma_R < x\} \\ &\approx \Pr\left\{\frac{E_s \gamma_{SR}}{E_R \gamma_{RR}+1} < x\right\} \\ &\approx 1 - \left(1 - exp\left(-\frac{(1-\alpha)}{2x\eta\alpha \bar{\gamma}_{RR}}\right)\right). \end{aligned} \tag{14}$$

Substituting (6) into (4), the CDF in the second hop can be calculated as

$$\begin{aligned} F_{\gamma_D}(x) &= \Pr\{\gamma_D < x\} \\ &= \Pr\left(2\eta\alpha(1-\alpha)^{-1} E_s \gamma_{SR} \gamma_{RD} < x\right) \\ &= \int_{y=0}^{\infty} f_{\gamma_{SR}}(y) \left(1 - e^{\left(-\frac{x(1-\alpha)}{2\eta\alpha E_s y \bar{\gamma}_{RD}}\right)}\right) dy \\ &= 1 - \sqrt{\frac{2x(1-\alpha)}{(\eta\alpha E_s \bar{\gamma}_{SR} \bar{\gamma}_{RD})}} K_1\left(\sqrt{\frac{2x(1-\alpha)}{(\eta\alpha E_s \bar{\gamma}_{SR} \bar{\gamma}_{RD})}}\right), \end{aligned} \tag{15}$$

where the last equation is obtained in [[18], §3.324.1].

To this end, the CDF for e2e SNR can be given by

$$\begin{aligned} F_{\gamma_{eq}}(x) &= 1 - (1 - F_{\gamma_R}(x)) \times (1 - F_{\gamma_D}(x)) \\ &= 1 - Q_1(x) + Q_2(x), \end{aligned} \tag{16}$$

where $Q_1(x)$ and $Q_1(x)$ are defined in (11).

This ends the proof for this section.

3.2 Ergodic Capacity with the End-to-end SNR

We are able to obtain the PDF of the received e2e SNR after taking the partial derivative of the derived expression in (11) with respect to x as follows

$$f_{\gamma_{eq}}(x) = Q_3(x) + Q_4(x), \tag{17}$$

where

$$Q_3(x) = \frac{1}{2x}\left[1 + exp\left(-\frac{(1-\alpha)}{2\eta\alpha x \bar{\gamma}_{RR}}\right)\right] \Psi^2(x) K_0(\Psi(x)),$$

$$Q_4(x) = \frac{(1-\alpha)}{2\eta\alpha\bar{\gamma}_{RR}x^2} \exp\left(-\frac{(1-\alpha)}{2\eta\alpha x\bar{\gamma}_{RR}}\right) \Psi(x) K_1(\Psi(x)),$$

and using the property of Bessel function, $\frac{\partial}{\partial x}(z^v K_v(z)) = -z^v K_{v-1}(z)$ [[18], §8.486.18].

Thus, using (17), the ergodic capacity at high SNR is given by

$$C_{\gamma_{eq}} = \int_{x=0}^{\infty} (Q_3(x) + Q_3(x)) \log_2(1+x)\, dx. \tag{18}$$

3.3 Average Bit Error Probability (ABEP)

In this section, we are going to consider the average bit error probability (ABEP) which is a probability used to evaluate the performance of wireless communication applications. For different types of modulations, such as BPSK, BFSK with orthogonal signaling, and M-ary square QAM alphabet, which transmits data by changing the amplitude of two carrier signals, is widely used in wireless communication systems.

For the sake of simplicity, the ABEP, $P_b(e)$ at R can be expressed by

$$P_b = \int_{\gamma=0}^{\infty} P_b(e|\gamma) f_{\gamma_{eq}}(\gamma)\, d\gamma, \tag{19}$$

where $P_b(e|\gamma)$ is averaging the conditional BEP over the PDF.

Using BEP in an AWGN channel [[20], Eq.(14)], the k-th bit ABEP of Gray bit-mapped M-ary square QAM is given by

$$P_b(e|k) = \frac{2}{\sqrt{M}} \sum_{i=0}^{(1-2^{-k})\sqrt{M}-1} \left\{ (-1)^{\left\lfloor \frac{i.2^{k-1}}{\sqrt{M}} \right\rfloor} \times \left(2^{k-1} - \left\lfloor \frac{i.2^{k-1}}{\sqrt{M}} + \frac{1}{2} \right\rfloor\right) \right. \\ \left. \times \varepsilon\left[Q(2i+1)\sqrt{\frac{3\gamma}{M-1}}\right] \right\}, \tag{20}$$

where $\varepsilon[.]$ denotes the statistical expectation operator, and $Q(x)$ is the Gaussian Q-function defined as $Q(x) = \frac{1}{\sqrt{2\pi}}\int_x^{\infty} e^{-t^2/2} dt$ and $x \leq 0$.

Eventually, the ABEP of M-ary square QAM can be written as

$$P_b(e) = \frac{1}{\log_2\sqrt{M}} \sum_{k=0}^{\log_2\sqrt{M}} P_b(e|k). \tag{21}$$

In order to compute the ABEP as in [20], we have the following expression

$$\mathcal{J} = E\left|pQ\left(\sqrt{2q\gamma}\right)\right|, \tag{22}$$

where $(p,q) = (1,2)$ for BPSK, and $(p,q) = (1,1)$ for QPSK. As a result, before obtaining the ABEP, the distribution function of γ has to be expected.

Then, we rewrite the expression for ABEP derived in (22) over the CDF with the e2e SNR as follows

$$\mathcal{J} = \frac{p\sqrt{q}}{2\sqrt{\pi}} \int_{\gamma=0}^{\infty} \frac{e^{-q\gamma}}{\sqrt{\gamma}} F_{\gamma_{eq}}(\gamma)\, d\gamma. \tag{23}$$

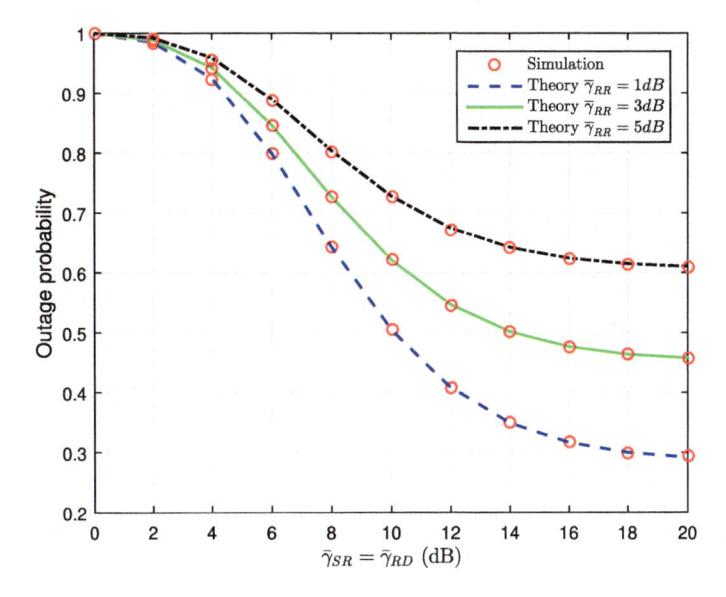

Fig. 2. The CDF of the received SNR versus the average SNR.

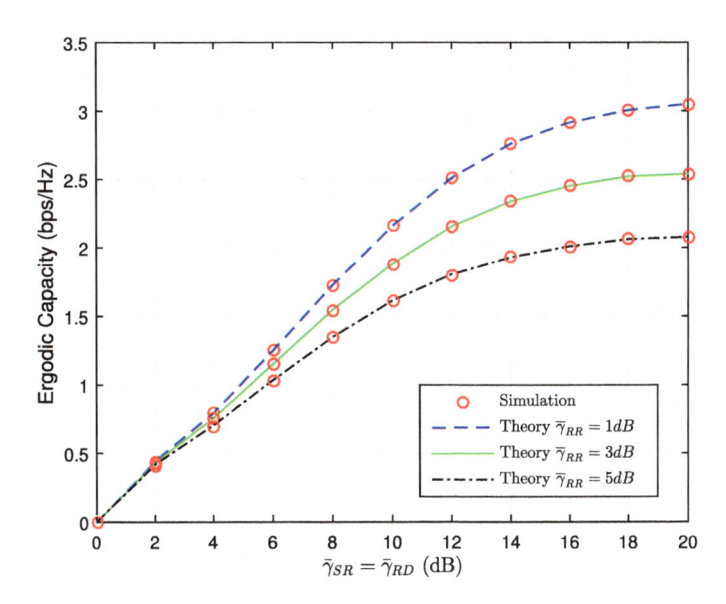

Fig. 3. The ergodic capacity with the e2e SNR versus the average SNR.

4 Numerical Results

In this section, we provide the simulation results based on Monte Carlo method to prove the robustness of the system performance. In particular, we assume that $\bar{\gamma}_{SR} = \bar{\gamma}_{RD}$. Besides that, we set $\eta = 0.4$, $\alpha = 0.2$, and $P_S = 1(W)$.

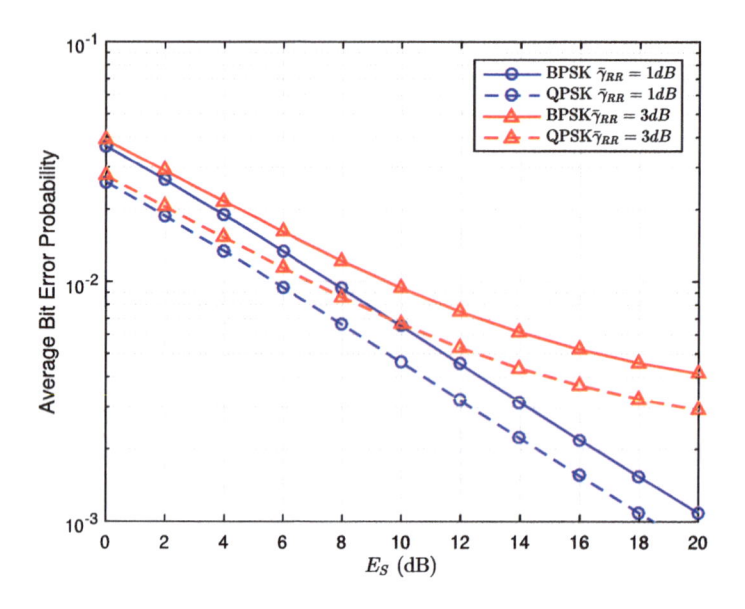

Fig. 4. The ABEP performance versus the transmit power at S.

In Fig. 2, it illustrates the CDF of the received SNR as a function of the average SNR in the S → R and R → D links in the presence of SI. It is shown that the CDF drops significantly when $x = 5dB$ as $\bar{\gamma}_{SR}$ increases. In case $\bar{\gamma}_{SR} = 12dB$, $\bar{\gamma}_{RR} = 5dB$, and $\bar{\gamma}_{RR} = 3dB$, the CDF of declines rapidly. It is observed that the ideal outage performance relies on the high level of $\bar{\gamma}_{SR}$ and the low level of $\bar{\gamma}_{RR}$.

The ergodic capacity versus the average SNR is depicted in Fig. 3 with three different values of $\bar{\gamma}_{RR}$. It is obvious that as SNR increases the ergodic capacity rises. In particular, in case $\bar{\gamma}_{RR} = 1$, it enjoys the best ergodic capacity compared to other two cases.

Figure 4 depicts the ABEP as a function of the transmit power at S in QPSK and BPSK modulations, i.e., ($p = q = 2$); ($p = 1, q = 2$)) under the impact of SI. It is noticeable that BER depends on the transmit power R. In HPSK scheme, the ABEP performance is better than that of others.

5 Conclusion

In this paper, we studied the system performance of a CRN considering TSR protocol to study EH at the relay node. In order to improve the system performance, expressions for outage probability at high e2e SNR, ergodic capacity, and the ABEP were obtained. Eventually, we validated our derivations by providing Monte Carlo simulations.

Acknowledgments. This research received funding from the grant No. SP2018/59 conducted by VSB-Technical University of Ostrava, Czech Republic and partially was supported by The Czech Ministry of Education, Youth and Sports from the Large Infrastructures for Research, Experimental Development and Innovations project No. LM2015070.

References

1. Kang, J.-M., Kim, I.-M., Kim, D.I.: Joint optimal mode switching and power adaptation for nonlinear energy harvesting SWIPT system over fading channel. IEEE Trans. Commun. **66**(4), 1817–1832 (2018). https://doi.org/10.1109/TCOMM.2017.2787568
2. De Rango, F., Gerla, M., Marano, S.: A scalable routing scheme with group motion support in large and dense wireless ad hoc networks. Comput. Electr. Eng. **32**(1–3), 224–240 (2006). https://doi.org/10.1016/j.compeleceng.2006.01.017
3. Fazio, P., De Rango, F., Sottile, C.: A new interference aware on demand routing protocol for vehicular networks. In: Proceedings of the 2011 International Symposium on Performance Evaluation of Computer and Telecommunication Systems, SPECTS 2011, pp. 98–103 (2011). art. no. 5984853
4. Nguyen, H.-S., Nguyen, T.-S., Voznak, M.: Relay selection for SWIPT: performance analysis of optimization problems and the trade-off between ergodic capacity and energy harvesting. AEU Int. J. Electron. Commun. **85**, 59–67 (2018)
5. De Rango, F., Lonetti, P., Marano, S.: MEA-DSR: a multipath energy-aware routing protocol for wireless Ad Hoc Networks. IFIP Int. Fed. Inf. Process. **265**, 215–225 (2008)
6. Rabie, K.M., Adebisi, B., Alouinik, M.-S.: Half-duplex and full-duplex AF and DF relaying with energy-harvesting in log-normal fading. IEEE Trans. Green Commun. Netw. **1**(4), 468–480 (2017). https://doi.org/10.1109/TGCN.2017.2740258
7. Zhao, Y., Adve, R.: Symbol error rate of selection amplify-and-forward relay systems. IEEE Commun. Lett. **10**(11), 757–759 (2016). https://doi.org/10.1109/LCOMM.2006.060774
8. Bai, X., Shao, J., Tian, J., Shi, L.: Power-splitting scheme for nonlinear energy harvesting AF relaying with direct link. Wirel. Commun. Mobile Comput. **2018** (2018). https://doi.org/10.1155/2018/7906957
9. Nasir, A.A., Zhou, X., Durrani, S., Kennedy, R.A.: Relaying protocols for wireless energy harvesting and information processing. IEEE Trans. Wireless Commun. **12**(7), 3622–3636 (2013). https://doi.org/10.1109/TWC.2013.062413.122042
10. Nasir, A.A., Tuan, H.D., Ngo, D.T., Duong, T.Q., Poor, H.V.: Beamforming design for wireless information and power transfer systems: receive power-splitting versus transmit time-switching. IEEE Trans. Commun. **65**(2), 876–889 (2017). https://doi.org/10.1109/TCOMM.2016.2631465

11. Liu, L., Zhang, R., Chua, K.-C.: Information, wireless, transfer, power: a dynamic power splitting approach. IEEE Trans. Commun. **61**(9), 3990–4001 (2013). https://doi.org/10.1109/TCOMM.2013.071813.130105
12. Nguyen, H.-S., Nguyen, T.-S., Nguyen, M.T., Voznak, M.: Optimal time switching-based policies for efficient transmit power in wireless energy harvesting small cell cognitive relaying networks. Wireless Pers. Commun. Int. J. **99**(4), 1605–1624 (2018). https://doi.org/10.1007/s11277-018-5296-2
13. Lu, G., Shi, L., Ye, Y.: Maximum throughput of TS/PS scheme in an AF relaying network with non-linear energy harvester. IEEE Access, 26617–26625 (2018). https://doi.org/10.1109/ACCESS.2018.2834225
14. Nguyen, H.-S., Voznak, M., Nguyen, M.-T., Sevcik, L.: Performance analysis with wireless power transfer constraint policies in full-duplex relaying networks. ELEKTRONIKA IR ELEKTROTECHNIKA **24**(4), 1215–1392 (2017). https://doi.org/10.5755/j01.eie.23.4.18725
15. Yan, Z., Chen, S., Zhang, X., Liu, H.-L.: Outage performance analysis of wireless energy harvesting relay-assisted random underlay cognitive networks. IEEE Internet Things J. (2018). https://doi.org/10.1109/JIOT.2018.2800716
16. Nguyen, H.-S., Nguyen, T.-S., Vo, V.-T., Voznak, M.: Hybrid full-duplex/half-duplex relay selection scheme with optimal power under individual power constraints and energy harvesting. Comput. Commun. **124**, 31–44 (2018)
17. Peng, C., Li, F., Liu, H.: Wireless energy harvesting two-way relay networks with hardware impairments. Sensors (2017). https://doi.org/10.3390/s17112604
18. Gradshtein, I.S., Ryzhik, I.M.: Table of Integrals, Series, and Products, 4th edn. Academic Press Inc., New York (1980)
19. Lou, Y., Qi-Yue, Y., Cheng, J., Zhao, H.-L.: Exact BER analysis of selection combining for differential SWIPT relaying systems. IEEE Signal Process. Lett. **24**(8), 1198–1202 (2017). https://doi.org/10.1109/LSP.2017.2705066
20. Cho, K., Yoon, D.: On the general BER expression of one and two dimensional amplitude modulations. IEEE Trans. Commun. **50**(7), 1074–1080 (2002). https://doi.org/10.1109/TCOMM.2002.800818

LCCT vs. LLC Converter - Analysis of Operational Characteristics During Critical Modes of Operation

Michal Pridala, Michal Frivaldsky[(✉)], and Pavol Spanik

University of Zilina, 010 26 Zilina, Slovakia
michal.frivaldsky@fel.uniza.sk

Abstract. Following the invention of resonant power converters, lots of new topologies with significant improvements considering increase of efficiency and power density are arising. The main procedure how to optimize operational characteristics is through the modification of converter's main circuit. In this paper, circuit topology of LLC resonant DC-DC converter and half-bridge LCCT resonant DC-DC converter are comparatively studied for target application, which is modular architecture of power supply. Simulation models of LLC and LCCT circuits are created using magnetic and thermal models of individual components using PLECS simulation software. Based on the circuit investigation within critical operational conditions, short-circuit and overload states are analysed, because of the selection procedure of the proper topology for the modular power supply system. For these purposes extended range of output power, flat characteristic of efficiency, and low ripple current and/or voltage will be necessary.

Keywords: Resonant converter · Efficiency · Redundancy · Critical operation modes

1 Introduction

Nowadays well know and popular solutions of resonant converters have been studied deeply, thus functionality and applications area are well covered even due to initial problems with their understanding [1–4]. But focusing on the expectation of next days they may be off their future use for custom specific or industrial applications. The main reasons are limiting values of regulation range, and low functionality from characteristic properties point of view.

In this article, initially two topologies (LLC/LCCT) are compared in terms of efficiency, cost and behavior in nominal and dynamic states. LCCT converter presents multi-tank resonant topology with one series LC tank and one parallel LC tank [5, 6]. The main purpose of this approach is to be able to establish operational regions and operational characteristics, what is necessary for more complex designs. It is expected that selected topology shall be used for modular solutions of power supply systems, thus redundancy, overloading capability, short-circuit prevention, no-load operation must be investigated in order to secure reliable and efficient operation [7–11].

The structure of the paper is as follows:

- Comparisons of the transfer characteristics of LLC and LCCT converter, description of operational regions and recommendations for optimal operation
- Input - output parameters and resonant components calculation of LLC and LCCT converter for selected application
- Steady state analysis and critical modes of operation (start-up, over-load, over-current).

2 Transfer Characteristics of LLC and LCCT Converter

The main difference of properties between those two converters is in the shape of the transfer characteristics (Fig. 1). These characteristics are considered for resonant tanks of the LLC and LCCT converter using 100 kHz as the main resonant frequency, nominal quality factor qN = 1 and input to output voltage transformation 400 Vdc/48 Vdc. Resonant tanks have constant transfer ratio at this frequency, thus input voltage is the same as output voltage for all load values. Steepness of the characteristics is depending on the components of resonant tanks, it mainly depends on the quality factor of resonant components.

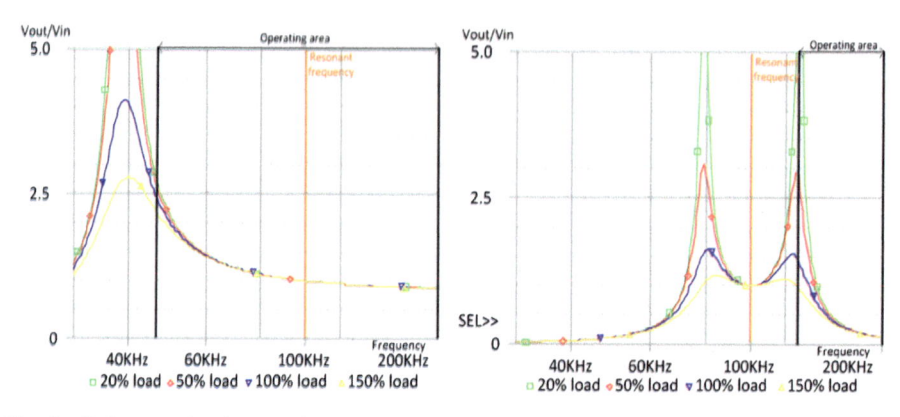

Fig. 1. Voltage gain characteristic of LLC converter (left) and LCCT converter (right) for similar input-output conditions.

As can be seen from Fig. 1, the transfer characteristic of LCCT has more steepness. This means it is possible to change the output voltage of power converter in wide range through the smaller change of switching frequency. It also has two gain peaks, but a feasible operating area of this type of converter is divided into two regions. First starts from the resonant frequency and continues to the right side what is similar to the LLC converter. The ZCS region should be avoided, while is bounded by the individual gain peaks relevant for different power load to the left bounded by the main resonant frequency. Second region starts from the resonant frequency to the left peak, while prohibited part is larger compared to right side from resonant frequency (regions are mirrored). The shape of LCCT transfer characteristic enables more control flexibility during various operation states, whereby compared to the LLC converter frequency range can be much smaller. This is also related to critical operational conditions, i.e. start-up process or safety – functionality (overload, short-circuit etc…). LLC resonant converter is usually triggered from very high switching frequencies, typically three-times higher than the resonant frequency of resonant tank. The same is valid when critical operation mode occurs. That is one of the reasons, why manufacturers are using lower operating frequencies (100 kHz–200 kHz) for medium power levels (500 W–3 kW) when using LLC topology. If switching frequency of 300 kHz has to be used within nominal operating state, for example, the start-up or protective frequencies must achieve ranges around 900 kHz. These values are problematic for the most of the digital controllers, while combined A/D circuits have also their limits.

3 Comparison of LLC and LCCT Converter – Steady State, Critical Operational Modes

For determination of the resonant circuit components of LLC and LCCT (Fig. 2) module the same resonant frequency was considered (100 kHz). Both converters have the same conversion ratio i.e. 400 V of input voltage and 48 V of output voltage, while their nominal output power is 500 W. From these parameters, the resonant circuit elements were determined (Table 1), whereby detailed derivation of the computational equations presented well in [12–14].

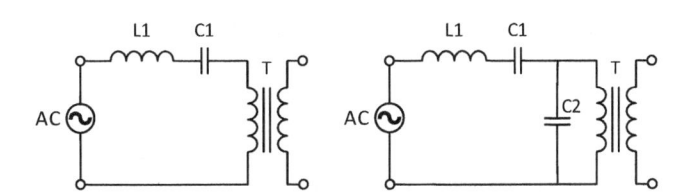

Fig. 2. The LLC (left) and LCCT (right) tank configuration.

Table 1. Resonant tank components of proposed converters.

Name of component	LLC	LCCT
Serial capacitor C1	120 nF	10 nF
Serial inductor L1	21 µH	254 µH
Parallel capacitor C2	n/a	39 nF
Magnetizing inductance of transformer	121 µH	64 µH
Turn-ration of transformer	8/2	8/2

In this chapter simulation models are presented for steady state analysis and investigation of critical modes of operation. The models are designed in PLECS software (Fig. 3), which is primarily suited for analysis of power electronic systems. It enables precise modeling from thermal performance point of view (semiconductor devices) as well as from precise modeling of high frequency inductive components. For proposed converters the RM12 core with 3F3 properties of ferrite material was selected.

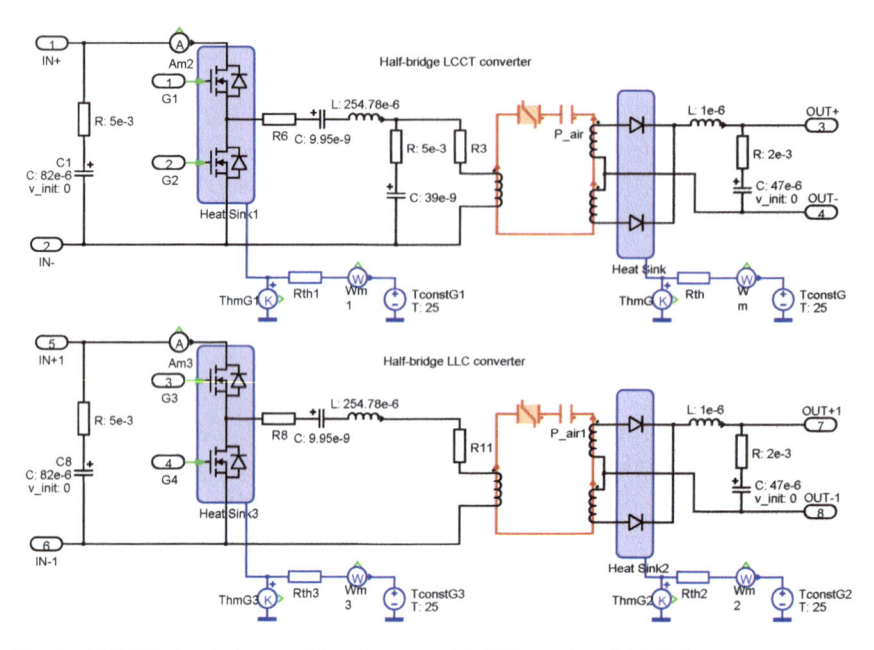

Fig. 3. PLECS simulation models of proposed LCCT (top) and LLC (bottom) converters.

3.1 Steady-State Operation

Control of the LCCT converter is similar to standard LLC converter, so pulse frequency modulation is used. Steady state operation is investigated for the full load operation (100%), light load operation (20%). The nominal output power is 500 W/48 Vdc. Switching frequencies for this operational point for both converters is 118 kHz, thus it is slightly shifted to the right from main resonant frequency. It is seen, that ZVS switching is valid for primary switches. Output diodes are switched within ZCS conditions (Fig. 4).

The series part of resonant tank (LCCT) is characterized by sinusoidal waveforms of voltage and current (L1 – C1), while voltage at the parallel resonant capacitor is limited to the voltage of power supply. Current waveform has pulsating character (LCCT) with higher peak value, anyway soft switching is achieved at the output rectifier.

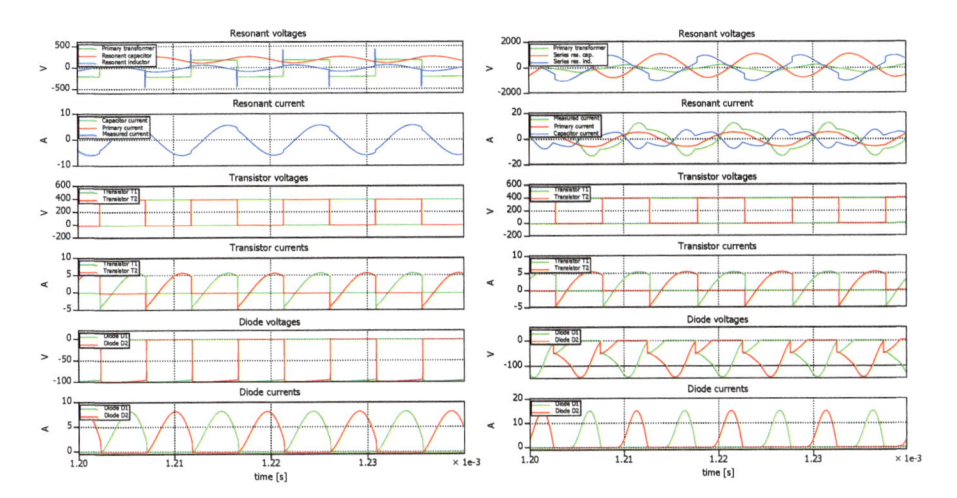

Fig. 4. Operational waveforms of LLC (left) and LCCT converter (right) for 100% load.

Light – load operation can be considered when different power load of the individual modules within modular power supply system occur. The output power for this operational point is 100 W/48 Vdc. The switching frequency remains at 118 kHz for LLC converter due to its flat transfer characteristic. There is no dramatic change within the operational waveforms, just the amplitudes of the currents decreased to the certain values (Fig. 5).

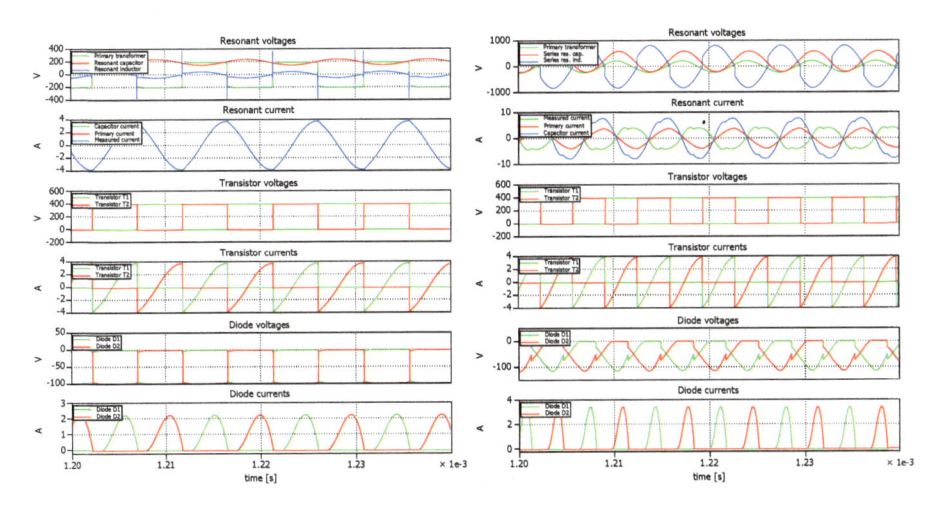

Fig. 5. Operational waveforms of LLC (left) and LCCT converter (right) for 20% of load

Because LCCT converter has much steeper characteristic when change of load is considered, the switching frequency was tuned to 147 kHz. Figure 5 shows most important waveforms of circuit components. As in previous case, ZVS/ZCS conditions are achieved at the primary/secondary side switches, sinusoidal character of voltage and currents of resonant tank is visible, and nominal voltage stress for resonant capacitors is also visible. Similarly to the previous situation mostly the amplitudes of the individual currents decreased.

Next experiment is focused on the investigation of the operational waveforms during overload condition. Such situation can occur, when power imbalance of individual converters during modular operation becomes valid. In such situation one module shall withstand higher power stress oppose to nominal operation, because of the failure of other one, or because of unexpected operation of whole modular system. It is considered that during such operation, one module shall withstand double of the nominal output power, so for this point Pout = 1 kW/48 Vdc. Looking at the Fig. 1, switching frequency will be closer to resonant frequency, thus it was tuned to 105 kHz for both converters. Almost for this case, ZVS conditions are achieved for primary switches, what is preferred for all of the conditions. Also sinusoidal character is visible at the waveforms of voltage and current of resonant tanks (LCCT). The output current is doubled oppose to nominal conditions, whereby ZCS conditions for secondary rectifier can be visible. Based on these results, it can be said the overload condition might not be critical, when safety margin for switches and designed circuit components will be considered (Fig. 6).

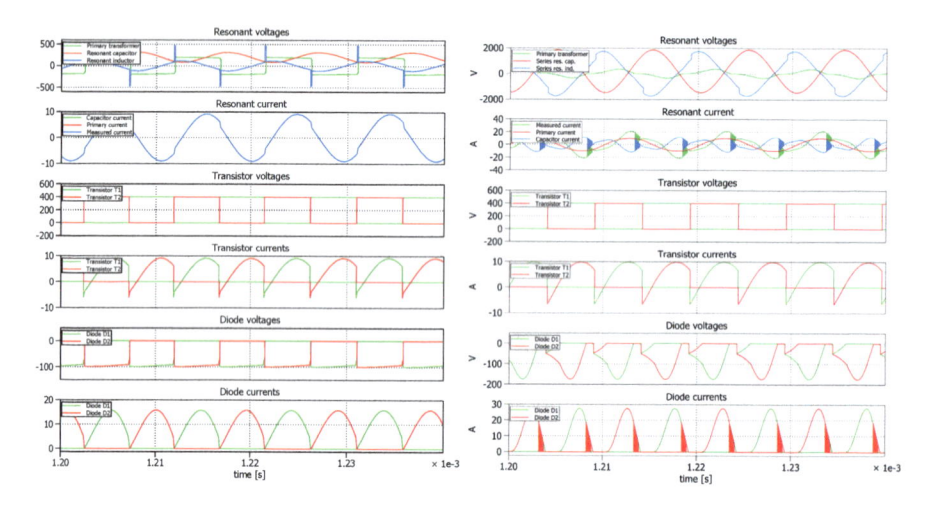

Fig. 6. Operational waveforms of LLC (left) and LCCT converter (right) for 200% of load

3.2 Critical Dynamic Operation

Start – up of the resonant converters is task that needs individual attention. It is related with issues that are current and voltage stress of resonant components, as well as current inrush of semiconductor devices, or output capacitors. Most common and reliable way how to eliminate negative impacts is the use of frequency ramp. LLC converter is starting – up with switching frequency that is several times higher (3–5 times) than nominal operating frequency [15, 16]. In this way, the gain of resonant circuit is limited, thus output voltage is not at its nominal value initially. The value of starting frequency is given by the character and shape of transfer characteristic. In this way, overregulation at the output as well as limitation of inrush current is secured. On the other side, dynamics must be satisfactory. From Fig. 1 is seen, that the shape of transfer characteristic of LCCT converter is not as flat as in the case of LLC converter. For the start-up sequence of the LCCT converter, the lower frequency will be required thus the voltage gain modification is more flexible compared to LLC. In this way the value of inrush current can be significantly lowered. Oppose to well-known LLC converter, this is big advantage, because much lower regulation range will be required.

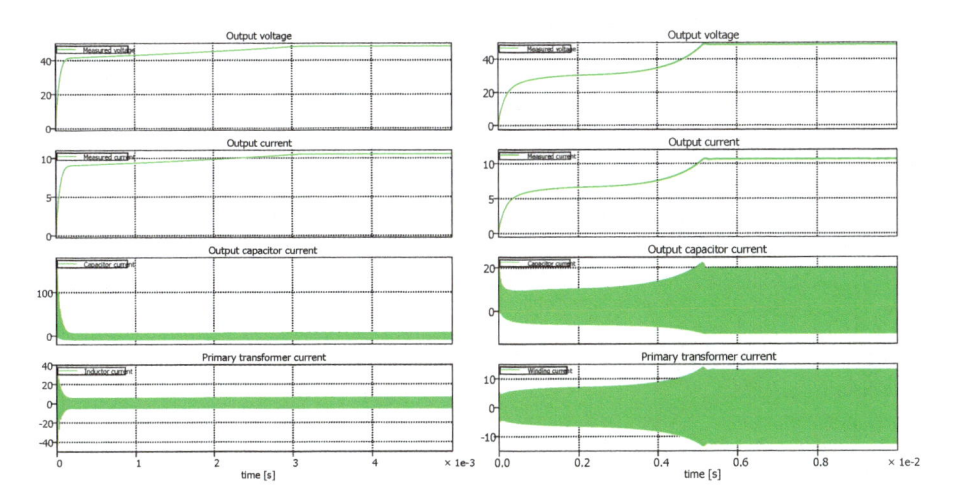

Fig. 7. Start-up waveforms of proposed LLC (left) and LCCT (right) converter

Figure 7 shows start – up of LLC converter for nominal power (100%), where operation with frequency ramp from initial 150 kHz to 105 kHz was applied. As can be seen from the waveforms the high value of inrush-current is presented, while peak value exceeds 140 A at the output filtering capacitor. Elimination is possible through the use of several times higher start-up frequency, or with the use of start-up circuit (relay + shunt resistor). In the case of LCCT converter, the inrush current of the output capacitor is comparable to the value of the nominal current. This ability is also valuable when lifetime of circuit components is considered, because electric - stresses during operation will be reduced.

Another critical situation is short-circuit. This situation is critical for any power electronic system. It was analysed from nominal operation of converter, while at certain time, short-circuit at the output terminal occurred. Figure 8 shows described situation. It can be seen that another advantage of proposed LCCT converter is clearly visible. It is its natural short-circuit prevention, while any increase of switching frequency, or skipping of pulses, or reduction of pulse width is not necessary. The resonant tank is acting as circuit with very high impedance, thus current flowing from the source to the output is naturally limited by its characteristic. The short-circuit situation reflects mainly in the resonant circuit. Voltage at series capacitor C1 increases, also with the increase of current flowing through inductor L1. These are the main components that can be harmed during short-circuit operation.

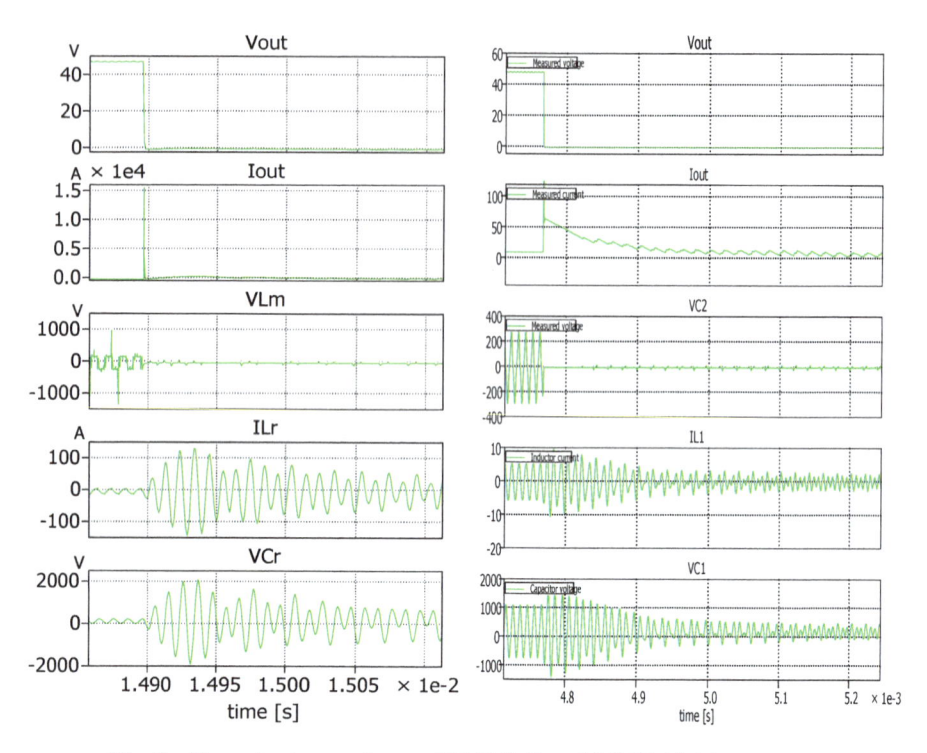

Fig. 8. Short-circuit waveforms of LLC (left) and LCCT (right) converter

4 Conclusion

In this paper LCCT multi-resonant converter with DC-output of voltage type is presented, while focus is given on the application for modular power supply architectures. For that purpose initial description of proposed converter is given, while isolated version is designed because of target application scope. Voltage transfer characteristic was designed in order to describe most proper operating regions of proposed converter. After it, simulation analysis was performed, whereby attention was paid to steady state operation as well as to the investigation of dynamic behaviour within critical and standard regimes. Well-known LLC converter was compared to proposed LCCT within these experiments. It was found that proposed LCCT converter has several advantages, which are suiting it for further perspective use in modular architectures of power supply. These are better dynamic performance like start-up and dynamic load regulation, redundancy, better short-circuit and overload capabilities. Mentioned properties are targeting characteristics which are required from modular power supply, i.e. reliability, high efficiency and long-life operation. Future works will be focused on the further investigation of imbalance and dynamic control during closed loop operation.

Acknowledgements. The authors wish to thank Slovak grant agency APVV for the project no. 0396-15 - Research of perspective high-frequency converter systems with GaN technology. Authors also would like to thank for the support from the national grant agency VEGA for the project no. 1/0479/17 – Research of the optimal methodologies for energy transfer within the systems with distributed storage systems.

References

1. Hu, Z., Qiu, Y., Wang, L., Liu, Y.F.: An interleaved LLC resonant converter operating at constant switching frequency. IEEE Trans. Power Electron. **29**(6), 2931–2943 (2014)
2. Prazenica, M., Dobrucky, B., Sekerak, P., Kalamen, L.: Design, modelling and simulation of two-phase two-stage electronic system with orthogonal output for supplying of two-phase ASM. Adv. Electr. Electron. Eng. **9**(1), 56–64 (2011)
3. Chen, S.-M., Haung, Y.-H., Chung, Y.-Y., Hsieh, Y.-H., Liang, T.-J.: A novel interleaved LLC resonant converter. In: IECON 2013 - 39th Annual Conference of the IEEE Industrial Electronics Society, Vienna, pp. 293–297 (2013)
4. Wu, L.M., Chen, P.S.: Interleaved three-level LLC resonant converter with fixed-frequency PWM control. In: IEEE 36th International Telecommunications Energy Conference (INTELEC), Vancouver, pp. 1–8 (2014)
5. Nobile, G., Cacciato, M., Scarcella, G., Scelba, G.: Multi – criteria experimental comparison of batteries circuital models for automotive applications. Komunikácie/Commun. **1**, 97–104 (2018). ISSN 2585-7878 (online), ISSN – 1335-4205 (print)
6. Dobrucky, B, Prazenica, M, Kascak, S, Kassa, J.: HF link LCTLC resonant converter with LF AC output. In: 38th Annual Conference on IEEE-Industrial-Electronics-Society (IECON), Canada, pp. 447–452 (2012). ISBN 978-1-4673-2421-2, ISSN 1553-572X
7. Dobrucky, B., Benova, M., Kascak, S.: Transient analysis and modelling of 2nd-and 4th-order LCLC filter under non-symmetrical control. Elektronika IE Elektrotechnika **111**(5), 89–94 (2011)
8. Dudrik, J., Pastor, M., Lacko, M.: High-frequency soft-switching PWM DC-DC converter with active output rectifier operating as a current source for arc welding applications. Electric Power Compon. Syst. **45**(6), 681–691 (2017)
9. Kindl, V., Kavalir, T., Pechanek, R., Skala, B., Sobra, J.: Key construction aspects of resonant wireless low power transfer system. In: ELEKTRO 2014, Rajecké Teplice, pp. 303–306, 19–20 May 2014
10. Kindl, V., Kavalir, T., Pechanek, R., et al.: Basic operating characteristics of wireless power transfer system for small portable devices. In: 40th Annual Conference of the IEEE-Industrial-Electronics-Society (IECON), Dallas, TX, pp. 3819–3823, October 2014
11. Zaskalicky, P.: Modelling of a serial wound DC motor supplied by a semicontrolled rectifier. Adv. Electr. Electron. Eng. **5**(1), 110–113 (2006)
12. Aiello, G., Cacciato, M., Messina, S., Torrisi, M.: A high efficiency interleaved PFC front-end converter for EV battery charger. Komunikácie/Commun. **1**, 86–92 (2018). ISSN 2585-7878 (online), ISSN 1335-4205 (print)

13. Dobrucky, B., Benova, M., Abdalmula, M., Kascak, S.: Design analysis of LCTLC resonant inverter for two-stage 2-phase supply system. Automatika **54**(3), 299–307 (2013). ISSN 0005-1144
14. Chlebis, P., Tvrdon, M., Baresova, K., et al.: The system of fast charging station for electric vehicles with minimal impact on the electrical grid. Adv. Electr. Electron. Eng. **14**(2), 89–94 (2016)
15. Tvrdon, M., Chlebis, P., Hromjak, M.: Design of power converters for renewable energy sources and electric vehicles charging. Adv. Electr. Electron. Eng. **11**(3), 204–209 (2013)
16. Stepins, D., Huang, J.: Optimization of modulation waveforms for improved EMI attenuation in switching frequency modulated power converters. Adv. Electr. Electron. Eng. **13**(1), 10–21 (2015)

Control Renewable Energy System and Optimize Performance by Using Weather Data

Duy Tan Nguyen, Duy Anh Nguyen[✉], and Lien Son Chau Hoang

Ho Chi Minh University of Technology,
268 Ly Thuong Kiet, Ho Chi Minh City, Vietnam
duyanhnguyen@hcmut.edu.vn

Abstract. There is no doubt that renewable energy (RE) is now playing a key role in our daily life and has become more popular thanks to many advantages such as environment friendly, unlimited sources and so on, variable of applications. Obviously, the only one obstacle we must face is the dependence upon weather conditions which have been, so far, the unpredictable variants of most RE systems. Hence, in this paper, we propose a new algorithm using weather forecast and historical weather data in order to optimize performance of RE systems. Basically, this algorithm is not only making full use of weather data base but also helping us decide whether to discharge the load or have it operated on high performance at least in the next three hours. We have already embedded it into our RE system combining with Darius wind turbine and photovoltaic (PV) system, the result is showed below.

Keywords: Weather database · Optimize resources · Wind power system · Solar system

1 Introduction

We have built an algorithm to make full use of database including historical weather data and live weather data to make decision should or not discharge loads in our renewable energy system. The live weather data we use is from Weather Underground (one of the weather forecast sources of Google) and web page Timeanddate while our historical weather data is just corrected the featured months in a year.

Hence, we analyze the data from the past and compare with the other one at the present. We start with simple techniques such as finding the correlation between wind energy, solar energy and weather conditions. Then we use Visual Studio plots for fine grained analysis. We try to find the relation between our renewable energy and weather attributes based on another system from [4] and [10].

After processing data and making decision, our system can work more fluently with a higher performance. Once the algorithm is embedded in the system, the power from wind turbine and PV panels will be used more efficiently and economically.

2 Basic Theory

2.1 Wind Turbine System

The first source of system is a vertical wind turbine operating well in any wind direction. In addition, the output power is ratio with the input velocity wind. In fact, wind turbine cannot collect 100% kinematic energy from the wind or converts this kind of energy to electrical energy. Thus, we needed to pay more attention to the coefficient of the system, this coefficient gives full effect on the wind turbine output power. 20%–25% is the normal average value of this coefficient. For more detail, the formula below shows us about relationship between wind speed with output power and wind energy with electrical energy [1].

$$P_{turbine} = \frac{1}{2} C \rho A v^3 \tag{1}$$

Where:

ρ – air density (kg/m^3)
C – coefficient of power
A – frontal area (m^2)
v – velocity of the wind (m/s)

Equation (1) show us how power output depends on the velocity wind. On the one hand, although a small change of wind speed can make the wind turbine power increases significantly, the wind is not stable during a day or at least in hours. To improve the reliability of our system, we built another power source, and it was photovoltaic system.

2.2 Photovoltaic System

The second source is PV panels from inductors, it is made from Silicate. Basing on follow equations, we can find out the output power and how temperature effect on our results. Notifying that the equivalent circuit and the basic equations of the PV cell/panel in Standard Test Conditions (STC) are shown, as well as the parameters extraction from data-sheet values.

$$V_{OC}(T) = V_{OC} + k_v(T - T_{STC}) \tag{2}$$

$$I_{SC}(T) = I_{SC}\left[1 + \frac{k_i}{100}(T - T_{STC})\right] \tag{3}$$

Where V_{OC} is open-circuit voltage in STC, I_{SC} is short-circuit current in STC, k_v is temperature coefficient of the open-circuit voltage, k_i is temperature coefficient of the open-circuit voltage, T is temperature at real time and T_{STC} is temperature in Standard Test Conditions. Both of these parameters V_{OC} and I_{SC} change with ambient temperature. As a result, the output power is not constant and we must consider other parameters like V_{MPP} and I_{MPP}. These are the voltage and current at the maximum power points in Standard Test Condition, all these technical characteristics have impact on the maximum power, the following equations will illustrate:

$$\eta = \frac{V_{OC} I_{SC} FF}{P_{in}} = \frac{P_{max}}{P_{in}} \qquad (4)$$

$$FF = \frac{V_{MP} I_{MP}}{V_{OC} I_{SC}} \qquad (5)$$

The fill factor, more commonly known by its abbreviation "FF", is a parameter which, in conjunction with V_{OC} and I_{SC} determines the maximum power from a solar cell. The FF is defined as the ratio of the maximum power from the solar cell to product of V_{OC} and I_{SC}.

In another way Eq. (4) can rewrite as below:

$$\eta_C = \frac{P_{max}}{P_{in}} = \frac{V_{max} I_{max}}{I(t) A_C} \qquad (6)$$

Where V_{max}, I_{max} is voltage and current at maximum power point, I(t) is solar intensity and A_C is area of solar cell.

2.3 Data Analyze

The main point of this work was how could we combine the historical weather data and live data from the weather forecast. We chose which months were the best for wind turbine operating well and which months were suitable for PV panels working with the highest performance.

Ho Chi Minh city is affected by two main wind direction and mainly monsoon winds in the west – southwest and north – northeast. Southwest wind from the Indian Ocean blew in the rainy seasons, from June to October with an average speed 3.6 m/s and the strongest winds in August, averaging 4.5 m/s. The Northeast wind blew from East Sea in every dry season, from November to February with an average speed of 2.4 m/s. In addition, there is another wind direction, south – southeast, from March to May, average speed of 3.7 m/s. Basically, Ho Chi Minh city is in the area without wind storms. From above information, we chose August is the month wind operates well.

The first figures belows is the graph illustrating annual average temperature in Ho Chi Minh City. From this point, we can see the different temperature between months is small and almost there is not change much between months. So, we can choose August for the month PV panels working well (Fig. 1).

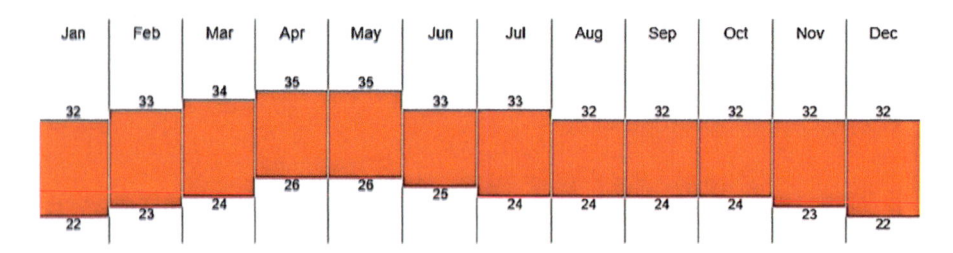

Fig. 1. Annual temperature average in Ho Chi Minh City

3 Methodology

3.1 Details About Renewable Energy System

Block diagram in the Fig. 2 illustrates all components in our system. In this figure, wind turbine and solar panel are the main sources of our system. A back up source is battery bank, after connecting with some converters to meet electrical characteristics of all system. Then there are four switches connect these components to the main DC bus. In addition, there are two MPPT algorithms which maintain the maximum output power as long as possible. The output is DC loads or energy can go through a converter to supply the AC loads.

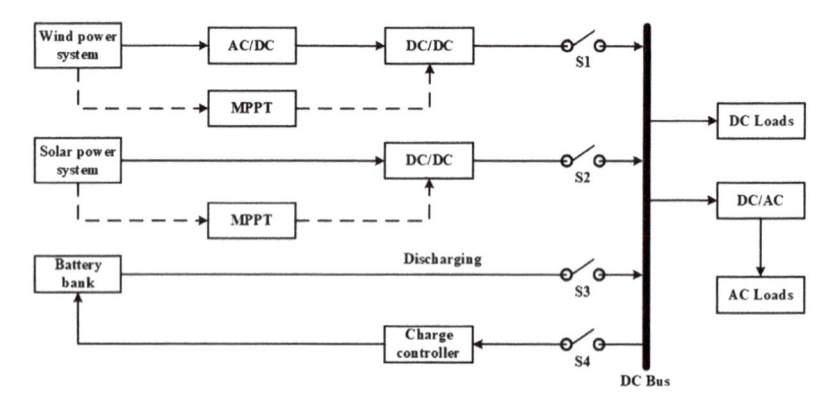

Fig. 2. Block diagram of system

In Fig. 2, the core of wind power system (the first block) is vertical wind turbine operating well in any wind direction but its power still depends on wind velocity. Equation (1) shows how big dependent is. Wind speed and power are third-order proportional, when the wind speed increase or decrease, the output power increase (decrease) in proportion to the cube. To solve this drawback, an algorithm MPPT is added into system. The algorithm helps turbines can operates at maximum output power as long as possible, even the wind speed changes or not. The output power is converted into DC current before providing all the loads.

The second core block is solar power system and the sources are solar panels had strong impact by temperatures and sky conditions. Equation (10) expresses the effect of environment temperature on cells temperature. Then, Eq. (11) illustrates how cells temperature carries out output performance solar system. Moreover, this system needs algorithm MPPT and some converter to convert and optimize output DC current.

The final core block is battery bank recommend as a backup source in system. This battery banks operates with an algorithm to control its charge or discharge action. However, we want this back up source to be used hardly, we still want turbine and solar panels to be the main sources for all the loads.

Next, the operational principle of system is described through each of steps below:

First, system updates the parameters and current status of the power supply of the wind turbine, the solar cell and the storage status of battery. Thereby, system sets a parameter that the available capacity is calculated by the total power of the turbine and the capacity of the solar cell.

Second, from the available capacity, system compares to the required power to deliver the loads. If available capacity is equal to the load power consumption, system close two switches S1 and S2 supply and disconnect switch S3, S4. Then system continues to update status of the initial parameters.

Third, If the available capacity is less than the consumption load, first of all, the current state of battery is checked, if the number is between 40–80%, system closes two switches S1 and S2 and disconnect switches S3, S4. In the case this figure less than 40%, no power source can meet the load capacity, disconnect all the loads.

Fourth, in case the available capacity is greater than the consumption demand, system priority the loads first which mean close switches S1 and S2.

Finally, considering the status of the battery, if the SOC status is less than 95%, system uses the extra power (in step 4) to charge the battery, this closes the S4 simultaneously. Disconnecting S3 to avoid battery backwards into the load system. If SOC is greater than 95%, we now connect excess energy to a dump load.

3.2 Weather Data Algorithm

Next step, we built an algorithm for our system. This algorithm used the data after analyzing and the status battery to calculate how many hours user can discharge all the loads. That means the system would announce when user could let all loads operate at 100% performance. For this purpose, we needed to know the status of all the load, we just recommend some representative loads. We present detail information about these loads in the next part, now we move to the condition to active mode which all loads work 100% performance. The Fig. 3 illustrates details about this algorithm.

More specifically, the algorithm works as 6 steps follow:

First, a variable i is set up. This variable represents for hours in a day. In the beginning of the day, value i is set to 0 and move on the second step.

Second, based on the current variable i, the weather data is got from database. From the present, system collects the 3-next-hour wind speed. For clearly understand, we name the wind speed in first hour is W_i, the second is W_{i+1}, the third is W_{i+2} and move on the next step.

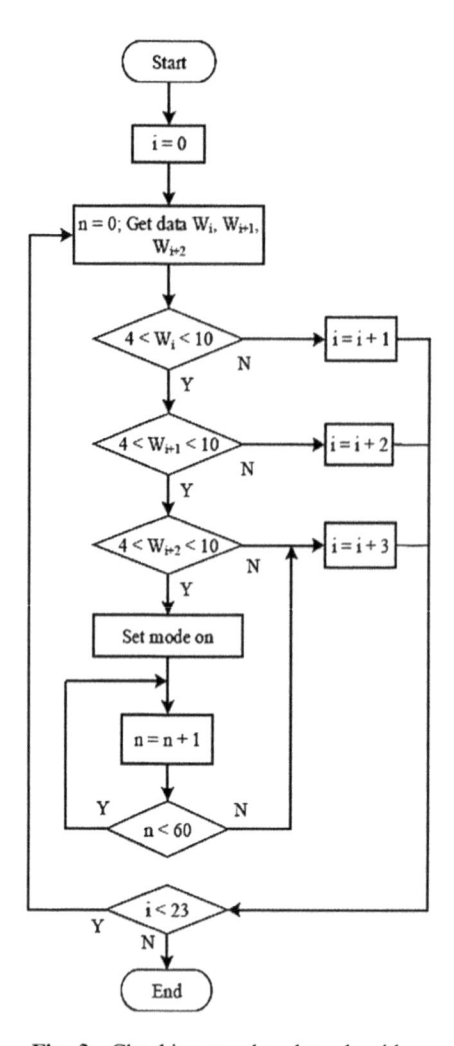

Fig. 3. Checking weather data algorithm

Third, the system sets up a range from 4 (m/s) to 10 (m/s), if the W_i value is in this range, the system moves on the fourth step. Otherwise, the variable i is added 1 unit. New variable i is checked, if its value smaller than 23, the system updates new values i and return the second step. In case variable i is bigger than 23, the system stops.

Fourth, the value W_{i+1} is has the same process with value W_i in the previous step. Instead of adding unit, variable i is added to 2 units if W_{i+1} is not in allowed range.

Fifth, this step almost has the same process with the fourth step. The value W_{i+2} is checked in allowed range, when its value still is in the range, the active mode is set on. Active mode means all the loads in system can operate with 100% performance (this mode is just only actives when all 3 values W_i, W_{i+1}, W_{i+2} meet fully requirement system). Otherwise, variable i is added to 3 units if W_{i+2} is not in allowed range.

Sixth, n is added 1 unit and checked it larger than 60 or not. If n is still smaller than 60, the system continues add 1 unit to n and check when it's larger than 60. When value n is beyond 60, variable i is added 3 units. If new variable i is smaller than 23, the system updates another new values i and return the second step. Or if not, the system stops.

4 Simulation Result

After building algorithms, we create a user interface based on WinForms C# from Visual Studio which illustrates how algorithm operate. Besides, this interface focus output power of all system and display how weather condition have impact on wind turbine and solar panels. Figure 4 exhibits all information such as weather data in recent day, graphics conveying the different of wind speed and temperature in a day, status of wind turbine, solar panel, battery bank and info loads.

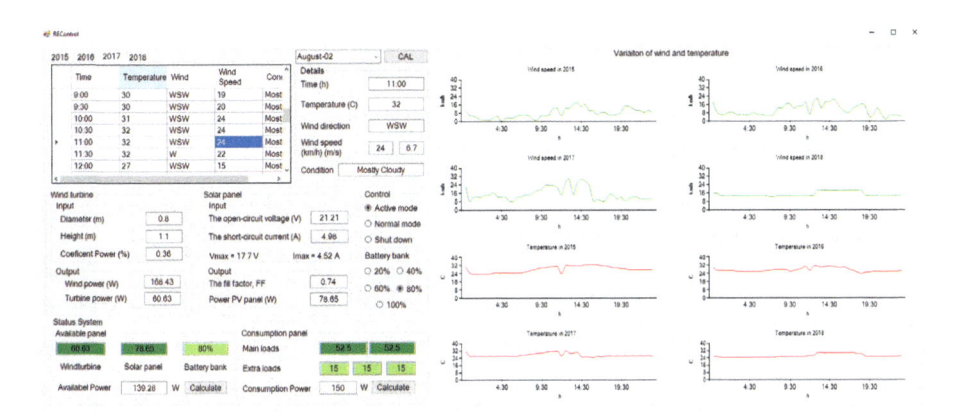

Fig. 4. Weather data interface

In Fig. 5, we have built a weather database whose source is Weather Underground. There are four tables relative four tabs 2015, 2016, 2017 and 2018, data tables are shown when user click on relative tab. In addition, user can pick a date and data changes accordingly. Besides, Details tab on the right of figure shows relative information when every cell on table is clicked. Once user have chosen day and time, user can move to the next tabs which have displayed status of wind turbine, solar panels and battery. Data table also describe the weather data in next hours, so user can notice that when they can let all the loads operate with 100% performance.

Figure 6 illustrates all parameters about the core components in system. In tab Wind turbine, there are two fixed figures which are diameter and height wind turbine (we choose from the practice model), user can set up coefficient power. That is input parameter and output are wind power and turbine power. Solar panel tab has similar information, the input includes specifications of a solar panel from Solageo (open-circuit voltage, short-circuit current and Vmax, Imax), the output tab embodies the fill

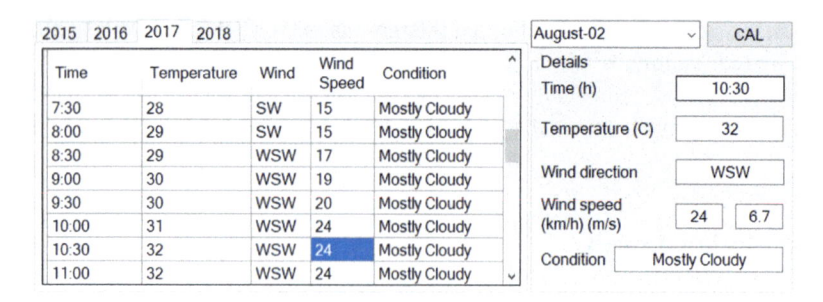

Fig. 5. Weather data tabs

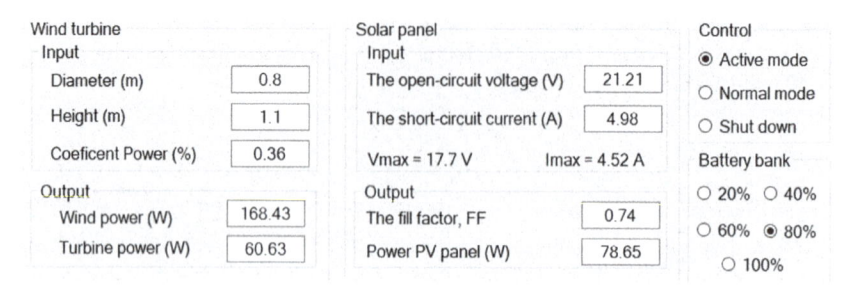

Fig. 6. Specifications tab

factor and power PV panel. The final tab in this figure is percentage of battery bank which is status of charge.

Based on the Bezt laws, the coefficient power cannot beyond 59,3%. The wind power is calculated by the follow equation [1]:

$$P_{wind} = \frac{1}{2}\rho A v^3 \qquad (7)$$

From (1) and (7) the equation for coefficient power is as follow [10]:

$$C_p = \frac{P_{turbine}}{P_{wind}} \qquad (8)$$

From two above equations, we see that how strong wind have impact on wind turbine. Besides, the output power from solar panel also is depend on temperature which is described by follow equations [2]:

$$V_{OC} = \frac{nkT}{q}\ln\left(\frac{I_L}{I_0} + 1\right) \qquad (9)$$

Where I_0 is dark saturation current, I_L is light generated current, n is ideal factor, T is temperature, V_{OC} is open circuit voltage. The short-circuit current, Isc, increases slightly with temperature [2]:

$$\frac{1}{I_{sc}}\frac{dI_{SC}}{dT} \approx 0{,}0006 \tag{10}$$

$$\frac{1}{FF}\frac{dFF}{dT} \approx \left(\frac{1}{V_{OC}}\frac{dV_{OC}}{dT} - \frac{1}{T}\right) \approx -0{,}0015 \tag{11}$$

In fact, the solar panels manufacture has published specific thermal parameters which shows how much this figure increase or decrease as the temperature rise or drop by one Celsius. So, does our solar panel, three parameters are current temperature coefficient $\alpha I_{SC} = 0.08\%/°C$, voltage temperature coefficient $\beta V_{OC} = -0.32\%/$, Power temperature coefficient $\gamma P_{max} = -0.38\%/°C$.

In tab control, there are some options, user can choose they use active mode or another mode. When all needed information is available, system actives Active mode which all the loads turn on green, if Normal mode is on, all the extra loads turn on yellow, that mean they are cut off to keep power supply for the main loads. The final option is shut down all the load, in case the weather condition is too bad to operate or cause damaged.

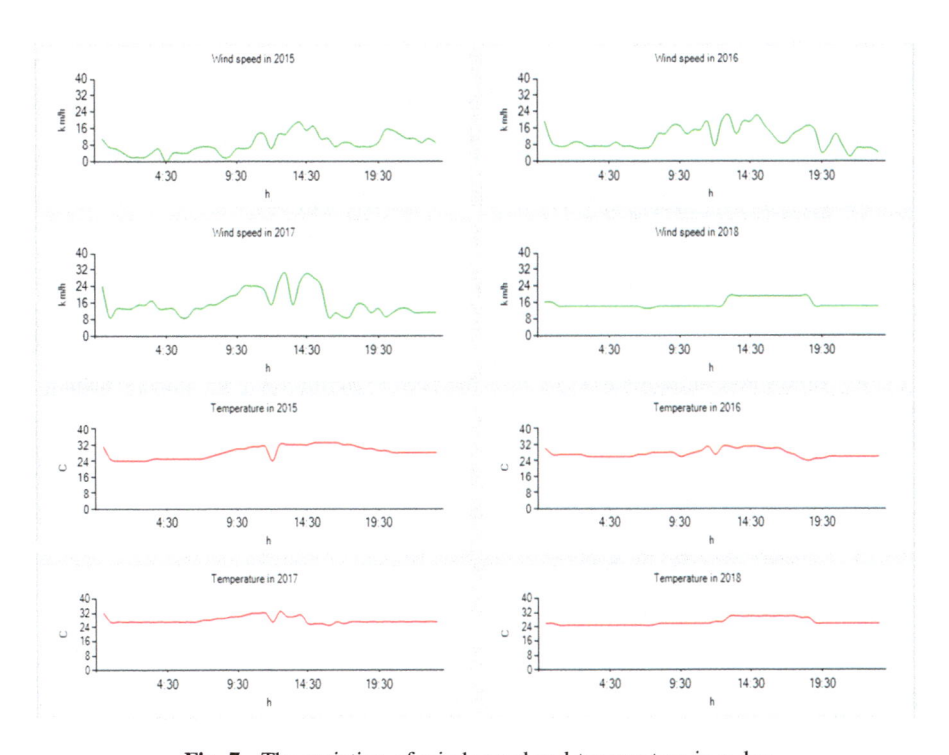

Fig. 7. The variation of wind speed and temperature in a day

In Fig. 7, there are eight charts illustrating the variation of wind and temperature in day which user have already chosen in weather data tab. More details, the first fourth chart displayed by green line are describing wind speed in 2015, 2016, 2017 and 2018. The red lines in four charts left are conveying temperature at the same time.

Figure 8 shows status of all system. After aggregation data from two resources and back up sources, the system shows power available and power consumption. In the available panel, the power of the wind turbine, solar panels and the status of the charge are displayed and combined with the last total cell after clicking the calculation button. The colors shown on the screen give user the ability to provide loads of power and battery charge status. Green is best ability to provide the loads with 100% capacity until red is not capable to sustain operation for loads. In addition, at the power consumption tab, there are two main loads and three sub-loads, the colors also representing the operating level corresponding to the respective power. Finally, battery is only available in the case the main sources are not enough capacity to maintain loads in system.

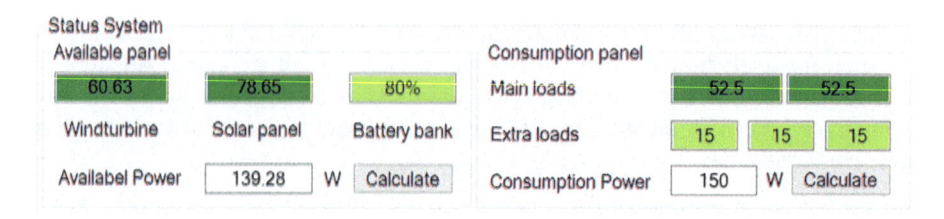

Fig. 8. Status system tab

In summary, we have built an interface that allows users to know the operating status of the system. With time-based predictions, users can actively activate the 100% performance mode at appropriate intervals, or they can choose to use the load at moderate power, or when a backup power supply is needed.

5 Conclusion

Obviously, no one can deny the key role which RE system play in the energy industry recently. With all the advantages RE have, human rate is now changing their mind to make full use of it as a Noah's ship for the sustainability of their civilization. The only problem we have here is their over dependence on weather characteristic that leading to differentiate between supply and demand in electricity market. That is the reason why we propose this algorithm so that we can optimize all of those systems to get a better balance operating in the utilities.

References

1. The Physics of Wind Turbines, Kira Grogg Carleton College, p. 8 (2005). Accessed 6 Nov 2013
2. Griffith, J.S., Rathod, N.S., Paslaski, J.: Some tests of flat plate photovoltaic module cell temperatures in simulated field conditions. In: Proceedings of the 15th IEEE Photovoltaic Specialists Conference, Kissimmee, FL, pp. 822–830 (1981)
3. Dubey, S., Sarvaiya, J.N., Seshadri, B.: Energy Research Institute, Nanyang Technological University Ankur Sahai, Renewable Energy Prediction using Weather Forecasts for Optimal Scheduling in HPC Systems, Published by Elsevier Ltd
4. Kaldellis, J.K.: Stand-alone and hybrid wind energy system. Woodhead Publishing Limited, Abington Hall, Granta Park, Great Abington, Cambridge CB21 6AH, UK (2010)
5. Islam, M., Ting, D.S.K., Fartaj, A.: Aerodynamic models for Darrieus type straight-bladed vertical axis wind turbines. Renew. Sustain. Energy Rev. **12**(4), 1087–1109 (2008)
6. Appleyard, D.: Solar Trackers: Facing the Sun. Renewable Energy World Magazine, UK: Ralph Boon, 1 June 2009
7. Sinton, R.A., Cuevas, A.: Contactless determination of current–voltage characteristics and minority-carrier lifetimes in semiconductors from quasi-steady-state photoconductance data. Appl. Phys. Lett. **69**, 2510–2512 (1996)
8. Baruch, P., De Vos, A., Landsberg, P.T., Parrott, J.E.: On some thermodynamic aspects of photovoltaic solar energy conversion. Solar Energy Mater. Solar Cells **36**, 201–222 (1995)
9. Levy, M.Y., Honsberg, C.B.: Rapid and precise calculations of energy and particle flux for detailed-balance photovoltaic applications. Solid-State Electron. **50**, 1400–1405 (2006)
10. Afonso, M., Pereira, P., Martins, J.: Weather monitoring system for renewable energy power production correlation. In: DoCEIS 2011. IFIP AICT, vol. 349, pp. 481–490 (2011)
11. Weather Underground Homepage. https://www.wunderground.com. Accessed 26 July 2018
12. Timeanddate Homepage. https://www.timeanddate.com. Accessed 26 July 2018

Analysis of Efficiency and THD in 7-Level Voltage Inverters with Reduced Number of Switches

Ales Havel[✉] [iD], Martin Sobek, and Petr Chamrad

VSB - Technical University of Ostrava,
17. listopadu 15, 708 33 Ostrava-Poruba, Czech Republic
{ales.havel,martin.sobek,petr.chamrad}@vsb.cz

Abstract. This paper focuses on analyzes of theoretical efficiencies and total harmonic distortions of modern topologies of 7-level voltage inverters with reduced number of switches. Three different topologies of 7-level voltage inverters controlled with subharmonic PWM methods were built and analyzed in the Matlab/Simulink environment. The results are focused mainly on determining the THD of the output phase voltage and total power losses of the presented topologies. These parameters are used to determine the optimal circuit topology and modulation technique for control of the 7-level voltage inverter.

Keywords: 7- level voltage inverter · Efficiency ·
Reduced number of switches · Subharmonic PWM · Total harmonic distortion ·
Total power losses

1 Introduction

Multi-level inverters create a higher number of voltage levels in the output (load) voltage. There is also a better distribution of the input voltage between the switches. Due to this fact, multi-level inverters are used especially in high-voltage areas, where conventional two-level inverters would be more stressed with overvoltage on switches. Due to the increasing number of applications of multi-level inverters, their modern topologies with reduced number of switching elements or sources are also emerging. These reductions lead to a decrease in the price of the converter and its size. With the increasing number of inverter levels, its control complexity also increases, but the total harmonic distortion of the output phase voltage is reduced. All methods for controlling multi-level inverters are based on known methods for two-level inverters. Some modern topologies may require special control requirements [1, 2]. For proper application, the suitability of the inverter is judged in terms of the number of components used, efficiency and total harmonic distortion [3, 4] and [5].

Three topologies of 7-level voltage inverters with reduced number of switches along with suitable modulation techniques will be described in Sects. 2 and 3. These topologies will be simulated in the Matlab/Simulink environment working to resistive and resistive inductive loads. The simulation output will focus mainly on determining the THD of the output phase voltage and total losses of the inverter. These parameters will be key data for determining optimal circuit topology and modulation technique.

2 7-Level Inverter Topologies with Reduced Number of Switches

The main purpose of these modern topologies of inverter power circuits is to reduce the number of switches or power supplies while maintaining the same number of output voltage levels. All this is done because of the reduction in the price of the inverter, its size and the complexity of the control algorithm. Most modern structures are based on the topology of the cascade interconnection of a multi-level inverter. The control of these inverters is most often done using subharmonic PWM methods.

2.1 Symmetrical Topology of 7-Level Inverter with 4 Sources and 6 Switches

An increase in the output voltage levels of the inverter is achieved by adding a DC source and one switch. Compared with the cascade connection, where it was necessary to add the whole bridge (four switches), this is a major advantage. Switches S_1 and S_6 are used to set the desired polarity of the voltage on the load. The zero voltage value is created by switching on the switch S_2 when the remaining switches are disconnected. Switches S_3, S_4 and S_5 generate individual voltage levels. Switches S_2, S_3, S_4 and S_5 must be bidirectional for correct operation of the inverter [1]. The number of output levels of the load voltage is calculated from the following formula:

$$N_{step} = 2 \cdot n - 1 \qquad (1)$$

Maximal voltage on the load is determined as:

$$U_{z\,max} = (n-1) * U_d \qquad (2)$$

Where n is the number of DC sources and U_d is the value of DC voltage.

Figure 1 shows the symmetrical topology of 7-level inverter with 4 sources and 6 switches. The output voltage levels are created by switching the switches S1 to S6 according to the above-described rules. The switching diagram of the switches is described above in the Table 1.

Table 1. Switching combinations of 7-level inverter with 4 sources and 6 switches.

S1	S2	S3	S4	S5	S6	U_z
0	0	0	0	1	1	U_d
0	0	0	1	0	1	$2U_d$
0	0	1	0	0	1	$3U_d$
0	1	0	0	0	0	0
1	0	1	0	0	0	$-U_d$
1	0	0	1	0	0	$-2U_d$
1	0	0	0	1	0	$-3U_d$

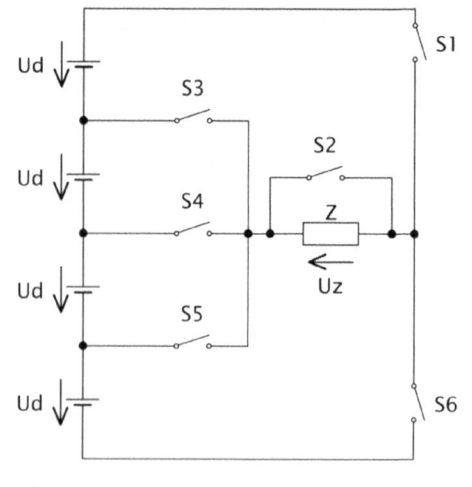

Fig. 1. Symmetrical topology of 7-level inverter with 4 sources and 6 switches.

2.2 Asymmetrical Topology of 7-Level Inverter with 2 Sources and 7 Switches

This inverter consists of n DC sources and S power switches, of which only two are bidirectional. The proposed topology is configured in such a way that the output voltage produced is created from U_{d1} and only one of the remaining DC sources. Due to this, only two bidirectional switches (S_3, S_4) are sufficient, which leads to reduction of switching power losses [1, 4]. DC sources are selected according to the equations below. They should also be arranged in such a way to avoid the loss of any level during output voltage generation, which may reduce the THD on the output. As with symmetric topology, even here it is necessary to add a power switch together with the DC source to increase the output voltage value.

$$U_{dX} = 2*(x-1)*U_{d1} \quad for\, x = 2,3,\ldots,n \tag{3}$$

The number of output voltage levels and the maximal load voltage are given by:

$$N_{step} = \frac{2*U_{dn}}{U_{d1}} + 3 \tag{4}$$

$$U_{z\,max} = U_{d1} + U_{dn} \tag{5}$$

Figure 2 shows the asymmetrical topology of 7-level inverter with 2 sources and 7 switches. The DC source U_{d2} is twice as large as U_{d1}. Against the previous symmetrical topology of 7-level inverter, asymmetrical connection requires one more switch, but only half of the DC power sources. The number of output voltage levels remains the same. The switching diagram of the switches is described above in the Table 2.

Table 2. Switching combinations of 7-level inverter with 2 sources and 7 switches.

S1	S2	S3	S4	S5	S6	S7	U_Z
0	1	1	0	0	0	0	U_{d1}
0	0	0	1	1	0	1	U_{d2}
0	1	0	0	1	0	1	$U_{d1} + U_{d2}$
0	0	0	0	1	1	0	0
1	0	0	1	0	0	0	$-U_{d1}$
0	0	1	0	0	1	1	$-U_{d2}$
1	0	0	0	0	1	1	$-(U_{d1} + U_{d2})$

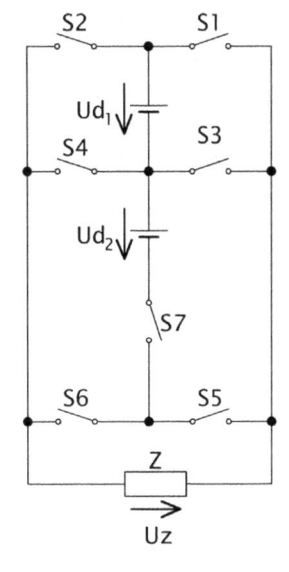

Fig. 2. Asymmetrical topology of 7-level inverter with 2 sources and 7 switches.

2.3 Symmetrical Topology of 7-Level Inverter with 3 Sources and 6 Switches

The structure of the inverter is created by the ideal combination of three existing topologies of 7-level inverters, which are composed from 9 switches and 3 sources, 7 switches and 3 sources, and 6 switches and 4 sources. The ideal combination of those will be an inverter, consisting of 6 switches and 3 DC sources [2]. Required number of switches and sources depends on the number of output voltage levels by equations:

$$S = \frac{N_{step} + 5}{2} \quad (6)$$

$$n = \frac{N_{step} - 1}{2} \quad (7)$$

Where S is the number of power switches and n is the number of power sources. Only two switches (S4, S6) set the polarity of the output voltage. The remaining four switches are used to generate the respective seven levels of output voltage, of which S1, S2 are bidirectional, see the Fig. 3. The corresponding output voltage levels are given by the switching combinations described in the Table 3.

Table 3. Switching combinations of 7-level inverter with 3 sources and 6 switches.

S1	S2	S3	S4	S5	S6	U_Z
0	1	0	0	0	1	U_d
1	0	0	0	0	1	$2U_d$
0	0	1	0	0	1	$3U_d$
0	0	0	0	0	0	0
1	0	0	1	0	0	$-U_d$
0	1	0	1	0	0	$-2U_d$
0	0	0	1	1	0	$-3U_d$

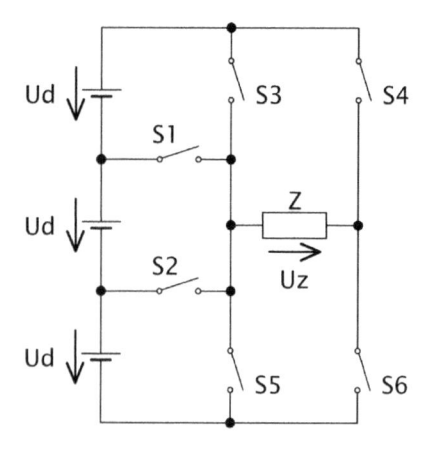

Fig. 3. Symmetrical topology of 7-level inverter with 3 sources and 6 switches.

The zero output voltage can be also obtained by other switching combinations. A suitable combination can be selected according to the type of load and the direction of the current flow, simultaneously with respect to the size of the switching losses.

3 Modulation Techniques for Multilevel Inverters

The two most commonly used modulation techniques for multi-level inverters are subharmonic PWM and vector modulation. Both types of modulations are based on two-level inverter variants. A suitable modulation technique is chosen based on the parameters: switching losses, amount of harmonics in the output signal, reaction time on change the input parameters and so on. Contrary to modulations of conventional multi-level inverters, the modulations of modern topologies with a reduced number of switches or sources may vary slightly. There is a larger number of unique topologies, which can have specific requirements on the inverter switching algorithms.

3.1 Subharmonic PWM

This method is based on pulse width modulation, which is a comparison of two signals: a reference signal, most commonly a sine wave, and a carrier signal, which is either saw or triangular wave. By two-level inverters, the carrier signal is always one. This differentiates the modulation of two-level from multi-level inverters in which the number of carrier signals is derived from the number of inverter levels. Generally, the required number of carrier signals can be expressed by the formula:

$$N_{signal} = N_{step} - 1 \qquad (8)$$

Where N_{signal} is the number of carrier signals and N_{step} is the number of voltage levels. The carrier signals are mutually shifted vertically or horizontally from each other. With a horizontal displacement, a constant number of pulses is generated over the period, regardless of a magnitude of the generated voltage. Modulation with horizontal displacement of carrier signals is called PSC PWM. However, the vertical displacement is easier to reach by the processor and occurs in three basic forms [3]:

(1) *PD* (Phase Disposition) = all carrier signals are in phase
(2) *POD* (Phase Opposition Disposition) = positive carrier signals are in phase, negative carrier signals are also in phase, but positively displaced by 180°
(3) *APOD* (Alternative Phase Opposition Disposition) = the carrier signal is always shifted by 180° to the adjacent signal

The different properties of each type of arrangement of carriers differ more markedly at low switching frequencies. It also depends on the type of the inverter and number of its levels. For the 3-phase inverter, the smallest THD of the output signal is with PD. The POD and APOD methods have the same distortion for 3-level inverters, while the 5-level inverter has worst performance with POD. The problem of selecting the modulation technique for each type of inverter is described in more detail in [5]. For multi-level inverters, the value of the modulation index M is determined as:

$$M = \frac{A_m}{(N_{step} - 1) * A_n} \qquad (9)$$

Where A_m is the amplitude of reference signal, A_n the amplitude of carrier signal and N_{step} is the number of voltage levels of the inverter.

4 Simulation of 7-Level Inverter Topologies

The simulation was performed in the Matlab/Simulink environment, which was chosen with respect to the relatively simple generation of the inverter's control algorithm. It also has the function of fast and easy evaluation of THD in the simulated circuit. However, the program does not work with advanced models of real semiconductor components, so circuit power losses cannot be easily determined. All the topologies described in Sect. 2 were selected for simulation. From Sect. 3, a subharmonic PWM was chosen as a modulation technique with PD, POD and APOD methods.

4.1 Simulation Setup and Creation of Power Circuit

A series combination of R = 10 Ω and L = 10 mH was used as a load for simulated inverters. Whereas the switching algorithm for only resistive load was described in the literature [1, 2, 4], it had to be modified to enable current switching even with inductive load. That modification was performed using the knowledge and operational principles of switched semiconductor components in a two-level inverter working with inductive load.

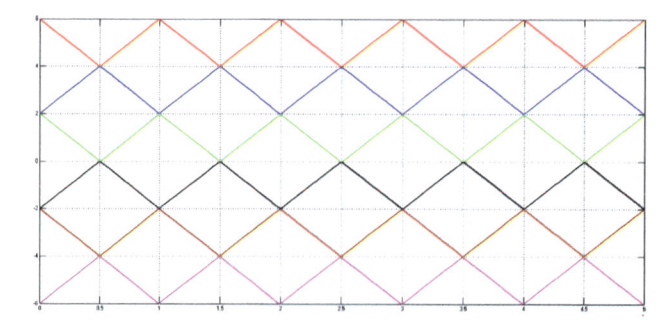

Fig. 4. Example of carrier signals layout for APOD subharmonic PWM method

The control circuit for modulation techniques of subharmonic PWM is composed from sine and triangular wave generators, constants, sum blocks, comparator blocks, and logical blocks. The sine wave generator works as a reference signal. Its amplitude is set according to the desired modulation index (for M = 1 the amplitude reaches 6 V). The carrier signal is created in six triangular waveform generators, the frequency of which determines the switching frequency of the inverter. The phase shift is selected either 0° or 180° according to the subharmonic PWM method. The amplitude of output signals from the triangular waveform generators is ±1 V, it is therefore necessary to change the offset of these signals in order to obtain their correct vertical arrangement - see Fig. 4. Carrier signals are then compared with the reference signal. Subsequently, the desired control algorithm with output signals in the form of switching pulses for individual IGBT transistors is created using the logical blocks.

The simulation time was set with the respect to the computational demands and the amount of needed output values to 1 s. Step size was fixed; always 100 times lower than the switching period. The selected simulation method was ode4 (Runge - Kutta).

4.2 Evaluation of Total Harmonic Distortion

THD and total power losses are used to compare the quality of simulated topologies and modulation methods. THD determines the total harmonic distortion of the output signal with respect to the sinusoidal waveform and is determined by the formula (10).

$$THD = \frac{\sqrt{\sum_{x=2}^{h} U_x^2}}{U_1} * 100 \, [\%] \tag{10}$$

Where U_1 is the voltage of 1st harmonics and h is the number of harmonics taken into the calculation. THD was evaluated for all simulated topologies and all types of modulations (PD, POD, APOD) for the range of switching frequencies 1 kHz to 5 kHz (step 1 kHz) and the range of modulation indexes M = 0,1–1,2 (step 0,1).

4.3 Evaluation of Power Losses and Efficiency

To calculate total power losses, it is necessary to observe what happens on particular switch during the period - the collector-emitter voltage U_{CE} before and after switching, the number of turn on times per period k, the current through the switch I_C, and the time of current conduction t_{COND} - see Fig. 5. All these values were measured by virtual oscilloscope during the simulation on all switches. Output power was calculated from RMS values of load voltages and currents. The input power could not be measured directly on DC sources, as they conducted alternately during the period (sometimes one or two sources conducted together) - therefore it was determined to add total losses to the output power. Due to high computational demands, the losses are determined for R-type loads only, with f_{sw} = 1 kHz and a modulation index M = 1. The difficulty of the calculations lies in the analytical determination of each voltage and current change associated with switching on/off each switch during one period.

Total power losses and inverter efficiencies can be calculated from above-mentioned values and parameters using the following Eqs. (11–16). Switching energy loss:

$$E_{SW} = E_{ON} + E_{OFF} = (t_{on} + t_{off}) * k * \frac{U_{CE}}{2} * \frac{I_C}{2} \ [J] \tag{11}$$

$$t_{on} = t_{d(ON)} + t_r \ [s] \tag{12}$$

$$t_{off} = t_{d(OFF)} + t_f \ [s] \tag{13}$$

Energy loss during the current conduction:

$$E_{COND} = (U_{CE0} * I_C + R_{CE} * I_C^2) * t_{COND} \ [J] \tag{14}$$

Total power losses:

$$P_{tot} = (E_{SW} + E_{COND}) * f_{out} \ [W] \tag{15}$$

Theoretical efficiency of the inverter:

$$\eta = \frac{U_{Zrms} * I_{Zrms}}{P_{tot} + (U_{Zrms} * I_{Zrms})} * 100 \ [\%] \tag{16}$$

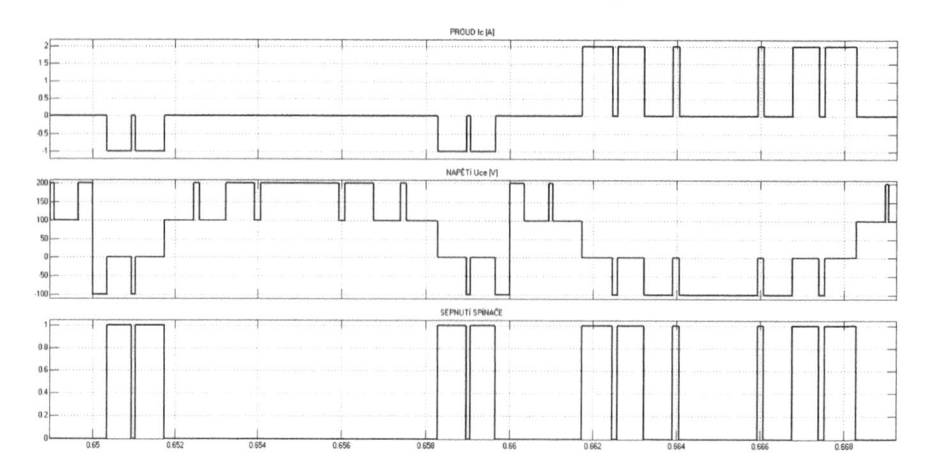

Fig. 5. Symmetrical topology of 7-level inverter with 3 sources and 6 switches. The waveforms of current (top), voltage (middle) and switch status (bottom) on the switch S1. R = 100 Ω, f_{SW} = 1 kHz, modulation index M = 1, modulation APOD.

5 Results

The data describing THD and total power losses of particular topologies are used to determine optimal inverter topology including modulation techniques. THD has been determined for the Nyquist frequency, so it is considered with the full bandwidth. Parameters of the load were R = 10 Ω, L = 10 mH. Figures 6, 7 and 8 are showing waveforms of load current and voltage on all tested topologies. Table 4 shows the results from efficiency calculations and Fig. 9 compares the dependencies of THD on the modulation index for 2-level and 7-level inverters with PD, POD and APOD methods.

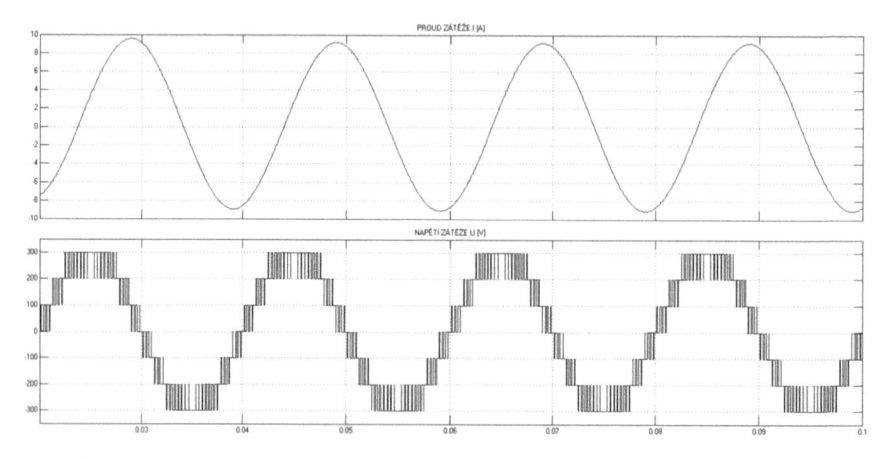

Fig. 6. Symmetrical topology of 7-level inverter with 4 sources and 6 switches. Waveforms of current (top) and voltage (bottom) on the RL load. f_{SW} = 3 kHz, M = 1, modulation APOD.

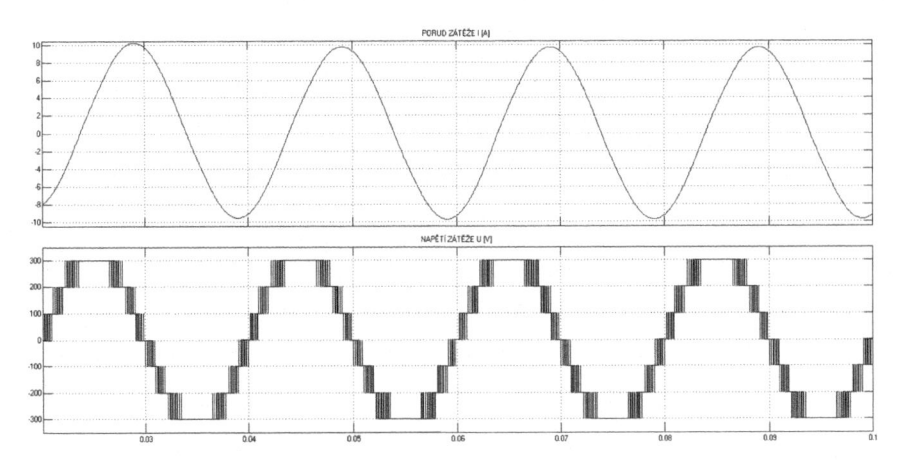

Fig. 7. Asymmetrical topology of 7-level inverter with 2 sources and 7 switches. Waveforms of current (top) and voltage (bottom) on the RL load: $f_{SW} = 5$ kHz, $M = 1.1$, modulation POD.

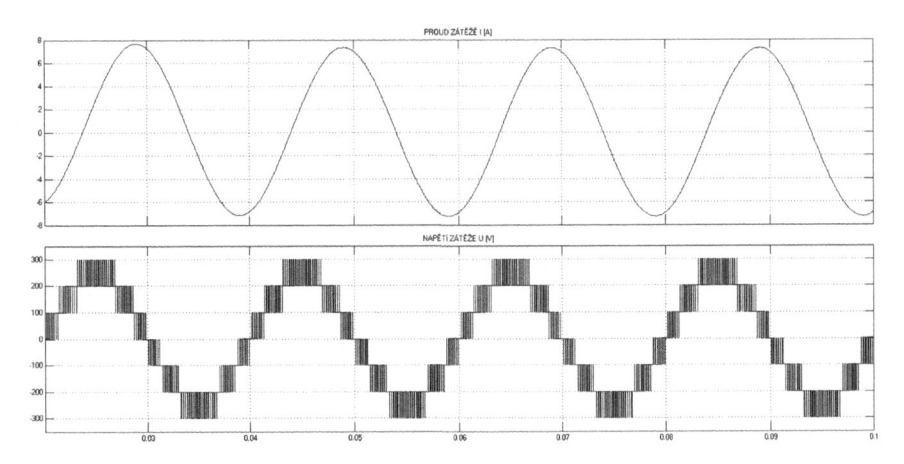

Fig. 8. Symmetrical topology of 7-level inverter with 3 sources and 6 switches. Waveforms of current (top) and voltage (bottom) on the RL load: $f_{SW} = 5$ kHz, $M = 0.8$, modulation PD.

Table 4. Comparison of power losses and theoretical efficiencies for analyzed 7-level inverters.

Inverter topology	Total power losses [W]	Output power [W]	Input power [W]	Efficiency [%]
7-level (4 voltage sources, 6 switches)	8,82	463,5	472,32	98,13
7-level (3 voltage sources, 6 switches)	11,91	463,5	475,41	97,49
7-level (2 voltage sources, 7 switches)	13,32	463,5	476,82	97,21
2-level (1 voltage source, 6 switches)	7,00	463,5	470,5	98,51

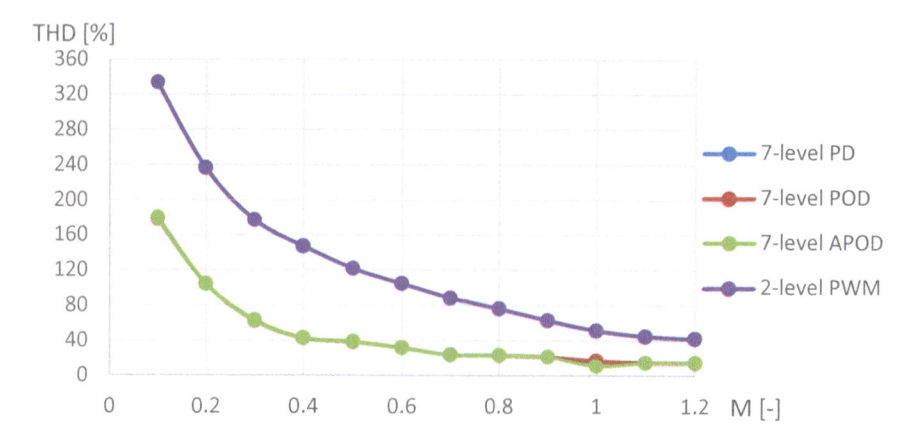

Fig. 9. Dependencies of THD on the modulation index (comparison of modulation methods).

6 Conclusion

The advantage of modern inverter topologies increases significantly with the increasing number of output voltage levels. The predicted trend of THD decrease with the increasing modulation index and the number of voltage levels of the inverter can be observed from the obtained dependency. From a THD perspective, it cannot be easily determined, which of the simulated 7-level inverter topologies is the most appropriate. In terms of modulation techniques, the best results were achieved with the APOD method. Highest theoretical efficiency was achieved with the symmetrical topology of 7-level inverter with 4 sources and 6 switches, while the efficiency of a classical 2-level inverter is about 0.38% higher due to lower value of switching power losses.

Acknowledgment. This paper was supported by the projects: Center for Intelligent Drives and Advanced Machine Control (CIDAM) project, Reg. No. TE02000103 funded by the Technology Agency of the Czech Republic and Project Reg. No. SP2018/162 funded by the Student Grant Competition of VSB – Technical University of Ostrava.

References

1. Gnana, P.M., Balamurugan, M., Umashankar, S.: A new multilevel inverter with reduced number of switches. IJPEDS **5**(1), 63–70 (2014)
2. Chintala, L.R., Peddapelli, S.K., Malaji, S.: Improvement in performance of cascaded multilevel inverter using triangular SVPWM. AEEE **14**(5), 562–570 (2016)
3. Komolafe, O.A., Olayiwola, O.I.: Gapped alternate phased opposite disposition pulse width modulation control for multilevel inverters. J. Eng. Appl. Sci. **9**(4), 560–567 (2014). ARPN
4. Omar, R., et al.: Optimization of a three phase cascaded H-bridge multilevel inverters for harmonic elimination. AJBAS **10**(14), 225–240 (2016)
5. Brandstetter, P., Kuchar, M., Skuta, O.: Implementation of RBF neural network in vector control structure of induction motor. IREE **9**(4), 749–756 (2014)

Waste Management - Weighing-Machine Automation

Zdenek Slanina[(✉)], Rostislav Pokorny, and Jan Dedek

Faculty of Electrical Engineering and Computer Science, VSB-TU Ostrava, 708 33 Ostrava, Czech Republic
zdenek.slanina@vsb.cz

Abstract. Nowadays the environment is polluted in different ways. One of these is the wrong disposal. A considerably ineffective method of disposal of waste can be the storage at a designated site of so-called landfills that have a significant adverse effect on the environment. Most of the larger cities suffer from these aspects of inefficient waste management, which is why the landfill is mostly on the periphery. Due to rapid population growth, this situation is only getting worse. A possible solution is reuse of waste by so-called recycling to a utility product. Effective and efficient recycling is required for many people. Which individual materials will be separated and further processed into a utility product. A possible solution is to automate these tasks using current information technologies.

Keywords: Waste management · Industry 4.0 · Latex · Linux · Embedded systems

1 Recycling of PET Material

The final stages of recycling are the effective weighing of the resulting product. At TOMA RECYCLING company dealing with the recycling of plastics technologies have been developed which enable the automation of the weighing of the output product (regranulate), the printing scales and the subsequent processing of this data into the company information systems.

The empty PET (Polyethylene terephthalate) packaging is discarded by the consumer, after use and becomes PET waste. When the PET bottles are returned to an authorized redemption center, or to the original seller in some jurisdictions, the deposit is partly or fully refunded to the redeemer. In both cases the collected post-consumer PET is taken to recycling centers, where it is sorted and separated from other materials such as metal, objects made out of other rigid plastics such as PVC, HDPE, polypropylene, flexible plastics such as those used for bags (generally low-density polyethylene), drink cartons, glass, and anything else which is not made out of PET. This is point one and two in Fig. 1.

Post-consumer PET is often sorted into different color fractions: transparent or uncoloured PET, blue and green colored PET, and the remainder into a mixed

```
Collecting and separating
      PET bottles                    Recycling              Regranulation

┌──────────────────┐         ┌──────────────────┐         ┌──────────────────┐
│ 1. Collection of │────────▶│ 3. Crushing PET  │────────▶│ 7. Processing of │
│    PET bottles   │         │     bottles      │         │ PET flakes for PET│
│                  │         │                  │         │     granulate    │
└────────┬─────────┘         └────────┬─────────┘         │  (regranulation) │
         │                            │                   └──────────────────┘
         ▼                            ▼
┌──────────────────┐         ┌──────────────────┐
│ 2. Classification│         │ 4. Washing and   │
│ and treatment of │         │ cleaning of PET  │
│   PET bottles,   │─────────│ bottoms, removal │
│   separation of  │         │     of glue      │
│   other wastes   │         │                  │
└──────────────────┘         └────────┬─────────┘
                                      ▼
                             ┌──────────────────┐
                             │ 5. Separation of │
                             │ impurities, labels│
                             │     and HDPE     │
                             └────────┬─────────┘
                                      ▼
                             ┌──────────────────┐
                             │ 6. Manufacture of│
                             │    PET flakes    │
                             └──────────────────┘
```

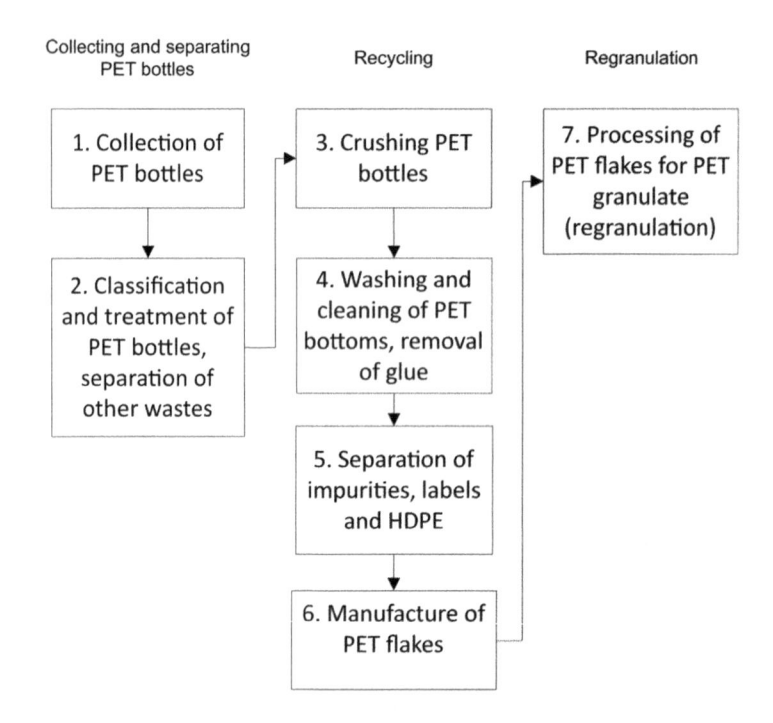

Fig. 1. PET recycling diagram.

colors fraction. The further treatment process includes crushing, washing, separating and drying. Recycling companies further treat the post-consumer PET by shredding the material into small fragments. These fragments still contain residues of the original content, shredded paper labels and plastic caps. These are removed by different processes, resulting in pure PET fragments, or "PET flakes".

PET flakes are used as the raw material for a range of products that would otherwise be made of polyester. Examples include polyester fibers (a base material to produce clothing, pillows, carpets, etc.), polyester sheets, strapping, or back into PET bottles. For these purposes, must be PET flakes firstly regranulated.

Recycling process consists from:

1. Collection of PET bottles.
2. Classification and treatment of PET bottles, separation of other wastes. Recycling line inputs are PET bottles molded into 100 - 450 kg bales. Bottles are purchased from sorting companies, which are an intermediate step between sorted waste containers and recycling itself. It is point 1 and 2 at Fig. 1 (Collection of PET bottles and Classification and treatment of PET bottles, separation of other wastes). PET bottles are sorted by color most often for clear, blue, green and multicolor.

3. Description of the recycling process:
   - Crushing PET bottles.
   - The packages are loaded into the conveyor, passes through the weighing conveyor where each package is individually weighed, and the weight is automatically recorded in the system.
   - The operator of the entrance section of the package releases the cutting of the binding wires, the package travels through the devices which ensure the mechanical separation.
   - The individual bottles then go to the optical separation and separation of non-ferrous metals, followed by manual puncturing and the separation of ferromagnetic metals. Next step is wet grinding.
4. Washing and cleaning of PET bottoms, removal of glue. After grinding the PET are washing, drainage, floatation to separate polymers of different densities (HDPE, PP, LDPE, PS vs. PET and PVC). After this separation the material passes through the drying section and the air separation of the impurities.
5. Separation of impurities, labels and HDPE followed by the end part of the line, where the cleaning of the output (removal of residual metal - feromag and not-feromag.)
6. Manufacture of PET flakes. The product are PET flakes and HDPE, PP flakes. Then there is refranulating process. PET flakes travel through the weighing stack into stockpiles, from where they are further forwarded by the airway for reprocessing to regranulate.
7. Processing of PET flakes for PET granulate (regranulation):
   - PET airflakes are dispensed into the pre-drying force
   - From there, they travel through the hopper onto the weighing belt and into the extruder
   - In the extruder, PET flakes melt (at the addition of a dye) at a predetermined temperature which must not exceed a limit at which molten PET would degrade (*approx. 280 °C*)
   - The PET melt passes through a perforated plate where undercutting occurs with the knife head for underwater regranulation
   - The regranulate then goes through the washing (dust removal), the drying zone by the drying and air cooling
   - The final step before shipping to customers is to dispense in bags of a total weight of 1200 kg in which the regranulat completely cool
   - The part of the line is, of course, a vacuum system that provides the extraction and subsequent condensation of the gaseous organic components that are produced by the melting of PET flakes.

2 Printing of Weights Label

The whole device is composed of generic devices that are capable of functioning independently, see [2,3,6,9–12]. All components are commonly available in distribution networks of electrotechnical merchandise shops. Because of the absence of atypical components, this solution is very economical, easy to reproduce, and easy to maintain in the future. Another advantage was the absence of a lengthy development and testing of a comprehensive hardware that could not be easily modified in the future according to the customer's requirements; in the case of a comprehensive solution this would require the development and testing of a new platform in case of hardware change. This is both time-consuming and costly. Using the individual components is shown in the Fig. 2.

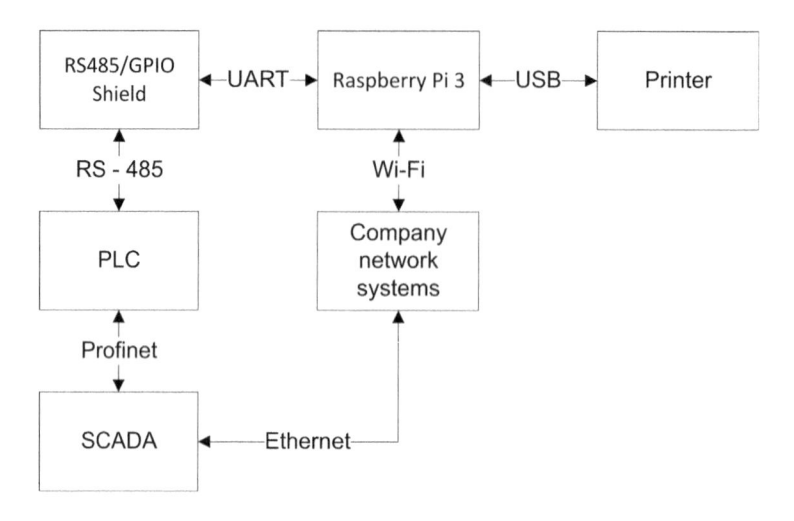

Fig. 2. Technology schematic.

2.1 Raspberry Pi 3

As the main board was used Raspberry Pi 3, see Fig. 3. This is a one-board computer sized like credit card developed by the British foundation Raspberry Pi Foundantion in 2012 to create a low cost mini-computer as a platform for learning and managing various devices. The cost of the equipment at the beginning of 2018 was around 55$.

Raspberry Pi 3 is built on a 1.2 GHz 64-bit ARM processor. The processor is based on the Cortex A53 quad core RISC architecture. Raspberry Pi, run on own Raspbian Linux operating system, which is derived from Debian. This system is officially supported and developed by the Raspberry Pi Foundantion Foundation, see [1,4,5,7,8]. This system is highly optimized for running on ARM processors.

2.2 RS485 GPIO Shield for Raspberry Pi V3.0

Due to the lack of direct RS-485 bus support needed to communicate with the PLC machines in production. The MAX485 was used on the Raspberry Pi 3 expansion board. The MAX485 is a low power integrated circuit for RS485 and

Fig. 3. Raspberry Pi 3.

RS422 communications. This driver is included on the PCB board which allows easy insertion into the Raspberry Pi control board where it is connected to the UART bus (Fig. 4).

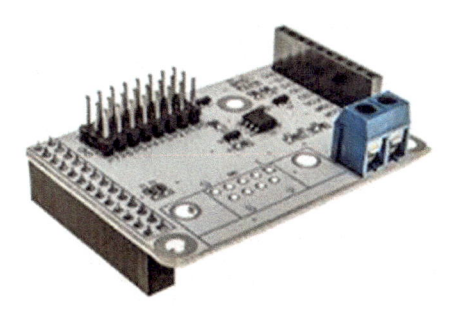

Fig. 4. RS 485 shield.

2.3 USB Office Printer

For the final output, a classic USB printer was chosen because of its cost and operating costs. The printer can print on a standard A5 office format, and it is not necessary to logically search for specialized labels printed on industrial printers. In this case is use laser printer Brother HL-L2340DW, because have native support in driver for Raspbian OS (Fig. 5).

2.4 Technology into Production

The device is running in regranulation, see Fig. 2, where it communicates with the PLC via the RS485 communication line, which is further connected through PROFINET to other PLCs and SCADA/MES servers. On these servers, production data is recorded and evaluated. In the future, it is intended to use Integrated Wi-Fi to back up the weighing machines (Fig. 6) to a remote server.

Fig. 5. Printer for final labels.

Fig. 6. Weighing the regranulate process.

3 Software and Sequence of Printing Processes

The sequence of processes is as follows (dataflow diagram is shown at Fig. 7):

- Receiving data from the PLC
- Check that the balance is zero
- Confirmation of data reception

- Load the template
- Modify the template based on the received data
- Complaints of LaTeX to a PDF file
- Print a PDF file through CUPS on a USB printer
- Save a PDF file to FTP storage (Currently not implemented).

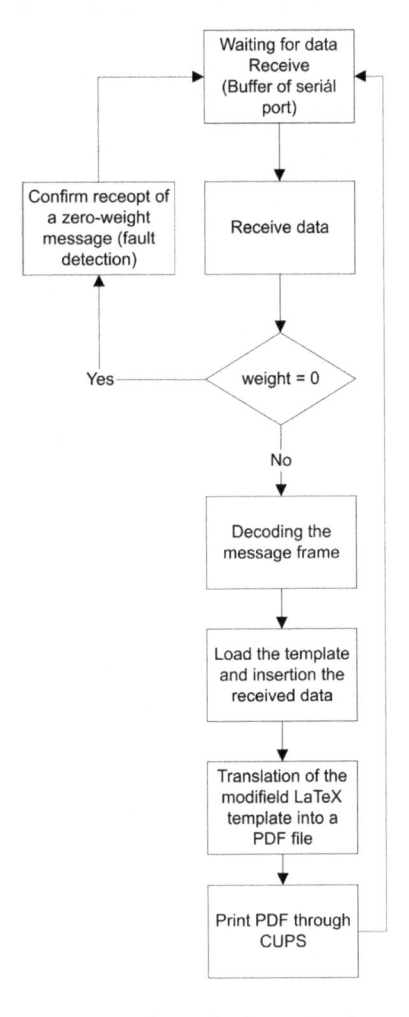

Fig. 7. Dataflow of print technology.

3.1 Serial Communication

The program uses system serial link service for Raspberry Pi, so it was not necessary to treat by interrupting the reading of each apartment if the received content in the buffer reaches the length of the expected message is read out and subsequently processed. The communication protocol between the PLC and RPi is as follows, the PLC sends a message at regular intervals, thus it is master on the bus and RPi accepts the receipt of this message. This communication frame has been chosen because of the possibility of detecting a RPI fault from the PLC and subsequently alerting the operator to an error.

The content of the message varies depending on whether the bag has been weighed (this operator is controlled by the PLC automaton) if yes the weighted weight is sent, otherwise the weight is zero and RPi ignores these messages only sends confirmation of its receipt.

3.2 Working with Text Templates in LaTex

Tex is a computer program. It was created by Professor Donald Ervin Knuth in the 1970s. The program provides a set of macros for creating complex documents, which makes it possible to create a pre-made text template that can be subsequently machine-operated. As a part of the project, this feature was used when replacing the keywords in the received document and then translating this source code through the pdfTeX tool into a PDF file that is then used for printing as well as a backup on the remote server. On Raspbian was installed these packages (Table 1):

Table 1. Message frame sent from PLC to RPi

Bit	Data
0	Address
1	Address
2	Color (0–9 i ASCII)
3	Number of bag (Year e.g. 1 in ASCII)
4	Number of bag (Year e.g. 8 in ASCII)
5	Number of bag (Production number from 1 to 8 in ASCII)
6	Number of bag (Serial number - 5. digit)
7	Number of bag (Serial number - 4. digit)
8	Number of bag (Serial number - 3. digit)
9	Number of bag (Serial number - 2. digit)
10	Number of bag (Serial number - 1. digit)
11	Weight (Weight - 5. digit)
12	Weight (Weight - 4. digit)
13	Weight (Weight - 3. digit)
14	Weight (Weight - 2. digit)
15	Weight (Weight - 1. digit)
16	CR
17	LF

- tex-common
- tex-gyre
- texlive-base
- texlive-binaries
- texlive-fonts-recommended
- texlive-latex-base.

3.3 Printing

Another process is to call the Common UNIX Printing System (CUPS) system service. It is a modular print system for Unix operating systems that allows you to act as a print server. CUPS consists of a spooler and print scheduler (Sheduler/Spooler), a converter to convert a file into a printer-compatible format, and then a backend that takes care of sending the data itself to the printer (see Fig. 8).

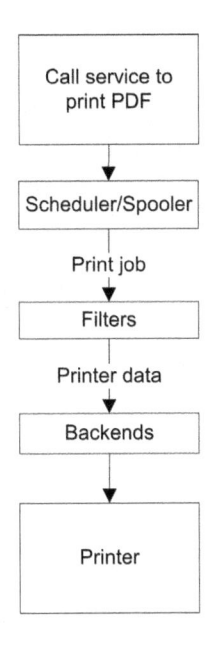

Fig. 8. Printing flow for final label.

3.4 Cron and Deamons

To ensure the correct functionality of the program, the Cron system service is used, it is a software deamon that automatically launches a defined process at a certain time. This is a specialized system process that serves as a task scheduler. In this application, Cron is set to cycle the script for a one minute. The script is a simple program that first detects whether the print process from the process listing runs if no action is taken. If the process is not found, the application runs and logs the current time and date that the application was started. This log can then be evaluated.

TOMA RECYCLING a.s. Jana Palacha 538 735 81, Bohumin Czech Republic	TOMA RECYCLING
(ID) 04074157, **(VAT)** CZ04074157	
(W) toma-recycling.cz, **(E)** info@toma-recycling.cz	

Product:	TOMAPET
Color:	Clear
Production:	18/7/00001
Gross Weight:	1220
Net Weight:	1200
Operator:	

Fig. 9. Final label attached to waste bale.

4 Conclusion

The total size of the application is about 2 GB of flash memory. In this case, there was a 16 GB card, so 85% of the capacity is still available. The main program that caters to system service calls. It communicates with the PLC through the serial line and creates the weight labels. It was developed in Python scripting language. Final label is shown at Fig. 9. There is a possibility to bring solution base on virtual instrumentation (see [13–15]) and extend tasks done by weighing machine control system, but described simple solution is used and successful now.

Acknowledgement. This work was supported by the project SP2018/160, "Development of algorithms and systems for control, measurement and safety applications IV" of Student Grant System, VSB-TU Ostrava.

References

1. Mikolajek, M., Otevrel, V., Koziorek, J., Slanina, Z.: Data trends in industry automation using.NET framework. IFAC-PapersOnLine **28**(4), 418–423 (2015)
2. Bradac, Z., Fiedler, P., Cach, P., Vrba, R.: Wireless communication in automation. In: Proceedings of the IEEE International Conference on Electronics, Circuits, and Systems, vol. 2, pp. 659–662 (2003). Art. no. 301871
3. Zezulka, F., Bradac, Z., Fiedler, P., Sir, M.: Trends in automation - investigation in network control systems and sensor networks. IFAC Proc. Vol. (IFAC-PapersOnline) **10**(PART 1), 109–113 (2010)
4. Havlikova, M., Sediva, S., Bradac, Z., Jirgl, M.: A man as the regulator in man-machine systems. Adv. Electr. Electron. Eng. **12**(5), 469–475 (2014)
5. Arm, J., Bradac, Z., Kaczmarczyk, V.: Real-time capabilities of Linux RTAI. IFAC-PapersOnLine **49**(25), 401–406 (2016)
6. Misik, S., Bradac, Z., Arm, J., Stastny, L.: Embedded telemetry system with data presentation using HTTP and data logging. IFAC-PapersOnLine **28**(4), 101–106 (2015)
7. Sir, M., Bradac, Z., Kaczmarczyk, V.: Ontology and automation technique. In: Recent Researches in Communications and IT - Proceedings of the 15th WSEAS International Conference on Communications, Part of the 15th WSEAS CSCC Multiconference, Proceedings of the 5th International Conference on CIT 2011, pp. 171–174 (2011)
8. Ozana, S., Pies, M., Hajovsky, R., Koziorek, J., Horacek, O.: Application of PIL approach for automated transportation center. In: Ozana, S., Pies, M., Hajovsky, R., Koziorek, J., Horacek, O. (eds.) LNCS (LNAI and LNBI), vol. 8838, pp. 501–513 (2014)
9. Pies, M., Hajovsky, R., Ozana, S., Haska, J.: Wireless sensory network based on IQRF technology. In: 2014 the 4th International Workshop on Computer Science and Engineering-Winter, WCSE 2014 (2014)
10. Pies, M., Hajovsky, R.: Using the IQRF technology for the internet of things: case studies. Lecture Notes in Electrical Engineering. vol. 425, pp. 274–283 (2018)
11. Jurenoks, A., Jokic, D.: Sensor network information flow control method with static coordinator within internet of things in smart house environment. Proc. Comput. Sci. **104**, 385–392 (2016)
12. Prauzek, M., Kromer, P., Rodway, J., Musilek, P.: Differential evolution of fuzzy controller for environmentally-powered wireless sensors. Appl. Soft Comput. J. **48**, 193–206 (2016)
13. Martinek, R., Kahankova, R., Bilik, P., Nedoma, J., Fajkus, M., Blaha, P.: Speech quality assessment based on virtual instrumentation. In: Proceedings of the 10th International Conference on Computer Modeling and Simulation, pp. 49–53. ACM (2018)
14. Martinek, R., Sincl, A., Vanus, J., Kelnar, M., Bilik, P., Machacek, Z., Zidek, J.: Modelling of fetal hypoxic conditions based on virtual instrumentation. In: Proceedings of the Second International Afro-European Conference for Industrial Advancement AECIA 2015, pp. 249–259. Springer, Cham (2016)
15. Martinek, R., Koudelka, P., Latal, J., Vitasek, J., Vanus, J., Wen, H., Nazeran, H.: Modelling of wireless fading channels with RF impairments using virtual instruments. In: 2016 IEEE 17th Annual Wireless and Microwave Technology Conference (WAMICON), pp. 1–6. IEEE (2016)

Optimization of Voltage Model for MRAS Based Sensorless Control of Induction Motor

Ondrej Lipcak$^{(\boxtimes)}$ and Jan Bauer

Faculty of Electrical Engineering, Czech Technical University in Prague,
Technicka 2, 166 27 Prague, Czech Republic
{lipcaond,bauerja2}@fel.cvut.cz

Abstract. Omission of the speed sensor in induction motor based electric drive is very actual topic because of cost savings. When omitting the speed sensor, induction machine mathematical model that does not require information about rotor speed has to be used, such as voltage model. The quality of control then depends on the model stability and accuracy of induction motor parameter identification. This paper strives to discuss most important quantities that influence stability and accuracy of the voltage model.

Keywords: Voltage model · Current model · Induction machine · Field oriented control · Sensorless control · MRAS

1 Introduction

Sensorless control strategies for AC drives are very popular nowadays [1–3]. They are already starting to spread from academic domain into industrial applications. Main reason is to minimize costs of the drive by reducing number of equipped sensors. Omitting the speed sensor increase drive reliability and decrease maintenance requirements. However, sensor reduction is connected with increased requirements on accuracy of the drive's parameters. Algorithms for speed estimation with respect to the used estimation technique can be divided as: techniques based on machine model and techniques based on open loop estimators, various observers, artificial intelligence methods or model reference adaptive systems (MRAS) [2, 3].

As induction motor (IM) belongs to the most widespread electro-mechanical converter in variable speed drives, most sensorless control strategies are based on field-oriented control algorithms (FOC), where measured parameters such as currents, voltages are firstly transformed to DC quantities and then separately regulated [1]. Transformation angle and other quantities required for FOC such as rotor flux are obtained from mathematical models of the IM [4, 5]. In conventional sensored FOC, two main types of mathematical models are mostly used to estimate flux space vector amplitude and position

$$u_1 = R_1 \underline{i}_1 + \frac{d\underline{\Psi}_1}{dt} + j\omega_k \underline{\Psi}_1$$
$$0 = R_2 \underline{i}_2 + \frac{d\underline{\Psi}_2}{dt} + j(\omega_k - \omega)\underline{\Psi}_2.$$
$$\underline{\Psi}_1 = L_1 \underline{i}_1 + L_m \underline{i}_2$$
$$\underline{\Psi}_2 = L_2 \underline{i}_2 + L_m \underline{i}_1. \quad (1)$$

where symbol_marks space vectors. Equation (2) is used for inner motor torque calculation

$$m_i = \frac{3}{2} p_p \frac{L_m}{L_2} \left(\Psi_{2\alpha} i_{1\beta} - \Psi_{2\beta} i_{1\alpha} \right), \quad (2)$$

where α denotes real and β imaginary axis of a stator-fixed coordinate system.

The combination of variables that are easy to measure like DC link voltage, stator phase currents or rotor angular speed are used to calculate the value and position of the rotor flux.

Model that uses measured values of stator currents and angular speed according to (3) is usually labeled as a current model:

$$\frac{d\Psi_{2\alpha}}{dt} = \frac{L_m R_2}{L_2} i_{1\alpha} - \frac{R_2}{L_2} \Psi_{2\alpha} - \omega \Psi_{2\beta}$$
$$\frac{d\Psi_{2\beta}}{dt} = \frac{L_m R_2}{L_2} i_{1\beta} - \frac{R_2}{L_2} \Psi_{2\beta} + \omega \Psi_{2\alpha}. \quad (3)$$

This model shows good and stable performance in low speed operation, the non-linearities of the inverter can be overcome by fast current control loop. However, its main disadvantage is a required knowledge of the rotor resistance R_2 which can´t be measured directly and can change its value with IM temperature in about 150%. The requirement of the speed information when talking about sensorless control is problematic too.

Second model is based on knowledge of voltages applied on IM terminals and stator currents (4):

$$\frac{d\Psi_{2\alpha}}{dt} = \frac{L_2}{L_m} \left(u_{1\alpha} - R_1 i_{1\alpha} - \sigma L_1 \frac{di_{1\alpha}}{dt} \right)$$
$$\frac{d\Psi_{2\beta}}{dt} = \frac{L_2}{L_m} \left(u_{1\beta} - R_1 i_{1\beta} - \sigma L_1 \frac{di_{1\beta}}{dt} \right). \quad (4)$$

This model is known as a voltage model, however, due to recalculation of stator flux space vector to rotor flux space vector, the derivative of the stator current $\underline{i}_1$ is introduced and thus stability of the model is reduced. Moreover, voltage on the machine terminals can't be measured directly because of its pulsating nature.

By fusing (3) and (4) together, the structure of MRAS system for control without speed sensor is obtained.

In this case, voltage model is used as a reference model and current model as an adaptive model. The angular speed of the IM $\widehat{\omega}_r$ is estimated by minimizing the error ε between flux vectors from reference and adaptive model (5).

$$\varepsilon = \widehat{\Psi}_{2\alpha}\Psi_{2\beta} - \widehat{\Psi}_{2\beta}\Psi_{2\alpha}. \tag{5}$$

The accuracy of the inverter's voltage estimation remains essential for the sensorless control functionality (Fig. 1).

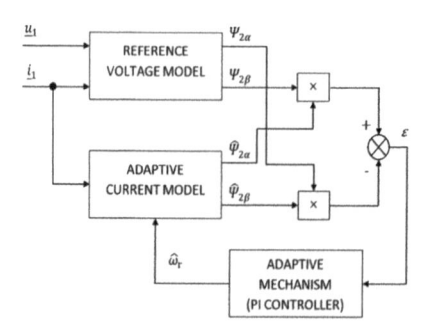

Fig. 1. Rotor flux MRAS

2 Flux Estimator with DC Offset Compensation

Equation (4) can rewritten into a different form:

$$\Psi_{2\alpha} = \frac{L_2}{L_m}\left(\int_0^t (u_{1\alpha} - R_1 i_{1\alpha})d\tau - \sigma L_1 i_{1\alpha}\right)$$
$$\Psi_{2\beta} = \frac{L_2}{L_m}\left(\int_0^t (u_{1\beta} - R_1 i_{1\beta})d\tau - \sigma L_1 i_{1\beta}\right). \tag{6}$$

The members within the integrals correspond to real and imaginary component, respectively of stator magnetic flux vector in stator-fixed coordinates:

$$\Psi_{1\alpha} = \int_0^t (u_{1\alpha} - R_1 i_{1\alpha})d\tau$$
$$\Psi_{1\beta} = \int_0^t (u_{1\beta} - R_1 i_{1\beta})d\tau. \tag{7}$$

It is therefore sufficient to solve Eq. (7) first and then substitute obtained results into (6), which will yield an algebraic expression for components of the rotor magnetic flux space vector.

Unfortunately, it is impossible to use pure integrator during practical implementation of Eq. (7). Two main reasons are unknown initial values and a DC offset which is superimposed on digital representation of voltage and current vectors due to drift and offset of analog hardware [7–10].

So far, multiple approaches have been proposed in literature on how to deal with this severe issue. These methods can be divided into three groups: methods using low-pass filters, integrators with amplitude saturation and adaptive flux observers [4].

The scheme proposed in this paper is depicted in Fig. 2. It is a direct modification of the one presented in [6]. Symbol $^\wedge$ represents estimated values, subscripts $_{vm}$ and $_{cm}$ stand for voltage model and current model, respectively.

In principle, the magnitude of the estimated stator magnetic flux vector is limited by a PI controller based on an error signal of the same phase but stable and/or constant or quasi constant amplitude. In other words, the stator flux vector is forced to stay on a circular trajectory in a complex plane around origin of coordinates (DC component causes the circle to drift away from the origin) [6].

A prerequisite for a successful operation of the proposed integrator lies within a well implemented current model. Cross members in Eq. (3) that represent current model ensure natural suppression of the arising DC component. A robust Fourth Order Runge-Kutta numerical method was selected in order to obtain stable and accurate approximation [7].

The flux error vector $\widehat{\underline{\Psi}}_{off}$ is then calculated as follows:

$$\widehat{\underline{\Psi}}_{off} = \widehat{\underline{\Psi}}_{1vm} \left(1 - \frac{|\widehat{\underline{\Psi}}_{1cm}|}{|\widehat{\underline{\Psi}}_{1vm}|}\right). \tag{8}$$

The error signal $\widehat{\underline{\Psi}}_{off}$ is then fed into a PI controller that, if properly tuned, adjusts the estimated offset voltage vector $\hat{\underline{u}}_{off}$ in such manner, that the endpoint of $\widehat{\underline{\Psi}}_{1vm}$ follows approximately a circular trajectory around origin. Recalculation of stator flux vector to rotor flux vector and vice versa can be easily done with help of (6) after substituting for the integral terms from (7).

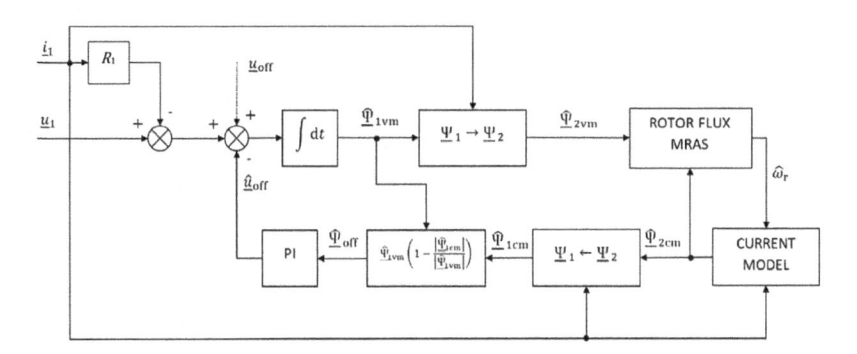

Fig. 2. Modified integrator with DC offset compensation based on current model

3 Compensation of VSI Nonlinearities

Another problem connected with voltage model is the acquisition of stator voltage. In most cases of vector control, a PWM controlled two level voltage source inverter (VSI) is used to supply the machine. Therefore, it becomes very difficult and hardware demanding to measure the stator voltage directly. Because of that, the voltage is mostly determined indirectly [10].

One method is to reconstruct the voltage from known gating signals of the VSI and measured DC link voltage. Another approach is based on using the desired voltage vector (i.e. input to the modulator) as an input to the voltage model [10]. This method is simple and can be quite efficient, but one must deal with a key problem and that is a nonlinear behavior of the inverter caused by semiconductor switches and necessary protective time (deadtime) inserted by microcontroller in order to prevent short-circuit of the DC link [10, 11].

3.1 Deadtime and IGBT Delay Compensation

The inserted deadtime causes a distortion of the inverter phase voltage u_{a0} which becomes significant in the area of small voltage vectors applied to the motor (i.e. during low-speed, low-torque operation). For initial analysis, let's omit non-ideal characteristics of the IGBTs and consider only the deadtime effects. In case of vector-controlled IM, the desired stator voltage vector is mostly created by space vector modulation (SVM) techniques. Inside the microcontroller, up-down counters are usually used for this purpose. The control signal for the upper switch is then symmetrical around the top of the counter and the signal for the lower switch is complementary. Furthermore, the control signal for switching on the respective IGBT is delayed by the deadtime, the control signal for turning off is sent without delay [9–11].

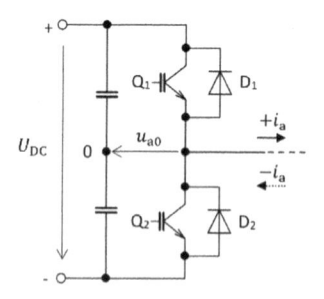

Fig. 3. One leg of voltage source inverter

It is obvious that during deadtime both switches are off and the polarity of the voltage u_{a0}, when considering inductive load, depends on the direction of the current i_a. If i_a is positive, then D_2 must conduct and u_{a0} is equal to $-U_{DC}/2$. If i_a is negative, then D_1 conducts and u_{a0} is equal to $+U_{DC}/2$.

Now, let's move on to real semiconductor switches represented by IGBTs. Key datasheet parameters necessary for further analysis are: turn-on delay time $T_{d(on)}$, turn-on rise time T_r, turn-off delay time $T_{d(off)}$ and turn-off fall time T_f. The interpretation and effects of $T_{d(off)}$ and $T_{d(off)}$ are obvious. Now we will look at role of T_r and T_f of IGBT modules connected in a topology of two-level VSI. Following considerations are made for IGBT modules without snubber.

Let's assume, that there is a positive current flowing to inductive load through D_2 (see Fig. 3). Shortly after a turn-on signal is applied to the gate of Q_1, its collector current starts to rise while collector-emitter voltage stays approximately at full DC link value. If we assume $i_a = \text{const.}$, the current through D_2 must descent in a complementary manner. When the forward current of D_2 reaches zero, an opposite reverse recovery current starts to flow through the diode adding to the collector current of Q_1. Around the peak of the D_2 reverse recovery current, the collector-emitter voltage of Q_1 starts to sharply fall to zero. The transient state is over and Q_1 conducts full load current.

Now, when a turn-off signal is applied to the gate of Q_1, after a delay, its collector-emitter voltage starts to build up to the full DC link value while maintaining approximately constant collector current. After the collector-emitter voltage reaches the DC link voltage, D_2 starts to take over the current. From the analysis performed in the last two paragraphs it can be noted that T_f doesn't distort the resulting phase voltage u_{a0}. Value of T_r should be theoretically considered, but its effect is in case of used IGBT modules (T_r around 10 ns) negligible (Fig. 4).

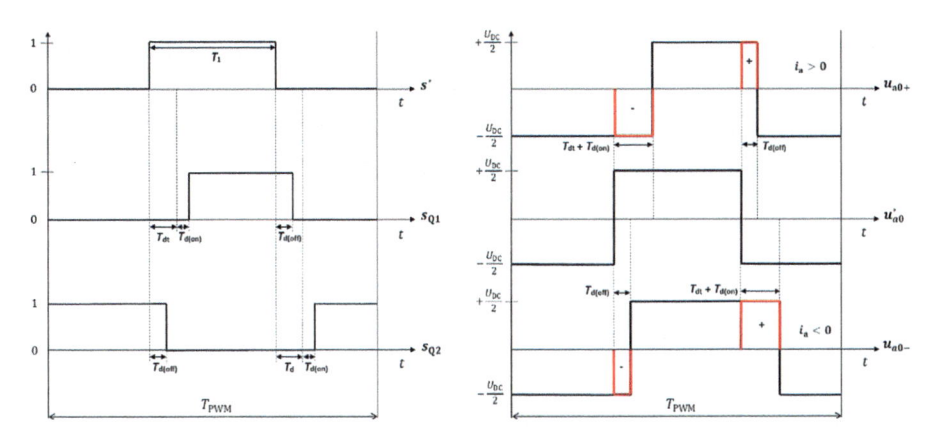

Fig. 4. Deadtime and IGBT delay effects on the resulting inverter phase voltage: s^* ideal gating pulse for Q_1 (for Q_2 negated), $s_{Q1}(s_{Q2})$ logical state (1 on, 0 off) of Q_1 (Q_2) considering deadtime and delay effects, u_{a0}^* desired inverter phase voltage, u_{a0+} (u_{a0-}) actual inverter phase voltages for $i_a > 0$ ($i_a < 0$).

The mean value of the ideal (desired) inverter phase voltage can be expressed as

$$U_{AV0} = \frac{1}{T_{PWM}} \int_0^{T_{PWM}} u_{a0}(t) dt = \frac{1}{T_{PWM}} \left(\frac{U_{DC}}{2} T_1 - \frac{U_{DC}}{2} (T_{PWM} - T_1) \right)$$
$$= \frac{U_{DC}}{2 T_{PWM}} (2T_1 - T_{PWM}). \tag{9}$$

Now we introduce an "effective deadtime" T_{eff} given by

$$T_{eff} = T_{dt} + T_{d(on)} - T_{d(off)}. \tag{10}$$

The mean value of distorted voltage for $i_a > 0$ can be then expressed as

$$U_{AV+} = \frac{1}{T_{PWM}} \int_0^{T_{PWM}} u_{a0+}(t) dt = \frac{1}{T_{PWM}} \left(\frac{U_{DC}}{2}(T_1 - T_{eff}) - \frac{U_{DC}}{2}(T_{PWM} - T_1 + T_{eff}) \right)$$
$$= \frac{U_{DC}}{2T_{PWM}}(2T_1 - T_{PWM}) - \frac{1}{T_{PWM}} U_{DC} T_{eff} = U_{AV0} - \Delta U_{DD}$$
(11)

and in a case of $i_a < 0$ as

$$U_{AV-} = \frac{1}{T_{PWM}} \int_0^{T_{PWM}} u_{a0-}(t) dt = \frac{1}{T_{PWM}} \left(\frac{U_{DC}}{2}(T_1 + T_{eff}) - \frac{U_{DC}}{2}(T_{PWM} - T_1 - T_{eff}) \right)$$
$$= \frac{U_{DC}}{2T_{PWM}}(2T_1 - T_{PWM}) + \frac{1}{T_{PWM}} U_{DC} T_{eff} = U_{AV0} + \Delta U_{DD}$$
(12)

It is apparent from the last two equations, that the actual mean value U_{AV} can be written as follows:

$$U_{AV} = U_{AV0} - \Delta U_{DD} = U_{AV0} - \frac{1}{T_{PWM}} U_{DC} T_{eff} \mathrm{sgn}(i_x) \quad x = a, b, c \quad (13)$$

Remark 1: It is possible to convert Eq. (13) into space vector notation and derive correction voltage vector that has to be added to the desired one before entering modulator [5, 10]. However, simpler and computationally less demanding approach is to adjust the desired output duty cycle from modulator. In a similar manner as in the case of the mean values of the inverter phase voltage, we can derive a relationship between the corrected duty cycle s_{cor} and the output duty cycle from modulator s_{mod} [11]

$$s_{cor} = s_{mod} + \frac{T_{eff}}{T_{PWM}} \mathrm{sgn}(i_x) \quad x = a, b, c. \quad (14)$$

Remark 2: Taking into consideration that the phase currents are formed by a PWM controlled VSI, it can be difficult to exactly detect the instant when they change their polarity. This problem can be further aggravated by imperfections of the sensing hardware. It has been experimentally observed, that if the current controllers are well tuned with good transient response, the actual values can be replaced by the desired values which are smoother. This means transforming the desired current vector from rotor coordinates to its three-phase components.

Remark 3: Unfortunately, the stated IGBT parameters are dependent on temperature, collector-emitter voltage, collector current, gate voltage and load type. For experimental results presented in this paper, a current based correction of IGBT switching characteristics has been implemented. The effective deadtime T_{eff} is being adjusted during motor operation according to actual collector currents by a correction function based on switching characteristics stated in datasheet.

Remark 4: It has been noticed that adding a full time effective compensation when the respective phase currents are lower than some threshold current I_{th} can lead to a bigger distortion of the current space vector. This was later confirmed by measurement on the VSI. The transient switching states during low or zero phase currents differ from the ones during higher current (Fig. 5). Therefore, every time when $i_x < |I_{th}|$ the correction duty cycle is calculated as

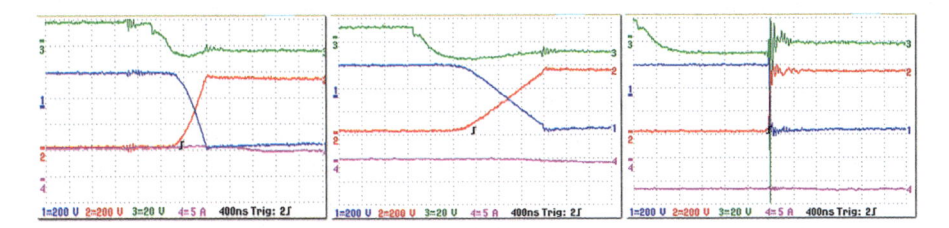

Fig. 5. Inverter leg switching detail ($u_{GEQ1} - 3$, $u_{a0} - 1$, $u_{Q1} - 2$, $i_a - 4$)

$$s_{cor} = s_{mod} + \frac{T_{eff}}{T_{PWM}} \frac{i_x}{I_{th}} \quad x = a, b, c. \tag{15}$$

Figure 5 shows VSI leg switching detail for one leg of the VSI for three states of phase current: $i_a > 0$, $i_a \approx 0$, $i_a < 0$. From the figure above it is obvious, that when the current is positive, the voltage u_{a0} changes polarity instantly after Q_1 is switched off and the transition takes approx. 400 ns. In case of small value of the phase current, the transition takes much longer time (1.2 μs) and for negative current, u_{a0} changes polarity instantly after Q_2 is switched on, e.g. after dead time which was set to 2 μs.

4 Optimization Results

As already mentioned at the beginning, this paper strives to analyze and optimize accuracy of so-called voltage model of the IM in order to increase stability of FOC with putting emphasis on sensorless control. For this purpose, direct FOC control with flux based MRAS speed estimation was implemented into a Texas Instrument TMS28F335 controller. As a drive, 12 kW induction motor with the following parameters (Table 1) coupled with DC motor has been used.

Table 1. IM and DC motor parameters

Induction motor				DC motor	
P_n	12 kW	R_1	0.358 Ω	P_n	8.8 kW
I_n	22 A	R_2'	0.354 Ω	I_{an}	38.3 A
U_n	380 V	$L_{\sigma 1}$	0.00227 H	U_{an}	230 V
f_n	50 Hz	$L_{\sigma 2}$	0.00227 H	I_{bn}	2.8 A
$\cos\varphi_n$	0.8	R_{Fe}	202.1 Ω	U_{bn}	110 V
n_n	1400 rev/min	L_h	0.0779 H	n_n	1400 rev/min

Behavior and influence of the accuracy of the voltage model has been tested during different conditions. Figure 6 shows acceleration of the IM drive under no-load condition (continuous line) and under load (dashed line).

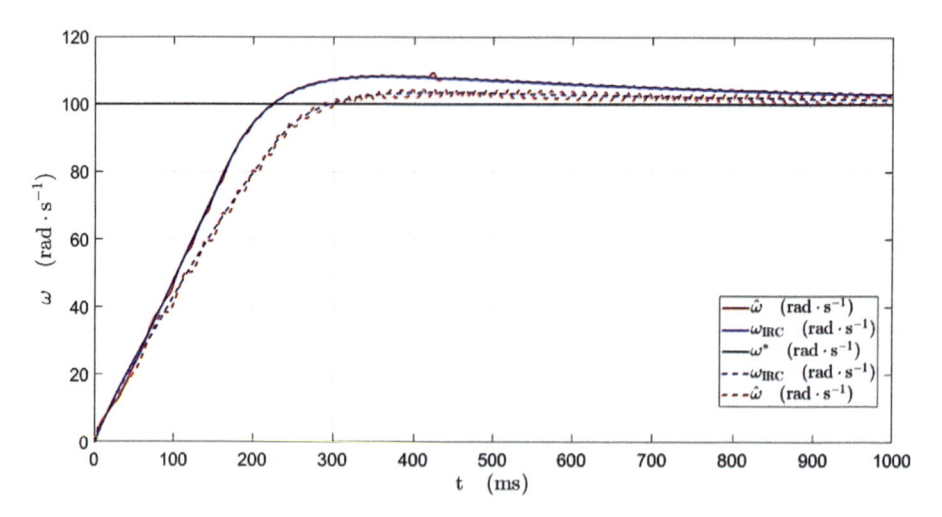

Fig. 6. IM drive acceleration from 0 to 100 rad s^{-1}

Figure 7 shows drive reversal from 20 rad s^{-1} to −20 rad s^{-1} with dead time compensation and without dead time compensation.

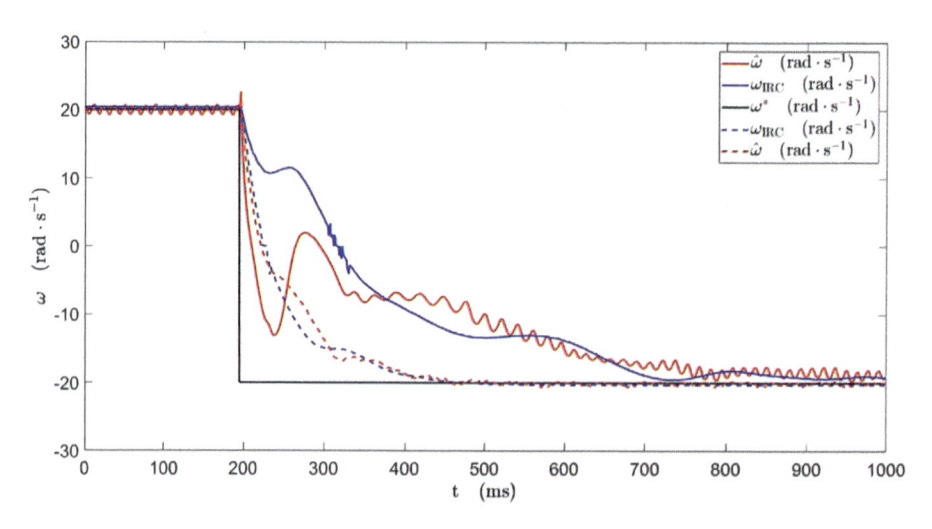

Fig. 7. IM reversal with (dashed lines) and without (continuous lines) deadtime compensation

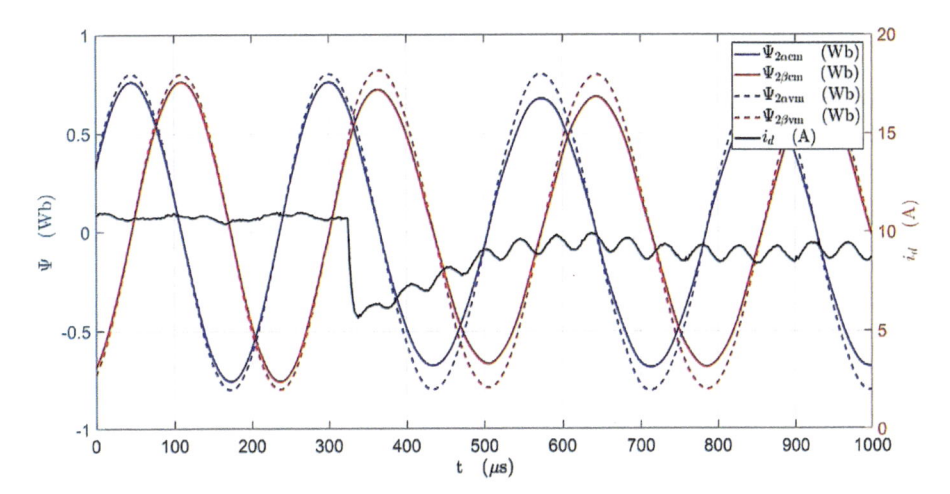

Fig. 8. Influence of VSI nonlinearities on flux estimation

In the Fig. 8, deadtime compensation was switched off at the instant $t = 300$ μs. It can be seen that the rotor flux components estimated by the current model (continuous lines) remain stable. However, flux components estimated by the voltage model change immediately their amplitude. This causes wrong reaction of i_d controller and wrong estimation of IM torque.

5 Conclusion

A proper operation of so-called voltage model is a key aspect when considering a practical implementation of sensorless control of IM based on rotor flux MRAS.

First, and perhaps the most serious problem one has to deal with, is the arising DC offset which is superimposed on the resulting components of the stator magnetic flux vector. It has been experimentally confirmed, that the proposed flux estimator, if properly tuned, successfully eliminates this negative phenomenon both in steady and transient states.

Furthermore, the voltage model requires accurate knowledge of the stator voltage vector. One can use desired voltage vector that is an input to the modulator when a proper compensation of the inverter most significant nonlinearities is handled. Experimental results show, that the proposed deadtime and IGBT delay compensation improves the IM performance in a low-speed area and ensures more consistent operation of the both used IM mathematical models.

References

1. Bose, B.K.: Modern Power Electronics and AC Drives. PHI Learning Private Limited, New Jersey (2013)
2. Rao, P., Nakka, J., Shekar, R.: Sensorless vector control of induction machine using MRAS techniques. In: International Conference on Circuits, Power and Computing Technology (2013)
3. Vo, H., Brandstetter, P., Dong, Ch., Tran, T.: Speed estimators using stator resistance adaption for sensorless induction motor drive. Power Eng. Electr. Eng. **14**(3) (2016)
4. Koteich, M.: Flux estimation algorithms for electric drives: a comparative study. In: 3rd International Conference on Renewable Energies for Developing Countries (REDEC) (2016)
5. Holtz, J., Quan, J.: Drift- and parameter-compensated flux estimator for persistent zero-stator-frequency operation of sensorless-controlled induction motors. IEEE Trans. Ind. Appl. **39**(4), 1052–1060 (2003)
6. Lascu, C., Andreescu, G.: Sliding-mode observer and improved integrator with DC-offset compensation for flux estimation in sensorless-controlled induction motors. IEEE Trans. Ind. Electron. **53**(3), 785–794 (2006)
7. Bauer, J., Lipcak, O., Kyncl, J.: Different approaches in numerical solution of continuous mathematical models of induction machine. In: 18th International Conference on Power Electronics and Motion Control (2018)
8. Brandstetter, P., Kuchar, M.: Rotor flux estimation using voltage model of induction motor. In: 16th International Scientific Conference on Electric Power Engineering (2015)
9. Quang, N.P., Dittrich, J.: Vector control of three-phase AC machines: system development in the practice (2008)
10. Sediki, H., Djennoune, I.S.: Compensation method eliminating voltage distortions in PWM inverter. Int. J. Electr. Comput. Eng. **3**(6) (2009)
11. Anuchin, A., Gulyaeva, M., Briz, F., Gulyaev, I.: Modeling of AC voltage source inverter with dead-time and voltage drop compensation for DPWM with switching losses minimization. In: International Conference on Modern Power Systems (MPS) (2017)

Capability of Predictive Torque Control Method to Control DC-Link Voltage Level in Small Autonomous Power System with Induction Generator

Pavel Karlovský(✉) and Jiri Lettl

Department of Electric Drives and Traction, Faculty of Electrical Engineering,
Czech Technical University in Prague, Technicka 2,
166 27 Prague, Czech Republic
{karlopav,lettl}@fel.cvut.cz

Abstract. In a small autonomous power system, the electric power is usually produced by means of an electric machine operated as a generator. The often used one is an induction generator. The generator supplies the DC-link through the controlled converter (voltage source inverter – VSI). The appliances connected to the same DC-link consume power and therefore the DC-link capacitor is discharged. To maintain the decreasing DC-link voltage level, the same power amount must be delivered by the generator. In case of the induction generator, its control is specific as no external source for machine excitation is present. To obtain the torque value desired by the DC link voltage PID controller, the generator is usually controlled by the field-oriented control (FOC) method or direct torque control (DTC) method. However, the model predictive control (MPC) method has been utilized for electric machine control in recent time. The paper explores the possibility of employing the predictive torque control (PTC), which is often used implementation of MPC in this type of applications. The PTC method ability to control the DC-link voltage in a small autonomous power system has been verified experimentally on a laboratory system with induction generator.

Keywords: Predictive torque control · Induction generator ·
DC-link voltage control · Small autonomous power system

1 Introduction

Many arrangements of small autonomous power systems are known. Such system can run completely in a standalone operational mode [1] or it can be partially connected to the grid [2]. Usually, the power sources are composed of a prime mover and a generator, or a photovoltaic cell as well as a battery. The power management of all devices is a key aspect for the proper operation of the autonomous power system [3]. As the prime mover can be used wind or water rotor or diesel engine. In this cases, constant rotational speed can but doesn't have to be maintained. The generator terminals are connected to the input of the converter with the common DC-link [4]. Another possible energy source in the system can be a photovoltaic cell. The photovoltaic cell provides DC voltage and is connected to the common DC-link via its own converter. This converter can be controlled to the maximum power or constant voltage [5]. Multiple appliances consuming power are connected to the DC-link. Typically, they are DC loads, AC loads, or the energy storage systems such as accumulators. All of them are usually connected to the DC-link via their own converters (AC or DC). The DC link energy is stored in the capacitor and the power consuming appliances reduce the charge of the capacitor and therefore reduce the DC-link voltage level. In order to maintain the constant voltage level, it is necessary to supply the DC link with the power same as is consumed at any time instant. The main requirement for the controller is to ensure the constant voltage [6]. The typical arrangement of small autonomous system is presented in Fig. 1.

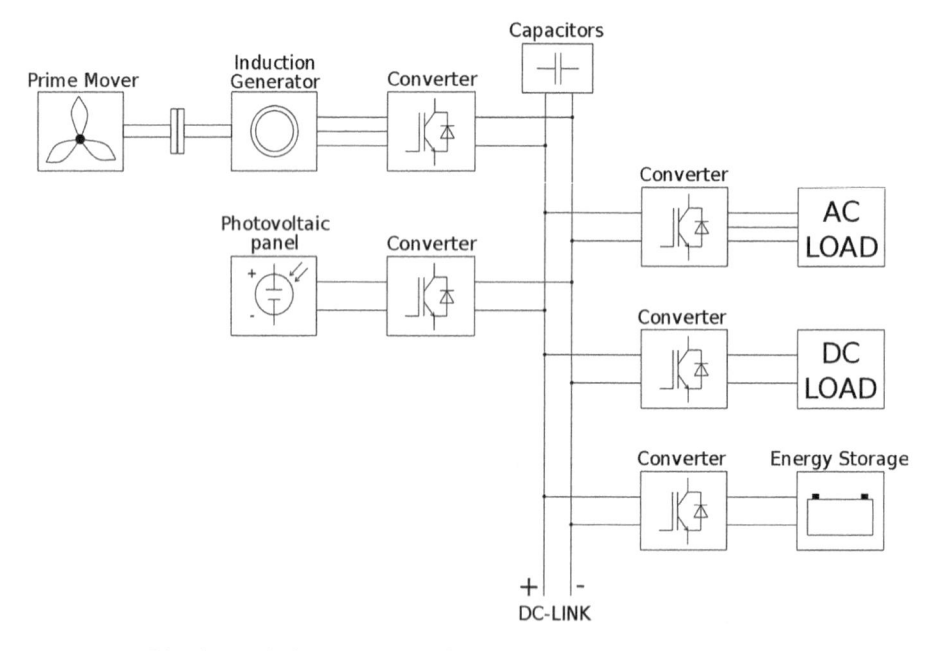

Fig. 1. Typical arrangement of a small autonomous power system

Many applications require the voltage controlled to constant value without bigger peaks. The magnetized rotating induction generator creates voltage at its terminals and this value can be adjusted by the torque produced by the generator. The common approach to the voltage level control is by means of the PID controller. The PID controller gives the torque reference for the generator. To reach the torque reference, a special control algorithm is used [7]. The commonly used algorithms to the machine control are the field oriented control (FOC) and direct torque control (DTC). Both can be used to control the machine working as a generator or as a motor. The use of fuzzy logic that eliminates the PID controller is also possible [8].

The model predictive approach (MPC) to control electric drives is nowadays a very actual topic [9, 10]. The performance comparison of the common industrial methods like DTC and FOC to MPC was analyzed mainly for motor applications [11]. The MPC used for electric drives can be distinguished as two strategies. The predictive torque control (PTC) method controls the torque and flux amplitude and corresponds to DTC. The predictive current control (PCC) method controls torque and flux components of the current and therefore corresponds to FOC. The results show that the MPC is a very promising strategy that eliminates some disadvantages of the previous strategies [9]. Very similar results were observed in generator applications. The application of MPC for doubly fed induction generator in a wind power system is often implemented [12, 13] with the speed encoder or as an encoderless variant [14] in order to control the system to the required power.

The paper applies the PTC method to keep the machine magnetized and to reach the torque reference of the induction generator and hence, control the voltage in the DC-link to its reference value. The performance of such solution is examined and verified experimentally on small laboratory power system with induction generator. The results show the capability of PTC to control the voltage level in the DC-link of a small autonomous power system with induction generator.

2 Method

The generator terminals are connected to the converter. The converter creates switch combinations that define eight usable voltage vectors. Each voltage vector influences the behavior of the generator in a specific way. The control aim of PTC is to minimize the error between the torque and flux values to their references. The same aim has the DTC and similarly the FOC. Hysteresis controller in case of DTC, PI controller in case of FOC, and predictive controller in case of PTC are used. The PTC is based on the idea, that the next time instant value of the torque and flux amplitude can be predicted for each voltage vector when the actual state of the generator is known [9]. As there are eight possible voltage vectors, all of them are predicted and each future state is evaluated. Then, the most suitable one, according to the defined criteria, is chosen. The principle of the PTC is shown in Fig. 2.

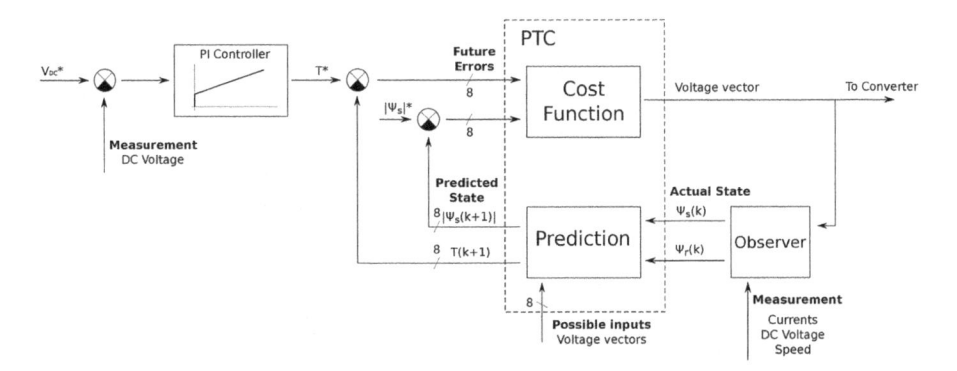

Fig. 2. Working principle of the DC-link voltage control using PTC

The stator flux vector of the next time instant $\psi_s(k + 1)$ is predicted for each of the eight possible voltage vectors $v_s(n)$ according to the voltage equation of the induction machine and the future torque $T(k + 1)$ as a cross product of the flux and current. After Euler discretization the torque and flux are obtained as (1) and (2)

$$\overrightarrow{\psi_s(k+1)} = \overrightarrow{\psi_s(k)} + T_s \cdot \overrightarrow{v_s(n)} - R_s \cdot \overrightarrow{i_s(k)}, \tag{1}$$

where R_s is stator resistance, i_s is the stator current vector and T_s is the sampling time and

$$T(k+1) = \frac{3}{2} \cdot p_p \cdot \left(\overrightarrow{\psi_s(k+1)} \times \overrightarrow{i_s(k+1)} \right), \tag{2}$$

where p_p is the number of pole pairs. In order to predict the torque (2), the prediction of the stator current $i_s(k + 1)$ must be known. The current can be predicted as (3)

$$\overrightarrow{i_s(k+1)} = \left(1 + \frac{T_s}{\tau_\sigma}\right) \cdot \overrightarrow{i_s(k)} + \frac{T_s}{\tau_\sigma + T_s} \cdot \frac{1}{R_\sigma} \cdot \left(\left(\frac{k_r}{\tau_r} - k_r \cdot j \cdot \omega(k)\right) \cdot \overrightarrow{\psi_r(k)} + v_s(n) \right), \tag{3}$$

where τ_σ is time constant $\tau_\sigma = \sigma * L_s/R_\sigma$, L_s is stator inductance, R_σ is resistance $R_\sigma = R_s + R_r^* k_r^2$, R_s is stator resistance, R_r is rotor resistance, k_r is coefficient $k_r = L_m/L_r$, L_m is mutual inductance, L_r is rotor inductance and τ_r is rotor time constant $\tau_r = L_r/R_r$. The ω is rotor electrical frequency and ψ_r is the rotor flux vector.

The individual states are evaluated according to the cost function. The cost function is defined as the sum of the errors between the references and the predicted values. We can also refer to it as to predictive controller. The chosen cost function minimizes the errors between the real and reference values of torque and flux amplitude.

$$g(n) = k \cdot \left| \left| \overrightarrow{\psi_s(k+1)} \right| - |\psi_s|^* \right| + \left| \overrightarrow{T(k+1)} - T^* \right| \tag{4}$$

where k is the weighting factor that ensures equal weight of torque and flux component of the cost function [15].

In the electric machine, the requirement for the constant amplitude of the flux is usually applied. The reference for the torque must be calculated. One possible way is the use of PI controller. The error for the controller is the difference between the actual and reference voltage levels in the DC-link. With the negative torque the machine works as generator and increases the voltage level. With the positive torque, the machine works as motor and decreases the voltage. Therefore, the negative result of the PI controller is the torque reference [7] (4).

$$T^* = -\left(K_p \cdot (V_{DC}^* - V_{DC}) + K_i \cdot \int_0^t (V_{DC}^* - V_{DC}) dt \right), \tag{5}$$

where the constants K_p and K_i are the PI controller proportional and integral coefficient and V_{DC} is the voltage in DC-link.

3 Results

The experiment was carried in order to verify the results. The model of the small autonomous system was created in the laboratory. The prime mover was simulated by the separately excited DC machine. Its shaft speed was controlled by the voltage source connected to the DC machine terminals. The induction generator was mechanically connected directly to the shaft of the DC machine and its terminals to the converter. The VSI of a three-phase full-bridge topology was utilized and connected to the DC-link with capacitor battery. The load was simulated by the resistance connected to the DC-link via transistor half-bridge topology and its power consumption was adjustable by the PWM. The important parameters of the laboratory power system are summarized in Table 1 and the experimental setup is depicted in Fig. 3. The dSPACE ds1103 platform was used for the algorithm implementation. The system was controlled with 50 μs time loop. No PWM was used for the VSI outputs. The change of transistor switch combination was possible only at the beginning of each time period, so the minimum duration of every transistor combination was 50 μs.

Table 1. Parameters of the system

Parameter	Value	Description
V_n	230 V	Nominal voltage of the generator
I_n	11.8 A	Nominal current of the generator
P_n	5.5 kW	Nominal power of the generator
ω_n	145 rad/s	Nominal speed of the generator
	Y	Star connected of the generator
p_p	2	Number of pole pairs of the generator
J	0.3 kg * m^2	Moment of inertia of the machines
C	2350 μF	Capacity of the DC link

In the first experiment, the constant voltage source was connected to the DC machine simulating the prime mover, so its shaft was rotating with constant speed. The DC-link was charged and was not connected to any external power supply. No power was consumed from the DC-link. The aim of the PTC algorithm was to reach the desired voltage reference in the DC-link. The Fig. 4 shows the DC-link voltage waveform following its reference (dashed curve) in the top of the figure, the torque created by the generator in order to reach the voltage reference in the middle part of the figure and the speed of the shaft in the figure bottom. The speed was set by the DC machine constant voltage, so it was changing only a little because of the slope of the DC machine speed characteristic. The presented results show quick response to the voltage reference step change and the desired voltage maintaining when no external power consumption is present.

In the second experiment, the voltage reference was constant, and the external power consumption was simulated by the resistor. As the load was increasing, the generator was producing greater torque in order to maintain the voltage reference. The

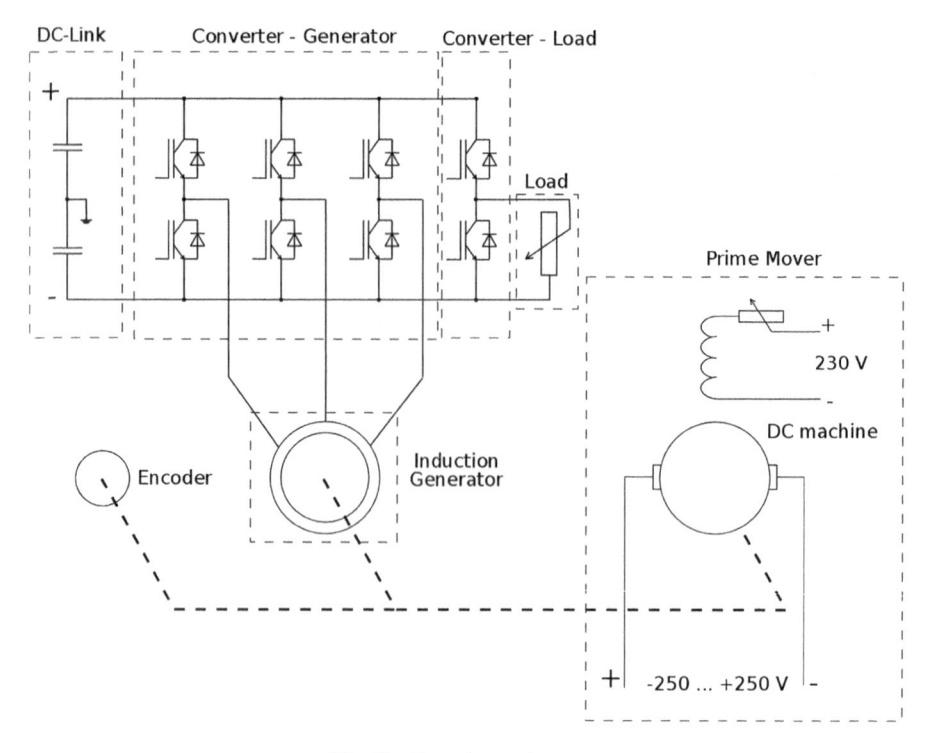

Fig. 3. Experimental setup.

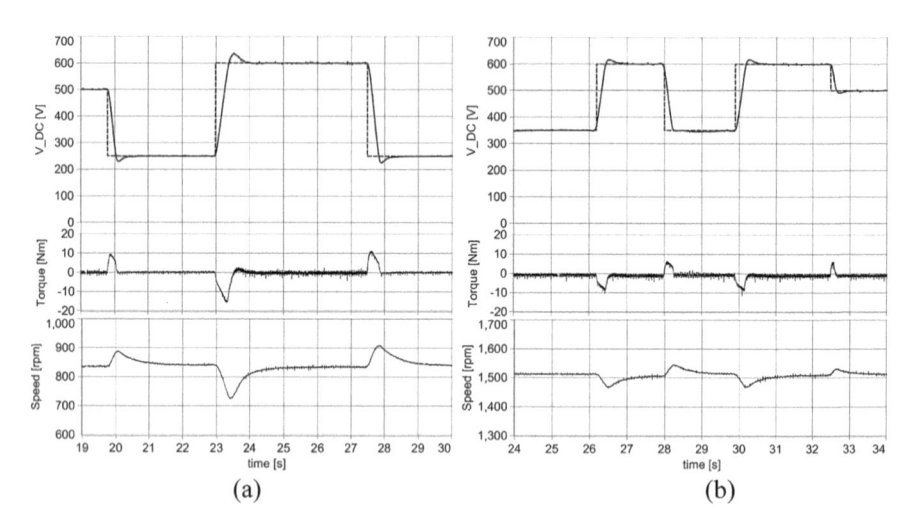

Fig. 4. DC-link voltage control at no load operation and voltage reference changes. The prime mover rotates at (a) 800 rpm. (b) 1500 rpm.

speed of the shaft decreased because of the DC machine speed characteristic. The experiment was done for different speeds of the prime mover. The Fig. 5 from the top shows the voltage in the DC-link, the torque produced by the generator, the current consumed by the load, and shaft speed. The presented results show capability of the proposed algorithm to maintain the constant DC-link voltage with the increasing power consumption.

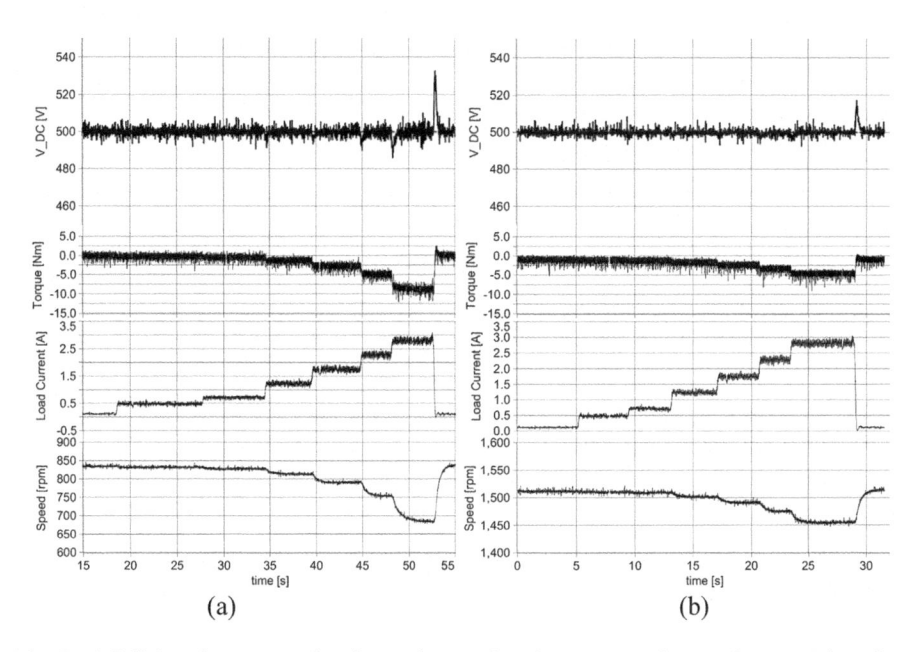

Fig. 5. DC-link voltage control at increasing load and constant voltage reference. The prime mover rotates at (a) 800 rpm. (b) 1500 rpm.

In the third experiment, the performance of the PTC and DC-link voltage waveform when the prime mover stops rotating. The energy stored in the rotating mass is utilized by the generator to supply the DC-link for a certain time period. In Fig. 6, the voltage value maintaining by utilizing the energy stored in the rotating mass at the power consuming is shown. From the top the figure depicts the controlled DC-link voltage waveform with its reference, the torque produced by the generator, the current consumed by the load, and the decreasing shaft speed. The presented results illustrate the PTC capability to supply power for a short time period in case of the prime mover fault situation.

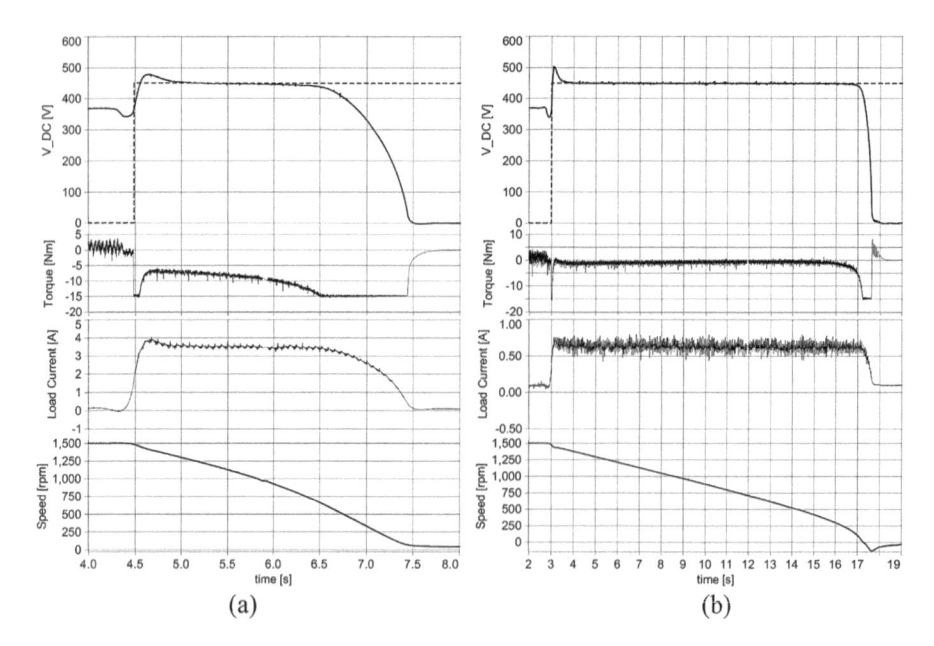

Fig. 6. The control algorithm ensures the constant voltage value even when the prime mover stops rotating. The current consumed by the load (a) 3.5 A (b) 0.7 A.

4 Conclusion

In the paper, the capability of the PTC to control the DC-link voltage of a small autonomous power system was investigated. The experiments were done in order to verify the functionality and performance of the proposed method. The simulated small autonomous power system was realized in the laboratory. The prime mover was simulated by the separately excited DC machine. The speed of the shaft was set by means of the voltage connected to the DC machine terminals. The load was simulated by the resistance with adjustable power consumption connected to the DC-link. The induction generator was connected to the common shaft and its terminals to the VSI of the transistor full-bridge topology. The voltage level of the DC-link was controlled by proposed algorithm. The experiment has verified that the voltage level is controllable at multiple prime mover speeds and at multiple load power consumption. The results prove the good capability of the PTC method to control the DC-link voltage level in a small autonomous power system with induction generator. The advantages and disadvantages of the proposed solution compared to other methods were not analyzed. This investigation will follow.

Acknowledgement. This material is based on the work supported by the Technology Agency of the Czech Republic under the grant for Competence Centers Program, project No. TE02000103, and on the work supported by the Student Grant Agency of the Czech Technical University in Prague under grant No. SGS18/132/OHK3/2T/13.

References

1. Ryvkin, S., Valiyev, M., Fligl, S., Bauer, J.: Control of island mode working induction generator based on state space controller. In: 16th International Power Electronics and Motion Control Conference and Exposition, Antalya, Turkey, pp. 527–532 (2014)
2. Wei, X., Xiangning, X., Pengwei, C.: Overview of key microgrid technologies. Int. Trans. Electr. Energy Syst. **28**(7) (2018)
3. Tazvinga, H., Zhu, B., Xia, X.: Energy dispatch strategy for a photovoltaic-wind-diesel-battery hybrid power system. Sol. Energy **108**, 412–420 (2014)
4. Kuo, S., Wang, L.: Analysis of isolated self-excited induction generator feeding a rectifier load. IEEE Proc. – Gener. Transm. Distrib. **149**(1), 90–97 (2002)
5. Pichlík, P., Zděnek, J.: Converter regulation of stand-alone photovoltaic system at low solar radiation. In: Proceedings of International Conference on Applied Electronics, Pilsen, Czech Republic, pp. 207–210 (2012)
6. Wang, L., Lee, D.: Coordination control of an AC-to-DC converter and a switched excitation capacitor bank for an autonomous self-excited induction generator in renewable-energy systems. IEEE Trans. Ind. Appl. **50**(4), 2828–2836 (2014)
7. Albu, M., Horga, V., Ratoi, M., Botan, C.: DC link voltage control of an induction generator/PWM converter system. In: Proceedings of International Aegean Conference on Electrical Machines and Power Electronics, Bodrum, Turkey, pp. 766–769 (2007)
8. Pena, R., Cardenas, R., Blasco-Gimenez, R., Henriquez, C.: DC link voltage control of a PWM excited induction generator. In: Proceedings of 28th Annual Conference of the Industrial Electronics Society, IECON 2002, Sevilla, Spain, vol. 1, pp. 247–250 (2002)
9. Rodriguez, J., Cortes, P.: Predictive Control of Power Converters and Electrical Drives. Wiley-IEEE Press, Hoboken (2012)
10. Wang, F., Li, S., Mei, X., Xie, W., Rodríguez, J., Kennel, R.: Model-based predictive direct control strategies for electrical drives: an experimental evaluation of PTC and PCC methods. IEEE Trans. Ind. Inf. **11**(3), 671–681 (2015)
11. Wang, F., Zhang, Z., Mei, X., Rodríguez, J., Kennel, R.: Advanced control strategies of induction machine: field oriented control, direct torque control and model predictive control. Energies **11**(1), 120 (2018)
12. Dirscherl, C., Hackl, C.M.: Model predictive current control with analytical solution and integral error feedback of doubly-fed induction generators with LC filter. In: Proceedings of IEEE International Symposium on Predictive Control of Electrical Drives and Power Electronics (PRECEDE), Pilsen, Czech Republic, pp. 25–30 (2017)
13. Yang, X., Liu, G., Li, A., Dai, L.V.: A predictive power control strategy for DFIGs based on a wind energy converter system. Energies **10**(8), 1098 (2017)
14. Bayhan, S., Abu-Rub, H., Ellabban, O.: Sensorless model predictive control scheme of wind-driven doubly fed induction generator in DC microgrid. IET Renew. Power Gener. **10**(4), 514–521 (2016)
15. Mamdouh, M., Abido, M.A., Hamouz, Z.: Weighting factor selection techniques for predictive torque control of induction motor drives: a comparison study. Arab. J. Sci. Eng. **43**(2), 433–445 (2018)

Feasibility Structural Analysis of Engineering Plastic Reel Module for Carrying Wound High-Voltage Electric Transmission Line

Jungyun Kim[1], Ho-Young Kang[2], Young-Geon Song[2], and Chan-Jung Kim[3](✉)

[1] School of Mechanical and Automotive Engineering,
Catholic University of Daegu, 13-13, Hayang-ro,
Gyeongsan-si, Gyeongbuk 38430, South Korea
kjungyun@cu.ac.kr
[2] Eco-Friendly Auto Parts Technology Institute, Gyeongbuk Technopark,
120, Gongdan 7-ro, Sinje-ri, Jillyang-eup, Gyeongsan-si,
Gyeongsangbuk-do 38463, South Korea
{hykang,younggeun}@gbtp.or.kr
[3] Mechanical Design Engineering, Pukyong National University,
365 Sinseon-ro, Nam-gu, Yongdang Campus, Busan 48547, South Korea
cjkim@pknu.ac.kr

Abstract. Current wooden reel modules are frequently used to carry high-voltage electric transmission lines by winding them around the reel module. But the wooden reel has been reported to have several problems for users and manufacturers, such as high manufacturing cost, heavy structure, and difficulty in recycling, so that it is necessary to make them from other light-weight materials with sufficient strength. One of the alternative materials is engineering plastic, and a new design process for reel modules should be developed for the engineering plastic material to minimize the total development duration. In this study, a numerical approach using a finite-element model of the reel module was used to ensure structural rigidity requirement over the maximum payload, 4,000 kgf, as well as the equivalent impact load from more than 10-cm free fall. The candidate finite-element model of the reel module was simulated for more than the maximum payload, and a structural static analysis was conducted for the equivalent impact force derived from free-fall motion with the center of the reel module fixed. Both simulation results revealed that the candidate reel model made from engineering plastic has satisfactory static rigidity compared with the current wooden reel module.

Keywords: Reel module · Engineering plastic material · Weight force · Equivalent impact force · Static rigidity

1 Introduction

A high-voltage electric transmission line has a cross section with a relatively large diameter, so the heavy electric line is wound around a reel module. A wooden reel module has been used for most winding devices for high-voltage electric transmission line because it provides sufficient structural rigidity and is convenient to dispose of after use. However, wooden reel modules have several problems for users and manufacturers, such as a heavy structure, difficulty in recycling, and high manufacturing cost. Manufacturers have been seeking other alternative lightweight materials suitable for reel modules instead of wood.

Nowadays, several light-weighted structural materials have been applied for the mechanical or electric industries to increase the fuel efficiency of transportation or enhance the dynamic behavior of interesting structures by reducing total weight under a promising similar rigidity [1, 2]. Nonsteel materials such as aluminum, magnesium, and titanium, have been widely used in transportation systems, and composite materials based on aluminum or other lightweight materials were also frequently considered in several field applications [3, 4]. Reinforced-plastic materials based on glass fiber or carbon fiber have been studied to understand their mechanical properties in different situations, and progress has been made in mass-production methods [5]. The engineering plastic materials, i.e., Polyethylene (PE), Polycarbonates (PC), and Acrylonitrile butadiene styrene (ABS), were also a sound solution by saving manufacturing effort and by providing both reasonable structural rigidity and excellence in recycling. Therefore, engineering plastic was selected as an alternative material to current wooden reel modules in this study.

The mechanical properties of engineering plastic and wood are different, so that the existing design for wooden material cannot be used for the candidate engineering plastic. The development of reel modules based on plastic material should begin by approaching it from the conceptual design of reel modules. The design requirement of reel modules is that they should carry the wound high-voltage electric transmission line up to a 4,000-kg payload, and no structural deformation or cracks should be allowed during the delivery service. In practice, few field claims are received during the delivery service of high-voltage transmission electric lines by means of the current wooden reel modules. The design requirement will be still valid for the proposed reel module made from engineering plastic, so that many rigorous development processes should be followed for new design model. First, there is concern regarding whether the reel module has enough structural rigidity to support the maximum payload from the high-voltage transmission line, up to 4,000 kg, and the load distribution should be able to change according to the condition of the electric transmission line wound around the reel module. The equivalent impact force during the collision between the loaded reel module and ground may also be a primary concern during the delivery service. Customers have demanded that the allowable height of a 4,000-kg-loaded reel module be more than 10 cm under free-fall motion.

This main focus of this study was new design model of reel module based on engineering plastic material suitable for carrying out the wounded heavy transmission line instead of current wooden reel module. So the feasibility of proposed reel model was verified for two static loading situations; one is structural rigidity over maximum payload and the other is maximum allowable height of free fall against rigid basement. Since both of design requirements had been applied for current wooden reel module, it can be possible to judge that the proposed reel model has enough structural rigidity to supersede it from current wooden one. The structural analysis focused on fast checking of the structural rigidity of the candidate model by applying a static analysis technique. So, the simulation was conducted for two cases of static conditions; one is structural analysis under a full-payload condition, more than 4,000 kg, and the other is structural analysis under equivalent impact force derived from free-fall motion at more than 10-cm height. The stresses at fragile locations were investigated for two cases to judge whether the structural rigidity of the engineering-plastic reel module is capable of support in each harsh situation. Deformations were also analyzed at clamped locations to determine whether the clamping condition using several bolts is sufficient to assemble all panels together. The direct comparison of reel models, both wood and plastic engineering, could not proceeded here because all quality check of wooden reel module had been done by experimental inspection without consideration of simulation results from finite element model.

2 Static Analysis Condition

The engineering plastic considered in this study was PE, and the mechanical properties were obtained using a sample specimen, as summarized in Table 1. Because all the stress should fall within the elastic zone of the engineering-plastic material, the maximum stress under all possible load conditions may be less than 11.9 MPa, which is 90% of the ultimate strength. In addition, an analytic method using a finite-element model of the reel module cannot consider uncertainties in the manufacturing process and the loading conditions, so the allowable stress was set far less than 11.9 MPa considering design safety factors.

Table 1. Mechanical property of applied engineering-plastic material

Item	Value
Elastic modulus	0.56 GPa
Density	1,020 kg/m^3
Poison's ratio	0.46
Ultimate strength	13.2 MPa

The reel body consisted of one cylinder component and two lateral-disk components, and eight bolts were used to clamp the cylinder component and disks together. All bolts were not considered as specific finite-element model components, but they were modeled as rigid body elements (RBEs) to represent the tight clamping of bolts, the indirect connection between two disk and one cylinder. The configuration of the target reel model and the RBEs used in this model is illustrated in Fig. 1. In addition, two kinds of analytic model of the reel module were considered according to different angles of the attached disks, 0° and 90°, because the disk model had an asymmetric structure with a hole in the circumferential direction. The asymmetric structure of the disk may result in the different stress distributions over the applied force in the loading cases.

The static analysis of the engineering-plastic reel model was conducted for two force-loading conditions according to the design requirement. First, the weight of the high-voltage electric transmission line was considered in this model to verify that the proposed model has the ability to support it under full payload. The direction of the weight force may act on the vertical direction, but the reaction force acting on the reel model can be changed according to winding condition around the cylinder component. Since the maximum payload was determined to be 4,000 kg by the customer, it can be assumed that the maximum vertical force from the weight of the electric transmission line may not exceed 4,000 kgf. However, in certain cases, the weight force can influence two disk components in the horizontal direction. So, a static structural analysis was done for bidirectional force loadings, 4,000 kgf in the vertical direction and 800 kgf for the horizontal direction, which were the most severe loading condition, as illustrated in Figs. 2 and 3. Here, the simulation was done for two cases of disk angles, 0° and 90°.

The second static analysis considered the unexpected impact force when the reel module collides with the ground with a free-fall motion. The heavy electric transmission line rested on the ground, and a rolling motion was carried out frequently by the operator to move the wound transmission line in the horizontal direction. If the wound reel module were dropped in the vertical direction, it would accelerate considerable damage to the reel module structure owing to the weight of the high-voltage transmission line. Therefore, the structural rigidity of the reel module against an unexpected impact force should be checked in the early stage of the design process. The minimum height of the free-fall motion was determined to be 10 cm by the customer, and the candidate model of the reel module should be robust enough to withstand more than the impact force from a 10-cm free fall. Transient analysis of the reel module model may provide the most reliable result for the free-fall motion of the reel module. However, the transient analysis of the free-fall motion required considerable calculation effort as well as proven dynamic material information, so it cannot be recommended for fast checking of structural rigidity over the impact force loading. The equivalent impact force was derived from the height of the free fall, and the calculated equivalent impact force was applied at the disk location where it was expected to face the ground at the time of the collision. In this case, all translational degrees of freedom of the mass center of the reel module were fixed in the vertical direction to provide the equivalent boundary condition of the free-fall motion. The configuration of the boundary condition, both constraints and loading forces, is illustrated in Fig. 4 for two cases of the simulation model, 0° and 90° disk angles, as shown in Fig. 5.

Unlike the weight force from the wound transmission line, the equivalent impact force can be changed over the assigned height (h). The theoretical impact force, $f_{imp}(h)$, can be calculated from the principle of impact and momentum in Eq. (1) from state i to state j, and the final velocity v_j can be derived from the linear energy equation in Eq. (2) [6]. Thus, the equivalent impact force can be expressed as the function of height

782 J. Kim et al.

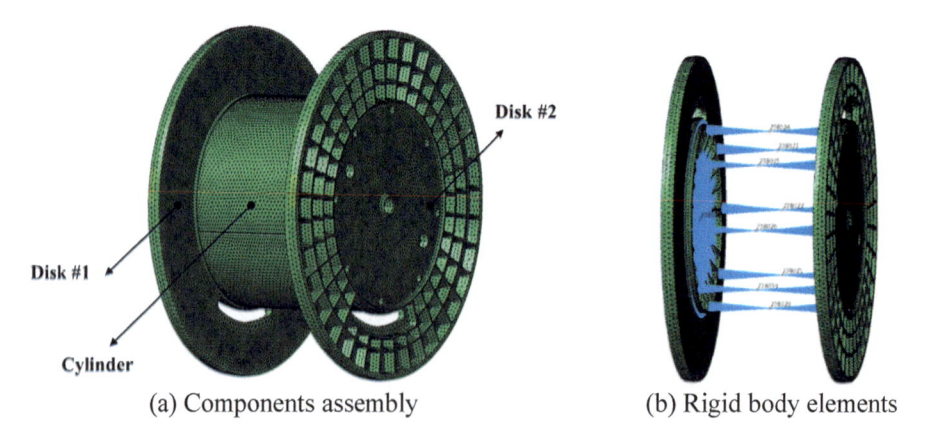

(a) Components assembly

(b) Rigid body elements

Fig. 1. Configuration of reel module with two disk components and one cylinder component

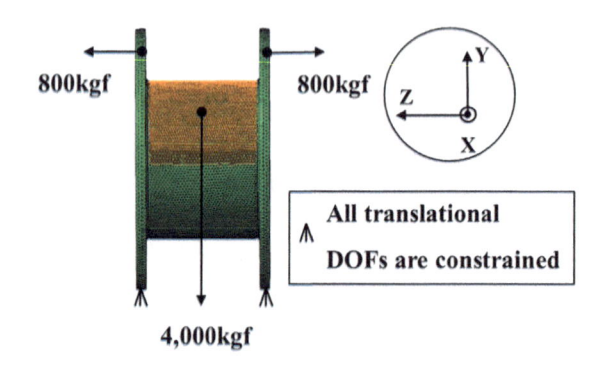

Fig. 2. Boundary condition of weight force analysis at front view: loading weight forces and constraints

(a) Angle of disk component: 0°

(b) Angle of disk component: 90°

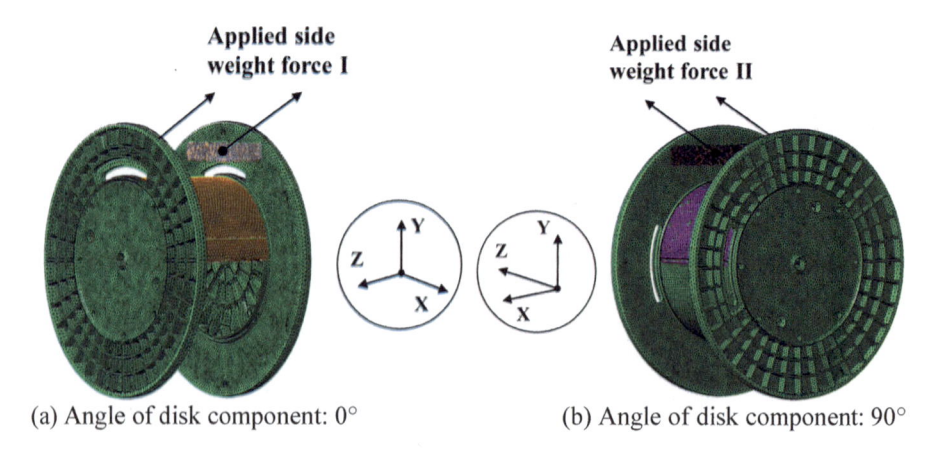

Fig. 3. Two weight force loading cases of reel module model

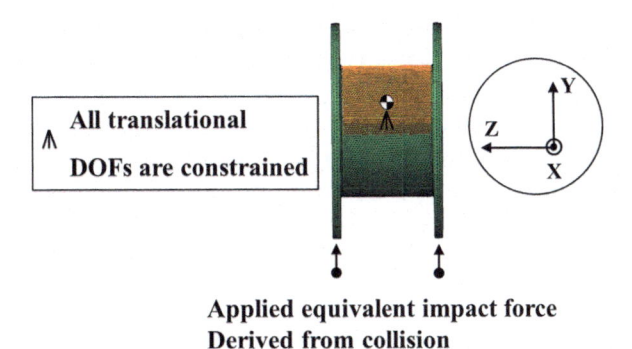

Fig. 4. Boundary condition of equivalent impact force analysis at front view: loading impact forces and constraints

(a) Angle of disk component: 0°

(b) Angle of disk component: 90°

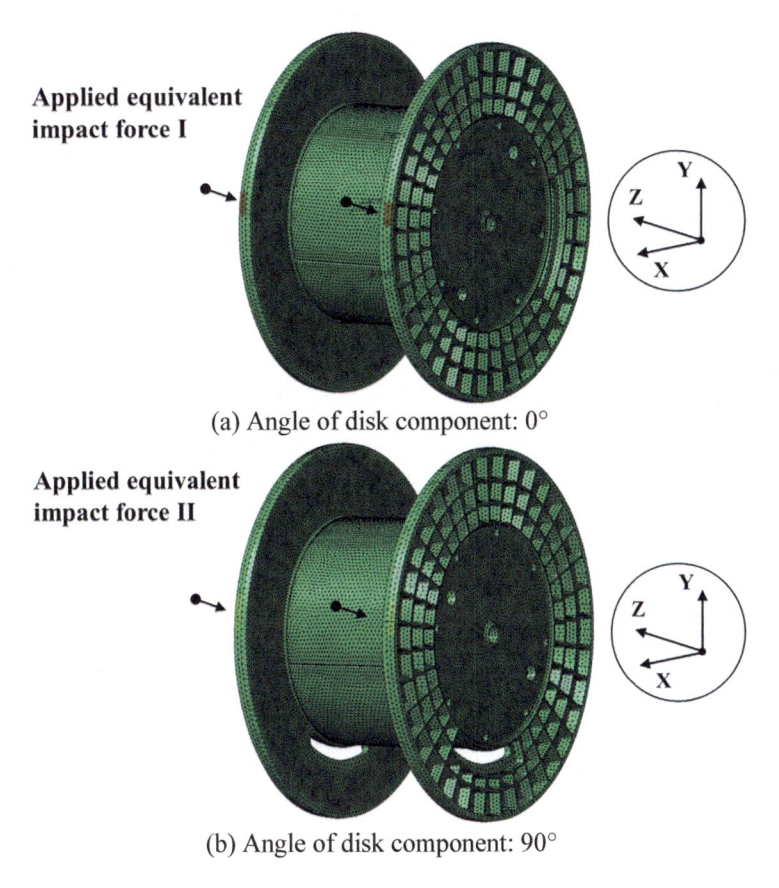

Fig. 5. Two equivalent impact force loading cases of reel module model

h, as shown in Eq. (3), and the time derivative Δt was set at 0.1 s after discussion with the customer. If the maximum payload of the wound electric transmission line was assumed to be 4,000 kg and the allowable height was set for 10 cm, as stated in the design requirement, the equivalent impact force was calculated to be 17,709 N. In this study, the maximum height was increased up to the allowable stress of 11.9 MPa, as discussed in Sect. 2.

$$m v_i + \sum \int_{t_1}^{t_2} (f_{imp}) dt = m v_j \tag{1}$$

$$\frac{1}{2} m (v_j)^2 = mgh \tag{2}$$

$$f_{imp}(h) = \frac{m\sqrt{2 \times 9.8 \times h}}{\Delta t} \tag{3}$$

3 Static Analysis Result

The static analysis was conducted for two cases of loading conditions, weight force from the electric transmission line itself, and equivalent impact force from the free-fall motion. The structural rigidity of each simulation result was determined from the maximum stress value exposed over the scheduled force loadings, and all maximum stresses should not exceed the design criterion of 11.9 MPa. If the maximum concentrated stresses derived from each simulation case shows less than 11.9 (MPa), which means the proposed reel module model has enough structural strength over the static loading case owing to all responding stress range falls within elastic region.

The static analysis of weight forces revealed that the maximum stresses were 9.4 MPa and 5.8 MPa, and the maximum deflections of the disk component facing the cylinder component were 9 mm and 8 mm, respectively, for different weight loading conditions. The maximum stress levels were all less than the stress criterion, 11.9 MPa, so that the candidate reel module model had enough structural rigidity over the maximum payload of the wound electric transmission line. The minimum safety margin, 1.26 (= 11.9/9.4), may not guarantee the proposed model clearly if we consider uncertainties at a field case but we should take into account the severe loading condition explained in Fig. 2. The sum of static weight force, 4,800 kg, is 120% of the design requirement, 4,000 kg, so that simulation results itself compensates uncertainty factors in experiments. The maximum deflection, 9 mm, was also kept at the acceptable level discussed with the customer. The limit of deflection between the disk component and the cylinder component was pre-determined as 12 mm by considering the size of cylinder seat at the two disk components as illustrated in Figs. 6(a) and 7(a). Simulation results are illustrated in Figs. 6 and 7 and summarized in Table 2.

The structural rigidity over two equivalent impact forces was simulated for the possible maximum height against the design criterion of stress concentration value, 11.9 MPa. The equivalent impact force was calculated with Eq. (3), and several loading cases were prepared to find the final simulation result approaches to the stress criterion, 11.9 MPa. For two impact loading cases, the maximum heights of the free-fall motion of the reel modules were 16.5 cm and 15 cm; all results were more than the design criterion, 10 cm. If the maximum height is converted to the impact force for two loading cases, 16.5 cm and 15 cm are equivalent to 22,747 N and 21,689 N, respectively, and both simulation results are all beyond the minimum design criterion, 17,709 N equivalent to 10 cm height. So the proposed reel module model has reliable capability to support the maximum pay-loaded electric transmission line over unexpected impact situation. Simulation results regarding impact force are illustrated in Figs. 8 and 9, respectively, and are also summarized in Table 2.

Both simulation results showed that the engineering-plastic reel module has suitable structural rigidity with respect to the design requirement currently applied for wooden reel modules. Therefore, the field problems regarding current wooden reel modules can be efficiently solved by replacing them with the proposed reel module made from the engineering-plastic material. In addition, loading case I (angle of disk:

(a) Maximum deflection locations

(b) Stress concentration locations

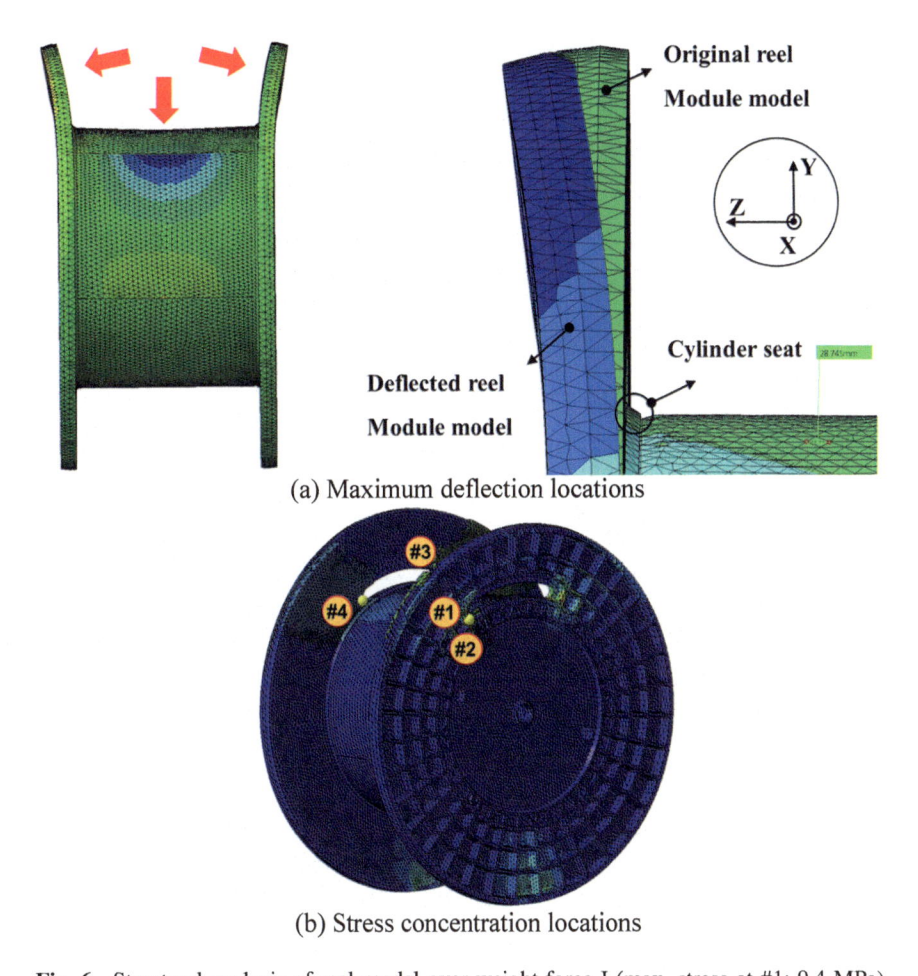

Fig. 6. Structural analysis of reel model over weight force I (max. stress at #1: 9.4 MPa)

(a) Maximum deflection locations

(b) Stress concentration locations

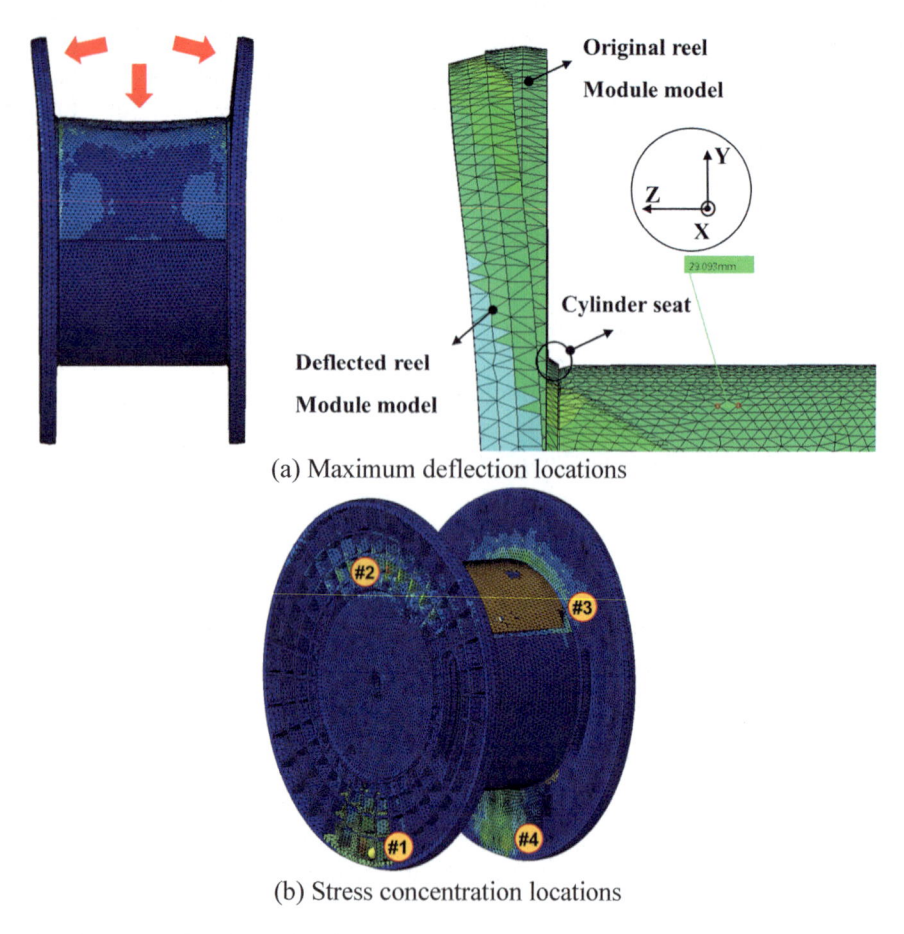

Fig. 7. Structural analysis of reel model over weight force II (max. stress at #1: 5.8 MPa)

(a) Maximum deflection location   (b) Stress concentration locations

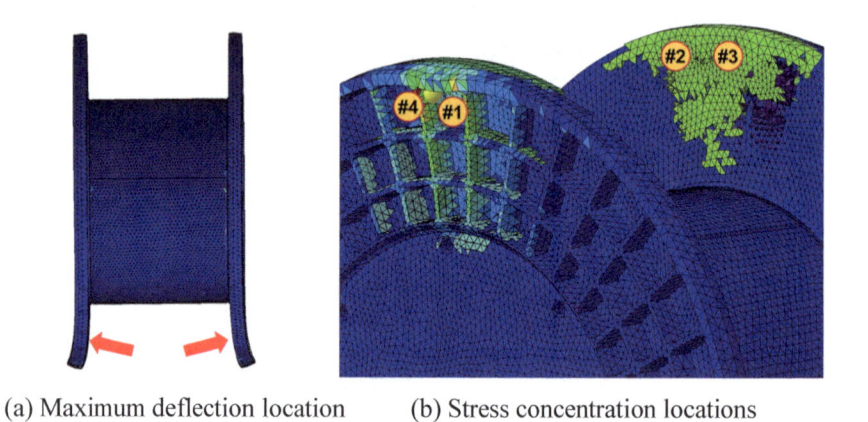

Fig. 8. Structural analysis of reel module model over equivalent impact force I

0°) is a more severe case than case II (angle of disk: 90°), because the position of the hole in the disk component is nearer the fragile area of the disk component at case I than in case II.

(a) Maximum deflection location (b) Stress concentration locations

Fig. 9. Structural analysis of reel module model over equivalent impact force II

Table 2. Summarization of simulation results

Simulation case	Result		
Weight force	Max. stress	Loading case I	9.4 MPa
		Loading case II	5.8 MPa
	Deflection	Loading case I	9 mm
		Loading case II	8 mm
Equivalent impact force	Max. height	Loading case I	16.5 cm
		Loading case II	15 cm

4 Conclusion

The static analysis of an engineering-plastic reel module model was conducted to determine the structural rigidity of the candidate model over the design requirements applied for current wooden reel modules. Both simulation results revealed that the maximum stress and deflection with respect to the harsh force loading cases, both more than 4,000-kg payload and 10-cm free-fall motion, were all within the design criteria, so that the plastic reel model can be used as an alternative to the current wooden reel module. Because the engineering-plastic reel module has several advantages over wooden reel modules, such as a light structure, low manufacturing cost, and excellence in recycling, the replacement of current reel modules may be a reasonable solution for manufacturers as well as customers.

Acknowledgement. This work were sponsored by both NRFK (Grant No. 2017R1D1A1B03034510) and Support Project for Base institute of System Industry (Grant No. P0002331), South Korea.

References

1. Eskandari, M., Najafizadeh, A., Kermanpur, A., Karimi, M.: Potential application of nanocrystalline 301 austenitic strainless steel in lightweight vehicle structures. Mater. Des. **30**, 3869–3872 (2009)
2. Kim, C.J., Kang, Y.J., Lee, B.H., Ahn, H.J.: Sensitivity analysis for reducing critical responses at the axle shaft of a lightweight vehicle. Int. J. Autom. Technol. **13**(3), 451–458 (2012)
3. Mamalis, A.G., Robinson, M., Manolakos, D.E., Demosthenous, G.A., Ioannidis, M.B., Carruthers, J.: Crashworthy capability of composite material structures. Compos. Struct. **37**, 109–134 (1997)
4. Ma, X.C., He, G.Q., He, D.H., Chen, C.S., Hu, Z.F.: Sliding wear behavior of copper-graphite composite material for use in maglev transportation system. Wear **265**, 1087–1092 (2008)
5. Davim, J.P., Reis, P., Antonio, C.C.: Experimental study of drilling glass fiber reinforced plastic (GFRP) manufactured by hand lay-up. Compos. Sci. Technol. **64**, 289–297 (2004)
6. Hibbeler, R.C., Yap, K.B.: Dynamics, 13th edn. Pearson, Singapore (2013)

Improving Fault Tolerant Control to the One Current Sensor Failures for Induction Motor Drives

Cuong Dinh Tran[1,2], Pavel Brandstetter[2], Sang Dang Ho[1,2], Thinh Cong Tran[1,2], Minh Chau Huu Nguyen[2], Huy Xuan Phan[3], and Bach Hoang Dinh[1(✉)]

[1] Faculty of Electrical and Electronics Engineering, Ton Duc Thang University, 19 Nguyen Huu Tho, Dist. 7, Ho Chi Minh City, Vietnam
{trandinhcuong, hodangsang, trancongthinh, dinhhoangbach}@tdtu.edu.vn
[2] Faculty of Electrical Engineering and Computer Science, VSB-Technical University of Ostrava,
17. listopadu 15/2172, 708 33 Ostrava-Poruba, Czech Republic
{pavel.brandstetter,huu.chau.minh.nguyen.st}@vsb.cz
[3] Long An Power Company,
168 Tuyen Tranh, Ward. 4, Tan An City, Long An Province, Vietnam
huypcla@gmail.com

Abstract. This paper presents a solution to deal with a fault of stator current sensors of induction motor (IM) drives by improving the fault tolerant control (FTC) method. The proposed FTC unit including the scheme of three current sensors and their associated observers is used to keep the IMs in the stable operation with any faulty sensor conditions. It consists of three parts: a fault detector, a fault locator, and a stator current reconfiguration. The paper has proposed a new model to estimate stator currents from the differential equations of rotor currents and stator currents for the fault detector. To verify the proposed approach, the simulation in MATLAB/Simulink has been applied and the result has demonstrated the effectiveness of our proposed method.

Keywords: Field Oriented Control (FOC) · Fault-tolerant control (FTC) · Current sensors

1 Introduction

Induction motors are used widely in various industrial applications due to their simplicity, ruggedness, reliability, low cost and volume manufacturing advantages [1]. However, in the past, the induction motors had been almost operated at a fixed speed (or uncontrollable speed modes) due to the complex nonlinear torque-speed characteristics. Nowadays, the combination of a high performance power converter and vector control techniques provides the excellent control capabilities for the induction motor drives. However, during operation, it should be considered the unexpected machinery failures that can deteriorate the performance of the IM drive as classified below [2, 3]:

- The mechanical faults such as the bent shaft, broken bar or cracked end-ring of the rotor.
- The electrical faults such as short circuits or open circuits of stator phase windings.
- The sensor faults such as current sensor, voltage sensor, and speed sensor failures.

In the control aspect, the sensor failure is one of the most common problems in controlled IM drives [4], where the improper of feedback signals leads to the incorrect control actions. Moreover, if the fault situation cannot be detected and isolated quickly, it will cause to breakdown the drive system [5, 6]. Furthermore, the measured stator currents have a very important role to estimate the state variables of the IM drive system, like the rotor flux, the electromagnetic torque, and the rotor speed. In practice, we can use either the schemes of two or three current sensors to provide the feedback current signals for the control model. Although using two sensors is more cost-effective than three sensors, however, the scheme of three current sensors can achieve the better control performance in cases of sensor faults [5]. Thus, there are many control modes of the drive system using the scheme of three current sensors in order to increase the reliability and stability in case of sensor failures. If occurring a single fault while the IM operating, the Fault Tolerant Control (FTC) unit will be activated to implement the fault detection, isolation, and compensation functions. The FTC unit comprises fault diagnosis and fault-tolerant control [6]. The fault diagnosis method for current sensors is based on Kirchhoff's law which checks whether the sum of three measured phase currents equals zero or not [2, 5–8], and then the FTC models will use an estimated current value to replace the faulty physical signal of that failure phase [6, 7].

In this paper, we concentrate on improving the fault tolerance algorithm against a single failure by one current sensor to keep IM drive stably. It uses the scheme of three stator current observers together with their stator current estimators for the fault tolerant control unit. First of all, the fault detection is checked by comparing the stator current values between the estimated model and physical sensors, then the observed errors among three observers are used to determine the failure phase location, and finally a new measurement configuration based on the remaining current sensors is suggested to implement the control task. This paper proposes a new model to estimate stator currents from the differential equations of rotor currents and stator currents in the fault detector. It can eliminate the confusion between the healthy and faulty sensor states when one of phase current sensors is the inaccurate gain.

2 Fault Tolerant Control for Current Sensors

In order to satisfy high performance and maintain stable operation even in faulty current sensor conditions, the IM drive system is controlled by Vector control (VC) method coupled with FTC unit. This section will introduce the mathematical model of the IM and the proposed FTC unit to control the IMs in the abnormal situations of current sensors faulty.

2.1 The Field Oriented Control Technique

The Vector Control, also called Field Oriented Control (FOC), is a popular control technique for IM drive systems, where they can be independently controlled by the torque and the flux variables as the same as the controllers of separated excited DC motors. In FOC, the stator current space vectors are decomposed into two perpendicular elements, i_{Sx} and i_{Sy} in the rotating reference frame [x, y], corresponding to the rotor flux space vector orientation to the x-axis as shown in Fig. 1. In this way, the component i_{Sx} is to maintain the identified amplitude of the flux rotor and the torque is controlled by adjusting the component i_{Sy} in the linear relationship characteristics [9, 10].

The stator current signals in the stationary reference frame [a, b, c] can be transformed into two-phase stationary reference frame [α, β] (Clarke transform) as follows:

$$\begin{aligned} i_{S\alpha} &= \frac{2i_a - i_b - i_c}{3}; \\ i_{S\beta} &= \frac{i_b - i_c}{\sqrt{3}} \end{aligned} \tag{1}$$

Generally, the dynamic model of an induction motor in [α, β] frame can be described as the below Eqs. [1, 3, 7]:

$$\boldsymbol{u}_S^S = R_S \boldsymbol{i}_S^S + \frac{d\boldsymbol{\Psi}_S^S}{dt} \tag{2}$$

$$0 = R_R \boldsymbol{i}_R^S + \frac{d\boldsymbol{\Psi}_R^S}{dt} - j\omega_R \boldsymbol{\Psi}_R^S \tag{3}$$

$$\boldsymbol{\Psi}_S^S = L_S \boldsymbol{i}_S^S + L_m \boldsymbol{i}_R^S \tag{4}$$

$$\boldsymbol{\Psi}_R^S = L_m \boldsymbol{i}_S^S + L_R \boldsymbol{i}_R^S \tag{5}$$

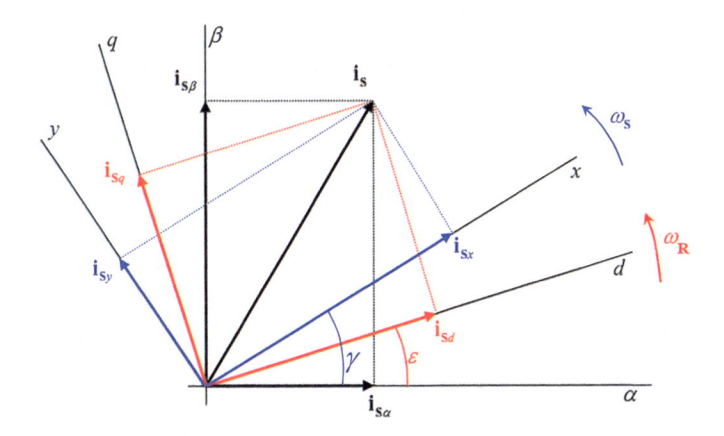

Fig. 1. Vector diagram of the induction motor - principle of vector control.

where R_S and R_R are the rotor and stator phase resistances, L_S and L_R are the rotor and stator inductances, L_m is the magnetizing inductance, Ψ_S^S and Ψ_R^S are the stator and rotor magnetic fluxes, ω_R is the rotor angular speed, and "θ" is the angle of the rotor flux. Using Park transform, we can obtain [i_{Sx}, i_{Sy}] from [$i_{S\alpha}$, $i_{S\beta}$] as follows:

$$i_{Sx} = i_{S\alpha} \cdot \cos\theta + i_{S\beta} \cdot \sin\theta;$$
$$i_{Sy} = -i_{S\alpha} \cdot \sin\theta + i_{S\beta} \cdot \cos\theta \qquad (6)$$

Hence, in normal condition, FOC models need to measure the three phase currents and transform to the two-phase stationary reference frame [α, β] as well as the rotating reference frame [x, y]. However, in practice, there would be some abnormal conditions that current sensors are possibly malfunctioning to make the failure occurrences. Therefore, it is necessary to apply the FTC unit into the control system of IM drives to detect any failure occurrence and reconfigure the control scheme to keep the system stably. The scheme of FOC structure, including FTC unit, used in this paper is presented in Fig. 2.

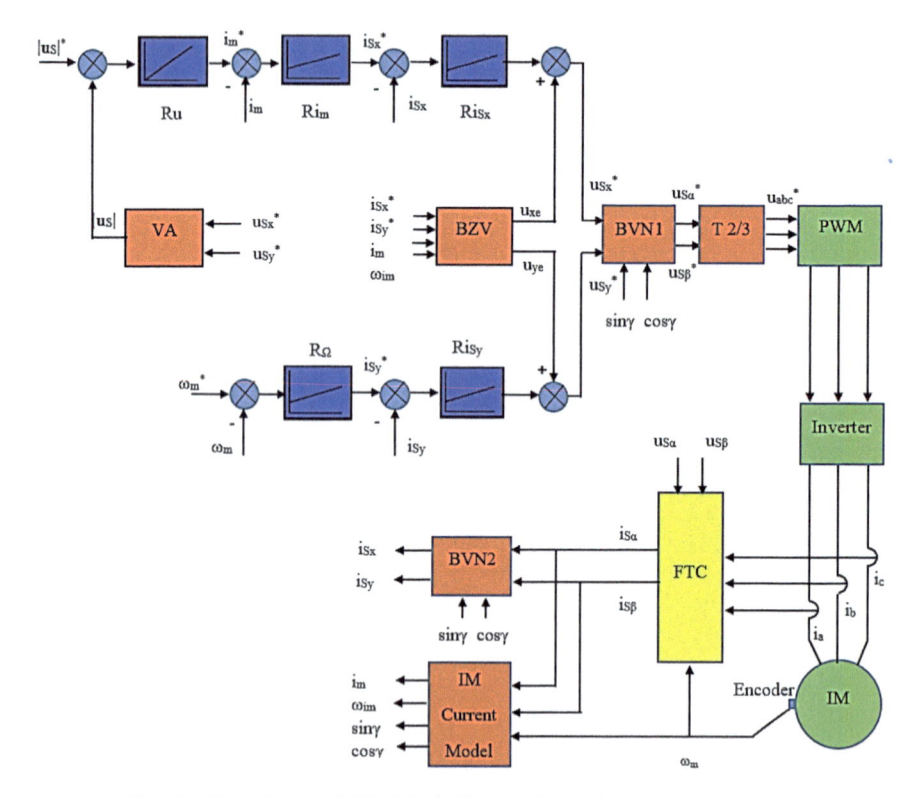

Fig. 2. The scheme of FOC including FTC unit for induction motors

2.2 FTC Unit in the FOC Induction Motor Drive Structure

The inputs of FTC unit comprise the stator voltages $[u_{S\alpha}, u_{S\beta}]$, the stator currents $[i_a, i_b, i_c]$ from current sensors, the rotor angular speed ω_R calculated from a speed encoder ω_m and the number of poles, 2p. The voltages "$u_{S\alpha}, u_{S\beta}$" can be obtained by one of two ways, transforming from three phase voltages "u_a, u_b, u_c" or calculating from PWM signals and DC voltage of the inverter [4]. The outputs, $i_{S\alpha}$ and $i_{S\beta}$, resulted by estimating, calculating and comparing expressions in the block diagram, Fig. 3, for controlling the IM drive.

The FTC algorithm comprises three parts: The fault detection, fault location, and reconfiguration.

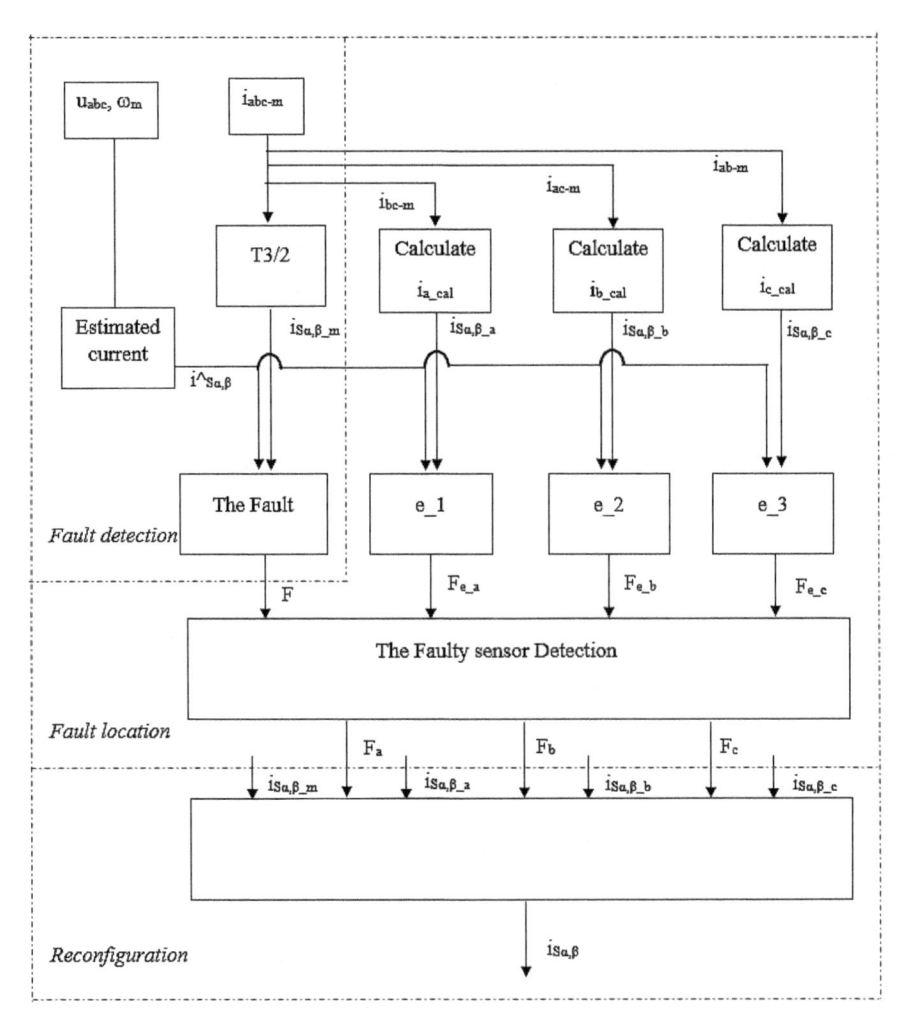

Fig. 3. Fault detection, location and reconfiguration block diagram.

Fault Detection Step: An error condition is detected by comparing the amplitude value between estimated stator currents and measured stator currents. In the stationary reference frame [α, β], the measured stator current [$i_{S\alpha\_m}$, $i_{S\beta\_m}$] are obtained from three current sensors as mentioned in Eq. (1).

As suggested in [9], from the current model of the IM, we can easily estimate the rotor flux from the voltages, the stator currents and the rotor speed. However, in the simulation, it is difficult to obtain the estimated stator currents from the differential equation of the rotor flux and stator currents. Therefore, a new model has been proposed to improve the estimated stator currents from the differential equation of the rotor currents and stator currents as well as using inputs from the voltages and the rotor speed.

From Eqs. (2), (3), (4), (5), we derive the differential equations of the stator and rotor currents in the stationary reference frame [α, β] as

$$\frac{di_{S\alpha}}{dt} = (\frac{L_R}{L_S L_R - L_m^2})[u_\alpha - R_S i_{S\alpha} + \frac{L_m^2}{L_R}\omega_R i_{S\beta} + \frac{L_m R_R}{L_R} i_{R\alpha} + L_m \omega_R i_{R\beta}] \quad (7)$$

$$\frac{di_{S\beta}}{dt} = (\frac{L_R}{L_S L_R - L_m^2})[u_\beta - R_S i_{S\beta} - \frac{L_m^2}{L_R}\omega_R i_{S\alpha} - L_m \omega_R i_{R\alpha} + \frac{L_m R_R}{L_R} i_{R\beta}] \quad (8)$$

$$\frac{di_{R\alpha}}{dt} = (\frac{L_S L_m}{L_m^2 - L_S L_R})[\frac{1}{L_S}u_\alpha - \frac{R_S}{L_S} i_{S\alpha} + \omega_R i_{S\beta} + \frac{R_R}{L_m} i_{R\alpha} + \frac{L_R}{L_m}\omega_R i_{R\beta}] \quad (9)$$

$$\frac{di_{R\beta}}{dt} = (\frac{L_S L_m}{L_m^2 - L_S L_R})[\frac{1}{L_S}u_\beta - \omega_R i_{S\alpha} - \frac{R_S}{L_S} i_{S\beta} - \frac{L_R}{L_m}\omega_R i_{R\alpha} + \frac{R_R}{L_m} i_{R\beta}] \quad (10)$$

Therefore, the estimated stator currents [$i\hat{}_{S\alpha}$, $i\hat{}_{S\beta}$] can be calculated from the voltage and speed rotor signals, and the criterion to determine a current sensor fault can be defined as:

$$IF: ||i^S_{S\_m}| - |\hat{i}^S_S|| \geq threshold$$
$$THEN: F = 1; ELSE: F = 0. \quad (11)$$

The reliability of the fault detection algorithm depends on the accuracy of inductive motor known parameters, especially, the value of stator resistance R_S. In practice, these parameters are changed during the IM's operation due to the temperature, magnetic saturation, etc. As a result, the mismatch of the algorithm can occur at the low-speed range operation.

In [9], it can be improved by combining a suitable threshold with other observers of the fault locator. By implementing various experiments about the influence of R_S into the estimated stator currents, a suitable threshold of 0.1 (based the nominal stator current) is proposed. Based on the Kirchhoff's law, the phase current observers can be calculated as follows in [10]

$$Observer_a : i_{a\_cal} = -i_b - i_c$$
$$Observer_b : i_{b\_cal} = -i_c - i_a \qquad (12)$$
$$Observer_c : i_{c\_cal} = -i_a - i_b$$

where [i_{a_cal}, i_{b_cal}, i_{c_cal}], [i_a, i_b, i_c] are the estimated currents and sensor currents, respectively.

In the stationary reference frame [α, β], the measured stator currents [$i_{S\alpha\_i}$, $i_{S\beta\_i}$] of each observer can be obtained from (i_a, i_b, i_c) using Eq. (1). As a result, the currents [$i_{S\alpha\_i}$, $i_{S\beta\_i}$] of each observer are independent from the respective current sensors. Thus, the currents [$i_{S\alpha\_i}$, $i_{S\beta\_i}$] of each observer will be compared with the estimated stator current [$\hat{i}_{S\alpha}$, $\hat{i}_{S\beta}$] by

$$e\_i = \left| |i_{S\alpha\_i} \times \hat{i}_{S\beta}| - |i_{S\beta\_i} \times \hat{i}_{S\alpha}| \right| \qquad (13)$$

When occurring the failure at a specific current sensor, it makes the error values increasing in two other observers. Thus, in this step, the faulty sensor will be located as the minimum value among the three errors by the proposed algorithm below

$$IF : (F_{e\_i} = \min\{e\_i\}), THEN : (F_{e\_i} = 1), ELSE : (F_{e\_i} = 0);$$
$$IF : (F = 1) \& (F_{e\_i} = 1), THEN : (F_i = 1), ELSE : (F_i = 0). \qquad (14)$$

After the faulty current sensor is determined, the FTC unit will replace the measured stator currents [$i_{S\alpha\_m}$, $i_{S\beta\_m}$] by the stator currents [$i_{S\alpha\_i}$, $i_{S\beta\_i}$] of the relative observers. This FTC unit will keep the IM drive to stably operate in the faulty conditions.

3 Simulation Results

This part presents the simulation of the proposed FTC unit in MATLAB/SIMULINK when there is a single current-sensor fault. The simulation parameters of IMs are listed as:

P_n = 2.2 kW, T_n = 14.8 Nm, ω_n = 1420 rpm, p = 2.
I_{Sn} = 4.85 A, I_{Sxn} = 3.96 A, I_{Syn} = 5.6 A.
U_{Sn} = 230/400 V, Ψ_{Sn} = 0.757 Wb.
R_S = 3.179 Ω, R_R = 2.118 Ω.
L_S = 0.209 H, L_R = 0.209 H, L_m = 0.192 H, T_R = 0.0987 s.

In this simulation, the induction motor has been operated at the normal reference speed areas where the FOC technique has applied in the control scheme with three current sensors. A constant load torque of 3 (N.m) has been applied during the time of 0.2 s. Due to the symmetric three phases, if any failure of a current sensor, i.e. phase A, occurs, the effect of this faulty sensor will be similar to that of other phases. Figure 4 shows the characteristics of the speed controller where the three phase stator currents have been fully measured by healthy sensors. Figure 5 depicts the performance of the controller with a current sensor fault at the time of 1.0 s without FTC.

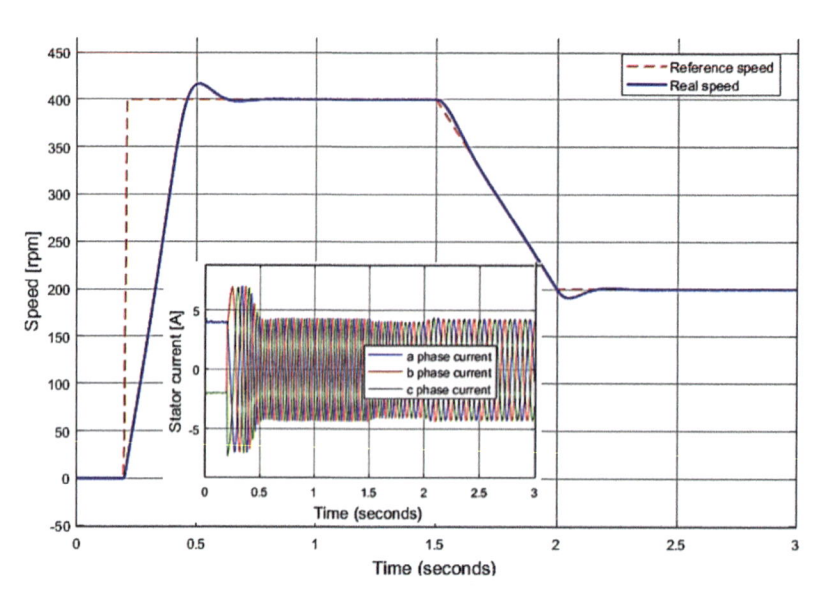

Fig. 4. FOC technique, reference speed ω_{m_ref} = 400 rpm, reference speed, real speed, and corresponding three phase currents in the healthy sensors – condition.

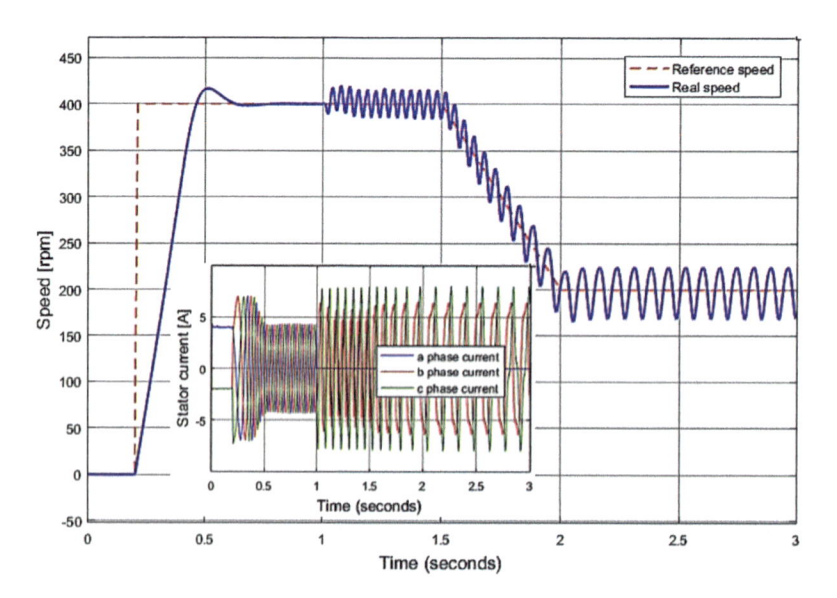

Fig. 5. FOC technique, reference speed ω_{m_ref} = 400 rpm, reference speed, real speed, and corresponding three phase currents with a phase current sensor fault.

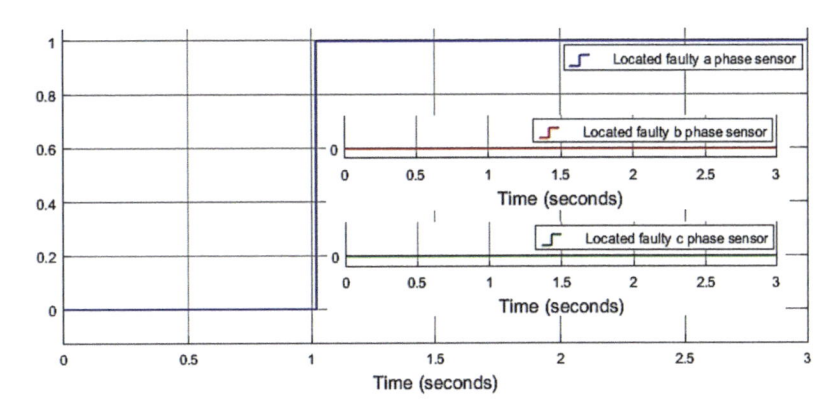

Fig. 6. Analysing fault current sensor signals

Figure 6 presents the functions of how to detect a sensor fault and locate exactly the phase failed of the FTC unit. The proposed model uses the minimum error among the three comparators to locate the faulty phase. In this case, the current sensor of phase A is the faulty device. Then, Fig. 7 described the performance of the proposed FTC unit in this faulty situation, i.e. one current sensor failed. As shown in the figure, the actual speed characteristics of the proposed FTC unit has smoothly kept the reference speed with a small deviation. The performance of the FTC unit is similar to that of the IM's controller with fully good condition sensors as mentioned in Fig. 4 and must be better

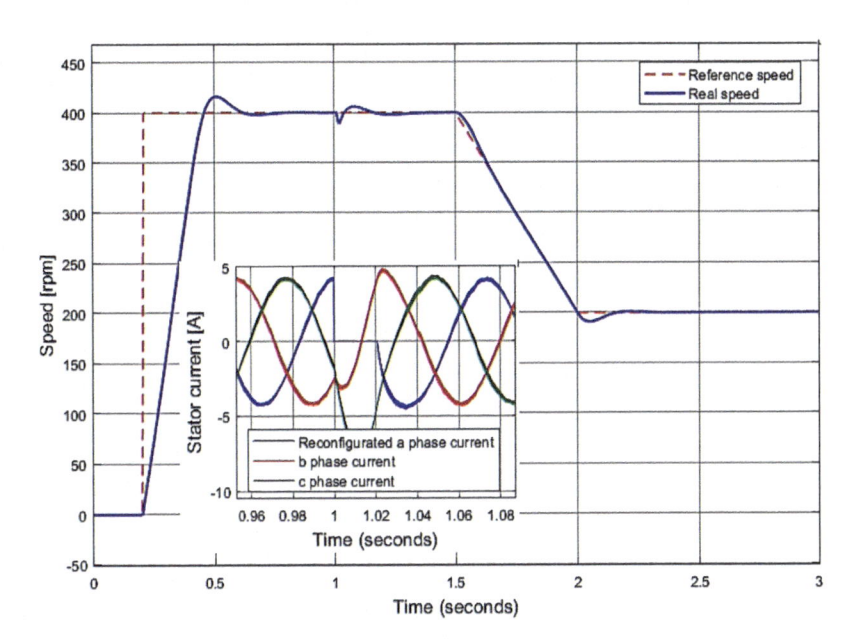

Fig. 7. FOC technique, reference speed $\omega_{m\_ref} = 400$ rpm, reference speed, real speed, and corresponding three phase currents with FTC technique when "a phase" current sensor is faulty.

than that of the IM's controller without FTC presented in Fig. 5. Therefore, we can see the effectiveness of the FTC unit to deal with the faulty condition of current sensors of IM's drives. If occurring any problem in current sensors, the rotor speed is disturbed in just a short time, but then quickly become stable again.

4 Conclusion

This paper has proposed a modified fault-tolerant control unit to deal with the abnormal condition of one phase current failure in IM's drives. This method includes three separate steps: detection, location, and reconfiguration, where a new model to estimate the stator currents from the differential equation of the rotor currents and the stator currents has been applied to detect the location of the faulty phase sensor. This new model is based on modifying the mathematical equations of the IM and the result is that it has high accuracy in estimating the stator currents. The simulation results have demonstrated the effectiveness of the proposed algorithm to maintain both speed and torque of IM drive systems in the case of occurring a single phase sensor fault.

Acknowledgement. The paper was supported by the Project reg. no. SP2018/162 - Student Grant Competition of VSB-Technical University of Ostrava, Research and development of advanced control methods of electrical controlled drives, member of research team, 2018.

References

1. Najafabadi, T.A., Salmasi, F.R., Jabehdar-Maralani, P.: Detection and isolation of speed-, DC-link voltage-, and current-sensor faults based on an adaptive observer in induction-motor drives. IEEE Trans. Ind. Electron. **58**(5), 1662–1672 (2011)
2. Lebaroud, A., Clerc, G.: Classification of induction machine faults by optimal time–frequency representations. IEEE Trans. Ind. Electron. **55**(12), 4290–4298 (2008)
3. Klimkowski, K., Dybkowski, M.: A fault tolerant control structure for an induction motor drive system. Automatika **57**(3), 638–647 (2016)
4. Yu, Y., Zhao, Y., Wang, B., Huang, X., Xu, D.: Current sensor fault diagnosis and tolerant control for VSI-based induction motor drives. IEEE Trans. Power Electron. **33**(5), 4238–4248 (2018)
5. Romero, M.E., Seron, M.M., De Dona, J.A.: Sensor fault-tolerant vector control of induction motors. IET Control Theory Appl. **4**(9), 1707–1724 (2010)
6. Rothenhagen, K., Fuchs, F.W.: Current sensor fault detection, isolation, and reconfiguration for doubly fed induction generators. IEEE Trans. Ind. Electron. **56**(10), 4239–4245 (2009)
7. Brandstetter, P.: Electrical Drive III. Ostrava Univ (2014)
8. Gaeid, K.S., Ping, H.W.: Fault tolerant control of induction motor. Modern Appl. Sci. **5**(4), 83 (2011)
9. Brandstetter, P., Kuchar, M.: Rotor flux estimation using voltage model of induction motor. In: 2015 16th International Scientific Conference on Electric Power Engineering (EPE), pp. 246–250. IEEE (2015)
10. Vasic, V., Vukosavic, S.N., Levi, E.: A stator resistance estimation scheme for speed sensorless rotor flux oriented induction motor drives. IEEE Trans. Energy Convers. **18**(4), 476–483 (2003)

Impact of Parameter Variation on Sensorless Indirect Field Oriented Control of Induction Machine

Andrej Kacenka[(✉)], Pavol Makys, and Lubos Struharnansky

Faculty of Electrical Engineering, University of Zilina, Žilina, Slovakia
andrej.kacenka@gmail.com,
{pavol.makys,struharnansky}@fel.uniza.sk

Abstract. A method is taken to the control of induction motor based electric drives without the aid of shaft mounted speed or position sensors. This method is called sensorless control. This paper describes a set of three observers used to estimate rotor speed and rotor flux angle which is needed to perform the Park transformation in Indirect Field Oriented Control (IFOC). The validity of the proposed method has been verified by simulation in Matlab and experimentation with frequency inverter controlled by digital signal controller (DSC). The accuracy of the speed estimate relies on the parameters of the machine, such as rotor time constant and stator resistance. These parameters variation causes an estimation error of the motor speed. This error could result in the decrease of control performance. This paper describes the effect of the motor parameters variation on angular velocity estimation.

Keywords: Induction machine · Sensorless indirect field oriented control · Pseudo sliding-mode observer

1 Introduction

Induction machines (IM) are widely used in industry and are adequate for almost any kind of environment. IM provide a wide speed range, high reliability, mechanically robustness, high efficiency and low price. However, induction machines have complex mathematical model, non-linear behaviour due to saturation and parameters variation which depends on the physical influence of temperature. For those reasons is the control of induction machines much more complex.

Since 1990 technology development in power semiconductors, microprocessors and logical programmable devices has made effective controls with adjustable speed feasible [1].

In general, two schemes have been proposed to control induction machines. The first scheme is called Scalar Control (SC). The constant Volt per Hertz ratio represents the scalar control. SC is relatively simple method mainly used in HVAC applications. The second scheme is called Field Oriented Control (FOC). The development of the FOC, induction machine drives became able to provide the same performance as the direct current machine drives. The FOC technique brings overall improvements in drive performance when compared to the scalar control (full torque control from zero to nominal speed, decoupled control of flux and torque, improved dynamics). With this technique, it is possible to uncouple the stator current components. Uncoupling establishes two independent and single controlled currents. FOC principle is to separate out the component of the stator currents responsible for producing torque and component responsible for producing magnetic flux. The torque component of stator current isq is similar to armature current of DC motor and the flux component isd is similar to field current of DC motor. Using these currents, the flux and torque can be independently controlled. In addition, a 90° electric angle is ensured between the uncoupled control currents. As a result, the model of the induction motor loses its complexity, and high drive performance can be realized [2]. The field oriented control ensures good and robust control in case of transients. The principle of field oriented control works with rotating vectors (or phasors) in a complex coordinate system. The magnitude and the phase of the controlled current change. FOC can be carried out by system and coordinate transformations of the basic equations of the motor. After applying the transformations, the alternating and sinusoidal quantities become non-alternating quantities. Due to uncoupling, the currents can be controlled, and then, after back-transformation it is possible to modify the output of the inverter with three-phase quantities. Subsequently, the magnitude and the phase of supplied voltage or current can be modified.

The most common topology for AC drives is based on a Voltage-Source Inverter (VSI) with a rectifier and a DC-link. Generally, the FOC for the induction motor drives can be divided into two categories: indirect and direct schemes. The field to be oriented can be stator, rotor or air-gap flux linkage. In the indirect field oriented control (IFOC) the slip velocity needs to be calculated and rotor speed can be either measured by sensor attached to the rotor or estimated to get synchronous speed. In IFOC scheme rotor flux estimation is not appearing. For the Direct Field Oriented Control (DFOC), the synchronous speed is calculated based on the flux angle, which is obtained from flux estimator or flux sensor.

If in IFOC estimators and/or observers are used to get rotor speed instead of speed sensor attached to the rotor, it can be spoken about sensorless indirect field oriented control. In recent years, a large number of sensorless field oriented control schemes for induction machine have been proposed. Where efficiency, low cost and control of the induction machine drive is a concern, the sensorless IFOC provides the best solution. The term sensorless does not represent the lack of sensors, but it means that speed or position sensor is missing. In IFOC, the speed or position are estimated using other parameters such as phase voltages, currents and IM equivalent circuit parameters. The needlessness of speed sensor decreases both cost and size of induction machine drive. Moreover, this technique is being deployed in hostile environments such as high temperature, corrosive contact, etc. During the last few decades, the sensorless controlled electrical drives has undergone rapid expansion due to the benefits of digital signal controllers. DSC become available allowing to obtain rotor speed by means of observers and estimators integrated with motor control.

In this paper is presented the sensorless Indirect Field Oriented Control of real induction machine. This control method is verified by simulation in Simulink and experimentation with three-phase frequency inverter controlled by DSC TMS320f28062. The effect of motor parameters variation on rotor speed is taken into account.

2 Mathematical Modelling of IM

The mathematical model of the induction motor is used with the consideration of an electromagnetic process in the simulation and experimentations. This mathematical model represents a multi—parameters, non—linear system with the following conclusions, for simplification of the mathematical description [3]:

- The induction machine is symmetrical.
- The air gap is considered to be constant.
- Iron losses can be neglected.
- Resistances and inductances are the same at all phases.

One of the motivations was to eliminate the on-line computation of time varying transformation matrices, as required in conventional field oriented control methods, and for this reason, the basis for the control system development is the following induction motor model, formulated in the stator-fixed (α, β) reference frame. The angular velocity, rotor flux and current differential equations may be expressed as [4]:

$$\dot{\omega}_r = \frac{1}{J}(\Gamma - \Gamma_L) = \frac{1}{J}\left(c_5 \Psi^T T^T I - \Gamma_L\right) \quad (1)$$

$$\dot{\Psi} = -P(\omega_r)\Psi + c_4 I \quad (2)$$

$$\dot{I} = c_1[c_2 P(\omega_r)\Psi - a_1 I + U] \quad (3)$$

where $\Psi^T = [\Psi_\alpha \Psi_\beta]$ is the rotor magnetic flux, $I^T = [i_\alpha i_\beta]$ is the stator current, $U^T = [u_\alpha u_\beta]$ is the stator voltage, Γ is the torque developed by the motor, ω_r is the mechanical rotor speed, $c_i, i = 1, 2, 3, 4, 5$ and a_1 are constants, given by: $c_1 = L_r/(L_s L_r - L_m^2)$, $c_2 = L_m/L_r$, $c_3 = R_r/T_r = 1/T_r$, $c_4 = L_m/T_r$, $c_5 = 3pL_m/2L_r$, $a_1 = R_s + (L_m^2 + L_r^2)R_r$, where L_s, L_r and L_m are, respectively, the stator and rotor resistances and p is the number of stator pole pairs [4]. Also,

$$\mathbf{P}(\omega_r) = \begin{bmatrix} c_3 & p\omega_r \\ -p\omega_r & c_3 \end{bmatrix} \quad (4)$$

3 State Estimation and Filtering

The rotor magnetic flux, rotor speed and load torque are produced by the following set of three observers. The first is rotor magnetic flux estimator. The second is a stator current vector pseudo sliding—mode observer formulated for generation of an unfiltered estimate of the rotor speed. The third observer provides filtered rotor speed and load torque estimates, a direct measurement of load torque being assumed to be unavailable [4].

3.1 The Rotor Magnetic Flux Estimator

For induction motors a means of estimating the rotor magnetic flux components may be devised by eliminating the rotor speed, ω_r, between Eqs. (2) and (3), yielding Eq. (5) [4].

$$\begin{bmatrix} \Psi_\alpha^* \\ \Psi_\beta^* \end{bmatrix} = \int \left[\left(c_4 - \frac{a_1}{c_2} \right) \begin{bmatrix} I_\alpha \\ I_\beta \end{bmatrix} + \left(\frac{1}{c_2} \right) \begin{bmatrix} U_\alpha \\ U_\beta \end{bmatrix} \right] dt - \left(\frac{1}{c_1 c_2} \right) \begin{bmatrix} I_\alpha \\ I_\beta \end{bmatrix} \quad (5)$$

With zero initial conditions, all the quantities on the right side of Eq. (5) are known, but the pure integration would be subject to drift in practice. This problem is overcomed here by noting that $\int_0^\infty \Psi(t)dt = 0$. Accordingly, if $\|\Psi\|$ exceeds $(1 + \lambda)\|\Psi\|_d$ where $0 < \lambda < 1$, the drift is prevented by replacing the integral of Eq. (5) by first order filter with a time constant, T_q such that $T_q \gg 1/\omega_{rmin}$ is the lowest angular velocity envisaged for the particular application [4].

3.2 The Pseudo Sliding Mode Observer and Angular Velocity Extractor

A stator current pseudo—sliding—mode observer is formulated for generation of an unfiltered estimate, $c_1 c_2 P(\widehat{\omega_r}) \Psi$, of the term, $c_1 c_2 P(\omega_r) \Psi$. The observer is therefore formed as a stator current real time model purposely omitted terms containing ω_r. Thus:

$$\dot{\mathbf{I}}^* = c_1[-a_1 \mathbf{I}^* + \mathbf{U}] - \mathbf{v} \quad (6)$$

$$\mathbf{v} = -v_{max} \cdot \text{sgn}[\mathbf{I}^* - \mathbf{I}] \quad (7)$$

Where $\mathbf{v}^T = [v_\alpha\ v_\beta]$ are the model corrections, i_α^* and i_β^* are estimates of i_α and i_β, as in conventional observers. The useful observer output here, however, are the continuous equivalent values $\mathbf{v}_{eq}$, of the rapidly switching $\mathbf{v}$. But Eq. (7) cannot directly generate the equivalent values. Instead, a pseudo—sliding—mode observer may be formed by replacing the signum function by high gain.

$$\mathbf{v}_{eq} = K_{SM}[\mathbf{I}^* - \mathbf{I}] \quad (8)$$

Where K_{SM} is a high gain so that $\mathbf{v}$ is continuous and closely approaches $\mathbf{v}_{eq}$ for sufficiently high gain, limited only by the non-zero iteration interval, h, of the digital implementation $K_{SM} < (2 - c_1 a_1 h)/h$ with Euler explicit numerical integration. The resulting approximation to $\mathbf{v}_{eq}$ is denoted $\mathbf{v}_{eq}^*$. Assuming that observer operates in the ideal sliding mode, $\mathbf{v} = \mathbf{v}_{eq}$ if $\mathbf{I}^* = \mathbf{I}$ and $\dot{\mathbf{I}}^* = \dot{\mathbf{I}}$. Then the right sides of Eqs. (3) and (6) may be equated, yielding (with $\mathbf{v}_{eq}$ replaced by $\mathbf{v}$) [4]:

$$c_1[-a_1 \mathbf{I}^* + \mathbf{U}] - \mathbf{v}_{eq}^* = c_1[c_2 P(\omega_r) \Psi - a_1 \mathbf{I} + \mathbf{U}] \quad (9)$$

Replacing Ψ and ω_r in Eq. (9) by their estimates Ψ^* and ω_r^* yields:

$$v_{eq}^* = -c_1 c_2 P(\omega_r^*) \Psi^* \tag{10}$$

The following formula for the desired angular velocity estimate ω_r^* is then derived from the components, v_{eq}^* of Eq. (10):

$$\omega_r^* = \frac{-v_{eq\,\alpha}^* \Psi_\beta^* + v_{eq\,\beta}^* \Psi_\alpha^*}{c_1 c_2 p \|\Psi^*\|} \tag{11}$$

3.3 Observer for Load Torque Estimation and Rotor Speed Estimate Filtering

The real time model of this observer is based on motor torque equation. The load torque is treated as a state variable whose differential equation augments the real time model. In this case, the load torque is assumed constant in the formulation of the real time model and so its state differential equation is simply $\dot{\Gamma}_L = 0$. The observer correction loop is actuated by the error between the rotor speed estimate, ω_r^*, and the filtered estimate, $\hat{\omega}_r$, from the real time model. Since, $\hat{\omega}_r$ is a filtered version of ω_r^*, it is used directly in the control loop instead of ω_r^*. The continuous time version of this observer is therefore [4]:

$$e_\omega = \omega_r^* - \hat{\omega}_r \tag{11}$$

$$\frac{d\hat{\omega}_r}{dt} = \frac{1}{J}\left[c_5(\Psi_\alpha i_\beta - \Psi_\beta i_\alpha) - \widehat{\Gamma}_L\right] + k_\omega e_\omega \tag{12}$$

$$-\frac{d\widehat{\Gamma}_L}{dt} = k_\Gamma e_\omega \tag{13}$$

The observer poles are both placed at $s = -1/T_f$ so that the filtering time constant, T_f, is a single design parameter for the gains, k_ω and k_Γ as it is shown in (15), where the right side of equation is the characteristic equation of the filtering observer characteristics polynomial [4]:

$$s^2 + \frac{2}{T_f}s + \frac{1}{T_f^2} = s^2 + k_\omega s + \frac{k_\Gamma}{J} \tag{15}$$

$$k_\omega = \frac{2}{T_f} \quad k_\Gamma = \frac{J}{T_f^2} \tag{16}$$

4 Simulation

Simulation has been realized using software product Matlab-Simulink to verify a validity of proposed sensorless IFOC method. Control structure is shown in Fig. 1. Set of three above mentioned observers has been used to get rotor velocity estimate.

Table 1. Parameters of the investigated IM

Sign	Definition	Value	Unit
P_N	Rated power	1.5	kW
U_N	Rated voltage	400	V
I_N	Rated current	3.4	A
J	Rotor moment of inertia	0.0035	$kg.m^{-2}$
R_S	Stator winding resistance	5.155	Ω
R_r	Rotor winding resistance	4.426	Ω
L_S	Stator winding inductance	0.291	H
L_r	Rotor winding inductance	0.291	H

This velocity estimate has been used as a feedback signal in speed control loop. The coefficients of PI controller have been tuned by trial and error method [5].

In simulation we use the induction machine parameters as shown in Table 1. The demanded magnetic flux is set 1.5 Wb and demanded angular velocity is set to 157 rad/s. The simulation results are shown in Fig. 2, 3 and 4.

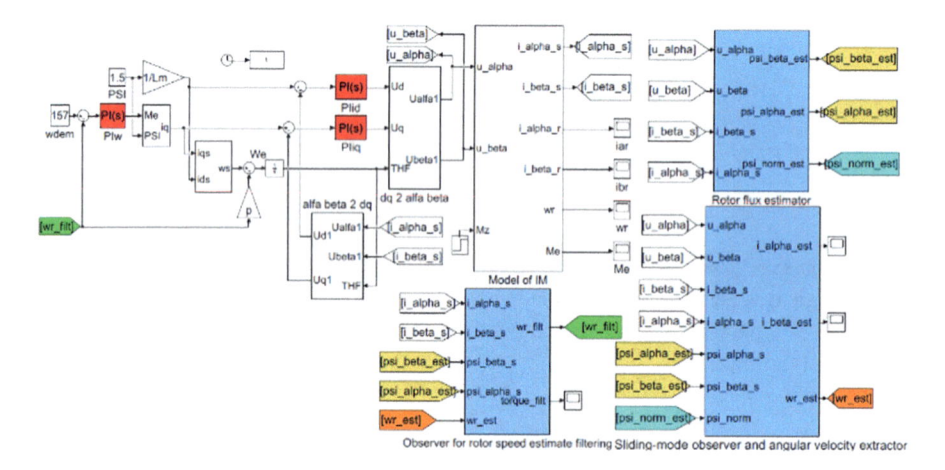

Fig. 1. A simulation model.

In Fig. 2, can be seen comparison of both rotor angular velocity and the rotor angular velocity estimate. The start up of IM has been simulated under no load conditions and since time 0.5 s load torque has been kept constant at rated value 10.1 Nm. In Fig. 2 can be also seen the difference between rotor angular velocity and rotor angular velocity estimate at steady state. It is approximately 1.9 rad/s at both no-load conditions and at rated load torque.

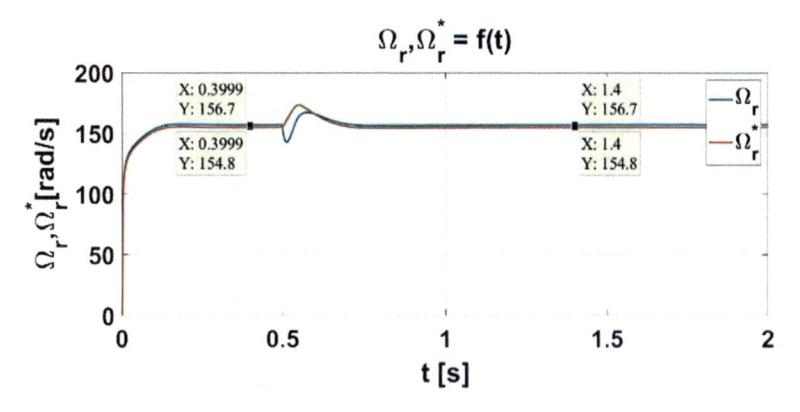

Impact of Parameter Variation on Sensorless Indirect Field Oriented Control of IM   805

Fig. 2. Simulated rotor speed and rotor speed estimate

Figure 3 shows rotor flux expressed as a square of rotor flux estimates magnitudes.

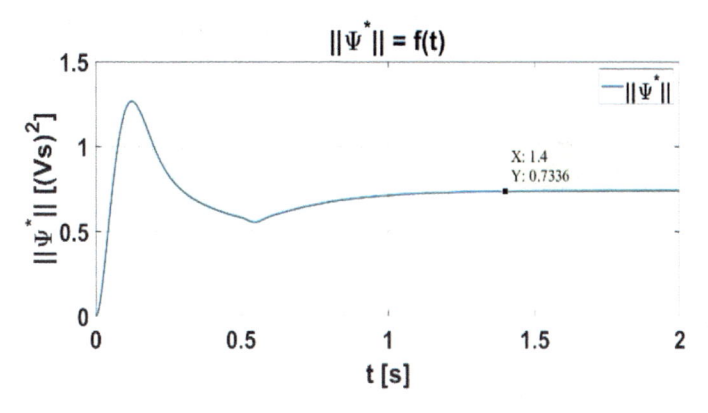

Fig. 3. Simulated rotor flux norm

In the Fig. 4, there are shown motor currents for start up and steady state. Current i_α is α-component of stator current and current i_α^* is its estimated value.

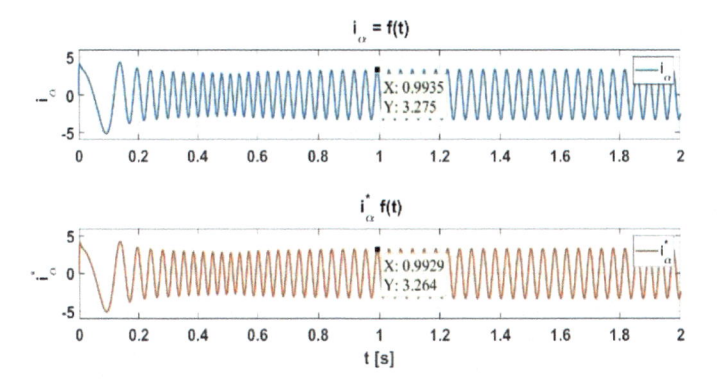

Fig. 4. Simulated α-component current and its estimate

5 Experimental Results

To verify the proposed method and simulation results an experimental stand with IM coupled with dynamometer as a load has been used. The experiments were performed with 1.5 kW induction machine powered by three-phase frequency inverter as shown in Fig. 5. Table 1 gives the machine parameters. The control law was implemented via Digital Signal Controller TMS320f28062 by Texas Instruments. The switching frequency was set as 10 kHz. The coefficients of PI controller have been tuned by trial and error method and by Ziegler-Nichols method.

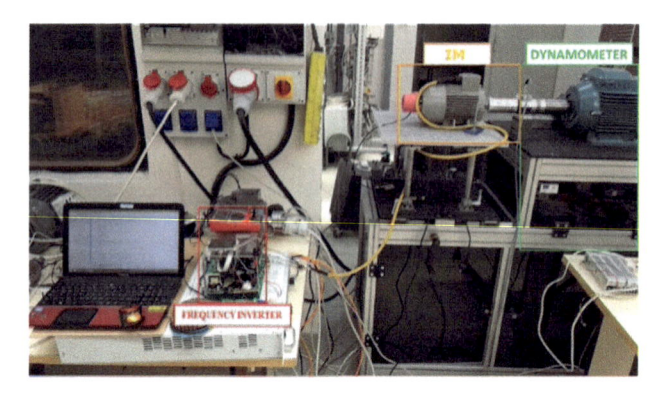

Fig. 5. Experimental stand equipped with IM and dynamometer

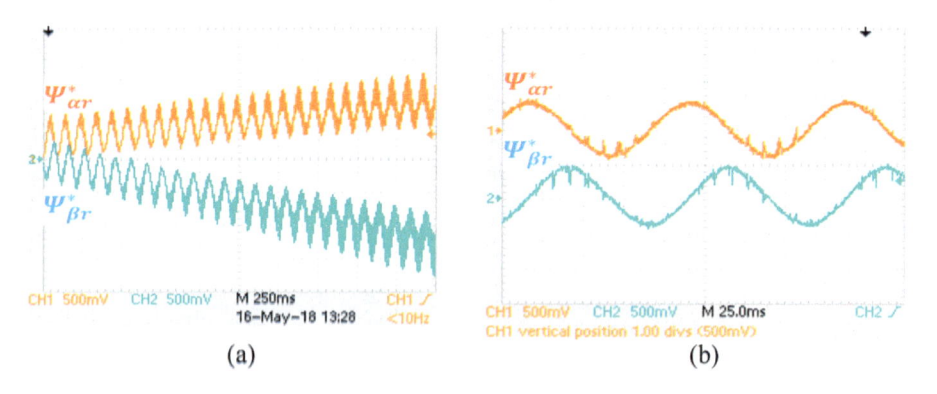

Fig. 6. The estimated rotor flux components before filtering (a) and after filtering (b)

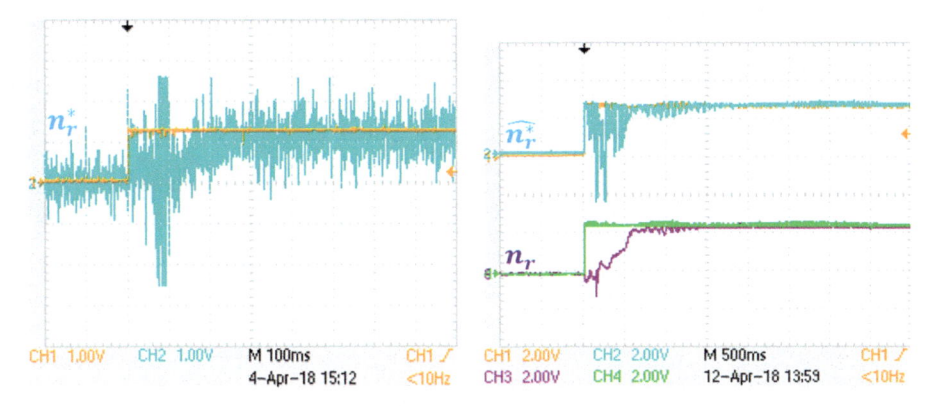

Fig. 7. The unfiltered rotor speed estimate

Fig. 8. The filtered rotor speed estimate and actual rotor speed

In the Fig. 6, there are shown estimated rotor flux components both before filtering (a) and after filtering (b). The pure integration of the Eq. (5) is a subject to drift in practice as shown in (a). To eliminate this drift the first order filter with time constant $t_q = 0.15$ s has been applied.

In the Fig. 7 is shown step response of unfiltered rotor speed estimate at start up, where demanded speed is 500 rpm. In the Fig. 8 is shown step response of filtered rotor speed estimate (blue colour) and actual rotor speed measured by encoder (purple colour) at start up, where demanded speed is 500 rpm.

6 The Effect of Machine Parameters Error on Rotor Speed Estimation

In Indirect Field Oriented Control and mainly in sensorless IFOC which utilizes a set of observers presented in this paper to get rotor speed estimate, equivalent circuit parameters need to be known.

These parameters are not constant and may differ in a wide range from operating point to another operating point i.e. these parameters are affected by temperature, saturation, frequency and skin effect [6]. The machine parameters used in sensorless IFOC includes the stator inductance, rotor inductance, stator resistance and rotor resistance, L_s, L_r, R_s, R_r, respectively.

Especially, a winding temperature can vary a lot during operation, resulting winding resistances mismatch. This mismatch influence on rotor speed estimation has been analysed in this paper. To emulate resistances mismatch, the incorrect parameters in digital signal controller have been set. One of the most important tasks of DSC is to calculate rotor speed both from the mathematical model of IM and from mathematical model of observers in a real time. Described models utilize winding resistance as a parameter and are highly sensitive to its error [7]. If the assumed values of resistances are not consistent with real ones, rotor speed is not correctly estimated.

6.1 The Effect of Rotor Resistance Error on Speed Estimation

First of all, the effect of rotor resistance error on rotor speed estimate has been investigated. Rotor resistance is a parameter which is important both in sensored IFOC and in sensorless IFOC in order to calculate slip frequency (Fig. 9).

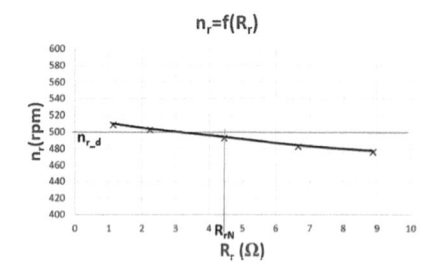

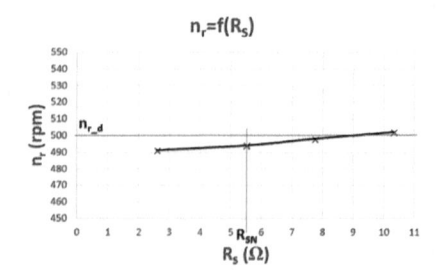

Fig. 9. The rotor speed versus rotor resistance

Fig. 10. The rotor speed versus stator resistance

As shown in Fig. 10, with the proposed sensorless IFOC scheme, the rotor speed decreases from 510 rpm to 477 rpm as rotor resistance increases from $0.25R_{rN}$ to $2R_{rN}$. The speed reference is set as 500 rpm. It can be seen, that there is also a deviation between reference speed and actual speed at rated rotor resistance due to incorrect equivalent circuit parameters determination.

6.2 The Effect of Stator Resistance Error on Speed Estimation

Another parameter being analysed is a stator resistance. It has a strong influence on speed estimation in sensorless IFOC drive but contrary to IFOC with sensor on the shaft its influence can be neglected [8]. As shown in Fig. 10, the rotor speed increases from 491 rpm to 502 rpm as stator resistance increases from $0.5R_{sN}$ to $2R_{sN}$. The speed reference is set as 500 rpm. It can be seen, that there is also a deviation between reference speed and actual speed at rated stator resistance due to incorrect equivalent circuit parameters determination.

7 Conclusion

This paper has presented the sensorless Indirect Field Oriented Control in Induction Machine drive. A set of three observers needed to estimate rotor velocity has been described. The validity of proposed method has been verified by simulation in Simulink and experimentation with frequency inverter controlled by DSC. Speed estimation applying set of described observers has led to inaccuracy due to motor parameters mismatch. The effect of rotor and stator resistance variation on the performance of the sensorless indirect field oriented control has been also investigated. The results indicate that when the parameters are inaccurate, the actual rotor speed diverges from the command. It is to be highlighted, that rotor resistance error has stronger influence on rotor speed variation.

Acknowledgment. This work was supported by Slovak Scientific Grant Agency VEGA No. 1/0774/18. and by project ITMS: 26220120046, co-funded from EU sources and European Regional Development Fund.

References

1. Rodríguez-Reséndiz, J, Rivas-Araiza E., Herrera-Ruiz, G.: Indirect field oriented control of an induction machine sensing DC-link current. In: IEEE Electronics, Robotics and Automotive Mechanics Conference (2008)
2. Kohlrusz, G., Fodor, D.: Comparison of scalar and vector control strategies of induction motors University of Pannonia, Faculty of Engineering (2011)
3. Brandstetter, P., Kuchar, M.: Rotor flux estimation using voltage model of induction motor. Technical University of Ostrava (2015)
4. Vittek, J., Dodds, J.S.: Forced dynamics control of electric drives (2003)
5. Kacenka, V., Rafajdus, P., Makys, P., Vavrus, V., Szabo, L.: Static and dynamic fault analysis of switched reluctance motor. University of Zilina, Faculty of Electrical Engineering (2012)
6. Soliman, H.M.: Effect of the parameters variation for induction motor on its performance characteristics with field oriented control compared to scalar control (2016)
7. Timer, J., Adzic, E., Porobic, V., Marcetic, D.: Influence of rotor time constant error on IFOC control structure (2009)
8. Dumnic, B., Ivanovic, Z., Katic, V., Vasic, V., Delimar, M.: Sensorless vector control and effects of machine parameters mismatch in variable speed wind turbines (2007)

Validation the FEM Model of Asynchronous Motor by Analysis of External Radial Stray Field

Petr Kacor[(✉)] and Petr Bernat

VŠB-TU Ostrava, FEECS, 17. listopadu 15,
708 32 Ostrava-Poruba, Czech Republic
petr.kacor@vsb.cz

Abstract. The paper deals with the validation of the FEM model of asynchronous motor by using the basic diagnostic method of an external magnetic field. The basic parameter to be analyzed is the induced voltage generated on the coil, which senses the external stray field of the asynchronous motor in its radial direction. Fast Fourier Transformation is used to evaluate the individual components of the spectrum according to common diagnostic procedures. Verification of the FEM model is performed by measuring the radial external field on a real asynchronous motor in the laboratory, where machine failures such as a broken rotor bar or eccentricity are also simulated. The electrical and mechanical parameters of the FEM model of motor correspond with the real machine in the laboratory.

Keywords: Asynchronous motor · FFT · FEM · Radial stray field · Rotor bar

1 Introduction

Operational diagnostics of electrical machines is a diagnostic field using information carried in operating parameters of the machine, most often voltage, current, temperature, mechanical vibration [1]. In this respect, the electric machine is a relatively complicated device containing both static and rotating mechanical elements and their mutual connection, whether fixed, rolling or sliding. Integral components include electrical components, including connecting terminals, stator and rotor windings, and magnetic circuit. Malfunctions may arise during operation of an electrical machine for various reasons [1, 2].

Statistically, the most frequently malfunctioning components of the machine are its electrical circuit, in particular the stator winding insulation system and the integrity of the rotor winding [2]. These malfunctions significantly disrupt the machine's operability and cause disruption of all machine's operating characteristics and variables (torque, noise, vibration, heating). Failures of the magnetic circuit are less frequent, usually only resulting as a consequence of other failures in the electrical or mechanical system of the machine. Mechanical defects typically have a dual origin, either of a purely mechanical origin, or are the direct or indirect consequence of other defects in the electrical circuit [2, 3].

The statistics further show that most of the malfunctions in a machine do not occur independently, but that a number of them are mutually interconnected. With electric rotary machines, the dominant energy flow is between the power supply and the machine shaft, including the air gap, and the electromagnetic field plays a key role in the transmission of energy between the stator and the rotor [4]. From this perspective, it can be concluded that the indication of any defect, failures, or asymmetry must necessarily become manifest in this electromagnetic field [4, 5].

Several diagnostic methods have been developed over the last 50 years to detect asynchronous machine failures, the most well-known of these probably being MCSA-Motor Current Signature Analysis or the scanning of an external stray field of asynchronous motor with a coil in a radial or axial motor direction [4–7]. Recently, the possibilities of placing a sensor directly into the machine space are being developed, which, moreover, can be powered by electromagnetic wireless transmission [8].

In our paper, we will deal with diagnostics from an external field, where the sensor is placed on the surface of the machine, which is reliably insulated from the perspective of dangerous contact with live parts. Measurement can even take place at a sufficient distance from the rotating parts of the machine, thereby eliminating the potential associated risks [4–6].

2 Finding of Motor Failures

Finding failures of the machine most often consists in evaluating and analyzing the FFT of the timeline of an operating variable, such as the power supply current or the induced voltage from the external coil [4–6]. When a failure occurs, a new field appears in the rotating magnetic field of the machine, whose rotation speed is no longer synchronous and cyclically deforms the main field. What actually occurs is an asymmetry in the magnetic field of the air gap, and this phenomenon is transmitted to the stator field and it also influences the flow of the current by which the machine is powered. As a result, characteristically additional frequencies appear in the power current, as well as in the stray field of the machine, which can be separated from the supply frequency by measurement and analysis and used for diagnostics. In the sides of the main harmonic field, lateral bands appear - mirror-arrayed frequency lines, corresponding to the slip frequency of the machine [2]:

$$f_v = f_1(1 - 2s) \quad (1)$$

Where is: f_X - the frequency of lateral bands expressing asymmetry, f_1 - the frequency of machine supply voltage ("dominant harmonic"), s - the rotor slip. It is further stated in the literature that the ratio of the amplitudes of these characteristic frequencies and the power supply frequency is directly proportional to the magnitude of the failure, and this ratio can thus be used as a diagnostic variable to determine the presence of asymmetry and, if a comparative measurement is available, to determine the degree of asymmetry [2–7]. Other frequencies occurring in the machine's current or stray field are frequencies generated by the slotting of the stator and rotor and possibly mechanical asymmetry - most often through eccentricity [2, 7]:

Stator slots frequency:

$$f_v = f_1 \left(\frac{Z_S}{p} \pm 1 \right) \tag{2}$$

Rotor slot frequency:

$$f_v = f_1 \left[\frac{Z_R(1-s)}{p} \pm 1 \right] \tag{3}$$

$$f_v = \frac{f_1}{p}(1-s)(Z_R \pm p) \tag{4}$$

Frequency generated by static and dynamic eccentricity:

$$f_v = f_1 \left[(k \cdot Z_R \pm n_d) \cdot \frac{1-s}{p} \pm n_x \right] \tag{5}$$

$$f_d = f_1 \left(\pm n \cdot \frac{1-s}{p} + 1 \right) \tag{6}$$

Where is: Z_S - the number of stator slots, Z_R - the number of rotor slots, f_1 - the basic harmonic frequency of supply current, p - the number of pole-pairs, s - the slip, k - the any integer (k = 1, 2, 3, etc.), n_x - the order of the stator time harmonics that are present in the power supply of motor (n_x = 1, 3, 5, etc.), n_d - the eccentricity order.

Knowing these frequencies, the frequency analysis of the current flow can be used to diagnose both rotor electric asymmetries and other defects, typically mechanical asymmetries, like rotor eccentricity and incorrect assembly [1, 2].

2.1 Identification of Motor – Determination the Number of Rotor Bars

As noted in several literature sources [9, 10], the knowledge of number of rotor bars Z_R, is very important because from it mainly derives the accuracy of determining the type of defect (broken bar, eccentricity). We do not usually have information on the number of rotor bars and they are not listed even on the machine identification table (name plate). In addition, in the recent years, smooth rotor technology has been used to produce cast rotors of high-efficiency asynchronous motors. So even when disassembling the machine, we are not able to physically determine the number of rotor bars - Z_R [11, 12].

At this stage, we can use measurement of the stray field in the radial direction and the subsequent FFT analysis. Despite the fact that this methodology initially appears to be easy, the scanned spectrum of the distribution field shows a strong effect of harmonics and finding significant harmonics, which are generated by the rotor teeth, proves difficult [13]. Thus, the number of rotor slots can generally be determined from the external radial field spectrum line at the machine's nominal load. The line usually appears relatively high above the base harmonic field ($f = 50$ Hz), and can be identified in hundreds of Hz to units of kHz [14].

2.2 Measurement of Rotor Bars Number by Slot Harmonic

To determine the number of bars (slots) in the rotor, the TiePie HandyScope HS3 Oscilloscope and the external coil with a shielding placed on the side of the asynchronous motor were used. The measured motor has the following parameters: 3-phase asynchronous motor, $P_N = 2.2$ kW, $U_N = 400/230$ V Y/D, $2p = 4$, $n_N = 1425$ rpm, $I_N = 4.9$ A, $\cos\phi = 0.81$. To record data from the oscilloscope and further analysis we used PC.

Figure 1 shows the measured motor and its coupling with dynamometer. The close-up shows the actual sensing coil is displayed, which is loosely placed close to the motor frame of the machine.

Fig. 1. Motor coupled with dynamometer and detail of sensing coil location

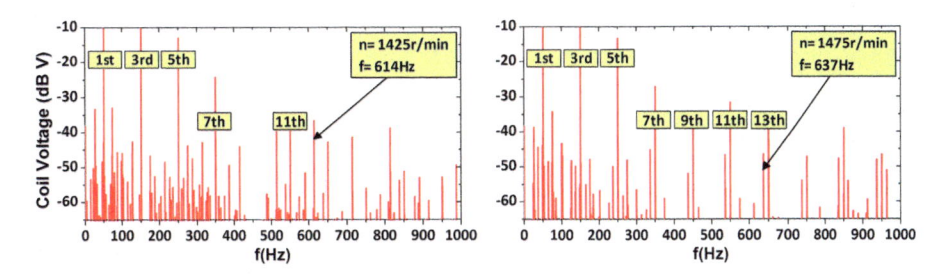

Fig. 2. FFT analysis of coil voltage at different speed of rotor, $n_1 = 1425$ rpm and $n_2 = 1475$ rpm

Figure 2 shows the FFT analysis of the stray field for two different motor loads, the nominal state - $n_1 = 1425$ rpm, and further, when decreasing the load the machine at a speed of $n_2 = 1475$ rpm. Figure 2 also highlights the dominant tooth harmonics and their corresponding frequencies. To calculate the number of rotor slots, or the number of rotor bars, respectively, we can use the formula (4) to determine the rotor slots frequency (7). The overall results for the 3 measured slip values are shown in Table 1. The stated method also requires, in addition to the FFT analysis, a relatively precise rotor rpm measurement, from which we eventually determine the slip - s.

Table 1. Estimation of number of rotor bars Z_R from direct measurement of slot harmonic

Revolution of rotor n(rpm)	Slip s(-)	Measured frequency f(Hz)	Number of rotor slots Z_R(-)	ROUNDED
1425	0.0500	614	27.96	28
1450	0.0333	625	27.93	28
1475	0.0166	637	27.94	28

The number of rotor slots can be computed by expressing of Z_R from the Eq. (3).

$$f_v = f_1 \left[\frac{Z_R(1-s)}{p} \pm 1 \right] \rightarrow Z_R = \frac{p(f_v \pm f_1)}{f_1(1-s)} = \frac{2(614+50)}{50(1-0.05)} = 27.96 \quad (7)$$

2.3 Determination of Rotor Slots Number by Method of rpm Difference

Another method that can be used to determine the number of rotor bars is more or less based on simple considerations. The rotor teeth generate a band of harmonics that shifts to lower or higher values with changes in rotor speeds. This shift can be traced relatively well during a gradual change in rotor speed. The generated harmonic band is directly proportional to the harmonics of the slots and the rate of decrease or increase of its main frequency, including the bands, must exactly correspond to the rate of decrease or increase in rotation speed of rotor. In the next steps we will only measure the difference in frequency of the selected harmonics. To eliminate the effect of additional harmonics generated primarily by the influence of the magnetic circuit saturation, we reduce the machine's supply voltage to approximately $U_{TEST} = 0.2 U_N$.

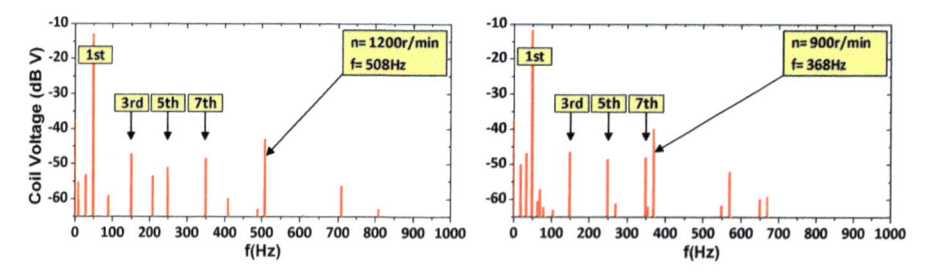

Fig. 3. FFT analysis of coil voltage at two different rotor speed, $n_1 = 1200$ rpm and $n_2 = 900$ rpm

The harmonics are captured for predefined speeds selected as $n_1 = 1200$ rpm, $n_2 = 900$ rpm, $n_3 = 600$ rpm. At synchronous speed $n_S = 1500$ rpm, the slot harmonics are negligible, since the currents in the rotor are close to $I_2 = 0$ A. Similarly, in the case of $n_0 = 0$ rpm, the rotor and stator harmonics bands merge.

On Fig. 3 it can be seen that during a decrease in rotor speed, the slot harmonics band gradually shifts towards lower values, and Table 2 then captures the measured frequency of the selected line of the rotor harmonics spectrum. As you can see, the number of rotor bars is determined by taking into consideration that difference in rotations $\Delta n = 300$ rpm decreases the frequency of slot harmonic by $\Delta f = 140$ Hz.

Table 2. Differential method of determinations of rotor slot number Z_R

Revolution of rotor n(rpm)	Shifting frequency f(Hz)	Different frequency Δf (Hz)	Number of rotor slots Z_R(-)
1200	508	140	28
900	368	140	28
600	228	140	28
300	88	–	–

The number of rotor bars (slots) - Z_R can be computed from simple derived equation:

$$Z_R = \frac{\Delta f}{\Delta n} = \frac{f_n - f_{n-1}}{n_n - n_{n-1}} = \frac{60(508 - 368)}{1200 - 900} = \frac{8400}{300} = 28 \qquad (8)$$

2.4 Measurement of Broken Rotor Bar

The asynchronous motor was further disassembled and a hole was made near the end ring simulating the cracked rotor bar, see Fig. 4. A section of the metal sheets of the rotor had to be removed by turning to be able to find the correct position of the rotor bar. Among other things, the Z_R - number of rotor bars that could be identified during this rotor adjustment was confirmed.

Fig. 4. Healthy rotor and rotor with broken bar made by drilling of selected bar

After reassembly, the motor was coupled with the dynamometer again and subjected to measurement. A nominal load corresponding to the motor name plate values was set and, in terms of diagnostics, FFT analysis of induced voltage from the radial stray field was performed. The results of these analyses are shown in Fig. 5. As assumed, in contrast to a healthy motor, a more significant sideband (mark as BB) appears near the basic harmonic, which at rotor slip - $s = 0.05$ corresponds to the frequencies $f_{X1} = 45$ Hz and $f_{X2} = 55$ Hz. These sidebands are distinctive and identify the broken of the rotor bar. Also other sidebands which identify the eccentricity are increased in amplitude. Typical frequency pairs are: f_{ECC1} = (73.75 and 26.25) Hz and the second pair f_{ECC2} = (97.5 and 2.5) Hz. These frequencies can be confirmed by a theoretical relationship from Eq. (5).

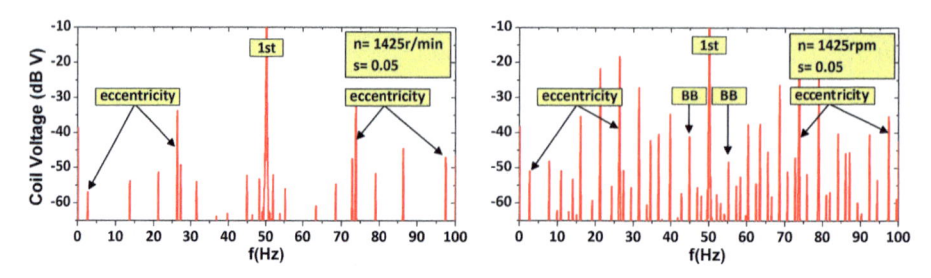

Fig. 5. FFT analysis of waveform of healthy motor and motor with broken bar

2.5 2D FEM Model of Asynchronous Motor

To validate the results of the performed analyses, a 2D FEM electromagnetic model of an asynchronous motor was created [15]. For electromagnetic analysis, Ansys Maxwell was used to solve 2D transient electromagnetic analysis including rotational mechanical motion. The model contains Z_S = 36 stator slots and Z_R = 28 rotor slots. The material property of the magnetic circuit has been selected from M850-50A and the BH curve is also entered. FEM model dimensions and parameters correspond to the measured machine in the laboratory, see Table 3.

Table 3. Comparison of main measured and simulated values of motor

Basic parameters	Real motor	FEM model
Power (W)	2200	2150
Speed (rpm)	1425	1428
Torque (Nm)	14.7	14.4
PF (-)	0.81	0.795
Current (A)	4.9	5.2
Efficiency (%)	81	79

The time step of electromagnetic analysis is set to small to capture the waveforms of the induced voltage with sufficient smoothness and for calculation with numerical stability. In Fig. 6, on the left, you can see the distribution of magnetic flux density in the FEM model without a failure at nominal load state of motor. The sensing coil is located on the left side of the motor. The FFT analysis of the induced voltage on sensing coil is shown in Fig. 6 in the range of frequency f = (0–1000) Hz. As can be seen, even in the numerical model a harmonic of rotor slots with the frequency f_{ZR} = 615 Hz, can be found, which confirms the previous theoretical facts and measurement.

There are also multiples of slot harmonic f_{ZR1} = 715 Hz, f_{ZR2} = 815 Hz and f_{ZR3} = 915 Hz. In the numerical model, the eccentricity harmonic f_{ECC1} = 25 Hz and f_{ECC2} = 75 Hz are absent. They are typical for technological eccentricity in the real machine, but in this step of simulation, we consider an ideal assembly of stator and rotor without eccentricity.

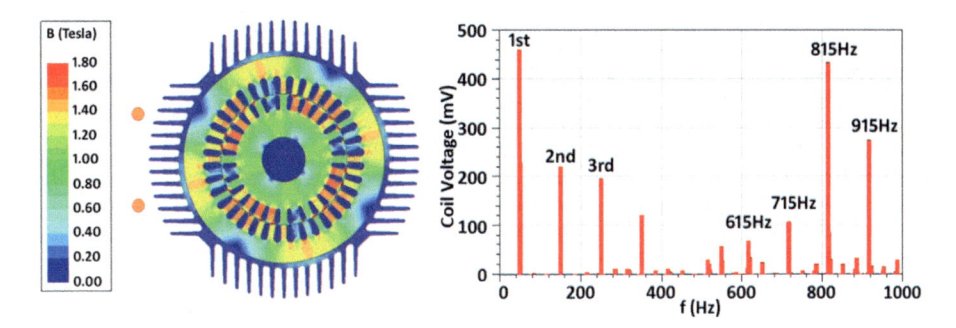

Fig. 6. FEM model of health asynchronous motor and FFT of coil voltage

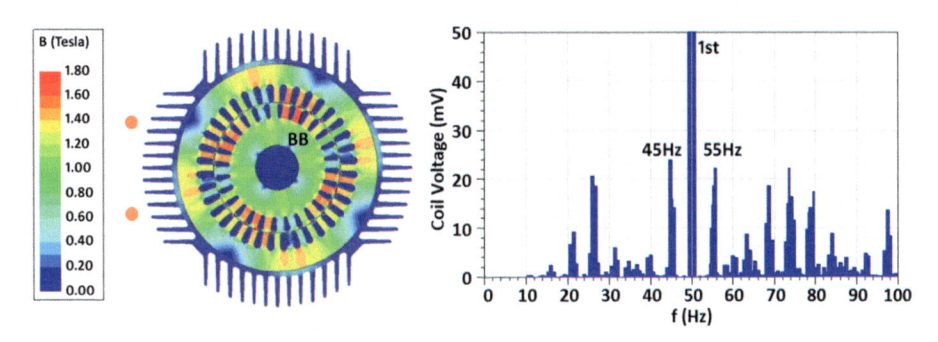

Fig. 7. FEM model of asynchronous motor with one broken bar and FFT of coil voltage

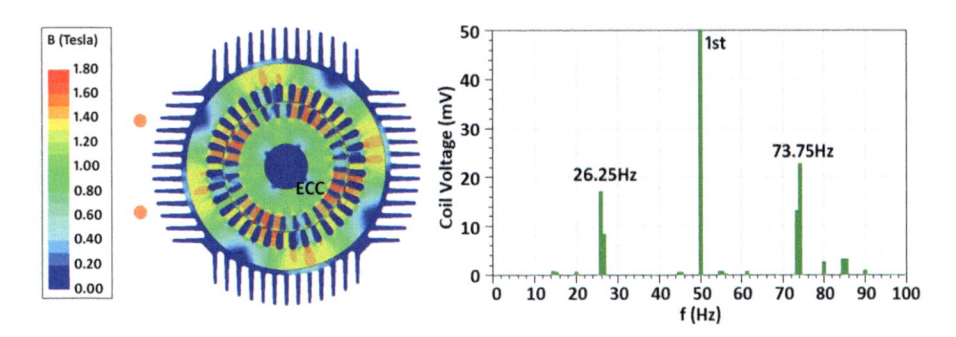

Fig. 8. FEM model of asynchronous motor with eccentricity and FFT of coil voltage

We also created the model of broken rotor bar by defining zero electrical conductivity for the selected bar. Load conditions are the same as for a failure-free motor. Figure 7 shows the magnetic flux density in the machine, which is deformed on one side (mark as BB) by an asymmetric distribution of rotor bars current densities. By FFT analysis, we get the image, which again confirms the theoretical assumptions and real measurement. Around the fundamental harmonic f_1 = 50 Hz, two spectral lines f_{X1} = 45 Hz and f_{X2} = 55 Hz appeared. This suggests that the rotor bar is broken. Other harmonic bands are represented only insubstantially.

2D FEM model is further usable for assessing eccentricities and the pure dynamic eccentricity can be seen in Fig. 8. Eccentricity was simulated by shifting the rotor sheet at *x*-axis (mark as ECC). Value of distance from axis of rotation is approximately Δecc = 0.02 mm. From the FFT analysis of the induced voltage can be seen the characteristic side harmonics with frequency f_{ECC1} = 26.25 Hz and f_{ECC2} = 73.75 Hz which is also in good match with theory and real measurement.

3 Conclusion

The article used a well-known diagnostic method for detecting rotor bar defects by measuring an external stray field of asynchronous motor. The analysis results were compared with a numerical 2D FEM model of the same design dimensions and parameters. A very good match was found between simulation results and real measurement of the asynchronous motor in the laboratory.

The article also used a simple comparative method for detecting the number of rotor bars, or of the rotor slots, respectively. The current form of the rotors of highly efficient asynchronous motors does not allow identification of the rotor even when it is disassembled. Disassembling also should be the last option to determine the number of rotor slots. The differential method used needs to be further developed and validated to be able to deploy it for a wider range of performance and design lines of asynchronous machines. As further follows from the paper, a motor can be diagnosed with relatively simple means with sufficient precision and the equipment necessary can be obtained self-sufficiently.

The advantage of the numerical model is the relative ease proceeds in which the individual faults can be simulated, but the paper also points to the importance of validation of simulated and measured data where the simple numerical FEM models of machines often lead to erroneous conclusions.

Acknowledgement. This research was supported by the SGS grant No. SP2018/61 from VSB - Technical University of Ostrava.

References

1. Henao, H., et al.: Trends in fault diagnosis for electrical machines: a review of diagnostic techniques. IEEE Ind. Electron. Mag. **8**(2), 31–42 (2014). https://doi.org/10.1109/MIE.2013.2287651
2. Nandi, S., Toliyat, H.A.: Condition monitoring and fault diagnosis of electrical machines - a review. In: Conference Record of the 1999 IEEE Industry Applications Conference. Thirty-Forth IAS Annual Meeting (Cat. No.99CH36370), Phoenix, AZ, vol. 1, pp. 197–204 (1999) https://doi.org/10.1109/IAS.1999.799956
3. Dorrell, D.G., Thomson, W.T., Roach, S.: Analysis of airgap flux, current, and vibration signals as a function of the combination of static and dynamic airgap eccentricity in 3-phase induction motors. IEEE Trans. Ind. Appl. **33**(1), 24–34 (1997). https://doi.org/10.1109/28.567073
4. Romary, R., Corton, R., Thailly, D., Brudny, J.F.: Induction machine fault diagnosis using an external radial flux sensor. EPJ. Appl. Phys. **32**(2), 125–132 (2005). https://doi.org/10.1051/epjap:2005079
5. Ceban, A., Pusca, R., Romary, R.: Study of rotor faults in induction motors using external magnetic field analysis. IEEE Trans. Industr. Electron. **59**(5), 2082–2093 (2012). https://doi.org/10.1109/TIE.2011.2163285
6. Cuevas, M., Romary, R., Lecointe, J.P., Jacq, T.: Non-invasive detection of rotor short-circuit fault in synchronous machines by analysis of stray magnetic field and frame vibrations. IEEE Trans. Magn. **52**(7), 1–4 (2016). https://doi.org/10.1109/TMAG.2016.2514406
7. Thomson, W.T., Fenger, M.: Current signature analysis to detect induction motor faults. IEEE Ind. Appl. Mag. **7**(4), 26–34 (2001). https://doi.org/10.1109/2943.930988
8. Kindl, V., Kavalir, T., Pechanek, R.: Key construction aspects of low frequency wireless power transfer system using parallel resonance. In: 2015 17th European Conference on Power Electronics and Applications (EPE 2015 ECCE-Europe), Geneva, pp. 1–5 (2015). https://doi.org/10.1109/EPE.2015.7311758
9. Keysan, O., Ertan, H.B.: Determination of rotor slot number of an induction motor using an external search coil. In: Facta Universitatis - Series: Electronics and Energetics, vol. 22, br. 2, pp. 227–234 (2009). https://doi.org/10.2298/FUEE0902227K
10. Elhaija, W.A., Ghorbanian, V., Faiz, J., Nejadi-Koti, H.: Significance of rotor slots number on induction motor operation under broken bars. In: 2017 IEEE International Electric Machines and Drives Conference (IEMDC), Miami, FL, pp. 1–8 (2017). https://doi.org/10.1109/IEMDC.2017.8001862
11. Kindl, V., Hruska, K., Pechanek, R., Sobra, J., Skala, B.: The effect of space harmonic components in the air gap magnetic flux density on torque characteristic of a squirrel-cage induction machine. In: 2015 17th European Conference on Power Electronics and Applications (EPE 2015 ECCE-Europe), Geneva, pp. 1–5 (2015). https://doi.org/10.1109/EPE.2015.7309221
12. Ferkova, Z., Kindl, V.: Influence of skewed squirrel cage rotor with intermediate ring on magnetic field of air gap in induction machine. Elektron. Elektrotechnika **23**(1), 26–30 (2017). https://doi.org/10.5755/j01.eee.23.1.17580
13. Barzegaran, M.R., Mohammed, O.A.: Condition monitoring of electrical machines for extreme environments using electromagnetic stray fields. In: 2014 International Conference on Electrical Machines (ICEM), Berlin, pp. 2479–2485 (2014) https://doi.org/10.1109/ICELMACH.2014.6960535

14. Gritli, Y.: Diagnosis and fault detection in electrical machines and drives based on advanced signal processing techniques. Dissertation Thesis, Department of Electrical, Electronics and Information Engineering "Guglielmo Marconi" (DEI) at University of Bologna, Bologna, ITALY, March 2014. http://www.die.ing.unibo.it/dottorato_it/Gritli/these-YASSER%20GRITLI.pdf. Accessed 24 July 2018
15. Ceban, A., Fireteanu V., Romary, R., Pusca, R., Taras, P.: Finite element diagnosis of rotor faults in induction motors based on low frequency harmonics of the near-magnetic field. In: 8th IEEE Symposium on Diagnostics for Electrical Machines, Power Electronics and Drives, Bologna, pp. 192–198 (2011). https://doi.org/10.1109/DEMPED.2011.6063623

Outage and Intercept Probability Analysis for Energy-Harvesting-Based Half-Duplex Relay Networks Assisted by Power Beacon Under the Existence of Eavesdropper

Tan N. Nguyen[1,2], Phuong T. Tran[1(✉)], Nguyen Dao[3], and Miroslav Voznak[1,2]

[1] Wireless Communications Research Group,
Faculty of Electrical and Electronics Engineering, Ton Duc Thang University,
No. 19, Nguyen Huu Tho Street, Tan Phong Ward, District 7,
Ho Chi Minh City, Vietnam
{nguyennhattan,tranthanhphuong}@tdtu.edu.vn
[2] VSB-Technical University of Ostrava,
17. listopadu 15/2172, 708 33 Ostrava - Poruba, Czech Republic
miroslav.voznak@vsb.cz
[3] Faculty of Electrical and Electronics Engineering, Ton Duc Thang University,
No. 19, Nguyen Huu Tho Street, Tan Phong Ward, District 7,
Ho Chi Minh City, Vietnam
nguyendao@tdtu.edu.vn
http://feee.tdtu.edu.vn, http://www.vsb.cz/en

Abstract. In this paper, we propose a half-duplex relaying schemes for wireless sensor networks, in which both source node and relay node are equipped with energy harvesting technology and assisted by a power beacon (PB) to supply energy for their transmission duty. In this network, the security and privacy issues are significant due to the possible eavesdropping by surrounding users. Motivated by this observation, we carefully investigate the security and reliability performance of the proposed systems under the existence of an eavesdropper in the nearby environment. Here, the security performance and the reliability performance are represented by outage probability (OP) and intercept probability (IP), respectively. The power-splitting energy harvesting protocol is applied in our analysis. We rigorously derive the closed-form expressions of both OP and IP of the system and study the effect of various parameters, including power-splitting factors, channel parameters, transmit power, and noise power, on these performance factors. Finally, Monte Carlo simulation results are also performed to confirm the correctness of all theoretical analysis derived.

Keywords: Energy harvesting · Physical layer security · Half-duplex · Power beacon · Intercept probability · Outage probability

1 Introduction

Nowadays, with the rapid growth of wireless services, energy consumption issues for wireless networks have become increasingly critical. Different energy-saving algorithms have been investigated for wireless communication networks [1–4]. Among them, energy harvesting (EH) from the available RF energy sources in the surrounding environment has attracted significant interest [5] because it can remove the burden of recharging or replacing the batteries frequently [6]. Another advantage of this method is that the RF signals can convey both energy and information simultaneously. Hence, a RF-based energy harvesting technique introduced in 2008 by Varshney [7], called simultaneous wireless information and power transfer (SWIPT), has quickly become a hot research topic recently [8–12]. After the seminal paper of Varshney [7], Zhang and Ho [13] proposed two practical receivers for EH, namely, time-switching and power-splitting protocols. Later, the throughput performance of wireless relay networks for both time-switching (TSR) and power-splitting (PSR) protocols have been studied rigorously in [14] and [15].

One of the main challenges to implement EH in practice is the vast gap between the operational sensitivity of the decoder and the energy harvester [16]. That makes this technology ineffective at medium and long distance. To resolve this problem, Huang and Lau [9] have proposed a novel network architecture where stations called power beacons (PBs) are deployed in an existing wireless network for recharging mobiles and sensors via microwave radiation. Since then, the effectiveness of this strategy has been investigated in several papers [16–18]. Le [16] evaluated the throughput performance of a multiple-access wireless networks with distributed PBs. Zhou et al. [17] considered the performance of PB implementation in milimeter-wave ad-hoc networks. Liang [18] considered a power-beacon-assisted two-way relay networks to maximize the minimum rate of two users under the constraint of user fairness.

Physical layer security has recently attracted a lot of attention to wireless communication community, especially when the traditional cryptography methods seems to be vulnerable to advanced attacks. Wyner introduced the idea of physical layer security in 1975 [19], but this topic has only been reconsidered recently [20–22]. The authors in [21] proposed an algorithm to maximize the secrecy sum rate of a two-way relay system by optimizing the relay beamforming vector and the transceivers' powers under total power constraint. A physical-layer-security scheme for an underlay relay-based cognitive radio network (CRN) that uses orthogonal frequency-division multiplexing (OFDM) as the medium access technique is proposed in [22], in which the secrecy rate is maximized subject to the constraints on transmission power and interference level at the primary users.

Physical layer security in power-beacon-assisted networks has just been addressed in the last few years. For example, the authors in [23] investigated the secrecy outage performance analysis of a power-beacon-assisted wireless network with multiple eavesdroppers. In [24], the secrecy performance of a wiretap channel with the assistance of PB was studied. The authors proposed and evaluated different jamming schemes to enhance the security of the systems. However, no relay node is involved in the above papers.

Motivated by these facts, in this paper, we investigate a power-beacon-assisted energy-harvesting-based half-duplex relay networks with an eavesdropper. Here, both reliability and security are taken into account. Different from work in [24], we use artificial noise method to enhance the protection of the system. We derive the closed-form expressions of the outage probability, intercept probability, and the likelihood of successful communication without an interception; then conduct the Monte Carlo simulation to verify the analysis.

The rest of the paper is organized as follows. The system model of our proposed system is described in the next section. Section 3 presents the mathematical analysis of the system performance. Numerical results obtained from Monte Carlo simulation are introduced in Sect. 4. Finally, Sect. 5 concludes the paper.

2 System Model

Figure 1 illustrates our proposed half-duplex relaying system model. In this model, the source S sends information to the destination D with the help of a decode-and-forward relay R. Both source and relay are equipped with energy harvesting technology, and their energy is supplied from a PB, located at some convenient place near the source and the relay. Due to broadcasting nature of wireless communications, the signals transmitted by S and R can also be received by an eavesdropper node E in the vicinity. We assume that D is located very far from S, hence the direct transmission from S to D is not feasible. The solid lines in Fig. 1 represent the main links (from S to R and R to D), while the dash lines represent the wiretap links (from S and R to E). In this communication scheme, the relay is assumed to have no own data to transmit and use only the energy from the PB to help forward the message from S to D. All communication nodes, as well as the PB, are equipped with a single antenna.

Regarding the channel models, we assume an independent Rayleigh fading channel model such that the channel gains remain unchanged during each transmission block but independently vary from one block to another block. Let h_{BS} and h_{BR} denote the channel gains from the PB to S and R, respectively; h_{SR} and h_{SD} denote the channel gains from the source to the relay and from the relay to the destination, respectively; and h_{SE} and h_{RE} denote the channel gains from S and R to E, respectively. Therefore, all link power gains are random variables (RVs) and subject to exponential distribution. Specifically, $|h_{AB}|^2$ is an exponential RV with parameter λ_{AB}, where h_{AB} is the channel gain from node 'A' to node 'B', with $A \in \{B, S, R\}$ and $B \in \{S, R, D, E\}$.

In this paper, the basic time-switching protocol for EH is applied as shown in Fig. 2 [14]. Let T denote the duration of one transmission block. The communication between S and D consists of two processes: the energy harvesting process and the transmission process. The first process is corresponding to time slot 1 in Fig. 2, which has the length of αT. The transmission process has the duration of $(1-\alpha)T$ and involves two phases with equal length. In the first phase, the

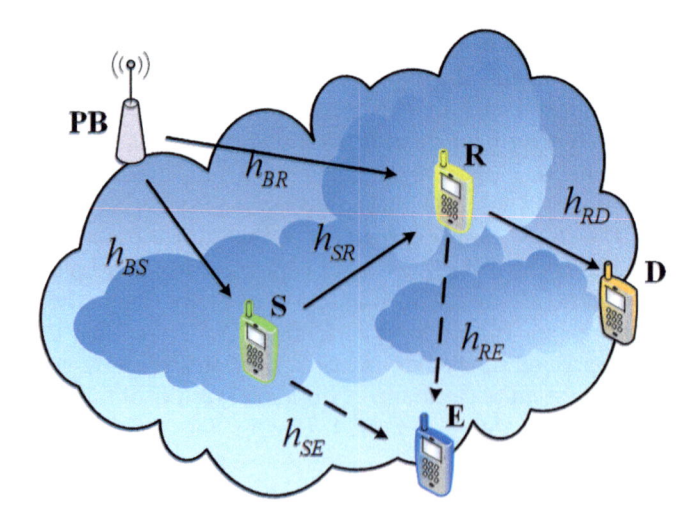

Fig. 1. Power-beacon-assisted energy-harvesting-based half-duplex relay networks model with eavesdropper.

source node sends its message to the relay node, and in the second phase, the relay node employs the decode-and-forward (DF) protocol to resend the decoded message to the destination node.

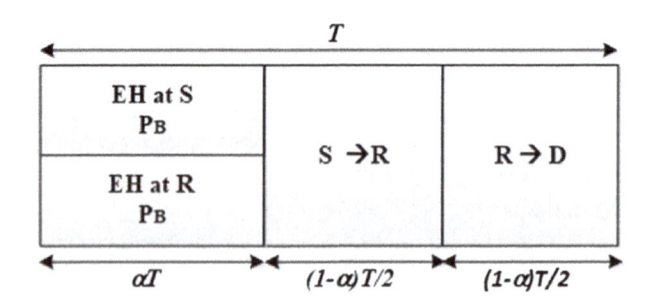

Fig. 2. TSR protocol for energy harvesting.

During the energy harvesting phase, both S and R harvest the energy from the PB. The amount of energy harvested by the source and the relay can be calculated as $E_S = \eta P_B.|h_{BS}|^2.\alpha T$ and $E_R = \eta P_B.|h_{BR}|^2.\alpha T$, respectively, where P_B is the transmit power of the PB and η is the energy harvesting efficiency (for simplicity, we assume the same efficiency at S and R).

According to the law of energy conservation, the energy used for transmission in the next process must be equal to the harvested energy mentioned above. Hence, the transmit power of S and R during the information transmission process can be found by

$$P_S = \frac{E_S}{(1-\alpha)T/2} = \frac{2\eta\alpha T P_B |h_{BS}|^2}{(1-\alpha)T} = \kappa P_B |h_{BS}|^2 \quad (1)$$

and

$$P_R = \frac{E_R}{(1-\alpha)T/2} = \frac{2\eta\alpha T P_B |h_{BR}|^2}{(1-\alpha)T} = \kappa P_B |h_{BR}|^2 \quad (2)$$

where $\kappa \triangleq \frac{2\eta\alpha}{1-\alpha}$.

Now, let's consider the information transmission process. During the first phase of this process, the source sends its message to the relay. The received signal at the relay can be expressed as

$$y_R = h_{SR} x_S + n_R \quad (3)$$

Here, x_S is the transmitted signal, which satisfies $E|x_S|^2 = P_S$, where $E[X]$ denotes the expectation of the random variable X. n_R is the zero mean additive white Gaussian noise (AWGN) with variance N_0. From (1) and (3), the signal-to-noise-ratio (SNR) at the relay can be calculated as

$$SNR_1 = \frac{\kappa P_B |h_{BS}|^2 |h_{SR}|^2}{N_0} \quad (4)$$

Similarly, the following signal is received at the destination:

$$y_D = h_{RD} x_R + n_D \quad (5)$$

where x_R is the forwarded message from the relay, which satisfies $E|x_R|^2 = P_R$, and n_D is the zero-mean AWGN at D with variance N_0. The SNR at D can be expressed as

$$SNR_2 = \frac{\kappa P_B |h_{BR}|^2 |h_{RD}|^2}{N_0} \quad (6)$$

3 Performance Analysis

In this section, the reliability and security performance of the proposed system is investigated. Reliability performance is represented by outage probability, while the intercept probability represents security performance. These two performance factors are studied at various configurations of the proposed scheme.

3.1 Outage Probability

An outage occurs if and only if at least one of the two links, source-relay link and relay-destination link, does not work. Assume that the source transmits at a constant rate R. Then by Shannon's theorem, $\gamma_{th} = 2^R - 1$ must be the lower threshold for SNR at both relay and destination to communicate normally. Hence, the outage probability of the proposed system is defined as

$$OP = \Pr\left\{\frac{\kappa P_B}{N_0} \min\left(|h_{BS}|^2|h_{SR}|^2, |h_{BR}|^2|h_{RD}|^2\right) < \gamma_{th}\right\} \qquad (7)$$

Let's denote $X = |h_{BS}|^2|h_{SR}|^2, Y = |h_{BR}|^2|h_{RD}|^2$. Note that X and Y are independent random variables, each of which is the product of two exponential random variables. The cumulative distribution function (CDF) of this kind of random variables can be found in some previous works, for example, in [25]. Specifically, the CDFs of X and Y are respectively given by

$$F_X(x) = 1 - 2\sqrt{\frac{x}{\lambda_{BS}\lambda_{SR}}} K_1\left(2\sqrt{\frac{x}{\lambda_{BS}\lambda_{SR}}}\right) \qquad (8)$$

and

$$F_X(x) = 1 - 2\sqrt{\frac{x}{\lambda_{BS}\lambda_{SR}}} K_1\left(2\sqrt{\frac{x}{\lambda_{BS}\lambda_{SR}}}\right) \qquad (9)$$

Now, the Eq. (7) can be rewritten as

$$OP = 1 - \Pr\left(\frac{\kappa P_B X}{N_0} \geq \gamma_{th}\right) \Pr\left(\frac{\kappa P_B Y}{N_0} \geq \gamma_{th}\right) \qquad (10)$$

By using (8) and (9), we obtain

$$\Pr\left(\frac{\kappa P_B X}{N_0} \geq \gamma_{th}\right) = 1 - F_X\left(\frac{\gamma_{th} N_0}{\kappa P_B}\right)$$
$$= 2\sqrt{\frac{\gamma_{th} N_0}{k P_B \lambda_{BS}\lambda_{SR}}} K_1\left(2\sqrt{\frac{\gamma_{th} N_0}{\kappa P_B \lambda_{BS}\lambda_{SR}}}\right) \qquad (11)$$

and similarly,

$$\Pr\left(\frac{\kappa P_B Y}{N_0} \geq \gamma_{th}\right) = 2\sqrt{\frac{\gamma_{th} N_0}{\kappa P_B \lambda_{BR}\lambda_{RD}}} K_1\left(2\sqrt{\frac{\gamma_{th} N_0}{\kappa P_B \lambda_{BR}\lambda_{RD}}}\right) \qquad (12)$$

Finally, from (10), (11), and (12), we obtain the closed-form formula for the outage probability as stated in the following theorem.

Theorem 1. *For the DF protocol, the outage probability of the power-beacon-assisted energy-harvesting-based half-duplex relay network as proposed in Fig. 1 is given as*

$$OP = 1 - \frac{4\gamma_{th}}{\kappa Q \sqrt{\lambda_{BS}\lambda_{SR}\lambda_{BR}\lambda_{RD}}} \times K_1\left(2\sqrt{\frac{\gamma_{th}}{\kappa Q \lambda_{BS}\lambda_{SR}}}\right) \times K_1\left(2\sqrt{\frac{\gamma_{th}}{\kappa Q \lambda_{BR}\lambda_{Rd}}}\right) \qquad (13)$$

where Q is defined as $Q \triangleq \frac{P_B}{N_0}$ *and* $K_1(\cdot)$ *is the first-order modified Bessel function of the second kind.*

Proof. The proof follow directly from (10), (11), and (12). □

3.2 Intercept Probability

Now, let's consider the case that there is an eavesdropper in the surrounding environment. This eavedropper can overhear the message sent from the source or from the relay. Therefore, the SNR at the eavesdropper for decoding the source message can be formulated by

$$SNR_E = \max(\varphi_{SE}, \varphi_{RE}) \qquad (14)$$

where φ_{SE} and φ_{RE} are instantaneous signal-to-noise ratios (SNRs) of the S-E link and R-E link, respectively.

To enhance the security of the proposed system, we apply the physical layer security strategy. The source node S and the relay node R can randomly generate codebooks to confuse E from combining the received data with MRC (see randomize-and-forward (RF) method [26,27]). The received signal at the eavesdropper during the first phase of transmission process is given by

$$y_E = \sqrt{(\rho)}h_{SE}x_S + \sqrt{1-\rho}h_{SE}x_{AN1} + n_E \qquad (15)$$

where x_{AN1} is the artificial noise generated by the source, which also satisfies $E|x_{AN1}|^2 = P_S$; x_S is the source message as described in (3); ρ is a power-splitting factor to divide the available power at the source into two parts: message and artificial noise; and n_E is the AWGN at the eavesdropper with zero mean and variance of N_0.

Similarly, the received signal at the eavesdropper during the second phase of the transmission process is expressed as

$$y_E = \sqrt{(\rho)}h_{RE}x_R + \sqrt{1-\rho}h_{RE}x_{AN2} + n_E \qquad (16)$$

where x_{AN1} is the artificial noise generated by the source, which also satisfies $E|x_{AN2}|^2 = P_R$. For simplicity, we assume that the same power-splitting factor is used at both the source and the relay.

Then the instantaneous SNRs at the eavesdropper are calculated as

$$\varphi_{SE} = \frac{\rho P_S |h_{SE}|^2}{N_0 + (1-\rho)P_S|h_{SE}|^2} = \frac{\rho}{\frac{N_0}{\kappa P_B |h_{BS}|^2 |h_{SE}|^2} + (1-\rho)}, \qquad (17)$$

$$\varphi_{RE} = \frac{\rho P_R |h_{RE}|^2}{N_0 + (1-\rho)P_R|h_{RE}|^2} = \frac{\rho}{\frac{N_0}{\kappa P_B |h_{BR}|^2 |h_{RE}|^2} + 1 - \rho}. \qquad (18)$$

The intercept probability is defined as the probability such that $SNR_E \geq \gamma_{th}$. By using (14), we have:

$$IP = \Pr(SNR_E \geqslant \gamma_{th}) = 1 - \Pr\{\max(\varphi_{SE}, \varphi_{RE}) < \gamma_{th}\} \qquad (19)$$

where IP denotes the intercept probability at the eavesdropper. Let's denote $Z = |h_{BS}|^2|h_{SE}|^2$ and $T = |h_{BR}|^2|h_{RE}|^2$. By substituting the expressions of the instantaneous SNRs at the eavesdropper into (19), we get

$$IP = 1 - \Pr\left(\frac{\rho}{\frac{1}{\kappa QZ}+1-\rho} < \gamma_{th}\right)\Pr\left(\frac{\rho}{\frac{1}{\kappa QT}+1-\rho} < \gamma_{th}\right)$$

$$= \begin{cases} 0 & \text{if } \zeta < 0 \\ 1 - \Pr\left(Z < \frac{\gamma_{th}}{\kappa Q\zeta(1-\rho)}\right)\Pr\left(T < \frac{\gamma_{th}}{\kappa Q\zeta(1-\rho)}\right) & \text{if } \zeta \geqslant 0 \end{cases} \quad (20)$$

where $Q = P_B/N_0$ and $\zeta = \frac{\rho}{1-\rho} - \gamma_{th}$. In the above equation, both Z and T are the products of two exponential random variables. Hence, their CDFs have the similar form to (8) and (9). By using the same approach as in the computing of outage probability, we have the following theorem.

Theorem 2. *For the DF protocol, the intercept probability of the power-beacon-assisted energy-harvesting-based half-duplex relay network with an eavesdropper is given as follows.*

– *Case 1:* $\frac{\rho}{1-\rho} \geq \gamma_{th}$:

$$IP = 1 - \left[1 - 2\sqrt{\frac{\gamma_{th}}{\kappa Q\zeta(1-\rho)\lambda_{SE}\lambda_{BS}}} K_1\left(2\sqrt{\frac{\gamma_{th}}{\kappa Q\zeta(1-\rho)\lambda_{SE}\lambda_{BS}}}\right)\right]$$
$$\times \left[1 - 2\sqrt{\frac{\gamma_{th}}{\kappa Q\zeta(1-\rho)\lambda_{RE}\lambda_{BR}}} K_1\left(2\sqrt{\frac{\gamma_{th}}{\kappa Q\zeta(1-\rho)\lambda_{RE}\lambda_{BR}}}\right)\right] \quad (21)$$

– *Case 2:* $\frac{\rho}{1-\rho} < \gamma_{th}$:

$$IP = 0 \quad (22)$$

where $\lambda_{SE}, \lambda_{BS}, \lambda_{RE}$, and λ_{BR} are the parameters of the distribution of $|h_{SE}|^2$, $|h_{BS}|^2, |h_{RE}|^2$, and $|h_{SE}|^2$, respectively, and $K_1(\cdot)$ is the first-order modified Bessel function of the second kind.

Proof. We have:

$$\Pr\left(Z < \frac{\gamma_{th}}{\kappa Q\zeta(1-\rho)}\right) = F_Z\left(\frac{\gamma_{th}}{\kappa Q\zeta(1-\rho)}\right)$$
$$= 1 - 2\sqrt{\frac{\gamma_{th}}{\kappa Q\zeta(1-\rho)\lambda_{SE}\lambda_{BS}}} K_1\left(2\sqrt{\frac{\gamma_{th}}{\kappa Q\zeta(1-\rho)\lambda_{SE}\lambda_{BS}}}\right) \quad (23)$$

and similarly,

$$\Pr\left(T > \frac{1-\rho}{\kappa Q\zeta}\right) = 1 - 2\sqrt{\frac{\gamma_{th}}{\kappa Q\zeta(1-\rho)\lambda_{RE}\lambda_{BR}}} K_1\left(2\sqrt{\frac{\gamma_{th}}{\kappa Q\zeta(1-\rho)\lambda_{RE}\lambda_{BR}}}\right) \quad (24)$$

By substituting (23) and (24) into (20), we obtain the final results. □

3.3 Reliable Communication Without Interception

In this section, we combine both reliability and security analysis to derive feasible configurations for our proposed scheme. It's easy to see that the outage event and the interception event are independent. So, by using the results obtained from the two previous sections with some adjusting in the algebras, we can claim the third theorem as following.

Theorem 3. *For the DF protocol, the probability of reliable communication without being intercepted of the power-beacon-assisted energy-harvesting-based half-duplex relay network with an eavesdropper is given as follows.*

- Case 1: $\frac{\rho}{1-\rho} \geq \gamma_{th}$:

$$P = \left[1 - 2\sqrt{\frac{\gamma_{th}}{\kappa Q \zeta (1-\rho)\lambda_{SE}\lambda_{BS}}} K_1\left(2\sqrt{\frac{\gamma_{th}}{\kappa Q \zeta (1-\rho)\lambda_{SE}\lambda_{BS}}}\right)\right]$$
$$\times \left[1 - 2\sqrt{\frac{\gamma_{th}}{\kappa Q \zeta (1-\rho)\lambda_{RE}\lambda_{BR}}} K_1\left(2\sqrt{\frac{\gamma_{th}}{\kappa Q \zeta (1-\rho)\lambda_{RE}\lambda_{BR}}}\right)\right]$$
$$\times \frac{4\gamma_{th}}{\kappa Q \rho \sqrt{\lambda_{BS}\lambda_{SR}\lambda_{BR}\lambda_{Rd}}} K_1\left(2\sqrt{\frac{\gamma_{th}}{\kappa Q \rho \lambda_{BS}\lambda_{SR}}}\right) K_1\left(2\sqrt{\frac{\gamma_{th}}{\kappa Q \rho \lambda_{BR}\lambda_{Rd}}}\right)$$
(25)

- Case 1: $\frac{\rho}{1-\rho} < \gamma_{th}$:

$$P = \frac{4\gamma_{th}}{\kappa Q \rho \sqrt{\lambda_{BS}\lambda_{SR}\lambda_{BR}\lambda_{Rd}}} K_1\left(2\sqrt{\frac{\gamma_{th}}{\kappa Q \rho \lambda_{BS}\lambda_{SR}}}\right) K_1\left(2\sqrt{\frac{\gamma_{th}}{\kappa Q \rho \lambda_{BR}\lambda_{Rd}}}\right)$$
(26)

4 Numerical Results

To verify the analytical results developed in the previous section, we conduct the Monte Carlo simulation for a network with one source, one relay, one destination, and one eaves dropper. The impact of the distance between nodes on the system performance has been represented by the parameters λ_{ab} of the channel gain distributions. So, in this limited space of the paper, we don't add the distance variable to our analysis. All simulation parameters are listed in Table 1.

Figures 3 and 4 shows the outage probability and intercept probability of the proposed system versus the transmitted power P_s in 3 cases, which are corresponding to 3 values of the time-switching factor α, i.e., 0.3, 0.5, and 0.7. The data rate is set to $R = 3$ bps/Hz. It's can be observed that the simulation curve and the analytical curve for each case overlap together. That confirms the correctness of our analysis. It's also easily seen that when the ratio P_B/N_0 increases, the outage probability decreases but the intercept probability increases.

The impact of time switching factor α on the outage probability and intercept probability is shown in Figs. 5 and 6. Here, the ratio P_B/N_0 is set to

Table 1. Simulation parameters

Symbol	Parameter name	Values
R	Source rate	3 bps/Hz
γ_{th}	SNR threshold	7
η	Energy harvesting efficiency	0.8
α	Time-switching factor	0.3
ρ	Power-splitting factor	0.7
λ_{ab}	Parameter of $|h_{ab}|^2$	0.5 (for all links)
P_B/N_0	Beacon-power-to-noise-ration	0–30 dB

20 dBW. Again, the analytical solutions are in exact agreement with the simulation results. The outage probability tens to decrease as time-switching factor increase. That is because when α increases, more power is used to transmit the message, and hence, the message is more likely to be detected as the receiver. In contrast, the intercept probability increases with α with the same reason.

Figure 7 illustrate the impact of the application of artificial noise to prevent eavesdropping. Again, we set $\frac{P_B}{N_0} = 20$ dB in this simulation. The effect of ρ depends on the selection of time-switching factor α. For small α such as 0.3 or 0.5, the probability of successful communication increases at low regime of ρ. However, after exceeding a threshold of ρ this probability is reduced. The reason for this phenomenon is that for low ρ, the intercept probability is equal to zero;

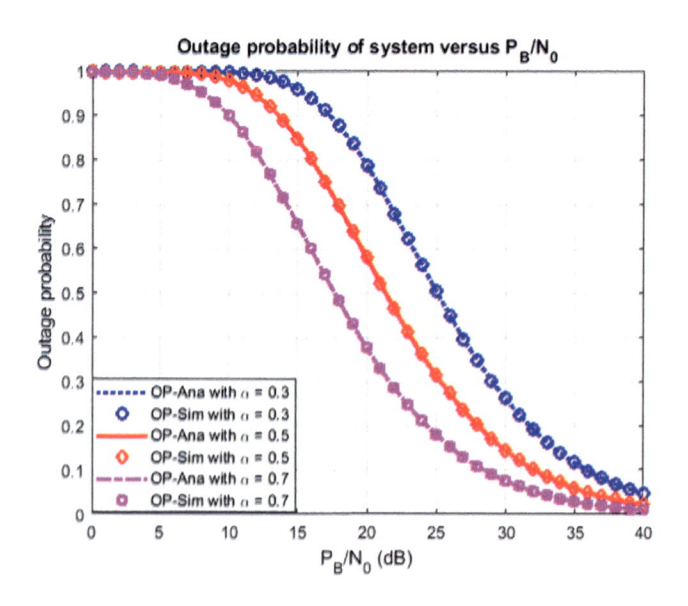

Fig. 3. Outage probability versus P_B/N_0.

but for high value of ρ, which satisfies the condition $\rho > \gamma_{th}(1-\rho)$, the intercept probability becomes positive and increasing. Another phenomenon occurs with $\alpha = 0.7$. In this case, the probability of successful communication decreases with the power-splitting factor for all values of ρ.

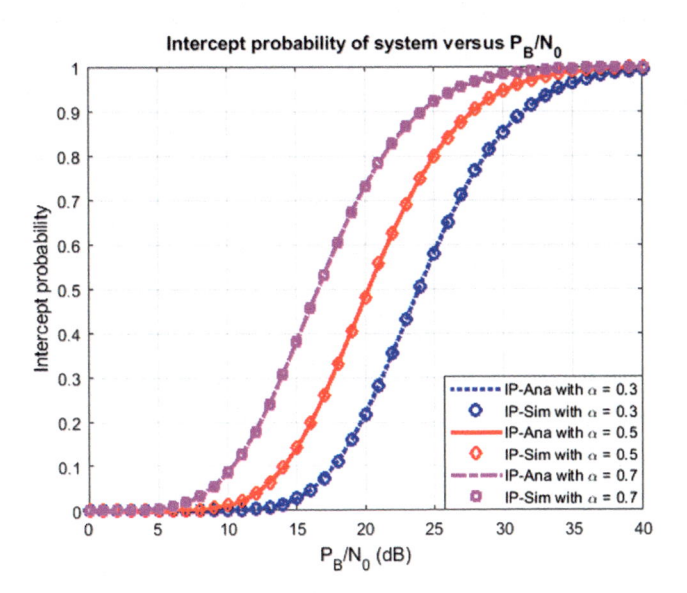

Fig. 4. Intercept probability versus P_B/N_0.

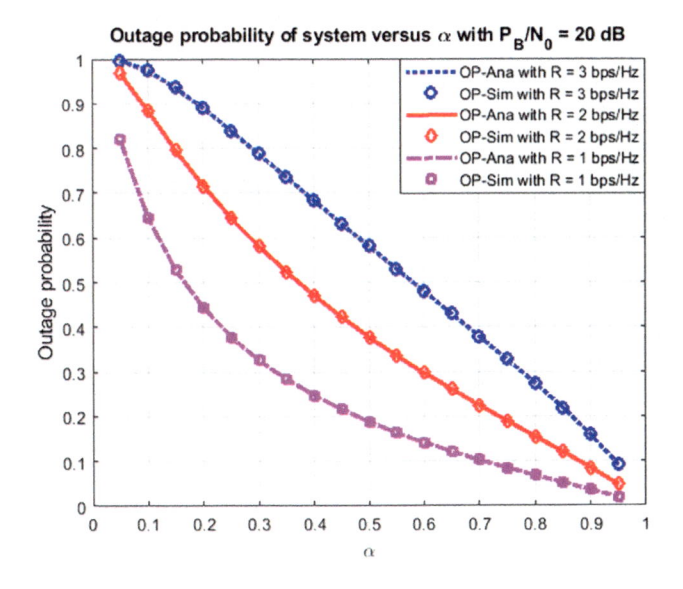

Fig. 5. Outage probability versus time switching factor.

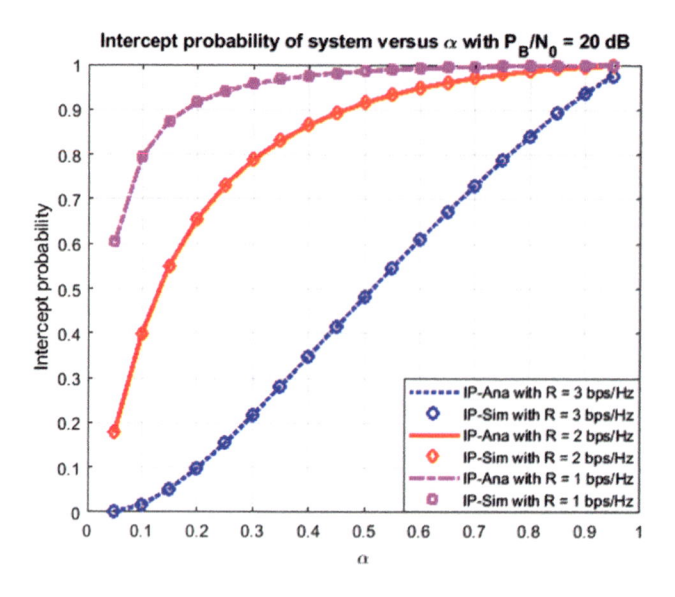

Fig. 6. Intercept probability versus time switching factor.

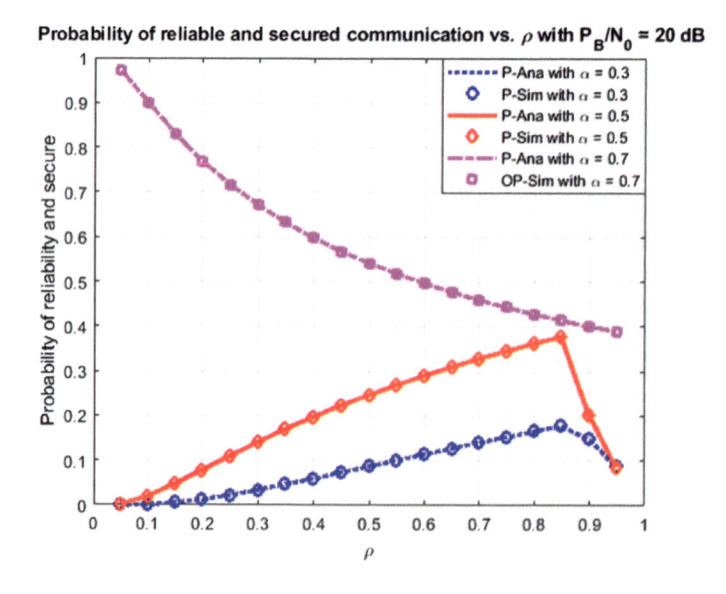

Fig. 7. Probability of reliable and secured communication versus power-splitting factor.

5 Conclusion

In this paper, we study the reliability and security performance of a dual-hop decode-and-forward half-duplex relay networks using RF energy harvesting with time-switching protocol and the assistant of a power beacon. The main contribution of this work is the rigorous analysis of the system performance to figure out the closed-form expressions for the outage probability, intercept probability, and finally, the probability of successful communication without being accepted. The analytical results are supported by Monte Carlo simulations as well. From both analytical results and simulation results, we can find the good configurations for the proposed system according to each specific condition of the network topology and channel gains. This can provide significant information to implement practical systems using the technologies mentioned in this paper.

Acknowledgment. This research received funding from the grant No. SP2018/59 conducted by VSB-Technical University of Ostrava, Czech Republic.

References

1. Alshaheen, H., Takruri-Rizk, H.: Energy saving and reliability for wireless body sensor networks (WBSN). IEEE Access **6**, 16678–16695 (2018)
2. Wei, Y., Ma, X., Yang, N., Chen, Y.: Energy-saving traffic scheduling in hybrid software defined wireless rechargeable sensor networks. Sensors **17**(9), 2126 (2017)
3. Lei, L., Yuan, D., Ho, C.K., Sun, S.: Optimal cell clustering and activation for energy saving in load-coupled wireless networks. IEEE Trans. Wirel. Commun. **14**(11), 6150–6163 (2015)
4. Ren, J., Zhang, Y., Zhang, N., Zhang, D., Shen, X.: Dynamic channel access to improve energy efficiency in cognitive radio sensor networks. IEEE Trans. Wirel. Commun. **15**(5), 3143–3156 (2016)
5. Ulukus, S., Yener, A., Erkip, E., Simeone, O., Zorzi, M., Grover, P., Huang, K.: Energy harvesting wireless communications: a review of recent advances. IEEE J. Sel. Areas Commun. **33**(3), 360–381 (2015)
6. Bi, S., Ho, C.K., Zhang, R.: Recent advances in joint wireless energy and information transfer. In: 2014 IEEE Information Theory Workshop (ITW), pp. 341–345, November 2014
7. Varshney, L.R.: Transporting information and energy simultaneously. In: 2008 IEEE International Symposium on Information Theory, pp. 1612–1616, July 2008
8. Zhou, X., Li, Q.: Energy efficiency for swipt in mimo two-way amplify-and-forward relay networks. IEEE Trans. Veh. Technol. **67**(6), 4910–4924 (2018)
9. Huang, K., Lau, V.K.N.: Enabling wireless power transfer in cellular networks: architecture, modeling and deployment. IEEE Trans. Wirel. Commun. **13**(2), 902–912 (2014)
10. Nguyen, T.N., Quang Minh, T.H., Tran, P.T., Voznák, M.: Energy harvesting over Rician fading channel: a performance analysis for half-duplex bidirectional sensor networks under hardware impairments. Sensors **18**(6), 1781 (2018)
11. Peng, H., Lin, Y., Lu, W., Xie, L., Liu, X., Hua, J.: Joint resource optimization for DF relaying SWIPT based cognitive sensor networks. Phys. Commun. **27**, 93–98 (2018). http://www.sciencedirect.com/science/article/pii/S1874490717306213

12. Grover, P., Sahai, A.: Shannon meets Tesla: wireless information and power transfer. In: 2010 IEEE International Symposium on Information Theory, pp. 2363–2367, June 2010
13. Zhang, R., Ho, C.K.: Mimo broadcasting for simultaneous wireless information and power transfer. In: Global Telecommunications Conference (GLOBECOM 2011), pp. 1–5. IEEE, December 2011
14. Nasir, A.A., Zhou, X., Durrani, S., Kennedy, R.A.: Relaying protocols for wireless energy harvesting and information processing. IEEE Trans. Wirel. Commun. **12**(7), 3622–3636 (2013)
15. Nasir, A.A., Zhou, X., Durrani, S., Kennedy, R.A.: Throughput and ergodic capacity of wireless energy harvesting based DF relaying network. In: 2014 IEEE International Conference on Communications (ICC), pp. 4066–4071, June 2014
16. Le, N.P.: Throughput analysis of power-beacon-assisted energy harvesting wireless systems over non-identical Nakagami- m fading channels. IEEE Commun. Lett. **22**(4), 840–843 (2018)
17. Zhou, X., Guo, J., Durrani, S., Renzo, M.D.: Power beacon-assisted millimeter wave Ad Hoc networks. IEEE Trans. Commun. **66**(2), 830–844 (2018)
18. Liang, H., Zhong, C., Lin, H., Suraweera, H.A., Qu, F., Zhang, Z.: Optimization of power beacon assisted wireless powered two-way relaying systems under user fairness. In: 2017 IEEE Global Communications Conference, GLOBECOM 2017, pp. 1–6, December 2017
19. Wyner, A.D.: The wire-tap channel. Bell Syst. Tech. J. **54**(8), 1355–1387 (1975)
20. Wang, L., Wong, K., Jin, S., Zheng, G., Heath Jr., R.W.: A new look at physical layer security, caching, and wireless energy harvesting for heterogeneous ultra-dense networks. IEEE Commun. Mag. **56**(6), 49–55 (2018). https://doi.org/10.1109/MCOM.2018.1700439
21. Obeed, M., Mesbah, W.: Efficient algorithms for physical layer security in two-way relay systems. Phys. Commun. **28**, 78–88 (2018). http://www.sciencedirect.com/science/article/pii/S1874490717304378
22. Shah, H.A., Koo, I.: A novel physical layer security scheme in OFDM-based cognitive radio networks. IEEE Access **6**, 29486–29498 (2018)
23. Huang, Y., Zhang, P., Wang, J., Wu, Q.: Secure transmission in power beacon assisted wireless communication networks. In: 2017 IEEE 28th Annual International Symposium on Personal, Indoor, and Mobile Radio Communications (PIMRC), pp. 1–6, October 2017
24. Jiang, X., Zhong, C., Zhang, Z., Karagiannidis, G.K.: Power beacon assisted wiretap channels with jamming. IEEE Trans. Wirel. Commun. **15**(12), 8353–8367 (2016)
25. Zhong, C., Jin, S., Wong, K.K., McKay, M.R.: Ergodic mutual information analysis for multi-keyhole MIMO channels. IEEE Trans. Wirel. Commun. **10**(6), 1754–1763 (2011)
26. Mo, J., Tao, M., Liu, Y.: Relay placement for physical layer security: a secure connection perspective. IEEE Commun. Lett. **16**(6), 878–881 (2012)
27. Duy, T.T., Son, P.N.: Secrecy performances of multicast underlay cognitive protocols with partial relay selection and without eavesdropper's information. KSII Trans. Internet Inf. Syst. **9**(11), 4623–4643 (2015)

Design of Electrical Regulated Drainage with Energy Harvesting

Vaclav Kolar[1(✉)], Roman Hrbac[1], Tomas Mlcak[1], and Jiri Placek[2]

[1] VSB-TU Ostrava, 17 listopadu 15, Ostrava, Czech Republic
vaclav.kolar@vsb.cz
[2] Dopravni podnik Ostrava, Podebradova 494/2, Ostrava, Czech Republic

Abstract. This paper deals with the process of designing electrical regulated drainage, which has been done by the authors in the past years. It is focused especially on the development of power sources. Electrical drainage is used for the protection of underground devices (pipelines) from the corrosive effects of stray currents. The drainage is connected between the pipeline and railway or tramway rails. This electrical regulated drainage can work without an external power source. Power supply from the power network or a battery is possible; however, users often come with the requirement that there should not be such a source. For this reason, energy harvesting systems were developed to get energy from current flowing through the drainage and voltage on the drainage, using them for the control circuits of the drainage. This paper describes the construction and laboratory testing of circuits developed for energy harvesting.

Keywords: Electrical drainage · Energy harvesting · Anticorrosive protection

1 Introduction to Electrical Drainage

In electrified railways or trams (electric traction), rails are used as conductors conducting electricity. Because rails are never perfectly isolated from the ground, current leaks and partly flows through it, thus giving rise to stray currents [1]. Trolleybuses are more convenient, as they create no stray currents [2]. If stray current flows through metal objects buried in the ground (e.g. pipelines), it contributes to the corrosion of these objects, which is both undesirable and dangerous. The corrosive effect of DC stray currents is much stronger than that of AC stray currents [3].

The formation mechanism of stray currents is shown in Fig. 1. The anode area of an underground device can be protected by electrical drainage, which is shown in Fig. 2.

The most basic and primitive drainage type is the so-called direct drainage. This is a simple conductive connection between an underground device and rails in the place where current usually leaves the underground device and returns to the rails. But the position of anode and cathode areas may change. In these cases, direct drainage might sometimes make the situation even more complicated.

An improved version is called "polarised drainage". In this version, the connection between the rail and the underground device is ensured by an element which makes it possible to make current flow only in one direction, e.g. a diode. This is a very widespread type of electric drainage.

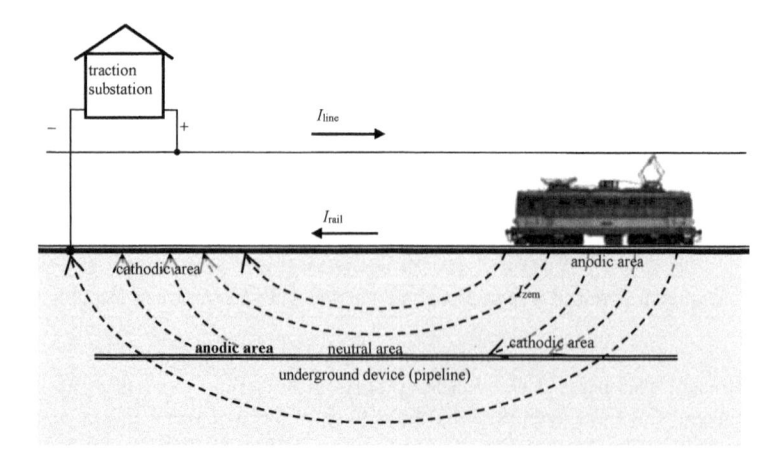

Fig. 1. Formation of stray currents caused by electric traction, and their flow through a underground device

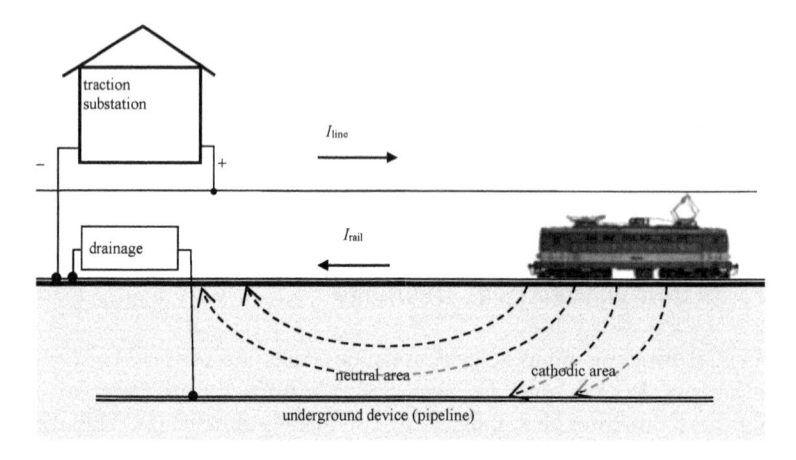

Fig. 2. Connection of electric drainage between rails and an underground device

Another development stage is the so-called regulated electric drainage, which has the same function as polarised drainage and in addition, this drainage can regulate the current flowing through it. This regulation becomes necessary when current values are too high and when the potential of the underground device reaches higher negative values. (This condition is never good for metals buried underground because it leads to the release of hydrogen and to embitterment of the metal). Regulation is usually controlled by a proportional or proportional-integral controller, but there are also adaptive regulation and self learning based solutions [4].

Regulated electric drainage is used especially in electrified railways.

Current regulation can be done either by means of switching resistors of different ohmic values (this solution was more common in the past), or using pulse width modulation (PWM), which is more frequent nowadays.

For the past two years, the authors of this paper have been working on the development of a new regulated electrical drainage system for tram tracks, which may also be used on railways in the future [5].

2 Energy Harvesting System

One of the requirements in designing the drainage was that it should be able to work without an external power source. Power supply from an external source or from a battery is possible, but the drainage also needs to be able to work without it, at least to a certain extent. In order to achieve this, alternative ways of energy harvesting from the rail - pipeline circuit were developed.

Figure 3 shows a simplified block diagram of the drainage. Some parts, such as overvoltage and overcurrent protection, are not shown in the diagram because they have nothing to do with the topic of this paper. This paper deals with an energy harvesting system. From this point of view, blocks 6, 7 and 8 are important. Block 8 works only as a simple charge pump consisting of resistors, condensers and a Zener diode voltage regulator. Its connection and function are not going to be described here. Blocks 6 and 7 are the key for energy harvesting.

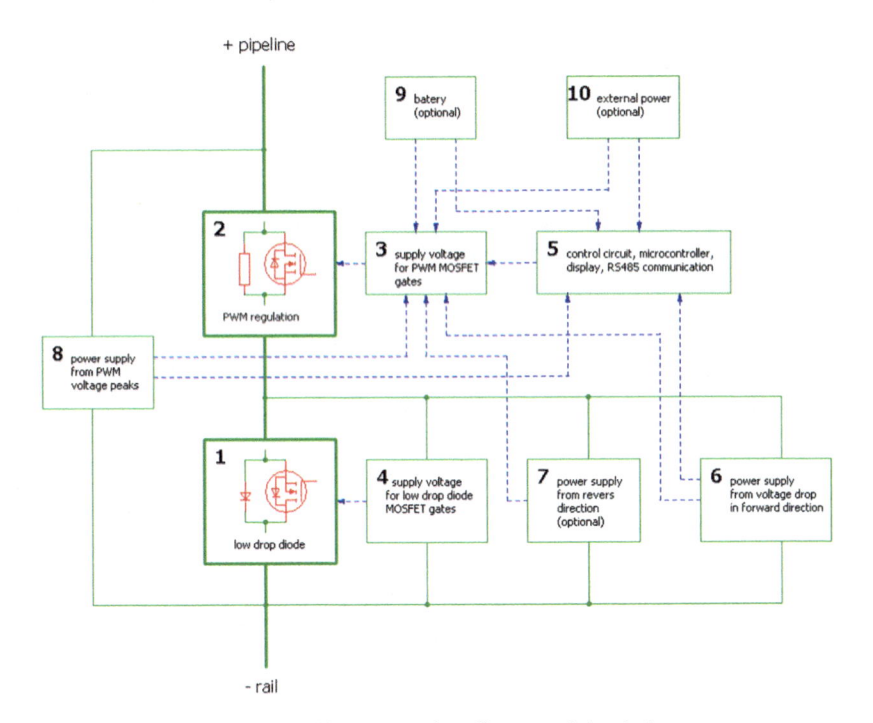

Fig. 3. Simplified connection diagram of the drainage

3 Block 6 – Power Supply from Voltage Drop in the Forward Direction

If there is current flowing through the drainage in the forward direction, voltage drop of a few mV to a few V occurs on the MOSFET transistors of the low drop diode. The resistance of the MOSFETs of the low drop diode is approximately 3 mΩ and the initial voltage drop for the MOSFETs to open is approximately 20 mV. The following applies to the voltage drop on the low drop diode:

$$V = 20 + 3 \cdot I \qquad (mV; A) \qquad (1)$$

During operation, current values are within ten to hundreds of A. For currents up to about 30 A, it is not necessary to regulate current, the drainage only works as polarised drainage and control block 5 does not have to work. This means that block 6, which supplies the control circuit, must be able to work with current from some 30 A, i.e. voltage starting on some 110 mV. Current taken by block 6 can be quite large; values within several amperes are acceptable. The diagram of the connection of block 6 is shown in Fig. 4.

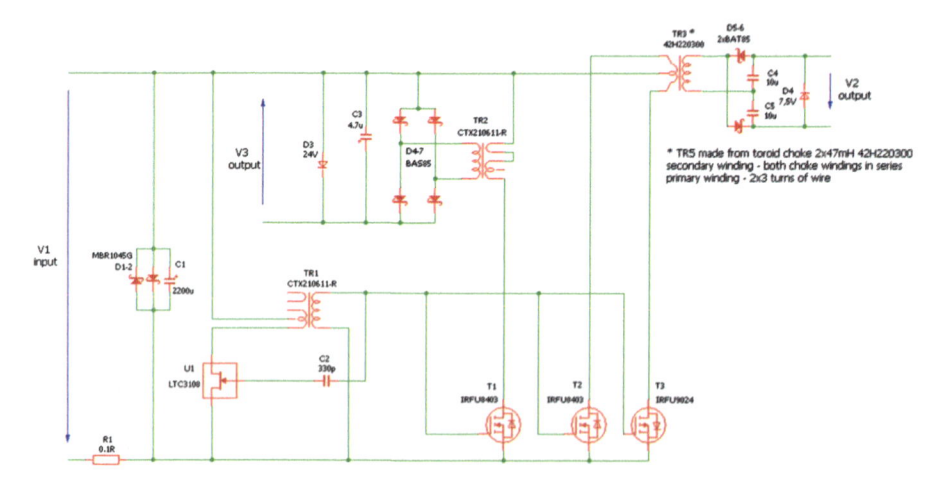

Fig. 4. Diagram of block 6 - supply from the voltage drop in the forward direction.

Block 6 supplies two galvanically separated output voltages. The first one, V2, is used to supply the control microcontroller. The microcontroller works at a supply voltage starting from 3.5 V, in two modes, which are switched automatically, based on how much energy is available:

1. Saving mode - consumed current is about 0.5 mA; the microcontroller controls the drainage, but neither the display nor outer communication works.

2. Standard mode - the microcontroller controls the drainage, after pressing a button it turns on the display and communicates with the user. The microcontroller and the display both consume approximately 2 mA together.

The second output voltage, V3, is used for supplying energy to MOSFET gates, which perform PWM. Here it is necessary for voltage to be at least 6 V, while consumption is lower than 1 µA. Block 6 is based on an integrated circuit LTC3108, which is originally an ultra-low voltage step-up converter, able to work with values starting at 20 mV. In this case, it is impossible to use the data sheet recommended application; because two galvanically separated output voltages and greater power are needed.

Only one JFET transistor of the LTC3108 integrated circuit is used, and it works together with the TR1 transformer as a very low voltage oscillator that starts to oscillate at about 20 mV. Output voltage from this oscillator drives MOSFETs T2 and T3, which are connected to transformer TR3, supplying voltage V2. The output of the oscillator also drives MOSFET T1, which is connected to transformer TR2, supplying voltage V3.

4 Block 7 – Power Supply in the Reverse Direction

If there is no current flowing through the drainage for a long time and no voltage is supplied by block 6, after a few tens of hours capacitors in block 3 become discharged, power MOSFETs in block 2 are closed and the drainage is turned off and does not even work as polarised (non-regulated) drainage. In this case, voltage of at least 6 V in the forward direction needs to be on the drainage while a train or tram is passing, so that the drainage can start working again. It was experimentally verified that most of the time, this "self-starting" occurs on the drainage when a tram passes by for the first time. It was also verified that capacitors in block 3 are not discharged even after the night break when trams are not running, so during normal operation, the above mentioned situation is not very likely to occur.

Nevertheless, in case it is needed, a solution was designed to charge capacitors on block 3 by drainage voltage in the reverse direction, where voltage on the drainage is almost all the time. In trams, it oscillates between 1 to 30 V, in trains it is approximately between 10 and 100 V. However, any current going through the drainage in the reverse direction is undesirable in terms of electro-chemical corrosion; this type of power supply has to work with very low input current values, within a few mA or even less.

A buck-boost converter was designed, based on a multivibrator built from a 74HC00 circuit. This circuit starts working at 1 V and when power values are low; its consumption is way under 1 mA. Figure 5 shows the diagram of Block 7.

Since only very little current may flow through the drainage in the reverse direction, block 7 does not supply the control circuit, whose consumption starts at 0.5 mA, but it only charges condensers in block 3, for which about 1 µA is needed.

Since voltage in the reverse direction is lower for trams than for trains, two different variants have been designed, one for trams and one for trains, differing in the value of

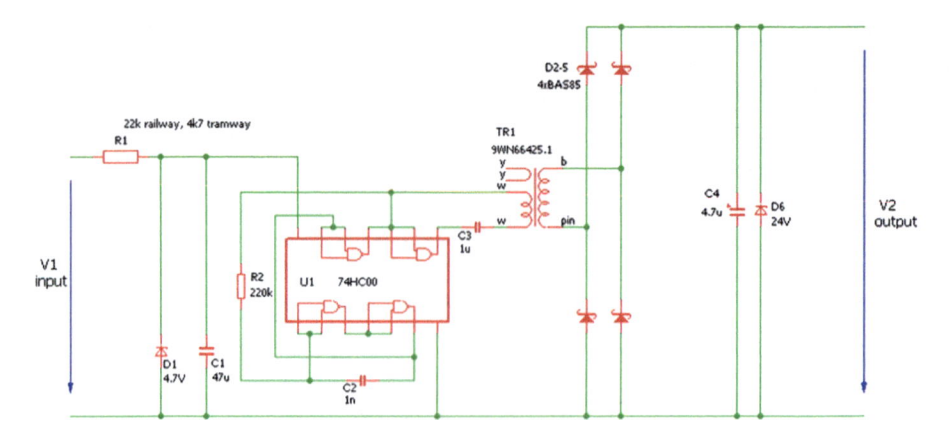

Fig. 5. Block 7 diagram - power supply from the reverse direction.

resistor R1. The tram version of the drainage works with the value 4.7 kΩ, while the railway version works with 22 kΩ.

5 Block 6 Parameters Based on Experiment Results

Functioning of block 6 was measured for supply voltage ranging from 50 to 350 mV, which corresponds to 10–110 A flowing through the drainage. Values of current flowing through the drainage were calculated based on relationship 1. Output voltage V2 was measured for two load values, 6.8 kΩ (corresponding to microcontroller consumption in saving mode) and 1.6 kΩ (microcontroller consumption in standard mode). Output voltage V3 was loaded with a 10 MΩ resistor, which is less favourable than the real situation. In reality, this voltage is used to supply the gates of MOSFET transistors, whose resistance is within GΩ. Measurement results are presented in Fig. 6.

Measurement results (see Fig. 6) show the following:

- Voltage V3 reaches the 6 V value (i.e. the value necessary to keep the drainage in the polarised non-regulated drainage state) when the value of current passing through the drainage is 15 A and more.
- Voltage V2 reaches 3.5 V (i.e. the value necessary for the microcontroller, so that the drainage may work as regulated drainage) when current passing through it is approximately 20 A for the saving mode and approximately 40 A for the standard mode.

These values are in compliance with the requirements concerning the operation of the drainage.

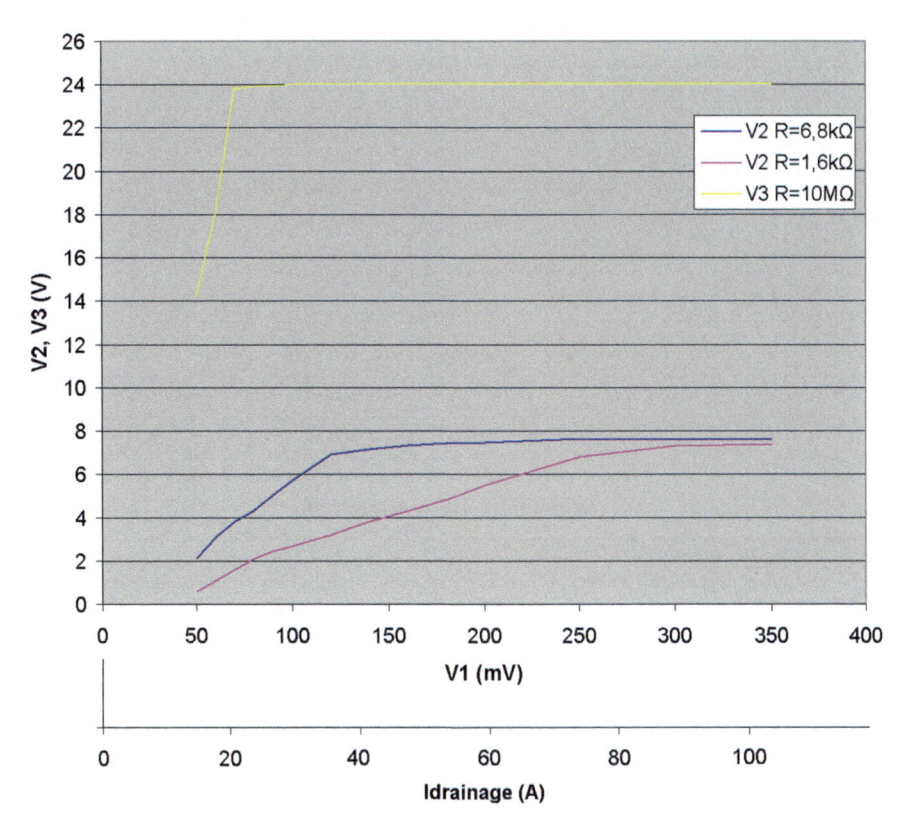

Fig. 6. Output voltages of block 6 as a function of drainage current and voltage drop on drainage

6 Block 7 Parameters Based on Experiment Results

Measurement results for the convertor supplied from the reverse direction are presented in Fig. 7.

During the experiment, the converter was loaded with a 10 MΩ resistor, which is less favourable than in reality.

The results show clearly that the converter reaches the necessary output voltage value of 6 V starting from approximately 2 V in the railway version of the drainage, and from 1.3 V for the tram version. These values are satisfactory in terms of the correct functioning of the drainage.

Current consumed by Block 7 when the drainage is reverse polarized, is within the mA range, which means that it is low enough.

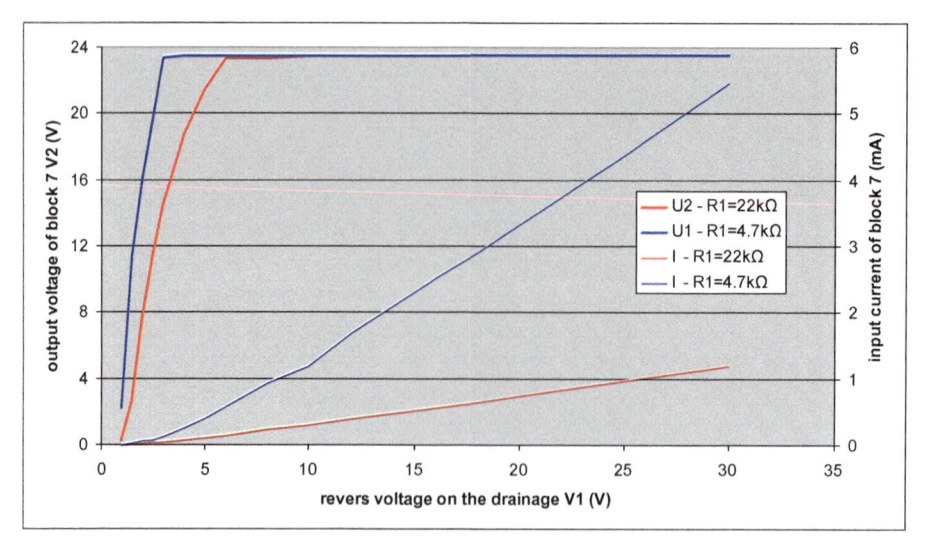

Fig. 7. Output voltage and input current of block 7 as a function of reverse voltage on the drainage.

7 Discussion

Converters developed to supply the drainage were selected as a compromise solution considering utility properties and construction complexity. Higher complexity would increase the price of the device, which would also become more susceptible to failure. In recent years, thermoelectric generators have been used to obtain energy necessary to supply devices like this drainage [6]. The authors have discarded this possibility, due to its complexity and little certainty of energy supply. Another popular form of energy harvesting, photovoltaic cells [7], was not used for similar reasons. After their previous experience, the users of the drainage were against using photovoltaic cells (installation outside the switchboard, necessity to keep the cells clean and remove snow, risk of damage by vandals). For this reasons, solutions relying only on energy coming from the rail-pipeline circuit were chosen. These solutions do not require extra work on the user side.

8 Conclusion

When developing electrical regulated drainage for trams and railways, converters were developed to supply the drainage by the current flowing through the drainage and from reverse polarised voltage on the drainage. The purpose of designing these converters is that the drainage does not have to be supplied from an external source or from a battery. When developing the converters, it was taken into consideration that power converters should be functional but also simple to install and that they should not require maintenance. All the solutions which were used were discussed with the contracting authority and the future users of the drainage. It was verified by experiment that the converters comply with the requirements related to drainage operation.

The tram version of the drainage was tested within the Ostrava tram network. It was verified on several locations that after installation, the drainage is able to start working even without a battery or an external power source. By the time this paper was written, it had been working well for over one year.

References

1. Kolar, V., Hrbac, R., Mlcak, T.: Investigating ground currents leaking from rails in DC and AC traction. In: Proceedings of the 8th International Scientific Symposium on Electrical Power Engineering, Elektroenergetika 2015, 16–18 September 2015. Technical University of Kosice, Stara Lesna (2015). ISBN 978-805532187-5
2. Bartlomiejczyk, M., Polom, M.: Spatial aspects of tram and trolleybus supply system. In: Proceedings of 8th International Scientific Symposium on Electrical Power Engineering (Elektroenergetika), Stara Lesna, pp. 223–227 (2015). ISBN 978-80-553-2187-5
3. Chen, Z., Koleva, D., Breugel, K.: A review on stray current-induced steel corrosion in infrastructure. Corros. Rev. **35**(6), 397–423 (2017)
4. Repka, M., Danel, R., Neustupa, Z.: Intelligent control of treatment technological processes in preparation plants using neural networks. In: SGEM2013 Conference Proceedings, 16–22 June 2013, pp. 441–447 (2013). ISBN 978-954-91818-9-0/ISSN 1314-2704
5. Hrbac, R., Kolar, V., Mlcak, T.: Limiting the influence of regulated electrical drainage on track circuits. Elektronika ir Elektrotechnika **2**(5), 99–110 (2016). ISSN 1392-1215
6. Yan, J., Liao, X., Yan, D.: Review of micro thermoelectric generator. J. Microelectromech. Syst. **27**(1), 1–18 (2018)
7. Ali, S.A., Ali, S.H., Khan, S.N., Aman, M.A.: Energy harvesting for remote wireless sensor network nodes. Int. J. Adv. Comput. Sci. Appl. **9**(4), 198–203 (2018)

Analysis of Appliance Impact on Total Harmonic Distortion in Off-Grid System

Michal Petružela[(✉)], Vojtěch Blažek, and Jan Vysocký

VŠB – Technical University of Ostrava,
17. listopadu 15/2172, 708 33 Ostrava – Poruba, Czech Republic
{michal.petruzela,vojtech.blazek,jan.vysocky}@vsb.cz

Abstract. Smart Grid is name for Off-Grid system controlled by intelligent data analysis, predictive models and real-time adjustments. They are one of the possible future options of power engineering with focus on energy supply decentralization at various scales from small configurations used in single house or cabin to the large systems operating on scale of small village or part of town. Power sources of this systems are predominantly renewable power sources which are lower and have stochastic nature. This leads to lower short-circuit power which is one of key characteristics of such system. Outcome of this is possible difficulties of sustainable power quality (PQ). In our, paper we present an analysis on the lowest possible level, where we examined impact of appliance towards response of PQ parameters in this case, total harmonic distortion of the voltage (THDV). From the results of analysis, we can say there are several types of appliances that affect THDV more and these appliances must be taken into account if we want to maintain certain level of THDV.

Keywords: Power quality · Total harmonic distortion of voltage · Smart Grid · Off-Grid system

1 Introduction

So-called Off-Grid systems are autonomous energy systems supplied from renewable energy sources (RESs) and utilise some sort of energy storage usually in form of battery bank. Compared to the standard distribution and transmission systems, application of such an energy source implies different grid parameters.

Stability of power quality (PQ) parameters in the Off-Grid systems are conjoined with short-circuit power and RES has stochastic and unstable character which causes lower short-circuit power and therefore lower stability of PQ parameters [1, 2]. Among the other parameters, THDV belongs to the most important PQ parameter [3]. According to electromagnetic compatibility (EMC) standards EN 61000, THDV has to be lower than 8%. Currently, there are no requested limits for the THDV in Off-Grid systems but ordinary appliances are designed for use and connection to On-Grid system where is keeping the requested limits expected [4]. For safe and reliable operation and to prevent premature ageing of appliances, it is better that we keep PQ parameters in limits even in the Off-Grid [5].

This raised an interest of the researchers in the development of optimization tools that can detect and backwards optimize PQ parameters to fulfil the limits of internationally defined standards and norms. Initial concept of load control named Active Demand Side Management (ADSM) was proposed and designed by [6] and [7] proposed a power quality forecasting module. Results of this measurements will be incorporated into ADSM's power quality forecasting module.

Our goal is to find out which appliances negatively influence PQ in our Off-Grid system. We had 10 different appliances that created 120 combinations of three and these combinations were switched on and off at defined intervals. We measured and stored systems THDV response for further evaluations. They revealed the appliances that affect THDV more than others. Measurements took two and a half days, as we focused on the effect of individual appliances on THDV in the off-grid system In the future, this measurement will be done over a longer time period and with more appliances. According to the negative effects of the appliance on the THDV, the appliance will be classified in the future. Individual groups will have their weight. The biggest weight will have the group of appliances with the worst impact on THDV. The ADSM algorithm will continue to work with these groups to avoid reaching the unwanted level of THDV. The article focuses on the issue of poor power quality from the point of view of electricity.

2 Experiment Description

Power quality measurement and evaluation is done in accordance with European standard EN 50160 and EN 61000. According to standards, short measuring interval of THDV is 10 min and long interval is 2 h. For the reason of feasibility of the experiment and because some appliances like microwave and kettle aren't designed to be working for longer periods of time, we chose to do short measuring interval.

As we mentioned, before ten appliances were combined into triples creating 120 combinations in total. At once, there are connected three appliances because in normal household it is quite uncommon to have only one appliance running. To measure system's THDV response in our Off-Grid, we turned on appliances for an equal time (12 min). To avoid stochasticity caused by RES, ensure repeatability of the experiment and keep short-circuit power low, system was supplied by charged batteries. After each measure of triplet, batteries were re-charged to the same defined state of charge (SoC). To achieve different SoCs, we had to adjust charging voltage of the hybrid inverter. To truly ensure that batteries are at desired SoCs, the stage of battery charging lasts 18 min. SoC is reached when the voltage of battery bank is approximately same as the charging voltage of the hybrid inverter.

Appliances in Table 1 were selected with the intention to incorporate all types of the most influencing appliances in a household. All measurements were conducted at two SoCs (50% and 100%) of the battery to examine if and how much is THDV affected by it. KMB SMC 144 device is used for THDV measurement with time step set to one minute. For every minute minimum, maximum and average values of THDV, load and power factor were stored.

Table 1. Characteristic features of used appliances

Appliance	Load (W)			Power factor (−)		Power Supply
	avg	min	max	avg	Characteristic	
Mower	537.6	532.1	549.27	0.52	Inductive	Continuous
Drill	157.1	149.5	167	0.49	Inductive	Continuous
Kettle	619.1	617	628.3	1	Resistive	Continuous
Fridge	207.6	195.5	219.5	0.72	Inductive	Continuous
Switched-mode power supply	410	409.7	420.2	0.78	Capacitive	Continuous
AC heating	880	852.5	910	0.91	Inductive	Continuous
Microwave	203	76.8	1348.3	0.84	Inductive	Switched
Boiler	307	305.8	346.5	0.99	Resistive	Continuous
TV	44	42.8	50.5	0.6	Capacitive	Continuous
Lights	156	152.5	165.1	0.84	Capacitive	Continuous

2.1 Platform Description

Figure 1 shows our Off-Grid system where under normal conditions are connected two photovoltaic panels 2 kWp each. But for reasons mentioned before, batteries supply all of the energy consumed by appliances.

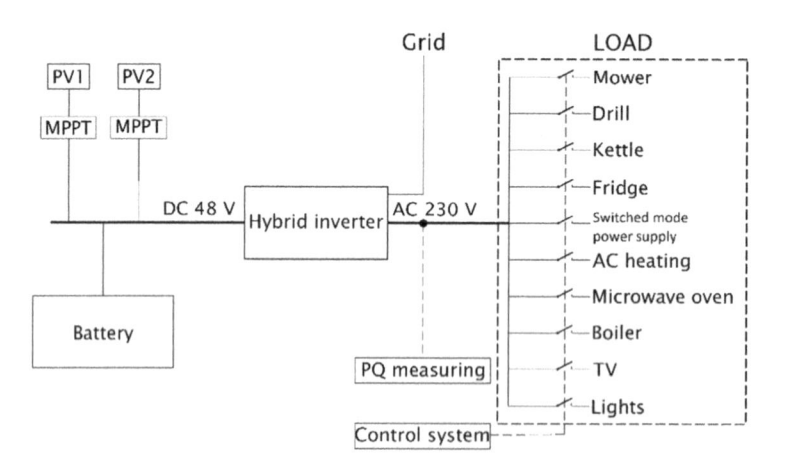

Fig. 1. Scheme of a testing platform

Off-Grid system in this experiment consists of hybrid inverter, battery bank and PQ measuring device. Hybrid inverter is Schneider Electric Conext XW+ 8545 with continuous output power 6.8 kW, this inverter also charges battery bank composed of 37 Ferrak 375 KPL batteries. Batteries are NiCd with 375 Ah capacity and 1.2 nominal voltage. AC distribution grid and appliances are connected to the hybrid inverter via switchboard controlled by the control system. This control system is automatically switching on appliances for measuring and alternating current (AC) grid for battery recharging according to schedule. Measured PQ values are stored in system's database for further analysis.

3 Testing and Results

As mentioned minimal, maximal and average values of each minute were measured and stored for 12 min period. First and last minute of measuring period contains transient phenomenon caused by switching and for this reason, they were discarded.

Figure 2 shows the measured full cycle (120 combinations), charging and discharging at both the SoCs. From these results, we can say that SoC of batteries has very little influence on THDV. Although, in all cases, THDV limit of 8% wasn't exceeded we examined all combinations that exceeded THDV 5%, Fig. 3 shows these 23 cases more in detail.

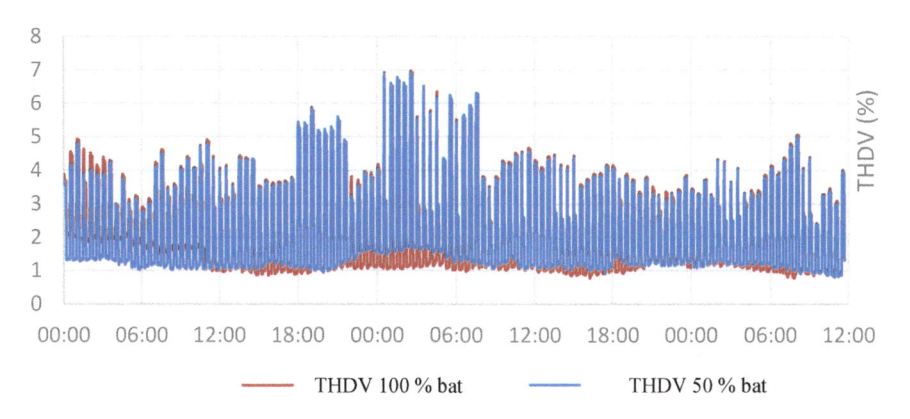

Fig. 2. Measured average values of THDV at both SoCs.

In all but one cases that exceeded THDV 5%, one of running appliances was drill. This is because drill utilises universal motor and most of distortion comes from sparking at the commutator. Another noticeable THDV change is in 6 cycles where is AC connected. At the start is THDV high and load isn't at it's maximum, after 3 min of standby mode AC starts cooling, load grows and THDV drops to values under 5%, for example, we can observe this situation on measure cycle starting after 120th min. Another troublesome appliance is microwave, not only load of microwave isn't continuous but switched as mentioned in Table 1. When running flux density in magnetic core is high and as a result there are higher harmonics of current that creates higher harmonics voltage drops that deforms fundamental voltage. Microwave was also in 6 cases that surpassed 5% THDV. Last but not least appliance that affected THDV was switched-mode power supply even though this was a laboratory power supply where better filtration is expected in comparison to the switched-mode power supply used for example in personal computer, still took part in 8 cases that had THDV higher than 5%.

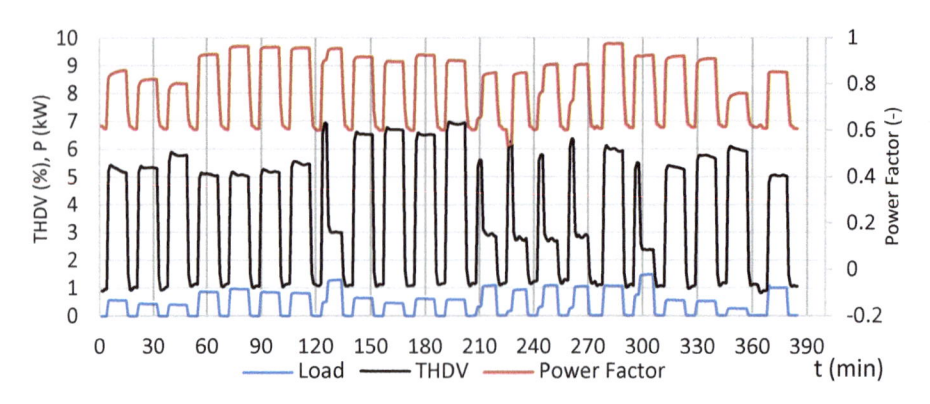

Fig. 3. Selected measure cycles of combinations that exceeded THDV 5%

4 Conclusion

Results and their examination presented in this paper show that SoC of batteries has negligible effect on THDV and confirmed that certain appliances have bigger impact on THDV than the simpler ones that have continuous load with power factor very close to the 1 in our case represented by boiler and kettle. These certain appliances are drill with universal motor and sparking on commutator, switched-mode power supply due it's way of operation, AC during standby mode where load is caused only by electronics and filter isn't presumably designed for AC in this mode, and the last one is microwave that has switched load and on top of that high voltage transformer inside microwave has very high saturation of magnetic core that causes higher harmonics of current.

Nevertheless, no combination of appliances caused THDV to grow past 8% limit given by the EN 61000. One of possible causes is that the utilised hybrid inverter has nominal power of 6.8 kW and largest load of connected appliances was 1.8 kW, so the inverter was operating in low load zone where it is able to sustain all PQ parameters in required limits.

Our future work is to connect load with power consumption closer to the nominal power of the hybrid inverter and add appliances of everyday use that have higher negative impact on THDV.

Acknowledgement. This paper was supported by the following projects: LO1404: Sustainable development of ENET Centre; CZ.1.05/2.1.00/19.0389 Development of the ENET Centre research infrastructure; SP2018/58 Students Grant Competition and TACR TH01020426, Czech Republic and the Project LTI17023 "Energy Research and Development Information Centre of the Czech Republic" funded by Ministry of Education, Youth and Sports of the Czech Republic, program INTER-EXCELLENCE, subprogram INTER-INFORM.

References

1. Eckbert, D., Larsson-Edefors, P.: Interconnect-driven short-circuit power modeling. In: Euromicro Symposium on Digital Systems Design, pp. 414–421. IEEE (2001)
2. Saradarzadeh, M., Farhangi, S., Schanen, J.-L., Jeanin, P.-O., Frey, D.: Combination of power flow controller and short-circuit limiter in distribution electrical network using a cascaded h-bridge distribution static synchronous series compensator. In: IET Generation, Transmission & Distribution, vol. 6, no. 11, pp. 1121–1131. IET Digital Library (2012)
3. Broshi, A.: Monitoring power quality beyond EN 50160 and IEC 61000-4-30. In: 2007 9th International Conference on Electrical Power Quality and Utilisation, EPQU, pp. 1–6. IEEE (2007)
4. Mišák, S., Stuchlý, J., Vramba, J., Prokop, L., Uher, M.: Power quality analysis in off-grid power platform. In: Advances in electrical and electronic engineering, vol. 12, no. 3, pp. 177–184. AEEE (2014)
5. Bhattacharyya, S., Myrzik, J., Kling, W.: Consequences of poor power quality – an overview. In: 42nd International Conference On Universities Power Engineering, UPEC 2007, pp. 651–656. IEEE (2007)
6. Mišák, S., Stuchlý, J., Platoš, J., Krömer, P.: A heuristic approach to active demand side management in off-grid systems operated in a smart-grid environment. Energy Build. **96**, 274–284 (2015)
7. Vantuch, T., Mišák, S., Ježowicz, T., Buriánek, T., Snášel, V.: The power quality forecasting model for off-grid system supported by multiobjective optimization. IEEE Trans. Industr. Electron. **64**(12), 9507–9516 (2017)

Influencing of Current Sensors by an External Magnetic Field of a Nearby Busbar

Tadeusz Sikora[(✉)] [iD] and Jan Hurta

VSB-Technical University of Ostrava, Ostrava, Czech Republic
tadeusz.sikora@vsb.cz

Abstract. Based on several identified erroneous current measurements using clamping current measuring transformers, a methodology for detecting the influence of near busbar magnetic field to current sensors was designed and tested. With a 600 A current in the busbar, the ammeters and current sensors showed a current of up to 11 A, even if they did not cover any conductor. The error was dependent on the position of the current sensor and also the distance. Measurement errors were detected by some sensors even more than 50 cm from the busbar. To compare individual ammeters and current sensor a method was proposed to measure and calculate relative error value in percent of current range per kiloampere of interfering busbar current.

Keywords: Current measurement error · Current sensor crosstalk error · Disturbing magnetic field · Busbar magnetic field · Current sensor

1 Introduction

Connecting temporary meters in low-voltage substations is routine for most technicians. Choosing the location of current sensors for measurements of just a short duration, usually leads to choosing the easiest place to connect. Sometimes this hastened process, when we are focusing only on the electrical (of conductivity) part of electrical engineering during installation, can backfire.

Let us now describe one such measurement, which basically led to further examination of the issue and to writing this contribution. The aim of the measurement was to verify, by measuring with an independent, calibrated device, the output of an approximately 1 MVA generator, during warranty testing. The entire duration of measurement was to take 2 h, so the measuring instruments were connected in the easiest way, see Fig. 1.

The input from the generator was from below, via the generator switch and then led to the busbars where the current transformers (CT 1 to CT 3, see picture) are installed. At the output from the metering transformers with a ratio of 2000/5 A, clamp current sensors were connected (essentially also current measuring transducers) of the MN71 [1] type with a 10 A RMS range.

MN71 clamp probes were connected to the calibrated measuring device, the voltage inputs of the instrument were connected directly to the 400 V busbar. Measuring commenced, the warranty test section was measured, and the Generator Output Report was processed and submitted.

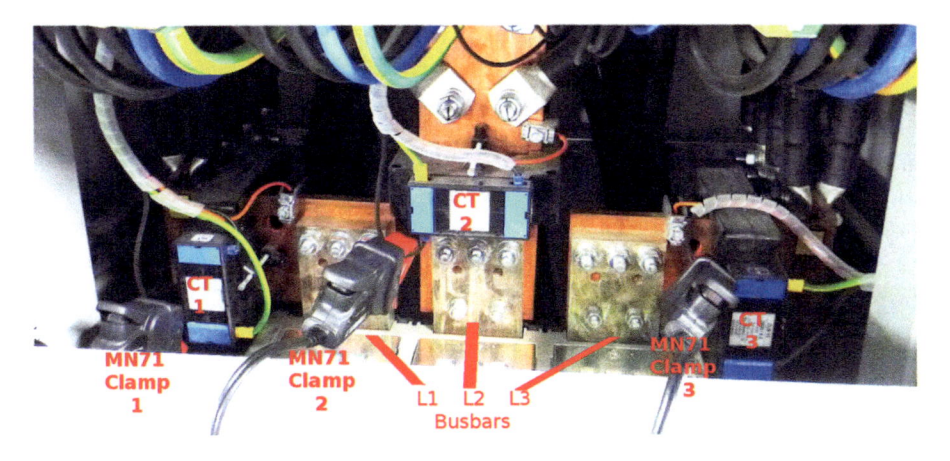

Fig. 1. Current sensors connection

However, the measured output of the generator significantly differed, compared with the operating measurements, which showed a value about 10% lower. Therefore, it was necessary to carry out a repeated warranty test and, above all, to determine who was to blame: the supplier of the cogeneration unit, of which the generator was a part, the customer and its operational measurement, or the supplier of the warranty tests.

2 Analysis of Erroneous Measurement

Through thorough analysis, it was found that the error is most likely on the part of the measuring described above, because the recorded output current of the three-phase synchronous generator was very unsymmetrical, see Fig. 2 (upper three waveforms).

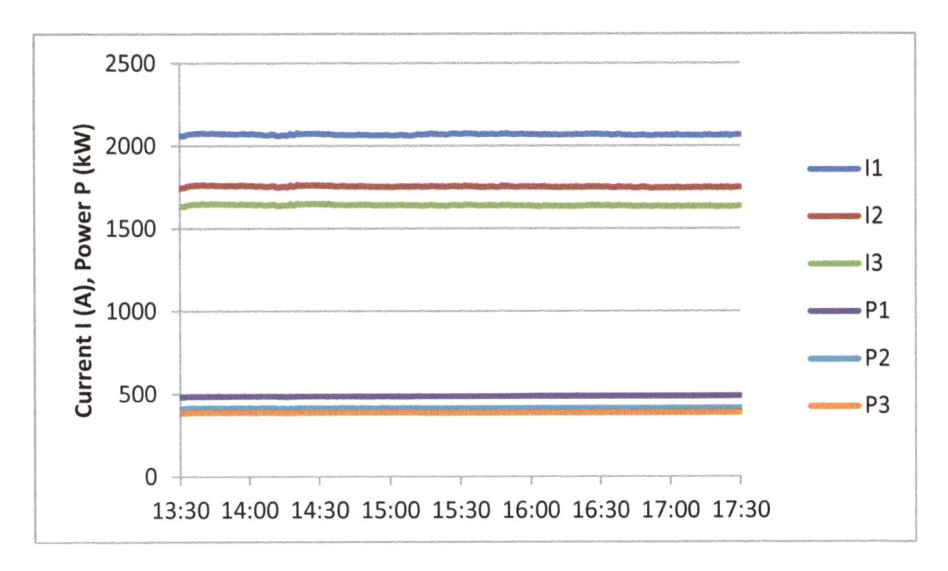

Fig. 2. Current and active power course during erroneous measurement

Equally unsymmetrical was the produced active power (lower three waveforms). Lack of symmetry or the non-zero sum of phase currents is sign of current sensor fault [2].

Thus, the suspicion arose that the MN71 current sensors, which measured the secondary current of the 2000/5 A current transformers (i.e., the current of several amperes) are influenced by the external magnetic field of the busbars as the busbars had a roughly 400× higher current flow. This suspicion proved to be correct at the next measurement, when the current probes were again connected in the same way, and with a slight change in the position of the current sensors, the measured current changed by percentage points.

Also interesting was the finding that the sensors were affected only by the own bus phase, but not by the next phase current. If this were the case, then there should be a phase shift, since the current of the secondary phase is shifted by 120°. Based on the measuring, however, the phase shift of all three phases approaches the same value – one, see Fig. 3.

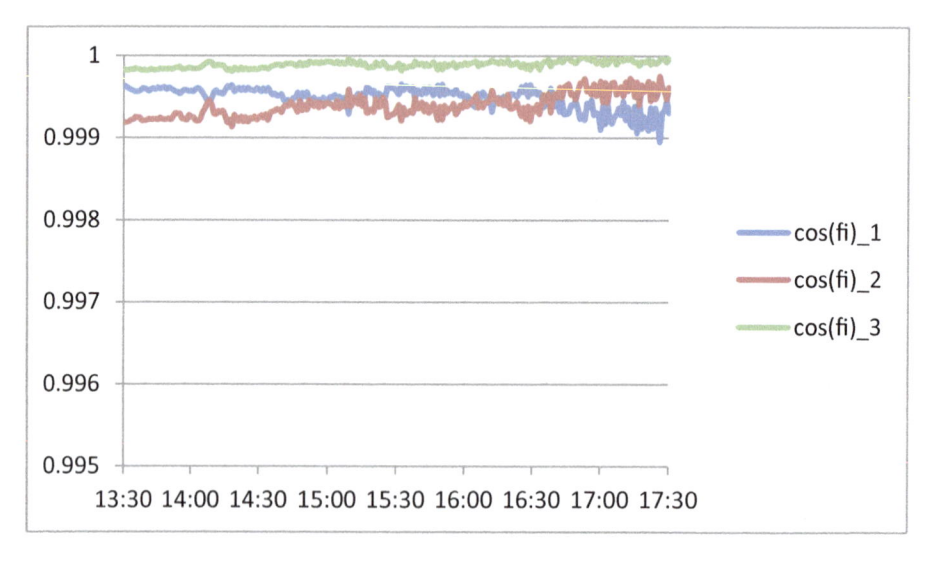

Fig. 3. First harmonic order cos(fi) course during erroneous measurement

The entire measurement therefore had to be repeated using Rogowski sensors (AmpFlex [3]), which were connected directly to the high-current busbar. With this measurement, no noticeable current asymmetry occurred.

So even with progress in field of electrical current sensing (e.g. non-invasive current sensing for multi-core cables [4, 5], optical current sensing [6], or coreless current transducers [7]), the influence of external fields must be taken into account.

Since many measurements are carried out within the VŠB-TU Ostrava, where temporary measurements using current sensors are installed in the switchboards, it was advisable to examine this issue more closely.

3 Method Used

A source with a maximum current of 600 A was used for the experiment. It was connected using four 150 cm cables to two connected copper buses with a cross-section of 2× (5×30) mm and 150 cm in length. The busbars were laid on a wooden veneer base and mounted. The wooden veneer table was placed on two plastic supports at a height of about 50 cm, since on the floor, or on laboratory tables, the rebar in the concrete or metal construction of the table could affect the measurements. A scale of distance from the edge of the busbar was marked on the wooden table (Fig. 4).

Fig. 4. Current source and test busbar

All devices were measured in three "axes" according to their mutual parallel, or the perpendicularities of the busbar axis, the longitudinal axis of the device and the loop of the current transformer. Individual axes are illustrated in the following figure.

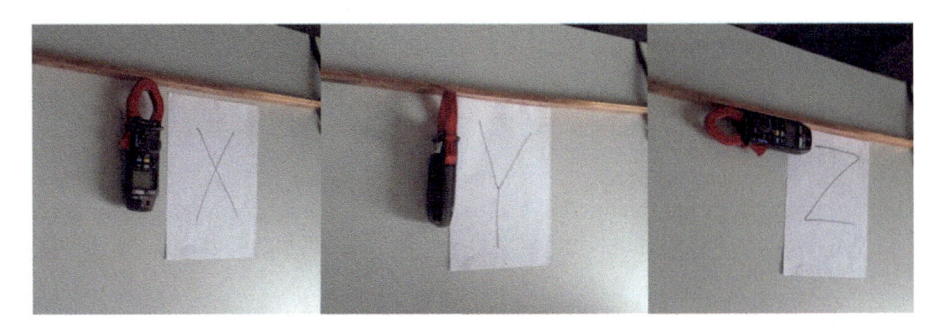

Fig. 5. Illustration of current meter's positions to the busbar (zero-distance)

The values shown by individual instruments in each of the axes (X, Y and Z) were read at 0, 0.5, 1, 2, 3, 5, 10, 20, 25, 50 and 100 cm distance from the busbar. The status was considered to be zero when the clamps directly touched the busbars, as illustrated in Fig. 5. In addition, one measurement was made in a special position where the center of the attachment gap was directly above the busbar.

The individual devices were measured at these current ranges (maximum range is always used): MN71 10 A, Mastech 1000 A, Range 200 A [8], Prova 100 A [9], CA F205 600 A [10]. The exception among the measuring instruments consists of MN 71 clamps which are only a current sensor, more precisely a current transformer with load resistance (10 A/1 V transmission). These clamps were connected to the ESCORT 3136 A Multimeter (milivoltmeter). The measurements were made at 100, 300 and 590 A current flowing through busbar.

4 Results and Analysis

The devices were measured from the position on the busbar and at distances from 0 to 100 cm from the busbar, each in three axes. If there is no value (or zero) from a certain distance, the other values were not measured.

To display all measured values in graphs with logarithmic coordinates, a fictitious distance was assigned to two special positions (no negative or zero values can be displayed in the logarithmic chart): 0.01 cm for position on the busbar, 0.1 cm for zero distance from the busbar.

In order to be able to compare the individual devices with each other and for extrapolation of the measured currents for other current values in the busbar, a relative error value in range was introduced, depending on the current of the busbar. The formula for this variable and an example of the calculation is given below:

$$I_{err} = \frac{I_m}{I_R \cdot I_B} \cdot 100\% = \frac{6,34}{600 \cdot 0,590} \cdot 100\% = 1,709 \quad (\%/kA; \ A, \ A, \ kA, \ \%) \quad (1)$$

In Eq. (1) I_{err} is relative current error value, I_m is measured erroneous current, I_R is current range of measuring device and I_B is current through busbar.

The resulting error is in percent of the range per 1 kA of current flowing through the busbar. Thus, this formula calculates the linear dependence of the wrong reading and the current of the busbar that generates a disturbing magnetic field. Later we will show that this assumption is of limited validity.

4.1 Testing of Clamp Ammeters

Below are the graphs of the measured values for all devices in their individual positions with the current of the bus being 590 A (the impedance of the current circuit was higher than the current source could provide; this value is the maximum of the source) (Figs. 6, 7 and 8).

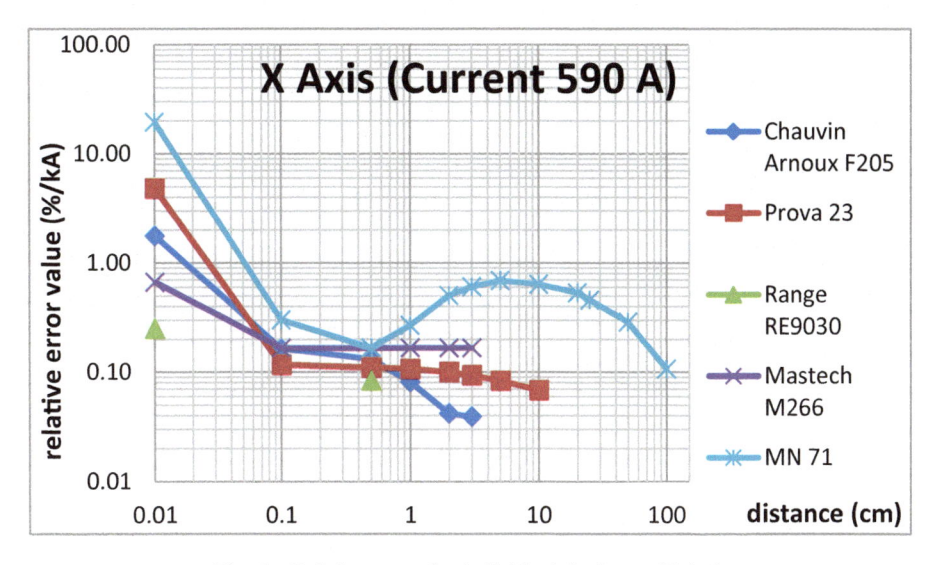

Fig. 6. Relative error for individual devices – X Axis

From the graphs, it follows that the position most susceptible to influence is that of position Y: the device axis and the level of the loop are perpendicular to the axis of the busbar.

Erroneous values were recorded for most devices in all positions up to a distance of 10 cm from the busbar, and two devices recorded an error current at a distance of about 20 cm but at a level of up to 0.1% of the range in kA of the busbar current.

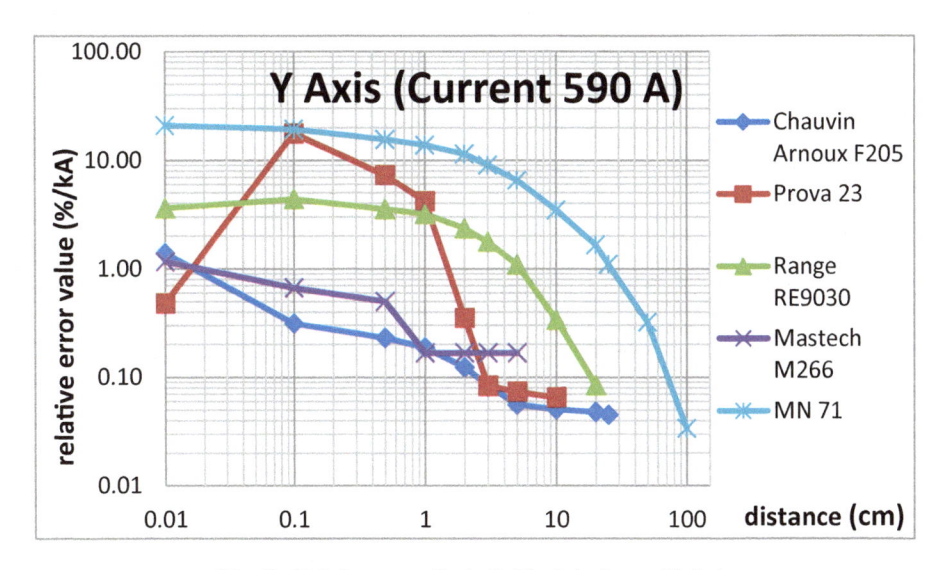

Fig. 7. Relative error for individual devices – Y Axis

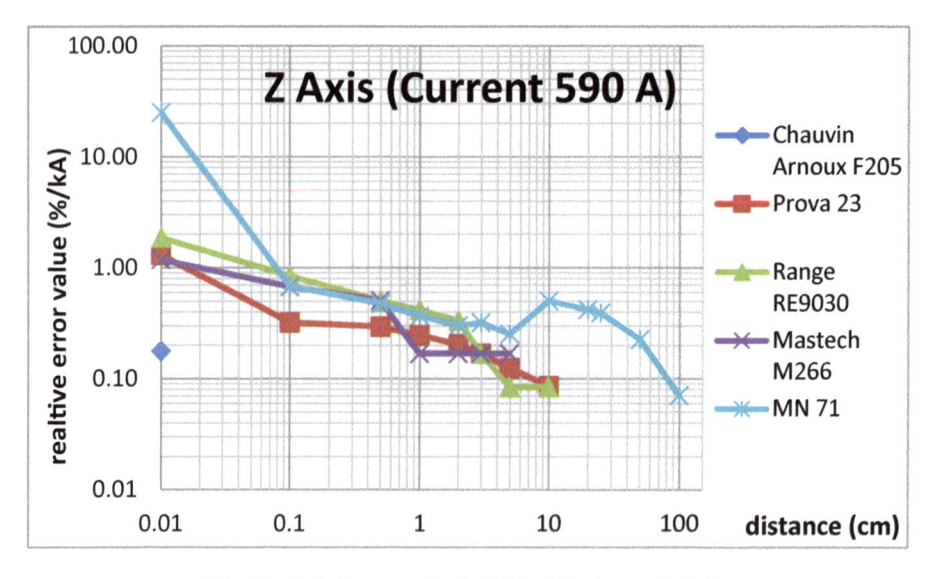

Fig. 8. Relative error for individual devices – Z Axis

The exception is the MN71 sensor, which shows the highest error in all positions, up to a maximum measured distance of 100 cm. This is a sensor similar to that which was mentioned in the introduction of this paper. This is just a sensor that is connected by a meter-long cable to the measuring device. This cable is a simple twin-cord, unshielded and uncrossed, and therefore very easily affected.

Thanks to this antenna, or more precisely, loop, through which the AC magnetic flux flows from the busbar, the interference range is much higher than for the clamp ammeters. The geometry of the measuring apparatus during the test is shown Fig. 9. The diagram is sketched approximately to scale. It shows a high current source,

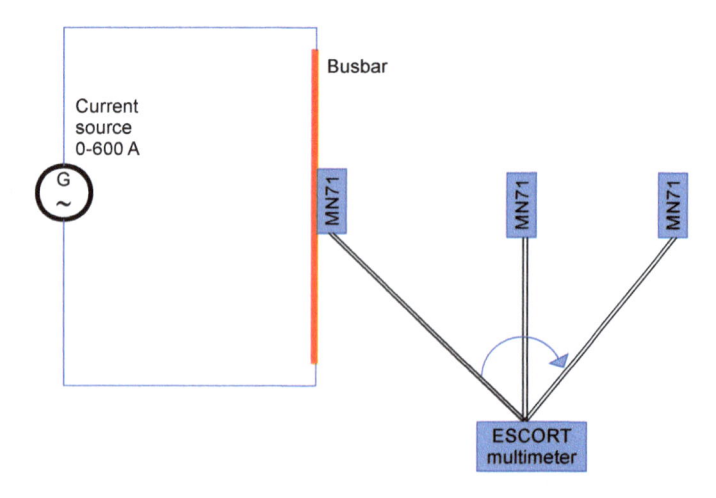

Fig. 9. Measurement geometry of influencing MN71 clamps with multimeter

a supply current line, a coupled busbar pair, and also MN71 clamps at various distances from the busbar, and finally an ESCORT multimeter.

4.2 Testing the Current Sensors of the Network Analyzer

In the next part, the current sensors of the network analyzer were tested, namely the ENA330 [11] model by Elcom, a.s. The standard accessories for measuring current are the MN71 clamps and Rogowski coil sensor AmpFlex A106. For the MN71 clamps, the same axes are applied as those for the measurement in the previous case. But the Rogowski sensor has a different geometry, so here is a photograph of the sensor in all three positions.

Both sensors have switchable ranges within the network analyzer: MN71 clamps have ranges of 2 and 10 A, Rogowski sensors 30, 300 and 3000 A (Fig. 10).

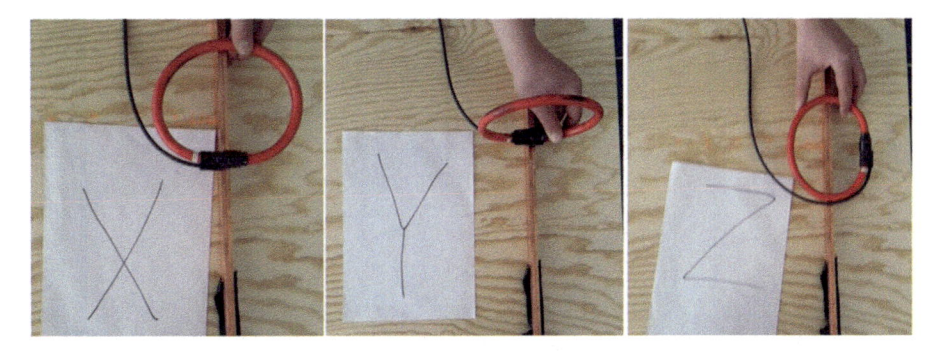

Fig. 10. Illustration of Rogowski coil's positions to the busbar (zero-distance)

The analyzer was tested at 10A range for MN71, or 30 A for AmpFlex A106, respectively, and all three axes and two busbar currents: 300 and approx. 550A. The values of the relative current errors calculated from the measured values are given in the following two graphs in Figs. 11 and 12. It is noteworthy that for both graphs and in all axes the proportional current errors related to the 1 kA current of the busbar were almost identical.

The Rogowski coil had a relative error of up to 70%. It should be borne in mind, however, that the measurements were made on a 30 A range, which is quite atypical for a sensor of this size. It is very likely that the 300A range would be an error of only 7%, and the 3000A range should already be below one percent of kA of the interfering current.

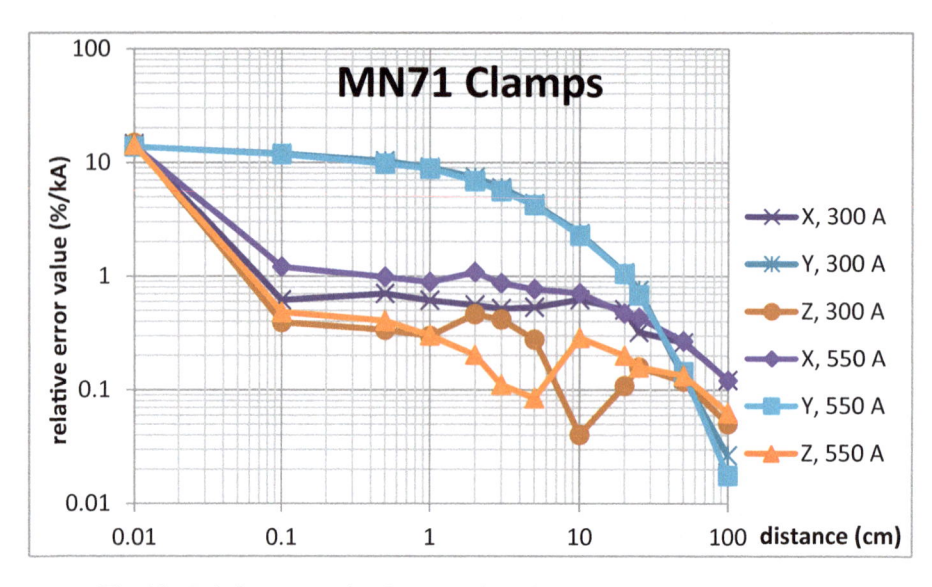

Fig. 11. Relative error value for network analyser with MN71 current clamps

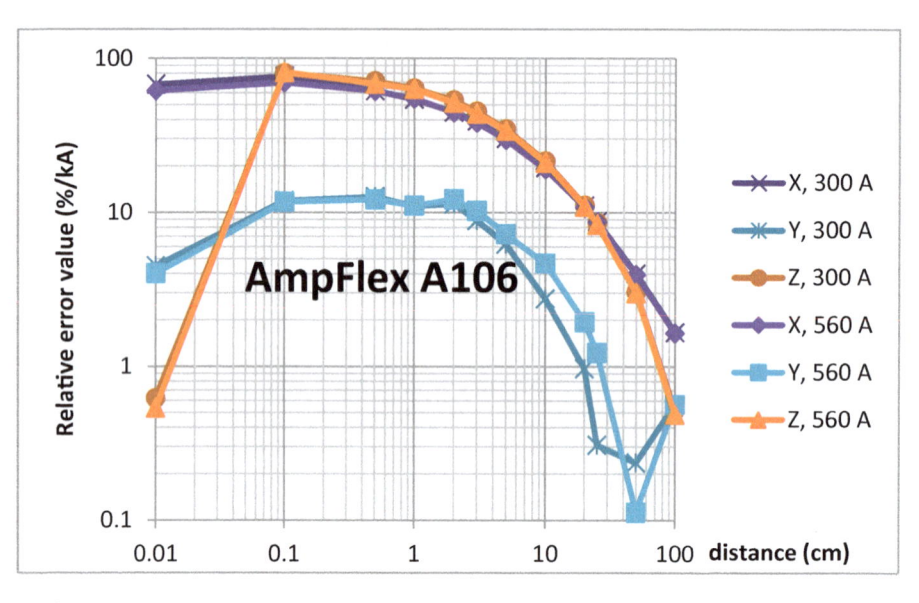

Fig. 12. Relative error value for network analyser with AmplFlex A106 Rogowski coil

5 Summary and Conclusion

All magnetically based current sensors are sensitive to external magnetic fields. All of them, including new types [12, 13], have to deal with mitigating magnetic crosstalk error. But when it gets to comparison of commercially sold ammeters there commonly used method to compare how prone the device is to external fields in comparison with others on market. This paper thus proposes a method how to measure and compare ammeters.

For all tested hand clamp ammeters, the measured error-current from a distance of 10 cm from the busbar was less than one percent of the range, based on the 1 kA interference current of the busbar. For current sensors that are connected to either a multimeter or a network analyzer, the error was more than 1% even at a significantly greater distance from the busbar.

Acknowledgement. This research was partially supported by the SGS grant from VSB - Technical University of Ostrava (No. SP2018/61) and by the project TUCENET (No. LO1404).

References

1. Chauvin Arnoux Metrix. http://www.chauvin-arnoux.com/en/produit/mn71.html. Accessed 31 Mar 2018
2. Bouakoura, M., Nait-Said, N., Nait-Said, M.S., Belbach, A.: Novel speed and current sensor FDI schemes with an improved AFTC for induction motor drives. Adv. Electr. Electron. Eng. **16**(1), 1–14 (2018)
3. Chauvin Arnoux Metrix. http://www.chauvin-arnoux.com/en/produit/a110-ampflex-sup-reg-sup.html. Accessed 31 Mar 2018
4. Geng, G., Yang, X., Gao, Y., Hammond, W., Xu, W.: Non-invasive current sensor for multi-core cables. IEEE Trans. Power Delivery (Early access), 1–8 (2017). https://doi.org/10.1109/TPWRD.2018.2813540
5. Itzke, A., Weiss, R., DiLeo, T., Weigel, R.: The influence of interference sources on a magnetic field-based current sensor for multiconductor measurement. IEEE Sens. J. **18**, 6782–6787 (2018)
6. Perciante, C.D., Ferrari, J.A.: Magnetic crosstalk minimization in optical current sensors. IEEE Trans. Instrum. Meas. **57**, 2304–2308 (2008)
7. Ripka, P., Chirtsov, A.: Busbar current transducer with suppression of external fields and gradients. IEEE Trans. Magn. 1–4 (2018)
8. RE9030, Range Digital Multimeter. http://www.range.com.hk/products/RE9030.html
9. Prova 23 Power Harmonics & Current Leakage Multifunction Clamp Meter. https://www.test-equipment.com.au/prova-23-power-harmonics-current-leakage-multifunction-clamp-meter/. Accessed 31 Mar 2018
10. Chauvin Arnoux Metrix. http://www.chauvin-arnoux.com/en/produit/f205.html. Accessed 31 Mar 2018
11. ENA330 (PQA) - Elcom, a.s. https://www.elcom.cz/en/product/test-and-measurement/electrical-network-analyzers/instruments/ena330. Accessed 31 Mar 2018
12. Weiss, R., Makuch, R., Itzke, A., Weigel, R.: Crosstalk in circular arrays of magnetic sensors for current measurement. IEEE Trans. Industr. Electron. **64**, 4903–4909 (2017)
13. Ripka, P., Chirtsov, A.: Influence of external current on yokeless electric current transducers. IEEE Trans. Magn. **53**, 1–4 (2017)

A Model for Predicting Energy Savings Attainable by Using Lighting Systems Dimmable to a Constant Illuminance Level

T. Novak, J. Sumpich, J. Vanus, K. Sokansky, R. Gono, J. Latal, and P. Valicek[✉]

Faculty of Electrical Engineering and Computer Science,
VSB TU Ostrava, Ostrava, Czech Republic
`pavel.valicek@vsb.cz`

Abstract. The objective of this paper is to demonstrate the feasibility of predicting electric energy savings achievable by using dimmable interior lighting systems. The prediction model uses an artificial interior lighting system which can be dimmed to a constant illuminance level and takes into account the contribution of daylight entering the space through side and top daylight openings, i.e. windows and skylights. The basic motivation for setting up this prediction model was the requirement raised by interior lighting system designers who need to know the net economic effect of dimmable lighting systems and to use the figures in discussions with the investor during the design project preparation stage. The prediction model developed gives transparent and unambiguous evidence of the economic benefit of using lighting systems.

Keywords: Daylight · Artificial light · Lighting systems · Energy requirements

1 Introduction

The prediction model developed gives transparent and unambiguous evidence of the economic benefit of using lighting systems that reduce the intensity of light emitted by them in dependence on the amount of daylight entering the rooms being illuminated.

The core of the prediction model is daylight dynamics. The uniformly overcast sky assumption was used in the model in order to allow the effect of the orientation of the daylight entrance areas with respect to the cardinal points to be disregarded and to use the least favorable daylight conditions. The uniformly overcast sky model uses incident radiation based on the position of Sun with respect Earth in the position in which the lighting system is evaluated. Conversion of the parameters of the uniformly overcast sky and of the height of the Sun is used to obtain information on the illuminance of a horizontal unshaded plane. Calculations the dynamically changing illuminance on an unshaded outdoor plane via a network of points of daylight factors calculated in the interior space analyzed are used to generate illuminance levels in those points due to daylight both in dependence on time and in the interior space of the building.

The daylight contributions so obtained (in time and space) are deducted from the illuminance level stipulated by the applicable standard to obtain the illuminance level to be generated by the artificial lighting system. The power requirement difference between the currently installed non-dimmable lighting system working at the 100% input and the dimmable lighting system represents the electric energy savings (during the time of expected use of the workplace in question), from which the period of return of the artificial lighting system upgrading cost can be derived.

2 Dynamic Model of Uniformly Overcast Sky

The modelling of the potential savings should be based on external illumination, which, however, varies appreciably. Currently, calculations of the daylight factor use a model of uniformly overcast sky with 5 klx or 20 klx illuminance. We designed a dynamic model based on a uniformly overcast sky with variable illuminance during the day. This model can be used throughout the year to calculate illuminance from daylight in an indoor setting without having to know which directions the windows face.

Although in fact, the sky is uniformly overcast a few times or never during the year, this is a tolerable simplification enabling us to set up a sky model which can be used in the calculations and provides a clear idea of how sky luminance changes when passing from the horizon to the zenith. The model of uniformly overcast sky and dark earth ground used by us postulates a horizon-to-zenith 1:3 luminance gradient (CIE). Luminance of any point in the sky can be then calculated as [1]:

$$L_\gamma = L_Z/3.(1 + 2.\sin \gamma) \ [cd.m^{-2}] \quad (1)$$

where L_γ is sky luminance at angle γ above the horizon, L_z is sky luminance in the zenith [1, 2].

This sky is then input to the daily illumination calculation software. The daylight factor is one more parameter which must be introduced to clearly explain the basis for the daily interior illumination calculations. The daylight factor, expressing the daylight contribution to illumination in buildings at sites where good vision is imperative, is the ratio between interior illuminance (from both direct and reflected light) and the illuminance of an outdoor unshaded plane under uniformly overcast sky. The contributions of direct sunrays are excluded from both illuminances. The daylight factor accounts for the effect of glazing, air pollution, indoor and/or outdoor shading, etc. The effect of the orientation of the windows and skylights, on the contrary, is not included owing to the postulated uniformly overcast sky. The daylight factor (in percent values) is determined from the relation:

$$D = \frac{E}{E_V}.100[\%] \quad (2)$$

where E is the illuminance (in lx) of the reference interior plane, E_V is reference illuminance (in lx) in a point of the outdoor unshaded plane (ČSN EN 73 0580-1:2007). However, it should be borne in mind that the illuminance level E_v on the outdoor unshaded plane is based on the above values of 5 and 20 klx. Hence, the daylight factor only tells us how many per cent outdoor illuminance will occur in the point of interest in the interior area. Information regarding the dynamics of this contribution due to outdoor illuminance changes vanishes altogether. For this reason, illuminance dynamics calculations must be performed for the unshaded outdoor plane during the day and throughout the year, respecting the luminosity distribution patterns of the uniformly overcast sky [4–6].

2.1 Diffuse Illuminance Calculation

Diffuse illuminance D_v, which is the illuminance of an unshaded outdoor plane under uniformly overcast sky can be determined, based on the knowledge of the particular geographical and time coordinates, by using the following Eq. (3).

In our model, horizontal illuminance of an unshaded outdoor plane was additionally corrected by means of the skylight transparency factors D_{Vm}/E_V, obtained by long-term measurements in Bratislava [1].

$$D_V = \left(\frac{D_{Vm}}{E_V}\right).E_{V0}.\varepsilon.\sin \gamma_S \; [\text{lx}], \qquad (3)$$

Where (D_{Vm}/E_V) is the skylight transparency factor. E_{V0} is the luminous solar constant = 133,334 lx, γ_S is solar elevation (degrees), ε is the eccentricity factor (dimensionless).

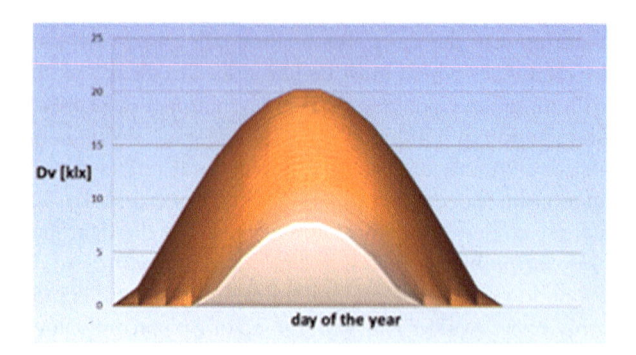

Fig. 1. Diffuse illuminance levels on an unshaded outdoor plane throughout the year

The relative values of ε and γ_s for different times throughout the year are inserted in Eq. (3) to obtain the illuminance levels on the unshaded outdoor plane (see Fig. 1).

3 Prediction Model

The illuminance levels D_v are converted and inserted in Eq. 3 to calculate the expected interior illumination levels contributed by daylight assuming uniformly overcast sky.

$$E_{x,y,t} = D_{x,y} \cdot D_{v,t}/100 \; [\text{lx}], \tag{4}$$

where $E_{x,y,t}$ is interior illuminance in time and space due the daylight contribution, $D_{x,y}$ is the daylight factor in a point of the area, $D_{v,t}$ is the variable outdoor illuminance in time. Since the goal of the calculation is to gain insight into the potential for power savings obtained by installation of a dimmable interior lighting system, the illuminance stipulated by the applicable standard must be subtracted from the expected daylight contribution to obtain the required contribution of the artificial interior lighting system to total illuminance.

$$E_{a,x,y,t} = E_{\text{norm}} - E_{x,y,t} \; [\text{lx}], \tag{5}$$

where $E_{a,x,y,t}$ is the resulting illuminance in a specific point of the interior space, E_{norm} is the illuminance required for the space by the standard, $E_{x,y,t}$ is internal illuminance (in space and time) due to the daylight component. It holds that $E_{x,y,t}$ must be larger than E_{norm} or else it is zero. Based on the technical design of the interior lighting system and cyclic calculations of the dimming levels for light lines or even for the individual lights so as to approach the required $E_{x,y,t}$ levels for all the time points considered, the dimming development patterns are obtained for the light lines/individual lights. Now the dimming figures are converted to electric power to enable the power used for the lighting system to be calculated in dependence on time in the uniformly overcast sky conditions [8].

Knowing the expected disposition of operations in the areas analyzed, the power consumption difference can be estimated between the designed interior lighting system operating at a full (constant) input power and the interior lighting system where the lights in the areas are dimmed to provide a constant illumination level. The difference can be looked upon as a prediction of the power savings potential of the controllable interior lighting system which can be dimmed to a constant illuminance level.

In order to make the model verifiable, its parts were implemented into the software for daylight calculations and into the software for artificial interior illumination calculations [3].

4 Verification of the Prediction Model

Apart from mathematical calculations of the daylight and artificial light, verification of the model required long-term measurements of the electrical parameters in selected interior areas equipped with lighting systems dimmable to a constant illumination level. External photographs of the selected rooms with dimmable lights are shown in Fig. 2. Long-term measurements of the lighting system input power were performed in both rooms. The rooms are located at the geographical coordinates 49.83 N and 18.16 E (a house in the city of Ostrava, Czech Republic).

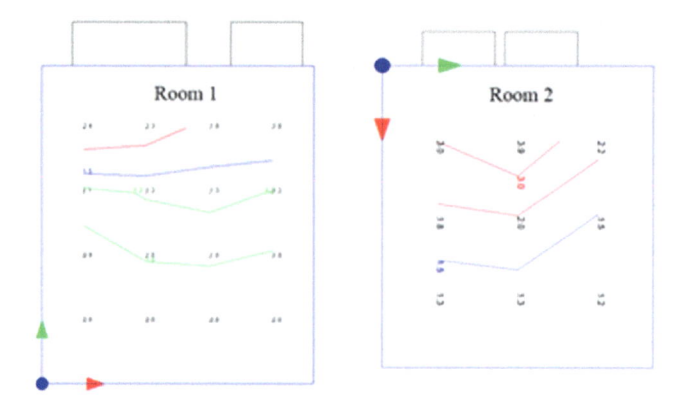

Fig. 2. Daylight factor calculations. Left: Room 1, right: Room 2

Room 1, 4300 × 5000 mm area and 2670 mm height, is located on the 3rd (above-ground) floor of an administrative building (the shading surrounding building and trees were included in the calculations). The shorter wall accommodates two windows which are 700 mm apart: a triple window, 1400 mm × 1800 mm area, at a height of 1030 mm above the floor and at a distance of 480 mm from the wall, and a double window, 1400 mm × 1140 mm area, at a height of 1030 mm above the floor and at a distance of 180 mm from the wall. Room 2 is located on the 2nd (above-ground) floor of a testing family house, built by the MSDK (Moravian-Silesian Wood Industry Cluster) for research purposes at the VŠB – Technical University of Ostrava. Once again, the shading surrounding buildings and trees were included in the calculation. The area of the room is 3975 mm × 3372 mm, height 2100 mm. The shorter wall accommodates two windows which are 120 mm apart: one window, 900 mm ×1000 mm area, at a height of 810 mm above the floor and 500 mm from the wall, and one window 900 mm × 1000 mm area, at a height of 810 mm above the floor and 1520 mm from the wall (Fig. 3).

Fig. 3. The model rooms viewed from the exterior. Left: Room 1, right: Room 2

Ground plants of the two rooms (also indicating the locations of the windows) obtained as outputs of the software for daylight calculations (in accordance with the applicable ČSN standard) are shown in Fig. 2. The ground plan in Fig. 2 includes points for daylight illuminance calculations (for uniformly overcast sky) and the daylight factor isophotes at the height of the reference plane of the expected visual procedure sites (i.e. desktop), viz. 0.75 m. As mentioned above, outdoor shading objects such as buildings and trees were included in the calculations. Taking into account the outdoor illuminance data for the unshaded plane at the building's geographical coordinates, the calculated daylight factor values are converted to the expected daylight contributions (in lx) to the daily illumination in the rooms examined. The pictures in Fig. 6 (daylight factor conversion) show an example of conversion for external illuminance for the two rooms.

4.1 Artificial Lighting System Calculations

Now, the artificial lighting systems for the two rooms examined must be designed to enable the power savings to be calculated. The rooms are normal office rooms (Fig. 4), for which the standard (i.e. EN 12464-1) sets lower maintained illuminance limits of 500 lx and 300 lx for Rooms 1 and 2, respectively. Since no particular vision procedure sites have been specified for either room, the calculation was conducted so that the 500 lx/300 lx requirement be met in any calculation point. The design of the artificial lighting system was created in the software based on the precise room dimensions, whereby implementation of the calculation cycles was made possible [3].

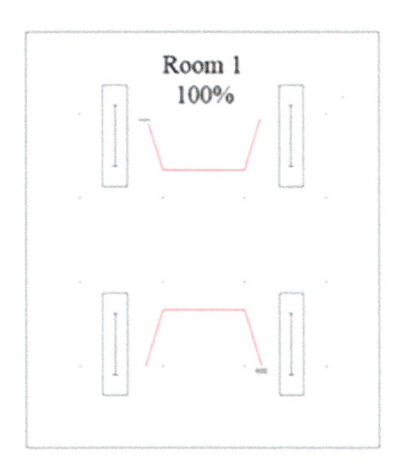

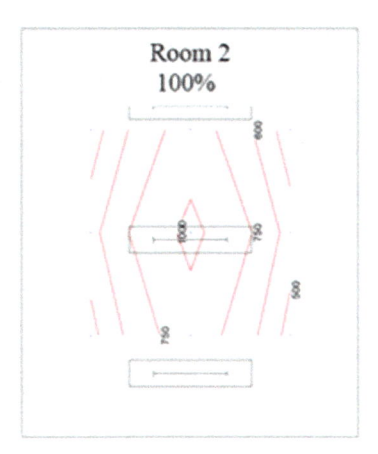

Fig. 4. Artificial lighting system design to indoor illuminance (lx). Left: Room 1, right: Room 2

Room 1 is equipped with four ceiling-mounted QUATTROC BODY 2X36 W TC-L HFI WL6 L840 fluorescent light luminaires (Fig. 7) on the right side. Room 2 is equipped with three ceiling-mounted ZC T5 228/12LOS ZK, 2×28 W/840 fluorescent lights luminaires (Fig. 5) on the left side.

The lighting system calculations for the areas under study provide the maximum power input of the lighting system, from which the power consumption by the non-dimmable lighting system can be obtained for the postulated use of the area.

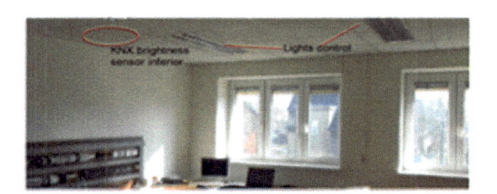

Fig. 5. The model rooms viewed from the exterior. Left: Room 1, right: Room 2

4.2 Prediction Model Application

Once the time dependences of the illuminances from daylight (in the uniformly overcast sky conditions) are known for the space treated, a cycle of dimming calculations for the parts of the lighting systems (i.e. individual lights or light lines) can be launched based on the requirement that the minimum regulatory illuminance levels in the area are met. So the illumination levels (or permissible degrees of dimming) of the lighting system are obtained. An example of the artificial lighting system dimming level is shown in Fig. 7 for the daily illuminance level on the unshaded outdoor plane. As to the conversion of the dimming level to input power, the software uses a finite dependence, assuming that the light's specific power will not vary with power regulation.

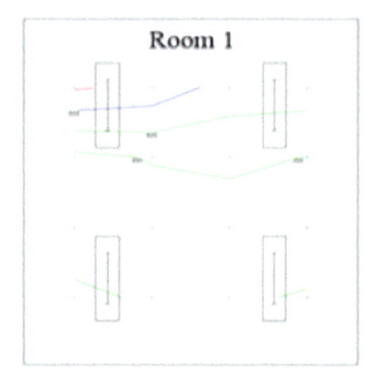

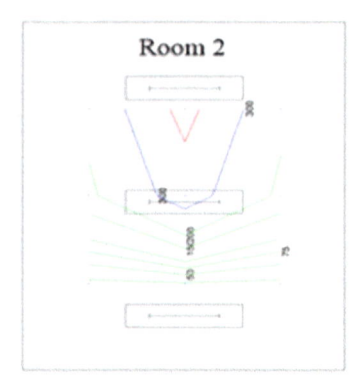

Fig. 6. Daylight factor conversion to indoor illuminance (lx) for Room 1 (left) and Room 2 (right)

Figure 6 shows examples of daylight factor values converted to illuminance levels for a specific time and day of the year. Examples of daylight/artificial lighting combination calculations for one specific time of the year are shown in the next figure (Fig. 7). Following the basic illumination system calculations for the daylight/artificial lighting combinations, the potential of regulation of the artificial lighting systems can be calculated by using a modified version of the WILS application software. The lighting system regulation calculations are performed in 10% steps (from 0% to 100%). For this example, the linear relationship between luminous flux and wattage was used. When the required luminous flux provided by the lighting system is < 10%, the system is virtually disconnected by the software. For practical reasons, it is advisable to limit the number of independently controlled light lines in the power savings calculations to four because the computation time increases substantially if more complex lighting systems are treated [3, 7].

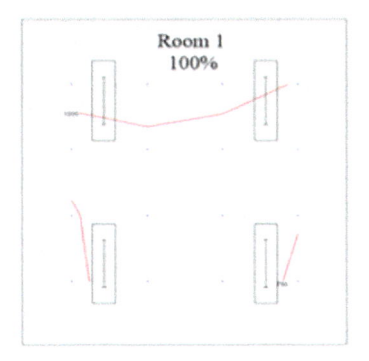

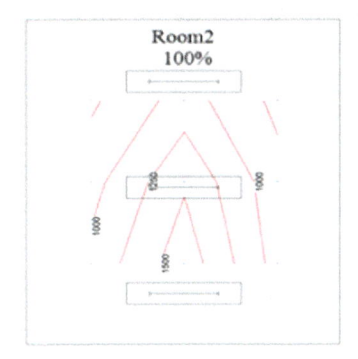

Fig. 7. Sum of the artificial lighting system at 100% and the daylight illuminance (lx) component. Left: Room 1, right: Room 2

In addition to a detailed list of power taken by the lighting system, the computation output may include a short (simplified) table demonstrating the prediction of the dimming levels for the individual lights or light lines in dependence on the levels of illuminance of the unshaded outdoor plane at different illuminance levels (Tables 1 and 2). Table 1 demonstrates the levels of regulation of two lighting systems, R_1 and R_2, at different luminance on the unshaded outdoor plane [6].

Table 1. Regulation of the light lines (%) for Room 1

Model Room 1		
D_V [lx]	R_1[%]	R_2[%]
1000	100	100
5000	90	90
7000	90	90
10000	80	90
20000	70	60
25000	60	40

Table 2. Regulation of the light lines (%) for Room 2

Model Room 2			
D_V [lx]	R_1[%]	R_2[%]	R_3[%]
1000	20	40	100
5000	0	30	90
7000	0	30	70
10000	0	30	60
20000	0	0	30
25000	0	0	0

For example, if the horizontal illuminance on the unshaded outdoor plane is 20 klx, the first and second lines of the artificial lighting system will illuminate to 60% and 70%, respectively (Fig. 7). The same dependence is included in Table 2 for Room 2, which is equipped with 3 lighting systems, R_1, R_2 and R_3. For example, if, as above, the horizontal illuminance on the unshaded outdoor plane is 20 klx, neither the first line nor the second line of the artificial lighting system will illuminate and the third line will illuminate to mere 30% [6].

5 Conclusion

Data obtained by testing the prediction model in real conditions gave evidence that the assumptions underlying the model of power consumption savings by an interior lighting system which is dimmable to a constant illuminance level are correct. The sky conditions simplification to uniformly overcast sky, owing to which the effect of spatial orientation of the windows can be disregarded, proved to be acceptable. The differences between the observed and calculated data lies within 7%, which is very good. Because of unpredictable behavior of normal sky.

Within subsequent investigations into the topic of dimming interior lighting systems dimmable to a constant illuminance, the response of the sensors to changes in the outdoor conditions will be examined and the question will be answered as to in which real conditions the control system is actually capable of regulating the lighting system to a constant illuminance level.

Acknowledgments. This paper has been elaborated in the framework of the project SP2018/117 of Student Grant System, VSB-TU Ostrava.

References

1. Kittler, R., Darula, S.: Parametrization problems of the very bright cloudy sky conditions. Sol. Energy **62**(2), 93–100 (1998)
2. CIE. www.cie.co.at
3. Software WDLS a WILS (2013). http://www.astrasw.cz Accessed 25 July 2013
4. Sumpich, J., Novak, T., Sokansky, K., Carbol, Z.: Calculation of saving possibilities in interior lighting system using both daylight and artificial light. In: Przeglad Elektrotechniczny, pp. 345–347 (2013)
5. Sumpich, J., Sokansky, K., Novak, T., Carbol, Z.: Electrical energy saving possibilities in lightening of indoor workplace while using both daylight and artificial light. In: 13th International Scientific Conference Electric Power Engineering 2012, EPE 2012, pp. 1165–1168 (2011). ISBN: 978-80-248-2480-2
6. Novak, T., Vanus, J., Sumpich, J., Koziorek, J., Sokansky, K., Hrbac R.: Possibility to achieve the energy savings by the light control in smart home. In: Proceedings of the 7th International Scientific Symposium on Electrical Power Engineering, Stará Lesná, Slovakia, 18 September–20 September 2013, pp. 260–263. Košice: Elektroenergetika 2013 (2013)

7. Vanus, J., Novak, T., Koziorek, J., Konecny, J., Hrbac, R.: The proposal model of energy savings of lighting systems in the smart home care. In: IFAC Proceedings Volumes (IFAC-PapersOnline) 2013, pp. 411–415 (2013)
8. Parise, G., Martirano, L.: Daylight impact on energy performance of internal lighting. IEEE Trans. Ind. Appl. **49**(1), 242–249 (2013).art. no. 6363598

Strategy of Metropolis Electrical Energy Supply

Valery Beley, Andrey Nikishin(✉), and Dmitriy Gorbatov

Kaliningrad State Technical University, Kaliningrad 236034, Russia
andrey.nikishin@outlook.com

Abstract. It is shown that the globalization of the electrical energy systems does not provide a reliable power supply for built environment in case of natural and man-made disasters, which can cause massive damage. The article propose the strategy of building a metropolis power supply on a "Smart City" concept. Implementation of this strategy through the use of decentralized power supply based on renewable energy sources, new high-reliability approach provided by the principle of (n-2) and the use of storage devices will ensure reliable power supply of the metropolis, and improve the ecology of the built environment.

Keywords: Renewable energy sources · Power engineering · Electrical energy supply system · Distributed generation · Metropolis

1 Introduction and Problem Definition

Power industry development trends show globalization of the energy sector, especially - the electrical energy industry [1]. Several advantages of electrical energy are the reasons of its wide use:

- versatility: it can be produced relatively easily by the conversion from any forms of energy and re-converted into thermal, mechanical, light and other forms of energy;
- using the electric power systems, power plants built in the world areas with easy access to primary energy (fuel) can be connected to electric power consumers.

Figure 1a shows (in different colors) the European interconnected power systems working independently on alternating current, and respective power systems managed by control centers [2].

These power systems are connected through direct current (DC) lines and DC-links (Fig. 1b). DC-lines and DC-links are unique: they only transport the active power and they do not disturb the synchronous mode of interconnected power systems; if the DC-line is built on the principle of superconductivity it has no energy losses [3].

Along with the undoubted advantages, globalization of energy systems has a number of disadvantages, the main ones are: a substantial increase of number and complexity of interconnections in the power system and system control; vulnerability to natural and manmade disasters.

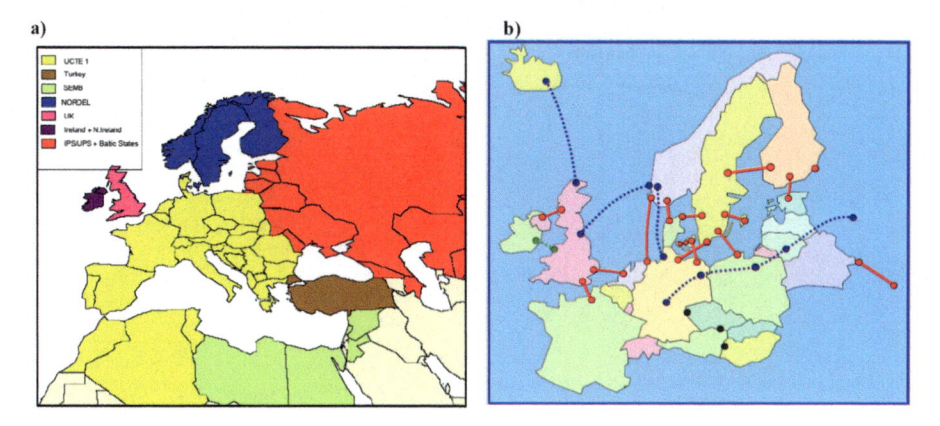

Fig. 1. European interconnected power systems (a) and direct current interconnections between them (b) [2]

It means that in case of the loss of external power supply, situation close to catastrophic is possible. It is not only about the comfort, which is lost at the electricity cut at metropolis; it puts under the risk all the citizens and makes their lives much more complicated. One of the biggest failures of this kind took place in US energy system on August 14, 2003 [4]. Due to the cascade of short-circuit faults, eight US states and part of Canada were disconnected from the electrical system, and it affected more than 50 million people. The largest cities of the region lost electricity: Detroit, New York, Cleveland, Ottawa, and Toronto. The economic damage of such accidents is enormous; in that case it was estimated as much as 6 billion dollars. Taking this into consideration, the organization of rational, economic, environmental friendly and reliable metropolis electrical energy supply system is one of the important scientific and technical problem.

2 Authors Proposals and Discussion

Considering the structure of metropolis energy supply system [5–8] as the most vulnerable part of built environment, it should be noted that they mainly receive power from the power system. Metropolis backbone electrical grid is a network of power transmission lines with the voltage level of 220 kV and more. The following is organized by means of this network: interconnection to electrical power system, power output from the local power stations and electrical energy supply of the consumers through the electrical network on the second and third hierarchical levels of metropolis power supply system (Fig. 2).

In case of system accidents in the power system, this leads to catastrophic consequences. In order to increase the reliability of metropolis electricity supply, the authors propose an integrated strategic approach to solving this problem (Fig. 3), which consist of few keypoints.

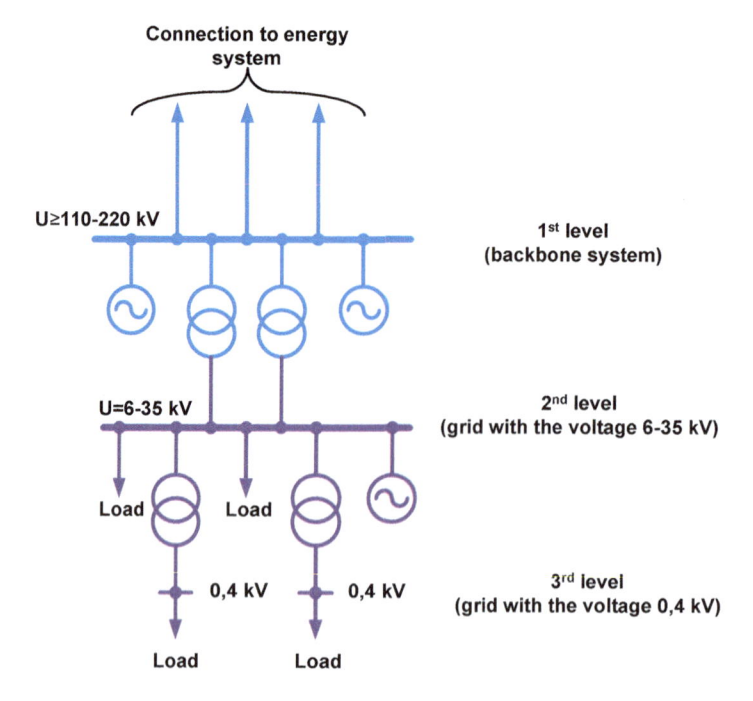

Fig. 2. The hierarchical system of metropolis power supply

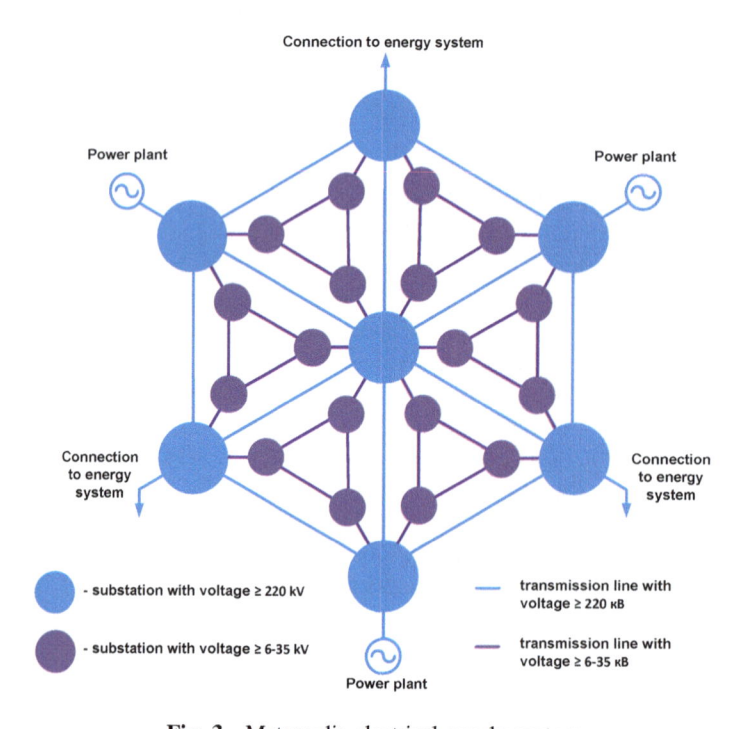

Fig. 3. Metropolis electrical supply system

- Increase the number of energy supply lines for each metropolis substation of at least 3, which will allow to fulfill the principle (n-2).

Reliability of power supply on 1st and 2nd hierarchical levels of metropolis power supply system should be provided by the principle of (n-2): in case of a transmission line maintenance and the simultaneous failure on the second line, the third transmission line remains in operation and provides a connection to the electrical system. Reliability of metropolis electrical energy supply is also provided by distributed power plants connected to the grid on the 1st and 2nd hierarchical levels of supply system (Fig. 2).

- In order to ensure metropolis reliable energy supply, the concept of "Smart City" with the maximum use of local renewable energy sources (Fig. 4) should be used.

With the loss of connection with the power system, the metropolis electrical energy supply system switches to stand-alone (isolated) operation. It is reasonable to use the "Smart City" concept as part of general "Smart Grids" concept (Fig. 4) when implementing the stand-alone operation of metropolis electrical energy supply system. Development of "Smart Grids" is a key objective in developed countries. In smart metropolis energy supply system along with the power network (Fig. 4) there is an information networks (Fig. 4), which collects information on power flows, generation and consumption of electricity, commands from the control center to control the operating modes of power plants and metropolis consumers [9].

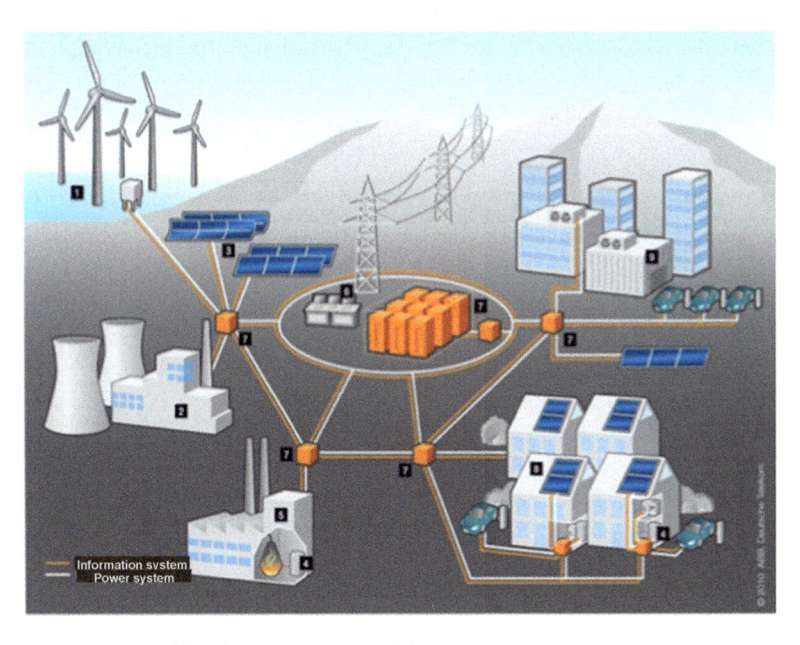

Fig. 4. Smart metropolis energy supply system

As the power source for the metropolis the priority should be given to the maximum use of renewable energy sources: thermal, hydro, wind and solar power. The metropolis with a population of 1 million on average produces about 600 thousand tons of municipal solid wastes (MSW), which can be used as fuel for thermal power plants. MSW have high calorific value and can be compared with brown coal and peat (Table 1) [10].

Table 1. Calorific values of various types of fuel [10]

Fuel	Calorific values, MJ/kg
Fuel oil	40
Natural gas	36
Coal	25
Brown coal	13
Peat	12
Unsorted municipal waste	6,27–8,36
Solid combustible waste	8,36–12,54
Waste in tailings	4,18–6,27

The share of electricity generation through the use of MSW energy in 2016 was around 2.5% of the total amount of generated electricity. Foreign thermal power plants comply with the requirements of the European Directive on the Incineration of Waste 2000/76/EC cleaning emissions of the flue gas [11]. Table 2 shows some of the most effective types of these stations [12].

Table 2. Effective thermal power plants that use MSW as fuel [12]

TPP	Commissioning	Capacity, tons of MSW per year	Electrical capacity, MW	Price, millions of EUR
Rudersdorf Germany	2008	260000	35,5	110
EBSKnapsack Germany	2009	240000	33,5	105
Spreerecycling Germany	2013	255000	17	138
Gaoantun China	2008	300000	25	123

For a metropolis with 1 million citizens, capacity of thermal power stations using MSW as a fuel will be around 50–60 MW, which will provide around 10–15% of the electricity needs of the metropolis and a certain part of the heat supply, as well as it will solve the problem of solid waste disposal.

If average wind speed is over 6.4 m/s at the wind turbine hub height (Table 3) it is efficient to install wind turbines, which are connected to 2nd levels metropolis energy supply system (Fig. 3) [12].

Table 3. Roughness class of the area for wind energy installation (height 50 m) [12]

№	Characteristic	Specific output, W/m²	Annual wind speed, m/s
1	Poor	0–200	0,0–5,6
2	Low-profit	200–300	5,6–6,4
3	Average	300–400	6,4–7,0
4	Good	400–500	7,0–7,5
5	Excellent	500–600	7,5–8,0
6	Great	600–800	8,0–8,8
7	Outstanding	>800	>8,8

Decrease in the cost of solar generated electricity (solar power), makes the use of solar power plants for urban consumers a cost-effective solution (Table 4).

Table 4. The normalized production cost of solar electricity in the world by 2015 [9]

Type of the system	Price for 1 MW·h without government subsidies		Price for 1 MW·h with government subsidies	
	From	To	From	To
Solar PV - Rooftop residential	$180	$265	$138	$203
Solar PV - Rooftop commercial buildings	$126	$177	$96	$135
Solar PV - Utility scale	$72	$86	$56	$66
Solar Thermal with storage	$118	$130	$96	$105

However, the design and operation of solar power plants in urban areas has a number of features. First of all, it is the low density of solar radiation, which depends on the geographic city location and weather conditions (200–250 W/m² on average on the earth's surface), as well as shading created by buildings and other infrastructure objects. As a result, the effective utilization of solar radiation demand special planning of residential areas taking into account daily and annual sun cycles.

The strategy of urban solar power development involves their usage on two levels. There are solar photovoltaic power plants (Solar PV) for individual buildings with solar cells located on the roof or on the walls, so called rooftop solar PV. They provide the power supply of building consumers and infrastructure (Smart Home). They connect on the third level of metropolis power supply system (Fig. 2) to provide power reserve in case solar PV does not generate enough electricity or for the output of the surplus of energy into the urban network in case it generates more than consumed.

Utility scale solar PV are designed for centralized power generation, have a capacity of tens, hundreds, and, in good conditions, thousands of kW. They are connected to the first or second levels of urban power supply system (Fig. 3). They are usually built at the city outskirts, because they require the alienation of large areas.

As for solar PV electricity production control, despite the largely probabilistic nature of solar electricity generation, the advantage is its overlap with the daily cycle of domestic consumers load, and, for a large proportion of air-conditioning loads - annual cycles.

One of the best example of special planning with the use of "Smart City" concept is the Japanese city of Fujisawa that was built in 2014 and there the main part of the energy consumption covered with solar PV and thermal collectors (Fig. 5).

Fig. 5. Energy efficient residential buildings in the city of Fujisawa

In the case of an emergency and disconnection of the city from external supply it can fully provide its population with electricity, communications, and even hot water off-line for three days.

- In order to ensure reliable operation in stand-alone mode metropolis power grid, power storage is urgent problem.

One of the most important directions of technological breakthroughs, especially caused by the wide distribution of renewable energy sources, is a radical reduction in the costs of stationary and mobile technologies for accumulating electricity on the basis of storage batteries (Fig. 6) [13].

Currently, the only well-mastered technology for accumulating electricity is pumped storage power plants (PSPs). Like any accumulation system, PSP is a pure consumer of electricity, as it uses 20–25% more electricity than it gives back to the energy system [14].

They are large system-oriented storage systems, whereas for a new generation of production technologies and small electricity consumers, distributed storage systems are have to comply with new calls of the future:

- provide controlled output of power from flickering generation (wind and solar power stations) taking into account the needs of the energy system, which will allow optimizing of the loading of generators and grids in the power system, and also reduce the necessary inputs of new generation units capacity;
- expand the ability of consumers to price-dependent demand management (demand response), allowing them to actively influence the price equilibrium on the electricity market.

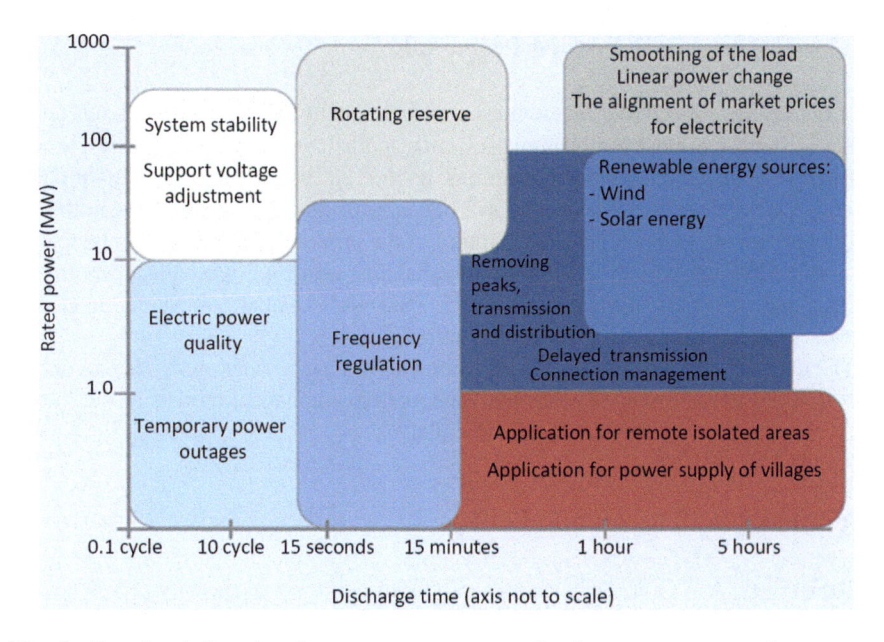

Fig. 6. Functional diversity of power storage systems for the power system and consumers

Batteries storage systems can be connected in parallel to the renewable energy power stations (Fig. 3) [15].

- Use the especially designed algorithms of consumption control of metropolis consumers during stand-alone operation.

In case of to stand-alone operation mode it is impossible to supply all consumers with energy (Fig. 7, curve 1), the management system of the power system must provide power to vitally important consumers of the metropolis (Fig. 7, curve 2).

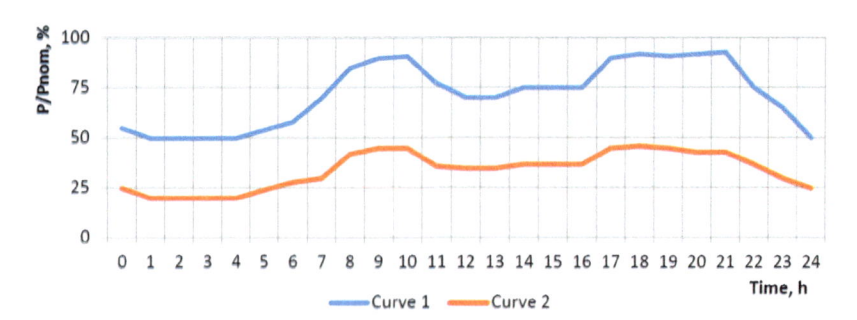

Fig. 7. Metropolis electrical loads

3 Conclusions and Future Proposals

The organization of reliable, economic, environmentally friendly and rational energy supply system for electricity supply of metropolis consumer, taking into account increasing dependence of such consumers on the electricity supply is an important problem for developed countries and will be such for developing countries in the near future. Authors are given general description of the proposals, which should be realized for discussed systems, such as increase of reliability using n-2 principle, "Smart City" concept, use of power storage of different kind and adequate consumer load management systems during stand-alone operation.

The detailed assessment of the possibilities of the use of such methods can only be done with the use of especially designed adequate mathematical models of the system, which is a question of the future development.

References

1. Belogor'yev, A.M.: i dr. Trendy i stsenarii razvitiya mirovoy energetiki v pervoy polovine XXI veka (2011)
2. SIEMENS: Challenge and Solutions for high voltage transmission. Sector Energy
3. Posse, A.V.: Skhemy i rezhimy elektroperedach postoyannogo toka. - Energiya. Leningr. otd-niye (1973)
4. Andersson, G., et al.: Causes of the 2003 major grid blackouts in North America and Europe, and recommended means to improve system dynamic performance. IEEE Trans. Power Syst. **20**(4), 1922–1928 (2005)
5. Shvedov, G.V.: Elektrosnabzheniye gorodov: elektropotrebleniye, raschetnyye nagruzki, raspredelitel'nyye seti. MEI (2012)
6. Kozlov, V.A., Bilik, N.I., David, L.F.: Spravochnik po proyektirovaniyu elektrosnabzheniya gorodov. - Energoatomizdat. Leningr. otd-niye (1986)
7. Kozlov, V.A.: Elektrosnabzheniye gorodov (1977)
8. Karapetyan, I.G., Faybisovich, D.L., Shapiro, I.M.: Spravochnik po proyektirovaniyu elektricheskikh setey (2009)
9. Beley, V.F., Selin, V.V., Zadorozhnyy, A.O., Nikishin, A.YU., Yelagin, N.N., Solovey, A. I.: Spravochnik modulya: vozobnovlyayemyye istochniki energii: spravochnoye izdaniye. Kaliningrad: Izd-vo OOO "TESK" (2015)

10. McKendry, P.: Energy production from biomass (part 1): overview of biomass. Bioresour. Technol. **83**(1), 37–46 (2002)
11. Directive 2000/76/ec of the European Parliament and of the Council of 4 December 2000 on the incineration of waste
12. http://www.eew-energyfromwaste.com/de/standorte/knapsack.html
13. Makarov, A.A.: Prognoz razvitiya energetiki mira i Rossii 2016. INEI RAN (2016)
14. Beley, V.F., Gorbatov, D.S.: Avtonomnyy rezhim Kaliningradskoy energosistemy, postroyennaya na raspredelennoy generatsii i ispol'zovanii vozobnovlyayemykh istochnikov energii. Novyye tekhnologii gazovoy, neftyanoy promyshlennosti, energetiki i svyazi: XXII Mezhdunarodnyy kongressa: sb. trudov.-Moskva: Ekonomika, 22, 328–337 (2015)
15. Bezrukih, P.P., Bezrukih, P.P. (ml.), Gribkov, S.V.: Vetroehnergetika: Spravochno-metodicheskoe izdanie. Pod obshchej redakciej Bezrukih, P.P. — « Intekhehnergo-Izdat » , « Teploehnergetik » , 304 s (2014)

Robotics

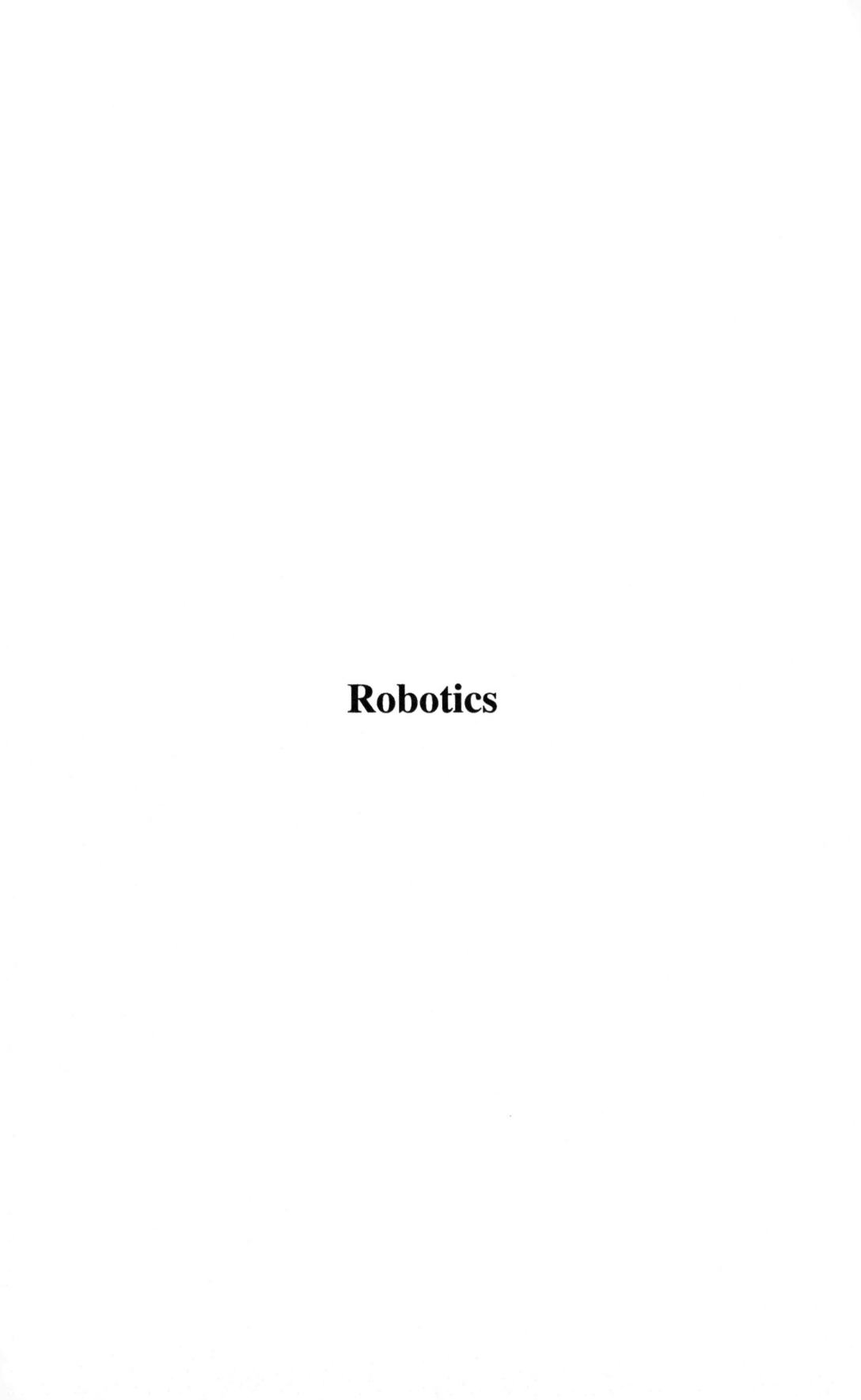

Attitude Control of Jumping Robot with Bending-Stretching Mechanism

Chea Xin Ong, Yurika Nomura, and Jun Ishikawa[✉]

Tokyo Denki University, Tokyo, Japan
ishikawa@fr.dendai.ac.jp

Abstract. This article proposes an attitude control method to stabilize the posture of a jumping robot with bending-stretching mechanism as its leg mechanism. The proposed method is based on nonlinear out-put zeroing control, and the motion of the trunk part is used to control the posture of the leg mechanism. To show the validity of the proposed method, three computer simulations have been carried out. One is to see the response to an impulse-like disturbance while the robot is stabilizing its posture, another is to see the tracking performance of the angle of the leg mechanism to the time-varying reference, and the last one is to see the ability for the angle of leg mechanism to be stabilized at the upright posture during bending-stretching motion. As the results, it is proved that the proposed method will effectively help the robot to ensure its stability in all three situations.

Keywords: Jumping robot · Attitude control · Output zeroing control · Under-actuated system

1 Introduction

Recently, robots for exploration in environment where human being could not enter such as disaster site have been studied and developed. There are various types of locomotive mechanism which allow robots to travel along rough terrain such as, wheeled robot [1], quadruped robot [2, 3], hexapod walking robot [4], blade type crawler robot [5], and so on. One category of these robots is robots which can jump over obstacles larger than the robot's size and move along rough terrain by jumping [6–11]. Among jumping robot, there are also different types of jumping mechanism has been developed. For example, robot that jump using a CO_2-powered piston [6], robot with cam system and spring [7], robot with four-bar spring linkage [8, 9], and jumping monopod with various types of leg mechanism [10, 11] From these previous research, it could be found that a typical method to realize jumping motion is using bending-stretching mechanism such as Salto 1p robot [11], the 7g miniature jumping robot [7], and the jumping-crawling robot [8].

The objective of this research is to achieve stable jumps by legs with bending-stretching mechanism, mainly focusing on the stabilization of robot's attitude. For the first step of the development, basic performance of the output zeroing controller [6], which is typically using to stabilize states of under-actuated system, is examined. In the controller design, a 3-link mechanism that consists of a torso, a thigh and a lower leg moving in the sagittal plane is modeled as an under-actuated system of a simple 2-link mechanism. Specifically, the lower-leg link connected to ground by a freely-rotating joint is fixed to the thigh link by setting a bending-stretching knee joint angle to be a constant value. Then, the combined leg mechanism is considered to be the first link, and the torso part driven by an actuator is the second link. Based on the under-actuated 2-link model, a nonlinear output zeroing controller is designed to stabilizes the attitude of the jumping robot. Although there are also other methods to stabilize the robot attitude by using axially symmetric reaction wheels such as reported in [12], a method of balancing using an asymmetrical torso part is proposed. The validity of the proposed method is evaluated by computer simulations.

2 Modelling of Jumping Robot

In this section, the modelling of the controlled plant, which is the jumping robot, will be describe. Figure 1 shows a schematic diagram of modeling for the jumping robot, and the definitions and values of parameters in the model diagram are listed in Table 1. Since the link model of the leg mechanism driven by the servo motor is not an essential part in the simulation study of this article, the first prototype is adopting the same design as Salto [11]. The representation with (θ_m) in Table 1 indicates that the parameter is a function of the angle θ_m of the servo motor related to the angular displacement of the knee joint.

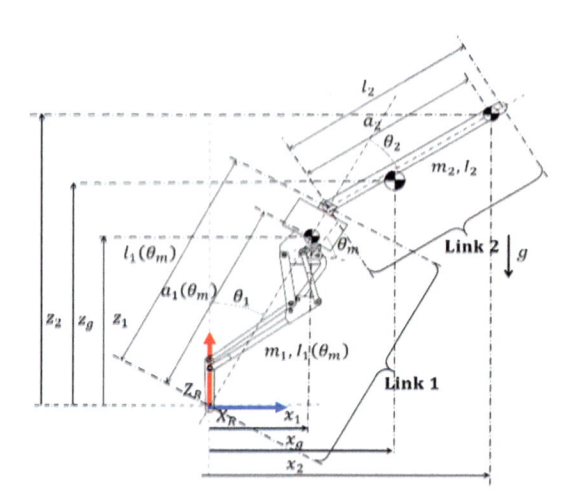

Fig. 1. Schematic diagram for modeling of jumping robot.

Since the first joint of the robot is a joint in contact with the ground, it is modeled as a freely rotatable joint, and the corresponding generalized torque is set to 0. The second joint is driven by a motor, and the robot is controlled to be in an inverted state by utilizing the movement of the connected torso part. The length of link 1, l_1, and the distance to the center of gravity of link 1, a_1, which vary along with bending and

Table 1. Definition and value of parameters.

Definition	Symbol	Value [Unit]
Acceleration of gravity	g	9.81 [m/s^2]
Length of link 1	$l_1(\theta_m)$	2.50×10^{-1}–4.00×10^{-1} [m]
Length of link 2	l_2	2.40×10^{-1} [m]
Distance to the center of gravity of link 1	$a_1(\theta_m)$	1.50×10^{-1}–3.00×10^{-1} [m]
Distance to the center of gravity of link 2	a_2	1.92×10^{-1} [m]
Mass of link 1	m_1	4.00×10^{-1} [kg]
Mass of link 2	m_2	1.50×10^{-1} [kg]
Moment of Inertia of link 1	I_1	2.00×10^{-3} [kg·m^2]
Moment of Inertia of link 2	I_2	0 [kg·m^2]
Viscosity coefficient of link 1	c_1	0 [N·m/(rad/s)]
Viscosity coefficient of link 2	c_2	1.40×10^{-4} [N·m/(rad/s)]

stretching movement are to be functions of the angle θ_m of the servo motor as described above. Note that θ_m is not the bending and stretching angle of the knee joint itself, and the knee angle, θ_b, will be described in Sect. 4. The moment of inertia around the center of gravity of link 1, I1, also varies as a function of θ_m, but as the mass at the upper end of the leg mechanism (waist part) is assumed to be sufficiently heavier than the rest other parts of the leg mechanism, I1 was set to be constant value, considering the variation of I1 be negligibly small.

Equation of motion of the robot derived by Lagrange's method can be obtained as

$$\boldsymbol{M} \cdot \begin{bmatrix} \ddot{\theta}_1 \\ \ddot{\theta}_2 \end{bmatrix} + \boldsymbol{H} \cdot \begin{bmatrix} \dot{\theta}_1 \\ \dot{\theta}_2 \end{bmatrix} + \boldsymbol{G} = \begin{bmatrix} \tau_1 \\ \tau_2 \end{bmatrix}$$

$$\boldsymbol{M} := \begin{bmatrix} \alpha + \gamma + 2\beta\cos\theta_2 & \gamma + 2\beta\cos\theta_2 \\ \gamma + 2\beta\cos\theta_2 & \gamma \end{bmatrix}$$

$$\boldsymbol{H} := \begin{bmatrix} -\beta\dot{\theta}_2(2\dot{\theta}_1 + \dot{\theta}_2)\sin\theta_2 \\ \beta\dot{\theta}^2\sin\theta_2 \end{bmatrix} \quad (1)$$

$$\boldsymbol{G} := \begin{bmatrix} -k_1\sin\theta_1 - k_2\sin(\theta_1 + \theta_2) \\ -k_2\sin(\theta_1 + \theta_2) \end{bmatrix}$$

where the coefficients in the equation are

$$\alpha := m_1 a_1^2 + m_2 l_1^2 + I_1$$
$$\beta := m_2 l_1 a_2$$
$$\gamma := m_2 a_2^2 + I_2$$
$$k_1 := (m_1 a_1 + m_2 l_1)g$$
$$k_2 := m_2 a_2 g$$

respectively.

3 Output Zeroing Control

An output zeroing control law is designed base on the equation of motion (1) derived in the previous chapter. The angular momentum P of the robot system is given by

$$P = (\alpha + \gamma + 2\beta\cos\theta_2)\dot{\theta}_1 + (\gamma + \beta\cos\theta_2)\dot{\theta}_2. \tag{2}$$

In the proposed method, this angular momentum P is chosen to be the output function of the system and stabilized by using the method described in the reference [13]. For this purpose, the derivative of P is derived as

$$\dot{P} = k_1\sin\theta_1 + k_2\sin(\theta_1 + \theta_2) \tag{3}$$

from the gravitational term of the equation of motion (1). Solving (2) for $\dot{\theta}_1$ yields

$$\dot{\theta}_1 = \frac{P - (\gamma + \beta\cos\theta_2)\dot{\theta}_2}{\alpha + \gamma + 2\beta\cos\theta_2}. \tag{4}$$

Then, concidering $\dot{\theta}_2$ as a system input, v, and defining the state variable as $x = [\theta_1, \theta_2, P]^T$, a nonlinear state equation of the system can be obtained as

$$\begin{bmatrix} \dot{\theta}_1 \\ \dot{\theta}_2 \\ \dot{P} \end{bmatrix} = \begin{bmatrix} \frac{P}{\alpha + \gamma + 2\beta\cos\theta_2} \\ 0 \\ k_1\sin\theta_1 + k_2\sin(\theta_1 + \theta_2) \end{bmatrix} + \begin{bmatrix} -\frac{\gamma + \beta\cos\theta_2}{\alpha + \gamma + 2\beta\cos\theta_2} \\ 1 \\ 0 \end{bmatrix} v. \tag{5}$$

By differentiating Eq. (3), the term $\dot{\theta}_2$ will appear explicitly as follows:

$$\ddot{P} = (k_1\cos\theta_1 + k_2\cos(\theta_1 + \theta_2)) \cdot \dot{\theta}_1 + k_2\cos(\theta_1 + \theta_2) \cdot \dot{\theta}_2. \tag{6}$$

Substituting (2), (3) and (6) into a stable characteristic equation,

$$\ddot{P} + f_2\dot{P} + f_1 P = 0, \tag{7}$$

which the angular momentum, P, should be satisfied, yield

$$[k_1\cos\theta_1 + k_1\cos(\theta_1 + \theta_2) + f_1(\alpha + \gamma + 2\beta\cos\theta_2)]\dot{\theta}_1 + [k_2\cos(\theta_1 + \theta_2)$$
$$+ f_1(\gamma + \beta\cos\theta_2)]\dot{\theta}_2 + f_2(k_1\sin\theta_1 + k_1\sin(\theta_1 + \theta_2)) = 0 \tag{8}$$

Thus, the velocity of the driven joint, $\dot{\theta}_2$, as the control law, v, can be derived as

$$v = -\frac{[k_1\cos\theta_1 + k_1\cos(\theta_1 + \theta_2) + f_1(\alpha + \gamma + 2\beta\cos\theta_2)]\dot{\theta}_1 + f_2(k_1\sin\theta_1 + k_1\sin(\theta_1 + \theta_2))}{\{k_2\cos(\theta_1 + \theta_2) + f_1(\gamma + \beta\cos(\theta_2))\}}. \tag{9}$$

Both f_1 and f_2 in the Eq. (9) are positive design parameters for adjusting the response. The attitude of the robot can be stabilized by controlling the angular velocity of the driven joint, $\dot{\theta}_2$, to track the desired one, v. Therefore, the torque input, τ_2, can be designed as

$$\tau_2 = k_v\left(u - \dot{\theta}_2\right) + k_p(\theta_d - \theta_1) + c_2\dot{\theta}_2, \tag{10}$$

where k_v and k_p is the feedback gain and the third term is a compensation term of viscous friction. In the next chapter, the stability of the system using this control law is verified by computer simulations.

4 Computer Simulations

In this section, the results of computer simulations to show the validity of the proposed control system will be describe.

The value of the control gains in Eq. (10) obtained by trial and error are summarized in Table 2. f_c and ζ are parameters that determine the gains f_1 and f_2 shown in Eq. (8), which is the equation that the output function P must satisfied. From f_c and ζ, f_1 and f_2 can be obtained as

$$f_1 = (2\pi f_c)^2 \tag{11}$$

$$f_2 = 2\zeta(2\pi f_c). \tag{12}$$

The simulation was performed under the following three conditions.

1. Input an impulse like disturbance to link 1 to evaluate the robustness of the system.
2. Apply a sine wave with frequency of 0.2 Hz and amplitude of $\pm 10°$ as the target value, θ_d to the angle of link 1 to evaluate the tracking performance
3. Apply bending and stretching motion to the knee joint to evaluate the attitude maintaining ability during bending-stretching motion

In the simulation, to represent the jumping mechanism shown in Fig. 1, instead of using the displacement of the servomotor that drives the knee joint θ_m, as a configuration capable of directly controlling the bending and stretching angle θ_b of the knee, the robot is solved as a 3-link mechanism by forward dynamics, as shown in Fig. 2. That is, the relational expression between the displacements of the first link (the leg mechanism considered as one rigid body), θ_1, the second link, θ_2 and the displacements θ_a, θ_b, θ_c of the 3-link mechanism on the simulator are

$$\theta_1 = \theta_a + \frac{1}{2}\theta_b, \tag{13}$$

$$\theta_2 = \theta_c + \frac{1}{2}\theta_b \tag{14}$$

respectively.

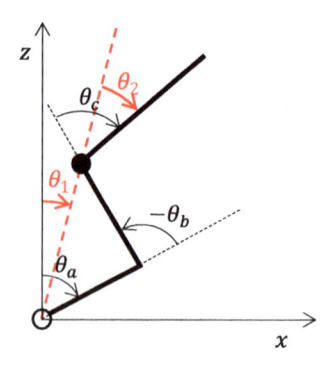

Fig. 2. Schematic diagram to show the relationship between the simplified two link model and the actual three link model.

Table 2. Definition and value of parameters used in simulation.

Definition	Symbol	Value [Unit]
Position feedback gain	k_p	10 [N·m/rad]
Velocity feedback gain	k_v	−8 [N·m/(rad/s)]
Resonant frequency	f_c	0.8 [Hz]
Attenuation rate	ζ	1.2 [-]
Gain 1	f_1	25.3 [(rad/s)^2]
Gain 2	f_2	12.1 [rad/s]

4.1 Evaluation of Robustness Against Torque Disturbance

A simulation was performed to evaluate the response when an impulse like disturbance of 0.1 s width is inputted to the first joint at time 1 s. The knee joint displacement θ_b is fixed to −90° by using a phase-lead compensator. The time response of θ_1, θ_2, θ_a, θ_b, θ_c is shown in Fig. 3, and the sequence photography captured every 1 s of the animation generated from the time response is shown in Fig. 4. The blue point, red point and green point in the figure represent the position of center of gravity of the leg mechanism, torso part and entire robot, respectively.

θ_1 and θ_2 in Fig. 3 were calculated by Eqs. (11) and (12) using the values of θ_a, θ_b, θ_c obtained from simulation. From Fig. 3, it can be seen that θ_1 and θ_2 converge to 0° after a certain period of time elapses when the torque disturbance is applied. Also, from Fig. 4, it is found that the balance of the robot was broken at 1.5 s due to impulsive disturbance, but it was able to recover to the inverted state at 3 s. Therefore, it was confirmed that the attitude of the leg mechanism can be maintained when torque disturbance is applied.

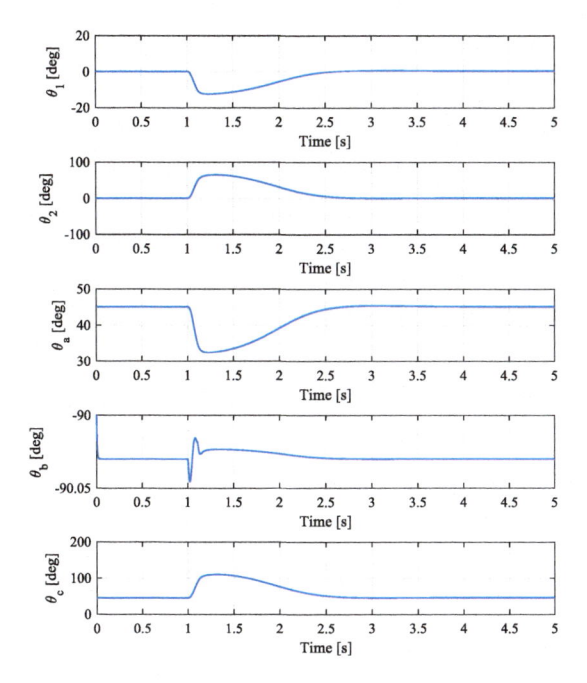

Fig. 3. Impulse response of jumping robot under output zeroing controller on simulation.

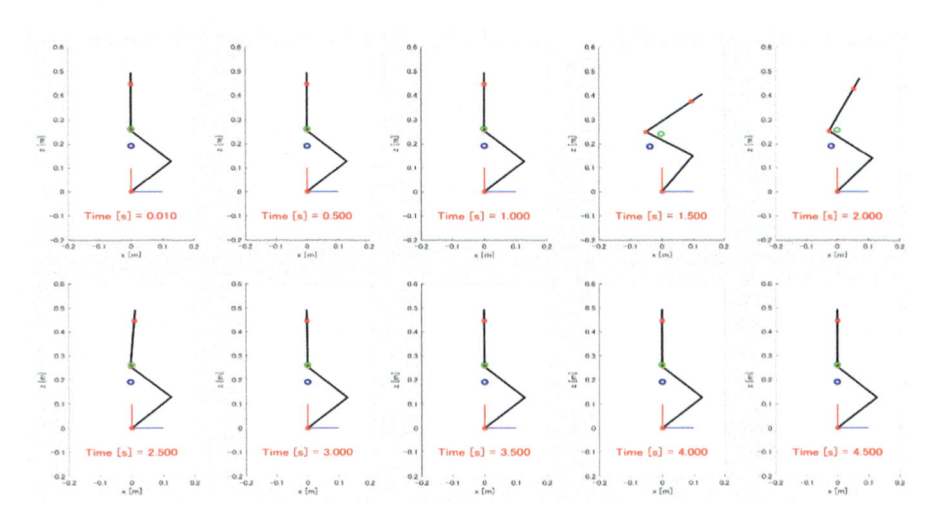

Fig. 4. Sequence photography captured every 0.5 s of impulse response of jumping robot under output zeroing controller on simulation.

4.2 Evaluation of Tracking Performance of Posture Angle θ_1 to Target Value θ_d

Another simulation was performed to evaluate the tracking performance of the angle of link 1, θ_1 to the applied sine wave with frequency of 0.2 Hz and amplitude of $\pm 10°$ as the target value. Similarly, in the previous section, the knee joint displacement θ_b is fixed to $-90°$ by using a phase-lead compensator. The time response of θ_1, θ_2, θ_a, θ_b, θ_c is shown in Fig. 5, and the sequence photography captured every 1 s of the animation generated from the time response is shown in Fig. 6. The blue point, red point and green point in the figure represent the position of center of gravity of the leg mechanism, torso part and entire robot, respectively, as in the previous section.

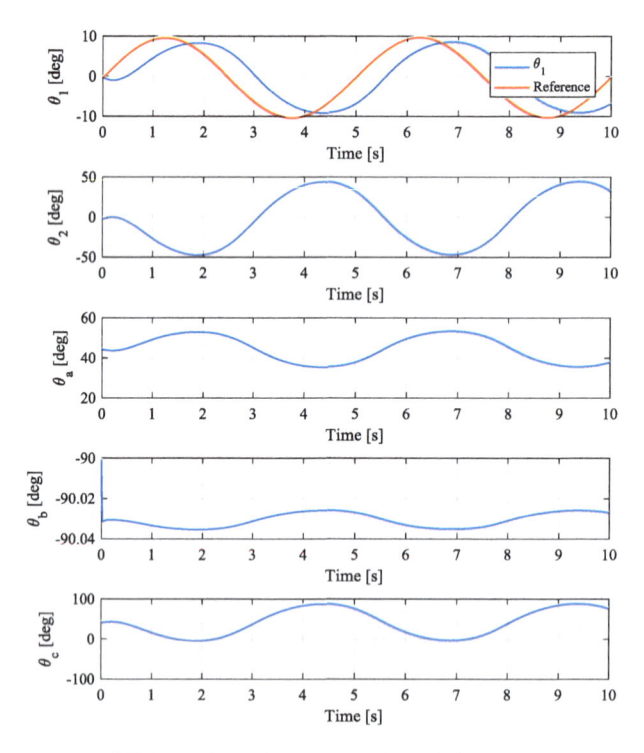

Fig. 5. Time response of link angle while target value of θ_1 is apply under output zeroing controller on simulation.

From Fig. 5, although there is a delay of about 0.5 s, θ_1 follows the target value of the sinusoidal wave, and accordingly θ_2 moves at an amplitude of 45° to $-45°$ to balance the robot. Further, from Fig. 6, it is understood that the inverted state can be maintained while the first link and the second link are moved back and forth. Therefore, in this simulation, although the control band is narrow, it was confirmed that the tracking control of the attitude angle can be realized by the control rule shown in Eq. (10).

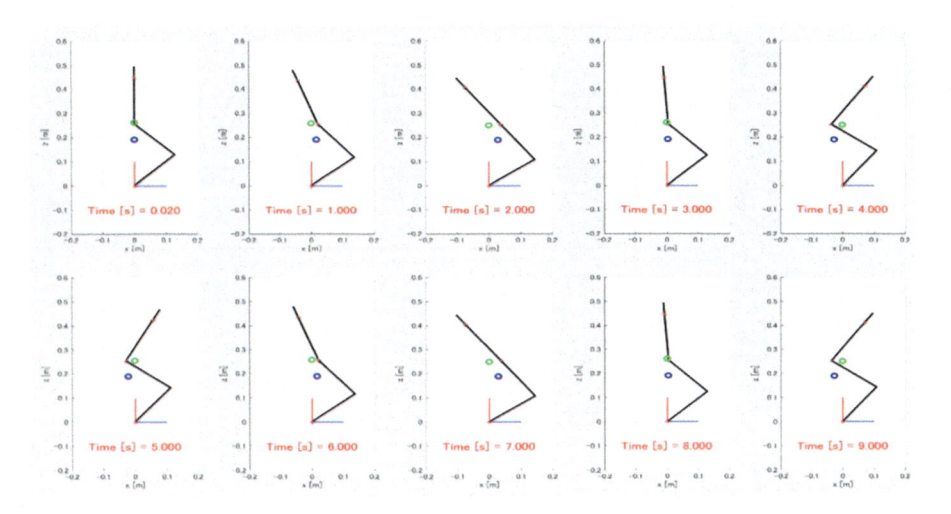

Fig. 6. Sequence photography captured every 1 s of link angle while target value of θ_1 is given under output zeroing controller on simulation.

4.3 Evaluation of Attitude Maintaining Ability During Bending Stretching Motion

Lastly, a simulation was performed to evaluate the attitude maintaining ability during bending stretching motion. In this simulation, the knee joint θ_b is controlled to follow a sinusoidal wave with frequency of 1.0 Hz and amplitude of $\pm 15°$ and a vertical offset of $-90°$ as the target value by using the phase lead compensator to perform the bending and stretching motion of the leg mechanism, while controlling the attitude of the robot. The time response of $\theta_1, \theta_2, \theta_a, \theta_b, \theta_c$ is shown in Fig. 7, and the sequence photography captured every 0.2 s of the animation generated from the time response is shown in Fig. 8. The definition of the center of gravity position in the figure is the same as in the previous section.

From Fig. 7, both θ_1 and θ_2 are stabilized although small angles of oscillation occur due to the bending and stretching motion of the knee joint. And from Fig. 8, it can be seen that the first link and the second link are maintained in an inverted state while bending and stretching motion applied. Therefore, the attitude maintaining ability during bending stretching motion also been verified.

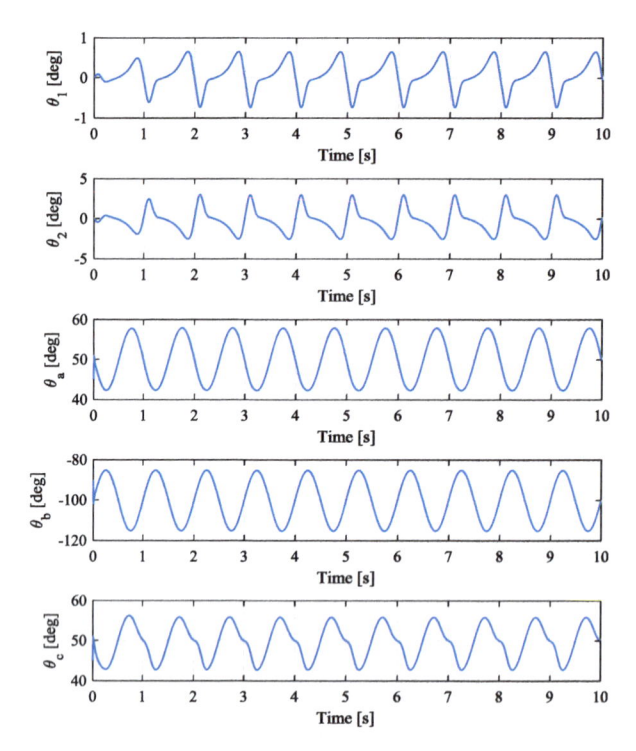

Fig. 7. Time response of link angle when bending-stretching movement is generated under output zeroing controller on simulation.

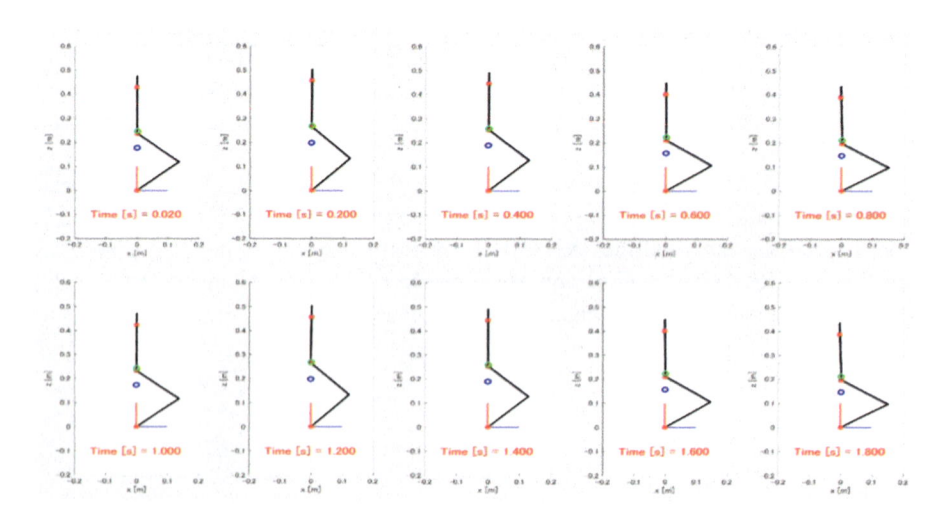

Fig. 8. Sequence photography captured every 0.2 s of link angle during bending-stretching motion under output zeroing controller on simulation (0 s–1.8 s).

5 Conclusion

The objective of this study is to realize stable leap with jumping robot which has bending and stretching mechanism as its leg mechanism. As the first step of it, in this paper, computer simulations to evaluate the performance of the proposed output zeroing control to maintain stable attitude of the robot has been carried out. As the result, it was confirmed that constant performance can be realized when impulse-like disturbance is applied, when the leg mechanism is tracking a sine wave target value, and during bending-stretching motion.

In the future, actual machine experiments will be conducted to confirm the effectiveness of the purposed controller in real machine.

References

1. Zhu, J.H., et al.: Design of a wheel-type mobile robot for rough terrain. In: Proceedings of the 2012 IEEE/ROBIO International Conference, December 2012
2. Raibert, M., Blankespoor, K., et al.: BigDog, the rough-terrain quadruped robot. In: Proceedings of the 17th IFAC World Congress, July 2008
3. Li, X., Gao, J.Y., et al.: A new control method of quadruped robot walking on rough terrain based on linear inverted pendulum method. In: Proceedings of the 2014 IEEE/ROBIO International Conference, December 2014
4. Huang, K.J., Chen, S.C., et al.: A bio-inspired hexapod robot with noncircular gear transmission system. In: Proceedings of the 2012 IEEE/ASME International Conference, July 2012
5. Yamada, Y., Endo, G., et al.: Blade-type crawler vehicle bio-inspired by a wharf roach. In: Proceedings of the 2014 IEEE/ICRA International Conference, June 2014
6. Ackerman, E.: Boston dynamics sand flea robot demonstrates astonishing jumping skills. IEEE Spectrum March 2012
7. Kovac, M., Fuchs, M., et al.: A miniature 7g jumping robot. In: Proceedings of the 2008 IEEE/ICRA International Conference, May 2008
8. Jung, G.P., et al.: An integrated jumping-crawling robot using height-adjustable jumping module. In: Proceedings of the 2014 IEEE/ICRA International Conference on Robotics and Automation, June 2016
9. Reddy, N.S., Ray, R., et al.: Modeling and simulation of a jumping frog robot. In: Proceedings of the 2011 IEEE/ICMA International Conference, August 2011
10. Zhao, J.G., et al.: A Miniature 25 grams running and jumping robot. In: Proceedings of the 2014 IEEE/ICRA International Conference on Robotics and Automation, June 2014
11. Haldane, D.W., Plecnik, M.M., Yim, J.K., Fearing, R.S.: Repetitive extreme-acceleration (14-g) spatial jumping with Salto-1P. In: Proceedings of the 2017 IEEE/RSJ International Conference on Intelligent Robots and Systems (IROS2017), September 2017
12. Nomura, Y., Ishikawa, J.: Aerial attitude control of hopping robots using reaction wheels: evaluation of prototype II in the air. In: Proceedings of the 4th International Conference on Advanced Engineering – Theory and Applications (AETA 2017), pp. 361–372, December 2017
13. Mita, T.: Introduction to nonlinear control theory: skill control of underactuated robots, Chap. 7, Shokodo (2000). (in Japanese)

Geometric Foot Location Determination Algorithm for Façade Cleaning Robot

Shunsuke Nansai[✉] and Hiroshi Itoh

Tokyo Denki University, 5 Senju-asahi-cho, Adachi, Tokyo 120-8551, Japan
{nansai,itoh}@mail.dendai.ac.jp

Abstract. The purpose of this paper is to construct a foot location determination algorithm for façade cleaning robot. To accomplish the façade cleaning, some geometrical constraints are considered: The module robot moves along with the line of a window frame. Distance of one step has to be shorter than the robot's motion range. The cleaning units cover all area from top edge to bottom of the window frame. These conditions are discussed in this paper. After that, the algorithm satisfying these conditions are constructed by geometrical equations with respect to each situation. In particular, the situations fall into four scenes with initial location, middle area, final location as well as left foot location. The effectiveness of the algorithm is verified via a numerical simulation.

Keywords: Façade cleaning robot · Foot location algorithm · Geometry

1 Introduction

In recent decades, a lot of skyscrapers have been built by their cutting-edge construction technologies and processes. Even in such modern skyscrapers, outdated methods were used for glass façade cleaning and maintenance. The cleaning practice possess high chances of accidents and concomitant casualties, which may result in the loss of life. Numerous accidents have been reported even with the use of a gondola for façade cleaning jobs. For instance, the sudden blow of storm at Shanghai World Financial Center [1] resulted in the loss of control of gondola. In another example, the gondola got suspended at a height of 240 m during the maintenance of the World Trade Center in New York city [2]. The involvement of robots in scenarios like glass façade cleaning can potentially minimize the risk of accidents and maximize the productivity.

We have identified plenty of research works on the robot involved solution for the façad cleaning, which have sprouted up in a decade. A series of sky-cleaners that are driven by pneumatic actuators for the concurrent locomotion and cleaning of a glass wall by using the vacuum suction cups are discussed in [3]. The works mentioned above explore sky-cleaners with suction cups on the both ends of actuators in the X–Y stage and suction cups attached to a vacuum pump.

Seo et al. has developed ROPE RIDE [4], which can climb vertical surfaces by utilizing a rope dropped down from the top of the building. The ROPE RIDE is installed with two additional propeller thrusters to ensure a strong attachment to the wall. Furthermore, Fraunhofer IFF, a research institute, developed a wall cleaning robot named SIRIUS [5]. The SIRIUS robot is capable of performing an up-and-down motion on a vertical surface, using a crane installed on buildings for façade maintenance. The robot adheres to the wall by the suction system and moves with the help of linear actuators [6]. Even though the existing robot-aided solutions for façade cleaning each excel in their effectiveness and experimental results, the performance of the current robots are constrained to their target buildings. Precisely, the robots require additional equipment such as a crane, —i.e. the system is still requires significant development to adapt to various building architecture. Hence, adaptability is a factor that lacks in the present façade cleaning scenario.

Our ultimate goal is to develop a glass façade cleaning robot capable of adapting to any skyscrapers. To work a robot system at different shaped buildings, to possess some different morphologies is required. A self-reconfigurable robot is one of potent solution to realize high adaptability. Reconfigurable modular robot system allows to realize different morphology through assembling/disassembling/organizing system of each module. This research aims to implement the glass façade cleaning robot through development of a module robot possessing cleaning ability and design of both its locomotion control system and a assembling/disassembling/organizing system. The modular robot strategy accomplishing the façade cleaning on any skyscrapers has been proposed [7]. And, the modular robot based on the strategy has been developed in our previous study [8]. However, the module robot is still operated the target foot location manually, it is necessary to design an autonomous control system to determine its foot location appropriately.

This paper aims to construct the foot location determination algorithm. The most right module on a glass façade is considered in order to limit complexity as initial stage of this study. Key points of this algorithm is geometrical constraints of accomplishing the façade cleaning. The module robot is required to move its cleaning unit along the line with the window frame. In addition, the module always have a constraint of motion range. Since these two conditions are able to be considered geometrically, the appropriate foot locations are able to be calculated by solving the geometrical equations. In this paper, we describe challenges for the algorithm based on the façade cleaning method, first. After that, the algorithm is constructed with respect to each situations. Finally, effectiveness of the algorithm is verified via a numerical simulation.

This paper is organized as follows: Sect. 2 described the façade cleaning method, and introduce our module robot. Also, challenges for the algorithm based on the façade cleaning method is described. The foot location determination algorithm is constructed in Sect. 3. The algorithm consists of equations with respect to each situations. In Sect. 4, effectiveness of the constructed algorithm is verified via the numerical simulation. Section 5 concludes this paper.

2 Challenges for Façade Cleaning

In the case of glass façade cleaning, the robot design must satisfy the two factors we mentioned below.

- terrain adaptability:
  - adaptability to window shapes
  - ability to avoid obstacles (window frames)
- cleaning ability:
  - effective cleaning mechanisms
  - effective cleaning strategy

The execution of cleaning tasks varies according to several external factors associated with the surfaces to be cleaned. "Terrain adaptability" refers to the locomotion capability of the robot on glass frames having different shapes. Typically, the glass panels can be observed in five different shapes including horizontally long, vertically long, parallelogram, trapezoidal, and curvilinear. Moreover, the shape of the glass panels can be a combination of these five shapes to form a complex shape such as a horizontally-long-parallelogram, vertically-long-trapezoidal, etc. Hence, the robot must adapt to the glass panel having a complex geometry to execute a locomotion independent of the architecture of a building. Even though there exist glass panels with numerous shapes, all of them can be categorized into two. From the close observation on the geometry of various glass panels, we can infer that the window size and the window shape are the two key parameters that characterize all the glass panels. In this context, we define the window size as ratio of horizontal and vertical dimensions of the glass window to be cleaned. The horizontally-long-parallelogram and vertically-long-parallelogram are different descriptions belonging to the same category: the parallelogram. Hence the robot design should be capable of adapting to different window size and window shape to achieve the terrain adaptability factor.

Cleaning ability, the second factor to be considered for the design of the proposed system refers to the identification of the best method to achieve effective cleaning. To identify the efficient cleaning mechanism for the robot, we observed the way of glass façade cleaning done by experts. For manual cleaning, the scrubber and squeegee are the primary tools used by experts. The experts clean glass walls by approaching the glass surface to be cleaned, washing the dirt off by using the scrubber and cleaning liquid, and finally wipe the dust residue using a squeegee (see Fig. 1). We identified that to ensure an effective cleaning, experts clean the window from top to bottom and ensure the movement of the squeegee such that it traces the surface covered by the scrubber. Besides, they ensure the complete contact of the squeegee over the glass surface during cleaning and cleaning is done towards a single direction (see Fig. 1 (**b**)→(**c**)→(**d**)).

Figure 2 shows our developed module robot. The developed module possesses a morphology like a biped robot. Each foot part as foot unit has equivalent elements/abilities: It includes the cleaning unit, a suction cup, a vacuum pump, a seal mechanism, a mechanical valve to break vacuum as well as servo motors for joint actuate.

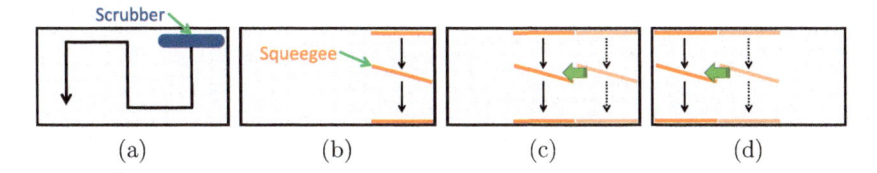

Fig. 1. The cleaning strategy adopted by the experts. Firstly, the experts wash the dirt off by using the scrubber and cleaning liquid (**a**). Secondly, the dust residue is wiped by squeegee which is manipulated from top to bottom of the window (**b**). The wiping is done towards a single direction with respect to each column as shown as (**b**)→(**c**)→(**d**).

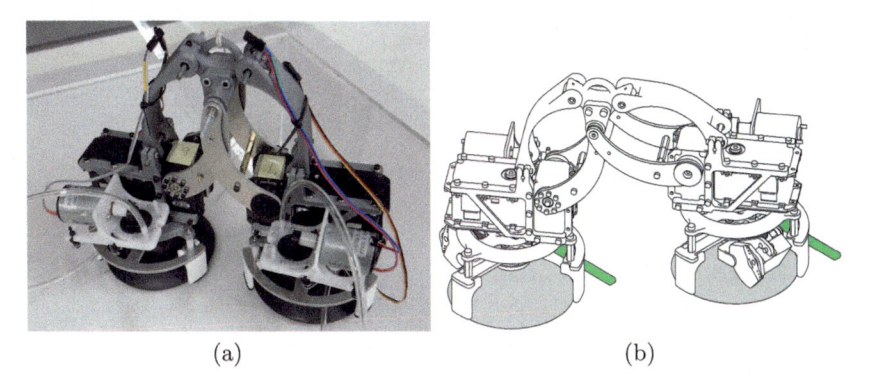

Fig. 2. The developed module robot. (a) Picture of the module. (b) Skeleton model of the module. The gray area represents the suction cups. The green area represents the cleaning units.

Figure 3 shows a solution for window size and window shape adaptation that belongs to the factor "terrain adaptability". Using a fixed number of robot modules, controlling the foot location of individual modules helps the robot system to adapt to various window shapes. On the other hand, a different sized glass window can be cleaned by adapting the number of modules. The solution we identified is a modular system. The local control of each module helps the robot to adapt to various window shapes. Similarly, the system becomes adaptable to window size by varying the number of robot modules to be deployed.

Figure 4 shows the identified cleaning mechanism and cleaning strategy for achieving another factor "cleaning ability". The developed module equipped with a squeegee next to the scrubber can execute the cleaning of the window from top to bottom. The scrubber can trace the movement of the squeegee and it can clean toward the one direction as well.

Our strategy accomplishing the window cleaning has been shown as Figs. 3 and 4. Therefore, the developed module robot is capable of executing the cleaning

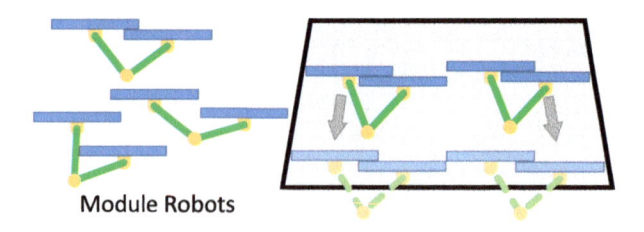

Fig. 3. Modular robot system for the glass façade cleaning robot. Controlling the foot location of individual modules helps adapting to window shapes with different shapes. The robot system adapts to various window shapes by varying the number of modules deployed. The arrows in the right figure and the dotted lines represent the movement flow.

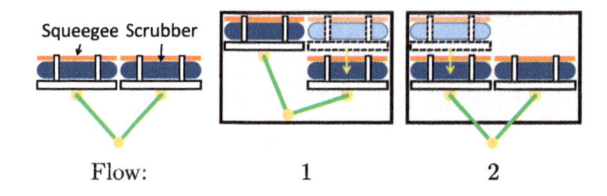

Fig. 4. Module's strategy of transition to accomplish the cleaning task. Since the squeegee is installed next to the scrubber, and the glass surface is cleaned by the movement from top to bottom.

tasks by designing autonomous locomotion control system. From Figs. 3 and 4, challenges for the control system are listed as follows:

- The module moves its one cleaning unit along the line with the window frame.
- The cleaning unit covers all area from top edge to bottom.
- Covering area of right-and-left cleaning unit overlaps a little in order to wipe away the cleaning liquid perfectly.

3 Foot Location Algorithm

The foot location determination algorithm satisfying the challenges discussed in Sect. 2 is constructed in this section. Since the module robot equips the vacuum suction cup as its leg, the robot is capable of sticking on vertical surface certainly without stabilization control such as zero moment point. Thus, the robot is allowed to locomote with the longest step within the limitation of motion range. In addition, the robot walks with a step at a time, since the general walking locomotion moving forward left and right leg alternatively is not effective to satisfy our proposed strategy (see Fig. 4). As an early stage of the research, we set three assumptions as below:

- The window frame is arbitrary quadrangle.
- The coordinates of the window frame's vertex are known.
- The window frame is enough larger than the module size.

Figure 5 shows a schematic figure of the system. Physical parameters and function variables are defined as Table 1. In Fig. 5 and Table 1, in order to interpolate between θ_{to} and θ_{bo} smoothly, $\theta(s)$ is formulated by utilizing fifth polynomial interpolation as follow:

$$\theta(s) = \sum_{i=0}^{5} a_i t^i,$$

where a_i ($i = 1, \cdots, 5$) are denoted as coefficients of the fifth polynomial interpolation.

From Fig. 5, the foot locations are formulated as follows:

$$\boldsymbol{P}_r = \boldsymbol{p}_{op} + \boldsymbol{p}_{pr} = s \begin{bmatrix} \cos\theta_0 \\ \sin\theta_0 \end{bmatrix} - h_r \begin{bmatrix} \cos(\theta(s) + \theta_{hr}) \\ \sin(\theta(s) + \theta_{hr}) \end{bmatrix}, \quad (1)$$

$$\boldsymbol{P}_l = \boldsymbol{p}_{oq} + \boldsymbol{p}_{ql} = (s - \Delta s) \begin{bmatrix} \cos\theta_0 \\ \sin\theta_0 \end{bmatrix} - h_l \begin{bmatrix} \cos(\theta(s - \Delta s) + \theta_{hr}) \\ \sin\theta(\theta(s - \Delta s) + \theta_{hr}) \end{bmatrix}, \quad (2)$$

where $\boldsymbol{p}_{mn}$ represents vector from $\boldsymbol{P}_m$ to $\boldsymbol{P}_n$. Hence, by determining appropriate s and ds situationally, the foot location of both left and right leg is calculatable. And, determination algorithms of s and ds fall into four patterns (initial location, middle area, final location, and left foot location) with corresponding to differences of each property.

3.1 Initial Location

At the initial location, edge of the right cleaning unit is on the right-top vertex, —i.e., $s = 0$, $\Delta s = 0$. Thus, by substituting $s = 0$, $\Delta s = 0$ to (1) and (2), we obtain

$$\boldsymbol{P}_r = -h_r \begin{bmatrix} \cos(\theta_{to} + \theta_{hr}) \\ \sin(\theta_{to} + \theta_{hr}) \end{bmatrix}, \qquad \boldsymbol{P}_l = -h_l \begin{bmatrix} \cos(\theta_{to} + \theta_{hl}) \\ \sin(\theta_{to} + \theta_{hl}) \end{bmatrix}. \quad (3)$$

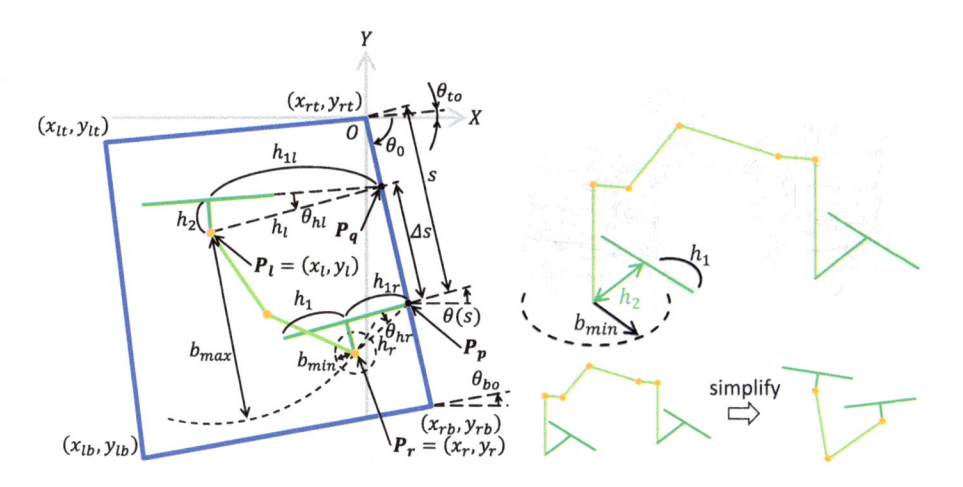

Fig. 5. The schematic figure of the algorithm. And, kinematic model of the robot.

Table 1. Physical parameters and function variables

Parameters	Notation	Value
Half length of the cleaning unit	h_1 [m]	0.5
Length from window frame to the cleaning unit	$\begin{cases} h_{1r} \\ h_{1l} \end{cases}$ [m]	$\begin{cases} h_1 \\ 2.8h_1 \end{cases}$
Length from the cleaning unit to joint	h_2 [m]	0.15
Length from window frame to foot location	$\begin{cases} h_r \\ h_l \end{cases}$ [m]	$\begin{cases} \sqrt{h_{1r}^2 + h_2^2} \\ \sqrt{h_{1l}^2 + h_2^2} \end{cases}$
Maximum step length of the module	b_{max} [m]	0.9
Safety margin to avoid collisions	b_{min} [m]	0.15
Coordinate of window frame's vertex	$\begin{cases} (x_{rt}, y_{rt}) \\ (x_{lt}, y_{lt}) \\ (x_{rb}, y_{rb}) \\ (x_{lb}, y_{lb}) \end{cases}$ [m]	$\begin{cases} (0.0, 0.0) \\ (-2.06, 0.16) \\ (0.28, -1.65) \\ (-1.65, -2.46) \end{cases}$
Angle of top window frame	θ_{to} [rad]	$\tan^{-1} y_{lt}/x_{lt}$
Angle of bottom window frame	θ_{bo} [rad]	$\tan^{-1} \frac{y_{rb}-y_{lb}}{x_{rb}-x_{lb}}$
Angle of right window frame	θ_0 [rad]	$\tan^{-1} y_{rb}/x_{rb}$
Angle between window frame and right cleaning unit	θ_{hr} [rad]	$\tan^{-1} h_2/h_1$
Angle between window frame and left cleaning unit	θ_{hl} [rad]	$\tan^{-1} h_2/2.8h_1$
Distance to edge of right cleaning unit	s [m]	variable
Angle of right cleaning unit	$\theta(s)$ [rad]	variable
Gap between edge of left and right cleaning unit	Δs [m]	variable
Foot location of both leg	$\begin{cases} P_r \\ P_l \end{cases}$ [m]	variable
Coordinate of edge of right cleaning unit	P_p [m]	variable
Coordinate of edge of left cleaning unit	P_q [m]	variable

3.2 Middle Area

At the middle area, the right foot location is on maximum step from the left foot location. From (1) and (2),

$$\boldsymbol{p}_{lr} = -\boldsymbol{p}_{ol} + \boldsymbol{p}_{or} = \begin{bmatrix} s\cos\theta_0 - h_r\cos(\theta(s) + \theta_{hr}) - l\cos\theta_l \\ s\sin\theta_0 - h_r\sin(\theta(s) + \theta_{hr}) - l\sin\theta_l \end{bmatrix}, \quad (4)$$

where $l = \|\boldsymbol{p}_{ol}\|$ and $\theta_l = \angle\boldsymbol{p}_{ol}$. When $\|\boldsymbol{p}_{lr}\| = b_{max}$, it is the maximum step length. Thus, s satisfying the maximum step length is calculated by solving following equation:

$$f_1(s) = 0, \quad (5)$$
$$f_1(s) = \|\boldsymbol{p}_{lr}\|^2 - b_{max}^2,$$
$$= s^2 + h_r^2 + l^2 - b_{max}^2 - 2sh_r\cos(\theta(s) + \theta_{hr} - \theta_0) - 2ls\cos(\theta_0 - \theta_l)$$
$$+ 2h_r l\cos(\theta(s) + \theta_h - \theta_l).$$

3.3 Final Location

At the final location, to avoid collision against bottom window frame, the leg has to take a distance from the bottom window frame as well as the right window frame. Thus, the foot location is determined so that distance between P_r to the bottom window frame equals to b_{min}. The bottom window frame is able to be formulated by a linear equation as follow:

$$y = ax + b,$$
$$a = \tan\theta_{bo}, \qquad b = s_f(\sin\theta_0 + \tan\theta_{bo}\cos\theta_0).$$

Thus, s for the final location is calculated by solving following equation:

$$f_2(s) = 0, \tag{6}$$
$$f_2(s) = \frac{-ax_r + y_r - b}{\sqrt{a^2 + b^2}},$$
$$= s\sin(\theta_0 - \theta_{bo}) + h_r\sin(\theta_{bo} - \theta(s) - \theta_{hr}) + s_f\sin(\theta_{bo} - \theta_0) - b_{min}.$$

3.4 Left Foot Location

For the left foot location after calculating s for determining the right foot location, it has to satisfy two requirements: The cleaning area of the left leg overlaps the one of the right a little. Both legs have to avoid collision. To satisfy these requirements, the left leg is located at 10% right of the cleaning unit's length than standard distance from the left leg, —i.e. $h_{1l} = 2.8h_1$. And, in order to avoid collision, edge of the one cleaning unit possesses a distance from another cleaning unit as shown in Fig. 6. In Fig. 6, distance from edge of the right and left cleaning unit denoted as d_r and d_l have to possess minimum margin defined as b_d. That is, Δs is determined so that shorter way of either d_r or d_l equals to b_d. Both right and left cleaning units are able to formulated by linear equations as follows:

$$y_r = a_r x + b_r,$$
$$a_r = \tan\theta(s), \qquad b_r = s(\sin\theta_0 - \tan\theta(s)\cos\theta_0),$$
$$y_l = a_l x + b_l,$$
$$a_l = \tan\theta(s - \Delta s), \qquad b_l = (s - \Delta s)(\sin\theta_0 - \tan\theta(s - \Delta s)\cos\theta_0).$$

Note that s is addressed as a constant value, since s has already been determined by (3), (5) and (6). Thus, Δs for the left foot location is calculated by solving following equation:

$$g(\Delta s) = 0, \tag{7}$$

$$\begin{cases} g(\Delta s) &= \frac{-ax_r + y_r - b}{\sqrt{a^2+b^2}}, \qquad (\theta(s) \geq 0) \\ &= -\Delta s\sin(\theta_0 - \theta(s)) + 1.8h_1\sin(\theta(s) - \theta(s - \Delta s)) - b_d, \\ g(\Delta s) &= \frac{-ax_r + y_r - b}{\sqrt{a^2+b^2}}, \qquad (\theta(s) < 0) \\ &= 2h_1\sin(\theta(s - \Delta s) - \theta(s)) - \Delta s\sin(\theta(s - \Delta s) - \theta(s)) + b_d, \end{cases}$$

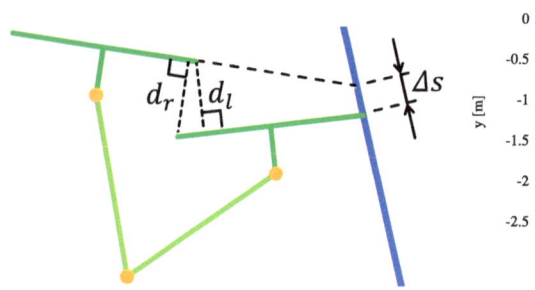

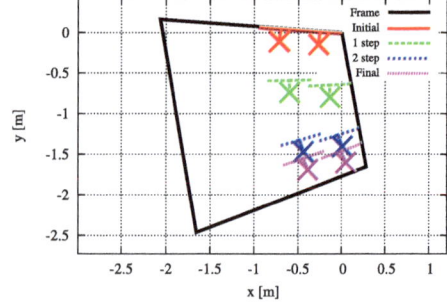

Fig. 6. The location of left foot with collision avoidance.

Fig. 7. The simulation result of the algorithm. "×" marks represent the determined foot locations by solving (3), (5), (6), and (7). And, "T"-like shape represents the foot unit of the module robot including the cleaning unit.

where b_d represents minimum distance as margin for collision avoidance.

The foot locations for both left and right leg are determined by solving (3), (5), (6), and (7) corresponding to the situations. And in this paper, Newton-Raphson method is utilized to solve these equations.

4 Numerical Simulation

The effectiveness of the algorithm is verified in this section via a numerical simulation. The physical parameters are set as Table 1. The simulation is performed by MaTX VC version 5.3.45. The simulation result is shown in Fig. 7. In Fig. 7, "×" marks represent the determined foot locations by solving (3), (5), (6), and (7). And, "T"-like shape represents the foot unit of the module robot including the cleaning unit.

Figure 7 shows that the algorithm calculates the foot locations along the right window frame. And, the area of both left and right cleaning unit are overlapped a little. Thus, the most right column is fully covered by the cleaning unit of the module robot. Hence, the algorithm determines appropriate foot locations satisfying challenges mentioned in Sect. 2.

5 Conclusion

This paper has constructed the foot determination algorithm for façade cleaning robot in order to develop the autonomous façade cleaning robot. First, requirements for the façade cleaning and our developed robot have been described. Second, the challenges for autonomous façade cleaning robot have been listed up: The module moves its one cleaning unit along the line with the window frame. The cleaning unit covers all area from top edge to bottom. Covering area of right-and-left cleaning unit overlaps a little in order to wipe away the cleaning liquid perfectly. Next, the foot determination algorithm satisfying the challenges has been constructed. The algorithm has based on the geometry of the module robot and the widow frame. Finally, the effectiveness of the algorithm has been verified via the numerical simulation. As the result, the algorithm has successfully determined appropriate foot locations satisfying the challenges.

As our future works, in order to realize the modular robot system for the façade cleaning, extended foot location algorithms for another column including the most left column will be constructed, since the algorithm for first column has been addressed in this paper. And, the experiment system will be built and the effectiveness of the algorithm will be verified via experiments.

Acknowledgement. This research is supported by Research Institute for Science and Technology of Tokyo Denki University Grant Number Q18X-00/Japan and The Precise Measurement Technology Promotion Foundation.

References

1. BBC: Shanghai window cleaning cradle swings out of control. http://www.bbc.com/news/world-asia-china-32176401
2. BBC: Window washers rescued from high up world trade center. http://www.bbc.com/news/world-us-canada-30028969
3. Zhang, H., Zhang, J., Zong, G., Wang, W., Liu, R.: Sky cleaner 3: a real pneumatic climbing robot for glass-wall cleaning. IEEE Robot. Autom. Mag. **13**(1), 32–41 (2006)
4. Seo, K., Cho, S., Kim, T., Kim, H.S., Kim, J.: Design and stability analysis of a novel wall-climbing robotic platform (rope ride). Mech. Mach. Theory **70**, 189–208 (2013)
5. Elkmann, N., Hortig, J., Fritzsche, M.: Cleaning automation. In: Springer Handbook of Automation, pp. 1253–1264. Springer, Heidelberg (2009)
6. Elkmann, N., Lucke, M., Krüger, T., Kunst, D., Stürze, T., Hortig, J.: Kinematics, sensors and control of the fully automated facade-cleaning robot siriusc for the fraunhofer headquarters building, Munich. Ind. Robot Int. J. **35**(3), 224–227 (2008)
7. Nansai, S., Elara, M.R., Tun, T.T., Veerajagadheswar, P., Pathmakumar, T.: A novel nested reconfigurable approach for a glass façade cleaning robot. Inventions **2**(3), 18 (2017)
8. Nansai, S., Onodera, K., Elara, M.R.: Development of a module robot for glass façade cleaning robot. In: International Conference on Advanced Engineering Theory and Applications, pp. 704–714. Springer, Cham (2017)

Smart Manipulation Approach for Assistant Robot

Yeyson Becerra[1(✉)], Jaime Leon[1], Santiago Orjuela[1], Mario Arbulu[1], Fernando Matinez[2], and Fredy Martinez[2]

[1] Corporacion Unificada Nacional de Educacion Superior, Bogota, Colombia
{jaime.leonbaq,yeyson.becerra,mario.arbulu}@cun.edu.co
[2] Universidad Distrital Francisco Jose de Caldas, Bogota, Colombia
{fmartinezs,fhmartinezs}@udistrital.edu.co

Abstract. This work deals with a smart visual assisted manipulation algorithm, for an assistant robot. The algorithm is divided in two parts: Object visual position feedback, and whole body motion computation. The object position feedback is based on image analysis to define coordinates in R3 space, and provide a reference to the robot about the object localization. The whole body motion computation deals with hand motion planning, and body motion to improve the reach of the hand, and broaden arm workspace. Consequently, grasping the object, picking it up, and bringing it to a goal position. An image analysis method is proposed, by using triangle similarity, knowing in advance object geometric conditions, and distance from camera to the object. A novel motion modified D-H parameters were used to build the workspace. Simulation and experimental results are discussed in order to validate our proposal.

Keywords: Robotics · Modeling · Assistance · Workspace · Computer vision · Planning

1 Introduction

Object manipulation is a necessary task in humans, it requires a process to be effective, this process is simpler in humans. In contrast, robot manipulation is a more complex process as it involves several factors, that does not affect human object manipulation, such as manipulation of nonrigid objects and substances or variation in the structure of the environment [1].

As a result, previous advances in local manipulation and object approach pursue a variety of approaches, this to overcome the current limitations of autonomous robot manipulation in human environments. As a part of robot modeling, this work proposes an integrated model of the robot, by not dividing the robot end-effector as a separate tool, as other previous works [9]. Therefore, as an advantage the analysis is simpler and more accurate by also proposing a Cartesian workspace not published before in previous works.

Robotic assistance can benefit people with movement disabilities, specially, robot manipulators can benefit people, with limited upper body mobility [2]; as well as, it can be useful for everyday tasks [4], office tasks, physical rehabilitation, and therapy, for children with ASD (Autism Spectrum Disorder). Therefore, human-robot interaction is crucial, to achieve a good object approach process [8]. Thus, safety and robustness issues are common, in the focus of human-robot interaction strategies [5]. In robot object manipulation, there is a close interaction between robot and human partner [7], hence, a coordination strategy is employed, to specify the reactive/compliant behavior of the robot [3].

This paper addresses a manipulation strategy performed by an assistant robot. This process was carried out, by modeling the robot arms and legs, and by identifying joint initial configuration space. In the same way, a local object approach, without collisions was achieved, considering a local vision algorithm.

Theoretical framework is addressed in Sect. 2, considering a brief robot description, robot workspace and kinematics modeling, as well as local vision algorithms. Section 3 presents results and discussion. Finally, conclusions are presented in Sect. 4.

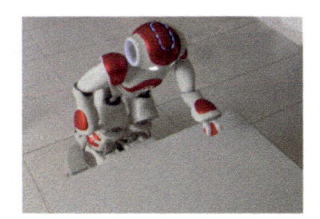

Fig. 1. Assistant robot

1.1 Contributions

Work contributions are presented as follows:

- The proposed humanoid robot model of 5 DOF (Degrees of Freedom) arms, and 6 DOF legs, is an important contribution as it differs from previous works [9]. Because, the proposed D-H modified model, does not imply dividing the hand end-effector as a separate tool. We are using it for defining the workspace.

- The assistant robot was provided with a computer vision system, that let it identify an object over a white surface, recognizing the exact object position. The computer vision system works with a reference image, that is compared with subsequent images, through triangle similarity. In this way **z** coordinate is identified; **x** and **y** coordinates are known, comparing object dimensions with captured image dimensions.

2 Theoretical Framework

In this section, both robot modeling and local vision algorithms are presented.

2.1 Robot Modeling

Left Arm and Left Leg Forward Kinematics Modeling Using D-H Parameters. A forward kinematic solution is proposed, for computing the robot workspace, [11], thus, left arm modeling is achieved by modified D-H (Denavit-Hartenberg) parameters (see Table 1), and left leg by standard D-H parameters (Table 2), [6]. In the same way, for right humanoid limbs.

Table 1. Left arm D-H modified parameters for assistant robot

i	α (Alpha)	a	θ (Theta)	d (distance)
1	−π/2	0	q1	0
2	π/2	0	q2	0
3	−π/2	0	q3	d3
4	π/2	0	q4	0
5	−π/2	0	q5	d5

Table 2. Left leg D-H parameters for assistant robot (modified)

i	α (Alpha)	a	θ (Theta)	d (distance)
1	−3π/4	0	q1	0
2	−π/2	0	q2	0
3	π/2	0	q3	0
4	0	a4	q4	0
5	0	a5	q5	0
6	−π/2	0	q6	0

In order to compute the upper limbs workspace, the end effector hand position equations (see Eqs. 1, 2, 3), were obtained from previously modified D-H parameters, as following:

$$x = (d4*sin(q1)*sin(q3)*sin(q4) - cos(q1)*(sin(q2-\pi/2)*$$
$$((d4*cos(q4))+l2)+(d4*cos(q3)*cos(q2-pi/2)*sin(q4))) \quad (1)$$
$$y = cos(q2-\pi/2)*((d4*cos(q4))+d2)-(d4*cos(q3)*sin(q4)*sin(q2-\pi/2)) \quad (2)$$
$$z = sin(q1)*(sin(q2-\pi/2)*((d4*cos(q4))+d2)+(d4*cos(q3)*cos(q2$$
$$-\pi/2)*sin(q4)))+(d4*cos(q1)*sin(q3)*sin(q4)) \quad (3)$$

Robot Description. Robot structure consists on a 85 cm of height and 27.5 cm of width and 10 kg of weight, it has a total of 25 Degrees of Freedom for manipulation tasks, and omni-directional wheeled locomotion. It has Two HD cameras, four microphones, two sonar range finders to detect obstacles, two infrared emitters and receivers, an inertial board, nine tactile sensors and eight pressure sensors. Each arm consists on a 5 Degree of freedom set of joints and each arm is 21.8 cm length (see Fig. 1).

Augmented Workspace. When the robot is performing object manipulation, its body must lean towards the object, in order to perform a better approach and reach it successfully. As a result, when the robot torso leans towards the object, the workspace increases, allowing the arm to extend to an augmented distance (see Fig. 6).

2.2 Local Vision Algorithms

The problem of vision was based on identifying an object (red ball) on a white surface and knowing its position in a 3D space, in which this was found, having as framework the assistant robot's upper front camera. To develop this task, the geometric characteristics of the object were defined, for this particular case, the object is 3.8 cm of diameter.

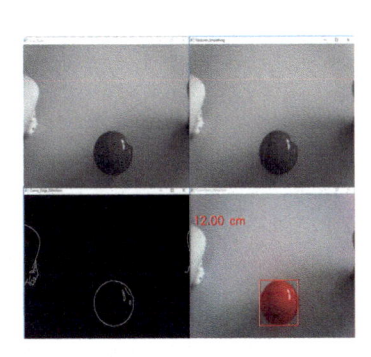

```
                        START
                              DEFINE distance from camera to object
                              DEFINE width of the object
                              DEFINE area (x,y) of the image
                              TAKE picture_1 of the object

                                    APPLY gray scale
                                    APPLY filtering
                                    APPLY edge detection
                                    APPLY contours

                                    D = distance from camera to object
                                    W = width of the object
                                    P = apparent width in pixels
                                    F = focal length
                                    COMPUTE F = (P*D)/W

                              for PICTURE_1 to PICTURE_n do

                                    APPLY gray scale
                                    APPLY filtering
                                    APPLY edge detection
                                    APPLY contours

                                    C = center of the object
                                    COMPUTE C = contours/2
                                    D' = new distance from camera to object
                                    COMPUTE D' = (F*W)/P

                                    COMPUTE Y = (C/(P/W))-(area(x)/2)
                                    COMPUTE Z = (area(y)/2)-(C/(P/W))

                              PRINT X coordiante = D'
                              PRINT Y coordiante = Y
                              PRINT Z coordiante = Z
                        END
```

Fig. 2. Image processing **Fig. 3.** Vision algorithm pseudocode

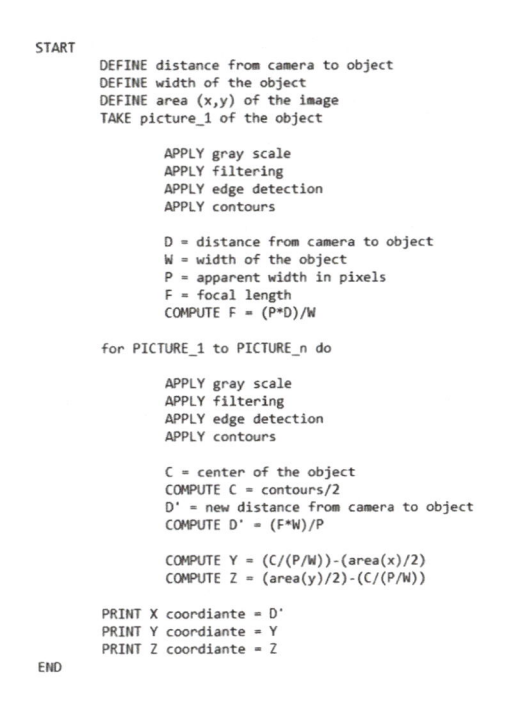

Once the geometric conditions of the object were identified, the next step was to proceed to capture an image of the environment in which the object was localized to perform image processing. The captured image was converted to gray-scale, then, 2D Gaussian Filter was applied (Eq. 4) to remove high frequency noise. It is important to notice that edge detection will work properly if a filtering stage was implemented previously. Canny edge detection was utilized [15], threshold values were defined by a minimum value and a maximum value, edges with intensity gradient bigger than maximum value were sure to be edges, and edges lower than minimum value were sure to be non-edges, edges between maximum and minimum values depend on their connectivity (Eq. 5).

$$G_2(x,y) = \frac{1}{2\pi\sigma^2}e^{-\frac{x^2+y^2}{2\sigma^2}} \quad (4) \quad G_n = \frac{\partial G_2(x,y)}{\partial n} = n \cdot \nabla G_2(x,y) \quad (5)$$

Image processing after applying gray-scale, filtering, edge detection and contour that represents the object of interest is shown (see Fig. 2). It was assumed that the contour with the largest area in the picture was the object. The twelve centimeter seen in Fig. 2 is from robot's upper front camera to the object (D).

Having defined image processing, it was proceed to know the distance from camera to object. The method used to perform this was triangle similarity, consisted of knowing geometric conditions of the object, measuring the actual distance from camera to object and introducing it as a reference in the algorithm, and finally, applying some trigonometry.

Summarizing, triangle similarity depended on knowing width of the object (W), a reference distance from camera to object (D) and the apparent width of the object given in pixels (P). The result was the focal length of camera (F) as represented in Eq. 6. Focal length parameter was taken from a reference picture that later was used to compare with subsequent pictures and in this way estimating different distances from camera to object.

$$F = \frac{P \times D}{W} \quad (6) \qquad D' = \frac{F \times W}{P} \quad (7)$$

Subsequent pictures taken after reference picture utilized triangle similarity to determine a new distance (D') from object to camera. Every time the distance between camera and object was changed, D' took a new value according to constants (F) and (W) and variable (P) as seen in Eq. 7.

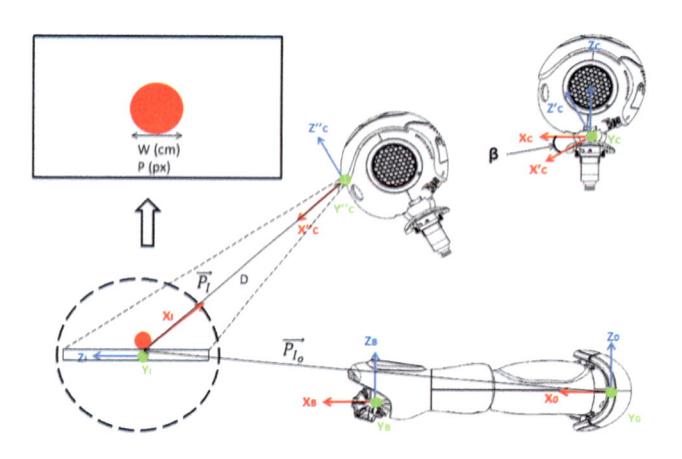

Fig. 4. Frames vision - manipulation system

Once D was known, named x coordinate, (see Fig. 4) it was necessary to determine the position of object in a coordinate plane, located in the camera of the robot. To do this, it was needed to know the number of pixels of the object and making a comparison among physical world and digital world. In consequence, it was possible to learn the dimension in centimeters of the image, which was useful to know the position of the object over a y,z plane (see Fig. 4). Vision algorithm pseudocode is shown below (see Fig. 3)

Cartesian coordinates that define localization of the object were already established. The next step was to relate camera frame and robot's arm frame in order to generate communication and understanding between vision algorithm and manipulation algorithm. The next matrix (see Eq. 8) show the relation among 3 different frames located at camera, object and arm's base (see Fig. 3). The angle β defined pitch in the head of robot, $\vec{P}_{I_0} = (x_{I_0}, y_{I_0}, z_{I_0})$ defined position of arm's end effector, which was a 3D vector, $\vec{P}_C = (x_C, y_C, z_C)$ defined camera position, which was a 3D vector and $\vec{P}_I = (x_I, y_I, z_I)$ defined image position, which was a 3D vector. Camera position was related to origin frame through a translation, a rotation and a translation; these 3 movements were grouped in a homogeneous transformation matrix.

$$\begin{bmatrix} x_{I_0} \\ y_{I_0} \\ z_{I_0} \\ 1 \end{bmatrix} = \begin{bmatrix} 1 & 0 & 0 & x''_C \\ 0 & 1 & 0 & y''_C \\ 0 & 0 & 1 & z''_C \\ 0 & 0 & 0 & 1 \end{bmatrix} \begin{bmatrix} cos(\beta) & 0 & sen(\beta) & 0 \\ 0 & 1 & 0 & 0 \\ -sen(\beta) & 0 & cos(\beta) & 0 \\ 0 & 0 & 0 & 1 \end{bmatrix} \begin{bmatrix} 1 & 0 & 0 & x_C \\ 0 & 1 & 0 & y_C \\ 0 & 0 & 1 & z_C \\ 0 & 0 & 0 & 1 \end{bmatrix} \begin{bmatrix} x_I \\ y_I \\ z_I \\ 1 \end{bmatrix} \quad (8)$$

3 Results

3.1 Vision Experimental Results

The vision algorithm took into account the distance from camera to object as a reference. The distance was 12 cm and measurement of object was 3.8 cm of diameter. Having these 2 measurements allowed to estimate missing coordinates (y,z) as it was explained in Sect. 2.2. Light condition must to be considered to improve the response of the algorithm as well as an appropriate measurement from camera to object. It was also needed determining properly parameters for Gaussian Filter and Canny edge detection for its relevance of identification of the object over a coordinate plane.

Before testing with a real object, it was decided to perform test with a simulated object that was in front of the head of the robot, in order to know the reach of robot arm, predict collisions in a real environment and compare results with the workspace of the robot. The description of this experiment was detailed below.

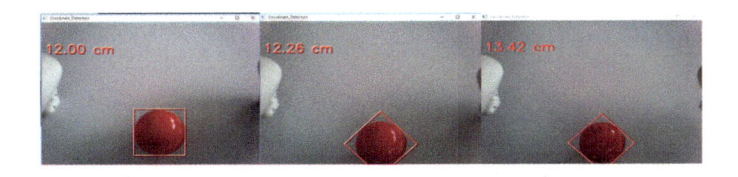

Fig. 5. Vision experimental results

The head of the robot was positioned at 90° and the simulated object in front of it, more exactly, in front of the upper camera. It was used both, the left robot arm and the right, to develop this experiment. The utilized distances (D) were 5 cm, 10 cm and 15 cm in between head of the robot and the simulated object. The measurements for the image were 21 cm height and 27.9 cm width or 480 pixels height and 640 pixels width.

The first position for the simulated object was at the center of image taken by camera of the robot (y,z coordinates) and 5 cm (x coordinate), so the first attempt was to reach that position; the subsequent attempts were at corners of the image (y,z coordinates) and varying distances (x coordinate), it was supposed the simulated object was positioned in all these points, which in total were 15. The result of this test was summarized (see Table 3).

Table 3. Vision experimental results.

Distance	Center	Upper right corner	Upper left corner	Lower right corner	Lower left corner
5 cm	✓	✗	✗	✗	✗
10 cm	✓	✗	✗	✓	✓
15 cm	✗	✗	✗	✓	✓

Results of the test showed the following, 9 positions out of 15 were errors, thus, the percentage of success was 40%. Nevertheless, 8 positions were in an extreme part of the image that rarely were used in a real environment (cornes). It could also be appreciated that the most successful area to manipulate an object was located from the center to the bottom of the image; moreover it should be avoided to attempt to grab the object when it is located from the top to the center of the image, therefore the robot should be repositioned to find a new perspective of the object.

As it was stated before, the simulated object was positioned parallel to the head of robot, which in that case was 90°. But in practice the real object had to be over a white surface, reason why, it was chosen 3 different positions to the body of robot, tilting it forward to 73°, 75° and 78° in order to observe the object at different distances (D) and assessing the vision algorithm response. The results are shown (see Fig. 5).

3.2 Visual Assisted Manipulation Simulation and Experimental Results

The object manipulation simulation was carried out considering vision algorithm and motion planning. The goal was to approach and reach the object (red ball) in order to grasp it and pick it up. It was used a simulator of robots to test the vision algorithm before it was tested in the physical robot.

Manipulation experimental results showed robot-object approaching and object manipulation, these were based on the vision algorithm and the manipulation algorithm. As it was stated before, the vision algorithm provides information of Cartesian coordinates of the object that are then sent to the manipulation algorithm, as a result, the robot approaches to the object, open its hand, moves the arm towards the object and once the hand is located over the object, it grabs it and picks it up (see Fig. 6).

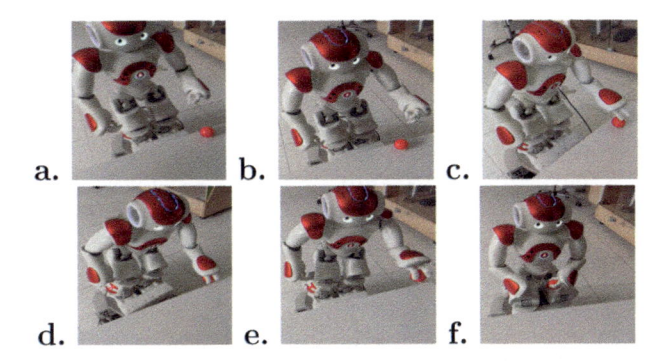

Fig. 6. Object manipulation experimental results

The workspace helped to validate what was found in the visual experiment. As it was stated before, there were 9 positions out of 15 that the robot arm could not reach, the distances (D) between camera and object were selected randomly without taking into account the workspace of the robot arm and therefore, it was necessary to know more precisely what the unreachable positions were for the robot arm. The 15 positions were examined and validated by the workspace of left arm and right arm, black lines in Fig. 6 show the workspace path defined by ShoulderRoll, ShoulderPitch and ElbowRoll while red line shows workspace path defined by ShoulderRoll and ElbowYaw. Dashed blue line refers to the augmented workspace (see Fig. 7).

3.3 Discussion

The main issue of this investigation was to develop a method to identify, to localize and to manipulate an object over a surface through an assistant robot. The results show that the utilized object is properly identified by the vision algorithm, allowing knowing the Cartesian coordinates of the object, moreover, the object might take some other positions over the surface and the vision algorithm will still identify the localization of the object, thus, the grasping and picking tasks are effectively carried out.

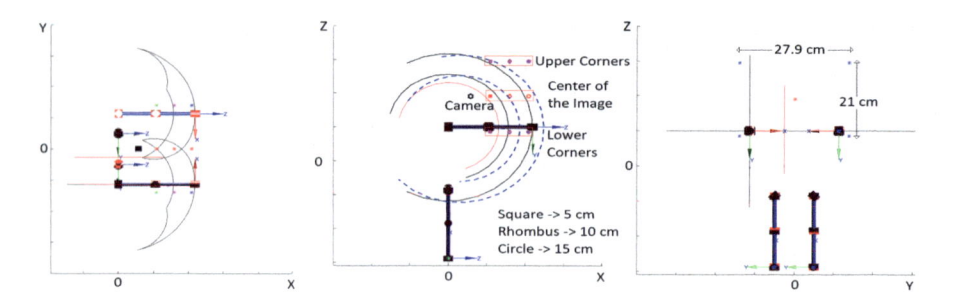

Fig. 7. Views of workspace

On the other hand, a workspace study of the robot arm is developed to know the points that can be reached by its end effector, besides; an augmented workspace is defined to represent the reach of the robot including not only arms but legs also. The augmented workspace provides information to know if the position of the object is reachable by its end effector.

The vision algorithm allows developing a smart manipulation task; it is just necessary a robot arm and a camera to perform the grasping and picking tasks, besides, this work is complemented by the definition and use of the augmented work space that allows knowing the reaching and constraints that the robot has when grasping and picking the object.

Before this work, it was already used a camera and robot to perform manipulation task [2] but with an extra element that was a kinect to detect the depth of the object. It was also aimed to people with disability in their upper body.

4 Conclusion

In the experimental setup, the robot accomplished object approach, manipulation and picking. It is important to remark, that the performed manipulation tasks were achieved, by robotic perception through vision algorithm.

Gaussian filter as Canny Edge Detection have to be defined for specific light conditions; because, it was noticed that vision's algorithm variated its response in not identifying the object properly. Furthermore, it was considered specific geometric conditions for the object (red ball) so, if the object changes, the algorithm has to be updated. Finally, the algorithm is suitable to use different kind of filters to improve its response in dynamic environments and different kind of objects.

Further research may be oriented to develop a path planning algorithm to the robot arm as well as to the mobile robot in order to take the picked objects to a new destination. Additionally, it is also possible to improve the vision algorithm in order to identify more than an object or even to optimize the image filtering. Finally, it is recommended to use a different gripper to avoid the issue to deal with just small objects.

References

1. Kemp, C., Edsinger, A.: Challenges for robot manipulation in human environments grand challenges of robotics. IEEE Robot. Autom. Mag. **14**(1), 20–29 (2007)
2. Quintero, C., Ramirez, O., Jägersand, M.: VIBI: assistive vision-based interface for robot manipulation. In: IEEE International Conference on Robotics and Automation (ICRA), pp. 4458–4463. Seattle, WA (2015)
3. Cehajic, D., Erhart, S., Hirche, S.: Grasp pose estimation in human-robot manipulation tasks using wearable motion sensors. In: IEEE/RSJ International Conference on Intelligent Robots and Systems (IROS), pp. 1031–1036. Hamburg, Germany (2015)
4. Kruse, D., Radke, R., Wen, J.: Collaborative human-robot manipulation of highly deformable materials. In: 2015 IEEE International Conference on Robotics and Automation (ICRA), pp. 3782–3787. Seattle, WA (2015)
5. Hayne, R., Luo, R., Berenson, D.: Considering avoidance and consistency in motion planning for human-robot manipulation in a shared workspace. In: IEEE International Conference on Robotics and Automation (ICRA), pp. 3948–3954. Stockholm, Sweden (2016)
6. Corke, P.: A simple and systematic approach to assigning Denavit-Hartenberg parameters. IEEE Trans. Robot. **23**(3), 590–594 (2007)
7. Yue, C., Guo, S., Li, Y., Li, M.: Bio-inspired robot launching system for a motherson underwater manipulation task. In: 2014 IEEE International Conference on Mechatronics and Automation, pp. 174–179. Tianjin, China (2014)
8. Young, M.: The Technical Writer's Handbook. University Science, Mill Valley (1989)
9. Kofinas, N.: Forward and Inverse Kinematics for the assistant Robot. Technical University of Crete, Greece Department of Electronic and Computer Engineering, Chania, Greece, pp. 27–48 (2012)
10. Hashemi, E., Ghaffari, M.: Dynamic modeling and control study of the NAO biped robot with improved trajectory planning. In: Materials with Complex Behavior, pp. 671–688 (2012)
11. Li, J., Zhang, R., Cui, F., Zhang, Y.: Analysis on the workspace of six-degreesof-freedom industrial robot based on AutoCAD. In: 2017 the Second International Conference on Mechanical, Manufacturing, Modeling and Mechatronics, pp. 1–6. Kortrijk, Belgium (2017)
12. Cianchetti, M., Calisti, M., Margheri, L., Kuba, M., Laschi, C.: Bioinspired locomotion and grasping in water: the soft eight-arm OCTOPUS robot. Bioinspiration Biomim. **10**(3C), 035003 (2015)
13. Franchi, A., Petitti, A., Rizzo, A.: Distributed estimation of the inertial parameters of an unknown load via multi-robot manipulation. In: 53rd IEEE Conference on Decision and Control, pp. 6111–6116. Los Angeles, CA (2014)
14. Canny, J.: A computational approach to edge detection. IEEE Trans. Pattern Anal. Mach. Intell. **8**, 679–698 (1986)
15. Caban, J., Rosebrock, A., Yoo, T.: Automatic identification of prescription drugs using shape distribution models. In: 19th IEEE International Conference on Image Processing, pp. 1005–1008. Orlando, FL (2012)
16. OpenCV home page, Canny Edge Detector Documentation. https://docs.opencv.org/2.4/doc/tutorials/imgproc/imgtrans/cannydetector/cannydetector.html?highlight=canny20edge. Accessed 28 May 2018

Computational Study on Upward Force Generation of Gymnotiform Undulating Fin

Van Hien Nguyen[1(✉)], Canh An Tien Pham[2], Van Dong Nguyen[2], Hoang Long Phan[2], and Tan Tien Nguyen[2]

[1] PetroVietnam Camau Fertilizer Joint Stock Company, Camau City, Vietnam
hiennv@pvcfc.com.vn
[2] Ho Chi Minh City University of Technology (HCMUT), Ho Chi Minh City, Vietnam
antienpham@gmail.com, nguyendongnavy@gmail.com, phlong65@yahoo.com, nttien@hcmut.edu.vn

Abstract. Gymnotiform is a type of fish swimming which uses only a long anal fin as the main propeller. In the real gymnotiform fish, the fin undulates with the counter-propagating motion for the movement of upward and downward while the body still rigid. There are some studies about this type of motion which apply the experiment and the digital particle image velocimetry (DPIV) method. In the robotics field, computational fluid dynamics (CFD) is extremely outstanding with the accuracy, fast and thorough display the impact of fluid on objects. This paper proposes a model of CFD to determine about 2 types of counter-propagating motion.

Keywords: Gymnotiform · Propeller · Counter-propagating · Upward motion · CFD

1 Introduction

Autonomous Underwater Vehicles (AUVs) is a feasible approach for discovering underwater environment. Besides, it has a variety of applications in terms of transportation and scout. Gymnotiform Undulating Fin Robot is a biomimetic robot which benefits the characteristics of natural systems such as high efficiency, undisturbed environment, flexible movement. With these advantages, this field has been studied over the world in the effort to improve this type of biomimetic robot. Many concepts are introduced with a various mechanical design [2, 12, 15, 16, 21]. To thoroughly understand the process of undulating and generating force, there are three approaches. The first method basing on the fluid dynamics, the relative equation between force and others parameters is establish to consider the influence of this parameters on force generated [3, 6, 7, 10, 21]. Although this solution is visual and strongly bases on mathematical foundation, there are still some obstacles while modeling the fin and the impact of fluid on this, especially the impact of vortexes and turbulent flow. The second method basing on experiment, after measuring the force in various cases, the data was analyzed to identify the relation between force and others parameters in a reasonable law [8, 17, 24, 25]. The third approach is computational fluid dynamics (CFD) method which benefits the rapid calculation process on computer and gives the more accurate results [9, 11, 13, 15, 18–20, 23, 24]. In CFD, the domain is divided into a number of cells and the calculation process is accomplished on each cell, so that all the results are received with the help of computers.

In our prior study, we have determined the number of fin-rays and the force generated in the direction along the fin [1, 5]. For movement in the other directions, the fin has to have another undulating law. This study focuses on the upward movement of the fin by applying the 3D CFD method.

2 Problem Setting

In the typical fish of Gymnotiform branch – black ghost knife fish which use only a caudal fin as a main propeller while the body is rigid, the undulating law of the fin decides the direction of movement (Fig. 1).

Fig. 1. Expected directions of fin movement

The forward and backward movements are the main directions in terms of robotics which have been studied in the majority of researches [7, 17, 19, 22, 24]. For these directions, the fin undulates with the sinusoidal law. In addition, upward movement enables black ghost knife fish to rise up (if the fin in the bottom) or sink down without the forward or backward movement. To accomplish this movement, knife fish will undulates with counter-propagating law, which has been studied by Curet et al. [14] using experimental method with digital particle image velocimetry (DPIV). However, they only focus on upward force in inward counter-propagating and outward counter-propagating cases. This study creates a simulation model of gymnotiform undulating fin and employs CFD method to survey these two cases and measure forces in 3 directions and compares the efficiency of upward force production in 2 cases.

3 Model of Simulation

3.1 Model of Gymnotiform Undulating Fin

In our previous study [5], the fin with 16 fin-rays in one wavelength is superior. The model has 2 full wavelengths with counter-propagating motion. The fin's motion is defined by (1) and (2) where $\theta(n,t)$ is the angle of n-th ray at moment t and their specifications shown in Table 1.

$$\theta(n,t) = \theta_{max} \sin\left(2\pi ft - k\frac{2\pi n}{N-1}\frac{L}{\lambda}\right), n = \overline{0,15} \quad (1)$$

$$\theta(n,t) = \theta_{max} \sin\left(2\pi ft + k\frac{2\pi n}{N-1}\frac{L}{\lambda}\right), n = \overline{16,31} \quad (2)$$

Table 1. Specifications of fin

f	Frequency	1 Hz, 1.2 Hz, 1.5 Hz, 1.7 Hz, 2 Hz
θ_{max}	Angle amplitude	$20°, 30°, 40°$
λ	Wavelength	0.3 m
L	Length of fin	0.6 m
N	Number of fin-rays	16
k	Motion factor	1 for inward counter-propagating motion
		-1 for outward counter-propagating motion

3.2 Method

This study applies the Finite Volume Method (FVM) provided in ANSYS FLUENT to calculate the force exerting to fin in its undulating process. The fluid is supposed to be ideal with the constants of properties and incompressible. Therefore, the calculation bases on pressure with the fluid's properties in fluent database. The Reynolds-averaged Navier-Stokes (RANS) equations in Cartesian tensor form are expressed by [4]:

$$\frac{\partial \rho}{\partial t} + \frac{\partial}{\partial x_i}(\rho u_i) = 0 \quad (3)$$

$$\frac{\partial}{\partial t}(\rho u_i) + \frac{\partial}{\partial x_j}(\rho u_i u_j) = -\frac{\partial p}{\partial x_i} + \frac{\partial}{\partial x_i}\left[\mu\left(\frac{\partial u_i}{\partial x_j} + \frac{\partial u_j}{\partial x_i} - \frac{2}{3}\delta_{ij}\frac{\partial u_l}{\partial x_l}\right)\right] + \frac{\partial}{\partial x_j}\left(-\rho\overline{u_i'u_j'}\right) \quad (4)$$

Where velocity components $u_i = \overline{u_i} + u_i'$, $\overline{u_i}$ and u_i' are the mean and fluctuating velocity components ($(i,j,l = 1,2,3)$, ρ is fluid density, and μ is dynamic viscosity.

To reduce the computational cost, The Boussinesq hypothesis is applied in the $k - \varepsilon$ models with the presence of turbulence kinetic energy k and turbulence dissipation rate ε, the Reynolds-averaged equation and the transport equations for the Standard $k - \varepsilon$ Model is [4]:

$$-\overline{\rho u_i' u_j'} = \mu_t \left(\frac{\partial u_i}{\partial x_j} + \frac{\partial u_j}{\partial x_i} \right) - \frac{2}{3} \left(\rho k + \mu_t \frac{\partial u_k}{\partial x_k} \right) \delta_{ij} \quad (5)$$

$$\frac{\partial}{\partial t}(\rho k) + \frac{\partial}{\partial x_i}(\rho k u_i) = \frac{\partial}{\partial x_j} \left[\left(\mu + \frac{\mu_t}{\sigma_k} \right) \frac{\partial k}{\partial x_j} \right] + G_k + G_b - \rho \varepsilon - Y_M + S_k \quad (6)$$

$$\frac{\partial}{\partial t}(\rho \varepsilon) + \frac{\partial}{\partial x_i}(\rho \varepsilon u_i) = \frac{\partial}{\partial x_j} \left[\left(\mu + \frac{\mu_t}{\sigma_\varepsilon} \right) \frac{\partial \varepsilon}{\partial x_j} \right] + C_{1\varepsilon} \frac{\varepsilon}{k}(G_k + C_{3\varepsilon} G_b) - C_{2\varepsilon} \rho \frac{\varepsilon^2}{k} + S_\varepsilon \quad (7)$$

Where μ_t is turbulent viscosity. G_k and G_b demonstrate the production of turbulent kinetic energy of the mean velocity gradients and buoyancy respectively. Y_M illustrates the influence of fluctuating dilatation on dissipation rate. $C_{1\varepsilon}$, $C_{2\varepsilon}$ and $C_{3\varepsilon}$ are model constants. σ_k and σ_ε are the turbulent numbers for k and ε. S_k and S_ε are source terms.

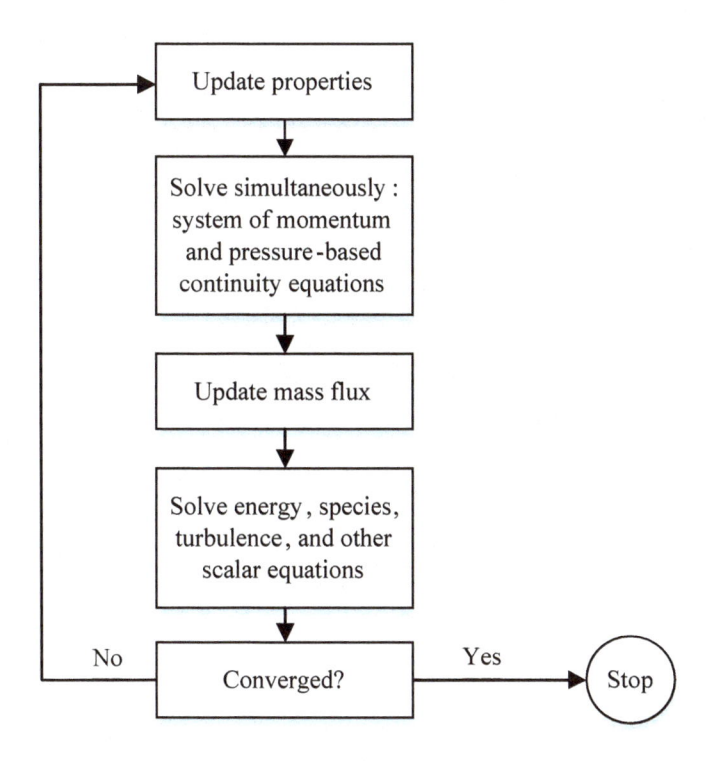

Fig. 2. Overview of the pressure-based coupled algorithm method [4].

For the solver theory, the pressure-based coupled algorithm is used to solve the system of momentum and continuity equations in a closely coupled manner. In Fig. 2, the properties need for calculating process are used to solve momentum and pressure-based continuity equations (Eqs. 5, 6, 7). The output of this step is velocity components of fluid and the value of k and ε. After that, these variables is used to calculate the residual functions to check the convergence of calculating process. If these residuals satisfy the convergence criteria, the calculation pass to next time step. If not, the calculation will use these variables received in this process and re-calculate until the convergence criteria are satisfied or maximum repeat times are reached.

For convergence criteria, residuals of continuity, x-velocity, y-velocity, z-velocity, k and ε have to smaller than 0.001. To achieve it, each time step includes 20 interactions for computational process.

3.3 Simulation Model

In this study, a geometry of fin and domain are created with specifications shown in Fig. 4. When the fin undulates, the force generated is exerted on the fin and pushes the fin to rise up. Therefore, at the top of the fin, water is always statics and turbulent in the other direction. It means the boundary in the top of the fin is inlet, and that in others direction are outlet (Fig. 3).

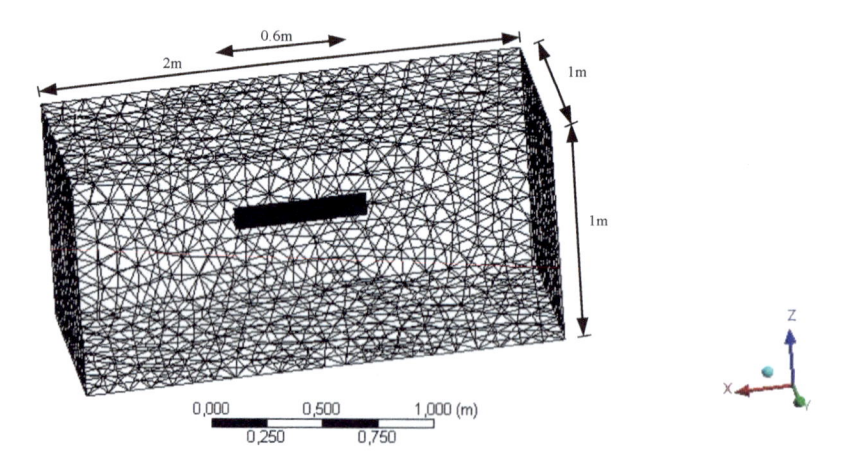

Fig. 3. Mesh model

The domain is divided into 775530 tetrahedron elements with 145024 nodes. Unstructured elements is used because of the severity of re-meshing in the undulating process. That is to say, when the fin undulates, the elements near the fin have to be skew remarkably to meet the motion of the fin. Therefore, unstructured elements are proper to facilitate their re-meshing. In addition, the elements near the fin are small to improve the accuracy of result with the boundary layer effect (see in Fig. 4).

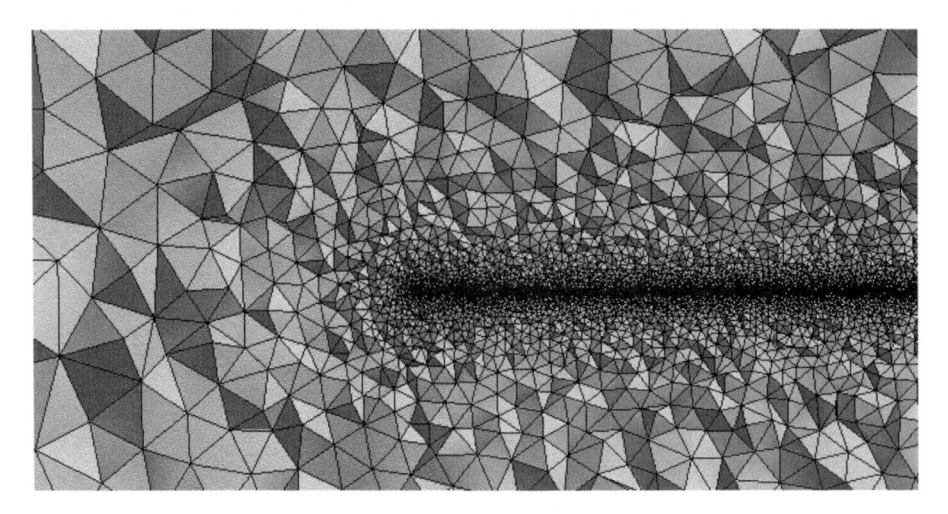

Fig. 4. Mesh near fin.

The undulating motion of the fin is created by using User-Define Function (UDF). This function allows to define the motion of each node on the fin, which bring about the motion of the fin. The nodes between 2 fin-rays are interpolated as long as nodes which are equidistant from the x-axis are aligned. The fin's motion is divided into 2 processes. Initially, the fin is flat in order to create a geometry and re-meshing model straightforwardly. After that, the fin move to form a sinusoid profile until this profile has a specific amplitude. In the second process, the fin undulates with a specific frequency.

4 Results

Calculation process is accomplished in 2000 steps with time step $\Delta t = 0.005$. Force exerting on fin is divided into 3 components associated with 3 directions. In this simulation, the upward force is F_z (see Fig. 5.).

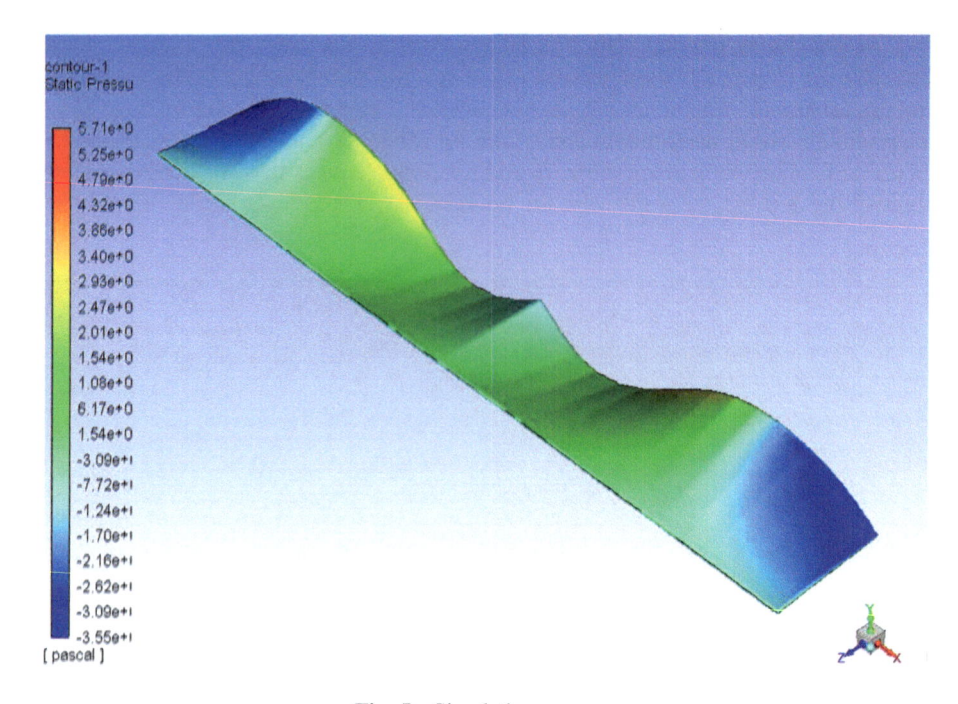

Fig. 5. Simulation process

The results of upward force are shown in Fig. 6(a). The simulation includes 3 amplitudes $(20°, 30°, 40°)$ and arranges between 1 Hz and 2 Hz. As can be seen in Fig. 6. The upward force increase with the rise of amplitude (proper with previous study). And there is insignificant difference between 2 types of counter-propagating motion.

In addition, the efficiency of force generated which is defined in Eq. (8) is shown in Fig. 6(b).

$$efficiency = \frac{|F_z|}{\sqrt{F_x^2 + F_y^2 + F_z^2}} \quad (8)$$

According to this results, although the value of force generated is similar in 2 types of counter-propagating motion, the efficiency of inward motion is higher than that of outward motion. That is to say, with the same specifications of motion, inward motion produce less force in other unexpected directions. This is extremely important because the less force in other unexpected directions means the more stability of the fin.

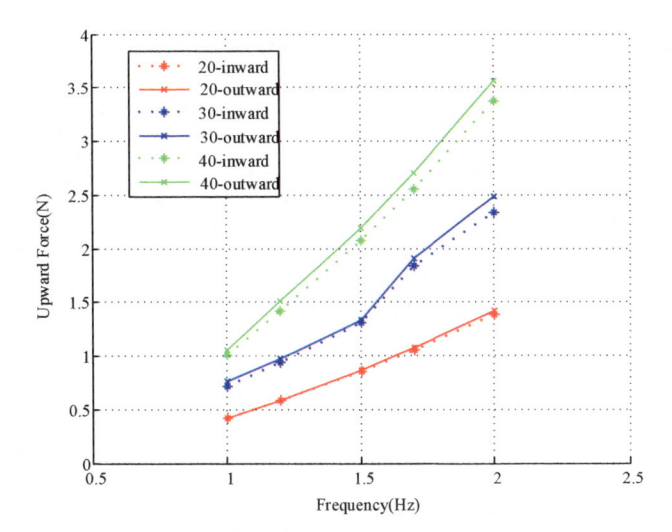

a) Upward force generated with various frequency

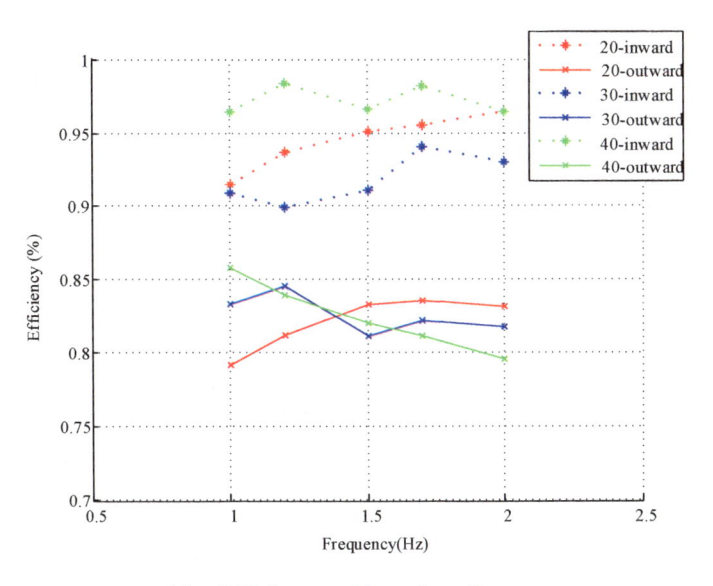

b) Efficiency with various frequency

Fig. 6. Simulation results

5 Conclusion

This study proposes a model of CFD simulation and applies this for a case of upward force generation. 2 types of counter-propagating motion are compared in both value of force generated and the efficiency of force generating process. The results suggest that the inward counter-propagating motion produces force more stably. For further study, the model should apply the Fluid Structure Interaction (FSI) to gain more accurate and realistic because the fin had better move in the fluid.

References

1. Nguyen, V.H., Pham, C.A.T., Nguyen, V.D., Kim, D.H., Nguyen, T.T.: A study on force generated by gymnotiform undulating fin. In: 15th International Conference on Ubiquitous Robots, pp. 247–252, 27–30 June 2018
2. Wang, S., Wang, Y., Wei, Q., Tan, M., Yu, J.: A bio-inspired robot with undulatory fins and its control methods. IEEEASME Trans. Mechatron. **22**, 206–216 (2017)
3. Liu, H., Curet, O.M.: Propulsive performance of an under-actuated robotic ribbon fin. Bioinspir. Biomim. **12**, 036015 (2017)
4. ANSYS Fluent Theory Guide, Release 18.1, April 2017
5. Nguyen, V.D., Phan, D.K., Pham, C.A.T., Kim, D.H., Dinh, V.T., Nguyen, T.T.: Study on determining the number of fin-rays of a gymnotiform undulating fin robot. In: Lecture Note in Electrical Engineering, vol. 465, pp. 745–752. Springer, 7–9 December 2017. ISSN 1876-1100
6. Sfakiotakis, M., Fasoulas, J., Gliva, R.: Dynamic modeling and experimental analysis of a two-ray undulatory fin robot. In: 2015 IEEE/RSJ International Conference on Intelligent Robots and Systems (IROS), pp. 339–346. IEEE (2015)
7. Bale, R., Shirgaonkar, A.A., Neveln, I.D., Bhalla, A.P.S., MacIver, M.A., Patankar, N.A.: Separability of drag and thrust in undulatory animals and machines. Sci. Rep. **4**, 7329 (2015)
8. Sfakiotakis, M., Fasoulas, J.: Development and experimental validation of a model for the membrane restoring torques in undulatory fin mechanisms. In: 2014 22nd Mediterranean Conference of Control and Automation (MED), pp. 1540–1546. IEEE (2014)
9. Maddalena, L., Vergine, F., Crisanti, M.: Vortex dynamics studies in supersonic flow: merging of co-rotating streamwise vortices. Phys. Fluids **26**, 046101 (2014)
10. Neveln, I.D., Bai, Y., Snyder, J.B., Solberg, J.R., Curet, O.M., Lynch, K.M., MacIver, M.A.: Biomimetic and bio-inspired robotics in electric fish research. J. Exp. Biol. **216**, 2501–2514 (2013)
11. Yong-Hua, Z., Jian-Hui, H., Kin-Huat, L.: Numeric simulation on the performance of an undulating fin in the wake of a periodic oscillating plate. Int. J. Adv. Robot. Syst. **10**, 352 (2013)
12. Curet, O.M., Patankar, N.A., Lauder, G.V., MacIver, M.A.: Mechanical properties of a bio-inspired robotic knifefish with an undulatory propulsor. Bioinspir. Biomim. **6**, 026004 (2011)
13. Zhang, Y.H., He, J.H., Zhang, G.Q.: Influence of the obliquity of fin ray on propulsion performance for biorobotic underwater undulating propulsor. Appl. Mech. Mater. **52–54**, 267–272 (2011)

14. Curet, O.M., Patankar, N.A., Lauder, G.V., MacIver, M.A.: Aquatic manoeuvering with counter-propagating waves: a novel locomotive strategy. J. R. Soc. Interface **8**, 1041–1050 (2011)
15. Liu, F., Yang, C.-J., Lee, K.-M.: Hydrodynamic modeling of an undulating fin for robotic fish design. In: 2010 IEEE/ASME International Conference on Advanced Intelligent Mechatronics (AIM), pp. 55–60. IEEE (2010)
16. Chen, J., Hu, T., Lin, L., Xie, H., Shen, L.: Learning control for biomimetic undulating fins: an experimental study. J. Bionic Eng. **7**, S191–S198 (2010)
17. Peter, B., Ratnaweera, R., Fischer, W., Pradalier, C., Siegwart, R.Y.: Design and evaluation of a fin-based underwater propulsion system. In: 2010 IEEE International Conference on Robotics and Automation (ICRA), pp. 3751–3756. IEEE (2010)
18. Zhou, H., Hu, T., Xie, H., Zhang, D., Shen, L.: Computational and experimental study on dynamic behavior of underwater robots propelled by bionic undulating fins. Sci. China Technol. Sci. **53**, 2966–2971 (2010)
19. Zhou, H., Hu, T., Xie, H., Zhang, D., Shen, L.: Computational hydrodynamics and statistical modeling on biologically inspired undulating robotic fins: a two-dimensional study. J. Bionic Eng. **7**, 66–76 (2010)
20. Zhang, Y., Jia, L., Zhang, S., Yang, J., Low, K.H.: Computational research on modular undulating fin for biorobotic underwater propulsor. J. Bionic Eng. **4**, 25–32 (2007)
21. Low, K.H.: Locomotion and depth control of robotic fish with modular undulating fins. Int. J. Autom. Comput. **3**, 348–357 (2006)
22. Epstein, M., Colgate, J.E., MacIver, M.A.: Generating thrust with a biologically-inspired robotic ribbon fin. In: 2006 IEEE/RSJ International Conference on Intelligent Robots and Systems, pp. 2412–2417. IEEE (2006)
23. Kern, S., Koumoutsakos, P.: Simulations of optimized anguilliform swimming. J. Exp. Biol. **209**, 4841–4857 (2006)
24. Epstein, M., Colgate, J.E., MacIver, M.A.: A biologically inspired robotic ribbon fin. In: IEEE/RSJ International Conference on Intelligent Robots and Systems, workshop on Morphology, Control, and Passive Dynamics, pp. 2412–2417 (2005)
25. Willy, A., Low, K.H.: Initial experimental investigation of undulating fin. In: 2005 IEEE/RSJ International Conference on Intelligent Robots and Systems, 2005 (IROS 2005), pp. 1600–1605. IEEE (2005)

Modular Design of Gymnotiform Undulating Fin

Van Dong Nguyen[1], Canh An Tien Pham[1], Van Hien Nguyen[2], Thien Phuc Tran[1], and Tan Tien Nguyen[1(✉)]

[1] Ho Chi Minh City University of Technology (HCMUT),
Ho Chi Minh City, Vietnam
nguyendongnavy@gmail.com, antienpham@gmail.com,
{ttphuc.rectie,nttien}@hcmut.edu.vn
[2] PetroVietnam Camau Fertilizer Joint Stock Company, Camau City, Vietnam
hiennv@pvcfc.com.vn

Abstract. In the field of underwater robot, gymnotiform undulating fin inspired a number of research group over the world. A variety of design has been consequently created and developed with the design of multi-actuators. However, this type of design has the remarkable downsides. That is to say, the complex mechanical design and control, the deterioration of motors' quality should not be underestimated. In this study, a modular design of gymnotiform undulating fin is created in order to avoid the aforementioned problem. This design aims to simplify the mechanism of the fin, reduce the number of actuators and the size of the fin. For this reason, a camshaft is designed to transmit momentum of a motor to all fin-rays in this model.

Keywords: Gymnotiform · Modular design · Camshaft

1 Introduction

Biomimetic robot is a potential approach for autonomous underwater vehicles (AUVs) which benefits the outstanding characteristics of natural species. Gymnotiform is a type of undulating motion which uses a long anal fin as a main propeller while its body is rigid. This is a remarkable advantage for robotic field because of its stability. Besides, its noise could be mixed with noise from fish, so that it could not be detected in military field or not disturb the underwater environment.

The typical gymnotiform fish is black ghost knife fish which inspire for modelling and design process. In the real fish, the long anal fin locates below the body and 2 pectoral fins balancing for swimming. However, the design of robot can be adjusted by changing the combination of fins. Therefore, an improved fin should be developed before designing the AUVs.

Recently, this type of fin's motion attracts a couple of research and a variety of models are created [2, 10–13, 16–18]. In the most models, each fin-ray is transmitted by a motor in order to form the undulating process of the fin. In our previous studies [1, 4], 2 models are designed (see Fig. 1), the factors of the fin such as frequency, amplitude and wavelength are straightforwardly adjusted. However, weight of models, complicated design and control are the obstacles of the fin because of the excessive number of motors. Besides, motors' overheating results from continuous reverse of motors to transmitted torque to fin-rays. This paper focuses on designing a simple modular fin's model which has a minimal size and one motor transmitting force to all fin rays of model and restricts the reversion of motor.

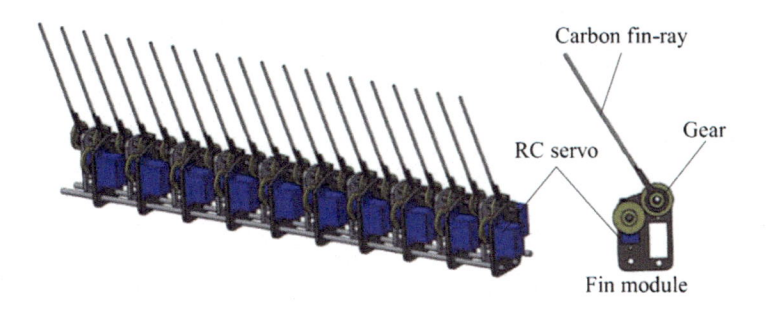

a) Model – 1

b) Model – 2

Fig. 1. Previous models. (a) Model – 1 is assembled by various fin module which fix on 3 thread rods in order to adjust the distance between 2 fin modules. (b) Model – 2 is assembled with 16 DC motors transmitting force to 16 fin-rays.

2 Principle of Design

2.1 Principle of Gymnotiform Undulating Fin

The undulating process of gymnotiform is introduced in our previous study [1], the undulating motion is transmitted through the motion of fin-ray which is defined by [20]:

$$\theta(n,t) = \theta_{max} \sin\left(2\pi f t + \phi_0 - \frac{2\pi n}{N-1}\frac{L}{\lambda}\right) \quad (1)$$

Where $\theta(n, t)$ is the angle of n-th ray at moment t, f is the frequency of fin's motion, ϕ_0 is initial phase, in this paper, $\phi_0 = 0$, θ_{max} is amplitude of n-th ray, λ is wavelength, L is the length of fin and N is number of fin-rays in one wavelength (Fig. 2).

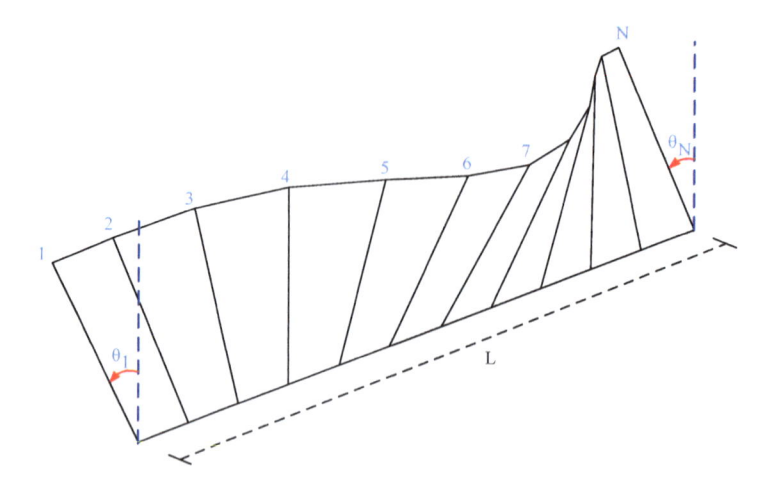

Fig. 2. Modelling of gymnotiform undulating fin. The fin-rays are equidistant from each other nearby. The position of fin-rays is defined by the angular between fin-ray and the vertical axis.

According to study [4], 16 is the most proper number of fin-rays in one wavelength. In addition, initial phase ϕ_0 is assumed to 0, and the fin has one full wavelength. Therefore, Eq. (1) is transferred to:

$$\theta(n, t) = \theta_{max} \sin\left(2\pi ft - \frac{2\pi n}{15}\right) \quad (2)$$

2.2 Concept of Modular Design

In the field of gymnotiform undulating fin robot, there are a variety of designs which have the various combinations of fin. The important and complex problem is that how many fin in a robot and the type of combination of fins are proper. Although the combination of fin could be different in a variety of designs, these fins are based on the unit of one full wavelength because of the stability of model. That is to say, the force generated by the integer number of full wavelength is more stable, because the force generating process at the different moments is almost similar without the effect of water flow. Initially, the unit of one full wavelength needs to be improved and determined thoroughly.

In an effort to determine the force generated in undulating process, the relation between force and factors such as frequency, amplitude, etc. are discovered.

$$\vec{F}_{unit} = f(f, \theta_{max}, \ldots) \quad (3)$$

The force for movement in water depends on specific model and requirement. Therefore, the number of modular fins could be combined until the force required for movement achieved. Approximately, the total force is a sum of forces generated by modular fins:

$$\vec{F}_{total} = \sum \vec{F}_{unit} \qquad (4)$$

In terms of combination of modular fins, it is divided into 3 main types: serial, parallel and complex combination (Fig. 3).

a) Serial combination

b) Parallel combination

c) Complex combination

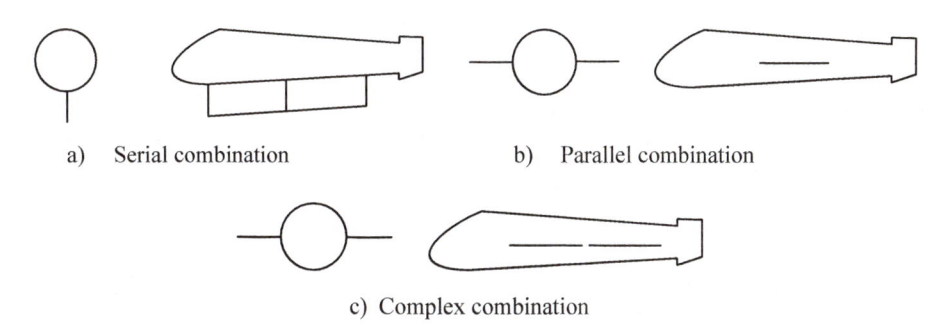

Fig. 3. Types of combination

2.3 Design of Power Transmission

To simplify the mechanism of model, a Grooved Cam System is designed to transmit the momentum from a motor to all fin-rays. In this Grooved Cam System, the distance between 2 fin-rays is reduced significantly. In addition, there is only one motor for a modular design and this motor do not reverse continuously (Fig. 4).

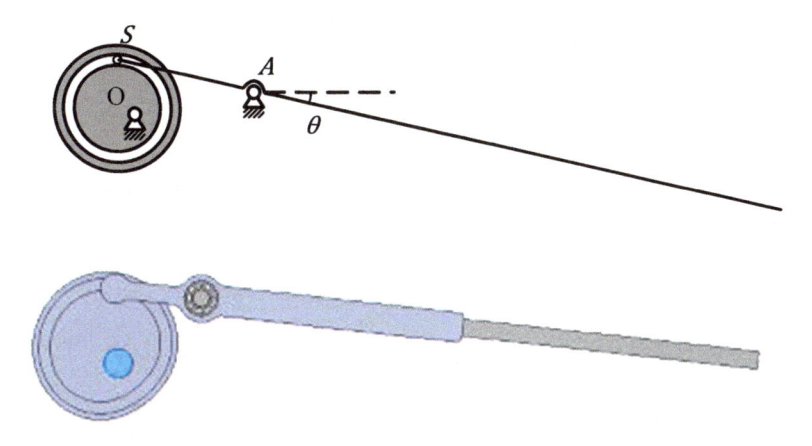

Fig. 4. Grooved Cam with the swing fin-ray. A follower in one top of the fin-rays move in the groove of cam.

3 Model and Results

The modular model includes 16 cam pieces corresponding to 16 fin-rays. This cam series is equidistantly attached in a shaft and the angular deflection of 2 adjacent pieces is 24°. Each cam piece transmits momentum to a fin-ray via the contact of a small bolt and the groove. The shaft is transmitted by a motor via a couple of gears. The specifications of design are shown in Table 1.

Table 1. Specifications of modular design

Specifications	Value
Fin	198 mm × 80 mm
Cam system	ϕ40 mm × 232 mm
Amplitude	15°
Frequency	2 hz, 2.5 hz, 3 hz, 3.5 hz
Weight	282 g

In this design, the size groove and bolt has to fit accurately together to avoid the vibration of module of fin-ray. Besides, the position of cam modules in a shaft need to be fixed but the fin-ray still have a capability to undulate. So that ball-bearing are used to contact the fin-rays and spacers help fix the position of ball-bearing (Fig. 5).

a) Frame of model b) Assembled model

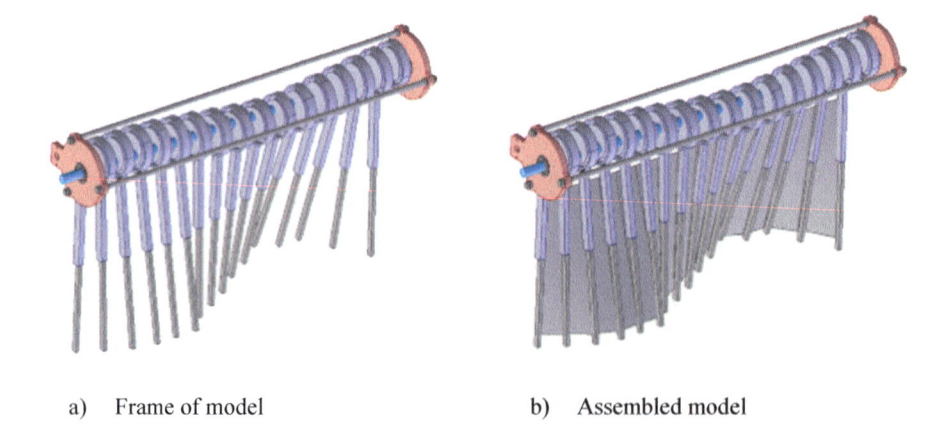

Fig. 5. Model of Grooved Cam System. (a) Frame of model is a set of 16 grooved cams which are deflected to create a sine profile of the fin. (b) An fin membrane is attached to fin-rays.

An experiment is conducted to verify the undulating process of this modular design. The testing tank's dimension is 1.2 m × 0.6 m × 0.6 m. The model slides in a slide-rail to free in a direction along to the fin and fix the others direction. A load cell is fixed on a slide-rail to measure the force generated by the fin. The experiment model and result are shown in the Fig. 6.

Because of the limit of size and torque of motor, the experiment is only conducted with the frequency varying from 2 hz to 3.5 hz. However, the force generated does not increase while the frequency increases from 3 hz to 3.5 hz. Therefore, 3 hz is the most proper frequency for this modular design.

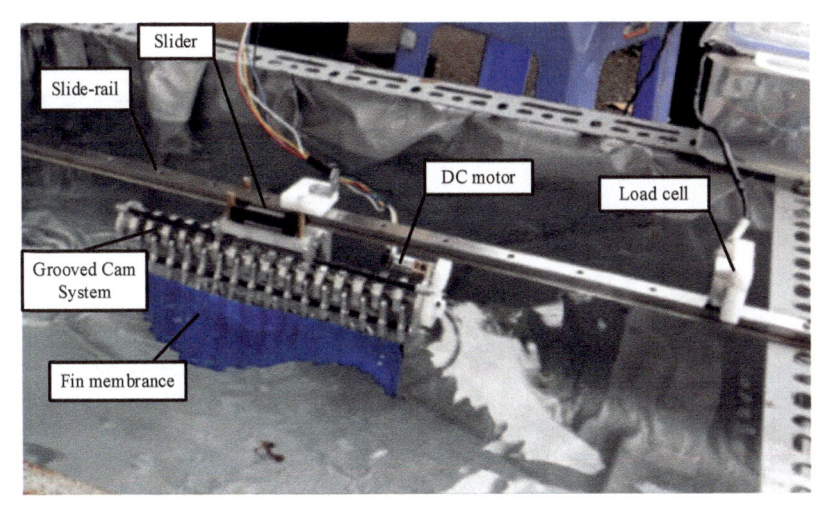

a) Experiment model

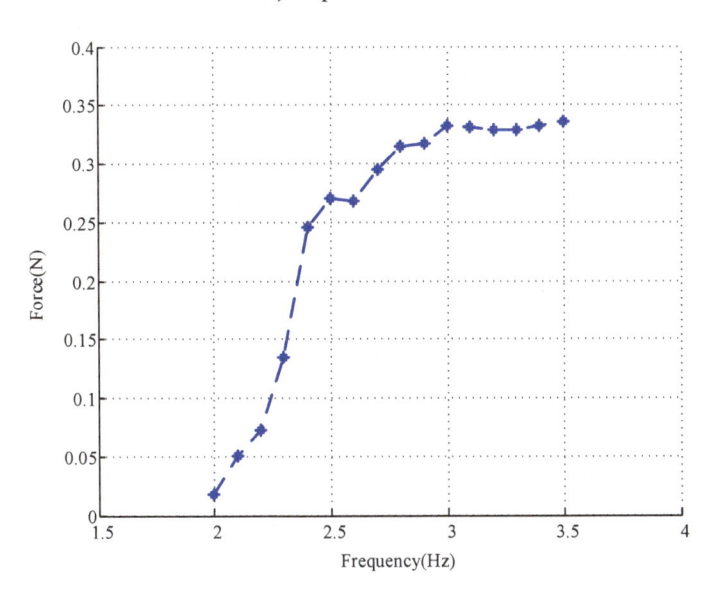

b) Experiment result

Fig. 6. Experiment model and result. (a) The setting of experiment model to measure the force generated in the forward direction in the undulating process. (b) Result of force measured in experiment with the frequencies varying from 2 hz to 3.5 hz

4 Conclusion

To simplify the mechanical structure of gymnotiform undulating fin and avoid the drawback of multi-actuator design, a modular prototype is created. The Grooved Cam System is designed to transmit the momentum of a motor to all fin-rays. Besides, to minimum the size of this prototype, the amplitude and frequency are investigated to accomplish the undulating process. For the further study, this prototype should be improved by waterproof cover and examining the undulating process with 6 DoF. In addition, the encompassing model of a biomimetic robot could be created by apply in this modular design.

References

1. Nguyen, V.H., Pham, C.A.T., Nguyen, V.D., Kim, D.H., Nguyen, T.T.: A study on force generated by gymnotiform undulating fin. In: 15th International Conference on Ubiquitous Robots, pp. 247–252, 27–30 June 2018
2. Wang, S., Wang, Y., Wei, Q., Tan, M., Yu, J.: A bio-inspired robot with undulatory fins and its control methods. IEEEASME Trans. Mechatron. **22**, 206–216 (2017)
3. Liu, H., Curet, O.M.: Propulsive performance of an under-actuated robotic ribbon fin. Bioinspir. Biomim. **12**, 036015 (2017)
4. Nguyen, V.D., Phan, D.K., Pham, C.A.T., Kim, D.H., Dinh, V.T., Nguyen, T.T.: Study on Determining the Number of Fin-Rays of a Gymnotiform Undulating Fin Robot. Springer - Lecture Note in Electrical Engineering, vol. 465, pp. 745–752, 7–9 December 2017. (ISSN 1876-1100)
5. Sfakiotakis, M., Fasoulas, J., Gliva, R.: Dynamic modeling and experimental analysis of a two-ray undulatory fin robot. In: 2015 IEEE/RSJ International Conference on Intelligent Robots and Systems (IROS), pp. 339–346. IEEE (2015)
6. Bale, R., Shirgaonkar, A.A., Neveln, I.D., Bhalla, A.P.S., MacIver, M.A., Patankar, N.A.: Separability of drag and thrust in undulatory animals and machines. Sci. Rep. **4** (2015)
7. Maddalena, L., Vergine, F., Crisanti, M.: Vortex dynamics studies in supersonic flow: merging of co-rotating streamwise vortices. Phys. Fluids **26**, 046101 (2014)
8. Neveln, I.D., Bai, Y., Snyder, J.B., Solberg, J.R., Curet, O.M., Lynch, K.M., MacIver, M.A.: Biomimetic and bio-inspired robotics in electric fish research. J. Exp. Biol. **216**, 2501–2514 (2013)
9. Yong-Hua, Z., Jian-Hui, H., Kin-Huat, L.: Numeric simulation on the performance of an undulating fin in the wake of a periodic oscillating plate. Int. J. Adv. Robot. Syst. **10**, 352 (2013)
10. Curet, O.M., Patankar, N.A., Lauder, G.V., MacIver, M.A.: Mechanical properties of a bio-inspired robotic knifefish with an undulatory propulsor. Bioinspir. Biomim. **6**, 026004 (2011)
11. Peter, B., Ratnaweera, R., Fischer, W., Pradalier, C., Siegwart, R.Y.: Design and evaluation of a fin-based underwater propulsion system. In: 2010 IEEE International Conference on Robotics and Automation (ICRA), pp. 3751–3756. IEEE (2010)
12. Liu, F., Yang, C.-J., Lee, K.-M.: Hydrodynamic modeling of an undulating fin for robotic fish design. In: 2010 IEEE/ASME International Conference on Advanced Intelligent Mechatronics (AIM), pp. 55–60. IEEE (2010)

13. Chen, J., Hu, T., Lin, L., Xie, H., Shen, L.: Learning control for biomimetic undulating fins: an experimental study. J. Bionic Eng. **7**, S191–S198 (2010)
14. Zhou, H., Hu, T., Xie, H., Zhang, D., Shen, L.: Computational hydrodynamics and statistical modeling on biologically inspired undulating robotic fins: a two-dimensional study. J. Bionic Eng. **7**, 66–76 (2010)
15. Zhou, H., Hu, T., Xie, H., Zhang, D., Shen, L.: Computational and experimental study on dynamic behavior of underwater robots propelled by bionic undulating fins. Sci. China Technol. Sci. **53**, 2966–2971 (2010)
16. Hu, T., Shen, L., Low, K.H.: Bionic asymmetry: from amiiform fish to undulating robotic fins. Chin. Sci. Bull. **54**, 562–568 (2009)
17. Low, K.H.: Mechatronics and buoyancy implementation of robotic fish swimming with modular fin mechanisms. Proc. Inst. Mech. Eng. Part J. Syst. Control Eng. **221**, 295–309 (2007)
18. Low, K.H.: Locomotion and depth control of robotic fish with modular undulating fins. Int. J. Autom. Comput. **3**, 348–357 (2006)
19. Epstein, M., Colgate, J.E., MacIver, M.A.: Generating thrust with a biologically-inspired robotic ribbon fin. In: 2006 IEEE/RSJ International Conference on Intelligent Robots and Systems, pp. 2412–2417. IEEE (2006)
20. Sfakiotakis, M., Fasoulas, J.: Development and experimental validation of a model for the membrane restoring torques in undulatory fin mechanisms. In: 2014 22nd Mediterranean Conference of Control and Automation (MED), pp. 1540–1546 (204)

Path Following Control of Automated Guide Vehicle Using Camera Sensor

Dae Hwan Kim[(✉)] and Sang Bong Kim

Department of Mechanical Design Engineering, College of Engineering,
Pukyong National University, Busan, Korea
kimdh2599@pknu.ac.kr

Abstract. The development of techniques for path following control of vehicles has become an important and active research topic in the face of emerging markets for advanced autonomous guided vehicles (AGVs). This paper presents a two-layer control architecture for path following of a AGV using camera sensor. The AGV is a tricycle wheeled mobile robot with three wheels, two fixed wheels and one driving steering wheel. Camera sensor is used to measure the tracking position error and heading angle error. Based on these errors, a controller that integrates two control loops, inner loop and outer loop, is designed. The outer loop control is based on fuzzy logic framework and the inner loop control is based on two conventional PID controllers. The effectiveness of the proposed control system is demonstrated through simulations and experiments.

Keywords: AGV · Automated Guide Vehicle · Path following · PID · Fuzzy

1 Introduction

Automated Guided Vehicle (AGV) is a transportation vehicle automatically traveling on a predefined route. AGV is most often used to deliver materials around a manufacturing facility or a warehouse. With lots advantages such as reduction in the number of worker, improvement of productivity and quality, improvement of work environment and safety, less damage on transporting goods, real-time control of material flow and improved management on product, AGV has attracted the attention to many researchers as well as manufacturers [1–9].

This paper is about path following control of Automated Guided Vehicle (AGV) using camera sensor. To do this task, the followings are considered. First, modeling of AGV is derived. The AGV is a tricycle wheeled mobile robot with two fixed passive wheels and one steering driving wheel. Second, tracking errors consisting of position error and heading angle error are defined. These errors can be measured using camera sensor. Based on the tracking errors, a control algorithm for path-following of AGV is presented. The controller has two loops: outer loop control and inner loop control. The outer loop calculates the command angular velocity and steering angle of the driving steering wheel. The outer loop control is based on Fuzzy logic framework. The inner loop control uses two PID controllers to control driving motor and steering motor in the fashion that the driving steering wheel follows the desired command from the outer loop control. Hardware and software for building the experimental system are described. Simulation and experimental results has been done to evaluate the effectiveness of the proposed controller.

2 System Description

To The AGV has three wheels, two fixed wheels and one steering wheel. The steering wheel is directly driven by two DC motors, the first one for its orientation and the second one for its rotation. Figure 1(a) shows the AGV system.

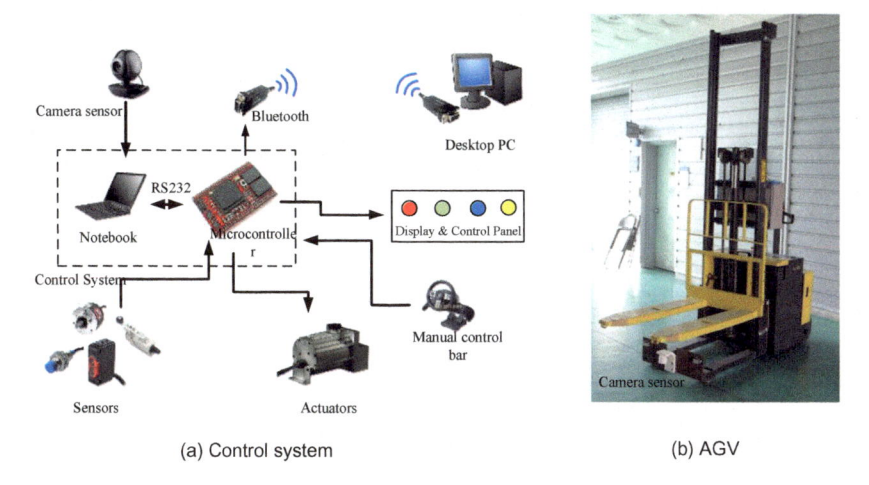

Fig. 1. Structure of the AGV system.

Figure 1(b) shows the control system for AGV. The control system is composed of two parts: high level computer control and low level microprocessor control. The former is used for image processing and control algorithm. The latter is used for motors and peripheral control. Computer and microprocessor communicate each other through RS232 protocol.

3 System Modeling

Figure 2 shows the configuration of the AGV's in the coordinate system where XOY is the global coordinate frame and $X_Q Q Y_Q$ is the moving coordinate frame. A reference point Q is an intersection point between the common axis of two fixed wheels and the symmetric axis through steering wheel of the AGV. β is steering angle and ω_s is angular velocity of the driving steering wheel. θ is heading angle of AGV. v is linear velocity of AGV at point $Q(x,y)$ and $\omega = \dot{\theta}$ is angular velocity of AGV at Q. r_s is driving steering wheel radius and l is the distance from the reference point Q to the driving steering wheel.

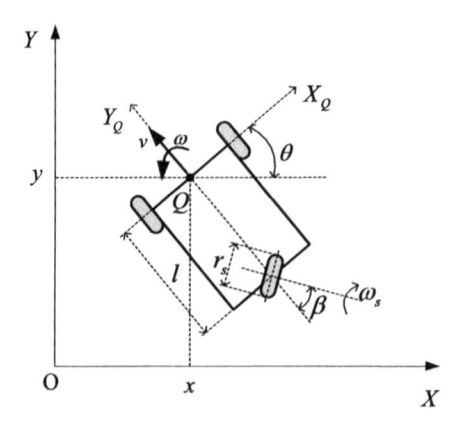

Fig. 2. AGV's configuration in the coordinate system

Because the linear velocity vector, v, is in the same direction with Y_Q axis of the moving coordinate, the following is obtained.

$$v = -\dot{x}\sin\theta + \dot{y}\cos\theta \tag{1}$$

$$\begin{cases} v = -r_s\omega_s \sin\beta \\ \omega = \dot{\theta} = -\frac{r_s}{l}\omega_s\cos\beta \end{cases} \tag{2}$$

Form Eq. (2), the followings are obtained.

$$\begin{cases} \begin{cases} \beta = \arctan\left(\frac{v}{\omega l}\right) & \text{if } \omega \neq 0 \\ \beta = \frac{\pi}{2} & \text{if } \omega = 0 \end{cases} \\ \omega_s = -\frac{1}{r_s}(v\sin\beta + \omega l\cos\beta) \end{cases} \tag{3}$$

4 Tracking Error and Error Measuring Using Camera Sensor

Figure 3 shows schematic of the tracking error of AGV. e_{pos} is the position error between the position of the reference point Q on the AGV and the tracking point on the reference path. e_θ is the angle error between the Y_Q axis and the tangent line of the path at the tracking point. B denoted as the reference point on the path is the intersection between QX_Q axis and the desired path. e_{pos} and e_θ can be defined as shown in Fig. 3.

$$e_{pos} = x_B \tag{4}$$

$$e_\theta = arc\tan\left(\frac{x_C - x_A}{y_A - y_C}\right) \tag{5}$$

Figure 3 shows schematic of the tracking error of AGV. e_{pos} is the position error between the position of the reference point Q on the AGV

(a) Schematic of tracking error

(b) Schematic for measuring the error tracking using camera sensor

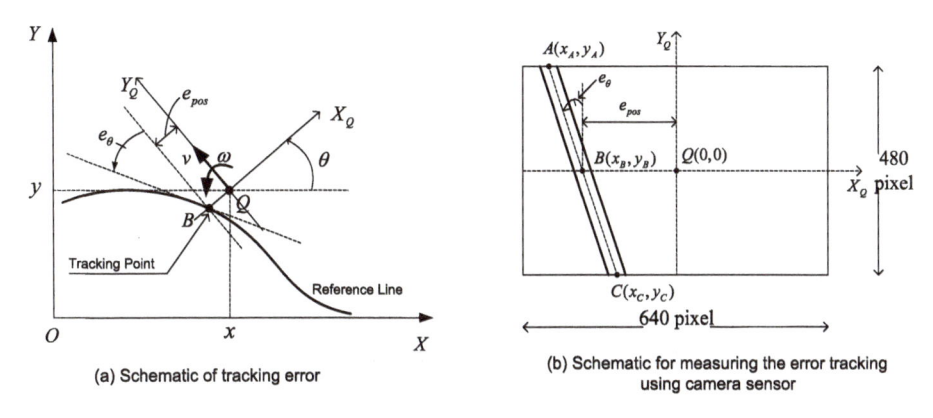

Fig. 3. Schematic tracking error of AGV

5 Controller Design

Figure 4 shows the structure of the proposed path-following algorithm for AGV. The controller integrates two control loops, inner control loop and outer control loop. The superscript $(*)$ denotes the command value from the outer loop control.

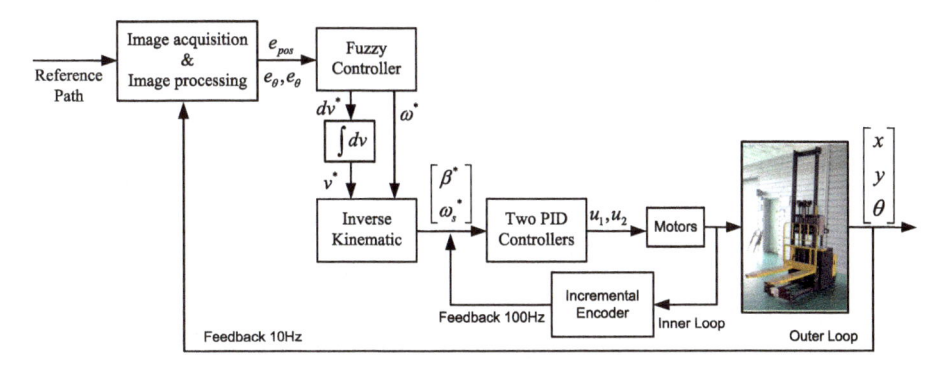

Fig. 4. Structure of the proposed path-following algorithm for AGV

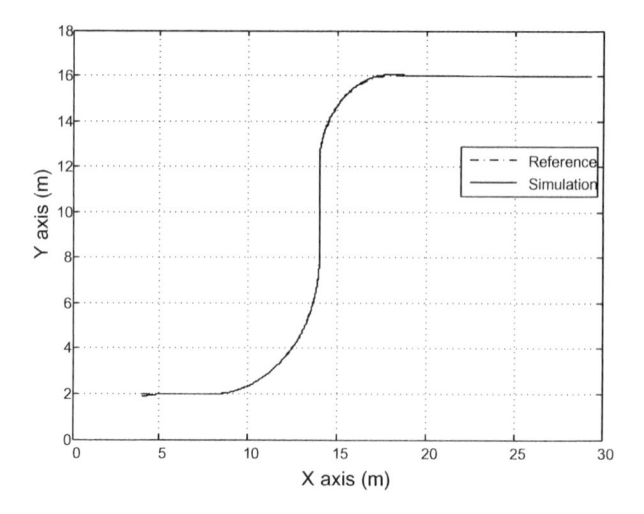

Fig. 5. Tracking the reference path in simulation

The outer loop calculates the command angular velocity and steering angle of the driving steering wheel. The outer loop control is based on Fuzzy logic framework. The fuzzy controller has three inputs, position error, e_{pos}, angle error, e_θ, and derivative of angle error, $\dot{e}_\theta$. In addition to the position error and angle error, derivative of angle error is also very important, especially when the AGV passes through sharp and narrow turns. The outputs of fuzzy controller are derivative of linear velocity, dv, and angular velocity, ω, of the AGV at Q point. The membership functions are given as shown in Fig. 4.

The inner loop control uses two PID controllers to control driving motor and steering motor in the fashion that the driving steering wheel follows the desired command from the outer loop control.

k_{p1}, k_{i1} and k_{d1} are the coefficients of the PID controller 1 used for steering motor control. e_β is the steering angle error and u_1 is PID controller 1 output as follows:

$$e_\beta = \beta^* - \beta \tag{5}$$

$$u_1 = k_{p1}e_\beta + k_{i1}\int e_\beta + k_{d1}\frac{de_\beta}{dt} \tag{6}$$

k_{p2}, k_{i2} and k_{d2} are the coefficients of the PID controller 2 used for driving motor control. e_β is the driving angular velocity error and u_2 is PID controller 2 output as follows:

$$e_\beta = \beta^* - \beta \tag{7}$$

$$u_2 = k_{p2}e_{\omega_s} + k_{i2}\int e_{\omega_s} + k_{d2}\frac{de_{\omega_s}}{dt} \tag{8}$$

6 Simulation and Experimental Results

Simulation is done on industrial AGV system with parameters $r_s = 0.125$ m and $l = 0.3$ m. Initial conditions are $x(0) = 4$ m, $y(0) = 1.9$ m, $\theta(0) = -90°$, $\omega_s(0)=0$ rad/s and $\beta(0) = 0°$.

Figure 5 shows the path following result in simulation and experiment. The reference path has five segments with three straight line segments and two curved line segments. The radius of the first curve is 6 m and the radius of the second curve is 4 m. Figure 6(a) shows the position error that becomes zero when the path is straight and small enough when the path is curve. Figure 6(b) shows angle error that is bounded by 6°.

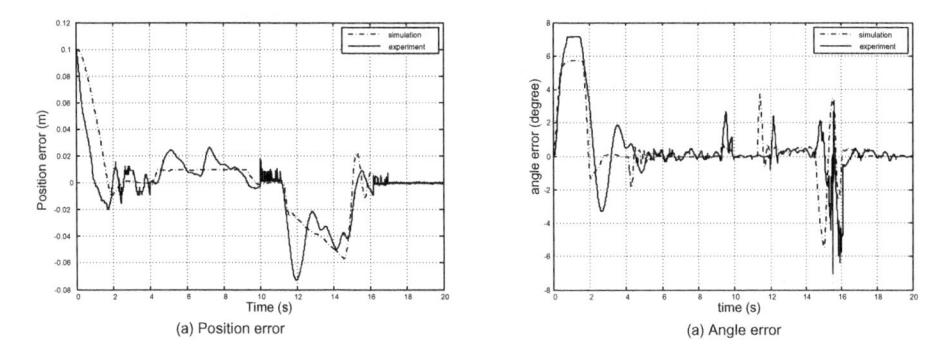

Fig. 6. Tracking errors in simulation and experiment

7 Conclusions

In this paper, hardware and controller system for the AGV was introduced. After that, modeling of AGV and tracking error using camera sensor was given. Based on the tracking error, a control algorithm based on Fuzzy and two PID controllers for path following of AGV was proposed. The simulation and experimental results presented that the AGV could track the reference path very well. The position error converged to zero for the straight line and small value enough for curved path. The angle error was bounded by 6°. The effectiveness of the proposed system was shown through the simulation and experimental results.

Acknowledgement. This research was conducted under the Pukyong National University Research Park(PKURP) for Industry-Academic Convergence R&D support program, which is funded by the Busan Metropolitan City, Korea.

References

1. Abdalla, T.Y., Abdulkarem, A.A.: PSO-based optimum design of PID controller for mobile robot trajectory tracking. Int. J. Comput. Appl. **47**(23), 30–35 (2012)
2. Campion, G., Bastin, G., Novel, B.D.A.: Structural properties and classification of kinematic and dynamic models of wheeled mobile robots. IEEE Trans. Robot. Autom. **12**(1), 47–62 (1996)
3. Gracia, L., Tornero, J.: Kinematic control of wheeled mobile robots. Latin Am. Appl. Res. **38**, 7–16 (2008)
4. Bloch, A., Reyhanoglu, M., McClamroch, N.: Control and stabilization of nonholonomic dynamic systems. IEEE Trans. Autom. Control **37**(11), 1746–1756 (1992)
5. Kamga, A., Rachid, A.: Speed, steering angle and path tracking controls for a tricycle robot. In: Proceedings IEEE International Symposium Computer-Aided Control System Design, pp. 56–61 (1997)
6. Fierro, R., Lewis, F.L.: Control of a nonholonomic mobile robot: backstepping kinematics into dynamics. Elsevier J. Robot. Syst. **14**(3), 149–163 (1997)
7. Eghtesad, M., Necsulescu, D.S.: Experimental study of the dynamic based feedback linearization of an autonomous wheeled ground vehicle. Elsevier J. Robot. Auton. Syst. **47**, 47–63 (2004)
8. Chakraborty, N., Ghosal, A.: Dynamic modeling and simulation of a wheeled mobile robot for traversing uneven terrain without slip. J. Mech. Des. **127**(5), 901–910 (2004)
9. Jeon, Y.B., Kim, S.B.: Modeling and motion control of mobile robot for lattice type welding. KSME Int. J. **16**(1), 83–93 (2002)

Binary Classification of Terrains Using Energy Consumption of Hexapod Robots

Valeriia Iegorova[✉] and Sebastián Basterrech

Department of Computer Science, Faculty of Electrical Engineering and Computer Science, Czech Technical University, Prague, Czech Republic
{Iegorval,Sebastian.Basterrech}@fel.cvut.cz

Abstract. The terrain classification problem is a relevant task in autonomous robots, which can help in the control locomotion and motion planning of autonomous robots. We conduct several experiments in different environments, where a hexapod walking robot covers some specific terrains. In this paper, we present an experimental analysis of the binary terrain classification problem using the most important variable (current signal) related to the energy consumption of the robot. The current signal is a sequential data that evolves in time, therefore our problem is limited to develop a machine learning method for classifying this signal according to the terrain. We analyze the problem using the Long Short-Term Memory (LSTM) model, which is a Recurrent Neural Networks that has obtained good performance for time-series classification. We evaluated several binary scenarios, where each scenario presents two different types of terrains. Our results show that the LSTM model trained only with information related to the current signal is able to distinguish binary situations of terrain.

Keywords: Control locomotion · Terrain classification · Time-series classification · Neural networks · Long short-term memory

1 Introduction

There is an increasing demand in developing self-navigating robots that are able to properly operate in unknown environments. A robot has to be able to quickly analyse the surroundings and make real-time decisions regarding its operations. One of important environmental factors that influences the behaviour of a robot is the terrain on which it moves. The terrain classification task is an important problem for the self-navigating robotics, to have knowledge about the terrain can be useful for improving the motion planning. It is usually associated with mobile robots, navigating in the priorly unknown surroundings. Such robots have to modify their behaviour based on the current state of the environment. The type of terrain, on which robot operates, is one of the important factors for the real-time decision-making. Based on the input data, which is usually the readings from several sensors of a robot, a model has to determine the surface on which the robot is located. Different approaches have been explored in the robotic community. Considering recent success of Neural Network (NN) models, several works analyzed the power of NN for solving the terrain classification problem. In [1], the authors studied the frequency response method with training data obtained by taking the Fast Fourier Transform of the vibration measurement. The authors used a probabilistic NN classifier, and they obtained an accuracy larger than 73.3%, which depended on speed of a robot and terrain type. Another contribution in the area was done by [2], the data gathered from multiple sensors located on a mobile robot is used for the multi-class terrain classification problem. The authors applied feedforward NN, and they achieved a success rate between 70% and 90% varying between different types of terrains. Different types of Recurrent Neural Networks (RNNs) were used for terrain classification using visual information of the environment [3]. The authors analysed visual inputs regarding to the terrain using the standard RNN, Dynamic Cortex Memories (DCMs), and Long Short-Term Memory (LSTM) models. In [4], audio data was used for the multi-class terrain classification problem. It has been used Convolutional Neural Network (CNN) and LSTM, where the CNN creates a feature map of the audio, then the LSTM produces the predictions. Authors combine both types of the NNs in a single pipeline. CNN was also applied in [5], where the authors classify the energy signals according to the terrains.

In this work, we analyze the performance of LSTM for classifying the terrain. The contribution of our work is based in the fact that we predict the terrain type using only information related to the energy consumption of the robot. Instead of using visual inputs or other type of input information from the environment, that can be computationally expensive for the robot, we analyze only sequential data produced by the energy consumption. Instead of using data from multiple sensors or images, as in other works on this topic, we limit ourselves with using only the values of the current signal (which is one important variable of the energy consumption). This approach allows us to minimize the required computational operations, as well as the size of the dataset, therefore is more suitable for embedded systems. The input signals are obtained by a walking multi-legged robot over different terrains. Since this data contains temporal dependencies, we explore the performance of the LSTM model, which is a popular method for time-series classification. Besides, this type of NNs have been shown to correctly learn from temporal dependencies in many different settings [6,7].

This article is structured as follows: Sect. 2 provides a concise theoretical overview of LSTM model, also briefly describing its basis, a regular Feed-Forward Neural Network (FFNN). Section 3 describes into more details LSTM applied to the problem of time-series classification, as well as related work in the area of terrain classification. In Sect. 3.1, we introduce the available dataset, specifications of the classification problems addressed and the network architecture used, along with the obtained results.

2 Long Short-Term Memory

A Neural Network (NN) is a complex system of parametric functions interconnected among them. Each neuron receives information from the environment, processes the information using an activation function and sends the results to its neighbours and/or to the environment. The model has an abstraction of the synaptic weight which is a real number that weights the edges among the neurons connections. When a NN is used for solving a ML problem the weights are the main parameters which are adjusted according to the specific task. If the information flow inside the NN is in one direction only, such a network is called a feedforward NN (FFNN). There are no feedback connections in FFNNs [8]. In the case of FFNN the model is a parametric mapping from an input to an output space, and the parameters are the weights.

Another family of NNs, which is called a Recurrent Neural Network (RNN), introduces a feedback connection. Developed in the 1980s, RNNs constitute a special family of NNs that are able to deal with time dependencies presented in the data. While the FFNNs treat their inputs as if they were independent from each other, the RNNs do not make such an assumption. They introduce the concept of memory which is represented as an internal (hidden) state of a RNN cell, which is dynamic and changes as the time-series passes through the cell. This property makes the RNN model suitable for solving problems with sequential data.

In the following we formalize the RNN model. Let denote by $\mathbf{x}_t$ the input pattern at time t. At each time t, a RNN cell is characterized by its state $\mathbf{h}_t$ which depends on the hidden state at time $t-1$, $\mathbf{h}_{t-1}$, and the input at time t, $\mathbf{x}_t$. Then the hidden state at time t depends is defined as:

$$\mathbf{h}_t = \sigma(\mathbf{W}^{\text{in}}\mathbf{x}_t + \mathbf{W}^{\text{rec}}\mathbf{h}_{t-1}), \tag{1}$$

where $\mathbf{W}^{\text{in}}$ is a matrix collecting the input weights, and $\mathbf{W}^{\text{rec}}$ is a matrix containing the recurrent weights and $\sigma(\cdot)$ is the activation function. The model output is computed as

$$\mathbf{y}_t = \mathbf{W}^{\text{out}}\mathbf{h}_t, \tag{2}$$

where the vector h_t denotes the hidden state at time t and y_t is the output of the model at time t. Due to the recurrent nature of the Vanilla RNNs, it is hard to adjust the hidden weights int he learning process. The problem is often studied in the literature as the vanishing-exploding gradient [9]. A possible solution for this issue was solved by so-called gated RNNs, in which a simple RNN unit is substituted by a more complicated gated unit. The LSTM model is one of such gated units, introduced in 1997 by Hochreiter and Schmidhuber [10]. It consists of three internal gates that govern the cell operations. All these gates can be concisely described by the following expressions [6]:

$$\mathbf{f}_t = \sigma(\mathbf{W}_f^{\text{rec}}\mathbf{y}_{t-1} + \mathbf{W}_f^{\text{in}}\mathbf{x}_t + \mathbf{b}_f), \tag{3}$$

$$\mathbf{i}_t = \sigma(\mathbf{W}_i^{\text{rec}}\mathbf{y}_{t-1} + \mathbf{W}_i^{\text{in}}\mathbf{x}_t + \mathbf{b}_i), \tag{4}$$

$$\mathbf{o}_t = \sigma(\mathbf{W}_o^{\text{rec}}\mathbf{y}_{t-1} + \mathbf{W}_o^{\text{in}}\mathbf{x}_t + \mathbf{b}_o), \tag{5}$$

where $\mathbf{f}_t$ is a value of the forget gate, $\mathbf{i}_t$ is a value of the input gate and $\mathbf{o}_t$ is a value of the output gate all of them at time t. The parameters of these equations are:

- $\mathbf{W}_f^{\text{in}}$, $\mathbf{W}_i^{\text{in}}$, and $\mathbf{W}_o^{\text{in}}$ are the input weights of expressions (3), (4) and (5).
- $\mathbf{W}_f^{\text{rec}}$, $\mathbf{W}_i^{\text{rec}}$, and $\mathbf{W}_o^{\text{rec}}$ are the recurrent weights of expressions (3), (4) and (5).
- $\mathbf{b}_f$, $\mathbf{b}_i$, and $\mathbf{b}_o$ are the biases of the corresponding gates.
- $\sigma(\cdot)$ is an activation function.

In the standard LSTM, all the gates are represented by the sigmoid functions, deciding which information to forget or update at the current time step. Let denote the Hadamard product (the entry-wise product) with the symbol $\odot$. The output at any time step is controlled by the output gate and depends upon the hidden state of the LSTM cell:

$$\mathbf{y}_t = \mathbf{o}_t \odot \tanh(\mathbf{c}_t), \tag{6}$$

where

$$\mathbf{c}_t = \mathbf{f}_t \odot \mathbf{c}_{t-1} + \mathbf{i}_t \odot \mathbf{h}_t, \tag{7}$$

$$\mathbf{h}_t = \tanh(\mathbf{W}_h^{\text{rec}}\mathbf{y}_{t-1} + \mathbf{W}_h^{\text{in}}\mathbf{x}_t + \mathbf{b}_h), \tag{8}$$

where $\mathbf{h}_t$ is the hidden state.

3 Methodology

LSTM networks are versatile and can be used for different applications. In this work, we focus on LSTM used for solving a time-series classification problem. In this problem the learning dataset is composed by an input-output pair where the inputs are a time-series and the outputs are classes. LSTM can be applied for binary and multivariate classification problems.

The dataset is obtained from the experiments with a low-cost hexapod walking robot. The robot has six legs, each with 3 joints attached to the trunk which hosts the electronics and sensory equipment [11]. The robot dimensions are approximately 45×40 cm. A hall-effect current sensor was used for estimating the instantaneous robot power consumption, which is provided with a 62 Hz frequency. Altogether six type of terrains were evaluated, for each terrain three experiments were conducted. The considered terrains sorted according to difficulty were: PVC flooring, turf-like carpet and semi-transparent soft black fabric represent different flat terrains. Then, wooden cubes covered with turf-like carpet, wooden cubes covered with the black fabric and bare wooden cubes, which are irregular height and sloped 10×10 cm wooden blocks, are representatives of rough terrain scenarios. We used the same turf-like carpet and black-fabric have been used for rough as-well-as plain ground setup. We denote the six different classes of the terrain with the following keywords: "grass flat", "grass rough", "cubes", "flat", "black flat" and "black rough". There are two main types of the terrains: flat and rough, the latter being made of the wooden cubes with a side of 10 cm. In addition, there are two types of covers (soft black cover and a grass-like mat). All these types of terrains and covers combined, in additional to the uncovered surfaces, gives us the six classes mentioned above. Figure 1 shows an example of the current signal in four terrains during 100 time steps. All the experiments has been conducted in a Robotic Computational Laboratory[1]. We derive several separate problems from the available data. In this work, we combine the terrain classes, which are presented in the raw dataset, in several distinct ways so as to obtain different instances of a binary terrain classification problem. In one of these problems, the model tackles the problem of distinguishing between two separate classes, as presented in the dataset. In the other problems, several of six classes are grouped together. These problems are described in details in Table 1. We denote by A the problem of distinguish between a flat terrain and a rough terrain composed by wooden cubes. The problem B refers to the binary classification between a surface covered by a black textile versus a a surface covered by artificial grass. The experiments were made over two type of surfaces a flat one and a surface with a slope. The problem C defines the

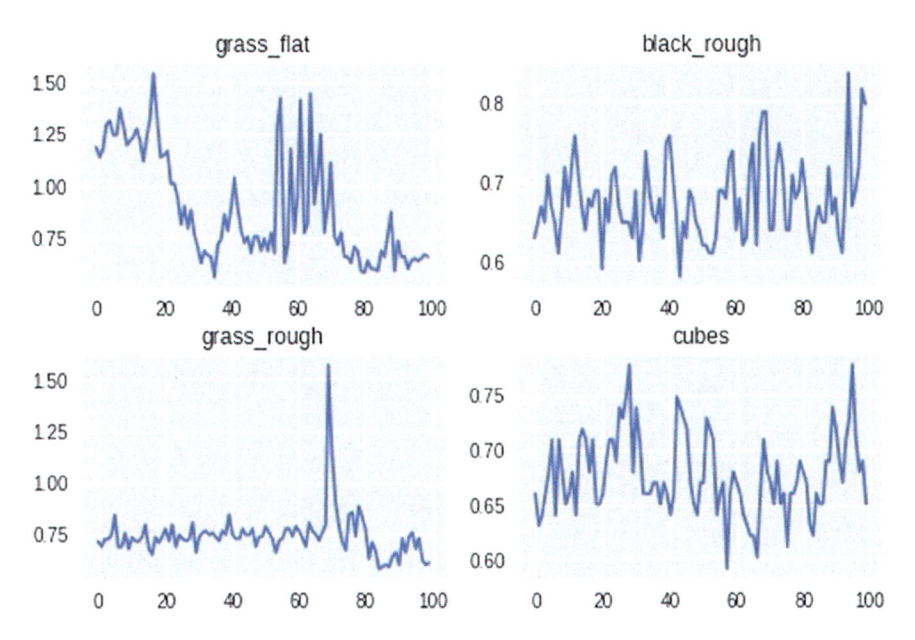

Fig. 1. Examples of sequences used for training of the model (only 100 time steps for each of the sequences). The signals show the measured currency of the robot in four different terrains.

---

[1] Center for Robotics and Autonomous Systems, Faculty of Electrical Engineering, Czech Technical University: http://robotics.fel.cvut.cz/cras/.

classification between a flat surface and surface with a slope. The flat surface was covered by the textile and artificial grass, and the same was made over the slope. The dataset consists of several recordings of sensor data from a hexapod. In this experiment, we use only the values of the current, as our goal is to use the least amount of the data possible for a fast classification of the terrain. The whole dataset of current values is stored as several long sequences of current values, recorded during several experiments for the different types of terrains. Since we are looking towards the real-time terrain classification, as required for many practical applications, we need significantly shorted sequences of input data than those presented in the unedited dataset. As the preprocessing step, we cut all the dataset sequences into the smaller sequences of 100 time steps. We discard resulting sequences whose length is smaller than 100. We randomly shuffled the data and divide it into the training and testing sets, with 85% of data used for training, and 15% used for testing.

Table 1. Description of the binary classification problems.

Problem identification	Description
A	"flat" versus "cubes"
B	"black rough" and "black flat" versus "grass rough" and "grass flat"
C	"black flat" and "grass flat" versus "black rough" and "grass rough"

Table 2. Evaluated architectures of the LSTM model.

Architecture	Description
1	Input $->$ LSTM32 $->$ Dense $->$ Activation
2	Input $->$ LSTM64 $->$ Dense $->$ Activation
3	Input $->$ LSTM32 $->$ LSTM32 $->$ Dense $->$ Activation
4	Input $->$ LSTM64 $->$ LSTM64 $->$ Dense $->$ Activation
5	Input $->$ LSTM32 $->$ Dropout40 $->$ LSTM32 $->$ Dropout40 $->$ Dense $->$ Activation
6	Input $->$ LSTM64 $->$ Dropout40 $->$ LSTM64 $->$ Dropout40 $->$ Dense $->$ Activation

3.1 Results

We evaluated six different architectures of the LSTM model. They differ in the number of hidden units, amount of layers and in the regularization parameters. Table 2 presents the different model architectures, where LSTM32/64 refers to the amount of units used in a specific layer, and Dropout40 refers to the dropout parameter with a probability of 40%. More information about Dropout technique can be seen in [12].

In addition to the different architectures, two distinct approaches to training the models was used. A relevant control parameter of the training algorithms is the learning rate. This parameter controls the variation update of the weights in each iteration. In the first case, we train the model for 500 epochs with the fixed learning rate of $\eta = 0.001$, where epoch is defined as a single pass through the whole training set. In the second case, we used the decaying learning rate, which starts at $\eta_0 = 0.01$ and decays as $\eta = \eta_0 e^{-0.1t}$, where t is the epoch number, and the network is trained for 100 iterations. Due to the randomness of initial parameters initialization, inherent to the training of the NNs, one run of the training and testing cycle is not enough to judge the performance of a given model. Therefore, we evaluate the architecture using the same architecture with different initialization several times. First, we estimate which of the combinations of introduced models and training processes seem to be give better results. We decide this based on five runs of training/testing evaluations, performed by each LSTM model over each of the problems with different setting of the learning rate, as discussed above. These results are shown in Table 3. For comparing the LSTM model, we use their averaged accuracy (ACC) for different problems:

$$Av(ACC) = \frac{1}{3}ACC_A + \frac{1}{3}ACC_B + \frac{1}{3}ACC_C,$$

where ACC_i is the accuracy, as averaged from 5 runs, of a given combination of model and training procedure, for the problem identified as i. From the results, presented in Table 3, the architectures 2 and 4 obtained better average accuracies (73.7% and 76.3%, correspondingly). Therefore, these two LSTM models have been evaluated in more details. Here it is important to note that, though the models with fixed learning rate have shown better average performance, the exponentially decaying learning rate had significantly reduced length of the training phase, while achieving comparable results. However, for the purpose of this paper we do not orient on the speed of learning, and only use average testing set accuracy as the measure for evaluating the different models. Now, we focus on the two most promising architectures. To get more accurate estimates of the expected performance, we repeated the training and testing cycle. This time, the results presented are averaged from 30 runs. The mean and the standard deviation as measured from the 30 runs, are shown in Table 4. Also, the 95% Confidence Interval (CI) is computed to show, in which range the accuracy of the model is likely to fall. The CI for the averaged accuracy value was computed using an approximation of the Gaussian distribution. In the case of problem A, the CI of the accuracies of architectures 2 and 4 don't overlap, therefore we can affirm that the architecture 4 is more suitable for solving the problem A. On the other hand, the CIs computed in the problems B and C show that we can not distinguish what architecture obtains better accuracy. From these results, we can note that adding the second LSTM layers tends to significantly modify the accuracy on a given task. While we cannot state that this step improves accuracy values for all the problems, we can note the trend in standard deviation values: for all the tasks, it tends to be significantly smaller for the model with two LSTM layers. This means that, if we are only able to perform one run of the model, we are more likely to end up with the estimated mean accuracy than in the case of the other model.

Table 3. Average of the accuracy obtained among five trials.

Problem	Architecture	Learning rate with exponential decay	Fixed learning rate
A	1	80.2	76.2
	2	70.6	81.2
	3	74.8	80.0
	4	71.8	87.2
	5	69.2	82.6
	6	69.0	76.4
B	1	74.0	62.8
	2	68.4	64.0
	3	71.2	61.4
	4	66.2	64.4
	5	66.2	60.6
	6	62.0	63.6
C	1	60.6	75.6
	2	62.0	76.0
	3	64.2	75.2
	4	63.0	77.4
	5	61.2	64.6
	6	66.2	77.8

Table 4. Accuracy obtained among 30 trials.

Problem	Architecture	Mean	Standard deviation	95% Confidence Interval
A	2	68.60	15.86	(62.92, 74.28)
	4	86.17	7.90	(83.34, 89.00)
B	2	61.97	5.41	(60.03, 63.91)
	4	63.40	4.77	(61.69, 65.11)
C	2	77.37	13.03	(72.71, 82.03)
	4	76.73	6.93	(74.25, 79.21)

4 Conclusions and Future Work

The terrain classification problem is a relevant task in autonomous robots, to obtain information from the environment and predict the terrain can improve the motion planning of autonomous robots. In this paper, we explored the problem of terrain classification, as an instance of the time-series classification problem. We analyzed the problem using Long Short-Term Memory (LSTM) networks in a binary scenario. We focused on several instances of binary classification and compared the performance of different LSTM architectures for each of them. We assume that there are two types of terrains, and we use only information from the energy consumption of the robot for making the prediction. This last assumption was considered in order of reducing the computational cost of the robotic learning system. We make the predictions using limited amount of the input data: namely, readings of the power consumption. This approach can proof beneficial for the self-navigating robots because using the logs only from one sensor can significantly reduce the amount of hardware on a robot, as well as save the space required to store the input data. For the approach presented in this paper, we make an assumption that a robot is traversing only one type of surface during a given time frame. We have conducted several experiments using a hexapod robots, and we have shown that the LSTM is able of obtaining good predictions in a binary situation when the types of terrains are very different among them.

The obtained classifier does not obtain acceptable accuracy when the problem is extended to several classes of terrains. Therefore, as future work we plan to evaluate the LSTM with more input variables, we plan to use the information captured by other sensors instead of having only the current signals.

Acknowledgment. This work has been supported by the Czech Science Foundation (GAČR) under research project No. 18-18858S, and the authors acknowledge the support of the OP VVV MEYS funded project CZ.02.1.01/0.0/0.0/16_019/0000765 "Research Center for Informatics".

References

1. DuPont, E.M., Moore, C.A., Collins, E.G., Coyle, E.: Frequency response method for terrain classification in autonomous ground vehicles. Auton. Robots **24**(4), 337–347 (2008). https://doi.org/10.1007/s10514-007-9077-0
2. Lauro, O., Johann, B., Gary, W., Robert, K.: Terrain characterization and classification with a mobile robot. J, Field Robot. **23**(2), 103–122 (2006). https://onlinelibrary.wiley.com/doi/abs/10.1002/rob.20113
3. Otte, S., Laible, S., Hanten, R., Zell, A.: Robust visual terrain classification with recurrent neural networks. In: European Symposium in Artificial Neural Networks (ESANN), pp. 451–457, Jan 2015
4. Valada, A., Spinello, L., Burgard, W.: Deep feature learning for acoustics-based terrain classification. In: International Symposium on Robotics Research (ISRR) (2015)

5. Falck, R.H., Čížek, P., Basterrech, S.: Recurrence plot and convolutional neural networks for terrain classification using energy consumption of multi-legged robots. In: International Conference on Soft Computing MENDEL 2018. Czech Republic, Brno (2018)
6. Gers, F.A., Schmidhuber, J.A., Cummins, F.A.: Learning to forget: continual prediction with LSTM. Neural Comput. **12**(10), 2451–2471 (2000). https://doi.org/10.1162/089976600300015015
7. Greff, K., Srivastava, R.K., Koutník, J., Steunebrink, B.R., Schmidhuber, J.: LSTM: A search space odyssey (2015). *CoRR*, vol. abs/1503.04069. http://arxiv.org/abs/1503.04069
8. Goodfellow, I., Bengio, Y., Courville, A.: Deep Learning. MIT Press (2016). http://www.deeplearningbook.org
9. Bengio, Y., Frasconi, P., Simard, P.: The problem of learning long-term dependencies in recurrent networks. In: IEEE International Conference on Neural Networks, vol. 3(2), pp. 1183–1188 (1993)
10. Hochreiter, S., Schmidhuber, J.: Long short-term memory. Neural Comput. **9**(8), 1735–1780 (1997)
11. Čížek, P., Faigl, J.: On localization and mapping with RGB-D sensor and hexapod walking robot in rough terrains. In: IEEE International Conference on Conference Systems, Man, and Cybernetics (SMC), pp. 2273–2278 (2016)
12. Srivastava, N., Hinton, G., Krizhevsky, A., Sutskever, I., Salakhutdinov, R.: Dropout: a simple way to prevent neural networks from overfitting. J. Mach. Learn. Res. **15**, 1929–1958 (2014). http://jmlr.org/papers/v15/srivastava14a.html

The Movement of Swarm Robots in an Unknown Complex Environment

Quoc Bao Diep[(✉)] and Ivan Zelinka

Faculty of Electrical Engineering and Computer Science,
Technical University of Ostrava, 17. Listopadu 15 Ostrava, Czech Republic
zelinkaivan65@gmail.com, ivan.zelinka@vsb.cz

Abstract. This paper presents a method for swarm robots catching multiple moving targets without colliding any dynamic obstacles and other robots in an unknown complex environment. An imaginary map, including multi-layers corresponding to the number of robots, is built in which the starting position, the target, the obstacles, and the robot denoted by the highest position, the lowest position, the small hills, and the spherical ball on the map. The PSO algorithm was proposed to lead the robot to move on the map toward the given targets safely. Simulation results are also presented to show the feasibility of the method.

Keywords: Swarm robot · Particle swarm optimization
Obstacle avoidance · Path planning

1 Introduction

Avoiding collisions with dynamic obstacles and other robots to catch the given target is one of the most issues in the field of swarm robotics. That is, in an unknown environment where many robots are active, each robot not only catches its moving target but also must be able to avoid each other and avoid dynamic obstacles that the robot does not know about them until the sensors detect them. Various methods have been proposed to solve this problem such as [4,6,8]. Each method has its advantages in some specific environments. There is no doubt that considering the movement of the robot as an optimization problem is one of the most common methods.

The particle swarm optimization (PSO) algorithm was first introduced in 1995 [5] and became popular and effective in solving optimal problems. It is widely used for optimal path planning for robots as well as avoiding obstacles such as [1,9].

In this paper, the author presents a method of constructing an imaginary map that the start position, the target, the obstacles, and the robot, in turn, are considered as the highest position, the lowest position, the small hills and the spherical ball. This ball can be moved on the imaginary map due to gravity, corresponding to the movement of the robot in the actual environment. The PSO algorithm, which acts as gravity on this map, is used to solve the problem.

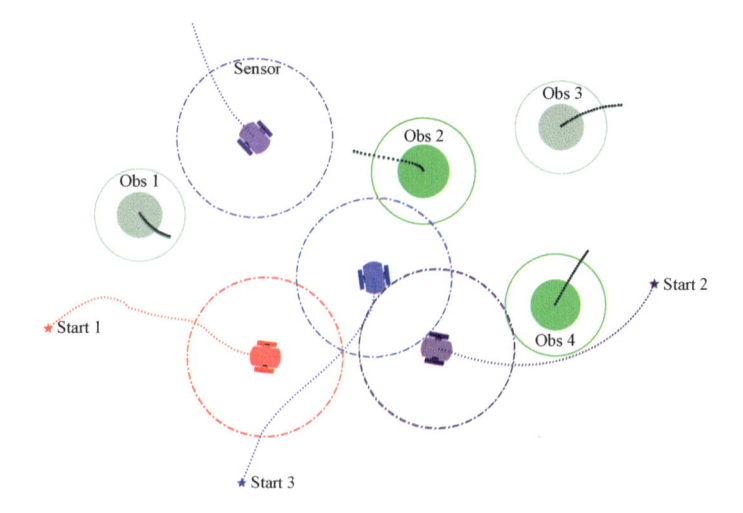

Fig. 1. The model of the obstacles and the robots.

2 The Method

2.1 The Movement of Swarm Robots

Catching the given moving target without colliding any dynamic obstacles and other robots in an environment that robots do not know their position until detecting them by sensors is the final goal of swarm robots. Robots can only operate under certain conditions. Therefore, to ensure this task can be achieved, some assumptions about robots, targets, and obstacles are described below.

The robots can move easily from the current position to the farthest position that given before without any dilemmas. The distance between that two points is called the moving step of the robot, which depends on the physical structure of the robot and does not affect the quality of the algorithm. The sensors are placed on the robot to accurately measure the distance to the obstacles and other robots in their range, see Fig. 1.

The obstacles are considered as circles of radius r_{obs} containing all their physical dimensions. They move at the speed less than the speed of the robot with unspecified trajectory, and robots do not know their position until detecting them.

The target moves at the speed less than the speed of the robot, so the robot is able to catch it. Its position is provided to the robot.

2.2 The Imaginary Environment

The movement of the robot in the environment mentioned above is considered as the movement in an imaginary map with different heights. This map is constructed based on the following principles (Fig. 2):

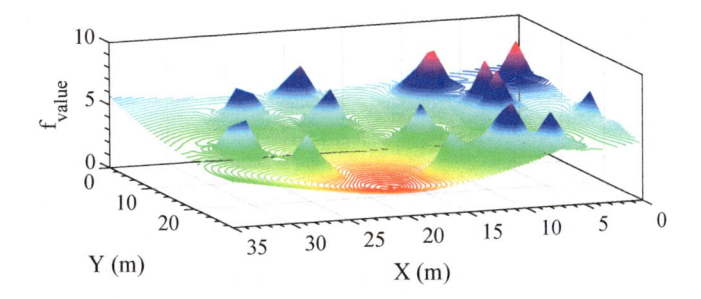

Fig. 2. The imaginary map.

- The starting position of the robot is the highest point and the target position is the lowest point on the map,
- The robot is considered as a spherical ball,
- The obstacles and other robots are considered as small hills,
- The PSO algorithm is considered to be gravity pulling the spherical ball that rolls down from higher to lower without moving back up the hills, i.e. not colliding with any obstacles and other robots.

To build this map, Eq. 1 was proposed by Diep and Ivan in [2]. The first component in the right-hand side creates the lowest point on the map, corresponding to the target position, and the second component creates the small hills on the map, corresponding to the obstacles and other robots position.

$$f_{value} = \gamma \, d_{target} + \sum_{n=0}^{number_{obs}} \frac{\alpha + r_{obs}}{\beta + d_{obs}^2} \quad (1)$$

where:

f_{value}: the fitness value,
γ: the coefficient of the target,
β: the coefficient of the obstacles and other robots,
α: the dynamic coefficient of the small hills,
$number_{obs}$: the number of detected obstacles,
r_{obs}: the radius of the obstacles,
d_{target}: the distance from the robot to the target,
d_{obs}: the distance from the robot to each detected obstacle.

The movement problem of swarm robots becomes an optimization problem, and the robot trajectory is the rolling way of the spherical ball on the map. When multiple robots move together in the same environment, each robot movement is considered the movement of the spherical ball on a separate map, corresponding to a layer in multiple layers.

During the move, if a local minimum occurs, i.e. the height of the spherical ball does not change, corresponding to the robot being held by the surrounding obstacles, the size of the small hills will increase accordingly by increasing the α in Eq. 1 until the robot leaves from the hold, as shown in Fig. 3.

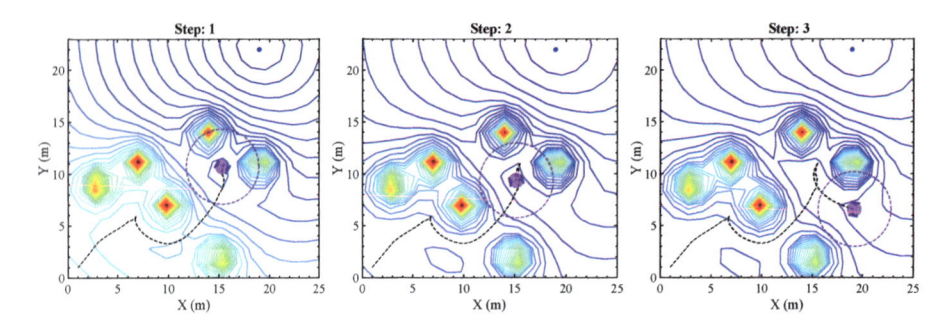

Fig. 3. The robot is trapped between obstacles.

3 Particle Swarm Optimization Algorithm

The PSO algorithm, first introduced by Eberhart and Kennedy in 1995, was inspired by the movement of fish and birds, which each individual is considered to be a particle in a swarm [5].

In the moving and avoiding obstacles issue, each particle represents a virtual position of the robot on the map (see Sect. 2). A swarm consisting of particles is randomly generated around the actual position of the robot at the start of the algorithm. These particles are evaluated based on the fitness function (see Eq. 1) to choose two particles that have the best fitness value in the swarm and in its trajectory, respectively corresponding to the global best and the local best. These two particles are used to calculate the velocity based on Eq. 2.

$$v_{ij}^{t+1} = w\ v_{ij}^{t} + c_1\ rand\ (pbest_{ij} - x_{ij}^{t}) + c_2\ rand\ (gbest_j - x_{ij}^{t}), \quad (2)$$

where:

v_{ij}^{t+1}: the velocity of the ith particle in the $(t+1)^{th}$ iteration,
v_{ij}^{t}: the velocity of the ith particle in the tth iteration,
w: the inertia weight,
c_1, c_2: the priority factors,
$rand$: the random number from 0 to 1,
$pbest_{ij}$: the best value in the ith particle trajectory,
$gbest_j$: the best value in the swarm,
x_{ij}^{t}: the current position of the ith particle in the tth iteration.

The new position of the particle is obtained based on the current position and velocity as shown in Eq. 3, where x_i^t and x_i^{t+1} are the current and the new position of the particle. This process is repeated to get the best new position.

$$x_i^{t+1} = x_i^t + v_i^{t+1}. \quad (3)$$

The linear decreasing inertia weight strategy is used to influence the velocity of the particles in each iteration [7].

$$w = w_{max} - \frac{(w_{max} - w_{min})\, t}{t_{max}}, \qquad (4)$$

where:

w_{min}, w_{max}: the minimum and the maximum inertia weight given before,
t, t_{max}: the current iteration and the maximum iteration.

The next position that the robot will move to not only has the best value in the swarm but also has the distance to the current position smaller than the moving step mentioned in Sect. 2.

4 Simulation Results

4.1 Setup Environment

Three different maps were built to simulate the proposed method using Matlab. In the first map, there were ten static obstacles, a robot, and a standing target. The second map consisted of five static obstacles, five standing targets, and five robots. Four robots, corresponding to four moving targets, and seven dynamic obstacles that moved at independent speeds and trajectories were given in the last map.

The starting position of all robots and targets are given in Table 1. The size and position of the obstacles of the three maps are given in Table 2. The robots used in all simulations have a radius $r_{robot} = 0.8$ m. The robot sensors have an active range with a radius $r_{sensor} = 3.6$ m. The moving step of the robots is $r_{movingstep} = 0.5$ m.

4.2 The Results and Discussions

Figure 4 which captured at the 17th, 54th, and 122nd steps shows the movement of the robot in form of the actual environment and the imaginary map.

An imaginary map was built based on the principle in Sect. 2. In this map, the 3rd obstacle did not appear in the entire robot movement, i.e. it was not detected by the robot sensor and completely did not affect the robot trajectory, even though it did exist in reality.

Table 1. The starting positions and the target positions of the robots.

Map		1st	2nd					3rd			
Position		Ro1	Ro1	Ro2	Ro3	Ro4	Ro5	Ro1	Ro2	Ro3	Ro4
Start	x (m)	01	16	05	02	32	07	01	29	10	05
	y (m)	01	23	03	14	05	23	07	09	00	25
Target	x (m)	19	17	25	30	05	27	30	02	20	22
	y (m)	22	04	25	15	25	10	15	16	25	02

Table 2. The starting position of the obstacles.

Map	$Obstacle_i$	1	2	3	4	5	6	7	8	9	10
1st	x_i (m)	10	14	09	03	07	26	24	15	14	19
	y_i (m)	07	19	16	09	11	07	19	02	14	11
	r_i (m)	1.9	2.4	2.1	2.0	2.1	2.3	2.1	2.0	2.4	2.3
2nd	x_i (m)	26	07	23	20	10	-	-	-	-	-
	y_i (m)	19	09	04	14	16	-	-	-	-	-
	r_i (m)	1.9	2.0	2.1	2.2	2.3	-	-	-	-	-
3rd	x_i (m)	10	14	25	03	07	26	14	-	-	-
	y_i (m)	07	15	20	03	11	07	23	-	-	-
	r_i (m)	1.9	2.4	2.1	2.0	2.1	2.3	2.2	-	-	-

The local minima problem has appeared twice, which means that the surrounding obstacles have held the robot, the 1st, 4th, and 5th obstacles for the first time, and the 9th, and 10th obstacles for the second. At this time, the dynamic coefficient α in Eq. 1 increases, causing the size of the small hills to increase correspondingly, leading the robot out of the hold.

In the first map, the robot reached the target with 122 moving steps without colliding with any obstacles.

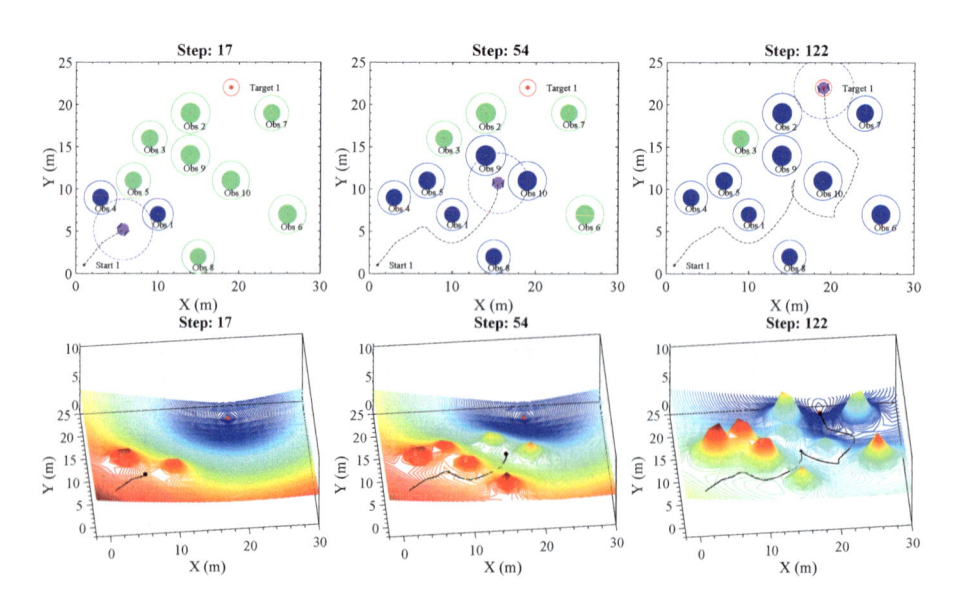

Fig. 4. The movement of the robot in an unknown static environment.

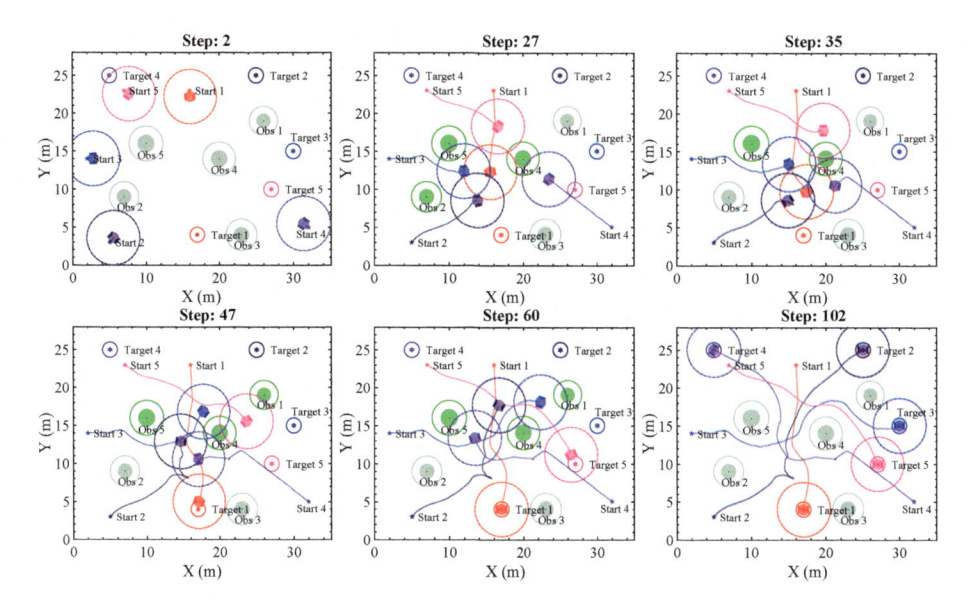

Fig. 5. The movement of the multiple robot in an unknown static environment.

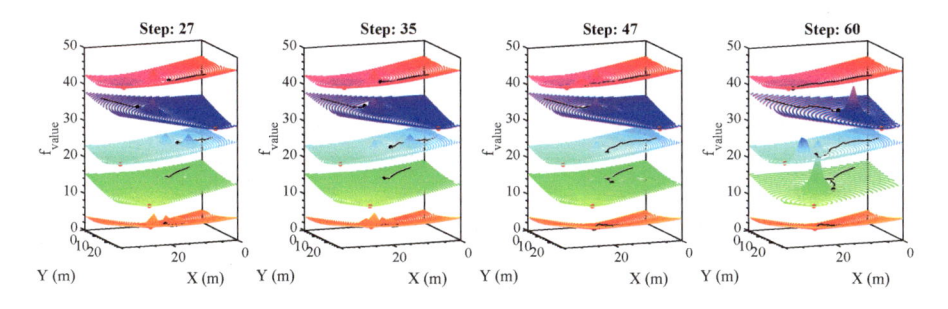

Fig. 6. The detailed trajectory of the robot in the imaginary map.

At the 27th and 35th moving steps, the figure shows that the robots must avoid each other robots when they are close together besides avoid surrounding obstacles. The 1st robot detected the 4th obstacle and the 2nd, the 3rd robots, and all of them are considered as dynamic obstacles, leading to three small hills around the 1st robot and it moved away from them to reach the target, as shown in the 27th moving step in the lowest layer of Fig. 6. This happened in the same way for the other robots until all of them reached their target. The 4th robot took the longest distance to reach its target, which was 102 moving steps, and the 1st robot with the shortest distance took 50 moving steps. The 2nd, the 3rd, and the 5th robots reach their targets at the 90th, the 84th, and the 64th moving steps respectively (Fig. 5).

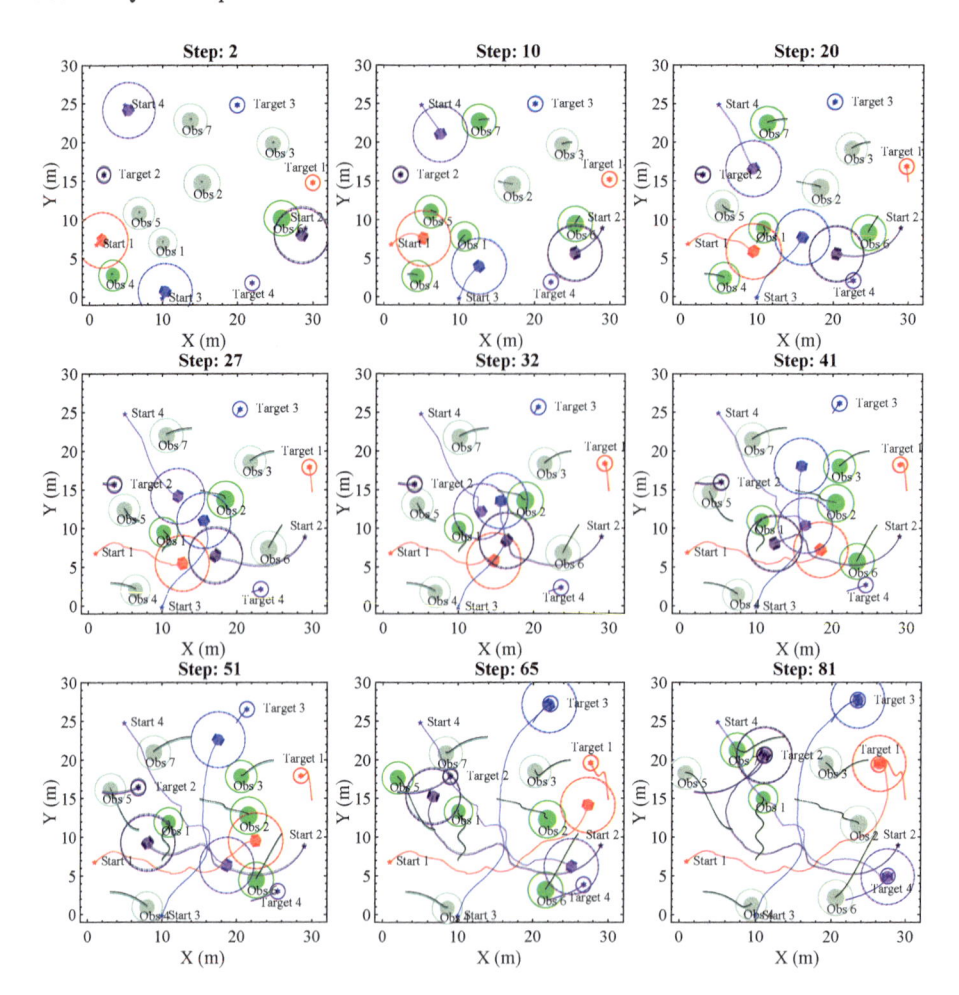

Fig. 7. The moving process of the robots in the complex environment.

Figure 6 shows the movement of the robot in form the imaginary map, with five layers corresponding to five robots, the lowest layer indicating the 1st robot, the highest layer indicating the 5th robot respectively. When other robots or obstacles are detected, small hills pop up in the corresponding layer. As can be seen in these layers, the robot is like a spherical ball rolling from the highest position, around these small hills, to the lowest position, as well as the robot reached its target.

Thus, all robots have completed their tasks without colliding with each other as well as colliding with obstacles in the second map.

In the last map, a complex environment was built in which all objects moved, as shown in Fig. 7. During the 27th to 41st moving steps, the movement of the robots became complicated as the robots were close together and surrounded by the 1st, 2nd, 3rd, and 6th obstacles. However, the robots did not collide with other robots and obstacles to move away towards their target safely.

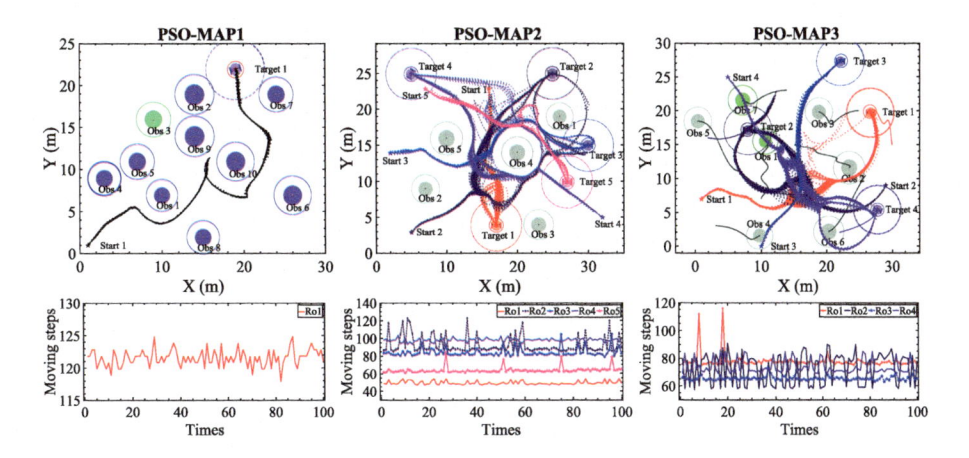

Fig. 8. Simulations are repeated 100 times for each map.

The simulations were repeated 100 times for each map. In Fig. 8, the trajectories of the robots were recorded and plotted in the same actual map to easily distinguish the differences between them, and the number of moving steps that the robot performed at each loop to catch its target was recorded simultaneously to determine the stability of the algorithm. The statistical values of the moving steps of the robots in each map are given in Table 3. The distance from each robot to the rest were also determined in the simulations to confirm that the robot did not collide with any obstacles and other robots during the move.

Unlike the simplicity of the trajectories in the first map, the trajectories of the robots in the second and third maps are complex and different because of the movement of obstacles or other robots. However, the task of catching the target has been completed without colliding with each other or colliding with any obstacles.

Table 3. The statistical values of the moving steps of the robots.

Map	1st	2nd					3rd			
Values	Ro1	Ro1	Ro2	Ro3	Ro4	Ro5	Ro1	Ro2	Ro3	Ro4
Min	118	48	86	79	92	62	74	57	63	69
Max	125	55	123	105	104	85	116	91	73	86
Mean	121.7	50.2	93.2	84.4	98.4	64.6	77.6	71.8	65.3	73.6
Median	122	49	89	83	99	64	77	78	65	71
Mode	122	49	88	82	99	63	77	59	65	71

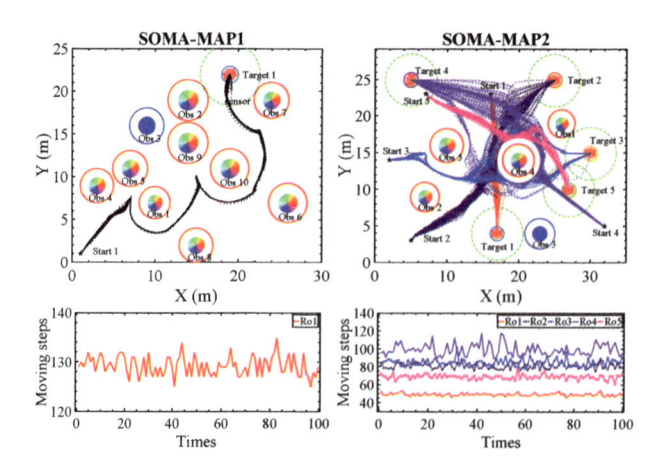

Fig. 9. The movement of the robots using SOMA.

The method proposed in [3] has been used in comparison with this method. The authors have performed 100 simulations for the 1st and 2nd map that used the same parameters of position, size, moving step of robots and obstacles as shown in Fig. 9.

Figures 8 and 9 show that the proposed method is more efficient as the robots move with more stable trajectories and fewer steps.

5 Conclusion

The paper fully addresses the problem of the movement of swarm robots in a complex environment, including moving targets and unknown dynamic obstacles, in which the movement of the robot is viewed as an optimization problem. The PSO algorithm was used to solve this. The simulation results in different environments were presented to demonstrate the feasibility of the method.

Acknowledgment. The following grants are acknowledged for the financial support provided for this research: Grant of SGS 2018/177, VSB-Technical University of Ostrava.

References

1. Deepak, B., Parhi, D.R., Raju, B.: Advance particle swarm optimization-based navigational controller for mobile robot. Arab. J. Sci. Eng. **39**(8), 6477–6487 (2014)
2. Diep, Q.B., Zelinka, I.: An algorithm for swarm robot to avoid multiple dynamic obstacles and to catch the moving target. In: XIIIth International Symposium Intelligent Systems, INTELS18, St. Petersburg, Russia (2018, in print, accepted)
3. Diep, Q.B., Zelinka, I.: Obstacle avoidance for swarm robot based on self-organizing migrating algorithm. In: The 2018 IEEE Symposium Series on Computational Intelligence (2018, in print, accepted)

4. Hossain, M.A., Ferdous, I.: Autonomous robot path planning in dynamic environment using a new optimization technique inspired by bacterial foraging technique. Robot. Auton. Syst. **64**, 137–141 (2015)
5. Kennedy, J., Eberhart, R.C.: Particle swarm optimization. In: Proceedings of the 1995 IEEE International Conference on Neural Networks, vol. 4, pp. 1942–1948. Perth, Australia. IEEE Service Center, Piscataway (1995)
6. Montiel, O., Orozco-Rosas, U., Sepúlveda, R.: Path planning for mobile robots using bacterial potential field for avoiding static and dynamic obstacles. Expert Syst. Appl. **42**(12), 5177–5191 (2015)
7. Nickabadi, A., Ebadzadeh, M.M., Safabakhsh, R.: A novel particle swarm optimization algorithm with adaptive inertia weight. Appl. Soft Comput. **11**(4), 3658–3670 (2011)
8. Rashid, A.T., Ali, A.A., Frasca, M., Fortuna, L.: Path planning with obstacle avoidance based on visibility binary tree algorithm. Robot. Auton. Syst. **61**(12), 1440–1449 (2013)
9. Zhang, Y., Gong, D.W., Zhang, J.H.: Robot path planning in uncertain environment using multi-objective particle swarm optimization. Neurocomputing **103**, 172–185 (2013)

Image Processing

Contour Detection Method of 3D Fish Using a Local Kernel Regression Method

Jong Min Oh[1], Sung Rak Kim[1], Sung Won Kim[1], Nam Soo Jeong[2], Min Saeng Shin[2], Hak Kyeong Kim[1], and Sang Bong Kim[1(✉)]

[1] Department of Mechanical Design Engineering, Pukyong National University, Busan 48549, South Korea
{Ohmin100, cdxv77, blckl022}@naver.com,
{hakkyeong, kimsb}@pknu.ac.kr
[2] Department of Mechanical System Engineering,
Dongwon Institute of Science and Technology, Yangsan, Republic of Korea
{nsj, nssin}@dist.ac.kr

Abstract. This paper proposes a fish contour detection method using a local kernel regression method. To do this task, the followings are done. Firstly, 3D depth map of a fish is obtained by using Kinect camera sensor. Secondly, edge of the fish in 3D depth map is transformed into numerous points. Thirdly, all information is removed except the edge points. However, recognition points for recognizing the fish are appropriately left because there are only points left to distinguish between background and fish. Fourthly, the points are recognized as contour of the fish by using the local kernel regression method. Finally, experiment results using Kinect camera sensor are shown to verify the validity of the proposed method compared to Canny method.

Keywords: Local kernel regression · Edge point · Contour of fish · Kinect camera sensor

1 Introduction

Measuring the wound of a fish technique needs fish detection and recognition. Shape, contour and volume of fish is very important for fish detection/recognition. In [1], a new rectangle representation algorithm based on the Kinect camera was proposed to localize the grasping point of the fish. For 3D point cloud image, Papazov et al. [2] utilized a Kinect stereo camera sensor to acquire depth images of the scene. Bley et al. [3] proposed another approach of grasp selection by fitting learned generic fish models to point cloud data. Canny [4] proposed a computational approach to edge detection. In most of 3D point cloud processing, the fish are blurry and invisible, so image processing is needed separately.

To solve this problems, this paper proposes a fish contour detection method using local kernel a regression method [5]. Recognition of the fish is obtained by using Kinect camera sensor based on Euclidean distance. A contour detection method is proposed using a local kernel regression method. Finally, the effectiveness of the proposed method is verified by experiment. The experimental results show that the contour of the fish is detected well compared to Canny method.

2 Contour Detection Method of 3D Fish Using a Local Kernel Regression Method

In this chapter, a contour detection method of 3D fish using a local kernel regression method is presented to extract a 3D point cloud of depth map of 3D fish image and the points is transformed into a contour of the 3D fish. The method is processed as follows:

2.1 3D Point Cloud

In this section, a 3D point cloud is extracted by using neighborhood-based filtering techniques. The neighborhood-based filtering method determines the filtered position of a point using similarity measures between a point and its neighborhood which has a strong influence on the efficiency and effectiveness of filtering approach. As described in the following methods, the similarity can be defined by position of points, normal vectors or regions.

A general filter can be expressed into [6]:

$$I'_f(x,y) = \frac{1}{w_p} \sum_{i,j \in \Omega} w(i,j) * I(i,j) \quad (1)$$

$$w_p = \sum_{i,j \in \Omega} w(i,j) \quad (2)$$

where $I_f(x,y)$ is a filtered image of pixel (x,y), w_p is a normalizing factor. Ω is the neighborhood range of pixel (x,y) as the center of the rectangular area called "window", (i,j) is one pixel coordinate of neighboring pixels, (x,y) is a coordinate of current pixel to be filtered, $w(i,j)$ is the weighting value of the point (i,j) to be filtered, and $I(i,j)$ is the original depth image in the point (i,j).

A weight value w in the bilateral filter is defined using the spatial closeness w_s and the intensity difference w_r as follows [8]:

$$w(i,j) = w_s(i,j) \times w_r(i,j) \quad (3)$$

$$w_s(i,j) = e^{\left(-\frac{(i-x)^2 + (j-y)^2}{2\sigma_s^2}\right)} \quad (4)$$

$$w_r(i,j) = e^{\left(-\frac{(I(i,j)-I(x,y))^2}{2\sigma_r^2}\right)} \quad (5)$$

where (i,j) is the neighborhood of a center point (x,y), $I(i,j)$ presents the intensity at (x,y), and σ_s and σ_r are the standard deviations of Gaussian functions.

In order to reduce time complexity, the weight of gray domain in the bilateral filters are replaced with a binary function (6) to achieve a better performance [8]. However, this kind of filters deals with the point cloud containing intensity components.

As a consequence, normal vector as one of the important attributes of point cloud is considered in the process of bilateral filters of which the weight is defined as a function of spatial location and normal information of points shown in Eq. (7) [7].

$$w_r = \begin{cases} 1 & |I(x,y) - I(i,j)| \leq 3, I(x,y) \neq 0 \\ 0 & |I(x,y) - I(i,j)| > 3, I(x,y) \neq 0 \end{cases} \quad (6)$$

$$w = w_s \times w_r = f[d(p,q)] \times g[c(n_p, n_q)] = \frac{e^{-(p-q)^2}}{2\sigma_s^2} \frac{e^{-(n_p-n_q)^2}}{2\sigma_r^2} \quad (7)$$

where f and g are the Gaussian function with σ_s and σ_r. σ_s and σ_r are the standard deviations of f and g. $d(p,q)$ is defined as certain distance(Euclidean distance) between point p and its neighbor q. $c(n_p, n_q)$ indicates the normal relation at points p and q.

2.2 Canny Edge Point Detection

Canny Edge Detection is a popular edge detection method. It is a multi-stage algorithm as follows [4]:

Step 1 Noise Reduction: Since edge detection is susceptible to noise in the image remove the noise in the image with a 5×5 Gaussian filter. Equation (8) is a Gaussian filter kernel of size $(2k+1) \times (2k+1)$ as follows:

$$H_{ij} = \frac{1}{2\pi\sigma^2} exp\left(\frac{(i-(k+1))^2 + (j-(k+1))^2}{2\sigma^2}\right); 1 \leq i, j \leq (2k+1) \quad (8)$$

Equation (9) is an example of a 5×5 Gaussian filter used to create the adjacent image with σ = 1.4. The asterisk denotes a convolution operation.

$$I_f = \frac{1}{159} \begin{bmatrix} 2 & 4 & 5 & 4 & 2 \\ 4 & 9 & 12 & 9 & 4 \\ 5 & 12 & 15 & 12 & 5 \\ 4 & 9 & 12 & 9 & 4 \\ 2 & 4 & 5 & 4 & 2 \end{bmatrix} * I \quad (9)$$

Step 2 Finding Intensity Gradient of the Image: Smoothened image is then filtered with a Sobel kernel in both horizontal and vertical direction to get the first derivative in horizontal direction (G_x) and vertical direction (G_y). From these two images, edge gradient and direction for each pixel can be obtained as follows:

$$Gradient\ magnitude = Edge_{Gradient(G)} = \sqrt{G_x^2 + G_y^2} \quad (10)$$

$$Gradient\ direction = Angle(\theta) = \tan^{-1}\left(\frac{G_y}{G_x}\right) \quad (11)$$

Gradient direction is always perpendicular to edges. It is rounded to one of four angles representing vertical, horizontal and two diagonal directions.

Step 3 Non-maximum Suppression: After getting gradient magnitude and direction, a full scan of image is done to remove any unwanted pixels which may not constitute the edge. For this, at every pixel, pixel is checked if it is a local maximum in its neighborhood in the direction of gradient as shown in Fig. 1.

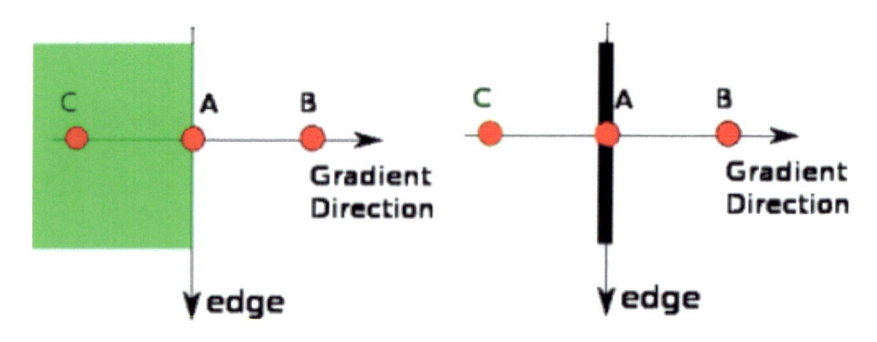

Fig. 1. Non-maximum suppression

Point A is on the edge (in vertical direction). Gradient direction is normal to the edge. Point B and point C are in gradient directions. So the point A is checked with the point B and the point C to see if it forms a local maximum. If so, it is considered for next stage, otherwise, it is suppressed (put to zero).

Step 4 Hysteresis Thresholding: This stage decides whether all edges are really edges or not. For this, we need two threshold values, minVal and maxVal as shown in Fig. 2. Any edges with intensity gradient more than maxVal are sure to be edges. Those below minVal are sure to be non-edges, and discarded. Those which lie between these two thresholds are classified edges or non-edges based on their connectivity. If they are connected to "sure-edge" pixels, they are considered to be part of edges. Otherwise, they are also discarded.

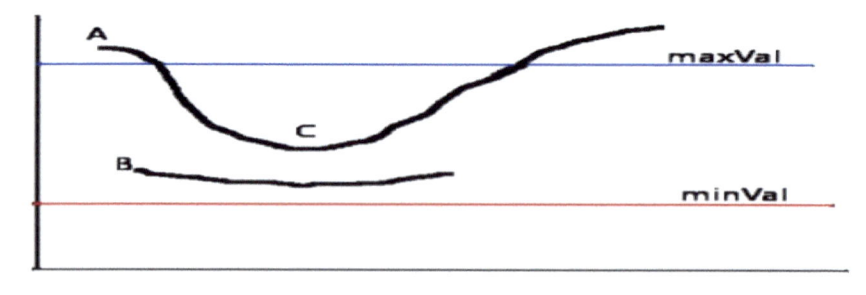

Fig. 2. Hysteresis thresholding

Because the edge A is above the maxVal, it is considered as 'sure edge'. Although the edge C is below maxVal, it is connected to the edge A. So the edge C is also considered as valid edge and the full curve is 'sure edge'. Although the edge B is above minVal and is in the same region with the edge C, it is not connected to any 'sure edge'. Therefore, the edge B is discarded. So it is very important that minVal and maxVal must be selected accordingly to get the correct result. Figure 3 shows a block diagram of Canny edge detector.

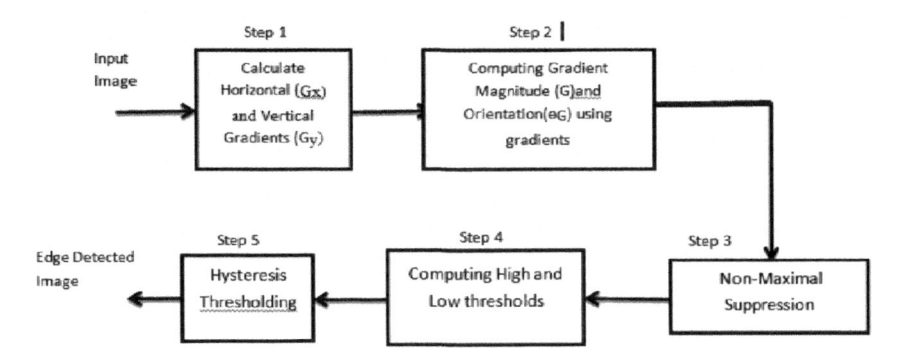

Fig. 3. Block diagram of Canny edge detector

2.3 Local Kernel Regression

In most of 3D point cloud processing, the fish are blurry and invisible, so image processing is needed separately. The Canny method is specialized for edge detection of 2D image. Therefore, the Canny method is not suitable for edge detection of 3D point cloud. To solve this problems, this paper proposes a fish contour detection method using local kernel a regression method as follows [9]:

A function used for general kernel regression is represented as

$$f(x) = \frac{\sum y_i \emptyset_i(x)}{\sum \emptyset_i(x)} \tag{12}$$

$$\emptyset_i(x) = \emptyset(\|x - x_i\|)$$

Local kernel regression is a supervised regression method to approximate a function $f(x) : \mathbb{R}^d \to \mathbb{R}$ given its output image $y_i \in \mathbb{R}$ at its sampled points x_i, and $x \in \mathbb{R}^d$ is a set of x_i. The input data might be corrupted with noise such that $y_i = f(x_i) + \mathcal{E}$, where $\mathcal{E}$ is a random noise with zero mean, and $\emptyset_i(x)$ is the kernel function.

The essence of the method is to approximate the unknown function $f(x_i)$ around the evaluation point x in terms of a Taylor expansion:

$$f(x_i) = f(x) + (x_i - x)^T \nabla f(x) + \frac{1}{2}(x_i - x)^T H f(x)(x_i - x) + \ldots \tag{13}$$

where $Hf(x)$ denotes the Hessian matrix of $f(x)$. The order o of the expansion is defined as the number of terms used in the Taylor expansion minus one.
For $o = 0$, $f(x_i) = f(x)$ and Eq. (13) can be reformulated as follows:

$$f(x_i) = s_0 + a_i^T s_1 + b_i^T s_s + \ldots \quad (14)$$

where $s_0 = f(x)$, $s_1 = \nabla f(x) = \left[\frac{\partial f}{\partial x_1}, \frac{\partial f}{\partial x_2}, \ldots, \frac{\partial f}{\partial x_d}\right]$, $s_2 = \left[\frac{\partial f(x)}{\partial x_i \partial x_j}\right] \in R^{d \times d}$
for $i, j = 1, \cdots, d$, $a_i = (x_i - x)$, $b_i = \text{vech}\left((x_i - x)(x_i - x)^T\right)$ and s_0, s_1, s_2 are function itself, gradient and Hesian of f, respectively.
Vech(A) is the half-vectorization of $A \in R^{n \times n}$ as follows:

$$\text{vech}(A) = [A_{11}, A_{21}, \ldots, A_{n1}, A_{22}, A_{32}, \ldots, A_{n2}, \ldots, A_{n-1n-1}, A_{nn-1}, A_{nn}]^T \quad (15)$$

For a symmetric matrix $A \in R^{3 \times 3}$, $\text{vech}(A) = [A_{11} A_{12} A_{13} A_{22} A_{23} A_{33}]^T$.
Since Eq. (14) uses finite number of terms, Eq. (14) will be most accurate at the samples around the point x.

This suggests using a weighted least squares approach to find the unknown parameters.

Moving least square (MLS) reconstruction of a function f for a sample set $\{x_i, y_i\}$ is defined as follows:

$$J \equiv \min f_{MLS}(x) = \min_f \sum \varepsilon_i^T \varepsilon_i \varnothing_i(x) \quad (16)$$

where $\varnothing$ is a decreasing weight function to give samples close to the evaluation point x more weight in the minimization.

$$J = \min_{s_k} \sum \left(y_i - \left(s_0 + a_i^T s_1 + b_i^T s_2 + \ldots\right)\right)^2 \varnothing_i(x) \quad (17)$$

Equation (17) is equivalent to minimizing the following expression

$$J = \min(y - Xs)^T \Phi(y - Xs) \quad (18)$$

$$y = [y_1, y_2, \cdots, y_n]^T \quad (19)$$

$$X = \begin{bmatrix} 1 & (x_1 - x)^T & \text{vech}\left((x_1 - x)(x_1 - x)^T\right)^T & \cdots \\ 1 & (x_2 - x)^T & \text{vech}\left((x_2 - x)(x_2 - x)^T\right)^T & \cdots \\ \vdots & \vdots & \vdots & \vdots \\ 1 & (x_n - x)^T & \text{vech}\left((x_n - x)(x_n - x)^T\right)^T & \cdots \end{bmatrix} \quad (20)$$

$$s = [s_1 s_2^T \cdots s_{o+1}^T]^T \quad (21)$$

$$\Phi = \begin{bmatrix} \phi_1(x) & 0 & 0 & \cdots \\ 0 & \phi_2(x) & 0 & \cdots \\ \vdots & \vdots & \ddots & \vdots \\ 0 & 0 & \cdots & \phi_n(x) \end{bmatrix} \qquad (22)$$

For o = 0, Eq. (17) is reduced into:

$$J = \min_{s_i}(y_i - s_0)^2 \varnothing_i(x) = \min \sum (y_i - s_0)^2 \varnothing_i x \qquad (23)$$

The solution of Eq. (23) is simply given as follows:

$$f(x) = s_0 = \sum \left(n_i^T(x - x_i) - s_0\right)^2 \varnothing_i(x) \qquad (24)$$

From Eq. (12) and Eq. (24), the following is obtained as

$$y_i = f(x_i) = n_i^T(x - x_i) \qquad (25)$$

where n_i is the normal in the sample point x_i defined as the normalized gradient as follows:

$$n = \frac{\nabla f}{\|\nabla f\|} \text{ for } \|\nabla f\| \neq 0 \qquad (26)$$

For o = 1 and $y_i = 0$, the followings are obtained for $\nabla f(x_i) = n_i = s_i$.

$$\min \sum (y_i - s_0 + a_i^T s_1) \varnothing_i(x) = \min_{s_0} \sum (s_0 + a_i^T \nabla f(x_i))^2 \varnothing_i(x)$$
$$= \min_{s_i} \sum (s_0 + a_i^T n_i)^2 \varnothing_i(x) \qquad (27)$$

For $\|s_1\| = \|\nabla f(x)\| = 1$ and o = 1, s_0, s_1 are obtained as [9]:

$$s_0 = s_1^T(x - c) \qquad (28)$$

$$(C - cc^T) s_1 = 0 \qquad (29)$$

$$c = \frac{\sum x_i \varnothing_i(x)}{\sum \varnothing_i(x)} \text{ and } C = \frac{\sum x_i x_i^T \varnothing_i(x)}{\sum \varnothing_i(x)} \qquad (30)$$

Therefore, the solution s_1 of Eq. (29) is the eigenvector of matrix $(C - cc^T)$ and corresponds to the normal of $(C - cc^T)$.

3 Experiment Result

3.1 Kinect Camera

In this paper, Kinect camera is selected as shown in Fig. 4.

Fig. 4. Kinect camera

The specifications of the Kinect camera are shown in Table 1.

Table 1. Specifications of the Kinect camera

RGB pixel	640 × 480
Depth pixel	640 × 480
FPS(frame/second)	30 fps
Minimum depth sensor range	800 mm
Maximum depth sensor range	4000 mm

3.2 3D Point Cloud

Figure 5 shows 3D point cloud image of a fish using the bilateral filter with $\sigma_s = 3$ and $\sigma_r = 3$ using Kinect camera. As shown in Fig. 5, there are a lot of blurry fish.

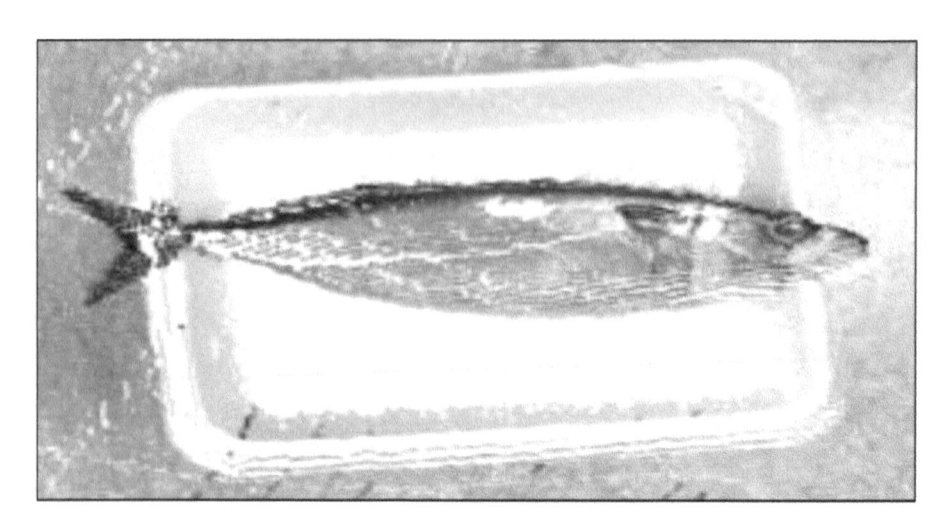

Fig. 5. 3D point cloud image using the bilateral filter

3.3 Canny Edge Points Detection

Figure 6 is the edge point extracted from the 3D point cloud picture in Fig. 5 using Canny method with $\sigma = 1.4$. mentioned in the Sect. 2.2. Since the 3D point is an array of points, the resulted edge image included a large amount of edge points extracted at the time of edge point extraction. Therefore, the full edge extraction time is about 3 s.

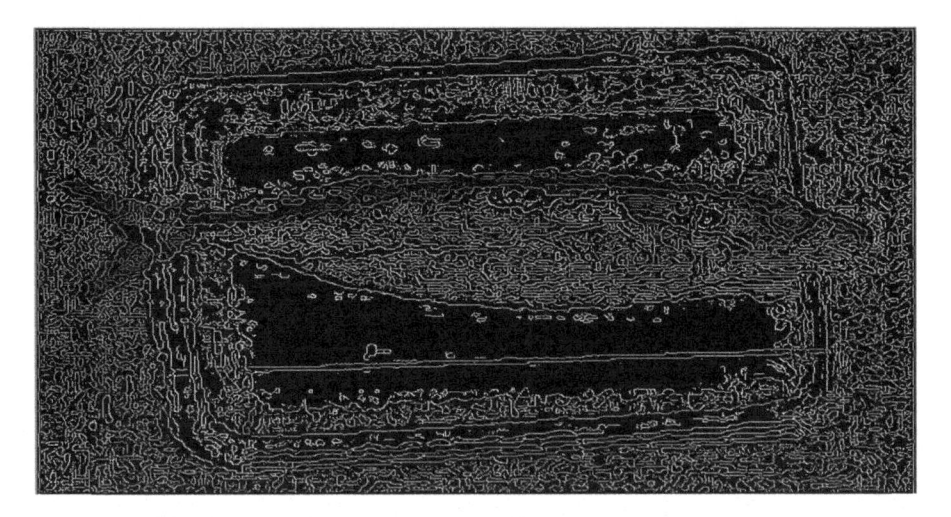

Fig. 6. Edge points extraction using Canny method

3.4 Local Kernel Regression

Figure 7 shows the edge point extracted from the 3D point cloud picture in Fig. 5 using local kernel regression mentioned in the Sect. 2.3. $\varnothing$ uses bilateral filter of Eq. (3). Nevertheless, the 3D point is an array of points, the resulted edge image using local kernel regression includes less amount of edge points extracted at the time of edge point extraction than that using Canny edge detection method. Therefore, the full edge extraction time is about 2 s.

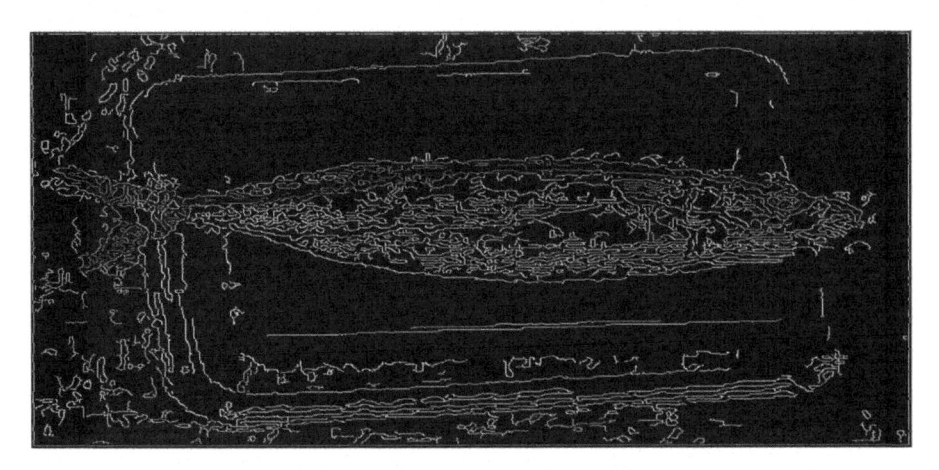

Fig. 7. Contour detection using local kernel regression

4 Conclusions

In this paper, a contour detect method of 3D fish image by local kernel regression using Kinect camera was proposed and compared to that by Canny method. The validity of the proposed method was verified by experiments. The conclusions of this paper are as follows. Firstly, 3D cloud point of the 3D fish image taken by Kinect camera was obtained in the Sect. 2.1 and was blurry. Secondly, the Canny method extracted numerous edges from 3D point cloud. Therefore, the numerous edges were not recognized as real edges of the 3D fish and edge detection time was about 3 s. Therefore, the Canny method was not suitable as edge detection for 3D point cloud image. Thirdly, the resulted edge image for 3D point cloud image by the proposed local kernel regression method included less edge points than that by the Canny method and edge detection time was about 2 s. Finally, the experimental results were shown to verify the validity of the proposed method. The local kernel regression method is better and faster in performance for edge detection of 3D point cloud image than the Canny method.

Acknowledgments. This research was a part of the project titled "Localization of unloading automation system related to Korean type of fish pump (20150446)", funded by the Ministry of Oceans and Fisheries, Korea.

References

1. Nguyen, T.H., Jeong, S.K., Kim, H.K., Kim, S.B.: A method for localizing and grasping objects in a picking robot system using kinect camera. In: Proceedings of 2016 International Symposium on Advanced Mechanical and Power Engineering (ISAMPE), pp. 178–180 (2016)
2. Papazov, C., Haddadin, S., Parusel, S., Krieger, K., Burschka, D.: Rigid 3D geometry matching for grasping of known objects in cluttered scenes. Int. J. Robot. Res. **31**(4), 538–553 (2012)
3. Bley, F., Schmirgel, V., Kraiss, K.F.: Mobile manipulation based on generic object knowledge. In: Proceedings of the IEEE International Symposium on Robot and Human Interactive Communication (2006)
4. Canny, J.F.: A computational approach to edge detection. IEEE Trans. Pattern Anal. Mach. Intell. **8**(6), 679–698 (1986)
5. Seber, G.A.F., Wild, C.J.: Nonlinear Regression. Wiley, New York (1989)
6. Han, X.F., Jin, J.S.: A review of algorithms for filtering the 3D point cloud. Sig. Process.: Image Commun. **57**, 103–112 (2017)
7. Wand, M., Berner, A., Bokeloh, M., Jenke, P., Fleck, A., Hoffmann, M., Maier, B., Staneker, D., Schiling, A., Seidel, H.: Processing and interactive editing of huge point clouds from 3D scanners. Comput. Graph. Sci. Direct **32**, 204–220 (2008)
8. Gao, T.S.: 3D image reconstruction algorithm based on depth map information of a Kinect camera sensor, Thesis, Pukyong National University (2017)
9. Oztireli, A.C., Guennebaud, G., Gros, M.: Feature preserving point set surfaces based on Non-Linear Kernel Regression. EUROGRAPHICS **28**(2)

Camera Based Tests of Dimensions, Shapes and Presence Based on Virtual Instrumentation

Lukas Soustek[(✉)], Radek Martinek, Lukas Snajdr, and Petr Bilik

Department of Cybernetics and Biomedical Engineering, Faculty of Electrical Engineering and Computer Science, VSB–Technical University of Ostrava, Ostrava, Czech Republic
{lukas.soustek, radek.martinek, lukas.snajdr, petr.bilik}@vsb.cz

Abstract. This article focuses on the use of machine vision for the needs of automated inspections based on virtual instrumentation using a visualization tool called Vision Builder for Automated Inspection (VBAI) - National Instruments (NI). The experimental part presents a real application for camera tests of dimensions, shapes and presence. The application deals with sorting of nuts and washers M6 to M12, wherein the result is a list of the total number of nuts, washers and the number of individual indexes of nuts, washers and faulty objects and inspection status. The original contribution of the article is to demonstrate the use of machine vision based on virtual instrumentation by means of VBAI on a real industrial machine vision application.

Keywords: Camera · Vision Builder for Automated Inspection (VBAI) · Machine vision system · Virtual instrumentation

1 Introduction

The person as an operator of the device can work effectively only for a short period of time until they can concentrate on the issue being dealt with. This time is affected, among other things, by the number of repetitions of the same steps; our attention decreases over time and the error rate rises. At the same time, the person has limited possibilities to recognize the degree of damage and the speed of the evaluation. For these reasons, an automated machine vision system is the best choice for visual inspection of products when producing multiple pieces, see [1–3].

Machine vision is the use of camera systems in industry, for image information processing by the superior system in which an algorithm is created and its outcome then affects the automated decision making of the machine or the information system, see [4, 5]. Typically, it is about checking the presence, dimensions, completeness or defects of the product within the production line. Based on the image information, the system evaluates which products are correct and which are not. The huge advantage of automated machine vision systems is still the same quality of inspection without limitation to the number of pieces checked or to the system operating time, see [6, 7].

The basic HW for machine vision systems is an optical sensor - a camera with a specific optical system (lens, filters) required for the task. The camera itself is often inadequate due to varying lighting parameters in the workplace. Therefore, another important part of the optical inspection system is the illumination device for illuminating the scene being scanned and increasing the accuracy and thus the quality of the optical inspection system, see [8, 9]. Machine vision systems today are also used in robotics for object detection and recognition tasks [11]. This paper focuses on dimensions, shape and presence tests, namely the sorting of nuts and washers M6 to M12. There are more programs for machine vision software development; in this article, LabVIEW or Vision Builder for Automated Inspection by National Instruments are used, see [12–14].

2 Automated Image Acquisition and Processing - HW for Machine Vision

At the Department of Cybernetics and Biomedical Engineering, Faculty of Electrical Engineering and Computer Science, VSB - Technical University of Ostrava, Czech Republic, there is a unique machine vision laboratory based on virtual instrumentation in which the experiments were carried out. Cameras with a planar image sensor (with data nut output) in industrial designs (such as the Basler aca1300-30gm camera) as well as commonly available standard cameras (Microsoft LifeCam Studio) are available for different types of inspections.

A standard matrix camera (Fig. 1a) comprises a planar sensor comprised of light-sensitive cells located in the matrix. Thanks to this, the camera can capture a larger image area, but the quality depends on the features such as the sensor type (CCD or CMOS), sensor size and resolution, scanning speed, colour range, etc. The size of the sensor is indicated in inches and determines the size of the photocell as well, so it means that a larger sensor with the same resolution has a larger pixel size, lower noise and a better image, the disadvantage is the larger focal length and larger, heavier and more expensive lenses. Sensor resolution indicates the number of pixels in the image and, in general, sensor resolution must be at least twice as large as the smallest measurable dimension required. Sensor resolution is specified in the Megapixel unit and, in industry, a resolution of 2 Mpx is usually sufficient. Scanning speed is indicated in frames per second (fps) and is especially important for selecting a suitable camera for an application where process speed is important.

Another specific type of camera is a line scan camera (Fig. 1b). Line scan cameras are a bit more demanding as for data processing because we only get 1 line of the image to the sensor, so, to compose the image of the entire scene, cooperation of the camera with the sliding system is necessary. Line scan cameras bring many advantages such as better resolution in the X axis of the image due to the large number of light-sensitive points (pixels), a practically unlimited image length in the Y axis (we create the Y axis of the image ourselves by putting the individual lines behind each other). Line scan cameras are, on principle, functions suitable for applications where the object or the camera is moving, such as your photocopy.

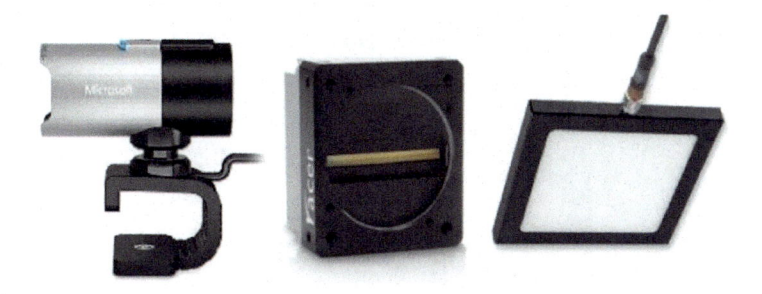

Fig. 1. (a) Microsoft LifeCam Studio Webcam, (b) Basler Racer Line Scan Camera, (c) Back Light BL-130 W Illuminator.

For the experiments performed, a commercially available camera with a planar sensor for static scene analysis was used due to simplicity. Classification of products, specifically their presence and dimensions, was chosen for the experiments. Steel nuts and washers of different sizes were selected as objects for the classification.

For machine vision tasks, additional illumination of the scene being examined is often useful for increasing the accuracy. There are several types of illumination with specific parameters and outcomes for the illuminated object; such an overview would be too extensive, so for the task of classifying tiny glossy objects, I chose the back light type of planar illumination (Fig. 1c). The light shines evenly from the back of the scene, so the resulting objects will be dark on the light background when the objects contours are highlighted.

3 Testing the Possibilities of an Automated System for Optical Inspection of Products

The actual experimental part deals with sorting of nuts and washers M6 to M12, wherein the result is a list of the total number of nuts, washers and the number of individual indexes of nuts, washers and faulty objects and inspection status.

3.1 Implementation

The hardware was assembled as shown in Fig. 2. The camera captures the image from above, and, underneath it, there is a back light BL-130 W illuminator on the active area of which objects for scanning were placed. Appropriate distance means that the camera can record all objects placed on the back light illuminator.

3.2 Calibration

Because we want to measure the dimensions of the parts, it is necessary to calibrate the system regularly (in our case, calibration of the gauge) after intervening in the optical system or in operation. Using a familiar length gauge, we will teach the system what

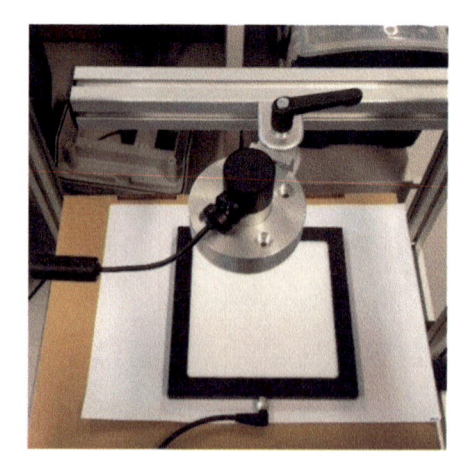

Fig. 2. Hardware assembly.

lengths the resolution has in the X and Y axes. Calibration is very simple. After creating the image of the calibration pattern, (in our case a ruler picture), we will run the Image Calibration Setup function, place the cursors on the points of the calibration pattern having familiar dimensions, and enter the real distance of these 2 points into the calibration function, see Fig. 3. The program will internally calculate the resolution of the image in the particular axis.

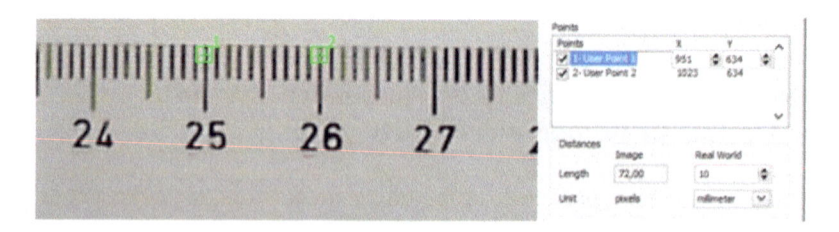

Fig. 3. Calibration pattern - selecting the points and entering the actual dimensions.

3.3 The Actual Image Processing

We work with most of the VBAI functions in Wizard mode when we select general function and inside it we edit internal functions and parameters where we immediately see the effect of the operation on the image being processed in the preview screen. A function parametrized this way is inserted into the image inspection (analysis) diagram by the user, when, after starting the program, each function returns the image processing status was successful/failed, see Fig. 4.

As mentioned before, each scene-specific analysis requires different image processing techniques. If we want to measure the dimensions of the parts, it is necessary to convert the image taken from the camera to a black and white image, only this way the PC detects the edges of the parts and can assign their coordinates after the subsequent

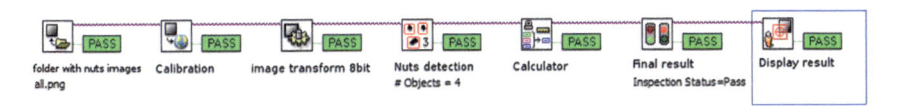

Fig. 4. VBAI inspection diagram.

recalculation of the dimensions. Using function Vision Assistant and selecting Threshold image processing with the correct settings do this transfer. The goal of the experiments conducted is to detect the presence and dimensions of steel nuts of different types. After converting the image looks like Fig. 5.

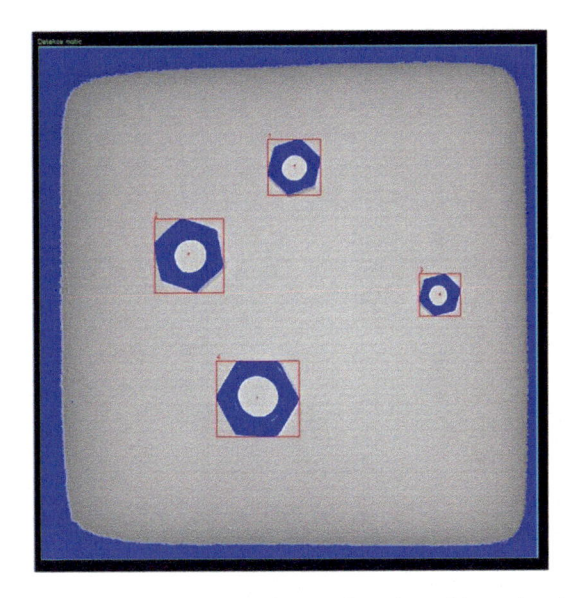

Fig. 5. Analysis of the presence of a part based on object edge detection.

By appropriate setting of the search parameters and the size limits of the parts found, we specify the use of the application for the analysis of specific components, in our case, steel nuts.

The next step is mathematical processing of the parameters of the components found in the function of the calculator, namely the calculation of the nut surface value and its assignment to the correct group of parts, see Fig. 6. Each area is compared to the minimum and maximum area value according to Table 1. If the area of the object is within the minimum and maximum area of the individual nut, add 1 to the number of the individual nut, thus meeting the condition for writing the current index to the specified nut, which is then converted to the text format and the separator between the indexes is a "comma". Objects that are not within the minimum and maximum areas of the individual nuts are evaluated as faulty objects. In the case of faulty objects, their number and indexes are also dealt with, as with the individual types of nuts. The whole

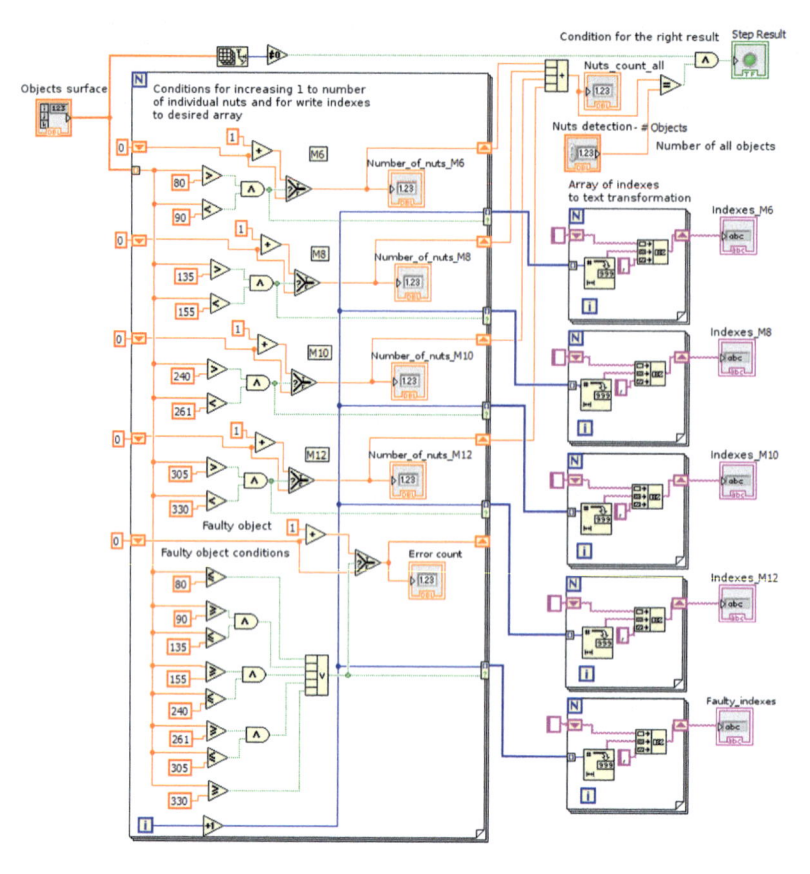

Fig. 6. Solution of the calculator function for laboratory task 1

step evaluation is correct if the number of detected objects is greater than 1 and, at the same time, if the sum of all individual numbers of nuts is equal to the total number of objects.

Table 1. Database of nut types and their dimensions.

	M6	M8	M10	M12
s (mm)	10.0	13.0	17.0	19.0
e (mm)	11.1	14.4	18.9	21.1
Area (mm^2)	86.60	146.36	250.28	312.64
Minimum (mm^2)	80.00	135.00	240.00	305.00
Maximum (mm^2)	90.00	155.00	261.00	330.00

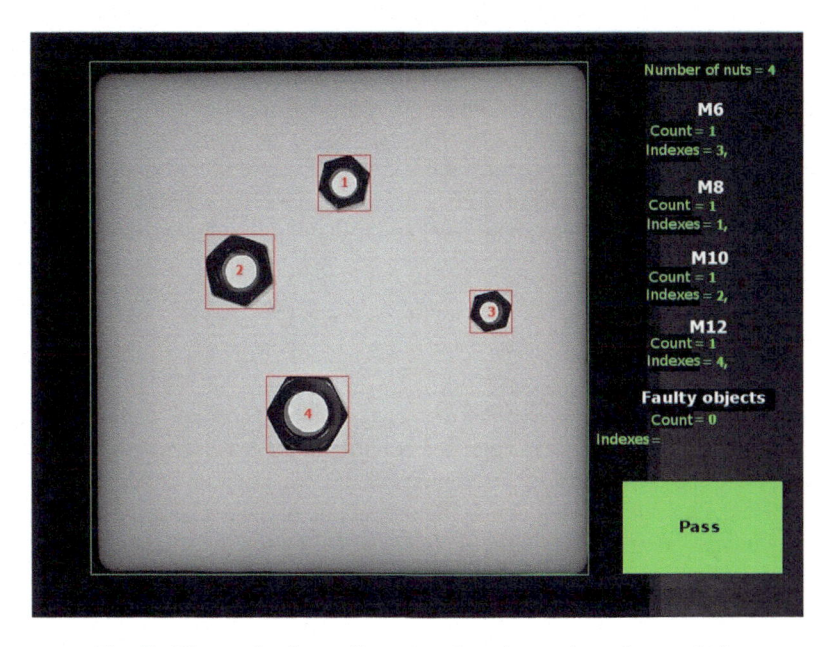

Fig. 7. The result of nut, dimension detection and number analysis.

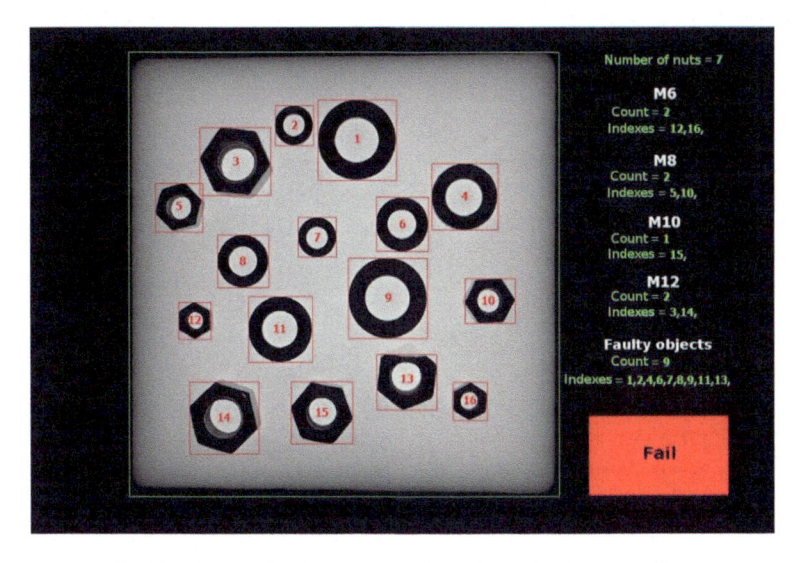

Fig. 8. The result of a mixture of good and bad parts analysis

Once we know which part falls into which group, it is easy to count the number of elements in the given group. The result of the analysis can be written in the form of text into the actual analysed image, see Fig. 7.

When the application is presented with parts different than it expected or damaged parts that do not meet one of the analysis conditions, the application declares the overall inspection to be wrong. Of course, you can set the behaviour of the application to the status of each analysis step as needed and choose the appropriate way of passing the information to the superior system or the human operator. An example of good and bad parts analysis is shown in Fig. 8. In a text statement, the application informs the operator of the number of defective pieces, their labelling (indexes), and displays the positions of the parts.

4 Conclusion

The possibility of using the machine vision system is inexhaustible, there are basically two types of analysis techniques - edge detection/finding a match between the image of the pattern and the area in the analysed image. Edge detection tasks are usually faster, computationally less demanding due to the conversion of the analysed image to a black and white image, but this reduces the accuracy by losing the information on the colour brightness differences of the objects.

The application presented deals with sorting of nuts M6 to M12 using the areas of the object detected. After sorting the known parts comparing to the total number of parts found, the application evaluates whether it has detected the presence of bad parts (The inspection has not been passed, wrong). Generally, the accuracy of the application always depends on the tolerance settings of the partial functions results. By correct setting of the product quality evaluation application, it is possible to achieve up to 0% rejects occurrence in finished products.

Acknowledgment. This article was supported by the Ministry of Education of the Czech Republic (Project No. SP2018/170). This work was supported by the European Regional Development Fund in the Research Centre of Advanced Mechatronic Systems project, project number CZ.02.1.01/0.0/0.0/16_019/0000867 within the Operational Programme Research, Development and Education.

References

1. Sonka, M., Hlavac, V., Boyle, R.: Image processing, analysis, and machine vision. Cengage Learning (2014)
2. Tanimoto, S. (ed.): Structured Computer Vision: Machine Perception Through Hierarchical Computation Structures. Elsevier (2014)
3. Kuhajda, B., et al.: New strategies for frequency measurement using high-speed video camera system. In: 2015 International Conference on Applied Electronics (AE), pp. 125–130. IEEE (2015)
4. Jain, R., Kasturi, R., Schunck, B.G.: Machine Vision. McGraw-Hill, New York (1995)
5. Davies, E.R.: Machine Vision: Theory, Algorithms, Practicalities. Elsevier (2004)
6. Vernon, D.: Machine vision-Automated visual inspection and robot vision. NASA STI/Recon Technical Report A, p. 92 (1991)

7. Steger, C., Ulrich, M., Wiedemann, C.: Machine Vision Algorithms and Applications. Wiley (2018)
8. Beck, J., Hope, B., Rosenfeld, A. (ed.): Human and Machine Vision. Academic Press (2014)
9. Sarkar, N.R.: Machine vision for quality control in the food industry. In: Instrumental Methods for Quality Assurance in Foods. Routledge, pp. 177–198 (2017)
10. Cubero, S., et al.: Automated systems based on machine vision for inspecting citrus fruits from the field to postharvest—a review. Food Bioprocess Technol., 1623–1639 (2016)
11. Li, X., et al.: Fault-tolerant control method of robotic arm based on machine vision. In: 2018 Chinese Control and Decision Conference (CCDC). IEEE (2018)
12. Hryniewicz, P., et al.: Technological process supervising using vision systems cooperating with the LabVIEW vision builder. In: IOP Conference Series: Materials Science and Engineering. IOP Publishing, p. 012086 (2015)
13. Joshi, K.D., Surgenor, B.W., Chauhan, V.D.: Analysis of methods for the recognition of Indian coins: a challenging application of machine vision to automated inspection. In: 2016 23rd International Conference on Mechatronics and Machine Vision in Practice (M2VIP), pp. 1–6. IEEE (2016)
14. Ding, Z., Zhang, R., Kan, Z.: Quality and safety inspection of food and agricultural products by LabVIEW IMAQ vision. Food Anal. Methods, 290–301 (2015)

A 3D Scanner Based on Virtual Instrumentation Implemented by a 1D Laser Triangulation Method

Jindrich Brablik[✉], Radek Martinek, Marek Haluska, and Petr Bilík

Department of Cybernetics and Biomedical Engineering,
Faculty of Electrical Engineering and Computer Science,
VSB–Technical University of Ostrava, Ostrava, Czech Republic
{jindrich.brablik, radek.martinek, marek.haluska,
petr.bilik}@vsb.cz

Abstract. This paper deals with the design and implementation of a 3D scanner based on 1D laser triangulation method. Virtual instrumentation principles and tools were used in the design. NI myRIO embedded device was utilized to control the scanner hardware and graphical development environment LabVIEW was used to develop a control application for the scanner and to develop GUI for a PC platform. Purpose of implemented scanner is to demonstrate the principle of 3D scanners and 1D laser triangulation method to students of virtual instrumentation class. First part of this paper provides concise theoretical information about 3D scanners, 1D laser triangulation method and virtual instrumentation principles. Second part focuses on mechanical, hardware and software design and implementation of realized scanner.

Keywords: 3D scanner · Virtual instrumentation · 1D laser triangulation · NI myRIO · LabVIEW

1 Introduction

The 3D scanner is a HW input device that has been designed for shape, or possible appearance (only certain types of scanners), analysis of real objects or environments and acquisition of their digital form. The principle of the 3D scanner function is based on scanning the surface of the analyzed object and determining the distance of its individual points from the reference point (e.g. the scanner). This way, the information about the 3D coordinates (X, Y, Z) of the object being scanned in space is gained. In most cases, only one scan is not enough to acquire a complete model of the object scanned. Multiple scanning from different directions is usually required to obtain detailed information on all the sides and dimensions of the object. The output of the analysis is the so-called point cloud, which is then converted to a geometric model using a suitable polygonal network. Using the network, the image pattern is reconstructed, and a 3D model of the object scanned is obtained.

The beginnings of the development of 3D scanners date back to the 1960s [2, 20]. Primary scanners used lights, cameras and projectors for their functions. Since then an enormous progress was made in development of 3D scanners [12]. Nowadays, there are several numbers of scanners based on different physical principles (optical, magnetic, x-ray, ultrasonic, etc.) [2, 11, 12, 20]. The application area of 3D scanners is very wide and is still growing [1]. The examples of application areas are as follows:

- Medical industry (orthopedics, prosthetics, dentistry, diagnostics).
- Film and game industry (figure scanning).
- Product quality control (dimension check).
- Automotive industry (automated driving system).
- Surface mapping.

Thanks to enormous progress in development of 3D scanners and advances in software and hardware, broad spectrum of industries is nowadays interesting in digitizing their products [3]. This fact leads to higher demand for people with knowledge of this area in labor market. This paper is presented as a result of a design and implementation of relatively low-cost 3D scanner intended for demonstration of 3D scanner principles to students of virtual instrumentation class. 1D laser triangulation technique was selected, because of its simple principle and implementation [12]. This technique also provides relatively high precision in comparison with for example time of flight or phase shift techniques [7]. Results obtained from implemented scanner are compared to two different low-cost scanners based on time of flight and structured light techniques [16, 21]. Comparison with commercially available scanner based on laser triangulation principle was performed as well [12].

1.1 1D Laser Triangulation

Triangulation is currently the most widely used method of digitization in optical methods. 1D laser triangulation falls into a group of non-contact active optical methods whose principle consists in reconstructing the object scanned by irradiating its surface with a light source (light beam in the case of 1D triangulation) and its simultaneous scanning, for example, with a camera. The point scanned on the object, the scanner and the light source together form a triangulation triangle. The triangulation base (B) marks the connecting line of the light source and the scanner. The angle between the base and the side of the light source does not change. The angle between the triangulation base (β) and the scanner is variable according to the point of the object scanned. Distance between scanned point and scanner can be determined based on the known value of angle (β) and the size of triangulation base (B) according to the following Eq. (1).

$$L = \frac{B}{\tan(\beta)} \quad (1)$$

The triangulation measurement principle of modern laser scanners is based on the fact that position of the beam reflected on the scanner front end is proportional to distance of scanned object from the scanner. The beam must reflect from the scanned object at a constant angle. Point where the reflected beam appears on the scanner is evaluated instead of the intensity of the incidental beam or the time of its flight (other non-contact methods). This makes the detection much more accurate (up to tens of micrometers) reliable and more resistant to interference. Drawback of this method are disadvantages associated with the inherent properties of lasers and limited scanning range of several meters [2, 3, 17, 19].

1.2 Virtual Instrumentation

Virtual instrumentation falls to the field of measurement, testing and automation. Its principle is replacing a single-purpose device (oscilloscope, DMM and others) with industrial or desktop PC (eventually an industrial controllers or modular systems e.g. PXI, VXI) and general HW for measurement and control. The function of this connection is defined by a SW application that is running on a PC and controls the connected HW, collects and processes data and interprets it to the user. The entire user interface of the created system is defined in the custom software application. The main advantage of this approach is to easily change the functionality of the device by creating a new SW application and possibly replacing or extending the HW [3].

LabVIEW – (Laboratory Virtual Instrumentation Engineering Workbench) is a graphical development environment developed by National Instruments, which is primarily dedicated to measurement, control and virtual instrumentation. Its purpose is to provide a programming tool to engineers in various technical branches without having to study in detail computer science and software engineering. The LabVIEW code is created by merging of so-called virtual instruments, or VIs, which are equivalent to the functions in classic text programming languages. For the 3D scanner implementation, the LabVIEW development environment and its expansion modules (Real-Time and FPGA) were used.

2 Design and Principles of Implemented 3D Scanner

The designed scanner uses the above-described 1D laser triangulation method. The schematic arrangement of the scanner is shown in Fig. 1. The object scanned is fixed to a rotating platform (carousel) rotated by the first stepper motor (NI ISM-7400E) by the desired angle of rotation α. The second stepper motor (the same model) moves the linear shift (SLW-BB-PT-1040-200 made by Igus), thus defining the Z coordinate. The laser scanner (OptoNCDT 1320 made by Mikro-Epsilon) that scans the analyzed object is attached to the shift. The information about the radius of the scanned object is determined based on the known distance of the center of the carousel from the laser distance scanner and the distance measured by it. It is necessary to know the SMR, which is the start of measuring range of the laser distance scanner. An infrared barrier, or distance, scanner was used to detect the linear shift run-out. IBM FRU P/N 49G2196 was selected as the HW component power supply. The entire scanner control unit consists of the NI myRIO-1900 student platform, including a gate array (FPGA) and a microprocessor with a real-time operating system.

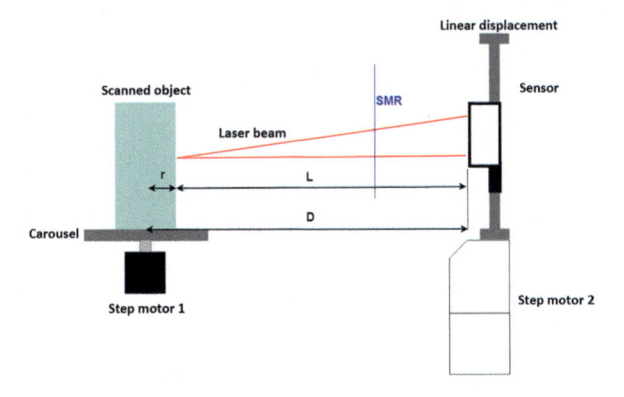

Fig. 1. Schematic arrangement of designed 3D scanner.

2.1 Mechanical Structure

The designed scanner is placed into a mechanical structure having dimensions of 37 × 27 × 47 cm made of 30 mm ITEM aluminum profiles. The metal structure and the remaining structural components were designed using the INVENTOR 2017 program. The rotating platform that serves for anchoring the scanned object and its rotation during the actual scanning was an important design element. The structure designed is shown in Fig. 2.

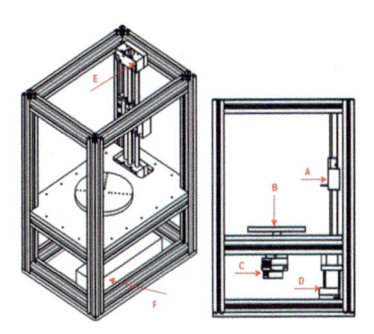

Fig. 2. Mechanical structure design

The structure consists of the rotary carousel (B) rotated by the first stepper motor (C). The linear shift (E) is moved by the second stepper motor (D). The optical sensor (A) is fixed to the linear shift. The voltage source (F) is also part of the structure.

3 3D Scanner Implementation

For the high-quality interconnection of the NI myRIO control unit with individual scanner components, a simple PCB was designed. It ensures a fixed connection, space saving and clarity, as opposed to the alternative considered in the form of a bread board. The complete structure of the scanner for educational purposes is shown in the following Fig. 3.

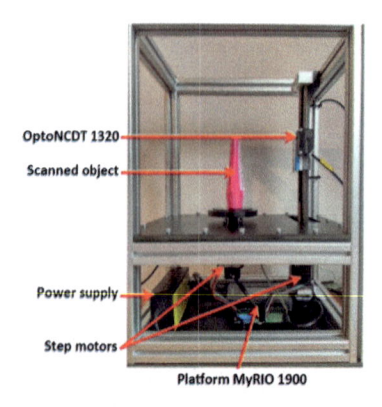

Fig. 3. Complete structure and HW components of implemented 3D scanner.

Figure 4 shows the block diagram of the hardware components connection to the NI myRIO-1900 platform. The communication between the PC and the myRIO control unit is primarily conducted via the 802.11 b,g,n (WI-FI) interface.

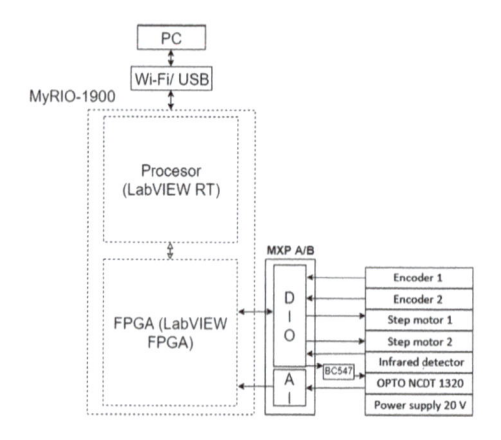

Fig. 4. Schematic design of HW interconnection.

3.1 SW Application

The scanner function is controlled by a three-layer SW application created in the LabVIEW development environment and its expansion modules (Real-Time and FPGA). The lowest SW layer is designed for programmable gate array (FPGA) and implements direct control of both stepper motors, reading data from their encoders, laser scanner and infrared barrier scanner. The interface with next SW layer that is designed for the NI Linux Real-Time operating system (RTOS) is also part of the FPGA layer. RTOS layer serves as a bridge between the PC and the FPGA, it implements data forwarding and scaling. The top SW layer is designed for the PC and includes a graphical user interface (Fig. 5). GUI allows user to control the scanner and displays scanned data in form of XY graph and 3D graph.

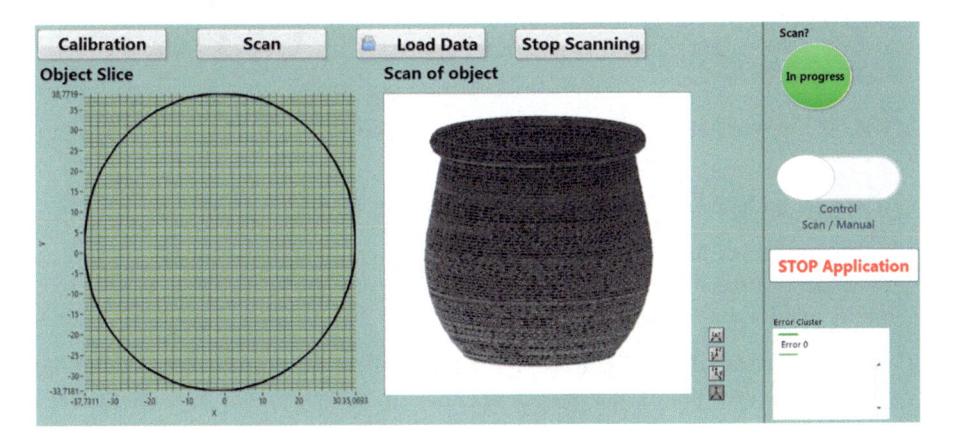

Fig. 5. Graphical user interface of PC SW layer.

The PC layer also allows the user to run the scanner calibration, the result of which is positioning of the laser scanner on top of the carousel and the preparation for the scan itself. After the calibration, the user can run an auto scan or switch to the manual mode and perform controlled configuration and scanning. The PC layer includes storing and retrieving the data scanned as well. The .xlsx format, which provides clear display and processing in Microsoft Office was used for storing. Also, .stl format was used, because it is supported by the CAD system and allows, for example, the subsequent 3D print of the scanned object.

4 Results

Data of a total of three objects with its scanned copies was compared to test the accuracy of the 3D scanner created. The first object was measured using a caliper, the second object's exact diameter was known, and the last one was scanned with Handy-Scan - EXAscan scanner made by SolidVision. The resulting data was compared to the 3D scanner created. First, the arithmetic mean of the data measured (both original and the scan) was calculated. The absolute and relative errors were calculated from the average. Figure 6 shows the original of the object and the scan created using the implemented scanner.

Fig. 6. The original of the object and the scan created using the implemented scanner.

The scan parameters (for all three objects were the same):

- rotation speed of the circular platform - half a revolution per second,
- movement of the linear shift - one millimeter,
- one revolution per scanning.

Slight waviness of the digitized object caused by its gloss can be seen. At the beginning (measured point), the diameter of the original object was measured using a calliper. Only this point, which is the most appropriate for accuracy because of zero height, was measured. The resulting data was averaged and compared to the object scanned. The relative error when measuring the point average was around 0.13% with corresponding absolute error around 0.07 mm. Also, the height of scanned object was measured with relative error of 0.26% and corresponding absolute error of 0.21 mm. The measurement results of the first object are shown in the following Table 1 (altogether 5 measurements were conducted).

Table 1. Comparison of the diameter and the height of the first object to its scan.

	Original Ø [mm]	Scan Ø [mm]	Original h [mm]	Scan h [mm]
1	53,1	53,20	79,51	79,35
2	53,11	53,18	79,57	79,65
3	53,09	53,11	79,53	79,15
4	53,11	53,11	79,52	79,45
5	53,07	53,21	79,56	79,05
$\overline{xr}, s[mm]$	53,09	53,16	79,54	79,33
Δx [mm]	0,07		0,21	
δ[%]	0,13		0,26	

Accuracy of implemented scanner is slightly better in comparison with low-cost 3D scanner based on structured light technique [16], where the absolute error of 0.3 mm was reached. However, in comparison with low-cost time of flight 3D scanner [21], where the absolute error was in interval from 1 cm to 2.5 cm, the accuracy of our scanner is superior. Achieved accuracy is comparable or slightly worse than the accuracy of commercially available low end to mid end 3D scanners (steintek 20–300 μm, ShapeGrabber 25–200 μm, Cyberware 50–300 μm) using laser triangulation technique. Since the implemented scanner is intended for teaching purposes, the achieved accuracy is more than satisfying.

5 Conclusion

This paper described design and implementation of 3D scanner based on 1D laser triangulation method and virtual instrumentation principles. Scanner is controlled by SW application developed in LabVIEW and consists of mechanical structure, rotary carousel, two step motors, linear shift, laser scanner, obstruction sensor and NI myRIO embedded controller. Point clouds of three objects were acquired, analyzed and compared to reference measurements (caliper, declared dimensions and commercial scanner) to determine accuracy of realize scanner. Accuracy was determined as absolute and relative error and comparison with other techniques and commercially available scanners was performed.

Implemented scanner has all the advantages and drawbacks associated with used technique. It has good accuracy of 100 um to 200 um and is resistant to noise. Its main disadvantage is that it can scan objects only small objects from short distance about 25 cm, also the object must be static during the scanning process. There are planned improvements to present software application that will facilitate real time display of scanning process, volume calculation and display of error plot based on reference scan.

References

1. Daneshmand, M., Helmi, A., Avots, E. et al.: 3D scanning: a comprehensive survey. Comput.Res. Repository, abs/1801.08863 (2018)
2. Ebrahim, M.: 3D laser scanners' techniques overview. Int. J. Sci. Res. **4**(10), 5–611 (2015)
3. Montilla, M., Orjuela-Vargas, S.S., Philips, W.: State of the art of 3D scanning systems and inspection of textile surfaces. In: Proceedings of SPIE - The International Society for Optical Engineering, vol. 9018 (2014). https://doi.org/10.1117/12.2042552. author, F.: Contribution title. In: 9th International Proceedings on Proceedings, pp. 1–2. Publisher, Location (2010)
4. Chromy, A.: Application of high-resolution 3D scanning in medical volumetry. Int. J. Electron. Telecommun. **62**(1), 23–31 (2016)
5. Acost, D., Garcia, O., Aponte, J.: Laser triangulation for shape acquisition in a 3D scanner plus scan. In: Electronics, Robotics and Automotive Mechanics Conference (CERMA 2006), Cuernavaca, pp. 14–19 (2006). https://doi.org/10.1109/cerma.2006.54
6. Jezersek, M., Flezar, M., Mozina, J.: Laser multiple line triangulation system for real-time 3-D monitoring of chest wall during breathing. Strojniski Vestnik **54**, 503–506 (2008)

7. Ebrahim, M.: 3D Laser Scanners: History, Applications, and Future. https://doi.org/10.13140/2.1.3331.3284
8. Wulf, O., Wagnr, B.: Fast 3D Scanning Methods for Laser Measurement Systems (2003)
9. Al-Sharif, L.: Three dimensional scanner using laser triangulation technology. In: 23rd Canadian Congress of Applied Mechanics (2011)
10. Mikulski, S.: Laser triangulation in three-dimensional scanners. Comput. Appl. Electr. Eng. **11** (2013). https://doi.org/10.21008/j.1508-4248, ISSN: 1508-4248
11. Hao, L.: 3D Scanning USA: University of Southern California Department of Computer Science. http://www.hao-li.com/cs599-ss2015/slides/Lecture04.2.pdf
12. Blais, F.: Review of 20 years of range sensor development. J. Electron. Imaging **13**, 231–243 (2004). https://doi.org/10.1117/1.1631921
13. Obrenovic, Z., Starcvic, D., Jovanov, E.: Virtual Instrumentation. Wiley encyclopedia of Biomedical Engineering (2006)
14. Dorsh, R.G., Husler, G., Herrman, J.R.: Laser triangulation: fundamental uncertainty in distance measurement. Appl. Opt. **33**(7), 1306 (1994). https://doi.org/10.1364/A0.33.001306
15. Reyes, A.L., Cervantes, J.M., Gutirrez, N.C.: Low cost 3D scanner by means of 1D optical sensor. Procedia Technol. **7**, 223–230 (2013)
16. Rocchini, C., Cignoni, P., Montani, C., Pingi, P., Scopigno, R.: A low cost 3D scanner based on structured light. Comput. Graph. Forum **20**, 299–308 (2001). https://doi.org/10.1111/1467-8659.00522
17. Amann, M.C., Bosch, T., Lescure, M., Myllylä, R., Rioux, M.: Laser ranging: a critical review of usual techniques for distance measurement. Opt. Eng. **40**, 10–19 (2001). https://doi.org/10.1117/1.1330700
18. Xianyu, S., Qican, Z.: Dynamic 3-D shape measurement method: a review. Opt. Lasers Eng. **48**, 191–204 (2009). https://doi.org/10.1016/joptlasng.2009.03.012
19. Chen, F., Brown, G.M., Song, M.: Overview of three-dimensional shape measurement using optical methods. Opt. Eng. **39**, 10–22 (2000). https://doi.org/10.1117/1.602438
20. Sansoni, G., Trebeschi, M., Docchio, F.: State-of-the-art and applications of 3d imaging sensors in industry, cultural heritage, medicine, and criminal investigation. Sensors (Basel, Switzerland) **9**(1), 568–601 (2009). https://doi.org/10.3390/s90100568
21. Cui, Y., Schuon, S., Chan, D., Thrun, S., Theobalt, C.: 3D shape scanning with a time-of-flight camera. In: 2010 IEEE Computer Society Conference on Computer Vision and Pattern Recognition, San Francisco, CA, pp. 1173–1180 (2010). https://doi.org/10.1109/CVPR.2010.5540082

Author Index

A
Aouaouda, Sabrina, 548
Arbulu, Mario, 904
Augustynek, Martin, 45

B
Baca, Jakub, 683
Basterrech, Sebastián, 939
Bauer, Jan, 758
Becerra, Yeyson, 904
Behan, Ladislav, 409
Beley, Valery, 870
Béreš, Michal, 175
Bernat, Petr, 810
Bilík, Petr, 973, 982
Blažek, Vojtěch, 844
Boulkaibet, Ilyes, 67
Bououden, Sofiane, 67
Brablik, Jindrich, 982
Brandejsky, Tomas, 294
Brandstetter, Pavel, 261, 580, 789
Butakova, Maria A., 475

C
Chadli, Mohammed, 67, 548
Chamrád, Petr, 514, 736
Chatterton, Steven, 647, 658
Chen, Chih-Keng, 78
Chernov, Andrey V., 475
Chueh, Chin-mei, 134
Cortes, Darío, 673

D
Dang, Dac Chi, 533
Dang, Phuoc Vinh, 647, 658
Dang, Truong Giang, 283
Dao, Nguyen, 821
Davendra, Donald, 134, 239
Dedek, Jan, 747
Diep, Quoc Bao, 949
Dinh, Bach Hoang, 78, 261, 303, 561, 789
Dinh, Cuong Tran, 261
Dobeš, Petr, 207
Dolezel, Petr, 101
Dolgiy, Alexander I., 24
Domesová, Simona, 228
Duy, Tran Trung, 371
Duy, Vo Hoang, 441
Duží, Marie, 55, 162

E
Efimov, Nikolay N., 628

F
Fait, Michal, 55
Fiedorova, Klara, 45
Florez, David, 673
Frivaldsky, Michal, 715
Fusek, Radovan, 207

G
Gavlas, Antonin, 390
Genkin, Arkady, 185

Gomez-Casseres, Andrés, 673
Gono, Radomir, 88, 860
Gorbatov, Dmitriy, 870
Gresak, Erik, 400

H
Ha, Dac-Binh, 693
Ha, Duy-Hung, 693
Haluska, Marek, 982
Hamel, Emmanuel, 134
Hasal, Martin, 313
Havel, Ales, 736
Hendrych, Jakub, 484
Ho, Sang Dang, 261, 580, 789
Hoang, Lien Son Chau, 725
Honc, Daniel, 101
Horák, David, 13
Hrbac, Roman, 835
Hurta, Jan, 850

I
Iegorova, Valeriia, 939
Ishikawa, Jun, 569, 883
Itoh, Hiroshi, 894

J
Jalowiczor, Jakub, 420
Jasek, Roman, 197
Jeong, Nam Soo, 963
Jeong, Sang Kwun, 605
Jespersen, Bjørn, 162
Jung, Suk Ho, 533

K
Kacenka, Andrej, 799
Kacor, Petr, 810
Kadavy, Tomas, 197, 272
Kahankova, Radana, 125
Kang, Ho-Young, 778
Karlovsky, Pavel, 769
Kazikova, Anezka, 272
Kieu, Tam Nguyen, 359
Kim, Chan-Jung, 778
Kim, Dae Hwan, 494, 521, 932
Kim, Hak Kyeong, 521, 605, 963
Kim, Jungyun, 3, 778
Kim, Sang Bong, 450, 494, 521, 605, 932, 963
Kim, Sung Rak, 605, 963
Kim, Sung Won, 521, 963
Kim, Young Bok, 533
Kolar, Vaclav, 835

Kolarik, Jakub, 429
Kolpakhchyan, Pavel G., 628
Kopitza, Vadim V., 628
Kouřil, Daniel, 514
Kovalev, Sergey M., 24, 475
Koziorek, Jiri, 390
Krömer, Pavel, 313
Krutas, Alexander D., 154
Kuchar, Martin, 683
Kudryashova, E. V., 639
Kuncicky, Lumir, 429
Kuncicky, Radim, 429
Kuznetsov, N. V., 639
Kuznetsova, O. A., 639

L
Lapkova, Dora, 144
Latal, J., 860
Le, Kim Hung, 34
Le, Nhat Binh, 533
Lee, Choong Hwan, 521
Leon, Jaime, 904
Leonowicz, Zbigniew, 88
Lettl, Jiri, 769
Ličev, Lačezar, 484
Lipcak, Ondrej, 758
Lobov, Boris N., 628
Lycka, Patrik, 335

M
Majdisova, Zuzana, 325
Makys, Pavol, 799
Martinek, Petr, 239
Martinek, Radek, 125, 429, 973, 982
Martinez, Fredy, 904
Matinez, Fernando, 904
Mehic, Miralem, 114
Melnikov, Boris, 185
Mlcak, Tomas, 835
Mokaev, R. N., 639
Mokaev, T. N., 639
Mrovec, Martin, 461

N
Nam, Pham Minh, 371
Nam, Tran Thanh, 441
Nansai, Shunsuke, 894
Ngo, An H., 303
Ngoc, Long Nguyen, 359
Nguyen Kieu, Tam, 371
Nguyen, Duy Anh, 251, 283, 618, 725

Nguyen, Duy Tan, 725
Nguyen, Hoang Linh, 450
Nguyen, Hoang-Sy, 705
Nguyen, Hong Nhu, 359
Nguyen, Hung, 34, 505
Nguyen, Huu Vinh, 34
Nguyen, Huy Hung, 533
Nguyen, Minh Chau Huu, 261, 580, 789
Nguyen, Phung Hung, 618
Nguyen, Tan N., 693, 705, 821
Nguyen, Tan Tien, 450, 914, 924
Nguyen, Thang T., 303
Nguyen, Thanh-Sang, 705
Nguyen, Trong Hai, 505
Nguyen, Tu-Trinh Thi, 359
Nguyen, Van Dong, 914, 924
Nguyen, Van Hien, 914, 924
Nguyen, Van Lanh, 521
Nguyen, Van Vu, 450
Nguyen, Viet Duong, 251
Niemiec, Marcin, 114
Nikishin, Andrey, 870
Nomura, Yurika, 883
Novak, T., 860
Nowaková, Jana, 313

O

Oh, Jong Min, 605, 963
Oh, Sea June, 605
Ong, Chea Xin, 883
Orjuela, Santiago, 904

P

Palacky, Petr, 683
Parshukov, Vladimir I., 628
Pecha, Marek, 13
Pennacchi, Paolo, 647, 658
Petružela, Michal, 844
Pham, Canh An Tien, 914, 924
Pham, Nguyen Nhut-Thanh, 592
Pham, Truong Thinh, 251
Pham, Viet Anh, 618
Phan, Hoang Long, 914
Phan, Huy Xuan, 789
Phong, Tran Thanh, 561
Phuong, Nguyen Thanh, 505
Phuong, Pham Nhat, 580
Placek, Jiri, 835
Platoš, Jan, 313
Pluhacek, Michal, 197, 272
Pokorny, Rostislav, 747
Pridala, Michal, 715

Q

Quynh, Truong Duc, 441

R

Rakay, Robert, 390
Rantung, Jotje, 605

S

Sayankina, Maria, 154
Senkerik, Roman, 197, 272, 344
Sevcik, Lukas, 409
Shin, Min Saeng, 963
Sikora, Tadeusz, 850
Skala, Vaclav, 325
Skanderova, Lenka, 217
Slanina, Zdenek, 747
Smaglichenko, Alexander, 154, 185
Smaglichenko, Tatyana A., 154, 185
Smolik, Michal, 325
Snajdr, Lukas, 973
Sobek, Martin, 514, 736
Sokansky, K., 860
Song, Young-Geon, 778
Soustek, Lukas, 429, 973
Spanik, Pavol, 715
Struharnansky, Lubos, 799
Styskala, Vitezslav, 24
Sukhanov, Andrey V., 24, 475
Sumpich, J., 860

T

Tam, Nguyen Kieu, 441
Tien, Dung Vo, 88
Tin, Phu Tran, 371
Tomaszek, Lukas, 335
Tran, Anh Minh Duc, 533
Tran, Anh-Minh Duc, 78
Tran, Cuong Dinh, 580, 789
Tran, Manh Son, 533
Tran, Ngoc-Huy, 592
Tran, Phuong T., 371, 821
Tran, Thien Phuc, 450, 924
Tran, Thinh Cong, 261, 580, 789
Truong, Ngoc Cuong, 283
Tuan, Nguyen Manh, 441

V

Valicek, P., 860
Van Dung, Mai, 441
Van Huynh, Van, 561
Van Yen, Nguyen, 441
Vanus, J., 860
Vargovsky, Jan, 217

Vestenický, Martin, 382
Vestenický, Peter, 382
Viktorin, Adam, 197
Vinh, Le The, 441
Vo, Dinh Due, 618
Voznak, Miroslav, 114, 359, 371, 400, 409, 420, 693, 705, 821
Vu, Tri-Vien, 78
Vu, Trung-Hieu, 78
Vysocký, Jan, 844

Y
Yamada, Shihono, 569
Yuldashev, M. V., 639
Yuldashev, R. V., 639

Z
Zajaczek, Stanislav, 475
Zdralek, Jaroslav, 359, 693
Zelinka, Ivan, 67, 335, 949

Printed by Printforce, the Netherlands